Adult Learning Theories in Post Compulsory and Vocational Education and Training

Modern Interpretations and Applications

Richard Skiba

AFTER MIDNIGHT
PUBLISHING

Skiba, Richard (Author)

Adult Learning Theories in Post Compulsory and Vocational Education and Training: Modern Interpretations and Applications

ISBN 978-1-7647987-2-3 (Paperback) 978-1-7647987-3-0 (eBook)

Non-fiction

Contents

Introduction

Education has always been fundamental to human development, but never before has the need for continuous learning been as significant as it is today. Rapid technological advancement, globalisation, demographic change, evolving labour markets, artificial intelligence, environmental sustainability, and changing societal expectations are transforming the knowledge, skills, and capabilities required to participate effectively in work and society. In this environment, learning is no longer confined to childhood or early adulthood. Instead, it has become a lifelong process that enables individuals to adapt, grow, innovate, and contribute throughout their personal and professional lives.

Adult education and vocational education and training occupy a central position within this lifelong learning landscape. Across the world, millions of adults return to education each year to gain qualifications, develop new occupational skills, transition into new careers, maintain professional competence, or pursue learning for personal fulfilment. Whether studying at a university, undertaking an apprenticeship, completing workplace training, participating in professional development, earning industry certifications, or engaging in self-directed online learning, adult learners represent one of the largest and most diverse educational populations. Understanding how these learners acquire knowledge, develop expertise, and transform their understanding has therefore become essential for educators, trainers, instructional designers, policymakers, employers, and learners themselves.

The study of adult learning has evolved considerably over the past century. Early educational theories sought to explain learning primarily through observable behaviour or cognitive processes, while later perspectives increasingly recognised the importance of experience, reflection, social interaction, culture, motivation, identity, and critical inquiry. Rather than replacing one another, these theories have progressively expanded our understanding of learning, each providing valuable insights into different aspects of the adult learning process. Contemporary educational practice draws upon multiple theoretical perspectives, recognising that no single theory can adequately explain the complexity of learning across the diverse settings in which adults live, work, and study.

This book examines adult learning from an international perspective, bringing together the major theories that have shaped adult education while exploring their relevance within contemporary post-compulsory education and vocational education and training systems. It moves beyond presenting theories as isolated historical concepts by demonstrating how they continue to inform modern educational practice across higher education, vocational education, workplace learning, professional development, online education, community education, and lifelong learning. Throughout the book, classical theories are considered alongside contemporary developments, including digital learning, artificial intelligence, competency-based education, micro-credentials, sustainability, learning analytics, immersive technologies, and the future of work.

While many texts examine individual learning theories in isolation, this book adopts a broader, integrative approach. Adult learning is explored through psychological, sociological, philosophical, technological, and workplace perspectives, recognising that effective educational practice often draws upon multiple theoretical traditions simultaneously. Behaviourism, cognitivism, constructivism, humanism, experiential learning, transformative learning, critical pedagogy, sociocultural theory, connectivism, and other contemporary perspectives are examined not as competing explanations but as complementary frameworks that illuminate different dimensions of adult learning. The book also demonstrates how these theories underpin instructional design, curriculum development, assessment, learner engagement, educational technology, workplace capability development, and lifelong learning policy.

A distinguishing feature of this book is its emphasis on practical application. Each theoretical perspective is connected to authentic educational and workplace contexts, illustrating how theory informs the design and delivery of effective learning experiences. Examples are drawn from vocational education, higher education, professional training, apprenticeships, industry certification, online learning, and workplace learning environments, enabling readers to appreciate how educational theory influences everyday practice. This applied focus recognises that theory and practice are mutually reinforcing; educational theory provides a framework for understanding learning, while practical experience continually refines and extends theoretical understanding.

A further deliberate feature of this book is its emphasis on contemporary scholarship. While many of the foundational theories of adult learning were developed during the twentieth century, they continue to evolve through ongoing research, technological innovation, and changing educational practice. Wherever possible, this book draws upon recent peer-reviewed literature to provide modern interpretations of well-established concepts, demonstrating how classical theories continue to inform contemporary adult education and vocational education and training. Rather than presenting these theories solely in their historical context, the discussion integrates current research with practical examples from higher education, vocational education and training, workplace learning, professional development, online learning, artificial intelligence, digital technologies, competency-based education, and lifelong learning. This approach enables readers to appreciate both the enduring value of foundational theories and the ways in which they continue to be refined, challenged, and applied within today's rapidly evolving educational landscape. By connecting established theoretical principles with contemporary evidence and real-world practice, the book seeks to bridge the gap between educational theory and implementation, ensuring that readers gain not only an understanding of how adult learning theories developed but also how they remain relevant for designing effective learning experiences in the twenty-first century.

The book also recognises that adult learning takes place within rapidly changing social and technological environments. Digital technologies have transformed access to education, enabling learning to occur across geographical boundaries and throughout every stage of life. Artificial intelligence is reshaping the ways learners access information, receive feedback, develop skills, and solve complex problems. At the same time, employers increasingly seek adaptable, reflective, collaborative, and digitally capable workers who can continue learning throughout their careers. These developments require educators to reconsider not only what adults learn, but also how learning is designed, facilitated, assessed, and recognised.

Although this text has broad international relevance, it gives particular attention to post-compulsory education and vocational education and training. Vocational systems occupy a critical position in developing occupational competence, supporting economic productivity, promoting workforce mobility, and enabling lifelong learning. Increasingly, the boundaries between vocational education, higher education, workplace learning, and professional certification are becoming more permeable, creating integrated lifelong learning pathways that support individuals across multiple careers and changing occupational demands. Understanding adult learning theories provides an essential foundation for designing these educational systems and ensuring they remain responsive to learners, employers, and society.

This book is intended for a wide audience, including undergraduate and postgraduate students, vocational education teachers and trainers, university academics, instructional designers, workplace educators, learning and development professionals, researchers, policymakers, and anyone with an interest in how adults learn. It is equally relevant to practitioners seeking to improve their teaching and to scholars wishing to understand the theoretical foundations of contemporary adult education.

Adult learning is about far more than acquiring knowledge or mastering workplace skills. It is about enabling individuals to think critically, adapt to change, solve complex problems, participate meaningfully in society, and continue developing throughout their lives. As the pace of technological, economic, and social change continues to accelerate, understanding how adults learn has never been more important. It is hoped that this book will provide readers with both a comprehensive understanding of adult learning theory and a practical appreciation of how these theories can be applied to create engaging, inclusive, evidence-informed, and effective learning experiences for adults across the diverse educational contexts of the twenty-first century.

Chapter 1
Foundations of Adult Learning

Learning does not stop when formal schooling ends. In today's rapidly changing world, adults are continually required to acquire new knowledge, develop new skills, adapt to emerging technologies, and respond to evolving personal, professional, and societal demands. Whether pursuing a new qualification, undertaking workplace training, changing careers, maintaining professional competence, or learning for personal enrichment, adults engage in learning throughout their lives. As a result, understanding how adults learn has become increasingly important for educators, trainers, employers, policymakers, and learners themselves.

Adult learning is a unique and complex field that differs in important ways from traditional approaches to teaching children and adolescents. Adults bring a wealth of life experience, established beliefs, professional expertise, and personal responsibilities to learning environments. These characteristics influence their motivations, expectations, learning preferences, and approaches to acquiring new knowledge. Consequently, educational practices designed for adult learners must recognise and respond to these distinctive characteristics if they are to be effective.

This chapter provides the foundation for understanding adult learning within the broader context of adult education and vocational education and training (VET). It explores what adult learning is, how it has evolved as a field of study, and why it has become central to lifelong learning and workforce development. The chapter examines the historical development of adult learning theory, introduces the major theorists and concepts that have shaped contemporary practice, and explores the key characteristics that distinguish adult learners from younger learners. It also considers the role of lifelong learning, vocational education, professional development, and continuing education in supporting individuals throughout their personal and professional lives.

By establishing the theoretical, historical, and practical foundations of adult learning, this chapter provides a framework for understanding the learning processes, teaching approaches, and educational systems that will be examined throughout the remainder of this book. An appreciation of these foundations is essential for anyone seeking to design, deliver, facilitate, or participate in effective adult learning experiences in an increasingly complex and rapidly changing world.

Defining Adult Learning

Adult learning refers to the process through which adults acquire, develop, and transform knowledge, skills, attitudes, values, behaviours, and competencies throughout their lives. It encompasses learning that occurs in formal educational institutions, vocational education and training (VET) systems, workplaces, community settings, and everyday life experiences. Adult learning has become an increasingly important field of study in response to social, economic, and technological changes that require individuals to engage in continuous learning across the lifespan. As a distinct area of educational theory and practice, adult learning recognises that adults bring different experiences, motivations, responsibilities, and expectations to learning than children and adolescents, necessitating approaches that acknowledge these unique characteristics.

The study of adult learning emerged as a significant field during the twentieth century, drawing upon psychology, sociology, philosophy, and education. Early theorists sought to understand whether adults learn differently from children and how educational practices should be adapted to meet adult learners' needs. This inquiry led to the development of several influential theories that continue to shape adult education and vocational training internationally. Among the most influential contributors are Malcolm Knowles (Knowles, 1980, 1984), whose work on andragogy helped establish adult education as a distinct field of practice; Jack Mezirow, who developed transformative learning theory (Mezirow, 1991); Paulo Freire, whose critical pedagogy emphasised education as a means of empowerment and social change (Paulo Freire, 1970); and David Kolb, whose experiential learning theory highlighted the central role of experience in adult learning (Kolb, 1984b).

One of the defining characteristics of adult learning is the importance of prior experience. Adults enter learning environments with accumulated knowledge derived from work, family life, social participation, and previous educational experiences. Knowles (1980) argued that these experiences represent a rich resource for learning and should be actively incorporated into educational processes. Unlike traditional pedagogical models that position learners as relatively dependent recipients of knowledge, andragogical approaches recognise that adult learners possess substantial experiential foundations upon which new learning can be constructed. Adult learners frequently interpret new information through existing cognitive frameworks, comparing new concepts with prior knowledge and evaluating their relevance to real-world situations.

Adult learning is also commonly characterised by goal orientation and purposeful engagement. Many adults participate in education and training to achieve specific personal, professional, or social objectives. These objectives may include gaining employment, advancing careers, obtaining professional credentials, changing occupations, improving workplace performance, or pursuing personal interests. Knowles (1984) suggested that adults are more likely to engage with learning when they perceive a clear relationship between educational activities and their immediate needs or life circumstances. Consequently, adult education programs often emphasise practical application, problem-solving, and authentic learning experiences that demonstrate clear relevance to learners' goals.

Closely related to goal orientation is the concept of self-direction. Knowles (1975) described self-directed learning as a process in which individuals take initiative in diagnosing their learning needs, formulating learning goals, identifying resources, implementing learning strategies, and evaluating outcomes. While not all adults exhibit the same degree of self-direction, research suggests that many adult learners value autonomy and expect opportunities to influence the content, pace, and methods of their learning. Contemporary adult education therefore frequently incorporates learner-centred approaches that encourage participation, collaboration, and shared responsibility for learning outcomes.

The importance of reflection represents another key feature of adult learning. Reflection enables learners to examine their assumptions, evaluate experiences, and integrate new knowledge into existing perspectives. This process forms the foundation of transformative learning theory described by Mezirow (1991), which proposes that adult learning can involve profound changes in the way individuals understand themselves and the world around them. According to Mezirow, transformative learning occurs when individuals critically reflect upon previously unquestioned assumptions, leading to the development of new perspectives and more inclusive ways of thinking. Such transformation is particularly relevant in contexts involving career transitions, significant life events, professional development, and social change.

The role of experience in adult learning is further emphasised in Kolb's experiential learning theory (Kolb, 1984b). Kolb conceptualised learning as a cyclical process involving concrete experience, reflective observation, abstract conceptualisation, and active experimentation. According to this model, learning is not simply the acquisition of information but the transformation of experience into knowledge through reflection and application. Experiential learning has had a profound influence on vocational education, workplace training, apprenticeships, internships, and professional development programs because it aligns closely with the practical and applied nature of adult learning.

Adult learning is also shaped by social, cultural, and political contexts. Paulo Freire (1970) challenged traditional models of education that treated learners as passive recipients of information, describing such approaches as the "banking model" of education. Instead, Freire advocated a dialogical approach in which learners and educators collaboratively construct knowledge through critical inquiry and reflection. His concept of conscientisation emphasised the development of critical awareness of social, political, and economic realities, enabling learners to become active participants in transforming their circumstances. Freire's work has been particularly influential in community education, literacy programs, workforce development initiatives, and adult education programs aimed at promoting social justice and empowerment.

Contemporary understandings of adult learning are closely linked to the concept of lifelong learning. Lifelong learning recognises that education extends beyond childhood and formal schooling and continues throughout an individual's life. International organisations such as the UNESCO and the Organisation for Economic Co-operation and Development have promoted lifelong learning as a means of supporting economic competitiveness, social inclusion, active citizenship, and personal development. In an era characterised by rapid technological change, globalisation, demographic shifts, and evolving labour markets, individuals increasingly need to update their knowledge and skills throughout their careers and lives (UNESCO, 2021a).

Adult learning occurs within formal, non-formal, and informal contexts. Formal learning typically takes place within structured educational systems and leads to recognised qualifications, such as university degrees, vocational qualifications, apprenticeships, and professional certifications. Non-formal learning occurs outside traditional educational institutions but remains organised and intentional, including workplace training, professional development workshops, and community education programs. Informal learning occurs through everyday experiences and activities, often without explicit educational objectives. Adults learn informally through workplace interactions, problem-solving activities, self-directed study, online engagement, social networks, and participation in communities of practice. Research consistently demonstrates that a substantial proportion of adult learning occurs informally, particularly within workplace environments where learning is embedded in daily activities and professional practice.

Within vocational education and training, adult learning is primarily concerned with the development of occupational competence and workforce capability. Vocational learners often seek practical knowledge and skills that can be directly applied to workplace contexts. Consequently, vocational education systems around the world frequently incorporate competency-based training, work-integrated learning, apprenticeships, simulations, workplace assessment, and recognition of prior learning. These approaches reflect many of the principles identified by adult learning theorists, including experiential learning, learner autonomy, practical relevance, and the value of prior experience.

Adult learning theories have therefore played a significant role in shaping contemporary educational practice. Knowles' andragogy highlights the importance of self-direction and experience; Mezirow's transformative learning theory emphasises critical reflection and perspective transformation; Kolb's experiential learning theory demonstrates how experience becomes knowledge through reflection and application; and Freire's critical pedagogy illustrates how learning can promote empowerment and social change. Together, these perspectives provide a comprehensive framework for understanding how adults learn across diverse educational, vocational, and social contexts.

A contemporary definition of adult learning may therefore be understood as the lifelong process through which adults intentionally or unintentionally acquire, develop, and transform knowledge, skills, attitudes, values, and competencies through formal, non-formal, and informal experiences. This process supports personal growth, professional development, social participation, critical awareness, and adaptation to changing life circumstances. Such a definition reflects the multidimensional nature of adult learning and provides a foundation for examining the theories, practices, and policies that shape adult education and vocational education and training systems worldwide.

Adult learning is widely recognized as a lifelong process encompassing formal, non-formal, and informal dimensions. The European Centre for the Development of Vocational Training (CEDEFOP) defines lifelong learning as "all learning activity undertaken throughout life, which results in improving knowledge, know-how, skills, competencies and/or qualifications for personal, social, and/or professional reasons" (Grigoroviča et al., 2022, p. 755). Similarly, lifelong education refers to the ongoing pursuit of learning throughout an individual's lifespan, encompassing activities that enhance knowledge, skills, competencies, and understanding to support personal growth, employability, social inclusion, and active participation in civic life (Butcher & Bruce, 2015).

UNESCO conceptualises adult education as a comprehensive and lifelong process through which adults engage in formal, non-formal, and informal learning to expand their knowledge, develop their capabilities, enhance professional and technical competencies, and contribute more effectively to personal development, social inclusion, and active participation in society (Egetenmeyer & Fedeli, 2017a). These learning processes occur across formal, non-formal, and informal settings (Dobrich, 2022; Sofo, 2011), supporting both economic goals and humanistic aims including critical awareness and social transformation (Roumell et al., 2018). The multidimensional nature of adult learning, spanning personal growth, professional development, and active citizenship, is reflected in policy frameworks worldwide that shape adult education and vocational training systems (Krishna, 2019).

Structured Adult Educational Systems

University degrees, vocational qualifications, apprenticeships, and professional certifications all involve learning and credentialing, but they differ significantly in their purpose, structure, learning approach, assessment methods, and relationship to employment. University degrees emphasize general education and deep theoretical knowledge, whereas vocational qualifications and apprenticeships focus on occupation-specific practical skills (Bol & Weeden, 2014; Krishna, 2019). Certifications validate competency in specific professional domains through standardized assessments, functioning as non-statutory credentials distinct from licensure (Albert, 2017; Phillips, 2011). Apprenticeships combine workplace-based learning with formal instruction, culminating in employer-recognized qualifications (Bol & Weeden, 2014; Ding, 2023), while degrees follow primarily classroom-based curricula (Decker, 2019). Assessment methods range from academic examinations in universities to competency-based practical evaluations in vocational and certification contexts (Krishna, 2019; Loquinte, 2016). Regarding employment, certifications and vocational credentials demonstrate occupational specificity that signals direct job readiness (Albert, 2017; Baird et al., 2022), whereas degrees signal broader analytical capabilities. These pathways can be complementary rather than competing; frameworks integrating all credential types, from diplomas to degrees, apprenticeships, and certifications, support diverse routes to professional success (Decker, 2019; Krishna, 2019).

University degrees, vocational qualifications, apprenticeships, and professional certifications are all important forms of education and credentialing, yet they differ significantly in their purpose, structure, learning approaches, assessment methods, and relationship to employment. While each pathway supports the development of knowledge and skills, they are designed to meet different educational, occupational, and professional objectives.

Table 1: Structured Educational Systems Comparison,

Feature	University Degrees	Vocational Qualifications	Apprenticeships	Professional Certifications
Primary Purpose	Develop broad theoretical and disciplinary knowledge	Develop occupational competence	Develop occupational competence through work and training	Verify competence or specialised knowledge in a profession
Learning Focus	Academic and theoretical	Practical and occupational	Workplace-based practical learning	Professional standards and specialised skills
Delivery	Universities and higher education institutions	Vocational colleges, training providers, technical institutes	Combination of workplace and formal training	Professional bodies, industry associations, certification organisations
Duration	Typically 3–8 years	Several months to 2–4 years	Usually 1–4 years	Days to months of preparation
Assessment	Exams, essays, projects, research	Demonstration of workplace skills and knowledge	Workplace performance and competency assessment	Examination and/or portfolio assessment
Outcome	Degree (Bachelor, Master, Doctorate)	Certificate, Diploma, Advanced Diploma, National Qualification	Qualification plus occupational experience	Industry-recognised certification or designation
Employment Relationship	Not usually required	May or may not include workplace experience	Requires employment relationship with an employer	Usually undertaken by existing practitioners
Orientation	Education-focused	Training-focused	Employment-focused	Credential-focused

University degrees are primarily intended to develop broad theoretical knowledge, critical thinking, analytical abilities, and disciplinary expertise (Durrans et al., 2020; Klassen et al., 2020). Offered by universities and other higher education institutions, degree programs include qualifications such as bachelor's, master's, and doctoral degrees (Liễu et al., 2018; Shay et al., 2016). These programs typically require several years of study and emphasise conceptual understanding, research, analysis, and the application of knowledge to complex problems (Guerra, 2017; Klassen et al., 2020). For example, a Bachelor of Engineering program may include studies in mathematics, physics, engineering science, design principles, research methods, and professional practice (Durrans et al., 2020; Long, 2020). Students learn not only how engineering systems function but also the scientific and theoretical principles that underpin them (Durrans et al., 2020; Klassen et al., 2020). University education is generally grounded in academic disciplines and is designed to prepare graduates for a wide range of career opportunities, further study, research, leadership roles, and professional practice (Guerra, 2017; Klassen et al., 2020; Shay et al., 2016).

In contrast, vocational qualifications focus on developing the practical knowledge, skills, and competencies required to perform specific occupations . These qualifications are typically delivered through vocational education and training (VET) systems, technical institutes, colleges, and industry training providers. Vocational programs are commonly aligned with industry competency standards and are designed to ensure that learners can perform workplace tasks to an acceptable professional standard (Koca & Ortakaya, 2023; Moodie, 2008; Sule et al., 2024; Wheelahan, 2015). Examples include qualifications in construction, automotive technology, plumbing, aged care, hospitality, logistics, and information technology. Vocational education places a strong emphasis on practical application, workplace relevance, and occupational competence, reflecting its primary purpose of preparing learners for employment and workforce participation (Billett, 2011). Assessment is often competency-based and may involve practical demonstrations, workplace projects, simulations, portfolios, and direct observation of performance. Competency-based approaches seek to determine whether learners can consistently perform workplace activities to prescribed industry standards rather than primarily assessing theoretical knowledge alone (Hodge, 2007; Mulcahy, 2000). Although vocational education incorporates underpinning theoretical knowledge, its principal focus is on ensuring that learners can competently perform the tasks, responsibilities, and problem-solving activities associated with a particular occupation (Billett, 2001).

Apprenticeships represent a distinct form of vocational learning that combines paid employment with structured education and training. They are among the oldest forms of occupational preparation and have historically served as a primary mechanism for transmitting trade and occupational skills across generations (Fuller & Unwin, 2011). Apprenticeships remain widely used in trades, technical occupations, and increasingly in professional and higher-skilled industries. An apprentice is employed by an organisation while simultaneously undertaking formal training that leads to a recognised qualification. Learning occurs primarily through participation in authentic workplace activities under the supervision of experienced practitioners, supported by classroom instruction, workshops, online learning, and other structured educational experiences. This integration of work and learning reflects the principles of situated learning and communities of practice, which emphasise that occupational competence develops through participation in real workplace environments (Lave & Wenger, 1991).

For example, an electrical apprentice develops practical skills by working alongside qualified electricians while simultaneously studying electrical theory, regulations, safety requirements, and installation practices through a registered training provider or educational institution. Unlike vocational qualifications undertaken independently of employment, apprenticeships require a genuine workplace, an employer, ongoing supervision, and progressive skill development over an extended period. (Fuller & Unwin, 2003, 2011) argue that effective apprenticeships provide expansive learning environments that enable learners to develop not only technical competence but also occupational identity, professional judgement, and broader workplace capabilities. Apprentices benefit from earning wages while learning and typically complete their apprenticeship with both a recognised qualification and substantial industry experience. Consequently, apprenticeships are widely regarded as one of the most effective mechanisms for developing occupational competence because they integrate theoretical learning, practical skill development, and workplace socialisation within a single educational pathway (Billett, 2001; Fuller & Unwin, 2011).

Professional certifications serve to validate competence and specialized knowledge within specific professional domains, distinguishing them from broader academic qualifications (Muzhanova et al., 2022; Tapis et al., 2020). These credentials are awarded by professional associations and industry bodies such as ISACA, which offers the Certified Information Systems Auditor (CISA) designation (Merhout & Cothran, 2006; Tapis et al., 2020), CompTIA, which provides vendor-neutral security certifications (Alaryani et al., 2024; Muzhanova et al., 2022), and the Project Management Institute (PMI), which administers the Project Management Professional (PMP) credential (Nance & Hay, 2007; Remer & Martin, 2009). Accounting designations such as CPA and CA further exemplify professional certifications (Borisova & Bekhteneva, 2015; Mucenski-Keck et al., 2012), while Cisco offers vendor-specific networking certifications (Alaryani et al., 2024).

Professional certifications typically validate existing knowledge rather than providing comprehensive education. Candidates generally possess relevant qualifications and work experience prior to seeking certification (Boyd et al., 2009; Merhout & Cothran, 2006). The certification process commonly involves examinations, verification of professional experience, adherence to codes of ethics, and continuing professional development requirements (Nance & Hay, 2007; Remer & Martin, 2009). For instance, CISA certification requires passing an examination, five years of professional experience, and ongoing continuing education (Merhout & Cothran, 2006; Remer & Martin, 2009). Similarly, PMP certification demands demonstrated education and experience, examination, adherence to professional conduct standards, and continuing education (Nance & Hay, 2007; Remer & Martin, 2009).

These four educational pathways reflect different educational philosophies and approaches to learning. University education is primarily concerned with developing broad knowledge, intellectual capability, critical thinking, disciplinary expertise, and the capacity to engage with complex theoretical and practical problems (Barnett, 1990; Marginson, 2016). Vocational education focuses on developing occupational competence and work readiness through practical, applied, and industry-relevant learning experiences that prepare individuals for participation in specific occupations and sectors of the labour market (Billett, 2011; Wheelahan, 2015). Apprenticeships emphasise learning through participation in authentic work activities under the guidance of experienced practitioners, integrating employment and education into a single developmental process. This approach reflects theories of situated learning and communities of practice, which view occupational competence as emerging through engagement in real workplace environments and social participation within occupational communities (Fuller & Unwin, 2011; Lave & Wenger, 1991). Professional certifications, meanwhile, are principally concerned with validating competence against recognised professional, regulatory, or industry standards and often serve as mechanisms for assuring quality, maintaining professional credibility, and supporting continuing professional development (Eraut, 1994; Young & Allais, 2013).

Despite their differences, the boundaries between these pathways are becoming increasingly blurred. Educational systems in many countries have developed articulation arrangements and qualifications frameworks that enable learners to move between vocational and higher education sectors, allowing vocational graduates to progress into university degree programs and university graduates to undertake vocational or professional training (Moodie, 2008; OECD, 2021i). This trend reflects broader efforts to create more flexible and integrated education systems capable of supporting lifelong learning and workforce mobility. Apprenticeships have also evolved significantly, with higher apprenticeships, degree apprenticeships, and professional apprenticeships emerging in many jurisdictions as mechanisms for combining advanced academic study with workplace learning (Fuller & Unwin, 2011; OECD, 2018a). Similarly, many professions now encourage or require practitioners to combine academic qualifications with industry certifications, professional registration, and ongoing professional development in order to maintain competence in rapidly changing occupational environments (Eraut, 1994).

As economies become increasingly knowledge-based, technology-driven, and globally interconnected, lifelong learning has become essential for workforce participation, career progression, and social inclusion (Field, 2006; UNESCO, 2021a). Rapid technological change, evolving occupational requirements, and the increasing complexity of work mean that initial education alone is rarely sufficient to sustain an individual's career over a lifetime. Consequently, individuals are increasingly likely to engage in a combination of university education, vocational qualifications, apprenticeships, workplace learning, and professional certification throughout their lives. International organisations such as the UNESCO and the Organisation for Economic Co-operation and Development have argued that lifelong learning systems should facilitate movement between different forms of education and training while supporting continuous skills development and personal growth (OECD, 2021i; UNESCO, 2021a). Rather than representing competing forms of education, these pathways are increasingly viewed as complementary components of an integrated lifelong learning ecosystem that supports personal development, occupational competence, professional recognition, employability, and career advancement in a rapidly changing world.

Adult Education and Vocational Education and Training in a Global Context

Adult education and vocational education and training (VET) play a critical role in economic development, workforce preparation, social inclusion, and lifelong learning throughout the world. Although educational systems differ considerably between countries, adult education and VET share the common purpose of supporting individuals to acquire the knowledge, skills, competencies, and qualifications needed to participate effectively in society and the labour market (UNESCO, 2016). In an era characterised by rapid technological change, globalisation, demographic shifts, and evolving employment patterns, adult learning has become increasingly important for both individuals and nations. International organisations, governments, and employers increasingly recognise that continuous learning is essential for economic competitiveness, social cohesion, and individual wellbeing (OECD, 2021i).

Historically, adult education emerged as a means of promoting literacy, citizenship, and personal development among populations that had limited access to formal schooling (Gonzales, 2020; Mikulec & Kovšca, 2023). During the nineteenth and twentieth centuries, many countries established adult education programs to improve literacy rates, expand educational opportunities, and support social and economic development (Gonzales, 2020; Mikulec & Kovšca, 2023; Reid et al., 2021). Influenced by democratic, humanistic, and social reform movements, adult education gradually expanded beyond basic literacy and civic education to encompass professional development, workplace learning, community education, higher education access, and lifelong learning initiatives (Gonzales, 2020; Mikulec & Kovšca, 2023; Reid et al., 2021). At the same time, vocational education developed as a mechanism for preparing workers with the practical skills and occupational competencies required by industry and commerce, particularly during periods of industrialisation and economic transformation (Amponsah, 2020).

Today, adult education encompasses a broad range of learning activities, including literacy and numeracy programs, community education, continuing professional development, workplace learning, university study, online learning, and personal enrichment courses (Kuczera, 2018; Markowitsch & Bjørnåvold, 2022). Vocational education and training focuses more specifically on developing occupational competence and workforce capabilities through structured learning programs aligned with industry needs (Eichhörst et al., 2015). While adult education has traditionally emphasised personal, social, and civic development, VET has been primarily concerned with employability, occupational skills, and economic participation (Alvunger, 2024; Oecd, 2022b). However, contemporary education systems increasingly recognise that these objectives are interconnected, and many policies now seek to integrate personal development, employability, citizenship, and lifelong learning outcomes within a unified framework (Markowitsch & Bjørnåvold, 2022; Κατσικοπούλου, 2016).

Today, adult education encompasses a broad range of learning activities, including literacy and numeracy programs, community education, continuing professional development, workplace learning, university study, online learning, and personal enrichment courses. Vocational education and training focuses more specifically on developing occupational competence and workforce capabilities through structured learning programs aligned with industry needs (Moodie, 2008). While adult education has traditionally emphasised personal, social, and civic development, VET has been primarily concerned with employability, occupational skills, and economic participation. However, contemporary education systems increasingly recognise that these objectives are interconnected, and many policies now seek to integrate personal development, employability, citizenship, and lifelong learning outcomes within a unified framework (Field, 2006; UNESCO, 2021a).

International organisations have played a significant role in shaping global perspectives on adult education and VET. UNESCO has long advocated lifelong learning as a fundamental human right and a cornerstone of sustainable development. Through initiatives such as the CONFINTEA conferences and its lifelong learning agenda, UNESCO has promoted adult learning as a means of advancing social justice, active citizenship, inclusion, cultural development, and personal empowerment (UNESCO, 2016, 2021a). Similarly, the Organisation for Economic Co-operation and Development (OECD) has emphasised the importance of skills development, workforce adaptability, and lifelong learning in supporting economic growth, innovation, and labour market resilience (OECD, 2021i). The OECD's work has highlighted the need for education and training systems that can respond effectively to technological disruption, demographic change, and evolving workforce requirements.

Across the world, VET systems vary significantly in structure, governance, funding, and delivery. Countries such as Germany, Switzerland, and Austria are widely recognised for their dual-system apprenticeship models, which combine workplace training with formal education (Fazio et al., 2016; Lassnigg, 2024; Martínez-Izquierdo & Sánchez, 2022). In these systems, learners divide their time between employment and structured study, allowing them to develop both theoretical knowledge and practical workplace skills (Fazio et al., 2016; Kuczera & Field, 2018). These apprenticeship systems are frequently cited as examples of effective collaboration between governments, employers, industry associations, trade unions, and educational institutions (Bergseng, 2019; Emmenegger & Seitzl, 2020;

Martínez-Izquierdo & Sánchez, 2022), contributing to strong employment outcomes and relatively smooth transitions from education to work (Fazio et al., 2016).

In contrast, countries such as Australia, New Zealand, and the United Kingdom have developed competency-based VET systems that focus on demonstrating workplace competence against nationally recognised standards (Deißinger, 2012; Fazio et al., 2016). These systems typically utilise qualifications frameworks that facilitate progression between qualification levels and support pathways between vocational and higher education sectors (Deißinger, 2012). Competency-based approaches seek to ensure that qualifications accurately reflect workplace requirements while providing flexibility for learners and employers. Such systems have become influential internationally and have been adopted, adapted, or referenced in numerous countries seeking to strengthen workforce development and skills recognition (Deißinger, 2012; Fazio et al., 2016).

In many developing countries, adult education and VET are closely linked to national development priorities. Programs frequently focus on literacy, basic education, agricultural development, entrepreneurship, technical trades, and workforce capability building. Adult learning initiatives may also address broader social objectives such as poverty reduction, gender equality, public health, environmental sustainability, and community development (UNESCO, 2016). In these contexts, adult education is often viewed not only as a means of improving employability but also as a mechanism for empowering individuals and communities, reducing social inequalities, and supporting sustainable economic development.

Globalisation has significantly influenced both adult education and VET. Increasing international mobility of workers, multinational industries, and global supply chains have created growing demand for internationally recognised qualifications and transferable skills. Consequently, many countries have developed national qualifications frameworks and competency standards that support the recognition and portability of skills across jurisdictions (Pan et al., 2021; Šarčević, 2024). The quality of developed skills is assured by mechanisms including qualifications frameworks, occupational standards, and certification (Schmoelz, 2023). NQFs are expected to promote inter-sectoral mobility, connect education and training to the labour market, and improve the status of the VET sector (Perla et al., 2023). International comparisons of qualifications and efforts to improve credential recognition have become increasingly important in facilitating workforce mobility, supporting migration, and enhancing economic competitiveness in a global labour market (Pan et al., 2021; Secretariat, 2022). The European Qualifications Framework, for instance, groups qualifications into eight reference levels to ease portability and transfer across systems and learning contexts (Krishna, 2019). RVA has also been shown to play a role in helping youth and adults create employment opportunities abroad through skills recognition across national borders (Secretariat, 2022).

Technological change has further transformed adult education and VET. The expansion of digital learning technologies, online courses, virtual classrooms, simulations, micro-credentials, and artificial intelligence has created new opportunities for flexible and accessible learning (OECD, 2021b). Adults are now able to participate in education regardless of geographical location, often balancing study with work and family commitments. The growth of online and blended learning has been particularly significant in increasing access for rural populations, working adults, and learners who may have previously faced barriers to participation in traditional educational settings (Merriam & Bierema, 2014).

The emergence of the knowledge economy has also reshaped the role of adult education and VET. Traditional assumptions that individuals complete their education early in life and rely upon that knowledge throughout their careers are increasingly outdated. Rapid technological advancement, organisational change, and evolving occupational requirements mean that workers must continually update their knowledge and skills to remain employable and productive (Field, 2006). The rise of the knowledge economy has created new and higher skill demands on the labour market, posing particular challenges for collective skill formation systems traditionally catering to manual, mid-level skills (Abrassart & Wolter, 2019). Consequently, lifelong learning has become a central policy objective in many countries, with governments, employers, and educational institutions promoting ongoing education, reskilling, and upskilling throughout working life (OECD, 2021b).

Despite their importance, adult education and VET systems face numerous challenges. These include funding constraints, unequal access to learning opportunities, digital exclusion, changing labour market demands, demographic ageing, and the need to balance employer requirements with broader educational and social objectives (UNESCO, 2021a). Many countries report that attracting students to VET is difficult, as VET is considered less prestigious than higher education by both students and their parents (Reisel et al., 2015; Remington, 2017). Many countries also continue to struggle with status hierarchies that position academic education above vocational pathways (Mugler, 2025; Smak, 2022; Usman, 2026), despite evidence that vocational qualifications often provide strong employment outcomes and contribute significantly to economic productivity (Wheelahan, 2015). Indeed, the desirability of occupations requiring vocational education depends significantly on the level of education required rather than the skill content of occupations themselves (Benke & Rachwał, 2022).

Increasingly, policymakers and educators recognise that adult education and VET should not be viewed as separate or competing sectors. Rather, they are increasingly understood as interconnected components of lifelong learning systems that support personal development, social participation, workforce capability, and economic prosperity (Gordon, 2015; Markowitsch & Bjørnåvold, 2022; Salajan & Roumell, 2021). Contemporary education policies frequently emphasise flexible learning pathways, recognition of prior learning, credit transfer arrangements, and opportunities for learners to move between vocational and higher education throughout their lives (Gordon, 2015; Markowitsch & Bjørnåvold, 2022; McGrath et al., 2023). Such approaches reflect the growing recognition that learning is a continuous process that extends across the lifespan and occurs in a variety of formal, non-formal, and informal settings (Gordon, 2015; Markowitsch & Bjørnåvold, 2022; Salajan & Roumell, 2021).

In a global context, adult education and vocational education and training have become essential mechanisms for helping individuals, communities, and nations respond to economic, technological, and social change (Gordon, 2015; Markowitsch & Bjørnåvold, 2022; McGrath et al., 2023). They contribute not only to workforce development and economic competitiveness but also to personal fulfilment, social inclusion, democratic participation, and sustainable development (McGrath et al., 2023; Salajan & Roumell, 2021). As societies continue to evolve, the importance of accessible, flexible, and high-quality adult learning opportunities is likely to increase, reinforcing the central role of adult education and vocational education and training within lifelong learning systems worldwide (Gordon, 2015; Markowitsch & Bjørnåvold, 2022; McGrath et al., 2023; Salajan & Roumell, 2021).

Historical Development of Adult Learning Theory

The development of adult learning theory reflects evolving understandings of education, human development, work, citizenship, and lifelong learning. Although adults have always engaged in learning throughout their lives, the systematic study of how adults learn emerged relatively recently. Adult learning theory has developed through contributions from philosophy, psychology, sociology, and education, with each theoretical tradition reflecting the social, economic, and intellectual contexts of its time. Contemporary adult learning theory is therefore the product of more than a century of scholarship and practice, encompassing diverse perspectives on how adults acquire knowledge, develop skills, construct meaning, and transform their understanding of themselves and the world (Jarvis, 2004; Merriam & Bierema, 2014).

The roots of adult learning can be traced to ancient civilisations, where education often focused on civic participation, moral development, religious instruction, and vocational preparation. In ancient Greece, Socrates emphasised dialogue, questioning, and critical reflection as means of developing understanding and wisdom. The Socratic method, based on guided inquiry rather than direct instruction, remains influential in adult education because it encourages learners to examine assumptions, challenge beliefs, and construct knowledge through dialogue (Knowles et al., 2015).

Throughout the Middle Ages, adult learning occurred primarily through apprenticeships, religious institutions, and guild systems. Learning was largely practical and workplace-oriented, with knowledge transmitted through observation, imitation, participation, and mentorship. These early apprenticeship models established many of the principles that continue to underpin vocational education and workplace learning, including experiential learning, situated learning, and the development of occupational competence through practice (Billett, 2001; Lave & Wenger, 1991).

During the Enlightenment of the seventeenth and eighteenth centuries, education became increasingly associated with reason, individual development, and social progress. Philosophers such as John Locke and Jean-Jacques Rousseau emphasised the role of experience, observation, and human development in learning. Although their work focused primarily on childhood education, their ideas influenced later educational theorists who viewed learning as an active process shaped by experience and interaction with the environment (Merriam & Bierema, 2014).

The nineteenth century witnessed significant expansion in adult education as a result of industrialisation, urbanisation, and democratic reform movements. Growing literacy demands and economic change led to the establishment of workers' education programs, mechanics' institutes, extension courses, public libraries, and community education initiatives across Europe, North America, and other regions. Adult education during this period was closely associated with social reform, citizenship, and self-improvement. Educational opportunities were increasingly viewed as essential for economic advancement and participation in democratic societies, although adult education remained largely practice-oriented and lacked a coherent theoretical foundation (Field, 2006; Jarvis, 2004).

The early twentieth century marked the emergence of adult education as a distinct field of study. Researchers and educators began to question whether adults learned differently from children and whether pedagogical methods developed for schools were appropriate for adult learners. These questions laid the groundwork for the development of formal adult learning theories and the establishment of adult education as a recognised area of academic inquiry (Merriam & Bierema, 2014).

One of the most influential contributors to the early development of adult learning theory was John Dewey. Dewey's philosophy of progressive education emphasised the central role of experience in learning and argued that education should be grounded in real-world problems and active inquiry. He rejected traditional approaches that positioned learners as passive recipients of information and instead proposed that knowledge is constructed through interaction with the environment, reflection, and problem-solving. Dewey's work provided a foundation for experiential learning theory and continues to influence adult education, vocational training, workplace learning, and professional development (Dewey, 1938; Kolb, 1984b).

The mid-twentieth century saw significant advances in psychology that shaped adult learning theory. Humanistic psychologists such as Abraham Maslow and Carl Rogers emphasised personal growth, self-actualisation, autonomy, and learner-centred education. Maslow's hierarchy of needs suggested that individuals are motivated by a progression of needs culminating in self-actualisation, while Rogers argued that meaningful learning occurs when learners are actively involved, perceive learning as relevant, and participate in supportive environments characterised by trust and respect (Maslow, 1954; Rogers, 1969). Humanistic psychology profoundly influenced adult education by promoting learner-centred practices, self-directed learning, and recognition of the importance of individual experience and motivation.

The most influential figure in modern adult learning theory is generally considered to be Malcolm Knowles, who popularised the concept of andragogy during the 1960s and 1970s. Knowles defined andragogy as the art and science of helping adults learn and proposed that adult learners differ from children in several important respects. According to Knowles, adults need to understand why they are learning something, bring extensive experience to learning situations, prefer learning that is immediately relevant to their lives, are motivated by practical problem-solving, and desire increasing levels of self-direction and autonomy (Knowles, 1980; Knowles et al., 2015). Although andragogy has been criticised for overstating differences between adult and child learners, it remains one of the most influential frameworks in adult education and vocational training internationally.

During the same period, adult learning theory was increasingly influenced by critical social theory. Paulo Freire challenged traditional educational models that treated learners as passive recipients of knowledge, which he described as the "banking model" of education. Instead, Freire advocated a dialogical approach in which learners and educators collaboratively construct knowledge through critical reflection and discussion. Central to his work was the concept of conscientisation, or the development of critical consciousness, through which learners become aware of social, political, and economic inequalities and are empowered to challenge oppressive structures (Paulo Freire, 1970). Freire's critical pedagogy has had a profound influence on literacy education, community development, social justice initiatives, and adult education worldwide.

The 1970s and 1980s also saw increasing interest in self-directed learning. Research by scholars such as Allen Tough demonstrated that adults frequently engage in independent learning projects outside formal educational settings. Adults were found to take responsibility for identifying learning needs, locating resources, implementing learning strategies, and evaluating outcomes across a wide range of personal and professional contexts (Tough, 1979). This research expanded understanding of informal learning and reinforced the view that much adult learning occurs beyond traditional educational institutions.

A major theoretical development occurred with the publication of David Kolb's Experiential Learning Theory in 1984. Building on the work of Dewey, Kurt Lewin, and Jean Piaget, Kolb conceptualised learning as a cyclical process involving concrete experience, reflective observation, abstract conceptualisation, and active experimentation. Learning, according to Kolb, occurs through the transformation of experience into knowledge through reflection and application (Kolb, 1984b). Experiential Learning Theory became highly influential in adult education, workplace learning, leadership development, and vocational training because it provided a practical framework for understanding how experience contributes to learning and professional competence.

Another major contribution emerged through Jack Mezirow's Transformative Learning Theory. Mezirow argued that adult learning can involve profound changes in the way individuals understand themselves and their world. Through critical reflection, adults may challenge previously held assumptions, beliefs, and values, resulting in what Mezirow termed perspective transformation (Mezirow, 1991). Transformative learning shifted attention from the acquisition of knowledge to changes in meaning, identity, and worldview and remains one of the most influential contemporary theories in adult education.

During the 1990s, sociocultural perspectives became increasingly influential in adult learning research. Jean Lave and Etienne Wenger proposed that learning is fundamentally social and occurs through participation in communities of practice. Their theory of situated learning argued that knowledge cannot be separated from the social and cultural contexts in which it is developed and used. Learning occurs through participation in authentic activities and interaction with others rather than solely through formal instruction (Lave & Wenger, 1991). These ideas have significantly influenced understandings of workplace learning, apprenticeships, professional development, and vocational education.

In the twenty-first century, adult learning theory has expanded to address the challenges and opportunities created by globalisation, digital technologies, demographic change, and rapidly evolving labour markets. New perspectives such as connectivism, networked learning, digital learning theory, and workplace learning research have emerged in response to technological and social transformation (Jarvis, 2004; Siemens, 2005b). Contemporary theories increasingly recognise that adult learning occurs across formal, non-formal, and informal settings and is shaped by social, cultural, technological, and economic influences. There is growing emphasis on lifelong learning, employability, digital literacy, sustainability, and continuous professional development as essential components of modern adult education systems.

Figure 1: Summary historical development of Adult Learning Theory.

The historical development of adult learning theory therefore demonstrates a progression from early philosophical ideas concerning experience, inquiry, and dialogue to sophisticated contemporary theories that recognise the complexity of adult learning across diverse contexts. Contributions from scholars such as Dewey, Knowles, Freire, Kolb, Mezirow, Lave, and Wenger have profoundly shaped contemporary understandings of how adults learn. Together, these theories provide the intellectual foundation for modern adult education, vocational education and training, workplace learning, and lifelong learning policy and practice throughout the world.

Characteristics of Adult Learners

Adult learners differ from children and adolescents in several important ways that influence how they approach, engage in, and benefit from learning. Although adults are a highly diverse group with varying backgrounds, experiences, motivations, and learning needs, adult learning theory identifies several common characteristics that distinguish adult learners from younger learners. Understanding these characteristics is essential for designing effective educational programs, vocational training, workplace learning initiatives, and professional development activities (Knowles et al., 2015; Merriam & Bierema, 2014).

One of the most widely recognised characteristics of adult learners is their increasing desire for self-direction and autonomy. As individuals mature, they develop a stronger sense of independence and personal responsibility. Adult learners generally prefer to participate actively in decisions about what they learn, how they learn, and how their learning is assessed. Rather than relying entirely on teacher-directed instruction, adults often seek opportunities to take ownership of their learning process, establish learning goals, and select learning strategies that align with their needs and circumstances (Knowles, 1980). This characteristic forms a central principle of andragogy and is reflected in contemporary approaches that emphasise learner-centred education.

A second defining characteristic of adult learners is the extensive range of life experiences they bring to learning environments. Unlike children, adults possess accumulated experiences from employment, family life, community involvement, cultural participation, and previous education. These experiences serve as valuable resources that influence how new information is interpreted and understood. Adults often relate new learning to prior knowledge and use personal experiences to evaluate ideas, solve problems, and construct meaning. Consequently, effective adult education frequently incorporates discussion, reflection, case studies, simulations, and workplace examples that enable learners to draw upon and share their experiences (Kolb, 1984b; Merriam & Bierema, 2014).

Adult learners are also characterised by a strong emphasis on relevance and practical application. Adults generally want to understand why they are learning something and how it will benefit them in their personal, professional, or social lives. Learning that appears disconnected from real-world needs is often perceived as less valuable. As a result, adults tend to be motivated by educational activities that address immediate problems, support career advancement, improve workplace performance, or contribute to personal goals. This preference for practical and applicable learning explains why workplace-based learning, vocational education, and problem-centred instructional approaches are often particularly effective with adult learners (Knowles et al., 2015).

A related characteristic is that adults are typically problem-centred rather than subject-centred in their orientation to learning. While younger learners may study subjects primarily because they form part of a curriculum, adults often approach learning with specific goals or challenges in mind. They seek knowledge and skills that help them solve workplace issues, manage life transitions, meet professional requirements, or address personal interests. Consequently, adult learning programs are often most effective when organised around authentic tasks, real-world scenarios, and practical outcomes rather than abstract disciplinary content alone (Knowles, 1980).

Another important characteristic is readiness to learn, which is often linked to developmental tasks and changing life circumstances. Adults frequently become motivated to learn when faced with new roles, responsibilities, or challenges. Changes such as entering employment, changing careers, becoming a parent, assuming leadership responsibilities, or adapting to technological innovations can create a need for new knowledge and skills. Adult learning is therefore often closely connected to immediate social, occupational, or personal demands rather than predetermined educational sequences (Merriam & Bierema, 2014).

Adult learners are generally motivated by a combination of intrinsic and extrinsic factors. Extrinsic motivations may include obtaining qualifications, improving employment prospects, increasing income, meeting regulatory requirements, or gaining promotion opportunities. Intrinsic motivations often include personal satisfaction, self-improvement, intellectual curiosity, increased confidence, and the desire for personal growth. Humanistic theorists such as (Maslow, 1954) and (Rogers, 1969) emphasised the importance of intrinsic motivation, arguing that adults possess an inherent desire for growth, competence, and self-actualisation. Research consistently suggests that intrinsic motivation can be particularly powerful in sustaining engagement and persistence in adult learning activities.

Many adult learners demonstrate a strong capacity for self-directed learning. (Tough, 1979) found that adults frequently initiate, organise, and evaluate their own learning projects outside formal educational settings. Self-directed learners identify learning needs, locate resources, select learning strategies, and assess their progress independently. Advances in digital technologies have further expanded opportunities for self-directed learning by providing access to online courses, digital libraries, professional networks, webinars, podcasts, and other educational resources. However, the degree of self-direction varies among individuals and may depend on factors such as prior educational experience, confidence, and learning context.

Adult learners are also influenced by the multiple roles and responsibilities they manage in their daily lives. Unlike many younger learners, adults often balance education with employment, family obligations, financial commitments, and community responsibilities. These competing demands can affect the time, energy, and resources available for learning. Consequently, flexibility in course delivery, scheduling, assessment methods, and learning support services is often critical to successful participation and completion (Jarvis, 2004).

Another significant characteristic is the diversity that exists among adult learners. Adults vary considerably in age, culture, language, educational background, employment experience, socioeconomic status, learning preferences, and personal goals. This diversity enriches learning environments by introducing multiple perspectives and experiences, but it also requires educators to adopt inclusive and adaptable teaching approaches. Contemporary adult education increasingly recognises the importance of culturally responsive pedagogy, accessibility, and equitable participation in supporting diverse learner populations (Merriam & Bierema, 2014).

Adult learners are often capable of engaging in critical reflection, a process through which individuals examine assumptions, beliefs, values, and experiences. (Mezirow, 1991) argued that critical reflection is central to transformative learning, enabling adults to reassess existing perspectives and develop new ways of understanding themselves and the world. Through reflection, adults can challenge established beliefs, evaluate evidence, and make informed decisions about future actions. This capacity for reflective thinking is particularly important in professional practice, leadership development, and lifelong learning.

Social interaction also plays an important role in adult learning. Although adult learners are often characterised as independent, learning frequently occurs through collaboration, discussion, mentoring, and participation in communities of practice. (Lave & Wenger, 1991) argued that learning is fundamentally social and occurs through participation in authentic activities within specific social and cultural contexts. Workplace learning, professional networks, peer support groups, and collaborative learning environments all demonstrate the importance of social engagement in adult learning processes.

Contemporary adult learners increasingly require digital literacy and adaptability. Rapid technological change means that adults must continually update their knowledge and skills throughout their lives. The rise of online learning, artificial intelligence, digital collaboration tools, and virtual workplaces has increased the importance of technological competence. Consequently, adaptability, continuous learning, and lifelong skill development have become defining characteristics of successful adult learners in the twenty-first century (OECD, 2021i).

Overall, adult learners are characterised by self-direction, accumulated experience, practical orientation, readiness to learn, diverse motivations, self-directed learning capabilities, multiple life responsibilities, diversity, reflective capacity, social engagement, and adaptability. These characteristics distinguish adult learners from younger learners and provide the foundation for adult learning theories such as andragogy, experiential learning, transformative learning, self-directed learning, and situated learning. Understanding these characteristics enables educators, trainers, and organisations to design learning experiences that are relevant, engaging, and effective for adult learners across educational, vocational, workplace, and community contexts.

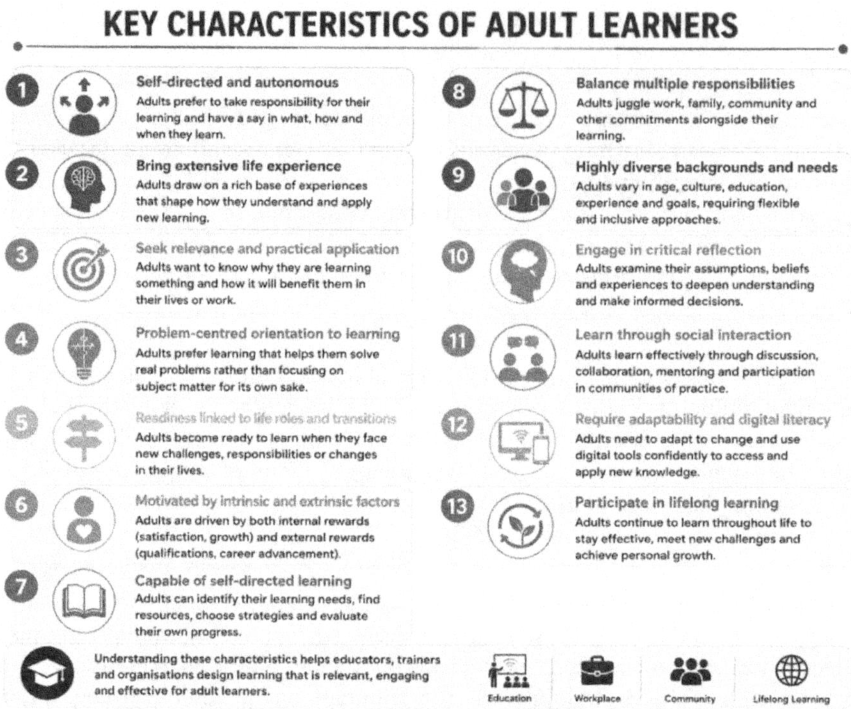

Figure 2: Summary of characteristics of adult learners.

Adult learning theories seek to explain why adults learn differently from children and adolescents, while teaching practices translate these theoretical insights into approaches for designing, delivering, and evaluating educational experiences. The characteristics commonly associated with adult learners have significantly influenced the development of adult learning theories and have shaped contemporary approaches to adult education, vocational education and training, workplace learning, and professional development. Collectively, these theories advocate educational practices that recognise adults as experienced, motivated, autonomous, and goal-oriented learners whose learning needs differ from those of younger students.

One of the most significant contributions of adult learning theory has been the movement away from traditional teacher-centred approaches toward learner-centred education. Knowles' theory of andragogy emphasises that adults generally desire greater autonomy and self-direction in their learning than children (Conway, 2022; Loeng, 2023; Purwati et al., 2022). As individuals mature, they increasingly seek opportunities to participate in decisions about what they learn, how they learn, and how their learning is assessed (Blaschke & Hase, 2019; Loeng, 2023). Knowles described self-directed learning as a process in which individuals take initiative in diagnosing needs, formulating goals, and evaluating outcomes (Loeng, 2020; Samaroo et al., 2013). Consequently, contemporary teaching practices increasingly position educators as facilitators of learning rather than transmitters of knowledge (Akintolu & Letseka, 2021; Caruso et al., 2020; Mbatha, 2024). McGrath noted that acknowledging adult learners as self-directed beings shifts the focus from teacher-centred to learner-centred approaches (Mbatha, 2024), while andragogic teaching reimagines the teacher as a mentoring facilitator guiding students toward collaborative assignments (Caruso et al., 2020). Learner-centred approaches encourage collaboration, learner choice, self-assessment, goal setting, and independent inquiry (Blaschke & Hase, 2019; Caruso et al., 2020), enabling adults to assume greater responsibility for their own learning processes (Ghamrawi et al., 2025; Loeng, 2020).

The recognition that adults bring extensive life and work experiences to learning environments is a fundamental principle in adult learning theory. Experiential learning theorists argue that knowledge is constructed through reflection on experience (Camps & Blancafort-Masriera, 2019; Darzón et al., 2021; Tinoco et al., 2022; Tolosa-Guzman et al., 2021). Kolb proposed that learning occurs through a cycle involving concrete experience, reflective observation, abstract conceptualisation, and active experimentation (Camps & Blancafort-Masriera, 2019; Darzón et al., 2021; Tinoco et al., 2022; Tolosa-Guzman et al., 2021). This cyclical model defines learning as "the process whereby knowledge is created through the transformation of experience" (Darzón et al., 2021; Tinoco et al., 2022). Billett highlighted the importance of workplace experience in developing vocational competence (Camps & Torres-Carballo, 2010).

These perspectives suggest that adult learners should not be treated as blank slates; rather, their prior experiences constitute valuable resources for learning (Ledezma-Ramírez, 2024; Mamani et al., 2024). Effective teaching practices incorporate learners' experiences through discussion, case studies, reflective activities, simulations, workplace projects, and collaborative problem-solving (Bazán et al., 2024; Darzón et al., 2021; Vivar et al., 2023). Learning thus becomes a process of integrating new knowledge with existing understanding rather than simply acquiring information (Ledezma-Ramírez, 2024).

Adult learners' preference for relevant and practical learning is well-established in andragogical theory. Knowles identified that adults' orientation to learning is problem-centred rather than subject-centred, and that they engage more readily when learning is directly applicable to their circumstances (Grus et al., 2025; Rahman et al., 2022). Consequently, educational programs for adults frequently incorporate workplace scenarios, project-based learning, case studies, simulations, and practical tasks (Coelho et al., 2024; Nesamany & Subasini, 2024). Problem-based learning and simulation provide opportunities for learners to apply knowledge in "real-world" environments (Weissova et al., 2024), while case studies help bridge theory and practice through analysis of real-life situations (Coelho et al., 2024).

Closely related is the concept of readiness to learn, one of Knowles's six assumptions of andragogy (Mukhalalati & Taylor, 2019; Rahman et al., 2022). Adults' learning needs often emerge from life transitions, changing responsibilities, or new occupational demands (Corbetta et al., 2019; Grus et al., 2025). This understanding has driven the development of lifelong learning and continuing professional development initiatives that provide flexible, timely learning opportunities (Corbetta et al., 2019; Tong-Battung, 2021). Effective CPD stresses learner-centred approaches that are responsive to individual needs and accessible when most relevant to learners' immediate goals (Hanshaw & Hanson, 2019; Tong-Battung, 2021).

Motivation is a central consideration in adult learning theory. Humanistic theorists such as Maslow and Rogers argued that adults possess an inherent desire for growth and self-actualisation (Moll, 2023, 2024; Wang & Sarbo, 2004). Knowles was strongly influenced by Rogers' view that formal education denied self-actualisation (Moll, 2023, 2024), and this humanistic philosophy underpins andragogical approaches that position learners as autonomous agents (Dean & Parson, 2021; Portugal, 2015). While adults may be motivated by external factors, intrinsic motivations, including personal satisfaction and intellectual stimulation, are equally important (Nihishova & Kryvonosova, 2020; Storey & Wang, 2016). Lindeman identified that adults are motivated to learn as they experience needs and interests that learning will satisfy (Storey & Wang, 2016). This understanding has influenced practices focused on creating supportive, non-threatening learning environments and providing meaningful feedback (Buckingham, 2000; Lipscomb & Ishmael, 2009).

The capacity for self-directed learning has profoundly influenced adult learning theory. Tough's research demonstrated that adults frequently engage in independent learning projects outside formal settings (Loeng, 2020; Morris, 2023). Tough concluded that adults spend remarkable amounts of time on such projects for acquiring and maintaining knowledge (Loeng, 2020). This contributed to self-directed learning theory, which views learners as active participants capable of identifying needs, locating resources, and evaluating outcomes (Loeng, 2020; Morris, 2023. Self-directed learning is grounded in humanistic assumptions (Loeng, 2020), and contemporary practices increasingly support it through flexible and online learning environments (Morris, 2023; Sufirmansyah, 2019).

Transformative Learning Theory further extends understanding of adult learning by recognising adults' capacity for critical reflection. Mezirow argued that adults can critically examine existing assumptions, beliefs, and values, leading to significant changes in perspective and understanding. Learning is therefore not limited to the acquisition of knowledge and skills but may involve profound personal and professional transformation. Teaching practices informed by transformative learning theory often incorporate reflective journals, critical discussions, ethical dilemmas, debates, and opportunities for learners to challenge their own assumptions.

Although adult learners are frequently characterised as independent, sociocultural theories emphasise that learning remains fundamentally social. Lave and Wenger proposed that learning occurs through participation in communities of practice, where individuals develop knowledge and skills through engagement with others in authentic social and professional contexts (Dahl, 2020; Sayer, 2013; Wastesson, 2024). This situated learning perspective posits that learning is "an integral and inseparable aspect of social practice" rather than merely individual cognition (Barry et al., 2016; Dahl, 2020). Knowledge is acquired through legitimate peripheral participation, whereby newcomers gradually move toward full participation through social interaction (Harris, 2008; Sayer, 2013; Usman et al., 2021). This perspective highlights the value of collaborative learning, peer mentoring, workplace learning, and professional networks (Barry et al., 2016; Endersby et al., 2019; Sayer, 2013). Learning is viewed as participation in shared practices and communities that shape knowledge, identity, and professional competence (Sayer, 2013; Usman et al., 2021; Wastesson, 2024).

The diversity of adult learners necessitates inclusive and responsive pedagogical approaches. Contemporary adult education must address differences in cultural background, language, educational experience, and learning preferences through differentiated instruction, culturally responsive pedagogy, and flexible assessment (Halder et al., 2024; Kavashev, 2024; Panda, 2020). Effective practices include multiple pathways for participation, accessible learning designs, and recognition that diverse perspectives enrich learning environments (Brodén et al., 2024; Scoppio & Covell, 2016; Zhang & Wang, 2025). Educators are called upon to implement strategies addressing equity, diversity, and inclusion as foundational elements of quality adult education (Brodén et al., 2024; Kayi, 2024).

Technological change and lifelong learning imperatives have further expanded adult learning theory. Digital technologies, AI, and online platforms create new opportunities requiring educators to support digital literacy and continuous learning (Jaldemark et al., 2021; Mozelius et al., 2024; Shohel et al., 2024). Contemporary teaching incorporates blended learning models, virtual collaboration tools, and multimedia resources (Kayi, 2024; Scoppio & Covell, 2016; Shohel et al., 2024). Learning increasingly occurs through networks and digital connections, demanding that educators prepare learners for adaptability across their lives and careers (Jaldemark et al., 2021; Lock et al., 2021; Mozelius et al., 2024).

Taken together, these theoretical perspectives suggest that effective adult education should be learner-centred, experience-based, relevant, flexible, reflective, collaborative, inclusive, and responsive to technological change. Adult learning theories consistently emphasise the importance of recognising learners' experiences, supporting self-direction, promoting critical reflection, fostering social interaction, and facilitating the practical application of learning. These principles have shaped contemporary educational practice across adult education, vocational education and training, workplace learning, and professional development.

The characteristics of adult learners therefore underpin virtually all major adult learning theories, including andragogy, experiential learning, transformative learning, self-directed learning, critical pedagogy, and situated learning. Collectively, these theories suggest that adults learn most effectively when learning is meaningful, relevant, experience-based, collaborative, reflective, and immediately applicable. Consequently, effective teaching practice must be designed around the distinctive characteristics and needs of adult learners rather than relying solely on approaches originally developed for children and adolescents. In doing so, educators can create learning environments that maximise engagement, support achievement, and promote lifelong learning in an increasingly complex and rapidly changing world.

Lifelong Learning and Continuing Education

Lifelong learning and continuing education have become central concepts in contemporary education, workforce development, and social policy. In an increasingly interconnected and rapidly changing world, individuals must continually acquire new knowledge, skills, and competencies to adapt to technological innovation, economic transformation, demographic change, and evolving social expectations (Chen & Murphy, 2022; Sung et al., 2022). Consequently, learning is no longer viewed as an activity confined to childhood or early adulthood but as a continuous process that extends throughout an individual's life (Egetenmeyer & Fedeli, 2017a; Pata et al., 2021). The concept has existed under various names, continuing education, recurrent education, adult education, but has gained particular policy prominence since the 1990s (Abukari, 2005; Regmi, 2020). International organizations such as the EU, UNESCO, and OECD have been instrumental in promoting lifelong learning as essential for economic competitiveness, social cohesion, and active citizenship (Guimarães, 2017; Jõgi & Gross, 2009; Milana & Holford, 2014). However, implementation varies significantly across national contexts, shaped by political, economic, and historical factors (Koulaouzides & Romano, 2022; Prokou, 2008). Critics note that policy discourse increasingly emphasizes economic utility and employability over humanistic and emancipatory aims of education (Guimarães, 2017; Koulaouzides & Romano, 2022; Singh et al., 2020).

The concept of lifelong learning emerged during the twentieth century as educators and policymakers began to recognise that formal schooling alone could not adequately prepare individuals for the complexities of modern society. One of the earliest advocates of lifelong education was Hutchins (1968), who argued that education should not be regarded as preparation for life but as an integral part of life itself. He proposed the idea of a "learning society" in which opportunities for learning would be available throughout adulthood and embedded within social institutions. This vision challenged traditional educational models that separated education from the remainder of an individual's life course and laid important foundations for later lifelong learning policies.

The modern concept of lifelong learning gained international prominence through the work of UNESCO. The influential Faure Report, *Learning to Be* (Faure et al., 1972), argued that education should be organised around the principle of lifelong learning and that societies should provide continuous opportunities for individuals to learn, develop, and participate fully in social and economic life. The report emphasised personal development, democratic participation, social justice, and human fulfilment as key purposes of lifelong learning. Rather than viewing education solely as a mechanism for workforce preparation, the Faure Report positioned learning as a lifelong process essential to individual and societal development.

This vision was further developed in the Delors Report, *Learning: The Treasure Within* (Delors et al., 1996), which remains one of the most influential documents in lifelong learning policy. The report identified four fundamental pillars of education: learning to know, learning to do, learning to live together, and learning to be. These pillars emphasise that lifelong learning should support not only the acquisition of knowledge and occupational skills but also personal growth, social cohesion, intercultural understanding, and active citizenship. The Delors Report significantly broadened international understanding of lifelong learning by highlighting its humanistic and social dimensions alongside its economic benefits.

Theoretical perspectives on lifelong learning also evolved during this period. Lengrand (1975) argued that education should be viewed as a continuous process that accompanies individuals throughout their lives and across diverse contexts. He emphasised that learning occurs not only within formal educational institutions but also through work, family life, community participation, and cultural engagement. This perspective contributed to contemporary understandings of lifelong learning as encompassing formal, non-formal, and informal learning experiences.

Today, lifelong learning is generally understood as the ongoing pursuit of knowledge, skills, values, and competencies throughout life for personal, social, civic, and professional purposes. Contemporary frameworks recognise three interconnected forms of learning. Formal learning occurs within structured educational institutions such as schools, universities, and vocational education and training providers and usually leads to recognised qualifications. Non-formal learning occurs through organised educational activities outside formal systems, including professional development programs, workplace training, community education, and short courses. Informal learning occurs through everyday experiences, social interactions, self-directed study, problem-solving, and participation in work and community life (Merriam & Bierema, 2014). Together, these forms of learning contribute to the development of individuals across the lifespan.

Continuing education represents a significant component of lifelong learning and typically refers to structured learning undertaken after initial education has been completed. Continuing education may include professional development, postgraduate study, vocational upskilling, industry certification, workshops, conferences, and retraining programs. While lifelong learning encompasses all learning throughout life, continuing education focuses specifically on organised educational activities that enable individuals to maintain, update, or expand their knowledge and skills (Jarvis, 2004). In many professions, continuing education has become essential for maintaining professional competence and responding to changing regulatory, technological, and industry requirements.

The increasing importance of lifelong learning is closely linked to the emergence of the knowledge economy. Economic activity is increasingly dependent upon knowledge, innovation, technology, and the capacity of workers to adapt to change. Skills that were once sufficient for an entire career may become obsolete within a relatively short period due to technological advances and organisational transformation. Field (2006) argues that lifelong learning has become a central policy response to these changes, with governments and employers encouraging individuals to continually develop their capabilities throughout their working lives. Similarly, (OECD, 2021f) emphasises that lifelong learning is critical for employability, workforce resilience, productivity, and economic competitiveness in rapidly changing labour markets.

Workplace learning has become one of the most important contexts for lifelong learning. Billett (2001) argues that much learning occurs through participation in everyday work activities and that workplaces function as important sites of knowledge development and skill acquisition. Similarly, (Eraut, 2004) highlights the significance of informal learning in professional practice, noting that a substantial proportion of workplace learning occurs outside formal training programs through problem-solving, collaboration, observation, and experience. These perspectives have expanded understandings of continuing education beyond classroom-based instruction to include learning embedded within work itself.

Lifelong learning also contributes significantly to personal development and social inclusion. Participation in learning can enhance self-confidence, wellbeing, health literacy, civic engagement, and social connectedness. Educational opportunities enable individuals to pursue personal interests, develop new skills, engage with their communities, and participate more effectively in democratic processes. The humanistic tradition underpinning lifelong learning emphasises that education should support the development of the whole person rather than focusing solely on economic outcomes (Delors et al., 1996; Faure et al., 1972). Consequently, lifelong learning is increasingly viewed as a mechanism for promoting social cohesion, cultural participation, and individual fulfilment.

Contemporary adult learning theories provide important foundations for understanding lifelong learning. Knowles' theory of andragogy highlights the importance of self-direction, autonomy, and learner responsibility, all of which are essential for sustained learning throughout adulthood (Knowles, 1980). Tough (1979) research demonstrated that adults frequently initiate and manage their own learning projects, while Mezirow (1991) Transformative Learning Theory emphasises the role of critical reflection in enabling adults to adapt to changing circumstances and develop new perspectives. These theories collectively suggest that lifelong learning depends not only on access to educational opportunities but also on the development of learners' capacity to direct, evaluate, and sustain their own learning.

Technological change has further transformed lifelong learning opportunities. Digital technologies, online learning platforms, open educational resources, micro-credentials, webinars, virtual classrooms, and massive open online courses have expanded access to education for millions of adults worldwide. Siemens (2005b) theory of connectivism reflects this changing landscape by suggesting that learning increasingly occurs through digital networks, online communities, and access to distributed sources of information. The growth of technology-enhanced learning has enabled adults to engage in flexible, self-paced learning while balancing employment, family responsibilities, and other commitments.

Despite its many benefits, lifelong learning remains unevenly distributed across populations. Participation rates are often influenced by educational background, socioeconomic status, age, employment conditions, geographical location, and digital access. Alheit and Dausien (2002) note that lifelong learning possesses a "double face": while it can empower individuals and expand opportunities, it can also reinforce inequalities when access to learning is uneven. Policymakers therefore face ongoing challenges in ensuring that lifelong learning opportunities are accessible, inclusive, and equitable.

In response to these challenges, governments and international organisations have increasingly developed policies designed to support learning throughout life. Qualifications frameworks, recognition of prior learning systems, flexible pathways between vocational and higher education, workplace learning initiatives, and lifelong learning strategies seek to create educational systems that accommodate learners at different stages of life. UNESCO (2021b) argues that future education systems must move beyond traditional models of initial education and embrace lifelong learning as a guiding principle for social, economic, and environmental sustainability.

Lifelong learning and continuing education represent a fundamental transformation in how education is understood. Rather than viewing learning as a finite process completed during youth, contemporary perspectives recognise that learning is continuous, lifelong, and essential for personal fulfilment, active citizenship, professional competence, and social participation. As technological, economic, and social change continues to accelerate, lifelong learning is likely to become even more important for individuals, organisations, and societies. Consequently, lifelong learning and continuing education have become foundational principles underpinning adult education, vocational education and training, workforce development, and educational policy throughout the world.

Formal, Non-Formal, and Informal Learning

One of the most important concepts within adult education and lifelong learning is the recognition that learning occurs in many different settings and through a variety of experiences. Traditionally, education was often associated exclusively with schools, colleges, and universities. However, lifelong learning scholars increasingly recognise that individuals continue to learn throughout their lives in workplaces, communities, families, and everyday activities. To describe these different forms of learning, educational researchers commonly distinguish between formal learning, non-formal learning, and informal learning (Coombs & Ahmed, 1974).

These categories help educators, policymakers, and employers understand the diverse ways in which knowledge, skills, attitudes, and competencies are acquired across the lifespan. Although the boundaries between the categories are not always clear-cut, they provide a useful framework for understanding lifelong learning.

Formal Learning

Formal learning refers to structured and organised learning that takes place within recognised educational institutions and leads to officially recognised qualifications or credentials. It is typically characterised by a planned curriculum, defined learning outcomes, qualified instructors, assessment processes, and certification upon successful completion.

Examples of formal learning include primary and secondary schooling, university degrees, vocational education and training qualifications, apprenticeships, diplomas, professional certifications, and postgraduate studies. Formal learning is usually delivered within institutions that have legal authority to award qualifications recognised by governments, industry bodies, or professional associations.

Formal learning is often highly structured, with learners following prescribed programs of study and progressing through clearly defined educational pathways. Assessment plays an important role in determining whether learners have achieved specified learning outcomes. Because of its structured nature, formal learning provides a recognised mechanism for credentialing knowledge and skills and often serves as a pathway to employment, career advancement, and professional recognition.

Despite its importance, formal learning represents only one component of an individual's overall learning experience. Many valuable skills and competencies are developed outside formal educational institutions, leading researchers to recognise the importance of non-formal and informal learning.

Formal learning occurs in virtually every country and remains the primary mechanism through which societies educate citizens, develop workforce skills, transmit cultural knowledge, and prepare individuals for employment and civic participation. Although formal learning is characterised by structured curricula, recognised qualifications, assessment systems, and accredited educational institutions, its organisation, delivery, and objectives vary considerably between countries and regions (Huang, 2024; ÜZÜM & Duruhan, 2020). These differences reflect historical traditions, cultural values, economic priorities, political systems, and labour market needs (Below et al., 2013; ÜZÜM & Duruhan, 2020).

One of the most familiar examples of formal learning is compulsory school education. Most countries provide primary and secondary education through nationally regulated systems that follow prescribed curricula and lead to recognised qualifications. However, the structure of schooling differs substantially. In contrast to systems like Australia's, Germany's education system separates students into different academic and vocational pathways at an early age, typically around age ten, through differentiated secondary schools (Maaz et al., 2008; Puhani & Weber, 2005). Students may attend Gymnasium, Realschule, Hauptschule, or other pathways depending on their academic performance and career aspirations (Güllüpınar, 2012; Huang, 2024; ÜZÜM & Duruhan, 2020). This reflects Germany's strong tradition of vocational specialisation and workforce preparation, exemplified by its dual system combining workplace training with vocational schooling (Chürr, 2015; Huang, 2024; ÜZÜM & Duruhan, 2020).

Higher education provides another important example of formal learning. Universities throughout the world offer structured degree programs leading to recognised qualifications. However, the organisation and accessibility of higher education vary significantly. In countries such as Germany, public universities have historically provided free education, reflecting a view of higher education as a public good (Below et al., 2013). These differences influence participation rates, educational equity, and access to advanced learning opportunities (Below et al., 2013; Huang, 2024).

Vocational education and training (VET) systems demonstrate important variations in formal learning across nations. Germany, Switzerland, and Austria are internationally recognised for their dual apprenticeship systems, which combine classroom-based instruction with substantial workplace training (Brunetti & Corsini, 2019; Kuczera, 2017). In these countries, apprentices typically spend roughly three-quarters of their time training in companies, with the remainder in vocational schools (Kuczera, 2017). These systems facilitate effective school-to-work transitions and are associated with lower youth unemployment (Brunetti & Corsini, 2019; Kuczera, 2017).

Australia's VET system differs structurally, being competency-based and organised around nationally endorsed training packages comprising units of competency (Schmidt et al., 2025; Smith, 2010; Wheelahan & Carter, 2001). Qualifications range from Certificates through Diplomas and are delivered by Registered Training Organisations (RTOs) and Technical and Further Education (TAFE) institutes (Jones, 2006; Polesel, 2010; Smith, 2010). Recognition of Prior Learning (RPL) is a significant feature, though state agencies have shown reluctance to fund it due to concerns about assessment processes (Dunn, 2004). Learners are assessed against industry-defined performance standards rather than comparative examinations (Hager & Smith, 2004; Smith, 2010). However, critics have noted that the competency-based approach can result in narrow pedagogy and may disadvantage some learner groups (Schmidt, 2020; Smith, 2010).

Germany, Switzerland, and Austria are internationally recognised for their dual apprenticeship systems, which combine classroom-based instruction with substantial workplace training (Graf, 2013; Moretti et al., 2017; Vujkov, 2024). Students typically divide their time between vocational schools and employers—for instance, apprentices in Switzerland spend three to four days per week in the firm and the remainder in vocational school (Graf, 2013). These systems are highly integrated into national labour markets and are widely regarded as models of effective school-to-work transition (Eichhörst et al., 2013; Moretti et al., 2017). Countries maintaining substantial dual apprenticeship systems exhibit smoother transitions from school to work and lower youth unemployment (Eichhörst et al., 2013; OECD, 2019).

Australia also maintains a well-developed VET system, but its structure differs significantly. Australian VET is competency-based, organised around nationally endorsed training packages and industry-defined performance standards rather than comparative academic examinations (Deißinger, 2012). Recognition of Prior Learning (RPL) is a significant feature, enabling formal recognition for skills acquired through work and life experience (Deißinger, 2012). This competency-based approach contrasts with the input-steered, curriculum-based qualifications characteristic of German-speaking countries (Deißinger, 2012).

In many East Asian countries, formal learning is shaped by cultural values emphasising academic achievement and examination performance. In China, the Gaokao heavily influences higher education admission, while South Korea and Japan similarly emphasise competitive examinations for university access (Vicente et al., 2024). Japan's education system relies on standardised national exams, and both Japan and Singapore achieve high PISA scores, though Japan faces concerns about rote learning and student stress (Vicente et al., 2024). Educational success is closely linked to social mobility, resulting in highly competitive environments and extensive participation in supplementary tutoring and private education (Kalashnyk et al., 2024).

Singapore's formal education system combines strong academic standards with government planning aligned to economic development (Liu, 2022; Vicente et al., 2024). The system emphasises mathematics, science, technology, and bilingualism (Singh, 2025; Vicente et al., 2024). Singapore's bilingual education policy requires students to learn English alongside their mother tongue (Singh, 2025; Vicente et al., 2024). More recently, Singapore has expanded lifelong learning through SkillsFuture, which encourages citizens to continually update their skills throughout their careers (Costa et al., 2020; Z. Y. Lim et al., 2024; Liu, 2022). SkillsFuture promotes "a mindset of continually striving towards greater excellence" so individuals "can own a better future" (Costa et al., 2020). From 2015 to 2020, Singapore spent an average of S$1 billion annually on continuing education and training (Liu, 2022). This reflects a broader shift toward viewing formal education as part of a lifelong learning continuum (Z. Y. Lim et al., 2024; Sung et al., 2022).

In developing countries, formal learning systems often face challenges associated with access, infrastructure, funding, and educational equity. While many countries have expanded participation in primary and secondary education, significant disparities may remain between urban and rural populations, genders, socioeconomic groups, and linguistic communities. In India, for instance, access, equity, and quality have been persistent objectives of elementary education reform, with policies such as DPEP, SSA, and RTE seeking to universalise schooling while addressing inequalities across states and populations (Huda et al., 2023). International organisations such as UNESCO continue to support efforts aimed at improving educational access, quality, and inclusion worldwide.

Religious and cultural traditions also influence formal learning in some regions. Islamic education systems in countries such as Saudi Arabia, Egypt, and Indonesia integrate religious studies into formal curricula. In Saudi Arabia, core subjects are dominated by Islamic sciences such as tajwid, tafsir, hadith, fiqh, and tawhid (Yoyo et al., 2023), while in Indonesia, Islamic boarding schools and madrasas have long served as models of religious education, with some integrating Saudi-oriented curricula alongside the national education curriculum (Febriansyah & El-Alami, 2021; Jamhari & Asrori, 2022; Siswanto & Ngadri, 2022). Egypt's Al-Azhar University has historically been a centre of Islamic education with branches and affiliated schools operating internationally, including in Indonesia (Cahyo et al., 2023; Mas'ud et al., 2019). In India, formal education includes a diverse mixture of government schools, private schools, technical institutes, and universities, with reform policies iteratively addressing the balance between access, equity, and quality across this vast system (Huda et al., 2023). Educational policy often seeks to balance traditional knowledge systems with the demands of economic modernisation and technological development.

Several key differences can be observed in formal learning systems around the world. Educational structures and funding models vary significantly, with countries differing in how they distribute funding across sectors and in the division between public and private sources (Ho et al., 2026; OECD, 2023d). Assessment methods range from traditional examinations to competency-based approaches, with some systems incorporating recognition of non-formal and informal learning (Jayaraman, 2021; OECD, 2021g). Vocational pathways differ considerably, particularly regarding integration between education and employment, as seen in Germany's dual system (Zaber et al., 2019) and broader TVET reforms globally (Secretariat, 2022). Curriculum priorities vary, with an emergent global curriculum encompassing citizenship education, technological skills, and lifelong learning (OECD, 2023). Access and equity remain significant concerns, with participation rates varying among population groups (Spiel et al., n.d.; , OECD, 2023). The role of government and industry in curriculum design differs, from centralized to decentralized models (Ding, 2023; , ("Unleashing the potential: transforming technical and vocational education and training", 2015; .

Despite these differences, formal learning systems share common characteristics: they are structured, curriculum-based, and lead to recognised qualifications (Secretariat, 2022). Increasingly, systems are adapting to support lifelong learning, digital education, and flexible delivery (OECD, 2021g, 2023d; Secretariat, 2022). Contemporary education policies globally advocate shifts "towards flexible and non-formal education, towards digital literacy" (OECD, 2023d), while qualification frameworks facilitate transparency and mobility (Secretariat, 2022). As global economies evolve, formal learning is becoming more diverse and responsive to changing workforce needs (Ho et al., 2026; OECD, 2021g).

Non-Formal Learning

Non-formal learning refers to organised and purposeful learning activities that occur outside the formal education system. Like formal learning, non-formal learning is planned and structured, but it does not usually result in a formally recognised qualification.

Non-formal learning can occur in workplaces, community organisations, professional associations, religious institutions, sporting clubs, and other social settings. Examples include workplace training programs, professional development workshops, industry seminars, short courses, community education classes, leadership programs, first aid training, mentoring schemes, and continuing professional education activities.

Non-formal learning is often designed to meet specific learning needs and may be more flexible than formal education. Programs are frequently shorter, more targeted, and more responsive to immediate personal, occupational, or community requirements. Participants typically enrol voluntarily because they wish to acquire particular knowledge or skills rather than pursue a formal qualification.

For example, an engineer attending a one-day workshop on new construction technologies, a nurse completing continuing professional development activities, or a manager participating in a leadership development program are all engaging in non-formal learning. Although these activities may not lead to formal qualifications, they contribute significantly to professional competence and lifelong learning.

Informal Learning

Informal learning refers to learning that occurs naturally through everyday life experiences rather than through structured educational activities. It is generally unplanned, unintentional, and often unconscious. Individuals may not even recognise that learning is taking place until they reflect upon their experiences.

Informal learning occurs continuously through work activities, family interactions, social relationships, hobbies, volunteering, problem-solving, observation, experimentation, and participation in community life. Examples include learning to use a new software application by experimenting with it, developing communication skills through workplace interactions, acquiring parenting skills through experience, or learning about history through personal reading and travel.

Research suggests that a substantial proportion of adult learning occurs informally. Eraut (2004) argued that much workplace learning takes place outside formal training programs through observation, collaboration, trial and error, problem-solving, and participation in daily work activities. Similarly, (Tough, 1979) found that adults frequently undertake self-directed learning projects independently, often without enrolling in formal educational programs.

Informal learning is particularly important because it allows individuals to continuously adapt to changing circumstances and acquire practical knowledge in authentic contexts. However, because it is often undocumented and uncertified, informal learning may be overlooked despite its significant contribution to competence and expertise.

Comparison of Formal, Non-Formal, and Informal Learning

The three forms of learning differ primarily in their degree of structure, organisation, and recognition.

Table 2: Differences in forms of learning.

Characteristic	Formal Learning	Non-Formal Learning	Informal Learning
Structured and planned	Yes	Yes	Usually No
Occurs in educational institutions	Usually	Not necessarily	No
Leads to recognised qualifications	Yes	Usually No	No
Assessment included	Yes	Sometimes	Rarely
Curriculum defined	Yes	Usually	No
Voluntary participation	Sometimes	Usually	Generally
Examples	Degrees, diplomas, apprenticeships	Workshops, professional development, community courses	Workplace experience, self-study, hobbies, daily activities

Recognition of Prior Learning

The distinction between formal, non-formal, and informal learning has important implications for educational policy and workforce development. Many individuals acquire substantial knowledge and skills through work experience, volunteering, military service, community involvement, and self-directed learning (Singh, 2015b, 2015c). Recognition of Prior Learning (RPL) processes seek to identify, assess, and formally recognise competencies acquired through non-formal and informal learning experiences (Guimarães & Mikulec, 2020b; Meghnagi & Tuccio, 2022).

RPL reflects the understanding that valuable learning occurs in many contexts and that competence should not be judged solely on the basis of participation in formal education (Sanséau & Ansart, 2013; Singh, 2015g). This principle has become increasingly important within vocational education and training systems worldwide (Singh, 2015h). RPL has been implemented under various names across countries, including APEL in the UK, PLAR in Canada, and VAE in France, with objectives ranging from improving employability to widening access to qualifications (Sanséau & Ansart, 2013; Singh, 2015a). The European Union has been particularly influential, issuing recommendations urging member states to establish validation arrangements for non-formal and informal learning (Guimarães & Mikulec, 2020b; Meghnagi & Tuccio, 2022). In developing countries, where the majority of workers acquire skills through informal training, RPL serves as a critical tool for formalising competencies and enabling workforce participation (Singh, 2015b, 2015e).

The Role of Formal, Non-Formal, and Informal Learning in Lifelong Learning

Modern lifelong learning frameworks recognise that individuals move continuously between formal, non-formal, and informal learning environments throughout their lives (Ioannou, 2024; Owusu-Agyeman, 2024; Singh, 2015e). A person may complete a university degree through formal learning, attend professional development workshops through non-formal learning, and acquire additional skills through workplace experience and self-directed study as forms of informal learning (Owusu-Agyeman, 2024). Formal learning occurs in educational institutions and leads to recognised qualifications, while non-formal learning represents systematic educational activity carried out outside formal frameworks, and informal learning encompasses self-directed learning and experiential workplace learning (Owusu-Agyeman, 2024; Pires et al., 2021). Together, these learning experiences contribute to personal development, professional competence, social participation, and adaptability (Ioannou, 2024; Mikulec & Kovšca, 2023).

Consequently, contemporary education systems increasingly seek to create pathways that connect these different forms of learning and recognise their collective contribution to individual and societal development (Ambrósio et al., 2014; Singh, 2015e; Роляк et al., 2020). Legislative frameworks in countries such as Denmark, Norway, and Thailand explicitly identify lifelong education as the integration of formal, non-formal, and informal learning (Singh, 2015e). Understanding formal, non-formal, and informal learning therefore provides a foundation for understanding lifelong learning and the diverse ways in which adults continue to learn throughout their lives (Ioannou, 2024; Owusu-Agyeman, 2024; Singh, 2015e).

Forms of Learning and Learning Theories

The distinction between formal, non-formal, and informal learning has become a foundational concept in adult education and lifelong learning because it recognises that adults acquire knowledge, skills, values, and competencies through a wide range of learning experiences that extend far beyond traditional educational institutions. Early educational systems tended to equate learning with formal schooling, viewing education as a discrete phase of life completed during childhood or early adulthood. However, researchers in adult education increasingly observed that significant learning continues throughout adulthood and frequently occurs in workplaces, communities, families, professional networks, and everyday life situations (Coombs & Ahmed, 1974; Jarvis, 2004; UNESCO, 2021b). This recognition challenged conventional understandings of education and contributed to the emergence of lifelong learning as a central principle within adult education theory and practice. Consequently, adult learning theories seek not only to explain learning within classrooms but also to understand how adults learn through experience, social participation, work, reflection, and self-directed inquiry.

The distinction between formal, non-formal, and informal learning is particularly relevant to Knowles' theory of andragogy. Knowles (1980) argued that adults differ from children in several important respects, including their desire for autonomy, self-direction, and practical application of learning. Adult learners are often motivated by immediate personal or professional needs and frequently engage in learning activities outside formal educational institutions. Professional development programs, workplace training, mentoring relationships, community education initiatives, and self-directed study are all examples of learning environments that reflect the principles of andragogy. The existence and prevalence of non-formal and informal learning provide support for Knowles' contention that adults are capable of taking responsibility for their own learning and that learning is most effective when it addresses real-life problems and situations (Knowles et al., 2015).

Experiential Learning Theory provides another important connection between adult learning theory and the distinction between formal, non-formal, and informal learning. (Kolb, 1984b) proposed that learning occurs through the transformation of experience into knowledge and that individuals learn through a cyclical process involving experience, reflection, conceptualisation, and experimentation. Informal learning environments are particularly significant from this perspective because they provide continual opportunities for adults to engage with new experiences and reflect upon their meaning. Workplace activities, family responsibilities, community participation, and everyday problem-solving all generate experiences that can contribute to learning. Billett (2001) similarly emphasised the importance of workplace experiences as sources of learning and competence development. These perspectives suggest that adults often develop expertise not primarily through formal instruction but through engagement with authentic situations that require the application and refinement of knowledge and skills.

The concept of self-directed learning is also closely associated with informal and non-formal learning environments. Research by Tough (1979) demonstrated that adults frequently undertake substantial learning projects independently, often without enrolling in formal educational programs. Self-directed learners identify their own learning needs, seek out resources, select learning strategies, and evaluate their progress. Much of this learning occurs outside institutional settings and is driven by personal, occupational, or social goals. The widespread occurrence of self-directed learning provides empirical support for theories that emphasise learner autonomy and personal responsibility, reinforcing the view that significant learning can occur without formal curricula, classrooms, or instructors.

The distinction between formal, non-formal, and informal learning is also highly relevant to Situated Learning Theory. (Lave & Wenger, 1991) argued that learning is fundamentally social and occurs through participation in communities of practice. According to this perspective, knowledge is developed through engagement in authentic activities alongside more experienced members of a community. Workplaces, professional associations, volunteer organisations, and community groups therefore become important learning environments because they enable individuals to acquire knowledge and skills through observation, collaboration, mentoring, and participation. The concept of communities of practice illustrates how non-formal and informal learning environments can support learning that is highly contextualised, socially situated, and directly relevant to professional and personal development.

Transformative Learning Theory further highlights the importance of learning beyond formal educational settings. (Mezirow, 1991) argued that adults frequently experience transformative learning when they critically reflect upon assumptions, beliefs, and perspectives that have previously guided their understanding of the world. Such transformative experiences often arise from life events, workplace challenges, personal relationships, social interactions, and other informal learning contexts rather than from structured educational programs. Informal learning environments frequently provide the disorienting dilemmas and reflective opportunities that stimulate perspective transformation, personal growth, and changes in behaviour. Consequently, transformative learning theory reinforces the importance of recognising learning that occurs outside formal educational institutions.

The growing recognition of non-formal and informal learning has also influenced educational policy and practice through the development of Recognition of Prior Learning (RPL) systems. Traditional education systems historically privileged learning acquired through formal education and qualifications. However, adult learning theories recognise that substantial competence can be developed through employment, volunteering, military service, community participation, family responsibilities, and self-directed learning. RPL processes seek to identify, assess, and formally recognise learning acquired through these experiences, thereby acknowledging that valuable learning can occur in many contexts (Guimarães & Mikulec, 2020a; Singh, 2015f). The development of RPL reflects a key principle of adult learning theory: that learning should be valued according to demonstrated competence rather than solely according to where or how it was acquired.

Perhaps the most significant connection between formal, non-formal, and informal learning and adult learning theory is found in the concept of lifelong learning. Lifelong learning frameworks recognise that adults move continuously between different learning environments throughout their lives, engaging in formal education, workplace learning, community activities, professional development, and self-directed learning as circumstances require (Delors et al., 1996; Faure et al., 1972; UNESCO, 2021b). Adult learning theories provide explanations for how learning occurs across these diverse contexts and why adults continue to learn throughout their personal and professional lives. Rather than viewing education as a finite process completed during youth, lifelong learning perspectives recognise learning as a continuous, multidimensional process embedded within everyday experience.

The distinction between formal, non-formal, and informal learning broadens understanding of adult learning by demonstrating that learning occurs through education, work, social participation, reflection, experience, and self-directed inquiry. Adult learning theories emerged largely in response to the recognition that meaningful learning extends beyond classrooms and formal qualifications. Consequently, understanding these different forms of learning provides an essential foundation for understanding andragogy, experiential learning, self-directed learning, situated learning, transformative learning, lifelong learning, workplace learning, and contemporary adult education practice. Together, these perspectives highlight the diverse and interconnected ways in which adults continue to learn, adapt, and develop throughout their lives.

Global Trends Shaping Adult Education

Adult education is being transformed by a range of global social, economic, technological, demographic, and environmental changes. These trends are influencing why adults learn, what they learn, how learning is delivered, and the role that education plays throughout the lifespan. As societies become increasingly interconnected and knowledge-intensive, adult education has expanded beyond its traditional focus on literacy and basic skills to encompass workforce development, lifelong learning, digital competence, social inclusion, and sustainable development. Understanding these global trends is essential for educators, policymakers, employers, and learners seeking to respond effectively to changing educational needs.

Lifelong Learning as a Global Policy Framework

One of the most significant developments in contemporary adult education is the widespread adoption of lifelong learning as a guiding educational principle. International organisations such as UNESCO, the OECD, and the European Union increasingly promote lifelong learning as essential for economic competitiveness, social inclusion, and personal development (Salas, 2017). The Faure Commission's *Learning to Be* advocated lifelong education as the "master concept" for both developed and developing countries (Carlsen & Haddad, 2013), while the Delors Commission's *Learning: The Treasure Within* reinforced this vision for the twenty-first century (Carlsen & Haddad, 2013; Milana, 2012). The EU declared 1996 the European Year of Lifelong Learning, further institutionalising the concept (Barros, 2019). The OECD's ministerial commitment to "making lifelong learning a reality for all" marked a decisive shift from earlier, narrower conceptions of recurrent education (Roy, 2003). More recently, UNESCO has continued reinforcing lifelong learning's importance in addressing global challenges (Auclair, 2023).

As a result, many countries have developed national lifelong learning strategies encouraging adults to continually update their knowledge and skills (Roy, 2003). With the adoption of lifelong learning by OECD countries, adult learning has become a central aspect of education policy, with all examined countries adopting reforms in this area (Roy, 2003). Adult education is increasingly viewed not as remedial but as a normal component of modern citizenship and workforce participation (Evans, 2015).

Technological Change and Digital Transformation

Rapid technological advancement is a primary driver of adult learning, as digital technologies, AI, automation, and data analytics transform workplaces and demand continuous skill updating (Barus & Bontisesari, 2023; Saragih et al., 2023; Sikora et al., 2023). Digital transformation has fundamentally altered how adult education is delivered, with online platforms, virtual classrooms, learning management systems, and mobile applications expanding access to learning (Bennett & McWhorter, 2022; Meaney, 2024). MOOCs and micro-credential programs enable learners to acquire targeted skills outside traditional degree programs, though evidence suggests they primarily serve already-educated populations (Meaney, 2024).

Connectivism, as a learning theory, posits that learning increasingly occurs through digital networks and online communities rather than solely through traditional institutions (Serim & Bilgili, 2023). Digital literacy has become a fundamental competency for participation in contemporary society (Almagarbeh, 2022), requiring adult education providers to help learners develop technological skills, information literacy, and the capacity to learn effectively in online environments (Saragih et al., 2023). The COVID-19 pandemic significantly accelerated this digital transformation across higher and adult education globally (Bennett & McWhorter, 2022; Roy & Al-Absy, 2022; Serim & Bilgili, 2023).

Technological Change and Digital Transformation

Rapid technological advancement constitutes a primary driver of adult learning globally. Digital technologies, automation, and data analytics are transforming workplaces, requiring workers to continually update their skills (Duplàa et al., 2023; OECD, 2020c). The fourth industrial revolution demands competencies such as critical thinking, continuous learning capacity, and adaptation to new tools (Duplàa et al., 2023).

Digital transformation has fundamentally altered adult education delivery. Online learning platforms, virtual classrooms, and learning management systems have expanded access to learning (Bates & Glikman, 2021; Poëllhuber et al., 2017). MOOCs and micro-credential programs enable learners to acquire targeted skills outside traditional degree programs (Bakki et al., 2020; Bruillard et al., 2021; Poëllhuber et al., 2017). As Bates and Glikman note, the first connectivist MOOC was created in 2008 by Siemens, Downes, and Cormier (Bates & Glikman, 2021). Siemens' (2005) theory of connectivism posits that learning increasingly occurs through digital networks rather than solely through traditional institutions (Bakki et al., 2020; Duplàa et al., 2023; Glikman, 2021). This theory emphasizes that connections and information location matter more than content itself (Duplàa et al., 2023).

Consequently, digital literacy has become fundamental for societal participation (Michelot, 2022; Toro & Kisi, 2022). Adult education providers increasingly focus on developing technological skills, information literacy, and digital communication abilities (Michelot, 2022; Toro & Kisi, 2022; إبراهيم, 2024). Competencies in digital, informational, and media literacies are recognized as central twenty-first-century skills (Michelot, 2022).

Workforce Transformation and Skills Development

The transition towards knowledge-based economies has elevated the significance of adult education for workforce development. Employers increasingly seek workers possessing not only technical expertise but also adaptability, critical thinking, communication, collaboration, and problem-solving skills (Marszowski, 2019; Wanyu Wang et al., 2024). As industries evolve and new technologies emerge, workers must continuously update their skills to remain competitive (Wanyu Wang et al., 2024), and the challenge of encouraging individuals to continue their education to upgrade vocational and career skills persists (Osborne, 2017).

International organisations emphasise continuous skills development, with adult education understood as promoting employment opportunities while supporting competence development required by the labour market (Egetenmeyer & Fedeli, 2017b). Continuing vocational education and training is defined as a means to "improve participation of adults in lifelong learning, reinforce their employability and increase employment" (Egetenmeyer & Fedeli, 2017b). Adult education systems are increasingly aligned with workforce development strategies and labour market needs (Alvunger, 2024; Stephen & Festus, 2022). Vocational education and training, professional development programs, and workplace learning initiatives have become important mechanisms for supporting lifelong employability (Egetenmeyer & Fedeli, 2017b; Stephen & Festus, 2022). Micro-credentials have emerged as flexible tools enabling individuals to reskill and upskill, representing "a response to the need to upskill existing employees and reskill people looking for work" (Diab & Zhang, 2023), and are increasingly recognised as supporting lifelong learning across academic and business sectors (Brown, Mhichíl, et al., 2021; Темјановски et al., 2023).

The Growth of Micro-credentials and Flexible Learning Pathways

Traditional educational models are increasingly complemented by shorter, flexible learning formats. Micro-credentials, digital badges, industry certifications, and short courses enable learners to acquire specific competencies without committing to full qualifications (Antonaci, Henderikx, et al., 2021; Brown, Mhichíl, et al., 2021; Темјановски et al., 2023). Brown et al. characterize micro-credentials as "smaller units of study, which are usually shorter than traditional forms of accredited learning" (Brown, Mhichíl, et al., 2021), while UNESCO notes the term encompasses "nano-degrees, micro-masters credentials, certificates, badges, licences and endorsements" (Brown, Mhichíl, et al., 2021).

This trend reflects changing workforce requirements and learner expectations. Adults increasingly seek targeted, immediately applicable learning (Diab & Zhang, 2023; Phelan & Glackin, 2020). Jacobson attributes the rise of micro-credentials to increasing college costs, employer-education skills gaps, and nontraditional students who "prefer to finish college as quickly, and as flexibly as possible" (Phelan & Glackin, 2020). Micro-credentials support individuals balancing education with employment and other commitments through flexible, competency-based formats (Antonaci, Henderikx, et al., 2021; Thorpe et al., 2024). Antonaci et al. note that "continuing education programs are not flexibly organised, and most people cannot invest years in a bachelor's or master's degree" (Antonaci, Henderikx, et al., 2021).

Consequently, educational systems increasingly recognize smaller learning units and create stackable pathways (Ashizawa et al., 2024; Brown, Mhichíl, et al., 2021; Selvaratnam & Sankey, 2021). Governments are aligning micro-credentials with national qualification frameworks (Ashizawa et al., 2024; Brown, Mhichíl, et al., 2021), and institutions are establishing governance models enabling learners to accumulate credentials over time (Porto & Presant, 2023; Selvaratnam & Sankey, 2021).

Demographic Change and Ageing Populations

Many countries are experiencing significant demographic shifts, including ageing populations and increased life expectancy. Average life expectancy across OECD countries rose from 69 years in 1970 to 79.7 years in 2010 (Goodwill & Chen, 2021). As people live longer and remain economically active for extended periods, adult education plays an increasingly important role in supporting career transitions, ongoing employability (Merriam & Kee, 2014; OECD, 2013), health literacy (Cummins & Kunkel, 2015; Goodwill & Chen, 2021), and social participation (Findsen, 2016; Martin, 2018).

The concept of lifelong learning has become particularly relevant in ageing societies because individuals may need to adapt to multiple career changes during their working lives (Millar et al., 2020; OECD, 2013; Sørensen, 2024). With impending changes to the employment landscape and nonlinear career pathways, educational institutions and individuals need proactive approaches towards lifelong learning (Moroz, 2020; . Adult education is also increasingly recognised as contributing to healthy ageing, cognitive wellbeing, social engagement, and quality of life among older adults (Findsen, 2016; Martin, 2018; Moroz, 2020; Staudinger & Bowen, 2011). Research demonstrates that participation in non-formal lifelong learning promotes psychological wellbeing and may protect against cognitive decline (Mason & Carr, 2021; Moroz, 2020), while education later in life stimulates cognitive skills and provides social exchange (Goodwill & Chen, 2021).

Globalisation and International Mobility

Globalisation has increased the movement of people, knowledge, and labour across national borders (María Amor Barros del et al., 2023; OECD, 2017). The number of international migrants worldwide has grown substantially, creating increasingly diverse communities (OECD, 2017). Workers increasingly operate within international labour markets, while organisations compete globally, demanding internationally recognised qualifications, transferable skills, intercultural competence, and multilingual communication abilities (Egetenmeyer et al., 2017; María Amor Barros del et al., 2023; Saraiva et al., 2022).

Adult education systems are responding by promoting global citizenship, intercultural understanding, and international credential recognition (Egetenmeyer et al., 2017; Oliver, 2010). The development of global citizenship has become an important objective, implying deeper understanding of international interconnections, comparable skills, and increased civic engagement (Egetenmeyer et al., 2017). Qualifications frameworks such as the European Qualifications Framework (EQF) provide shared reference points for defining competence profiles across countries (Oliver, 2010), while credit transfer systems like ECTS and mutual recognition agreements facilitate educational and occupational mobility (Egetenmeyer et al., 2017; María Amor Barros del et al., 2023; Saraiva et al., 2022). UNESCO's Global Convention on the Recognition of Higher Education Qualifications further supports fair and transparent evaluation of qualifications across borders (Reghenzani-Kearns, 2021). Initiatives coordinated by organisations such as the European Training Foundation aim to develop global framework descriptors, positioning education systems for increased knowledge sharing and global mobility of lifelong learners (Reghenzani-Kearns, 2021). Adult learners are therefore encouraged to develop competencies, including plurilingual and intercultural skills, that enable participation in diverse and interconnected global environments (María Amor Barros del et al., 2023; OECD, 2017).

Recognition of Prior Learning and Alternative Learning Pathways

There is growing recognition that valuable learning occurs outside formal educational institutions. Many adults acquire substantial knowledge and skills through employment, volunteering, military service, community participation, and self-directed learning (Bhandari, 2024; Roy & Marsafawy, 2021). As a result, Recognition of Prior Learning (RPL) and validation of non-formal and informal learning have become increasingly important international trends (Alves et al., 2020; Guimarães & Mikulec, 2020b).

Countries around the world have developed systems that assess and formally recognise competencies acquired through experience, commonly referred to as "identifying, documenting, assessing and certifying formal, non-formal and informal learning outcomes against standards used in formal education and training" (Bhandari, 2024). Different terminologies exist across jurisdictions: PLA in the USA, PLAR in Canada, APL/APEL in the UK, and RPL in Australia, New Zealand, and South Africa (Alves et al., 2020; Bhandari, 2024). International organisations also use varying terms, with the EU referring to "validation of non-formal and informal learning" and UNESCO using "recognition, validation and accreditation" (Bhandari, 2024; Guimarães & Mikulec, 2020b). These systems help reduce barriers to further education, improve workforce mobility, and acknowledge the value of lifelong learning (Rius et al., 2021).

RPL has proven beneficial for improving employability, enabling further education and training, and promoting social inclusion (Bhandari, 2024; Jentsch et al., 2022). The growing acceptance of RPL reflects broader recognition that learning occurs across multiple contexts, including workplaces, communities, and self-directed settings (Singh, 2015c), and should not be judged solely on the basis of formal qualifications (Priadi et al., 2019; Roy & Marsafawy, 2021).

Equity, Inclusion, and Social Justice

Contemporary adult education increasingly emphasises equity, inclusion, and access (Brian, 2021; Hill et al., 2025; Lacher, 2024). Despite advances in educational participation, significant disparities remain based on socioeconomic status, gender, ethnicity, disability, geographic location, and digital access (Heymann et al., 2020; Hill et al., 2025). Adults from marginalized communities encounter persistent barriers including financial instability, childcare demands, and limited awareness of available programs (Hill et al., 2025). International organisations and governments continue to promote policies aimed at widening participation and reducing educational inequalities, as exemplified by SDG 4, which aims to "ensure inclusive and equitable quality education and promote lifelong learning opportunities for all" (Lacher, 2024; Spiel et al., 2018).

Adult education is increasingly viewed as a tool for social inclusion, community development, and democratic participation (Gravani et al., 2019; Hill et al., 2025; Lacher, 2024). Programs designed to improve literacy, language proficiency, digital skills, and employability often target disadvantaged populations, including migrants, refugees, and low-skilled adults, and seek to address broader social inequalities (Gravani et al., 2019; Hutfilter et al., 2018; Oliver, 2010; Ugurel-Kamisli, 2022). Government financial support for lifelong learning is most impactful when targeted to adults with the lowest qualifications (Hutfilter et al., 2018). Inclusive education, culturally responsive teaching, and accessible learning environments have therefore become important priorities (Brian, 2021; Xhomaki et al., 2019). However, equal access to technology does not automatically imply equal learning opportunities, as digital divides persist alongside gaps in basic skills (Barger, 2020; Brian, 2021; Xhomaki et al., 2019).

Sustainability and Education for Sustainable Development

Environmental challenges including climate change, biodiversity loss, and resource depletion are increasingly shaping adult education agendas. UNESCO's Education for Sustainable Development (ESD) framework emphasises learning that supports sustainable lifestyles, environmental responsibility, and informed citizenship (Essomba et al., 2022; Glavič, 2020; Kefalaki, 2023). ESD is defined as empowering learners "to take informed decisions and responsible actions for environmental integrity, economic viability and a just society" (Ang, 2021), encompassing key issues such as "climate change, disaster risk reduction, biodiversity, poverty reduction, and sustainable consumption" (Glavič, 2020).

Adult education programs increasingly incorporate sustainability concepts and green skills. Vocational education and training (VET) offers opportunities for integrating ESD, equipping workers with competencies related to renewable energy, sustainable resource management, and environmental compliance (Lambini et al., 2021). The need for continuous adult learning on sustainability topics is recognised, though a critical gap persists in education for adult learners on climate action and biodiversity conservation (Vaughter et al., 2022). Funding challenges further complicate sustained adult ESD programming, as project-based financing creates instability for long-term implementation (Košmerl et al., 2024). Sustainability has thus emerged as a significant focus within lifelong learning and workforce development, with ESD positioned as both a target and critical enabler of sustainable development (Lambini et al., 2021; Vaughter et al., 2022).

Health, Wellbeing, and Resilience

The COVID-19 pandemic underscored the critical relationship between health literacy, resilience, and wellbeing across populations. Research demonstrated that community resilience and sense of community mitigated pandemic impacts on psychological wellbeing (Awiphan et al., 2026), while individual resilience mediated the negative effects of COVID-19-related stress (Mannarini et al., 2021; Versteeg & Kappe, 2021). Lower socioeconomic position was associated with reduced psychological resilience and poorer wellbeing during the pandemic (Bono et al., 2020).

The pandemic accelerated recognition that education serves purposes beyond occupational training. Higher education institutions were urged to promote resilience and support services to enhance student wellbeing (Şoitu et al., 2022; Versteeg & Kappe, 2021). Educational activities stimulate cognitive resilience and brain plasticity, with continuous learning improving mental agility and adaptability (Mahmood et al., 2022). Employers increasingly value adaptability, resilience, emotional intelligence, and wellbeing alongside technical competencies (Neilsen-Hewett et al., 2022).

Community organisations and educational institutions play vital roles in supporting social connectedness, which proved protective during the pandemic (Awiphan et al., 2026; Bozkurt, 2022).

The broader understanding that learning contributes to physical health, mental wellbeing, and social engagement reflects multisystem approaches to human development (Mahmood et al., 2022; Sharpe et al., 2021), where resilience is shaped by individual, social, cultural, and familial factors.

Implications for Adult Learning Theory

These global trends have reinforced many of the central assumptions of adult learning theory. The need for continual adaptation supports lifelong learning frameworks developed by Hutchins (1968), Lengrand (1975), Faure et al. (1972), and Delors et al. (1996). The growth of workplace learning aligns with experiential learning theory (Kolb, 1984) and situated learning theory (Lave & Wenger, 1991). Increased learner autonomy and flexible learning pathways reflect principles of andragogy (Knowles, 1980) and self-directed learning (Tough, 1979). Digital transformation has contributed to new theories such as connectivism (Siemens, 2005), while social and environmental challenges reinforce the importance of transformative learning (Mezirow, 1991).

Global trends are reshaping adult education in profound ways. Technological innovation, workforce transformation, demographic change, globalisation, sustainability challenges, and the expansion of lifelong learning are creating new opportunities and demands for adult learners. These developments are influencing both educational policy and practice, encouraging more flexible, inclusive, technology-enabled, and learner-centred approaches to education. As societies continue to evolve, adult education will play an increasingly important role in supporting employability, social participation, personal development, and sustainable futures. The ability to learn continuously throughout life is therefore becoming one of the most important capabilities for individuals and societies in the twenty-first century.

Adult Learning in Developed and Developing Economies

Adult learning plays a vital role in economic development, workforce capability, social inclusion, and lifelong learning across both developed and developing economies (Maruatona, 2014; Msiska, 2015; Spatafora, 2021). Developed economies, having already acquired the infrastructure of lifelong learning, focus on workforce adaptability, technological skills, and continuing education outside formal institutions (Msiska, 2015). In contrast, developing economies prioritize literacy, basic education, and poverty reduction, as adult literacy remains a prerequisite for socioeconomic development (Awgichew & Seyoum, 2017; Fute et al., 2022). For instance, Tanzania has used adult education as an instrument to tackle poverty, unemployment, and illiteracy (Fute et al., 2022), while Ethiopia's Integrated Functional Adult Literacy program links literacy with community development activities (Awgichew & Seyoum, 2017).

Challenges in developing contexts include inadequate infrastructure, poor institutionalization, and insufficient funding (Awgichew & Seyoum, 2017; Mehta, 2024). Meanwhile, social inclusion through adult learning remains critical, as minority communities in countries like Botswana face exclusion due to decontextualized curricula and unfamiliar languages (Maruatona, 2014). Despite these differences, investment in education and human capital yields substantial returns across all settings, expanding economic opportunities and promoting social mobility (Mehta, 2024; Spatafora, 2021). The UN's SDG

4 commitment to "ensure inclusive and equitable quality education and promote lifelong learning opportunities for all" underscores this universal importance (Hinton & Kazmi, 2022).

Historically, adult education in developing economies has been closely associated with literacy campaigns, community development initiatives, agricultural extension programs, health education, and poverty alleviation. Many developing countries inherited education systems that were unable to provide universal schooling, resulting in significant adult populations with limited literacy and numeracy skills. Consequently, adult education programs often emerged as mechanisms for addressing educational disadvantage and supporting social and economic development. Coombs and Ahmed (1974) argued that non-formal education could play a crucial role in reducing rural poverty by providing practical knowledge and skills relevant to local economic and social needs. Their work helped establish non-formal education as an important strategy within international development policy.

In many developing economies, adult learning remains closely linked to basic education and literacy. International organisations such as UNESCO continue to support adult literacy initiatives as part of broader efforts to achieve educational equity and sustainable development. Literacy programs frequently address not only reading and writing but also numeracy, financial literacy, health education, agricultural productivity, civic participation, and gender equality. In many regions, adult education serves as a mechanism for empowering marginalised populations and improving access to economic opportunities.

Workforce development is also an important objective of adult learning in developing economies. However, unlike many developed countries where training often focuses on advanced technical skills and professional development, adult education in developing contexts may prioritise employability, entrepreneurship, vocational skills, and income generation. Informal economies often employ a substantial proportion of the workforce, meaning that learning frequently occurs through apprenticeships, workplace experience, family enterprises, and community-based training rather than through formal vocational education systems. As a result, non-formal and informal learning play particularly important roles in skill development.

Recognition of Prior Learning (RPL) has become increasingly significant in developing economies because large numbers of workers acquire occupational skills outside formal education systems. (Singh, 2015f) notes that validation and recognition systems can help workers gain formal recognition for competencies developed through employment, apprenticeships, and informal learning. Such recognition can improve employment opportunities, facilitate access to further education, and support workforce mobility.

In contrast, adult learning in developed economies is often driven by the demands of knowledge-based economies, technological innovation, and changing labour markets. Most developed countries have achieved near-universal literacy and broad access to formal education, meaning that adult education is less focused on basic skills acquisition and more concerned with lifelong learning, professional development, reskilling, and upskilling. The rapid pace of technological change means that workers must continually update their knowledge and competencies throughout their careers.

The concept of lifelong learning has become particularly influential in developed economies. Reports such as *Learning to Be* (Faure et al., 1972) and *Learning: The Treasure Within* (Delors et al., 1996) promoted the idea that learning should continue throughout life rather than ending upon completion of formal schooling. This perspective has been widely adopted by governments and educational institutions in Europe, North America, Australia, New Zealand, Japan, South Korea, and Singapore. Adult learning is increasingly viewed as essential for maintaining employability, supporting career transitions, and enabling participation in rapidly changing societies.

Workplace learning occupies a particularly prominent position in developed economies. Billett (2001) argues that contemporary workplaces have become important sites of learning where employees continually develop new knowledge and skills through participation in work activities. Professional development, industry certification, postgraduate education, leadership programs, and continuing professional education are common forms of adult learning. Many professions now require ongoing learning as a condition of professional registration and accreditation.

Digital transformation has further expanded adult learning opportunities in developed economies. Online learning platforms, micro-credentials, virtual classrooms, open educational resources, and digital learning networks provide flexible pathways for adults seeking to update their skills. Siemens (2005b) theory of connectivism reflects the growing importance of digital networks as sources of learning. Adults increasingly engage in self-directed learning through online courses, webinars, professional communities, podcasts, and digital knowledge-sharing platforms.

Despite these differences, developed and developing economies face several common challenges. Globalisation, automation, artificial intelligence, climate change, and demographic shifts are creating new demands for learning across all regions. Both developed and developing countries must support adults in acquiring new skills, adapting to changing employment conditions, and participating effectively in increasingly complex societies. Lifelong learning has therefore become a shared policy objective across diverse economic contexts.

Significant differences remain, however, in access to learning opportunities. Developed economies generally possess more extensive educational infrastructure, higher participation rates in continuing education, greater availability of digital technologies, and stronger systems for recognising non-formal and informal learning. Developing economies may face challenges related to funding, infrastructure, educational access, digital connectivity, and workforce participation. Rural populations, women, minority groups, and disadvantaged communities often experience barriers to participation in adult education.

The role of technology also differs between contexts. In developed economies, technology is often used to enhance learning through sophisticated online platforms and digital learning ecosystems. In developing economies, technology may serve a more fundamental role in expanding access to education where traditional infrastructure is limited. Mobile learning, radio education, community technology centres, and low-cost digital platforms have become important mechanisms for delivering adult education in many regions.

Adult learning theories provide valuable frameworks for understanding learning in both developed and developing economies. Knowles' andragogy highlights the importance of self-directed and relevant learning, regardless of economic context. Kolb's Experiential Learning Theory explains how adults learn through workplace and life experiences. Freire's critical pedagogy has been particularly influential in developing countries, where adult education is often linked to empowerment, social justice, and community development. Mezirow's Transformative Learning Theory helps explain how adults adapt to social, economic, and technological change through critical reflection and perspective transformation.

Increasingly, the distinction between developed and developing economies is becoming less pronounced in some areas of adult learning. Globalisation, digital technologies, international qualifications frameworks, and lifelong learning policies are creating greater convergence in educational approaches. Many developing countries are expanding higher education, vocational education, and lifelong learning systems, while developed countries are increasingly recognising the importance of community-based learning, social inclusion, and recognition of informal learning.

Ultimately, adult learning in both developed and developing economies serves the common purpose of enabling individuals to improve their knowledge, skills, capabilities, and quality of life. The specific priorities may differ, from literacy and poverty reduction in some contexts to workforce adaptability and professional development in others, but the underlying principle remains the same: learning is a lifelong process that contributes to individual empowerment, economic development, social inclusion, and societal wellbeing. Understanding these similarities and differences provides valuable insight into the diverse ways adult education supports development across the world.

The Role of Education in Workforce Development

Education serves as a foundational mechanism for workforce development, functioning across multiple dimensions including formal schooling, vocational training, skills development, and lifelong learning. The relationship between education and workforce development has been shaped by evolving policy frameworks, economic transformations, and changing labour market demands.

A critical policy framework governing the intersection of adult education and workforce development in the United States is the Workforce Innovation and Opportunity Act (WIOA), which became law in 2014. WIOA explicitly frames adult basic education (ABE) as serving economic purposes and mandates coordination between ABE providers and workforce development partners (Hou et al., 2024). Under WIOA, there is "increased accountability requirements that narrowly focus on economic outcomes" and a deliberate "framing adult education for economic purposes" (Hou et al., 2024). This policy architecture illustrates how governments increasingly view education not merely as a public good but as an instrument for labour market integration.

WIOA emphasizes "greater involvement by adult education providers with workforce partners" and requires local workforce development boards (WDBs) to maintain a majority of business members, reflecting the growing influence of employers as stakeholders in educational programming (Hou et al., 2024). However, this alignment has created tensions, as "the relationship with adult education providers is often strained, reflecting the long-standing tension between workforce and adult education systems regarding core purposes and competition for students" (Hou et al., 2024). This tension highlights a fundamental debate about whether education should primarily serve economic functions or broader human development goals.

Vocational education and apprenticeship systems represent a direct pathway connecting education to workforce development. Modern apprenticeship is defined as "a talent training mode that focuses on skills training for students through in-depth cooperation between schools and enterprises, joint training of teachers and mentors" (Lin & Tian, 2020). Research on apprenticeship systems demonstrates significant economic returns: "participating in apprenticeship programs can yield significant, statistically significant benefits" (Lin & Tian, 2020). Furthermore, apprenticeships serve as effective workforce pipelines, as "providing apprenticeship positions is a strategy for attracting their future skilled labour force" and apprentices "accumulated firm-specific human capital, thereby reducing the firm's training and recruitment costs" (Lin & Tian, 2020).

The registered apprenticeship model in the United States exemplifies how education and workforce development converge. Registered apprenticeships "provide structured career advancement from low-wage entry-level positions to technical jobs" and must include both "classroom-based relevant technical instruction (RTI) and on-the-job training (OJT)" (Lin & Tian, 2020). This dual structure ensures that educational content is directly aligned with workplace requirements.

In China's manufacturing sector, the demand for highly skilled talent has intensified as "industry competitiveness has shifted from simply relying on resources and cheap labour to technological innovation" (王, 2016). The concept of "integration of industry and education" (产教融合) has been promoted to address the disconnect between educational outputs and industrial needs, aiming to "organically coordinate the education chain, talent chain, and industry chain" (王, 2016). However, significant challenges persist, including the cultural bias of "valuing academic credentials over skills" (重学历轻技能), which has contributed to shortages of highly skilled workers in manufacturing (王, 2016).

Education level significantly affects employment outcomes and labour mobility. Research on China's floating population demonstrates that "education level, marital status, type of registered place, registered permanent residence, inflow area and flow range" all have significant effects on unemployment (W. Li, 2018). Specifically, "low education levels limit employment choices" for migrant workers, contributing to structural unemployment (W. Li, 2018). This finding underscores the role of education in expanding employment options and reducing vulnerability to joblessness.

In the context of poverty alleviation and resettlement programs, vocational skills training has been identified as a key mechanism for promoting non-agricultural employment. Research on concentrated resettlement communities shows that "providing vocational skills training can significantly promote labour force outward employment" and that improving public services including education and vocational training enables resettled populations to "achieve livelihood transformation at lower non-agricultural employment costs" (Wang et al., 2021). Human capital endowment is considered "a primary factor determining the non-agricultural employment of resettled populations" (Wang et al., 2021).

Similarly, in resource-based cities undergoing economic transformation, education plays a critical role in workforce adaptation. Research indicates that cities with higher education levels experience different labour market outcomes during industrial transitions, and policy recommendations emphasize the need to "strengthen employment guidance and training, promote the improvement of local labour force quality, and enable the labour force to break free from existing path dependence to meet the new requirements of successor industries" (吴, 2017).

Education also contributes to workforce development through entrepreneurship education. In Chinese higher education, entrepreneurship education was introduced in response to increasing graduate employment difficulties, with the goal of cultivating students' entrepreneurial spirit, creating suitable jobs for the development of society and individuals, enabling smooth self-employment and creating new jobs, increasing employment opportunities for others (Zhan et al., 2022). Effective entrepreneurship education requires faculty who possess both theoretical knowledge and practical experience, and institutions are encouraged to "invite alumni, entrepreneurs, famous innovators, social institutions and other experts" to provide diverse perspectives (Zhan et al., 2022).

Education is equally vital for workforce development in agricultural sectors. The cultivation of "new-type professional farmers" requires individuals who possess "knowledge, technology, management skills, responsibility and social status" (Peng & Peng, 2017). Supply-side reforms in farmer education emphasize building comprehensive training systems, including "online and offline learning support services and teaching quality supervision and evaluation systems" (Peng & Peng, 2017). International experience, such as the United Kingdom's "National Training Award" for outstanding contributions to agricultural technical training, provides models for incentivizing quality in agricultural workforce education (Peng & Peng, 2017).

For ecological immigrants and ethnic minority populations, education and vocational training serve dual purposes of economic integration and cultural adaptation. Governments are urged to "accelerate the development of vocational and technical education, carry out multi-dimensional, multi-level technical training for immigrants, and achieve the alignment of education and training with job requirements" (张, 2018).

Several persistent challenges characterize the education-workforce development nexus. First, funding limitations constrain program quality and reach; in the ABE context, there is ongoing concern about "limited ABE funding and the need to use funds efficiently" (Hou et al., 2024). Second, structural mismatches between educational outputs and labour market demands remain prevalent, as evidenced by the simultaneous existence of graduate unemployment and skilled worker shortages in multiple national contexts (张, 2018; 王, 2016). Third, the instrumentalization of education for purely economic purposes risks neglecting broader educational goals, a tension that scholars and practitioners continue to debate (Hou et al., 2024).

Education functions as a multifaceted driver of workforce development, operating through formal schooling, vocational training, apprenticeships, entrepreneurship education, and continuing professional development. Effective workforce development requires alignment between educational institutions, employers, and government policy, though achieving this alignment involves navigating fundamental tensions about the purposes of education. The evidence suggests that investments in education, particularly vocational training and skills development, yield significant returns for individuals, employers, and economies, but success depends on contextual factors including institutional coordination, adequate funding, and responsiveness to evolving labour market demands.

Contemporary Challenges in Adult Learning

Adult learning has become increasingly important in the twenty-first century as individuals, organisations, and societies respond to rapid economic, technological, social, and environmental change. While opportunities for learning have expanded significantly through advances in technology, greater access to education, and the widespread adoption of lifelong learning policies, adult learners continue to face numerous challenges that can affect participation, engagement, and successful learning outcomes. These challenges are complex and interconnected, reflecting broader transformations in labour markets, demographic patterns, digital technologies, and social structures. Understanding contemporary challenges in adult learning is essential for educators, policymakers, employers, and learning organisations seeking to create effective and inclusive learning opportunities.

One of the most significant challenges is the accelerating pace of technological change. Advances in automation, artificial intelligence, robotics, data analytics, and digital communication technologies are continually reshaping workplaces and occupational requirements. Skills that were once sufficient for long-term employment may become obsolete within relatively short periods, requiring workers to engage in continual reskilling and upskilling. The challenge for adult learners is not merely acquiring new knowledge but developing the capacity to adapt to ongoing change throughout their working lives. Educational institutions similarly face difficulties in ensuring that curricula remain relevant in rapidly evolving industries. The need for continuous learning has therefore become both an opportunity and a burden for many adults who must balance learning demands with employment and personal responsibilities.

Digital transformation has created both new opportunities and new barriers to participation. Online learning, virtual classrooms, learning management systems, mobile learning applications, and digital resources have dramatically increased access to education. However, these developments have also highlighted the issue of the digital divide. Not all adults possess the technological skills, equipment, internet access, or digital confidence required to participate effectively in online learning environments. Older learners, individuals from disadvantaged socioeconomic backgrounds, rural communities, and those with limited prior exposure to technology may experience particular difficulties. Digital literacy has consequently emerged as a fundamental requirement for educational participation, workforce engagement, and social inclusion (OECD, 2021f).

Time constraints represent another major challenge for adult learners. Unlike younger students who may be able to devote substantial time to education, adults often balance multiple responsibilities simultaneously. Employment commitments, family obligations, childcare responsibilities, financial pressures, community involvement, and caregiving duties can limit the time available for learning. (Knowles, 1980) recognised that adults approach learning within the context of competing life roles and responsibilities. These competing demands frequently contribute to lower participation rates, delayed completion of programs, and difficulties maintaining engagement in long-term educational activities.

Financial barriers continue to affect access to adult learning opportunities. Although many countries promote lifelong learning, participation often requires significant financial investment. Tuition fees, course materials, technology requirements, travel costs, and lost income associated with study can discourage participation. Individuals with lower incomes are often less able to invest in education despite potentially benefiting most from additional learning opportunities. This creates a paradox in which those with the greatest educational needs may face the greatest barriers to participation. The challenge of ensuring equitable access to lifelong learning remains a significant policy concern in many countries.

Educational inequality more broadly remains a persistent challenge within adult learning. Participation rates in adult education are often strongly associated with prior educational attainment. Individuals who have already achieved higher levels of education are generally more likely to engage in further learning than those with lower educational qualifications (Field, 2006). This phenomenon, sometimes referred to as the "Matthew Effect" in education, can reinforce existing social and economic inequalities. Adults with limited literacy, poor educational experiences, language barriers, disabilities, or experiences of social exclusion may be particularly underrepresented in adult education programs. Consequently, achieving equitable participation remains a central challenge for adult education providers and policymakers.

The increasing diversity of adult learners presents both opportunities and challenges. Contemporary classrooms and learning environments often include learners from diverse cultural, linguistic, educational, occupational, and generational backgrounds. Globalisation and migration have contributed to increasingly multicultural learning environments, requiring educators to adopt inclusive and culturally responsive teaching practices. Adult educators must accommodate varying levels of prior knowledge, learning preferences, language proficiency, and educational experience while ensuring that learning remains relevant and accessible to all participants. This diversity enriches learning environments but also increases the complexity of educational design and delivery.

Motivation and learner engagement present ongoing challenges within adult education. Adult learners typically choose to participate in education voluntarily and often have specific goals related to employment, career advancement, personal development, or social participation. However, maintaining motivation can be difficult, particularly when learning programs are lengthy, demanding, or poorly aligned with learners' immediate needs. Knowles' theory of andragogy emphasises that adults are most motivated when learning is relevant, practical, and applicable to real-life situations (Knowles et al., 2015). Educational programs that fail to demonstrate relevance may struggle to maintain learner engagement and completion rates.

The recognition and validation of non-formal and informal learning remains another important challenge. Significant learning occurs outside formal educational institutions through employment, volunteering, community participation, military service, family responsibilities, and self-directed learning. Although many countries have introduced Recognition of Prior Learning (RPL) systems, the implementation of such systems remains uneven. Assessing and validating experiential learning can be complex, resource-intensive, and inconsistent across sectors and jurisdictions. As lifelong learning becomes increasingly important, educational systems face growing pressure to recognise learning acquired through diverse pathways rather than relying solely on formal qualifications (Singh, 2015f).

Workforce transformation continues to create challenges for adult learners and educational providers alike. Globalisation, technological disruption, changing employment patterns, and the growth of non-standard forms of work have increased uncertainty within labour markets. Many workers face the prospect of career transitions, occupational change, and the need to acquire entirely new skill sets during their working lives. Educational systems must therefore provide flexible and responsive learning opportunities that support reskilling and career mobility. At the same time, learners must navigate increasingly complex labour markets while making decisions about educational investment and career development.

The growth of micro-credentials and alternative learning pathways presents both opportunities and challenges. Short courses, digital badges, industry certifications, and micro-credentials provide flexible opportunities for skill acquisition. However, questions remain regarding quality assurance, recognition, transferability, and employer acceptance. The proliferation of credentials can create confusion for learners and employers regarding the value and relevance of different qualifications. Educational systems are increasingly required to balance flexibility with quality and consistency.

Mental health and wellbeing have emerged as significant concerns within adult learning. Economic uncertainty, workplace pressures, social change, and global events such as the COVID-19 pandemic have increased stress and anxiety for many learners. Adult learners may experience challenges related to confidence, self-efficacy, fear of failure, and educational anxiety, particularly if they have been absent from formal education for extended periods. Educational institutions increasingly recognise the importance of providing learner support services, counselling, mentoring, and wellbeing initiatives to support successful participation and completion.

Environmental and sustainability challenges are also influencing adult learning. Climate change, environmental degradation, resource scarcity, and the transition to more sustainable economies require workers and citizens to acquire new forms of knowledge and competence. Many industries are undergoing significant transformation as governments and organisations respond to environmental challenges. Adult education therefore plays an important role in developing sustainability literacy, environmental awareness, and the green skills required for emerging occupations and industries. Educational providers must continually adapt curricula to address these evolving needs.

The increasing emphasis on lifelong learning itself presents a paradoxical challenge. While lifelong learning is widely promoted as essential for employability, adaptability, and personal development, the expectation that individuals continuously update their skills throughout life can create pressure and uncertainty. Some scholars have argued that lifelong learning policies may place excessive responsibility on individuals to remain employable in changing labour markets rather than addressing broader structural and economic issues (Field, 2006). This tension highlights ongoing debates regarding the purposes of adult education and the balance between individual responsibility and societal support.

Adult learning theories provide valuable frameworks for understanding and addressing many of these challenges. Andragogy highlights the importance of relevance, autonomy, and practical application. Experiential Learning Theory emphasises the value of workplace and life experiences as sources of knowledge (Kolb, 1984b). Situated Learning Theory recognises the role of social participation and communities of practice in learning (Lave & Wenger, 1991). Transformative Learning Theory provides insight into how adults adapt to changing circumstances through critical reflection and perspective transformation (Mezirow, 1991). More recently, connectivism has sought to explain learning within increasingly digital and networked environments (Siemens, 2005b).

Key Concepts and Conclusions

Adult learning is a lifelong process through which individuals acquire, develop, and transform knowledge, skills, attitudes, values, and competencies throughout their lives. Unlike children, adults bring extensive personal, social, educational, and workplace experiences to learning environments, and these experiences significantly influence how they interpret, apply, and construct new knowledge. Adult learning occurs across formal, non-formal, and informal settings, including universities, vocational education and training (VET), workplaces, professional development programs, community organisations, and everyday life experiences. As societies become increasingly complex, technology-driven, and interconnected, adult learning has become essential for personal development, workforce participation, social inclusion, and active citizenship.

A central theme throughout this chapter is that adults learn differently from younger learners. Adult learners tend to be more self-directed, goal-oriented, and motivated by learning that has immediate relevance to their personal, professional, or social circumstances. They generally prefer practical, problem-centred learning that can be applied to real-world situations rather than abstract or purely theoretical instruction. Adults also possess a strong capacity for reflection and frequently evaluate new knowledge in relation to their existing experiences, beliefs, and values. These characteristics have shaped the development of adult learning theories and continue to influence contemporary educational practice.

The chapter has demonstrated that no single theory fully explains how adults learn. Instead, contemporary understanding draws upon multiple complementary perspectives. Knowles' theory of andragogy emphasises self-direction, experience, readiness to learn, and practical application. Kolb's experiential learning theory highlights the role of experience, reflection, conceptualisation, and experimentation in the learning process. Mezirow's transformative learning theory explains how critical reflection can lead to significant changes in perspective and understanding. Freire's critical pedagogy emphasises dialogue, empowerment, and social transformation, while situated learning theory highlights the importance of social participation and communities of practice in the development of knowledge and competence. Together, these theories provide a comprehensive framework for understanding adult learning across a wide variety of educational and workplace contexts.

The chapter has also explored the diverse educational pathways available to adult learners, including university degrees, vocational qualifications, apprenticeships, and professional certifications. While these pathways differ in purpose, structure, and delivery, they all contribute to the development of knowledge, skills, and professional competence. Increasingly, these pathways are viewed as complementary components of broader lifelong learning systems that enable individuals to move between academic, vocational, and professional learning throughout their careers. This flexibility reflects the growing recognition that learning is not confined to a single stage of life but continues throughout adulthood.

Adult education and vocational education and training play a critical role in supporting economic development, workforce capability, social inclusion, and personal growth. Around the world, governments, employers, and international organisations recognise that lifelong learning is essential for responding to technological change, globalisation, demographic shifts, and evolving labour market demands. Effective adult learning systems therefore support not only employability and productivity but also wellbeing, civic participation, social mobility, and community development.

Effective adult education must be learner-centred, relevant, flexible, and responsive to the unique characteristics of adult learners. Educational experiences should acknowledge prior knowledge and experience, support autonomy and self-direction, encourage critical reflection, promote collaboration, and provide opportunities for practical application. By recognising these principles, educators, trainers, employers, and policymakers can create learning environments that maximise engagement, enhance learning outcomes, and support lifelong learning in an increasingly complex and rapidly changing world.

Chapter 2
Theoretical Foundations of Adult Learning

Learning is at the heart of education, training, professional development, and personal growth. Yet despite its importance, learning is a complex phenomenon that cannot be explained through a single perspective. Adults learn in different ways depending on their experiences, motivations, goals, social environments, cultural contexts, and access to information. Over the past century, researchers and educators have developed a range of theories to explain how learning occurs, why people learn differently, and how educational experiences can be designed to support effective learning.

This chapter explores the major theoretical foundations of adult learning and examines how these theories have shaped contemporary educational practice. Beginning with Behaviourism and Cognitivism, the chapter traces the evolution of learning theory through Constructivism, Humanism, Social Learning Theory, Experiential Learning Theory, Critical Theory, Sociocultural Theory, and Connectivism. Each perspective offers unique insights into the learning process, highlighting different aspects of how adults acquire knowledge, develop skills, construct meaning, interact with others, and adapt to changing environments.

As the chapter demonstrates, the development of learning theory reflects an expanding understanding of human learning. Early theories focused primarily on observable behaviour and measurable outcomes, while later approaches increasingly recognised the importance of cognition, experience, social interaction, personal growth, culture, and technology. This progression has led to a more comprehensive view of learning as a dynamic, multidimensional process that extends well beyond the classroom and continues throughout an individual's life.

Understanding these theories is essential for educators, trainers, instructional designers, workplace supervisors, and anyone involved in supporting adult learning. Learning theories provide the conceptual foundations that underpin teaching strategies, assessment practices, curriculum design, learner support, and professional development. By understanding the strengths, limitations, and practical applications of each theory, educators can make informed decisions about how best to engage learners and achieve desired learning outcomes.

Rather than viewing these theories as competing explanations of learning, this chapter encourages readers to see them as complementary perspectives that illuminate different dimensions of the adult learning experience. Together, they provide a rich framework for understanding how adults learn and offer valuable guidance for designing learning experiences that are meaningful, relevant, engaging, and effective in an increasingly complex and interconnected world.

Introduction to Major Theories of Adult Learning

Understanding how adults learn has been a central concern of educators, psychologists, and researchers for more than a century. Over time, a range of learning theories have emerged to explain the processes through which people acquire knowledge, develop skills, change behaviours, and construct meaning from their experiences. Each theory reflects particular assumptions about the nature of learning, the role of the learner, the role of the educator, and the factors that influence learning outcomes.

No single theory fully explains all aspects of adult learning. Instead, each provides a different perspective on how learning occurs and offers valuable insights for educational practice. Some theories focus on observable behaviour and the effects of reinforcement, while others examine cognitive processes, personal development, social interaction, experience, culture, or technology. Together, these theories provide a rich foundation for understanding the complexity of adult learning and for designing effective educational experiences across higher education, vocational education and training (VET), workplace learning, community education, and professional development.

Historically, learning theories can be viewed as an evolution in thinking about learning. Early theories concentrated on observable behaviour and measurable outcomes, while later perspectives increasingly recognised the importance of cognition, experience, social interaction, personal growth, cultural influences, and digital technologies. This progression reflects a broadening understanding of learning as a multifaceted process that extends beyond simple behaviour change.

This evolution can be summarised as follows:

- **Behaviourism** – focuses on what people do and how behaviour can be shaped through reinforcement and conditioning.
- **Cognitivism** – examines how people think, process information, remember, and solve problems.
- **Constructivism** – explores how people actively create meaning by connecting new knowledge with prior experience.
- **Humanism** – emphasises why people learn, focusing on personal growth, self-direction, and motivation.
- **Social Learning Theory** – explains how people learn through observing and interacting with others.
- **Experiential Learning Theory** – highlights how learning develops through experience, reflection, and application.
- **Critical Theory** – investigates how learners question assumptions, challenge power structures, and transform their understanding.
- **Sociocultural Theory** – examines how culture, language, communities, and social participation influence learning.

- **Connectivism** – explains how learning occurs within digital networks and interconnected information systems.

The following sections examine each of these theories in greater detail, exploring their origins, key concepts, major contributors, practical applications, strengths, and limitations. Understanding these perspectives enables educators to make informed decisions about teaching strategies, learning activities, assessment methods, and learner support.

Modern adult education rarely relies on a single learning theory. Effective teaching typically combines elements from multiple theoretical perspectives to address diverse learner needs, learning objectives, and educational contexts. For example, a vocational training program may use behaviourist techniques to develop technical skills, cognitivist strategies to promote understanding, experiential activities to support workplace learning, and social learning approaches to encourage collaboration and mentoring. By drawing on the strengths of different theories, educators can create more engaging, effective, and meaningful learning experiences for adult learners.

Behaviourism and Adult Learning

Behaviourism is one of the earliest and most influential learning theories, focusing on observable behaviour rather than internal mental processes. Behaviourist theorists argue that learning occurs when an individual's behaviour changes as a result of interactions with the environment. From a behaviourist perspective, learning is demonstrated by a measurable change in behaviour that results from experience, practice, reinforcement, or conditioning.

The foundations of behaviourism were established by several pioneering researchers. Ivan Pavlov, a Russian physiologist, introduced the concept of classical conditioning through his experiments with dogs, demonstrating how responses such as salivation could be associated with initially neutral stimuli (Chirimuuta, 2021; Guercio, 2018; Morris et al., 1990). Pavlov discovered that dogs exhibited salivation not only at the sight of food but also at stimuli associated with food delivery, such as the footsteps of the servant bringing food (Staddon, 2024). This established the principle that a temporary relationship could be formed between a neutral stimulus and an unconditioned response through repeated association (Chirimuuta, 2021).

John B. Watson extended these principles to human behaviour and is widely regarded as a foundational figure in behaviourism (Guercio, 2018). Watson firmly believed that psychology should focus on the objective examination of behaviour, arguing that eradicating references to consciousness would yield a better formulation of psychological processes (Guercio, 2018). As Watson declared, psychology's "theoretical goal is the prediction and control of behavior" (Watson, 1913, p. 158). Watson's methodological behaviourism set consciousness aside, insisting that psychology should concern itself only with observable actions,

B.F. Skinner further developed behaviourism through the concept of operant conditioning, which differs from Pavlov's classical conditioning in that the response precedes rather than follows the critical stimulus (Buser, 2006; İNankul, 2016). Skinner's approach was based on the observation that behaviour "operates" on the environment to produce advantageous outcomes (Guercio, 2018). In operant conditioning, a reward or punishment follows after the organism has emitted a specified pattern of behaviour (Buser, 2006; Chirimuuta, 2021). Behaviours followed by positive reinforcement (reward) are repeated and learned, whereas behaviours followed by negative reinforcement or punishment tend to disappear (Staddon, 2024), (Lecas, 2006). Skinner studied these principles extensively through experiments with rats and pigeons in what became known as the "Skinner box" (İNankul, 2016), and his work on schedules of reinforcement contributed significantly to understanding human behaviour (Guercio, 2018).

Within educational contexts, behaviourism views learning as the acquisition of new behaviours through reinforcement and repetition. Classical and operant conditioning principles are both integrated into classroom management and instructional systems. Behavioural strategies such as positive reinforcement, token economies, and behaviour contracts have been shown to be effective in managing behaviour, increasing student engagement, and improving academic performance. According to Skinner's planned instruction model, the material is structured in short teaching units and presented according to the learner's rhythms, with the student's participation being active (Staddon, 2024). Each teaching step consists of information presentation, a question, a space for the learner to respond, and the correct answer as feedback (Staddon, 2024)

The educator's role in behaviourist frameworks centres on structuring instruction, providing opportunities for practice, and delivering reinforcement. Classical conditioning allows educators to shape behaviours by exposing learners to specific stimuli, while operant conditioning enables the systematic reinforcement of desired responses,. Learning objectives are clearly defined, and performance is measured against predetermined standards, consistent with behaviourism's emphasis on observable and measurable outcomes (Lecas, 2006).

Behaviourist Approaches in Vocational and Workplace Education

Behaviourist approaches are particularly prevalent in vocational education and training (VET), workplace training, compliance programs, and competency-based education. In the medical education context, behaviourism is closely linked to competency-based education, where learning is "anchored to behaviorally developed objectives, which clearly define what and how a given task should be performed and lay down the criteria of its acceptable performance" (Omer, 2020a, p. 197). Competency-based training more broadly aims to ensure that the skills delivered by the training systems match the skills needed by industry (Carolino & Rouco, 2022), and is fundamentally performance- and standards-based, involving pre-specification of learning outcomes in a realistic workplace practices (Carolino & Rouco, 2022). Many occupational skills require learners to perform tasks consistently and accurately according to established procedures, whether in healthcare, trades, or industrial settings. Behaviourism's core proposition is that stimuli for learning are embedded in the environment and that learning is manifested as a change in behaviour (Omer, 2020b), making it well-suited for contexts where observable, measurable performance is paramount.

Instructional Techniques Rooted in Behaviourism

Several instructional techniques commonly associated with behaviourism continue to be widely used in adult and vocational education. Behaviourism stands on two legs—"deliberate practice and feedback" (Omer, 2020b, p. 391), and learners are expected to engage in deliberate practice of what their teachers do 'drill and skill' to master a given task or competency" (Omer, 2020a). In workplace safety training, behaviour-based approaches use immediate feedback: for example, "each time a worker would lift or transfer a patient, they would receive immediate feedback from a trainer on actions performed correctly or in need of improvement. This process was repeated several times over several days until the workers were observed to be performing safe lifts in a rapid, fluid way over 90% of the time" (Howard et al., 2004, p. 34). Research with miners similarly found an "overwhelming preference for training that would allow them to try the new skill, including hands-on practice in classroom or lab (61.9%), practice at worksite (48.8%), and simulation or drill (41.7%) (Eckes et al., 2016).

Assessment strategies aligned with behaviourist principles rely on direct observation and structured tools. In medical education, assessment is based on direct observation of performance, which can be run through an objective structured clinical examination (OSCE), rubrics, or checklists (Omer, 2020b). Performance checklists and competency assessments are also central in emergency medicine simulation, where "objective checklist and expert scoring systems must be created" and "each step must be identified, and the proper sequence of those steps must be established" (Brnich et al., 2002, p. 47). In healthcare training for infectious disease care, "repetitive training, practice, and feedback are critical components for ensuring the safety of staff" (Sartoni et al., 2025, p. 11), with documentation of training incorporated into personnel files. Simulations, drills, and step-by-step demonstrations are used extensively across vocational contexts, including firefighter training where trainees are submitted to drills, simulations and operational exercises (Fuchs et al., 2025), and food safety training where interactive approaches including videos, simulations, and demonstrations complement traditional methods (Bond et al., 2007).

Consistent research demonstrates the importance of feedback in learning outcomes (Azaria et al., 2017). Nikitara et al. (2025) identify formative assessment as an effective practice in vocational education, as it provides continuous information about learner progress that can be used to refine both teaching methods and learning activities.

Reinforcement in Workplace Settings

In workplace settings, reinforcement extends beyond immediate corrective feedback and can include a range of incentives designed to encourage desired behaviours. Behaviour-based safety approaches advocate recognising and rewarding safe work practices through mechanisms such as praise, recognition programs, and financial incentives. Proponents argue that the costs associated with these rewards may be offset by reductions in workplace injuries and associated expenses (Howard et al., 2004). Behaviourist theory also emphasises the importance of contingency, whereby learning is strengthened when reinforcement or feedback follows performance promptly and consistently (Omer, 2020b).

Within vocational education and training (VET), reinforcement can take additional forms, including the awarding of qualifications, certificates, licences, and formal recognition of achievement. Evidence of training and assessment outcomes is commonly retained within personnel and professional development records, supporting career progression and ongoing employability (Sartoni et al., 2025). Competency-based assessment systems further reinforce learning by evaluating learners against clearly defined industry standards and classifying performance according to competency outcomes. This creates a structured relationship between successful performance, recognition of competence, and access to professional qualifications and employment opportunities (Carolino & Rouco, 2022).

Strengths and Limitations of Behaviourism in Adult Learning

Behaviourist theories have historically provided the foundation for numerous competency-based curricula and training programmes, where observable changes in behaviour result from specific stimuli in the learning environment (Buhrig, 2023). This approach is particularly effective for developing technical and procedural skills. For instance, in police recruit training, physical aspects such as firearms, use of force, and tactics may be more effectively taught through behavioural methods where skills are repeated over time and subsequently integrated into experiential activities such as scenarios or simulations (Dean & Parson, 2021). Similarly, in health sciences education, clinical simulation has been found effective for developing technical skills in controlled environments (Obashoro-John & Findsen, 2023). The emphasis on measurable performance outcomes aligns well with competency-based frameworks, where learners must demonstrate the acquisition of knowledge and skills that can be repeated and retrieved reliably. In vocational and professional contexts, behaviourist approaches support the development of both task-specific technical competencies and the demonstration of competence through objective assessment (Berríos-Arvey et al., 2024).

Despite these strengths, behaviourism has been subject to considerable criticism from adult learning theorists. A central concern is that behaviourist approaches place insufficient emphasis on learners' experiences, motivations, and capacity for self-direction. Knowles's theory of andragogy, by contrast, posits that adults are self-directed learners whose varied life experiences serve as rich resources in the learning environment, and who are motivated by internal rather than external drives,,. Transformative learning theory, developed by Mezirow, emphasises that adults learn by reflecting on and adjusting their perspectives, a process that behaviourism largely overlooks (Jain & Madaan, 2020; Obashoro-John & Findsen, 2023). Adult learning theorists within the humanistic tradition argue that learning is fundamentally about becoming a person in society and transforming experience into knowledge, skills, and attitudes to support individual development. Behaviourism's focus on external stimuli and observable responses tends to treat learners as passive recipients rather than active constructors of meaning, neglecting critical thinking, creativity, and reflective practice (Dean & Parson, 2021; Matorevhu, 2022). Furthermore, the absence of attention to self-directed learning opportunities diminishes learners' agency and their capacity to continuously evolve in professional practice (Matorevhu, 2022).

Contemporary adult learning practice increasingly combines behaviourist methods with other theoretical frameworks to address these limitations. For example, the Justice Institute of British Columbia revised its police recruit training programme to move away from a lecture-based and behaviourist model, incorporating self-directed learning, case studies, simulation and scenario training, experiential activities, and reflection alongside skill-based repetition (Dean & Parson, 2021). In health sciences education, problem-based learning is used to develop transversal skills such as critical thinking and collaboration, while clinical simulation develops technical skills, suggesting that a hybrid approach balancing different methodologies is ideal (Obashoro-John & Findsen, 2023). Experiential learning theory, rooted in the work of Kolb, Dewey, and others, emphasises that knowledge is created through the transformation of experience and that time and space for reflection help learners absorb and make sense of their experiences (Jain & Madaan, 2020; Rismiyanto, 2017). Vocational training contexts similarly recommend combining competency-based approaches with direct experiential learning in workplace or community settings (Berríos-Arvey et al., 2024). Creative and play-based pedagogies, transformative learning approaches, and collaborative problem-solving are increasingly sought after alongside traditional skill-development methods in professional education (Ghamrawi et al., 2025). This integrated perspective recognises that while behaviourist strategies remain highly effective for developing specific, measurable competencies, adult learning also fundamentally involves experience, reflection, social interaction, and personal meaning-making (Dean & Parson, 2021; Jain & Madaan, 2020).

Practical Applications of Behaviourism in Adult Learning

Behaviourist principles continue to be widely applied across higher education, vocational education and training (VET), workplace learning, and English as a Second Language (ESL) education. These applications typically involve clearly defined learning objectives, structured instruction, deliberate practice, repetition, reinforcement, and measurable performance outcomes (Abidin, 2016; Edgar, 2012; İNankul, 2016; LaMonica, 2000; Nalliah, 2014; Sani & Mohamad, 2023; Utami, 2021; Zhang, 2024).

In higher education, behaviourist approaches are commonly used when students must master foundational knowledge, technical procedures, or professional competencies. For example, medical students may repeatedly practise clinical procedures such as taking blood pressure, administering injections, or performing cardiopulmonary resuscitation (CPR). Instructors provide immediate feedback and assess performance using structured checklists and objective criteria.

Similarly, engineering students may complete repeated calculations and problem-solving exercises to develop proficiency in applying formulas and design standards. In nursing education, simulation laboratories often require students to perform specific clinical tasks repeatedly until competence is demonstrated. Online learning platforms frequently incorporate behaviourist principles through automated quizzes, instant feedback, progress tracking, and mastery-based progression systems.

Behaviourism is particularly evident within competency-based training systems. Apprentices learning welding, carpentry, automotive repair, or electrical installation often begin by observing a demonstration before repeatedly practising the task under supervision. Trainers provide corrective feedback and reinforcement until the learner can consistently perform the task to the required industry standard.

In heavy vehicle driver training, learners repeatedly practise vehicle inspections, coupling and uncoupling procedures, and safe driving techniques. In hospitality training, students may repeatedly practise food preparation, customer service interactions, or coffee-making procedures. Assessment commonly involves direct observation against performance criteria, with learners required to demonstrate competency through repeated successful performance.

Safety and compliance training also frequently adopts behaviourist approaches. Workers may complete drills, emergency response exercises, lockout-tagout procedures, or manual handling activities where correct behaviours are reinforced through coaching, feedback, and formal assessment.

Many workplace induction and compliance programs are behaviourist in design. New employees are taught specific procedures, complete practice activities, receive feedback, and are assessed against predetermined standards. Examples include workplace health and safety training, infection control procedures, cybersecurity awareness programs, and equipment operation training.

Behaviour-based safety programs are another common application. Employees are observed performing tasks, provided with immediate feedback, and recognised for demonstrating safe work practices. Organisations may reinforce desired behaviours through recognition programs, performance incentives, certificates of achievement, or promotion opportunities.

In language education, behaviourist theory has had a particularly profound influence. Behaviourism conceptualizes language learning as occurring through environmental stimulation, where learners acquire language through stimulus-response associations and habit formation (Abidin, 2016) . The Audio-Lingual Method, widely used in ESL contexts, is directly grounded in Skinnerian operant conditioning principles, emphasizing drilling, repetition, and reinforcement of correct responses to build linguistic habits (Utami, 2021). Mobile web-based learning platforms for ESL learners similarly incorporate behaviourist principles, creating environments that promote active engagement, continuous practice, and positive learning behaviours (Sani & Mohamad, 2023).

For example, an ESL teacher may introduce a grammatical structure and then engage learners in repeated drills:

- Teacher: "I am going to work."
- Student: "I am going to work."
- Teacher: "She is going to school."
- Student: "She is going to school."

Pronunciation training often involves repeated modelling and imitation, with immediate corrective feedback provided when errors occur. Vocabulary acquisition may be reinforced through flashcards, repetition exercises, matching activities, and computer-assisted language learning programs that reward correct responses and provide instant feedback.

Modern language-learning applications such as gamified mobile platforms also incorporate behaviourist elements through repetition, reinforcement, progress indicators, achievement badges, and reward systems that encourage continued practice.

However, critics have noted that commercial ESL software has "departed little from past behaviorist products," remaining locked in drill-and-practice paradigms rather than embracing communicative approaches (LaMonica, 2000).

Many contemporary learning technologies incorporate behaviourist principles regardless of the educational sector. Learning management systems often use quizzes, badges, completion tracking, automated feedback, and competency dashboards to reinforce learning behaviours. Adaptive learning platforms may require learners to demonstrate mastery before progressing to more advanced content, reflecting behaviourism's emphasis on observable performance and measurable achievement.

Across higher education, vocational education, workplace learning, and ESL instruction, behaviourist approaches are most effective when learners must acquire specific knowledge, procedural skills, or observable competencies. Demonstration, repetition, deliberate practice, feedback, reinforcement, and objective assessment remain central features of many contemporary adult learning environments, even when combined with more learner-centred approaches such as experiential, constructivist, or transformative learning.

Cognitivism and Information Processing

Cognitivism emerged during the mid-twentieth century as a response to the limitations of behaviourism (An, 2020; Dakich, 2014; Edgar, 2012). While behaviourism focused on observable behaviours and external stimuli, cognitivism shifted attention to the internal mental processes involved in learning (Dakich, 2014; Yeo, 2024). Cognitivist theorists argue that learning is not simply a change in behaviour but rather a change in knowledge, understanding, memory, and thinking (Tehlan, 2023; Yeo, 2024). Learning occurs when individuals acquire, organise, store, retrieve, and apply information (Yeo, 2024).

The rise of cognitivism was influenced by developments in psychology, linguistics, computer science, and neuroscience (Dakich, 2014; Tehlan, 2023). Researchers began to recognise that behaviour alone could not fully explain complex human learning processes such as reasoning, problem-solving, language acquisition, and decision-making (An, 2020; Edgar, 2012). Cognitivism therefore seeks to understand what happens inside the learner's mind during learning rather than focusing solely on observable actions (Dakich, 2014; Tehlan, 2023; Yeo, 2024).

Information Processing Theory

One of the most influential branches of cognitivism is Information Processing Theory. This theory likens the human mind to an information-processing system, suggesting that information is received, processed, stored, and retrieved through a series of interconnected mental processes. While the comparison between the mind and a computer has limitations (Torres et al., 2023), it provides a useful framework for understanding how learning occurs and how information is transformed into knowledge (Khalil & Elkhider, 2016; Torres et al., 2023).

According to information processing theory, learning involves three primary memory systems: sensory memory, working memory, and long-term memory (Banikowski & Mehring, 2017; Khalil & Elkhider, 2016; Torres et al., 2023; Yeo, 2024). Information first enters the learner's sensory memory through the senses, including sight, hearing, touch, smell, and taste. Sensory memory retains incoming information for only a brief moment Torres et al., 2023), Yeo, 2024). Unless the learner directs attention towards the information, it quickly disappears (Khalil & Elkhider, 2016; Torres et al., 2023; Yeo, 2024). For example, a student may briefly see a diagram displayed during a lecture. If the student pays attention to the diagram, the information is transferred to working memory for further processing; if not, it is rapidly forgotten (Khalil & Elkhider, 2016; Torres et al., 2023).

Working memory, sometimes referred to as short-term memory, is the stage where conscious thinking and active processing occur (Lemieux, 2004; Yeo, 2024). Information is temporarily held while learners interpret, analyse, compare, and manipulate it Cowan, 2013). However, working memory has a limited capacity and can only process a relatively small amount of information at any one time (Cowan, 2013; El-Kishawi et al., 2021; Khalil & Elkhider, 2016). When learners are presented with excessive information or overly complex tasks, cognitive overload may occur, reducing their ability to learn effectively (Khalil & Elkhider, 2016; ,. This concept has significant implications for instructional design, particularly in adult education and vocational training, where information must often be presented in manageable stages (El-Kishawi et al., 2021; Khalil & Elkhider, 2016). Activities that rely on working memory include solving mathematical problems, following instructions, comparing concepts, participating in discussions, and learning new workplace procedures (Cowan, 2013).

Information that is effectively processed within working memory can be transferred to long-term memory (Banikowski & Mehring, 2017; Khalil & Elkhider, 2016; Torres et al., 2023). Long-term memory serves as the brain's primary storage system and is capable of retaining information for extended periods, potentially throughout an individual's lifetime (Yeo, 2024). Information is more likely to be stored in long-term memory when it is meaningfully understood, rehearsed, organised, and connected to existing knowledge and experiences (Banikowski & Mehring, 2017; Khalil & Elkhider, 2016; Torres et al., 2023). Once stored, knowledge can be retrieved and applied when required (Banikowski & Mehring, 2017; Zainuddin et al., 2024). Frequent use and reinforcement strengthen these memory pathways, making recall easier and more reliable (Banikowski & Mehring, 2017; Stokes & Berlacher, 2022; Zainuddin et al., 2024).

Examples of information stored in long-term memory include remembering how to operate machinery, recalling professional regulations and procedures, applying engineering formulas, speaking a second language, or performing a clinical procedure (Cowan, 2013; Ketev et al., 2023). The ability to retrieve and apply such knowledge in new situations is a key indicator that meaningful learning has occurred (Torres et al., 2023; Zainuddin et al., 2024). Information Processing Theory therefore provides valuable insights into how educators can structure learning experiences to support attention, understanding, memory retention, and knowledge transfer (Khalil & Elkhider, 2016; Yeo, 2024).

Key Cognitive Processes

Information Processing Theory explains the process of acquiring, processing, storing, and retrieving information from memory and provides guidance on how memory can be enhanced (Zhang et al., 2023). The theory was initially presented by Atkinson and Shiffrin as a cognitive theory of learning that addresses the processing, storage, and retrieval of knowledge in the mind (Metsämuuronen & Räsänen, 2018). Research on the relationship between cognition, knowledge, and education has confirmed the positive impacts of instructional design on each learning cycle phase, including concentration, preparation, encoding, retrieval, and attention (Tangen & Borders, 2017).

Attention determines which information from the surrounding environment enters working memory for further processing. Selective attention is crucial to most types of information processing. Attention and awareness play a major role in understanding learning and retention; attended learning is far superior, and voluntary orienting vastly improves encoding (He & Chen, 2026). Information received from the environment via sensory memory that is within the focus of attention is processed consciously in working memory for possible storage in long-term memory (Terrell, 2006). Educators often employ strategies such as questioning, visual aids, demonstrations, storytelling, multimedia presentations, and practical activities to attract and sustain learner attention. Expressiveness in teaching, through physical movement, voice intonation, and eye contact, primes students' information processing by facilitating their selective attention. A learner can retain knowledge better when information is encoded together with visuals, and through utilisation of visualisation during lessons, students' attention can be drawn (Metsämuuronen & Räsänen, 2018).

Encoding is the process of transforming information into a form that can be retained within long-term memory. Effective encoding occurs when learners actively engage with the material and connect new information to existing knowledge. Elaboration is the process of linking new information with previous information (Zhang et al., 2023). The more meaningful the analyses of stimuli at hand are, the higher the levels of subsequent retention will be (Sarfo, 2016). Multimodule sensory-motor encoding of information can facilitate memory storage and retrieval (Meylani, 2024). Clarity in instruction plays an important role in the encoding and storage stages of information processing, which enhance and facilitate retrieval. Learning activities that encourage discussion, reflection, summarisation, and application strengthen the encoding process; instructional applications that enhance emotional arousal and call for reflection upon previous experiences can indeed enhance retention in learning experiences (Hagen & Park, 2016). Drawing, for instance, improves memory by encouraging a seamless integration of semantic, visual, and motor aspects of encoding (Leong & Raja Lailatul Zuraida Raja Maamor, 2023).

Storage involves maintaining information within long-term memory so that it can be accessed in the future. Information is often organised into interconnected mental frameworks known as schemas. Schema is described as a memory structure that contains the records of human experiences or information, consisting of a network of interrelations among its constituent parts (Sanatullova-Allison, 2014). Schemas provide a cognitive framework that helps individuals organise and retrieve information, promoting efficient memory storage and retrieval processes. A schema is an organised framework of long-term knowledge that, when activated, facilitates the encoding and learning of new, related concepts, promotes inference, reasoning, and problem solving, and guides recall of relevant information (Yaşar, 2016). As learners acquire additional knowledge, their schemas become more complex; schema-consistent information is more easily integrated into existing schemas, promoting efficient encoding and storage. The highest learning efficiency occurs when short-term memory is transformed into long-term memory with lasting schemas (Abeysekera et al., 2024). Information previously stored in long-term memory is triggered and retrieved when processing new information, which substantially reduces learners' working memory load and increases learning efficiency (Meylani, 2024).

Retrieval involves accessing information from long-term memory when it is required. The knowledge retention process entails three phases: information encoding, information storage, and information retrieval; a student's failure in any phase will result in forgetting or incorrect memory formation (Metsämuuronen & Räsänen, 2018). Memory retrieval plays an important role in the functioning of working memory (Sarfo, 2016). By providing frequent distributed practice, the learner can retrieve and utilise information stored in long-term memory (Metsämuuronen & Räsänen, 2018). Activities that require learners to actively retrieve information help reinforce learning; distributed practice with increasing intervals after correct retrievals generates a much higher retention rate than massed practice (He & Chen, 2026). Spaced practices result in a more robust enhancement of long-term memory encoding, recall, and recollection through both direct and indirect effects of memory processing characteristics. Retrieval practice and spaced repetition have been shown to reinforce memory encoding and retrieval processes (Schönwetter, 1993).

Together, attention, encoding, storage, and retrieval form the core cognitive processes that underpin learning. By understanding how these processes operate, educators can develop instructional strategies that maximise learner engagement, improve memory retention, and support the effective application of knowledge and skills. Teaching students in brain-friendly ways means implementing principles that align with how the human brain encodes, consolidates, and retrieves information (Chew, 2021). Neuroscience-informed educational design can promote learning motivation, high-level information processing, and knowledge retention (Chew, 2021).

Major Contributors to Cognitivism

Several influential theorists contributed to the development of cognitivism and significantly shaped contemporary understandings of learning. Jean Piaget proposed that learners actively construct knowledge through interaction with their environment. His concepts of assimilation and accommodation explain how individuals incorporate new information into existing mental structures and modify those structures when new experiences cannot be readily understood within existing frameworks (Piaget, 1952, 1970).

Jerome Bruner further advanced cognitivist thinking through his emphasis on discovery learning, arguing that learners should actively engage with ideas and concepts rather than simply receive information from instructors. Bruner also introduced the concept of scaffolding, whereby support is provided to learners during the learning process and gradually withdrawn as competence and independence develop (Bruner, 1961; Wood et al., 1976).

Robert Gagné contributed to instructional design through his theory of learning and his identification of different types of learning outcomes. He proposed a systematic model of instruction that outlined specific events of instruction designed to support attention, learning, retention, and transfer of knowledge and skills (Gagné, 1965; Gagné et al., 1992).

Finally, Richard Atkinson and Richard Shiffrin developed the influential multi-store model of memory, which describes information processing as a movement between sensory memory, short-term memory, and long-term memory. Their model provided a foundational framework for understanding memory processes and remains highly influential in educational psychology and instructional design (Atkinson & Shiffrin, 1968).

Cognitivism in Adult Learning

Cognitivism aligns closely with many of the characteristics associated with adult learners. Adults typically bring substantial prior knowledge, life experiences, and professional expertise to learning situations (Al-Mosawi, 2022; Figurska & Sokół, 2016; Rubaia'an, 2023). Cognitivist theories recognise that learning does not occur in isolation; rather, new knowledge is interpreted and understood through the lens of what learners already know (Al-Mosawi, 2023; Badyal & Singh, 2017; Brieger et al., 2020). The organisation of existing knowledge within memory influences how new information is processed, understood, and retained (Al-Mosawi, 2019; Badyal & Singh, 2017; Moll, 2024).

Specifically, cognitivism focuses on how information is acquired, organised, stored, and retrieved within the mind, viewing learning as a process of information processing and knowledge construction (Al-Mosawi, 2019, 2022). Cognitivist approaches emphasise the importance of helping learners understand the meaning of new information and connect it to their existing knowledge structures. Learning is therefore seen as an active mental process in which learners interpret, organise, and integrate information to develop understanding and support long-term retention (Brieger et al., 2020).

Within this framework, learners acquire knowledge through activities such as listening, observing, reading, and direct experience. However, learning does not occur simply through exposure to information; it requires learners to actively process, interpret, and remember what they encounter (Al-

Mosawi, 2019, 2023). The effectiveness of learning is influenced by how well new information can be connected to existing cognitive structures and prior experiences.

Constructivism, which shares many principles with cognitivism, places particular emphasis on the role of prior knowledge in the learning process. Constructivist theorists argue that learners build new understanding by drawing upon and extending what they already know. As a result, learning experiences that acknowledge and utilise learners' existing knowledge and experiences are generally associated with deeper understanding and more effective learning outcomes (Al-Mosawi, 2022, 2023).

Adult learners are typically characterised by a high degree of self-direction, a wealth of accumulated life and professional experience, a readiness to learn when knowledge is relevant to their needs, a preference for practical and problem-centred learning, and strong intrinsic motivation (Al-Mosawi, 2022; Figurska & Sokół, 2016; Silva, 2018). These characteristics align closely with cognitivist principles because adults draw upon existing cognitive structures developed through prior learning and experience when acquiring new knowledge. From a cognitivist perspective, learning occurs by building upon these established mental frameworks, allowing learners to interpret, organise, and integrate new information with what they already know (Badyal & Singh, 2017; Bigdeli et al., 2023).

Cognitivism is particularly relevant in situations where learners must process information, retrieve previously acquired knowledge, develop understanding, and refine existing mental models. Learning is viewed as an active process in which new knowledge is incorporated into and reshapes existing cognitive structures (Badyal & Singh, 2017). Consequently, prior knowledge plays a critical role in determining how learners interpret new information, respond to learning activities, and construct meaning from educational experiences. Existing knowledge and experiences influence attention, comprehension, and problem-solving, thereby shaping the effectiveness of learning (Brieger et al., 2020).

The importance of prior knowledge can also be understood through cognitive models of memory. The tripartite model of memory proposes that learning involves interactions between sensory memory, working memory, and long-term memory. Working memory functions as the central processing system responsible for attention, reasoning, and higher-order cognitive tasks, while long-term memory stores existing knowledge and experiences. New information is interpreted and integrated through these existing knowledge structures, providing a cognitive explanation for how prior learning influences the acquisition and application of new knowledge (Moll, 2024).

Adult learners are frequently required to analyse complex information, solve problems, make decisions, and apply knowledge in practical contexts. They must often integrate new learning with existing expertise and transfer knowledge and skills to workplace situations. As a result, cognitivist approaches focus on helping learners understand concepts, organise information, identify relationships, and apply knowledge effectively, rather than simply memorising facts. Learning activities are designed to promote understanding, reasoning, and the development of mental frameworks that support long-term retention and transfer of learning.

Cognitivist principles are widely applied in higher education, where learning frequently involves complex reasoning, critical analysis, and problem-solving. Teaching methods such as lectures, tutorials, case studies, concept mapping, reflective activities, and collaborative discussions are designed to help students process information and develop deeper understanding.

For example, engineering students may analyse design problems and evaluate alternative solutions using established principles and formulas. Business students often examine case studies to identify organisational issues, assess possible responses, and justify their recommendations. Nursing students apply clinical reasoning to patient scenarios, while law students interpret legislation, legal principles, and precedents to construct legal arguments. In each case, learners are required to process information, identify patterns and relationships, and apply their knowledge to new and often complex situations.

In vocational education and training (VET), cognitivism complements competency-based approaches by helping learners understand not only how to perform a task, but also why it is performed in a particular manner. While behaviourist approaches may focus on demonstrating and practising procedures, cognitivist approaches encourage learners to understand the principles, reasoning, and decision-making processes that underpin competent performance.

Examples include understanding the principles behind electrical safety procedures, analysing load restraint requirements in transport operations, interpreting engineering drawings, diagnosing faults in mechanical systems, or planning construction activities. To support cognitive processing, vocational training often incorporates diagrams, flowcharts, troubleshooting exercises, simulations, workplace scenarios, and problem-based learning activities. These strategies help learners organise information, develop mental models, and apply knowledge in realistic workplace contexts.

Cognitivist theories have also had a significant influence on modern English as a Second Language (ESL) and language education more broadly. Unlike traditional behaviourist approaches that rely heavily on repetition and drills, cognitivist approaches encourage learners to understand language structures, recognise patterns, and apply language rules in meaningful contexts.

Learners may analyse grammar rules, explore relationships between sentence structures, use vocabulary in authentic situations, engage in reading comprehension activities, and complete writing tasks that require planning and organisation. Reflection on language-learning strategies is also encouraged, helping learners become more aware of how they learn and how they can improve their language skills. Through these approaches, learners actively construct understanding of the language system rather than simply imitating language patterns, leading to deeper comprehension and more flexible language use.

Strengths and Limitations of Cognitivism

Cognitivism has made a significant contribution to understanding how learning occurs and remains one of the most influential learning theories in education (Angulo-Alvarez & López, 2024; Beatriz et al.; Mantuano et al., 2021). One of its major strengths is its recognition of the importance of internal mental processes such as thinking, memory, perception, reasoning, and problem-solving (Meza, 2022; Pástor et al., 2021; Ríos & Rodríguez, 2022). Unlike behaviourism, which focuses primarily on observable behaviour, cognitivism seeks to explain how learners acquire, organise, process, and apply knowledge (Angulo-Alvarez & López, 2024; Meza, 2022; Pástor et al., 2021). This enables the theory to explain more complex forms of learning that involve understanding, analysis, decision-making, and the transfer of knowledge to new situations (Brítez, 2023; Meza, 2022).

Another strength of cognitivism is its acknowledgement of the role of prior knowledge and experience in learning. The theory recognises that learners do not approach new information as blank slates; rather, they interpret and integrate new knowledge based on existing mental structures and experiences (Brítez, 2023; Cortés & Polo, 2021; Uribe-Canónigo, 2017). Cognitivist principles have also provided valuable guidance for instructional design by emphasising the organisation of content, the management of cognitive load, and the use of strategies that support memory and understanding (Meza, 2022). Many contemporary educational technologies, online learning systems, adaptive learning platforms, and instructional design models are grounded in cognitivist principles and information-processing concepts (Douieb, 2021; Pástor et al., 2021).

Despite these strengths, cognitivism is not without limitations. One common criticism is that it may underestimate the influence of social, cultural, and environmental factors on learning (Espinoza, 2023). By focusing primarily on individual mental processes, cognitivism can overlook the ways in which learning is shaped through interaction with others, participation in communities, and engagement with cultural practices (Espinoza, 2023; Uribe-Canónigo, 2017). This has led some researchers to argue that cognitivist approaches provide only a partial explanation of learning (Espinoza, 2023).

Critics have also suggested that information-processing models may oversimplify the complexity of human thinking by comparing the mind to a computer (Pástor et al., 2021; Torres et al., 2022). While such models offer useful frameworks for understanding memory and learning, they may not fully capture the dynamic and interconnected nature of human cognition (Molina, 2022; Torres et al., 2022). Additionally, emotional, motivational, and contextual influences on learning may receive insufficient attention within purely cognitivist approaches (Espinoza, 2023; Ríos & Rodríguez, 2022). Factors such as learner confidence, personal values, social relationships, and workplace culture can significantly affect learning outcomes but are not always adequately explained by information-processing models (Espinoza, 2023; Ríos & Rodríguez, 2022).

Overall, cognitivism views learning as a process of acquiring, organising, storing, and retrieving knowledge (Angulo-Alvarez & López, 2024; Brítez, 2023). Information Processing Theory explains this process through the movement of information between sensory memory, working memory, and long-term memory (Meza, 2022; Torres et al., 2022). Whereas behaviourism focuses on observable actions and behavioural change, cognitivism seeks to understand the mental processes that underpin learning and performance (Mantuano et al., 2021; Meza, 2022; Torres et al., 2021). Its influence can be seen across higher education, vocational education and training, workplace learning, and language education, where learners are encouraged to actively process information, develop knowledge structures, solve problems, and apply understanding in meaningful and practical contexts (Brítez, 2023; E. A. Ortiz et al., 2021; Parra et al., 2023).

Constructivism and Social Constructivism

Constructivism is a learning theory that proposes learners actively construct knowledge rather than passively receive it (Ardiansyah & Ujihanti, 2017; Malik et al., 2019; Sulistyowati, 2019). According to constructivist perspectives, learning occurs when individuals interpret new information and experiences in light of their existing knowledge, beliefs, and experiences (Bell, 2021; Sulistyowati, 2019). Knowledge is therefore not simply transmitted from teacher to learner; instead, it is actively built by the learner through engagement with ideas, problems, experiences, and reflection (Ardiansyah & Ujihanti, 2017; Bell, 2021; Burkšaitienė, 2006)

Constructivism emerged largely as a response to behaviourist and traditional instructional approaches that viewed learning as the accumulation of facts or the acquisition of conditioned responses (Malik et al., 2019; Vásquez et al., 2023). Constructivist theorists argue that learners are active participants in the learning process who continuously create, modify, and reorganise their understanding of the world (Ardiansyah & Ujihanti, 2017; Malik et al., 2019). Learning is viewed as a process of meaning-making rather than memorisation (Ardiansyah & Ujihanti, 2017; Botross, 2024; Doolittle et al., 2023).

Foundations of Constructivism

One of the most influential contributors to constructivist learning theory was Jean Piaget. Piaget proposed that learners actively construct knowledge through their interactions with the environment rather than passively receiving information from others. Central to his theory are the processes of assimilation and accommodation, which explain how individuals develop and modify their understanding of the world. Assimilation occurs when new information is integrated into existing cognitive structures or schemas, while accommodation occurs when those schemas must be altered or reorganised to account for new experiences that cannot be adequately explained by existing knowledge (Piaget, 1952, 1970).

According to Piaget, learning is driven by a desire to achieve cognitive equilibrium, a state in which an individual's understanding of the world is internally consistent. When learners encounter information that conflicts with their existing knowledge, a state of disequilibrium occurs, motivating them to adapt their thinking through assimilation or accommodation. Learning is therefore viewed as an active and continuous process of exploration, questioning, experimentation, and reflection, through which individuals gradually develop more sophisticated understandings of concepts and phenomena (Piaget, 1952; Schunk, 2020).

Constructivism more broadly is founded on the principle that learners actively construct their own understanding of reality. Knowledge is not viewed as an objective body of information that can simply be transmitted from teacher to learner; rather, it is created through the learner's interpretation of experiences and interactions with the environment. Prior knowledge plays a critical role in this process because new learning is understood and interpreted through the lens of existing cognitive structures. Consequently, learners may construct different understandings of the same experience based on their individual backgrounds, experiences, and perspectives (Bruner, 1966; Fosnot, 2013).

From a constructivist perspective, learning involves the active creation of meaning, with understanding developing and evolving over time as learners encounter new experiences and information. Knowledge is therefore seen as dynamic rather than fixed, and learning is understood as a process of continuous refinement and reconstruction of understanding. This perspective acknowledges that multiple interpretations of reality may exist and that learners develop knowledge through personal engagement, reflection, and experience rather than through passive reception of facts (Fosnot, 2013; Schunk, 2020).

Social Constructivism

While constructivism primarily focuses on how individuals construct knowledge through personal experience, social constructivism emphasises the importance of social interaction, language, culture, and collaboration in the learning process. Social constructivists argue that learning is not solely an individual cognitive activity but is fundamentally shaped by participation in social and cultural contexts.

Knowledge develops through interactions with others and is influenced by the values, beliefs, practices, and language of the communities in which learners participate.

The most influential contributor to social constructivism was Lev Vygotsky. Vygotsky argued that learning is inherently social and that cognitive development occurs through interaction with more knowledgeable others, such as teachers, mentors, peers, supervisors, and community members (Vygotsky, 1978). From this perspective, learning begins in social contexts before being internalised by the individual. Through dialogue, observation, collaboration, and shared activities, learners acquire knowledge and skills that they gradually incorporate into their own understanding.

Language occupies a central role within social constructivist theory because it serves as the primary tool through which knowledge is communicated, negotiated, and constructed. Through language, learners ask questions, share ideas, clarify misunderstandings, and engage in collaborative problem-solving. Vygotsky argued that language not only facilitates communication with others but also supports the development of internal thought processes and higher-order cognitive functions (Vygotsky, 1978).

Social constructivism therefore views learning as a socially mediated process in which knowledge is co-constructed through interaction with others. Understanding is shaped by cultural influences, social relationships, and participation in communities of practice. Collaboration is considered essential because it exposes learners to diverse perspectives and supports the development of deeper understanding. Learning is not viewed as the acquisition of objective knowledge alone but as participation in shared meaning-making processes that occur within social and cultural environments (Schunk, 2020).

One of Vygotsky's most influential contributions to educational theory is the concept of the Zone of Proximal Development (ZPD). The ZPD refers to the difference between what a learner can accomplish independently and what they can achieve with appropriate guidance and support from a more knowledgeable individual (Vygotsky, 1978). This concept highlights the idea that learning is often most effective when learners are challenged beyond their current level of independent competence but are provided with sufficient support to succeed.

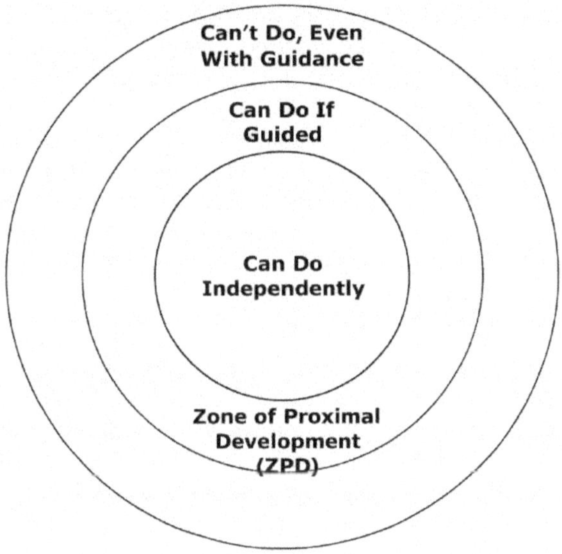

Figure 3: Zone of Proximal Development (ZPD).

Within the Zone of Proximal Development, learners can perform tasks and solve problems that would otherwise be beyond their capabilities. Teachers, mentors, supervisors, and experienced peers play an important role in providing assistance, modelling strategies, asking guiding questions, and offering feedback. As learners gain knowledge, skills, and confidence, they become increasingly capable of performing these tasks independently.

The concept of the ZPD gave rise to the educational practice known as scaffolding. Scaffolding involves providing temporary instructional support to help learners complete tasks that they cannot yet perform on their own. This support may take many forms, including demonstrations, hints, explanations, questioning, modelling, feedback, or collaborative activities. As learners develop competence and confidence, the level of support is gradually reduced until they are able to perform independently. Scaffolding has become a widely used instructional strategy across higher education, vocational education and training, workplace learning, and professional development because it enables learners to progressively develop knowledge and skills within a supportive learning environment (Schunk, 2020; Wood et al., 1976).

Constructivism in Adult Learning

Constructivism aligns strongly with the principles of adult learning because adults bring a wealth of prior knowledge, life experiences, workplace expertise, values, and beliefs to educational settings (Ardiansyah & Ujihanti, 2017; Woodward et al., 2021). Rather than passively receiving information, adult learners actively interpret, evaluate, and integrate new knowledge based on what they already know (Mishra, 2023; O'Donnell, 2012). Learning is therefore viewed as a process of constructing meaning, where individuals connect new ideas with existing understandings and experiences to develop deeper and more personally relevant knowledge (O'Donnell, 2012; Woodward et al., 2021).

From a constructivist perspective, effective adult learning builds upon learners' prior experiences and encourages active participation in the learning process (Vaishali & Mısra, 2020; Woodward et al., 2021). Learners are encouraged to question, investigate, analyse, discuss, and reflect upon information rather than simply memorising facts (Megawati et al., 2023; Mishra, 2023; Mugambi, 2018). Learning activities are designed to promote problem-solving, critical thinking, collaboration, and the application of knowledge to authentic situations (Megawati et al., 2023; Mugambi, 2018; Saldivar, 2025; Uzorka et al., 2020). Reflection is particularly important because it enables learners to examine their assumptions, evaluate their experiences, and refine their understanding (Uzorka et al., 2020; Woodward et al., 2021). Through these processes, learners construct personal meaning and develop knowledge that can be applied in real-world contexts (Mishra, 2023; Shah, 2019).

Adult learners are therefore particularly likely to benefit from educational approaches that involve investigation, discussion, collaboration, and practical application (Megawati et al., 2023; Uzorka et al., 2020; Woodward et al., 2021). Constructivist learning environments typically position the educator as a facilitator who guides inquiry, supports exploration, and encourages learners to develop their own understanding rather than simply transmitting information (Ardiansyah & Ujihanti, 2017; Mishra, 2023; Mugambi, 2018; Shah, 2019; Thoron & Myers, 2012; Vaishali & Mısra, 2020).

Constructivist approaches are widely used throughout higher education, particularly in disciplines that require critical thinking, analysis, professional judgement, and the application of knowledge to complex situations (AlAfnan & Dishari, 2024; Roxas et al., 2023). Rather than focusing solely on lectures and information delivery, constructivist teaching strategies encourage students to actively engage with concepts, problems, and real-world scenarios (Adhikari & Kunwar, 2023; Bell & Liu, 2019; Doolittle et al., 2023; Saldivar, 2025).

For example, engineering students may be required to analyse complex design challenges, evaluate alternative solutions, and justify their recommendations based on technical and practical considerations. Nursing students often work through patient case studies that require them to assess symptoms, interpret clinical information, and determine appropriate interventions. Law students may analyse legislation, case law, and legal precedents to construct arguments and resolve hypothetical legal disputes. Similarly, business students frequently examine organisational case studies and develop strategies for addressing management, financial, or operational challenges.

Constructivist approaches in higher education commonly include problem-based learning, inquiry-based learning, research projects, collaborative discussions, simulations, reflective journals, capstone projects, and industry-based projects. In these environments, learning occurs through investigation, reasoning, dialogue, and reflection, enabling students to develop a deeper understanding of concepts and their practical applications.

Constructivism has become increasingly influential within vocational education and training (VET), where learners must often apply theoretical knowledge to authentic workplace situations (Hébrard, 2013; Margeviča-Grīnberga et al., 2023; Ruslin, 2017). While competency-based training focuses on demonstrating workplace skills and standards (Hébrard, 2013; Qing, 2023; Ruslin, 2017), constructivist approaches help learners understand the reasoning and decision-making processes that underpin competent performance (Luaensutthi, 2023; Sarfo, 2016).

For example, apprentice electricians may investigate electrical faults within simulated systems and develop diagnostic strategies to identify and rectify problems. Construction students may plan, schedule, and manage building projects that mirror real workplace conditions. Heavy vehicle driver trainees may analyse route-planning scenarios involving fatigue management, load restraint, and regulatory compliance. Hospitality students may respond to customer service situations that require problem-solving and decision-making rather than simply following scripted procedures.

Workplace projects, simulated work environments, troubleshooting exercises, apprenticeships, traineeships, workplace learning placements, and scenario-based activities are common examples of constructivist learning within VET. These approaches encourage learners to integrate theory and practice while developing the judgement and adaptability required in real workplaces.

Constructivist principles are also widely applied in workplace training and professional development. Rather than relying solely on formal instruction, organisations increasingly use learning activities that encourage employees to solve authentic problems, collaborate with colleagues, and reflect on their experiences (Armson, 2009; Evanick, 2024; Mathews, 2007).

For example, leadership development programs may involve participants analysing organisational challenges and developing strategic solutions. Project management training often requires learners to plan and manage realistic projects while responding to changing circumstances. Healthcare professionals may participate in multidisciplinary case reviews where they collaboratively evaluate patient outcomes and identify opportunities for improvement. Emergency services personnel frequently engage in scenario-based exercises that replicate real incidents and require teams to make decisions under pressure.

Mentoring programs, communities of practice, action learning projects, workplace investigations, and reflective practice activities all reflect constructivist principles by encouraging employees to construct knowledge through experience, collaboration, and reflection.

Constructivist principles have significantly influenced modern English as a Second Language (ESL) and language education more broadly (Jumaah, 2024; Sulindra et al., 2024). Rather than focusing primarily on memorisation, grammar drills, and repetition, constructivist approaches encourage learners to actively use language in meaningful and authentic contexts (Camara-Walker & Irizarry, 2013; Eslit, 2023; Jumaah, 2024).

For example, learners may participate in group discussions, collaborative projects, role plays, workplace communication simulations, and problem-solving activities that require the practical use of language. Adult migrants learning English may engage in activities such as completing government forms, participating in job interviews, communicating with service providers, or planning community events. Business English learners may participate in meetings, presentations, negotiations, and workplace communication exercises that mirror real professional situations.

Project-based learning, authentic communication tasks, reflective language journals, peer collaboration, and task-based learning activities enable learners to construct understanding of language through use and interaction. In this way, language is learned as a tool for communication and meaning-making rather than as a collection of isolated grammatical rules.

Constructivist approaches are also common in community education and adult learning programs. Adult literacy classes, financial literacy programs, parenting education, health education, and community development initiatives often draw upon learners' existing experiences as a foundation for new learning (Ardiansyah & Ujihanti, 2017; L. R. Brown et al., 2023; Saldivar, 2025; Woodward et al., 2021).

In the domain of literacy education, constructivist frameworks emphasize that learners construct knowledge by merging prior knowledge and experience with current learning (Anna et al., 2023). Social constructivism specifically posits that school literacy learning among students of diverse backgrounds improves when educators address instructional goals, the role of home language, instructional materials, and relationships with the community (Ardiansyah & Ujihanti, 2017). From a social constructivist perspective, literacy learning outcomes are viewed as the result of interactions among students, teachers, families, communities, and educational institutions. Successes and challenges in literacy development are therefore not seen solely as individual learner achievements or shortcomings but as outcomes influenced by the broader social and educational environment (Ardiansyah & Ujihanti, 2017). Effective teacher training in constructivist approaches can help educators create engaging, supportive, and collaborative learning environments that promote the development of literacy and numeracy skills (Saldivar, 2025).

Financial and digital literacy programs similarly benefit from constructivist principles. Constructivism centres on the learner's characteristics of self-direction and experience to create new knowledge, reflecting real-life problem-solving and supporting the general principles of adult education and lifelong learning (L. R. Brown et al., 2023). In corporate settings, a constructivist adult education philosophical approach leveraging Kolb's experiential learning theory enables employees to engage in hands-on learning, reflect on their experience, build conceptual knowledge, and incorporate new approaches, an effective strategy for building digital literacy at an individualized pace (L. R. Brown et al., 2023).

Parenting education represents another domain where constructivism has demonstrated influence. Shannon, working in the space of pregnancy, childbirth, and early parenting education, has noted that constructivist theory significantly influenced her views on learning and teaching, as students construct their understanding based on their own unique experiences and reflection on those experiences (Woodward et al., 2021). As a practitioner and advocate for constructivist methods in adult education, she highly values the "learning by doing" approach that experiential learning activities embrace (Woodward et al., 2021).

Health education programs also draw on constructivist principles. Simulation methodology in nursing and healthcare education is rooted in adult learning theory, with constructivist theory explaining how learners gain knowledge through simulation experiences (Rutherford-Hemming, 2012). Experiential learning, such as simulations, allows students to engage directly with concepts, and research has shown that participation in simulated experiences promotes deliberation and deep learning (Rutherford-Hemming, 2012). Community-based adult learners' wellbeing can be enhanced through participation in adult and community education, resulting in an increased intention toward continual learning (Chau & Ren, 2024).

Community development initiatives more broadly align with constructivist pedagogy through their emphasis on building upon participants' lived experiences. Constructivism emphasizes active, experiential, and student-based learning as it facilitates critical thinking, collaboration, and the construction of knowledge on the basis of lived experiences (Saldivar, 2025). Participatory learning methods such as simulations, group discussions, and case studies have been shown to increase participants' understanding and awareness of social issues (Yasmiatti, 2024). Value-based education within constructivist frameworks helps learners understand core values, adopt them as profound principles, and act on those values in their own lives (Yasmiatti, 2024). Experiential learning, defined as "the process by which learners construct knowledge, skills, and values from direct experience," encompasses all aspects of human beings, not just cognitive aspects (Yasmiatti, 2024).

For example, participants in financial literacy programs may analyse their own budgeting challenges and develop practical solutions. Community leadership programs may require learners to identify local issues and design projects that address community needs. Adult literacy programs frequently use authentic documents, workplace materials, and real-life communication tasks to ensure learning remains relevant and meaningful.

These approaches recognise that adult learners bring valuable knowledge and experiences to the learning environment and that learning is most effective when it is connected to real-life challenges and personal goals.

Constructivism is particularly well suited to adult education because it recognises the importance of prior experience, active participation, reflection, problem-solving, and real-world application. Across higher education, vocational education and training, workplace learning, professional development, ESL education, and community learning, constructivist approaches encourage learners to actively engage with knowledge, collaborate with others, and construct meaning from their experiences. By connecting learning to authentic contexts and encouraging learners to take an active role in the learning process, constructivism supports the development of deep understanding, critical thinking, and lifelong learning capabilities.

Strengths and Limitations of Constructivism

Constructivism offers several significant advantages for adult learning and education. One of its primary strengths is its emphasis on developing deep understanding rather than relying on rote memorisation (Krishna et al., 2024). By encouraging learners to actively engage with ideas, analyse information, and connect new knowledge to prior experiences, constructivist approaches promote meaningful learning that is more likely to be retained and applied in new situations (Damoah, 2024; Mishra, 2023). The theory recognises that learners bring existing knowledge, experiences, and perspectives to the learning environment, and that these form the foundation upon which new understanding is built (Andersen, 2018; Ardiansyah & Ujihanti, 2017; Okochi et al., 2023).

Constructivist learning environments also support the development of critical thinking, problem-solving, and decision-making skills (Mishra, 2023; Mohamad & Romli, 2021). Learners are encouraged to investigate problems, evaluate evidence, consider alternative viewpoints, and construct their own understanding rather than simply accepting information provided by an instructor (Mishra, 2023; Saldivar, 2025) This active involvement in the learning process often increases learner engagement, motivation, and ownership of learning (Saldivar, 2025; Sioukas, 2022). Furthermore, because learning activities are frequently based on authentic tasks and real-world situations, constructivist approaches can facilitate the transfer of knowledge and skills to workplace, professional, and everyday contexts (Huang, 2021; Okochi et al., 2023). These characteristics align particularly well with adult learning principles, which emphasise relevance, experience, autonomy, and practical application (Krishna et al., 2024; Woodward et al., 2021).

Social constructivism extends these benefits by highlighting the importance of collaboration, discussion, mentoring, and participation in learning communities (Bhattarai & Basnet, 2022; Damoah, 2024; Mishra, 2023). Through interaction with peers, educators, mentors, and other members of a community, learners are exposed to diverse perspectives and opportunities to negotiate meaning (Huang, 2021; Lin et al., 2025). Collaborative learning activities can deepen understanding, enhance communication skills, and support the development of professional and social competencies (Huang,

2021; Stafford, 2021). The emphasis on social interaction reflects the reality that much adult learning occurs within workplaces, communities, and professional networks rather than solely through individual study (Deulen, 2013; Okochi et al., 2023).

Despite its strengths, constructivism also presents several challenges. Constructivist learning activities often require substantial time for investigation, discussion, reflection, and collaboration, which can make them more resource-intensive than traditional instructional approaches (Ameri, 2020; Sioukas, 2022). Because learners are encouraged to develop their own understanding, there is also a risk that misconceptions may arise or persist if appropriate guidance and feedback are not provided (Y. Kim, 2024; Woodward et al., 2021). Effective facilitation is therefore essential to ensure that learners construct accurate and meaningful knowledge (Y. Kim, 2024; Sioukas, 2022).

Assessment can also be more complex within constructivist learning environments. Evaluating learners' understanding, reasoning processes, and application of knowledge may require more sophisticated assessment methods than traditional tests that focus primarily on factual recall (Y. Kim, 2024). Additionally, some learners may prefer more structured forms of instruction, particularly when they are unfamiliar with a subject area or lack confidence in their ability to learn independently (Sioukas, 2022). Discovery-based learning approaches can place significant cognitive demands on novice learners who may require greater guidance and support before they are able to engage effectively in open-ended inquiry (Maimunah et al., 2023; Woodward et al., 2021).

For these reasons, many educators adopt a balanced approach that combines constructivist principles with more structured instructional methods. Direct instruction may be used to introduce foundational concepts and skills, while constructivist activities provide opportunities for learners to apply, explore, and extend their understanding (Krishna et al., 2024; Woodward et al., 2021). This combination allows educators to benefit from the strengths of constructivism while addressing some of its potential limitations.

Constructivism views learning as an active process in which learners construct knowledge by connecting new information to existing experiences, understanding, and beliefs (Ardiansyah & Ujihanti, 2017; Mishra, 2023; Yasmin & Islam, 2021). Social constructivism expands this perspective by emphasising the role of social interaction, language, culture, and collaboration in the development of knowledge (Huang, 2021; Lin et al., 2025; Maimunah et al., 2023). Together, these theories have had a profound influence on contemporary adult education, higher education, vocational education and training, workplace learning, and language education (Y. Kim, 2024; Maimunah et al., 2023; Woodward et al., 2021). They encourage educators to move beyond the simple transmission of information and instead create learning environments in which learners actively explore, discuss, reflect, collaborate, and develop meaningful understanding through engagement with both ideas and other people (Mishra, 2023; Okochi et al., 2023; Saldivar, 2025).

Humanism and Learner-Centred Education

Humanism is a learning theory that places the learner at the centre of the educational process. Unlike behaviourism, which focuses on observable behaviour, or cognitivism, which focuses on mental processes, humanism emphasises the development of the whole person (Arghode et al., 2017; Badyal & Singh, 2017; Waseem & Aslam, 2020). Humanist educators are concerned not only with what learners know and can do, but also with their personal growth, self-awareness, motivation, emotions, values, creativity, and potential (Arghode et al., 2017; Bell, 2021). According to humanistic learning theory, education should support both professional competence and personal development, recognising that learning contributes to an individual's growth as both a practitioner and a person (Badyal & Singh, 2017). Humanistic approaches emphasise learner autonomy and personal responsibility, viewing learners as active participants who take ownership of their educational journeys. Within this framework, instructors act primarily as facilitators, mentors, and guides who create supportive learning environments that encourage self-directed learning, reflection, and personal growth (Arghode et al., 2017; Ashfia & Septianti, 2024).

Humanism emerged during the mid-twentieth century as a response to both behaviourist and psychoanalytic approaches, which many educators believed did not adequately recognise human agency, free will, and the capacity for personal growth (Omer, 2020b; Waseem & Aslam, 2020). Behaviourism had dominated educational psychology until the middle of the last century, measuring human behaviour in terms of reflex responses to stimuli (Waseem & Aslam, 2020), while cognitivism subsequently challenged behaviourism by emphasising internal mental processes (Nalliah, 2014; Waseem & Aslam, 2020). Humanist theorists argued that learning is most effective when individuals are intrinsically motivated, actively involved in their learning, and working towards personally meaningful goals (Arghode et al., 2017; Bell, 2021). Education should therefore support learners in becoming autonomous, self-directed, and fulfilled individuals. Humanistic learning theory views education as a means of fostering personal growth, independence, self-reliance, and self-awareness. Rather than focusing solely on the acquisition of knowledge and skills, humanism emphasises the development of the whole person and encourages learners to become autonomous, reflective, and capable of directing their own learning and personal development (Arghode et al., 2017).

Humanism has had a profound influence on adult education because it aligns closely with the characteristics of adult learners (Arghode et al., 2017; Omer, 2020b). Adults typically bring established experiences, personal goals, values, and motivations to educational settings (Figurska & Sokół, 2016; Waseem & Aslam, 2020). Humanist approaches recognise that adults are capable of directing their own learning and are often motivated by personal development, career advancement, self-improvement, or the desire to achieve meaningful life goals (Arghode et al., 2017; Waseem & Aslam, 2020). Knowles's theory of andragogy shares many characteristics with the humanistic perspective, particularly its emphasis on learner autonomy, self-direction, and intrinsic motivation. Andragogy proposes that adult learners prefer learning experiences that are relevant to real-life problems and practical situations rather than being organised solely around subject content. It also recognises that adults are motivated largely by internal factors such as personal growth, professional development, achievement, and self-fulfilment, reinforcing the strong alignment between andragogical and humanistic approaches to learning (Omer, 2020b).

Humanistic learning theory is based on several interconnected principles. First, learning is viewed as a personal process that involves emotional, social, intellectual, and psychological dimensions (Bell, 2021). Education should therefore address the whole learner rather than focusing solely on academic achievement (Arghode et al., 2017; Badyal & Singh, 2017).

Second, learners are viewed as inherently capable of growth and self-improvement. Humanist educators believe that individuals naturally strive to learn, develop, and fulfil their potential when provided with supportive environments and meaningful opportunities. Humanistic education therefore places a strong emphasis on fostering personal development, self-awareness, and the realisation of individual potential, recognising that education should contribute not only to knowledge acquisition but also to the broader growth of the learner as a person (Arghode et al., 2017).

Third, personal meaning is considered essential to effective learning. Learners are more likely to engage deeply with learning activities when they perceive them as relevant to their own interests, goals, and experiences (Bell, 2021). Learning that lacks personal significance is often less motivating and less effective.

Fourth, learner autonomy and self-direction are highly valued. Humanistic approaches encourage learners to take responsibility for their learning, set goals, evaluate their progress, and make decisions about their educational pathways (Arghode et al., 2017; Omer, 2020b). In this framework, behaviourism is teacher-centred, while cognitivism, humanism, and constructivism theory are learner-centred (Badyal & Singh, 2017).

Finally, positive relationships are viewed as fundamental to learning. Respect, trust, empathy, and psychological safety create conditions that support exploration, reflection, creativity, and growth (Bigdeli et al., 2023). As noted in the context of adult learning, learners actively participate in the learning process most effectively in a safe and supportive setting (Bigdeli et al., 2023).

Abraham Maslow and the Hierarchy of Needs

One of the most influential humanist theorists was Abraham Maslow (Clark, 2016; Omodan & Mtshatsha, 2022; Ursula, 2024). Maslow proposed that human behaviour is influenced by a hierarchy of needs, ranging from basic physiological requirements through to self-actualisation (Qin, 2022; Thoresen et al., 2016; Ursula, 2024). Humanistic psychology, in which Maslow and Rogers are well-known pioneers, emerged as a complement to psychoanalysis and behaviourism (Clark, 2016; Schjetne & Мельникова, 2020), and emphasised the holistic development of the individual (Omodan & Mtshatsha, 2022).

According to Maslow, individuals must have their fundamental needs reasonably satisfied before they can focus fully on learning and personal development (Qin, 2022; Simanjuntak, 2019; Ursula, 2024). These needs include physiological requirements such as food and shelter, safety and security, belonging and social connection, and self-esteem (Thoresen et al., 2016; Ursula, 2024). Maslow's hierarchy consists of five levels: physiological needs, safety needs, social needs (belonging and love), esteem needs, and self-actualisation needs (Simanjuntak, 2019; Thoresen et al., 2016; Ursula, 2024). At the highest level of the hierarchy is self-actualisation, which refers to the desire to achieve one's full potential and become the best version of oneself (Clark, 2016; Ursula, 2024). As the theory maintains, a person's basic human needs must be satisfied before he/she reaches ultimate performance (Qin, 2022).

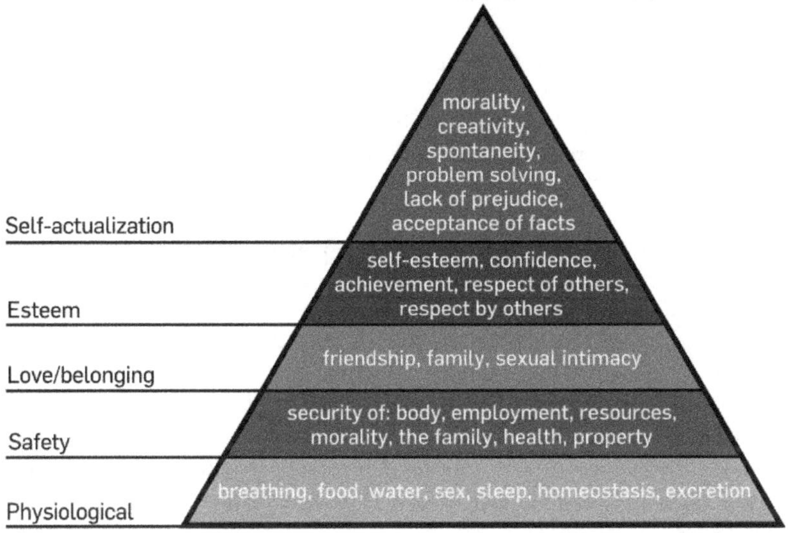

Figure 4: Maslow's Hierarchy of Needs.Factoryjoe, CC BY-SA 3.0, via Wikimedia Commons.

Maslow's theory has important implications for education (Omodan & Mtshatsha, 2022; Simanjuntak, 2019; Ursula, 2024). Literature studies on Maslow's hierarchy show that fulfilling basic needs such as safety and esteem greatly influences students' desire to learn (Ursula, 2024) . Learners who feel unsafe, unsupported, excluded, or undervalued may struggle to engage effectively with learning (Omodan & Mtshatsha, 2022; Simel, 2020; Ursula, 2024). Educators therefore have a responsibility to create learning environments that foster inclusion, respect, support, and psychological wellbeing (Gerdin et al., 2021; Omodan & Mtshatsha, 2022; Simel, 2020).

Humanistic pedagogy places strong emphasis on the emotional well-being of learners and views education as a process that nurtures personal growth, self-worth, and positive human development. Humanistic educators generally maintain that individuals possess an inherent capacity for growth, learning, and positive development, and that supportive educational environments can help learners realise their potential and develop into confident, self-directed individuals (Omodan & Mtshatsha, 2022). Humanistic educators emphasise the importance of creating learning environments in which learners feel safe, respected, supported, and free to express themselves without fear of judgement. Such environments are believed to encourage creativity, exploration, personal growth, and meaningful learning by enabling learners to participate openly and confidently in the educational process (Simel, 2020).

A humanistic approach to education also recognises the importance of understanding and valuing learners' perspectives, experiences, emotions, and opinions. Educators are encouraged to foster inclusive classrooms that demonstrate respect for individual differences and actively support learner participation, belonging, and personal development. By acknowledging learners as whole individuals with unique needs and experiences, humanistic education seeks to create conditions that support both academic achievement and personal growth (Omodan & Mtshatsha, 2022).

Carl Rogers and Learner-Centred Education

Another major contributor to humanistic education was Carl Rogers. Rogers argued that meaningful learning occurs when learners are actively involved in the learning process and when education addresses issues that are personally relevant to them (deCarvalho, 1991; Khatib et al., 2013; Štěrba, 2018). He believed that learners should have considerable control over their learning experiences and that the role of the educator is to facilitate rather than direct learning (Budd et al., 2015; deCarvalho, 1991; Štěrba, 2018).

Rogers proposed that effective educators demonstrate empathy, authenticity, and unconditional positive regard towards learners (Bermea, 2022; Budd et al., 2015; Štěrba, 2018). Rather than acting as the sole source of knowledge, educators create supportive environments that encourage exploration, self-discovery, and personal growth (deCarvalho, 1991; Khatib et al., 2013; Štěrba, 2018).

Learner-centred education emerged directly from Rogers' work (Budd et al., 2015; deCarvalho, 1991). In learner-centred classrooms, learners participate actively in setting goals, selecting learning activities, evaluating progress, and reflecting on their learning experiences (Khatib et al., 2013; Msonde, 2021). The educator acts as a guide, mentor, coach, or facilitator who supports learners in achieving their goals (Budd et al., 2015; Foster et al., 2021; Khatib et al., 2013).

Humanism in Adult Learning

Humanism aligns particularly well with adult learning because adults are often self-directed and goal-oriented learners (Grabarski et al., 2023). Many adults choose to participate in education voluntarily and are motivated by personal aspirations, professional development, career advancement, or life transitions (Barath & Ross, 2024; Gardner et al., 2021; Greer, 2013).

Humanistic approaches encourage adults to identify their own learning needs, establish personal goals, and take responsibility for their progress. Reflection is an important component of humanistic learning because it helps learners evaluate their experiences, identify strengths and weaknesses, and develop greater self-awareness.

In adult education, learning is often most effective when learners perceive a clear connection between educational activities and their personal or professional objectives. Humanistic educators therefore seek to create flexible, supportive learning environments that acknowledge individual differences and encourage personal growth.

Humanistic principles are widely applied in higher education, particularly in programs that emphasise professional development, critical reflection, and personal growth.

For example, university students may develop individual learning plans that allow them to pursue areas of personal interest within a broader curriculum. Reflective journals are commonly used to encourage students to examine their experiences, beliefs, and learning processes. Nursing, teaching, counselling, and social work programs frequently incorporate reflective practice activities that promote self-awareness and professional identity development.

Student-centred learning approaches, self-directed research projects, mentoring programs, and portfolio assessments all reflect humanistic principles by giving learners greater ownership of their educational experiences.

Humanistic approaches are also evident within vocational education and training. While VET focuses on developing workplace competence, learner-centred practices can support engagement, motivation, and personal responsibility for learning.

For example, apprentices may participate in developing their own training plans and identifying areas where they require additional practice or support. Learners undertaking management qualifications may select workplace projects that align with their career aspirations and organisational needs. Recognition of Prior Learning (RPL) processes also reflect humanistic principles by recognising the value of learners' existing knowledge and experiences.

Workplace coaching, mentoring, self-assessment activities, reflective journals, and individual development plans are common examples of humanistic approaches within VET environments.

Humanistic principles have become increasingly important in workplace learning and professional development. Organisations recognise that employee development extends beyond technical skills and includes leadership, communication, emotional intelligence, resilience, and personal effectiveness.

Leadership development programs often encourage participants to reflect on their values, strengths, goals, and leadership styles. Coaching programs support employees in identifying personal objectives and developing strategies to achieve them. Professional development plans frequently allow individuals to choose learning opportunities that align with their interests and career aspirations.

These approaches recognise that employees are motivated not only by organisational requirements but also by personal growth, achievement, and self-fulfilment.

Humanistic approaches have also influenced language education. Humanistic language teaching seeks to create supportive, low-anxiety environments in which learners feel comfortable taking risks and communicating in a new language.

Teachers often encourage learners to discuss topics that are personally meaningful and relevant to their lives. Activities may include personal storytelling, reflective writing, collaborative projects, discussions about life experiences, and goal-setting exercises. By connecting language learning to learners' interests and experiences, educators can increase motivation and engagement.

Example of Humanism in Higher Education

Scenario: Bachelor of Nursing Program – Developing Professional Identity and Self-Awareness

A university lecturer teaching a final-year nursing subject adopts a humanistic approach to support both the professional and personal development of students. Rather than relying solely on lectures and examinations, the lecturer begins the semester by asking students to develop individual learning goals based on their career aspirations, perceived strengths, and areas for improvement. Some students wish to improve their leadership skills, while others focus on communication, cultural competence, or clinical decision-making.

Throughout the semester, students maintain reflective journals in which they analyse their experiences during clinical placements. They are encouraged to examine not only what occurred during patient interactions but also how they felt, how their values influenced their decisions, and what they learned about themselves as future healthcare professionals. The lecturer provides supportive feedback that encourages deeper reflection rather than simply judging whether answers are correct or incorrect.

The classroom environment is designed to promote psychological safety and mutual respect. Students participate in facilitated discussions where they can openly discuss challenges, mistakes, ethical dilemmas, and personal experiences without fear of embarrassment. The lecturer acts as a mentor and facilitator, encouraging students to explore different perspectives and develop their own professional identities.

Assessment includes a professional development portfolio in which students document their growth throughout the course, identify future learning goals, and reflect on how their values and experiences shape their nursing practice. Success is measured not only through academic achievement but also through evidence of personal growth, self-awareness, reflective capacity, and professional development.

This example demonstrates several humanistic principles, including learner autonomy, personal meaning, self-reflection, self-directed learning, supportive educator-learner relationships, and the development of the whole person rather than simply the acquisition of technical knowledge.

Examples of Humanism in Vocational Education and Training (VET)

Scenario: Licensed Practical Nurse (LPN) Program at a Community College

A community college in the United States delivers a Licensed Practical Nurse (LPN) program for adult learners from diverse backgrounds. Many students are mature-age learners balancing employment, family responsibilities, and study. Recognising these challenges, instructors adopt a humanistic, learner-centred approach that focuses not only on developing clinical competence but also on supporting students' personal growth, confidence, and professional identity.

At the beginning of the program, students complete a self-assessment identifying their strengths, previous healthcare experiences, personal learning goals, and areas where they would like additional support. Some students may be returning to education after many years, while others may already work as nursing assistants and wish to progress their careers. Instructors meet individually with students to discuss their aspirations and help them develop personalised learning plans.

Throughout the course, students participate in reflective practice activities following clinical simulations and hospital placements. After caring for patients, they are encouraged to reflect on questions such as:

- What aspects of patient care did I perform well?
- What challenged me during this interaction?
- How did I respond emotionally to the situation?
- What did I learn about myself as a healthcare professional?
- What skills would I like to strengthen moving forward?

Rather than focusing exclusively on technical performance, instructors encourage students to consider empathy, communication, professionalism, cultural awareness, and ethical decision-making. Feedback sessions are conducted as coaching conversations in which instructors help learners identify strengths and opportunities for growth rather than simply pointing out deficiencies.

The learning environment is intentionally designed to foster respect, inclusion, and psychological safety. Students are encouraged to share personal experiences, ask questions freely, and support one another during challenging learning activities. Peer mentoring programs pair new students with more experienced learners, helping to build confidence and a sense of belonging within the program.

Students are also given flexibility to pursue projects aligned with their interests. For example, one student interested in geriatric care may complete a project on dementia support strategies, while another interested in community health may investigate health education programs for underserved populations. This allows learners to connect their studies with personally meaningful goals and career aspirations.

Assessment includes traditional competency evaluations required for licensure, but also incorporates reflective portfolios documenting professional growth, learning achievements, clinical experiences, and future career objectives. Success is measured not only by mastery of nursing skills but also by evidence of increasing self-awareness, confidence, professionalism, and commitment to lifelong learning.

This example illustrates how humanistic principles can be integrated into a highly regulated vocational program by recognising learners as whole individuals, supporting personal development alongside technical competence, promoting learner autonomy, and creating a supportive environment that encourages growth, reflection, and self-directed learning.

Scenario: Certificate III in Carpentry – Individual Learning and Personal Development in Australia

A trainer delivering a Certificate III in Carpentry recognises that learners enter the program with varying levels of experience, confidence, and career goals. At the commencement of training, each apprentice meets with the trainer to discuss their previous experience, learning preferences, strengths, and professional aspirations. Together, they develop an individual learning plan that identifies specific goals for the apprenticeship period.

One apprentice may wish to improve leadership skills in preparation for becoming a leading hand, while another may focus on developing confidence when interpreting construction drawings. The trainer encourages learners to take responsibility for monitoring their own progress and regularly reflect on their development.

During practical activities, apprentices are given opportunities to make decisions, solve problems, and evaluate their own work. For example, rather than simply following step-by-step instructions to construct a timber frame, learners are encouraged to plan the sequence of work, identify potential challenges, and determine the most effective construction methods. After completing the task, they participate in reflective discussions about what they learned, what difficulties they encountered, and how they could improve future performance.

The trainer creates a supportive environment where mistakes are viewed as learning opportunities rather than failures. Apprentices are encouraged to ask questions, seek assistance, and share experiences from their workplaces. Regular coaching conversations focus not only on technical competence but also on confidence, communication skills, teamwork, workplace relationships, and career aspirations.

Assessment includes a reflective skills development log in which apprentices record workplace experiences, identify lessons learned, evaluate their performance, and establish future development goals. Recognition of Prior Learning (RPL) opportunities are also discussed, acknowledging the value of learners' existing skills and experiences.

This example demonstrates how humanistic principles can be applied in VET by promoting learner ownership, self-direction, reflection, personal growth, workplace relevance, and supportive trainer-learner relationships while still achieving competency-based outcomes.

Strengths and Limitations of Humanism

Humanism offers several important strengths. It promotes learner motivation, engagement, self-confidence, autonomy, and personal responsibility (Houssami & Benattabou, 2024; Neufeld & Kassam, 2025). It recognises the importance of emotions, relationships, and personal meaning in learning, making it particularly relevant to adult education (Kálmán, 2019; Neufeld & Kassam, 2025). Humanistic approaches can foster lifelong learning habits by helping learners develop self-awareness and self-directed learning skills (Houssami & Benattabou, 2024; Kálmán, 2019).

However, humanism also has limitations. Some learners may require more structure and guidance than humanistic approaches typically provide (Dachner et al., 2017; Neufeld & Kassam, 2025). Learner-centred approaches can be difficult to implement in large classes or highly regulated training environments (Neufeld & Kassam, 2025). Measuring outcomes related to personal growth and self-development can also be challenging (Orsini et al., 2017). Additionally, excessive emphasis on individual choice may sometimes result in insufficient attention to essential content or competency requirements (George, 2010; Neufeld & Kassam, 2025).

Humanism views education as a process of supporting the development of the whole person. It emphasises personal growth, self-direction, intrinsic motivation, meaningful learning, and the achievement of human potential (Houssami & Benattabou, 2024; Rovai et al., 2020). Through the work of theorists such as Abraham Maslow and Carl Rogers, humanism gave rise to learner-centred education, which positions learners as active participants in their own learning journeys (Kálmán, 2019; Neufeld & Kassam, 2025). Humanistic principles continue to influence adult education, higher education, vocational training, workplace learning, and language education by encouraging educators to create supportive environments that foster autonomy, reflection, personal growth, and lifelong learning (Houssami & Benattabou, 2024; Kálmán, 2019; Rovai et al., 2020).

Social Learning Theory

Social Learning Theory explains learning as a process that occurs through observing, modelling, and interacting with others. Unlike behaviourism, which focuses primarily on direct reinforcement and punishment, Social Learning Theory proposes that people can learn new behaviours, skills, attitudes, and knowledge simply by watching others and observing the consequences of their actions (Carcea & Froemke, 2019; Lyons & Berge, 2012).

The theory was developed primarily by Albert Bandura, who argued that learning is a social process influenced by the interaction of personal factors, behaviour, and the environment (Lyons & Berge, 2012; Yilmaz & Yılmaz, 2019). Bandura challenged the idea that learning must result from direct experience, demonstrating that individuals can acquire new knowledge and behaviours indirectly through observation, (Carcea & Froemke, 2019). His work represented a bridge between behaviourist and cognitive approaches because it recognised both external influences and internal mental processes (Fithriani et al., 2021; Lyons & Berge, 2012).

One of Bandura's most famous studies was the Bobo Doll Experiment, which demonstrated that children who observed adults behaving aggressively towards a toy were more likely to imitate that behaviour themselves (Galanaki & Malafantis, 2022; Yilmaz & Yılmaz, 2019). The study showed that people can learn simply by observing others, even without receiving direct rewards or punishments (Carcea & Froemke, 2019; Galanaki & Malafantis, 2022).

Observational Learning

At the centre of Social Learning Theory is the concept of observational learning, also known as modelling, which proposes that individuals can acquire new knowledge, skills, attitudes, and behaviours by observing others rather than solely through direct experience (Bandura, 1977, 1986; Schunk, 2020). Bandura argued that much human learning occurs in social environments where individuals are exposed to the actions of parents, teachers, peers, supervisors, mentors, and other role models. Through observation, learners develop an understanding of how behaviours are performed and the consequences associated with those behaviours, enabling them to adopt similar actions in appropriate situations (Bandura, 1986; Illeris, 2018). Observational learning is particularly important in educational settings, workplaces, families, and communities, where individuals continually learn from the examples provided by others (Bandura, 1977; Schunk, 2020).

Bandura (1977) proposed that observational learning involves four interconnected cognitive processes: attention, retention, reproduction, and motivation. These processes determine whether an observed behaviour is successfully learned and subsequently performed.

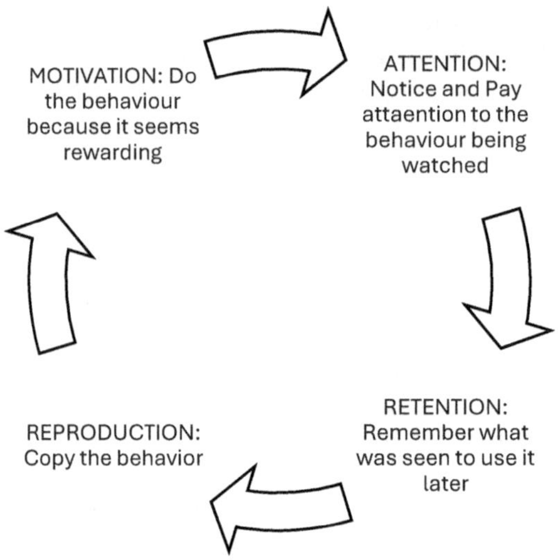

Figure 5: Observational learning.

The first process, attention, requires learners to focus on the model and the behaviour being demonstrated. Not all observed behaviours are learned; learners are more likely to attend to models who possess characteristics such as expertise, credibility, authority, status, competence, or social attractiveness. Environmental factors, the complexity of the behaviour, and the learner's interests and needs also influence the degree of attention given to a model (Bandura, 1977; Schunk, 2020).

The second process, retention, involves storing information about the observed behaviour in memory so that it can be recalled later. Learners may retain information in the form of mental images, verbal descriptions, concepts, procedures, or symbolic representations. Effective retention enables individuals to remember the key elements of a behaviour after the model is no longer present and provides the foundation for future performance (Bandura, 1977; Illeris, 2018).

The third process, reproduction, refers to the learner's ability to translate remembered observations into actual behaviour. Even when a learner has paid attention to and retained information about a behaviour, successful performance requires the necessary physical, cognitive, and behavioural capabilities. Practice, feedback, coaching, and repeated opportunities to perform the behaviour help learners refine their skills and improve performance over time (Bandura, 1977; Schunk, 2020).

The final process, motivation, determines whether the learner chooses to perform the observed behaviour. Individuals are more likely to imitate behaviours when they expect positive outcomes, such as rewards, recognition, achievement, social approval, career advancement, or personal satisfaction. Motivation may also arise through observing others receive positive consequences for their actions, a phenomenon known as vicarious reinforcement. Conversely, observing negative consequences may discourage the adoption of particular behaviours (Bandura, 1977, 1986; Schunk, 2020).

Together, these four processes explain how individuals learn through observation and why some observed behaviours are adopted while others are ignored. The model highlights that learning is not a passive process of imitation but rather involves active cognitive engagement, memory, practice, and motivation. As a result, Social Learning Theory provides a comprehensive explanation of how people learn within social environments and remains highly influential in education, workplace training, professional development, and adult learning (Bandura, 1986; Illeris, 2018; Schunk, 2020).

Vicarious Learning and Reinforcement

A key contribution of Social Learning Theory is the concept of vicarious learning, whereby individuals learn from observing the consequences experienced by others. At the core of this process, the observer forms a cognitive or mental representation of the modelled behaviour, which then serves as a guide for the observer's own performance. Vicarious learning occurs not only through direct observation of actual models' behaviours but also through covert visualization, whereby imagination can produce scenes having the same potential to influence behaviour as would actual events.

For example, an apprentice may observe a colleague receiving praise for following safe work procedures and become more likely to adopt similar behaviours. Indeed, research in the informal sector has documented that apprentices and work-and-pay artisans who follow others are likely to observe the safety protocols at work, while master artisans practice safety protocols to serve as an illustration to their apprentices (Ishak & Obeng-Ofori, 2022). Conversely, observing another worker receive disciplinary action for breaching safety rules may discourage unsafe behaviour. As one supervisor explained, apprentices often need ongoing reinforcement regarding the serious risks associated with workplace injuries and the potentially severe consequences that can result from unsafe work practices (Ishak & Obeng-Ofori, 2022).

The transmission of workplace knowledge and behaviours across generations of workers is a well-established aspect of occupational learning. For example, within the electrical industry, safe work practices have traditionally been learned through observation, mentoring, and modelling, with experienced tradespeople demonstrating appropriate techniques and behaviours that apprentices subsequently adopt and apply in their own work. This process highlights the important role of observational learning in the development of workplace competence and safety culture.

This process, known as vicarious reinforcement, allows learning to occur without direct personal experience. Research applying social learning theory to inter-organisational contexts has demonstrated that vicarious learning encompasses four key processes, attention, retention, motor reproduction, and motivation, through which organisations and individuals learn from the consequences experienced by others without undergoing those consequences themselves.

Self-Efficacy

Bandura later introduced the concept of self-efficacy, which refers to an individual's belief in their ability to successfully perform a task or achieve a goal (Guerrin, 2012; Nagels et al., 2018; Par et al., 2023; Safourcade & Alava, 2009; Soussi et al., 2021). More precisely, self-efficacy is defined as the belief of the individual in his or her capacity to organize and execute the course of action required to produce desired results (Guerrin, 2012; Nagels et al., 2018; Safourcade & Alava, 2009).

Individuals with high self-efficacy are generally more willing to attempt challenging tasks, persist when difficulties arise, and recover from setbacks (Leroy et al., 2013; Nagels et al., 2018) . As Bandura's social cognitive theory posits, self-efficacy helps individuals orient their behaviours, choose their goals, persist in the face of obstacles, and manage efforts and emotional reactions when confronting difficulties (Nagels et al., 2018). Conversely, those with low self-efficacy may avoid challenges, give up more easily, or doubt their capabilities (Par et al., 2023). Students who doubt their capacity to use educational resources effectively become rapidly overwhelmed and experience higher stress, while individuals with learning difficulties who have experienced few successes tend to exhibit low self-efficacy (Par et al., 2023).

Educational practices that provide successful experiences, constructive feedback, mentoring, and role models can strengthen learners' self-efficacy and improve learning outcomes (Achille, 2024; Leroy et al., 2013; Nagels et al., 2018; Par et al., 2023). Mastery experiences constitute a primary source of self-efficacy construction (Nagels et al., 2018; Par et al., 2023), as success achieved through effort and perseverance consolidates beliefs in one's own capacities (Par et al., 2023). Vicarious experiences, observing others, including peers and expert models, successfully perform tasks, also contribute to self-efficacy development (Par et al., 2023). Furthermore, through encouragement and feedback on progress, teachers can influence students' self-efficacy beliefs (Leroy et al., 2013), and these beliefs are positively associated with academic achievement (Achille, 2024; Leroy et al., 2013). Teachers' own efficacy beliefs in their capacity to motivate and foster learning also play a critical role in shaping students' self-evaluations of their intellectual capabilities (Achille, 2024).

Social Learning Theory in Adult Education

Social Learning Theory is particularly relevant to adult education because adults frequently learn through observing and interacting with colleagues, mentors, supervisors, industry experts, and peers (Chao, 2005; Chuang, 2021). Bandura's social learning theory focuses on an individual's learning through vicarious experiences, such as observing a role model or hearing from others, within a social context (Chuang, 2021). While formal training programs play an important role in adult learning, a substantial proportion of workplace learning occurs informally through observation, participation, and social interaction (Jeong et al., 2018; Scott, 2022b). Indeed, informal learning has been identified as the most dominant type of learning found in the workplace, with early estimates suggesting approximately 80% of workplace learning is informal compared to approximately 20% formal learning (Scott, 2022b).

Adults often acquire new knowledge and skills by watching experienced practitioners perform tasks, solve problems, communicate with clients, manage projects, and respond to unexpected challenges (Chao, 2005; Smith, 2003). Studies on how managers develop have found that work relationships are essential for their development, as managers learn from seasoned managers through direct interactions, including advice seeking and feedback, and indirectly by observation (Chao, 2005). Through these observations, learners develop an understanding of both technical procedures and the professional behaviours required for effective performance in their occupational roles (Chuang, 2021; Moon, 2018).

In social learning theory, the instructor serves as a role model to learners, and social interaction and human relations are major components in applying this theory in adult learning and vocational training settings (Chuang, 2021). Sources of learning in the workplace can include loosely organized mentoring or coaching, or new knowledge and skills acquired in the pursuit of work itself, and participation in groups or circumstances that enable learning through interaction with a more expert other is particularly satisfactory for adult learners (Scott, 2022b; Smith, 2003).

The theory aligns closely with many contemporary approaches to adult learning, including communities of practice, mentoring relationships, coaching programs, peer learning activities, apprenticeships, internships, and professional networking. In these environments, learning occurs not only through direct instruction but also through participation in shared activities and exposure to the knowledge, skills, attitudes, and behaviours demonstrated by more experienced practitioners. As a result, Social Learning Theory provides a valuable framework for understanding how adults develop competence and professional identity throughout their careers.

Social Learning Theory is widely applied in higher education, particularly in disciplines where professional practice and collaboration are important components of learning. Students frequently learn through observing demonstrations, participating in collaborative activities, and interacting with experienced practitioners. Common applications include clinical demonstrations in nursing and medicine (Esterhazy et al., 2020; Muthaura et al., 2015; Shikuku et al., 2024), laboratory demonstrations in science and engineering (Qafisheh, 2012), moot courts in law programs (Esterhazy et al., 2020; Fourie & Coetzee, 2017; Gautam, 2017; Parsons, 2016; Quinot & Tonder, 2015; Scott, 2012; Yun-bo & Zhang, 2025), peer-assisted learning schemes (Marchi et al., 2025; Shikuku et al., 2024), group projects (Mitchell et al., 2011), research supervision (Baetens & Cheah, 2013; Ghori, 2011), mentoring arrangements (Baetens & Cheah, 2013; Marchi et al., 2025; Shikuku et al., 2024), professional placements (Ghori, 2011; Marchi et al., 2025; Pascoe, 2017; Tully & Avramenko, 2015), and internships (Mitchell et al., 2011; Muthaura et al., 2015; Patel et al., 2021; Singh, 2021; Yun-bo & Zhang, 2025).

For example, engineering students may accompany experienced engineers during site inspections, design reviews, or project meetings. By observing how engineers assess risks, interpret technical information, communicate with stakeholders, and make design decisions, students develop practical insights that extend beyond what can be learned through textbooks alone. Similarly, Master of Laws students may observe practising lawyers during court proceedings, client consultations, or negotiations. These experiences allow students to model professional behaviours, advocacy techniques, ethical decision-making processes, and communication strategies that they can later apply during moot court exercises, internships, or legal practice.

Social Learning Theory is particularly influential within vocational education and training because many trades, technical occupations, and service industries rely heavily on observation, modelling, and guided practice (MacMahon et al., 2024; Zhao & Ko, 2018). Learners commonly acquire skills by watching experienced workers perform tasks before attempting those tasks themselves under supervision (Aakernes, 2020; Aheisibwe & Kintu, 2019; Schmid et al., 2020). This approach allows learners to observe not only technical procedures but also workplace expectations, communication styles, safety practices, and professional conduct (MacMahon et al., 2024).

Examples can be found across a wide range of vocational fields. Apprentices often observe qualified tradespeople carrying out construction, electrical, or mechanical tasks before practising the skills themselves. Hairdressing learners watch demonstrations of cutting, colouring, and styling techniques before applying them to clients. Heavy vehicle drivers learn by observing experienced operators conduct vehicle inspections, manage fatigue risks, and navigate challenging driving conditions. Hospitality learners observe customer service interactions, while mining trainees may shadow supervisors during inspections and operational activities. Beauty therapy learners similarly observe client consultations, treatment procedures, and professional communication techniques before undertaking supervised practice. In each case, learners develop competence through a combination of observation, imitation, practice, and feedback.

Many workplace learning and professional development activities are founded on the principles of Social Learning Theory. Organisations frequently use mentoring programs, coaching relationships, job shadowing, on-the-job training, communities of practice, peer learning groups, and leadership development initiatives to facilitate learning through social interaction and observation. These approaches recognise that employees often learn most effectively when they can observe experienced practitioners performing real workplace tasks and discuss their experiences with others.

For example, a newly appointed underground coal mine manager may accompany an experienced manager during inspections, operational meetings, emergency exercises, and incident investigations. Through observing how the experienced manager identifies hazards, communicates with personnel, interprets monitoring data, and makes operational decisions, the learner gains both technical knowledge and professional judgement. Over time, opportunities to practise these activities independently, combined with feedback and reflection, enable the learner to develop the competencies required for effective management and leadership within the mining environment.

Social Learning Theory therefore provides a powerful explanation of how professional knowledge, workplace culture, technical skills, and leadership capabilities are developed across higher education, vocational education, and workplace settings. By emphasising observation, modelling, interaction, and feedback, the theory highlights the importance of social environments in supporting adult learning and professional growth.

Strengths and Limitations of Social Learning Theory

Social Learning Theory offers several important strengths that help explain its enduring influence in education and training. One of its greatest contributions is its recognition of the central role that social interaction plays in learning. The theory explains how individuals can acquire complex knowledge, skills, attitudes, and behaviours through observation without necessarily having to experience situations directly themselves (Crocco & Cseh, 2020; Middleton et al., 2018; Mwikonji, 2021). This makes it particularly useful for understanding workplace learning, where employees frequently learn by observing colleagues, supervisors, mentors, and industry experts (Chuang, 2019; Johnson et al., 2018; Zhao & Ko, 2018). Social Learning Theory also incorporates important cognitive processes such as attention, memory, and motivation, acknowledging that learners actively interpret and process what they observe rather than simply imitating behaviour automatically (Liu et al., 2022; Middleton et al., 2018). Furthermore, the theory highlights the significant influence of role models and mentors in shaping learning and professional development (Hussain & Tasnim, 2023; Miao et al., 2019; Mohammed et al., 2026) and emphasises the importance of self-efficacy, or an individual's belief in their own capabilities, as a key factor influencing performance, persistence, and achievement (Middleton et al., 2018; Sanders et al., 2015; Ugheoke, 2024).

Despite these strengths, Social Learning Theory also has several limitations. Because learning occurs through observation, learners may adopt undesirable behaviours as readily as desirable ones if inappropriate conduct is modelled by influential individuals (Johnson et al., 2018; Miao et al., 2019). Observation alone may also be insufficient to develop deep conceptual understanding, particularly in areas requiring complex theoretical knowledge or critical analysis (Setiyo et al., 2018). The effectiveness of social learning is heavily dependent on the quality and competence of the role models being observed, meaning poor examples can result in poor learning outcomes (Johnson et al., 2018; Setiyo et al., 2018). Additionally, individual differences among learners may influence what they notice, remember, and ultimately reproduce, leading to variations in learning outcomes even when individuals are exposed to the same models (Howardson & Behrend, 2015; Middleton et al., 2018). Finally, while observation is an important component of learning, many skills require substantial hands-on practice, feedback, and experience before proficiency can be achieved (Fang, 2025; Setiyo et al., 2018).

Social Learning Theory views learning as a fundamentally social process in which individuals acquire knowledge, skills, attitudes, and behaviours by observing others and the consequences of their actions (Crocco & Cseh, 2020; Khanyase, 2022; Zhao & Ko, 2018). Through concepts such as observational learning, modelling, vicarious reinforcement, and self-efficacy, Bandura demonstrated that learning extends well beyond direct instruction and personal trial-and-error experiences (Khanyase, 2022; Mohammed et al., 2026; Mwikonji, 2021). The theory continues to exert a significant influence on adult education, higher education, vocational training, workplace learning, mentoring programs, apprenticeships, and professional development (Chuang, 2019, 2021; Fang, 2025; Zhao & Ko, 2018) because it reflects the way people commonly learn in real-world social and professional environments (Crocco & Cseh, 2020; Hussain & Tasnim, 2023; Johnson et al., 2018).

Experiential Learning Theory

Experiential learning theory explains learning as a process through which individuals develop knowledge, skills, and understanding by engaging in and reflecting upon experiences. Rather than viewing learning as the passive reception of information from an instructor, experiential learning emphasises active participation, reflection, interpretation of experiences, and the application of new insights to future situations (Baker et al., 2012; Hedin, 2010; Hill, 2017; Ilyas et al., 2020; Matsuo, 2015a; Salinas-Navarro et al., 2024; Xu et al., 2023; Zavadskienė & Bubnys, 2018). From this perspective, learning occurs when experiences are examined, interpreted, and transformed into knowledge that can guide future actions and decision-making (Baker et al., 2012; Ilyas et al., 2020; Nguyen et al., 2020; Salinas-Navarro et al., 2024).

Kolb's experiential learning model describes this process as a continuous cycle involving four interconnected stages: concrete experience, reflective observation, abstract conceptualisation, and active experimentation (Dillard et al., 2024; Ganira & Odundo, 2023; Matsuo, 2015a; Salinas-Navarro et al., 2024; Voon et al., 2022). Learners first engage in an experience, then reflect on that experience, develop concepts or generalisations from their reflections, and subsequently test these ideas through new experiences. Through repeated cycles of action and reflection, understanding becomes progressively refined and expanded.

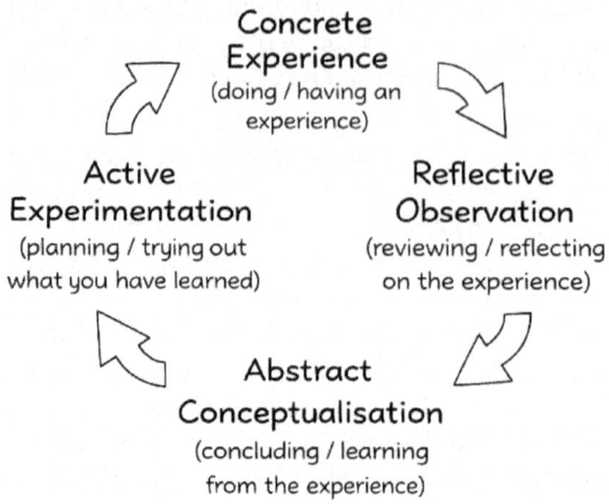

Figure 6: Kolb's learning cycle.

In the first stage, the learner has a direct experience. This may involve completing a task, participating in a simulation, undertaking workplace practice, solving a problem, or engaging in a real-life activity. In the second stage, the learner reflects on the experience by considering what happened, what worked well, what did not work, and why. In the third stage, the learner develops concepts, principles, or general lessons from the experience. In the final stage, the learner applies those lessons to a new situation, testing and refining their understanding through further action.

Experiential learning is grounded in the work of influential scholars such as John Dewey, Kurt Lewin, and Jean Piaget, whose theories emphasised the importance of experience, reflection, interaction with the environment, and cognitive development in the learning process (Hedin, 2010; Hill, 2017; Newman et al., 2017). Consistent with these traditions, experiential learning proposes that learners construct their own understanding by actively engaging with experiences and reflecting on their meaning rather than simply absorbing information provided by others (Hill, 2017; Xu et al., 2023). As a result, experiential learning has become one of the most influential and widely cited approaches to understanding how individuals learn from experience across educational, workplace, and professional contexts.

Experiential learning is particularly relevant to adult education because adults bring existing life and work experience to learning. They often prefer learning that is practical, relevant, problem-based, and connected to real-world tasks. It is also highly relevant to vocational education and training, workplace learning, professional education, apprenticeships, simulations, clinical placements, fieldwork, internships, and project-based learning.

For example, in higher education, nursing students may participate in a clinical simulation, reflect on their communication and decision-making, connect the experience to clinical theory, and then apply the learning during placement. In engineering, students may design and test a structure, evaluate its performance, revise their assumptions, and improve the design.

In vocational education, an apprentice mechanic may diagnose a vehicle fault, reflect on the diagnostic process, connect the experience to electrical or mechanical principles, and apply the learning to future faults. A construction learner may participate in a site safety inspection, identify hazards, reflect on risk controls, and then apply improved safety practices on another worksite.

Experiential learning has several strengths. It supports practical skill development, problem-solving, confidence, reflection, and transfer of learning to real situations. It also encourages learners to take an active role in their learning rather than remaining passive recipients of information.

However, experiential learning also has limitations. Experience alone does not always lead to effective learning. Learners may repeat mistakes or develop incorrect assumptions if reflection, feedback, and guidance are weak. It can also be time-consuming and may require access to suitable workplaces, simulations, equipment, supervision, and assessment conditions.

Overall, experiential learning theory is highly influential in adult education because it recognises that meaningful learning often occurs through active engagement with real or realistic experiences. It encourages educators to design learning that combines practice, reflection, theory, feedback, and application.

Application Examples

The following examples demonstrate how Kolb's experiential learning cycle can be implemented in different educational contexts.

Higher Education Example: Bachelor of Civil Engineering

A final-year civil engineering subject requires students to design a pedestrian bridge for a local council. Rather than learning solely through lectures and textbook examples, students work in project teams to develop a real-world design solution.

The concrete experience occurs when students analyse the site, collect survey data, identify design constraints, and prepare preliminary bridge designs. They then construct scale models or use computer simulations to test their designs under various loading conditions.

The reflective observation stage occurs when students review the performance of their designs. They examine why certain design elements performed well while others failed, discuss challenges encountered during modelling and testing, and consider the assumptions they made during the design process.

During abstract conceptualisation, students connect their observations to engineering theories and principles such as structural analysis, material properties, load distribution, fatigue, and safety factors. They identify broader lessons that can be applied to future engineering projects.

The active experimentation stage occurs when students revise their designs based on their findings and test the improved solution. The cycle may be repeated several times throughout the project as students progressively refine their understanding and design capability.

This approach mirrors professional engineering practice, where engineers continuously learn through design, testing, evaluation, and improvement.

Higher Education Example: Master of Laws (LLM)

A Master of Laws program incorporates experiential learning through a simulated commercial litigation exercise.

Students are presented with a complex contractual dispute involving multiple parties. Acting as legal practitioners, they review evidence, interview simulated clients, draft legal advice, negotiate settlements, and participate in a mock court hearing.

The concrete experience involves undertaking the role of a practising lawyer and managing the case from commencement to resolution.

The reflective observation stage occurs after the hearing, where students analyse their advocacy performance, communication skills, legal reasoning, negotiation strategies, and ethical decision-making. They consider what influenced the outcome and identify areas for improvement.

During abstract conceptualisation, students connect their experiences to legal doctrine, procedural rules, dispute resolution theory, professional ethics, and case law. They develop broader insights into effective legal practice and client representation.

The active experimentation stage may involve applying these lessons in subsequent moot courts, negotiations, workplace placements, or professional practice simulations.

Experiential learning helps law students bridge the gap between legal theory and professional practice by allowing them to experience the realities of legal decision-making and client advocacy.

Vocational Education Example: Diploma of Beauty Therapy

A beauty therapy learner is undertaking advanced facial treatment training within a simulated salon environment.

The concrete experience occurs when the learner conducts a client consultation, assesses skin conditions, recommends treatments, and performs a facial treatment on a client.

Following the treatment, the learner enters the reflective observation stage by evaluating the effectiveness of the consultation, treatment outcomes, client communication, time management, and technical performance. Feedback is obtained from both the trainer and the client.

During abstract conceptualisation, the learner links the experience to underlying theories of skin anatomy, dermatology, product chemistry, infection control, and client care. They identify why particular treatment techniques were successful and how client characteristics influenced treatment decisions.

The active experimentation stage occurs when the learner applies these insights during future client appointments, adjusting treatment plans, communication techniques, and product selection to improve outcomes.

Through repeated cycles of client interaction, reflection, and improvement, learners develop both technical competence and professional judgement required within the beauty industry.

Vocational Education Example: Underground Coal Mining Management

A learner undertaking a Diploma of Underground Coal Mining Management participates in a simulated emergency response exercise involving a spontaneous combustion event in an underground mine.

The concrete experience begins when the learner assumes the role of an underground mine manager during the simulated emergency. They receive reports from control room personnel, assess gas monitoring data, activate emergency procedures, communicate with statutory officials, and coordinate evacuation and response activities.

Following the exercise, the learner engages in reflective observation by reviewing the decisions made throughout the incident. They evaluate the effectiveness of communication systems, emergency procedures, risk assessments, and leadership actions. They consider what information was available at the time, how decisions were made, and what could have been improved.

During abstract conceptualisation, the learner links the experience to mine safety legislation, risk management principles, ventilation engineering, spontaneous combustion management plans, emergency preparedness requirements, and human factors affecting decision-making during emergencies.

The active experimentation stage occurs when the learner participates in subsequent emergency simulations or workplace exercises, applying lessons learned to improve incident command, risk assessment processes, communication protocols, and emergency response strategies.

This approach is particularly valuable because many of the most critical competencies required of underground coal mining managers involve high-consequence decision-making that cannot be safely learned through actual emergencies. Experiential learning allows realistic practice in a controlled environment while developing the judgement, confidence, and leadership capabilities required for statutory management roles.

These examples demonstrate that experiential learning can be applied across both academic and vocational contexts. Whether designing infrastructure, conducting legal proceedings, treating clients, or managing mining emergencies, learners develop deeper understanding when they actively engage in authentic experiences, reflect on those experiences, derive lessons from them, and apply those lessons in future situations. This continuous cycle of experience, reflection, conceptualisation, and experimentation remains one of the most effective approaches for developing professional competence and lifelong learning capability.

Critical Theory and Emancipatory Education

Critical Theory is a perspective on learning and society that seeks to understand, question, and transform social structures that create inequality, oppression, and injustice. Unlike learning theories that focus primarily on how individuals acquire knowledge and skills, Critical Theory is concerned with the relationship between education, power, culture, ideology, and social change. It argues that education should do more than transmit knowledge; it should help learners critically examine the social, political, economic, and cultural forces that shape their lives and empower them to challenge inequitable conditions (Ndimande, 2018; Neupane, 2021; Rasmussen, 2024).

The origins of Critical Theory can be traced to the Frankfurt School, a group of German scholars including Max Horkheimer, Theodor Adorno, Herbert Marcuse, and later Jürgen Habermas (Bachmann & Moisio, 2022; Chrismastianto et al., 2024; Macdonald, 2014; Rasmussen, 2024). These theorists sought to understand why social inequalities and forms of domination persisted despite advances in education, democracy, and economic development (Bachmann & Moisio, 2022; Rasmussen, 2024). They argued that social institutions, including education, media, government, and workplaces, can reinforce existing power structures by shaping beliefs, values, and assumptions in ways that appear natural or unquestionable (Bachmann & Moisio, 2022; Charles, 2018; Ndimande, 2018). The Frankfurt School historically emerged as a critique influenced by Hegel's philosophical idealism, Karl Marx's thought, and Freud's psychoanalysis, and is classified as a school of Neo-Marxism (Chrismastianto et al., 2024). Notably, the very concept of "critical theory" was first coined not in Frankfurt but during the Institute's exile period in the United States, in essays by Horkheimer and Marcuse published in 1937 (Reitz, 2009). Critical theory thereby provides the intellectual basis for confronting processes of political and economic reification and for making citizens aware of subtle mechanisms of domination (Bachmann & Moisio, 2022).

In educational contexts, Critical Theory asks questions such as:

- Whose knowledge is represented in the curriculum? (Batz et al., 2025; Ndimande, 2018; Wright et al., 2023)
- Who benefits from existing educational structures? (Batz et al., 2025; Ndimande, 2018)
- Which voices are excluded or marginalised? (Batz et al., 2025; Ndimande, 2018; Wright et al., 2023)
- How do social class, gender, race, culture, language, and economic circumstances influence educational opportunities? (Atehortúa, 2020; Mustafaa & Martinez, 2022; Ndimande, 2018)
- How can education contribute to social justice and democratic participation? (Mustafaa & Martinez, 2022; Ndimande, 2018; Neupane, 2021; Rasmussen, 2024)

Rather than accepting knowledge as neutral or objective, Critical Theory encourages learners to critically examine the assumptions, interests, and power relationships that underpin knowledge and social practices (Charles, 2018; Ndimande, 2018; Neupane, 2021). As Freire argued, the curriculum of the oppressed is characterised by biased knowledge oriented toward the ruling class and dominant groups who control curriculum content (Ndimande, 2018; Neupane, 2021). Critical pedagogy, which draws directly from the principles of critical theory developed by the Frankfurt School, aims to emancipate oppressed or marginalised people by challenging domination, subordination, and oppression, and by developing critical consciousness through emancipatory education (Chrismastianto et al., 2024; Neupane, 2021). The goal of critical theory, in this educational sense, is emancipation from oppression through the arising of critical consciousness based on conscientisation (Mustafaa & Martinez, 2022; Neupane, 2021).

Paulo Freire and Emancipatory Education

The educational application of Critical Theory is most strongly associated with the Brazilian educator Paulo Freire, whose work laid the foundation for emancipatory education and critical pedagogy (Ardill, 2019; Biesta, 2017; Salinas-Atausinchi & Huaman-Lucana, 2021; Vlieghe, 2013). Freire argued that traditional education often positions learners as passive recipients of information (Bello, 2023; Bezerra, 2021; Mokotso, 2024). He described this approach as the banking model of education, in which teachers "deposit" knowledge into learners who are expected to receive, memorise, and reproduce it (Bello, 2023; Bezerra, 2021; Biesta, 2017; Mokotso, 2024; Morais et al., 2023).

Freire believed that this approach can reinforce existing power structures by discouraging critical thinking and limiting learners' ability to question their circumstances (Barros et al., 2024; Bello, 2023; Bezerra, 2021; Morais et al., 2023). Instead, he proposed a more democratic and participatory approach to education based on dialogue, reflection, and action (Barros et al., 2024; Bello, 2023; Zanchetta et al., 2016).

Central to Freire's work is the concept of conscientisation (critical consciousness) (Coon, 2019; Mokotso, 2024; Zanchetta et al., 2016). Conscientisation refers to the process through which individuals become aware of the social, political, economic, and cultural forces that influence their lives (Mokotso, 2024; Zanchetta et al., 2016). Through critical reflection and discussion, learners develop the ability to recognise inequality, challenge assumptions, and take informed action to improve their circumstances and communities (Omodan, 2022; Salinas-Atausinchi & Huaman-Lucana, 2021; Zanchetta et al., 2016). For Freire, education should not simply prepare individuals to adapt to society; it should help them understand society and participate in transforming it (Bello, 2023; Lima, 2024; Valenzano, 2021).

Dialogue and Problem-Posing Education

Freire proposed a model known as problem-posing education, which differs significantly from traditional teacher-centred instruction (Ignacio, 2020; Pouwels, 2019; Uddin, 2019). In problem-posing education, learners and educators work together to investigate real-world issues and problems (Ignacio, 2020; Melo, 2019; Torre et al., 2017). Rather than presenting predetermined answers, educators facilitate discussion, questioning, and critical reflection (Bernard, 2024; Torre et al., 2017; Uddin, 2019). Learners draw upon their own experiences and knowledge while exploring social realities and identifying possible solutions (Melo, 2019; Roka, 2023; Uddin, 2019).

Dialogue plays a central role in this process (Chambers, 2019; Ignacio, 2020; Pouwels, 2019). Freire viewed dialogue as a collaborative exchange in which learners and educators learn from one another (Ignacio, 2020; Nweke & Owoh, 2020; Zhao, 2013). Through dialogue, learners develop critical thinking skills, examine multiple perspectives, and construct deeper understandings of complex issues (Barbosa & Wang, 2020; Pouwels, 2019; Torre et al., 2017).

This approach transforms the educator from an authority figure who transmits knowledge into a facilitator who supports inquiry, reflection, and collective learning (Dias & Boulder, 2023; Torre et al., 2017; Zhao, 2013).

Emancipatory Learning

Emancipatory learning refers to a form of learning that enables individuals to critically examine their assumptions, question established beliefs, and gain greater agency over their personal, professional, and social circumstances. Rather than focusing solely on the acquisition of knowledge and skills, emancipatory learning emphasises the development of critical awareness, reflective thinking, and the capacity to take informed action (Fleming, 2018; Preece, 2003). Through this process, learners move beyond simply accepting information and instead evaluate the social, cultural, political, and economic influences that shape their perspectives and experiences (Babacan & Babacan, 2012; Bergman & Hessel, 2023).

A central aim of emancipatory learning is to help individuals recognise assumptions, biases, and taken-for-granted beliefs that may influence their thinking and behaviour (Fleming, 2022b; Kloubert, 2019). Learners are encouraged to critically evaluate accepted practices, explore alternative viewpoints, and develop a deeper understanding of the power relationships that exist within organisations, communities, and society (Babacan & Babacan, 2012; Buttigieg & Calleja, 2020a; Hoggan et al., 2016). This process can increase awareness of issues such as discrimination, inequality, exclusion, and social injustice (Fleming, 2024; Formenti & Hoggan-Kloubert, 2023), while also fostering independent thinking and informed decision-making (Bergman & Hessel, 2023; Wu & Crocco, 2019).

Emancipatory learning also promotes active participation in democratic and civic life by encouraging learners to engage with social issues, contribute to discussions, and participate in efforts to improve their communities and workplaces (Hoggan-Kloubert & Mabrey, 2022; Landry & Lieres, 2022). The ultimate goal is not merely personal understanding but the development of the knowledge, confidence, and critical capacity needed to contribute to positive social change (Formenti & Hoggan-Kloubert, 2023; Hoggan et al., 2016; Kattari, 2025). By becoming more aware of the forces that shape their lives, learners are better positioned to make informed choices and influence their environments in meaningful ways (Bergman & Hessel, 2023; Rennick, 2015).

This perspective has had a significant influence on adult learning theory, particularly through the work of Jack Mezirow (Fleming, 2018; Wang et al., 2019). Mezirow's theory of transformative learning builds upon many of the principles of emancipatory learning by emphasising the importance of critically reflecting on previously held assumptions, beliefs, and perspectives (Buttigieg & Calleja, 2020a; Fleming, 2022b; Kloubert, 2019). Through such reflection, learners may revise their understanding of themselves and the world, leading to more inclusive, informed, and autonomous ways of thinking and acting (Babacan & Babacan, 2012; Hoggan et al., 2016; Schnepfleitner & Ferreira, 2021).

Critical Theory in Adult Education

Critical Theory has had a significant influence on adult education because adult learners bring diverse life experiences, professional backgrounds, and social perspectives to the learning environment. Rather than treating education as the simple transfer of knowledge from teacher to learner, critical approaches encourage adults to examine their experiences within broader social, economic, political, cultural, and historical contexts. Learners are encouraged to reflect on how institutions, policies, traditions, and power structures shape individual opportunities and outcomes, and to consider how these structures might be challenged or improved.

Rather than focusing exclusively on vocational skills or academic content, critical adult education promotes inquiry, reflection, dialogue, and action. Learners are encouraged to ask why particular systems operate as they do, whose interests are served by existing arrangements, and how alternative approaches might create more equitable outcomes. Learning therefore becomes a process of developing critical awareness and informed participation rather than simply acquiring information. This approach is commonly reflected in community education programs addressing social issues, trade union and worker education initiatives, environmental sustainability programs, citizenship and civic participation courses, human rights education, and social inclusion projects throughout Europe and other parts of the world.

Critical Theory is frequently applied in higher education through activities that encourage students to analyse complex issues from multiple perspectives and consider the broader social implications of professional practice (Wallace, 2020; West, 2023). Students are encouraged to move beyond technical competence and develop an understanding of the ethical, political, economic, and cultural factors that influence decision-making within their disciplines (Fields et al., 2022; Lawrence & Maphalala, 2021; Wallace, 2020).

Applications of critical theory and critical pedagogy within higher education include transformation from systemic pedagogy to a culture that encourages student self-transformation and empowerment (West, 2023). Critical pedagogy, rooted in critical theory, seeks to analyse and transform educational contexts by applying principles that critique and change power relations in society (Allen, 2011; Fields et al., 2022). This educational philosophy aims to raise the consciousness of, empower, and liberate students, moving beyond the mere transmission of knowledge to enable individuals to critically probe and challenge social disparities and injustices (Diko, 2023; Fields et al., 2022). Core philosophical principles from critical pedagogy that inform higher education include historicity of knowledge, cultural politics, political economy, and dialectical theory (Wallace, 2020). Dialectical theory, in particular, contests traditional theories of education that tend to reinforce positivism, certainty, and technical control of knowledge and power (Wallace, 2020).

In practice, critical pedagogy encourages dialogic relationships, promotes student agency and critical social consciousness, and facilitates the development of critical thinking skills (Diko, 2023; Nomdo, 2023). Students are supported in developing awareness of their environment and society, equipped with the ability to address imbalances in power and justice (Chrismastianto et al., 2024). Higher education institutions are considered spaces of possibility where social justice and equity can be promoted through socially just pedagogical practices (Lawrence & Maphalala, 2021). Teaching and learning processes remain a platform from which to promote social justice, with curricula designed to help students articulate practice as emancipatory (Gormally et al., 2021; Lawrence & Maphalala, 2021). This approach connects social and political change directly to education, helping learners understand the importance of critically engaging with their reality, identifying social issues, and taking proactive steps toward creating a just society (Diko, 2023).

For example, students enrolled in a Master's degree in Public Policy may critically examine the impacts of housing, migration, welfare, or labour market policies across different European countries. Rather than simply studying policy frameworks, they may investigate how policy decisions affect various population groups, explore competing political perspectives, and consider how policy reforms might improve social outcomes.

In environmental science and sustainability programs, students may analyse the social and economic implications of the European Green Deal, renewable energy transitions, or climate adaptation strategies. They may examine how environmental policies affect different industries, communities, and regions, and evaluate the challenges involved in balancing environmental objectives with economic and social priorities.

Similarly, students undertaking urban planning or architecture programs may critically evaluate urban regeneration projects, public transport systems, or housing developments. They may investigate how planning decisions influence social inclusion, accessibility, affordability, and community wellbeing, while considering the perspectives of governments, businesses, residents, and other stakeholders. These learning activities encourage students to recognise that professional decisions often have significant social consequences and require careful ethical consideration.

Application in Vocational Education and Training

Although vocational education and training is often associated with the development of occupational competence, Critical Theory can also contribute to the development of reflective and socially aware practitioners. By encouraging learners to examine the broader context of their work, critical approaches help learners understand not only how tasks are performed but also why particular practices exist and how they affect individuals, organisations, and communities (Bamber & Crowther, 2011; Majola et al., 2024; Schmidt, 2020; Wight, 2021).

For example, learners undertaking hospitality management qualifications in Europe may examine issues such as labour conditions, seasonal employment, workforce diversity, and the social impacts of tourism. Rather than focusing solely on operational procedures, learners may analyse how management decisions influence employee wellbeing, customer experiences, and local communities.

In agricultural and agribusiness training programs, learners may critically examine topics such as food security, sustainable farming practices, environmental regulations, and the economic pressures faced by farming communities. They may explore how agricultural policies influence producers, consumers, rural economies, and environmental outcomes, developing a broader understanding of the role agriculture plays within society.

Similarly, learners undertaking qualifications in logistics and supply chain management may investigate issues such as globalisation, ethical sourcing, labour standards, and environmental sustainability. Through critical analysis, learners can develop a greater appreciation of how supply chain decisions affect workers, communities, businesses, and the environment, while identifying opportunities for more responsible and sustainable practices.

These applications demonstrate how Critical Theory can complement vocational and professional education by encouraging learners to become reflective practitioners who understand the wider social, economic, and ethical dimensions of their occupations. Rather than viewing work solely as a technical activity, learners develop the capacity to critically evaluate practices, challenge assumptions, and contribute to positive change within their industries and communities.

Strengths and Limitations of Critical Theory

Critical Theory promotes critical thinking, reflection, social awareness, and democratic participation. Critical pedagogy, which emerged from Critical Theory, encourages learners to examine assumptions, question accepted knowledge, and analyse the power relationships that influence social institutions, cultural practices, and educational systems (T. H. Chiang, 2021; Fenwick, 2005). Rather than viewing knowledge as neutral or value-free, Critical Theory emphasises that knowledge is shaped by social, political, economic, and cultural contexts. Learners are therefore encouraged to investigate how certain perspectives become dominant, whose interests are served by existing arrangements, and how alternative viewpoints may contribute to greater equity and inclusion (T. H. Chiang, 2021).

A central contribution of Critical Theory is its emphasis on developing critical consciousness. Through reflection and dialogue, learners become more aware of the social forces, structures, and relationships that shape their experiences and opportunities. This increased awareness can help individuals recognise inequality, discrimination, and exclusion while developing the capacity to participate more actively in social, civic, and professional life. Critical pedagogy seeks to empower learners by fostering independent thinking, encouraging informed participation in democratic processes, and supporting efforts to address social injustice and promote positive change (Achilleos et al., 2021).

Despite its strengths, Critical Theory has several limitations. Some critics argue that it can place a strong emphasis on social and political issues at the expense of disciplinary knowledge and technical competence. Concerns have been raised that educational outcomes and social mobility are also heavily influenced by learners' mastery of academic content and professional knowledge, and that these areas may receive insufficient attention in some critical pedagogical approaches (T. H. Chiang, 2021).

Others note that concepts such as power, ideology, oppression, and social justice can be interpreted differently by different individuals and groups, making consistent implementation challenging. In practice, creating genuinely democratic learning environments can be difficult because power relationships continue to exist between educators and learners, as well as among learners themselves. Furthermore, rational discussion and critical analysis may not always address deeper personal beliefs, interests, values, or emotional influences that shape behaviour and decision-making (Todić & Christensen, 2022).

Critical pedagogy has also been criticised for operating at a relatively high level of abstraction and for being easier to describe in theory than to implement consistently in practice. Educational institutions often face practical constraints such as curriculum requirements, assessment frameworks, time limitations, regulatory obligations, and organisational expectations, which can restrict opportunities for extensive dialogue, reflection, and critical inquiry. These challenges may be particularly evident in highly regulated, competency-based, or compliance-focused learning environments where the primary focus is on achieving prescribed learning outcomes and demonstrating occupational competence (Lorenzetti et al., 2016).

Sociocultural Perspectives on Learning

Sociocultural perspectives on learning view learning as a fundamentally social and cultural process rather than an activity that occurs solely within an individual's mind. These perspectives emphasise that knowledge, thinking, language, and learning are shaped through interactions with other people and participation in social, cultural, and historical contexts (Kandaurova & Milovanova, 2023; Rezvanova et al., 2019). Learning is therefore understood not simply as the acquisition of information but as a process of becoming a participant in the practices, values, and ways of thinking of a particular community (Rezvanova et al., 2019).

The sociocultural perspective is most strongly associated with the work of the Russian psychologist Lev Vygotsky, whose theories have had a profound influence on education, psychology, and adult learning (Kandaurova & Milovanova, 2023; Margolis, 2020a; Nguyen, 2017). Vygotsky argued that cognitive development occurs through social interaction and that learning is mediated by cultural tools such as language, symbols, technologies, and shared practices (Nguyen, 2017; Rezvanova et al., 2019). According to this view, individuals develop understanding through participation in social activities and interactions with more knowledgeable members of their communities (Bovshik & Gaivoronskaya, 2022; Nguyen, 2017).

Vygotsky proposed that learning and development are inseparable from the social and cultural environments in which individuals live (Kandaurova & Milovanova, 2023; , (Rezvanova et al., 2019). Rather than viewing learning as an isolated cognitive process, he argued that knowledge is first constructed through interaction with others before becoming internalised by the learner (Bednarskaya, 2022; Rezvanova et al., 2019). In this sense, learning moves from the social level to the individual level (Rezvanova et al., 2019).

For example, a learner initially develops understanding through discussions with teachers, peers, mentors, or supervisors. Through repeated participation in these interactions, the learner gradually internalises concepts, strategies, and ways of thinking until they can perform tasks independently (Bovshik & Gaivoronskaya, 2022; Nguyen, 2017). This process highlights the importance of communication, collaboration, and guided participation in learning (Bovshik & Gaivoronskaya, 2022; Nguyen, 2017).

One of Vygotsky's most influential concepts is the Zone of Proximal Development (ZPD) (Margolis, 2020a; Nguyen, 2017). The ZPD refers to the gap between what a learner can accomplish independently and what they can achieve with guidance and support from a more knowledgeable person (Abakumova et al., 2012; Nguyen, 2017). As Vygotsky defined it, the ZPD represents the distance between the actual level of development, determined by independent problem-solving, and the level of potential development, determined by problem-solving under the guidance of an adult or in collaboration with more capable peers (Nguyen, 2017).

Tasks that fall within the learner's current capabilities may require little assistance, while tasks beyond the learner's current level may be impossible without support (Abakumova et al., 2012; Bednarskaya, 2022). Learning is believed to occur most effectively when instruction targets the area between these two points (Bednarskaya, 2022; Margolis, 2020a). Within the ZPD, learners can successfully complete more challenging tasks with appropriate guidance, gradually developing the skills and knowledge needed to perform independently (Abakumova et al., 2012; Bovshik & Gaivoronskaya, 2022). The concept highlights the importance of matching learning activities to the learner's developmental stage and providing appropriate levels of support to facilitate growth (Bednarskaya, 2022; Bovshik & Gaivoronskaya, 2022; Kalimullina, 2023).

Scaffolding

Although the term scaffolding was developed by later researchers, it is closely associated with Vygotsky's work (Javeed, 2019; Polly & Byker, 2020; Ries et al., 2023). Scaffolding refers to the temporary support provided by educators, mentors, supervisors, or peers to help learners complete tasks that they could not yet perform independently (Le et al., 2026; Polly & Byker, 2020; Ries et al., 2023).

As learners gain competence, the support is gradually reduced until they can perform the task on their own (Le et al., 2026). Scaffolding may involve demonstrations, questioning, coaching, prompts, feedback, modelling, or collaborative problem-solving activities (Holcomb, 2025; Le et al., 2026; Polly et al., 2013; Yi-Cline, 2020).

For example, an experienced electrician may initially guide an apprentice through fault-finding procedures, explaining each step and providing feedback. As the apprentice develops competence, the mentor gradually reduces assistance, allowing the apprentice to assume increasing responsibility for diagnosing and resolving faults independently.

Language as a Tool for Learning

A central feature of sociocultural theory is the importance of language in learning and cognitive development. Vygotsky viewed language as the primary tool through which individuals communicate, share knowledge, and construct understanding.

Through discussions, questioning, explanations, and collaborative activities, learners develop new ways of thinking and interpreting the world. Language enables learners to access the knowledge of others, participate in social activities, and internalise concepts and problem-solving strategies. In educational settings, dialogue therefore becomes an essential mechanism for learning rather than merely a means of transmitting information.

Communities of Practice

A central feature of sociocultural theory is the importance of language in learning and cognitive development. Vygotsky viewed language as the primary tool through which individuals communicate, share knowledge, and construct understanding (Córdoba, 2020; García, 2022; Maigua et al., 2024). For Vygotsky, language is the most important psychological tool, as through it individuals develop and create their reality (Maigua et al., 2024), and it constitutes the preferred sign system for human psychological development, promoting mediation of the person's relationships with themselves through internal dialogues and with others (Córdoba, 2020). Language serves as the tool that enables the process of internalisation of social activity (interpersonal) to internal activity (intrapersonal), thereby achieving the development of higher functions (Hine & María Nuria Rodríguez de, 2021).

Through discussions, questioning, explanations, and collaborative activities, learners develop new ways of thinking and interpreting the world. Language enables learners to access the knowledge of others, participate in social activities, and internalise concepts and problem-solving strategies (Cruz & Cruz,

2023; García, 2022). The sociocultural theory posits that mental activity results from socio-cultural learning, implying the internalisation of elements acquired through social interaction (Maigua et al., 2024), and that collaboration within the Zone of Proximal Development, through language, allows direct contact among students who participate in collaborative work (García, 2022). Language permits the restructuring of thought with meaning, and when the subject has become involved in the world of words and symbols, the construction of concepts and dissemination of new knowledge becomes more significant. Dialogue related to reading texts acquires fundamental importance, as it not only drives shared understanding but also contributes to the development of discursive communicative skills (Cruz & Cruz, 2023). In educational settings, dialogue therefore becomes an essential mechanism for learning rather than merely a means of transmitting information (Córdoba, 2020; Cruz & Cruz, 2023; García, 2022). Within this framework, the teacher uses language as a psychological tool to mediate and manage information so that the student constructs knowledge, with the student becoming the principal active agent in the pedagogical process (García, 2022). The sociocultural linguistic perspective seeks to integrate language and social processes in the classroom to understand the processes of teaching and learning (Córdoba, 2020).

Situated Learning

Sociocultural perspectives also emphasise the concept of situated learning, which suggests that learning is inseparable from the context in which it occurs (Beneito-Montagut et al., 2018; S. L. Espinoza, 2020; Llorente & Iglesias, 2018). Knowledge is not viewed as abstract information that can simply be transferred between situations (Dujo & García, 2009; Vargas, 2024). Instead, learning is shaped by the activities, tools, social relationships, and environments in which it takes place (Franky, 2009).

This perspective suggests that authentic learning experiences are often more effective than decontextualised instruction because learners can directly connect knowledge to real-world practice. Consequently, internships, apprenticeships, simulations, workplace learning, and project-based learning are all strongly aligned with sociocultural principles (Picón & Correa, 2022).

From a situated learning perspective, knowledge is inherently linked to the contexts in which it is developed and applied. Learning is not viewed as the acquisition of abstract information that can be transferred unchanged from one setting to another. Instead, understanding is shaped by the activities, social interactions, cultural practices, tools, and environments that surround the learning process (Brown et al., 1989; Lave & Wenger, 1991). What individuals know and how they use that knowledge are therefore influenced by the situations in which learning occurs. For example, understanding workplace safety procedures in a classroom differs significantly from applying those procedures on an active construction site, in a hospital ward, or within a manufacturing facility. Learners must not only understand formal rules and procedures but also develop an appreciation of how those procedures are interpreted, adapted, and enacted within real operational environments. As a result, learning involves acquiring both explicit knowledge and the tacit knowledge that emerges through participation in professional practice (Eraut, 2004; Lave & Wenger, 1991).

A central concept within situated learning is legitimate peripheral participation, which describes the process through which newcomers gradually become members of a professional, occupational, or social community (Lave & Wenger, 1991). New learners typically begin by engaging in relatively simple, low-risk, or supportive activities while observing more experienced members of the community. Through increasing participation, responsibility, and interaction, they progressively acquire the knowledge, skills, values, behaviours, and identities associated with competent membership. Learning is therefore viewed not only as acquiring knowledge but also as becoming a participant in the social practices of a community (Wenger, 1998).

For example, an apprentice carpenter may initially assist with measuring materials, preparing work areas, and handling tools while observing qualified tradespeople undertake more complex construction activities. As competence develops, the apprentice assumes increasing responsibility under supervision until they are able to perform tasks independently. Throughout this process, learning occurs through observation, discussion, coaching, problem-solving, and participation in authentic workplace activities rather than through formal instruction alone. Knowledge is therefore developed through engagement in practice and interaction with experienced members of the trade community (Billett, 2001; Lave & Wenger, 1991).

Situated learning also highlights the important role played by the tools, technologies, language, symbols, and artefacts used within particular communities of practice. Competence involves more than understanding theoretical concepts; it requires learners to become familiar with the instruments, systems, procedures, communication methods, and cultural norms that characterise professional practice (Vygotsky, 1978; Wenger, 1998). For example, engineering students must learn not only engineering principles but also how to use industry-standard design software, interpret technical drawings, apply professional standards, participate in project meetings, and communicate using the specialised language of the profession. These tools and practices help shape how knowledge is generated, shared, and applied within professional communities.

Situated learning suggests that authentic learning experiences are often more effective than decontextualised instruction because learners can directly connect knowledge to real-world practice (Brown et al., 1989; Herrington & Oliver, 2000). When learning occurs within realistic contexts, learners are better able to understand the relevance of knowledge, recognise how concepts interrelate, and appreciate the complexities of professional decision-making. Authentic learning environments also expose learners to the uncertainties, constraints, and social interactions that characterise real-world practice, helping them develop judgement, adaptability, and professional competence alongside technical knowledge.

Consequently, many contemporary educational approaches reflect the principles of situated learning. Internships, apprenticeships, clinical placements, workplace learning programs, project-based learning, simulations, fieldwork, and professional practicums all provide opportunities for learners to engage in authentic activities within realistic contexts (Billett, 2001; Lave & Wenger, 1991). Medical students develop clinical competence through participation in healthcare settings, engineering students gain practical experience through industry projects and site-based learning, and vocational learners develop occupational skills through apprenticeships and workplace training. In each case, learning occurs through participation in the activities and practices of a professional community rather than solely through classroom-based instruction.

Situated learning therefore provides a powerful explanation for the development of professional expertise and occupational competence. By emphasising participation, context, social interaction, and authentic practice, it demonstrates that learning involves more than acquiring information. It is a process through which individuals become increasingly capable participants in communities of practice, developing both the knowledge and professional identity required for effective performance within real-world environments (Lave & Wenger, 1991; Wenger, 1998).

Sociocultural Perspectives in Adult Education

Sociocultural perspectives are highly relevant to adult education because adults frequently learn through participation in workplaces, professional networks, community groups, and social organisations. Adult learners bring existing knowledge, experiences, cultural backgrounds, and professional identities into learning situations, all of which influence how new knowledge is interpreted and applied.

Learning often occurs through collaboration, mentoring, coaching, discussion, observation, and shared problem-solving rather than through individual study alone. Professional learning communities, peer mentoring programs, workplace training, and collaborative projects all reflect sociocultural principles by recognising that learning is embedded within social relationships and professional practice.

In higher education, sociocultural perspectives can be seen in collaborative learning activities, group projects, peer-assisted learning programs, laboratory work, clinical placements, and research supervision. These activities encourage students to construct knowledge through interaction with peers, academics, and professionals.

As an example, students in a civil engineering program may work together on the design of a transport infrastructure project. Through collaboration, discussion, and feedback, they develop both technical knowledge and professional skills while learning to operate within the norms and practices of the engineering profession.

Similarly, medical students learn through clinical placements where they interact with experienced practitioners, patients, and multidisciplinary healthcare teams. These experiences allow them to develop professional competence through participation in authentic healthcare environments.

Sociocultural perspectives are particularly evident in vocational education and training because many occupational skills are learned through participation in workplace activities. Apprenticeships, traineeships, workplace learning, and practical training all rely heavily on interaction with experienced practitioners.

For example, apprentice carpenters learn by working alongside qualified tradespeople on real construction projects. Through observation, discussion, coaching, and practice, they gradually develop the technical skills, safety knowledge, and workplace behaviours required for competent performance.

Similarly, learners undertaking qualifications in hospitality, logistics, healthcare, or manufacturing acquire much of their knowledge through participation in authentic workplace activities where learning occurs within communities of practice.

Strengths and Limitations of Sociocultural Perspectives

Sociocultural perspectives offer several important strengths. They recognise the importance of social interaction, culture, language, and context in learning (Korte & Lin, 2012; Wasim et al., 2023). They provide a strong explanation for workplace learning, apprenticeship models, mentoring relationships, and professional development (Derwik & Hellström, 2023; Garza & Harter, 2014; Wakahiu, 2017). These perspectives also highlight the value of collaboration, authentic learning experiences, and participation in real-world activities (Bukko & Dhesi, 2021; Różewski et al., 2015; Wasim et al., 2023). By recognising that learning occurs through engagement with communities and social practices, sociocultural theories offer a more holistic understanding of how knowledge and skills develop (Brødsjø et al., 2023; Brown, 2017; Korte & Lin, 2012).

Beside their strengths, sociocultural perspectives have limitations. Some critics argue that these theories place less emphasis on individual cognitive processes and personal differences in learning (Benoliel & Schechter, 2017; Różewski et al., 2015). The concepts can also be difficult to operationalise in large-scale educational settings where individual support and meaningful social interaction may be limited (Gould et al., 2022; Różewski et al., 2015). Furthermore, access to effective communities of practice, mentors, and supportive learning environments is not always equally available to all learners (Derwik & Hellström, 2023; Gendera et al., 2025; Liden et al., 2016).

Connectivism and Networked Learning

Connectivism is a contemporary learning theory that seeks to explain how learning occurs in a digitally connected world. Developed primarily by George Siemens (Siemens, 2005a) and later expanded by Stephen Downes (Downes, 2007, 2012), Connectivism emerged in response to the limitations of traditional learning theories in explaining how people learn in environments characterised by rapid technological change, abundant information, and extensive digital networks.

Unlike behaviourism, cognitivism, constructivism, or sociocultural theories, which were largely developed before the rise of the internet and digital communication technologies, Connectivism argues that learning increasingly occurs through connections between people, information sources, technologies, and networks. Knowledge is no longer viewed as something that resides solely within an individual's mind. Instead, knowledge is distributed across networks of individuals, organisations, databases, digital platforms, and technological systems.

According to Connectivism, the ability to locate, evaluate, connect, and apply information is often more important than memorising information itself. In a world where knowledge changes rapidly, learners must continually update their understanding by accessing and participating in networks that provide current information and expertise.

Origins of Connectivism

Connectivism emerged during a period of rapid technological advancement, increasing internet access, and growing reliance on digital communication. Siemens argued that many existing learning theories focused primarily on individual cognition and did not adequately account for the ways technology was transforming learning (Chen & Chan, 2024; Lamtara, 2023a). Connectivism is often described as a significant departure from traditional learning theories such as behaviourism, cognitivism, and constructivism because it extends the focus of learning beyond the individual learner. Rather than viewing learning primarily as an internal cognitive process, Connectivism emphasises the role of networks, relationships, digital technologies, and distributed sources of knowledge in the learning process (Chen & Chan, 2024). Learning is understood as occurring through interactions among people, information systems, organisations, and technological platforms, reflecting the increasingly interconnected nature of knowledge in contemporary society.

Similarly, proponents of Connectivism argue that many traditional learning theories do not adequately account for the influence of technology and organisational networks on learning. While earlier theories largely concentrate on how individuals acquire, process, and construct knowledge, Connectivism highlights the importance of external knowledge sources, digital environments, and networked communities in supporting learning and professional development. This perspective recognises that learning increasingly occurs through engagement with technological systems and participation in broader information networks that extend beyond the individual learner (Lamtara, 2023a).

Traditional learning theories generally view learning as a process that occurs within the individual through experience, reflection, instruction, or social interaction (Sa'adi, 2016). Behaviourist and cognitivist perspectives typically conceptualise knowledge as existing outside the learner and becoming internalised through learning processes, while constructivist theories emphasise the learner's active role in constructing knowledge through the interpretation of experiences and interactions with the environment (Sa'adi, 2016).

Connectivism extends these perspectives by proposing that learning can occur not only within the individual but also through connections to external sources of knowledge, including digital technologies, databases, information repositories, organisations, and social networks (Chen & Chan, 2024; Kim et al., 2011; Yalmeh et al., 2020). From a connectivist perspective, knowledge is distributed across networks that include both human and technological elements. Learning therefore involves developing the ability to establish, maintain, and navigate connections between these various sources of information and expertise (Aluko et al., 2022; Kim et al., 2011).

Within this framework, knowledge is not confined to what an individual personally knows. Instead, valuable knowledge may reside in professional networks, online communities, databases, digital platforms, artificial intelligence systems, and other information resources that can be accessed when needed (Yalmeh et al., 2020). Consequently, an individual's capacity to learn depends not only on their existing knowledge and skills but also on their ability to identify relevant information sources, evaluate their credibility, and effectively connect with networks that support ongoing learning and knowledge development (Aluko et al., 2022; Chen & Chan, 2024). In this way, Connectivism emphasises the importance of network formation, information literacy, and technological engagement as central components of learning in a digitally connected world.

This perspective reflects the reality that modern professionals increasingly rely on digital technologies and networked sources of information to solve problems, make decisions, and acquire new knowledge. Search engines, online databases, artificial intelligence systems, professional networks, social media platforms, and collaborative technologies have become integral components of learning and professional practice (Keller et al., 2020; Yan & Gašević, 2026). Connectivism emerged in response to these developments, recognising that knowledge is increasingly distributed across complex networks of people, organisations, information resources, and technological systems rather than residing solely within individuals (Yan & Gašević, 2026).

Contemporary learners frequently access, share, and construct knowledge through digital platforms such as social media, online communities, learning management systems, professional forums, and collaborative workspaces (Chen & Chan, 2024). These environments enable learners to engage with diverse sources of expertise, participate in discussions, and contribute to collective knowledge creation. As a result, learning increasingly involves the ability to build and maintain connections within digital networks and to effectively utilise technology to access relevant information and expertise (Keller et al., 2020).

The growing use of artificial intelligence has further expanded the networked nature of learning. AI systems are increasingly capable of generating content, interpreting user needs, recommending resources, supporting decision-making, and personalising learning experiences (Yan & Gašević, 2026). These technologies influence how learners access information, engage with learning tasks, and navigate knowledge networks. Consequently, the ability to work effectively with digital technologies and critically evaluate information sources has become an increasingly important component of learning in contemporary educational and professional environments.

Core Principles of Connectivism

Connectivism is founded on several principles that reflect the nature of learning in a digitally connected and information-rich world. A central principle is that learning and knowledge exist within networks rather than solely within individual learners (Chatti ́ et al., 2010; Corbett & Spinello, 2020; Lajmiri, 2016). Knowledge is distributed across connections between people, organisations, databases, technologies, and information sources (Benjamin et al., 2024; Chatti et al., 2010; Kotzee & Palermos, 2021). Learning therefore involves creating, maintaining, and strengthening these connections in order to access relevant information and expertise when required (Downes, 2020; Wang et al., 2014; Zhu et al., 2024). From this perspective, the ability to participate effectively in knowledge networks becomes a critical component of learning and professional practice (Chen & Chan, 2024; Corbett & Spinello, 2020; Lee & Wong, 2025).

Another important principle is that the capacity to acquire new knowledge is often more valuable than the knowledge an individual currently possesses (Yalmeh et al., 2020). In rapidly changing fields, information can quickly become outdated, making it impossible for individuals to rely solely on what they already know (Chen & Chan, 2024; Yalmeh et al., 2020). As a result, learners benefit from developing the skills needed to locate, evaluate, and access current information as circumstances change (Benjamin et al., 2024; Tınmaz, 2012). Learning is therefore viewed as an ongoing process of adaptation and continuous knowledge renewal rather than the accumulation of fixed bodies of information (Chen & Chan, 2024; Yalmeh et al., 2020).

Connectivism also recognises that learning frequently involves connecting specialised sources of knowledge (Chen & Chan, 2024; Corbett & Spinello, 2020; Lajmiri, 2016). Modern professional and academic fields are increasingly complex, and no individual can reasonably possess all the knowledge required to address every problem or situation (Benjamin et al., 2024). Effective learners understand where expertise resides and how to access it through professional networks, digital resources, communities of practice, and technological systems (Chen & Chan, 2024; Lee & Wong, 2025; Tınmaz, 2012). The ability to identify appropriate sources of information and connect with knowledgeable individuals becomes an important aspect of competence in many disciplines (Benjamin et al., 2024; Corbett & Spinello, 2020).

Maintaining these connections is equally important because learning is viewed as a continuous and dynamic process (Downes, 2020; Kotzee & Palermos, 2021; Zhu et al., 2024). Professional networks, industry associations, online communities, social media groups, collaborative platforms, and communities of practice all contribute to ongoing knowledge development (Benjamin et al., 2024; Corbett & Spinello, 2020). By remaining connected to these networks, learners can stay informed about emerging developments, new research, evolving practices, and changing professional requirements (Benjamin et al., 2024; Chen & Chan, 2024). Learning therefore extends beyond formal education and becomes an ongoing activity embedded within professional and social networks (Corbett & Spinello, 2020; Downes, 2020).

A further principle of Connectivism is that decision-making itself constitutes a form of learning (Černý, 2020; Lajmiri, 2016). Learners must continually evaluate the relevance, reliability, accuracy, and credibility of information obtained from multiple sources (Benjamin et al., 2024; Tınmaz, 2012; Yalmeh et al., 2020). Because knowledge and information environments are constantly evolving, learners are required to make informed judgements about which information is most useful and appropriate in a given context (Chen & Chan, 2024; Yalmeh et al., 2020). This process of selecting, interpreting, and applying information contributes to learning and highlights the importance of critical thinking, information literacy, and reflective decision-making within networked environments (Benjamin et al., 2024; Tınmaz, 2012; Yalmeh et al., 2020).

Learning in Networks

A central concept in Connectivism is the idea of the network. Networks consist of nodes and connections. Nodes may include individuals, organisations, websites, databases, learning management systems, artificial intelligence tools, professional communities, research publications, or social media groups. Learning occurs as individuals create, strengthen, and navigate connections between these nodes.

For example, an engineer seeking information about sustainable infrastructure may draw upon academic journals, industry standards, government publications, professional associations, online discussion forums, webinars, and colleagues working in related fields. Learning occurs through interacting with and synthesising information from multiple sources rather than relying on a single authority. The quality and diversity of these connections influence the learner's ability to acquire accurate, current, and useful knowledge.

Digital Technology and Learning

Connectivism places significant emphasis on technology as an active component of learning. Digital technologies are not viewed merely as tools for delivering instruction but as integral parts of learning networks. Search engines, online databases, collaborative platforms, artificial intelligence systems, learning management systems, and social networking tools all contribute to knowledge creation and dissemination.

This perspective recognises that modern learners frequently access information on demand rather than relying solely on previously acquired knowledge. Professionals often solve problems by consulting digital resources, collaborating with geographically dispersed colleagues, and participating in online communities. As a result, technological literacy becomes an important component of effective learning.

Networked Learning

Closely related to Connectivism is the concept of networked learning. Networked learning focuses on how technology facilitates connections between learners, educators, resources, and communities. While Connectivism is often presented as a learning theory, networked learning is generally viewed as an educational approach or framework that emphasises learning through connected relationships.

Networked learning recognises that meaningful learning occurs through interactions among people, digital technologies, and learning resources. It emphasises collaboration, communication, knowledge sharing, and collective problem-solving.

The concept emerged from the work of researchers such as Chris Jones, Peter Goodyear, and others who examined how digital technologies support collaborative learning and knowledge construction. Networked learning environments encourage learners to engage with others, contribute to shared knowledge, and participate in learning communities that extend beyond traditional classroom boundaries.

Personal Learning Networks

A practical application of Connectivism is the development and use of Personal Learning Networks (PLNs). A Personal Learning Network refers to the collection of people, organisations, communities, technologies, and information sources that an individual draws upon to support ongoing learning and professional development. Unlike traditional learning environments that rely primarily on formal instruction, PLNs enable individuals to access knowledge from a diverse range of sources and to engage in continuous, self-directed learning throughout their careers.

A PLN may include professional colleagues, mentors, industry experts, academic researchers, industry associations, online communities, and communities of practice. It can also incorporate a variety of digital resources and technologies such as professional journals, digital libraries, podcasts, webinars, social media platforms, learning management systems, government and regulatory websites, and artificial intelligence tools. Together, these interconnected sources provide access to specialised knowledge, emerging trends, expert advice, and professional discussions that may not be available through formal educational programs alone.

The composition of a Personal Learning Network often reflects an individual's professional role, interests, and learning goals. For example, a civil engineer may follow professional engineering bodies, subscribe to technical journals, participate in infrastructure-focused online forums, attend industry webinars, and engage with colleagues through professional networking platforms. Similarly, a healthcare professional may rely on clinical practice databases, professional associations, research publications, specialist discussion groups, and continuing professional development resources to remain informed about advances in healthcare practice.

Through active participation in Personal Learning Networks, individuals can continuously access current information, share experiences, seek advice, collaborate with others, and contribute to collective knowledge creation. PLNs therefore support lifelong learning by enabling learners to remain current within their fields, respond to changing professional requirements, and develop new knowledge and skills as needed. In a rapidly evolving digital environment, the ability to establish and maintain effective learning networks has become an increasingly important capability for both personal and professional development.

Connectivism in Adult Education

Connectivism is particularly relevant to adult education because adults frequently engage in self-directed learning and ongoing professional development throughout their careers (Benjamin et al., 2024; Conceição & Biniecki, 2024). Rapid technological advancement, changing workplace requirements, evolving regulations, and emerging industry practices require many adult learners to continually update their knowledge and skills in order to remain effective in their professional roles (Conceição & Biniecki, 2024; Dai, 2024). As a result, learning is increasingly viewed as a continuous process that extends beyond formal education and occurs throughout an individual's personal and professional life.

Rather than relying exclusively on structured courses or traditional educational institutions, adult learners often acquire knowledge through professional networks, online communities, webinars, digital resources, collaborative platforms, and social learning environments (Benjamin et al., 2024; Conceição & Biniecki, 2024). Digital technologies provide access to a wide range of learning opportunities, including virtual learning environments, professional networking platforms, social media communities, and collaborative tools that facilitate communication and knowledge sharing. Adults also frequently engage with experts from diverse disciplines and utilise resources such as blogs, podcasts, online discussions, professional publications, and other digital media to support their learning and professional growth (Benjamin et al., 2024).

Connectivism provides a useful framework for understanding how adults acquire, maintain, and apply knowledge in rapidly changing environments. The theory emphasises the importance of developing and maintaining connections with people, information sources, organisations, and technologies that support ongoing learning (Benjamin et al., 2024; Conceição & Biniecki, 2024). Through these networks, learners can access current information, exchange ideas, identify emerging trends, and continually update their knowledge and skills. This capacity to connect with and navigate knowledge networks is increasingly important in professional contexts where information changes rapidly and expertise is distributed across multiple sources.

The theory also aligns closely with the principles of lifelong learning. Connectivism recognises that learning is not confined to formal education or specific stages of life but occurs continuously across a range of personal, social, and professional contexts (Benjamin et al., 2024; Conceição & Biniecki, 2024). As technological innovation continues to reshape workplaces and societies, individuals must continually adapt and develop new capabilities. Lifelong learning therefore becomes essential for maintaining professional competence, responding to change, and participating effectively in contemporary knowledge-based economies (Dai, 2024). Connectivism provides a framework for understanding how this ongoing learning occurs through participation in networks, engagement with digital technologies, and interaction with diverse sources of knowledge and expertise.

In higher education, Connectivism has influenced online learning, blended learning, open educational resources, and massive open online courses (MOOCs). Students are increasingly encouraged to engage with multiple information sources, collaborate across digital platforms, and develop digital literacy skills.

For example, students in a Master's program in Data Science may access research databases, participate in online coding communities, collaborate through cloud-based platforms, attend virtual conferences, and engage with practitioners from around the world. Learning occurs through interaction with a broad network of people and resources rather than solely through lectures and textbooks.

Similarly, students undertaking research degrees frequently rely on international networks of scholars, online repositories, collaborative software, and professional communities to develop and share knowledge.

Connectivism is increasingly relevant in vocational education and training as industries adopt digital technologies, automation, data-driven systems, and networked forms of work. Learners are frequently required to access online resources, industry databases, technical documentation, digital manuals, and professional networks to remain current with changing technologies, regulations, and workplace practices. As vocational occupations become more technologically sophisticated, the ability to locate, evaluate, and apply information from multiple sources has become an important component of occupational competence (Rodríguez, 2021).

For example, apprentices in advanced manufacturing may use digital manuals, online troubleshooting platforms, remote support systems, and professional forums to solve technical problems. Logistics and supply chain learners may access real-time tracking systems, regulatory databases, and global industry networks. Healthcare workers increasingly rely on online clinical resources, evidence-based practice databases, and professional communities to support decision-making and continuing professional development. These examples illustrate how vocational competence increasingly depends on the ability to access and utilise distributed sources of knowledge.

Strengths and Limitations of Connectivism and Networked Learning

Connectivism offers several important strengths. It addresses the realities of learning in a digital age, recognises the importance of technology in knowledge creation and access, and provides a framework for understanding lifelong learning in rapidly changing environments (Chen & Chan, 2024; Conceição & Biniecki, 2024; Jeny, 2024). The theory highlights the importance of information literacy, digital literacy, networking, collaboration, and continuous learning (Benjamin et al., 2024; Jeny, 2024) . It also reflects the increasingly distributed nature of knowledge in modern professional practice (Chen & Chan, 2024; Cvetković, 2022; Ossiannilsson, 2022).

Networked learning similarly supports collaboration, knowledge sharing, interdisciplinary learning, and access to diverse perspectives (Benjamin et al., 2024; Chen & Chan, 2024; Ossiannilsson, 2022). It enables learners to participate in communities that extend beyond geographical and institutional boundaries (Arniza, 2024; Cvetković, 2022; Ossiannilsson, 2022).

Despite its influence, Connectivism has attracted criticism. Some scholars argue that it is more accurately described as a pedagogical perspective than a fully developed learning theory because it does not explain cognitive processes in the same depth as established theories such as cognitivism or constructivism (Arniza, 2024; Benjamin et al., 2024; Chen & Chan, 2024). Others argue that many of its principles can be explained through existing sociocultural and constructivist perspectives (Conceição & Biniecki, 2024; Isidori et al., 2023).

Additionally, the effectiveness of networked learning depends on learners' ability to evaluate information critically. The abundance of online information can create challenges related to misinformation, information overload, digital inequality, and varying levels of technological access and competence (Benjamin et al., 2024; Jeny, 2024; Odularu et al., 2023). Learners must therefore develop strong critical thinking and information literacy skills to navigate complex information environments effectively (Arniza, 2024; Dai, 2024; Jeny, 2024).

Relationship Between Adult Learning Theories

The various adult learning theories are best understood not as competing explanations of learning but as complementary perspectives that focus on different dimensions of the learning process. Each theory highlights a particular aspect of how adults learn, and together they provide a comprehensive framework for understanding learning in contemporary educational and workplace settings.

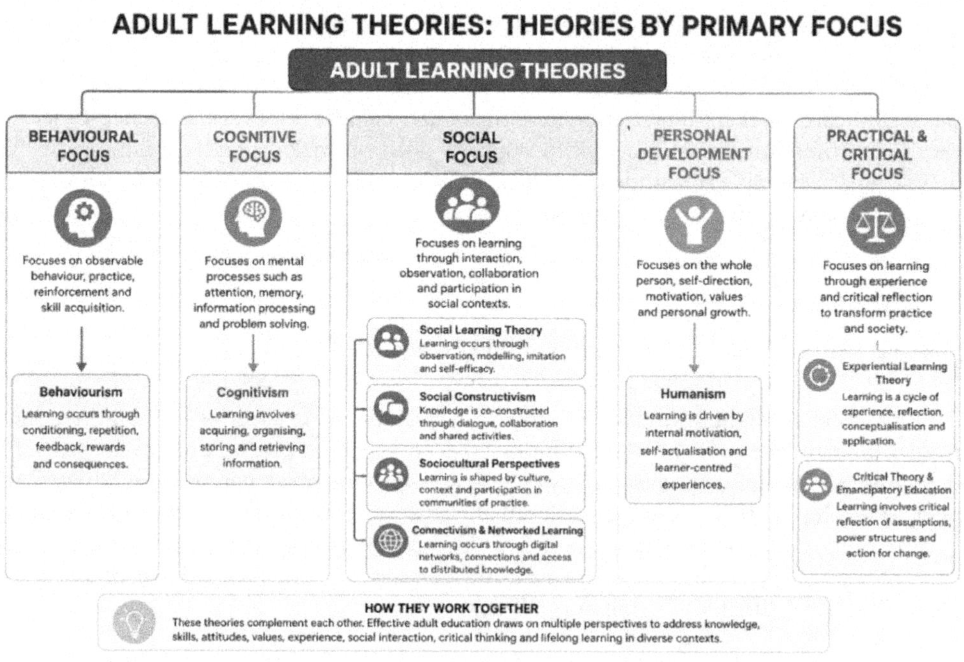

Figure 7: Adult Learning Theories by primary focus.

At the most basic level, Behaviourism focuses on observable behaviour and skill acquisition. It explains how adults learn through practice, repetition, feedback, reinforcement, and consequences. Behaviourism is particularly useful for developing procedural skills, compliance requirements, safety behaviours, and routine workplace tasks.

Cognitivism builds upon this foundation by examining the internal mental processes involved in learning. It focuses on how adults perceive, process, organise, store, and retrieve information. While behaviourism emphasises what learners do, cognitivism seeks to explain what happens within the learner's mind during learning.

Constructivism extends cognitive perspectives by proposing that learners actively construct knowledge rather than simply receiving information. Adults interpret new experiences through the lens of their existing knowledge, beliefs, and experiences. Learning therefore becomes a process of meaning-making rather than information transmission.

Social Constructivism further develops this idea by recognising that knowledge construction often occurs through interaction with others. Discussions, collaboration, mentoring, and shared problem-solving help learners develop deeper understanding. This theory highlights the social nature of learning and the importance of dialogue in knowledge construction.

Humanism shifts attention toward the learner as a whole person. It emphasises learner autonomy, personal growth, self-direction, intrinsic motivation, and self-actualisation. Humanistic approaches recognise that adult learners bring goals, aspirations, emotions, values, and life experiences that influence learning.

Social Learning Theory complements these perspectives by explaining how adults learn through observation, modelling, imitation, and self-efficacy. It highlights the importance of role models, mentors, supervisors, and peers in shaping learning and professional development.

Experiential Learning Theory focuses on the role of experience as the basis of learning. Adults learn by engaging in activities, reflecting on their experiences, developing concepts, and applying their learning in new situations. Experiential learning bridges the gap between theory and practice.

Critical Theory and Emancipatory Education introduce a broader social dimension by encouraging learners to critically examine assumptions, power structures, social systems, and inequalities. Learning becomes not only a process of acquiring knowledge but also of developing critical awareness and the capacity for informed action.

Sociocultural Perspectives further expand the understanding of learning by emphasising that learning is embedded within social, cultural, historical, and professional contexts. Knowledge is shaped by participation in communities of practice, workplace cultures, and social interactions.

Finally, Connectivism and Networked Learning address learning in the digital age. They recognise that knowledge is increasingly distributed across networks of people, organisations, databases, technologies, and digital platforms. Learning involves developing the ability to access, evaluate, connect, and apply knowledge from diverse sources.

These theories can be viewed as a progression from simple to increasingly complex understandings of learning:

Table 3: Progression of learning theories.

Theory	Primary Focus
Behaviourism	Behaviour and skill acquisition

Theory	Primary Focus
Cognitivism	Mental processes and information processing
Constructivism	Individual knowledge construction
Social Constructivism	Collaborative knowledge construction
Humanism	Personal growth and learner autonomy
Social Learning Theory	Observation and modelling
Experiential Learning	Learning through experience
Critical Theory	Critical reflection and social awareness
Sociocultural Theory	Learning within social and cultural contexts
Connectivism	Learning through networks and technology

Rather than replacing one another, each theory adds another layer of understanding. Together they move from understanding learning as a behavioural response to understanding learning as a cognitive, social, experiential, cultural, critical, and networked process.

Comparing and Evaluating Learning Theories

The study of learning theories provides educators with different perspectives on how learning occurs, why learners behave as they do, and how effective educational experiences can be designed. No single learning theory fully explains all aspects of human learning. Instead, each theory offers unique insights into particular dimensions of the learning process and is most effective when applied in appropriate contexts. Comparing and evaluating learning theories allows educators, trainers, and instructional designers to select teaching approaches that best align with learning objectives, learner characteristics, and educational environments.

One of the most significant differences between learning theories concerns how they view the learner and the learning process. Behaviourism focuses on observable behaviour and views learning as a change in behaviour resulting from environmental stimuli and reinforcement. Learning is achieved through repetition, practice, rewards, and consequences. Behaviourist approaches are particularly effective for developing routine skills, procedural knowledge, and compliance-based learning where consistency and accuracy are important. However, behaviourism has been criticised for paying limited attention to internal cognitive processes, learner motivation, and the social dimensions of learning.

In contrast, Cognitivism emphasises the mental processes involved in learning, including perception, memory, information processing, and problem-solving. Learning is viewed as the acquisition and organisation of knowledge within the learner's cognitive structures. Cognitivist approaches are useful for teaching complex concepts, developing critical thinking skills, and helping learners organise and retain information. However, critics argue that cognitivism may underestimate the influence of social, cultural, and environmental factors on learning.

Constructivism builds upon cognitive theories by proposing that learners actively construct knowledge through their experiences and interactions with the world. Rather than receiving knowledge passively, learners interpret information based on their existing understanding and experiences. Constructivist approaches encourage inquiry, exploration, problem-solving, and reflection. These approaches are particularly effective for developing higher-order thinking skills and supporting meaningful learning. However, constructivist learning environments can be resource-intensive and may require substantial learner independence and educator facilitation.

Social Constructivism extends constructivist principles by emphasising the importance of social interaction in knowledge construction. Influenced by Vygotsky's work, social constructivism argues that learning occurs through collaboration, dialogue, and participation in social activities. Concepts such as scaffolding and the Zone of Proximal Development highlight the role of more knowledgeable others in supporting learning. Social constructivist approaches are widely used in collaborative learning, group projects, mentoring, and workplace training. Their effectiveness, however, depends heavily on the quality of social interactions and the availability of appropriate support.

Humanism takes a different perspective by focusing on personal growth, self-development, and learner autonomy. Humanistic theories view learners as capable individuals who are motivated by the desire for self-actualisation and personal fulfilment. Learning environments based on humanistic principles emphasise respect, empathy, learner choice, and intrinsic motivation. These approaches can be highly effective for adult learners and professional development contexts but may be less suitable when highly structured instruction or standardised outcomes are required.

Experiential Learning Theory, particularly as developed by Kolb, emphasises learning through experience, reflection, conceptualisation, and experimentation. Experiential approaches recognise that knowledge is often developed through direct engagement with real-world situations. This perspective is widely used in professional education, workplace learning, simulations, fieldwork, and practical training. While experiential learning provides valuable opportunities for applying theory to practice, the quality of learning depends on the learner's ability to reflect upon and interpret their experiences effectively.

Social Learning Theory, developed by Bandura, highlights the importance of observation, modelling, and self-efficacy in learning. Individuals learn by observing others, evaluating the consequences of behaviours, and developing confidence in their own abilities. Social Learning Theory provides a strong explanation for workplace learning, mentoring, apprenticeships, and professional development. However, observational learning alone may be insufficient for developing deep conceptual understanding without opportunities for practice and reflection.

Critical Theory and Emancipatory Education focus on the social, political, and cultural dimensions of learning. These perspectives encourage learners to question assumptions, examine power relationships, and develop critical awareness of social structures. Critical approaches are particularly valuable for promoting critical thinking, social justice, and democratic participation. However, they can be challenging to implement in highly structured educational environments and may sometimes place less emphasis on technical or disciplinary knowledge.

Sociocultural Perspectives emphasise that learning is shaped by social interaction, cultural tools, language, and participation in communities of practice. Learning is viewed as a process of becoming an active participant within a particular social or professional community. These perspectives are especially effective for explaining workplace learning, apprenticeships, mentoring, and professional identity development. Their primary limitation is that they may not fully explain individual cognitive differences or learning processes that occur independently of social interaction.

Connectivism and Networked Learning address the realities of learning in a digitally connected world. These perspectives recognise that knowledge is distributed across networks of people, organisations, technologies, and information sources. Learning involves developing the ability to access, evaluate, and connect information from multiple sources. Connectivism is particularly relevant to lifelong learning, professional development, and digital learning environments. However, some scholars argue that it is better viewed as a framework for understanding learning in digital environments than as a fully developed learning theory.

When comparing these theories, it becomes clear that they often focus on different aspects of learning rather than directly competing with one another. Behaviourism emphasises behaviour, cognitivism emphasises mental processes, constructivism emphasises knowledge construction, sociocultural theories emphasise social participation, humanism emphasises personal growth, and connectivism emphasises networked knowledge. Each perspective provides valuable insights into specific learning situations and educational challenges.

Modern educational practice increasingly adopts an eclectic or integrated approach that draws upon multiple theories rather than relying exclusively on a single perspective. For example, a vocational training program may use behaviourist techniques to develop safety procedures, cognitivist strategies to teach technical concepts, constructivist activities to solve workplace problems, sociocultural approaches to support apprenticeships, and connectivist tools to facilitate ongoing professional learning. By combining insights from multiple theories, educators can design learning experiences that address the cognitive, social, emotional, practical, and technological dimensions of learning.

Ultimately, the value of any learning theory lies not in its ability to explain all learning but in its usefulness for understanding particular learning situations and informing effective educational practice. Comparing and evaluating learning theories enables educators to make informed decisions about teaching, learning, assessment, and curriculum design while recognising the complexity and diversity of human learning.

Using the Theories Collectively in Adult Education

Effective adult education rarely relies on a single learning theory. Adult learners bring diverse experiences, motivations, learning preferences, and professional goals to the learning environment, requiring educators to draw upon multiple theoretical perspectives to support effective learning. Each learning theory contributes unique insights into how adults acquire knowledge, develop skills, construct meaning, interact with others, and apply learning in real-world contexts. Consequently, contemporary adult education often adopts an integrated approach that combines elements from several theories to address a broad range of learning needs and outcomes.

For example, in a professional development program for project managers, behaviourist principles may be used to teach mandatory safety procedures, organisational policies, regulatory requirements, or software applications through structured practice, repetition, feedback, and reinforcement. At the same time, cognitivist approaches may be employed to help learners understand project management methodologies, risk management frameworks, scheduling techniques, and decision-making models that require information processing, comprehension, and memory.

Constructivist approaches can be incorporated through case studies, scenario-based activities, and problem-solving exercises that require learners to apply concepts to realistic workplace situations. Social constructivist principles may further enhance learning through group discussions, collaborative projects, peer learning activities, and team-based problem-solving, allowing participants to share experiences and co-construct knowledge through dialogue and interaction.

Humanistic principles may guide the overall design of the learning program by encouraging learner autonomy, self-direction, and personal growth. Learners may be given opportunities to identify their own learning goals, select project topics relevant to their professional interests, and reflect on their personal and career development. Such approaches recognise the importance of motivation, individual needs, and the learner's sense of purpose in the learning process.

Social Learning Theory can contribute through mentoring programs, workplace coaching, demonstrations, guest presentations, and opportunities to observe experienced project managers in practice. By observing the behaviours, strategies, and decision-making processes of others, learners can develop confidence, self-efficacy, and professional competence. Experiential Learning Theory further complements these approaches by providing opportunities for learners to engage in workplace projects, simulations, role plays, and reflective practice activities that transform experience into learning.

Critical Theory and emancipatory approaches encourage learners to examine broader organisational, social, and ethical dimensions of professional practice. Learners may be challenged to analyse stakeholder interests, organisational culture, ethical dilemmas, and the social consequences of project decisions. Such activities promote critical thinking and encourage learners to question assumptions and consider alternative perspectives.

Sociocultural perspectives can be reflected through workplace-based learning, participation in professional communities of practice, and collaboration with experienced practitioners. These approaches recognise that learning occurs within social and cultural contexts and that professional competence develops through participation in authentic professional activities. Connectivist approaches further extend learning opportunities by encouraging learners to build professional networks, engage with industry associations, access digital resources, participate in online communities, and utilise emerging technologies to support continuous learning and professional development.

An Integrated Model of Adult Learning

When viewed collectively, these theories suggest that effective adult learning is a multidimensional process that involves several interconnected elements. Adults acquire knowledge, skills, and competencies through instruction, practice, and information processing, as described by behaviourist and cognitivist perspectives. They construct personal understanding by connecting new information with prior knowledge and experiences, consistent with constructivist principles. Learning is also enhanced through interaction with others, collaboration, observation, and participation in social activities, reflecting social constructivist and social learning perspectives.

Adult learning is further strengthened through experience, reflection, and application, which are central to experiential learning theory. Personal growth, self-direction, intrinsic motivation, and the pursuit of meaningful goals are emphasised within humanistic approaches. Critical theory highlights the importance of questioning assumptions, examining social realities, and developing critical awareness, while sociocultural perspectives emphasise participation in professional, cultural, and workplace communities. Connectivism recognises the increasing importance of digital technologies, information networks, and distributed knowledge sources in supporting lifelong learning.

Taken together, these theories provide a holistic understanding of adult learning that recognises learners as cognitive, social, emotional, experiential, cultural, and technologically connected individuals. Rather than viewing learning through a single lens, contemporary adult education benefits from integrating multiple theoretical perspectives to create learning experiences that are engaging, relevant, collaborative, reflective, practical, and adaptable to the complex needs of modern learners and workplaces.

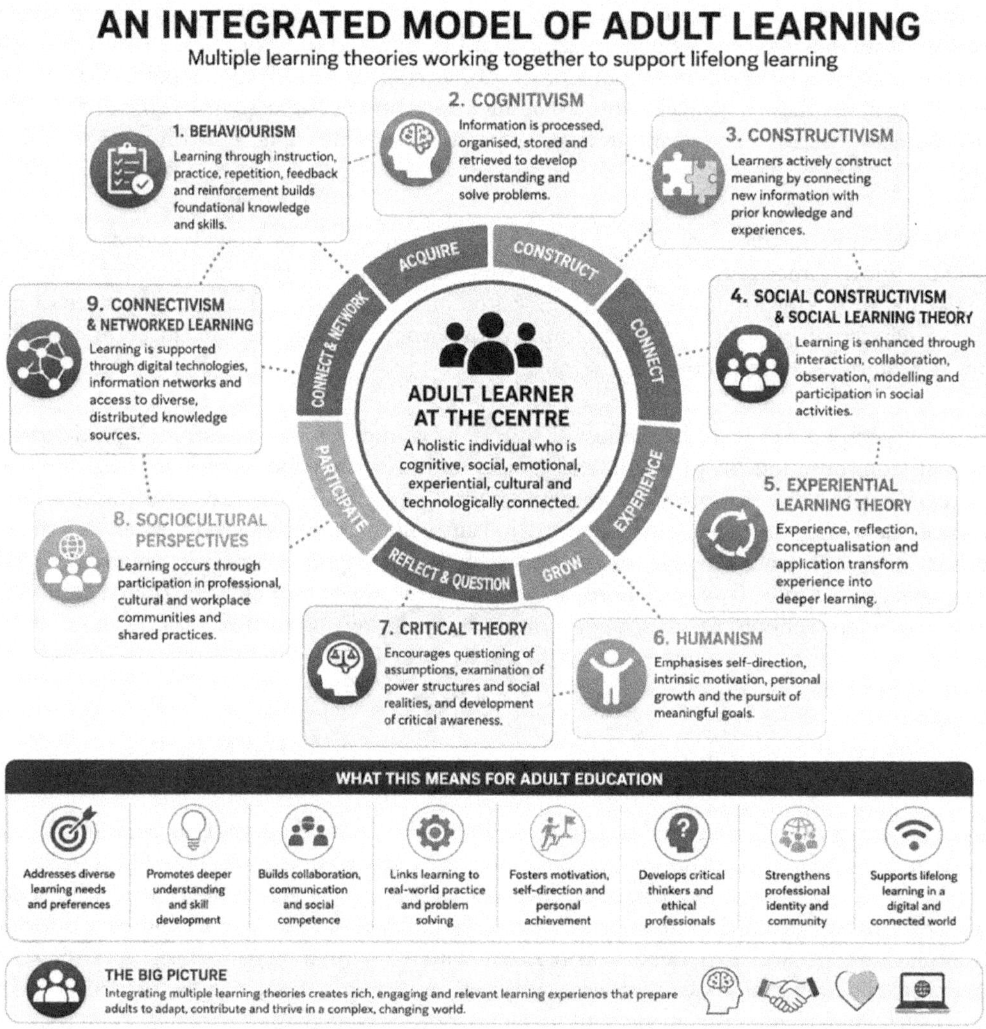

Figure 8: Integrated model of adult learning.

This model presents adult learning as a holistic and interconnected process rather than a collection of separate and competing theories. The central premise is that no single learning theory can fully explain how adults learn. Instead, each theory offers a distinct perspective on the learning process, highlighting particular aspects of how adults acquire knowledge, develop skills, construct understanding, interact with others, and adapt to changing environments. Effective adult education therefore draws upon multiple theoretical perspectives to support knowledge acquisition, skill development, personal growth, professional competence, and lifelong learning.

At the centre of the model is the adult learner, represented as a cognitive, social, emotional, experiential, cultural, and technologically connected individual. This reflects the understanding that adult learners do not enter educational environments as blank slates. Rather, they bring with them a wealth of prior knowledge, life experiences, professional backgrounds, personal motivations, cultural influences, and established social networks. These factors shape how they interpret new information, engage with learning activities, and apply learning in practice.

Surrounding the learner is a circular framework that represents the major processes involved in adult learning. These processes include acquiring knowledge and information, constructing understanding and meaning, connecting and collaborating with others, experiencing and applying learning, growing personally and professionally, reflecting upon and questioning assumptions, participating in professional and social communities, and building networks through digital and professional connections. Together, these processes illustrate the dynamic and lifelong nature of adult learning, highlighting that learning extends beyond formal education and continues throughout an individual's personal and professional life.

Behaviourism forms one of the foundational layers of the model by focusing on observable behaviours and the development of skills through instruction, practice, repetition, feedback, reinforcement, and assessment. In adult education, behaviourist approaches are particularly effective for teaching procedures, compliance requirements, safety practices, technical competencies, and routine workplace tasks. Activities such as learning to operate equipment, following workplace procedures, or mastering software applications rely heavily on behaviourist principles. Within the model, behaviourism contributes primarily to the learner's ability to acquire foundational knowledge and skills.

Cognitivism complements this perspective by focusing on the mental processes involved in learning. It examines how information is perceived, organised, stored, and retrieved, helping to explain how learners develop understanding, solve problems, and make decisions. Adult learners engage cognitive processes when studying theories, analysing information, planning projects, or evaluating complex situations. Cognitivism supports the development of knowledge structures that enable learners to organise information effectively and apply it in meaningful ways. Within the model, it contributes to the learner's capacity to acquire, organise, and process understanding.

Constructivism builds upon cognitive perspectives by proposing that learners actively construct meaning by connecting new information with their existing knowledge and experiences. Learning is therefore not simply the transmission of information from teacher to learner but a process of interpretation and meaning-making. Adult learners evaluate new ideas through the lens of their previous experiences, making learning highly personal and context-dependent. Within the model, constructivism contributes to the development of understanding through the active construction of knowledge.

Social Constructivism and Social Learning Theory further extend the learning process by emphasising the importance of social interaction. Social Constructivism highlights the role of discussion, collaboration, and shared meaning-making in learning, while Social Learning Theory focuses on observation, modelling, imitation, and the development of self-efficacy. Adults frequently learn through interactions with colleagues, mentors, supervisors, peers, and professional communities. Activities such as teamwork, coaching, mentoring, collaborative projects, and professional networking demonstrate the social nature of learning. Within the model, these theories support the learner's ability to connect with others and learn through social engagement.

Experiential Learning Theory highlights the importance of experience as a source of learning. According to this perspective, learning occurs through a continuous cycle of experience, reflection, conceptualisation, and application. Adults often learn most effectively when they can apply knowledge in realistic situations and reflect on the outcomes of their actions. Workplace projects, simulations, internships, fieldwork, and practical exercises all provide opportunities for experiential learning. Within the model, experiential learning contributes to the learner's capacity to learn through direct experience and practical application.

Humanism shifts attention to the learner as a whole person and emphasises self-direction, intrinsic motivation, personal growth, self-awareness, and the pursuit of meaningful goals. Humanistic approaches recognise that adult learners are autonomous individuals capable of directing their own learning and making decisions about their personal and professional development. Adults are often motivated by a desire to improve their careers, enhance their quality of life, or achieve personal fulfilment. Within the model, humanism supports personal and professional growth by recognising the importance of motivation, purpose, and self-development.

Critical Theory introduces a reflective and transformative dimension to learning by encouraging learners to question assumptions, examine power structures, analyse social realities, and develop critical awareness. Rather than accepting information uncritically, learners are encouraged to reflect on the broader social, organisational, ethical, and political influences that shape knowledge and practice. This perspective is particularly relevant in areas such as leadership, education, healthcare, policy development, and professional ethics, where critical thinking and informed judgement are essential. Within the model, Critical Theory contributes to reflection, questioning, and transformative learning.

Sociocultural perspectives emphasise that learning is shaped by culture, context, language, and participation in communities of practice. Knowledge develops through engagement with workplaces, professional groups, organisations, and social communities. Learning therefore becomes a process of participation, social interaction, and identity formation. Apprenticeships, workplace learning, professional associations, and industry communities all illustrate how learning occurs within social and cultural contexts. Within the model, sociocultural theories support the learner's participation in communities and professional practice.

Connectivism and Networked Learning reflect the realities of learning in a digitally connected world. These perspectives recognise that knowledge is increasingly distributed across networks of people, organisations, databases, digital platforms, and technological systems. Learning involves the ability to locate, evaluate, connect, and apply information from diverse sources. Professional networks, online communities, social media platforms, webinars, digital libraries, learning management systems, and artificial intelligence tools all contribute to contemporary learning environments. Within the model, Connectivism supports the learner's ability to connect, network, and engage with distributed sources of knowledge within digital environments.

Collectively, these theories provide a comprehensive framework for understanding adult learning as a multidimensional process. They recognise that adults learn through a combination of instruction, cognition, experience, social interaction, reflection, participation, personal growth, and technological engagement. By integrating these perspectives, educators can design learning experiences that are engaging, relevant, collaborative, reflective, practical, and adaptable to the diverse needs of contemporary learners and workplaces. The model therefore provides a holistic view of adult learning that acknowledges the complexity of human learning and the importance of drawing upon multiple theoretical perspectives to support lifelong learning and professional development.

The lower section of the model highlights the benefits that can be achieved when multiple learning theories are integrated within adult education and training. Rather than relying on a single approach, educators can draw upon the strengths of different theories to create more effective and inclusive learning experiences. This integrated approach enables educators to address the diverse learning needs, backgrounds, motivations, and preferences that adult learners bring to the learning environment. By incorporating multiple perspectives, learning experiences can be designed to accommodate a wider range of learners while supporting both individual and collective learning outcomes.

The integration of multiple theories also promotes deeper understanding and more comprehensive skill development. Behaviourist and cognitivist approaches help learners acquire knowledge and develop competence, while constructivist, experiential, and social learning approaches encourage learners to apply, analyse, and extend that knowledge in meaningful ways. Through collaboration, discussion, observation, and participation in shared activities, learners develop communication skills, social competence, teamwork capabilities, and professional relationships that support both learning and workplace performance.

An integrated approach further strengthens the connection between learning and real-world practice. Experiential learning, sociocultural participation, and constructivist problem-solving activities help learners apply knowledge to authentic situations and workplace challenges. At the same time, humanistic approaches encourage motivation, self-direction, confidence, and personal achievement by recognising learners as autonomous individuals with unique goals and aspirations. Critical theory contributes by fostering critical thinking, ethical reasoning, and the ability to examine assumptions and evaluate complex social, organisational, and professional issues.

The model also recognises the importance of professional identity, community participation, and lifelong learning. Sociocultural perspectives highlight the role of professional communities and workplace participation in developing competence and belonging, while connectivist approaches emphasise the importance of digital technologies, professional networks, and distributed knowledge sources. Together, these perspectives support learners in maintaining current knowledge, adapting to change, and engaging in ongoing professional development throughout their lives.

Ultimately, the model argues that adult learning is inherently multidimensional. Adults learn not only through formal instruction but also through thinking, experience, reflection, social interaction, professional participation, personal growth, and engagement with digital networks. Learning is therefore best understood as a complex process involving cognitive, behavioural, social, emotional, cultural, experiential, and technological dimensions that interact and influence one another.

Rather than seeking to identify a single "best" learning theory, the model suggests that effective adult education draws upon multiple theories simultaneously. For example, a learning program may use behaviourist techniques to teach procedures and foundational skills, cognitivist strategies to develop understanding, constructivist activities to support problem-solving, experiential learning to build practical competence, social learning approaches to encourage collaboration, humanistic principles to promote self-direction, critical reflection to challenge assumptions, sociocultural participation to develop professional identity, and connectivist tools to support ongoing learning through professional and digital networks.

Viewed in this way, the model provides a comprehensive framework for designing adult learning experiences that are engaging, relevant, practical, collaborative, reflective, and adaptable. By integrating multiple theoretical perspectives, educators can create learning environments that better reflect the complexity of adult learning and prepare learners to succeed in the dynamic and continually evolving contexts of modern life, work, and society.

The greatest strength of contemporary adult education lies in its ability to draw upon multiple learning theories simultaneously. In practice, effective learning experiences rarely align with a single theoretical perspective. Instead, educators combine approaches to address the cognitive, social, experiential, emotional, cultural, and technological dimensions of learning. The following examples demonstrate how multiple theories can be integrated into adult teaching and learning practice.

Example 1: Vocational Education – Heavy Vehicle Driver Training

A heavy vehicle driver training program provides a clear example of how multiple theories can operate together.

Behaviourist approaches are used when learners practise vehicle inspections, coupling procedures, and defensive driving techniques. Repetition, feedback, and assessment help learners develop safe and consistent behaviours. Cognitivist principles are applied when learners study road transport regulations, fatigue management requirements, and vehicle systems, requiring them to process, organise, and recall information.

Constructivist learning occurs when learners analyse accident case studies and develop solutions to realistic transport scenarios. Social constructivism is incorporated through group discussions examining Chain of Responsibility obligations and fatigue-related incidents. Humanistic principles are reflected when learners identify personal driving goals and reflect on their own attitudes toward safety and professional responsibility.

Social Learning Theory is evident when learners observe experienced drivers demonstrating best-practice techniques. Experiential Learning Theory is central to practical driving exercises, where learners gain experience, reflect on performance, and apply lessons learned during subsequent driving activities. Critical Theory may be used to examine broader issues such as driver wellbeing, workplace pressures, and industry safety culture. Sociocultural learning occurs through participation in the professional transport community, while Connectivism supports ongoing learning through telematics systems, online compliance resources, industry forums, and digital fatigue management tools.

Example 2: Higher Education – Civil Engineering Project

In a university civil engineering course, students may be required to design a bridge for a regional community.

Behaviourism contributes through repeated practice of engineering calculations and design standards. Cognitivism supports the development of technical knowledge related to structural analysis, materials science, and risk management. Constructivist learning occurs as students develop their own design solutions to complex engineering challenges.

Social constructivist principles are evident when students work collaboratively in design teams, sharing ideas and solving problems together. Humanism is reflected when students are encouraged to pursue design approaches aligned with their professional interests and career aspirations.

Social Learning Theory is incorporated through interactions with experienced engineers, guest lectures, and industry mentors. Experiential Learning Theory is applied through site visits, laboratory testing, design projects, and simulations. Critical Theory may encourage students to examine environmental sustainability, social equity, and the ethical implications of infrastructure decisions. Sociocultural perspectives emerge through participation in professional engineering communities and industry engagement activities. Connectivism supports learning through digital modelling software, online engineering databases, professional networks, and emerging artificial intelligence design tools.

Example 3: Healthcare Professional Development

A hospital implementing a new patient safety initiative provides another example of integrated learning.

Behaviourist approaches are used to ensure staff consistently follow new clinical procedures. Cognitivist strategies help staff understand the underlying medical rationale, risk factors, and decision-making processes associated with patient safety.

Constructivist learning occurs when healthcare workers analyse clinical incidents and develop strategies to prevent recurrence. Social constructivism supports interdisciplinary collaboration between nurses, doctors, pharmacists, and allied health professionals. Humanistic approaches recognise healthcare workers' personal commitment to patient wellbeing and professional growth.

Social Learning Theory is evident when staff observe experienced practitioners and participate in mentoring programs. Experiential learning occurs through simulations, case reviews, and workplace practice. Critical Theory may encourage examination of organisational culture, communication failures, and systemic causes of patient harm. Sociocultural perspectives recognise the influence of professional norms and healthcare culture. Connectivism supports learning through clinical decision-support systems, online medical databases, professional networks, and continuing professional education platforms.

Example 4: Workplace Leadership Development

A leadership development program for middle managers often incorporates numerous learning theories.

Behaviourist approaches may be used to develop specific leadership behaviours through practice and feedback. Cognitivist principles support understanding of leadership models, organisational strategy, and decision-making frameworks. Constructivism allows participants to interpret leadership concepts through their own workplace experiences.

Social constructivism is evident in peer coaching, group discussions, and collaborative leadership challenges. Humanism encourages self-awareness, personal growth, and reflection on leadership values and goals. Social Learning Theory is incorporated through executive mentoring and observation of effective leaders.

Experiential Learning Theory underpins workplace projects, action learning sets, and reflective journals. Critical Theory encourages participants to question organisational assumptions, power structures, and workplace inequalities. Sociocultural perspectives recognise leadership as a socially situated practice influenced by organisational culture. Connectivism supports ongoing development through professional networks, leadership communities, webinars, podcasts, and digital learning platforms.

Example 5: Community Adult Education

A community-based financial literacy program demonstrates how learning theories can be integrated outside formal education.

Behaviourism helps learners practise budgeting techniques and financial calculations. Cognitivism supports understanding of financial concepts such as interest, debt, investment, and risk. Constructivism enables participants to connect new knowledge to their own financial experiences.

Social constructivism is reflected in group discussions where participants share strategies and challenges. Humanistic approaches recognise learners' personal financial goals and life circumstances. Social Learning Theory occurs when participants learn from facilitators and peers who model effective financial behaviours.

Experiential learning involves applying budgeting tools to real-life situations. Critical Theory may encourage examination of social and economic structures that influence financial wellbeing. Sociocultural perspectives recognise the role of family, community, and culture in shaping financial decision-making. Connectivism supports continued learning through online financial tools, government resources, digital calculators, and financial education websites.

Example 6: Apprenticeship and Trades Training

An electrical apprentice learning to install and maintain electrical systems experiences learning through multiple theoretical perspectives simultaneously.

Behaviourism supports the development of safe work practices and technical procedures. Cognitivism assists with understanding electrical theory and regulations. Constructivism occurs as apprentices apply theory to practical workplace situations.

Social constructivism is evident through interaction with tradespeople and supervisors. Humanism supports growing professional confidence and identity. Social Learning Theory is highly visible as apprentices observe experienced electricians and model their practices.

Experiential learning occurs daily through workplace activities. Critical Theory may encourage reflection on workplace safety, industry practices, and sustainability issues. Sociocultural perspectives explain how apprentices gradually become members of the electrical trade community. Connectivism supports access to digital manuals, online training resources, technical support systems, and professional industry networks.

Case Study: Integrated Learning in a Project Management Course

Consider a twelve-week project management course designed for working professionals.

During the course, learners study project management methodologies and risk management frameworks (Cognitivism). They complete quizzes and practical exercises to reinforce key concepts (Behaviourism). Learners analyse workplace case studies and develop solutions to realistic project challenges (Constructivism).

Participants work in teams to develop project plans and share professional experiences (Social Constructivism). They observe guest presentations from experienced project managers and participate in mentoring activities (Social Learning Theory). Learners undertake workplace-based projects and maintain reflective journals documenting lessons learned (Experiential Learning Theory).

The course encourages learners to identify personal development goals and select project topics relevant to their careers (Humanism). Ethical decision-making, stakeholder management, and organisational power dynamics are examined through critical discussions (Critical Theory). Participants engage with professional associations and industry communities (Sociocultural Theory) while using online resources, webinars, project management software, and professional networking platforms (Connectivism).

The result is a rich learning environment that addresses knowledge acquisition, practical skill development, critical thinking, collaboration, professional identity, lifelong learning, and workplace application simultaneously.

Key Principle

The most effective adult learning environments do not ask, *"Which learning theory is correct?"* Instead, they ask, *"Which combination of learning theories will best support the desired learning outcomes?"* By integrating multiple theoretical perspectives, educators can create learning experiences that address the full complexity of adult learning and prepare learners to succeed in dynamic professional, social, and technological environments.

Key Concepts and Conclusions

Adult learning is a complex and multifaceted process that cannot be fully explained by any single theory. Adults learn through a combination of cognitive, behavioural, social, emotional, experiential, cultural, and technological influences. As a result, effective adult education requires an understanding of multiple perspectives on learning and the ability to apply these perspectives appropriately within different educational, workplace, and community contexts.

Behaviourism highlights the role of observable behaviour in learning and emphasises the importance of practice, repetition, reinforcement, and feedback. This perspective is particularly valuable when developing procedural knowledge, technical skills, compliance behaviours, and workplace routines where consistent performance and skill mastery are required.

Cognitivism focuses on the mental processes involved in learning, including perception, information processing, memory, problem-solving, and decision-making. Learning involves the development of knowledge structures that enable learners to organise information, develop understanding, and apply knowledge effectively in a range of situations.

Constructivism views learning as an active process in which individuals construct meaning by connecting new information to their existing knowledge and experiences. Learning is therefore influenced by prior understanding, personal interpretation, and contextual factors, making it both individual and experiential in nature.

Social Constructivism and Social Learning Theory extend learning beyond the individual by recognising the importance of social interaction. Learning occurs through observation, modelling, collaboration, mentoring, discussion, and participation in shared activities. These perspectives emphasise that knowledge is often developed and refined through engagement with others and participation in social environments.

Experiential Learning Theory highlights the importance of experience as a source of learning. Adults frequently learn most effectively when they are able to apply knowledge in authentic situations, reflect upon their experiences, develop conceptual understanding, and apply their learning in new contexts. Experience, reflection, and action therefore form an important cycle of adult learning.

Humanistic approaches place the learner at the centre of the educational process. Humanism emphasises self-direction, intrinsic motivation, personal growth, self-awareness, and self-actualisation. Adult learners are viewed as autonomous individuals who bring unique goals, experiences, values, and aspirations to the learning environment.

Critical Theory and Emancipatory Education emphasise the importance of critical reflection and transformative learning. Learners are encouraged to question assumptions, examine power relationships, analyse social realities, and develop greater awareness of the forces that shape knowledge and practice. Learning is therefore viewed not only as the acquisition of knowledge but also as a means of promoting critical consciousness and meaningful change.

Sociocultural perspectives emphasise that learning is shaped by culture, language, context, and participation in communities of practice. Knowledge develops through engagement with workplaces, professions, organisations, and social groups, making learning both socially situated and culturally influenced.

Connectivism reflects the realities of learning in a digitally connected world. Knowledge is increasingly distributed across networks of people, organisations, databases, technologies, and digital platforms. Learning involves developing the ability to locate, evaluate, connect, and apply information from multiple sources while participating in professional and digital networks.

A recurring theme across contemporary learning theories is the importance of lifelong learning. Rapid technological, social, and economic change requires adults to continually update their knowledge, skills, and capabilities. Learning increasingly occurs beyond formal educational institutions through workplaces, professional communities, online environments, and digital networks.

Collectively, these perspectives demonstrate that effective adult education integrates multiple learning theories rather than relying on a single approach. Different theories contribute unique insights into how adults learn, and combining these perspectives allows educators to address the diverse needs, experiences, motivations, and goals of adult learners.

A major conclusion emerging from the study of adult learning theories is that no single theory can fully explain the complexity of adult learning. Behaviourism, Cognitivism, Constructivism, Humanism, Social Learning Theory, Experiential Learning Theory, Critical Theory, Sociocultural Theory, and Connectivism each illuminate different aspects of the learning process. A comprehensive understanding of adult learning therefore requires consideration of multiple theoretical perspectives.

Another important conclusion is that adult learners bring significant prior experience to the learning environment. Unlike children, adults possess extensive life, work, and educational experiences that shape how they interpret new information and engage with learning activities. Effective adult education recognises, values, and builds upon these existing experiences.

The chapter also demonstrates that learning extends well beyond the acquisition of information. Adult learning involves developing skills, solving problems, reflecting on experiences, constructing meaning, forming professional identities, collaborating with others, and adapting to changing personal and professional circumstances. Learning is therefore both developmental and transformative in nature.

Social interaction emerges as a fundamental component of learning. Adults learn through collaboration, mentoring, observation, discussion, participation, and engagement with professional and social communities. Learning is not solely an individual activity but is enhanced through interaction with others and participation in shared practices.

The importance of reflection is another key conclusion. Reflection enables learners to examine experiences, challenge assumptions, evaluate evidence, and improve future performance. Reflective practice is central to experiential, transformative, and professional learning and supports continuous personal and professional development.

The chapter further highlights the value of authentic learning contexts. Learning is more meaningful and effective when it is connected to realistic situations, workplace challenges, professional practice, and genuine problems. Authentic learning environments improve learners' ability to transfer knowledge and skills to real-world settings.

Technology has also transformed the nature of adult learning. Digital technologies, online communities, professional networks, artificial intelligence, and information systems have expanded access to knowledge and learning opportunities. Contemporary adult learning increasingly occurs within networked and technology-rich environments that extend beyond traditional classrooms.

The need for lifelong learning is another significant conclusion. The pace of change within modern workplaces and society requires individuals to continually develop new knowledge, skills, and competencies. Lifelong learning has become an essential component of professional success, personal development, and social participation.

The chapter concludes that educators should adopt an integrated approach to adult learning. The most effective educational programs combine behavioural, cognitive, social, experiential, humanistic, critical, sociocultural, and connectivist approaches to create richer and more effective learning experiences. Such integration enables educators to address the diverse and complex needs of adult learners.

Ultimately, adult learning should be viewed as a holistic process that encompasses cognitive, behavioural, social, emotional, experiential, cultural, ethical, and technological dimensions. Effective learning environments recognise this complexity and support the development of knowledgeable, skilled, reflective, self-directed, socially engaged, and lifelong learners.

Adult learning is a dynamic and lifelong process shaped by experience, cognition, social interaction, culture, reflection, motivation, and technology. Theories of adult learning provide valuable frameworks for understanding different aspects of this process, but no single theory offers a complete explanation of how adults learn. Each theory contributes important insights into the factors that influence learning and the conditions that support effective educational practice.

By integrating the insights of Behaviourism, Cognitivism, Constructivism, Humanism, Social Learning Theory, Experiential Learning Theory, Critical Theory, Sociocultural Theory, and Connectivism, educators can develop learning experiences that are meaningful, relevant, engaging, and effective. Such an integrated approach recognises the diverse characteristics, experiences, motivations, and aspirations that adult learners bring to the learning environment.

As society becomes increasingly interconnected, knowledge-driven, and technologically advanced, the importance of effective adult education continues to grow. Understanding the complementary contributions of different learning theories enables educators to design learning environments that not only support the acquisition of knowledge and skills but also foster critical thinking, collaboration, adaptability, professional identity, and lifelong learning. Through this holistic approach, adult education can better prepare individuals to thrive in the complex and continually evolving personal, professional, and social environments of the twenty-first century.

Chapter 3
Andragogy and Adult Learning Principles

Learning is a lifelong process that extends far beyond the years of formal schooling. In contemporary societies, adults continually engage in learning to adapt to changing workplace requirements, develop new skills, pursue career advancement, respond to technological innovation, fulfil professional obligations, and achieve personal goals. As economies become increasingly knowledge-based and industries evolve at an accelerating pace, the ability to learn effectively throughout life has become an essential capability for individuals, organisations, and communities. Understanding how adults learn is therefore a fundamental concern for educators, trainers, workplace supervisors, instructional designers, and policymakers.

One of the most influential frameworks for understanding adult learning is andragogy, a concept most closely associated with Malcolm Knowles. Andragogy emerged from the recognition that adults often approach learning differently from children and that educational methods designed for school-aged learners may not always be appropriate for mature learners. Unlike traditional pedagogical approaches, which are often characterised by teacher-directed instruction, andragogy emphasises learner autonomy, prior experience, practical relevance, problem-solving, and self-direction. These principles have profoundly influenced adult education, vocational education and training (VET), workplace learning, higher education, professional development, and continuing education initiatives around the world.

This chapter explores the origins, development, and contemporary application of andragogy and adult learning principles. It begins by examining the historical development of andragogy and the contributions of Malcolm Knowles to the modern adult learning movement. The chapter then explores Knowles's six assumptions of adult learning, including the need to know, self-concept, prior experience, readiness to learn, orientation to learning, and motivation to learn. These assumptions provide a foundation for understanding the distinctive characteristics of adult learners and the implications for educational practice.

The chapter also examines key concepts closely associated with adult learning, including self-directed learning, learner readiness and motivation, the role of experience as a learning resource, experiential learning, problem-centred learning, and the importance of social and collaborative learning. Contemporary developments such as transformative learning, technology-enhanced learning, lifelong learning, and heutagogy are also considered, highlighting how adult learning theory has evolved in response to changing educational, social, and technological contexts. In addition, the chapter explores the increasing recognition of diversity, inclusion, and cultural responsiveness within adult education and examines how contemporary educators integrate multiple learning theories to support effective learning outcomes.

By the end of this chapter, readers will have developed an understanding of the foundational principles of adult learning and the ways in which these principles can be applied in educational, workplace, and professional settings. The chapter demonstrates that effective adult education involves more than the transmission of knowledge; it requires recognising learners as active participants who bring valuable experiences, motivations, and perspectives to the learning process. Understanding these principles enables educators and trainers to design learning experiences that are relevant, engaging, practical, and responsive to the needs of adult learners in an increasingly complex and rapidly changing world.

Origins of Andragogy

Andragogy is a theory and practice of adult learning that focuses on the unique characteristics, needs, and motivations of adult learners. The term is commonly used to distinguish adult learning from pedagogy, which traditionally refers to the education of children (Akintolu & Letseka, 2021; Budiarto et al., 2023; Ghamrawi et al., 2025; Khan, 2024; Kiwelu & Ogbonna, 2020; Loeng, 2023; Moll, 2024; Samaroo et al., 2013). Andragogy is based on the premise that adults learn differently from children because they possess greater life experience, a stronger sense of self-direction, and a preference for learning that is relevant to their personal, professional, or social needs (Budiarto et al., 2023; French et al., 2020; Ghamrawi et al., 2025; Khan, 2024; Purwati et al., 2022).

At its core, andragogy views adult learners as autonomous individuals who take responsibility for their own learning. Adults are typically motivated to learn when they perceive a clear purpose or benefit from the learning experience, such as improving job performance, solving practical problems, advancing their careers, or pursuing personal interests. Consequently, adult learning is often characterised by self-direction, practical application, problem-solving, and the integration of prior experience with new knowledge.

Malcolm Knowles, the scholar most closely associated with andragogy, identified several assumptions about adult learners. He argued that adults become increasingly self-directed as they mature, bring a rich reservoir of life experiences to learning situations, are ready to learn when they encounter new roles or responsibilities, prefer learning that addresses real-life problems, and are motivated primarily by internal factors such as personal growth, achievement, and self-fulfilment. These assumptions have had a profound influence on adult education, workplace training, higher education, and professional development.

Andragogy remains one of the most influential frameworks in adult education. Its principles are widely applied in vocational education and training (VET), workplace learning, professional development programs, community education, and university teaching, where learners are encouraged to actively participate in the learning process, draw upon their experiences, and apply learning to authentic situations.

Although Malcolm Knowles is often credited with popularising andragogy, the concept has a much longer history. The term andragogy was first introduced in 1833 by the German educator Alexander Kapp (Clair & Käpplinger, 2021; Loeng, 2023; Moll, 2023; Onwuatuegwu, 2023). Kapp used the term to describe educational approaches appropriate for adults, particularly in relation to the lifelong learning and self-development of individuals (Clair & Käpplinger, 2021; Loeng, 2023). However, the concept received little attention at the time and remained largely absent from educational discourse for many decades (Moll, 2024).

Interest in adult learning increased during the late nineteenth and early twentieth centuries as industrialisation, social change, and expanding educational opportunities created a growing demand for adult education (Loeng, 2023; Tymchuk et al., 2021). During this period, educators and social reformers began recognising that adults often approached learning differently from children and that educational methods designed for schools were not always suitable for mature learners (Loeng, 2023).

The modern development of andragogy emerged during the mid-twentieth century. European educators, particularly in Germany and Eastern Europe, began using the term more frequently to describe the theory and practice of adult education (Clair & Käpplinger, 2021; Moll, 2024). However, it was Malcolm Knowles who introduced and popularised andragogy in North America during the 1960s and 1970s (Clair & Käpplinger, 2021; Moll, 2024; Onwuatuegwu, 2023).

Knowles initially described andragogy as the "art and science of helping adults learn" (Clair, 2024; Moll, 2023) . He contrasted it with pedagogy, which he characterised as the art and science of teaching children (Caruso et al., 2020; Moll, 2024). While this distinction has sometimes been criticised as overly simplistic (Clair, 2024; Moll, 2024), it drew attention to the need for educational approaches that recognised the unique characteristics of adult learners (Moll, 2023).

Knowles's work was influenced by several intellectual traditions and scholars. Humanistic psychology, particularly the work of Abraham Maslow and Carl Rogers, shaped his emphasis on self-direction, personal growth, intrinsic motivation, and learner-centred education (Clair, 2024; Moll, 2024). Progressive educational thinkers such as John Dewey influenced the importance placed on experience, reflection, and practical application (Παπαθανασίου, 2023). Adult education pioneers such as Eduard Lindeman also contributed to the view that adult learning should be grounded in learners' experiences and life situations rather than predetermined subject matter (Loeng, 2023; Moll, 2024).

Throughout the 1970s and 1980s, Knowles refined his theory and identified six core assumptions about adult learners: the need to know, self-concept, prior experience, readiness to learn, orientation to learning, and motivation to learn (Bergkvist et al., 2023; Onwuatuegwu, 2023; Sasere & Makhasane, 2023). These assumptions became the foundation of modern andragogical practice and continue to influence adult education worldwide (Bergkvist et al., 2023; Clair, 2024).

Although andragogy has been widely adopted, it has also attracted criticism. Some researchers argue that the characteristics identified by Knowles are not unique to adults and may also apply to children in certain circumstances (Grus et al., 2025; Moll, 2024). Others suggest that cultural, social, and contextual factors can influence learning preferences more strongly than age alone (Moll, 2023). Nevertheless, andragogy remains a valuable framework for understanding adult learning because it highlights the importance of learner autonomy, experience, relevance, and practical application (Clair, 2024).

Today, andragogy is generally viewed not as a complete theory of learning but as a set of principles that guide effective adult education (Clair, 2024; Παπαθανασίου, 2023). It complements other learning theories such as Humanism, Constructivism, Experiential Learning Theory, Social Learning Theory, and Transformative Learning Theory, providing a practical framework for designing learning experiences that recognise the distinctive characteristics and needs of adult learners (Clair, 2024; Παπαθανασίου, 2023).

Malcolm Knowles and the Modern Adult Learning Movement

Malcolm Shepherd Knowles (1913–1997) is widely recognised as one of the most influential figures in the field of adult education and is often associated with the development and popularisation of andragogy in North America (Knowles, 1973, 1980; Merriam & Bierema, 2014). Although the concept of andragogy predates his work, Knowles was instrumental in refining its principles and applying them to educational practice, thereby shaping contemporary understandings of how adults learn. His contributions significantly influenced adult education, workplace learning, vocational training, higher education, and professional development throughout the latter half of the twentieth century (Knowles et al., 2020; Merriam & Bierema, 2014).

Knowles's work emerged during a period of considerable social, economic, and educational transformation following the Second World War. Expanding access to higher education, increasing workforce specialisation, rapid technological change, and growing participation in continuing education highlighted the need for educational approaches specifically suited to adult learners (Merriam & Bierema, 2014). At the time, many educational practices were based on pedagogical assumptions developed primarily for children and adolescents. Knowles argued that these approaches did not adequately reflect the characteristics, experiences, motivations, and learning needs of adults (Knowles, 1973, 1980).

A central feature of Knowles's work was his rejection of the notion that adults should be treated as passive recipients of knowledge. Instead, he proposed that adults are capable of directing their own learning and possess a wealth of experiences that influence how they interpret, evaluate, and apply new information (Knowles, 1984). This perspective contributed to a shift away from teacher-centred instruction towards learner-centred education, where educators function as facilitators, guides, coaches, and resources rather than simply transmitters of information (Merriam & Bierema, 2014; Smith, 2002).

Knowles defined andragogy as the art and science of helping adults learn (Knowles, 1980). Initially, he presented andragogy as a distinct alternative to pedagogy, arguing that adults and children differ in important ways that require different educational approaches. However, his views evolved over time, and he later acknowledged that pedagogical and andragogical approaches exist along a continuum rather than representing mutually exclusive models. He recognised that both approaches may be appropriate depending on the learner, the subject matter, and the educational context (Knowles, 1984; Knowles et al., 2020).

Among Knowles's most influential contributions was the articulation of a set of assumptions about adult learners. He proposed that adults need to understand why they are learning something before committing to the learning process. Adults tend to view themselves as self-directing individuals who seek a degree of control over their learning. They bring a rich reservoir of experiences that can serve as valuable learning resources. Their readiness to learn is often associated with changing life roles and responsibilities, while their orientation to learning is generally problem-centred and focused on immediate application. Furthermore, adults are often motivated by internal factors such as personal growth, self-esteem, achievement, and job satisfaction rather than solely by external rewards (Knowles, 1980, 1984).

These assumptions have had a profound influence on educational practice across a wide range of settings. Adult learning programs increasingly emphasise practical application, problem-solving, case studies, workplace projects, self-directed learning opportunities, collaborative activities, and the recognition of learners' prior experiences (Merriam & Bierema, 2014; Taylor & Hamdy, 2013). Such approaches have become particularly prominent in vocational education and training (VET), workplace learning, healthcare education, professional development, and continuing education.

The modern adult learning movement extends beyond Knowles's contributions, but his work provided much of its conceptual foundation. This movement emphasises the importance of learner autonomy, relevance, experience, and lifelong learning. It recognises that adults are active participants in their own education and that learning is most effective when it is meaningful, practical, and connected to real-world challenges (Merriam & Bierema, 2014). Furthermore, the movement acknowledges that learning occurs across multiple contexts, including formal education, workplaces, communities, professional networks, and informal environments.

Knowles's ideas were influenced by several broader intellectual traditions. Humanistic psychologists such as Abraham Maslow and Carl Rogers contributed to his emphasis on personal growth, self-actualisation, learner autonomy, and learner-centred education (Maslow, 1970; Rogers, 1969). His work also reflected the influence of Eduard Lindeman, who argued that adult education should be grounded in learners' experiences and life situations rather than organised around predetermined subject matter (Lindeman, 1926). These influences helped shape a view of adult learning that prioritises personal meaning, relevance, and active participation.

Following Knowles, the field of adult learning continued to expand through the contributions of numerous scholars. Jack Mezirow developed Transformative Learning Theory, emphasising critical reflection and perspective transformation (Mezirow, 1991). David Kolb advanced Experiential Learning Theory, highlighting the role of experience and reflection in learning (Kolb, 1984a). Albert Bandura's Social Learning Theory emphasised observation, modelling, and self-efficacy (Bandura, 1977), while Etienne Wenger's work on Communities of Practice explored learning through participation in professional and social groups (Wenger, 1998). Scholars such as Stephen Brookfield and Patricia Cranton further developed the role of critical reflection and transformative learning in adult education (Brookfield, 2017; Cranton, 2016). Together, these perspectives expanded and enriched the foundations established by Knowles.

Despite its enduring influence, Knowles's theory has been subject to criticism. Some researchers argue that the characteristics he identified are not exclusive to adults and may also apply to children and adolescents in certain circumstances (Pratt, 1993). Others suggest that andragogy places insufficient emphasis on social, cultural, political, and organisational influences on learning (Hartree, 1984; Merriam & Bierema, 2014). Critics have also noted that not all adults are self-directed learners and that learning preferences vary according to factors such as educational background, confidence, culture, and prior experience (Merriam & Bierema, 2014).

Nevertheless, Knowles's contribution remains foundational to contemporary adult education. His work helped shift educational thinking from a primary focus on teaching towards a greater focus on learning and the learner. By emphasising experience, relevance, self-direction, and practical application, Knowles provided a framework that continues to influence the design and delivery of adult learning programs around the world (Knowles et al., 2020; Merriam & Bierema, 2014).

The influence of Malcolm Knowles remains evident throughout higher education, workplace training, vocational education and training, professional development, and lifelong learning initiatives worldwide. His ideas continue to shape how educators design learning experiences for adults and remain a cornerstone of contemporary adult learning theory and practice.

Assumptions of Andragogy

A central feature of Malcolm Knowles's theory of andragogy is a set of assumptions about the characteristics of adult learners. These assumptions describe how adults typically approach learning and provide guidance for designing effective adult education programs. Knowles originally proposed four assumptions in the 1970s and later expanded the framework to six assumptions, which remain widely used in adult education, workplace learning, vocational education and training (VET), higher education, and professional development (Knowles, 1980, 1984).

Figure 9: Assumptions of Andragogy.

Need to Know

Knowles argued that adults need to understand why they need to learn something before they are willing to invest time and effort in learning it. Unlike children, who may accept learning because it is required by a teacher or curriculum, adults typically want to know the purpose, relevance, and benefits of learning activities (Khan, 2024; Onwuatuegwu, 2023). They are more likely to engage in learning when they can see how it will help them solve problems, improve performance, achieve goals, or address real-life challenges (Khan, 2024; Kiwelu & Ogbonna, 2020; Onwuatuegwu, 2023). Knowles identified that adult learning is distinguished by being more internally motivated, and that adults are stimulated to learn when learning is connected to their personal needs and interests (Clausen, 2023; Morris, 2019). The relationship between adult learning and the relevance of the learning to tasks, problems to solve, and the entire outcome of the learning is close (Kiwelu & Ogbonna, 2020). Adults become ready to learn things they need to know in order to cope effectively with real-life situations, reinforcing the necessity of relevant educational provision according to the roles and challenges learners face in their current lives (Khan, 2024). Furthermore, adults are motivated to learn as they experience the need or interest in learning information or skills that will help them perform tasks or deal with problems in their life situations (Khan, 2024).

For example, employees undertaking workplace training are often more motivated when they understand how the training will improve safety, increase productivity, support career advancement, or assist them in performing their roles more effectively. Adult learning is more goal-oriented and self-sought, meaning the person wants to learn to achieve a certain goal such as acquiring a new career, satisfying a curiosity, raising self-esteem, or improving performance at the place of work (Kiwelu & Ogbonna, 2020). Adults are motivated by both external factors (such as better jobs, promotions, and higher salaries) and internal factors (such as the desire for increased job satisfaction, self-esteem, and quality of life) (Aljohani & Alajlan, 2020). Educators can address this need by clearly explaining learning outcomes, demonstrating relevance, and connecting learning activities to practical applications. The andragogical approach emphasizes the learner's role in shaping the learning process and focuses on practical applications and problem-solving (Onwuatuegwu, 2023). If students do not have a strong motivation to learn, they will not learn and will not be able to learn (Aljohani & Alajlan, 2020).

Self-Concept

As individuals mature, they typically develop a stronger sense of personal responsibility and independence in managing their lives and decisions. Knowles argued that adult learners generally view themselves as self-directing individuals who prefer to exercise autonomy and control over their learning processes (Storey & Wang, 2016). According to this perspective, adults have a psychological need to be recognised as capable of making their own decisions and managing their own learning. As a result, educational approaches that treat adults as dependent learners may create resistance or reduce engagement (Storey & Wang, 2016).

The principle of self-direction is a core assumption of andragogy. Proponents argue that as people mature, they tend to move from dependence on others towards greater autonomy and self-management (Itasanmi et al., 2023). Consequently, adult learners often respond positively to learning environments that provide opportunities for choice, self-directed learning, goal setting, and active participation in educational decision-making (Clair, 2024; Purwati et al., 2022). Many adult education scholars regard the promotion of learner autonomy as a fundamental objective of adult learning, with self-direction widely recognised as a defining characteristic of effective adult education practice (Clair, 2024).

This assumption has significant implications for teaching practice. Adult educators frequently involve learners in setting learning goals, selecting learning activities, identifying learning resources, and evaluating their own progress (Samaroo et al., 2013). Within the andragogical framework, self-direction is viewed as a process in which learners take an active role in identifying their learning needs, establishing goals, locating appropriate resources, selecting learning strategies, and assessing learning outcomes. Rather than relying entirely on the educator to direct the learning process, adults are encouraged to assume increasing responsibility for planning, managing, and evaluating their own learning (Samaroo et al., 2013).

Providing opportunities for learners to contribute to the design and direction of learning experiences is therefore an important element of adult education. Involving adults in decisions about course content, learning activities, and expected outcomes helps ensure that learning remains relevant to their needs, interests, and goals. This approach reflects the concept of self-directed learning, in which adult learners actively initiate and contribute to the learning process rather than simply receiving information from an instructor (Purwati et al., 2022).

As a consequence, the educator's role shifts from that of a traditional authority figure to that of a facilitator who supports, guides, and enables learning (Loeng, 2020; Yusuf, 2021). Rather than controlling the learning process or determining all aspects of what is to be learned, the educator creates conditions that encourage participation, reflection, inquiry, and independent learning. Contemporary adult education therefore emphasises facilitation, collaboration, and learner empowerment, recognising that effective learning often occurs when adults are actively engaged in directing their own educational experiences (Loeng, 2020; Yusuf, 2021).

Prior Experience

One of the most distinctive characteristics of adult learners is the extensive range of experiences they bring to the learning environment. These experiences may include work history, family responsibilities, community involvement, previous education, and life events. Knowles viewed this reservoir of experience as a valuable resource that can enrich learning for both the individual and other learners (Delahunty & Kimbell, 2021; Okorie & Uzoaru, 2021). Knowles critically emphasized the necessity of considering adults' self-concept, experience, readiness to learn, and orientation to learning tasks in facilitating effective adult learning (Acton, 2023). As Knowles proposed, adults learn differently from young people, motivated by their experience and need for immediate application.

Prior experience influences how adults interpret new information, solve problems, and construct meaning. Learning activities that draw upon learners' experiences, such as case studies, discussions, simulations, reflective exercises, and workplace projects, are often particularly effective because they enable learners to connect new knowledge with existing understanding. Research in healthcare professional education has identified several common andragogical practices, including self-assessment and reflection on personal experiences, discussion of work-related scenarios with peers, and learning facilitated through coaching and mentoring relationships (Mukhalalati & Taylor, 2019). Similarly, Taylor identified six key elements that support adult learning in practice: individual experience, awareness of context, dialogue, authentic relationships, holistic orientation, and critical reflection (Blaschke & Hase, 2019). These approaches recognise that learning is enhanced when adults can draw upon and critically examine their existing knowledge and experiences.

The importance of experience is further reflected in experiential learning theories, which emphasise that meaningful learning occurs through a cycle of experience, reflection, conceptualisation, and application. Through reflective observation and collaborative learning processes, adults can deepen their understanding, test assumptions, and integrate new knowledge into their existing frameworks of understanding (Mishra & Raina, 2021). Mezirow also emphasised the central role of experience in learning, arguing that adult learning is fundamentally shaped by how individuals interpret and make sense of their experiences. He viewed adult education as a process that supports the development of increasingly independent, self-directed, and reflective learners (Wang et al., 2013).

However, prior experience can also present challenges. Long-held beliefs, assumptions, values, and habits may sometimes make it difficult for learners to accept new ideas or alternative perspectives. Recognising this, Mezirow developed Transformative Learning Theory, which focuses on how adults critically examine and revise their existing frames of reference. Transformative learning often begins when individuals encounter situations or experiences that challenge their current assumptions, prompting reflection and consideration of alternative viewpoints. Through this process, learners may modify their beliefs, perspectives, and ways of understanding the world (Blaschke & Hase, 2019).

Although andragogy and transformative learning theory share several common principles, including self-directed learning and the importance of experience, transformative learning extends these concepts by placing greater emphasis on cognition, meaning-making, perspective transformation, and critical self-reflection (Blaschke & Hase, 2019). Effective adult education therefore not only draws upon learners' experiences as valuable learning resources but also encourages learners to critically examine those experiences. Reflective learning is influenced by an individual's capacity to analyse experiences, consider alternative interpretations, and evaluate practice within its broader context. Research suggests that reflective thinking develops progressively over time and can be strengthened through supportive learning environments, constructive feedback, and skilled facilitation by educators (Wang et al., 2013). Consequently, critical reflection remains a cornerstone of adult learning theory, enabling learners to integrate abstract concepts with practical experience and to address increasingly complex personal and professional challenges.

Readiness to Learn

Knowles proposed that adults become ready to learn when they encounter situations that require new knowledge or skills. Their readiness to learn is closely related to their social roles and responsibilities (Hagen & Park, 2016; Khan, 2024; Purwati et al., 2022). Changes in employment, family circumstances, professional requirements, or personal goals often create a need for learning (Henschke; Moll, 2024).

For example, a newly promoted supervisor may seek leadership training (Hagen & Park, 2016), a healthcare worker may need to learn new clinical procedures, or a parent may participate in parenting education programs. Adults are generally most receptive to learning when they perceive an immediate need for the knowledge or skills being taught (Clair, 2024; Dachner & Polin, 2015; Ghamrawi et al., 2025).

This assumption highlights the importance of timing and relevance in adult education. Learning is often most effective when it is linked to current challenges, responsibilities, or life transitions (Clair & Käpplinger, 2021; Purwati et al., 2022; Sasere & Makhasane, 2023).

Orientation to Learning

Knowles argued that adults tend to adopt a problem-centred rather than a subject-centred orientation to learning (Clair, 2024; Loeng, 2023). As individuals mature, their learning increasingly becomes focused on performing tasks, solving problems, and addressing challenges encountered in their personal and professional lives. Unlike children, who are often required to learn content that may have relevance at some future point, adults generally prefer learning that can be applied immediately to real-world situations and current responsibilities (Akintolu & Letseka, 2021; Itasanmi et al., 2023; Moll, 2023). Adult learners are therefore more likely to engage with learning when they can clearly see its practical value and understand how it will help them address existing needs, improve performance, or achieve specific goals. Their readiness to learn is often closely linked to the perceived usefulness and immediate applicability of new knowledge and skills in everyday situations (Itasanmi et al., 2023).

As a result, adult learners are typically more engaged when educational activities focus on practical problem-solving, workplace challenges, and authentic performance outcomes (Khan, 2024; Knapke et al., 2024; Yusuf, 2021). Research has shown that learners respond positively to educational experiences that require them to address realistic problems and apply newly acquired skills in meaningful contexts (Knapke et al., 2024). Adult education therefore frequently emphasises learning activities that involve investigation, analysis, decision-making, and the application of knowledge to real-world scenarios. This preference for practical application over abstract subject matter is reflected in the widespread use of case studies, simulations, workplace projects, scenario-based learning, and other experiential activities that connect theory with practice (Bergkvist et al., 2023; Dale et al., 2008; Hyden & Cherrstrom, 2025; Khan, 2024).

Educational approaches that incorporate authentic learning experiences provide opportunities for learners to apply theoretical concepts in realistic contexts. Activities such as internships, workplace projects, collaborative investigations, community-based learning, reflective journals, interviews, observations, and applied research projects help learners bridge the gap between academic knowledge and professional practice (Hyden & Cherrstrom, 2025; Khan, 2024). Research suggests that adult learners benefit most when learning activities are contextualised and designed around realistic problems that mirror those encountered in everyday life and work environments (Bergkvist et al., 2023; Dale et al., 2008). Such approaches enable learners to test ideas, develop practical skills, and gain confidence in applying new knowledge to complex situations.

This assumption has significantly influenced the adoption of experiential learning and problem-based learning approaches throughout adult education (Bergkvist et al., 2023; Dale et al., 2008; Khan, 2024). These approaches are designed to encourage active participation, critical thinking, collaboration, and the practical application of knowledge. By engaging learners in authentic tasks and realistic problem-solving activities, they help develop higher-order cognitive skills while supporting the transfer of learning to professional and personal contexts (Dale et al., 2008; Khan, 2024). In professional development settings, learning activities are often structured around workplace problems and role-specific challenges to ensure that new knowledge can be directly integrated into practice. This emphasis on relevance, application, and performance remains one of the defining characteristics of adult learning and a central principle of andragogical practice (Bergkvist et al., 2023).

Motivation to Learn

Although external factors such as promotions, qualifications, salary increases, and professional requirements can motivate adults to learn, Knowles argued that internal motivations are often more powerful (Ilie, 2019; Pang & Lee, 2013; Purwati et al., 2022). Adults are frequently motivated by personal growth, self-esteem, achievement, quality of life, professional satisfaction, curiosity, and the desire to reach personal goals (Ilie, 2019; Pang & Lee, 2013; Rothes et al., 2016).

Intrinsic motivation tends to support deeper engagement and persistence in learning (Pang & Lee, 2013; Rothes et al., 2016). Adult educators therefore seek to create learning environments that foster autonomy, competence, confidence, and personal relevance (Kellenberg et al., 2017; Ray & Nelson, 2025). When learners perceive learning as meaningful and beneficial to their own goals, motivation is generally enhanced (Åström et al., 2025; Rothes et al., 2016; Schumann et al., 2025).

Significance of the Assumptions

Together, these six assumptions provide a framework for understanding how adults typically approach learning. They emphasise the importance of relevance, experience, autonomy, practical application, and intrinsic motivation. The assumptions have significantly influenced educational practice by encouraging learner-centred approaches that recognise adults as active participants in their own learning.

These principles have shaped the design of contemporary adult education programs across higher education, vocational education and training, workplace learning, professional development, healthcare education, and community education. Learning activities such as workplace projects, problem-based learning, reflective practice, mentoring, collaborative learning, and self-directed study all reflect the influence of Knowles's assumptions.

While researchers have debated whether these characteristics apply universally to all adults, the assumptions of andragogy remain one of the most influential frameworks in adult learning theory and continue to provide valuable guidance for educators seeking to design effective learning experiences for adult learners.

Self-Directed Learning

Self-directed learning (SDL) is a learning process in which individuals take primary responsibility for planning, managing, and evaluating their own learning. Rather than relying entirely on an instructor to determine what, how, and when learning occurs, self-directed learners actively identify their learning needs, establish goals, locate resources, select learning strategies, and assess their progress and outcomes (Lemmetty & Collin, 2020; Loeng, 2020; Samaroo et al., 2013).

The concept occupies a central place in adult learning theory and is closely associated with Malcolm Knowles's theory of andragogy. Knowles viewed self-direction as one of the defining characteristics of adult learners, arguing that as individuals mature they increasingly seek autonomy and control over their learning (Calabrese & Capraro, 2021; Clair, 2024; Onwuatuegwu, 2023). He described self-directed learning as a process in which learners take the initiative, with or without assistance from others, to diagnose learning needs, formulate learning objectives, identify resources, implement learning strategies, and evaluate results (Lemmetty & Collin, 2020; Loeng, 2020; Samaroo et al., 2013).

Although Knowles is most commonly associated with self-directed learning, the concept has deeper roots in adult education. Early adult education theorists such as Eduard Lindeman emphasised the importance of learner autonomy and experience-based learning (Loeng, 2020, 2023; Malinen & Piirainen, 2023). Lindeman claimed that adults learn best when actively involved in determining what, how, and when to learn (Loeng, 2023). During the 1970s and 1980s, self-directed learning became a major area of research in adult education, with scholars including Allen Tough, Huey Long, Lucy Guglielmino, and Paul Candy examining how adults learn independently outside formal educational settings (B. E. Ahmad et al., 2023; Henschke, 2010a).

Research by Tough revealed that many adults regularly undertake substantial learning projects on their own initiative, often without enrolling in formal courses (B. E. Ahmad et al., 2023; Loeng, 2020). These projects may involve learning new workplace skills, developing hobbies, managing health conditions, acquiring technological competencies, or pursuing personal interests.

Self-directed learners typically demonstrate several characteristics that support independent learning. They are proactive in identifying learning needs, motivated to achieve personal goals, and willing to take responsibility for their own development (Lemmetty & Collin, 2020). They often display curiosity, initiative, self-discipline, and persistence when faced with challenges. Knowles himself identified the ability to develop and be in touch with curiosities as a crucial dimension of self-directed learning skills (Henschke, 2010a).

Self-directed learners are also capable of monitoring and regulating their learning. They assess their progress, modify strategies when necessary, seek feedback, and reflect on their performance (Lemmetty & Collin, 2020; Loeng, 2020). This process of self-regulation helps learners adapt to changing circumstances and improve learning effectiveness over time. The stages of a self-directed process may include planning, monitoring, managing, and reflecting on learning (Lemmetty & Collin, 2020).

However, self-direction should not be confused with learning in isolation. Self-directed learners frequently seek support from instructors, mentors, colleagues, professional networks, online communities, and other resources (B. E. Ahmad et al., 2023; Loeng, 2020; Samaroo et al., 2013). The distinguishing feature is not the absence of assistance but the learner's primary responsibility for directing the learning process. As Knowles's definition explicitly states, individuals take the initiative "with or without the help of others" (Samaroo et al., 2013).

Components of Self-Directed Learning

Most models of self-directed learning describe a process in which learners assume primary responsibility for managing their own learning (Hezlett, 2016;, (Jimoyiannis & Tsiotakis, 2016;, Lubbe et al., 2016;. Although different models vary in their terminology and emphasis, they generally include a series of interconnected stages that guide learners from identifying a learning need through to evaluating the outcomes achieved (Chaiyasit et al., 2023; Cremers et al., 2016). Knowles's widely cited definition characterizes self-directed learning as a process in which individuals take the initiative to diagnose their learning needs, formulate learning goals, identify resources, choose and implement appropriate learning strategies, and evaluate learning outcomes (Bou et al., 2024; Chaiyasit et al., 2023; Gupta et al., 2024; Jimoyiannis & Tsiotakis, 2016; Lubbe et al., 2016; Yang et al., 2021). These stages reflect the active and autonomous nature of self-directed learning, where individuals take responsibility for planning, implementing, and assessing their own educational development (Hezlett, 2016; Laubscher, 2019; Lubbe et al., 2016; Poudel, 2024).

The process typically begins when learners identify a gap between their current knowledge, skills, or abilities and the level of competence they wish to attain. This recognition may arise from workplace requirements, professional development needs, personal interests, career aspirations, technological advancements, or significant life transitions. Awareness of this gap creates the motivation and rationale for learning, providing a foundation for subsequent learning activities.

Once a learning need has been identified, learners establish specific goals and objectives that define what they hope to achieve. Effective learning goals provide direction, clarify expectations, and establish criteria against which progress can later be measured. Goal setting helps learners focus their efforts and make informed decisions about the resources and strategies that will best support their learning.

The next stage involves identifying and accessing appropriate learning resources. Self-directed learners actively seek out information, support, and learning opportunities from a variety of sources. These resources may include books, academic journals, online courses, instructional videos, workplace training materials, professional associations, mentors, colleagues, coaches, or subject matter experts. The ability to locate and evaluate relevant resources is an important component of effective self-directed learning.

After identifying suitable resources, learners select learning strategies that align with their goals and preferred approaches to learning. Depending on the nature of the learning objective, these strategies may include reading, observation, practice, experimentation, discussion, reflection, coaching, simulation exercises, workplace projects, or formal study. Self-directed learners often combine multiple strategies to maximise learning effectiveness and adapt their approach as their understanding develops.

An essential aspect of self-directed learning is the evaluation of learning outcomes. Learners assess whether their goals have been achieved by reviewing their performance, seeking feedback from others, reflecting on their experiences, and measuring progress against established objectives. This evaluation process helps learners identify strengths, recognise areas requiring further development, and determine whether additional learning is necessary. Through ongoing reflection and self-assessment, learners continuously refine their knowledge, skills, and learning strategies, supporting lifelong learning and professional growth.

Self-Directed Learning in Practice

Self-directed learning occurs in a wide variety of contexts. In the workplace, employees often identify skill gaps and independently pursue learning opportunities to improve performance or prepare for career advancement. Professionals may undertake continuing professional development activities, attend conferences, complete online courses, or participate in industry networks (Sun, 2023; Yu et al., 2023). Continuing Professional Development (CPD) activities play a central role in professional learning, as they are often learner-initiated and tailored to personal development needs, reflecting the hallmarks of self-directed learning and promoting lifelong learning behaviours by encouraging individuals to identify learning gaps and seek out opportunities for growth. In Kenya, for example, the 'Teacher Education in Sub-Saharan Africa Massive Open Online Course ' (TESSA MOOC) was used to promote teachers' lifelong and professional development, offering self-regulated learning with flexible commencement dates (Maphosa & Maphosa, 2023). MOOCs have been recognized as useful for professional development, with learners reporting that certification can add value to their curriculum and career trajectories.

In higher education, students engage in self-directed learning when conducting independent research projects, selecting elective subjects, managing study schedules, and pursuing individual learning interests (S. Wang et al., 2024). The emergence of open educational resources, massive open online courses, and digital learning management systems has resulted in a sharp increase in student autonomy, as these tools provide external prerequisites for fostering learner independence. Students' self-directed learning encompasses the abilities to independently set goals, select strategies, manage the learning process, and assess outcomes (Selçuk et al., 2026). Similarly, vocational education and training increasingly incorporates self-directed learning through workplace projects, portfolio development, independent research tasks, and competency-based assessment,. The dual system of vocational education, for instance, combines parallel training on the job and in vocational schools, while TVET institutions are increasingly integrating digital tools including MOOCs and Open Educational Resources (OER) to facilitate self-directed professional learning.

The rise of digital technologies has further expanded opportunities for self-directed learning. Online learning platforms, open educational resources, podcasts, webinars, professional communities, and artificial intelligence tools provide learners with unprecedented access to information and learning opportunities (Maphosa & Maphosa, 2023; Selçuk et al., 2026; Yu et al., 2023). Open Educational Resources are digitized materials offered freely and openly for educators, students, and self-learners to use and reuse for teaching, learning, and research. MOOCs support self-directed learning by enabling learners worldwide to access open-access educational resources at convenient times and at optimal paces (D. Griffiths et al., 2024; Maphosa & Maphosa, 2023). The proportion of self-directed learning supported by online learning spaces is getting increasingly higher relative to face-to-face instruction (Yu et al., 2023). AI technology platforms can achieve comprehensive collection and management of teaching information, providing powerful support for subsequent autonomous learning (Lazorak et al., 2021), while AI-enhanced platforms such as Duolingo use automatic speech recognition and natural language processing techniques to facilitate adaptive, self-paced learning within networked communities. Digital education supports informal education and autonomous learning, driving vocational and continuing education innovations and enhancing the education system's openness and connectivity, paving the way for universal lifelong learning in the knowledge economy era (Selçuk et al., 2026).

Benefits of Self-Directed Learning

Self-directed learning offers several advantages. It allows learners to tailor learning experiences to their individual needs, interests, and goals (Kiwelu & Ogbonna, 2020; Neimeyer et al., 2010). Because learning is often intrinsically motivated and personally relevant, engagement and persistence may be enhanced (Khurma et al., 2024; Kiwelu & Ogbonna, 2020). Self-directed learning also develops important lifelong learning skills such as problem-solving, critical thinking, information literacy, self-management, and adaptability (Kiwelu & Ogbonna, 2020; Kultaieva & Sahuichenko, 2019).

In rapidly changing industries, self-directed learning has become increasingly important because employees must continually update their knowledge and skills (Hutasuhut et al., 2021; Šatienė, 2017). The ability to independently identify learning needs and acquire new competencies is widely regarded as a critical capability for the modern workforce (Hutasuhut et al., 2021; Neimeyer et al., 2010; Stoten, 2021a).

Challenges and Criticisms

Notwithstanding its benefits, self-directed learning is not without challenges. Not all learners possess the confidence, experience, motivation, or skills required to effectively manage their own learning. Some individuals may struggle with goal setting, time management, self-assessment, or identifying reliable learning resources. Research has confirmed that students report lacking confidence in their abilities to engage in self-directed learning activities, often due to a lack of information on possible learning resources or an inability to use resources effectively (Blaschke & Marín, 2020; Kharroubi & ElMediouni, 2024). Students may lack a sophisticated understanding of the variety of resources available and how to use them effectively for learning (Kharroubi & ElMediouni, 2024). Furthermore, challenges related to functioning as independent learners and developing self-regulated skills have been widely documented, with teachers noting the difficulties students face in maintaining routines for continuous learning (Blaschke, 2021; Lai, 2015).

Researchers have also noted that self-direction exists on a continuum rather than as an all-or-nothing characteristic. Learners may require varying levels of support depending on their prior knowledge, experience, confidence, and the complexity of the subject matter. This continuum perspective is well established in the literature; for instance, the pedagogy-andragogy-heutagogy (PAH) continuum describes learners moving from traditional, teacher-directed instruction to more self-directed and ultimately self-determined learning, requiring less instructor control and more learner autonomy as they progress, (Oates, 2019). Similarly, self-directed learning has been characterised as a short-term, largely self-directed continuum. Highly technical or unfamiliar topics may require more structured guidance before learners can become increasingly self-directed. Research supports this, showing that lower-capacity learners struggle with open-ended tasks and benefit significantly from structured guidance, whereas more capable learners benefit from reduced instructional guidance and can engage in deeper self-directed learning. Scaffolding techniques that provide targeted support help accelerate learners' development and foster autonomy as control is gradually handed over to the learner.

Critics have further argued that self-directed learning can sometimes overemphasise individual responsibility while underestimating the influence of social, cultural, organisational, and economic factors on learning opportunities and outcomes. Modern self-regulation theory acknowledges that learning is shaped by cognitive, motivational, social, and behavioural elements while also considering cultural, organisational, and contextual influences (Güler & Esen, 2021). Bandura's reciprocal determinism framework similarly posits that learning results from the interaction of personal, environmental, and behavioural factors, where environmental factors include quality of instruction, access to information, and help from peers (Schraw et al., 2006). Emergency remote teaching during the pandemic further exposed how socio-economic gaps, cultural isolation, and motivational challenges constrain learners' capacity for autonomous learning. The Person-Process-Context model of self-directed learning explicitly incorporates contextual factors including culture, power dynamics, and organisational policies as determinants of learning outcomes.

Self-Directed Learning and Adult Education

Self-directed learning remains one of the most influential concepts in adult education (Carré, 1993; Cosnefroy, 2011; Jézégou, 2008). It aligns closely with the principles of andragogy by recognising adults as autonomous learners who bring experience, motivation, and personal goals to the learning process (Carré, 1993; Cosnefroy, 2011; Samaroo et al., 2013). Knowles (1975) described self-direction as a process "in which individuals take the initiative, with or without the help of others, in diagnosing their needs, formulating learning goals, identifying human and material resources for learning, choosing and implementing appropriate learning strategies, and evaluating learning outcomes" (Samaroo et al., 2013). The concept of self-directed learning emerged from adult education research, notably following the foundational works of Rogers (1969), Tough (1979), and Knowles (1975), and proposes a conceptualisation based on the degree of learner control over the educational process (Carré, 1993). Contemporary adult education increasingly seeks to develop learners' capacity for self-direction while providing appropriate support, guidance, and learning resources (Candas, 2012; Carré, 1993). This includes metacognition, open learning, individualisation, experiential learning, and the andragogical approach as essential ramifications of self-directed learning in pedagogical practice (Carré, 1993). The capacity to direct one's own learning articulates three notions within a dynamic process: self-determination, self-regulation, and self-efficacy (Verzat et al., 2016). Mezirow further contended that adult education must address the learning needs of adults and release their potential to develop into "independent, self-directed, and reflective learners" (Samaroo et al., 2013).

As workplaces, technologies, and societies continue to evolve, self-directed learning is likely to remain a cornerstone of lifelong learning (Candas, 2012; Carré, 2015; Marsick & Watkins, 2007). The culture of lifelong learning has gradually come to compete with the older culture of training, with international organisations proclaiming the absolute necessity of developing permanent, everyday learning (Carré, 2015). Contemporary society envisions the citizen as an individual capable of learning throughout life (Candas, 2012), and the ability to independently acquire, apply, and evaluate new knowledge has become an essential capability for individuals seeking to adapt, grow, and succeed in a rapidly changing world (Carré, 2015; Marsick & Watkins, 2007). The economy of knowledge demands an increasingly educated, qualified, and competent workforce (Marsick & Watkins, 2007), while self-directed learning has been shown to function across formal, non-formal, and informal contexts, including distance education and workplace settings (Jézégou, 2008; Verzat et al., 2016).

Learner Readiness and Motivation

Learner readiness and motivation are two closely related assumptions within Malcolm Knowles's theory of andragogy. Together, they help explain why adults choose to engage in learning and what factors influence their willingness to participate, persist, and succeed in educational activities. Knowles argued that adult learners are generally most receptive to learning when they perceive a clear need for new knowledge or skills and when learning is relevant to their current circumstances, responsibilities, or goals.

Learner Readiness

Learner readiness refers to the extent to which an individual is prepared and willing to engage in learning. According to Knowles, adults become ready to learn when they encounter situations that require new knowledge, skills, or competencies. Unlike children, whose learning is often organised according to age-based curricula, adult learning readiness is frequently linked to developmental tasks, social roles, and life circumstances.

Changes in employment, career progression, family responsibilities, technological developments, health concerns, or personal interests can all create a need for learning. For example, a newly promoted supervisor may seek leadership training, a healthcare professional may need to learn updated clinical procedures, or an engineer may need to develop expertise in emerging technologies. In each case, learning is driven by an immediate and identifiable need rather than by a prescribed curriculum.

The concept of readiness is closely linked to relevance. Adults are generally more motivated to learn when they can see a direct connection between learning activities and the challenges they face in their personal or professional lives. Educational programs that align with learners' current roles, responsibilities, and aspirations are therefore more likely to generate engagement and commitment).

Research has shown that readiness to learn is influenced by a range of factors, including prior knowledge, confidence, self-efficacy, organisational support, access to learning opportunities, and perceived usefulness of the learning activity. Adults who believe that learning will help them solve problems, improve performance, or achieve desired outcomes are typically more willing to invest effort in the learning process.

Motivation to Learn

Motivation refers to the forces that initiate, direct, and sustain learning behaviour. Knowles acknowledged that adults may be influenced by external motivators such as promotions, salary increases, professional requirements, qualifications, or organisational expectations. However, he argued that internal motivators often become increasingly important as individuals mature.

Internal or intrinsic motivations may include personal growth, self-esteem, job satisfaction, curiosity, achievement, increased confidence, improved quality of life, and the desire to realise personal potential. Adults frequently engage in learning because they want to become more effective in their professional roles, solve problems they are experiencing, or pursue interests that are personally meaningful.

The importance of intrinsic motivation is supported by humanistic learning theories, particularly the work of Abraham Maslow and Carl Rogers. Humanistic theorists argue that individuals are naturally motivated to learn when learning contributes to personal development, self-actualisation, and the achievement of meaningful goals (Maslow, 1970; Rogers, 1969). From this perspective, motivation is strengthened when learners perceive learning as personally relevant, valuable, and aligned with their aspirations.

Self-Determination Theory, developed by Ryan and Deci, provides further insight into adult motivation by suggesting that motivation is enhanced when three fundamental psychological needs are met: autonomy, competence, and relatedness (Ryan & Deci, 2020). Adults are more likely to engage deeply in learning when they have some control over their learning, feel capable of succeeding, and experience supportive relationships within the learning environment.

The Relationship Between Readiness and Motivation

Readiness and motivation are closely interconnected. Adults who perceive an immediate need for learning are often more motivated to engage in educational activities, while motivated learners are more likely to seek out learning opportunities and persist when challenges arise. A change in circumstances may create readiness to learn, but motivation influences whether the individual actually commits to the learning process.

For example, the introduction of new technology in a workplace may create a need for employees to develop new skills. Employees who recognise the importance of the technology and believe that learning it will improve their performance are likely to be highly motivated. Conversely, if learners do not perceive the learning as relevant or beneficial, readiness and motivation may be low despite organisational requirements.

Implications for Adult Education

The concepts of readiness and motivation have important implications for educational design and delivery. Adult educators can enhance readiness by ensuring that learning activities are relevant, practical, and connected to learners' real-world needs. Conducting needs analyses, involving learners in planning, and demonstrating the value of learning outcomes can help increase learner readiness.

Similarly, motivation can be strengthened by creating supportive learning environments that foster autonomy, competence, and meaningful participation. Learning activities that provide opportunities for success, recognise prior experience, encourage collaboration, and demonstrate practical application are often more effective in sustaining learner motivation.

Many contemporary approaches to adult education, including workplace learning, professional development, competency-based training, and problem-based learning, are built upon the recognition that adults learn most effectively when they are both ready and motivated to learn. By aligning learning with learners' needs, goals, and experiences, educators can create conditions that encourage engagement, persistence, and meaningful learning outcomes.

The concepts of learner readiness and motivation are particularly important in adult and vocational education because most adult learners participate voluntarily and typically have clear reasons for undertaking training. Unlike school-aged learners who may follow a prescribed curriculum, adult learners are often balancing employment, family responsibilities, financial commitments, and personal goals. Consequently, they are more likely to engage with learning when they perceive an immediate need for new knowledge or skills and when they can see a clear connection between learning activities and desired outcomes.

Example 1: Heavy Vehicle Driver Training

Consider a learner undertaking training to obtain a Heavy Combination (HC) or Multi Combination (MC) licence. The learner's readiness to learn is often driven by a specific career goal, such as securing employment, obtaining a promotion, or meeting industry requirements. Because the learner can immediately see how the training will improve employment opportunities and earning potential, motivation is generally high.

An instructor applying andragogical principles would explain how each learning activity relates directly to safe vehicle operation and licensing requirements. Rather than focusing solely on theoretical concepts, the trainer would use realistic driving scenarios, fatigue management exercises, load restraint activities, and route planning tasks that reflect the learner's future work environment. The learner's readiness and motivation are strengthened because the learning has immediate practical relevance.

Example 2: Apprentice Electrician Learning New Technologies

An experienced electrician may be required to complete professional development in solar photovoltaic systems or battery storage technologies. The learner is ready to learn because changes in technology and market demand have created a need for new competencies. Motivation may stem from career advancement, increased employability, business growth opportunities, or compliance with industry standards.

Training programs that demonstrate how the new technology can be applied on actual job sites are likely to be more effective than purely theoretical instruction. Practical workshops, site visits, fault-finding exercises, and installation projects allow learners to directly connect learning to workplace performance.

Example 3: Healthcare Professional Development

A registered nurse undertaking training in new clinical procedures provides another example of learner readiness. Changes in healthcare practices, equipment, or regulatory requirements create an immediate need for learning. The nurse understands that the knowledge gained will directly affect patient outcomes and professional competence.

Motivation is reinforced when training incorporates realistic clinical simulations, case studies, and hands-on practice rather than relying exclusively on lectures. The nurse is more likely to engage actively because the learning addresses current workplace challenges and contributes to professional effectiveness.

Example 4: Leadership Training for New Supervisors

An employee promoted to a supervisory role may suddenly discover that technical expertise alone is insufficient for managing people, resolving conflict, conducting performance reviews, and leading teams. This transition creates a strong readiness to learn because the learner encounters new responsibilities requiring additional skills.

A vocational trainer can leverage this readiness by using authentic workplace scenarios, role-playing difficult conversations, analysing leadership case studies, and encouraging reflection on real management challenges. The learner's motivation remains high because the skills being developed can be applied immediately within their new role.

Example 5: Adult Literacy and Numeracy Programs

Adults returning to education to improve literacy, numeracy, or digital skills often have highly personal motivations. Some may wish to gain employment, support their children's education, manage finances more effectively, or participate more confidently in society.

Readiness to learn often emerges from recognising the limitations created by existing skill gaps. Effective educators connect learning activities to everyday situations such as completing workplace forms, reading safety instructions, calculating household budgets, or using online services. By demonstrating practical benefits, educators strengthen both readiness and motivation.

Example 6: Construction Industry Safety Training

Workers undertaking Work Health and Safety (WHS) training may initially view training as a compliance requirement. However, readiness and motivation increase significantly when learners understand how the training helps prevent injuries, protects co-workers, and improves workplace safety outcomes.

Instead of relying solely on legislation and policy documents, effective trainers use incident investigations, hazard identification exercises, site inspections, and risk assessment activities. These practical applications help learners appreciate the relevance of the training and encourage greater engagement.

Example 7: Continuing Professional Development (CPD)

Engineers, accountants, teachers, and other professionals frequently participate in CPD activities to maintain registration or professional membership. While compliance may initially motivate participation, adult learners are generally more engaged when they can see how the learning contributes to improved performance, career advancement, or professional competence.

For example, a civil engineer attending a seminar on sustainable infrastructure design may be motivated by the opportunity to apply new approaches to upcoming projects. A teacher completing training in educational technology may be motivated by the potential to improve student engagement and learning outcomes.

Understanding learner readiness and motivation enables vocational educators to design more effective and engaging learning experiences. Rather than assuming that learners are motivated simply because they have enrolled in a course, trainers should seek to understand learners' goals, aspirations, prior experiences, workplace requirements, and reasons for undertaking training. By identifying what learners hope to achieve, educators can better align learning activities with learners' needs and expectations, increasing the relevance and value of the learning experience.

This understanding can be used to tailor teaching strategies, examples, case studies, practical activities, and assessment tasks to reflect authentic workplace situations and learner interests. Adult learners are generally more engaged when they understand the purpose of learning activities and can clearly see how the knowledge and skills being developed will benefit them in their current or future roles (Adami et al., 2021; Grabarski et al., 2023; Moll, 2023). Knowles emphasised that adults are interested in "just-in-time" learning immediately related to their lives, in contrast to the "just-in-case" orientation of childhood education (Moll, 2023, 2024). Explaining the practical applications of learning, linking learning outcomes to workplace performance, and demonstrating how training supports career progression can help strengthen learner motivation and commitment (Beer et al., 2019; Malinga, 2024;

Ray & Nelson, 2025). Adult learners are frequently more motivated to perform well when they can see clear connections between training and workplace or career outcomes, (Ray & Nelson, 2025).

Effective vocational educators also create opportunities for learners to contribute their own experiences and knowledge to the learning process (Burns et al., 2020; Watkins & Marsick, 2014). Knowles recognised that adults accumulate a growing reservoir of experience that serves as a rich resource for learning Watkins & Marsick, 2014; ,. Drawing upon workplace examples provided by learners encourages participation and helps connect new learning to existing understanding (Adami et al., 2021; Burns et al., 2020). Providing opportunities for learners to immediately apply newly acquired skills through practical exercises, simulations, workplace projects, or problem-solving activities further reinforces the relevance of learning and supports knowledge transfer (Adami et al., 2021; Santos, 2019). The experiential learning theory, closely related to andragogy, states that the essence of adult learning is making sense of experiences and that adults learn best by doing (Burns et al., 2020).

Supporting learner autonomy is another important strategy. Adult learners often value having some degree of choice and control over their learning, including opportunities to influence learning activities, select project topics, or determine how they will demonstrate competency (Grabarski et al., 2023; Rahman et al., 2022; Ray & Nelson, 2025). Andragogy emphasises that the learner is autonomous and self-directed (Grabarski et al., 2023; Watkins & Marsick, 2014), and self-directed learning places greater emphasis on learners' autonomy and self-determination (Ray & Nelson, 2025). When learners are given responsibility for aspects of their learning and can see clear links between training and improved workplace performance, professional development, or career advancement, motivation is often enhanced (Malinga, 2024; Rahman et al., 2022). Knowles argued that the most persuasive motivators for adults are internal, including self-esteem and self-actualisation, though external motivators also play a role (Moll, 2023; Rahman et al., 2022).

When learners are both ready and motivated, they are more likely to engage actively in learning activities, persist when challenges arise, apply new knowledge and skills in workplace settings, and achieve meaningful learning outcomes (Rahman et al., 2022). Readiness to learn is influenced by the "need to know" assumption, and learners with high readiness are able to learn independently and perform successfully (Rahman et al., 2022). For this reason, learner readiness and motivation remain fundamental considerations in adult education, vocational education and training, workplace learning, and continuing professional development programs (Beer et al., 2019; Santos, 2019; Watkins & Marsick, 2014). By designing learning experiences that are relevant, practical, and aligned with learners' goals, educators can create conditions that support both learner engagement and long-term success (Adami et al., 2021; Malinga, 2024; Nebeker, 2014).

Experience as a Learning Resource

One of the most important assumptions of Malcolm Knowles's theory of andragogy is that adults bring a substantial reservoir of experience to the learning environment. Unlike children, whose life experiences are generally more limited, adult learners have accumulated knowledge, skills, beliefs, values, and perspectives through their employment, family responsibilities, education, community participation, and broader life experiences (Berry, 2021; Conway, 2022; Knapke et al., 2024; Smith, 2019). These experiences shape how adults interpret new information, approach problem-solving, and make sense of learning situations.

Knowles argued that prior experience represents one of the most valuable resources available in adult education and should be actively recognised and incorporated into the learning process (Desta & Gugssa, 2022; Henschke, 2011a; Thomas & Glazzard, 2025). Similarly, Lindeman viewed learners' experiences as the primary resource upon which adult education should be built, emphasising that adult learning differs from traditional schooling because it begins with the realities of learners' lives rather than predetermined subject matter (Henschke, 2011a). Adult learners often identify strongly with their accumulated experiences, which become an important part of their personal and professional identities (Conway, 2022). Consequently, effective adult education acknowledges and utilises these experiences as foundations for learning, reflection, discussion, and the development of new knowledge and skills.

Within the andragogical framework, experience serves both as a foundation for new learning and as a resource that can be shared with others (Henning, 2012; Henschke, 2011a). Adult learners do not enter educational settings as blank slates (Berry, 2021). Instead, they interpret new information through the lens of their existing experiences and use those experiences to construct meaning, solve problems, and evaluate new ideas (Chang, 2022; Desta & Gugssa, 2022; Henning, 2012). Knowles himself encouraged learners to analyse their experiences, stating that the core methodology of adult education is the analysis of experience (Henschke, 2011a). Consequently, effective adult education recognises and values learners' prior knowledge rather than treating all learners as if they are beginning from the same starting point (Smith, 2019).

The importance of experience is one of the key features that distinguishes adult learning from many traditional pedagogical approaches (Berry, 2021; Syaifar & S.Pd, 2017; Thomas & Glazzard, 2025). Traditional teacher-centred models often emphasise the transmission of knowledge from educator to learner, with limited recognition of the learner's existing knowledge and life experiences. These approaches have historically been based on assumptions that younger learners possess relatively little prior experience upon which to build new learning (Berry, 2021).

In contrast, andragogical approaches recognise that adult learners bring a wealth of personal and professional experience that can significantly enhance the learning process (Chang, 2022; Desta & Gugssa, 2022; Thomas & Glazzard, 2025). Learning is viewed as a collaborative endeavour in which educators and learners each contribute valuable knowledge, perspectives, and insights. Rather than serving solely as a source of information, the educator facilitates learning by drawing upon learners' experiences and helping them connect new concepts to real-world situations (Conway, 2022; Ngo et al., 2022).

This collaborative approach encourages learners to share experiences, compare perspectives, and collectively explore solutions to problems. Through discussion, reflection, and the exchange of ideas, learners contribute to the creation of knowledge rather than simply receiving it from an instructor. Critical adult education scholars have similarly emphasised the value of collaborative learning environments in which diverse experiences and viewpoints are acknowledged and used to deepen understanding (Chang, 2022). As a result, adult learning environments often place greater emphasis on dialogue, reflection, peer learning, and experiential activities that leverage the rich experiences learners bring to the educational setting.

Experience as a Foundation for Learning

Experience influences how adults perceive, interpret, and apply new information. When new knowledge can be connected to existing experiences, learning is often more meaningful and easier to understand. Adults frequently evaluate new concepts by comparing them with situations they have previously encountered and by considering how the information might apply in familiar contexts (Chuang, 2021; Gewurtz et al., 2016; Uzorka et al., 2020). Knowles's andragogical framework emphasises that adult learners bring an extensive and continually expanding body of experience to the learning environment, which serves as a valuable foundation for acquiring and applying new knowledge (Al-Mosawi, 2019). Because adult learners are often motivated by practical outcomes, they generally respond well to learning experiences that are directly connected to real-world tasks and workplace situations (Hix, 2013). Prior knowledge, skills, and experiences therefore become important resources that educators can use to facilitate learning and help learners build upon what they already know (Teszenyi et al., 2019; Thiel et al., 2022).

For example, an experienced heavy vehicle driver undertaking fatigue management training is likely to relate theoretical concepts to situations encountered during long-distance journeys. A construction supervisor participating in safety training may draw upon previous incidents or workplace observations when discussing risk management. Similarly, an enrolled nurse studying advanced clinical procedures may compare new techniques with prior patient care experiences. In each case, existing experience provides a framework through which new learning can be understood and applied.

Comparable examples can be found across many vocational disciplines. In printing and graphic arts, an experienced printing operator learning about colour management systems may relate the theory to previous situations where printed colours failed to match customer expectations. Concepts such as colour calibration, ICC profiles, and quality assurance become more meaningful because the learner can connect them to actual production challenges they have encountered. Likewise, a graphic designer learning advanced pre-press workflows may compare new digital processes with traditional proofing methods they have used in the past, enabling them to understand the practical benefits and limitations of emerging technologies.

In the Chemical, Hydrocarbons, and Refining industries, a process operator learning about process safety management may connect new concepts to previous experiences involving plant shutdowns, equipment failures, process deviations, or near-miss incidents. Understanding hazard controls, permit-to-work systems, and process risk management becomes easier when learners can relate the theory to events they have witnessed in operational environments. Similarly, a refinery maintenance technician studying corrosion management may interpret theoretical explanations of corrosion mechanisms through observations made during inspections of pipelines, pressure vessels, or processing equipment. Existing workplace experience helps transform abstract technical concepts into practical knowledge that can be applied directly in operational settings.

The value of connecting learning to experience is well documented across professional education and training. Research in healthcare education has demonstrated that learners are more likely to assimilate and retain knowledge when it is linked to authentic situations and practical applications (Berríos-Arvey et al., 2024). Similarly, clinical education research suggests that combining classroom learning with exposure to real or simulated practice environments helps learners develop the knowledge, judgement, and skills required for professional practice (Barrett et al., 2018; Torre et al., 2021). Learning activities situated within realistic contexts enable learners to bridge the gap between theory and practice, thereby improving understanding and transfer of learning.

The importance of experience in adult learning aligns closely with constructivist theories of learning, which propose that learners actively construct knowledge by building upon existing cognitive frameworks (Chuang, 2021; Gewurtz et al., 2016; Uzorka et al., 2020). Constructivist perspectives suggest that learning becomes meaningful when learners actively interpret new information, organise it within existing knowledge structures, and integrate it with prior understanding (Holmes, 2019). Existing knowledge and experience therefore provide the foundation upon which new learning is developed (Al-Mosawi, 2019).

Rather than passively receiving information, adult learners actively interpret, evaluate, and integrate new concepts into their existing understanding of the world (Bell & Bell, 2020; Chuang, 2021). Constructivist scholars argue that learners create new knowledge by filtering information through their previous experiences and existing understandings, gradually developing more sophisticated mental frameworks (Bell & Bell, 2020). Learners apply constructivist principles whenever they connect new knowledge to personal experiences, draw upon prior learning, and use their understanding to solve problems independently (Uzorka, 2021; Uzorka et al., 2020).

The relationship between andragogy and constructivism is particularly evident in their shared emphasis on experience, learner participation, self-direction, and problem-solving. Both perspectives recognise that learning is most effective when learners are actively engaged in making sense of information through reflection, discussion, and application. The experiential foundations of constructivist learning complement Knowles's view that adult learners bring valuable life experiences to the learning environment, while both approaches support learner-centred, problem-based learning in interactive and authentic contexts (Uzorka et al., 2020).

In terms of application examples, a printing and graphic arts worker learning colour management principles may relate the theory to previous experiences where printed colours did not match the client's expectations. Concepts such as colour profiles, calibration, and colour consistency become easier to understand because the learner can connect the new information to real production challenges they have encountered. Similarly, a pre-press operator learning about digital workflow automation may compare the new processes with traditional manual proofing and file preparation methods they have used in the past, allowing them to better understand both the advantages and limitations of the new technology.

A printing machine operator undertaking training in quality assurance may draw upon previous experiences of paper jams, ink density variations, registration problems, or finishing defects. Rather than viewing quality control as an abstract concept, the learner can immediately relate theoretical quality management principles to situations they have experienced on the production floor. This connection between experience and learning helps the learner understand why particular procedures are required and how they contribute to improved production outcomes.

In the Chemical, Hydrocarbons, and Refining industries, an experienced process operator learning about process safety management may connect new information about hazard identification and risk controls to previous plant shutdowns, equipment failures, process upsets, or near-miss incidents. Concepts such as layers of protection, permit-to-work systems, and process hazard analysis become more meaningful because the learner can visualise actual situations where these controls either succeeded or failed. Existing experience provides a practical framework for understanding the importance of safety systems and operational discipline.

Similarly, a refinery technician learning about corrosion management may compare theoretical corrosion mechanisms with equipment deterioration they have observed during inspections or maintenance activities. Understanding the causes of corrosion becomes easier when learners can relate the concepts to specific pipelines, vessels, heat exchangers, or processing equipment they have encountered in their workplace. The theory is no longer simply a technical explanation but a practical interpretation of familiar workplace conditions.

An experienced laboratory technician undertaking training in advanced analytical techniques may draw upon previous testing experiences when learning new methods. Concepts such as calibration, sampling integrity, measurement uncertainty, and quality assurance can be understood more deeply because they build upon an existing foundation of laboratory practice. The learner is able to integrate new techniques into an already established understanding of scientific testing and analysis.

Likewise, a control room operator learning about new process control systems may compare the features of modern digital control platforms with older systems they have operated. Their previous experiences with alarms, process deviations, and system responses provide a context for understanding how the new technology improves monitoring, control, and operational efficiency.

These examples demonstrate how adult learners actively interpret and integrate new concepts into their existing understanding of the world. Rather than simply receiving information from an instructor, they construct meaning by connecting new knowledge to previous experiences. This process reflects the constructivist view of learning, where knowledge is built upon existing cognitive structures rather than acquired in isolation. In vocational education, recognising and utilising learners' prior experiences allows educators to make learning more relevant, practical, and effective while helping learners transfer new knowledge directly to workplace performance.

Experience as a Resource for Others

Experience not only benefits individual learners but also contributes to the learning of others. Adult learning environments often include participants with diverse backgrounds, occupations, and expertise. Discussions, group activities, collaborative projects, and peer learning opportunities allow learners to share their experiences and perspectives, creating a rich learning environment that extends beyond the knowledge of the educator alone (Gerlach & Reinagel, 2016; Lakin & Crossman, 2004; Sumner, 2018).

For example, in a vocational training classroom, learners may have experience working in different industries, organisations, or roles. During discussions, they can share examples of workplace practices, challenges, and solutions that help others understand how concepts apply in different contexts (Lakin & Crossman, 2004; Zhao & Ko, 2024). This exchange of experience often makes learning more authentic, practical, and relevant (Gerlach & Reinagel, 2016; Sithole & Shokane, 2023).

Communities of practice, mentoring relationships, workplace coaching, and collaborative learning activities all rely heavily on the sharing of experience as a mechanism for learning (Sithole & Shokane, 2023; Sumner, 2018). In many professional settings, some of the most valuable learning occurs through conversations with experienced colleagues rather than through formal instruction alone (Sithole & Shokane, 2023; Williams, 2010).

As application examples, in an Animal Care and Management qualification, learners often bring experience from a variety of settings such as veterinary clinics, animal shelters, boarding kennels, wildlife rehabilitation centres, pet grooming businesses, zoos, or agricultural enterprises. During a class on animal behaviour, a learner employed in a veterinary clinic may share experiences managing anxious or aggressive animals during treatment procedures, while another learner working in a wildlife rescue organisation may discuss techniques used to minimise stress in injured native animals. These different perspectives help all learners develop a broader understanding of animal behaviour and handling practices across diverse environments. Similarly, when studying animal health and welfare, learners may contribute real-world examples of disease outbreaks, nutrition issues, or animal enrichment programs that illustrate how theoretical concepts are applied in practice.

In Food and Beverage vocational training, learners frequently possess varying levels of industry experience gained from cafés, restaurants, bars, catering operations, hotels, or event venues. During a lesson on customer service, one learner may describe strategies used to manage difficult customers during peak service periods, while another may explain how their workplace handles dietary requirements and allergen management. When studying food safety, learners can share experiences of inspections, contamination incidents, or workplace quality assurance procedures. These discussions allow learners to compare different operational approaches and gain practical insights that extend beyond textbook examples. A learner who has worked in a busy restaurant kitchen may provide valuable perspectives on workflow efficiency and communication during service, while another with experience in event catering may explain the challenges associated with large-scale food production and service delivery.

Group activities can further enhance learning by encouraging learners to collaboratively solve realistic workplace problems. For example, a team of food and beverage learners might be asked to develop a service plan for a large function. Learners with experience in front-of-house operations may contribute customer service strategies, while those with kitchen experience may identify production constraints and food safety considerations. The resulting discussion exposes all participants to aspects of the industry that they may not have previously encountered.

At the postgraduate level, Hotel Management students often bring substantial professional experience from a range of hospitality sectors, including hotels, resorts, restaurants, tourism operations, event management companies, airlines, and customer service organisations. This diversity of experience can significantly enrich classroom discussions. For example, when studying revenue management, a student who has worked in hotel reservations may share practical insights into occupancy forecasting and pricing strategies, while another with experience in airline yield management may introduce alternative approaches to capacity optimisation. Through discussion and comparison, learners develop a deeper understanding of how revenue management principles can be applied across different service industries.

Similarly, in a postgraduate hospitality leadership subject, a student who has managed a luxury hotel may discuss strategies for maintaining service quality and employee engagement, while another who has supervised a regional accommodation business may highlight the challenges of operating with limited staffing and resources. These varied perspectives allow learners to examine management theories through the lens of real organisational experiences and encourage critical evaluation of different approaches.

Communities of practice and mentoring relationships are particularly valuable in these industries. In animal care workplaces, experienced animal attendants, veterinary nurses, and supervisors often mentor new employees by sharing practical knowledge about animal handling, health monitoring, and workplace procedures. In food and beverage operations, much learning occurs through observation, coaching, and informal discussions between experienced chefs, supervisors, bartenders, and service staff. Similarly, in hotel management, emerging leaders often learn through conversations with experienced managers who share lessons learned from operational challenges, customer service failures, staff management issues, and strategic decision-making.

These examples demonstrate that adult learning is not solely dependent on the educator. The collective experiences of learners themselves become a valuable educational resource. By sharing workplace knowledge, discussing challenges, and reflecting on practical experiences, learners contribute to the learning of their peers and create richer, more authentic learning environments. This collaborative exchange of experience helps bridge the gap between theory and practice and reflects a core principle of adult learning and andragogy.

Experience and Experiential Learning

The importance of experience in andragogy is closely linked to experiential learning theories, particularly the work of David Kolb. Kolb (1984b) proposed that learning occurs through a continuous cycle involving concrete experience, reflective observation, abstract conceptualisation, and active experimentation (Baumgartner & Carr-Chellman, 2024; Friberg & Hamilton, 2025; Khadka, 2020; Susilo & Nugroho, 2019). According to this model, experience provides the raw material from which learning emerges (Khadka, 2020; Samaroo et al., 2013).

Experiential learning activities such as workplace projects, simulations, role plays, practical demonstrations, field exercises, internships, and case studies are widely used in adult education because they allow learners to connect theory with practice (Ekung & Odesola, 2018; Khadka, 2020; Knapke et al., 2024). These approaches acknowledge that adults often learn most effectively when they can apply knowledge in realistic situations and reflect on the outcomes (Berríos-Arvey et al., 2024; Friberg & Hamilton, 2025; Susilo & Nugroho, 2019).

For example, a trainee project manager may learn more from planning and delivering an actual project than from reading about project management principles alone. Similarly, an apprentice carpenter develops competence through hands-on construction activities that are subsequently analysed and refined through reflection and feedback (Ekung & Odesola, 2018; Khadka, 2020; Susilo & Nugroho, 2019).

Further practical examples of Kolb's experiential learning cycle can be found across many vocational and higher education contexts. In Community Services and Welfare training, learners often develop a deeper understanding of counselling, case management, and client support through direct engagement with realistic scenarios. For example, a learner studying case management may participate in a role-play involving a client experiencing homelessness. The role-play provides a concrete experience in which the learner practises communication, needs assessment, and referral skills. Following the activity, the learner reflects on what occurred, considering which communication strategies were effective and where improvements could be made. The learner then connects these observations to theories of client-centred practice, trauma-informed care, or strengths-based approaches studied in class. Finally, the learner applies these insights during future role-plays, workplace placements, or interactions with actual clients. Through this cycle of experience, reflection, conceptualisation, and application, professional competence gradually develops.

Similarly, a Community Services learner undertaking a work placement at a youth support organisation may observe how experienced workers engage with vulnerable clients. The learner reflects on these observations, compares them with the principles discussed in training, and develops a deeper understanding of professional boundaries, ethical decision-making, and effective communication. When faced with similar situations during subsequent placement activities, the learner is able to test and refine their developing skills. In this way, practical experience becomes the foundation for ongoing learning and professional growth.

Experiential learning is equally important in welfare and community development contexts. For example, learners may be required to design and deliver a small community engagement project. Through planning and implementing the activity, they encounter real-world challenges such as stakeholder engagement, resource constraints, cultural considerations, and participant needs. Reflecting on these experiences allows learners to better understand community development theories and improve their ability to design effective programs in the future.

At the bachelor degree level, Accounting students often participate in experiential learning activities that connect theoretical concepts to professional practice. For example, students studying financial accounting may complete a simulated business project in which they record transactions, prepare financial statements, and analyse business performance. The simulation provides a concrete experience that requires students to apply accounting standards and reporting principles in a realistic setting. After completing the exercise, students reflect on the challenges encountered, such as classifying transactions correctly or identifying errors in financial records. They then connect these observations to accounting theories, standards, and professional frameworks before applying the lessons learned to more complex scenarios.

Work-integrated learning and internships provide further examples of experiential learning in accounting education. A student undertaking an internship with an accounting firm may assist with preparing tax returns, conducting audits, or analysing financial data. Through these activities, the student gains direct exposure to professional accounting practice. Reflecting on workplace experiences helps the student understand how accounting principles are applied in real organisations and highlights the practical considerations that may not be fully apparent in classroom discussions. This reflection enables the learner to integrate theoretical knowledge with professional judgement and workplace expectations.

Case studies are another common experiential learning strategy in accounting education. For instance, students may analyse a company's financial statements and identify potential financial risks or compliance issues. By working through authentic business scenarios, students gain experience in applying accounting concepts to complex situations. Subsequent reflection and discussion help them understand not only the technical aspects of accounting but also the ethical, strategic, and regulatory considerations that influence financial decision-making.

These examples demonstrate how experiential learning allows learners to move beyond passive knowledge acquisition. Whether supporting vulnerable clients in community services or analysing financial performance in accounting, learners develop deeper understanding when they engage in authentic activities, reflect on their experiences, connect those experiences to theory, and apply their learning in new situations. This process reflects both Kolb's experiential learning cycle and the andragogical principle that experience serves as one of the most valuable resources for adult learning (Kolb, 1984b).

Critical Reflection and Transformative Learning

While experience is a valuable learning resource, experience alone does not guarantee learning. Adult learning theorists emphasise the importance of reflection in transforming experience into knowledge (Harper et al., 2024; Mohammed, 2020; Stawnychko, 2025). Through reflection, learners analyse experiences, identify patterns, evaluate assumptions, and develop new insights (Harper et al., 2024; Luning et al., 2022).

Jack Mezirow's Transformative Learning Theory extends this idea by suggesting that adults can critically examine and revise deeply held beliefs and assumptions through reflective processes (Stawnychko, 2025). According to Mezirow, significant learning often occurs when individuals encounter experiences that challenge their existing perspectives (Stawnychko, 2025). Transformative learning involves not just acquiring new knowledge but engaging in a fundamental shift in how individuals perceive themselves and their roles within their organisations (Stawnychko, 2025) . Through critical reflection, learners may develop new ways of understanding themselves, others, and the world around them (Mohammed, 2020; Stawnychko, 2025).

For example, a manager attending leadership training may realise that long-held assumptions about employee motivation are ineffective. Through reflection on workplace experiences and exposure to alternative perspectives, the manager may adopt new leadership approaches that improve team performance (Mohammed, 2020; Stawnychko, 2025). In this way, experience becomes not only a source of knowledge but also a catalyst for personal and professional transformation (Harper et al., 2024; Mohammed, 2020; Stawnychko, 2025).

Challenges Associated with Experience

Although experience is a powerful learning resource, it can also present challenges. Previous experiences may lead learners to develop fixed beliefs, habits, or assumptions that limit openness to new ideas (Acton, 2023; Sebova, 2020). Knowles et al. (2015) stressed that experiences, biases, and existing mental models can both foster and inhibit new learning (Sebova, 2020). Adults may sometimes interpret new information through existing biases or resist learning that contradicts established practices (Holladay, 2024; Lewis & Novak, 2023). Mental models are self-reinforcing and can only be changed if the person has new experiences that are markedly different from the past experiences that have shaped their unconscious mental model (Holladay, 2024). Growth was sometimes limited due to teacher resistance and lack of buy-in, as personal bias kept new approaches from being fully implemented (Lewis & Novak, 2023).

For example, an experienced worker may be reluctant to adopt new technologies because previous methods have been successful in the past. Similarly, long-standing workplace practices may persist even when more effective approaches become available (Sandberg, 2024; Sebova, 2020). In leadership development, transformative learning can be fostered by introducing challenging experiences significant enough to result in a shift in the individual's mental models (Sebova, 2020). From this perspective, development entails challenging mental models and questioning established behavioural patterns (Sandberg, 2024). Effective adult education therefore encourages learners not only to draw upon their experiences but also to critically examine them (Acton, 2023; Mukhalalati & Taylor, 2019).

Educators play an important role in helping learners question assumptions, consider alternative viewpoints, and evaluate whether previous experiences remain relevant in changing circumstances (Ghamrawi et al., 2025; Sebova, 2020) expressed the importance of adult learners developing critical awareness of their personal beliefs, understandings, and biases as one of the first steps in the learning process (Paja, 2013). Self-critical reflection denotes the ability to dispute one's own assumptions and stereotypes, described as "a uniquely adult form of metacognitive reasoning" (Slettli, 2019). Transformative learning involves experiencing a confusing issue, engaging in critical evaluation and self-reflection, and then taking action based on that reflection, which leads to a transformation of meaning and long-standing propositions (Mukhalalati & Taylor, 2019; Rubaia'an, 2023). Reflection, discussion, case analysis, and problem-based learning can all support this process (Acton, 2023; Buhrig, 2023; Sebova, 2020). Case-based learning requires students to construct their own understanding by integrating new problems with past experiences and knowledge rather than relying on memorization (Buhrig, 2023). The adult learning paradigm has influenced education strategies that include problem-based learning, action learning, reflective learning, and experiential approaches (Sebova, 2020). Educators must be intentional about having learners make their learning personal by thinking about learning experiences, and then thinking about how they thought about those experiences and why (Paja, 2013).

Implications for Adult and Vocational Education

Recognising experience as a learning resource has significant implications for teaching practice. Adult educators should create opportunities for learners to share experiences, reflect on practice, and apply learning to real-world situations. Learning activities such as discussions, workplace projects, simulations, case studies, portfolios, reflective journals, mentoring, and peer learning all leverage the value of experience within the learning process.

In vocational education and training (VET), this principle is particularly important because many learners bring substantial workplace experience to their studies. Trainers who acknowledge and incorporate this experience can make learning more relevant, engaging, and meaningful. Competency-based training, recognition of prior learning (RPL), workplace assessments, and work-integrated learning all reflect the recognition that experience is a valuable source of knowledge and evidence of competence.

Ultimately, the assumption that experience is a learning resource lies at the heart of andragogy. It recognises that adult learners arrive with rich and diverse backgrounds that can enhance both individual and collective learning. By valuing, utilising, and critically examining experience, adult education can create deeper, more relevant, and more transformative learning experiences that support lifelong learning and professional growth.

Problem-Centred Learning Approaches

Problem-centred learning is a fundamental principle of adult learning theory and one of the key assumptions of Malcolm Knowles's andragogical model. Unlike traditional educational approaches that are often organised around subject matter and the future application of knowledge, adult learners tend to prefer learning that is organised around real-life problems, tasks, and situations they currently face in their personal or professional lives (Knowles, 1980, 1984). Adults are generally less interested in learning information solely for its own sake and more interested in acquiring knowledge and skills that can help them address immediate challenges and improve their performance.

According to Knowles, as individuals mature, their orientation to learning shifts from being subject-centred to being problem-centred. Children often learn subjects because they are part of a prescribed curriculum and may not see an immediate application for the knowledge. Adults, however, typically want to understand how learning can help them solve practical problems, perform workplace tasks, make decisions, or achieve personal goals. As a result, adult learning is often most effective when it is organised around authentic situations rather than academic disciplines alone.

Problem-centred learning places learners in realistic situations that require them to apply knowledge, analyse information, make decisions, and evaluate outcomes. Rather than memorising facts or procedures in isolation, learners engage with complex issues that resemble those encountered in everyday life and professional practice. The focus shifts from acquiring knowledge to using knowledge.

This approach aligns closely with constructivist learning theory, which suggests that learners construct understanding through active engagement with meaningful tasks and experiences. It also aligns with experiential learning theories, particularly Kolb's experiential learning cycle, where learners develop understanding through experience, reflection, conceptualisation, and application (Kolb, 1984b).

Characteristics of Problem-Centred Learning

Problem-centred learning is characterised by learning activities that are organised around authentic problems, tasks, and situations that learners are likely to encounter in real-world settings. Problem-Based Learning (PBL) originated at McMaster University in Canada in the 1960s as a solution to traditional education problems, with Barrows and Tamblyn defining it as "the learning that results from the process of working toward the understanding or resolution of a problem",. Rather than focusing solely on the recall of information, learners actively engage in applying knowledge to analyse issues, evaluate options, and develop solutions. PBL stimulates students to analyse problems, draw answers, search for data, analyse data, and deduce answers to problems (Kulo & Cestone, 2021), and promotes learning in which students are not only passive recipients of information but also actively involved in the process of discovery and problem-solving (Gao et al., 2020).

These learning experiences are designed to mirror workplace and life situations, enabling learners to see the practical relevance of what they are studying. PBL is characterised as a student-centred pedagogy in which students learn about a subject through the experience of solving open-ended problems found in trigger material, and learners can actively involve themselves in problem solving through collaboration with peers and exchange different opinions in complex issues related to real life. Reflection plays an important role in helping learners evaluate their decisions, improve their understanding, and refine their performance. Students generally need to go through the cycle of "trying-negating-trying again," requiring continuous interaction with teachers to revise their cognition of the problem.

Learners are encouraged to draw upon their prior knowledge and experiences, using these as foundations for interpreting new information and addressing complex challenges. In the PBL process, students need to talk about the given problem based on their prior knowledge and experience with group members. Collaboration, discussion, and the sharing of diverse perspectives are often integral components of the learning process, allowing learners to benefit from the experiences and insights of others. PBL can be characterised by learning in small groups, with a medical educator facilitating learning in the group rather than acting in the "expert" role, and learning by using problems or cases that direct discussion among group members as the case unfolds (Nurdauletova et al., 2024). The social learning process organised in problem-based project teamwork provides a participant-directed approach through which learners practise self-directedness (Sorte & Kim, 2023).

Importantly, problem-centred learning recognises that many real-world problems do not have a single correct answer, and learners may explore multiple possible solutions based on available evidence and contextual factors. Group members review potential solutions and their pieces of evidence to determine a best-fit solution. Within this approach, the educator's role shifts from that of a transmitter of knowledge to that of a facilitator who guides inquiry, supports problem-solving, encourages reflection, and assists learners as they construct understanding through engagement with meaningful problems. Teachers play the role of guidance and cohesion throughout the PBL process, and the teacher is positioned as an organiser and guide while the student assumes the subjectivity of learning. PBL is a problem-oriented and student-centred pedagogy that usually involves learning in small groups to discuss certain topics or clinical cases and solve specific problems while supervised and guided by a teacher.

Problem-Based Learning

One of the most widely recognised forms of problem-centred learning is Problem-Based Learning (PBL). In PBL, learners are presented with a realistic problem before receiving detailed instruction. Working individually or in groups, they identify what they already know, determine what they need to learn, research relevant information, and develop solutions.

For example, nursing students may be presented with a patient displaying symptoms of a particular illness. Rather than beginning with a lecture on the condition, learners investigate the symptoms, explore possible diagnoses, identify required assessments, and recommend treatment options. Through the process of solving the problem, they acquire and apply relevant knowledge.

Similarly, business students may be asked to develop a strategy to improve a company's declining profitability. The problem becomes the vehicle through which concepts such as financial analysis, marketing, leadership, and strategic planning are learned.

In vocational Community Services training, learners may be presented with a scenario involving a young person experiencing homelessness, unemployment, and family conflict. Rather than simply studying theories of case management, learners must assess the client's needs, identify appropriate services, develop a support plan, and consider legal and ethical responsibilities.

As learners work through the scenario, they apply communication skills, referral processes, crisis intervention strategies, and professional judgement. The learning becomes highly relevant because it reflects situations they may encounter in the workplace. Discussions among learners often reveal different approaches based on their own experiences, further enriching the learning process.

In Animal Care and Management training, learners may be presented with a situation involving an animal displaying signs of stress, illness, or behavioural problems. Rather than studying animal welfare principles in isolation, learners investigate possible causes, assess environmental factors, recommend interventions, and develop monitoring plans.

This approach encourages learners to integrate knowledge of animal behaviour, nutrition, health, welfare, and workplace procedures while developing practical problem-solving skills required in animal care settings.

In Food and Beverage training, learners might be asked to manage a simulated service failure during a busy dinner period. The scenario could involve customer complaints, staffing shortages, dietary requirements, and food safety concerns occurring simultaneously.

To resolve the situation, learners must prioritise tasks, communicate effectively, apply customer service principles, maintain food safety requirements, and coordinate team activities. Through solving the problem, learners develop a deeper understanding of hospitality operations than they would through lectures alone.

At the bachelor degree level, accounting students may analyse a company's financial statements to identify why profitability is declining. Rather than simply calculating financial ratios, learners investigate operational issues, identify financial risks, evaluate management decisions, and recommend corrective actions.

This mirrors the work of professional accountants who rarely encounter isolated textbook problems. Instead, they must interpret complex information, exercise professional judgement, and develop practical recommendations based on available evidence.

At the postgraduate level, Hotel Management students may be tasked with responding to a sudden decline in occupancy rates at a resort. Learners analyse market conditions, competitor activity, customer feedback, pricing strategies, staffing levels, and operational costs.

Through collaborative discussion and analysis, students apply concepts from revenue management, marketing, leadership, and strategic management to develop evidence-based solutions. Because many postgraduate learners already possess industry experience, classroom discussions often become highly valuable opportunities for peer learning and professional reflection.

Benefits of Problem-Centred Learning

Problem-centred learning offers several advantages for adult learners. It increases learner motivation by demonstrating the relevance of learning to real-life situations (Alvarez, 2023; Jácome & Martinez, 2024; Mamani et al., 2024). It encourages critical thinking, decision-making, and problem-solving skills that are essential in professional practice (Jácome & Martinez, 2024; Mamani et al., 2024; Velásquez, 2021). It also promotes knowledge transfer because learners practise applying concepts in contexts similar to those they will encounter outside the classroom (García et al., 2024; Mamani et al., 2024; Velásquez, 2021).

Furthermore, problem-centred learning supports many of the core assumptions of andragogy. It recognises learners' prior experiences (Alvarez, 2023; Velásquez, 2021), encourages self-direction (Velásquez, 2021; , Alvarez, 2023), focuses on immediate application (Alvarez, 2023; Mamani et al., 2024), and addresses adults' preference for practical and meaningful learning experiences (Alvarez, 2023; Jácome & Martinez, 2024; Mamani et al., 2024).

Challenges and Considerations

Despite its advantages, problem-centred learning requires careful design and facilitation. Problems must be realistic, appropriately challenging, and aligned with learning outcomes (Gewurtz et al., 2016; Rubaia'an, 2023). Designing problems in an authentic context is a key factor that distinguishes problem-based learning from other instructional methods (Kanyesigye et al., 2022). The facilitator must create formative and summative assessments to define the knowledge gained and determine whether the project contains a problem-solving component (Yadav, 2017). Learners may require support when developing problem-solving skills, particularly if they are accustomed to more traditional forms of instruction (Alexander et al., 2024; Austin). Barriers to implementation and student acceptance of active learning include student preferences for passive, traditional lecture, and online learning may require increased facilitator training, administration time, and time to provide feedback (Alexander et al., 2024). Conceptual clarity of what self-directed learning entails and guidance for both teachers and students can help problem-based learning to bring forth self-directed learners (Loyens et al., 2008). Educators must also ensure that learners acquire the underlying theoretical knowledge needed to develop effective solutions (Kirch & Sadofsky, 2021; Sahin et al., 2025). The primary objective of problem-based learning is to accumulate the concepts of the discipline in the context of a clinical or authentic problem (Kirch & Sadofsky, 2021), and educators must critically evaluate content in terms of learning outcomes and support students' independent thinking and reasoning skills within the curriculum (Sahin et al., 2025).

When implemented effectively, however, problem-centred learning provides a powerful framework for adult education because it mirrors the way adults naturally learn and perform in professional environments (Clausen, 2023; Foster et al., 2021). Adults possess a life-centred orientation to learning and desire instrumental, applied learning that addresses the problems learners experience in real life (Foster et al., 2021). This assumption resonates prominently in contemporary educational frameworks related to problem-based learning (Clausen, 2023). Problem-based learning serves as a model for adult learning because it mobilizes students' prior knowledge to help them absorb and understand new knowledge, inculcates clinical reasoning, and facilitates interdisciplinary integration relevant to real-life situations (Dale et al., 2008). By focusing on authentic problems rather than abstract subject matter, educators can create learning experiences that are engaging, relevant, and directly applicable to workplace and life situations (Gewurtz et al., 2016; Nutwell et al., 2023). PBL strategies align well with adult learner characteristics such as self-direction, highly developed experiences, a desire to engage with learning to better cope with real-life tasks and problems, and the focus on performance rather than acquiring knowledge (Nutwell et al., 2023). Adult learning principles, including self-direction, relevant material, problem-based learning, self-motivation, and application of life experiences, are effectively met through the implementation of problem-based learning (Alexander et al., 2024). The student-centred nature of PBL, the fact that students start working on a problem before they have received other curriculum inputs, and the identification of their knowledge deficits are all crucial features that foster self-directed learning (Loyens et al., 2008).

Applications in Adult Education and Vocational Education and Training (VET)

The principle that experience is a valuable learning resource has significant implications for the design and delivery of adult education and vocational education and training (VET). Adult learners enter educational programs with accumulated knowledge gained through employment, family responsibilities, community involvement, and previous educational experiences. Effective adult education recognises these experiences as assets that can be used to support learning rather than treating all learners as though they are starting from the same point. As a result, adult education programs often emphasise practical application, authentic learning experiences, reflection, problem-solving, and learner participation.

In VET, this principle is reflected in the widespread use of workplace-based learning activities. Case studies, simulations, role plays, practical demonstrations, workplace projects, and scenario-based assessments allow learners to connect new knowledge to situations they have already encountered in their workplaces. For example, an experienced aged care worker studying a higher-level community services qualification may draw upon previous interactions with clients when learning advanced case management techniques. Similarly, an apprentice electrician learning fault-finding procedures can relate new diagnostic methods to electrical problems previously encountered on job sites. By linking learning to existing experience, learners are able to develop deeper understanding and improve their ability to transfer learning to workplace performance.

Recognition of Prior Learning (RPL) is another important application of this principle within VET. RPL acknowledges that valuable learning occurs outside formal educational settings and allows learners to receive credit for competencies developed through work experience, military service, volunteering, community participation, or previous training. Rather than requiring learners to repeat learning they have already mastered, RPL recognises existing competence and focuses educational effort on addressing identified skill gaps. This approach aligns strongly with andragogical assumptions regarding the value of prior experience and learner autonomy.

Adult education programs also frequently incorporate reflective practice activities that encourage learners to examine and learn from their experiences. Reflective journals, learning portfolios, professional discussions, mentoring conversations, and workplace debriefings provide opportunities for learners to analyse their experiences, identify lessons learned, and connect practical situations with theoretical concepts. This is particularly important in professions such as healthcare, education, community services, engineering, and management, where professional judgement develops through both experience and reflection.

Collaborative learning strategies further demonstrate the application of experience-based learning principles. Adult learners often possess diverse backgrounds and expertise that can enrich the learning environment. Group discussions, peer learning activities, communities of practice, and workplace learning teams allow learners to share experiences, compare approaches, and collectively solve problems. In many adult learning settings, valuable learning occurs through interaction with peers who have faced similar workplace challenges or who bring different perspectives to a problem.

The influence of experience is also evident in competency-based training, which forms the foundation of many VET systems. Competency-based approaches focus on the application of knowledge and skills in authentic workplace contexts rather than solely on theoretical understanding. Assessment tasks are therefore often designed to replicate workplace conditions and require learners to demonstrate competence through practical performance, workplace evidence, projects, simulations, and portfolios. Such approaches recognise that meaningful learning occurs when learners can apply new knowledge to realistic situations and build upon their existing experiences.

Consequently, the recognition of experience as a learning resource has shaped many contemporary adult education and VET practices, including workplace learning, RPL, experiential learning, problem-based learning, reflective practice, mentoring, coaching, collaborative learning, and competency-based assessment. These approaches acknowledge that adults bring valuable knowledge and experience to the learning environment and that learning is often most effective when new knowledge can be connected to existing understanding and applied to authentic situations.

Criticisms and Limitations of Andragogy

Despite its widespread influence and enduring popularity in adult education, andragogy has been subject to considerable criticism since its introduction. While many educators regard it as a useful framework for understanding adult learners, critics argue that it has limitations as a comprehensive theory of learning and may oversimplify the complexities of how adults learn (Clair, 2024; Clair & Käpplinger, 2021). Several criticisms focus on the assumptions underlying andragogy, its applicability across different contexts, and its theoretical foundations (Hagen & Park, 2016; Moll, 2023, 2024).

One of the most common criticisms of andragogy is that the characteristics identified by Knowles are not exclusive to adult learners. Knowles argued that adults are generally self-directed, experience-rich, problem-centred, and internally motivated (Akintolu & Letseka, 2021; Clair & Käpplinger, 2021). However, researchers have noted that many children and adolescents can also display these characteristics under appropriate circumstances, while some adults may not exhibit them at all (Clair, 2024; Moll, 2024).

For example, highly motivated secondary school students may demonstrate considerable autonomy, initiative, and self-directed learning behaviours. Conversely, adults entering unfamiliar learning environments may require substantial guidance, structure, and support (Volchenkova, 2015). Critics therefore argue that the distinction between pedagogy and andragogy is often less clear-cut than Knowles originally suggested (Clair, 2024; Moll, 2024), and that learning preferences may be influenced by factors other than age alone. Indeed, Knowles himself came to present andragogy and pedagogy as complementary approaches rather than mutually exclusive and age-specific (Clair, 2024).

Andragogy assumes that adults generally prefer and are capable of directing their own learning (Clair & Käpplinger, 2021; Volchenkova, 2015). While self-direction is undoubtedly an important characteristic of many adult learners, critics argue that not all adults possess the confidence, knowledge, skills, or motivation required to manage their own learning effectively (Conway, 2022; Moll, 2023).

Learners entering a new profession, studying a highly technical subject, or returning to education after a long absence may initially require substantial instructor guidance. In such situations, more structured teaching approaches may be appropriate until learners develop the competence and confidence needed for greater autonomy. Critics therefore suggest that self-direction should be viewed as a continuum rather than an inherent characteristic of all adult learners (Baumgartner & Carr-Chellman, 2024; Clair, 2024).

Another significant criticism is that andragogy focuses primarily on the individual learner while paying relatively little attention to broader social, cultural, political, and organisational influences on learning (Clair & Käpplinger, 2021; Conway, 2022; Moll, 2023).

Knowles's framework emerged largely from Western, individualistic cultural traditions that value independence, autonomy, and self-direction (Conway, 2022; Moll, 2023). However, learning preferences and educational expectations vary considerably across cultures. In some cultural contexts, learners may prefer more structured, teacher-directed approaches and may view educators as authoritative sources of knowledge rather than facilitators. Brookfield has regarded the theory as "culture blind," arguing that its key notion of self-directed learning marginalises people with racial and cultural identities that value the teacher as the source of knowledge and guidance (Moll, 2023). Similarly, Pratt described the neglect of cultural context in andragogy as the learner operating as if he/she has risen above the web of social structures (Moll, 2023).

Furthermore, factors such as socioeconomic status, access to education, workplace conditions, organisational culture, gender, language, and power relationships can significantly influence learning opportunities and outcomes (Baumgartner & Carr-Chellman, 2024; Clair & Käpplinger, 2021; Moll, 2023). Critics argue that andragogy does not adequately address these broader contextual influences. As scholars have noted, andragogy's assumptions were developed based on white, male, middle-class norms of the 1960s, and its bias toward the representation of highly educated, skilled men has been identified as a significant limitation (Baumgartner & Carr-Chellman, 2024; Conway, 2022; Moll, 2023).

Critical adult education theorists have argued that andragogy pays insufficient attention to issues of power, inequality, and social justice (Clair & Käpplinger, 2021; Conway, 2022). While Knowles focused primarily on improving learning effectiveness, scholars such as Paulo Freire emphasised that education also occurs within social and political systems that influence access to knowledge and opportunities (Harris & Roter, 2024).

From this perspective, learning is not simply a matter of individual motivation and self-direction. Educational experiences are shaped by workplace hierarchies, institutional structures, social norms, economic circumstances, and power relationships (Clair & Käpplinger, 2021; Moll, 2023). From a 21st-century viewpoint, andragogy appears to depoliticize and individualize learning and education, isolating them from social context and approaching them as psychological matters (Clair & Käpplinger, 2021). Critics suggest that andragogy tends to assume learners have equal opportunities to participate in learning and exercise autonomy, which may not always be the case (Conway, 2022; Duff, 2019).

Some scholars argue that andragogy is better understood as a set of instructional principles or assumptions about adult learners rather than a complete theory of learning (Clair, 2024; Hagen & Park, 2016). Unlike theories such as Behaviourism, Constructivism, Experiential Learning Theory, or Transformative Learning Theory, andragogy does not fully explain the cognitive, psychological, or social processes through which learning occurs (Hagen & Park, 2016; Moll, 2024). Andragogy has been variously defined as a theory of adult learning, a theory of adult education, a technique for adult education, and a set of assumptions, and this lack of a strong definition has paralleled critiques that it is not a theory at all (Hagen & Park, 2016). (Hartree, 1984)suggested that andragogy lacked definition and theoretical grounding, while others noted it was "under-theorized" (Clair, 2024; Hagen & Park, 2016).

Instead, it primarily provides guidance for designing learning environments and educational practices that are likely to be effective for adult learners (Clair, 2024). For this reason, many contemporary scholars view andragogy as a practical framework rather than a comprehensive explanatory theory (Clair, 2024; Hagen & Park, 2016; Moll, 2024). As Moll has argued, rather than a purported cleavage between the learning of adults and children, Knowles may actually have been driving at a distinction between non-formal and formal teaching methods (Moll, 2024).

Another limitation is the assumption that adults share common learning characteristics. In reality, adult learners are highly diverse. Differences in educational background, culture, occupation, age, life experiences, confidence, literacy levels, learning preferences, and previous educational experiences can significantly influence how adults approach learning (Clair, 2024; Cocquyt et al., 2018). Research has shown that "the andragogical assumptions seem to reflect preferences that are more aligned with males, higher levels of education, and highly skilled occupations, characteristics that are not necessarily representative of adults in need of adult and continuing education" (Roessger et al., 2018, p. 2) . Furthermore, studies have identified the need to "better understand adult learner characteristics with respect to different cultures and learner demographics" (Clair, 2024, p. 11).

For example, an experienced manager undertaking executive education may be highly self-directed and comfortable with independent learning, whereas an adult returning to study after many years may prefer more structure and support (Darden, 2014). Knowles' theory "promotes self-directing and independence in the adult learner, but not all adult learners embrace these ideals especially if the learner lacks self-confidence" (Darden, 2014, p. 809). Similarly, learners in highly regulated industries such as aviation, healthcare, or nuclear operations may require more directive instruction because of the safety-critical nature of the knowledge being taught (Clair, 2024).

This diversity suggests that no single educational approach is appropriate for all adult learners or all learning situations (Clair, 2024; Nassir, 2020). As Nassir notes, andragogical principles "can't be considered as recipes, which are suitable for all adults, regardless of their particularities and those of their contexts" (Nassir, 2020, p. 8). Knowles himself designed the Model of Andragogy in Practice to allow "the adaptation of andragogical principles" through "the analysis of contextual factors (individual and situational differences & goals and objectives of learning" (Nassir, 2020, p. 8).

Knowles initially presented andragogy as an alternative to pedagogy, creating a distinction between child and adult learning (Loeng, 2023; Newton, 1995). Critics argued that this dichotomy was overly simplistic because many educational approaches exist along a continuum rather than fitting into two distinct categories (Loeng, 2023). As Loeng observes, "the either-or perspective may be too restrictive, thereby representing a false dichotomy" (Loeng, 2023, p. 46).

In response to these criticisms, Knowles later revised his position and acknowledged that pedagogical and andragogical methods should not be viewed as mutually exclusive or as opposing approaches to education. Rather than maintaining a strict distinction between child and adult learning, he came to recognise that educational approaches exist along a continuum and that both pedagogical and andragogical strategies may be appropriate depending on the learner, the learning situation, and the educational objectives (Halupa, 2015). This shift in thinking was reflected in later editions of his work, where the relationship between pedagogy and andragogy was presented as a matter of degree rather than as fundamentally different forms of learning (Loeng, 2023).

This revised perspective suggests that educators should select teaching approaches based on factors such as learner characteristics, prior knowledge, experience, subject matter, learning goals, and contextual requirements rather than assuming that all adults prefer andragogical approaches or that all children require pedagogical approaches (Nassir, 2020). For example, adult learners entering a completely unfamiliar field may initially benefit from highly structured instruction and guidance, while younger learners with strong interest and capability may successfully engage in self-directed and problem-based learning activities.

Knowles also emphasised that andragogy should not be regarded as a rigid doctrine or universal prescription for teaching adults. Instead, he viewed it as a set of assumptions about learners that educators should consider, evaluate, and adapt according to the specific characteristics of learners and the circumstances in which learning occurs (Pembridge, 2014). This more flexible interpretation has contributed to the continuing relevance of andragogy by allowing educators to apply its principles selectively and in combination with other learning theories and instructional approaches.

For example, a novice adult learner studying an unfamiliar technical subject may initially benefit from structured instruction, while an experienced professional undertaking advanced training may thrive in a highly self-directed learning environment (Darden, 2014; Halupa, 2015). It has been noted that some adults are highly dependent on teachers for structure, while some children are independent learners (Halupa, 2015). The determination of teaching styles "should be situational, based on the connection between the teacher and learner" (Darden, 2014, p. 811).

Within vocational education and training (VET), the assumptions of andragogy can sometimes be challenging to apply consistently because adult learners enter training with diverse backgrounds, capabilities, and learning needs (Cocquyt et al., 2018; Moll, 2023). Learners may vary considerably in their levels of workplace experience, literacy, confidence, educational attainment, and readiness for self-directed learning. While some learners are comfortable taking responsibility for their own learning and participating in learner-centred activities, others may prefer more structured learning environments that provide clear instructions, organised learning materials, and direct guidance from trainers (Moll, 2023).

Despite these challenges, andragogical principles have been widely adopted within technical and vocational education and training (TVET), where they have influenced the design of occupational assessment, vocational training, job placement programs, and other forms of adult workforce development (Moll, 2023). The emphasis on practical application, workplace relevance, learner experience, and problem-solving aligns closely with the objectives of vocational education and training.

However, andragogy has also been criticised for reflecting assumptions that emerged from predominantly Western, individualistic educational traditions. Some scholars argue that the framework was developed within a social context largely shaped by white, middle-class, male perspectives and therefore may not adequately reflect the experiences, values, and learning preferences of all learner groups (Moll, 2023). Critics suggest that assumptions regarding autonomy, self-direction, and individual responsibility for learning may not align with the cultural norms of all societies and communities. Furthermore, concerns have been raised that a universal application of andragogical assumptions may overlook the influence of race, gender, culture, socioeconomic background, and other social factors that shape learning opportunities and educational experiences (Moll, 2023). These criticisms highlight the importance of recognising diversity among adult learners and adapting educational approaches to suit the needs of different individuals, cultures, and learning contexts.

Furthermore, competency-based training systems often require learners to achieve specific performance outcomes within prescribed frameworks (Moll, 2023). While andragogical principles such as relevance, practical application, and learner participation remain valuable, educators must balance learner autonomy with regulatory requirements, training package specifications, workplace standards, and assessment obligations (Clair, 2024; Moll, 2023). As noted in the literature, "not all adult learners fit the assumptions of andragogy, and that different approaches may be more effective for different individuals" (Onwuatuegwu, 2023).

Despite these criticisms, andragogy remains one of the most influential frameworks in adult education (Clair, 2024). Many criticisms focus not on whether the principles are useful, but on the extent to which they can be universally applied (Clair, 2024; Pembridge, 2014). Concepts such as learner experience, practical relevance, self-direction, problem-centred learning, and intrinsic motivation continue to underpin many contemporary approaches to adult learning (Clair, 2024; Volchenkova, 2015). Current research supports the view that "andragogy is a useful model for designing instruction in numerous fields of practice" (Clair, 2024, p. 7).

Most modern adult learning scholars view andragogy as one component of a broader body of adult learning theory (Clair, 2024; Halupa, 2015). Rather than treating it as a complete explanation of learning, educators often combine andragogical principles with insights from constructivism, experiential learning, transformative learning, social learning theory, humanism, and self-directed learning (Cocquyt et al., 2018; Halupa, 2015). Contemporary discussions of adult learning increasingly suggest that pedagogy, andragogy, and heutagogy should not be viewed as rigidly separate approaches but rather as complementary perspectives that may be applied according to the needs of learners and the demands of particular learning situations (Halupa, 2015). Learners may move between more structured, teacher-directed approaches and more autonomous, self-directed or self-determined forms of learning as their knowledge, confidence, experience, and learning capabilities develop. This perspective recognises that different instructional approaches may be appropriate at different stages of the learning process rather than being determined solely by the learner's age.

Similarly, research has highlighted the value of integrating andragogical principles with other educational perspectives, particularly social constructivism. Combining these approaches acknowledges both the importance of learner autonomy and experience, and the role of collaboration, interaction, and social participation in the construction of knowledge (Cocquyt et al., 2018). Such integration is particularly relevant in contemporary learning environments, including blended and online learning, where learners often engage in both independent study and collaborative activities.

Consequently, andragogy continues to provide a valuable framework for understanding many aspects of adult learning while recognising that learning is shaped by a complex interaction of individual characteristics, social relationships, cultural influences, educational contexts, and environmental factors (Clair, 2024; Cocquyt et al., 2018; Nassir, 2020). Rather than serving as a complete explanation of adult learning, andragogy is increasingly viewed as one of several complementary perspectives that can inform the design and delivery of effective learning experiences for adults.

Contemporary Interpretations of Adult Learning Principles

Since Malcolm Knowles first popularised andragogy during the twentieth century, understandings of adult learning have continued to evolve (Clair & Käpplinger, 2021; Henschke, 2010a). While the core principles of andragogy remain influential, contemporary scholars generally view adult learning as more complex and context-dependent than originally proposed (Clair, 2024; Clair & Käpplinger, 2021). Modern interpretations recognise that adult learning is shaped not only by individual characteristics such as experience, self-direction, and motivation (Abedini et al., 2021; Ghamrawi et al., 2025), but also by social, cultural, technological, organisational, and environmental influences (Conway, 2022; Moll, 2023).

Contemporary adult learning theory tends to move beyond the question of whether adults learn differently from children and instead focuses on understanding the conditions that support effective learning across diverse contexts (Berríos-Arvey et al., 2024; Clair, 2024; Hagen & Park, 2016). Rather than treating andragogy as a universal theory, many scholars now regard it as one of several complementary frameworks that can inform educational practice (Abedini et al., 2021; Henschke, 2011a; Peace, 2006). This perspective recognises that adults are a highly diverse population (Moll, 2023) and that effective learning strategies may vary according to learners' backgrounds, experiences, cultures, professions, and learning goals (Bakare, 2020; Conway, 2022; Moll, 2023)/

Learner-Centred Education

One of the most enduring contemporary interpretations of adult learning is the emphasis on learner-centred education. Modern adult education places the learner at the centre of the educational process, recognising learners as active participants rather than passive recipients of information (Ghamrawi et al., 2025; Jain & Madaan, 2020; Sasere & Makhasane, 2023). Educators increasingly design learning experiences that encourage participation, collaboration, inquiry, reflection, and problem-solving (Ghamrawi et al., 2025; Halpern & Tucker, 2015).

Learner-centred approaches emphasise flexibility, responsiveness, and personalisation. Rather than delivering identical learning experiences to all learners, educators seek to accommodate different levels of experience, prior knowledge, learning preferences, and professional goals (Caruso et al., 2020; Sasere & Makhasane, 2023). This approach reflects contemporary understandings that learning is most effective when learners can actively engage with content and connect learning to their own circumstances, (Castanha et al., 2025; Govender & Booyse, 2024).

Lifelong Learning

Contemporary interpretations of adult learning are closely associated with the concept of lifelong learning. Rapid technological change, globalisation, evolving labour markets, and changing social conditions have increased the importance of continuous learning throughout life (Desmarchelier & Cary, 2022; Rachel Fitzgerald & Henk Huijser, 2021; J. Lang, 2023; Темјановски et al., 2023). The Fourth Industrial Revolution is primarily knowledge-based and continuously requires new competencies, and thus education policies and systems need to be sufficiently agile to react to knowledge developments and redesign programmes to incorporate lifelong learning. Learning is no longer viewed as an activity confined to childhood or formal education but as an ongoing process that occurs across the lifespan (Parsons et al., 2023).

Adults are increasingly expected to update their knowledge and skills throughout their careers. As a result, adult learning principles are frequently applied in continuing professional development, workplace learning, community education, micro-credential programs (Diab & Zhang, 2023; Rachel Fitzgerald & Henk Huijser, 2021; J. Lang, 2023; Темјановски et al., 2023), and online learning environments (Sharma et al., 2024; Swanepoel, 2021). Micro-credentials, for instance, support lifelong learning and flexible opportunities for adults to upskill and reskill, while offering a flexible, advanced way to help people develop their knowledge, skills and competencies (Темјановски et al., 2023). Lifelong learning emphasises adaptability, continuous improvement, and the capacity to acquire new competencies in response to changing circumstances (Parsons et al., 2023).

The growing pace of technological, economic, and workforce change has increased the importance of adaptability as a key capability for individuals and organisations. As industries evolve, occupations change, and new technologies emerge, workers are increasingly required to update their knowledge and develop new competencies throughout their careers. In response to these demands, micro-credentials have gained prominence as flexible learning pathways that enable individuals to acquire targeted skills and knowledge, respond to changes in professional practice, and maintain employability in competitive labour markets. Their increasing adoption reflects the broader emphasis on lifelong learning within contemporary adult education, where continuous skill development and ongoing professional growth are viewed as essential for career sustainability and workforce resilience.

Self-Directed and Self-Determined Learning

While self-directed learning remains an important component of adult learning, contemporary interpretations extend beyond Knowles's original conception (Blaschke & Hase, 2019). Heutagogy, often described as self-determined learning, builds upon andragogy by placing even greater emphasis on learner autonomy, capability development, and learner control (Blaschke, 2021; Blaschke & Hase, 2019; Morrison, 2020).

In self-directed learning, learners generally assume responsibility for selecting learning strategies and managing their progress towards learning outcomes that have been established by an educator, training organisation, or curriculum framework. Self-determined learning, or heutagogy, extends this level of autonomy by giving learners greater control over the entire learning process, including the identification of learning needs, the development of learning objectives, the selection of learning activities, and the methods used to demonstrate achievement (Blaschke, 2021; Ishaq et al., 2024). Heutagogy places a strong emphasis on learner agency, recognising learners as active participants who shape both the direction and evaluation of their learning.

The distinction between self-directed and self-determined learning lies primarily in the degree of learner control. While self-directed learning often operates within educator-defined goals and outcomes, self-determined learning allows learners to negotiate or establish their own goals, identify relevant learning resources, and participate in the design of assessment approaches (Ishaq et al., 2024). This reflects contemporary educational environments in which learners have unprecedented access to information through digital technologies and are increasingly expected to take responsibility for planning, managing, and evaluating their own learning pathways (Blaschke & Hase, 2019; Morrison, 2020).

Heutagogical approaches are particularly evident in professional development, workplace learning, online education, and higher education settings where learners pursue personalised learning goals, respond to emerging professional needs, and develop the adaptability, self-regulation, and lifelong learning capabilities required in rapidly changing environments (Blaschke & Hase, 2019; Lock et al., 2021; Morrison, 2020; Stoten, 2021b). Through these approaches, learners develop not only specific knowledge and skills but also the capacity to independently identify future learning needs and continue learning throughout their lives.

Social and Collaborative Learning

Contemporary adult learning theory increasingly recognises that learning is not solely an individual activity. Social constructivist perspectives emphasise that knowledge is often developed through interaction, collaboration, discussion, and participation in communities (Abbott & Lee, 2022; Chuang, 2021). Social constructivism contends that knowledge is co-constructed as individuals interact with their culture and society and that learning occurs through a collaborative, negotiated process (Abbott & Lee, 2022).

The work of Etienne Wenger on Communities of Practice has been highly influential in demonstrating that adult learning frequently occurs through participation in professional, occupational, and social communities rather than solely through formal instruction (Bentley et al., 2010; Gray, 2018; Wareing, 2014). Communities of practice consist of individuals who share common interests, professional responsibilities, challenges, or areas of expertise and who develop their knowledge and skills through ongoing interaction, collaboration, and shared problem-solving (Gray, 2018). Within these communities, learning occurs as members engage with colleagues, mentors, professional networks, and workplace groups, exchanging experiences, sharing knowledge, and collectively addressing practical challenges (Bentley et al., 2010; Mesquita et al., 2014; Webster-Wright, 2009).

This perspective emphasises that learning is embedded within practice and social participation. Knowledge develops through engagement in authentic activities, observation of experienced practitioners, discussion, collaboration, and progressively increasing participation in professional tasks. Learning and application therefore occur simultaneously, enabling individuals to develop both competence and professional identity within their communities of practice (Bentley et al., 2010; Mesquita et al., 2014). Research has shown that effective professional learning is often sustained over extended periods and is strengthened when individuals participate in supportive communities that encourage collaboration, reflection, and engagement with authentic workplace problems (Webster-Wright, 2009).

The concept of communities of practice has influenced a wide range of contemporary educational and professional development strategies, including peer learning, collaborative projects, workplace mentoring, coaching, discussion forums, professional networks, and structured communities of practice (Abblitt et al., 2023; Abbott & Lee, 2022; Christiansen & Bell, 2010; Gray et al., 2022; Nebeker, 2014; Teszenyi et al., 2019). These approaches recognise that much professional learning occurs through interaction with others and participation in shared activities rather than through individual study alone. Peer learning, for example, draws upon the understanding that learning is situated within social contexts and develops through participation in professional communities (Christiansen & Bell, 2010).

Within professional environments, individuals often develop expertise through gradual participation in workplace activities and professional networks. Newcomers typically begin by observing and participating in simpler tasks before progressively assuming greater responsibility and becoming more central members of the professional community. Through this process, learners develop not only technical knowledge and skills but also an understanding of professional norms, values, practices, and identities (Abblitt et al., 2023). Contemporary perspectives therefore view adult learning as both an individual and a social process, recognising that learners frequently participate in multiple interconnected communities that collectively contribute to their ongoing professional and personal development (Teszenyi et al., 2019).

Experiential and Authentic Learning

Experience remains central to contemporary interpretations of adult learning. However, modern approaches place greater emphasis on authentic learning environments that closely resemble real-world situations (Baldacci, 2022; Duchatelet et al., 2023; Evanick, 2024). Learning activities are often designed to replicate workplace tasks, professional challenges, and practical problems that learners are likely to encounter outside the classroom (Baldacci, 2022; Evanick, 2024; Hovington et al., 2020).

Problem-based learning, project-based learning, simulations, workplace learning, internships, clinical placements, and scenario-based activities all reflect this emphasis on authentic learning (Baldacci, 2022; Duchatelet et al., 2023; Evanick, 2024; Huggins, 2017). These approaches recognise that adults frequently learn most effectively when they can apply knowledge in realistic contexts and immediately see the relevance of what they are learning (Duchatelet et al., 2023; Evanick, 2024; Hovington et al., 2020; X. Li, 2018).

Transformative Learning

One of the most significant developments beyond andragogy has been the emergence of Transformative Learning Theory, primarily associated with Jack Mezirow. Transformative learning theory was first articulated by Mezirow in the 1970s, originating from his study of the learning experiences of women returning to post-secondary education or professional workplaces after an extended time away (Biasin, 2018b; Yunga, 2019). Mezirow desired to formulate an alternative, more critical vision of Knowles's andragogical model, enhancing the function of perspective transformation through critical reflection on adults' assumptions (Biasin, 2018b). Transformative learning extends the role of experience by emphasising critical reflection and perspective transformation. Specifically, Mezirow uses the term "critical reflection" as a means of addressing the base presuppositions upon which a problem is built, also describing this sub-process as "premise reflection" (Yunga, 2019). The theory posits that perspective transformation occurs when adult learners undergo the process of revising their "meaning structures", culturally-defined frames of reference composed of "meaning schemes" and "meaning perspectives" (Njiro, 2014; Yunga, 2019).

Contemporary interpretations recognise that adult learning is not limited to acquiring new knowledge and skills. Learning can also involve questioning assumptions, re-evaluating beliefs, and developing new ways of understanding the world. There are three possible dimensions to such transformation: psychological (changes in understanding of the self), convictional (revision of belief systems), and behavioural (changes in lifestyle) (Fejes & Nicoll, 2013). Transformative learning occurs when individuals critically examine existing perspectives and construct new interpretations based on reflection and experience. As Mezirow delineates, the process begins with a "disorienting dilemma," after which people engage in critical reflection and re-evaluate the assumptions they have made about themselves and their world (Ballout & DePace, 2025; Njiro, 2014). Reflections on their meaning perspectives or meaning schemes can result in a perspective transformation or change in world view, and people then engage in "reflective discourse" to obtain consensual validation (Njiro, 2014). Critical reflection allows individuals to recognise how much their own biography has limited them, a capacity that develops fully in adulthood (Kušić et al., 2020). Over time, transformative learning theory has taken the place of andragogy and behaviourism as the dominant mainstream in adult education (Biasin, 2018b; Ludovico, 2017).

This perspective has become particularly influential in leadership development, professional education, social justice education, healthcare education, and community learning programs. In leadership development, transformative learning occurs when individuals critically reflect upon their environment and modify their frame of reference and view of the world, with critical reflection facilitating an adult's re-examination of basic beliefs (Mohammed, 2020). In professional education, including teacher preparation, integrating critical reflection journals and problem-based learning can foster perspective transformation central to transformative learning (Ghamrawi et al., 2025).

In social justice education, transformative learning theory provides a framework through which nursing students, for example, learn through communicative and reflective experiences that lead to changes in their "frame of reference," influencing social justice practices (Ballout & DePace, 2025; Elliott, 2024). In healthcare education, Mezirow's transformative learning theory has been applied to guide curriculum design, program evaluation, and the development of leadership skills and professional identity among health professionals (Unger et al., 2025). In community learning programs, andragogues serve as change agents who can carry out transformative learning in which critical reflection is the central category, helping marginalised groups through democratic discourse (Kušić et al., 2020). The theory's influence extends across educational leadership preparation programs designed to foster transformational learning outcomes that enable learners to critically reflect upon and shift pre-existing assumptions, biases, and paradigms (Young et al., 2006).

Technology-Enhanced Adult Learning

Digital technologies have significantly reshaped contemporary interpretations of adult learning. Online learning platforms, learning management systems, webinars, podcasts, virtual simulations, mobile learning applications, artificial intelligence, and open educational resources have expanded access to learning opportunities (Analyti et al., 2024; Cohen & Cohen, 2023). Cohen and Cohen specifically identify OERs as including "MOOCs, webinars/videoconferences, podcasts, videos, recorded lectures, papers, books, Wikipedia, social networks, interactive apps, and learning websites" (Cohen & Cohen, 2023, p. 40). Digital learning resources, repositories, and learning management systems have become important components of technology-enhanced learning environments (Analyti et al., 2024).

Technology allows learners to access information on demand, learn at their own pace, collaborate across geographic boundaries, and personalise their learning experiences (Letseka & Seeletso, 2021; Papadakis, 2023). Openness "brings flexibility which helps students to learn what they want, how they want to learn and where they want to learn it" (Letseka & Seeletso, 2021, p. 114). Modern adult learning principles increasingly emphasise flexibility, accessibility, digital literacy, and learner control over the time, place, and pace of learning (Gil-Jaurena, 2020; Panda, 2023; Papadakis, 2023).

Digital technologies have significantly expanded opportunities for adult learning while also introducing new challenges and expectations for learners. Contemporary understandings of digital literacy extend beyond the technical ability to use digital tools and resources, recognising that digital competence involves social, cultural, critical, and communicative dimensions that develop over time (Gil-Jaurena, 2020). Adult learners therefore need to be understood within broader learning environments that increasingly emphasise flexibility, autonomy, and continuous personal and professional development (Panda, 2023). As educational systems become more flexible and responsive to diverse learner needs, e-learning and blended learning approaches have become increasingly important for improving access to education, particularly for adult learners and those whose work, family, or geographic circumstances limit participation in traditional classroom-based programs (Papadimitriou, 2015).

The growth of Massive Open Online Courses (MOOCs), micro-credentials, digital badges, and online professional development programs illustrates the growing role of technology in supporting contemporary adult learning (Papadakis, 2023; Yousef & Sumner, 2020; Zuhaırı et al., 2020). These learning opportunities provide flexible pathways for adults to acquire new knowledge and skills, update professional competencies, and engage in lifelong learning. MOOCs, in particular, have emerged as an important component of the modern learning landscape by providing accessible educational opportunities to learners across diverse locations and circumstances (Yousef & Sumner, 2020). Their ability to deliver learning at scale has enabled educational institutions, professional bodies, and training providers to reach large numbers of learners without many of the traditional constraints associated with face-to-face education (Papadakis, 2023).

More broadly, developments in open educational resources (OER), open educational practices (OEP), and online learning platforms have expanded opportunities for lifelong learning and are continuing to reshape the provision of distance and online education (Zuhaırı et al., 2020). These technologies support learner autonomy by allowing individuals to access learning resources at times, locations, and paces that suit their personal and professional circumstances. They also provide new opportunities for professional development, workforce upskilling, career transitions, and continuing education. Increasingly, higher education institutions, professional organisations, and training providers are incorporating online courses, open educational resources, and micro-credential programs into their educational offerings to support ongoing learning and enhance access to education throughout the lifespan (Papadakis, 2023; Yousef & Sumner, 2020). As a result, technology-enabled learning has become a significant component of contemporary adult education, reinforcing the principles of flexibility, accessibility, lifelong learning, and learner autonomy.

Competency and Capability Development

Contemporary adult learning increasingly distinguishes between competency and capability. Traditional vocational education has often focused on developing specific competencies that can be demonstrated against established standards (Moodie, 2011). While competency remains important, modern approaches also emphasise broader capabilities such as critical thinking, adaptability, creativity, collaboration, communication, and lifelong learning (Choi-Lundberg, 2024; Moodie, 2011).

This shift reflects recognition that modern workplaces require individuals not only to perform current tasks effectively but also to adapt to future changes and solve novel problems (Grabarski et al., 2023; Harreveld & Singh, 2009). Consequently, adult learning programs increasingly focus on developing both technical competence and transferable capabilities (Choi-Lundberg, 2024; Moodie, 2011; Winberg & Hollis-Turner, 2021).

Diversity, Inclusion, and Cultural Context

Modern interpretations of adult learning place greater emphasis on diversity, inclusion, and cultural responsiveness than early andragogical models (Gallo et al., 2022; Housel, 2019; Lūka, 2019). Contemporary scholars recognise that learning is influenced by culture, language, identity, socioeconomic circumstances, organisational environments, and social relationships (Gravani et al., 2019; Hunter-Johnson, 2021; Lūka, 2019).

Rather than assuming a universal adult learner, educators increasingly acknowledge that adult learners bring diverse experiences, values, expectations, and learning preferences (Housel, 2019; Lūka, 2019; Weissova et al., 2024). This recognition has encouraged more inclusive and culturally responsive approaches to adult education that accommodate different ways of knowing, learning, and participating (Gallo et al., 2022; Gravani et al., 2019; Housel, 2019; Medina & Effrosyni, 2024).

Contemporary adult learning theory places considerable emphasis on diversity, inclusion, and cultural responsiveness. Increasingly, educators recognise that effective learning environments must acknowledge and value the cultural backgrounds, experiences, perspectives, and identities that learners bring to the educational setting. As a result, culturally responsive pedagogy (CRP) has been integrated with andragogical principles to support more inclusive teaching practices and to enhance educators' ability to differentiate instruction according to the needs of diverse learner populations (Housel, 2019). Rather than applying a uniform approach to all learners, culturally responsive adult education seeks to draw upon learners' cultural knowledge and experiences as resources that enrich learning and improve educational outcomes.

Research in nursing education has similarly highlighted the importance of inclusive andragogical practices that consider factors such as cultural background, language proficiency, academic preparation, learning preferences, and prior educational experiences when designing learning activities and support strategies (Gallo et al., 2022). Recognising these differences allows educators to create learning environments that are more accessible, equitable, and responsive to the needs of diverse learners.

The importance of culture in adult learning is also evident in language education. Research has demonstrated that effective adult language learning programs must take account of learners' social and cultural contexts, prior experiences, motivations, goals, learning strategies, and individual circumstances (Lūka, 2019). Learning is increasingly understood as a process that is influenced by cognitive, emotional, social, and cultural factors, rather than solely by individual learner characteristics. Educational experiences therefore occur within broader social and cultural environments that shape how learners interpret information, engage with learning activities, and construct meaning (Lūka, 2019)

These considerations are particularly significant in the context of migration, refugee education, and multicultural learning environments. Intercultural approaches to adult education seek to promote equity, inclusion, participation, and empowerment while recognising the linguistic, cultural, and social diversity of learners (Gravani et al., 2019). Such approaches often involve adapting curricula, learning activities, assessment methods, and teaching practices to better support learners from diverse cultural and educational backgrounds. Research has shown that educators working with migrant, refugee, and culturally diverse learners benefit from developing intercultural competence, which supports the creation of inclusive and supportive learning environments that facilitate learner engagement and success (Medina & Effrosyni, 2024).

Similarly, the increasing internationalisation of education has highlighted the need for teaching strategies that accommodate culturally diverse student populations. International learners frequently encounter unfamiliar educational expectations, communication styles, and classroom practices. Consequently, educators are often required to adapt their teaching approaches to create learning environments that support participation and success across diverse cultural groups (Hunter-Johnson, 2021).

These developments have also contributed to critiques of early andragogical models. Some scholars have argued that traditional interpretations of andragogy placed excessive emphasis on individual autonomy and self-direction while giving insufficient attention to social, cultural, political, and contextual influences on learning (Henschke, 2010b). Critics suggest that adult learning cannot be fully understood without considering issues such as culture, identity, power relationships, social structures, and opportunities for transformation. Furthermore, attempts to categorise learners according to broad cultural or national characteristics have been criticised for oversimplifying the diversity that exists within learner populations, particularly among adults whose experiences, identities, and backgrounds are often highly varied (Gazi, 2014).

Contemporary adult education increasingly recognises diversity as a valuable educational resource. Educators working in internationalised and multicultural learning environments acknowledge that learners bring a wide range of professional, cultural, and life experiences that can contribute to richer learning experiences for all participants (Weissova et al., 2024). Collaborative and intercultural learning initiatives have further demonstrated the potential of adult education to develop intercultural competence, cultural humility, and the skills required for effective interaction across cultural boundaries (Anderson & Or, 2023). These developments reflect a broader shift towards understanding adult learning as a socially situated, culturally influenced, and collaborative process that extends beyond the individualistic assumptions associated with some earlier interpretations of andragogy.

Integration of Multiple Learning Theories

Perhaps the most significant contemporary interpretation is the recognition that no single theory fully explains adult learning. Modern adult education increasingly draws upon multiple theoretical perspectives, including andragogy, humanism, constructivism, experiential learning, transformative learning, social learning theory, self-directed learning, and heutagogy. As Carré (2015) notes, the theories of learning developed over the twentieth century have constructed, through progressive sedimentation, a body of knowledge on how adults learn, with each theory emphasizing a specific dimension of the process, whether the importance of reinforcements (behaviourism), the helping relationship (humanism), the role of activity (constructivism), social interactions (socioconstructivism), or agency (Carré, 2015).

Contemporary perspectives on adult learning increasingly recognise a progression from pedagogy, through andragogy, to heutagogy, reflecting expanding levels of learner autonomy and responsibility. Within this continuum, learning approaches move from teacher-directed models, where educators determine learning goals and processes, towards self-directed and ultimately self-determined forms of learning in which learners assume increasing control over what, how, and why they learn (Blaschke, 2021; Saint-Luc, 2018). This progression highlights the growing importance of learner agency, self-direction, and learner-centred educational practices in contemporary adult education. As learners develop greater confidence, experience, and capability, they become increasingly able to manage their own learning, identify learning needs, establish goals, and evaluate outcomes, supporting lifelong learning and adaptability in rapidly changing personal and professional environments (Blaschke, 2021; Ishaq et al., 2024; Saint-Luc, 2018). The multiplicity of relevant theoretical frameworks is further evidenced by scholarship examining self-directed learning (Carré, 2015), transformative learning, and social constructivism (Saint-Luc, 2018), as distinct yet complementary lenses for understanding adult learning processes.

Rather than treating these theories as competing alternatives, contemporary educators often integrate principles from multiple perspectives to design learning experiences that address the diverse needs of adult learners. Carré observes that theories of learning are multiple both in general psychology and in the domain of adult education, yet give rise to few integrative syntheses (Carré, 2015). Saint-Luc demonstrates this integrative approach in practice, drawing simultaneously upon constructivism (Piaget), socioconstructivism (Vygotsky), social learning (Bandura), and transformative learning to develop cooperative training methodologies connecting education, training, and research. Bernardin similarly illustrates how andragogical principles can be combined with constructivist approaches and Mezirow's transformative learning in e-learning design for adults. This integrative approach reflects the growing understanding that adult learning is a complex, multidimensional process influenced by cognitive, emotional, social, cultural, and environmental factors. As Saint-Luc (2018) emphasizes, the affective dimension is critically important in cooperative learning frameworks, where feelings of connectedness and the psychological security provided by the cooperative group enable expression, communication, and productive confrontation through mediation and the development of intersubjectivity. Alhadeff-Jones further underscores this complexity by demonstrating that transformative learning involves temporal dimensions, both continuous and discontinuous processes, that must be understood rhythmically across the lifespan.

Significance for Adult Education and VET

Contemporary interpretations of adult learning have had a profound influence on vocational education and training (VET), higher education, workplace learning, and professional development (Burlakova et al., 2019; Burns et al., 2020; Harden et al., 2018). Learning is increasingly designed around authentic workplace tasks, learner participation, collaboration, reflection, digital technologies, and ongoing professional development (Cocquyt et al., 2018; Harden et al., 2018; Rodrigo et al., 2024). The integration of social-constructivist and andragogical principles has led to learning environments characterised by interactive learning, collaborative learning, authentic learning, and learner-centred approaches (Cocquyt et al., 2018). Digital technologies, including blended and online learning, have been widely adopted in adult education to support these goals (Burns et al., 2020; Cocquyt et al., 2018; Rodrigo et al., 2024). Educators are encouraged to act as facilitators, coaches, mentors, and learning partners rather than simply transmitters of information (Griffiths, 2015; Hoffert et al., 2021; Wáng & Millward, 2014). Coaching, for instance, has been shown to align well with adult and lifelong learning theory, moving beyond conventional transmission models toward active, collaborative, and engaging learning (Griffiths, 2015). In medical education, programs have been developed that explicitly train physicians in facilitation, mentoring, and coaching rather than didactic instruction (Hoffert et al., 2021). Similarly, in school leadership preparation, faculty are encouraged to adopt roles as mentors and coaches supporting candidates' self-directed growth (Cunningham et al., 2018).

While the principles of relevance, experience, self-direction, and practical application remain central (Cunningham et al., 2018; Rodrigo et al., 2024), contemporary adult learning theory recognises that effective learning also depends upon social interaction, critical reflection, cultural context, technological access, and organisational support (Burns et al., 2020; Cocquyt et al., 2018; Erol & Altunay, 2023; Harden et al., 2018). Research has demonstrated that the social nature of face-to-face tutorials, peer learning, workshops, and in-class support has the most positive impact on professional development (Harden et al., 2018). The quality of online interaction has been found to be positively related to adults' social capital in blended learning environments (Cocquyt et al., 2018). Cultural and organisational context matters significantly; challenges such as resource constraints, resistance to change, and inadequate institutional support can hinder effective implementation of adult learning programs (Burns et al., 2020; Erol & Altunay, 2023). Self-directed learning, while central to adult education theory, requires literacy, language proficiency, ICT skills, and a supportive mentor to guide learners through the process (Orn, 2023). As a result, modern adult education is characterised by a more flexible, inclusive, and holistic understanding of how adults learn throughout their lives (Burlakova et al., 2019; Cocquyt et al., 2018; Dosunmu & Dichaba, 2020). Lifelong learning has become a conceptual framework through which all structures of education for individuals are organised (Dosunmu & Dichaba, 2020), encompassing formal, non-formal, and informal learning opportunities (Burlakova et al., 2019; Pīgozne et al., 2019), and recognising that continuous personal and professional growth is vital across the life course (Harden et al., 2018; Orn, 2023).

A useful way to illustrate the progression from pedagogy to andragogy and ultimately to heutagogy is through the professional development of a railway infrastructure engineer who must learn to use a newly implemented digital track monitoring system. A rail network operator introduces an advanced asset monitoring platform that combines sensors, data analytics, and artificial intelligence to identify track defects and predict maintenance requirements. Engineers and maintenance personnel must develop the knowledge and skills necessary to use the system effectively, but the learning experience can differ significantly depending on the educational approach adopted.

Under a pedagogical approach, the training is primarily instructor-directed. The organisation develops a structured training program with predetermined learning objectives, content, learning activities, and assessments. Participants attend formal workshops where an instructor explains the system's functions, demonstrates software features, and provides step-by-step guidance on how to use the platform. Learners complete prescribed exercises and assessments designed by the trainer, with the educator determining what knowledge must be acquired, how learning will occur, and how competence will be measured. The learner's role is largely to absorb information, follow instructions, and demonstrate proficiency against established criteria.

An andragogical approach recognises that the engineer is an experienced professional who already possesses substantial knowledge of rail infrastructure, maintenance planning, and asset management. Rather than focusing solely on software functions, the learning process is centred on practical workplace problems and operational challenges. Discussions may explore how predictive maintenance can reduce track failures, how sensor data can improve maintenance scheduling, or how the new system can enhance rail safety and asset reliability. The educator acts as a facilitator rather than a lecturer, encouraging participants to draw upon their previous experience, share workplace examples, discuss implementation challenges, and collaboratively solve realistic maintenance scenarios. Although the organisation may still establish the overall learning outcomes, learners have greater influence over how they engage with the content and how the new knowledge is applied within their particular work environments.

A heutagogical approach extends learner autonomy even further by placing primary responsibility for learning with the engineer. After developing a foundational understanding of the monitoring system, the engineer identifies a personal learning objective related to their professional interests and workplace needs. For example, the engineer may decide to investigate how machine learning algorithms within the platform can be used to improve rail defect prediction on regional freight corridors. To achieve this goal, the engineer independently identifies learning needs, selects appropriate learning resources, engages with professional networks and industry forums, consults software vendors and subject matter experts, accesses technical documentation and online learning materials, and conducts workplace investigations using operational data.

Assessment within a heutagogical framework may also differ substantially from traditional approaches. Rather than completing a standardised examination or prescribed practical exercise, the engineer may negotiate a project-based assessment with their manager, such as developing and trialling a predictive maintenance model for a section of railway track. Success is evaluated through demonstrated workplace outcomes, problem-solving capability, and the learner's ability to apply knowledge in a meaningful operational context. In this situation, the learner determines not only how learning will occur, but also what will be learned, why it is important, which resources will be used, and how achievement will be demonstrated.

This example highlights the increasing level of learner autonomy across the pedagogy–andragogy–heutagogy continuum. In pedagogy, the organisation and educator exercise primary control over the learning process. In andragogy, learning becomes more learner-centred, drawing upon the learner's experience and focusing on practical application and problem-solving. In heutagogy, the learner assumes responsibility for identifying learning needs, designing learning pathways, selecting learning resources, and evaluating outcomes. The progression reflects contemporary adult learning theory's growing emphasis on learner agency, adaptability, self-direction, capability development, and lifelong learning in rapidly changing professional environments.

Key Concepts and Conclusions

Andragogy is the theory and practice of adult learning and is based on the understanding that adults differ from children in their learning needs, motivations, experiences, and expectations. Adult learners are generally self-directed, bring substantial life and work experience to learning situations, and prefer learning that is relevant, practical, and immediately applicable to their personal or professional circumstances. The concept has become one of the most influential frameworks in adult education and has shaped approaches to vocational education and training, workplace learning, higher education, and professional development.

The modern understanding of andragogy is closely associated with Malcolm Knowles, whose work popularised and refined adult learning theory during the twentieth century. Knowles helped shift educational thinking away from teacher-centred instruction and towards learner-centred education. In this approach, educators act as facilitators, mentors, coaches, and guides who support learning rather than simply transmitting information. His work emphasised the importance of learner autonomy, experience, relevance, and practical application in adult education.

Central to Knowles's theory are six assumptions about adult learners. Adults need to understand why learning is important before they are willing to invest effort in it. They generally view themselves as autonomous and self-directing individuals who prefer to exercise control over their learning. Adults bring a wealth of personal and professional experiences that influence how they interpret and apply new knowledge. Their readiness to learn is often linked to new roles, responsibilities, or challenges that require additional knowledge or skills. Adults tend to adopt a problem-centred orientation to learning, preferring activities that address real-world situations and can be applied immediately. Finally, adults are frequently motivated by internal factors such as personal growth, achievement, professional satisfaction, and self-improvement rather than by external rewards alone.

Self-directed learning represents an important extension of these principles. In self-directed learning, individuals assume responsibility for identifying learning needs, establishing goals, locating learning resources, selecting appropriate learning strategies, and evaluating outcomes. This process reflects the adult learner's desire for autonomy and control and has become increasingly important in contemporary learning environments where learners are expected to manage their own professional and personal development.

Learner readiness and motivation are also fundamental concepts in adult learning. Adults are generally most receptive to learning when they perceive a clear need for new knowledge or skills and can see a direct connection between learning activities and their personal or professional goals. Readiness and motivation are closely linked and significantly influence learner engagement, persistence, and achievement. Educational programs that are relevant, practical, and aligned with learner goals are therefore more likely to be effective.

Experience is one of the most valuable resources available in adult education. Adult learners interpret new information through the lens of their previous experiences and use those experiences to construct meaning, solve problems, and evaluate new ideas. Effective adult education not only draws upon learners' experiences but also encourages critical reflection on those experiences. Closely related to this concept is experiential learning, which emphasises learning through experience, reflection, conceptualisation, and application. Authentic activities, practical exercises, simulations, and workplace projects help learners connect theory with practice and facilitate meaningful learning.

Adult learning is also characterised by a preference for problem-centred learning. Rather than focusing solely on abstract subject matter, adults generally prefer learning that addresses real-world problems and workplace challenges. Consequently, adult education often incorporates case studies, simulations, workplace projects, scenario-based learning, and problem-solving activities that mirror the situations learners encounter in their professional and personal lives.

Contemporary adult learning theory further recognises the importance of social and collaborative learning. Learning is not solely an individual process but is often developed through interaction with others. Communities of practice, mentoring relationships, coaching, peer learning, professional networks, and collaborative projects all contribute to the development of knowledge, skills, professional identity, and workplace competence. These approaches acknowledge that learning is frequently embedded within social and professional contexts.

Another important development in adult learning theory is transformative learning. This perspective suggests that adult learning can involve more than simply acquiring new knowledge and skills. Through critical reflection, adults may challenge existing assumptions, reconsider long-held beliefs, and develop new perspectives about themselves, their professions, and the world around them. Transformative learning emphasises personal growth, perspective transformation, and critical self-awareness.

Technological developments have significantly influenced contemporary adult learning. Online learning platforms, learning management systems, MOOCs, micro-credentials, digital badges, artificial intelligence, and open educational resources have expanded access to flexible and accessible learning opportunities. These technologies support learner autonomy, self-directed learning, and lifelong learning by enabling learners to access education at times, locations, and paces that suit their individual circumstances.

Lifelong learning has become a central concept within contemporary adult education. Learning is increasingly viewed as a continuous process that extends throughout life rather than being confined to formal schooling. Rapid technological, economic, and social change requires individuals to continually update their knowledge and skills through formal, non-formal, and informal learning opportunities. Lifelong learning supports adaptability, employability, personal development, and active participation in society.

Contemporary theory has also extended beyond self-directed learning to self-determined learning, or heutagogy. In heutagogical approaches, learners exercise greater control not only over how they learn but also over what they learn, why they learn it, and how learning is assessed. This approach reflects increasing recognition of learner agency and the need to develop adaptable, capable individuals who can independently manage their learning throughout life.

Modern adult learning theory also places considerable emphasis on diversity, inclusion, and cultural responsiveness. Learning is influenced by culture, language, identity, socioeconomic circumstances, prior experiences, and social relationships. Effective adult education therefore requires flexible, inclusive, and culturally responsive approaches that acknowledge and value learner diversity while creating equitable opportunities for participation and success.

Finally, contemporary adult education increasingly integrates multiple theoretical perspectives. Rather than relying on a single explanation of learning, educators draw upon andragogy, humanism, constructivism, experiential learning, transformative learning, social learning theory, self-directed learning, and heutagogy to create rich and effective learning environments. These theories are viewed as complementary perspectives that collectively contribute to a more comprehensive understanding of how adults learn.

Andragogy remains one of the most influential frameworks for understanding adult learning and continues to shape educational practice across vocational education and training, workplace learning, higher education, professional development, and community education. Its emphasis on learner autonomy, experience, relevance, readiness, practical application, and intrinsic motivation provides a valuable foundation for designing effective learning experiences for adults.

The chapter demonstrates that adult learners are not passive recipients of information but active participants in the learning process. Adults learn most effectively when they understand the purpose of learning, can draw upon their existing experiences, participate in meaningful and relevant activities, and apply learning to authentic personal and professional situations. Learning is therefore most successful when it recognises and builds upon the knowledge, experiences, motivations, and goals that learners bring to the educational environment.

While Knowles's theory remains highly influential, contemporary understandings of adult learning extend beyond its original formulation. Modern perspectives recognise that learning is shaped not only by individual characteristics but also by social, cultural, technological, organisational, and environmental influences. Adult learning is increasingly understood as a complex and multidimensional process involving cognitive, emotional, social, and contextual factors.

The growing importance of lifelong learning, self-directed learning, workplace learning, professional development, and digital technologies has further expanded the relevance of adult learning theory. Individuals are increasingly expected to adapt to changing technologies, occupations, industries, and social conditions by continually acquiring new knowledge and skills throughout their lives.

The chapter also highlights the importance of collaboration, communities of practice, critical reflection, transformative learning, cultural responsiveness, and learner-centred educational approaches. These developments demonstrate that adult learning involves more than the acquisition of knowledge and skills; it also supports the development of professional identity, critical thinking, adaptability, self-awareness, and the capacity for continuous personal and professional growth.

Ultimately, no single theory fully explains adult learning. Contemporary adult education draws upon multiple complementary theoretical perspectives to create learning environments that are flexible, inclusive, relevant, and effective. The enduring significance of andragogy lies in its recognition that adult learners bring valuable experiences, unique motivations, and diverse needs to the learning process, and that education is most effective when it respects, engages, and builds upon these characteristics.

Chapter 4

Transformative Learning and Critical Reflection

Learning is often associated with acquiring new knowledge, developing skills, or improving performance. However, some learning experiences produce changes that extend far beyond the accumulation of information. Adults may encounter situations that challenge long-held assumptions, reshape their understanding of themselves and others, alter their professional identities, and fundamentally transform the ways they interpret the world. These deeper forms of learning are the focus of transformative learning theory, one of the most influential perspectives in adult education.

Transformative learning explores how adults make meaning from experience and how that meaning can change over time. Rather than viewing learning as simply the transfer of knowledge, transformative learning emphasises the processes through which individuals critically examine assumptions, question established beliefs, and develop more inclusive and adaptable perspectives. Such changes often occur in response to significant life experiences, professional challenges, educational opportunities, or encounters with alternative viewpoints that disrupt existing ways of thinking and create opportunities for growth.

Central to this chapter is the work of Jack Mezirow, whose theory of perspective transformation has shaped contemporary understandings of adult learning for several decades. Mezirow proposed that adults develop frames of reference through their experiences and that meaningful learning occurs when these frames of reference are critically examined and revised. Through critical reflection, dialogue, and engagement with challenging experiences, learners can transform their assumptions and develop new ways of understanding themselves, their professions, and the societies in which they live.

The chapter also examines the broader processes that support transformative learning, including reflective practice, meaning-making, identity development, learning through life transitions, critical consciousness, empowerment, dialogue, and discourse. These concepts highlight that learning is not solely an individual cognitive activity but is also shaped by social relationships, cultural contexts, power structures, and participation in communities. As learners engage with new experiences and perspectives, they continuously construct and reconstruct meanings that influence their identities, decisions, and actions.

Understanding transformative learning is increasingly important in contemporary society. Rapid technological change, evolving workplace expectations, social complexity, global interconnectedness, and ongoing personal and professional transitions require individuals to adapt continually throughout their lives. The capacity to critically examine assumptions, embrace new perspectives, and respond thoughtfully to changing circumstances has become an essential capability for lifelong learning, professional development, and active citizenship.

This chapter explores the theoretical foundations of transformative learning and critical reflection, examines the major contributions of Mezirow and other scholars, and considers how transformative learning operates across educational, professional, and community contexts. By understanding how adults construct meaning, develop identities, and transform perspectives, educators and learners can better appreciate the profound ways in which learning contributes to personal growth, professional development, social participation, and lifelong adaptation in a rapidly changing world.

Foundations of Transformative Learning

Transformative Learning Theory is one of the most influential perspectives in adult education and was developed primarily by Jack Mezirow during the late twentieth century (Schalkwyk et al., 2019; Vipler, 2021; Yunga, 2019). The theory focuses on how adults change the way they understand themselves, others, and the world around them. Unlike learning theories that emphasise the acquisition of knowledge and skills, transformative learning is concerned with fundamental changes in perspective, assumptions, beliefs, values, and ways of thinking (Kuechler & Stedham, 2017; Vipler, 2021). It proposes that learning can lead not only to increased knowledge but also to profound shifts in how individuals interpret their experiences and make meaning of their lives (Bennett et al., 2022).

The foundations of transformative learning are grounded in the idea that individuals develop frames of reference through their life experiences (K. Jones, 2019; Yunga, 2019). These frames of reference consist of assumptions, beliefs, expectations, values, cultural norms, and habits of mind that influence how people perceive and interpret information (Khuzwayo, 2022; Yunga, 2019). Over time, these perspectives become deeply embedded and often operate unconsciously, shaping decisions, behaviours, and understandings of the world (Wang et al., 2020; Yunga, 2019). While these frameworks help individuals make sense of experience, they can also limit understanding by reinforcing unexamined assumptions and biases (Khuzwayo, 2022; Samin, 2026; Schalkwyk et al., 2019).

A central concept within transformative learning is critical reflection. Mezirow argued that adults can learn to critically examine the assumptions underlying their beliefs and actions (Mbokota et al., 2022; Nyamunda & Westhuizen, 2020). Critical reflection involves questioning why certain beliefs are held, evaluating the evidence that supports them, and considering alternative viewpoints (Khokhotva & Albizuri, 2019; Nyamunda & Westhuizen, 2020). Through this process, learners become aware of assumptions that may previously have been accepted without question (Briciu et al., 2024; Hill et al., 2023). Critical reflection is central to transformative learning, and without it, transformative learning is unlikely to occur (Nyamunda & Westhuizen, 2020). According to Mezirow, there are three layers of critical reflection: reflecting on content (what people perceive, feel, think and act on), reflecting on process (how people perform the function of perceiving), and reflecting on premise (why people do what they do) (Mbokota et al., 2022). Critical reflection encourages individuals to move beyond habitual ways of thinking and develop more inclusive, open, and flexible perspectives (Horton et al., 2023; Marotta, 2024).

Transformative learning is often triggered by what Mezirow described as a disorienting dilemma (DeAngelis, 2021; Lombaerts et al., 2019; Mbokota et al., 2022). A disorienting dilemma is an experience, event, or situation that challenges existing assumptions and creates a sense of uncertainty or discomfort (DeAngelis, 2021; Mälkki & Green, 2018). A disorienting dilemma is considered a prerequisite for transformation because, when it occurs, people are more likely to critically reflect on their frame of mind (Mbokota et al., 2022). Such experiences may include career changes, organisational restructuring, technological disruption, migration, illness, personal loss (Shaik, 2022), exposure to different cultures, or participation in educational programs that challenge established beliefs (Póvoa et al., 2025; Wangding et al., 2024). A disorienting dilemma can be a slow process or sudden, triggered by a crisis (Nyamunda & Westhuizen, 2020). When individuals encounter situations that cannot easily be explained using their existing frames of reference, they may be prompted to critically examine their assumptions and reconsider their perspectives (Cox & John, 2016; Mbokota et al., 2022).

Dialogue and discourse are also fundamental to transformative learning. Mezirow believed that individuals refine and test their new understandings through discussion with others (Mbokota et al., 2022; Shaik, 2022). Dialogue becomes "the medium for critical reflection to be put into action, where experience is reflected on, assumptions and beliefs are questioned and habits of minds are ultimately transformed" (Marotta, 2024). Engaging in open, respectful, and rational dialogue allows learners to compare perspectives, evaluate evidence, challenge assumptions, and explore alternative interpretations (Horton et al., 2023; Uthman, 2023). According to Mezirow, dialogue enables learners to engage in reflective discourse, critically examine their existing beliefs, and construct new meanings through open and respectful conversations with others (Uthman, 2023). Through interaction with others, learners develop greater insight into their own thinking and become more capable of adopting perspectives that are broader and more inclusive (Pittman & Gioia, 2020; Shaik, 2022).

The theory is influenced by several intellectual traditions. Humanistic psychology contributed ideas about personal growth, self-actualisation, and individual development. Constructivism provided the view that individuals actively construct meaning from their experiences rather than passively receiving knowledge. Critical social theory, particularly the work of Jürgen Habermas, influenced Mezirow's emphasis on critical reflection, questioning assumptions, and emancipatory learning (Guo & Sheffield, 2006; Ibsen, 2023). Habermas understood the function of learning as providing individuals with the skills necessary to reflect critically on the ways in which ideology shapes interpretation of reality, and to thus bring about a process of change. Through emancipation, individuals become aware of how ideologies create and contribute to dependency on reified powers. Habermas identified three knowledge-constitutive interests, technical, practical, and emancipatory, which frame different orientations to knowledge and action (Guo & Sheffield, 2006; Thapaliya & Luitel, 2024). The emancipatory interest is grounded in achieving freedom from dogma through self-reflection and gives rise to critically reflective inquiry. Habermas's emancipation is also connected with notions of social justice, equality, autonomy, and responsibility (Thapaliya & Luitel, 2024). These influences collectively shaped a theory that views learning as a process of meaning-making, reflection, and personal transformation,

Transformative learning differs from many traditional educational approaches because its primary goal is not simply the mastery of content. Instead, it seeks to develop learners who are more reflective, autonomous, and capable of critically examining their assumptions. Mezirow's theory draws on constructivist epistemology for critical reflection, the way adult learners make sense of their experiences, and on constructionist ontology for critical discourse, the process of social interaction that influences how a person construes experiences. Perspective transformation involves the learner engaging in critical self-reflection of their own assumptions and the ways in which the learner's views of the world may be distorted or limited. Learning is viewed as a process through which individuals become increasingly aware of how their beliefs and perspectives have been shaped and how those perspectives can be revised in light of new experiences and evidence. As Mezirow's theory articulates, transformative learning is a meta-cognitive process that connects critical reflection, dialectical discourse, reason, discernment, insight, context, relevance, action, and outcomes in service of transformation (Nelson, 2012). This type of learning may occur either in group interaction or independently due to critical reflection on assumptions.

In adult education, transformative learning is frequently associated with activities that encourage reflection and critical inquiry. Reflective journals, case studies, workplace projects, discussions, simulations, mentoring relationships, and experiential learning activities can all support transformative learning by encouraging learners to examine their experiences and consider alternative viewpoints. Critical reflection is considered a key enabler for emancipatory learning, involving increased awareness and questioning of the assumptions under which we and others think and act, and being ready to act differently. Coaching and mentoring approaches that foster critical consciousness are capable of helping individuals think differently about themselves and their society. These approaches are particularly valuable in professional education, leadership development, healthcare, community services, and vocational education, where learners often encounter complex situations that require them to question assumptions and adapt their thinking. In healthcare contexts, critical reflective inquiry has been proposed as a method encompassing descriptive, reflective, and critical/emancipatory phases oriented toward understanding practice, correcting it through self-reflection, and generating models of good practice. In legal education, transformative pedagogies in work-integrated learning can foster personal and social transformation in addition to enhancing professional skills.

The foundation of transformative learning lies in the belief that adults are capable of fundamentally changing how they understand themselves and their world. Through critical reflection, dialogue, experience, and the examination of assumptions, learners can move beyond existing perspectives and develop more informed, inclusive, and adaptable ways of thinking. The emancipatory interest and transformative learning theories promote a vision of teaching and learning as a socio-political act that goes beyond conventional settings. Experience cannot be disconnected from social contexts, and transformative learning can be reinterpreted as the transformation of experience with both personal and social dimensions (Lofgreen, 2023). Transformative learning therefore represents one of the deepest forms of adult learning, involving not only the acquisition of knowledge but also the transformation of meaning, identity, and perspective.

Jack Mezirow and Perspective Transformation

Jack Mezirow (1923–2014) was one of the most influential scholars in adult education and the founder of Transformative Learning Theory (Fleming, 2022a). His work fundamentally changed how educators understand adult learning by proposing that learning involves more than acquiring knowledge and skills (Helskog, 2014; Tatarinceva & Gode, 2016). Mezirow (1991) argued that some learning experiences lead individuals to critically examine and transform the assumptions, beliefs, values, and perspectives that shape how they understand themselves and the world. This process, which he termed perspective transformation, became the central concept of Transformative Learning Theory.

Mezirow developed the theory following research into women returning to higher education in the 1970s (Biasin, 2018b; Brookfield, 2018; Hoggan & Higgins, 2023; Kokkos, 2022; McClain, 2024). He observed that participation in education often led learners to reassess long-held beliefs about themselves, their roles, and their capabilities (McClain, 2024; Παπαθανασίου, 2023). Rather than simply gaining new information, many learners experienced profound changes in their self-understanding and worldview (Bezard & Shaw, 2017; Sifakis, 2007). These observations led Mezirow to conclude that adult learning can result in a transformation of meaning structures rather than merely an expansion of knowledge (Biasin, 2018b; Hoggan & Higgins, 2023; Kokkos, 2022).

Meaning Perspectives and Frames of Reference

At the heart of Mezirow's theory is the concept of frames of reference. These are the assumptions, expectations, values, beliefs, and habits of mind that individuals develop through their life experiences (Fischbach & Guerrero, 2020; Şahin et al., 2016). Frames of reference influence how people interpret events, make decisions, solve problems, and interact with others (Ananmalay et al., 2022; Parker & Myrick, 2010). They are shaped by family, culture, education, religion, social institutions, and personal experiences (Bergman & Hessel, 2023; Jansen, 2023; Resch, 2022).

Because these assumptions often operate unconsciously (Ananmalay et al., 2022; Stapleton, 2020), individuals may not recognise how strongly they influence their thinking. Mezirow argued that many adults accept their frames of reference without question, assuming that their perspectives accurately reflect reality (Giambo & Garrido, 2024; Şahin et al., 2016). However, these perspectives can sometimes be incomplete, distorted, or no longer appropriate for changing circumstances (Akinsooto et al., 2020; Giambo & Garrido, 2024; Roux, 2024).

Perspective Transformation

Perspective transformation occurs when individuals critically examine their existing assumptions and develop new ways of understanding themselves and their experiences. It involves a fundamental shift in meaning perspectives, resulting in more inclusive, open, reflective, and adaptable ways of thinking (Coryell, 2013; Eltayb & Valcke, 2024). Transformative learning theory describes this process as one that "transforms frames of reference, such as fixed assumptions and expectations, to make them more inclusive, discriminating, open, reflective, and emotionally capable of change" (Eltayb & Valcke, 2024, p. 107).

Rather than simply learning new facts, perspective transformation changes the lens through which individuals view the world. The learner begins to question previously accepted beliefs, considers alternative viewpoints, and develops a more critical understanding of assumptions that may have guided their actions for many years. This process involves "developing autonomous thinking and bringing about a deep, structural shift in the basic premises of thought, feelings, and actions" (Eltayb & Valcke, 2024, p. 107). Critical reflection is central to this process, as it encourages individuals to reflect deeply on their perspectives and assumptions, leading to profound personal and professional growth (Coryell, 2013; Eltayb & Valcke, 2024). Research has confirmed that transformative learning facilitates "perspective transformation through critical reflection, social learning, and experiential knowledge, fostering profound shifts in beliefs and perspectives" (Eltayb & Valcke, 2024, p. 108). Collaborative inquiry frameworks can support transformative learning by providing structured opportunities for learners to explore ideas, challenge assumptions, and engage in meaningful dialogue with others. Such frameworks help learners navigate both the cognitive and emotional dimensions of examining deeply held beliefs and perspectives. By encouraging discussion, reflection, and shared problem-solving, they create supportive environments in which learners can critically evaluate their assumptions and consider alternative viewpoints.

This perspective highlights that transformative learning is not solely an individual process of reflection and personal change. It also occurs through social interaction, where learners engage with diverse experiences, perspectives, and interpretations. Through collaborative inquiry, individuals can test their ideas, receive feedback, and develop new understandings that may not emerge through solitary reflection alone. Consequently, transformative learning is increasingly viewed as both a personal and a social process, shaped by dialogue, relationships, and participation in learning communities (Coryell, 2013).

For example, a manager who has always believed that effective leadership requires strict control and authority may, through education and reflection, come to recognise the value of collaboration, empowerment, and shared decision-making. The change is not simply acquiring new management techniques; it is a transformation in the manager's understanding of leadership itself. Such shifts parallel findings in educational contexts where participants experienced transformative changes in their perspectives, attitudes, values, and beliefs about themselves and their professional practice (Coryell, 2013; Eltayb & Valcke, 2024).

The Disorienting Dilemma

According to Mezirow, perspective transformation often begins with what he termed a disorienting dilemma (Alhadeff-Jones, 2018; Breton, 2018; Florey & Émery--Bruneau, 2023; Kaçar, 2021; Wangding et al., 2024). A disorienting dilemma is an experience or event that challenges an individual's existing assumptions, beliefs, values, or ways of understanding the world (Florey & Émery--Bruneau, 2023; Wangding et al., 2024). When people encounter situations that cannot be easily explained using their current perspectives, they may experience uncertainty, discomfort, or confusion (Biasin, 2018a). This disruption encourages them to question previously accepted assumptions and consider alternative ways of interpreting their experiences (Alhadeff-Jones, 2018; Kaçar, 2021; Sifakis, 2007).

Disorienting dilemmas can arise from a wide range of personal, professional, educational, and social circumstances (Alhadeff-Jones, 2018; DeCapua et al., 2017; Dubouloz, 2014; Müller et al., 2020). Significant career changes, organisational restructuring, technological advancements, exposure to different cultures, major life events, professional development programs, higher education experiences, personal crises, or even unexpected successes can all challenge established patterns of thinking (DeCapua et al., 2017; Kaçar, 2021; Müller et al., 2020). These experiences often require individuals to re-evaluate existing beliefs and adapt to new realities, creating conditions that support transformative learning (Chasserio & Bacha, 2023; Florey & Émery--Bruneau, 2023; Sifakis, 2007).

For example, an experienced engineer may encounter new sustainability requirements that fundamentally challenge traditional approaches to infrastructure design. Practices that were once considered effective and acceptable may no longer align with contemporary environmental expectations, regulatory requirements, or community values. Initially, the engineer may struggle to reconcile these new expectations with long-established professional practices and assumptions. Through reflection, discussion, and further learning, however, the engineer may begin to develop a new understanding of engineering practice that places greater emphasis on sustainability, environmental responsibility, and long-term social outcomes.

Mezirow argued that while disorienting dilemmas can be uncomfortable, they play a crucial role in transformative learning (Alhadeff-Jones, 2018; Dubouloz, 2014; Wangding et al., 2024). By disrupting habitual ways of thinking, they create opportunities for critical reflection and personal growth (Clarke, 2022; DeCapua et al., 2017; Wangding et al., 2024). As individuals examine the assumptions underlying their beliefs and behaviours, they may develop more inclusive, flexible, and informed perspectives (Alhadeff-Jones, 2018; Clarke, 2022; Sifakis, 2007). In this way, disorienting dilemmas serve as catalysts for perspective transformation, enabling learners to move beyond existing frames of reference and develop new ways of understanding themselves and the world around them (Alhadeff-Jones, 2018; Biasin, 2018b; Breton, 2018; Florey & Émery--Bruneau, 2023).

Critical Reflection

Critical reflection is the central mechanism through which transformative learning occurs. Mezirow argued that meaningful transformation requires learners to critically examine the assumptions, beliefs, values, and expectations that shape their understanding of themselves and the world around them (Babacan & Babacan, 2012; Kokkos, 2022; Walters et al., 2016). According to Mezirow, critical reflection involves "a critical review of presuppositions from conscious and unconscious prior learning and their consequence" and results in a change in one's firmly held beliefs (Walters et al., 2016) . Many of these assumptions are developed over time through family influences, education, culture, professional experiences, and social interactions. Frames of reference are defined as the summation of one's intellectual capacity through which all thoughts, experiences, learning, and actions are filtered and processed, based on socialization, past experiences, and prior knowledge (Walters et al., 2016). Because they often become deeply ingrained, individuals may accept them as unquestioned truths rather than recognising them as perspectives that can be examined and challenged (Birdsall, 2022; Closs & Antonello, 2011).

The process of critical reflection involves deliberately questioning the foundations of one's thinking. Mezirow clarified that critical reflection goes beyond the thoughtful processes of judgments, interpretations, analyses, generalizations, and evaluations; it additionally includes deliberately reviewing and questioning the validity of assumptions (Kokkos, 2022). Learners are encouraged to explore why they hold particular beliefs, where those beliefs originated, and whether they remain valid in light of new experiences or evidence. As Mezirow articulated, critical reflection means reflecting upon "why we perceive, think, feel, or act as we do," or upon the causes which gave rise to problematic perspectives in the past (Kokkos, 2022). This process of questioning distinguishes critical reflection from other forms of reflection, as what differentiates critical reflection from any other reflection is that the first implies an evaluation of what is being reflected upon (Closs & Antonello, 2011). Through this process, learners move beyond simply accepting information and begin to evaluate the reasoning that underpins their own viewpoints (Bae, 2018; Woerkom, 2010).

Critical reflection also encourages learners to consider alternative ways of understanding complex issues. By engaging with different perspectives, experiences, and sources of knowledge, individuals become more aware of the limitations of their existing assumptions and more open to new possibilities (Crane & Sosulski, 2020; Holdo, 2022). Transformative learning challenges students to assess their values and perspectives through critical reflection and requires both intentional cognitive and affective action (Nolan & Roberts, 2023). This does not necessarily mean abandoning previous beliefs, but rather developing a more informed, balanced, and flexible understanding of a situation. Such reflection can reveal biases, misconceptions, or outdated ways of thinking that may no longer be appropriate in changing personal, professional, or social contexts (Closs & Antonello, 2011; Hanson, 2013). Indeed, critical reflection that is conceptualised as static or detached has been challenged by scholars who place more attention on critical reflection in praxis through a reflexive position (Hanson, 2013). Furthermore, experiential learning that supports only self-referenced reflection rarely involves an analysis of social and political processes, whereas critical reflection stresses the importance of considering the socially situated nature of experience (Closs & Antonello, 2011).

Through critical reflection, learners become aware of assumptions that were previously taken for granted and begin to assess their validity (Birdsall, 2022; Woerkom, 2010). By critical reflection, people become critically aware of how and why the structure of psychocultural assumptions has come to constrain the way they see themselves and their relationships (Woerkom, 2010). This process of questioning and evaluation creates opportunities for perspective transformation, enabling individuals to develop new insights, adopt more inclusive viewpoints, and construct more sophisticated ways of understanding their experiences (Babacan & Babacan, 2012; Walters et al., 2016; Wylie et al., 2020). "Transformative learning is a process of meaning becoming clarified through expanded awareness, critical reflection, validating discourse and reflective action, moving toward a fuller realization of agency" (Mezirow, 2000, p. 25). In transformative learning, critical reflection therefore serves as the bridge between encountering a challenging experience and achieving meaningful personal and intellectual growth (Holdo, 2022; Nyaradi et al., 2025; Wylie et al., 2020).

Dialogue and Rational Discourse

Mezirow emphasised that dialogue plays a crucial role in transformative learning (Babacan & Babacan, 2012; Mei et al., 2022; Schnepfleitner & Ferreira, 2021). While critical reflection is important, transformation does not occur solely through individual introspection (Jurkova & Guo, 2018; Schnepfleitner & Ferreira, 2021). Learners often develop new understandings by engaging with others, discussing experiences, and exploring different perspectives (Pittman & Gioia, 2020; Uthman, 2023). Through interaction and conversation, individuals have opportunities to examine the assumptions that shape their thinking and to consider viewpoints that may differ from their own (Babacan & Babacan, 2012; Hoggan-Kloubert & Mabrey, 2022).

Open, respectful, and reflective dialogue enables learners to compare experiences, evaluate evidence, and challenge existing beliefs in a supportive environment (Hoggan-Kloubert & Mabrey, 2022; Uthman, 2023). By listening to the perspectives of others, learners may become aware of assumptions they had previously taken for granted and gain insight into alternative ways of interpreting situations (Gunnlaugson, 2007; Jurkova & Guo, 2018). Such discussions encourage critical examination of ideas and help individuals assess the validity and limitations of their own reasoning (Babacan & Babacan, 2012; Hoggan-Kloubert & Mabrey, 2022).

Dialogue also provides a mechanism for refining emerging understandings (Grzegorczyk, 2018; Schnepfleitner & Ferreira, 2021). As learners articulate their thoughts, respond to questions, and engage with differing opinions, they are able to test new ideas and strengthen their interpretations (Grzegorczyk, 2018; Gunnlaugson, 2007). This process of shared inquiry helps learners move beyond narrow or unexamined perspectives and develop more nuanced understandings of complex issues (Helskog, 2014; Hoggan-Kloubert & Mabrey, 2022).

Through meaningful dialogue, transformative learning becomes both an individual and a social process (Jurkova & Guo, 2018). Reflection is enriched by interaction with others, allowing learners to construct perspectives that are more informed, balanced, and inclusive (Kloubert, 2019; Schnepfleitner & Ferreira, 2021). As a result, dialogue serves as an essential component of perspective transformation, supporting the development of greater self-awareness, critical thinking, and openness to alternative viewpoints (Pittman & Gioia, 2020; Uthman, 2023).

The Process of Perspective Transformation

Mezirow described perspective transformation as a developmental process through which individuals gradually revise their existing assumptions, beliefs, and ways of understanding the world (Mello et al., 2023; Schnepfleitner & Ferreira, 2021). Rather than occurring as a single event, transformation typically unfolds over time as learners encounter experiences that challenge their existing perspectives and prompt critical reflection (Gunnlaugson, 2007; Schnepfleitner & Ferreira, 2021). Although individuals may experience the process differently, Mezirow identified a series of stages that commonly characterise transformative learning (Mello et al., 2023; Parzen & Janzen, 2019).

The process often begins with a disorienting dilemma, an event or experience that challenges existing assumptions and creates uncertainty about previously accepted beliefs (Buttigieg & Calleja, 2020a; Mello et al., 2023; Parzen & Janzen, 2019). This is frequently followed by a period of self-examination, during which individuals reflect on their reactions, feelings, and existing perspectives (Mello et al., 2023; Parzen & Janzen, 2019). As learners engage in deeper reflection, they begin a critical assessment of their assumptions, questioning beliefs and values that may previously have been accepted without scrutiny (Mello et al., 2023; Parzen & Janzen, 2019).

As transformation progresses, learners often come to recognise that others have experienced similar challenges and uncertainties (Mello et al., 2023; Parzen & Janzen, 2019). This realisation can reduce feelings of isolation and encourage further exploration of new ways of thinking (Mello et al., 2023; Schnepfleitner & Ferreira, 2021). Learners then begin to explore alternative perspectives, considering different interpretations, viewpoints, and possibilities that may provide more effective ways of understanding their experiences and circumstances (Mello et al., 2023; Parzen & Janzen, 2019).

The next stages involve translating these new insights into action. Learners may plan a new course of action and identify the changes required to support their evolving perspectives (Mello et al., 2023; Parzen & Janzen, 2019). This often involves acquiring new knowledge and skills that enable them to operate effectively within their revised understanding of the world (Mello et al., 2023; Parzen & Janzen, 2019). As confidence develops, individuals begin trying out new roles, behaviours, or ways of thinking, testing their emerging perspectives in real-life situations (Mello et al., 2023; Parzen & Janzen, 2019).

Through continued practice and experience, learners gradually build competence and confidence in their new roles and perspectives (Mello et al., 2023; Parzen & Janzen, 2019). Over time, these changes become integrated into their everyday lives, leading to a process of reintegration in which they function with a transformed frame of reference (Mello et al., 2023; Parzen & Janzen, 2019; Schnepfleitner & Ferreira, 2021). The individual no longer simply possesses new knowledge but has developed a fundamentally different way of interpreting and responding to experiences (Bergman & Hessel, 2023; Kloubert, 2019).

Although not every transformative learning experience follows these stages in a precise sequence, Mezirow's framework provides a useful model for understanding how significant changes in perspective can occur (Mello et al., 2023; Parzen & Janzen, 2019; Schnepfleitner & Ferreira, 2021). It illustrates that transformative learning is often a gradual and reflective process involving critical examination, exploration of alternatives, personal growth, and the development of new ways of thinking and acting (Closs & Antonello, 2011; Mello et al., 2023; Schnepfleitner & Ferreira, 2021).

Examples of Perspective Transformation

Examples of perspective transformation can be found across a wide range of professional and educational contexts. In vocational education and training, a tradesperson undertaking leadership development may initially view supervision primarily as enforcing rules, monitoring compliance, and directing the work of others. Through critical reflection, training, and practical experience, the individual may come to recognise that effective leadership also involves mentoring employees, supporting professional growth, fostering teamwork, and creating environments that encourage high performance and continuous improvement. This shift represents a transformation in how leadership and workplace relationships are understood.

In healthcare settings, transformative learning may occur when a nurse reflects on patient experiences and outcomes and begins to question an exclusively clinical approach to care. Rather than focusing solely on procedures, diagnoses, and treatment protocols, the nurse may develop a broader appreciation of the emotional, social, cultural, and psychological factors that influence patient wellbeing. As a result, patient care becomes more holistic and person-centred, reflecting a transformed understanding of professional practice.

Higher education provides many opportunities for perspective transformation as students encounter new ideas and alternative viewpoints. For example, a business student may begin their studies with the assumption that organisational success is measured primarily through profitability and financial performance. Exposure to concepts such as sustainability, corporate governance, ethics, and social responsibility may challenge this perspective and encourage the student to adopt a more comprehensive understanding of organisational performance that considers environmental, social, and stakeholder outcomes alongside financial results.

Similarly, professionals working in community services may experience transformative learning through reflective practice and engagement with diverse communities. Exposure to different cultural perspectives and lived experiences can prompt practitioners to examine assumptions about disadvantaged or marginalised groups. Through critical reflection and dialogue, they may develop more empathetic, inclusive, and culturally responsive approaches to practice, resulting in improved relationships with clients and more effective service delivery. These examples illustrate how transformative learning can reshape professional perspectives, behaviours, and identities across a variety of fields.

Significance for Adult Education

Mezirow's theory has had a profound influence on adult education because it highlights learning as a process of meaning-making and personal growth rather than simply knowledge acquisition (Helskog, 2014; Kallou & Kikilia, 2021; Qiu, 2023). Mezirow's transformative learning theory suggests that "learning is not merely the acquisition of knowledge and skills but a process of critically reflecting on and questioning one's assumptions, beliefs, and values" (Qiu, 2023, p. 52). The theory is widely recognized as "the most integrated and investigated theory in the field of adult education during the last decades" (Kallou & Kikilia, 2021, p. 38), and no other theory since Knowles' principles of adult learning has triggered more interest and research in the field of adult education (Wang et al., 2013). Transformative learning emphasises that learning is fundamentally concerned with how individuals interpret and make meaning from their experiences, rather than simply the accumulation of knowledge. Unlike traditional approaches that focus primarily on what learners know, transformative learning examines the processes through which people construct, evaluate, and revise their understanding of themselves and the world. Learning is therefore viewed as an active process of meaning-making in which individuals critically reflect on their assumptions and perspectives in response to new experiences and information (Matthew-Maich et al., 2010).

Within this framework, Mezirow argued that the central purpose of adult learning is to facilitate transformation in the way individuals think, interpret experiences, and understand their place in the world. Through critical reflection and perspective transformation, adult learners can develop greater self-awareness, autonomy, and capacity for informed decision-making. This process not only promotes personal growth but also empowers individuals to engage more effectively with the challenges and opportunities they encounter in their personal, professional, and social lives (Matthew-Maich et al., 2010).

It encourages educators to create learning environments that promote reflection, dialogue, questioning, and critical thinking (Acton, 2023; Addleman et al., 2011; Carlisle & McCloskey, 2023; Tzovla & Kedraka, 2023). According to Mezirow's framework, the goal of adult education "is not merely to help students acquire information but rather to facilitate critical reflection and analysis of assumptions through discourse" (Addleman et al., 2011, p. 4). Transformative learning theory incorporates several interrelated elements that support deep and meaningful changes in perspective. Central among these are critical reflection, collaborative dialogue, and an awareness of the social, cultural, and organisational contexts in which learning occurs. Rather than focusing solely on the acquisition of knowledge, transformative learning seeks to help individuals examine the assumptions, beliefs, and values that shape their interpretations of experience and influence their actions (Acton, 2023).

Within this framework, the role of the adult educator extends beyond the transmission of information. Educators are expected to facilitate meaningful engagement with ideas, encourage participation and dialogue, support positive change, and create learning environments in which learners feel safe to question existing assumptions. At the same time, educators must avoid imposing predetermined viewpoints and instead foster independent thinking, enabling learners to develop greater autonomy and ownership of their learning (Tzovla & Kedraka, 2023).

A range of pedagogical strategies are commonly associated with transformative learning. These include critical reflection, rational discourse, and forms of practice that connect learning with social action and real-world application. Such approaches encourage learners to analyse their experiences, evaluate evidence, consider alternative perspectives, and apply new understandings to personal, professional, or community contexts (Brown, 2004).

Dialogue plays a particularly important role in this process. Through discussion and interaction with others, learners are able to reflect on their experiences, test their assumptions, examine differing viewpoints, and refine emerging understandings. Dialogue therefore serves as a key mechanism through which critical reflection is translated into learning and action, supporting the gradual transformation of habits of mind, beliefs, and perspectives (Marotta, 2024).

Transformative learning contributes to the development of learners who are more self-aware, reflective, adaptable, and capable of responding effectively to complex personal and professional challenges (Akinsooto et al., 2020; Sukri, 2024; Sweet, 2022). By encouraging individuals to critically examine their assumptions, beliefs, and experiences, transformative learning promotes deeper levels of understanding and supports meaningful changes in perspective. Research has shown that learners engaged in transformative learning often develop greater self-awareness, reassess previously held beliefs, experience shifts in their thinking, and become motivated to take informed action based on their new understandings (Sweet, 2022).

Education plays an important role in this process by fostering critical thinking and reflective capabilities that support personal growth and development. Through transformative learning experiences, adults are encouraged to re-evaluate their beliefs, values, and life goals, leading to changes in how they understand themselves and their relationships with others and the broader world (Sukri, 2024). This process of reflection extends beyond individual awareness and includes consideration of social, cultural, and community contexts. As learners critically examine their experiences and assumptions, they develop a deeper understanding of both their own capacities and the systems and environments in which they operate (Akinsooto et al., 2020).

The importance of these capabilities has increased in an era characterised by rapid technological advancement, social change, and organisational transformation (Fedeli & Olatunji, 2024; Kallou & Kikilia, 2021; Sukri, 2024). Individuals are increasingly required to adapt to new technologies, evolving workplace expectations, changing social conditions, and emerging forms of knowledge. Transformative learning provides a framework for developing the flexibility, critical thinking, and lifelong learning capabilities needed to navigate these challenges effectively.

The field of transformative learning has demonstrated considerable adaptability across a wide range of educational, professional, and cultural contexts, allowing it to remain relevant in addressing contemporary challenges (Fedeli & Olatunji, 2024). As educational institutions respond to technological innovation and societal change, there is growing recognition of the need to reconsider learning environments, teaching approaches, and curriculum design to better prepare learners for increasingly complex and uncertain futures (Kallou & Kikilia, 2021).

Through participation in transformative learning experiences, adults engage in critical reflection that can lead to significant personal and professional development. This process often results in greater self-understanding, revised perspectives, and a redefinition of personal and professional roles and responsibilities (Sukri, 2024). These outcomes are further strengthened by the collaborative and interactive nature of contemporary adult education, which encourages dialogue, shared inquiry, reflection, and the development of emotional and social competencies. Consequently, transformative learning continues to provide a valuable framework for supporting meaningful learning, personal growth, and adaptive capacity in a rapidly changing world.

Criticisms of Mezirow's Theory

Although Transformative Learning Theory has had a profound influence on adult education, it has also attracted significant criticism. One of the most common critiques is that Mezirow's original formulation places a strong emphasis on rational reflection and cognitive processes while giving comparatively less attention to emotions, intuition, relationships, spirituality, and broader social influences (Cappiali, 2024; McClain, 2024; Naik & Chattaraj, 2025; Ntseane, 2011). In Mezirow's framework, transformation is largely understood as a process through which individuals critically examine and revise their assumptions through reflection and rational discourse. While this emphasis on critical thinking has been widely recognised as a strength of the theory, some scholars argue that it provides only a partial explanation of how transformative learning occurs (Ntseane, 2011).

Critics contend that transformative experiences are often shaped not only by reasoned analysis but also by powerful emotional responses, personal relationships, intuitive insights, and deeply felt experiences. Research suggests that emotions frequently act as catalysts for transformation, influencing how individuals interpret experiences and motivating them to reconsider existing beliefs and assumptions (Cappiali, 2024; Perry, 2021). Consequently, scholars have argued for a more comprehensive understanding of the relationship between affective and cognitive dimensions of learning, recognising that reflection, emotion, and meaning-making are often closely interconnected rather than separate processes.

Related critiques have challenged the assumption that critical reflection and rational discourse are the sole or primary pathways to transformation. Alternative perspectives suggest that transformative learning may also emerge through artistic expression, spiritual practice, cultural engagement, intuitive understanding, embodied experience, and other forms of deep personal reflection (Naik & Chattaraj, 2025; A. Romano et al., 2024; Schnepfleitner & Ferreira, 2021; Yazdani et al., 2017). These approaches emphasise the imaginative, symbolic, emotional, aesthetic, and spiritual dimensions of human experience, arguing that meaningful learning can occur through ways of knowing that extend beyond analytical reasoning. From this perspective, an exclusive focus on critical thinking may undervalue other legitimate forms of learning and understanding, including embodied, intuitive, and aesthetic knowledge (Cappiali, 2024).

Another significant criticism concerns the cultural assumptions underlying Transformative Learning Theory. Several scholars have argued that the theory reflects Western educational traditions that prioritise individual autonomy, personal agency, and rational analysis (Cappiali, 2024; McClain, 2024; Naik & Chattaraj, 2025; Ntseane, 2011). While these values may align with many Western educational contexts, they may not fully reflect the ways learning and transformation are understood in other cultural settings. Critics suggest that an emphasis on individual perspective transformation can overlook the influence of social structures, cultural traditions, power relationships, collective identities, and community-based forms of learning.

These concerns are particularly evident in cultures where knowledge, identity, and learning are understood as fundamentally interconnected with family, community, culture, and social relationships. In such contexts, transformation may be influenced by factors such as gender, ethnicity, social class, cultural expectations, and power dynamics, rather than being viewed primarily as an individual cognitive process (Ntseane, 2011). For example, some non-Western perspectives emphasise the interconnectedness of individuals with their communities, cultures, and natural environments, highlighting collective growth and social responsibility alongside personal development (Naik & Chattaraj, 2025).

In response to these criticisms, contemporary scholars have expanded Transformative Learning Theory to incorporate a broader range of influences on learning and transformation. Recent developments recognise the importance of emotional, relational, cultural, moral, and social dimensions of learning and seek to provide a more holistic understanding of how transformation occurs (Cappiali, 2024; McClain, 2024; Perry, 2021; Schnepfleitner & Ferreira, 2021). These perspectives acknowledge that transformative learning can be shaped by interpersonal relationships, cultural contexts, emotional experiences, and social environments, as well as by critical reflection and rational discourse.

As a result, contemporary interpretations of transformative learning present it as a multidimensional process that integrates cognitive, emotional, social, cultural, and relational factors. Rather than rejecting Mezirow's original contributions, these developments build upon and extend the theory, providing a more comprehensive framework for understanding the diverse ways adults experience learning, growth, and transformation in contemporary educational and professional settings.

Legacy

Despite these criticisms, Mezirow's concept of perspective transformation remains one of the most influential ideas in adult learning theory (Brookfield, 2018; Hoggan et al., 2016; Park, 2026). It provides a powerful explanation of how adults can move beyond existing assumptions, develop more sophisticated ways of understanding the world, and adapt to changing personal and professional circumstances (Hoem & Ursin, 2025; Nichols et al., 2020; Pittman & Gioia, 2020). By emphasising critical reflection, dialogue, and the examination of assumptions (Brookfield, 2018; Hoem & Ursin, 2025; Khabanyane et al., 2014). Mezirow demonstrated that adult learning can involve profound changes in identity, meaning, and worldview (Buttigieg & Calleja, 2020b; Park, 2026; Pittman & Gioia, 2020), making transformative learning one of the deepest and most significant forms of adult education (Hoggan & Higgins, 2023; Hoggan et al., 2016).

Critical Reflection and Reflective Practice

Critical reflection and reflective practice are closely related concepts that play a central role in adult learning, professional development, and transformative learning. Both involve examining experiences to improve understanding and future performance, but they differ in their depth and purpose. Reflective practice generally focuses on improving actions and decision-making, whereas critical reflection extends further by questioning the assumptions, beliefs, values, and perspectives that shape those actions.

Reflective Practice

Reflective practice refers to the deliberate process of thinking about experiences in order to learn from them and improve future performance. The concept is most commonly associated with the work of Donald Schön (1992), who argued that professionals develop expertise not only through formal knowledge but also through reflection on their experiences in practice (Cadorette et al., 2023; Fragkos, 2016; Franco-Tantuico, 2021; Hosoda, 2006; Kruger, 2015).

Reflective practice involves examining what happened, why it happened, what was done well, what could have been improved, and how future situations might be approached differently. Through this process, learners and professionals develop greater awareness of their actions and improve their ability to respond effectively to complex situations (Turns et al., 2014; Williamson, 1999).

Schön distinguished between two forms of reflection (Cadorette et al., 2023; Kruger, 2015):

Reflection-in-action occurs while an activity is taking place. Individuals monitor situations as they unfold, assess emerging problems, and adjust their actions in real time (Cadorette et al., 2023). For example, a teacher may modify a lesson when students appear confused, or a nurse may adapt a treatment approach based on a patient's changing condition (Cadorette et al., 2023; Sylte, 2020).

Reflection-on-action occurs after an event has taken place. Individuals review the experience, analyse decisions and outcomes, and identify lessons for future practice (Cadorette et al., 2023; Sylte, 2020). For example, a project manager may review a completed project to determine what contributed to successes or challenges.

Reflective practice has become a fundamental component of many professions, including healthcare, education, engineering, business, social work, emergency services, and vocational trades (Fragkos, 2016; Kruger, 2015; Purvis, 2025; Turns et al., 2014; Williamson, 1999).

Critical Reflection

Critical reflection extends beyond reviewing actions and outcomes. It involves questioning the assumptions, beliefs, values, and social influences that shape how individuals interpret experiences and make decisions (Akinsooto et al., 2020; Closs & Antonello, 2011; Matthew-Maich et al., 2010) . Rather than simply asking whether an action was effective, critical reflection asks why certain approaches were chosen and whether the underlying assumptions remain valid (Sweet, 2022; Wu & Crocco, 2019).

The concept is strongly associated with Jack Mezirow's Transformative Learning Theory (Akinsooto et al., 2020; Buttigieg & Calleja, 2020b; Closs & Antonello, 2011). Mezirow argued that adults often operate according to assumptions developed through culture, education, family influences, and prior experiences (Buttigieg & Calleja, 2020b; Şahin et al., 2016). Many of these assumptions become so familiar that they are accepted without question (Akinsooto et al., 2020; Matthew-Maich et al., 2010). Critical reflection encourages individuals to examine these assumptions and determine whether they continue to be appropriate (Buttigieg & Calleja, 2020b; Closs & Antonello, 2011; Sweet, 2022).

Questions commonly associated with critical reflection include:

- Why do I think this way? (Akinsooto et al., 2020; Buttigieg & Calleja, 2020b)
- What assumptions am I making? (Matthew-Maich et al., 2010; Sweet, 2022)
- How did these beliefs develop? (Akinsooto et al., 2020; Buttigieg & Calleja, 2020b)
- What evidence supports or challenges my perspective? (Sweet, 2022; Wu & Crocco, 2019)
- How might others view this situation differently? (Chang et al., 2023; Coryell, 2013)
- What alternative interpretations exist? (Brown, 2004; Şahin et al., 2016)

Through this process, learners may discover that previously accepted beliefs are incomplete, biased, or no longer suitable (Closs & Antonello, 2011; Coryell, 2013; Şahin et al., 2016). This awareness can lead to significant changes in perspective and behaviour (Akinsooto et al., 2020; Buttigieg & Calleja, 2020b).

Relationship to Transformative Learning

Critical reflection is considered one of the primary mechanisms through which transformative learning occurs (Mohammed, 2020; Sebastien, 2017; Sweetman, 2018). According to Mezirow, transformation happens when individuals critically examine their assumptions and develop new ways of understanding themselves and the world (Mohammed, 2020; Sweetman, 2018). Mezirow's theory posits that transformative learning is "based on the principle that adult learners experience a transformation in the way they view experiences and formulate perspectives through critical reflection on the assumptions upon which one's interpretations, beliefs, and habits of mind or points of view are based" (Sweetman, 2018, p. 6). Critical reflection involves challenging one's assumptions and understandings in the effort to find new meaning (Sebastien, 2017), and through this process, individuals modify their frame of reference and view of the world (Mohammed, 2020).

For example, a manager may initially believe that effective leadership requires strict control and close supervision of staff. Research indicates that a perception of leadership as primarily about "productivity, efficiency, control and power" is increasingly being challenged (Preece, 2003), and that authoritarian and controlling leadership has a negative impact on learning (Lemmetty & Brem, 2023). Through critical reflection on workplace experiences and feedback from employees, the manager may come to recognise the value of collaboration, empowerment, and employee development. This shift aligns with what scholars describe as moving away from "prescriptive management practices to practices that are developmental in nature" that "empower individuals to enhance transformative learning" (Lemmetty & Brem, 2023, p. 8). Such a shift from hierarchical control to empowerment and shared power represents a broader trend in leadership thinking (Preece, 2003). This represents a transformation in perspective rather than simply the acquisition of new management techniques, what Mezirow described as a change in one's frame of reference rather than merely informational learning (Sebastien, 2017; Sweetman, 2018).

Differences Between Reflective Practice and Critical Reflection

Although the two concepts overlap, they differ in focus and depth.

Reflective practice primarily concentrates on improving professional performance and effectiveness. It focuses on evaluating actions, decisions, and outcomes to identify opportunities for improvement (Edelist et al., 2024; Rambaree & Nässén, 2021; Susler & Babacan, 2021). Reflective practice is widely recognised as an important professional learning process that supports the development of self-awareness and encourages individuals to consider alternative approaches to action and decision-making (Rambaree & Nässén, 2021). Through deliberate reflection on experiences, professionals are able to examine their actions, evaluate outcomes, and identify opportunities for improvement in future practice.

More broadly, reflective practice is understood as a process of learning from experience that promotes deeper insight into both oneself and professional practice. It involves the critical examination of everyday activities, decisions, behaviours, and underlying assumptions, enabling individuals to better understand how their beliefs, values, and actions influence professional outcomes. By engaging in this ongoing process of reflection and evaluation, practitioners can enhance their professional competence, improve decision-making, and support continuous personal and professional development (Susler & Babacan, 2021).

Critical reflection goes further by examining the assumptions and belief systems that underpin actions and decisions. It explores broader questions about values, power relationships, cultural influences, and ways of knowing (Almengor, 2018; Boyd et al., 2022; Edelist et al., 2024; Mohamed et al., 2022; Patel, 2023; Susler & Babacan, 2021). Critical reflection extends beyond personal reflection by encouraging individuals to examine the broader social, cultural, historical, and political influences that shape their beliefs, assumptions, and actions. While reflection often focuses on an individual's experiences, thoughts, feelings, and behaviours, critical reflection places greater emphasis on questioning societal norms, dominant assumptions, and the power structures that influence how knowledge, values, and practices are constructed and maintained (Edelist et al., 2024).

Several characteristics distinguish critical reflection from more general forms of reflection. It involves questioning assumptions that are often accepted without scrutiny, examining issues from a social as well as an individual perspective, analysing the influence of power relationships, and seeking greater equity, understanding, and empowerment through the reflective process (Susler & Babacan, 2021). Rather than focusing solely on improving performance or solving immediate problems, critical reflection encourages learners to consider how broader social systems and structures influence professional practice and personal experiences.

Critical reflection also requires consideration of the ethical, cultural, historical, political, and social values that underpin decision-making and behaviour (Patel, 2023). Through this process, individuals examine how their beliefs and assumptions have been shaped by their environments and experiences, and how these influences affect their interpretation of situations and interactions with others. Such reflection encourages a deeper understanding of the contexts in which professional and personal actions occur.

A key aspect of critical reflection is the examination of how individual and societal assumptions, together with existing power relationships, influence practice and decision-making (Boyd et al., 2022). This involves exploring the ways in which social norms, cultural expectations, institutional structures, and historical conditions shape perspectives and contribute to particular ways of thinking and acting. By recognising these influences, individuals are better positioned to challenge assumptions that may limit understanding or contribute to inequitable outcomes.

Critical reflection further encourages individuals to consider practical problems within their wider social and cultural contexts rather than viewing them as isolated events (Mohamed et al., 2022). Through this broader perspective, learners can develop a more comprehensive understanding of the factors influencing situations and identify alternative approaches that may be more effective, inclusive, or equitable.

From an adult learning perspective, critical reflection can be understood as a process through which individuals identify the assumptions underlying their actions, explore the historical and cultural origins of those assumptions, evaluate their continuing relevance, and develop alternative ways of thinking and acting (Almengor, 2018). In this way, critical reflection serves as a powerful mechanism for personal growth, professional development, and transformative learning, enabling individuals to move beyond unquestioned beliefs and engage more thoughtfully with the complexities of contemporary life and work.

A reflective practitioner may ask, "How could I have handled this situation more effectively?" A critically reflective practitioner may also ask, "What assumptions influenced my response, and are those assumptions justified?" Critical reflection and critically reflective practice are closely related concepts, but they differ in their application. Critical reflection involves developing a way of thinking that questions assumptions, examines social and organisational influences, and considers how power relationships shape beliefs, decisions, and professional practice. It encourages individuals to analyse issues through ethical, social, cultural, and political lenses, leading to a deeper understanding of the factors that influence both personal and professional actions (Boyd et al., 2022).

Critically reflective practice builds upon this foundation by translating critical awareness into action. Rather than stopping at reflection and analysis, practitioners use the insights gained through critical reflection to inform their everyday decisions, behaviours, and professional activities. In this sense, critical reflection provides a perspective for understanding practice, while critically reflective practice involves applying that understanding to improve professional performance and promote more informed, ethical, and equitable outcomes (Boyd et al., 2022).

A critically reflective professional therefore develops a way of seeing and engaging with the world that is guided by a commitment to ethical practice, social responsibility, and justice (Ng et al., 2022). Such practitioners continually examine the assumptions that influence their work, consider the broader implications of their actions, and remain open to alternative perspectives and approaches. This orientation encourages ongoing learning and supports more thoughtful and responsive professional practice.

When critical reflection is applied specifically to the improvement of professional practice, it becomes a form of reflective practice that places particular emphasis on examining the assumptions underlying decisions and actions, especially those influenced by power relationships and social structures (Bachore et al., 2024). By recognising and questioning these influences, practitioners can develop more inclusive, equitable, and effective approaches to their work while contributing to continuous personal and professional development.

Applications in Professional Practice

Reflective practice and critical reflection are applied across many professions as mechanisms for learning, improvement, and professional development. In healthcare, reflective practice enables clinicians to evaluate patient care decisions, assess outcomes, and identify opportunities to improve clinical performance. Critical reflection extends this process by encouraging practitioners to examine the assumptions that influence their decisions, including beliefs about patient needs, cultural backgrounds, communication practices, and models of care. This deeper level of analysis can contribute to more holistic, patient-centred, and culturally responsive healthcare delivery.

In vocational education and training, apprentices and trainees frequently engage in reflective practice to evaluate completed tasks, identify errors, and improve technical performance. Through reflection, learners can develop greater competence and confidence in applying workplace skills. Critical reflection, however, encourages learners to move beyond technical considerations and examine broader influences on practice, such as workplace culture, attitudes towards safety, organisational norms, and assumptions about what constitutes best practice. This process supports the development of adaptable and critically aware professionals who are better equipped to respond to changing workplace demands.

Engineering provides another context in which reflective practice and critical reflection play important roles. Reflective practice supports continuous improvement by encouraging engineers to review project outcomes, analyse technical decisions, and identify opportunities to enhance future performance. Critical reflection expands this process by prompting consideration of wider issues, including organisational practices, ethical responsibilities, sustainability objectives, stakeholder interests, and the social and environmental consequences of engineering decisions. Such reflection encourages engineers to consider not only technical effectiveness but also the broader impacts of their work.

In community services, critical reflection is particularly important for examining assumptions about clients, social disadvantage, and service delivery. Practitioners are encouraged to reflect on how their beliefs, values, and professional practices influence interactions with individuals and communities. By critically examining these assumptions and considering the effects of social, cultural, and structural factors, practitioners can develop more inclusive, equitable, and culturally responsive approaches to supporting clients and addressing community needs.

The importance of reflection is widely recognised within adult learning theory because adult learners bring substantial life experiences, prior knowledge, beliefs, and values to learning environments. Reflection enables learners to connect new knowledge with existing experiences, facilitating deeper understanding and more meaningful learning. Critical reflection takes this process further by encouraging learners to question and evaluate the assumptions that shape their perspectives, creating opportunities for personal growth and perspective transformation.

Together, reflection and critical reflection support lifelong learning, professional development, adaptability, and informed decision-making. They help individuals develop greater self-awareness, emotional intelligence, and the capacity to respond effectively to complex and uncertain situations. As a result, reflective practice and critical reflection have become essential components of contemporary adult education and professional learning. While reflective practice focuses on learning from experience and improving performance, critical reflection promotes deeper examination of assumptions, values, and perspectives. Combined, these processes support professional competence, personal growth, and transformative learning by encouraging individuals not only to improve what they do, but also to reconsider how and why they think and act in particular ways.

Meaning-Making Processes

Meaning-making refers to the process through which individuals interpret experiences, construct understanding, and develop coherent explanations of themselves, others, and the world around them (Bhang et al., 2025; Calvard, 2025; Piercy, 2013). Rather than passively receiving information, learners actively assign meaning to experiences by connecting new information with their existing knowledge, beliefs, values, emotions, and social contexts (Kiely et al., 2004; Park, 2026; Piercy, 2013); . Meaning-making is therefore a fundamental process in adult learning (Kiely et al., 2004; Park, 2026; Piercy, 2013), influencing how individuals understand experiences, solve problems, make decisions, and construct personal and professional identities (Bhang et al., 2025; Calvard, 2025; Kiely et al., 2004; Park, 2026).

The concept of meaning-making is closely associated with constructivist learning theories, which propose that knowledge is actively constructed rather than simply transmitted from teacher to learner. From this perspective, learning occurs when individuals interpret new experiences in relation to their existing mental frameworks. As learners encounter new information, they seek to integrate it into their current understanding, revise existing beliefs, or develop entirely new ways of interpreting their experiences.

Meaning-making is particularly important in adult learning because adults bring extensive prior experiences, knowledge, and assumptions to educational settings. These experiences provide the foundation upon which new learning is interpreted. When new information aligns with existing beliefs, learners may simply reinforce their current understanding. However, when experiences challenge existing assumptions, learners may need to engage in deeper reflection and reconstruction of meaning. This process can lead to significant changes in perspective and understanding.

Several adult learning theorists have emphasised the importance of meaning-making. Mezirow's Transformative Learning Theory, for example, describes learning as a process of revising meaning perspectives through critical reflection. According to Mezirow, individuals develop frames of reference that shape how they interpret experiences. Transformative learning occurs when learners critically examine these frames of reference and construct more inclusive, discriminating, and integrative understandings of their experiences. In this sense, meaning-making lies at the heart of perspective transformation.

Similarly, Robert Kegan's Constructive-Developmental Theory views human development as a process of increasingly complex meaning-making. Kegan argued that individuals continuously construct meaning from their experiences (Castegnaro et al., 1970; Gutiérrez-Gómez, 2023; Vaca, 2017) and that personal development involves shifts in the structures through which meaning is organised (Castegnaro et al., 1970; Santibáñez, 2016). Kegan's theory, described as "Neo-Piagetian," is founded on two key psychological ideas: that people actively construct the meaning of their experience, and that qualitatively distinct developmental stages can be identified in how individuals construct those experiences (Castegnaro et al., 1970). As Kegan articulated, there is no feeling, experience, thought, or perception independent of the meaning-making context in which it becomes such, because individuals *are* that meaning-making context (Castegnaro et al., 1970). The theory conceptualises individual differences as a function of how people construct and organise experiences related to themselves and their social environments, employing subjective structures (the lens through which the world is perceived) and objective structures (the content of experience that is organised) (Vaca, 2017).

Kegan describes five stages of development, beginning from an egocentric perspective and progressing through different levels of consciousness and complexity (Gutiérrez-Gómez, 2023), with each stage characterised by a particular way of perceiving the world and relating to oneself and others (Gutiérrez-Gómez, 2023; Santibáñez, 2016). As individuals mature, they become capable of interpreting experiences from more complex, flexible, and integrated perspectives (Gutiérrez-Gómez, 2023; Santibáñez, 2016), (Contreras, 2006). This developmental progression integrates three fundamental areas, the dynamic (inner world), the cognitive (thinking about the objective world), and the moral (social thinking), drawing from distinct psychological traditions (Santibáñez, 2016). The theory further suggests that meaning-making and the organisation of meanings constitute a fundamental function of the human organism (Contreras, 2006).

Meaning-making is influenced by both cognitive and emotional processes. Individuals do not interpret experiences solely through logical reasoning; emotions, values, relationships, culture, and identity also shape how meaning is constructed. Significant life events, career transitions, educational experiences, personal successes, failures, and interactions with others can all influence the meanings individuals assign to experiences. Consequently, meaning-making is often both an intellectual and emotional process.

Social interaction also plays an important role in meaning-making. Through dialogue, collaboration, and participation in communities of practice, learners encounter alternative perspectives that may confirm, challenge, or expand their existing understandings. Discussions with peers, mentors, educators, and colleagues can help learners test assumptions, consider different interpretations, and refine their perspectives. Meaning is therefore often constructed through interaction with others as well as through individual reflection.

In professional contexts, meaning-making supports learning and adaptation in complex environments. For example, a newly promoted manager may initially interpret leadership as directing and controlling staff. Through experience, feedback, and reflection, the manager may develop a more sophisticated understanding of leadership as facilitating collaboration, supporting employee development, and creating organisational culture. Similarly, a healthcare professional may come to understand patient care not only as clinical treatment but also as encompassing emotional, cultural, and social dimensions of wellbeing. In each case, learning involves changes in the meaning assigned to professional roles and responsibilities.

Meaning-making is also central to lifelong learning. As individuals encounter new technologies, changing workplace requirements, social developments, and personal challenges, they continually reinterpret experiences and revise their understanding of the world. This ongoing process enables adaptation, resilience, and personal growth throughout life.

Within adult education, educators can support meaning-making by creating opportunities for reflection, discussion, problem-solving, experiential learning, and critical inquiry. Activities such as case studies, simulations, workplace projects, journalling, collaborative learning, and reflective discussions encourage learners to connect new knowledge with prior experience and examine the assumptions that shape their understanding. These approaches help learners move beyond memorisation towards deeper learning and more meaningful knowledge construction.

Meaning-making is the process through which learning becomes personally significant. It enables individuals to integrate new experiences into their existing understanding, revise perspectives when necessary, and construct coherent explanations that guide future thinking and action. As a result, meaning-making is widely regarded as a foundational concept in adult learning, transformative learning, constructivism, and lifelong learning.

Identity Development and Learning

Identity development and learning are closely interconnected processes. Learning is not simply the acquisition of knowledge and skills; it also involves changes in how individuals understand themselves, their roles, their capabilities, and their place within communities and society. As people learn, they often develop new perspectives, adopt new values, acquire new competencies, and reshape their sense of self (Egetenmeyer & Fedeli, 2017b; Zhao et al., 2024). Consequently, learning can contribute to the ongoing development of personal, professional, social, and cultural identities (Egetenmeyer & Fedeli, 2017b).

Identity can be understood as an individual's sense of who they are, including their beliefs, values, experiences, relationships, and membership in various social and professional groups. Rather than being fixed, identity is dynamic and continually shaped by experiences, interactions, and learning throughout life, (Budwig & Alexander, 2021). Adult learning scholars increasingly recognise that many significant learning experiences involve not only changes in knowledge but also changes in identity (Egetenmeyer & Fedeli, 2017b).

From a constructivist perspective, identity develops through the process of meaning-making. As individuals interpret experiences and construct understanding, they simultaneously construct narratives about themselves. Learning therefore influences how people see themselves and how they are recognised by others (Bridgman et al., 2020). For example, a person undertaking a qualification in community services may gradually move from viewing themselves as a student to identifying as a professional practitioner (Egetenmeyer & Fedeli, 2017b; Eyal & Berkovich, 2024). This shift involves more than mastering technical knowledge; it reflects a transformation in self-perception and professional identity (Egetenmeyer & Fedeli, 2017b; Zhao et al., 2024).

Adult learners typically enter educational settings with well-established identities that have been shaped by their careers, family responsibilities, cultural backgrounds, life experiences, and social relationships (Jõgi, 2023; Park, 2026). Unlike younger learners who may still be forming their sense of self, adults often bring complex personal and professional identities into the learning environment. Research suggests that adult learners draw upon a rich range of life experiences when engaging in education, while simultaneously considering how learning may influence their future aspirations and opportunities (Karlsson, 2023). From an adult learning perspective, these experiences are not merely background characteristics but important resources that contribute to both learning and identity development (Jõgi, 2023). Adult learners frequently balance multiple identities, such as employee, parent, partner, community member, and student, as they participate in educational activities.

Learning experiences can reinforce existing identities or challenge learners to reconsider how they see themselves and their roles in society. When learners encounter new ideas, expectations, or experiences that conflict with their existing self-concepts, they may engage in reflection, adaptation, and identity reconstruction. Research in vocational education highlights that learning a vocation involves more than acquiring technical skills; it also involves developing an identity associated with that occupation (Karlsson, 2023). Identity formation is often a dynamic process in which individuals negotiate between their past experiences and future aspirations. Periods of transition and role change may create tensions or identity struggles that prompt learners to re-evaluate who they are and who they wish to become (Kriner et al., 2015). Transformative learning theory provides a useful framework for understanding these processes, as critical reflection on challenging experiences can lead to both perspective transformation and identity change.

For example, an experienced tradesperson undertaking supervisory or management training may initially view themselves primarily as a skilled technical practitioner. Through learning, workplace experience, and increased responsibility, they may gradually begin to identify as a leader responsible for supporting staff, managing teams, and contributing to organisational success. This transition involves more than acquiring leadership knowledge and skills; it reflects the development of a new professional identity. Research indicates that professional identity formation emerges through the interaction between educational experiences and workplace practice, and that technical competence alone is insufficient for the development of a strong vocational identity. Learners often need to develop new ways of understanding themselves as both professionals and lifelong learners (Jõgi, 2023). Professional identity continues to evolve throughout education and employment, reflecting an ongoing process of self-development and adaptation. Studies of adult educators similarly demonstrate that professional roles can be understood in different ways, such as technical expert, competent practitioner, or reflective professional, each associated with different forms of identity and practice (Ranieri & Pachler, 2014).

Identity development is also evident among mature-age students entering higher education after substantial workforce and life experience. Such learners may initially define themselves primarily through employment, family roles, or community responsibilities. Participation in higher education can introduce new identities, including those of university student, researcher, scholar, or future professional. Research suggests that identity formation in adult education extends beyond vocational development and involves broader personal identity trajectories that influence how individuals understand themselves and their place in society (Karlsson et al., 2021). Higher education can provide significant opportunities for identity development by exposing learners to new ideas, communities, and ways of thinking. Some scholars argue that higher education should be viewed not only as preparation for employment but also as a process of identity formation that supports personal growth and self-understanding.

Participation in academic and professional communities can further support identity development. Studies of postgraduate and doctoral students, for example, demonstrate how engagement with scholarly communities can facilitate transitions from seeing oneself as a student to identifying as a researcher or scholar (Kriner et al., 2015). At the same time, adult learners must often manage the interaction between multiple identities, including those associated with study, work, and family life. Successfully balancing these competing demands can lead to greater integration of personal and professional identities, while difficulties in managing these pressures may contribute to disengagement or withdrawal from education. Research indicates that successful participation in education can positively influence how adults perceive themselves as learners, professionals, and active members of society, contributing to increased confidence, self-efficacy, and a stronger sense of identity.

Transformative Learning and Identity

Identity development is a central feature of transformative learning. Mezirow's Transformative Learning Theory suggests that learners may experience profound changes in their frames of reference through critical reflection and perspective transformation (Hess, 2019; Leshem et al., 2015). Frames of reference are "the structures of assumptions through which we understand our experiences," encompassing cognitive, conative, and emotional components, and individuals "transform our frames of reference through critical reflection on the assumptions upon which our interpretations are based" (Leshem et al., 2015, p. 2). More specifically, "perspective transformation" occurs when adults reconsider their internal cognitive frames composed of "habits of mind" in light of biographical, social, and cultural experiences (Hess, 2019, p. 54). As assumptions, beliefs, and values are reconsidered, individuals often develop new understandings of themselves and their relationships with others (Hess, 2019; Leshem et al., 2015). Self-reflection can lead to significant personal transformations that are "more inclusive, differentiating, permeable, critically reflective and integrative of experience" (Leshem et al., 2015, p. 2).

For example, a healthcare professional who adopts a more holistic and patient-centred approach to care may not only change their professional practice but also their identity as a practitioner. Research on nurses in telehealth-based palliative care illustrates how adopting patient-centred holistic care models can reshape professional identity, as "the professional identity of nurses is also established in the trust-based nurse-patient relationship" and new technologies and approaches "transform the nurse-patient interaction" (Ma et al., 2025, p. 13). Similarly, positive clinical outcomes can reinforce practitioners' identity and sense of professional value, "serving as a key psychological resource that motivated continuous learning and engagement" (Liu et al., 2026, p. 8). A manager who moves from an authoritarian leadership style to a collaborative leadership approach may develop a new understanding of what it means to be a leader, as collaborative management structures are recognized as "critical to transforming care delivery" (Aldawsari et al., 2014, p. 495).

Transformative learning therefore involves changes in both knowledge and identity. Learners do not simply learn new information; they may come to see themselves differently as a result of the learning process. Research on academic librarians has demonstrated that individuals experience "perspective transformation around their teaching identities," with external experiences, relationships, and environments affecting "individuals' self-concepts in different ways" (Hess, 2019, p. 56). Similarly, studies of student teachers show that professional identity development is shaped through experiential learning "whereby knowledge is created through the transformation of experience" (Leshem et al., 2015, p. 2). Health care educators have likewise observed that student learning "becomes transformative when theoretical concepts and multicultural knowledge are situated in experiential learning environments" (Duffy, 2008, p. 96).

Robert Kegan and Identity Development

One of the most influential theorists linking identity and learning is Robert Kegan. His Constructive-Developmental Theory proposes that human development involves increasingly complex ways of making meaning and understanding oneself in relation to the world (Alhadeff-Jones, 2018; Éneau, 2016). Kegan argued that individuals progress through developmental stages in which they become capable of reflecting on aspects of themselves that were previously taken for granted. As development occurs, people gain greater capacity for self-authorship, the capacity to internally define one's beliefs, identities, and social relations rather than relying solely on external expectations. Self-authorship, as built upon Kegan's earlier works (e.g., 1982, 1994), reflects the use of an internal orientation to critically choose from multiple possibilities. This theory describes how one's fundamental assumptions evolve from being driven by external authorities to being driven by an internal identity.

From a constructive-developmental perspective, learning contributes to identity development by enabling individuals to develop increasingly sophisticated, integrated, and self-directed understandings of themselves and their relationships with the world. As learners encounter new experiences, perspectives, and challenges, they refine their beliefs, values, and ways of making meaning, leading to more complex forms of personal and professional identity. Professional education, leadership development programs, and higher education frequently provide opportunities for this type of growth by encouraging learners to examine their assumptions, engage with diverse perspectives, and develop greater capacity for independent judgement and self-authorship.

Within higher education, the development of self-authorship is widely regarded as an important educational outcome. Educational approaches informed by constructive-developmental theory seek to support learners in becoming active participants in the construction of knowledge rather than passive recipients of information. Such approaches recognise learners as capable contributors to the learning process, encourage them to draw upon their own experiences, and view learning as a collaborative process of meaning-making. By engaging learners in reflection, dialogue, and inquiry, these educational practices can foster greater autonomy, critical thinking, and identity development.

Transformative learning theory builds upon these developmental foundations by highlighting the relationships between experience, reflection, identity construction, and adult development (Alhadeff-Jones, 2018). Through critical examination of experiences and assumptions, learners may revise existing perspectives and develop new ways of understanding themselves and their roles in society. Identity development is therefore closely connected to processes of meaning-making and transformative learning, as individuals continually reconstruct their sense of self in response to new knowledge and experiences.

Research suggests that a wide range of educational experiences can support the development of self-authorship and identity growth. Programs that involve experiential learning, engagement with diverse perspectives, collaborative dialogue, and opportunities for personal reflection have been shown to encourage learners to question existing assumptions and develop more complex understandings of themselves and others. Examples include outdoor education programs, intergroup dialogue initiatives, and liberal arts education, all of which provide environments that challenge learners to engage with unfamiliar ideas and experiences. Such educational contexts can act as catalysts for identity development by encouraging learners to take greater responsibility for their own meaning-making and to construct more integrated and self-authored perspectives.

Social Identity and Communities of Practice

Identity development is not solely an individual process. Social learning theorists emphasise that identities are formed through participation in communities and social groups (Cheng, 2014; Doan, 2019; Jørgensen & Hadders, 2015). Etienne Wenger's theory of Communities of Practice suggests that learning involves becoming a participant in a particular community and gradually developing the identity associated with that community (Ataş & Daloğlu, 2024; McDermott et al., 2023). As Wenger argued, the development of knowledgeable skills and identity in a community of practice are not two separate processes; through participating in community activities, individuals develop a sense of "self" during engagement (Cheng, 2014). Wenger addresses multiple dimensions of identity formation dependent on macro, micro, individual, and group dynamics, including identity as negotiated experience, community membership, and learning trajectory (Ataş & Daloğlu, 2024).

For example, apprentices entering a trade initially participate at the periphery of the professional community. As they gain experience, knowledge, and acceptance, they move towards full participation and increasingly identify as members of that profession (Guile & Lahiff, 2016; Jørgensen & Hadders, 2015; Lensjø, 2020). Lave and Wenger's concept of legitimate peripheral participation illuminates how newcomers to occupational communities learn the expertise and identity associated with a community of practice as they progress from being novices to becoming experts (Guile & Lahiff, 2016). In the context of plumbing apprenticeships, for instance, apprentices involve their whole person in the social community, and these systems of relations not only define learning a craft but also construct an identity (Lensjø, 2020). Learning therefore involves both acquiring competence and developing a professional identity (Doan, 2019; Li et al., 2009b).

This process can be observed across many fields. Student nurses become nurses—as Lave and Wenger's framework of legitimate peripheral participation has been promoted as a theory for students to gain skills and professional identity in clinical placements (Berragan, 2011; Jørgensen & Hadders, 2015; Li et al., 2009b). In nursing education, learning is seen as an integral and inseparable aspect of social practice, with the traditional apprenticeship model being supplanted by a more contemporary view based on communities of practice (Berragan, 2011). Trainee teachers become teachers, with professional development understood through the lens of identity development within communities of practice (Ataş & Daloğlu, 2024; Cheng, 2014). Engineering graduates become engineers, and apprentice chefs become chefs. The use of communities of practice and situated learning to understand identity formation and socialisation has been explored across pharmacy, medicine, and nursing (McDermott et al., 2023). In each case, learning is closely linked to identity formation and professional socialisation (Abrahams et al., 2022; McDermott et al., 2023), as professional socialisation, the process of learning skills, attitudes, values, and behaviours of the profession, plays a key role in individuals acquiring the culture, norms, and values of the profession (Abrahams et al., 2022).

Professional Identity Development

Professional identity refers to the values, beliefs, behaviours, responsibilities, and self-perceptions associated with a particular occupation or profession. It encompasses how individuals understand their role within a profession, how they perceive their responsibilities to others, and how they align their actions with professional standards and expectations. As a result, professional education typically seeks to develop not only technical competence but also the professional identities that support ethical, responsible, and effective practice (Smith et al., 2015).

In engineering education, for example, the goal extends beyond developing technical knowledge and problem-solving skills. Engineering programs also seek to cultivate professional identities grounded in ethical responsibility, public safety, accountability, sustainability, and service to society. Research suggests that values play an important role in shaping professional identity during the educational process, influencing how students understand their future roles and responsibilities as professionals (DeFoor et al., 2019). Studies of engineering ethics have demonstrated that personal and professional values significantly influence ethical decision-making in practice. Consequently, contemporary engineering education increasingly recognises the importance of incorporating values, ethics, and emotional awareness into professional identity development. Educational approaches such as service learning, ethics instruction, and opportunities for reflection help students connect technical knowledge with broader professional responsibilities and societal impacts.

Healthcare education similarly aims to develop professional identities that combine clinical competence with ethical practice, compassion, professionalism, and patient-centred care. In nursing, professional identity development is recognised as a critical aspect of professional preparation and ongoing development. Continuing education, mentoring, and professional socialisation contribute to strengthening nurses' sense of professional identity and shared professional purpose (Keeler, 2024). Professional identity is also embedded within preceptor and mentorship programs, reflecting the expectation that experienced practitioners play a role in supporting the development of future members of the profession. In medical education, interprofessional learning experiences have been shown to influence how learners perceive their own professional roles and those of other health professionals. Through participation in clinical environments and collaborative learning activities, students gradually develop the attitudes, values, behaviours, and professional identities associated with healthcare practice.

Professional identity development is typically shaped through a combination of formal education, workplace experience, mentoring relationships, reflective practice, and participation in professional communities. Educational strategies that encourage reflection and self-examination can help learners connect professional knowledge with their emerging professional identities. For example, reflective writing activities, mentoring programs, and portfolio-based learning have been used to support professional identity development in healthcare education. Similarly, collaborative learning experiences, peer teaching, and interprofessional education initiatives help learners develop professional values related to teamwork, communication, and collaboration (Burgess & McGregor, 2018).

Ethics education also plays an important role in professional identity formation across many disciplines. Case-based discussions, collaborative problem-solving activities, and interprofessional learning experiences provide opportunities for learners to explore ethical issues, examine professional responsibilities, and consider the values that underpin professional practice. These experiences support the development of professional judgement, ethical reasoning, and a deeper understanding of the responsibilities associated with professional roles.

Across professions, professional identity development is increasingly viewed as an ongoing process that continues throughout an individual's career. As learners gain experience, engage in reflection, and participate in professional communities, they continually refine their understanding of what it means to be a member of their profession. Consequently, professional education is concerned not only with what learners know and can do, but also with who they become as professionals and how they embody the values, responsibilities, and standards of their chosen field

Identity, Lifelong Learning, and Career Change

Identity development becomes particularly significant during periods of transition, when individuals are required to adapt to changing circumstances and reconsider how they understand themselves and their roles in society. Career changes, promotions, technological developments, retirement, migration, and major personal life events can all disrupt established patterns of identity and prompt individuals to reassess their beliefs, capabilities, responsibilities, and future aspirations (Cohen-Scali, 2012; Orfanidou & Kedraka, 2022; Sukri, 2024). Such transitions often involve adjustments not only to external circumstances but also to an individual's sense of self and place within professional, social, and personal contexts.

Work-related transitions are particularly influential because they frequently alter professional responsibilities, social relationships, and expectations. Changes in employment, occupational roles, organisational structures, or retirement can require individuals to develop new understandings of who they are and how they contribute to their workplaces and communities (Cohen-Scali, 2012). These transitions may be temporary or permanent, voluntary or involuntary, but all have the potential to influence identity development by challenging existing self-concepts and encouraging the formation of new ones.

Adult education professionals often work with individuals experiencing such transitions and play an important role in supporting adaptation and personal growth throughout different stages of life (Cohen-Scali, 2012). Educational experiences can provide opportunities for reflection, skill development, confidence building, and exploration of new possibilities, helping learners navigate uncertainty and prepare for changing personal and professional circumstances. As adults encounter career shifts, social changes, and evolving life roles, education can become an important resource for understanding and managing these transitions (Sukri, 2024).

Learning during periods of transition frequently contributes to identity transformation. As individuals acquire new knowledge, develop new competencies, and engage in critical reflection, they may begin to change how they think, feel, and act in relation to themselves and others (Orfanidou & Kedraka, 2022). These changes can result in revised perspectives, new professional or personal identities, and greater capacity to adapt to future challenges. Consequently, identity development and learning are closely interconnected, particularly during times of change, when education can support both practical adaptation and deeper personal transformation.

For example, an experienced railway worker transitioning into an engineering management role may need to move from identifying primarily as a technical specialist to identifying as a leader, strategist, and decision-maker. The problem of professional identity formation in engineering fields is recognized as particularly pressing in the context of continuing engineering education (Кубрушко & Kozlenkova, 2019). Similarly, an individual retraining for a new career may need to reconstruct aspects of their identity as they adapt to new professional expectations and environments. The development of professional identity, encompassing beliefs, behaviours, values, and motives through which a person defines himself or herself in a professional role, is central to such transitions (Egetenmeyer & Fedeli, 2017b). Research on lifelong learners has demonstrated that learning experiences gradually develop identity, with significant influence on the development and transformation of other identities as well (Orfanidou & Kedraka, 2022). Marrying, setting up a family, practicing a profession, or changing a craft are key moments that transform and crystallize adult identities (Ilie, 2019).

Lifelong learning supports these transitions by providing opportunities for individuals to acquire new knowledge, develop new competencies, and construct new identities that align with changing circumstances (Ceschi et al., 2021; Sukri, 2024; Zhao et al., 2024; Кубрушко & Kozlenkova, 2019). Lifelong learning is considered an appropriate response to changes and a key lever for resilience, adaptation, and development of both individuals and organizations (Ceschi et al., 2021). Education is not only a means of acquiring new skills but also a key factor in reshaping personal identity and fostering adaptability in a rapidly changing world (Sukri, 2024). This identity shift is crucial in a world characterized by constant technological and social evolution, where adaptability and continuous learning are paramount (Sukri, 2024). In the context of continuing education, lifelong learning is necessary for both personal and professional development (Кубрушко & Kozlenkova, 2019), and the continuous acquisition of competences is perceived as determinant for professional success and career advancement (Ceschi et al., 2021). The ongoing updating of knowledge highlights the significance of lifelong learning and the need for education programs to integrate mechanisms that support ongoing professional growth (Zhao et al., 2024).

Significance for Adult Education

Understanding the relationship between identity development and learning has important implications for adult education. Educators increasingly recognise that learning involves more than transmitting information; it also involves supporting learners as they develop new ways of understanding themselves and their roles within society.

Learning activities that encourage reflection, dialogue, mentoring, experiential learning, and participation in professional communities can help facilitate identity development. Such approaches recognise that education contributes not only to what learners know, but also to who they become. Consequently, identity development is now widely regarded as a central dimension of adult learning. Through learning, individuals continuously construct, reconstruct, and refine their personal, social, and professional identities, enabling them to adapt to changing environments, engage in lifelong learning, and participate more effectively in their communities and professions.

The relationship between learning and identity development can be observed across a wide range of adult education contexts. Understanding this relationship enables educators to design learning experiences that support not only the acquisition of knowledge and skills but also the development of professional confidence, self-efficacy, and professional identity.

Consider a facilitator delivering a professional development program for newly appointed managers. Many participants may enter the course identifying primarily as technical specialists whose expertise is based on their occupational knowledge and individual performance. As they engage with leadership theories, workplace scenarios, reflective activities, and discussions with peers, they may begin to reconsider their understanding of leadership. Over time, learners may shift from viewing themselves as technical experts who supervise staff to seeing themselves as leaders who coach employees, build teams, manage organisational change, and influence workplace culture. The learning experience therefore contributes not only to leadership competence but also to the development of a new professional identity.

This transition from technical specialist to leader is well documented in the literature. Moving into formal leadership roles often brings a "disorienting shift in identity and perspective that can be overwhelming and difficult to manage" (Hinck, 2022). A leader must develop a "provisional self" to test new skills, practices, and behaviours associated with the expectations of the new leadership role, before ultimately incorporating this new leadership identity into their professional and personal roles (Turner et al., 2018).

Research on coach development programs has similarly demonstrated that participation in developmental learning experiences can lead to significant changes in how individuals understand themselves and their professional roles (Burt, 2024; Dhar, 2024; Mukherjee et al., 2024). Participants often report that the learning process provides new ways of thinking, communicating, and approaching leadership, resulting in changes to their professional practice and self-perception (Fang, 2023; Mukherjee et al., 2024). These changes frequently extend beyond the acquisition of specific coaching or leadership skills, influencing broader aspects of personal and professional identity (Burt, 2024; Dhar,

2024; Kragt & Day, 2020). In some cases, participants describe the emergence of a new professional identity as an unexpected outcome of the learning experience (Burt, 2024; Nakamura, 2025), highlighting how developmental programs can contribute to profound shifts in how individuals see themselves as leaders and practitioners (Fang, 2023).

Leadership development programs that incorporate engagement with leadership theories are essential to this identity shift. Contingency-based and relational leadership theories help participants move beyond simplistic views of leadership as supervision toward understanding leadership as a multifaceted, relational process (Rhodes & Fletcher, 2013). In professional sport contexts, for example, coaches and athlete leaders highlighted how appointment to leadership roles "required players to reach a new, more nuanced understanding of leadership and to reconceptualise their role and identity within the context of the team" (Çobanoğlu, 2019). Similarly, in military education, a leadership development course framework explicitly includes identity theory as one of six foundational theoretical frames for developing leaders (Haddad & O'Connor, 2022).

Workplace scenarios, reflective activities, and peer discussions serve as critical pedagogical mechanisms in this transformation. Reflective practice has been identified as playing a central role in leadership development, enhancing learning and transfer from development initiatives (Çobanoğlu, 2019). Coaching and mentoring relationships within such programs support self-discovery, self-inquiry, and self-action, which are fundamental to leadership growth (Haddad & O'Connor, 2022). Research on leadership development in community colleges found that mentoring and coaching relationships, alongside belonging to a culture of leadership, were key findings related to participants' growth and identity change (Hinck, 2022). Furthermore, characteristics such as identity clarity can be developed through specific cultural experiences, knowledge and skill building resources, as well as developing an informed view of self and identity clarity through reflective activities.

The development of leadership competencies, including coaching employees, building teams, managing change, and influencing culture, is intertwined with this identity development. Leadership development programs need to address capabilities including self-management, social skills, and work facilitation, while the networked domain of development involves culture, diversity, identity, networking, and team-building skills. Organisations should focus on developing leaders' work identities within their leadership development programs, taking into consideration the multiple complexities associated with leadership roles, as strengthening leader role identity ultimately enhances overall work identity and leadership effectiveness (Meadows & Braine, 2022). The construction of a professional identity implies an interaction between person and context as individuals adopt and adapt professional characteristics to the necessities of their immediate context (Oosten, 2014). Through processes of self-discovery and developing deeper connections with their authentic selves, participants in development programs become more self-assured, increasing their confidence in their roles as managers and leaders in their organisations,

Identity development is equally important in adult language, literacy and numeracy (LLN) education. Adult learners enrolling in LLN programs often bring prior educational experiences that have shaped their self-perceptions as learners (Fletcher et al., 2011; Gal et al., 2020). Some may have experienced repeated academic difficulties and consequently view themselves as incapable of succeeding in formal education (Fletcher et al., 2011; Mellard et al., 2012; Saal & Dowell, 2014). Through supportive instruction, achievable learning goals, constructive feedback, and successful learning experiences (Grummell, 2023; Paris & Parecki, 1995; Windisch, 2015), these individuals can gradually develop a more positive learner identity (Grummell, 2023; Keating, 2012; Muscat et al., 2020). Rather than seeing themselves as poor readers, writers, or learners, they may begin to view themselves as capable and successful participants in education (Paris & Parecki, 1995; Saal & Dowell, 2014). This shift in identity can significantly influence motivation, persistence, confidence, and willingness to engage in further learning opportunities (Gal et al., 2020; Mellard et al., 2012; Muscat et al., 2020).

Private pilot flight training provides another example of how learning contributes to identity development. Students commonly commence flight instruction identifying as enthusiasts or aspiring pilots. During training, they acquire technical flying skills, aeronautical knowledge, and regulatory understanding. However, they also begin to internalise the responsibilities, attitudes, and behaviours associated with aviation professionalism. The aviation industry possesses a strong professional culture with its own norms and values, and being exposed to this professional culture enables students to develop their own professional identity and transition into the industry (Lyle & Siao, 2021). Pilots have a strong sense of professional identity and take great pride in their work, being highly motivated to perform at the highest level of skill and expertise (Davey, 2004). Through flight planning, risk assessment, decision-making exercises, and interactions with instructors, learners gradually develop an identity as pilots rather than simply individuals learning to fly. Flight instructors play a significant role in this process; as both pilots and trainers of pilots, they have a vested interest in maintaining the status of commercial airline pilots as a professional group, and their interactions with students shape the professional attitudes and behaviours that trainees come to adopt.

The construction of training practices consistent with professional aviation values is not simply impression management but reflects a deeper process of identity formation. As they assume greater responsibility for flight operations and safety, they begin to think and act in ways consistent with the expectations of the aviation profession. Professional pilot culture embodies the attitudes and values linked with the specific profession (Davey, 2004), and collegiate aviation training serves the purpose of preparing students to enter the professional aviation workforce. The childhood dream of flying can serve as an identity anchor that sustains occupational identity throughout a pilot's career, illustrating how deeply this identity transformation can take root during initial training (Fraher & Gabriel, 2014). The transition from student pilot to licensed pilot therefore represents both a competency-based achievement and a significant identity transformation. Knowledge of concepts such as socialisation, group membership, and social interdependence can be of great value in understanding the dynamics of professional identity in aviation (Vermeulen, 2009).

These examples illustrate that adult learning frequently involves changes in how individuals understand themselves and their roles within professional and social contexts. Whether developing leadership capability, rebuilding confidence through literacy and numeracy education, or becoming a competent and responsible pilot, learners are often engaged in processes of identity construction alongside the acquisition of knowledge and skills. Recognising this broader developmental dimension of learning enables educators to create experiences that support not only competence but also personal growth, professional formation, and lifelong learning.

Learning Through Life Transitions

Learning through life transitions is a significant area of adult learning theory that examines how individuals learn, adapt, and develop as they navigate major changes in their personal, social, and professional lives. Unlike formal education that is often planned and structured, learning through transitions frequently emerges in response to changing circumstances that require individuals to acquire new knowledge, skills, behaviours, perspectives, and identities (Carter & Nicolaides, 2023; Crane & Hartwell, 2018; Vogelsang, 1993).

Adult life is characterised by a series of transitions that may include entering the workforce, changing careers, promotion to leadership roles, becoming a parent, migration, retirement, relationship changes, illness, bereavement, technological change, or returning to education (Allcock, 2018; Wyn, 2014). These events often disrupt established routines and assumptions, creating situations in which existing knowledge or ways of understanding are no longer sufficient (Carter & Nicolaides, 2023; Vogelsang, 1993). As a result, individuals are required to learn in order to adapt successfully to new circumstances (Parzen & Janzen, 2019; Vogelsang, 1993). Negotiating life transitions requires individuals to reflect critically on taken-for-granted perspectives and to restructure them in ways necessary to deal with new life situations (Parzen & Janzen, 2019; Vogelsang, 1993). The complexity of transitional processes means that transitions can be challenging periods requiring significant adaptation (Allcock, 2018), and new patterns of transition have been recognised as increasingly extended and multifaceted (Wyn, 2014).

Many adult learning scholars view transitions as powerful catalysts for learning because they create a gap between an individual's current capabilities and the demands of a new situation (Carter & Nicolaides, 2023; Crane & Hartwell, 2018). This gap often stimulates reflection, problem-solving, experimentation, and the acquisition of new competencies (Carter & Nicolaides, 2023; Parzen & Janzen, 2019; Vogelsang, 1993). Disorienting dilemmas triggered by life changes bring to conscious awareness conflicts that stimulate meaning making (Carter & Nicolaides, 2023), and leaders or educators can act as catalysts during transitions by providing intellectual stimulation to help individuals make developmental shifts Crane & Hartwell, 2018; . In this sense, learning becomes a mechanism for managing change and restoring a sense of competence and stability (Bøje et al., 2022; Carter & Nicolaides, 2023; Vogelsang, 1993).

Theoretical Foundations

The study of learning through life transitions draws upon several influential theoretical perspectives that help explain how adults learn, adapt, and develop when confronted with significant changes in their lives. These theories provide insight into the cognitive, social, emotional, and developmental processes that occur as individuals navigate new circumstances, roles, and identities.

Mezirow's Transformative Learning Theory (Mezirow, 1991) suggests that major life events and transitions can act as catalysts for profound learning. Significant changes such as career shifts, migration, retirement, illness, or returning to education may challenge existing assumptions and ways of understanding the world. These experiences can create a sense of uncertainty or disequilibrium that encourages critical reflection. Through examining their beliefs, values, and assumptions, learners may develop new perspectives and reconstruct their understanding of themselves and their place within society. As a result, transitions can lead not only to new knowledge but also to changes in identity and worldview,

Kegan's Constructive-Developmental Theory views life transitions as opportunities for individuals to develop increasingly sophisticated ways of making meaning. According to this perspective, adults continually evolve in their capacity to understand themselves, others, and complex social environments. When confronted with new challenges and responsibilities, individuals may move towards greater self-authorship, becoming less dependent on external expectations and more capable of defining their own beliefs, values, and goals. Transitions therefore provide opportunities for personal growth and the development of greater autonomy and responsibility.

Life Course Theory offers a broader perspective by emphasising that learning occurs throughout the lifespan and is shaped by social, cultural, historical, and personal influences. From this viewpoint, transitions are considered normal developmental events that occur across different stages of life (Kubberød & Pettersen, 2018; Young & Bunting, 2024). Experiences such as entering the workforce, becoming a parent, changing careers, or retiring are understood as important life events that influence both learning and identity formation (Donaldson & Villagrasa, 2024; Kubberød & Pettersen, 2018). The timing and impact of these transitions may vary between individuals, but each has the potential to contribute to ongoing development and adaptation.

A lifespan development perspective suggests that adults exercise considerable agency in determining when, how, and why they engage in learning throughout their lives. Educational participation is often influenced by changing personal, professional, and social circumstances, with individuals making decisions about learning in ways that help them balance competing demands and responsibilities. Choices regarding the timing of study, participation in professional development, or engagement in further education are frequently shaped by factors such as employment, family commitments, financial circumstances, and broader life goals (Kubberød & Pettersen, 2018).

Life transitions are therefore best understood from both lifelong and life-wide perspectives. Learning does not occur in isolation but is influenced by multiple contexts that interact throughout an individual's life. Significant transitions often involve simultaneous changes across several domains, including work, family, education, community participation, and personal relationships (Donaldson & Villagrasa, 2024). As a result, learning during transitions is often complex, requiring individuals to adapt to interconnected changes occurring across different areas of their lives.

Identity development is a particularly important aspect of these transitional experiences. As individuals move through significant life events, they frequently engage in processes of identity negotiation in which they reassess their roles, relationships, values, and future aspirations. These negotiations are shaped not only by personal reflection and individual choices but also by interactions with social institutions, cultural expectations, workplace structures, technologies, and other people (Donaldson & Villagrasa, 2024). Furthermore, identity development during transitions occurs within broader social contexts that include differing levels of influence, authority, and power. Consequently, learning through life transitions involves both personal meaning-making and ongoing engagement with the social, cultural, and institutional environments in which individuals live and work.

Situated Learning Theory and Wenger's concept of Communities of Practice highlight the social dimensions of learning during transitions. These perspectives suggest that learning occurs through participation in social and professional communities rather than solely through individual cognitive processes (Hicks, 2020; Karlsson, 2023).

Lave and Wenger (1991) conceptualised learning as a process of increasing participation within a community of practice. From this perspective, learning involves more than acquiring knowledge and skills; it is a process through which individuals become active and competent members of a social and professional community. Learning is therefore closely connected to participation, identity formation, and community membership, influencing how individuals understand themselves, what they are capable of doing, and how they interact with others within their social and professional environments.

As individuals enter new roles, organisations, occupations, or educational settings, they gradually acquire the knowledge, behaviours, values, expectations, and cultural norms associated with those communities (Hicks, 2020; Karlsson, 2023; Tobbell et al., 2010). Through participation in everyday activities and interaction with more experienced members, learners develop both practical competence and an understanding of what it means to belong to the community. Learning is therefore not solely concerned with mastering tasks but also with developing the identities associated with particular roles and professions (Hicks, 2020; Reddy & Shaw, 2019).

From a communities of practice perspective, learning can be understood as a transformative process that opens possibilities for new ways of thinking, acting, and being. As individuals engage with new communities, they often encounter opportunities to reconsider existing assumptions and develop new understandings of themselves. The transition into a new community of practice typically involves a process through which newcomers seek acceptance, legitimacy, and recognition as valued members of the group. This occurs as they interact with colleagues, learn workplace routines, acquire new skills, and adapt to the norms, expectations, and culture of the community (Tobbell et al., 2010). Through this process, the relationship between the individual and the community becomes integrated into their developing identity, shaping how they understand themselves and their professional role (Hicks, 2020).

Research across a range of professional and educational settings demonstrates the importance of communities of practice in supporting transitions and identity development. Studies involving vocational teachers, nurses, emergency medical personnel, and medical supervisors have consistently shown that learning during transitions involves both becoming a member of a new community and developing a corresponding sense of self within that community (Garth et al., 2024; Hicks, 2020; Kane, 2017; Karlsson, 2023; Reddy & Shaw, 2019; Tobbell et al., 2010). Similar processes can be observed in educational contexts, including children commencing school, mature-age learners entering university, professional students developing disciplinary expertise, and doctoral candidates becoming members of academic communities (Karlsson, 2023).

Importantly, transitions are not viewed merely as periods of induction or orientation. Rather, they are understood as ongoing processes of becoming, in which individuals continuously develop through participation, experience, and interaction with others. This process is often complex and non-linear, involving multiple pathways, relationships, and learning experiences that contribute to the gradual formation of competence, identity, and community membership. Through participation in communities of practice, learners not only acquire knowledge and skills but also develop the identities that enable them to function effectively within their chosen professions and social groups.

Together, these theoretical perspectives demonstrate that learning through life transitions is a multidimensional process involving changes in knowledge, meaning-making, participation, and identity. They highlight how significant life events can become powerful opportunities for personal growth, professional development, and lifelong learning (Tobbell et al., 2010; Young & Bunting, 2024).

Identity and Transition

One of the most significant aspects of learning through transitions is identity development. Major life changes often require individuals to reconsider who they are and how they relate to others. Learning therefore involves more than acquiring knowledge or skills; it frequently involves developing new identities (Chen, 2011; Valadas et al., 2019; Warhurst & Black, 2024).

For example, a tradesperson promoted to a supervisory position must learn not only management skills but also what it means to be a leader. Research on leader identity formation demonstrates that transitioning into a leadership role involves deeply personal processes of learning and growth, where formal authority alone does not make managers take on their role as leaders; rather, identity work becomes critical during transitions into new senior roles (Jog, 2021). Similarly, the process of professional socialization into a senior organizational role requires the new leader to reach an accommodation with the work role, the people with whom they interact, and the existing organizational culture (Male, 2006). A newly retired worker may need to develop a new sense of purpose and identity after leaving a long-established occupation.

The traditional understanding of retirement from an identity perspective involves identity loss, where the retiree can feel displaced, losing a valued identity while gaining a less desirable one (Warhurst & Black, 2024). Retirement requires individuals to encounter the challenge of role change from a paid worker to a retiree, which necessitates a significant change in life contents (Chen, 2011). Work functions as a powerful organizer of life dimensions, and identity and social adjustments are strongly connected to it, making the transition to retirement particularly consequential (Valadas et al., 2019). Research with retired Big 4 accounting partners illustrates how highly socialized professionals find it difficult to experience identity transition after retirement, since departure can heighten rather than diminish professional identity (Azambuja et al., 2025). A migrant entering a new country may need to negotiate cultural identities while learning new social norms and systems. Cultural determinants such as race/ethnicity and gender influence career decisions, and transitions in life are associated with individuals entering and exiting various roles as they move from one life stage to the next (Lytle et al., 2015).

Transitions often involve a process of identity reconstruction in which individuals integrate past experiences with future aspirations. Research on retired athletes demonstrates that identity reconstruction is essential for achieving post-sport identity synthesis, with retirement understood as a liminal phase marked by identity crisis and emotional turbulence (Rocco et al., 2026). Identity continuity, achieved through strengthening connections with past group identities and forming new identities related to previous ones, becomes critical to adjustment in new roles (Haslam et al., 2025). Retirement may also constitute a macro-career transition that triggers the re-evaluation of the meaning of a good life, motivating individuals to rediscover once-valued identities. Learning supports this process by providing opportunities to explore new roles, develop competence, and build confidence in unfamiliar situations. The counselling encounter itself represents a learning experience, empowering individuals to continue expanding on their lifelong learning experiences during transitions (Chen, 2011). Education and learning have an immense impact on the lives of older adults during the transition to retirement (Valadas et al., 2019), and career-crafting, involving questioning traditional notions of career progression, can support identity restoration and rejuvenation in later career (Warhurst & Black, 2024).

Types of Transitional Learning

Learning through transitions can occur across a wide range of personal, educational, and professional contexts. One of the most common forms is career transition, where individuals move into new occupations, accept promotions, change industries, or assume different workplace responsibilities. These transitions often require the development of new technical skills, knowledge, workplace behaviours, and professional identities (Federighi, 2018; Hawse & Wood, 2019). Research on non-linear career paths has demonstrated that professional growth is the result of accumulated learning outcomes across diverse life and work experiences, and that the development of professional identity is a key factor driving occupational mobility (Federighi, 2018). The transition from education to professional practice is "highly complex and critically important" (Baytiyeh and Naja (2012, p. 12) cited in Hawse and Wood (2019, p. 12), requiring not only discipline-specific competencies but also professional socialisation, defined as "the process of learning a professional role and emerging as a member of an occupational culture" (Melrose et al. (2012, p. 7) cited in (Hawse & Wood, 2019, p. 7). As individuals adapt to new roles, they must learn not only how to perform new tasks but also how to understand themselves within a different professional context, a process that involves identity matching and experimentation with possible selves (Killingberg et al., 2020). Career transitions are further complicated for specific populations such as qualified immigrants, who often must rebuild career capital while simultaneously adjusting their professional identity (Farashah & Blomquist, 2021), and industry professionals transitioning into academia, who experience ambiguity and flux as they negotiate contested spaces of identity (Myers et al., 2024).

Educational transitions are another important source of learning. Adults may return to study after many years away from formal education, commence higher education, undertake vocational training, or participate in professional development programs. These experiences frequently require learners to adapt to new learning environments, expectations, technologies, and ways of thinking (Baan, 2026; Wang & Shan, 2023). Research on students returning from military service highlights that the identity shift of "student-soldier-student" creates significant learning adaptation challenges, with students scoring poorly on adaptability to course difficulty, teaching styles, and learning methods (Wang & Shan, 2023). Higher education institutions have recognised the importance of supporting transitions through coaching programs that encourage learning and reflection, with employability development serving as a foundation for both smooth workplace entry and sustainable career development (Baan, 2026). In many cases, educational transitions involve the development of a learner identity alongside the acquisition of new knowledge and skills, as students' career aspirations are frequently expressed in terms of what they want to become, reflecting the centrality of identity capital (Benati & Fischer, 2020)

Learning also occurs during family-related transitions. Becoming a parent, partner, caregiver, grandparent, or assuming other family responsibilities often requires individuals to acquire new knowledge, develop practical skills, and adjust their perspectives and priorities. The menopause transition, for instance, has been recognised as a biopsychosocial life-course event that often coincides with significant role changes including caregiving responsibilities and evolving family dynamics, requiring identity renegotiation and shifts in relational roles (Cronin et al., 2026). These transitions can significantly influence personal identity as individuals adapt to changing family roles and relationships.

Health-related transitions provide another important context for adult learning. Individuals experiencing illness, injury, disability, rehabilitation, or age-related changes in physical capability frequently need to learn new ways of managing their health and daily activities. Research on transitions into mental health nursing illustrates how individuals entering health-related roles experience the shock of transitioning and must grapple with the realities of new work, with most adjusting over time through the development of professional identity, confidence, and coping self-efficacy (Foster et al., 2025). Occupational therapists in mental health practice similarly demonstrate that professional resilience is sustained by strong professional identity and that transitions from education into practice require strategies reinforcing identity through professional socialisation (Ashby et al., 2012). These experiences may involve acquiring medical knowledge, developing coping strategies, adapting routines, and redefining personal goals and expectations. Such transitions often require significant emotional adjustment as well as practical learning (Aryeequaye et al., 2025; Foster et al., 2025).

Retirement represents a major life transition that can involve substantial learning and identity change. For many individuals, employment has been a central source of purpose, routine, social interaction, and self-definition (Hodkinson, 2010; Murakami, 2021; Valadas et al., 2019). Retirement represents a significant life transition that often involves substantial learning, adaptation, and identity change.

Contemporary research views retirement not as a single event that occurs on a specific date but as an ongoing adjustment process that unfolds over time (Hodkinson, 2010). From a life course perspective, retirement involves major changes in social roles, daily routines, responsibilities, and patterns of social interaction, requiring individuals to adapt to new circumstances and redefine aspects of their identity (Murakami, 2021).

For many individuals, employment has provided structure, purpose, social connection, financial security, and a sense of personal identity for much of their adult lives. Consequently, transitioning into retirement may require the development of new routines, interests, social networks, and sources of meaning. Learning during this period often focuses on adapting to a different lifestyle while maintaining a sense of purpose, wellbeing, and personal fulfilment. As retirees navigate these changes, they frequently draw upon social support networks, personal experiences, and symbolic resources to build resilience and construct new understandings of themselves and their place within society (Murakami, 2021).

Research examining retirement experiences has highlighted the central role that work plays in shaping identity and life transitions. Studies of older adults have demonstrated that occupational roles remain an important influence on how individuals experience retirement and adapt to life after paid employment (Valadas et al., 2019). Education and learning can play a particularly important role during this period by supporting personal growth, social participation, confidence, and engagement with new activities and communities. Learning opportunities may help retirees develop new interests, maintain cognitive engagement, strengthen social connections, and create meaningful post-retirement identities.

Importantly, much of the learning associated with retirement occurs outside formal educational settings. Learning during this stage of life is often informal, experiential, and embedded within everyday activities and social interactions (Hodkinson, 2010). Retirees may learn through volunteering, community participation, hobbies, travel, family relationships, mentoring, and engagement with new technologies. These experiences contribute to ongoing personal development and support adaptation before, during, and after the transition from paid employment. Consequently, retirement can be understood not as the end of learning but as a period in which new forms of lifelong learning become increasingly important for wellbeing, identity development, and continued participation in society.

Technological transitions have become increasingly important in contemporary society. Rapid advances in digital technologies, automation, artificial intelligence, and workplace systems require adults to continually update their knowledge and competencies (Milosheva et al., 2023). Information literacy competencies have been identified as critical for navigating career transitions in the digital age, as career transitions are "complex endeavours, in which impactful decisions need to be made through the utilisation of appropriate information behaviour" (Milosheva et al., 2023, p. 11) Whether learning to use new software, adapting to digital communication platforms, or acquiring skills associated with emerging technologies, adults must engage in ongoing learning to remain effective in both their professional and personal lives.

Across all of these contexts, transitions create opportunities for growth, adaptation, and identity development. A comprehensive model of learning for career and labour market transitions has demonstrated that learning supports career transitions across the life-course, where these transitions may involve shifts away from initial occupational identities (Brown & Bimrose, 2014). The learning that occurs during periods of change extends beyond the acquisition of new knowledge and skills and often contributes to broader changes in how individuals understand themselves, their capabilities, and their roles within society (Brown & Bimrose, 2014; Federighi, 2018; Murakami, 2021).

Learning Processes During Transitions

Adults experiencing transitions often engage in a range of interconnected learning processes as they attempt to understand, adapt to, and manage changing circumstances. A common starting point involves reflecting on previous experiences and considering how past knowledge, skills, and assumptions relate to new situations. Theories of adult learning posit that adults have mental models constructed from previous experiences that serve as a growing resource for learning, and that they regularly exercise analogical reasoning in learning and practice (Shoaib et al., 2025). Through this reflection, individuals often evaluate existing beliefs and perspectives, questioning whether previously successful approaches remain effective or appropriate in their current context. Self-reflection coupled with a more critical understanding of one's context can turn adversity into an opportunity for learning and growth (Mishra & Ray, 2025). This process can encourage greater self-awareness and help learners identify areas where change or further development may be required. Research on internationally educated health professionals, for instance, found that self-awareness was one of four main outcomes of transformative education showing a positive impact on resilience during professional transitions (Mishra & Ray, 2025).

As individuals navigate unfamiliar situations, they frequently seek information, guidance, and support from a variety of sources, including formal education, professional networks, mentors, colleagues, family members, and community groups. Structured mentorship programs can mitigate transition anxiety and improve clinical confidence (Sommer et al., 2022), and mentorship has been shown to strengthen professional identity and decrease stress while fostering a greater sense of confidence, competence, and satisfaction (Oh et al., 2025; Upenieks, 2021). Motivation during transitions is shaped by personal aspirations, family support, faculty support, and self-confidence (Sommer et al., 2022). Accessing new information enables learners to better understand their changing circumstances and develop strategies for responding effectively. At the same time, they often work to develop new competencies that will help them meet emerging personal, social, or professional demands. This may involve acquiring technical skills, learning new ways of communicating, developing leadership capabilities, or building confidence in unfamiliar environments. As knowledge and experience are gained, self-doubt decreases, self-efficacy emerges, and learners report increased comfort, confidence, and competence.

Learning during transitions also involves experimentation and adaptation. Adults frequently test new behaviours, approaches, and ways of thinking as they attempt to find effective responses to new challenges. Research on novice teachers demonstrates how individuals initially perceive challenges as threats but gradually develop coping mechanisms and reframe difficulties as opportunities (Briciu et al., 2025). Through trial and error, feedback, and ongoing reflection, they gradually refine their actions and develop greater competence. Reflective learning has been recognized as a valuable tool for improving confidence and academic performance, helping individuals navigate uncertainties and enhance their learning experiences (Sommer et al., 2022). Participation in new workplaces, educational settings, social groups, or professional communities can further support learning by providing opportunities to build relationships, observe others, and gain access to shared knowledge and experience. Nurse practitioners transitioning into new roles, for example, build competence and confidence through trusting relationships and structured fellowship environments (Upenieks, 2021).

Over time, these learning processes often contribute to the construction of new personal and professional identities. As individuals acquire new knowledge, develop new skills, and participate in new communities, they may begin to see themselves differently and adopt new understandings of their roles, responsibilities, and capabilities. Research on nursing transfer students found that clinical experiences served as turning points in professional identity development, with participants noting that "accepting mistakes as part of the learning process has been key to my growth" (Oh et al., 2025, p. 6). Personal adaptability, the capacity to develop diverse role behaviours to respond effectively to changing environmental conditions, is closely linked to identity development and personal learning through mentoring relationships (Aller & Almrwani, 2022). Learning therefore becomes closely connected to identity development, helping individuals adapt not only to changing circumstances but also to evolving conceptions of who they are. Medical students during the COVID-19 pandemic, for instance, exhibited self-care, growth, and perspective formation even amid uncertainty (Yusof, 2025).

Importantly, these processes rarely occur in a straightforward or predictable manner. Learning through transitions is often characterised by periods of uncertainty, anxiety, self-doubt, and resistance, particularly when individuals are required to leave familiar roles or challenge long-held assumptions. This transitional period is characterised by psychological challenges, uncertainty, self-doubt, and struggle to align one's real identity with future aspirations (Otaki et al., 2025). Nurses facing redeployment reported emotional ambivalence reflecting the nonlinear and dynamic nature of transitions, underscoring how uncertainty can emerge early in the transition process. Students experiencing educational transitions displayed sustained fluctuation between confidence and doubt, oscillating between adjustment and renewed tension rather than moving smoothly across phases (Lang et al., 2024). Progress may involve setbacks as well as achievements, and learners may move back and forth between confidence and uncertainty as they adapt to change. Young adults navigating the pandemic found themselves "stalling, regressing, pivoting, reflecting, and ultimately, following new and revised pathways". However, these challenges can also create opportunities for growth. Overcoming fear, letting go of control, and being willing to be exposed to discomfort and failure opens the way to new insights and new understandings of self (Marangon et al., 2025). Certain negative emotions can actually act as catalysts for inner growth and role adaptation (Stephens et al., 2023), with moments of insight, increased confidence, and personal breakthrough contributing to deeper learning and successful adaptation to new circumstances.

Examples in Adult Education

The concepts associated with learning through life transitions can be observed across a wide range of educational settings. Adult learners frequently enter educational programs because they are experiencing significant personal, professional, or social change. In these situations, learning supports not only the acquisition of knowledge and skills but also adaptation, identity development, and successful participation in new communities and roles.

At a higher education level, students, for example, enrolled in political studies may experience learning as part of a broader process of intellectual and personal transformation. Consider a mature-age student who has spent twenty years working in local government and decides to undertake a Bachelor of Politics and International Relations. Initially, the student may identify primarily as a public servant whose understanding of political processes is based on practical workplace experience. As they engage with political theory, public policy analysis, international relations, and critical debates about governance, they may begin to question previously held assumptions about power, democracy, citizenship, and social change. Participation in academic discussions, research projects, and scholarly communities can encourage the development of a new identity as a political thinker, researcher, or policy analyst. Learning therefore involves more than mastering disciplinary content; it involves becoming a member of an academic community and developing new ways of understanding both society and oneself. Through this transition, the learner may move from relying primarily on organisational practices and established procedures to adopting a more analytical, critical, and evidence-based approach to understanding political issues.

A similar process can be observed in vocational education within the business services sector. Consider an experienced administrative assistant undertaking a Certificate IV or Diploma in Business (Australian qualification example) with the goal of moving into a supervisory role. The learner enters the program with a well-established identity as a support worker responsible for implementing instructions and maintaining organisational processes. Through studying leadership, project coordination, business communication, workplace planning, and team management, the learner gradually develops new competencies and begins to view themselves differently. Workplace projects, mentoring opportunities, and interaction with managers allow them to participate in a new community of practice associated with leadership and business management. Over time, they may begin to identify as a coordinator, supervisor, or emerging manager rather than solely as an administrative employee. The transition therefore involves both competence development and identity formation as the learner adapts to new responsibilities and expectations.

Learning through transitions can also be seen when individuals move between sectors or occupations. For example, a small business owner undertaking vocational training in business services may need to adopt new approaches to strategic planning, financial management, and human resource management as their business expands. The transition from self-employed operator to business manager requires not only new technical knowledge but also the development of a leadership identity. Similarly, a public servant studying political science may transition from seeing policy as something implemented within organisational frameworks to understanding themselves as someone capable of shaping policy through research, advocacy, or political engagement. In both cases, learning supports adaptation to changing professional roles while facilitating broader changes in identity and self-concept.

These examples demonstrate that learning through life transitions extends beyond the acquisition of new information. Whether in higher education or vocational education, learners often engage in processes of reflection, experimentation, participation, and identity reconstruction. As they navigate new roles, communities, and expectations, learning becomes a mechanism through which they make sense of change, develop confidence, and construct new personal and professional identities. In this way, education serves as both a source of knowledge and a means of supporting successful transitions throughout adult life.

Critical Consciousness and Empowerment

Critical consciousness and empowerment are closely related concepts in adult learning that focus on helping individuals understand, question, and transform the social, political, economic, and cultural conditions that shape their lives. Rather than viewing learning solely as the acquisition of knowledge and skills, these concepts emphasise learning as a process of developing awareness, agency, and the capacity to act upon the world (Chundur, 2020; Kastner & Motschilnig, 2021; Khandekar, 2021). They are particularly associated with critical pedagogy and the work of Brazilian educator Paulo Freire, whose ideas have had a profound influence on adult education, community development, and social justice education (Carnicelli et al., 2022; Grummell, 2023; Majola et al., 2024; Suarez-Balcazar, 2020).

Freire's concept of *conscientização* (P. Freire, 1970), or critical consciousness, refers to an engaged learning process that provides learners with the capacity to perceive social, political, and economic contradictions and to take action against oppressive elements of reality (Christens et al., 2015; Chundur, 2020). P. Freire (1970) described critical consciousness as consisting of both reflection and action geared toward the transformation of social systems and conditions (Christens et al., 2015), a unity he termed *praxis* (Majola et al., 2024; Nayak, 2026). This process involves transitioning from a state of adapted consciousness to critical consciousness through active dialogical education (Khandekar, 2021). Freire identified three levels of conscientisation: intransitive, semi-transitive, and critical consciousness (Seyi-Oderinde, 2021).

Central to Freire's pedagogy is the distinction between "banking" education, where learners passively receive knowledge, and "problem-posing" education, which encourages learners to critically analyse and understand their social and economic conditions (Carnicelli et al., 2022; Khandekar, 2021; Lalendle et al., 2020). Problem-posing education facilitates critical learning by engaging learners in dialogue about issues drawn from their everyday contexts (Schreiber-Barsch et al., 2023; Seyi-Oderinde, 2021). Freire's emancipatory model emphasises the development of "critical consciousness; the creative capacities; and the confidence, skills, and attitudes to intervene in the transformation of the social world" (Schugurensky (2011, p. 72) cited in Grummell (2023, p. 150).

Empowerment is intimately connected to critical consciousness. Both concepts share roots in attempts to overcome oppression and foster human development, community participation, and wellbeing (Christens et al., 2015). Freire referred to adult education as "democratic empowerment," and his work with the poor in Brazil promoted critical awareness about people's social and economic contexts and actions to generate social change (Suarez-Balcazar, 2020). Critical pedagogy implies that education's main purpose should be to convert existing society into one where equality, freedom, and empowerment exist. Education becomes critical pedagogy when educators, through dialogue, encourage learners to acknowledge constraints that impact their lives and recognise that some restrictions can be overcome (Nöel et al., 2019).

The critical approach to literacy exemplifies this integration of consciousness and empowerment. Rather than defining literacy from a purely economic or functional perspective, the critical approach emphasises the social and political context, viewing literacy as contributing to social transformation and empowering people to participate actively in the democratic process (Erbay, 2020). For Freire, literacy was not just about reading and writing skills but also about mobilising social resources in communities to confront and analyse surrounding social, political, and economic structures (Sykes & Wills, 2019). Reading, in Freire's pedagogy, is a process of interpreting, reflecting on, interrogating, and theorising, while writing is an acting on and dialogically transforming movement (Erbay, 2020).

These concepts have been applied across diverse educational contexts. In adult basic education, the framework of transformative learning intersects with critical consciousness to establish democratic spaces of learning that act as resources for self-empowerment, emancipation, participation, and collective action toward humanisation and social justice (Kastner & Motschilnig, 2021). In community development, participatory research methods drawing on Freire's critical consciousness have been used to understand and address disparities experienced by vulnerable populations (Chundur, 2020; Suarez-Balcazar, 2020). Community-based participatory research combines social investigation, educational work, and social action, favouring the oppressed and marginalised, with wider aims including emancipation, democratisation, and social justice (Kastner & Motschilnig, 2021).

The Irish adult literacy movement provides a concrete example of Freire's influence, where adult education was shaped by popular education and Freire's work brought to Ireland primarily through religious who had been involved in basic education in Latin America and the Philippines (Grummell, 2023). The movement was founded on an approach based on social justice and empowerment of learners stemming from Freire's philosophy (Grummell, 2023). Similarly, in South Africa, critical and radical traditions of adult education, including feminist adult education, anti-racist education, and decolonising methodologies, reflect the importance of self-determination, self-regulation, and self-reflection for successful adult learning (Lalendle et al., 2020).

Critical consciousness has also been extended beyond traditionally marginalised populations. Scholars have argued that conscientisation is critical in pedagogies where learners are not essentially marginalised or oppressed but may be considered privileged (Seyi-Oderinde, 2021). Freire's transformative conceptualisations have been shown to impact not only the "oppressed" but also the "privileged" who choose to work with them (Khandekar, 2021). In vocational education, Freirean concepts including generative themes, conscientisation, and dialogical learning have been applied to nurture critical consciousness, transformative action, and learner empowerment among students (Majola et al., 2024).

The relationship between empowerment and critical consciousness has been further elaborated in psychological research. Sociopolitical development, rooted in Black liberation psychology, builds upon critical consciousness by considering the sociocultural skills and contexts that influence and support it (Hope et al., 2023). While critical consciousness focuses primarily on cognitive shifts that motivate action for social justice, empowerment encompasses the emotional dimensions of civic development (Christens et al., 2015). Both frameworks contend that social change begins with problematising inequitable systems and that those in positions of subjugation have the power to effect change (Hope et al., 2023).

Contemporary applications continue to demonstrate the relevance of these concepts. In digital education, applying Freire's concept of critical consciousness recognises a dynamic and continuing cycle of action and reflection, enabling learners to understand how changes in technology impact their lives and to identify and overcome oppressive elements (Chundur, 2020). In health education, critical health literacy has built on Freire's emancipatory philosophy, with studies showing that programmes most successfully developing understanding of social determinants and political action are those based on a critical pedagogical cycle of identifying issues, reflection and dialogue on causes, and promotion of social action (Sykes & Wills, 2019).

Reflection and Action

A central feature of critical consciousness is the relationship between reflection and action. Freire argued that critical reflection alone is insufficient if it does not lead to meaningful action, while action without reflection may fail to address underlying causes of problems (Aguiar, 2023; Joerke et al., 2023; Lopes & Abílio, 2023). Freire emphasised that critical consciousness develops through the integration of reflection and action rather than through contemplation alone. From this perspective, individuals develop deeper awareness by critically examining their experiences, questioning assumptions, and acting upon the insights gained through that reflection. Learning is therefore understood as a dynamic process in which reflection and action continually inform one another, enabling individuals to develop new understandings of themselves and the world around them.

This relationship between reflection and action is commonly referred to as *praxis*. Freire conceptualised praxis as a process through which people critically engage with their social reality and take informed action to transform it (Lopes & Abílio, 2022). Through praxis, learners analyse their circumstances, develop alternative perspectives, and work towards improving their situations or contributing to broader social change (Joerke et al., 2023; Lopes & Abílio, 2023; Rodrigues, 2024). In this view, learning extends beyond acquiring knowledge and becomes a means of fostering personal agency and social transformation.

Scholars have argued that both reflection and action are necessary components of meaningful learning. Action that occurs without critical reflection may lack direction or understanding, while reflection that does not lead to action may fail to produce meaningful change (Lopes & Abílio, 2023). Consequently, transformative learning requires a balance between thoughtful analysis and practical engagement. Through this ongoing cycle of reflection and action, learners actively participate in shaping their own development and influencing the environments in which they live and work. Learning therefore becomes a process of engaging with and responding to the world, rather than simply acquiring information about it (Lopes & Abílio, 2022).

Empowerment Through Learning

Empowerment refers to the process through which individuals develop greater control over their lives, increase confidence in their abilities, and strengthen their capacity to make informed decisions and influence outcomes. Within adult education, empowerment is often viewed as both a process that occurs during learning and an outcome that emerges from successful educational experiences (Abbajobir, 2026; Cho, 2025; Finkel, 2014). As learners acquire knowledge, skills, and new perspectives, they frequently become better equipped to understand their circumstances, pursue their goals, and participate more actively in personal, professional, and community life (Abbajobir, 2026; Finkel, 2014; Hassi & Laursen, 2015).

Empowered learners typically demonstrate increased self-efficacy, confidence, autonomy, and personal agency (Hassi & Laursen, 2015; F. Wang et al., 2024). They become more capable of identifying and pursuing goals, making decisions, solving problems, and responding effectively to challenges (Hassi & Laursen, 2015) . Empowerment enables individuals to move beyond dependence on others for direction and develop a stronger sense of responsibility for their own learning, development, and future opportunities (Abbajobir, 2026; Cho, 2025). As a result, learners often become more willing to advocate for themselves, take initiative, and contribute to decision-making processes within their workplaces, communities, and social networks (Abbajobir, 2026; Finkel, 2014) .

Importantly, empowerment extends beyond personal confidence and individual achievement. It also encompasses the capacity to influence social, organisational, and political environments, (Finkel, 2014; F. Wang et al., 2024). Empowered individuals are better positioned to participate in collective decision-making, challenge inequitable practices, contribute to organisational improvement, and engage in civic and community activities (Abbajobir, 2026; Finkel, 2014; Shileche et al., 1970). Civic empowerment refers to an individual's belief that their participation and engagement can contribute to shaping civic and societal decisions. It reflects the extent to which people perceive themselves as capable of influencing issues that affect their communities and broader society. From this perspective, empowerment encompasses not only personal growth and increased self-confidence but also the capacity to participate meaningfully in democratic, community, and civic processes (Cho, 2025; F. Wang et al., 2024). Through empowerment, individuals develop the knowledge, skills, confidence, and agency needed to engage with social issues, contribute to decision-making, and influence outcomes within their communities and institutions.

Adult learning can support empowerment in a variety of ways. Educational experiences can help learners develop critical thinking skills that enable them to analyse information, evaluate assumptions, and make informed judgements (Hassi & Laursen, 2015; F. Wang et al., 2024). Learning can also increase self-awareness by encouraging individuals to reflect on their experiences, values, strengths, and aspirations (Hassi & Laursen, 2015; F. Wang et al., 2024). Through successful participation in educational activities, learners often build confidence and self-efficacy, gaining a stronger belief in their capacity to learn, adapt, and succeed (Abbajobir, 2026; Hassi & Laursen, 2015) .

Access to knowledge, information, and resources further contributes to empowerment by providing individuals with the tools needed to make informed decisions and pursue opportunities (Cho, 2025; Finkel, 2014). Educational programs can also create opportunities for learners to participate in decision-making processes, collaborate with others, and take responsibility for aspects of their learning (Abbajobir, 2026; F. Wang et al., 2024). Experiential learning, critical reflection, and collaborative learning have been identified as key processes that accelerate empowerment practices (F. Wang et al., 2024). In addition, adult education can encourage learners to question limiting assumptions, stereotypes, and social expectations that may constrain personal or professional growth (Hassi & Laursen, 2015). By engaging critically with their experiences and environments, learners can develop greater agency and become more active participants in their communities, workplaces, and professions (Abbajobir, 2026; Finkel, 2014).

Consequently, empowerment is widely regarded as an important goal of adult education (Abbajobir, 2026; Cho, 2025; Finkel, 2014). Research on integrated functional adult education has demonstrated that such programs can serve as a strong pathway to empowerment when it is inclusive, practical, and closely connected to community daily life, helping learners move from limited participation to active involvement in shaping their own futures (Abbajobir, 2026). Through learning, individuals can develop not only new knowledge and skills but also the confidence, critical awareness, and capacity for action that enable them to shape their own lives and contribute meaningfully to the broader society (Abbajobir, 2026; Hassi & Laursen, 2015; F. Wang et al., 2024).

Critical Consciousness and Adult Learning

Critical consciousness aligns closely with many principles of adult learning. Adults bring rich life experiences that provide a foundation for reflection and analysis (Halman et al., 2017; Jones, 2020). Critical consciousness, as theorized by Paulo Freire, was designed to foster transformative knowledge that allows learners to connect with their own personal, cognitive, and emotional experiences through reflection and dialogue with others (Jones, 2020). Through dialogue, collaborative learning, and examination of real-world issues, learners can explore how personal experiences connect to broader social contexts (Achilleos et al., 2021; Egan-Simon, 2022; Majola et al., 2024). Dialogue has been described as a fundamental component of critical pedagogy and the basis of democratic education (Pollard, 2020a, 2020b), with Freire asserting that dialogue is the encounter between men, mediated by the world in order to name the world. The collaborative and dialogic nature of critical learning empowers participants to actively shape discussions, contributing to richer exploration of educational complexities and potential for transformation (Majola et al., 2024; Teemant et al., 2024).

Unlike approaches that focus solely on technical competence, critical adult education encourages learners to question assumptions, examine power relationships, and consider alternative perspectives (Askari & Behdarvandirad, 2025; Halman et al., 2017; Sim & Bierema, 2025). Critical pedagogy is distinguished from traditional approaches by its concern with an ongoing process of discovery where knowledge emerges within a community through dialogue rather than being deposited into learners through what Freire termed "banking education" (Halman et al., 2017; Ignacio, 2020; Jones, 2020). Halman et al. (2017) identify key themes of critical consciousness in education as including "appreciating context," "illuminating power structures," "moving beyond 'procedural,'" "enacting reflection," and "promoting equity and social justice". Learning thus becomes a means of understanding not only how things work but also why they work that way and how they might be improved (Adhikari & Kunwar, 2023; Diko, 2023). Critical pedagogy probes power determinants, social inequalities, and the effects of authoritative ideologies Diko, 2023), encouraging learners to function as active agents who develop the ability to evaluate the validity, fairness, and authority within their educational and daily life situations. This transformative orientation moves learners from being passive objects to active subjects who can critically develop knowledge, question the status quo, and make possible changes through their agency (Askari & Behdarvandirad, 2025).

This approach is particularly relevant in contexts where learners face social, economic, or educational disadvantage (Pollard, 2020b; Sim & Bierema, 2025; Xu & Knijnik, 2021). By understanding structural influences on their experiences, learners may develop greater confidence and capacity to advocate for themselves and others (Achilleos et al., 2021; Killam & Camargo-Plazas, 2021; Rondini, 2015). The task of critical pedagogy is to bring members of oppressed groups to a level of critical consciousness so they feel empowered to become active agents of change (Diko, 2023; Egan-Simon, 2022). Freirean pedagogical philosophy and the critical consciousness process can both support and empower students from disadvantaged backgrounds in their demands for quality education (Xu & Knijnik, 2021). Watts et al. (2011), cited in Achilleos et al. (2021), identify three key elements of critical consciousness, critical reflection, political efficacy, and critical action, where critical reflection analyses structural causes of social issues, political efficacy refers to awareness of one's capacity to effect socio-political change, and critical action involves actions taken to change society. Research demonstrates that individuals with high critical consciousness scores tend to be actively engaged in a wide range of activities to support social justice (Sim & Bierema, 2025), and service-learning contexts show that students can develop structural analyses of inequities alongside reflexive consideration of their own potential to effect lasting social change (Rondini, 2015).

Applications in Adult Education

In higher education, critical consciousness can be developed through activities that encourage students to analyse social issues, evaluate competing perspectives, and reflect on their own assumptions. For example, students in political studies may examine issues such as inequality, public policy, human rights, or democratic participation (Christens et al., 2015; Landig & Landig, 2024; Obiagu, 2025). Through critical engagement with evidence and debate, they learn to understand the relationships between individual experiences and broader political structures (Christens et al., 2015; Moore et al., 2016). Critical consciousness involves the ability to analyse inequities connected to one's social conditions, including "critical analysis of social, political, economic, and race or gender-based inequities" (Christens et al., 2015, p. 17). Higher education institutions promote civic development through various pedagogies including curricular practices, service-learning, and community engagement (Alcantar et al., 2024). Research has found that educators often seek to foster students' critical awareness of social justice issues affecting society and local communities while simultaneously strengthening their sense of civic responsibility, participation, and personal agency (Alcantar et al., 2024). Through educational experiences that encourage reflection, dialogue, and engagement with social issues, learners can develop a deeper understanding of societal challenges and a greater capacity to contribute to positive social and community outcomes.

A mature-age student studying political science may initially view unemployment, housing affordability, or healthcare access as individual problems. Through engagement with political theory, public policy analysis, and social research, the student may develop a deeper understanding of the economic, political, and institutional factors that influence these issues (Moore et al., 2016; Obiagu, 2025). A systems worldview, grounded in Freire's concept of critical consciousness, involves recognising that social inequalities are often shaped by broader structural and institutional factors rather than solely by individual choices or circumstances (Moore et al., 2016). From this perspective, issues such as racism, sexism, and socioeconomic disadvantage are understood as products of historical and contemporary social systems that influence opportunities, resources, and outcomes for different groups within society.

Developing a critical understanding of these systemic influences enables individuals to examine the underlying causes of social inequities and to recognise how social, political, economic, and cultural structures contribute to unequal outcomes (Moore et al., 2016). Rather than viewing social conditions as inevitable or solely the result of personal responsibility, learners are encouraged to analyse the broader contexts in which those conditions emerge and persist.

This deeper awareness can contribute to both intellectual development and civic engagement. Research suggests that individuals who understand the structural dimensions of social inequality are more likely to participate in civic and community activities and to engage in efforts aimed at addressing social issues (Christens et al., 2015; Moore et al., 2016). Through the development of critical consciousness, learners can become better equipped to understand complex social challenges and to participate more actively in democratic and community processes.

Critical consciousness and empowerment are also relevant within vocational education and training. Although vocational programs often focus on practical skills, they can also encourage learners to understand the broader context of their work (Obiagu, 2025; Ray & Zarestky, 2021). (Obiagu, 2025) argues that vocational education has traditionally focused on developing skills and competencies associated with employment and economic productivity. While these outcomes remain important, education also has a broader role in supporting the overall development and wellbeing of individuals. From this perspective, educational programs should promote not only occupational competence but also the knowledge, values, and capabilities required for meaningful participation in social, cultural, and civic life.

The sociopolitical development model reflects this broader educational purpose by encouraging learners to develop an understanding of the political, economic, cultural, and systemic factors that shape society and influence individual opportunities and experiences (Obiagu, 2025). Through this approach, learners are encouraged to examine how social structures affect their lives and communities, while developing the critical awareness and agency needed to participate effectively in addressing social challenges. Consequently, vocational education can contribute not only to workforce preparation but also to the development of informed, engaged, and socially responsible citizens capable of contributing to both economic and societal wellbeing.

For example, a learner undertaking a Diploma of Business may study workplace communication, leadership, and organisational behaviour. Beyond developing management skills, learners can examine issues such as workplace culture, diversity, inclusion, ethical decision-making, and organisational power structures (Ray & Zarestky, 2021). VET programmes have the potential to create inclusive environments that challenge workplace norms and support equitable practices (Ray & Zarestky, 2021). This broader understanding helps future managers recognise how their decisions affect employees, customers, and organisational outcomes.

Similarly, workers undertaking leadership development programs may progress from viewing leadership primarily as a position of authority to understanding it as a process that involves enabling others, encouraging participation, supporting collaboration, and fostering positive workplace cultures (Christens et al., 2015; Sjöberg et al., 2016). Within this perspective, effective leadership is closely associated with creating opportunities for individuals and groups to participate in decisions that affect their work and professional lives, thereby increasing their capacity to influence outcomes and contribute meaningfully to organisational success (Sjöberg et al., 2016).

Learning in this context contributes not only to the development of managerial and leadership competencies but also to the formation of more reflective, ethical, and socially responsible professional identities (Christens et al., 2015; Obiagu, 2025). As learners develop greater awareness of the relationships between individual actions, organisational structures, and broader social contexts, they are better positioned to exercise leadership in ways that support both individual and collective development.

Research on critical consciousness and empowerment highlights that these concepts share a common concern with promoting human development, participation, wellbeing, and the capacity of individuals and communities to influence the conditions that affect their lives (Christens et al., 2015; Obiagu, 2025). Consequently, leadership education informed by these principles encourages learners to view leadership not merely as directing others but as creating environments in which people can participate, contribute, and thrive.

Critical Consciousness, Identity, and Transformation

Critical consciousness is closely connected to transformative learning. As individuals critically examine assumptions and develop new understandings of themselves and society, they may experience changes in identity and worldview (Barjesteh et al., 2026; Maison, 2022). Learning can lead individuals to reconsider previously accepted beliefs, adopt new perspectives, and develop a stronger sense of agency (Barjesteh et al., 2026; Filipiak & Caraballo, 2019).

For example, an adult literacy learner who previously interpreted educational difficulties as a personal shortcoming may, through reflection and dialogue, develop an understanding of the broader educational, social, and economic factors that can influence learning opportunities and outcomes. This expanded perspective can strengthen confidence, increase motivation, and encourage continued participation in education and lifelong learning (Barjesteh et al., 2026; Filipiak & Caraballo, 2019). Research on critical literacy suggests that such learning can enhance learners' capacity to question existing assumptions, challenge inequitable structures, and engage in actions aimed at creating positive change within their communities and broader society (Filipiak & Caraballo, 2019).

Similar outcomes have been observed in technology-enhanced learning environments. Studies examining the integration of critical pedagogical approaches within AI-supported education have found that these approaches can reshape traditional understandings of both teaching and learning by strengthening learner agency and encouraging more active participation in knowledge creation (Barjesteh et al., 2026). Through collaborative inquiry, reflection, and critical engagement with technology, learners and educators can develop new forms of digital citizenship that emphasise responsibility, participation, and informed decision-making.

Likewise, a professional undertaking leadership development may begin to question traditional assumptions about authority, hierarchy, and decision-making, leading to the development of a more collaborative, inclusive, and participatory leadership identity. Learning in this context extends beyond acquiring managerial skills and contributes to broader changes in how individuals understand their roles and responsibilities within organisations and communities.

Indigenous educational approaches provide further examples of the relationship between learning, identity, and empowerment. Educational environments that connect learning to lived experience, community relationships, and cultural knowledge can strengthen learners' sense of belonging and reinforce their identities as active members of their communities (Maison, 2022). Furthermore, educational approaches that encourage critical examination of underlying values, assumptions, and social narratives can support both transformative and decolonising forms of learning, enabling individuals to develop deeper understandings of sustainability, culture, and social responsibility (Maison, 2022). Through these processes, learning becomes a means of fostering personal growth, community engagement, and broader social transformation.

Criticisms and Contemporary Perspectives

Although critical consciousness and empowerment remain influential concepts within adult learning, they have also attracted a range of criticisms. Some scholars argue that critical pedagogy places substantial emphasis on structural explanations of social issues while giving insufficient attention to individual differences, personal agency, and the subjective dimensions of learning (Fenwick, 2005; Manojan, 2019). Earlier forms of critical pedagogy, particularly those influenced by orthodox Marxist perspectives, have been criticised for focusing primarily on class-based analyses while providing limited consideration of issues such as gender, race, identity, subjectivity, and the diverse ways individuals experience and interpret social realities (Manojan, 2019). As a result, contemporary scholars have sought to broaden critical pedagogy by incorporating a wider range of perspectives and experiences.

Questions have also been raised regarding the practical implementation of critical pedagogy. While dialogue is often presented as a democratic and emancipatory educational process, some researchers argue that power relations remain present within educational settings and can influence whose voices are heard and whose perspectives are privileged (Fenwick, 2005). In some circumstances, the educator's role as facilitator may not reduce power imbalances but instead create new forms of authority that continue to shape the direction and outcomes of learning. These concerns highlight the complexity of creating genuinely participatory learning environments and the challenges associated with balancing guidance, expertise, and learner autonomy.

Empowerment has faced similar conceptual challenges. Scholars have noted that the concept is difficult to define consistently and can encompass a wide range of meanings across different disciplines and contexts (Mohajer & Earnest, 2009). Definitions frequently include elements such as participation, autonomy, self-efficacy, agency, social change, and community development, making the concept broad and sometimes difficult to operationalise. The absence of clear agreement regarding the goals and outcomes of empowerment has made it challenging to evaluate empowerment initiatives and assess their effectiveness. Furthermore, some researchers have observed that programs claiming to promote empowerment do not always incorporate the critical, transformative, or social action dimensions that were central to Freire's original conception of empowerment and social change (Mohajer & Earnest, 2009).

Contemporary perspectives generally continue to recognise critical consciousness and empowerment as valuable components of adult learning while acknowledging the importance of adapting these concepts to diverse cultural, social, organisational, and educational contexts (Askari & Behdarvandirad, 2025; French et al., 2021; McClain, 2024). Rather than applying critical pedagogy as a fixed framework, many educators adopt flexible approaches that consider local conditions, learner backgrounds, and contextual factors. This approach recognises that strategies which are effective in one setting may require modification in another to remain relevant and meaningful.

As a result, contemporary adult education often combines elements of critical pedagogy with other learning theories and practices. Critical reflection, dialogue, social learning, and transformative learning are frequently integrated to support both personal development and broader social understanding (Babacan & Babacan, 2012; French et al., 2021; McClain, 2024). Educational approaches such as intergroup dialogue encourage learners to examine social and political issues from multiple perspectives while developing greater awareness of the factors that shape social relationships and inequalities (French et al., 2021). Similarly, transformative learning approaches emphasise the importance of experience, critical reflection, dialogue, trust, openness, and the influence of social and cultural contexts on learning and identity development (McClain, 2024).

Across a range of educational settings, educators employ strategies such as collaborative learning, simulations, reflective journals, debates, role plays, case studies, and problem-based learning to foster critical thinking, self-reflection, and social awareness (Babacan & Babacan, 2012). These approaches seek to support learners in examining assumptions, understanding complex social issues, and developing the knowledge, skills, and dispositions required for meaningful participation in professional, civic, and community life. Consequently, while debates regarding critical consciousness and empowerment continue, both concepts remain important influences on contemporary adult learning theory and practice.

Significance for Adult Education

Critical consciousness and empowerment highlight the transformative potential of adult learning. They emphasise that education can do more than develop knowledge and skills; it can help individuals better understand themselves, their communities, and the forces that shape their lives. Through critical reflection, dialogue, and action, learners can develop greater agency, confidence, and capacity to participate effectively in society.

Consequently, critical consciousness and empowerment remain important concepts in adult education because they position learning as a means of personal growth, professional development, democratic participation, and social change. They remind educators that learning is not only about what individuals know, but also about how they understand the world and their ability to influence it.

Dialogue and Discourse in Adult Learning

Dialogue and discourse are central concepts in adult learning theory because they emphasise learning as a social process rather than an exclusively individual cognitive activity. Through conversation, questioning, reflection, debate, and the exchange of perspectives, adults construct meaning, challenge assumptions, develop new understandings, and negotiate knowledge with others (Brown, 2006; Sherlock & Nathan, 2008; Wilhelmson, 2006). Dialogue and discourse are therefore closely associated with constructivist, transformative, critical, and sociocultural approaches to learning (Jenkins, 2021; Koulaouzidis et al., 2024).

Although the terms are sometimes used interchangeably, dialogue and discourse refer to related but distinct concepts Wilhelmson, 2006). Dialogue generally refers to purposeful communication between individuals or groups that promotes understanding, reflection, and collaborative meaning-making (Jenkins, 2021; Wilhelmson, 2006). It involves active listening, questioning, openness to alternative viewpoints, and mutual respect among participants (Wilhelmson, 2006). Dialogue seeks not merely to exchange information but to deepen understanding through interaction (Jenkins, 2021; Sherlock & Nathan, 2008).

Discourse refers more broadly to the language, ideas, assumptions, values, and patterns of communication that shape how people understand and discuss particular topics (Nomlomo, 2021). Discourses influence what is considered legitimate knowledge, how problems are defined, and which perspectives are privileged or marginalised (Callahan & Sandlin, 2007; Nomlomo, 2021). In educational settings, discourse shapes how learners interpret experiences, engage with knowledge, and develop professional identities (Nomlomo, 2021).

The importance of dialogue in adult learning is strongly associated with Paulo Freire's critical pedagogy (Cennamo et al., 2020; Jenkins, 2021; Παπαθανασίου, 2023). Freire viewed dialogue as a process through which learners and educators collaboratively investigate reality, question assumptions, and develop critical consciousness (Godoy & J., 2024; Jenkins, 2021). Rather than positioning teachers as the sole source of knowledge, dialogue creates opportunities for learners to contribute their experiences and participate actively in knowledge construction (Carlisle & McCloskey, 2023; Cennamo et al., 2020).

Dialogue also occupies a central place in Mezirow's Transformative Learning Theory (Brown, 2006; Carlisle & McCloskey, 2023; Kallou & Kikilia, 2021; Sherlock & Nathan, 2008; Wilhelmson, 2006; Παπαθανασίου, 2023). Mezirow argued that transformative learning often occurs through critical reflection combined with rational discourse (Brown, 2006; Carlisle & McCloskey, 2023; Παπαθανασίου, 2023). When learners encounter experiences that challenge their existing assumptions, dialogue allows them to examine evidence, consider alternative perspectives, and test the validity of their beliefs (Sherlock & Nathan, 2008; Wilhelmson, 2006). Through discussion and critical examination, learners may revise previously held assumptions and develop more inclusive and informed perspectives (Brown, 2006; Jenkins, 2021; Kallou & Kikilia, 2021).

Sociocultural theories of learning similarly emphasise the role of discourse in knowledge construction (Merzliakova et al., 2022; Nomlomo, 2021). From this perspective, learning occurs through participation in social activities and communities where language serves as a primary tool for thinking, communication, and learning (Merzliakova et al., 2022; Nomlomo, 2021). Knowledge is not simply transmitted from one person to another but is co-constructed through interaction and shared meaning-making (Merzliakova et al., 2022; Nomlomo, 2021)

Dialogue as a Process of Meaning-Making

Adults bring extensive experiences, beliefs, values, and assumptions to learning situations. Dialogue provides opportunities for learners to connect these experiences with new information and alternative perspectives (Muhtar et al., 2022; Odutayo, 2023). Through discussion, learners compare viewpoints, identify similarities and differences, and develop deeper understandings of complex issues (Drury et al., 2016; Egan-Simon, 2022; Wahlström, 2022).

For example, students in a political studies course may engage in discussions about democracy, public policy, or social justice (Barnes et al., 2022; Muhtar et al., 2022). Learners often enter these discussions with differing political experiences, ideological perspectives, and cultural backgrounds (Keegan, 2021; Muhtar et al., 2022). Through respectful dialogue, participants can explore alternative viewpoints, examine evidence, and develop more sophisticated understandings of political issues (Muhtar et al., 2022; Wahlström, 2022). The learning that occurs is not limited to acquiring factual knowledge but involves developing critical thinking, perspective-taking, and civic understanding (Barnes et al., 2022; Egan-Simon, 2022; Odutayo, 2023).

Similarly, learners in a vocational business services program may discuss workplace leadership, organisational culture, or ethical decision-making. Through dialogue, participants can draw upon workplace experiences, compare management approaches, and collaboratively explore solutions to organisational challenges (Drury et al., 2016; Odutayo, 2023). This process enables learners to connect theoretical concepts with practical workplace realities (Odutayo, 2023; Wahlström, 2022).

Dialogue and Critical Reflection

Dialogue plays an important role in promoting critical reflection. While individual reflection allows learners to examine their assumptions privately, dialogue introduces alternative viewpoints that may challenge existing ways of thinking (Frkal, 2018; Marotta, 2024; Nganga et al., 2025). Exposure to different perspectives can create opportunities for learners to reconsider previously unquestioned beliefs and develop more nuanced understandings (Frkal, 2018; Nganga et al., 2025; Ross-Gordon et al., 2015).

For example, a manager participating in leadership development may initially believe that effective leadership primarily involves directing and controlling employees. Through dialogue with colleagues, mentors, and educators, they may encounter alternative perspectives emphasising collaboration, empowerment, and shared decision-making (Gray, 2006; Lawrence, 2017; Spooner & John, 2020). These discussions can stimulate reflection and contribute to changes in leadership philosophy and practice (Frkal, 2018; Mbokota et al., 2022; Sweet, 2022).

In this way, dialogue functions as a catalyst for transformative learning by creating opportunities for learners to critically evaluate their assumptions and consider alternative ways of understanding themselves and the world (Boonyuen et al., 2016; Frkal, 2018; Marotta, 2024; Nganga et al., 2025).

Discourse and Power

Critical theorists emphasise that discourse is closely connected to power. The language used within institutions, professions, and societies influences how issues are understood and how individuals are positioned within social structures. Certain discourses may legitimise particular perspectives while marginalising others (Careless, 2015a, 2015b; Kamp, 2003). Critical discourse analysts aim to "reveal connections between language, power, and ideology, and to describe the way power and dominance are produced and reproduced in the discourse structures of generally unremarkable interactions" (Kamp, 2003, p. 139). Critical theorists seek to understand how inequity comes to be seen as "normal" in society, and believe that educational programmes and policies often serve to maintain the dominant ideology (Careless, 2015a, 2015b).

For example, workplace discussions about productivity, performance, efficiency, or professionalism often reflect broader organisational values and assumptions. Researchers have argued that contemporary education and training systems are increasingly influenced by economic and market-oriented priorities, with significant emphasis placed on productivity, competition, employability, credential attainment, and participation in a globalised economy (Careless, 2015b). As a result, the language and assumptions embedded within workplace and educational discourse can shape how success, competence, and professional achievement are defined and understood.

Similarly, public discussions concerning issues such as immigration, poverty, education, environmental sustainability, and social welfare influence how these topics are interpreted, debated, and addressed within society (Gouthro & Holloway, 2023; Wildemeersch et al., 2023). The ways in which these issues are framed can affect public perceptions, policy responses, and individual attitudes. Adult learning can support learners in critically examining such discourses by encouraging them to analyse how language, assumptions, values, and interests influence the presentation and interpretation of social issues. Through critical inquiry, learners can develop greater awareness of how media, institutions, and cultural narratives shape understandings of political and social realities (Sandlin et al., 2011).

Developing awareness of discourse enables learners to identify underlying assumptions, recognise alternative viewpoints, and critically evaluate dominant narratives and accepted forms of knowledge (Gouthro & Holloway, 2013; Gouthro & Holloway, 2023; Sumithran et al., 2022). Critical educators emphasise that learning does not occur in a neutral environment; rather, it is shaped by relationships of power that influence whose knowledge is valued, whose voices are heard, and which perspectives are considered legitimate (Gouthro & Holloway, 2013; Gouthro & Holloway, 2023). Consequently, critical pedagogy encourages learners to examine knowledge claims, question taken-for-granted assumptions, and investigate the social, cultural, political, and economic influences that shape understanding (Sumithran et al., 2022).

This process contributes to the development of critical consciousness by helping learners recognise the connections between knowledge, power, and social structures. Through critical reflection and dialogue, learners can develop a deeper understanding of social, political, and economic contradictions and become better equipped to evaluate how institutions, ideologies, and systems of power influence both individual experiences and broader societal outcomes (Gouthro & Holloway, 2013; Gouthro & Holloway, 2023; Mikelli, 2020). In this way, adult learning supports not only the acquisition of knowledge but also the development of the critical awareness necessary for informed participation in social, civic, and professional life.

Dialogue in Communities of Practice

Within communities of practice, dialogue serves as a mechanism through which newcomers learn the norms, values, language, and practices of a profession or community. Through interaction with more experienced members, learners gradually acquire both technical knowledge and an understanding of what it means to belong to the community (Burwell, 2012; Buus et al., 2022; Umbreş, 2021)

For example, apprentice tradespeople learn not only technical skills but also the language, safety practices, and professional expectations of their occupation through ongoing conversations with supervisors and colleagues (Howe et al., 2023; Sligo et al., 2011; Umbreş, 2021). Similarly, nursing students develop professional identities through dialogue with patients, educators, and experienced practitioners (Baker, 2019; Gallup, 2023). In both cases, learning occurs through participation in social interactions that support competence and identity development (Burwell, 2012; Buus et al., 2022; Ilomäki et al., 2022).

Technology and Online Dialogue

Digital technologies have expanded opportunities for dialogue beyond traditional classroom settings. Online discussion forums, virtual classrooms, social media, professional networks, and collaborative digital platforms allow learners to engage in dialogue across geographical and cultural boundaries (Mairs et al., 2013; Risner & Kumar, 2016).

These technologies can support collaborative learning (Fanguy et al., 2023; Mudau & Berg, 2023), peer feedback (Mohamed & Abdallah, 2021; Woo & Reeves, 2019), knowledge sharing (Mairs et al., 2013; Shanthi & Purwarno, 2022), and professional networking (Fayard & DeSanctis, 2005). However, they also present challenges, including misinformation (Mairs et al., 2013), unequal participation (Aderibigbe et al., 2021; Woo & Reeves, 2019), reduced social cues (Graham, 2019; Tarhouny et al., 2021), and difficulties establishing trust and meaningful engagement (Mudau & Berg, 2023; Shanthi & Purwarno, 2022). Effective online dialogue therefore requires careful facilitation (Aderibigbe et al., 2021; Woo & Reeves, 2019; Zhong & Norton, 2019) and the development of digital communication skills (Tarhouny et al., 2021; Yousaf & Masood, 2024).

Benefits of Dialogue and Discourse

Dialogue and discourse support adult learning in numerous ways by creating opportunities for learners to engage with ideas, experiences, and perspectives that may differ from their own. Through discussion, questioning, and collaborative inquiry, learners are encouraged to examine information critically, evaluate evidence, and reflect on their assumptions and beliefs (Mikulec, 2017; Scott, 2005). This process strengthens critical thinking and reflective practice, enabling individuals to develop deeper and more informed understandings of complex issues (Mikulec, 2017).

Engaging in dialogue also encourages perspective-taking and empathy. By listening to the experiences and viewpoints of others, learners gain insight into alternative ways of understanding the world (Coryell, 2013). Exposure to diverse perspectives can broaden understanding, challenge stereotypes, and promote greater appreciation of cultural, social, and professional differences (Akob, 2016;

Malilang, 2019). As learners consider viewpoints that differ from their own, they often develop a more nuanced and inclusive understanding of people and issues.

Dialogue and discourse further contribute to the collaborative construction of knowledge. Rather than viewing knowledge as something transmitted from expert to learner, many adult learning approaches recognise that understanding is often developed through interaction and shared meaning-making (Coryell, 2013; Schugurensky & Myers, 2008). Through discussion, debate, and collaborative problem-solving, learners build upon each other's experiences and ideas, generating new insights that may not emerge through individual study alone (Mikulec & Ramalho, 2022).

These processes can also promote transformative learning. Encountering alternative perspectives and engaging in critical discourse may challenge existing assumptions, prompting learners to reconsider previously held beliefs and develop new ways of understanding themselves and their environments (Coryell, 2013; Mikulec & Ramalho, 2022). Dialogue therefore serves as an important mechanism through which learners critically examine their perspectives and, where appropriate, revise them in light of new evidence and experiences (Mikulec, 2017).

Participation in dialogue also strengthens communication skills. Learners develop their ability to articulate ideas, present arguments, listen actively, provide constructive feedback, and engage respectfully with differing viewpoints (Akob, 2016; Ignatyeva et al., 2020). These skills are valuable not only within educational settings but also in workplaces, professional communities, and civic life, where effective communication is essential for collaboration and decision-making.

Dialogue and discourse play an important role in professional identity development. Through participation in professional conversations, learners acquire the language, values, norms, and expectations associated with particular occupations and disciplines (Schugurensky & Myers, 2008). Interaction with peers, mentors, educators, and experienced practitioners helps individuals develop an understanding of what it means to be a member of a professional community and supports the formation of professional identity (Coryell, 2013; Malilang, 2019).

Beyond individual and professional development, dialogue can enhance civic participation and democratic engagement. Discussing social, political, and community issues encourages learners to consider public concerns, evaluate competing viewpoints, and participate more actively in civic life (Martins, 2008; Mikulec, 2017). Through dialogue, individuals develop the knowledge and confidence needed to contribute to democratic processes, community decision-making, and public discourse (Martins, 2008; Mikulec, 2017).

Finally, dialogue and discourse encourage the critical examination of assumptions, values, and social structures. By exploring how language, culture, institutions, and power relationships influence understanding and behaviour, learners can develop greater awareness of the factors that shape individual and collective experiences (Scott, 2005). This critical awareness supports more informed decision-making and can contribute to personal growth, social responsibility, and meaningful participation in society (Martins, 2008).

Criticisms and Limitations

Despite their benefits, dialogue and discourse are not without limitations. Equal participation cannot always be assumed, as power relations, confidence levels, cultural norms, language proficiency, and organisational hierarchies may influence whose voices are heard (Dorsey, 2009; Kiramba & Smith, 2019; Skrefsrud, 2024; Sun, 2025). The culture of power that exists in classrooms and broader society, along with the hierarchies resulting from socialization, can serve as obstacles for participation in dialogue (Dorsey, 2009). In classroom settings, teachers may use their privileged position of hierarchy to restrict students' freedom of expression and overlook students' knowledge and experience (Kiramba & Smith, 2019).

Power dynamics represented by linguistic privilege, learning environment, and identity formation can influence students' language preferences and engagement (Sun, 2024). Furthermore, education is inherently non-neutral, often acting as a conduit for the transmission of dominant cultural norms and contributing to the perpetuation of societal power dynamics and hierarchies (Skrefsrud, 2024). Some participants may dominate discussions, while others may be reluctant to contribute (Bateman et al., 2025; Martin-Thomsen et al., 2021). Studies have shown that even in settings designed to encourage multilingual participation, students experienced hesitation and anxiety in formal classroom spaces where only the dominant language was typically sanctioned (Gorp, 2026). Discourse norms of classrooms can be misaligned with the norms of students' home communities, and imbalances in interpersonal power relations can limit participation (Bateman et al., 2025). Evaluation practices based on dominant culture language competency can function as gatekeepers, reinforcing constructs of cultural hegemony (Martin-Thomsen et al., 2021).

Additionally, dialogue does not automatically lead to transformation or consensus (Olatunji et al., 2025; Wade et al., 2008). Participants may interpret the same information differently, and deeply held beliefs may remain resistant to change (Wade et al., 2008). Research on peer-led dialogues among prospective teachers found that few students engaged in critical reflection, and many responses reflected deficit theories, stereotypical thinking, and technical-rational problem solving (Wade et al., 2008). Transformation manifests as a multifaceted process heavily influenced by the socio-cultural environment in which it takes place, and transformative conversations do not require a normative consensus (Olatunji et al., 2025). Educators therefore need to create environments that encourage respectful participation, critical inquiry, and openness to diverse perspectives (Evans et al., 2024; Parba, 2021; Skrefsrud, 2024).

Engaging students in critical dialogue requires creating spaces where diverse perspectives can be developed and discussed, contributing to culturally creative classroom practices in which the power of untold narratives enriches the overall classroom community (Skrefsrud, 2024). Teachers must move beyond traditional lessons to make learning experiences more socially engaged, enabling students to participate at higher levels of discourse interaction (Parba, 2021). Educators should foster inclusivity, cultural relevance, and linguistic equity, challenging monolingual norms while creating learning environments that are more inclusive, equitable, and socially responsible (Evans et al., 2024; Sun, 2024).

Significance for Adult Learning

Dialogue and discourse remain fundamental concepts in contemporary adult learning because they recognise learning as a social, collaborative, and communicative process. Through dialogue, learners construct meaning, examine assumptions, and develop new perspectives. Through discourse, they engage with the broader systems of language, knowledge, and power that shape how reality is understood. Together, dialogue and discourse support critical reflection, transformative learning, professional development, and democratic participation, making them essential components of effective adult education and lifelong learning.

As an example of application, the concepts of dialogue, discourse, power, and critical consciousness can be applied effectively within adult education programs in Agriculture, Horticulture, and Conservation and Land Management. Adult learners in these fields often bring extensive practical experience, strong occupational identities, and deeply held beliefs about land use, production systems, environmental management, and rural communities. Dialogue provides opportunities for learners to examine these experiences critically while engaging with alternative perspectives, new research, and evolving industry practices.

For example, a group of adult learners undertaking a Diploma of Agriculture may participate in a facilitated discussion about the use of chemical pesticides and herbicides. Some learners may draw upon decades of farming experience and argue that chemical controls are essential for maintaining productivity and profitability. Others may have experience with regenerative agriculture, integrated pest management, or organic farming systems and advocate for alternative approaches. Rather than presenting a single "correct" answer, the educator encourages learners to examine the evidence, question assumptions, and explore the environmental, economic, social, and regulatory implications of different management strategies. Through dialogue, learners gain a deeper understanding of the complexities of agricultural decision-making and develop the ability to evaluate competing perspectives.

Critical examination of discourse can further enhance learning. Learners might analyse how terms such as *sustainable agriculture*, *food security*, *environmental stewardship*, or *productive land use* are used by governments, industry groups, environmental organisations, and the media. They may discover that different stakeholders frame the same issue in very different ways depending on their values, priorities, and interests. This analysis helps learners recognise how language shapes understanding and influences agricultural policy, industry practices, and public opinion.

A similar approach can be applied in horticulture education. For instance, learners studying urban horticulture may explore debates surrounding water use, native vegetation, urban greening, and biodiversity conservation. Through structured dialogue, learners can examine the perspectives of local councils, developers, environmental groups, landscape designers, and community members. Such discussions encourage learners to move beyond technical horticultural knowledge and consider the broader social, environmental, and political contexts that influence horticultural practice.

In Conservation and Land Management, dialogue and discourse can be particularly valuable when examining land-use conflicts and environmental management issues. Consider a class discussing the management of a national park adjacent to agricultural land. Learners may represent different stakeholder groups, including farmers, Traditional Owners, conservation organisations, tourism operators, local residents, and government agencies. Through role plays, debates, and facilitated discussions, learners explore how competing interests influence decisions regarding biodiversity protection, cultural heritage, fire management, recreation, and economic development. This process enables learners to understand that environmental issues are rarely purely technical problems and often involve complex social, cultural, political, and economic considerations.

Dialogue can also support the examination of power and knowledge within land management. For example, learners may explore how Indigenous ecological knowledge has historically been marginalised within some land management systems and how contemporary conservation practices increasingly recognise the value of Indigenous land stewardship and cultural burning. Through critical discussion, learners can examine whose knowledge has traditionally been prioritised, how this has influenced environmental outcomes, and how more inclusive approaches to knowledge-sharing may contribute to improved land management practices.

In each of these examples, dialogue moves beyond the transmission of technical information. Learners are encouraged to critically examine assumptions, evaluate competing viewpoints, recognise the influence of language and power, and develop more sophisticated understandings of complex issues. As a result, learning contributes not only to technical competence but also to the development of reflective practitioners who are capable of making informed, ethical, and socially responsible decisions within agriculture, horticulture, and conservation and land management contexts.

Key Concepts and Conclusions

Transformative learning is a form of adult learning that involves fundamental changes in how individuals understand themselves, others, and the world around them. Unlike approaches that focus primarily on the acquisition of knowledge and skills, transformative learning is concerned with changes in perspectives, assumptions, values, beliefs, and meaning structures. It emphasises the capacity of learning to alter how individuals interpret experience and make sense of their lives.

Central to this perspective is Jack Mezirow's Transformative Learning Theory, which proposes that adults can transform their frames of reference through critical reflection and dialogue. Perspective transformation occurs when learners critically examine assumptions that have previously been accepted without question and develop more inclusive, reflective, and adaptable ways of understanding experience. Mezirow argued that transformative learning extends beyond informational learning by changing the lens through which individuals interpret the world.

Frames of reference are the assumptions, values, expectations, beliefs, and habits of mind developed through life experiences. These mental frameworks help individuals interpret information, make decisions, and navigate daily life. However, they can also limit understanding when they remain unexamined, causing individuals to interpret new experiences through assumptions that may no longer be appropriate or accurate.

Transformative learning is frequently initiated by a disorienting dilemma, an experience or situation that challenges existing assumptions and creates uncertainty. Such events disrupt established ways of thinking and encourage learners to question previously accepted beliefs. As individuals attempt to make sense of these experiences, they become more open to exploring alternative perspectives and developing new understandings.

Critical reflection is the primary mechanism through which transformative learning occurs. It involves examining the assumptions, values, beliefs, and perspectives that influence thinking and behaviour. Through critical reflection, learners question why they hold particular beliefs, evaluate the validity of those beliefs, and consider alternative viewpoints. This process enables individuals to move beyond habitual patterns of thinking and develop more informed and inclusive perspectives.

Dialogue and rational discourse play an essential role in supporting transformative learning. Through respectful discussion and engagement with alternative perspectives, learners compare experiences, challenge assumptions, evaluate evidence, and refine their interpretations of experience. Dialogue transforms reflection into a collaborative process, enabling individuals to construct new meanings and test emerging understandings in interaction with others.

Reflective practice is closely related to transformative learning and refers to the deliberate examination of experiences to improve future performance and decision-making. Drawing on the work of Donald Schön, reflective practice includes both reflection-in-action, which occurs during an activity, and reflection-on-action, which occurs after an event has taken place. Reflective practice is widely applied across professional and workplace learning contexts.

Although reflective practice and critical reflection are related, they differ in depth and purpose. Reflective practice focuses primarily on improving actions and outcomes, whereas critical reflection examines the assumptions, values, social influences, and power relationships that shape those actions. Consequently, critical reflection supports broader personal, professional, and social transformation.

Meaning-making is another foundational concept within transformative learning. It refers to the process through which individuals interpret experiences and construct understanding. Adult learners actively connect new information to their prior experiences, beliefs, emotions, and social contexts. Through this process, learning becomes personally significant and contributes to changes in understanding, perspective, and identity.

Identity development is closely connected to learning because educational experiences often influence how individuals understand themselves and their roles within society, communities, and professions. As learners acquire new knowledge, perspectives, and competencies, they frequently revise their sense of self and develop new personal and professional identities.

Robert Kegan's Constructive-Developmental Theory extends these ideas by explaining how individuals develop increasingly complex ways of making meaning throughout life. According to Kegan, learning involves not only acquiring knowledge but also developing more sophisticated structures for understanding experience, relationships, and social environments.

Transformative learning frequently occurs during major life transitions. Experiences such as career changes, promotions, retirement, migration, illness, or returning to education often challenge existing assumptions and require individuals to construct new meanings and develop new identities. These transitional periods provide fertile conditions for transformative learning and personal growth.

The chapter also explores critical consciousness and empowerment, drawing on the work of Paulo Freire. Critical consciousness involves recognising how social, political, economic, and cultural structures influence individual experiences and opportunities. Empowerment occurs when learners develop the knowledge, confidence, agency, and critical awareness necessary to influence their lives and participate effectively within society.

Dialogue and discourse further highlight the social dimensions of learning. Through interaction, discussion, and collaborative inquiry, learners construct knowledge, examine assumptions, consider multiple perspectives, and develop awareness of how language, power, and social structures shape understanding. Learning is therefore viewed as both an individual and a social process.

Together, these concepts contribute to professional growth and lifelong learning. Transformative learning encourages individuals to remain reflective, adaptable, self-aware, and open to new perspectives throughout their lives. By continually examining and revising their assumptions, learners develop the capacity to respond effectively to changing personal, professional, and social circumstances.

Transformative learning is one of the most influential theories in adult education because it explains how adults change not only what they know but also how they understand themselves and the world. Learning is fundamentally a process of meaning-making in which individuals construct, evaluate, and revise the assumptions that guide their thinking and behaviour. Through critical reflection, dialogue, experience, and engagement with alternative perspectives, adults can develop more inclusive, flexible, and informed ways of understanding complex personal, professional, and social realities.

The chapter demonstrates that transformative learning is often initiated by experiences that challenge existing assumptions. Disorienting dilemmas create opportunities for critical reflection, encouraging learners to question long-held beliefs and explore alternative ways of thinking. Through this process, individuals may experience perspective transformation, resulting in changes to identity, professional practice, values, and worldview.

Critical reflection emerges as the central mechanism of transformative learning. Unlike routine reflection, critical reflection requires learners to examine the assumptions, cultural influences, values, and power relationships that shape their interpretations of experience. This deeper level of inquiry promotes greater self-awareness, more informed decision-making, and increased personal agency.

Dialogue and discourse are equally important in supporting transformative learning. Through interaction with others, learners encounter alternative perspectives, test assumptions, refine their understanding, and construct new meanings. Learning therefore occurs not only through individual reflection but also through participation in social processes of inquiry, discussion, and collaboration.

The chapter further highlights the importance of meaning-making and identity development in adult learning. Learning experiences frequently reshape how individuals understand themselves, their capabilities, and their roles within communities and professions. Consequently, learning contributes not only to knowledge acquisition but also to ongoing personal, professional, and social transformation.

Although Transformative Learning Theory has been criticised for its original emphasis on rational reflection and Western individualism, contemporary interpretations have expanded the theory to incorporate emotional, relational, cultural, social, spiritual, and contextual dimensions of learning. These developments have broadened the theory's applicability and strengthened its relevance across diverse educational, cultural, and professional settings.

Ultimately, transformative learning positions education as a process of personal growth, critical inquiry, and lifelong development. By helping learners critically examine assumptions, construct new meanings, and adapt to changing circumstances, transformative learning supports the development of reflective, autonomous, socially aware, and adaptable individuals who are capable of navigating the complexities of contemporary life and work.

Chapter 5

Experiential and Workplace Learning

Much of what adults learn throughout their lives occurs not in classrooms but through participation in work, professional practice, community activities, and everyday experiences. Whether developing technical skills in a trade, managing complex workplace projects, caring for patients, leading teams, or adapting to new technologies, adults frequently learn by doing, reflecting, collaborating, and applying knowledge in real-world contexts. These forms of learning are often highly relevant, immediately applicable, and closely connected to the challenges and opportunities individuals encounter in their personal and professional lives. As a result, understanding how experience and work contribute to learning has become a central concern within adult education, vocational education and training, and professional development.

This chapter examines the theories and practices that explain learning through experience, participation, and workplace engagement. It begins by exploring the principles of experiential learning and David Kolb's influential Experiential Learning Theory, including the experiential learning cycle and its application across adult learning contexts. The chapter then examines major workplace learning theories, including Human Capital Theory, Situated Learning Theory, Communities of Practice, Informal and Incidental Learning, Organisational Learning Theory, Reflective Practice, and Activity Theory. These perspectives provide different but complementary explanations of how adults develop knowledge, skills, professional competence, and occupational identity through participation in authentic activities and social environments.

The chapter also explores apprenticeships, vocational learning, work-integrated learning, and Recognition of Prior Learning (RPL) as important mechanisms through which experiential and workplace learning are structured and recognised. These approaches demonstrate how theory and practice can be integrated to support skill development, professional growth, and lifelong learning. Collectively, the concepts explored in this chapter challenge traditional views of learning as something that occurs separately from practice and instead highlight the powerful role that experience, reflection, participation, and workplace engagement play in adult learning throughout life.

Principles of Experiential Learning

Experiential learning is based on the idea that learning occurs through experience and the reflection that follows experience. Rather than viewing learners as passive recipients of information, experiential learning positions them as active participants who construct knowledge through engagement with real-world situations, problem-solving activities, and practical application (Bell & Bell, 2020; Bell & Liu, 2019; Woodward et al., 2021). The approach recognises that adults bring prior experiences to learning environments and that these experiences provide a foundation for developing new knowledge, skills, attitudes, and understandings (Bell & Bell, 2020; Mueller & Anderson, 2014).

One of the fundamental principles of experiential learning is that learning is a continuous process grounded in experience. Individuals learn by interacting with their environments, encountering challenges, and reflecting on the outcomes of their actions (Y. Kim, 2024; Thomas et al., 2014). Experiences provide the raw material for learning, while reflection enables learners to interpret those experiences and extract meaning from them (Bell & Bell, 2020; Hedin, 2010). Learning therefore emerges through the interaction between action and reflection rather than through experience alone (Bell & Bell, 2020; Rutherford-Hemming, 2012).

A second principle is that learners actively construct knowledge rather than simply receive it. Experiential learning draws heavily on constructivist perspectives, which propose that individuals interpret new experiences through the lens of their existing knowledge, beliefs, and prior experiences (Saldivar, 2025; Splan et al., 2011). As learners encounter new situations, they compare these experiences with what they already know, revise their understandings, and develop more sophisticated ways of thinking about the world (Bell & Liu, 2019; Mueller & Anderson, 2014; Thomas et al., 2014).

Reflection is another essential principle. Experiential learning recognises that experience alone does not automatically result in learning. Learners must engage in deliberate reflection to examine what occurred, why it occurred, and what can be learned from the experience (Bell & Bell, 2020; Hedin, 2010; Yasmiatti, 2024). Through reflection, individuals identify successes, challenges, assumptions, and alternative approaches, enabling them to improve future performance and deepen their understanding, (Bell & Liu, 2019; Woodward et al., 2020).

Experiential learning also emphasises the integration of theory and practice. Knowledge is not viewed as separate from action but as something that is tested, refined, and applied in real-world contexts (Hashim et al., 2022; Mesuwini & Mokoena, 2023). Learners are encouraged to apply theoretical concepts to practical situations and then evaluate the outcomes of those applications (Chen, 2026; Elsayed et al., 2023). This relationship between theory and practice helps learners develop knowledge that is both meaningful and transferable to future situations (Močinić & Tatković, 2021; Shore & Dinning, 2023).

Another important principle is that learning involves the whole person. Experiential learning recognises that cognitive, emotional, social, physical, and behavioural dimensions all contribute to learning (Bwegyeme & Munene, 2021). Experiences often evoke emotions, challenge beliefs, influence attitudes, and affect identity (Onjewu et al., 2024). As a result, learning is understood as a holistic process that engages individuals intellectually, emotionally, and socially (Bwegyeme & Munene, 2021; Mesuwini & Mokoena, 2023).

Experiential learning is also learner-centred. Learners take an active role in directing their learning, setting goals, making decisions, solving problems, and evaluating outcomes (Abakah & Antwi, 2024; Elsayed et al., 2023). The educator's role shifts from transmitting information to facilitating experiences, supporting reflection, and creating opportunities for meaningful engagement (Farooqi & Scutt, 2026; Shore & Dinning, 2023). This aligns particularly well with adult learning principles, which emphasise autonomy, self-direction, and the importance of prior experience (Cornet et al., 2024; Mesuwini & Mokoena, 2023).

A further principle is that learning occurs through adaptation and experimentation. Learners continually test ideas, evaluate outcomes, and modify their actions in response to feedback (Bou-Ammar et al., 2025; Farooqi & Scutt, 2026). Through this cycle of action, reflection, conceptualisation, and experimentation, individuals develop increasingly effective ways of responding to challenges and opportunities (Chen, 2026; Elsayed et al., 2023). Learning is therefore viewed as an ongoing process of growth and adaptation rather than a finite outcome (Mesuwini & Mokoena, 2023; Močinić & Tatković, 2021).

Experiential learning also recognises the importance of context. Learning is influenced by the social, cultural, organisational, and environmental conditions in which experiences occur (Cornet et al., 2024; Shore & Dinning, 2023). What is learned and how it is interpreted may vary according to the setting, the participants involved, and the broader circumstances surrounding the experience (Onjewu et al., 2024). Consequently, experiential learning often emphasises authentic tasks and real-world situations that reflect the environments in which learners will apply their knowledge and skills (Mesuwini & Mokoena, 2023; Shore & Dinning, 2023).

These principles are most commonly associated with David Kolb's Experiential Learning Theory, which conceptualises learning as a cycle involving four interconnected stages: concrete experience, reflective observation, abstract conceptualisation, and active experimentation (Abakah & Antwi, 2024; Bwegyeme & Munene, 2021; Elsayed et al., 2023). According to Kolb, effective learning occurs when learners engage with all stages of this cycle, moving from experience to reflection, from reflection to conceptual understanding, and from understanding to further action and experimentation (Bou-Ammar et al., 2025; Chen, 2026; Farooqi & Scutt, 2026).

In practice, experiential learning can be observed across a wide range of adult learning contexts. Apprentice electricians learn through performing electrical installations under supervision and reflecting on their work. Nursing students develop competence through clinical placements where theoretical knowledge is applied to patient care. Agriculture students learn through fieldwork, analysing crop performance, testing management strategies, and evaluating outcomes. Leadership development programs often incorporate workplace projects, simulations, mentoring, and reflective activities that enable participants to learn through practice. In each case, learning emerges through the interaction of experience, reflection, knowledge construction, and application.

Collectively, these principles position experiential learning as one of the most influential approaches in adult education. By emphasising experience, reflection, active participation, and the integration of theory and practice, experiential learning provides a framework for understanding how adults develop knowledge, skills, professional competence, and personal insight throughout their lives.

David Kolb and the Experiential Learning Cycle

David A. Kolb is one of the most influential scholars in adult education and learning theory. His work on Experiential Learning Theory (ELT) provided a comprehensive framework for understanding how individuals learn through experience and has had a significant impact on adult education, higher education, vocational education and training (VET), professional development, management education, and workplace learning, (America & Mallon, 2020; Bertoni et al., 2024; Brabazon et al., 2020; Laidlaw & Johnson, 2025). Kolb's theory was first fully articulated in *Experiential Learning: Experience as the Source of Learning and Development* (1984), where he synthesised ideas from earlier scholars including John Dewey, Kurt Lewin, and Jean Piaget (Beck et al., 2017; Wuestewald, 2015).

Kolb argued that learning is not simply the acquisition of information but a continuous process in which knowledge is created through the transformation of experience (Beck et al., 2017; Fergusson & Laan, 2021). Rather than viewing learning as a product or outcome, he described it as a dynamic process involving interaction between experience, reflection, thinking, and action (W. S. Chiang, 2021; Zhao & Ko, 2024). Learning occurs as individuals engage with experiences, reflect upon them, develop conceptual understandings, and apply those understandings to new situations (Kennedy et al., 2022).

Kolb's theory draws upon several influential traditions in psychology and education. From John Dewey, he adopted the idea that education should be grounded in experience and that learning occurs through the interaction between individuals and their environments (America & Mallon, 2020; Wuestewald, 2015). From Kurt Lewin, he incorporated the concept of learning as a cyclical process involving action, observation, and feedback. From Jean Piaget, he drew on constructivist ideas concerning the active construction of knowledge through interaction with the environment (Bertoni et al., 2024).

These influences led Kolb to propose that effective learning requires the integration of concrete experience and abstract thinking, as well as reflection and active experimentation (Kennedy et al., 2022). Learning is therefore viewed as a holistic process involving cognitive, emotional, behavioural, and social dimensions, (Zhao & Ko, 2024).

The Experiential Learning Cycle

At the centre of Kolb's Experiential Learning Theory is the Experiential Learning Cycle, a model that describes learning as a continuous process involving four interconnected stages: concrete experience, reflective observation, abstract conceptualisation, and active experimentation (Esquivel et al., 2022; Pacas, 2021; Santiago et al., 2024; Tinoco et al., 2022; Tunno, 2023). Kolb argued that learning can begin at any point in the cycle (Horna & Reyes, 2019; ; however, effective learning occurs when individuals engage with all four stages (Gonzalez-Nuñez et al., 2024; Herrera & Hoz, 2014). Through this cyclical process, experiences are transformed into knowledge, understanding, and improved future performance.

The first stage, concrete experience, involves direct participation in an activity, event, or situation. Learners actively engage with an experience rather than simply observing it or studying it theoretically (Mariño et al., 2022; Villarroel et al., 2021). The experience provides the foundation for learning by exposing individuals to new challenges, observations, and opportunities for action (Esquivel et al., 2022). For example, an apprentice horticulturist may participate in planting and maintaining a new garden, a nursing student may provide patient care during a clinical placement, or a business student may undertake a workplace project focused on improving customer service. In each case, learning begins through active involvement in a real-world experience (Sánchez & Fernández, 2019; Villarroel et al., 2021).

Following the experience, learners move into the stage of reflective observation. During this phase, individuals think carefully about what occurred, examine the outcomes of their actions, identify successes and challenges, and consider their reactions to the experience (Diz-Comesaña et al., 2022; Tunno, 2023). Reflection encourages learners to ask questions such as what happened, why it happened, what worked well, what difficulties arose, and how they responded to the situation (Fernández et al., 2019). For example, after completing a conservation project, learners may analyse which aspects of the project were successful, what obstacles were encountered, and how environmental conditions influenced the results. Through reflection, experience becomes a source of learning as individuals identify patterns, recognise relationships, and gain insight into their actions and decisions (Lopes et al., 2024; Pacas, 2021).

The third stage, abstract conceptualisation, involves developing theories, principles, concepts, or explanations based on the reflections generated in the previous stage (Diz-Comesaña et al., 2022; Santiago et al., 2024). Learners connect their practical experiences to broader knowledge, frameworks, and ideas, transforming observations into more generalisable understanding, (Villarroel et al., 2021). During this process, learners consider what can be learned from the experience, what principles explain the outcomes, how the experience relates to existing theory, and whether alternative approaches might have been more effective. For example, a supervisor undertaking leadership training may reflect on a workplace conflict and connect the experience to theories of communication, motivation, or conflict resolution. Similarly, an agriculture student may interpret crop performance results using concepts from soil science, irrigation management, or plant nutrition. Through abstract conceptualisation, learners develop knowledge that extends beyond a single experience and can be applied to future situations (Carrasco-Walburg et al., 2021; Gonzalez-Nuñez et al., 2024).

The final stage, active experimentation, involves applying newly developed understanding to future experiences. Learners test ideas, modify behaviours, and try alternative approaches based on what they have learned (García-Romero & Lalueza, 2019; Horna & Reyes, 2019). Questions commonly considered during this stage include what can be done differently next time, how the new knowledge can be applied, and what outcomes might result from trying a different approach. For example, a conservation and land management student who identifies weaknesses in a weed control strategy may implement a revised method during a subsequent field activity. Likewise, a business manager may adopt new communication techniques after reflecting on a leadership challenge. The outcomes of these experiments create new experiences, which begin the learning cycle again and support ongoing development and improvement (Santiago et al., 2024).

Kolb's experiential learning cycle highlights that learning is not a single event but an ongoing process of experiencing, reflecting, thinking, and acting (Lopes et al., 2024; Mariño et al., 2022). By moving through each stage of the cycle, learners continually refine their understanding, improve their performance, and develop increasingly sophisticated ways of responding to new challenges and opportunities (Gonzalez-Nuñez et al., 2024; Villarroel et al., 2021).

Learning as a Continuous Cycle

A key contribution of Kolb's theory is the recognition that learning is continuous rather than linear. Each experience provides opportunities for reflection, conceptual understanding, and experimentation, which in turn generate new experiences and further learning (Jossou et al., 2022; Sharaf, 2021; Ziani et al., 2021). Learning therefore becomes an ongoing cycle of growth, adaptation, and development. As Kolb conceives it, learning is "a permanent process of knowledge creation from experiences" (Jossou et al., 2022), a process that operates in iterative loops where each cycle generates further learning (Sharaf, 2021; Ziani et al., 2021). This cyclical process transforms experience through reflection and conceptualization, with every learning episode leading to regulation of actions and planning of the next experience (Bourgain, 2022; Guillemette, 2020).

The cyclical nature of experiential learning makes it particularly relevant to adult learners, who frequently draw upon workplace experiences, professional practice, personal responsibilities, and significant life events as important sources of learning and development (Cristol & Müller, 2013; Dillard et al., 2024). Rather than separating learning from everyday experience, experiential learning recognises that adults continually construct knowledge by engaging with real-world situations, reflecting on those experiences, and applying new understandings to future challenges.

Experiential learning has become one of the foundational theories within adult education, adult learning, and andragogy because it emphasises active participation, reflection, and the transformation of experience into knowledge (Dillard et al., 2024). Through a continuous process of action and reflection, learners develop deeper understanding, improve professional practice, and enhance their capacity to respond effectively to changing circumstances.

The relationship between experience and learning has long been recognised as a central feature of adult education. Research has demonstrated the value of experiential approaches across a wide range of contexts, including professional practice, workplace learning, higher education, and collaborative learning environments (Cristol & Müller, 2013; Dillard et al., 2024). These applications highlight the importance of creating opportunities for learners to engage with authentic experiences, critically reflect on those experiences, and integrate their insights into future actions and decision-making.

Learning Styles and Later Critiques

Kolb later expanded Experiential Learning Theory by proposing that individuals may develop preferences for particular ways of learning within the experiential learning cycle. These preferences became known as learning styles and were based on how learners combine the processes of experiencing, reflecting, thinking, and acting (Daneil, 2021; Li et al., 2013). Kolb identified four primary learning styles: diverging, assimilating, converging, and accommodating (Sunggingwati & Haviluddin, 2019; Tian & Ng, 2024; Uysal, 2023). Each style reflects a different combination of the four stages of the learning cycle and suggests that learners may favour certain approaches when engaging with new experiences and solving problems (Li et al., 2013).

Individuals with a diverging style tend to emphasise concrete experience and reflective observation, preferring to view situations from multiple perspectives and generate ideas (Sunggingwati & Haviluddin, 2019; Tian & Ng, 2024; Yurt, 2025). Assimilating learners typically combine reflective observation with abstract conceptualisation, favouring logical analysis, theoretical understanding, and organised information (Tian & Ng, 2024; Williamson & Watson, 2006). Converging learners integrate abstract conceptualisation with active experimentation, focusing on problem-solving, decision-making, and the practical application of ideas (Diken & Özdemir, 2021; Sunggingwati & Haviluddin, 2019; Williamson & Watson, 2006). Accommodating learners emphasise concrete experience and active experimentation, preferring hands-on involvement, practical action, and learning through trial and error (Daneil, 2021; Sunggingwati & Haviluddin, 2019; Williamson & Watson, 2006).

Kolb's learning styles gained considerable popularity in education, training, and professional development because they provided a framework for understanding individual differences in learning preferences (Elsayed & Hassan, 2024; Li et al., 2013). Educators and trainers often used the model to design learning activities that catered to a variety of learner needs and encouraged engagement with different stages of the learning process (Daneil, 2021). The Learning Style Inventory (LSI) became one of the most commonly administered instruments for assessing learning styles in both research and teaching contexts (Elsayed & Hassan, 2024; Li et al., 2013; Yang et al., 2017)

Despite its widespread use, the learning styles component of Kolb's theory has attracted criticism. Researchers have questioned the empirical evidence supporting the classification of learners into distinct learning style categories and have expressed concerns that such classifications may oversimplify the complexity of learning. Critics argue that learners often adapt their approaches according to the context, task, and learning environment rather than consistently exhibiting a single preferred style (Lum et al., 2011). Indeed, research has demonstrated that students may indicate a preference for a certain learning style yet tend to adapt their learning strategies to the subject matter being studied.

More broadly, some scholars have suggested that Kolb's model places insufficient emphasis on the social, cultural, emotional, and political dimensions of learning. While the theory provides a valuable framework for understanding how individuals learn from experience, it tends to focus primarily on cognitive processes occurring within the individual learner (Li et al., 2013). Critics argue that learning is also shaped by relationships, cultural norms, organisational contexts, power structures, emotions, and social interactions, factors that may not be fully captured within the experiential learning cycle (Li et al., 2013). The role of feelings and emotions in determining whether and what people learn has been acknowledged, with positive feelings facilitating learning and negative emotions such as fear and anxiety blocking it (Li et al., 2013). Furthermore, culture has been identified as a powerful socialisation tool that may influence individuals' learning preferences.

Others have questioned the assumption that learning always follows a clear progression through the four stages of the cycle. In practice, learning is often messy, iterative, and non-linear. Individuals may move back and forth between stages, revisit earlier reflections, engage in multiple stages simultaneously, or experience learning in ways that do not conform neatly to a cyclical sequence, Some researchers have noted the practical impossibility of combining all four modes of learning because students have different preferences, and thus may choose between modes for perceiving and processing information. Consequently, while Kolb's model remains one of the most influential frameworks for understanding experiential learning (Elsayed & Hassan, 2024; Li et al., 2013), contemporary perspectives generally view learning as a more complex process that incorporates cognitive, emotional, social, cultural, and contextual influences (Li et al., 2013).

Significance of Kolb's Work

Despite these criticisms, Kolb's Experiential Learning Theory remains one of the most influential frameworks in adult education. Its enduring popularity stems from its practical relevance, intuitive structure, and emphasis on learning through experience. The model highlights the importance of combining action with reflection, theory with practice, and knowledge with application.

By demonstrating how experience can be transformed into learning, Kolb's work provides educators, trainers, and learners with a valuable framework for designing meaningful educational experiences and supporting lifelong learning, professional development, and continuous improvement across a wide range of contexts.

A useful example of Kolb's Experiential Learning Cycle can be found in a Level 3 Diploma in Adult Care programme delivered by a Further Education (FE) college in England. Many learners enrolled in these programmes may be adult students who are already employed as care workers in residential aged care facilities, supported living environments, or domiciliary care services. The qualification combines classroom learning with workplace practice, making it particularly suited to an experiential learning approach.

The learning cycle begins with a concrete experience. Imagine a learner working in a residential care home who is required to support an elderly resident experiencing confusion and distress due to dementia. During a morning shift, the resident becomes agitated and refuses assistance with personal care. The learner attempts to calm the resident and complete the required support tasks but finds the situation challenging. This real workplace event provides the experience from which learning can occur.

The learner then moves into reflective observation. Following the shift, the learner completes a reflective journal entry and discusses the incident with their workplace supervisor and college assessor. They consider what happened during the interaction, how they responded, what strategies appeared effective, and which actions may have increased the resident's distress. The learner may also reflect on their own emotional response, recognising feelings of uncertainty or frustration. Through reflection, the learner begins to identify factors that influenced the outcome and gains greater awareness of their own practice.

The next stage is abstract conceptualisation. During college workshops, the learner studies theories of dementia care, person-centred practice, communication techniques, and behaviour support. They connect these theoretical concepts to the experience they encountered in the workplace. The learner begins to understand that the resident's behaviour may have been influenced by confusion, environmental triggers, unmet needs, or communication barriers. They learn that person-centred approaches emphasise understanding the individual's perspective rather than focusing solely on task completion. Through this process, the learner develops broader principles that can guide future practice.

The learner then enters active experimentation. During subsequent shifts, they consciously apply the new knowledge gained through reflection and study. Instead of immediately encouraging the resident to complete personal care activities, the learner takes additional time to build rapport, uses simpler communication techniques, offers choices, and adjusts the environment to reduce potential sources of anxiety. The learner observes the results of these changes and evaluates whether the resident responds more positively. These actions create a new experience, which begins another cycle of reflection, conceptualisation, and experimentation.

Over time, repeated movement through the cycle helps the learner develop professional competence. Rather than simply memorising information about dementia care, they learn how to apply theory in authentic situations, evaluate outcomes, and continually improve their practice. The process also supports the development of critical thinking, self-awareness, confidence, and professional judgement.

This example illustrates why experiential learning can be particularly effective in UK adult education and vocational training. Many adult learners bring substantial workplace experience to their studies and are motivated by opportunities to apply learning directly to real-life situations. Programmes in health and social care, teaching assistance, management, construction, engineering, business administration, and community services frequently use workplace placements, practical activities, reflective journals, professional discussions, and work-based projects to facilitate movement through Kolb's learning cycle. By linking experience with reflection and theory, learners develop knowledge that is both meaningful and immediately relevant to their professional roles.

Workplace Learning Theories

Workplace learning theories seek to explain how individuals acquire knowledge, skills, attitudes, and professional competence through participation in work activities and organisational life. Unlike traditional educational approaches that often separate learning from practice, workplace learning theories emphasise that learning is embedded within everyday work, social interactions, problem-solving, and participation in professional communities (Andrew & Ferguson, 2008; Li et al., 2009a; Slade, 2013). The situated learning perspective, originally developed by Lave and Wenger, posits that knowledge is socially situated, practice-based, and fluid, and that meaningful learning involves individuals in practice even in peripheral roles within the workplace (Andrew & Ferguson, 2008; Price, 2020; Slade, 2013). As Wenger's social learning theory asserts, learning is fundamentally a social process whereby knowledge is socially co-constructed through participation in communities of practice (Li et al., 2009a; Shujahat & Wang, 2022; Wargent & Stocker, 2021). These theories have become increasingly important as organisations recognise that much of what employees learn occurs informally through experience rather than through formal training programs (Li et al., 2009a; Thompson, 2010). Indeed, the view of "learning on the job" emphasises that knowledge creation and sharing is embedded and embodied in social activity (Li et al., 2009a; Macpherson & Antonacopoulou, 2013).

Workplace learning is particularly significant in contemporary organisations because rapid technological advancement, changing job requirements, global competition, and increasing organisational complexity require workers to continually develop new knowledge and skills throughout their careers (Andrew & Ferguson, 2008; Li et al., 2009b). Organisations increasingly depend on employees who can respond effectively to complex and unpredictable challenges, adapt to new technologies, and contribute to innovation and continuous improvement. Consequently, the ongoing development of workforce capability has become a strategic priority across industries and professions.

This emphasis on continual learning is reflected in the concept of continuing professional development, which is widely understood as a structured and ongoing process through which professionals maintain and enhance their knowledge, skills, competence, and effectiveness throughout their working lives (Price, 2020). Although often associated with professions such as teaching, healthcare, and engineering, the principle of lifelong professional learning extends across virtually all occupational fields. Contemporary workers are expected not only to acquire new technical skills but also to develop critical thinking, adaptability, communication, and problem-solving capabilities in response to evolving workplace demands.

As a result, learning is increasingly viewed as an integral part of work rather than as an activity confined to classrooms, training centres, or formal educational programs (Andrew & Ferguson, 2008; Isah & Byström, 2015; Slade, 2013). Employees learn while performing tasks, solving problems, interacting with colleagues, accessing information, and responding to workplace challenges. Learning becomes embedded within daily work activities and develops through participation in professional practice rather than through formal instruction alone.

Research across a range of professions supports this understanding of workplace learning. Studies of medical practitioners have demonstrated that learning frequently occurs through routine engagement with information, decision-making processes, and patient care activities, with learning becoming intertwined with everyday professional practice (Isah & Byström, 2015). Similarly, research involving rural police officers has shown that professional learning emerges through direct engagement with the unique circumstances, environments, relationships, and challenges encountered in their daily work (Slade, 2013). These findings suggest that workplace learning is highly contextual and shaped by the specific situations in which work occurs.

Consequently, traditional views that separate learning from practice have been increasingly challenged by situated, experiential, and practice-based learning theories (Andrew & Ferguson, 2008; Barnard et al., 2022). These perspectives recognise that meaningful learning often occurs through participation in authentic work activities where theory and practice are closely integrated. Rather than viewing knowledge as something acquired independently of context and later applied, contemporary workplace learning theories emphasise that knowledge is developed through engagement in practice, collaboration with others, reflection on experience, and participation in professional communities. This shift has fundamentally influenced how organisations, educators, and researchers understand the relationship between learning, work, and professional development.

Human Capital Theory

One of the earliest perspectives on workplace learning is Human Capital Theory, which views education and training as investments that increase worker productivity and organisational performance. Associated with economists such as Theodore Schultz and Gary Becker (Fleming, 2017; Kuzminov et al., 2019), the theory proposes that individuals develop valuable knowledge and skills that enhance their capacity to contribute economically. Schultz introduced the concept of "human capital" in 1961, focusing on improving human resource productivity and analysing the link between socio-economic advancement and human capital (Baktymbet, 2021; Kasa et al., 2020). Becker subsequently systematised the theory, analysing how investment in education and training yields returns in the form of increased productivity and earnings (Fleming, 2017; Kuzminov et al., 2019). Becker distinguished between general human capital, skills transferable across organisations, and specific human capital, skills particular to a given firm (Fleming, 2017; Kuzminov et al., 2019).

From this perspective, workplace learning benefits both employees and employers. Workers increase their employability, career opportunities, and earning potential, while organisations benefit from improved productivity, innovation, and competitiveness (Auerbach & Green, 2024; Kasa et al., 2020). As Becker argued, each worker possesses qualities that may be enhanced through training and education, enabling calculation of rates of return on such investments (Baktymbet, 2021). Human

Capital Theory has strongly influenced government policies, vocational education systems, workforce development initiatives, and organisational training strategies (Auerbach & Green, 2024; Kasa et al., 2020). The theory became the basis for educational policy in many developed countries (Kuzminov et al., 2019) and provided a framework for justifying substantial investments in training and education at both national and organisational levels,, (Goodburn, 2020). As Kuzminov et al. note, by showing the mechanisms by which investment in education leads to economic growth, the model became the basis for new policies of increasing investment in education worldwide (Kuzminov et al., 2019).

While influential, critics argue that Human Capital Theory tends to focus primarily on economic outcomes and may underestimate the social, cultural, and relational aspects of workplace learning. Auerbach and Green contend that the embedding of the theory in methodological individualism makes impossible the incorporation of key non-peripheral factors that affect the demand for, and effects of, education, and that the resultant tensions from using an individualistic methodology to explain inherently social phenomena have been accommodated within the theory by ignoring them (Auerbach & Green, 2024). Fleming argues that the theory fundamentally individualises people, placing the costs of economic activity onto the employee and providing an ideological alibi for the radical responsibilisation of the workforce (Fleming, 2017). Critics from the radical economics school have challenged the theory for its failure to problematise the content and socioeconomic context of education (Auerbach & Green, 2024). Despite these critiques, human capital theory remains central to skills formation policies in most countries (Kasa et al., 2020).

Experiential Learning

David Kolb's Experiential Learning Theory has had a major influence on workplace learning. The theory proposes that learning occurs through a continuous cycle of experience, reflection, conceptualisation, and experimentation (Coffey & Davis, 2019; Dipitso, 2023; Goda et al., 2025; O'Connor et al., 2021; Santhi & Malathi, 2024). In workplace settings, employees learn by performing tasks, reflecting on outcomes, developing new understandings, and applying these insights to future situations (Dipitso, 2023; Goda et al., 2025; Matsuo, 2015b; Santhi & Malathi, 2024).

For example, a project manager may encounter difficulties coordinating a complex project. Through reflecting on the experience, analysing what contributed to the challenges, and implementing revised strategies, the manager develops greater competence and professional judgement (Dipitso, 2023; Goda et al., 2025; Matsuo, 2015b). Experiential learning highlights the importance of learning from both successes and failures and recognises that experience becomes valuable only when accompanied by reflection (Dipitso, 2023; Kambara, 2026; Matsuo, 2015b).

Situated Learning Theory

Situated Learning Theory was developed by Jean Lave and Etienne Wenger and argues that learning is inseparable from the social and cultural context in which it occurs (Amery & Griffin, 2020; Johnson, 2008; Stanley, 2017). According to this perspective, knowledge is not simply transferred from one person to another but is developed through participation in authentic activities within a community (Johnson, 2008; Schmid et al., 2020; Stanley, 2017).

Learning occurs through engagement in real work practices rather than through abstract instruction alone (Amery & Griffin, 2020; Mohammad, 2026; Schmid et al., 2020). New workers gradually move from peripheral participation to full participation as they gain experience and become accepted members of a professional community (Amery & Griffin, 2020; Holmes & Woodhams, 2013; Sayer, 2013; Schmid et al., 2020; Stanley, 2017).

For example, an apprentice electrician initially observes experienced tradespeople, assists with simple tasks, and gradually undertakes more complex work (Caldwell, 2011; Schmid et al., 2020). Over time, the apprentice develops not only technical skills but also professional norms, workplace values, and occupational identity (Holmes & Woodhams, 2013; Li et al., 2009b; Sayer, 2013; Schmid et al., 2020).

Communities of Practice

Building on situated learning, Wenger (1999) developed the concept of Communities of Practice (CoP). A community of practice consists of individuals who share a common profession, interest, or area of expertise and who learn through regular interaction and collaboration (Ralalatiana & Vatz-Laaroussi, 2016). Learning occurs through participation in shared activities, problem-solving discussions, mentoring relationships, and the exchange of knowledge (Ralalatiana & Vatz-Laaroussi, 2016). Knowledge is viewed as socially constructed rather than individually acquired, (Ralalatiana & Vatz-Laaroussi, 2016).

Examples include nursing teams discussing patient care, engineers collaborating on technical challenges, teachers sharing instructional strategies, or tradespeople mentoring apprentices. Communities of practice help organisations retain expertise, support innovation, and facilitate continuous learning (Ralalatiana & Vatz-Laaroussi, 2016). Ralalatiana and Vatz-Laaroussi (2016, p. 90), referring to Wenger (1999), describe the classroom as a "communauté de pratique" defining it as "un ensemble d'individus qui travaillent ensemble à l'atteinte d'un but" (Ralalatiana & Vatz-Laaroussi, 2016, p. 90). Impedovo highlights the role of communities of practice in supporting professional learning through collaboration, participation, and the exchange of knowledge among members (Ralalatiana & Vatz-Laaroussi, 2016). Within these communities, learning occurs as individuals engage with one another, share experiences, discuss challenges, and collectively develop new understandings. Knowledge is not simply transferred between participants but is co-constructed through ongoing interaction and engagement in shared professional activities.

This perspective emphasises the participatory and evolving nature of communities of practice, where learning emerges from active involvement in a community rather than from formal instruction alone. Through collaboration, reflection, and mutual support, members contribute to both individual and collective learning while strengthening professional identities and practices. Impedovo also draws on the work of (Wenger-Trayner & Wenger-Trayner, 2015) to highlight the growing importance of networked communities of practice, which enable learning and knowledge sharing across organisational, professional, and geographical boundaries (Ralalatiana & Vatz-Laaroussi, 2016). These networks create opportunities for practitioners to access diverse perspectives, develop new knowledge, and participate in broader professional learning communities.

Informal and Incidental Learning

Research by Marsick and Watkins (2018) significantly expanded understanding of workplace learning by highlighting the importance of informal and incidental learning (Barefield & Nicolaides, 2022; Watkins et al., 2018). Their work challenged the assumption that learning occurs primarily through structured training programs, formal courses, or organised professional development activities (Marsick & Watkins, 2018; Watkins et al., 2018). Instead, they argued that a substantial amount of workplace learning takes place during everyday work activities and often occurs without individuals consciously recognising that they are learning (Barefield & Nicolaides, 2022; Hunter, 2014).

Informal learning emerges naturally through participation in work and is typically self-directed, unstructured, and closely connected to immediate workplace needs (Anselmann, 2022; Marsick & Watkins, 2018). Employees may develop new knowledge and skills while solving problems, observing experienced colleagues, participating in meetings, receiving feedback, experimenting with different approaches, or engaging in discussions with co-workers (Berg & Chyung, 2008; Herrero et al., 2017; Wofford et al., 2013). Unlike formal training, informal learning is not usually planned in advance, assessed, or documented (Anselmann, 2022) yet it can make a significant contribution to professional competence and performance (Daryoush, 2013; Wofford et al., 2013).

Marsick and Watkins also distinguished incidental learning from informal learning (Herrero et al., 2017; Hunter, 2014; Shah et al., 2022). Incidental learning occurs as an unintended consequence of other activities rather than as a deliberate effort to learn (Daryoush, 2013; Shah et al., 2022). Individuals may acquire new knowledge, attitudes, skills, or insights while focusing on a different primary objective (Fee & Gray, 2020; Hunter, 2014). For example, a supervisor managing a difficult workplace conflict may develop leadership, communication, and conflict-resolution skills even though the immediate goal is to resolve the situation rather than to engage in leadership development. Similarly, an employee assigned to a new project may acquire project management skills, technical knowledge, or greater organisational awareness as a by-product of completing the task (Daryoush, 2013; Hunter, 2014).

The significance of informal and incidental learning lies in their close connection to authentic workplace experiences (Fee & Gray, 2020; Wofford et al., 2013). Because learning occurs within real situations and is immediately applied to practical problems, it is often highly relevant and transferable to future work activities (Jeong et al., 2018; Watkins et al., 2018). These forms of learning also enable workers to respond quickly to changing workplace demands, technological developments, and emerging challenges without waiting for formal training opportunities (Barefield & Nicolaides, 2022; Lokhtina & Faller, 2024).

Research consistently indicates that a substantial proportion of workplace learning occurs through informal and incidental processes rather than through structured educational programs (Ariffin & Hashim, 2010; Wofford et al., 2013). This finding highlights the importance of creating workplace environments that encourage collaboration, knowledge sharing, reflection, feedback, mentoring, and open communication (Ellinger & Cseh, 2007; Herrero et al., 2017; Kim et al., 2014). Organisational culture, leadership practices, and opportunities for social interaction therefore play a critical role in supporting learning and professional development (Berg & Chyung, 2008; Ellinger & Cseh, 2007). From this perspective, workplaces are not simply locations where tasks are performed but are also important environments in which continuous learning occurs through everyday participation in work (Kim et al., 2014; Lokhtina & Faller, 2024).

Organisational Learning Theory

Organisational Learning Theory extends the focus beyond individual learners to consider how entire organisations learn and adapt. Scholars such as Chris Argyris, Donald Schön, and Peter Senge argue that organisational success depends upon the collective capacity to acquire, share, and apply knowledge (Brătianu et al., 2020; Bui, 2020; Muthamia & Kilika, 2022).

Argyris and Schön introduced the concepts of single-loop and double-loop learning. Single-loop learning involves correcting errors without questioning underlying assumptions, while double-loop learning involves critically examining and potentially changing the assumptions, policies, and values that guide behaviour (Alaka et al., 2018). As Alaka et al. explain, single-loop learning "involves detecting and correcting errors (i.e performance gaps) enabling organizations to act upon present policies and objectives," whereas double-loop learning "occurs when the organization is willing to question long-held assumptions about its mission, customers, capabilities, or strategies" (Alaka et al., 2018, p. 75).

Peter Senge's concept of the learning organisation (Senge, 1990)emphasises continuous improvement, systems thinking, shared vision, team learning, and personal mastery (W. Ahmad et al., 2023; Bui, 2020; Jeong et al., 2007; Muthamia & Kilika, 2022; Silva et al., 2003). Senge developed this framework in *The Fifth Discipline*, proposing these five key disciplines as essential for organisational transformation (Bui, 2020; Park & Rojewski, 2006; Shahbaz, 2021). Systems thinking is considered the most important discipline and the "conceptual cornerstone" underlying all others (Bui, 2020; Warne, 2000), while personal mastery is regarded as the "spirit" of the learning organisation and team learning as its "foundation" (Bui, 2020). In learning organisations, knowledge is actively shared, innovation is encouraged, and employees are supported in ongoing professional development (Farrukh & Waheed, 2015; Muthamia & Kilika, 2022; Warne, 2000).

Peter Senge's concept of the learning organisation (Senge, 1990) also describes an organisation in which individuals and groups continuously develop their capabilities, expand their understanding, and work collaboratively to achieve shared goals (W. Ahmad et al., 2023; Muthamia & Kilika, 2022; Ng, 2020). Within such organisations, learning is embedded in everyday activities, and employees are encouraged to challenge assumptions, explore new ways of thinking, and engage in collective problem-solving. The emphasis is not only on individual development but also on creating organisational environments that support collaboration, innovation, and the ongoing generation of knowledge.

The concept of the learning organisation emerged from the broader field of organisational learning and has been refined through the contributions of numerous scholars over time (Brătianu et al., 2020; Evans et al., 2012). While Senge's work played a pivotal role in popularising the concept, subsequent research has expanded understanding of how organisations create, share, retain, and apply knowledge. Contemporary perspectives view learning organisations as dynamic systems that foster continuous learning, adaptability, and collective capacity building in response to changing internal and external environments.

Reflective Practice

Reflective Practice, particularly through the work of Donald Schön, has become a cornerstone of workplace learning (Kinsella, 2010; Nelson, 2012). Schön's theory, articulated in his influential books *The Reflective Practitioner* (1983) and *Educating the Reflective Practitioner* (1987), has had enormous impact across education and related fields (Kinsella, 2008, 2010). Over the past three decades, reflective practice has become a cornerstone of pedagogy in professional preparation in education, health sciences, management, and other fields, as well as a key concept in life-long learning and professional development (Nelson, 2012). Schön argued that professionals develop expertise through reflection both during and after action (Ahsanu et al., 2021; Cadorette et al., 2023; Nazir et al., 2022).

Reflection-in-action occurs while performing a task. Professionals think on their feet, adjust their actions, and respond to emerging situations in real time (Cadorette et al., 2023; Vera, 2023). Schön described reflection-in-action as "central to the art through which practitioners sometimes cope with the troublesome 'divergent' situations of practice" (Vera, 2023, p. 113). It involves observing one's thinking and action as they are occurring, to make adjustments in the moment (Ahsanu et al., 2021). Schön likened this process to a jazz pianist's improvisation or a basketball player's instant manoeuvring in response to a surprising move by an opponent (Kinsella, 2010).

Reflection-on-action occurs after an event or activity has taken place and involves a deliberate examination of performance, decisions, and outcomes in order to identify opportunities for improvement in future practice (Ahsanu et al., 2021; Cadorette et al., 2023). This process requires practitioners to revisit their actions, consider the reasoning behind their decisions, evaluate the effectiveness of their responses, and assess the consequences of those actions. By critically reviewing past experiences, individuals can develop a deeper understanding of their professional practice and identify strategies for enhancing future performance.

Unlike reflection that occurs during action, reflection-on-action involves a retrospective analysis of practice. Practitioners intentionally step back from the immediate demands of a situation to examine what occurred, why it occurred, and how different approaches may have produced alternative outcomes (Kinsella, 2010; Shah, 2022). This form of reflection encourages individuals to question assumptions, explore the factors that influenced their actions, and consider lessons that can be applied in similar situations in the future.

Schön argued that this reflective process contributes significantly to professional development and expertise. Through repeated reflection on past experiences, practitioners gradually refine their judgement, strengthen their understanding of complex situations, and improve their ability to anticipate the likely consequences of future actions (Malinin, 2018). Over time, reflection-on-action supports the development of specialised knowledge and professional competence by enabling individuals to learn systematically from experience and incorporate those insights into their ongoing practice.

This approach is widely used in professions such as teaching, nursing, social work, engineering, and management, where practitioners regularly encounter complex and unpredictable situations (Kinsella, 2010; Ramos-Rodríguez et al., 2016; Yanow & Tsoukas, 2009). Reflective practice has been widely adopted by nursing, health, and social care professions (Kinsella, 2010), and Schön's work has been hugely influential across disciplines dealing with professional work and expertise, including studies of nursing, teaching, planning, and management (Yanow & Tsoukas, 2009). Through ethnographic study of architects, engineers, town planners, managers, and psychotherapists, Schön described situated processes of professional thinking (Malinin, 2018). In the field of education, the reflective disposition highlighted by Schön has been especially influential (Ramos-Rodríguez et al., 2016), while in nursing, reflective practice stimulates the development of professional judgment (Atmani et al., 2023). The concept has also been applied in project management, where reflection-in-action manifests as managers "thinking on their feet" and "keeping their wits about them".

Activity Theory

The paragraph's claims are well supported by the literature. Activity Theory, rooted in Vygotsky's work and expanded by Engeström (2000), conceptualises human activity as object-oriented, mediated, collective, and social (Igira & Gregory, 2009; Kaatrakoski et al., 2016; Krichevsky, 2021; Roth & Lee, 2007; Tkachenko & Ardichvili, 2017). Cultural-historical activity theory views learning as embedded in sociocultural contexts, where behaviour cannot be understood independently of that context (Paavola et al., 2004; Sawchuk, 2007). The framework's second generation, credited to Engeström, emphasises the individual learning in interrelationship with the broader community, incorporating tools, rules, community, and division of labour as mediating elements of activity systems (Daniels & Warmington, 2007; Igira & Gregory, 2009; Krichevsky, 2021; Mentz & Beer, 2019).

Workplace learning, from this perspective, is shaped not only by individual cognition but also by organisational policies, technological resources, workplace relationships, and cultural norms (Kaatrakoski et al., 2016; Tkachenko & Ardichvili, 2017; Willocks, 2022a). Additional studies have identified the importance of workplace context in shaping learning behaviour (Kaatrakoski et al., 2016). Learning often emerges when contradictions or tensions within activity systems require workers to develop new ways of thinking and acting (Daniels & Warmington, 2007; Sawchuk, 2007; Taylor, 2014; Tkachenko & Ardichvili, 2017). Contradictions are defined as historically accumulating structural tensions within and between activity systems rather than mere problems or conflicts (Baldwin, 2020; Daniels & Warmington, 2007; Taylor, 2014). These contradictions manifest as conflicts, dilemmas, and disturbances in everyday work and serve as driving forces for change and development (Arinaitwe et al., 2022; Kaatrakoski et al., 2016; Tkachenko & Ardichvili, 2017; Willocks, 2022a). Expansive learning occurs through the collective identification and resolution of these contradictions, producing culturally new patterns of activity (Arinaitwe & Sannerud, 2019; Sawchuk, 2007; Tkachenko & Ardichvili, 2017).

When new tools or technologies are introduced into a workplace, they may generate secondary contradictions between existing elements and new objects within the activity system (Arinaitwe et al., 2022). Workers must collectively adapt by modifying existing practices, learning new skills, and collaborating differently (Arinaitwe & Sannerud, 2019; Kaatrakoski et al., 2016; Tkachenko & Ardichvili, 2017). Such tensions between emerging needs and organisational structures have been empirically identified in studies of educators adopting new digital practices (Kaatrakoski et al., 2016). The resolution of these contradictions through collective engagement constitutes the expansive, transformative type of learning central to the Cultural-Historical Activity Theory (CHAT) tradition (Mentz & Beer, 2019; Tkachenko & Ardichvili, 2017; Willocks, 2022b).

In CHAT, a contradiction is not simply a disagreement or conflict between people. Rather, it refers to a deeper tension, inconsistency, or mismatch within an activity system. These contradictions often emerge when existing ways of working are no longer adequate for achieving desired outcomes. For example, a hospital may introduce new digital health records while maintaining old paper-based procedures. The resulting tension between old and new systems creates inefficiencies, confusion, and frustration. Similarly, a vocational education provider may be expected to deliver highly flexible, individualised training while simultaneously complying with rigid regulatory requirements. These competing demands create contradictions within the system.

According to CHAT, contradictions are not viewed as problems to be eliminated but as important drivers of learning and change. When individuals and groups recognise these tensions, they begin to question existing practices, analyse underlying causes, and explore alternative ways of working. Through discussion, collaboration, experimentation, and reflection, participants collectively develop new approaches that better address the challenges they face. This process often leads to changes in tools, procedures, roles, relationships, or organisational structures.

The phrase "collective engagement" refers to the fact that this learning does not occur solely within individual minds. Instead, it is a social process involving multiple participants who work together to understand and resolve contradictions. Teachers, managers, workers, learners, or other stakeholders contribute different perspectives and expertise as they seek solutions. Learning therefore becomes a collaborative activity that transforms both individual understanding and organisational practice.

The term "expansive learning" is central to Engeström's interpretation of CHAT. Expansive learning occurs when people move beyond simply improving existing practices and instead create entirely new ways of working. Rather than learning how to perform current tasks more efficiently, participants redefine the object, purpose, or structure of the activity itself. Learning becomes transformative because it changes the activity system rather than merely improving performance within it.

For example, consider a UK Further Education college where teachers, administrators, and industry partners recognise that traditional classroom-based delivery is failing to meet the needs of adult learners who work full-time. The contradiction exists between the needs of learners and the existing delivery model. Through collaborative planning and experimentation, the college could develop a new blended learning approach that combines online learning, workplace projects, and flexible attendance. In this case, learning has occurred not simply because staff acquired new skills, but because the organisation collectively transformed how education is delivered. This represents expansive learning because the activity system itself has been redesigned.

Workplace Learning in Contemporary Organisations

Modern workplace learning increasingly combines elements from multiple theories. Formal training remains important, but organisations also recognise the value of mentoring, coaching, communities of practice, reflective practice, collaborative problem-solving, and experiential learning (Briggs, 2018; Harper et al., 2024). Workplace mentoring has been identified as particularly effective when it combines both clinical and management supervision, guiding learners through complex situations (Briggs, 2018). Leadership and management development programs increasingly emphasise the central role of learning facilitation and meta-learning processes in enabling capability development (Harper et al., 2024). Digital technologies have further expanded opportunities for workplace learning through online collaboration, knowledge-sharing platforms, virtual communities, and just-in-time learning resources (Mihalcea, 2017). In the digital age, organisations must shift from traditional workplace paradigms towards engagement, learning, and development of employees, with digital infrastructure creating cohesion among employees by promoting participatory organisational cultures (Mihalcea, 2017).

Contemporary approaches recognise that learning is not simply an individual process but is influenced by workplace culture, leadership, social relationships, organisational systems, and access to meaningful work experiences (Abid et al., 2023; Asriati, 2025; Harper et al., 2024; Mohammed et al., 2023). The workplace learning perspective is supported by situated learning theory, which assumes that learning goes beyond an individual process and includes learning conditions and the learning environment (Abid et al., 2023). Organisational learning culture has been shown to have a highly statistically significant positive correlation with workplace learning, alongside goal orientation and managerial effectiveness (Mohammed et al., 2023).

Various studies have demonstrated that culture and leadership have a critical influence on both organisational and individual learning processes and outcomes (Harper et al., 2024). Group cultures and developmental cultures, those that are inclusive, flexible, and open to change, are positively associated with organisational learning (Harper et al., 2024). Furthermore, psychological principles underscore the importance of autonomy, competence, and relatedness in enhancing motivation and well-being, while positive social relationships at work are associated with higher levels of job satisfaction, engagement, and organisational commitment (Asriati, 2025). Effective workplace learning environments therefore support both individual development and collective knowledge creation, with leadership playing a pivotal role in fostering cultures of continuous learning and growth (Asriati, 2025; Boverie et al., 2013; Harper et al., 2024). Leaders can create positive environments by promoting nurturing workplaces, encouraging meaningful work, and ensuring employees have opportunities for learning (Boverie et al., 2013).

Significance for Adult Education

Workplace learning theories have significantly influenced adult education, vocational education and training, and professional development (Carré, 2015; Evans, 2015). They highlight that adults learn most effectively when learning is relevant, contextualised, and connected to real-world practice (Coulombe, 2012; Evans, 2015). As Evans notes, the workplace as a learning environment has an important impact on motivating employees and facilitating the achievement of their goals across a broad range of professional and personal activities (Evans, 2015). These theories also demonstrate that valuable learning occurs not only through formal instruction but through participation in work, collaboration with others, reflection on experience, and engagement in professional communities (Carré, 2015; Coulombe, 2012; Cristol & Müller, 2013). Cristol and Müller identify that organisations stimulate informal learning through teamwork, problem-solving during work, and practice analysis (Cristol & Müller, 2013), while Coulombe emphasises that learning in work situations allows establishing links between theory and practice, work habits, collaborative spaces, and professional recognition (Coulombe, 2012). The development of work-study programmes, various forms of practice analysis, workplace training, and growing interest in informal professional learning all point toward this broader understanding (Carré, 2015).

Collectively, workplace learning theories provide a comprehensive understanding of how adults develop expertise, adapt to changing work environments, and engage in lifelong learning throughout their careers (Carré, 2015; Cristol & Müller, 2013; Evans, 2015). Evans highlights that at least six types of workplace learning occur naturally when people work collaboratively on tasks, including observation, peer support, focused discussions, and unsupervised practice (Evans, 2015). Carré further notes that international organisations from the European Union to UNESCO have proclaimed the absolute necessity of developing daily and permanent learning, formal and informal, as an essential ingredient of emerging cognitive societies (Carré, 2015). They reinforce the idea that workplaces are not merely locations where work is performed but are also powerful environments for learning, knowledge creation, and professional growth (Coulombe, 2012; Cristol & Müller, 2013; Evans, 2015). As Coulombe observes, the social nature of individuals, their capacity for dynamic engagement, and their desire for efficiency and understanding of work situations generate learning tied to those situations (Coulombe, 2012).

Apprenticeships and Vocational Learning

Apprenticeships and vocational learning represent some of the oldest and most enduring forms of workplace learning. Unlike traditional academic education, which often emphasises theoretical knowledge, apprenticeships focus on developing occupational competence through the integration of practical experience, workplace participation, and structured instruction. Learners acquire knowledge and skills by engaging in authentic work activities under the guidance of experienced practitioners while simultaneously developing an understanding of the theoretical principles that underpin professional practice (Li et al., 2023; Wolek, 1999).

Historically, apprenticeships emerged through guild systems in which novice workers learned directly from skilled craftspeople (Li et al., 2023; Patchett, 2017; Wolek, 1999). Apprentices gradually acquired the knowledge, skills, values, and professional standards of their trade through observation, practice, and participation in everyday work (Li et al., 2023; Patchett, 2017; Wolek, 1999). The craft guilds regulated and monitored apprenticeships, stipulating their duration, fees, the number of apprentices, their registration, and assessment (Li et al., 2023). The main mechanism through which tacit skills were transmitted across individuals was apprenticeship, a relation linking a skilled adult to a youngster whom he taught the trade. Although modern apprenticeship systems are more formally structured and regulated, the underlying principle remains largely unchanged: learning occurs through participation in real work activities supported by instruction, mentoring, and increasing levels of responsibility.

Apprenticeships differ from many other forms of adult learning because they are fundamentally centred on participation in productive work. While adult learning can occur in classrooms, online environments, community programs, or professional development courses, apprenticeships require learners to engage directly in real workplace activities. Learning occurs through performing occupational tasks under supervision, allowing apprentices to develop practical competence alongside theoretical knowledge. The workplace itself becomes the primary learning environment rather than simply a location where previously acquired knowledge is applied. As Schmid et al. describe, workplace learning can be understood as a sequenced pathway of activities towards vocational expertise, rooted in conceptions of learning as participation in communities of practice, describing learning as the gradual transition from marginal to core participation.

Another key difference is the close integration of theory and practice. In many adult education programs, theoretical concepts are taught first and practical application follows later. Apprenticeships, however, typically combine learning and application simultaneously, Apprentices learn concepts in formal training and immediately apply them in authentic workplace situations. In Norway, for example, two-year apprenticeships follow a dual-track model combining workplace training with school-based instruction. This continual movement between instruction and practice helps learners understand not only what to do, but why tasks are performed in particular ways and how theoretical principles influence workplace outcomes. Mentorship is innate to such training, as it stems from the apprenticeship-framed mentor/mentee relationship.

Apprenticeships also place a stronger emphasis on learning through observation, mentoring, and guided participation. Adult learners in other educational settings often work independently or interact primarily with instructors during scheduled learning activities. Apprentices, by contrast, spend significant time working alongside experienced practitioners who demonstrate skills, share tacit knowledge, provide feedback, and model professional behaviours (Patchett, 2017). The person-to-person transmission of tacit knowledge has been recognised as central to apprenticeship, as such knowledge is difficult to codify and must be conveyed through direct interaction. This mentoring relationship enables the transfer of practical expertise that is often difficult to teach through formal instruction alone. As Patchett argues, rather than passively absorbing the "secrets" of a workshop, the apprentice actively co-produces and reproduces its corporeal and material practices (Patchett, 2017).

A further distinction is the development of occupational identity. While many adult learning programs focus primarily on acquiring knowledge or specific skills, apprenticeships socialise learners into a profession or trade. Through participation in workplace communities, apprentices learn industry norms, workplace culture, professional values, communication practices, and standards of behaviour. It is part of a guide's role to organise learning opportunities and to monitor learners' participation, determining their readiness to engage in activities of increasing accountability and autonomy. As a result, apprenticeships contribute not only to skill development but also to the formation of a professional identity and a sense of belonging within an occupational community.

Finally, apprenticeships are typically competency-based and linked directly to workforce outcomes. Success is measured by the learner's ability to perform occupational tasks to industry standards rather than solely by academic achievement or knowledge acquisition. In Switzerland, for instance, the learning outcomes of two-year curricula are defined by occupational activities identified through analyses involving active workers and experts in the respective fields. Because apprenticeships are designed to prepare individuals for specific occupations, learning is highly contextualised, practical, and employment focused. Although other forms of adult learning may pursue broader personal, social, or intellectual goals, apprenticeships are distinguished by their direct connection to workplace competence, professional practice, and career entry or progression.

Scope of Apprenticeship Implementation

Apprenticeships are common in many countries, although their structure, status, and prevalence vary considerably. They are particularly prominent in countries with strong vocational education and training (VET) systems, where employers, industry bodies, and governments collaborate to develop occupational skills (Biavaschi et al., 2012; Koca & Ortakaya, 2023; Paudel & Eberhardt, 2023). In these countries, apprenticeships are often viewed as a mainstream pathway into skilled employment rather than an alternative to higher education (Graf, 2013; Lassnigg, 2024).

Germany is widely regarded as the international benchmark for apprenticeship systems (Lange, 2012; Lehmann, 2000). Its dual system combines paid employment with vocational school education, allowing apprentices to spend part of their time in the workplace and part in formal training (Locke & Jacoby, 2019; Paudel & Eberhardt, 2023). The content of training programs is determined in a tripartite manner at the federal level, with unions, employer associations, and the federal government negotiating the curriculum and the types of occupations to be covered (Locke & Jacoby, 2019). Apprenticeships are available across a broad range of occupations, including engineering, manufacturing, information technology, banking, healthcare, hospitality, and business administration, with service, construction, and industrial apprenticeships all represented (Biavaschi et al., 2012; Locke & Jacoby, 2019). Similar dual apprenticeship systems operate in Austria and Switzerland, where apprenticeships enjoy high levels of employer participation and social prestige (Graf, 2013; Lassnigg, 2024; Moretti et al., 2017). All three countries are classified as collective skill formation systems due to their historically evolved dual apprenticeship training systems (Ebner et al., 2013; Emmenegger & Seitzl, 2020; Graf, 2013). However, despite structural similarities, these three systems also show substantial differences in how they sustain upper secondary VET and cope with crises in the youth labor market (Lassnigg, 2024; Moretti et al., 2017).

Apprenticeships are also well established in the United Kingdom, where they are available from intermediate levels through to degree apprenticeships that combine university study with workplace learning (Lee, 2012). In recent years, the UK has expanded apprenticeship opportunities beyond traditional trades to include occupations such as digital technology, accountancy, nursing, project management, policing, and engineering. The introduction of Higher Apprenticeships has established synergies between vocational and higher education pathways (Lee, 2012). Ireland has similarly broadened its apprenticeship system, moving beyond construction trades into professional and service-sector occupations.

In the Nordic countries, including Denmark, Norway, and Finland, apprenticeships form an important part of vocational education. Denmark operates a strong dual training model similar to Germany, characterized by features such as a training levy and school-based training tailored to specific vocations completed prior to apprenticeship training (Emmenegger & Seitzl, 2020; Graf, 2013). Norway combines school-based vocational education with workplace apprenticeships; political reforms of the 1990s led to a hybrid system with a stronger apprenticeship component in Norway compared to Sweden's more school-based approach (Thunqvist et al., 2019). Finland utilizes an apprenticeship model in which school education incorporates compulsory elements of practical vocational training in the workplace (Marszowski, 2019). These countries generally place a high value on vocational pathways and maintain strong links between education providers and industry (Marszowski, 2019; Thunqvist et al., 2019).

Outside Europe, apprenticeships are common in Australia and New Zealand, where they are central components of vocational education and workforce development (Biavaschi et al., 2012). Traditionally associated with trades such as plumbing, electrical work, carpentry, automotive technology, and engineering, apprenticeships have expanded into areas such as business services, information technology, healthcare, logistics, agriculture, and community services. Australia also operates traineeships, which apply apprenticeship principles to a broader range of occupations.

In Canada, apprenticeships are particularly important in the skilled trades, including construction, manufacturing, transportation, and industrial maintenance (Esgandanian, 2023; Lehmann, 2000). Provincial governments regulate apprenticeship systems, and many occupations use the Red Seal Program to provide nationally recognised qualifications. Canadian apprenticeship models have been critiqued for not effectively matching young people to employment opportunities, with low completion rates relative to enrollments identified as a persistent challenge (Esgandanian, 2023). The German dual system is often considered a model for Canadian youth apprenticeship initiatives, though debates suggest that direct transfer of the German model requires careful consideration of institutional context (Lehmann, 2000). The United States also has registered apprenticeship programs, although participation rates are generally lower than in countries such as Germany or Australia (Biavaschi et al., 2012). Apprenticeships in the United States are common in construction, manufacturing, telecommunications, healthcare, and increasingly in information technology and cybersecurity.

Many developing and emerging economies also utilise apprenticeship models. India has expanded apprenticeship and work-based learning programs to address skills shortages and support economic development, with the Apprentices Act regulating training programs in industry under the Ministry of Skill Development and Entrepreneurship (Ravichandran, 2023). China has launched separate apprenticeship reforms through the Ministry of Human Resources and Social Security and the Ministry of Education, each geared towards enhancing the country's vocational education and training system, though neither model has delivered ideal training outcomes due to the absence of industry-level civil society governance (Zhang, 2021). Countries such as South Africa, Singapore, and Malaysia have similarly expanded work-based learning programs. In some countries, informal apprenticeship systems remain common, particularly in small businesses and traditional trades, where skills are transmitted directly from experienced workers to novices outside formal regulatory frameworks (Biavaschi et al., 2012).

Although apprenticeship systems differ significantly between countries, they share a common purpose: enabling learners to develop occupational competence through a combination of workplace experience, structured training, mentoring, and progressive responsibility (Biavaschi et al., 2012; Paudel & Eberhardt, 2023). Countries with strong apprenticeship traditions generally view work-based learning as a critical mechanism for workforce development, economic productivity, and the transition from education to employment (Biavaschi et al., 2012; Esgandanian, 2023; Koca & Ortakaya, 2023; Lassnigg, 2024).

Characteristics of Apprenticeship and Vocational Learning

A defining characteristic of apprenticeship and vocational learning is the emphasis on learning through participation in authentic occupational activities. Apprentices are not simply taught about work in a theoretical sense; they develop competence by actively engaging in workplace tasks and responsibilities (Aakernes, 2020; Rintala & Nokelainen, 2019). Learning occurs through direct involvement in real work environments where learners progressively acquire practical skills, knowledge, and professional judgement (Aakernes, 2020). Typically, apprentices begin with relatively simple tasks performed under close supervision. As their confidence and competence increase, they assume greater responsibility and undertake increasingly complex activities with greater independence (Dolenc et al., 2020; Li et al., 2009b). For example, an apprentice electrician may initially assist qualified tradespeople, identify tools and materials, and observe installations before progressing to installing wiring systems, diagnosing faults, interpreting plans, and coordinating work activities. This gradual progression reflects the principle that expertise develops through repeated exposure to authentic workplace situations rather than through theoretical instruction alone (Harnish & Wilke-Schnaufer, 1998; Ruiz et al., 2018).

The importance of participation is further explained by theories of situated learning developed by (Lave & Wenger, 1991). These theories propose that learning occurs through engagement in social and cultural practices rather than through the passive acquisition of abstract knowledge (Aakernes, 2020; Guile & Lahiff, 2016). New learners enter occupational communities as novices and participate in what Lave and Wenger termed legitimate peripheral participation (Aakernes, 2020; Rintala & Nokelainen, 2019). Initially, apprentices work at the periphery of the occupational community, undertaking lower-risk tasks while observing experienced practitioners (Li et al., 2009b; Rintala & Nokelainen, 2019). As they gain knowledge, confidence, and competence, they move towards fuller participation within the profession (Guile & Lahiff, 2016; Rintala & Nokelainen, 2019). Through this process, learners acquire not only technical skills but also the language, values, norms, behaviours, and professional identities associated with their occupation (Guile & Lahiff, 2016; Li et al., 2009b). For example, a nursing student undertaking clinical placement learns clinical procedures alongside professional communication, teamwork, ethical decision-making, and patient-centred care through participation in healthcare practice (Li et al., 2009b).

Apprenticeship learning also aligns closely with the concept of cognitive apprenticeship developed by (Collins et al., 1989) (Dolenc et al., 2020; Harnish & Wilke-Schnaufer, 1998; Stafeniak & Scott-Brown, 2016). Traditional apprenticeship models often focus on developing practical and physical skills; however, cognitive apprenticeship extends this approach by making expert thinking visible to learners (Dolenc et al., 2020; Girei et al., 2020; Napari, 2020). Experienced practitioners demonstrate not only what they do but also explain the reasoning, decision-making processes, and problem-solving strategies that underpin their actions (Dolenc et al., 2020; Tsukube & Matsuo, 2020). Learners observe these cognitive processes, receive coaching and feedback, and gradually assume greater responsibility for complex tasks (Bruin, 2019; Moon, 2018; Stafeniak & Scott-Brown, 2016). For example, an engineering supervisor may explain how risk assessments are conducted, why particular design decisions are made, or how faults are diagnosed and resolved. By exposing learners to expert thinking, cognitive apprenticeship helps develop the higher-order reasoning and judgement required for professional competence (Dolenc et al., 2020; Girei et al., 2020; Tsukube & Matsuo, 2020).

Another important characteristic of apprenticeship learning is its close relationship with experiential learning. Apprentices acquire knowledge through practical experiences, reflect on outcomes, develop conceptual understanding, and apply their learning to future situations. Workplace tasks provide the experiences that form the basis of learning, while mentoring, feedback, and formal instruction support reflection and knowledge development. For example, a carpentry apprentice may construct a roof frame, encounter difficulties during the process, discuss challenges with a supervisor, study relevant construction principles, and subsequently apply improved techniques to future projects. Through repeated cycles of action, reflection, and application, learners progressively refine their skills and deepen their understanding.

Apprenticeships are also strongly connected to Wenger's concept of communities of practice. Learning occurs within occupational communities where experienced and novice workers collaborate, share knowledge, solve problems, and engage in common activities (Varma & Malik, 2024; Warren, 2019). These communities provide opportunities for mentoring, observation, feedback, and collaborative learning that extend beyond formal instruction (Varma & Malik, 2024; Warren, 2019). Importantly, apprentices often learn from multiple individuals rather than from a single mentor. Supervisors, tradespeople, colleagues, managers, customers, and other stakeholders all contribute to the learning process (Warren, 2019). Participation in these social networks supports the development of both professional competence and occupational identity, helping learners become recognised members of their professional community (Varma & Malik, 2024; Warren, 2019).

While apprenticeships represent a significant form of vocational learning, vocational education encompasses a broader range of learning pathways. Vocational learning occurs in technical and further education institutions, training centres, workplaces, community organisations, and higher education settings (Rowe, 2019; Sanséau & Ansart, 2013). It may include traineeships, workplace-based learning, internships, clinical placements, industry projects, simulations, and professional development programs (Rowe, 2019; Sanséau & Ansart, 2013). Regardless of the delivery model, vocational learning is characterised by its focus on developing practical competence that can be applied directly within occupational contexts (Rowe, 2019; Sanséau & Ansart, 2013). Learners are encouraged to integrate technical knowledge, practical skills, problem-solving abilities, communication skills, and professional behaviours in ways that reflect real workplace requirements (Sanséau & Ansart, 2013; Tūtlys & Spöttl, 2017). Consequently, vocational learning plays a critical role across diverse industries including healthcare, information technology, engineering, agriculture, hospitality, aviation, logistics, and community services (Rowe, 2019).

The effectiveness of apprenticeship and vocational learning is heavily influenced by the quality of mentoring and supervision provided in the workplace. Experienced practitioners play a crucial role in demonstrating skills, monitoring performance, providing feedback, and supporting learner development (Warren, 2019). Effective mentors help learners bridge the gap between theoretical knowledge and practical application while fostering confidence, independence, and professional identity. Supportive supervision can accelerate skill acquisition, encourage reflective practice, and promote safe and effective workplace performance (Warren, 2019). Conversely, limited guidance or inadequate supervision may restrict learning opportunities and hinder professional growth (Warren, 2019).

Modern apprenticeship systems increasingly recognise that learning extends well beyond the completion of an initial qualification. Rapid technological advancement, changing industry requirements, and evolving workplace practices mean that workers must continue learning throughout their careers (Ceschi et al., 2021; Venkatraman et al., 2018). Apprenticeships therefore provide not only occupation-specific skills but also the foundations for lifelong learning, adaptability, and ongoing professional development (Ceschi et al., 2021; Venkatraman et al., 2018). Many occupations now offer pathways from apprenticeship qualifications to advanced vocational training, higher education programs, professional certifications, and specialist training opportunities (Ceschi et al., 2021; Venkatraman et al., 2018). As a result, apprenticeships are increasingly viewed as the beginning of a continuous process of learning and career development rather than simply preparation for entry into a particular occupation (Ceschi et al., 2021; Venkatraman et al., 2018).

Significance of Apprenticeships and Vocational Learning

Apprenticeships and vocational learning remain among the most effective methods for developing occupational competence because they integrate learning with authentic work practice. They provide opportunities for learners to acquire technical expertise, professional judgement, workplace behaviours, and occupational identities through participation in real-world activities. Supported by theories such as situated learning, cognitive apprenticeship, experiential learning, and communities of practice, apprenticeship systems demonstrate how learning emerges through experience, social interaction, mentoring, reflection, and participation in professional communities.

As economies continue to evolve and demand increasingly skilled workforces, apprenticeships and vocational learning will remain critical mechanisms for developing expertise, supporting workforce development, and enabling individuals to engage in lifelong learning and career progression.

Learning through participation is particularly effective in occupations where competence depends upon the ability to perform practical tasks in authentic workplace environments. Trades such as electrical work, plumbing, and carpentry, as well as professions such as nursing, aviation maintenance, hospitality, agriculture, and emergency services, rely heavily on this approach. An apprentice electrician develops competence by installing wiring, testing circuits, and troubleshooting faults under supervision, while a nursing student learns patient care through direct participation in clinical practice. In these contexts, active involvement allows learners to develop technical skills, situational awareness, professional judgement, and confidence that cannot be fully acquired through classroom instruction alone.

However, learning through participation may be less effective when the subject matter is highly theoretical or abstract and requires extensive foundational knowledge before meaningful application can occur. Disciplines such as advanced mathematics, theoretical physics, constitutional law, or philosophy often require learners to develop conceptual understanding before they can effectively engage in authentic activities. In such situations, participation alone may not provide sufficient opportunities to understand the underlying principles and theories that form the basis of professional practice.

Situated learning is particularly valuable when learners must become members of a professional or occupational community. Occupations such as policing, teaching, social work, healthcare, engineering, and the skilled trades require learners not only to acquire technical competence but also to understand professional norms, values, language, and workplace culture. A student teacher, for example, may initially observe experienced educators before gradually progressing to planning lessons, managing classrooms, and assessing student learning. Through participation in the professional community, learners develop both practical competence and a professional identity.

The effectiveness of situated learning can be reduced when learners have limited access to authentic communities of practice or when workplace cultures are dysfunctional. Organisations characterised by poor safety practices, ineffective leadership, or outdated procedures may inadvertently socialise learners into undesirable behaviours. Similarly, highly remote or dispersed workplaces may limit opportunities for observation, collaboration, and participation, making it more difficult for learners to become integrated into a professional community.

Cognitive apprenticeship is particularly effective in professions where expert thinking, decision-making, and problem-solving are central to performance. Fields such as medicine, engineering, law, project management, aviation, and software development require practitioners to apply complex reasoning processes in dynamic situations. When experienced professionals make their thinking visible by explaining how they analyse problems, assess risks, and make decisions, learners gain insight into the cognitive processes that underpin expert performance. For example, a senior engineer who explains the reasoning behind design decisions provides valuable learning opportunities that extend beyond technical procedures alone.

This approach may be less effective when experts are unable to articulate their thinking processes or when workplace demands leave little time for coaching and explanation. It may also be of limited value in highly routine occupations where decision-making requirements are minimal and tasks can be learned effectively through procedural instruction and repetition.

Experiential learning is highly effective when adult learners can apply knowledge directly to authentic problems and subsequently reflect on their experiences. Leadership development, management training, entrepreneurship, healthcare, community services, and emergency response training frequently benefit from experiential approaches. For example, a supervisor managing a difficult workplace conflict may reflect on the outcomes of their actions, analyse what contributed to the situation, and apply revised strategies in future interactions. This process promotes deep understanding, professional growth, and long-term retention of learning.

Experiential learning may be less effective when learners lack the prior knowledge necessary to interpret experiences accurately. Without appropriate guidance, reflection, and feedback, learners may reinforce misconceptions or develop ineffective practices. The approach may also be inappropriate in situations where mistakes involve unacceptable levels of risk, such as complex surgical procedures, nuclear power operations, or hazardous chemical processing. In such cases, simulation, structured instruction, and close supervision are often required before learners engage in authentic workplace activities.

Communities of practice are particularly effective in professions where knowledge is continually evolving and collaboration plays a central role in professional success. Fields such as teaching, healthcare, information technology, engineering, research, and management benefit from environments where practitioners regularly exchange knowledge, share experiences, and solve problems collectively. Teachers discussing instructional strategies, nurses sharing patient care insights, or software developers collaborating on technical challenges all contribute to the creation and dissemination of professional knowledge. These communities support ongoing professional development and foster collective expertise.

The effectiveness of communities of practice may be diminished in organisational cultures that discourage collaboration, knowledge sharing, or trust. Competitive workplaces where employees are reluctant to share expertise can limit opportunities for collective learning. In addition, communities of practice may unintentionally reinforce outdated practices if members become resistant to external perspectives, innovation, or change.

Vocational learning approaches such as traineeships, internships, workplace projects, clinical placements, and simulations are particularly effective when occupations require a combination of theoretical understanding and practical competence. Industries such as aged care, information technology, aviation, logistics, hospitality, engineering, and community services frequently utilise these approaches to help learners apply knowledge in realistic contexts. Simulated learning environments can be especially valuable when access to actual workplaces is limited or when risks must be carefully managed.

These approaches may be less effective when practical learning activities become disconnected from workplace realities. Simulations that fail to accurately replicate occupational conditions may not adequately prepare learners for employment. Similarly, vocational learning may be less suitable when educational goals focus primarily on personal enrichment, abstract theory, critical analysis, or intellectual exploration rather than occupational competence.

Mentoring is particularly effective when learners require guidance, feedback, encouragement, and professional socialisation. Apprentices, new managers, healthcare professionals, teachers, military personnel, and many other adult learners benefit from relationships with experienced practitioners who provide support and share expertise. Effective mentoring can accelerate skill development, build confidence, encourage reflective practice, and assist learners in navigating workplace challenges.

However, mentoring can be ineffective when mentors lack the necessary expertise, communication skills, or commitment to learner development. Poor mentoring relationships may reinforce ineffective practices, provide inconsistent guidance, or undermine learner confidence. Furthermore, highly experienced and self-directed adult learners may require less intensive mentoring than novices and may benefit more from peer collaboration or independent learning opportunities.

Lifelong learning and continuous professional development are particularly effective in occupations characterised by rapid technological, regulatory, or professional change. Fields such as cybersecurity, healthcare, engineering, aviation, education, and information technology require practitioners to continually update their knowledge and skills in order to remain competent and effective. Continuous learning enables professionals to adapt to new technologies, respond to emerging challenges, and maintain professional standards throughout their careers.

These initiatives may be less effective when they are poorly aligned with workplace needs, receive limited organisational support, or are viewed merely as compliance requirements. Adult learners are more likely to disengage when professional development activities are repetitive, irrelevant, or disconnected from their career goals and workplace realities. Meaningful lifelong learning requires relevance, practical application, and organisational commitment to continuous improvement.

The effectiveness of apprenticeship and vocational learning approaches depends largely on the relationship between the learning method, the learner's needs, and the occupational context. These approaches are particularly powerful when learning outcomes require practical competence, professional judgement, workplace socialisation, and the integration of theory and practice. They are generally less effective when learning objectives are highly abstract, when opportunities for authentic participation are unavailable, or when workplace environments fail to provide adequate support, mentoring, collaboration, and opportunities for reflection. Consequently, effective adult education often involves selecting and adapting learning approaches to suit both the nature of the learning objectives and the context in which learning occurs.

Work-Integrated Learning

Work-Integrated Learning (WIL) is an educational approach that intentionally combines academic learning with practical workplace experience (Abery et al., 2015; Choi-Lundberg, 2024; Curto-Reverte et al., 2025). It seeks to bridge the gap between theory and practice by providing learners with opportunities to apply knowledge, skills, and professional behaviours in authentic work environments (Bouwer et al., 2021b; Choi-Lundberg, 2024; Mesuwini et al., 2022). Unlike traditional classroom-based learning, which often separates learning from practice, WIL integrates the two, enabling learners to develop occupational competence while gaining an understanding of workplace expectations, culture, and professional identity (Abery et al., 2015; Blackford et al., 2024; Curto-Reverte et al., 2025).

WIL has been defined as an umbrella term for a range of approaches and strategies that integrate theory with the practice of work within a purposefully designed curriculum (Curto-Reverte et al., 2025; McNamara & Ruinard, 2016), and as "a curriculum design where individuals experience the professional context relevant to their study and work-related futures" (Bouwer et al., 2021a, p. 18). The International Journal of Work-Integrated Learning defines WIL as "an educational approach that uses relevant work-based experiences to integrate theory with meaningful workplace practice" (Mesuwini et al., 2022, p. 146).

The concept of work-integrated learning is based on the belief that learning is most effective when learners can connect theoretical knowledge with real-world application (Choi-Lundberg, 2024; Jackson, 2017; Scholtz & Bester, 2018). Through participation in workplace activities, learners gain opportunities to test ideas, solve practical problems, reflect on their experiences, and develop a deeper understanding of their field of study (Abery et al., 2015; Babacan & Babacan, 2015; Jackson, 2017). Reflection is recognised as integral to WIL, enabling students to develop both personally and professionally and to move beyond performative competence to understanding and judgement (Babacan & Babacan, 2015; Curto-Reverte et al., 2025).

Work-integrated learning therefore draws heavily upon experiential learning, situated learning, workplace learning, and constructivist theories, all of which emphasise the importance of authentic experience in the learning process (Bouwer et al., 2021a; Jooste & Shaikh, 2026; Mueller et al., 2022). WIL has been described as grounded in the theory of experiential learning (Jackson, 2017), while situated learning theory, particularly the work of Lave and Wenger, has been applied as a theoretical lens for understanding WIL experiences (Bouwer et al., 2021a). Constructivist perspectives similarly underpin WIL pedagogy, with scholars noting that WIL encompasses "a transformative pedagogy, which entails...personal development and experiential learning" (Abery et al., 2015, p. 88). Authentic and situated learning theories further inform WIL design by situating knowledge within real-world reasoning scenarios (Jooste & Shaikh, 2026).

Work-integrated learning can take many forms depending on the educational context and occupational field. Common examples include internships, practicums, clinical placements, apprenticeships, traineeships, cooperative education programs, industry projects, work placements, fieldwork, service learning, and workplace simulations. In higher education, nursing students undertake clinical placements in healthcare settings, engineering students participate in industry projects, and education students complete supervised teaching practicums. In vocational education and training, apprenticeships and traineeships represent well-established forms of work-integrated learning where structured workplace experience forms a central component of the qualification.

A key characteristic of work-integrated learning is the deliberate integration of workplace experiences with educational objectives. Learners are not simply placed in a workplace to gain experience; rather, the workplace activities are designed to support specific learning outcomes. Educational institutions, employers, supervisors, and learners typically collaborate to ensure that workplace experiences contribute to the development of required knowledge, skills, and competencies. Reflection is often incorporated through journals, portfolios, professional discussions, reports, or presentations to help learners connect their experiences with theoretical concepts.

Work-integrated learning provides several important benefits for adult learners. It enables learners to apply classroom knowledge in authentic settings, develop practical competence, and gain exposure to workplace technologies, systems, and procedures (OECD, 2024; Thistlethwaite, 2013). Practice-based learning facilitates the transfer of theory into the workplace through situated and experiential learning, and "is not solely about preparing for practice after qualification; rather it is about learning in and about practice through authentic experience and becoming part of a community of practice" (Thistlethwaite, 2013, p. 15). Participation in real work environments also helps learners develop employability skills such as communication, teamwork, problem-solving, time management, professionalism, and adaptability (OECD, 2024). The PISA-VET framework identifies employability skills among the key success outcomes for vocational learners, alongside occupational-specific learning outcomes and occupational identity (OECD, 2024). Furthermore, learners gain a clearer understanding of occupational expectations and workplace culture, which can ease the transition from education to employment (Dias et al., 2026; Thistlethwaite, 2013).

Work-integrated learning also supports the development of professional identity. Through interaction with experienced practitioners, learners begin to understand the values, norms, standards, and behaviours associated with their chosen profession (Juntunen et al., 2024; Thistlethwaite, 2013). Mentors serve as teachers, facilitators, and role models, supporting students "to develop a professional identity" and "to learn and safely achieve competence and autonomy in their professional role" (Juntunen et al., 2024, p. 703). A nursing student, for example, learns not only clinical procedures but also professional ethics, patient-centred care, communication practices, and teamwork (Dias et al., 2026; Thistlethwaite, 2013) Clinical mentorship in nursing has been shown to positively influence "students' confidence, workplace adaptation, clinical competence, and professional development" (Dias et al., 2026, p. 2). Similarly, an engineering student develops an understanding of professional responsibility, safety culture, and project management through participation in industry settings. Practice-based learning is indeed "a feature of professional and vocational training programmes such as teaching, the health professions, engineering and social work" (Thistlethwaite, 2013, p. 16).

The effectiveness of work-integrated learning is closely linked to the quality of workplace supervision and mentoring (Dias et al., 2026; Morley, 2018; Styles et al., 2022). Experienced supervisors play an important role in guiding learners, providing feedback, modelling professional practice, and helping learners make sense of their experiences (Juntunen et al., 2024; Morley, 2018; Styles et al., 2022). Supervision has been defined as "the provision of guidance and feedback on matters of personal, professional and educational development in the context of a trainee's experience of providing safe and appropriate patient care," and evidence suggests it "has a positive effect on patient outcomes and trainee development, and provides opportunities for focussed, constructive feedback" (Styles et al., 2022, p. 782). The quality of the mentor relationship affects "explicit support of learning, such as feedback, as well as influencing the motivation, self-efficacy and confidence of the student" (Morley, 2018, p. 3). Effective supervision encourages reflection, supports skill development, and assists learners in connecting workplace activities with educational objectives (Rees et al., 2019; Thistlethwaite, 2013).

Research suggests that work-integrated learning is most effective when learners are supported through a combination of active participation, experiential activities, collaborative learning, and dedicated time for learning and reflection. Such approaches have been shown to improve supervisory practices and enhance learner outcomes by creating opportunities for engagement, feedback, and the application of knowledge in authentic workplace contexts (Rees et al., 2019).

Conversely, the educational value of workplace learning can be diminished when learners receive inadequate supervision, have limited access to meaningful workplace experiences, or lack appropriate organisational support (Dias et al., 2026; Fordham, 2005). These challenges are particularly evident in professions that rely heavily on workplace placements and clinical supervision. For example, healthcare organisations often experience difficulties in recruiting and retaining sufficient numbers of qualified mentors to support growing student cohorts. At the same time, workforce shortages and staffing pressures can reduce the capacity of experienced practitioners to provide effective supervision, guidance, and learning opportunities. As a result, the quality of workplace learning experiences may be compromised, limiting opportunities for skill development, reflection, and professional growth (Fordham, 2005).

Work-integrated learning is particularly effective in professions and occupations where practical competence, professional judgement, and workplace socialisation are essential. Healthcare, teaching, engineering, aviation, agriculture, hospitality, construction, community services, and business are examples of fields that commonly utilise work-integrated learning (OECD, 2024; Thistlethwaite, 2013). Thistlethwaite notes that "a health professional student cannot fully understand the complexity of the working healthcare environment until immersed within it" (Thistlethwaite, 2013, p. 17). In these areas, authentic workplace participation allows learners to develop skills and knowledge that would be difficult to acquire solely through classroom instruction (OECD, 2024; Thistlethwaite, 2013). Work-based learning refers to "some combination of observing, undertaking, and reflecting on productive work in real workplaces" (OECD, 2024, p. 251).

However, work-integrated learning is not without challenges. Securing suitable workplace placements can be difficult, particularly in highly regulated professions or industries experiencing workforce shortages and increasing service demands (Coleman, 2022; Dias et al., 2026; Fordham, 2005). Learners may encounter considerable variation in the quality of workplace experiences, including differences in supervision, mentoring, learning opportunities, and access to meaningful tasks (Clarke, 2015; Dias et al., 2026; Morley, 2018). In some cases, limited contact between learners and supervisors can hinder the development of productive mentoring relationships, reducing opportunities for feedback, guidance, and professional support (Clarke, 2015). Additional challenges identified in workplace learning environments include communication difficulties, insufficient learner engagement, time pressures, and inadequate preparation of workplace mentors and supervisors (Dias et al., 2026).

Another challenge is ensuring that workplace experiences consistently align with educational goals and contemporary professional standards. Workplace practices may sometimes reflect local organisational norms rather than current best practice, creating tensions between educational expectations and workplace realities (Clarke, 2015; Thistlethwaite, 2013). Furthermore, participation in workplace learning does not automatically guarantee a smooth transition into professional employment. Some evidence suggests that learners may continue to experience difficulties when moving from supervised practice into independent professional roles, raising questions about how effectively workplace learning prepares learners for the full responsibilities of practice (Thistlethwaite, 2013). Consequently, successful work-integrated learning requires strong collaboration between educational institutions and employers, clearly defined learning outcomes, effective supervision, and structured opportunities for reflection and feedback (Fordham, 2005; Styles et al., 2022; Thistlethwaite, 2013). High-quality work-integrated learning depends upon the careful alignment of workplace activities, learning objectives, supervision processes, and assessment practices to ensure that workplace experiences contribute meaningfully to learner development (Thistlethwaite, 2013).

From an adult learning perspective, work-integrated learning aligns closely with principles of relevance, experience, and practical application (Dias et al., 2026; Thistlethwaite, 2013). Adult learners are generally more motivated when learning activities are directly connected to their professional aspirations and immediate workplace needs. Effective workplace learning environments therefore provide opportunities for learners to engage actively with authentic tasks, collaborate with experienced practitioners, and apply knowledge in meaningful contexts. Research suggests that adult learners benefit from learning environments characterised by interaction, mutual respect, trust, and opportunities for meaningful participation in workplace activities (Thistlethwaite, 2013). Similarly, learner engagement is strengthened when individuals perceive that they have ownership of their learning, autonomy in decision-making, and a clear understanding of the relevance of learning activities to their professional goals (Dias et al., 2026).

By integrating formal education with authentic workplace experiences, work-integrated learning provides a powerful mechanism for developing occupational competence, professional identity, and employability. It enables learners to connect theory with practice, develop workplace capabilities, and gain an understanding of professional expectations while participating in real-world activities. As a result, work-integrated learning strengthens the relationship between education and professional practice and remains a significant component of contemporary adult and vocational education (Dias et al., 2026; OECD, 2024; Thistlethwaite, 2013).

Work-Integrated Learning and Adult Learning Theories in Practice

One of the strengths of Work-Integrated Learning (WIL) is that it brings together many of the major adult learning theories within a single educational approach. Rather than relying on a single theory, WIL creates opportunities for learners to experience, reflect, construct knowledge, interact with professional communities, and develop occupational identities simultaneously. The following examples demonstrate how WIL aligns with key adult learning theories.

Andragogy (Knowles): Knowles' theory of andragogy proposes that adults are motivated when learning is relevant to their immediate needs, draws upon their experiences, and can be applied directly to real-life situations.

Consider an adult learner enrolled in a Diploma of Community Services who is undertaking a placement with a family support agency. Rather than studying client engagement techniques solely in a classroom, the learner conducts supervised client interviews, participates in case discussions, and develops support plans. The learner immediately sees the relevance of communication, ethics, and counselling concepts because they are being applied to real clients and real problems. The workplace experience provides a direct answer to the adult learner's common question: "Why do I need to learn this?" This alignment between learning and practical application reflects core principles of andragogy.

Experiential Learning (Kolb): Work-integrated learning strongly reflects Kolb's experiential learning cycle of experience, reflection, conceptualisation, and experimentation.

For example, a student teacher conducting a teaching practicum may deliver a lesson that does not engage students as effectively as expected. Following the lesson, the student reflects on what occurred, discusses the experience with a supervising teacher, reviews educational theories relating to student engagement, and modifies the lesson plan for the next class. The learner then applies these revised strategies and evaluates the results. This cycle of action, reflection, theory building, and reapplication closely mirrors Kolb's model of experiential learning.

Situated Learning (Lave and Wenger): Situated learning proposes that learning occurs through participation in authentic social and professional practices.

A nursing student undertaking a hospital placement begins by observing experienced nurses, assisting with basic patient care, and learning workplace routines. As competence develops, the student gradually assumes greater responsibility for patient assessments, documentation, medication administration, and communication with multidisciplinary teams. Learning occurs within the authentic context of healthcare practice rather than through abstract instruction alone. The student acquires not only technical skills but also professional language, workplace norms, and healthcare culture.

Legitimate Peripheral Participation: Lave and Wenger's concept of legitimate peripheral participation explains how newcomers gradually become full members of a professional community.

An apprentice carpenter may initially clean work areas, carry materials, and observe construction activities. Over time, the apprentice progresses to measuring materials, constructing wall frames, installing roofing systems, and eventually supervising portions of a project. The learner moves from peripheral participation to full participation within the occupational community. WIL creates the conditions that enable this progression to occur naturally.

Communities of Practice (Wenger): Wenger's theory suggests that learning occurs through participation in communities that share common goals, practices, and expertise.

An engineering student participating in an industry placement becomes part of a project team responsible for designing drainage infrastructure. Through meetings, site inspections, design reviews, and discussions with engineers, surveyors, and project managers, the student learns how professionals collaborate to solve complex problems. Knowledge is acquired not only through formal instruction but also through interaction with members of the professional community.

Constructivism: Constructivist theories propose that learners actively construct knowledge through experience and interpretation rather than simply receiving information.

Consider a logistics student undertaking a workplace project to improve warehouse efficiency. The learner analyses stock movements, identifies bottlenecks, proposes process improvements, and evaluates outcomes. Rather than being told the "correct" solution, the learner develops understanding through investigating problems, testing ideas, and constructing knowledge from experience. The workplace becomes an environment in which meaning is actively created.

Social Constructivism (Vygotsky): Social constructivism emphasises learning through interaction with more knowledgeable others.

A trainee flight instructor undertaking supervised flight instruction learns not only from personal experience but also through coaching from senior instructors. During lesson debriefings, experienced instructors explain teaching techniques, decision-making processes, and risk management strategies. Through guided participation, the trainee develops capabilities that would be difficult to achieve independently. This reflects Vygotsky's concept of the Zone of Proximal Development, where learners perform at higher levels with appropriate support.

Cognitive Apprenticeship: Cognitive apprenticeship focuses on making expert thinking visible.

An accounting student undertaking a professional placement may observe a senior accountant conducting a complex financial audit. Rather than simply demonstrating procedures, the accountant explains how risks are identified, why certain evidence is gathered, and how professional judgements are formed. By observing both actions and thought processes, the learner develops deeper understanding of professional reasoning. WIL provides rich opportunities for this type of cognitive modelling.

Transformative Learning (Mezirow): Work-integrated learning can create experiences that challenge existing assumptions and lead to perspective transformation.

For example, a social work student may enter a placement with preconceived beliefs about homelessness. Through direct engagement with clients experiencing housing insecurity, the student encounters complex social, economic, and personal factors that challenge these assumptions. Reflection, discussion, and critical analysis may lead the learner to develop a more sophisticated understanding of social disadvantage. The workplace experience becomes a disorienting dilemma that prompts transformative learning.

Humanism: Humanistic approaches emphasise personal growth, self-direction, and the development of individual potential.

A mature-age student completing a counselling placement may be encouraged to set personal learning goals, identify strengths and areas for improvement, and reflect on professional growth throughout the placement. The supervisor acts as a facilitator rather than a director of learning. The emphasis is placed not only on technical competence but also on self-awareness, empathy, confidence, and personal development.

Reflective Practice (Schön): Work-integrated learning provides extensive opportunities for reflective practice.

A graduate engineer participating in a construction project may encounter unexpected ground conditions that affect a design solution. After the project stage is completed, the engineer reflects on decisions made, evaluates alternative approaches, and considers lessons learned. This process of reflection-on-action contributes to professional expertise and ongoing development. WIL environments frequently incorporate reflective journals, portfolios, and debriefing sessions specifically to support this form of learning.

Professional Identity Formation: Many adult learning theorists recognise that learning involves becoming a particular type of professional rather than simply acquiring knowledge and skills.

A preservice teacher undertaking multiple school placements gradually begins to think, communicate, and behave as a teacher. Through interactions with students, colleagues, parents, and school leaders, the learner develops professional values, ethical standards, and a sense of belonging within the teaching profession. Similarly, apprentice electricians, nursing students, and trainee pilots develop occupational identities through participation in professional practice. WIL plays a critical role in this transition from learner to practitioner.

Integrating Multiple Theories in a Single Example: Consider a student nurse undertaking a hospital placement. The placement simultaneously demonstrates multiple adult learning theories. The learner applies classroom knowledge to real patient care situations, reflecting and refining practice through experience (Experiential Learning). Learning occurs within an authentic healthcare environment (Situated Learning), supported by experienced nurses who provide guidance and feedback (Social Constructivism and Cognitive Apprenticeship). The student gradually moves from observation to independent practice (Legitimate Peripheral Participation), becomes part of the nursing team (Communities of Practice), develops a professional nursing identity, and may even experience shifts in personal beliefs about healthcare through exposure to diverse patient experiences (Transformative Learning). At the same time, the learning remains highly relevant to career goals and immediate workplace needs (Andragogy).

This example illustrates why work-integrated learning is often regarded as one of the most powerful approaches in adult education. It provides a practical mechanism through which multiple adult learning theories operate simultaneously within authentic workplace contexts, creating rich opportunities for knowledge development, skill acquisition, professional socialisation, reflection, and personal transformation.

When Work-Integrated Learning Effectiveness

Work-integrated learning is particularly effective in fields where learners must apply theoretical knowledge within complex workplace environments (Cain et al., 2019; Ralph et al., 2009; Ralph et al., 2008; Thistlethwaite, 2013) . For example, student nurses undertaking clinical placements (Cain et al., 2019; Kaihlanen et al., 2021; Ralph et al., 2009), trainee teachers completing supervised practicums (Peiser et al., 2018; Ralph et al., 2009; Ralph et al., 2008), engineering students working on industry projects (Eisenstein, 2022; Ralph et al., 2009; Ralph et al., 2008), and aviation students participating in flight operations (Melrose et al., 2020; Roth et al., 2014) all benefit from opportunities to apply learning in authentic contexts (Cain et al., 2019; Thistlethwaite, 2013). In these situations, learners develop not only technical competence but also professional judgement, communication skills, and an understanding of workplace culture (Cain et al., 2019; Kaihlanen et al., 2021; Ralph et al., 2008; Thistlethwaite, 2013).

Work-integrated learning may be less effective when workplace experiences are poorly structured, unrelated to learning outcomes, or lack appropriate supervision. It may also be challenging in disciplines that are highly theoretical and where opportunities for authentic workplace participation are limited. For example, learners studying advanced theoretical mathematics or philosophy may gain less benefit from workplace placements than those preparing for practice-based professions. Additionally, workplaces with limited mentoring capacity, poor organisational culture, or inadequate learning opportunities may provide experiences that contribute little to learner development.

Research confirms that supervision is an essential factor in student success within work-based learning systems, and that good supervision requires instruction, modelling, supervised practice, and feedback (Morse, 2006). The quality of workplace placements is influenced by the capacity of supervisors and mentors; in Finland, for instance, lack of adequate staff in the workplace has been identified as a challenge in promoting WBL, and it may be unrealistic to expect extensive training for those who supervise trainees on work placements (Dogara, Saud, & Kamin, 2020). Similarly, employers have cited poor capacity to mentor or provide students with work exposure as a barrier to effective placements (Curto-Reverte et al., 2025). The challenge of matching the educative potential of the workplace with learners' needs has been recognised, with variables including students' competence, the demands of their tasks, their access to organisational and professional knowledge, and the facilitative capacity of workplace mentors or supervisors.

Regarding disciplinary suitability, a large-scale Australian scoping study found that traditional disciplines such as nursing, medicine, education and engineering have always had embedded work-integrated learning components, usually in the form of industry placements, but many disciplines in non-traditional areas such as the liberal arts were new to WIL and unsure of how to customise programs to meet their particular needs. This suggests that disciplines lacking established pathways for authentic workplace participation face greater challenges in implementing effective WIL. The importance of authentic, real-world work situations is well established; practice settings must offer experiences that are authentic in terms of acting out an occupation, with key strengths including access to authentic work activities, observation, access to experienced co-workers, and practice.

Obtaining and sustaining good quality WBL placements can be difficult, and there are non-financial costs and benefits that may not be widely recognised but which have significance when seeking placement opportunities. Challenges persist for students in balancing academic and professional responsibilities, and stronger cooperation between universities and organisations is critically needed, particularly through structured learning agreements and flexible curricula (Musset, 2019). The current research on WBL settings cannot sufficiently explain why employability in some settings is enhanced more than in others, highlighting the need to account for promotive factors from the university, the practical partner and the student perspective (Curto-Reverte et al., 2025; Dogara, Saud, & Kamin, 2020; Feldmann, 2016; Morse, 2006; Musset, 2019).

Reflective Practice in the Workplace

Reflective practice is the deliberate process of critically examining experiences, actions, decisions, and outcomes in order to improve future performance. It involves thinking carefully about what occurred, why it occurred, what was learned from the experience, and how that learning can be applied in future situations. Within workplace settings, reflective practice enables individuals to move beyond simply performing tasks to actively learning from their experiences and continuously improving their professional practice (Ingles, 2023; Woodward et al., 2023).

The concept of reflective practice is most closely associated with Schön (1983), who argued that professionals develop expertise not only through formal education and technical knowledge but also through systematic reflection on their work (Nzembayie et al., 2024). Schön proposed that professionals frequently encounter situations that cannot be addressed solely through predetermined procedures or rules. Instead, they must draw upon experience, judgement, and reflection to navigate complex and often unpredictable situations (Nzembayie et al., 2024). Reflective practice therefore helps professionals adapt to changing circumstances, solve problems, and refine their decision-making processes over time.

A key distinction within reflective practice is between reflection-in-action and reflection-on-action. Reflection-in-action occurs while an activity is taking place. Professionals monitor situations as they unfold, assess emerging information, and modify their actions in real time. For example, an experienced teacher may notice that students are struggling to understand a lesson and immediately change their instructional approach. A nurse may recognise a patient's condition is deteriorating and adjust care priorities accordingly. Reflection-in-action enables professionals to respond effectively to dynamic situations and supports adaptive expertise.

Reflection-on-action occurs after an event has taken place. Individuals review their actions, evaluate outcomes, and consider alternative approaches that might have produced better results. For example, an engineer may review the causes of a design problem after project completion, while a manager may analyse how a difficult conversation with an employee could have been handled more effectively. Reflection-on-action provides opportunities for deeper analysis and long-term learning because individuals have time to consider events in a broader context.

Reflective practice aligns strongly with experiential learning theory, particularly the Kolb (1984b) learning cycle. According to Kolb, learning begins with an experience, followed by reflection on that experience, the development of conceptual understanding, and the application of new insights to future situations (Coleman et al., 2025; Ingles, 2023; Jing et al., 2025). Kolb's Experiential Learning Theory posits that learning is a cyclical process comprising four stages: concrete experience, reflective observation, abstract conceptualization, and active experimentation (Coleman et al., 2025; Dillard et al., 2024; Jing et al., 2025). Workplace experiences provide the foundation for learning, while reflection enables individuals to extract meaning from those experiences and transform them into knowledge (Coleman et al., 2025; Ingles, 2023). Reflection is recognized as a critical learning process component for ensuring that learners' experiences are educative ones, as without it, experiences may remain isolated events rather than becoming opportunities for learning and improvement (Coleman et al., 2025). Indeed, reflective practice is considered "a pillar of quality educative experiences and a cornerstone of Experiential Learning Theory" (Ingles, 2023, p. 7), and researchers have emphasized that experiences lacking elements such as reflection and application may not be as educative (Coleman et al., 2025). Reflective capacity is viewed as central to experiential learning theory, whereby demonstrating reflective capacity is seen as evidence of transformative learning in which an individual shows the ability to reconcile concrete experiences within a field of theoretical concepts (Cathro et al., 2017).

Reflective practice is widely used across professional fields. In healthcare, nurses, doctors, and allied health professionals reflect on patient interactions, treatment decisions, and clinical outcomes to improve patient care and professional competence (H. Li, 2025; Sanchez-Ballado, 2025). In education, teachers reflect on lesson effectiveness, student engagement, and assessment outcomes to enhance instructional practice (Ann et al., 2018; H. Li, 2025; Lithoxoidou & Georgiadou, 2023). Reflective practice in education has been shown to foster deeper understanding of teaching methodologies and enhance teachers' abilities to apply strategies across diverse educational contexts (H. Li, 2025). Engineers may review project successes and failures to improve design processes and risk management. Managers and leaders use reflection to evaluate decisions, communication strategies, and team performance. In each case, reflection supports continuous improvement and professional growth, as reflective practice is recognised as an opportunity for professional growth that instructors may use to assess their teaching and advance professionally, (Yuan et al., 2025).

For adult learners, reflective practice is particularly valuable because it recognises experience as a rich source of learning. Adults bring extensive personal and professional experiences to learning situations. Reflection allows them to connect new knowledge with prior experiences, evaluate assumptions, and develop deeper understanding (Abri et al., 2024; Nurhidayah, 2024). Through reflection, adult learners become more aware of their strengths, limitations, values, and professional behaviours. Reflective practice nurtures self-awareness and professional development, enabling individuals to critically scrutinise the values, beliefs, and attitudes that shape their practice (H. Li, 2025). This increased self-awareness supports both personal development and professional competence, as teachers who engage in reflective practice develop informed decision-making skills and become proactive and confident (Abri et al., 2024; Aryal, 2024).

Reflective practice is also closely connected to transformative learning theory. Mezirow (1991) argued that critical reflection enables individuals to question previously held assumptions, beliefs, and perspectives. Workplace experiences may challenge existing ways of thinking and encourage professionals to reconsider how they understand situations, people, or professional responsibilities. Through critical reflection, individuals may experience significant shifts in perspective that lead to changes in behaviour and professional practice (Abri et al., 2024; Nurhidayah, 2024). Research confirms that reflection as a higher-order skill can help practitioners question aspects of teaching and learning, though such reflection may initially remain at a technical level before deepening (Khatter et al., 2024).

Several structured models have been developed to support reflective practice. Gibbs' Reflective Cycle encourages individuals to describe an experience, explore their feelings, evaluate what occurred, analyse contributing factors, draw conclusions, and develop action plans (Nurhidayah, 2024). Rolfe, Freshwater, and Jasper's model uses three simple questions: What? So what? Now what? These frameworks provide systematic approaches that help learners move beyond simple descriptions of events towards deeper analysis and learning (Ann et al., 2018). Additional approaches include writing reflective journals, peer observation, dialogue journals, and video-based analysis, all of which scaffold the reflective process (Abri et al., 2024; Grigg et al., 2024).

Reflective practice is most effective when supported by organisational cultures that encourage learning, openness, and continuous improvement (Aryal, 2024). Activities such as debriefings, supervision sessions, mentoring discussions, learning journals, peer reviews, incident analyses, and professional portfolios provide opportunities for structured reflection (Abri et al., 2024; H. Li, 2025). Peer observation, for instance, has been found to be a significant reflective tool for increasing professional development. In high-reliability industries such as aviation, healthcare, emergency services, and engineering, reflective practice is often embedded within organisational processes to support safety, performance improvement, and risk management.

Despite its benefits, reflective practice is not without challenges. Time pressures, heavy workloads, and organisational cultures that prioritise productivity over learning can limit opportunities for reflection

(Ann et al., 2018; Aryal, 2024). Some individuals may also focus only on successful outcomes or avoid examining mistakes critically; studies have found that pre-service practitioners' written reflections tend to be descriptive and lack critical judgement supported by evidence (Ann et al., 2018). Effective reflective practice requires honesty, self-awareness, and a willingness to consider alternative perspectives and areas for improvement (Khatter et al., 2024). Critics have also raised ethical concerns over "forced" reflection requiring individuals to be open and honest in assessed contexts (Khatter et al., 2024).

Reflective Practice and Adult Learning Theories

Reflective practice draws upon several adult learning theories simultaneously. It aligns closely with experiential learning because learning is understood to emerge through the interaction of experience and reflection. Within experiential learning theory, reflection is viewed as the mechanism through which experiences are examined, interpreted, and transformed into knowledge. Rather than learning solely from experience itself, individuals learn by analysing what occurred, considering its implications, and applying those insights to future situations. Research on experiential learning consistently highlights the importance of reflection in helping learners derive meaning from practical experiences and use those experiences as a foundation for ongoing development (Woerkom, 2010).

Reflective practice also supports constructivist perspectives on learning by enabling individuals to actively create meaning from workplace experiences. Constructivist theories propose that learners build knowledge by integrating new experiences with existing understandings, beliefs, and prior knowledge. Reflection plays a central role in this process because it encourages learners to examine their experiences critically, connect theory with practice, and develop new interpretations of events. Constructivist learning environments therefore place significant emphasis on self-reflection, critical thinking, and learner engagement as mechanisms for knowledge construction (Law, 2013; Waseem & Aslam, 2020). In professional education, reflective activities help learners connect classroom concepts with workplace realities and develop a deeper understanding of their professional practice (Hyde, 2021).

The relationship between reflective practice and humanistic learning theory is evident through the emphasis on self-awareness, personal growth, and self-directed development. Humanistic approaches view learning as a process that involves the whole person rather than simply the acquisition of knowledge or skills. Reflection encourages individuals to consider their strengths, limitations, values, motivations, and behaviours, thereby supporting personal and professional development. Increased self-awareness can enhance an individual's openness to feedback, willingness to improve, and capacity for self-directed learning. Consequently, reflective practice aligns closely with humanistic educational approaches that prioritise personal growth, autonomy, and lifelong self-improvement (Sharma & Sharma, 2021).

Reflective practice also contributes significantly to transformative learning by encouraging learners to critically examine their assumptions, beliefs, and perspectives. Through critical reflection, individuals become more aware of the factors that shape their thinking and behaviour, creating opportunities to reconsider established viewpoints and develop new ways of understanding themselves and the world around them. Constructivist and transformative learning approaches both emphasise the importance of reflection in developing autonomous, critical thinkers who are capable of questioning assumptions and engaging with complex issues. Reflection, awareness, and dialogue together provide a foundation for learning that extends beyond the acquisition of content knowledge and supports meaningful personal and professional transformation (Law, 2013; Phuong, 2018; Афанасенко et al., 2022).

Finally, reflective practice supports lifelong learning by fostering habits of continuous improvement and professional development. Many professions require practitioners to regularly evaluate their performance, identify areas for improvement, and adapt to changing workplace demands. Reflection enables individuals to monitor their learning, assess their effectiveness, and plan future development activities. The self-regulatory and metacognitive skills developed through reflective practice are particularly important for professionals who must engage in ongoing learning throughout their careers. By encouraging learners to take responsibility for their own development and continuously evaluate their practice, reflective practice contributes to the development of lifelong learners capable of adapting to evolving professional environments (Law, 2013; Woerkom, 2010).

Significance of Reflective Practice

Reflective practice is significant because it transforms workplace experience into professional learning. Rather than relying solely on formal education or accumulated experience, professionals actively examine and learn from their actions. Through systematic reflection, individuals develop deeper understanding, improve performance, strengthen professional judgement, and adapt to changing workplace demands. As a result, reflective practice is widely recognised as a cornerstone of professional competence, lifelong learning, and continuous improvement in modern workplaces.

The following examples demonstrate how reflective practice operates across three contrasting adult learning environments: military training, vocational fashion education in the United States, and bachelor-level mechanical engineering education in Sweden. Although these contexts differ in their purposes, levels of study, professional expectations, and forms of practical activity, each uses reflection to help learners interpret experience, evaluate performance, connect theory with practice, and improve future action. Collectively, the examples also show how reflective practice interacts with experiential learning, constructivism, cognitive apprenticeship, situated learning, communities of practice, transformative learning, professional identity development, and lifelong learning.

Reflective practice is widely used within military training because personnel frequently operate in complex, dynamic, and high-risk environments where continuous learning is essential (Bates, 2014; Johansen et al., 2021). Military contexts demand that individuals make critical decisions under stress, often with life-or-death consequences (Wilkes, 2021), and reflective approaches enable personnel to learn from experience and improve performance over time. Problem-based learning in military education, for instance, "offers opportunities for problem-solving in a collaborative setting and creates mental models for learning and forming self-directed learning habits through practice and reflection" (Johansen et al., 2021, p. 240). Similarly, experiential learning models emphasizing "the importance of action and reflection in a safe environment that can tolerate errors" have been applied to military medical training through virtual patient simulations (Stathakarou et al., 2024, p. 2).

Ethics and leadership training within military organisations increasingly recognise that effective learning extends beyond classroom-based instruction and requires opportunities for practical application and reflection (Bates, 2014). Reflective activities enable personnel to examine decisions, evaluate behaviour, and consider the implications of their actions in realistic operational contexts. Similarly, experiential learning approaches based on Kolb's learning cycle have been identified as particularly valuable for developing character, judgement, and leadership capabilities because they combine practical experience with systematic reflection, conceptual understanding, and the application of lessons learned to future situations (Gorey, 2021).

Military education programs frequently use field exercises, case studies, simulations, and guided discussions to encourage learners to compare observed actions and leadership behaviours with established theoretical frameworks. Through this process, learners develop reflective skills that support critical thinking, self-awareness, and professional growth (Melnychenko et al., 2024). Contemporary military training also incorporates reflective practice alongside a range of active learning strategies, including simulations, collaborative problem-solving activities, and structured discussions, to strengthen learning outcomes and decision-making capabilities (Осьодло et al., 2022). Collectively, these approaches demonstrate that reflection has become an integral component of military education because operational environments require personnel to continually evaluate their performance, adapt to changing circumstances, and learn from experience.

Consider a recruit undertaking infantry training. During a section attack exercise, the recruit may fail to maintain communication with other team members while moving through difficult terrain. Immediately following the exercise, the section conducts an after-action review. The recruit reflects on what occurred, why communication broke down, how the situation affected team performance, and what alternative actions could have been taken. During subsequent exercises, the recruit applies these lessons by improving communication procedures and situational awareness.

This example illustrates several adult learning theories simultaneously. The exercise itself represents experiential learning, as the recruit learns through direct experience. The after-action review demonstrates reflective practice, where experience is analysed and transformed into learning. The military section functions as a community of practice, with experienced soldiers modelling professional behaviours and sharing tacit knowledge. The instructor's guidance reflects cognitive apprenticeship, as expert decision-making processes are made visible to learners. Over time, the recruit develops not only technical military skills but also a professional military identity, illustrating aspects of situated learning and legitimate peripheral participation.

Reflective practice can also contribute to transformative learning in military contexts. A trainee officer may initially view leadership as issuing orders and enforcing discipline. Following a demanding field exercise in which morale and teamwork prove more important than authority alone, reflection may challenge this assumption. Through discussion, mentoring, and critical analysis, the officer may develop a more sophisticated understanding of leadership based on trust, communication, and influence rather than command alone.

In a vocational fashion design program in the United States, students often undertake industry placements, internships, or client-based design projects (Brudney, 2021; Dodge & McKeough, 2003; Valentino, 2010). Internships have become an increasingly prevalent form of experiential education across disciplines in the United States, with the number of students completing internships rising from 41% in 2004 to 61% in 2007 (Valentino, 2010). These experiential learning opportunities, which encompass cooperative education, internships, design projects, and practicums, serve as critical bridges between academic preparation and full participation in professional work (Dodge & McKeough, 2003; Johrendt et al., 2010). Internships are perceived as generally beneficial to students and employers as well as to educational institutions in the current professional world (Brudney, 2021), and industries such as media and fashion rely extensively on such placement structures (Brudney, 2021). Notably, the fashion industry is among the for-profit sectors where student internship and placement experiences are common, with enterprises in these industries required to provide "written assurance that the students will receive an educational experience" (Brudney, 2021, p. 180). However, concerns about equity persist, as wealthier students tend to reap a greater share of networking and CV-enhancing benefits from placements, particularly unpaid internships in industries such as fashion (Brudney, 2021). Cooperative education and internship models, which integrate learning with work, have existed in the United States since 1906 and remain foundational to vocational and professional preparation (Huiyan, 2019; Yin).

Imagine a student tasked with designing a sustainable clothing collection for a local fashion retailer. After presenting the designs, the student receives feedback that the garments are visually appealing but difficult and expensive to manufacture. The student reflects on the design process, evaluates the decisions made, reviews production constraints, and considers how sustainability, aesthetics, and manufacturability can be better balanced in future projects.

This example demonstrates constructivist learning, as the student actively builds new knowledge by integrating feedback with prior understanding. The workplace project provides a form of work-integrated learning, allowing theory learned in the classroom to be tested against real industry requirements. Reflection enables the student to move through Kolb's experiential learning cycle, transforming experience into improved design knowledge and practice.

The fashion industry placement also illustrates social constructivism. Through interaction with designers, production managers, buyers, and clients, the learner develops a deeper understanding of how design decisions are influenced by commercial realities. The workplace serves as a community of practice where professional knowledge is shared through collaboration and discussion. At the same time, the student develops a professional identity as a designer, learning industry standards, workplace expectations, and professional communication practices.

Reflective practice may also lead to transformative learning. A student who initially views fashion primarily as a creative activity may, through workplace experience and reflection, develop an understanding of broader issues such as sustainability, ethical sourcing, consumer behaviour, and global supply chains. Reflection encourages the learner to reconsider previous assumptions and adopt a more holistic view of the profession.

Some Swedish engineering programs incorporate industry projects, laboratory work, and workplace placements that encourage reflective learning. Consider a mechanical engineering student participating in a university-industry collaboration with a manufacturing company. The student is involved in redesigning a component to improve production efficiency. After testing the prototype, the design performs below expectations due to unexpected material fatigue. The student reviews the design assumptions, analyses test data, consults industry mentors, and reflects on how material selection, loading conditions, and manufacturing processes influenced the outcome.

This scenario strongly reflects experiential learning, as the student learns through direct engagement with an authentic engineering problem. Reflection transforms the project from a simple design failure into a valuable learning experience. Through this process, the student develops a deeper understanding of engineering principles and their practical application.

The project also demonstrates cognitive apprenticeship. Industry engineers explain how they analyse failures, interpret data, assess risks, and make design decisions. By observing expert thinking, the student learns not only technical procedures but also the professional judgement that underpins engineering practice. This exposure to expert reasoning is often difficult to replicate in classroom environments.

The workplace project further illustrates situated learning and communities of practice. The student participates in design meetings, manufacturing reviews, and technical discussions with engineers, technicians, and project managers. Through this participation, the learner acquires professional language, workplace norms, and engineering culture. Initially functioning as a novice contributor, the student gradually assumes greater responsibility, demonstrating legitimate peripheral participation.

Reflective practice may also foster lifelong learning skills. The student learns that engineering problems rarely have simple solutions and that continuous evaluation, adaptation, and learning are necessary throughout a professional career. Rather than viewing engineering knowledge as fixed, the learner develops habits of inquiry, reflection, and continuous improvement that support ongoing professional development.

Although military training, vocational fashion design, and university engineering education appear very different, reflective practice serves a similar purpose in each context. In the military, reflection improves operational effectiveness and leadership capability. In fashion design, it enhances creativity, problem-solving, and industry awareness. In engineering, it strengthens technical judgement and professional competence. Across all three settings, reflection transforms experience into learning, supports the construction of professional identity, encourages critical examination of assumptions, and develops the capacity for lifelong learning. These examples demonstrate why reflective practice is considered one of the most important mechanisms through which adult learners convert workplace experiences into meaningful professional growth.

Recognition of Prior Learning (RPL)

Recognition of Prior Learning (RPL) is an assessment process used to evaluate an individual's existing skills, knowledge, and experience against the requirements of a qualification, credential, competency standard, professional certification, or learning outcome (Conrad, 2023a; Graebe & McIntyre-Hite, 2024). Rather than requiring individuals to complete training for capabilities they have already developed, RPL enables prior learning to be formally recognised when sufficient evidence demonstrates that the required standards have been achieved (Garnett, 2015; Ginty et al., 2023).

RPL is founded on the principle that learning occurs throughout life and can be acquired through a wide range of formal, non-formal, and informal experiences (Garnett, 2015; Shaketange, 2018; Singh, 2015c). Individuals develop competencies through employment, military service, apprenticeships, volunteering, community participation, self-directed learning, professional development, hobbies, and everyday life experiences (Cameron, 2012; Conrad, 2023a; Garnett, 2015). When these experiences result in learning outcomes equivalent to those required by a qualification or professional standard, they may be recognised through an RPL process (Singh, 2015b, 2015c).

Although terminology varies internationally, similar recognition systems exist in many countries. These include Recognition of Prior Learning (RPL), Prior Learning Assessment and Recognition (PLAR), Accreditation of Prior Experiential Learning (APEL), Validation of Prior Learning (VPL), Recognition of Acquired Competencies, and Recognition of Current Competency (RCC) (Alves et al., 2020; Bhandari, 2024; Conrad, 2008; Conrad & McGreal, 2012; Werquin, 2021). Regardless of the terminology used, the underlying objective is the same: to recognise learning and competence regardless of where, when, or how that learning occurred (Alves et al., 2020; Singh, 2015b, 2015c).

RPL generally forms part of a broader framework for recognising learning. This framework may also include credit transfer, advanced standing, transfer of learning, or recognition of formal qualifications (Conrad, 2023a; Conrad & McGreal, 2012; Stenberg, 2022). While these mechanisms typically recognise previously completed courses or credentials, RPL focuses on learning and competence gained through experience, irrespective of whether it was acquired through formal education (Conrad, 2023a; Conrad & McGreal, 2012).

The RPL process typically begins with an initial discussion between the candidate and an assessor, evaluator, faculty member, or credentialing professional. During this stage, the requirements of the qualification or competency standard are explained, along with the evidence needed to demonstrate achievement. The candidate then gathers evidence that supports their claim of competence. Evidence may include employment records, résumés, position descriptions, performance evaluations, qualifications, licences, certifications, workplace documents, work samples, portfolios, logbooks, references, supervisor reports, professional registrations, photographs, videos, project reports, publications, or other relevant documentation.

Documentary evidence alone is often insufficient to establish competence. Consequently, RPL assessments frequently include additional methods such as professional interviews, competency conversations, workplace observations, practical demonstrations, simulations, challenge examinations, skills testing, portfolio reviews, or third-party verification. These assessment methods help confirm that the individual currently possesses the knowledge, skills, and capabilities required by the qualification or professional standard.

Effective RPL systems are guided by widely accepted assessment principles. Assessments should be valid, reliable, fair, transparent, and flexible. Similarly, evidence should be authentic, current, sufficient, and directly relevant to the competencies or learning outcomes being assessed. Assessors must be satisfied that the evidence demonstrates achievement of all required standards rather than only selected aspects of performance.

One of the most important principles of RPL is that recognition is granted against defined competency standards, learning outcomes, or performance requirements rather than job titles, years of experience, or qualifications alone. An individual may have extensive experience in an occupation, but recognition can only be awarded when sufficient evidence demonstrates that the required competencies have been achieved. Conversely, an individual with comparatively limited experience may still be granted recognition if they can demonstrate full competence against the relevant standards.

For example, a military logistics supervisor seeking recognition toward a qualification in supply chain management may provide evidence of workforce planning, inventory control, procurement activities, risk management, transportation coordination, and supervisory responsibilities acquired during military service. Assessors would review this evidence and compare it against the required learning outcomes or competency standards. Additional interviews, workplace observations, or practical assessments may be conducted to verify competence. Where sufficient evidence exists, the individual may receive full or partial recognition without undertaking additional training.

RPL provides significant benefits for learners, employers, educational institutions, and professional bodies. It reduces unnecessary duplication of learning, shortens the time required to obtain qualifications, lowers educational costs, supports workforce mobility, and formally recognises existing expertise. For employers, RPL can assist workforce development, succession planning, and employee retention by recognising the skills and knowledge already possessed by experienced personnel.

Despite these benefits, RPL requires rigorous assessment. Recognition decisions must be evidence-based, consistent, and defensible. The objective is not simply to reward experience but to confirm that the individual can demonstrate competence or achievement equivalent to that expected of someone who completed the traditional education or training pathway. As a result, effective RPL serves both as a mechanism for recognising lifelong learning and as a quality assurance process that maintains the integrity and credibility of qualifications, credentials, and professional standards while valuing the diverse ways adults acquire knowledge and skills throughout their lives.

Recognition of Prior Learning (RPL) and Adult Learning

Recognition of Prior Learning (RPL) is closely aligned with many of the core principles and theories of adult learning because it acknowledges that adults learn throughout their lives and that valuable learning occurs in a wide variety of settings beyond formal education (Dykes, 2014; Snyman & Berg, 2017). Rather than assuming that learning only takes place in classrooms, RPL recognises that adults develop knowledge, skills, and competencies through work, family responsibilities, military service, community involvement, professional practice, self-directed study, and everyday life experiences (Dykes, 2014; Snyman & Berg, 2017). In doing so, RPL reflects the fundamental adult learning principle that experience is a rich and legitimate source of learning (Dykes, 2014; Koutsoukos, 2022).

One of the strongest theoretical connections between RPL and adult learning can be found in Knowles' theory of andragogy. Knowles argued that adult learners bring extensive life and work experiences to learning situations and that these experiences form a valuable resource for future learning (Dykes, 2014; Koutsoukos, 2022; Mukhalalati & Taylor, 2019). RPL operationalises this principle by treating prior experience as evidence of learning rather than requiring adults to repeat training in areas where competence has already been achieved (Dykes, 2014). The RPL process therefore respects adult learners as knowledgeable individuals who possess existing capabilities and acknowledges that learning has occurred before entering a formal educational program (Dykes, 2014; Snyman & Berg, 2017). As Dennick notes, one of the principles of adult learning is that "the life experiences and background knowledge of adult learners becomes an important educational resource" (Dennick, 2016, p. 203).

RPL also aligns closely with experiential learning theory, particularly the work of Kolb (1984b). Experiential learning proposes that knowledge is created through the transformation of experience into understanding (Beaty & Cleary, 2025; Routh et al., 2022). Adults continually engage in cycles of experience, reflection, conceptualisation, and application throughout their personal and professional lives (Koutsoukos, 2022; Routh et al., 2022). The RPL process requires individuals to identify, analyse, and explain what they have learned from their experiences (Snyman & Berg, 2017). In developing portfolios, participating in competency conversations, or undertaking professional interviews, candidates engage in reflective activities that help make tacit knowledge explicit ((Routh et al., 2022; Snyman & Berg, 2017). In this sense, RPL is not merely an assessment process but also a learning process that encourages adults to reflect upon and articulate their learning (Koutsoukos, 2022; Snyman & Berg, 2017). As Dykes observes, RPL "underscores experiential learning and emphasises adult learning principles" (Dykes, 2014, p. 317). Notably, reflection is considered essential in the conceptual framework of RPL assessment, as without reflection, contextual learning remains trapped in the precise context in which it was learned (Routh et al., 2022; Snyman & Berg, 2017).

From a constructivist perspective, RPL recognises that individuals actively construct knowledge through interaction with their environments (L. R. Brown et al., 2023; Dennick, 2016). Constructivist theories suggest that learning occurs when people integrate new experiences with existing knowledge and understanding (Dennick, 2016; Waseem & Aslam, 2020). RPL acknowledges that adults have already constructed substantial knowledge through workplace practice, social interactions, and lived experiences (Dykes, 2014; Snyman & Berg, 2017). The assessment process requires learners to demonstrate how their experiences have contributed to the development of competencies, thereby recognising learning as a cumulative and personally constructed process rather than something acquired solely through formal instruction (Dennick, 2016; Woodward et al., 2020). As constructivism holds, "learning is always a building process whereby new knowledge can only be added on to and understood in terms of existing knowledge" (Dennick, 2016, p. 204).

RPL is also consistent with humanistic learning theory, which emphasises learner autonomy, self-direction, personal growth, and self-actualisation (Mukhalalati & Taylor, 2019; Waseem & Aslam, 2020; Woodward et al., 2020). Humanistic educators view learners as capable individuals who should be empowered to take responsibility for their own learning (Leong, 2022; Mukhalalati & Taylor, 2019). RPL supports this philosophy by allowing adults to take an active role in documenting and demonstrating their achievements (Snyman & Berg, 2017; Teele & Nkoane, 2024). Candidates must gather evidence, reflect on their experiences, and present a case for recognition, making them active participants rather than passive recipients of assessment (Snyman & Berg, 2017; Teele & Nkoane, 2024). The process can enhance self-confidence and self-awareness by helping individuals recognise the value of their own knowledge and accomplishments (Snyman & Berg, 2017; Teele & Nkoane, 2024). This aligns with the humanistic goal of contributing to the development of "energetic, positive, self-respecting, caring human beings who can meet all challenges encountered" (Woodward et al., 2020, p. 69).

The relationship between RPL and transformative learning theory is particularly significant. Mezirow argued that adults develop through critical reflection on their experiences and assumptions (Marotta, 2024; Paulos, 2018; Snyman & Berg, 2017). During the RPL process, candidates are often required to analyse their professional experiences, identify learning gained from practice, and explain how their understanding has developed over time (Day, 2016; Ions & Sutcliffe, 2019; Snyman & Berg, 2017). This reflection can lead individuals to recognise competencies they previously undervalued or failed to acknowledge (Conrad, 2011; Snyman & Berg, 2017). For some learners, the process can be transformative, changing how they perceive themselves, their professional capabilities, and their potential for further learning and career advancement (Authors, 2011; Conrad, 2011).

RPL also reflects the principles of lifelong learning, which emphasise that learning is a continuous process occurring across the lifespan (O'Leary, 2023; Sukri, 2024). Modern careers are increasingly characterised by changing occupations, evolving technologies, and ongoing professional development (Graebe & McIntyre-Hite, 2024; Sukri, 2024). Adults frequently acquire new competencies outside formal education systems (Authors, 2011; O'Leary, 2023). RPL supports lifelong learning by creating pathways through which previously acquired knowledge and skills can be recognised and connected to further educational and career opportunities (Graebe & McIntyre-Hite, 2024; Ions & Sutcliffe, 2019; O'Leary, 2023). Rather than viewing learning as something confined to schools, colleges, or universities, RPL acknowledges learning wherever it occurs (Authors, 2011; O'Leary, 2023).

The process also has strong links to workplace learning theories. Much adult learning occurs through participation in work activities, problem-solving, collaboration, mentoring, and communities of practice (Day, 2016; Paulos, 2018). Individuals often develop sophisticated knowledge and skills through everyday work experiences without receiving formal qualifications (Graebe & McIntyre-Hite, 2024; Snyman & Berg, 2017). RPL provides a mechanism for translating this workplace learning into recognised credentials (Day, 2016; Graebe & McIntyre-Hite, 2024; Ions & Sutcliffe, 2019). For example, a supervisor may have developed leadership, communication, planning, and risk management skills over many years of practice (Day, 2016; Graebe & McIntyre-Hite, 2024). Through RPL, these competencies can be assessed and formally recognised, creating a bridge between workplace learning and formal education systems (Day, 2016; Graebe & McIntyre-Hite, 2024; Ions & Sutcliffe, 2019).

From an adult education perspective, RPL serves both educational and social purposes. Educationally, it recognises the diverse ways adults learn and reduces unnecessary repetition of learning (Day, 2016; Ions & Sutcliffe, 2019). Socially, it promotes access, equity, and inclusion by valuing learning gained outside traditional educational pathways (Authors, 2011; Day et al., 2020; O'Leary, 2023). This is particularly important for mature-age learners, migrants, military veterans, Indigenous peoples, volunteers, and workers whose expertise may not be reflected in formal qualifications (Day et al., 2020; Hill et al., 2023; Snyman & Berg, 2017). By recognising existing competencies, RPL can increase participation in education, support career progression, and encourage further learning (Graebe & McIntyre-Hite, 2024; Ions & Sutcliffe, 2019; Sukri, 2024).

Informal and Incidental Workplace Learning

Informal and incidental learning are among the most common forms of adult learning in the workplace. Although organisations often focus on formal training programs, workshops, and professional development courses, research consistently shows that a significant proportion of workplace learning occurs outside structured educational activities (MacMahon et al., 2024; Manuti et al., 2015; Shah et al., 2022). Employees frequently develop knowledge, skills, attitudes, and professional judgement through everyday work experiences, interactions with colleagues, problem-solving activities, and participation in workplace practices (Coetzer et al., 2017; Møller, 2024).

The concepts of informal and incidental learning are most closely associated with the work of Marsick and Watkins, who argued that much adult learning occurs naturally as part of work rather than through planned instruction (Barefield & Nicolaides, 2022; Daryoush, 2013; Marsick & Watkins, 2018). They distinguished informal learning from formal learning by noting that informal learning is generally embedded within workplace activities, often initiated by learners themselves, and frequently driven by immediate practical needs rather than predetermined educational objectives (Han et al., 2022; Marsick & Watkins, 2018; Yoon et al., 2018). Marsick and Watkins initially conceptualised informal and incidental learning in contrast to structured, designed learning in formal settings in the workplace, focusing on how individuals learn by engaging an ongoing, interactive cycle of learning from experience that started with an unfamiliar challenge or problem at work (Marsick & Watkins, 2018).

Informal learning refers to learning that occurs outside structured educational programs but is often intentional (Han et al., 2022; Jeong et al., 2018). Individuals recognise a need to learn something and actively seek information, advice, or experience to address that need. Marsick and Watkins' model stresses that individuals largely learn from their experiences and that informal learning occurs when they have the need, motivation or opportunity for learning. For example, a maintenance technician encountering an unfamiliar equipment fault may consult a more experienced colleague, review technical manuals, search online resources, and experiment with solutions until the problem is resolved. Although no formal training course has occurred, significant learning has taken place through workplace activity (Rose & Chmutina, 2021; Wofford et al., 2013).

Incidental learning, by contrast, occurs unintentionally as a by-product of another activity (Coetzer et al., 2017; Jeong et al., 2018; Shah et al., 2022). The learner may not initially realise that learning is taking place (Coetzer et al., 2017; MacMahon et al., 2024). Marsick and Watkins described incidental learning as learning that occurs unintentionally while individuals are engaged in other workplace activities. Rather than being the primary objective, learning emerges as a secondary outcome of experiences such as observing organisational norms, adapting to workplace culture, solving problems, interacting with colleagues, or experimenting with different approaches to tasks and challenges (Manuti et al., 2015). Incidental learning is never planned, though it can sometimes be intentional, such as help constantly sought from a mentor (Rose & Chmutina, 2021). For example, a newly promoted supervisor participating in staff meetings may gradually develop negotiation, communication, and conflict management skills simply through repeated exposure to workplace situations. The primary purpose of attending the meetings is not learning, yet valuable learning occurs as a consequence of participation (MacMahon et al., 2024; Møller, 2024). Often, individuals only become aware of incidental learning when they later reflect on their experiences and recognise changes in their knowledge, behaviour, or understanding (Grandy & Mavin, 2020). Informal learning results in tacit knowledge acquisition and much of this type of learning happens outside an individual's conscious awareness (Grandy & Mavin, 2020; Shah et al., 2022).

Informal and incidental learning are closely connected to experiential learning theory because they emerge through direct engagement with workplace experiences (Barefield & Nicolaides, 2022; Kusaila, 2019; Nicolaides & Scully-Russ, 2018). Employees learn by performing tasks, solving problems, interacting with others, making decisions, and responding to challenges. The theoretical foundations of Marsick and Watkins' model drew upon Dewey's theory of experience, Argyris and Schön's action science, Lewin's field theory, Kolb's cycle of learning, and Simon's decision-making theory (Kusaila, 2019; Nicolaides & Scully-Russ, 2018). Reflection plays an important role in transforming these experiences into meaningful learning (Watkins et al., 2018; Wofford et al., 2013). When individuals analyse what occurred, why it occurred, and how similar situations might be handled in the future, workplace experiences become sources of ongoing professional development. The key components of informal and incidental learning theory include the integration of learning with daily routines, the inductive process of reflection and action, the linking of individual learning to that of others, and that the learning is not highly conscious (Kusaila, 2019; Watkins et al., 2018).

These forms of learning also align strongly with situated learning theory. Learning occurs within authentic workplace contexts rather than being separated from practice (Coetzer et al., 2017; Møller, 2024). Employees acquire knowledge while participating in the activities, routines, and social interactions of their occupational communities (Møller, 2024; Rose & Chmutina, 2021). A graduate nurse learns not only from formal instruction but also from observing experienced practitioners, participating in patient care, discussing cases with colleagues, and responding to unexpected clinical situations. Similarly, an apprentice electrician learns through daily interactions with tradespeople, exposure to workplace challenges, and participation in real work activities (MacMahon et al., 2024; Rose & Chmutina, 2021). As research in the mining industry has demonstrated, frontline workers conceptualise training and learning distinctively, with learning understood as something that occurs continuously on-the-job through both incidental and intentional means (MacMahon et al., 2024).

Communities of practice provide another important perspective on informal workplace learning (Watkins et al., 2018). Within communities of practice, employees share knowledge, discuss problems, exchange experiences, and collectively develop professional expertise (Coetzer et al., 2017; Watkins et al., 2018). Much of this learning occurs through conversations, mentoring relationships, collaborative problem-solving, and observation of more experienced colleagues (Ellinger & Cseh, 2007; Wofford et al., 2013). Newcomers gradually acquire the language, norms, values, and practices of their profession through participation in the community (Coetzer et al., 2017). In informal learning approaches such as communities of practice or action learning, development is more open-ended, focused on learning through challenging experiences, through interaction, and through solving organisational problems (Watkins et al., 2018).

Informal and incidental learning occur through a wide range of everyday workplace activities. Employees frequently learn while solving unexpected problems, observing experienced colleagues, participating in meetings and discussions, receiving feedback from supervisors and peers, and engaging in mentoring or coaching relationships (Schürmann & Beausaert, 2016; Scott, 2022a; Tan et al., 2026; Tannenbaum & Wolfson, 2022). Learning also emerges through trial-and-error experimentation, collaborative projects, workplace rotations, secondments, professional networking, and the use of workplace documents, manuals, and other resources (Margaryan et al., 2013; Scott, 2022a; Tan et al., 2026). Workplace learning research emphasises that informal training is inherently embedded in daily work and is influenced by interaction, problem-solving, and the broader social environment, rather than structured instruction (Tan et al., 2026). As organisations introduce new technologies, systems, and processes, employees often acquire new knowledge and skills through direct engagement with these changes (Crans, 2023; Scott, 2022a). Similarly, responding to workplace incidents, challenges, and unforeseen situations can generate valuable learning experiences that contribute to professional growth and competence. Individuals mainly learn on the spot, when facing the necessity to solve a problem or to respond to a novel situation (Crans, 2023), and incidental learning occurs as an unintended by-product of activities such as trial-and-error experimentation or interpersonal interaction (Dang & Chou, 2019; Tannenbaum & Wolfson, 2022).

For adult learners, informal and incidental learning are particularly valuable because they are highly relevant, contextualised, and immediately applicable to workplace practice (Crans, 2023; Kim et al., 2026). Learning is increasingly recognised as a context-embedded process rather than a series of discrete training activities (Kim et al., 2026). Adults are often motivated by practical problems and the need to perform effectively in their professional roles (Scott, 2022a). Learning that emerges directly from workplace experiences is therefore more likely to be viewed as meaningful and useful than learning that is disconnected from real-world applications. Informal learning also enables individuals to learn at the point of need, allowing them to address challenges as they arise rather than waiting for formal training opportunities (Shanks, 2023; Tannenbaum & Wolfson, 2022). Estimates suggest that between 70% and more than 90% of workplace learning occurs informally (Tan et al., 2026; Tannenbaum & Wolfson, 2022), underscoring its centrality to professional development.

The organisational environment plays a critical role in supporting informal and incidental learning. Workplaces that encourage collaboration, knowledge sharing, mentoring, reflection, and open communication create conditions in which learning can occur naturally and continuously (Jeong et al., 2018; Schürmann & Beausaert, 2016; Scott, 2022a). Four organisational factors that positively influence informal learning include learning-committed leadership and management, an internal culture committed to learning, work tools and resources, and people who form webs of relationships for learning (Schürmann & Beausaert, 2016). Employees are more likely to seek advice, share experiences, and engage in problem-solving when they feel supported by colleagues and supervisors (Crans, 2023; Schürmann & Beausaert, 2016; Tannenbaum & Wolfson, 2022). The meta-analysis reported by Tannenbaum and Wolfson found that personal support from managers and peers, formal organisational practices, and informal means of support such as climate were each positively related to informal learning (Tannenbaum & Wolfson, 2022). In contrast, highly hierarchical, competitive, or time-pressured workplaces may restrict opportunities for interaction, feedback, and knowledge exchange. Psychological safety, a shared belief that it is safe to take interpersonal risks, is highly relevant for informal learning; when psychological safety is low, it inhibits learning behaviours and information sharing (Opoku et al., 2019; Tannenbaum & Wolfson, 2022). Leadership support, psychological safety, and access to experienced colleagues are therefore important factors in fostering effective workplace learning (Crans, 2023; Shanks, 2023; Tannenbaum & Wolfson, 2022). Headteachers and managers have been found to serve as mentors or role models, encourage risk taking, give feedback, create informal learning opportunities, and enhance knowledge sharing (Schürmann & Beausaert, 2016; Shanks, 2023).

Despite its many advantages, informal and incidental learning can present challenges. Because it is often unstructured and undocumented, the quality and consistency of learning may vary considerably between individuals (Jeong et al., 2018; M. Singh, 2022). Employees may develop incomplete or inaccurate understandings if workplace practices are ineffective or if inappropriate behaviours are modelled by experienced workers. Furthermore, much workplace learning is tacit, meaning that individuals may find it difficult to explain, document, or demonstrate what they have learned (M. Singh, 2022). Structures and processes for helping workers identify their informal learning outcomes, competencies, or job descriptions are often lacking (M. Singh, 2022). This can create difficulties when seeking formal recognition of workplace-acquired skills and knowledge. Consequently, processes such as Recognition of Prior Learning (RPL) often require individuals to reflect upon, articulate, and provide evidence of learning gained through their workplace experiences (Pawlukewicz et al., 2013; Priadi et al., 2019; Singh, 2015c). RPL is defined as a judgment of the calibre of previous learning in the workplace, separate from formal learning, and whether to credit such learning (Pawlukewicz et al., 2013). An important goal of recognition processes should be to strengthen informal learning through the recognition of competences in the workplace (Singh, 2015c). Managers need to facilitate continuous professional development through portfolio compilation which acknowledges the learning opportunities within the workplace and can be used as recognition of prior learning (Jooste & Jasper, 2010).

Informal and Incidental Learning and Adult Learning Theory

Informal and incidental workplace learning embody many of the central principles of adult learning. They reflect andragogy by recognising the importance of experience and self-direction (Acosta-Leal et al., 2023; Pacavita & González, 2018; Pérez & Andrades, 2020). Andragogic theory, as developed by Knowles, identifies experience as the most important source of adult learning, alongside self-concept, readiness to learn, orientation toward learning, and motivation (Acosta-Leal et al., 2023; Pérez & Andrades, 2020). The adult learner's experience serves as "the book of learning," shaping new knowledge and creating the basis for identity (Cárdenas, 2021; Pacavita & González, 2018). Learning situations in adulthood depart from the learner's interests and are applicable to their work context, a fact not evident in pedagogy given children's lack of work experience (Duarte, 2021).

Informal and incidental workplace learning align with experiential learning through the transformation of experience into knowledge (Briceño et al., 2019; Vega & Domínguez, 2022). Experiential learning theory centres on the role that experience plays in the learning process, whereby knowledge is constructed through reflection and "giving meaning" to experiences (Briceño et al., 2019). This involves concrete experiences, observation and analysis, formation of abstract concepts based on reflection, and active experimentation with new concepts (Vega & Domínguez, 2022). The connection between practice and knowledge through reflection constitutes a fundamental learning process (García-Romero & Lalueza, 2019).

These forms of learning support constructivist views as individuals actively create meaning from workplace events (Briceño et al., 2019; Cárdenas, 2021). The adult consolidates learning through the experiential field, where context, work relationships, critical capacity, and human experience are indispensable elements (Cárdenas, 2021). The interaction between content and experience should achieve transformation, enabling the learner to reconstruct old ideas with new ones (Pacavita & González, 2018).

Informal and incidental workplace learning contribute to situated learning and communities of practice by embedding learning within authentic social and occupational contexts (Bermúdez, 2017; Vega & Domínguez, 2022). Situated learning, grounded in Vygotskian theory, positions context as fundamental for achieving meaningful learning (Acosta-Leal et al., 2023). Communities of practice represent spaces where collaborative learning occurs through shared experiences and knowledge directed toward common goals (Sánchez, 2016), and where participation in authentic practice fosters the development of professional competencies (Bermúdez, 2017).

Finally, these forms of learning may lead to transformative learning when workplace experiences challenge existing assumptions and encourage new ways of thinking (Contreras, 2006; Leupin, 2015; Prieto & Rojas, 2024). Transformative learning theory, developed by Mezirow, describes how adults integrate new information and perspectives into their worldviews by engaging in learning that transforms mental structures, termed perspectives and schemes of meaning, that condition experience (Leupin, 2015). Adults require "unlearning" simultaneously with learning to achieve transformational outcomes that stimulate behavioural modification through the revision of their own frames of reference (Contreras, 2006). The key factors of transformative learning comprise experience, critical reflection, and development (Guzón et al., 2021), and transformation occurs when subjects construct reflections from experience that change their perspectives of meaning and ways of acting (Prieto & Rojas, 2024).

Significance of Informal and Incidental Workplace Learning

Informal and incidental workplace learning are significant because they account for a substantial proportion of professional development throughout adult life. Studies suggest that approximately 70–90% of learning in the workplace derives from informal activities (McGill, 2024), with earlier estimates indicating roughly 80% of workplace learning is informal compared to 20% formal (McGill, 2024). While formal education provides foundational knowledge and credentials, much occupational expertise is developed through participation in everyday work (Hwang & Yoon, 2023; M. Singh, 2022; Yunga, 2019). Workplace learning constitutes activities enhancing employees' skills through professional development and problem-solving tactics to improve work performance, and is sometimes referred to as informal or non-formal learning (Scott, 2022a). Through problem-solving, collaboration, observation, reflection, and experience, workers continually expand their knowledge and capabilities (Kowalski & Russell, 2020; Lyons & Bandura, 2020). Informal learning occurs as individuals engage in daily work activities, networking, or performance planning, and is often unintentional and unplanned, involving interaction with others and consisting of action and reflection (Kowalski & Russell, 2020). Incidental learning, a related concept, takes place alongside tasks where workers concentrate on completing work but are not typically aware of the learning occurring (Bova & Kroth, 2001). Marsick and Watkins conceptualized informal and incidental learning as how individuals learn by engaging an ongoing, interactive cycle of learning from experience, triggered by unfamiliar challenges or problems at work (Lyons & Bandura, 2020). As a result, informal and incidental learning are widely recognised as essential components of lifelong learning, workforce development, and professional competence in contemporary organisations (Hwang & Yoon, 2023; M. Singh, 2022; Yunga, 2019). Non-formal and informal learning, as part of lifelong learning, are acknowledged as crucial for the constant development of professionals to cope with the ever-growing challenges of the world of work (Hwang & Yoon, 2023). Recognition of non-formal and informal learning is a core component of the concept of lifelong learning, which has significantly changed the ways in which learning and its key elements are understood (Bova & Kroth, 2001). Workplace learning is critical for an individual's capacity building in adult education and lifelong learning perspectives, and also for organizational capability building through training and development from a human resource development perspective (Lyons & Bandura, 2020).

Key Concepts and Conclusions

This chapter explored how adults learn through experience, work, participation, and professional practice. A central theme was that learning is not confined to classrooms or formal instruction but occurs continuously through engagement with real-world activities, workplace challenges, social interactions, and reflective practice. Experiential learning was presented as a learner-centred approach that views experience as the foundation of learning, with reflection acting as the mechanism through which experiences are transformed into knowledge, skills, and understanding. The chapter highlighted principles such as active participation, knowledge construction, the integration of theory and practice, learner autonomy, adaptation, experimentation, and the influence of context on learning. These principles culminated in David Kolb's Experiential Learning Theory, which conceptualises learning as a continuous cycle of concrete experience, reflective observation, abstract conceptualisation, and active experimentation.

The chapter examined the significance of workplace learning as a major source of adult learning throughout professional life. Workplace learning theories challenge traditional views that separate learning from practice and instead emphasise that learning is embedded within work activities, professional relationships, and organisational contexts. Human Capital Theory highlighted the economic value of education and training, while Experiential Learning Theory demonstrated how competence develops through reflection on experience. Situated Learning Theory and Communities of Practice emphasised that learning is inherently social and occurs through participation in authentic workplace activities and professional communities. Informal and incidental learning further illustrated how much workplace learning occurs naturally through everyday problem-solving, observation, collaboration, and workplace participation rather than through formal training programs.

The chapter also explored organisational perspectives on learning. Organisational Learning Theory demonstrated that learning extends beyond individuals to encompass collective learning, knowledge sharing, and organisational adaptation. Concepts such as single-loop learning, double-loop learning, learning organisations, systems thinking, and collective knowledge creation highlighted how organisations develop and apply knowledge to improve performance and respond to changing environments. Reflective Practice, particularly through the work of Donald Schön, was presented as a critical mechanism through which professionals develop expertise by reflecting both during and after action. Activity Theory further expanded understanding by illustrating how learning is shaped by social, cultural, organisational, and technological systems and how contradictions within these systems can drive transformative change and expansive learning.

The chapter also examined apprenticeships, vocational learning, work-integrated learning, and Recognition of Prior Learning (RPL) as important manifestations of experiential and workplace learning. Apprenticeships were shown to combine workplace participation, mentoring, practical experience, and formal instruction to develop occupational competence and professional identity. Work-integrated learning was presented as a structured approach that bridges theory and practice through placements, internships, practicums, industry projects, and workplace experiences. Recognition of Prior Learning highlighted how adults acquire valuable knowledge and skills through formal, non-formal, and informal experiences throughout their lives and how these competencies can be formally recognised. Collectively, these concepts reinforce the view that adult learning is a lifelong, experience-based process occurring across multiple contexts and supported by participation in authentic professional practice.

Experiential and workplace learning theories fundamentally challenge the assumption that learning occurs primarily through formal instruction. Instead, they demonstrate that adults develop much of their knowledge, competence, and professional expertise through engagement with authentic experiences, workplace participation, reflection, collaboration, and problem-solving. Experience alone, however, is insufficient. Learning occurs when individuals actively reflect upon their experiences, construct meaning from them, and apply their insights to future situations. This relationship between experience, reflection, and action is a recurring theme across many adult learning theories and remains central to understanding how adults learn.

The chapter also demonstrates that learning is both an individual and a social process. While individuals construct knowledge through personal experience and reflection, learning is strongly influenced by interactions with colleagues, mentors, supervisors, communities of practice, and organisational cultures. Knowledge is often developed collectively through participation in shared activities, collaborative problem-solving, mentoring relationships, and professional dialogue. Consequently, effective adult learning environments are those that encourage participation, communication, reflection, feedback, and knowledge sharing rather than relying solely on individual study or formal instruction.

A further conclusion is that workplaces should be understood not only as sites of production but also as powerful learning environments. Contemporary organisations increasingly depend upon employees who can learn continuously, adapt to change, solve complex problems, and contribute to innovation. As a result, workplace learning has become a strategic organisational capability rather than simply an individual responsibility. Learning organisations, communities of practice, mentoring systems, reflective practices, and opportunities for informal learning all contribute to workforce development and organisational effectiveness. The ability of organisations to create cultures that support learning has become increasingly important in environments characterised by technological change, global competition, and evolving workforce demands.

The chapter further highlights that effective adult learning requires the integration of theory and practice. Whether through apprenticeships, workplace projects, internships, clinical placements, simulations, or work-integrated learning, adults learn most effectively when they can connect abstract concepts to real-world situations. Authentic experiences allow learners to test ideas, evaluate outcomes, develop professional judgement, and refine their practice. Learning therefore becomes meaningful when knowledge is applied within the contexts in which it will ultimately be used.

Finally, the chapter reinforces the principle that learning is a lifelong process. Adults continue to develop knowledge and skills throughout their careers through formal education, workplace participation, professional development, reflection, and everyday experiences. Recognition of Prior Learning demonstrates that valuable learning occurs across multiple life contexts and that competence can be acquired through diverse pathways. Collectively, the theories and approaches explored in this chapter support a broader understanding of adult learning as a continuous, contextual, social, and experience-driven process that contributes not only to occupational competence but also to personal growth, professional identity, organisational capability, and lifelong development.

Chapter 6

Motivation, Engagement, and Learner Success

Adult learning begins with a decision to participate, however continuing that journey requires far more than simply enrolling in a course. Unlike many younger learners, adults must balance education alongside employment, family responsibilities, financial commitments, and an ever-changing range of personal and professional demands. Whether learning is undertaken to gain a qualification, advance a career, adapt to technological change, pursue personal interests, or respond to major life transitions, sustained participation depends on a complex interaction of motivation, engagement, confidence, persistence, and the ability to overcome barriers.

This chapter explores the factors that influence adult participation and achievement by examining why adults choose to learn, what encourages them to remain engaged, and what enables them to succeed. It reviews major theories of motivation, including intrinsic and extrinsic motivation, Self-Determination Theory, Knowles' andragogy, Expectancy Theory, Goal Theory, Achievement Motivation Theory, Maslow's Hierarchy of Needs, and Transformative Learning. The chapter also examines the roles of self-efficacy, learner persistence, retention, barriers to participation, learner engagement, wellbeing, resilience, and inclusive teaching practices in supporting successful educational outcomes.

Understanding these concepts is essential for educators, trainers, instructional designers, workplace learning professionals, and policymakers. By recognising the diverse motivations, challenges, and experiences that adult learners bring to education, learning environments can be designed to foster curiosity, build confidence, sustain engagement, remove barriers, and promote lifelong learning. Ultimately, effective adult education is not only about helping learners acquire knowledge and skills, but also about empowering them to persist, adapt, and thrive in an increasingly complex and rapidly changing world

Motivation Theories in Adult Learning

Motivation is one of the most important factors influencing adult learning. While children are often required to participate in education, adults generally make conscious decisions about whether, when, and how they will engage in learning (Abedini et al., 2021; Boeren & Holford, 2016; Yu, 2023). Adult learners frequently balance education with employment, family responsibilities, financial commitments, and other personal obligations (Leyretana & Trinidad, 2021; Storey et al., 2025). As a result, understanding what motivates adults to learn is essential for educators, trainers, employers, and policymakers seeking to design effective learning experiences (Lim et al., 2023; Sahoo et al., 2021).

Adult learning research has shown that motivation is complex and influenced by a combination of personal goals, workplace requirements, social expectations, self-beliefs, and environmental factors (Åström et al., 2025; Boeren & Holford, 2016; Broek et al., 2023; Millar et al., 2020). Several theories have been developed to explain why adults engage in learning and what sustains their participation over time (Åström et al., 2025; Sahoo et al., 2021; Zhao et al., 2024).

Intrinsic and Extrinsic Motivation

One of the most widely recognised distinctions is between intrinsic and extrinsic motivation. Intrinsic motivation refers to engaging in learning because the activity itself is interesting, enjoyable, or personally satisfying (Billsberry & Alony, 2023; M. Zhu et al., 2022). Adults who pursue a language course because they enjoy learning about other cultures or study history because of personal curiosity are primarily intrinsically motivated (Naumčiuk, 2023; M. Zhu et al., 2022).

Extrinsic motivation involves learning for external rewards or outcomes (Drew, 2008; Gardner et al., 2021). Adults may undertake training to gain a promotion, maintain professional registration, increase earnings, obtain a qualification, or satisfy employer requirements (Drew, 2008; Gardner et al., 2021; Kimmel et al., 2012). Although extrinsic motivators are often important, research suggests that intrinsic motivation tends to be associated with deeper engagement, persistence, and long-term learning outcomes (Billsberry & Alony, 2023; Rothes et al., 2016; Sethi et al., 2018).

For adult learners, intrinsic and extrinsic motivations frequently coexist (Drew, 2008; Sethi et al., 2018). A nurse undertaking postgraduate study may be motivated both by professional advancement and by a genuine desire to improve patient care (Jones et al., 2024; Sethi et al., 2018). Similarly, an apprentice electrician may seek certification for employment purposes while also enjoying the challenge of mastering technical skills (Gardner et al., 2021; Ostendorff).

Self-Determination Theory

Self-Determination Theory (SDT), developed by Edward Deci and Richard Ryan, is one of the most influential contemporary theories of motivation (Luo et al., 2021; Rubić & Matijević, 2019; Shelton-Strong, 2020; Turner, 2019). The theory proposes that individuals are more likely to be motivated, engaged, and persistent in learning when three fundamental psychological needs are satisfied: autonomy, competence, and relatedness (Jeno et al., 2017; Luo et al., 2021; Schweder & Raufelder, 2021; Turner, 2019). Autonomy refers to the need to have a sense of control over one's actions and decisions (Lee, 2016; Lou et al., 2017; Shelton-Strong, 2022). Competence involves feeling capable, effective, and able to successfully meet challenges (Alamer, 2021; Jeno et al., 2017; Lou et al., 2017). Relatedness reflects the need to feel connected to, supported by, and valued by others (Basma, 2015; Lou et al., 2017; Shelton-Strong, 2020). According to SDT, motivation is strengthened when learning environments support these needs and weakened when they are restricted or ignored (Basma, 2015; Jeno et al., 2017; Turner, 2019).

In adult learning contexts, autonomy is particularly important because adults generally prefer to take responsibility for their learning and make decisions about what, how, and when they learn (Lou et al., 2017). Learning experiences that provide choice, flexibility, and opportunities for self-direction are therefore more likely to generate engagement than highly controlled or prescriptive approaches (Bachman & Stewart, 2011; Evans & Boucher, 2015). For example, a professional development program that allows employees to select learning activities aligned with their career aspirations and workplace needs is likely to produce higher levels of motivation than a program in which all learning activities are predetermined (Evans & Boucher, 2015; Muñoz & Ramírez, 2015). Research grounded in SDT has consistently demonstrated that autonomy-supportive environments foster self-determined motivation and enhance intrinsic motivation (Bachman & Stewart, 2011; Roth et al., 2007; Siacor et al., 2024).

The need for competence is satisfied when learners experience success, receive constructive feedback, and develop confidence in their abilities (Alamer, 2021; Basten et al., 2014; Tanaka & Takeuchi, 2024). Adults are more likely to remain motivated when they believe they can successfully apply new knowledge and skills to real-world situations (Owen, 2020). Learning activities that are challenging yet achievable, supported by appropriate guidance and feedback, help learners develop a sense of mastery and effectiveness (Basten et al., 2014; Owen, 2020). As competence increases, so too does the willingness to engage in further learning and professional development (Jeno et al., 2017; Tanaka & Takeuchi, 2024).

Relatedness emphasises the social dimension of learning. Adult learners are more likely to be motivated when they feel respected, supported, and connected to instructors, mentors, colleagues, and fellow learners (Muñoz & Ramírez, 2015; Shelton-Strong, 2022). Collaborative learning activities, mentoring relationships, communities of practice, and supportive workplace cultures can strengthen feelings of belonging and encourage ongoing participation in learning (Jeno et al., 2017; Shelton-Strong, 2020). Positive social interactions not only support motivation but also contribute to knowledge sharing, professional growth, and learner confidence (Muñoz & Ramírez, 2015; Shelton-Strong, 2022).

Self-Determination Theory aligns closely with many principles of adult learning (Lou et al., 2017; Schweder & Raufelder, 2021). Adult learners typically value independence, personal responsibility, and learning opportunities that are meaningful and relevant to their lives and careers (Lou et al., 2017; Rubić & Matijević, 2019). By supporting autonomy, competence, and relatedness, learning environments can foster more sustained and self-directed forms of motivation, leading to greater engagement, persistence, and lifelong learning (Basma, 2015; Jeno et al., 2017; Luo et al., 2021).

Knowles' Andragogy and Motivation

Malcolm Knowles incorporated motivation directly into his theory of andragogy. He argued that adults are motivated primarily by internal factors rather than external pressures (Bergkvist et al., 2023; Conway, 2022; Khan, 2024; Purwati et al., 2022). While external incentives such as salary increases or promotions may influence participation (Bergkvist et al., 2023; Gardner et al., 2021), adults are often driven by personal growth, increased self-esteem, improved quality of life, professional competence, job satisfaction, and self-fulfilment (Gardner et al., 2021; Ibrahim et al., 2022).

Knowles also proposed that adults are most motivated when learning is relevant to their immediate needs and life situations (Darden, 2014; Khan, 2024). Learning that clearly addresses workplace challenges, personal goals, or practical problems is more likely to generate engagement than learning perceived as abstract or disconnected from real-world application (Darden, 2014; Onwuatuegwu, 2023).

The influence of internal motivation can be seen in many adult education settings. For example, an experienced enrolled nurse may choose to undertake a Bachelor of Nursing not because their employer requires it, but because they want to expand their professional knowledge, increase their confidence in clinical decision-making, and achieve personal career goals. Although obtaining a higher qualification may eventually lead to promotion or increased earnings, the learner is primarily motivated by professional growth and the desire to become a more capable healthcare practitioner. In this situation, the motivation comes largely from within the learner rather than from external rewards.

A similar example can be found in community education. An adult learner may enrol in a digital literacy course to gain confidence using computers and online services. Their motivation may stem from a desire to communicate with family members, access government services, manage personal finances, or participate more fully in modern society. The learner is not pursuing a qualification or financial reward but is seeking greater independence, self-esteem, and quality of life.

Knowles' emphasis on relevance can be observed in vocational education and training. Consider a supervisor in a warehouse who enrols in a leadership and team management course after being promoted to a supervisory role. The learner is motivated because the course addresses immediate workplace challenges such as conflict resolution, communication, staff supervision, and performance management. The direct connection between the learning content and the learner's daily responsibilities increases engagement because the knowledge can be applied immediately in the workplace.

In higher education, an adult student studying part-time while employed as a civil engineering technician may be highly motivated to complete an engineering degree because the learning directly supports their professional practice. Concepts learned in structural analysis, project management, or infrastructure design can be applied to projects they encounter at work. The immediate relevance of the content helps sustain motivation despite the challenges of balancing study, work, and family commitments.

A further example can be seen in agricultural education. A farmer attending workshops on drought management and sustainable farming practices is often motivated by the need to solve real problems affecting their property. Learning is valued because it provides practical solutions that can improve productivity, reduce costs, and support long-term business sustainability. If the training instead focused on highly theoretical concepts with little apparent application to the learner's circumstances, motivation and participation would likely be lower.

Similarly, an adult learner undertaking English language studies after migrating to a new country may be motivated by the desire to gain employment, communicate effectively within the community, support their children's education, and participate more fully in society. The learning is directly connected to immediate life needs and personal aspirations, making the educational experience meaningful and worthwhile.

These examples illustrate Knowles' argument that adult learners are most motivated when learning addresses real-life needs, supports personal or professional goals, and contributes to improved competence, confidence, self-esteem, and quality of life. While external rewards may influence participation, sustained engagement is often driven by the learner's desire to solve problems, achieve personal growth, and improve their ability to function effectively in their chosen roles and environments.

Expectancy Theory

Expectancy Theory, developed by Victor Vroom, proposes that motivation is influenced by an individual's expectations about effort, performance, and outcomes. According to the theory, people are more likely to invest effort in a task when they believe their effort will lead to successful performance, that successful performance will produce desired outcomes, and that those outcomes are personally valuable (Amali et al., 2023; Uriyo; Zheng et al., 2025). Vroom identified three key components that influence motivation: expectancy, instrumentality, and valence (Lokman et al., 2022; Swain et al., 2020; Zboja et al., 2020).

Expectancy refers to an individual's belief that their effort will result in successful performance (Moosa & Aloka, 2022; Swain et al., 2020; Yeheyis et al., 2016). In learning contexts, adults are more likely to engage in education or training when they feel confident that they can successfully complete the learning activities and achieve the required outcomes (Amali et al., 2023; Lokman et al., 2022). Factors such as prior experience, self-confidence, available support, and access to learning resources can influence these perceptions (Amali et al., 2023; Chopra, 2019). If learners believe a task is beyond their capabilities or that success is unlikely regardless of effort, motivation may decline (Amali et al., 2023; Swain et al., 2020).

Instrumentality refers to the belief that successful performance will lead to specific rewards or benefits (Yeheyis et al., 2016; Zboja et al., 2020; Zheng et al., 2025). Adult learners are more likely to be motivated when they can see a clear connection between learning achievement and desirable outcomes (Amali et al., 2023; J. J. Lim et al., 2024; Lokman et al., 2022). These outcomes may include career advancement, increased earnings, professional recognition, improved job performance, or access to new opportunities (Atiku, 2023; J. J. Lim et al., 2024). When learners perceive that completing a course or training program will have little impact on their personal or professional circumstances, their motivation to participate may be reduced (Amali et al., 2023; J. J. Lim et al., 2024).

Valence refers to the value or importance that individuals place on the expected outcomes (Savolainen, 2012; Wei Wang et al., 2024; Zheng et al., 2025). Even if learners believe they can succeed and that success will lead to rewards, motivation will remain low if those rewards are not personally meaningful (Moosa & Aloka, 2022; Swain et al., 2020). Different adults place different levels of importance on outcomes such as promotion, professional status, financial rewards, personal growth, or improved work-life balance (Atiku, 2023). The greater the perceived value of the outcome, the stronger the motivational influence (Moosa & Aloka, 2022; Wei Wang et al., 2024).

Expectancy Theory suggests that motivation is strongest when all three factors are present (John, 2023; Yeheyis et al., 2016). For example, an employee may be highly motivated to undertake leadership training if they believe they can successfully complete the program, that successful completion will improve their chances of promotion, and that career advancement is important to their personal and professional goals (Atiku, 2023; Lokman et al., 2022; Matthews et al., 2020). Conversely, motivation may decrease if learners perceive the learning task as excessively difficult, unlikely to result in meaningful benefits, or unrelated to their aspirations and needs (Amali et al., 2023; Swain et al., 2020). The multiplier effect in Vroom's equation means that if any one of the three factors is zero, the overall motivational force is zero (Savolainen, 2012; Yeheyis et al., 2016). The theory therefore highlights the importance of designing learning experiences that are achievable, clearly linked to desirable outcomes, and aligned with learners' personal goals and values (Lokman et al., 2022; Vries et al., 2019; Yeheyis et al., 2016).

Goal Theory

The theory proposes that individuals are more likely to remain motivated when they are working towards goals that are clear, meaningful, and appropriately challenging (O'Neill & Thomson, 2013; Schiller, 2022a). Goals provide a sense of purpose and direction, helping learners focus their attention, organise their efforts, and persist when difficulties arise (Mellard et al., 2012; Schiller, 2022a). According to Locke and Latham's goal setting theory, motivation is reflected in effortful behaviour directed toward achieving an objective, and having goals may have a stimulating effect on individual effort, persistence, and direction (Schiller, 2022a). When learners understand what they are trying to achieve and believe the goal is worthwhile, they are generally more willing to invest time and energy in the learning process (O'Neill & Thomson, 2013; Ong, 2014).

Adult learners often enter educational programs with specific objectives in mind. These objectives may include changing careers, gaining promotion, improving professional competence, obtaining a qualification, or pursuing personal interests and hobbies (Gardner et al., 2021; Ong, 2014). Generally, adult students engage in lifelong learning or continuing education for many reasons, including acquiring knowledge for personal development, career advancement, job promotion, and salary increment (Ong, 2014). Unlike many younger learners who may participate in education as part of a prescribed pathway, adults are frequently motivated by goals that are closely connected to their current life circumstances, professional responsibilities, or future aspirations (Ibrahim et al., 2012; O'Neill & Thomson, 2013). Adults tend to be more pragmatic, and their interest and engagement are more closely tied to relevance, relatedness, and practical application (O'Neill & Thomson, 2013). As many adults enrol in higher education for job-related reasons, they tend to demonstrate clear learning goals and greater intrinsic motivations than do younger students (Ibrahim et al., 2012). These goals provide a strong incentive to engage with learning and to persist despite competing demands such as work, family, and financial commitments (Mellard et al., 2012; O'Neill & Thomson, 2013). Low-skilled adults often have family and work responsibilities outside school and very little contact time with adult educators (O'Neill & Thomson, 2013).

The effectiveness of Goal Theory in adult learning is particularly evident when learning activities are explicitly linked to learner objectives. Adults are more likely to be motivated when they can clearly see how the knowledge and skills being developed will help them achieve their desired outcomes (Njenga, 2022; O'Neill & Thomson, 2013). Goal orientation remains a vital component that represents adult learners' direction in pursuing their educational programs, and by knowing their goal orientation, we are able to understand their learning strategies, achievement behaviour, and patterns of cognition (Ong,

2014). Adult Education Theory views adult learning as self-directed and goal-oriented activity (Njenga, 2022), and the key assumptions of effective adult learning include participants' internal motivation of learning and goal orientation in relation to acquiring new knowledge (Schiller, 2022a). For example, a healthcare worker studying for a nursing qualification may be motivated by the goal of expanding their scope of practice, while a tradesperson undertaking management training may be motivated by aspirations to move into a supervisory role. In both cases, the connection between the learning activities and the learner's goals enhances the perceived value of the educational experience (O'Neill & Thomson, 2013; Ong, 2014).

Goal Theory therefore highlights the importance of helping adult learners establish meaningful and achievable learning objectives (O'Neill & Thomson, 2013; Ong, 2014). Educators who align learning activities with learner goals, provide opportunities for progress toward those goals, and demonstrate the practical relevance of learning are more likely to foster motivation, persistence, and successful learning outcomes (Ibrahim et al., 2012; O'Neill & Thomson, 2013). Educational administrators are encouraged to design educational programs, teaching and learning strategies, and assessment methods that are congruent with adult learner goals, needs, and beliefs (Ibrahim et al., 2012). By connecting education to personal and professional aspirations, Goal Theory helps explain why purpose-driven learning is often particularly effective for adult learners (O'Neill & Thomson, 2013; Ong, 2014; Schiller, 2022a).

Achievement Motivation Theory

David McClelland's Achievement Motivation Theory proposes that individuals are motivated by varying combinations of three fundamental needs: the need for achievement, the need for affiliation, and the need for power (Aksah et al., 2023; Ilie, 2019; Siok et al., 2023). While all individuals possess each of these needs to some degree, one or more often become dominant and influence behaviour, decision-making, and engagement in learning activities (Chasaya, 2023; Denzinger et al., 2016). The theory suggests that understanding these motivational differences can help explain why adult learners respond differently to educational experiences and workplace development opportunities (Aksah et al., 2023; Ilie, 2019; Kichukova & Taneva, 2021).

Adults with a strong need for achievement are motivated by challenging tasks, opportunities for personal accomplishment, and the ability to demonstrate competence (Mohd Amirul Syafiq bin et al., 2023; Monk, 1996; Siok et al., 2023). These learners typically set high standards for themselves and are driven by a desire to improve performance and achieve success (Kichukova & Taneva, 2021; Monk, 1996). They often seek regular feedback to measure their progress and identify areas for improvement (Aishath et al., 2021). In educational settings, achievement-oriented learners may be particularly motivated by assessments, practical projects, professional certifications, and opportunities to apply new skills in ways that demonstrate their competence and expertise (Kichukova & Taneva, 2021).

Adults with a strong need for affiliation are motivated by social relationships, collaboration, and a sense of belonging (Aksah et al., 2023; Łukasiewicz et al., 2018; Tonnsen et al., 2017). They value positive interactions with others and often enjoy learning environments that encourage teamwork, discussion, and mutual support (Aksah et al., 2023; Brewer & Klein, 2006). For these learners, group activities, peer mentoring, collaborative projects, and communities of practice can enhance motivation by providing opportunities to build relationships and share experiences (Aksah et al., 2023; Brewer & Klein, 2006). Learning is often viewed not only as an individual pursuit but also as a social process that involves cooperation and connection with others (Aksah et al., 2023; Ilie, 2019).

Individuals with a strong need for power are motivated by opportunities to influence others, assume responsibility, exercise leadership, and contribute to organisational or community outcomes (Łukasiewicz et al., 2018; Rahim et al., 2023; Siok et al., 2023). In adult education, these learners may be attracted to leadership development programs, management training, mentoring roles, and learning experiences that prepare them for positions of greater authority or responsibility (Jain & Dutta, 2018; Tonnsen et al., 2017). Their motivation is often strengthened when learning is linked to career advancement, decision-making responsibilities, or opportunities to make a meaningful impact within their workplace or profession (Jain & Dutta, 2018).

Achievement Motivation Theory highlights the diversity of motivations that adult learners bring to educational environments (Aksah et al., 2023; Ilie, 2019). While some learners are driven primarily by personal accomplishment, others are motivated by social connection or opportunities for influence and leadership (Aksah et al., 2023; Ilie, 2019; Łukasiewicz et al., 2018). Recognising these differences allows educators, trainers, and workplace learning professionals to design learning experiences that cater to a range of motivational profiles (Aksah et al., 2023; Kichukova & Taneva, 2021). By incorporating opportunities for achievement, collaboration, and leadership, learning programs can better engage diverse groups of adult learners and support sustained participation and success (Aksah et al., 2023; Kichukova & Taneva, 2021; Siok et al., 2023).

Maslow's Hierarchy of Needs

Abraham Maslow's Hierarchy of Needs remains one of the most widely recognised theories of human motivation and continues to influence discussions of adult learning and development. Maslow proposed that human needs can be organised into a hierarchy, beginning with basic physiological and safety needs and progressing through social belonging and esteem needs to the highest level, self-actualisation (Malinga, 2024; Mensah et al., 2023; Shepherd, 2023; Wang & Poon, 2023). According to the theory, individuals are motivated to satisfy lower-level needs before focusing their attention on higher-order needs associated with personal growth and fulfilment.

At the foundation of the hierarchy are physiological needs, including food, water, shelter, and other requirements necessary for survival. Once these needs are reasonably satisfied, individuals seek safety and security through stable employment, financial security, health, and protection from harm. The next level involves social needs, including relationships, belonging, acceptance, and connection with family, friends, colleagues, and communities. Above these are esteem needs, which relate to self-respect, confidence, achievement, recognition, and the respect of others.

The highest level of the hierarchy, self-actualisation, represents the pursuit of personal growth, fulfilment, and the realisation of one's potential. At this level, individuals seek opportunities to develop their capabilities, achieve meaningful goals, and become the best version of themselves. In educational contexts, self-actualisation may be reflected in adults pursuing learning not solely for external rewards, but to expand their knowledge, develop expertise, explore personal interests, or contribute to society in meaningful ways.

Although aspects of Maslow's hierarchical structure have been questioned by later researchers, the theory remains influential because it highlights the relationship between personal circumstances and motivation. In adult learning, factors such as financial security, workplace stability, social support, and self-confidence can affect an individual's capacity and willingness to engage in education. The theory therefore suggests that effective learning environments should recognise and support the broader needs of adult learners while also providing opportunities for personal growth, achievement, and self-development.

Adult learning may support multiple levels of the hierarchy. Training that improves employability may contribute to financial security (Mensah et al., 2023). Professional qualifications may enhance esteem and recognition (Shepherd, 2023; Wang & Poon, 2023). Advanced study, creative pursuits, and personal development activities may contribute to self-actualisation by helping individuals realise their potential (Mensah et al., 2023; Sirko et al., 2020; Wang & Poon, 2023). Self-actualisation needs include continuous education and training, and the development and improvement of skills (Mensah et al., 2023). Research on recipients of second higher education has found that self-realisation and the realisation of potential opportunities are the primary motivational factors, corresponding to higher levels of needs in the hierarchy (Sirko et al., 2020). Similarly, higher education music teachers have demonstrated motivation toward lifelong learning and enhancing professional skills as expressions of self-actualisation (Wang & Poon, 2023).

Although the hierarchical structure has been criticised (Mensah et al., 2023), the theory highlights the importance of considering learners' broader personal circumstances and needs. Maslow's hierarchy lacks robust empirical validation as a strict sequential model (Malinga, 2024), and some research has found that individuals can simultaneously aspire to fulfil self-actualisation needs without following the particular sequence Maslow proposed (Malinga, n.d.). Other studies have suggested revising the hierarchy to fewer levels, as certain needs may be superfluous (Mensah et al., 2023). Nevertheless, the theory remains valuable as a framework for understanding learner motivation and identifying factors contributing to successful educational outcomes (Malinga, 2024; Shepherd, 2023).

Transformative Learning and Motivation

Transformative learning theory offers another perspective on motivation. Adults are often motivated to learn when experiences challenge existing assumptions, create uncertainty, or reveal gaps in knowledge. As Mezirow's theory posits, transformative learning involves a process of recognising and questioning personal frames of reference, which in turn stimulates changes in how individuals perceive and interact with the world (Prieto & Rojas, 2024). This transformation typically arises from critical dialogue and represents a deep level of learning that requires an act of consciousness on the part of the learner (Cela-Ranilla et al., 2017; Guzón et al., 2021). Significant life events such as career changes, promotions, migration, parenthood, military service, or organisational restructuring may stimulate learning by creating a need to adapt to new circumstances. Adults require "unlearning" simultaneously with learning to achieve transformational learning that allows them not only to acquire new knowledge but also to stimulate behavioural modification in light of reviewing their own frames of reference (Contreras, 2006). Experiences that confront individuals with situations outside their comfort zone can generate the "disorienting dilemma" that is central to transformative learning, compelling learners to adapt their mental schemas to a reality they cannot explain through existing perspectives.

In these situations, motivation emerges not simply from external rewards but from a desire to understand, cope with, and respond effectively to changing realities. This contrasts with extrinsic motivation; as research on transformative education notes, intrinsically motivated individuals care deeply about their inquiry and orient themselves toward making a real contribution (García-Romero & Lalueza, 2019). The transformative learning process challenges people to reflect critically on past experiences, leading to changes in feelings, beliefs, and attitudes. Adults are ready to learn when they see that what they will learn helps solve problems in their context, and the factors key to transformative learning comprise experience, critical reflection, and development. In this sense, transformative learning is understood as qualitatively different from everyday learning, set in motion by processes of cognitive dissonance when individuals encounter experiences that their existing frameworks cannot adequately explain.

Workplace Motivation and Adult Learning

In workplace settings, motivation is influenced not only by individual characteristics but also by a range of organisational factors. Opportunities for career progression, recognition and rewards, job security, leadership support, workplace culture, access to meaningful work, and opportunities for autonomy and decision-making can all shape an employee's willingness to engage in learning and development activities (Kittel et al., 2021; Ryu & Moon, 2019; Song et al., 2018). When employees perceive that their organisation values their contribution and invests in their professional growth, they are generally more likely to participate actively in learning opportunities (Abid et al., 2021; Ryu & Moon, 2019).

Career development opportunities are particularly significant motivators for many adult learners. Employees are often more willing to undertake training when they believe it will enhance their skills, improve their performance, or increase their prospects for promotion and advancement (Nafukho et al., 2017; Rijn et al., 2013). Career motivation has been shown to be positively related to informal learning activities such as keeping up-to-date, seeking feedback, and knowledge sharing (Rijn et al., 2013). Similarly, recognition for learning achievements, whether through formal qualifications, professional accreditation, salary progression, or acknowledgement from supervisors and peers, can reinforce motivation and encourage ongoing participation in professional development (Reichborn-Kjennerud, 2021; Singh, 2016). Research indicates that intrinsic motivators such as respect, recognition, trust, and autonomy play a particularly significant role in employee motivation (Singh, 2016).

Workplace culture also plays a critical role in shaping learning motivation. Organisations that encourage curiosity, innovation, knowledge sharing, and continuous improvement tend to create environments in which learning is viewed as a valued and expected part of professional practice (Gautam & Basnet, 2020; Kittel et al., 2021; Reichborn-Kjennerud, 2021). Organisational learning culture has been found to be connected with mastery-approach goal orientation, which in turn relates to self-regulated learning strategies (Kittel et al., 2021). Organisational culture dimensions such as job challenge, communication, trust, innovation, and social cohesion have been identified as essential for motivating employees toward training transfer (Gautam & Basnet, 2020). In contrast, workplaces that discourage questioning, penalise mistakes, or provide limited support for development may reduce employees' willingness to engage in learning activities (Abid et al., 2021; Reichborn-Kjennerud, 2021). Supportive leadership is particularly important, as managers and supervisors often influence whether employees feel encouraged to pursue new skills and apply what they have learned (Dewayani & Ferdinand, 2019; Na-Nan et al., 2017; Taheri et al., 2021). Research has demonstrated that supervisor support, peer support, and management support all positively influence motivation to transfer learning and the actual transfer of training to the workplace (Dewayani & Ferdinand, 2019; Na-Nan et al., 2017; Song et al., 2018).

The nature of the work itself can also affect motivation. Employees are generally more motivated to learn when they can see a clear connection between learning activities and their day-to-day responsibilities (Nafukho et al., 2017; Reichborn-Kjennerud, 2021). Training efficiency and relevance have been found to be critical factors in the transfer of learning, with employees more likely to apply knowledge when it is perceived as directly applicable to their work (Nafukho et al., 2017). Employees working in environments that prioritised competence development perceived training programmes to be better organised and more relevant to their work tasks (Reichborn-Kjennerud, 2021). Job characteristics such as task autonomy have been shown to have the greatest effect on adapting to organisational contexts (Song et al., 2018), while autonomy and feedback are related to self-efficacy, which in turn supports self-regulated learning strategies (Kittel et al., 2021). Learning that helps solve workplace problems, improve efficiency, enhance safety, or increase professional competence is often perceived as more valuable than learning that appears disconnected from practical workplace needs (Fatima, 2021; Nafukho et al., 2017). Opportunities to exercise autonomy, make decisions, and take responsibility for work outcomes can further strengthen motivation by fostering a sense of ownership and professional growth (Kittel et al., 2021; Torresan & Hinterhuber, 2023).

Employees are therefore more likely to participate in learning when they recognise clear benefits for job performance, professional development, and career advancement (Nafukho et al., 2017; Rijn et al., 2013; Ryu & Moon, 2019). Organisations that foster strong learning cultures, encourage experimentation and innovation, provide constructive feedback, and actively support employee development tend to generate higher levels of learning motivation (Gautam & Basnet, 2020; Kittel et al., 2021; Taheri et al., 2021). Workplace fun, for instance, has been found to have a positive indirect effect on informal learning by providing varied opportunities to learn, with management support mediating the relationship between workplace fun and learning opportunities (Taheri et al., 2021). By creating environments that value continuous learning and growth, organisations can enhance both individual performance and overall organisational capability (Fatima, 2021; Ryu & Moon, 2019; Torresan & Hinterhuber, 2023).

Applications in Adult Education

Motivation theories collectively demonstrate that adult learners are motivated by a combination of personal, professional, social, and psychological factors. Adults are more likely to engage in learning when it is relevant, meaningful, achievable, and connected to their goals and experiences. Effective adult education therefore requires more than the delivery of content; it requires learning environments that support autonomy, competence, relevance, achievement, reflection, and personal growth.

Understanding motivation helps educators design learning experiences that not only attract adult learners but also sustain engagement, persistence, and lifelong learning. As a result, motivation remains one of the most significant factors influencing participation, achievement, and success in adult education.

Self-Efficacy and Confidence in Adult Learning

Self-efficacy and confidence are important concepts in adult learning because they influence whether individuals choose to engage in learning, how much effort they invest, how they respond to challenges, and whether they persist when difficulties arise (Broek et al., 2023; Hammond & Feinstein, 2005). Although the terms are often used interchangeably, they have distinct meanings. Confidence generally refers to a person's overall belief in their abilities, whereas self-efficacy refers to an individual's belief that they can successfully perform a specific task or achieve a particular goal (Malureanu et al., 2021).

The concept of self-efficacy was developed by Albert Bandura as part of Social Cognitive Theory (Bi, 2023; Seevaratnam et al., 2023). Bandura (1977)argued that people's beliefs about their capabilities strongly influence their behaviour, motivation, and performance (Edokpolor & Omiunu, 2017; Studdard et al., 2013). Individuals with high self-efficacy are more likely to view difficult tasks as challenges to be mastered rather than threats to be avoided (Studdard et al., 2013). They are generally willing to attempt new activities, persist when faced with setbacks, and recover more quickly from failure (Bi, 2023). Conversely, individuals with low self-efficacy may doubt their abilities, avoid challenging tasks, and give up more easily when obstacles arise (Burnham & Ponton, 2021; Studdard et al., 2013).

In adult education, self-efficacy plays a significant role in determining participation and achievement (Hammond & Feinstein, 2005; Rothes et al., 2016). Many adult learners return to education after long periods away from formal study and may question their ability to succeed (Hammond & Feinstein, 2005). Concerns about academic skills, technology use, literacy, numeracy, assessment requirements, or balancing study with work and family responsibilities can affect their willingness to engage (Broek et al., 2023; Burnham & Ponton, 2021). Learners who believe they can successfully complete learning tasks are generally more motivated and more likely to persist than those who doubt their capabilities (Hammond & Feinstein, 2005; Reichenberg, 2021; Rothes et al., 2016).

Bandura identified four primary sources of self-efficacy (Bi, 2023; Elder, 2023; Howardson & Behrend, 2015; Mohammadi & Asadzadeh, 2011; Noel-Weiss et al., 2006). The first and most influential is mastery experiences, which involve successfully completing tasks or overcoming challenges (Bi, 2023; Elder, 2023),. Each successful learning experience strengthens a learner's belief in their ability to succeed in future situations (Jameson, 2019; Jameson & Fusco, 2014). For example, an adult learner who successfully completes an introductory computing course may become more confident about undertaking more advanced digital skills training (Calaguas & Consunji, 2022; Hyden & Cherrstrom, 2025).

The second source is vicarious experience, which occurs when individuals observe others similar to themselves successfully performing a task (Bi, 2023; Elder, 2023). Seeing peers overcome challenges and achieve success can increase a learner's belief that they too are capable of succeeding (Bi, 2023). In adult learning environments, peer mentoring, workplace coaching, and collaborative learning activities often provide opportunities for this type of learning and encouragement (Chauvin et al., 2020).

The third source is social persuasion, which involves encouragement, feedback, and support from instructors, supervisors, mentors, colleagues, or family members (Bi, 2023; Elder, 2023; Howardson & Behrend, 2015). Positive feedback can help learners develop confidence in their abilities and encourage them to persevere when facing difficulties (Hyden & Cherrstrom, 2025). Constructive guidance from educators can be particularly important for adults who may have had negative educational experiences in the past (Jameson, 2019; O'Neill & Thomson, 2013).

The fourth source is physiological and emotional states (Bi, 2023; Elder, 2023; Mohammadi & Asadzadeh, 2011; Noel-Weiss et al., 2006). Feelings such as anxiety, stress, fatigue, or fear can influence perceptions of competence (Huang et al., 2020; Jameson, 2019; Jameson & Fusco, 2014). Adult learners who experience high levels of stress may interpret these feelings as evidence that they are incapable of succeeding (Calaguas & Consunji, 2022; Jameson & Fusco, 2014). Conversely, positive emotional experiences and supportive learning environments can strengthen self-efficacy and encourage continued participation (Huang et al., 2020; O'Neill & Thomson, 2013).

Confidence is closely related to self-efficacy but is generally broader in scope. A learner may possess confidence in their overall abilities while still lacking self-efficacy in a specific area. For example, an experienced manager may be highly confident in leading teams but have low self-efficacy regarding the use of advanced data analysis software. Similarly, a skilled tradesperson may be confident in their practical abilities but uncertain about their capacity to succeed in academic study. Understanding this distinction helps educators identify areas where learners may require additional support and encouragement.

Self-efficacy and confidence have important implications for teaching practice. Adult educators can strengthen learners' self-beliefs by providing achievable challenges, clear instructions, constructive feedback, opportunities for success, and supportive learning environments. Breaking complex tasks into manageable stages, recognising progress, encouraging reflection on achievements, and facilitating peer support can all contribute to the development of stronger self-efficacy. Learning activities that allow adults to apply knowledge successfully in real-world contexts can further reinforce beliefs in their capabilities.

In workplace learning, self-efficacy influences employee willingness to undertake new responsibilities, adopt new technologies, participate in professional development, and respond to organisational change. Employees with high self-efficacy are generally more adaptable, resilient, and willing to engage in continuous learning. As workplaces become increasingly complex and technology-driven, fostering self-efficacy has become an important component of workforce development and lifelong learning.

Persistence and Learner Retention

Persistence and learner retention are critical concepts in adult education because they influence whether learners successfully complete their educational goals. While often used together, the two terms describe different aspects of participation. Persistence refers to a learner's ability and determination to continue progressing towards their educational objectives despite obstacles, interruptions, or challenges. Learner retention refers to an institution's ability to keep learners enrolled and engaged until they complete a course, program, or qualification. Together, these concepts are closely linked to learner success, educational quality, and workforce development outcomes.

Adult learners frequently face challenges that can affect their ability to persist in education. Unlike many younger students, adults often balance study with employment, family responsibilities, financial commitments, community obligations, and personal circumstances (Deng & Yang, 2021; Elkachradi et al., 2023; M. Wu et al., 2025). Unexpected events such as changes in employment, illness, caregiving responsibilities, or financial pressures can disrupt participation and increase the likelihood of withdrawal (Deng & Yang, 2021; M. Wu et al., 2025). Research on adult participation barriers has categorized these as situational barriers (such as cultural barriers, lack of time, and family responsibilities), institutional barriers (such as lack of infrastructure and high costs), and dispositional barriers (such as attitudes and willingness to participate) (M. Wu et al., 2025). As a result, persistence in adult learning often requires significant commitment, resilience, and effective support systems.

Research has shown that persistence is influenced by a combination of personal, social, academic, and institutional factors. Learners who have clear goals, strong motivation, positive self-efficacy, and confidence in their abilities are generally more likely to persist when faced with challenges (M. Wu et al., 2025). Some scholars consider adult learning motivation a prerequisite for individuals to sustain their engagement in lifelong learning, highlighting that adult learners with high motivation and a desire for success are more inclined to actively participate in learning activities (M. Wu et al., 2025). Goal-oriented learners often view difficulties as temporary obstacles rather than reasons to abandon their studies. Their commitment to achieving personal, professional, or educational aspirations provides a powerful source of motivation that supports continued engagement.

Social support also plays an important role in learner persistence. Encouragement from family members, friends, employers, colleagues, mentors, and fellow learners can help individuals manage competing demands and remain committed to their studies (Deng & Yang, 2021; Elkachradi et al., 2023). Learning environments that foster a sense of belonging and connection often contribute to higher levels of persistence because learners feel supported and valued. Conversely, feelings of isolation, disconnection, or lack of support can increase the likelihood of disengagement and withdrawal (Deng & Yang, 2021). Prior studies have found that a dearth of connections and support has become a significant stressor, especially when learners face the challenge of effectively navigating their educational environment (Deng & Yang, 2021).

Academic factors are equally important. Adult learners are more likely to persist when learning experiences are relevant, meaningful, and aligned with their goals (Elkachradi et al., 2023; M. Wu et al., 2025). Clear expectations, accessible learning materials, effective teaching practices, timely feedback, and opportunities for success help learners build confidence and maintain motivation. Learning motivation, learning obstacles, and learning engagement have been identified as independent factors influencing training effectiveness in adult learning contexts (M. Wu et al., 2025). Learning activities that allow adults to apply knowledge to real-world situations can reinforce the value of continued participation and strengthen commitment to learning (Elkachradi et al., 2023).

Institutional practices also have a significant influence on retention. Educational providers that offer flexible delivery options, responsive student support services, academic advising, mentoring programs, and accessible learning technologies are often more successful in retaining adult learners (Gibbs et al., 2023; Lynch & Lungrin, 2018). Flexibility is particularly important because adult learners frequently need to balance study with work and family responsibilities (Levy, 2017). Online learning, blended delivery, flexible assessment arrangements, and recognition of prior learning can help reduce barriers to participation and improve retention rates (Gibbs et al., 2023; Levy, 2017).

Several theoretical models have been developed to explain learner persistence and retention. Vincent Tinto's model of student retention emphasises the importance of academic and social integration, suggesting that learners are more likely to remain engaged when they feel connected to both the educational environment and the learning community (Lichtenstein et al., 2014). John Bean and Barbara Metzner later adapted retention theory to better reflect the experiences of non-traditional and adult learners, highlighting the influence of external factors such as employment, finances, family responsibilities, and life circumstances (Gibbs et al., 2023). Their work recognised that adult learners often make decisions about continuing education based on factors beyond the institution's direct control.

Persistence is particularly important in lifelong learning because adults frequently engage in multiple episodes of learning throughout their lives. Educational journeys may not always follow a continuous path. Some learners may temporarily pause their studies due to personal circumstances before returning later to continue their education (Gibbs et al., 2023) . From this perspective, persistence can be viewed not simply as continuous enrolment but as ongoing commitment to achieving educational goals over time, even when interruptions occur (Kuchumova et al., 2024).

In workplace learning, persistence is closely linked to professional development and career progression (Johnson et al., 2020; Kuchumova et al., 2024). Employees who persist with training programs, skill development initiatives, and continuing professional education are often better positioned to adapt to technological change, industry developments, and evolving workplace requirements (Efendi et al., 2025; Johnson et al., 2020). Organisations that support employee learning through mentoring, coaching, study assistance, and professional development opportunities can help foster persistence and improve learning outcomes (Jackson & Manley, 2021; Kuchumova et al., 2024).

Strategies for improving persistence and retention often focus on strengthening learner motivation, self-efficacy, and engagement while reducing barriers to participation (Engler et al., 2025; Lichtenstein et al., 2014). Effective approaches include providing orientation programs, establishing realistic expectations, offering academic and personal support services, creating opportunities for peer interaction, recognising learner achievements, and ensuring learning remains relevant to individual goals (Gibbs et al., 2023; Ranalli & Chappell, 2017). Early intervention when learners show signs of disengagement can also improve retention outcomes (Gibbs et al., 2023).

Ultimately, persistence and learner retention are essential for successful adult education. They reflect the complex interaction between learner motivation, personal circumstances, social support, educational quality, and institutional practices (Gibbs et al., 2023; Kuchumova et al., 2024). By understanding the factors that influence persistence and retention, educators and organisations can create learning environments that support adult learners in achieving their educational, professional, and personal goals (Oosten et al., 2017; Ranalli & Chappell, 2017).

Barriers to Participation in Adult Learning

Participation is a fundamental requirement for learning, yet many adults face obstacles that limit their ability to engage in education and training. Barriers to participation refer to the factors that prevent, discourage, delay, or interrupt involvement in learning activities. Understanding these barriers is important for educators, training providers, employers, and policymakers because reducing barriers can improve access, equity, learner retention, and educational outcomes. Adult learners often encounter multiple barriers simultaneously, and the impact of these barriers can vary depending on an individual's circumstances, life stage, and learning goals.

One of the most widely recognised frameworks for understanding barriers to participation was developed by Patricia Cross (1981), who categorised barriers as situational, institutional, and dispositional. This framework continues to be influential because it recognises that participation is shaped by both external circumstances and individual perceptions.

Situational Barriers

Situational barriers arise from an individual's personal circumstances and life responsibilities. These barriers are often beyond the learner's immediate control and are particularly common among adult learners who must balance education with other commitments. Employment obligations, family responsibilities, financial pressures, health concerns, and caregiving duties can all reduce the time, energy, and resources available for learning (Preussler et al., 2025; Swartzell et al., 2023). Unlike younger learners who may have fewer competing demands, adults frequently need to manage multiple roles simultaneously, making participation in education more challenging.

For example, a parent caring for young children may find it difficult to attend scheduled classes, while a full-time employee may struggle to participate in training that occurs during working hours. Financial challenges can further limit participation when learners are required to pay tuition fees, purchase learning materials, travel to educational facilities, or reduce paid working hours to accommodate study commitments (Gaster & Largent, 2024; Winslow, 2003). In many cases, the direct and indirect costs associated with education can create significant obstacles, particularly for learners with limited financial resources.

Research has also highlighted the impact of caregiving responsibilities on participation in education and other developmental activities. Caregivers often experience challenges such as limited family support, financial strain, difficulties accessing support services, emotional exhaustion, and the demands associated with caring for individuals with complex health needs (Uribe, 2022). These responsibilities can significantly reduce the time and capacity available for learning. Adult children caring for dependent older relatives, for example, frequently balance employment, family responsibilities, and caregiving obligations, creating substantial pressures on their personal time and wellbeing (Aaron, 2014). Caregivers may also face reduced income, increased healthcare expenses, transportation difficulties, housing pressures, and other practical challenges that further restrict their ability to engage in educational opportunities.

Situational barriers may be particularly pronounced among vulnerable or disadvantaged populations. Financial strain can be severe for some caregiving groups, with caregiving expenses consuming a substantial proportion of household income (Mage et al., 2025). Older adults may encounter additional challenges related to limited educational backgrounds, reduced access to support networks, restricted availability of services, or financial limitations that affect their capacity to prepare for and participate in learning opportunities. Geographic isolation, limited access to community resources, and a lack of available support services can further compound these difficulties.

Because situational barriers directly affect an individual's practical ability to participate, they are often regarded as some of the most significant obstacles to adult learning. Even highly motivated learners may struggle to engage in education when faced with competing demands on their time, finances, health, and personal responsibilities. Consequently, understanding and addressing situational barriers is essential for creating accessible and inclusive adult learning opportunities that accommodate the realities of adult learners' lives.

Institutional Barriers

Institutional barriers are created by the structures, policies, procedures, and practices of educational providers or training organisations (Ariwijaya & Ningsih, 2020; Jackson, 2014; Ostendorff, 2023). These barriers can make participation more difficult even when learners are highly motivated to engage in learning (Nieuwenhove & Wever, 2023; Surikova, 2024).

Examples include inflexible timetables, limited course availability, complex enrolment procedures, inaccessible learning materials, inadequate learner support services, restrictive admission requirements, and lack of recognition for prior learning. Specifically, institutional barriers have been described as encompassing "lack of services, lack of support groups, and the occurrence of those events, activities, or procedures and policies that unduly blocked access from admission to graduation" (Jackson, 2014, p. 46). Such barriers include lack of night and online classes, limited availability of faculty, and difficulties in reaching academic support (Ariwijaya & Ningsih, 2020). Further examples include inconvenient schedules, locations, and tuition fees, as well as issues navigating campus enrolment or financial systems and difficulty finding classes that fit adult learners' schedules (Ostendorff, 2023). Complex enrolment procedures and unclear policies have also been identified as significant institutional obstacles (Darko-Asumadu, 2023; Ostendorff, 2023).

The lack of recognition for prior learning represents another institutional barrier, with various methods being employed by higher education institutions to evaluate learning that has occurred outside traditional academic environments (Hearn & Long, 2023; Meghnagi & Tuccio, 2022). Adult learners may also encounter difficulties when programs are designed primarily for younger full-time students rather than individuals balancing multiple responsibilities (Ostendorff, 2023; Sualehi, 2023). As Ostendorff notes, institutional barriers "stem from the idea that adult learners are not like students fresh out of high school, and as such, the rules, which were written for those students, do not always fit an older student's needs" (Ostendorff, 2023, p. 42). Institutional support services with traditional business hours are typically difficult for adult learners to access (Ostendorff, 2023), and practices that discourage working professionals from participating in educational activities remain prevalent.

The increasing availability of online and blended learning has helped reduce some institutional barriers (Hearn & Long, 2023; Ostendorff, 2023), although access to technology and digital skills can create new challenges for some learners (Ngozwana, 2025; Sualehi, 2023). Online and hybrid degree programs appeal to adult learners because they reduce barriers of time and finances (Hearn & Long, 2023), and distance learning provides accessibility for those who may not manage brick-and-mortar classes due to family life, employment, or commuting distance (Ostendorff, 2023). However, online study requires study skills, self-discipline, and computer competence, which the most disadvantaged adult students are less likely to possess. Working learners from digitally disconnected communities face heightened institutional barriers even when digital literacy programming is offered (Sualehi, 2023), and learners with disabilities have experienced particular challenges with the shift to online learning, including lack of devices, inadequate technology skills, and unstable internet connectivity (Ngozwana, 2025). Furthermore, the shift to online learning during the COVID-19 pandemic intensified inequalities between participants, with some adults facing more constraints due to reduced supply of educational opportunities (Kalenda et al., 2023). Older learners and those with lower levels of education may also face challenges in accepting and using smart devices and online learning applications (M. Wu et al., 2025).

Dispositional Barriers

Dispositional barriers relate to an individual's attitudes, beliefs, perceptions, and self-concept. These barriers are often shaped by previous educational experiences, cultural expectations, and personal confidence levels (Porras-Hernández & Salinas-Amescua, 2011). Adults who had negative experiences at school may doubt their academic abilities or fear failure (Porras-Hernández & Salinas-Amescua, 2011). Others may question whether they are capable of learning new skills, particularly if they have been away from formal education for many years (Porras-Hernández & Salinas-Amescua, 2011).

Recent studies of adult participation in education have identified a range of factors that influence whether individuals choose to engage in learning opportunities. These factors include awareness of available educational options, personal motivation and goals, prior levels of educational attainment, previous learning experiences, and confidence in one's ability to succeed as a learner (Porras-Hernández & Salinas-Amescua, 2011). Adults who have experienced positive educational outcomes in the past are often more willing to re-engage in learning, whereas those who have encountered difficulties or negative experiences may be less likely to participate. Similarly, individuals who are unaware of available programs or who perceive limited value in education may be less inclined to enrol.

A particularly important factor influencing participation is self-efficacy, which refers to an individual's belief in their capacity to successfully perform the actions necessary to achieve desired outcomes (Porras-Hernández & Salinas-Amescua, 2011). Self-efficacy affects how people approach challenges, the effort they invest in learning, their persistence when difficulties arise, and their expectations of success. Adults with strong self-efficacy are generally more willing to undertake new learning experiences because they believe they possess the capability to overcome obstacles and achieve their goals.

The influence of self-efficacy extends beyond individual learning behaviour and can also operate within groups and communities. Individuals who believe in their collective ability to solve problems and achieve shared objectives are often more likely to participate in collaborative learning activities and community-based educational initiatives. Consequently, self-efficacy is considered a key factor in understanding adult participation, motivation, persistence, and achievement in educational settings.

These findings highlight the importance of fostering positive learning experiences, building learner confidence, and increasing awareness of educational opportunities. By addressing factors such as self-belief, motivation, and previous educational experiences, educators and institutions can help reduce barriers to participation and encourage greater engagement in lifelong learning.

Low self-confidence, fear of assessment, anxiety about technology, concerns about fitting in with younger students, and uncertainty about the value of education can all reduce participation. Research has demonstrated that computer anxiety, computer self-efficacy, and cognitive abilities are important predictors of technology adoption, with older adults being less likely to use technology, and the relationship between age and technology adoption being mediated by computer self-efficacy and computer anxiety (Czaja et al., 2006). Attitudinal obstacles in educational contexts include "anxiety of shame in front of classmates, percipience that technology does not improve learning change, percipience of computer as being difficult to use, lack of self-confidence in using technology, and undesirable experiences with using ICT in the past" (Mirzajani et al., 2015, p. 28). Undesirable past experiences affect perceptions of the importance of technology and ease of use, "decrease confidence and enhance anxiety" (Mirzajani et al., 2015). Dispositional barriers are particularly important because they can prevent individuals from enrolling in learning opportunities even when practical barriers are minimal. Indeed, research with poorly educated rural women in Mexico found that despite positive self-perceptions, sociocultural factors rather than dispositional barriers alone explained nonparticipation, demonstrating the complex interplay between dispositional and contextual factors (Porras-Hernández & Salinas-Amescua, 2011). Positive learning experiences, supportive educators, and early successes can help reduce these barriers by building confidence and self-efficacy. As Bandura's social cognitive theory supports, "strong perceived self-efficacy in the learning context leads the learner to self-regulate consistently at the affective, behavioural, and cognitive levels to improve learning conditions and performance" (Nouioua, 2018), and high levels of self-efficacy are associated with greater engagement and persistence in educational activities (Ann, 2023).

Financial Barriers

Financial barriers are frequently identified as a major obstacle to adult learning participation (Noya, 2018; OECD, 2018b). Tuition fees, textbooks, equipment, technology requirements, travel expenses, and lost income associated with study can create significant challenges (Noya, 2018). The financial constraints involved in pursuing education, including the lack of dedicated financial support to cover fees and other expenses, have been noted as significant obstacles to adult participation (Noya, 2018; OECD, 2018b). For many adults, particularly those supporting families or working in low-income occupations, the perceived cost of education may outweigh the anticipated benefits. Tuition fees may affect adult learners disproportionately, for example by worsening the opportunity cost of learning, and adults may also have limited access to financial aid and have less time to capture the returns in the labour market (Noya, 2018; OECD, 2018b).

Financial concerns can also influence decisions about the type of education pursued, the number of subjects undertaken, and whether learners persist to completion. Increasing tuition and private provision of education raise the issue of equity for students and workers who cannot afford the higher fees. Scholarships, government subsidies, employer-sponsored training, and flexible payment arrangements can help reduce the impact of financial barriers and improve access to learning opportunities (OECD, 2018b). Financial support for individuals may be needed to help address cost barriers (Noya, 2018; OECD, 2018b). Targeted financial aid to support adult learners, including grants, scholarships, tax subsidies, and student loans, has been recommended as a means to improve participation (Noya, 2018). Federal and state grant-aid programs, institutional aid, and private scholarships can further reduce the tuition and fee gap for adults in low-income situations.

Time Constraints

A lack of time is one of the most commonly reported barriers among adult learners (Catyong & Garciano, 2023; Darko-Asumadu, 2023; Omar & Rashid, 2019; M. Wu et al., 2025). Adults frequently juggle multiple responsibilities, including employment, family commitments, community involvement, and personal obligations (Barker et al., 2023; Darko-Asumadu, 2023; Omar & Rashid, 2019). Finding sufficient time for classes, study, assessments, and independent learning can be difficult (Lukács et al., 2023; Omar & Rashid, 2019; Shousha, 2021a).

Time constraints are particularly evident in professional education, where employees may be required to undertake training while maintaining full workloads (Ememe, 2020; Omar & Rashid, 2019; M. Wu et al., 2025). Flexible delivery models, self-paced learning, online education, and workplace-based learning arrangements can help address time-related barriers by allowing learners greater control over when and how they engage in study (Ememe, 2020; Lukács et al., 2023; Luu, 2021; Żammit, 2024).

Technological Barriers

As education becomes increasingly digital, technological barriers have become more significant. Limited access to computers, internet connectivity, software, or digital devices can restrict participation, particularly in remote or disadvantaged communities (Medhi & Baruah, 2025; Sualehi, 2023). The first digital divide concerns basic access: adult learners need the right hardware, software, and connectivity to develop basic digital literacy skills (Sualehi, 2023). Disproportionate access to information and communication technology by sub-populations such as older adults and adults with lower socioeconomic status has been a significant concern (Tamashita et al., 2024), and causes of the digital divide include education, income levels, geographical restrictions, motivation, and digital literacy (Jun, 2021). Some adult learners may also lack confidence in using technology or have limited digital literacy skills (Adomako et al., 2025; Sualehi, 2023). A "digital divide" exists characterized by lower levels of digital literacy and higher technology-use anxiety (McGinty et al., 2024), and individuals with lower education and income are less able to afford digital tools and may lack digital literacy and self-efficacy to use them.

These challenges became particularly evident during the widespread adoption of online learning during the COVID-19 pandemic. The COVID-19 pandemic induced digital exclusion and amplified the healthcare divide and poor mental health due to the worsening digital divide (Jinhyun Kim et al., 2023). During the pandemic, digital health literacy was considered a crucial ability to gain correct electronic health information and receive distance health care (Junghwan Kim et al., 2023). The pandemic transformed healthcare delivery with the rapid adoption of telehealth and digital technologies, creating a need to ensure that underserved communities do not face new technology-driven healthcare disparities (Rousseau et al., 2024). While technology can increase flexibility and access, it can also create new forms of exclusion if learners do not have the resources or skills required to participate effectively (Nguyen, 2020; Sualehi, 2023). The digital divide encompasses disparities in access to technology, internet connectivity, and digital literacy, and remains a significant barrier to equity, with older adults, individuals with lower income or educational attainment, and minoritized populations disproportionately affected (Adomako et al., 2025). Digitally disconnected working learners cannot gain the necessary skills to fully participate in 21st-century knowledge economies when trying to cross the first and second digital literacy divides (Sualehi, 2023).

Cultural and Social Barriers

Cultural and social factors can also influence participation in adult learning. Language barriers, cultural expectations, discrimination, social stereotypes, and lack of family or community support may affect learners' willingness or ability to engage in education (Al-Salem, 2020; Gravani et al., 2019; Obradović-Ratković et al., 2020). Immigrants and migrants commonly face "unexpected obstacles of poverty, discrimination, language, ambiguous immigration or legal status" that constrain their participation (Vaughn et al., 2009, p. 68). Language barriers in particular have been identified as preventing informed choices about education and other important areas (Hébert, 2016), while discrimination and social exclusion foster perceptions of intellectual inferiority and social insignificance among migrant learners (Maralani & Pfeiffer, 2024). Some individuals may come from backgrounds where participation in formal education was limited or where educational opportunities were not equally available (McBrien, 2022; Obradović-Ratković et al., 2020). Many refugee learners face challenges accessing education due to lack of academic transcripts, identity documents, insufficient funds, and lack of language skills required for engagement in educational institutions (Obradović-Ratković et al., 2020). Newcomer and refugee students are frequently marginalised due to their language, religion, ethnicity and culture (McBrien, 2022).

For migrant learners, refugees, Indigenous learners, and other underrepresented groups, cultural barriers may interact with financial, linguistic, and institutional challenges (Marcos et al., 2023;

Obradović-Ratković et al., 2020; Varsik & Gorochovskij, 2023). Refugees often encounter numerous barriers that can restrict access to education and training opportunities. These challenges may include limited access to information about available educational pathways, difficulties understanding admission and qualification requirements, challenges obtaining professional support services, and obstacles associated with the recognition and assessment of prior qualifications and credentials (Obradović-Ratković et al., 2020). Financial constraints can further limit participation, particularly when refugees face costs associated with tuition, learning materials, transportation, or the process of having overseas qualifications evaluated.

In addition to educational barriers, refugees frequently navigate complex immigration and settlement processes that can affect their capacity to engage in learning. Language differences, unfamiliarity with educational systems, employment pressures, housing insecurity, and limited social networks may further complicate participation. These challenges often interact with broader systemic factors, creating multiple layers of disadvantage that can make educational access and progression more difficult. As a result, refugees may require targeted information, guidance, financial support, and culturally responsive educational services to facilitate successful participation in learning and workforce development opportunities.

Indigenous students have been found more likely to be assessed as having learning difficulties, with their linguistic differences or effects of trauma interpreted by teachers as lack of ability (Tanyu et al., 2020). For LGBTQI+ refugees, cultural and language barriers may compound discrimination from both host and native cultures (Varsik & Gorochovskij, 2023). These intersecting challenges are intensified by cultural differences and discrimination that impede full engagement with educational environments (Young et al., 2024). Inclusive educational practices, culturally responsive teaching, and targeted support services can help create more equitable learning opportunities (Dediwalage et al., 2025; Gravani et al., 2019; Koukoulidis et al., 2024; Marcos et al., 2023). Culturally responsive pedagogy addresses structural barriers to learning for multicultural students (Dediwalage et al., 2025), while inclusive intercultural education involves adopting a pluralistic perspective that acknowledges diversity and benefits all students (Marcos et al., 2023). The intercultural approach in adult education encourages empowerment of migrant learners through adaptation of pedagogical approaches, provision of programs that stimulate intercultural interactions, and professional training of educators on intercultural competences (Gravani et al., 2019). Targeted support services, including language support, cultural mediators, cohort-based learning communities, and wrap-around services, have proven effective in creating more equitable access (Hamm et al., 2021; Oliveira et al., 2024; Young et al., 2024).

Workplace Barriers

Many adults participate in learning for employment-related reasons, yet workplace conditions can sometimes hinder participation. Lack of employer support, inflexible work schedules, excessive workloads, limited access to professional development opportunities, and organisational cultures that do not value learning can reduce engagement in education and training (Bailey et al., 2026; Jansen et al., 2025; Mlambo et al., 2021). Research has shown that workers most at risk of skill obsolescence are often least likely to engage in work-related learning, partly due to lower perceived access to supportive learning environments and fewer formal organisational resources for development (Jansen et al., 2025). Workload has been identified as a universal barrier to participation, alongside scheduling conflicts and time constraints (Bailey et al., 2026; Mlambo et al., 2021). Furthermore, organisational cultures that do not actively support professional development can leave employees feeling that learning is neither valued nor legitimised (Mlambo et al., 2021). The absence of supervisor support, unclear expectations regarding development, and limited funding for training further compound these barriers (Egetenmeyer & Fedeli, 2017b; Mlambo et al., 2021).

Employees are generally more likely to participate in learning when employers provide time, resources, encouragement, and recognition for professional development (Gerken, 2016; Jansen et al., 2025; Mlambo et al., 2021). Supervisors play a critical role by communicating the importance of ongoing development, acting as role models, and creating norms that encourage learning behaviours such as feedback seeking and information sharing (Gerken, 2016). Organisational support in the form of study time allocation, financial backing, and access to learning platforms can significantly enhance participation (Gerken, 2016; Mlambo et al., 2021; Windisch, 2015). Workplace learning cultures that support continuous improvement and lifelong learning can significantly reduce barriers to participation (Jansen et al., 2025; Windisch, 2015). A supportive learning climate, encompassing learning value of the job, supervisor support, and organisational support, has been positively associated with employability and engagement in both formal and informal learning (Heijden et al., 2009). Benefits of fostering such cultures extend beyond individual development to include improved employee retention, reduced error rates, and enhancement of the overall lifelong learning culture in the workplace (Windisch, 2015). Workers are more likely to engage in work-related learning when the learning culture in an organisation is supportive, and when formal organisational resources such as HR policies, recognition systems, and compensation systems are in place to stimulate learning (Jansen et al., 2025).

Psychological Barriers

Psychological barriers include factors such as fear of failure, anxiety, stress, low motivation, and lack of self-efficacy. Adults who have experienced educational difficulties in the past may worry about their ability to succeed (Goldman & Bradley, 2001; Thomson, 1994). Research has documented that adult learners frequently express concerns about rusty math and writing skills providing an inadequate base for successful college experiences (Thomson, 1994). Overcoming procrastination and public speaking anxiety were found to be the most prevalent personal concerns, with 52 percent and 42 percent of students respectively reporting high or moderate need for help (Thomson, 1994). Others may feel overwhelmed by competing responsibilities or uncertain about returning to study after a long absence (Goldman & Bradley, 2001; Thomson, 1994). Re-entry students, those who have been out of school for three years or more, face particular challenges because of their prolonged absence from academic settings (Thomson, 1994). Excessive demands upon time and poor study skills are also serious causes of concern for large numbers of re-entry students (Goldman & Bradley, 2001).

Psychological barriers can be particularly challenging because they may persist even when practical barriers have been removed. Research on self-efficacy demonstrates that strong self efficacy requires individuals to remain task-oriented when faced with obstacles or failure (Mungania, 2004), and that self-efficacy is inversely related to perceived barriers (Mungania, 2004). Enhancement of self-concept and self-esteem should therefore be a primary concern of institutions that educate adult students (Thomson, 1994). Supportive learning environments, positive feedback, mentoring, counselling services, and opportunities for early success can help learners overcome these challenges and build confidence (Goldman & Bradley, 2001; Thomson, 1994). Educators emphasize the important role instructors play as facilitators of self-concept and self-esteem in adult learners, fostering the development of potential and enhancing opportunities for success (Thomson, 1994). Peer mentoring and tutoring programs aimed at helping students feel needed, supported, and academically able have been shown to support student retention (Goldman & Bradley, 2001). Fennimore (1988) noted that the role of a mentor and the improvement of study skills, including active listening, problem solving, developing motivation, and clarifying goals, are essential for school success (Goldman & Bradley, 2001). Re-entry students themselves have suggested that greater financial, academic, and psychological support would increase both the number of returning students and their success rates (Goldman & Bradley, 2001). Counselling and advising services are also essential, as adult students fulfill many roles and experience stress in balancing work, home, and school responsibilities (Thomson, 1994).

Reducing Barriers to Participation

Addressing barriers to participation requires a learner-centred approach that recognises the diverse circumstances of adult learners (Singh, 2015h). Flexible learning options (Singh, 2015), financial assistance (OECD, 2018b), recognition of prior learning (Singh, 2015b, 2015c), accessible support services (Singh, 2015h), inclusive teaching practices (Singh, 2015h), and responsive educational policies can all improve participation rates (Misc, 2005),. Employers can contribute by supporting workplace learning and providing opportunities for professional development (Singh, 2015b), while educators can foster confidence, relevance, and learner engagement (Singh, 2015c, 2015h)

Flexible delivery approaches suited to the needs of adult learners can help encourage participation across all learning programmes (Singh, 2015b). Institutions of higher and further education are encouraged to provide more flexible, accessible and tailored courses designed with employers to equip students for the workplace (Singh, 2015b, 2015c). The lack of flexibility of delivery has been identified as complicating participation, alongside insufficient provision of dedicated academic support for adult students.

Regarding financial assistance, targeted financial aid to support adult learners may be needed, as tuition fees may affect adult learners disproportionately by worsening the opportunity cost of learning (OECD, 2018b). Adults may have limited access to financial aid and less time to capture returns in the labour market (OECD, 2018b). Research has recommended the creation of expanded financial supports through mechanisms such as the tax system to reduce cost barriers for adult learners and to provide stronger incentives to employers to invest in education and training for their employees (Singh, 2015c, 2015h). Providing targeted financial and non-financial support can improve the access and success of disadvantaged groups, alongside ensuring adequacy of broader support at the local level, such as transport and child care.

Recognition of prior learning (RPL) represents a critical mechanism for widening participation. RPL takes into account formal, non-formal and informal learning, and in the case of informal learning, recognition of existing competences may lead to an increased willingness among employees to take part in workplace training (Singh, 2015b). The portfolio process has been identified as a valuable support for those facing barriers of poverty, low formal schooling, social isolation, and lack of workplace experience (Singh, 2015c). Many universities in Europe are widening participation of non-traditional students by making pedagogical methods used for recognition an integral part of guidance and counselling activities (Singh, 2015c).

The role of employers is equally significant. There is indirect evidence that firms under-invest in workers' skills, including evidence of wage compression and the difficulty workers have internalising the benefits of training (OECD, 2005). Programmes designed from the start to support the participation and success of adult learners can leverage new funding opportunities (OECD, 2018b). Recognition is primarily about learner empowerment leading to personal development, employability and relevant qualifications, and it is in this wider context that lifelong learning has its true meaning and can open up a diversity of learning routes (Singh, 2015h).

Significance for Adult Education

Barriers to participation demonstrate that access to education is shaped by far more than individual motivation or willingness to learn. Adult learners exist within complex social, economic, cultural, and institutional contexts that can either facilitate or constrain their engagement in education and training (Windisch, 2015; M. Wu et al., 2025; Yu, 2023). Factors such as financial resources, family circumstances, workplace demands, cultural expectations, educational policies, and the availability of support services all influence whether adults are able to access and persist in learning opportunities. Recognising these influences enables educators, institutions, employers, and policymakers to develop more accessible and inclusive educational environments that support participation and lifelong learning (Sukri, 2024; Windisch, 2015; Yu, 2023).

A widely recognised framework for understanding barriers to adult learning categorises them as institutional, situational, and dispositional barriers (Shousha, 2021b; M. Wu et al., 2025). Situational barriers arise from an individual's personal circumstances and responsibilities. Common examples include limited time, family commitments, financial pressures, employment obligations, and competing life priorities (Elkachradi et al., 2023; Sukri, 2024; Teoh & Liau, 2023; Windisch, 2015; Yu, 2023). Research consistently identifies time constraints as one of the most significant challenges facing adult learners, as they often need to balance study with work, caregiving, and other responsibilities (Sukri, 2024; Teoh & Liau, 2023). Financial barriers can also be substantial, particularly when the costs of tuition, learning materials, transportation, or lost income create obstacles to participation (Jacobs et al., 2024; Sukri, 2024; Windisch, 2015; Yu, 2023).

Institutional barriers stem from the structures, policies, and practices of educational systems and organisations. These barriers may include inadequate infrastructure, inflexible scheduling arrangements, high costs, limited support services, complex administrative processes, and insufficient promotion of available learning opportunities (M. Wu et al., 2025; Yu, 2023). In some cases, potential learners may be unaware of educational programs because information is not effectively communicated or accessible. Policy and funding arrangements can also contribute to unequal access, particularly when eligibility requirements, program availability, or workforce-focused funding models fail to address the diverse needs of adult learners (Amoud & Raymer, 2025). Institutional barriers can disproportionately affect groups such as migrants, refugees, older adults, and individuals with lower levels of prior educational attainment (Hachem & Formosa, 2023; M. Wu et al., 2025; Καρανικόλα & Παλαιολόγου, 2021).

Dispositional barriers relate to learners' attitudes, beliefs, self-perceptions, and psychological readiness for learning. Many adults experience anxiety about returning to formal education, concerns about their academic abilities, fear of failure, or uncertainty about their capacity to succeed (Shousha, 2021b; Sukri, 2024). Previous negative educational experiences, low self-confidence, and limited self-efficacy can further reduce motivation and willingness to participate. These internal barriers may persist even when practical obstacles are removed, highlighting the importance of supportive learning environments that foster confidence, resilience, and positive learning identities.

The interaction of multiple barriers is particularly evident among disadvantaged and underrepresented populations. Individuals from vulnerable groups often experience overlapping challenges that compound their exclusion from educational opportunities. Refugees and migrants, for example, may face language barriers, difficulties obtaining recognition of prior learning and qualifications, limited access to literacy programs, and reduced access to educational funding and support services (Καρανικόλα & Παλαιολόγου, 2021). Older learners may encounter barriers related to curriculum design, accessibility, language of instruction, and assumptions about age and learning capacity (Hachem & Formosa, 2023). Distance learners can experience additional challenges associated with technology access, internet connectivity, digital literacy, and balancing study with other responsibilities (Teoh & Liau, 2023).

Support systems play an important role in helping adult learners overcome barriers to participation. Family encouragement, peer support, mentoring relationships, and institutional support services can significantly improve learner persistence and success (Karmelita, 2018). Adults who have access to strong support networks are generally better equipped to manage competing demands and cope with the challenges associated with learning. Guidance from educators, academic advisors, mentors, and employers can further assist learners in navigating educational pathways and maintaining motivation during periods of difficulty.

Research has also demonstrated that barriers can directly influence learning outcomes by reducing motivation, engagement, and training effectiveness (M. Wu et al., 2025). Learners who encounter significant obstacles may participate less actively, invest less effort in learning activities, or withdraw before completing their educational goals. Conversely, reducing barriers and strengthening support mechanisms can improve engagement, persistence, and overall educational achievement.

Addressing barriers to participation requires coordinated action at multiple levels. Governments can improve access through supportive policies, funding arrangements, and lifelong learning initiatives. Educational institutions can increase participation by offering flexible delivery options, reducing administrative complexity, providing preparatory and bridging programs, and ensuring that learning opportunities are widely promoted and accessible (Shousha, 2021b; Windisch, 2015). Financial assistance, subsidised education, and targeted support services can help reduce economic barriers, while inclusive teaching practices and learner support programs can address dispositional challenges (Jacobs et al., 2024; Sukri, 2024). The development of a skilled adult education workforce capable of supporting diverse learners is also an important component of improving participation and learner success (Windisch, 2015).

Barriers to participation are not simply individual problems but reflections of broader social, economic, and institutional conditions. Creating equitable access to adult learning requires sustained efforts from governments, educational providers, employers, communities, and learners themselves. By identifying and addressing the factors that restrict participation, educational systems can create pathways that enable more adults to engage in meaningful learning throughout their lives (Amoud & Raymer, 2025; Elkachradi et al., 2023; Saar et al., 2013).

Supporting Disadvantaged Learners

Supporting disadvantaged learners is a central objective of adult education because equitable access to learning contributes to social inclusion, workforce participation, economic development, and lifelong learning (Wilson et al., 2024). Adult learners come from diverse backgrounds and may experience educational disadvantage for a variety of social, economic, cultural, geographic, or personal reasons

(Wilson et al., 2024). Disadvantage does not reflect an individual's ability or motivation to learn; rather, it often results from unequal access to educational opportunities, resources, and support (Wilson et al., 2024). Effective adult education therefore seeks not only to provide access to learning but also to remove barriers that prevent learners from fully participating and succeeding (Wilson et al., 2024).

Disadvantaged learners may include individuals with low socioeconomic status Wilson et al. (2024),,, low levels of prior educational attainment (Shi et al., 2021; Wilson et al., 2024), people living in rural or remote communities,, migrants, refugees, Indigenous peoples, older adults (Shi et al., 2021), people with disabilities (Wilson et al., 2024), unemployed individuals (Wilson et al., 2024), people returning to education after long absences, and those with limited literacy, numeracy, or digital skills (Shi et al., 2021; Wilson et al., 2024). Many learners experience multiple forms of disadvantage simultaneously, creating complex barriers that require coordinated and flexible support (Trewartha et al., 2026). Understanding the diverse circumstances of adult learners enables educators to develop learning environments that are inclusive, responsive, and learner-centred (Wilson et al., 2024).

One of the most effective ways to support disadvantaged learners is by improving access to education. Flexible learning arrangements, including part-time study, evening classes, online and blended learning, workplace-based training, and self-paced programs, allow adults to balance education with employment, family responsibilities, and other commitments (Z. Y. Lim et al., 2024; Roumell & Jabarkhail, 2023; Wilbur & Taylor, 2018). Flexible assessment methods, multiple entry points throughout the year, and recognition of prior learning can further reduce barriers and encourage participation by acknowledging the knowledge and experience that adults already possess (Penland, 2015; Roumell & Jabarkhail, 2023).

Financial support is another important component of equitable adult education. Tuition fees, learning materials, transportation costs, technology requirements, and lost income can prevent adults from participating in education (Chowdhury & Behak, 2021; Leow et al., 2023). Scholarships, government subsidies, employer-funded training, flexible payment arrangements, and access to affordable learning resources can reduce financial barriers and make education more accessible (Leow et al., 2023; Z. Y. Lim et al., 2024; Manditereza & Chamboko-Mpotaringa, 2024). For many learners, financial assistance is the determining factor in whether participation is possible (Ndaimani, 2026).

Inclusive teaching practices are essential for supporting learners with diverse educational backgrounds and learning needs. Adult educators should recognise that learners enter educational programs with varying levels of prior knowledge, confidence, literacy, numeracy, digital capability, and study skills (Flore & Govender, 2021; Roumell & Jabarkhail, 2023). Differentiated instruction allows educators to adapt learning activities, teaching strategies, and assessment methods to accommodate individual needs while maintaining consistent learning outcomes. Providing clear explanations, practical examples, scaffolded learning activities, and multiple opportunities for practice can improve learner confidence and achievement (Flore & Govender, 2021).

Creating supportive learning environments also contributes significantly to learner success. Adults are more likely to engage in learning when they feel respected, valued, and psychologically safe (Flore & Govender, 2021; Norris & Lefrère, 2011). Positive relationships between educators and learners encourage participation, while collaborative learning activities help build peer support and a sense of belonging (Ndaimani, 2026; Norris & Lefrère, 2011). Learning environments that acknowledge diverse experiences and perspectives promote inclusion and encourage learners to contribute their own knowledge and expertise (Flore & Govender, 2021; Roumell & Jabarkhail, 2023).

Developing learners' confidence and self-efficacy is particularly important for adults who may have experienced previous educational failure or long periods away from formal study (Mahlangu, 2018; Roumell & Jabarkhail, 2023). Many disadvantaged learners question their academic abilities or fear that they will not succeed (Mahlangu, 2018). Educators can strengthen self-efficacy by setting achievable learning goals, providing regular constructive feedback, recognising progress, and celebrating success (Flore & Govender, 2021). Early positive learning experiences often increase motivation and encourage learners to persist through later challenges (Mahlangu, 2018).

Language and communication support are essential for learners from culturally and linguistically diverse backgrounds. Migrants, refugees, and learners for whom the language of instruction is not their first language may require additional assistance to understand course content, complete assessments, and participate confidently in learning activities (Shonfeld et al., 2021; Smythe, 2020). Providing plain language resources, bilingual support where available, language development programs, and culturally responsive teaching practices can improve participation and learning outcomes (Banegas, 2022; Smythe, 2020).

Digital inclusion has become increasingly important as education incorporates online learning technologies. Many disadvantaged learners experience limited access to computers, reliable internet connections, or digital devices, while others have limited digital literacy skills (Shonfeld et al., 2021). Providing access to technology, digital skills training, technical support, and alternative methods of accessing learning materials helps ensure that technological requirements do not become barriers to participation (Liu, 2021; Shonfeld et al., 2021).

Support services play an important role in promoting learner success. Academic advising, mentoring, counselling, disability support services, career guidance, study skills workshops, tutoring, and wellbeing services can assist learners in overcoming personal and educational challenges (Magro & Sfeir, 2023). Early identification of learners who may be at risk of disengagement allows institutions to provide timely interventions before difficulties lead to withdrawal from study. Strong support systems contribute to higher levels of persistence, retention, and successful course.

Employers also have an important role in supporting disadvantaged learners who participate in workplace education and training. Flexible work arrangements, paid study leave, mentoring, access to workplace learning opportunities, and encouragement from supervisors can make it easier for employees to undertake further education (Parr et al., 2022). Organisations that value continuous learning and professional development help create environments in which employees feel supported to develop new knowledge and skills throughout their careers (Parr et al., 2022; Shonfeld et al., 2021).

Culturally responsive education is particularly important when working with Indigenous learners, migrants, refugees, and learners from diverse cultural backgrounds (Bernard, 2024; Gouthro & Holloway, 2023). Effective educators recognise that learners bring different cultural perspectives, experiences, and ways of knowing to the learning environment (Cappiali, 2024; Ngubane & Makua, 2021). Incorporating culturally inclusive teaching practices, respecting cultural identities, and recognising the strengths that learners bring to education can improve engagement and foster a greater sense of belonging (Bernard, 2024; Botes, 2024; Jokikokko, 2021).

Universal Design for Learning (UDL) provides a useful framework for supporting diverse adult learners by designing learning experiences that are accessible from the outset rather than relying solely on individual accommodations (Liu, 2021). UDL encourages educators to present information in multiple ways, provide learners with different opportunities to demonstrate their knowledge and skills, and offer flexible methods of engagement that accommodate varying abilities, preferences, and learning styles (Liu, 2021). By anticipating learner diversity during course design, educators can reduce barriers and create more inclusive educational environments (Gouthro & Holloway, 2023; Liu, 2021).

Supporting disadvantaged learners also requires institutional and policy commitment. Educational providers need policies that promote equity, inclusion, accessibility, and lifelong learning (Gouthro & Holloway, 2023; Parr et al., 2022). Governments can contribute by funding adult education, supporting community learning programs, investing in literacy and numeracy initiatives, expanding digital inclusion, and ensuring that education remains accessible to underrepresented groups (Parr et al., 2022). Collaboration between educational institutions, employers, community organisations, and support agencies can provide coordinated assistance that addresses the complex needs of many adult learners (Bernard, 2024).

Contextualising Example: Supporting Disadvantaged Learners in a Correctional Education Environment

A vocational education teacher delivers a Certificate II in Construction Pathways within a correctional centre. The class consists of adult inmates with highly diverse educational backgrounds. Several learners have limited English language proficiency, some have interrupted schooling, others have low literacy and numeracy skills, and many have experienced long periods outside formal education. A number of learners also report previous negative educational experiences, resulting in low confidence and anxiety about returning to study. Rather than viewing these characteristics as deficits, the educator recognises that the learners bring valuable practical knowledge, life experiences, and a strong desire to improve their employment prospects upon release.

To support participation, the educator removes as many barriers to learning as possible. Written learning materials are rewritten using plain English, technical terminology is introduced gradually, and key construction vocabulary is supported with photographs, diagrams, demonstrations, and physical examples of tools and materials. Practical demonstrations precede written explanations so learners can first observe a task before reading about it. Instructions are broken into small, sequential steps, and learners are encouraged to ask questions without fear of embarrassment. Where appropriate, bilingual dictionaries and translated safety terminology are made available for learners with English as an additional language.

The educator also differentiates instruction to accommodate varying literacy levels while maintaining the same vocational competency standards. Learners demonstrate competence through practical tasks such as safely using hand tools, measuring timber, interpreting simple plans, or constructing basic framing components, rather than relying solely on written assessments. Oral questioning, workplace observations, photographs of completed work, and demonstrations of practical skills provide multiple ways for learners to demonstrate competency. Additional learning support is provided to those requiring assistance with reading assessment questions or completing workplace documentation.

Recognising that many learners have previously experienced educational failure, the educator deliberately builds confidence through achievable learning goals and regular positive feedback. Initial activities focus on simple tasks that allow learners to experience early success, such as correctly identifying construction tools, measuring materials accurately, or using personal protective equipment appropriately. As learners experience success, progressively more complex tasks are introduced. Constructive feedback focuses on improvement rather than mistakes, helping learners develop self-efficacy and confidence in their ability to learn new vocational skills.

A supportive classroom culture is also established to encourage engagement. Learners work collaboratively in small groups, pairing those with stronger English language skills or greater vocational experience with learners requiring additional support. Practical activities encourage peer learning, discussion, and shared problem-solving while maintaining individual accountability for competency. The educator actively promotes mutual respect, recognising that learners come from diverse cultural backgrounds and encouraging them to share previous work experiences that relate to the vocational tasks being undertaken.

The vocational program is designed to emphasise relevance to employment after release. Each practical activity is explicitly linked to workplace expectations, employability skills, and potential career pathways in the construction industry. Learners understand how mastering workplace safety procedures, communication skills, teamwork, and practical construction techniques will improve their employability and support successful reintegration into the community. This clear connection between learning and future employment increases motivation and persistence.

The correctional education provider further supports learners by collaborating with prison support services, literacy specialists, case managers, and post-release employment agencies. Learners requiring additional literacy or English language support are referred to foundation skills programs that complement their vocational training. Career planning sessions help learners identify employment pathways and prepare for transition into the community upon release. By coordinating educational, personal, and employment support, the institution addresses multiple barriers simultaneously and increases the likelihood that learners will complete their qualifications and successfully transition into employment.

This example illustrates how supporting disadvantaged learners extends beyond providing access to education. It requires educators to identify and reduce barriers, differentiate instruction, provide language and literacy support, foster confidence and self-efficacy, create psychologically safe learning environments, and connect learning to meaningful vocational outcomes. Within correctional education, these learner-centred strategies not only improve educational participation and achievement but also contribute to rehabilitation, employability, social inclusion, and reduced recidivism by equipping learners with the knowledge, skills, and confidence needed to successfully re-enter the workforce and wider community.

This scenario also demonstrates how multiple adult learning theories can be integrated to support disadvantaged learners in a correctional education environment. Rather than relying on a single teaching approach, the educator draws upon complementary theories that address the learners' experiences, motivations, confidence, and practical learning needs.

The educator's recognition that adult inmates bring valuable life experiences, despite interrupted education and limited literacy, reflects Knowles' theory of andragogy. Knowles argued that adults possess a wealth of prior experience that forms an important resource for learning and that education should acknowledge and build upon these experiences. By encouraging learners to relate previous employment, practical skills, and life experiences to construction tasks, the educator validates existing knowledge while creating meaningful connections between new learning and prior understanding. The emphasis on employability following release also reflects Knowles' principle that adults are most motivated when learning is immediately relevant to their personal circumstances and future goals.

The heavy emphasis on practical demonstrations before written instruction reflects Experiential Learning Theory, particularly Kolb's view that learning begins with concrete experience. Rather than introducing construction concepts through abstract written explanations, learners first observe demonstrations, handle tools, and perform practical tasks before reflecting on their experiences and applying newly acquired knowledge. This sequence is particularly appropriate for vocational learners with limited English language proficiency because it allows understanding to develop through direct experience before introducing technical language and theoretical concepts.

The educator's use of photographs, diagrams, demonstrations, scaffolded instruction, and gradual introduction of technical terminology reflects principles of Cognitivism. Cognitive learning theories recognise that learners process new information more effectively when it is organised into manageable components and connected to existing knowledge. Breaking complex construction procedures into sequential steps reduces cognitive load, while multiple forms of representation assist learners with differing literacy levels and language abilities in constructing accurate mental models of workplace procedures.

The collaborative classroom environment demonstrates Social Constructivism, particularly the work of Lev Vygotsky. Learners develop understanding through interaction with others, discussion, and shared problem-solving. Pairing learners with stronger English language skills or greater vocational experience alongside those requiring additional support creates opportunities for peer scaffolding within the Zone of Proximal Development. Through collaboration, learners gradually perform tasks they would have struggled to complete independently, while simultaneously developing communication and teamwork skills valued in the construction industry.

The educator's deliberate focus on early success and progressive skill development illustrates Bandura's Social Cognitive Theory, particularly the concept of self-efficacy. Many correctional learners have experienced repeated educational failure and may possess low confidence in their ability to succeed. By designing initial tasks that learners can complete successfully, providing constructive feedback, and gradually increasing task complexity, the educator creates mastery experiences that strengthen learners' beliefs in their own capabilities. Positive reinforcement from both the educator and peers further enhances confidence and encourages persistence when more challenging learning activities are introduced.

The supportive classroom climate also aligns with Humanistic learning theory. Humanistic educators argue that learning occurs most effectively when learners feel respected, accepted, and psychologically safe. Encouraging questions without fear of embarrassment, acknowledging cultural diversity, respecting learners' backgrounds, and focusing feedback on improvement rather than failure helps satisfy learners' emotional needs and creates conditions that support engagement and personal growth. This approach is particularly important in correctional education, where many learners have experienced stigma, exclusion, or repeated educational failure.

Several motivation theories are also evident throughout the example. The strong connection between vocational learning and future employment reflects Expectancy Theory, as learners understand that successful completion of training can lead to meaningful employment opportunities following release. The program also reflects Goal Theory, with learning activities clearly linked to specific objectives such as obtaining employment, developing construction skills, and successfully reintegrating into the community. The gradual development of competence, opportunities for learner success, and respectful learning environment also support Self-Determination Theory, as learners develop competence, experience supportive relationships, and gain increasing confidence in their ability to direct their own futures.

The educator's differentiated instruction and multiple methods of demonstrating competence illustrate principles of Universal Design for Learning (UDL) and inclusive adult education. Rather than expecting all learners to demonstrate knowledge through written assessments, competency is demonstrated through practical performance, oral questioning, workplace observation, and visual evidence. These varied assessment approaches reduce unnecessary barriers while maintaining the same vocational competency standards for all learners.

The collaboration between educators, literacy specialists, prison support services, case managers, and post-release employment agencies reflects the broader ecological perspective found in contemporary adult education. Adult learning is recognised as being influenced not only by classroom teaching but also by the social, institutional, and community supports surrounding the learner. By coordinating educational, personal, and employment services, the correctional education provider addresses the situational, institutional, and dispositional barriers that commonly affect disadvantaged adult learners.

Learner Engagement Strategies

Learner engagement is a fundamental component of successful adult education. It refers to the degree to which learners actively participate in, commit to, and invest effort in the learning process. Engaged learners are more likely to attend classes regularly, participate in discussions, complete learning activities, persist through challenges, and successfully achieve learning outcomes (Symonds et al., 2023). In adult education, engagement is particularly important because learners often balance education with employment, family responsibilities, and other competing commitments (Moquin et al., 2023). Maintaining engagement therefore requires learning experiences that are relevant, meaningful, and responsive to adult learners' needs (Miller, 2024).

Contemporary research recognises that learner engagement is multidimensional, encompassing behavioural, cognitive, emotional, and social dimensions (Petrescu et al., 2020; Symonds et al., 2023). Behavioural engagement refers to active participation in learning activities, attendance, and task completion (Symonds et al., 2023). Cognitive engagement involves the mental effort learners invest in understanding, analysing, and applying new knowledge (Petrescu et al., 2020; Symonds et al., 2023). Emotional engagement relates to learners' interest, enjoyment, confidence, and sense of belonging, while social engagement reflects interactions with educators and peers through collaboration, discussion, and shared learning experiences (Petrescu et al., 2020; Picca & Reed, 2023). Effective adult education seeks to foster all four dimensions simultaneously (Petrescu et al., 2020; Symonds et al., 2023)

Creating Relevant and Meaningful Learning

One of the most effective strategies for promoting engagement is ensuring that learning is directly relevant to adult learners' personal, professional, or community goals. Malcolm Knowles argued that adults are most motivated when they understand why they are learning something and when they can immediately apply new knowledge to real-life situations (Ghamrawi et al., 2025; Khan, 2024; Malinga). Knowles's andragogical model posits that adult learners "must know why they should learn something, before embarking on any learning journey" and that "relevance and context must be established in adult education programs" (Khan, 2024, p. 165). Adult learners are generally most engaged when they perceive that learning has immediate value and can be applied directly to their work, personal lives, or community roles (Ghamrawi et al., 2025; Khan, 2024; Malinga). This principle is central to andragogy, which proposes that adults are problem-centred rather than subject-centred and are motivated by learning that helps them address real-world challenges. Consequently, effective adult education connects new knowledge and skills to authentic situations that learners are likely to encounter in their daily lives or professional practice. Educators can strengthen engagement by clearly explaining the purpose of learning activities, identifying the intended learning outcomes, and demonstrating how the knowledge being developed contributes to workplace performance, career progression, or personal development (Grus et al., 2025; Khan, 2024).

Adult learning is most effective when learners actively participate in solving meaningful problems rather than simply acquiring theoretical knowledge. Learning activities that focus on realistic tasks, workplace scenarios, and practical applications help learners understand why the content is important and how it can be transferred to future situations (Nebeker, 2014). Providing clear learning objectives before, during, and after instruction further supports engagement by helping learners understand what they are expected to achieve and how individual learning activities contribute to broader educational goals (Mbatha, 2024). This emphasis on contextualisation, relevance, and practical application reflects contemporary understandings of meaningful adult learning, where knowledge is constructed through authentic experiences rather than memorised in isolation (Rodger Roberto Alves de, 2024).

For example, learners undertaking workplace health and safety training are more likely to remain engaged when they practise identifying hazards within their own workplace or a realistic simulated environment than when they study legislative requirements without practical application. By linking theory directly to workplace practice, learners can immediately see the purpose of the knowledge and develop skills that are directly transferable to their employment. Research has demonstrated that contextualising learning within authentic subject matter and workplace situations improves the transfer of knowledge and skills by helping learners understand how foundational concepts are used in practice (Perin, 2011). Rather than viewing learning as abstract or disconnected, learners recognise its immediate usefulness and are therefore more motivated to participate actively.

The relevance of learning is particularly important in professional education and workplace training, where the successful transfer of learning depends on the applicability of educational content to job requirements (Surikova, 2024). Adult learners typically approach education with clearly defined personal or professional goals and expect learning to assist them in fulfilling current social and occupational roles (Mbatha, 2024). Because adults are generally motivated by practical outcomes and meaningful objectives, learning activities that emphasise authentic problem-solving, workplace application, and immediate relevance are more likely to sustain engagement, improve knowledge transfer, and support long-term learning (Rodger Roberto Alves de, 2024; Surikova, 2024).

Building on Learners' Prior Experience

Adult learners bring extensive life and work experiences that provide valuable foundations for learning. Engagement increases when educators acknowledge these experiences and encourage learners to contribute their existing knowledge to classroom discussions and practical activities (Holdsworth et al., 2025; Shahid et al., 2023; Wolmarans, 2007). Adult learners are active learners who prefer hands-on learning opportunities and tend to focus on real-world situations, and their learning materials should offer explicit opportunities to relate to what they experience in practice (Wolmarans, 2007) . As Knowles argued, adults learn most effectively when they recognise a personal need for knowledge, often shaped by their life experiences, which serve as valuable learning resources (Holdsworth et al., 2025). The third aspect of adult learning theory holds that adult learners bring considerable information and life experiences to the journey of learning, possessing the ability to practice, link, and apply their learning to various situations and issues existing in the real world (Shahid et al., 2023). Rather than treating learners as passive recipients of information, educators create opportunities for them to compare experiences, solve problems collaboratively, and relate new concepts to familiar situations (Alghowinem et al., 2024; Mayombe, 2023; Wolmarans, 2007). Learning is more effective when adult learners are directly involved rather than passively receiving knowledge transmitted by facilitators (Mayombe, 2023). Case-based instruction, for instance, requires active learning and relies on a constructivist approach, making use of open-ended problems usually requiring small group collaboration, where learners solve problems by leveraging prior knowledge and past experience (Lyons & Bandura, 2020).

Discussion, reflection, case studies, workplace examples, and peer teaching all encourage learners to integrate new knowledge with prior experience (Alghowinem et al., 2024; Blackburne, 2024; Holdsworth et al., 2025). Case studies serve as effective means for providing learners with opportunities to become involved in phases of experiential learning, allowing integration of classroom learning with real-life experience (Blackburne, 2024; O'Flynn et al., 2023). Action learning draws on group interaction and dynamics to examine real-world problems, generally occurring within the actual workplace context. Hands-on activities and real-world case studies help participants understand new concepts, and the use of teamwork proves effective, aligning with adult learning theories for experiential and collaborative learning (Alghowinem et al., 2024). Deep learning is founded on reflection, using prior knowledge, and motivation, and methods that require learners to engage meaningfully with ambiguous and unfamiliar contexts facilitate learning by doing (Holdsworth et al., 2025). Recognising learners' expertise also demonstrates respect for their backgrounds, strengthening motivation and participation (Moll, 2023; Shahid et al., 2023). To fully engage in education, adult learners expect to be heard and given a platform to voice their thoughts and ideas, and their eagerness to learn can be stifled if educational programmes fail to take their assumptions or past educational practices into account (Shahid et al., 2023). Adults respond to some external motivators, but their most potent motivators are internal, driven by their understanding of the knowledge and skills necessary to achieve their potential (Moll, 2023).

Active Learning

Active learning strategies encourage learners to participate actively in constructing knowledge rather than passively receiving information from an instructor. Rather than relying solely on lectures or reading materials, active learning requires learners to engage with concepts through exploration, discussion, application, and reflection. This learner-centred approach recognises that adults develop deeper understanding when they are directly involved in the learning process and can connect new knowledge to their own experiences and real-world situations.

Research consistently demonstrates that adult learners achieve stronger learning outcomes when they actively engage with content through meaningful learning activities. Practical demonstrations, workplace simulations, group discussions, case studies, role-playing, project-based learning, problem-based learning, reflective activities, and hands-on practice all encourage learners to analyse information, solve problems, make decisions, and apply their knowledge in authentic contexts. These approaches move learning beyond memorisation by requiring learners to think critically, collaborate with others, and integrate theory with practice.

Active learning strategies are particularly valuable in adult education because they acknowledge the experience and expertise that learners bring to the classroom. Through discussion and collaborative activities, adults can compare perspectives, share workplace experiences, and contribute practical insights that enrich the learning experience for the entire group. Practical activities also provide opportunities to test ideas, receive feedback, and refine skills in realistic settings, helping learners develop both confidence and competence.

In vocational education and workplace training, active learning closely mirrors the environments in which learners will ultimately apply their knowledge and skills. For example, learners may practise operating equipment, participate in simulated workplace scenarios, analyse case studies based on real industry situations, or complete projects that reflect authentic workplace tasks. These experiences encourage learners to transfer theoretical knowledge into practical competence while developing problem-solving, communication, teamwork, and decision-making skills. By actively participating in meaningful learning experiences, adult learners are more likely to remain engaged, retain new knowledge, and successfully apply their learning in professional and everyday contexts.

Collaborative Learning

Social interaction is an important contributor to learner engagement. Collaborative learning allows adults to learn from one another by sharing experiences, discussing ideas, and solving problems collectively (Chuang, 2021; Smith et al., 2020; Smith, 2003). Group activities encourage learners to explain concepts, challenge assumptions, and consider alternative perspectives (Gauvain, 2018; Smith et al., 2020).

Collaborative learning is particularly valuable in vocational education, where teamwork mirrors workplace practice (Smith, 2003; Tynjälä & Häkkinen, 2005). Activities such as group projects, peer mentoring, workplace problem-solving exercises, and communities of practice encourage both knowledge development and interpersonal skill development (Cheng et al., 2011; Chuang, 2021; Smith, 2003; Stephens et al., 2022).

Goal Setting

Clearly defined learning goals help sustain engagement by providing learners with direction and purpose. Adults are more likely to remain motivated when they understand what they are working towards and can monitor their progress (Ortogero et al., 2024; Schiller, 2022b). As one participant in Schiller's study noted, "Well-defined goals are important for increasing motivation. You also need to be constantly open to new opportunities of learning" (Schiller, 2022b, p. 167). Educators can support engagement by establishing realistic learning objectives, breaking larger goals into manageable milestones (Schüll, 2024), and regularly reviewing learner progress (Mou, 2021). Progressive assessment and timely feedback have been identified as critical motivating factors for adult learners in higher education (Alhasov et al., 2020).

Goal setting also supports self-regulated learning by encouraging learners to take responsibility for planning, monitoring, and evaluating their own learning (Mou, 2021; Siregar et al., 2024). Self-regulated learning has been defined as an active, constructive process whereby learners set goals for their learning and then attempt to monitor, regulate, and control their cognition, motivation, and behavior, guided and constrained by their goals and the contextual features in the environment (Gregg & Nelson, 2017; Siregar et al., 2024). Research confirms that adult learners who are equipped with self-regulation skills can successfully navigate their academic paths by drawing on their prior experience and independence to attain significant learning outcomes (Siregar et al., 2024). Studies have further demonstrated that goal setting and self-monitoring are highly motivating for adult learners, with self-regulated learning models outlining key phases including forethought and planning, monitoring, control, and reaction and reflection (Ortogero et al., 2024). Adult learners have been shown to take responsibility for their learning, direction, and productivity to achieve study goals through better time and learning environment management (Lin, 2019).

Providing Choice and Autonomy

Self-Determination Theory suggests that autonomy is a major contributor to motivation and engagement (Lewis, 2024; O'Neill & Thomson, 2013; Rothes et al., 2016). Specifically, SDT posits three basic psychological needs, autonomy, competence, and relatedness, the fulfillment of which leads to autonomous motivation and optimal growth (Lewis, 2024; Wehmeyer et al., 2021). Adult learners generally value opportunities to exercise choice and control over their learning (Kálmán, 2019). Notably, a substantial 66.7% of respondents in one study recognized "more control over learning" as a pivotal advantage of self-directed learning (Chakraborty, 2024) . Whenever appropriate, educators can enhance engagement by allowing learners to select project topics, choose assessment formats, determine learning sequences, or identify workplace problems they wish to investigate (Chakraborty, 2024; Kálmán, 2019; Reddig et al., 2026). Research confirms that providing choices and tailor-making teaching in response to learners' preferences is among the most effective tools teachers possess to promote autonomous learning (Kálmán, 2019). AI-supported learning environments similarly emphasize that learners should be provided with choices in how they engage with content and control over how support is delivered (Reddig et al., 2026).

Providing autonomy demonstrates respect for adult learners as independent individuals while encouraging greater ownership of the learning process (Ghamrawi et al., 2025; Tracy, 2015; Wehmeyer et al., 2021). Knowles' Andragogical Model suggests that the self-concept of the adult learner drives the need for autonomy and self-directedness (Ghamrawi et al., 2025; Tracy, 2015). Many adult educators suggest the goal of all learning should be to increase personal autonomy, meaning to take control of the goals and purposes of learning and assume ownership of the learning process (Tracy, 2015). Empirical results confirm that adult learners with high autonomous motivation score higher in self-efficacy, behavioural engagement, and use of deep-learning strategies (Rothes et al., 2016). Autonomy-supportive teaching strategies align with adult learning approaches, suggesting that adult learners universally want to have control over their learning process (Tracy, 2015), and autonomy-supportive classrooms are learning environments in which students are motivated to act and engage in self-determined learning (Wehmeyer et al., 2021).

Scaffolding Learning

Many adult learners return to education after extended absences and may initially lack confidence. Scaffolding provides structured support that gradually decreases as learners develop competence and independence (Doo et al., 2020; Margolis, 2020b; Verenikina, 2004). Educators can scaffold learning by breaking complex tasks into smaller steps, modelling procedures, providing worked examples, offering guided practice, and progressively increasing task complexity (Ahmed et al., 2022; Carolyne, 2022; Verenikina, 2004).

Scaffolding helps maintain engagement because learners experience achievable success rather than becoming overwhelmed by tasks that appear beyond their capabilities (Ahmed et al., 2022; Koirala, 2019). Notably, meta-analytic evidence suggests that scaffolding's effects are particularly strong for adult learners, with effect sizes greater than those observed for younger populations (Belland, 2016; Doo et al., 2020). This is significant given that scaffolding was originally developed for use with young children (Belland, 2016; Doo et al., 2020). As Belland notes, "scaffolding's strongest effects are in populations the furthest from the target learner population in the original scaffolding definition" (Doo et al., 2020, p. 72). The application of scaffolding to adult learning contexts is well supported, as it helps develop adults as independent and lifelong learners (Ahmed et al., 2022), and educators can progressively withdraw support as learners move toward full participation and demonstrate greater autonomy (Koirala, 2019). Furthermore, scaffolding has been recognized as appropriate across diverse adult learning contexts, including adult literacy education (Ndoro, 2022; , Joyce et al., 1970), open and distance learning (Ahmed et al., 2022), and online higher education (Doo et al., 2020; Sanchez, 2020).

Providing Constructive Feedback

Timely, specific, and constructive feedback is essential for maintaining learner engagement. Effective feedback helps learners understand what they have achieved, identifies areas requiring improvement, and provides guidance for future learning. Research confirms that feedback enhances learning by informing learners about their performance in relation to learning outcomes and helps them redirect effort and activity to achieve those outcomes. Descriptive feedback, in particular, provides specific information about what students do well, what they need to improve, and how they can improve (Wati & Prawiro, 2019). Furthermore, feedback that incorporates known standards of performance with internal and external data improves a learner's ability to accurately self-assess and engage in reflection (Wisniewski & Zierer, 2021). Rather than focusing solely on errors, constructive feedback acknowledges progress, reinforces strengths, and encourages continued effort. Studies demonstrate that replacing grades with high-quality constructive feedback can impact students' level of engagement with the feedback offered, and that providing honest, explicit feedback helps students accurately calibrate their beliefs about their capabilities.

Regular, timely, and constructive feedback plays a vital role in developing learners' self-efficacy by reinforcing the understanding that improvement occurs through effort, practice, and persistence rather than fixed ability (Wati & Prawiro, 2019; Wisniewski & Zierer, 2021). Feedback helps learners recognise their progress, identify areas requiring further development, and understand the steps needed to improve future performance. As learners experience incremental success and receive guidance on how to refine their skills, their confidence in their ability to master increasingly complex tasks grows, encouraging continued engagement and persistence.

Research has consistently shown that learners with higher levels of self-efficacy are more likely to invest greater effort, persevere when encountering challenges, and achieve stronger learning outcomes (Artino, 2012). Feedback contributes to this process by helping learners develop accurate perceptions of their capabilities. Rather than relying solely on their own judgments, learners use feedback from educators, supervisors, and peers to evaluate their progress and adjust their learning strategies accordingly. This ongoing cycle of performance, feedback, reflection, and improvement enables learners to develop both competence and confidence.

Effective feedback is characterised by being specific, timely, constructive, and focused on learning rather than simply judging performance (Wisniewski & Zierer, 2021). Feedback that clearly identifies what has been achieved, explains where improvements can be made, and provides practical guidance for future learning is more likely to support learner development than feedback that merely indicates whether a task was correct or incorrect. Structuring learning activities so that learners receive frequent opportunities for meaningful feedback allows them to monitor their progress, refine their performance, and progressively develop expertise.

Feedback also benefits educators by informing instructional practice. Teachers who observe improvements in learner performance following the provision of effective feedback often report greater confidence in their teaching and stronger beliefs in their ability to facilitate learning successfully (Wisniewski & Zierer, 2021). Consequently, feedback functions as a reciprocal process that supports both learner development and teaching effectiveness. Within adult education, regular feedback helps maintain motivation, strengthens self-efficacy, promotes reflective practice, and contributes to sustained engagement and successful learning outcomes.

Building Self-Efficacy

Learners who believe they can succeed are more likely to participate actively and persist through challenges (Mellard et al., 2012; O'Neill & Thomson, 2013; Rothes et al., 2016). Adults with higher self-efficacy demonstrate greater engagement and are more likely to persist in and complete their courses (Rothes et al., 2016). Educators can strengthen self-efficacy by providing early opportunities for success, recognising learner achievements, offering encouragement, and gradually increasing task difficulty. Specifically, Bandura's theory advises that adult education programs should provide mastery experiences, verbal persuasion, and structured support to build self-efficacy (Comings et al., 1999; Orn, 2022). Feedback can boost self-efficacy in adults by reinforcing their abilities and encouraging persistence, building self-confidence, and increasing motivation and performance.

Additionally, self-efficacy can be cultivated indirectly when people witness the achievements of others they deem similar to themselves (Beauregard et al., 2015; Orn, 2022). Observing peers successfully complete learning activities enhances confidence through vicarious learning, as learners can relate to successful peers and think of themselves as competent to replicate the observed behaviour (Beauregard et al., 2015; Choo et al., 2019; Orn, 2022; Schunk, 1987). Vicarious learning is particularly influential when learners have little prior mastery experience with a skill, making them more reliant on observing others to develop self-efficacy (Choo et al., 2019).

Building confidence is particularly important for adults who have previously experienced educational failure or who have low literacy, numeracy, or digital skills. While adults may have a strong sense of self-efficacy in tasks related to their daily lives, feelings of academic self-efficacy are often weak for low-skilled adults based on past academic experiences (O'Neill & Thomson, 2013; Rolle, 2023). Adults with lower educational levels tend to show lower levels of perceived self-efficacy (Rothes et al., 2016). Many adults experience anxiety about returning to an academic environment, particularly if they have been away from formal education for a long period, and these psychological barriers can lead to lower levels of self-efficacy, which in turn diminishes motivation and persistence. Because sense of academic self-efficacy is predictive of persistence, building this up in adult learners is crucial (O'Neill &

Thomson, 2013). Dispositional variables relating to self-efficacy and resilience have been found to be moderate predictors of persistence in adult education programs (Mellard et al., 2012). Adult learners with low self-esteem also show less active participation in learning environments, and educators should pay attention to both self-efficacy and self-esteem to improve learning outcomes (Zhu, 2019).

Creating a Positive Learning Environment

Psychologically safe learning environments encourage learners to participate without fear of embarrassment or criticism (Haidari et al., 2020; Levesque, 2021; Wolcott et al., 2022). Adults are more willing to ask questions, contribute ideas, and attempt unfamiliar tasks when they feel respected and supported (Haidari et al., 2020; K. I. Hossain et al., 2023; W. H. Wang et al., 2025). Effective educators establish clear expectations, encourage respectful communication, value learner contributions, and foster inclusive classroom cultures Wang et al., 2025; , Wolcott et al., 2022; , Zhao, 2022).

Positive educator-learner relationships contribute significantly to engagement by creating trust, reducing anxiety, and promoting learner wellbeing (Haidari et al., 2020; Palalas & Doran, 2024; Roffey & Quinlan, 2021; W. H. Wang et al., 2025).

Using Authentic Learning Activities

Authentic learning involves educational activities that closely replicate the situations, tasks, and challenges learners are likely to encounter in real workplaces or community settings (Bartlett & Sims, 2021; Kittel et al., 2023). Rather than focusing on abstract knowledge alone, authentic learning places learners in realistic contexts where they apply knowledge and skills to solve genuine problems (Argüelles et al., 2022; Miller & Konstantinou, 2021). This approach helps learners understand not only what they are learning but also why it is important and how it will be used beyond the classroom (Hardin-Ramanan et al., 2018; Shouman et al., 2023). Adult learners are generally more engaged when learning reflects the realities of their professional and personal lives because they can immediately recognise its relevance and practical value (Bartlett & Sims, 2021; Black & Warhurst, 2025).

Authentic learning can take many forms, including workplace simulations, industry-based projects, case studies drawn from real organisations, practical demonstrations, fieldwork, work-integrated learning, and authentic assessment tasks (Ayuningrum et al., 2024; Gill, 2016; Kittel et al., 2023). These learning experiences encourage learners to analyse complex situations, make informed decisions, collaborate with others, and apply theoretical knowledge in ways that mirror professional practice (Argüelles et al., 2022; Hart & Bone, 2022). Rather than simply recalling information for assessment, learners develop the ability to use their knowledge effectively in realistic environments where multiple factors must be considered simultaneously (Akbari et al., 2021; Miller & Konstantinou, 2021).

In vocational education, authentic learning is particularly valuable because it closely aligns educational activities with workplace expectations (Ornellas et al., 2019),. Learners may operate industry-standard equipment, complete tasks using workplace procedures, solve realistic problems, or participate in simulated work environments that replicate actual job roles,. Authentic assessment further reinforces this approach by requiring learners to demonstrate competence through tasks that resemble those performed in the workplace, ensuring that assessment measures practical capability rather than theoretical knowledge alone (Akbari et al., 2021; Ornellas et al., 2019).

Authentic learning also supports the development of broader employability skills that extend beyond technical competence (Argüelles et al., 2022; Bartlett & Sims, 2021). As learners engage with realistic workplace scenarios, they develop communication, teamwork, critical thinking, problem-solving, time management, and decision-making skills alongside vocational knowledge (Black & Warhurst, 2025; Kittel et al., 2023; Miller & Konstantinou, 2021). By connecting learning to authentic contexts, educators help adult learners understand the immediate relevance of their studies, increase motivation and engagement, and better prepare them for successful participation in professional practice and lifelong learning (Bartlett & Sims, 2021; Hardin-Ramanan et al., 2018; Shouman et al., 2023).

Technology-Enhanced Engagement

Digital technologies can enhance learner engagement when they are used purposefully and appropriately. Online discussion forums have been widely used in various education settings to enable student-to-student collaboration and have been shown to build trust and a sense of learning community among participants. Interactive simulations, such as mixed reality environments like TeachLivE™, offer learners the opportunity to engage in practice sessions and learn new skills in environments specifically designed to enhance these skills. Educational videos, virtual classrooms, collaborative documents, mobile learning applications, and learning management systems can increase flexibility and provide additional opportunities for interaction (Shi & Lin, 2021).

However, technology should support rather than replace sound educational practice. As Moore and Miller emphasise, simply creating discussion forum assignments will not be enough to achieve deeper learning outcomes; instructors need to be intentional in how they design and structure their courses to ensure optimal engagement between learners and content. Educators must also recognise that some adult learners may require additional support to develop digital literacy skills before technology becomes an effective engagement tool. Cercone proposed designing online learning environments based on adult learners' limitations and considering their individual learning experiences in online instruction (Ng & Baharom, 2018). Similarly, the Understanding Dementia MOOC was designed specifically to support adult learners with limited educational backgrounds, demonstrating that purposeful design can accommodate diverse digital literacy levels.

Reflective Practice

Reflection encourages learners to think critically about their experiences, evaluate their progress, and identify opportunities for improvement. Learning journals, portfolios, reflective discussions, and self-assessment activities encourage adults to become more aware of their own learning processes and support the development of lifelong learning skills (Dison, 2016; Schedlitzki, 2019; Schumann & Baraka, 2024; Sultana, 2013). Portfolios, for instance, stimulate reflective peer and self-assessment, enabling learners to build capacity and confidence in making judgements about their own learning (Dison, 2016; Sultana, 2013). E-portfolios specifically support critical reflection and self-regulation, promoting deeper understanding of subject matter and encouraging learners to become independent, reflexive practitioners (Ouherrou et al., 2022; Schumann & Baraka, 2024; Useche et al., 2022). Learning journals and reflective writing similarly serve as instruments for systematising learning and fostering self-assessment in cognitive, psychomotor, and affective domains (Paraluppi et al., 2024). The learning portfolio, when used as a reflexive pedagogical tool, engages learners in ongoing, deeply situated learning practices that enable critical reflection on actions and decision-making processes (Schedlitzki, 2019). Self-assessment activities, including self-rating scales and reflective commentary, invite learners to evaluate their abilities and record their developmental progress over time (Neden, 2007; Vanhulle, 2005).

Reflection also strengthens engagement by helping learners recognise the value of their achievements and connect new learning with future goals. Portfolio-based approaches require learners to identify and reflect upon what motivates and satisfies them, documenting interests, skills, values, needs, goals, and strategies (Sultana, 2013). Teachers and counsellors perceive portfolios and personal development plans as instruments to stimulate self-responsibility, support students in reflecting on identity and future ambitions, and set up learning goals to achieve these (Sultana, 2013). The reflective process enables learners to revisit what was previously learned, rescuing both technical and emotional aspects of their development (Paraluppi et al., 2024). Furthermore, lifelong learning, understood as the continuous, self-motivated pursuit of knowledge, is enhanced through reflection activities that improve professional qualities and foster a sense of proactive growth (Luo et al., 2026; Said & Abdallah, 2024). Educators and learners with lifelong learning awareness more actively participate in reflective practice, continuously improving their professional development and adapting to evolving needs (Luo et al., 2026; Said & Abdallah, 2024).

Supporting Diverse Learners

Engagement strategies should recognise the considerable diversity that exists among adult learners. Adults enter education with varying educational experiences, cultural backgrounds, languages, abilities, learning preferences, employment histories, and personal responsibilities. Creating inclusive learning environments therefore requires educators to adopt teaching approaches that enable all learners to participate meaningfully and achieve success. Inclusive teaching practices, differentiated instruction, Universal Design for Learning (UDL), culturally responsive teaching, and accessible learning resources all contribute to reducing barriers and promoting equitable participation for learners with diverse needs (Housel, 2019; Rogers-Shaw et al., 2023; Ross-Gordon, 2024).

Universal Design for Learning has emerged as a comprehensive framework for designing learning environments that are accessible to the widest possible range of learners from the outset. Grounded in neuroscience, UDL promotes learner-centred education by emphasising accessibility, flexibility, collaboration, and learner engagement rather than relying on individual accommodations after barriers have been identified (Rogers-Shaw et al., 2023). Although UDL and differentiated instruction share the goal of meeting diverse learner needs, they differ in their approach. Differentiated instruction typically involves adapting teaching methods, learning activities, or content to suit individual learners, whereas UDL encourages educators to proactively design curriculum, learning resources, instructional practices, and assessment methods that provide flexible learning opportunities for everyone from the beginning (Sebti & Iqtadar, 2025). Together, these complementary approaches contribute to more accessible and inclusive educational environments (Tundaan, 2023).

Culturally responsive teaching further strengthens inclusive practice by recognising and valuing the diverse cultural experiences, perspectives, and identities that adult learners bring to education. Integrating culturally responsive pedagogy with UDL and principles of adult learning helps educators design learning experiences that acknowledge cultural diversity while challenging assumptions that may unintentionally disadvantage particular learner groups (Housel, 2019; Ross-Gordon, 2024). Providing learning materials in multiple formats, including text, graphics, audio, video, and other multimedia resources, enables learners from varied linguistic, educational, and cultural backgrounds to access content in ways that best support their learning (Rogers-Shaw et al., 2017). Increasingly, UDL, differentiated instruction, and culturally responsive teaching are recognised as complementary foundations for inclusive curriculum design and educational practice (Baldwin, 2021; Rabinowitz & Tondreau, 2021).

A key principle of inclusive education is providing learners with multiple ways to access information, participate in learning activities, and demonstrate their knowledge and skills. The three core principles of UDL, multiple means of representation, multiple means of action and expression, and multiple means of engagement, address the different ways learners perceive information, demonstrate learning, and remain motivated throughout the educational process (Dalton et al., 2012; Galkienė & Monkevičienė, 2021). In practice, this may involve presenting information through written text, visual diagrams, demonstrations, videos, or digital resources; allowing learners to demonstrate competence through written assignments, presentations, practical demonstrations, or multimedia projects; and offering a variety of collaborative and individual learning activities that encourage active participation (Sebti & Iqtadar, 2025). By embedding flexibility into curriculum design, educators reduce unnecessary barriers while maintaining consistent learning outcomes for all learners.

Contemporary perspectives increasingly view Universal Design for Learning not simply as a framework for providing accessibility accommodations, but as a broader educational philosophy that reshapes how adult learning environments are designed and delivered (Rogers-Shaw et al., 2017). This perspective is particularly important in adult education, where learners may have undiagnosed disabilities or may choose not to disclose disability-related needs. Rather than relying exclusively on individual adjustments, educators can create learning environments that are inherently flexible and accessible, thereby reducing barriers before they arise (Ross-Gordon, 2024). Designing learning environments to be accessible from the outset benefits all learners by increasing participation, improving engagement, and supporting more equitable educational outcomes across increasingly diverse adult learning populations (McClain & Mallary, 2025; Ross-Gordon, 2024).

Significance for Adult Education

Learner engagement is not achieved through a single teaching strategy but through the thoughtful integration of multiple learner-centred approaches. Effective engagement requires educators to create relevant learning experiences, build on learners' existing knowledge, encourage active participation, provide appropriate support, foster positive relationships, and develop learners' confidence and autonomy. These strategies draw upon many of the major adult learning theories, including andragogy, experiential learning, constructivism, self-determination theory, social cognitive theory, and humanistic education. When adult learners are actively engaged, they are more likely to persist, achieve meaningful learning outcomes, develop professional competence, and continue participating in lifelong learning.

The following examples illustrate how learner engagement strategies can be applied in different adult learning contexts. They demonstrate that while the underlying principles remain consistent, the implementation varies according to the learners' educational background, professional goals, and learning environment. These examples build directly on the learner engagement strategies discussed in this chapter.

Example 1: First-Year Architecture Student (United Kingdom)

A first-year architecture student at a university in the United Kingdom is studying a module on sustainable building design. Rather than relying solely on lectures about environmental design principles, the lecturer creates an authentic, learner-centred experience that integrates multiple engagement strategies.

The module begins by explaining why sustainable design is central to contemporary architectural practice and how the concepts being taught align with the requirements of professional bodies such as the Royal Institute of British Architects (RIBA). Understanding the relevance of the topic helps students appreciate how the knowledge will support both their academic success and future careers.

Students participate in active learning through studio-based design activities rather than passively listening to lectures. They work in small teams to analyse an existing university building, identify opportunities to improve its environmental performance, and develop redesign proposals that reduce energy consumption while maintaining functionality. These collaborative activities encourage discussion, critical thinking, and shared problem-solving as students compare ideas and justify design decisions.

The lecturer scaffolds learning by first demonstrating how to conduct a building performance analysis before guiding students through increasingly complex design tasks. Early activities include worked examples and structured worksheets, while later tasks require students to independently evaluate sustainability challenges and justify their own design solutions.

Authentic learning is embedded throughout the module. Students undertake site visits to local buildings, examine architectural drawings used by practising architects, and complete a design project based on a genuine urban redevelopment scenario. Assessment requires students to present their final design proposal to a panel simulating a client briefing, requiring them to communicate technical information clearly while responding to questions about their design choices.

Technology further enhances engagement through Building Information Modelling (BIM) software, digital modelling tools, virtual reality walkthroughs, and online collaboration platforms. Students receive regular formative feedback during studio critiques, allowing them to refine their designs before final submission. Feedback focuses not only on technical accuracy but also on creativity, environmental performance, and professional presentation, helping students build confidence and self-efficacy as developing architects.

Recognising the diversity of the cohort, the lecturer provides learning materials in multiple formats, including recorded lectures, annotated drawings, demonstration videos, written guides, and interactive design exercises. Students may demonstrate their understanding through architectural models, digital renderings, technical reports, or oral presentations, allowing different strengths to be recognised while maintaining consistent learning outcomes.

Example 2: Accountant Completing a Professional Development Course (United States)

An experienced accountant in the United States enrols in a professional development course on data analytics and artificial intelligence for financial reporting. Unlike undergraduate students, the participants already possess substantial accounting knowledge and industry experience. Consequently, the course is designed to build upon their professional expertise while introducing new technologies that are increasingly influencing accounting practice.

The instructor begins by demonstrating how artificial intelligence, data analytics, and automation are transforming auditing, financial reporting, fraud detection, and regulatory compliance. By linking course content directly to emerging workplace expectations and career advancement, participants immediately recognise the practical relevance of the learning.

Rather than relying on traditional lectures, participants complete authentic workplace activities using realistic financial datasets. They investigate unusual financial transactions, identify potential fraud indicators, develop data visualisations for management reporting, and use analytical software to support audit planning. These activities require participants to apply new technologies within familiar accounting contexts, enabling immediate transfer of learning to their workplaces.

Collaborative learning is incorporated through peer discussion groups where accountants from different industries compare how artificial intelligence and analytics are being implemented within their organisations. Participants share practical challenges, discuss ethical issues surrounding automated decision-making, and exchange strategies for improving financial reporting processes. These discussions allow learners to integrate new concepts with their existing professional knowledge.

The instructor scaffolds learning by first introducing basic analytical tools before progressing to more sophisticated predictive models and automated reporting techniques. Worked examples and guided exercises gradually give way to independent workplace projects in which participants analyse financial data drawn from scenarios similar to those encountered in professional practice.

Authentic assessment requires participants to complete a workplace improvement project rather than a traditional examination. Each learner identifies an opportunity within their own organisation to improve financial reporting or audit processes using data analytics, develops an implementation proposal, and presents recommendations supported by evidence generated during the course. This assessment directly mirrors the type of work participants are expected to perform in their professional roles.

Technology forms the foundation of the learning experience. Participants use cloud-based accounting software, business intelligence dashboards, spreadsheet automation tools, and collaborative online platforms that reflect contemporary accounting practice. Throughout the course, instructors provide timely, constructive feedback on practical tasks, enabling participants to refine their analytical techniques while strengthening their confidence in applying unfamiliar technologies.

Recognising that participants have varying levels of technological expertise, the course incorporates Universal Design for Learning principles by providing video demonstrations, written step-by-step guides, live workshops, downloadable reference materials, and optional advanced activities. Participants may demonstrate competence through written reports, recorded presentations, interactive dashboards, or workplace implementation plans, allowing flexibility while ensuring that all learners achieve the required professional outcomes.

These examples illustrate how learner engagement strategies can be adapted to different educational contexts while maintaining the same underlying principles. Whether supporting undergraduate architecture students or experienced accounting professionals, effective engagement is achieved by creating relevant learning experiences, encouraging active participation, providing authentic tasks, supporting collaboration, offering constructive feedback, scaffolding increasingly complex learning, and designing inclusive learning environments that accommodate diverse learner needs.

Wellbeing, Resilience and Success

Wellbeing, resilience, and success are closely interconnected concepts that play a significant role in adult learning. While knowledge, skills, and effective teaching strategies contribute to educational achievement, learners are also influenced by their physical health, psychological wellbeing, emotional resilience, and social circumstances (Cretton & Méndez, 2022; Hernández et al., 2019; Soroa et al., 2023). Research has demonstrated that resilience relates positively to optimism, life satisfaction, and perceived wellbeing, while also facilitating better coping with problems and relating negatively to depression, anxiety, and somatization (Cretton & Méndez, 2022). Furthermore, resilience has been identified as a capacity enabling individuals to anticipate, adapt, and successfully reorganize in the face of adversity, requiring physical, psychological, social, and cultural resources to sustain wellbeing (Soroa et al., 2023). Adult learners frequently balance study alongside employment, family responsibilities, financial commitments, and community obligations, making their educational experience substantially different from that of many younger learners (Cretton & Méndez, 2022; Hernández et al., 2019).

Non-traditional adult students who work and have family responsibilities have been found to report higher levels of resilience than traditional students, suggesting that assuming the challenge of studying at an older age with work and family responsibilities fosters a more solid construction of resilience (Cretton & Méndez, 2022). Similarly, emerging adults entering higher education experience opportunities and challenges such as combining school with work, greater independence, and establishing new social relationships (Hernández et al., 2019). Supporting learner wellbeing and resilience therefore contributes not only to academic achievement but also to persistence, motivation, engagement, and lifelong participation in education (Fernández et al., 2023; Hernández et al., 2019; Soroa et al., 2023).

Research confirms that students classified as resilient use adaptive coping strategies, know how to seek help, and perceive support from family and friends, while non-resilient students tend to procrastinate and have lower levels of psychological wellbeing (Hernández et al., 2019). Educational environments that foster supportive and affective relationships can contribute to resilience and psychological wellbeing (Soroa et al., 2023), and studies have identified significant links between psychological wellbeing, resilience, and academic performance (Fernández et al., 2023). Moreover, the perception of the learning environment as supportive of basic psychological needs is associated with increased psychological wellbeing and student recovery (Cretton & Méndez, 2022).

Wellbeing in Adult Learning

Wellbeing refers to an individual's overall physical, psychological, emotional, and social health (Chau & Ren, 2024; Lewis, 2020). Within education, learner wellbeing encompasses feeling safe, valued, supported, and capable of participating effectively in learning activities (MacDonald et al., 2018; Roffey & Quinlan, 2021). Positive wellbeing enables learners to concentrate, manage stress, regulate emotions, build relationships, and engage more fully with learning (Lewis, 2020; Zepke, 2013).

Adult learners often experience competing demands that can affect wellbeing, including employment pressures, financial stress, caring responsibilities, health concerns, and major life transitions Jokic et al., 2019; , Krikorian, 2020). Returning to education after a lengthy absence may also generate anxiety or uncertainty about academic ability (Jokic et al., 2019; Krikorian, 2020). When these pressures accumulate, learners may experience reduced motivation, lower confidence, diminished engagement, and an increased likelihood of withdrawing from study (Bartlett & Ross, 2022; Jokic et al., 2019).

Educational environments that promote wellbeing recognise these realities by creating supportive learning cultures characterised by respect, inclusion, flexibility, and open communication (Lewis, 2020; MacDonald et al., 2018). Educators contribute to learner wellbeing by establishing realistic expectations, providing timely feedback, encouraging collaboration, recognising learner achievements, and directing learners towards appropriate support services when required (Hassam et al., 2022; Jokic et al., 2019).

Understanding Resilience

Resilience is the capacity to adapt positively when encountering adversity, setbacks, stress, or change. Rather than avoiding difficulties, resilient learners develop the ability to recover from challenges, maintain motivation, and continue progressing towards their learning goals (Drewery et al., 2020; Kushch & Korolenko, 2024; Lee et al., 2024). The concept of resilience originates from the Latin *resilire*, meaning "to bounce back" (Kushch & Korolenko, 2024), and encompasses self-regulation, social interaction, problem-solving abilities, and motivation for learning and self-development (Kushch & Korolenko, 2024). Resilience is increasingly recognised as an essential capability for adult learners because educational journeys rarely progress without obstacles (Hall et al., 2015; Jansen & Wieland, 2024; Nygren et al., 2019). Educational resilience has been defined as "the heightened likelihood of success in school and other life accomplishments despite environmental adversities" (Gottron et al., 2024, p. 51), and it serves as a strong predictor of class participation and study motivation (Nygren et al., 2019).

Adult learners may encounter numerous challenges during their studies, including balancing competing priorities, adapting to unfamiliar technologies, managing academic workloads, coping with assessment pressures, or recovering from disappointing results (Hall et al., 2015; Jansen & Wieland, 2024; Liu & Li, 2023). The transition to higher education often brings significant challenges, such as becoming an independent learner or entering a new life phase (Jansen & Wieland, 2024). Learners with higher resilience are more likely to interpret setbacks as opportunities for learning rather than evidence of personal failure (Drewery et al., 2020; Liu & Li, 2023). A lifelong learning mindset, which incorporates curiosity, strategic thinking, and resilience, has been empirically associated with both objective career success (such as supervisor-rated performance and number of promotions) and subjective career success (including job satisfaction and work engagement) (Drewery et al., 2020). This growth-oriented perspective encourages persistence, problem-solving, and continuous improvement (Lee et al., 2024; Liu & Li, 2023). In transformative learning contexts, adult learners must develop resilience as "the ability to bounce back from adversity, and the determination to persevere" (Liu & Li, 2023, p. 8).

Importantly, resilience is not considered a fixed personal characteristic. It can be developed through supportive educational experiences, positive relationships, constructive feedback, reflective practice, and opportunities to experience incremental success (Hall et al., 2015; Jansen & Wieland, 2024; Kushch & Korolenko, 2024). Masten's work emphasises that resilience is a dynamic process, adaptable depending on circumstances, and requires intentional effort (Kushch & Korolenko, 2024). Research indicates that resilience training significantly improves emotional intelligence and fortitude among students (Jansen & Wieland, 2024), and higher education institutions, with their robust learning infrastructure and skilled educators, are considered ideal settings for delivering resilience training (Jansen & Wieland, 2024). Educational environments therefore play a critical role in helping learners strengthen resilience throughout their studies (Gottron et al., 2024; Hall et al., 2015; Jansen & Wieland, 2024). For Indigenous students, for example, resilience has been shown to be "a dynamic and multifaceted construct" that can be cultivated through pedagogies that value existing strengths and knowledge systems (Hall et al., 2015). Similarly, the development of information use practices during learning transitions has been found to support learning resilience by helping students navigate information and make informed decisions (Lee et al., 2024).

The Relationship Between Wellbeing and Resilience

Wellbeing and resilience reinforce one another. Positive wellbeing provides learners with the psychological resources needed to respond effectively to challenges, while resilience helps protect wellbeing during periods of stress or uncertainty (Molina-Montes et al., 2022; Omar, 2007). Learners who experience high levels of wellbeing generally demonstrate greater emotional regulation, confidence, motivation, and persistence (Noriega et al., 2020; Rico, 2022). Conversely, resilient learners are often better equipped to manage stress, recover from setbacks, and maintain positive wellbeing despite challenging circumstances (Freire et al., 2021; Molina-Montes et al., 2022; Moreno, 2023).

Research has demonstrated clear relationships between subjective wellbeing and resilience, with resilience components such as personal realisation, self-determination, and adaptation showing positive significant associations with specific wellbeing dimensions (Omar, 2007). The linkage between resilience and optimism can contribute to the design of effective programmes for stress prevention, and such programmes should include strategies that foster the development of psychological characteristics like resilience and optimism (Molina-Montes et al., 2022). Furthermore, school engagement has been identified as a factor promoting positive outcomes including student wellbeing and academic resilience (Gelpi-Trudo et al., 2021). Students who possess strong self-efficacy beliefs participate more confidently, work with greater rigour, persist longer, and exhibit fewer adverse emotional reactions when encountering difficulties (Rico, 2022). Educating under the pillars of positive psychology, together with elements constituting the theory of wellbeing, is considered essential for strengthening resilience and minimising emotional distress throughout life (Molina-Montes et al., 2022).

Educators contribute to both wellbeing and resilience by fostering psychologically safe learning environments where mistakes are viewed as normal components of learning (Bitrán, 2021; Iglesias-Díaz & Pérez, 2021; Salvo-Garrido et al., 2021). The creation of a climate of trust, respect, and reciprocity plays a fundamental role, as it allows students to perceive a sense of wellbeing and the psychosocial conditions necessary to engage with academic tasks (Rico, 2022). Positive psychology affirms that to facilitate the development and wellbeing of both students and teachers, it is necessary to generate climates of safety and positive emotions in the classroom (Molina-Montes et al., 2022). Educational centres can contribute to developing resilience, psychological wellbeing, and inclusion among the most vulnerable students through the learning environment (Iglesias-Díaz & Pérez, 2021). Encouraging reflection, celebrating progress, promoting realistic goal setting, and providing constructive support help learners develop confidence while reducing fear of failure. Teachers who demonstrate commitment, trust, motivation, orientation, and protective attitudes promote resilience and good academic performance through bonds of trust that enable personalised attention focused on students' socio-emotional wellbeing (Salvo-Garrido et al., 2021). Dialogic, cooperative, and participatory methodologies have been shown to enhance the emotional wellbeing of students by satisfying their basic psychological needs (Iglesias-Díaz & Pérez, 2021). It is the responsibility of teachers to detect students "at risk" and help them build their resilience, offering protective factors within schools (Noriega et al., 2020), while the school itself, after the family, constitutes a fundamental environment for students to acquire the competencies necessary to overcome adversity (Freire et al., 2021).

Strategies for Supporting Wellbeing and Resilience

Creating psychologically safe classrooms encourages learners to ask questions, express uncertainty, and participate without fear of embarrassment or criticism (Murray et al., 2025). Learning environments that demonstrate empathy, inclusiveness, and psychological safety have shown positive results in students' behaviour, self-esteem, motivation, and academic success (Murray et al., 2025). Respectful communication, inclusive teaching practices, and positive educator-learner relationships strengthen learners' sense of belonging and encourage active participation (Roffey & Quinlan, 2021). Supportive teachers have the power to foster a safe classroom climate and environment, set clear expectations, and open inclusive dialogue with students, while the day-to-day experience of teaching and learning within the classroom provides the greatest opportunity to contribute to student wellbeing (Pant & Rastogi, 2024). A positive school culture and climate fosters the wellbeing of all members and reflects an environment where everyone feels safe and experiences a sense of belonging (Pant & Rastogi, 2024).

Providing manageable learning challenges helps learners experience regular success while gradually increasing task complexity. Scaffolding difficult concepts enables learners to build competence progressively without becoming overwhelmed (Nohilly & Tynan, 2022). Constraints-informed pedagogies and learner-centred approaches can enhance learning by fostering autonomy through interactive engagement, whereby educators plan learning sequences informed by knowledge of learner constraints. Active learning pedagogies can benefit all participants when intentionally designed with wellbeing as a priority (Nohilly & Tynan, 2022).

Constructive feedback also strengthens resilience by helping learners view mistakes as opportunities for growth rather than evidence of failure (Murray et al., 2025). The social and cognitive safety net that enables learners to aim high while taking risks and to turn failure into resilience is rooted in safe relationships and in holding a growth mindset (Murray et al., 2025). Educators can help students recognise their strengths and build resilience through principles of titrated exposure, self-care, and peer support (Mabena-Segoe & Mabunda, 2025). Feedback that identifies strengths, explains areas for improvement, and provides clear guidance for future learning promotes self-efficacy and persistence. Positive education interventions that include strengths-based learning and growth mindset cultivation promote engagement, resilience, and intrinsic motivation.

Goal setting supports resilience by helping learners focus on achievable milestones rather than becoming discouraged by distant objectives. Given a comprehensive understanding of classroom climate significance, teacher educators can enhance learning by ensuring learners have the chance to manage their own learning with specific goals set (Sheinman & Russo-Netzer, 2021). Breaking larger tasks into smaller, manageable goals enables learners to experience continual progress, reinforcing motivation and confidence (Nohilly & Tynan, 2022).

Reflection encourages learners to recognise their own growth over time (Nohilly & Tynan, 2022; Roffey & Quinlan, 2021). Curriculum design should incorporate systematic resilience-building through evidence-based strategies including reflection, peer activities, experiential learning, and mindfulness (Nohilly & Tynan, 2022). Learning journals, portfolios, and self-assessment activities help learners appreciate improvements that may otherwise go unnoticed while developing greater awareness of effective learning strategies (Roffey & Quinlan, 2021). Collaborative activities, self-assessment and reflection, and combining subject learning with social-emotional learning abilities have been recommended as effective pedagogical approaches (Roffey & Quinlan, 2021).

Building supportive learning communities further enhances resilience. PBL tutorial groups and similar collaborative structures function as learning communities where participants negotiate meaning, develop shared practices, and construct professional identities, with the quality of these communities influencing psychological safety and wellbeing (Nohilly & Tynan, 2022). Peer collaboration, mentoring, discussion groups, and communities of practice reduce feelings of isolation and provide opportunities for learners to share experiences, exchange advice, and offer mutual encouragement (Nohilly & Tynan, 2022). Factors such as student agency and voice, school belonging, self and social awareness, and supportive peer relationships can support learners rather than further isolating them. Informal spaces of learning can foster student engagement and provide further opportunities for students to engage in social advocacy and promote a positive learning climate.

Wellbeing, Resilience, and Academic Success

A growing body of educational research demonstrates that learner wellbeing and resilience are associated with higher levels of engagement, persistence, academic achievement, and course completion. Learners who experience positive wellbeing are generally more capable of concentrating, regulating emotions, solving problems, and maintaining motivation (Balbín et al., 2022; Pilozo et al., 2024). Similarly, resilient learners are more likely to continue studying despite setbacks (Aguilar et al., 2023; Hinostroza, 2020), seek assistance when required (Aguilar et al., 2023), and adapt their learning strategies in response to feedback (Cerda et al., 2023).

Research consistently shows that resilient students maintain firm commitment to their academic tasks, evidenced by intrinsic motivation and confidence in their abilities and competencies (Hinostroza, 2020). Students with higher levels of resilience demonstrate lower intention to abandon their studies, while those with low resilience show low self-esteem, lack confidence in their possibilities, and experience anxiety regarding academic activities, which directly affects their academic performance (Aguilar et al., 2023). Indeed, resilience has been identified as having predictive value in relation to academic achievement, with evidence of its capacity to predict school engagement and academic performance (Revuelta et al., 2024). Emotional stability and the ability to control impulses, maintain a life plan, and remain optimistic in the face of problems are determining factors for developing resilience and achieving academic, personal, and social success (Balbín et al., 2022). Furthermore, students with resilient characteristics maintain significant positive associations with self-efficacy, self-regulation strategies, coping strategies, and motivational states of commitment (Cerda et al., 2023).

Academic success should therefore be viewed as extending beyond examination results or qualification completion. As (Kahu, 2013) argues, from a constructivist framework, universities should be more than obtaining qualifications, advocating for an approach that incorporates the notion of "becoming" a person as an educational objective, with student satisfaction considered as part of subjective wellbeing toward a broader consideration of academic success (Gutiérrez et al., 2018). Success also includes developing confidence, independence, adaptability, lifelong learning skills, professional competence, and the capacity to continue learning throughout changing personal and professional circumstances (Aguilar et al., 2023; Suárez et al., 2024). Working resilience as a component of the student's maturational process and as a cross-cutting content of teaching can lead to better emotional balance and better integral development (Aguilar et al., 2023). A positive school climate characterised by solid interpersonal relationships, a sense of belonging, and opportunities for participation promotes the development of resilience and emotional wellbeing (Suárez et al., 2024), while the promotion of resilience has important implications for preventing academic dropout and promoting students' emotional wellbeing (Suárez et al., 2024).

Wellbeing, Resilience, and Lifelong Learning

The importance of wellbeing and resilience extends well beyond formal education. Modern careers increasingly require individuals to adapt continuously to technological change, evolving workplace practices, and shifting labour market demands (Goodwill & Chen, 2021; Williams, 2024). Lifelong learning depends not only on acquiring new knowledge but also on maintaining the psychological capacity to embrace change, manage uncertainty, and recover from inevitable setbacks (Fang & Sim, 2024; Goodwill & Chen, 2021; Mengting, 2025).

Adult education therefore has a broader purpose than simply transmitting knowledge or developing vocational competence (Merriam & Kee, 2014; Narushima et al., 2016). By supporting learner wellbeing, fostering resilience, and encouraging reflective, self-directed learning, educators help learners develop the confidence, adaptability, and perseverance needed for continued personal, academic, and professional growth throughout their lives (Goodwill & Chen, 2021; Hameleers et al., 2022; Narushima et al., 2016; Paul et al., 2020).

Key Concepts and Conclusions

This chapter examined the major factors that influence adult learner motivation, engagement, persistence, and success. Adult learners differ from younger learners in that participation is generally voluntary and shaped by competing work, family, financial, and personal commitments. Consequently, motivation is influenced by a complex interaction of personal aspirations, workplace demands, psychological factors, social relationships, and the broader learning environment. Understanding these influences enables educators to design learning experiences that encourage participation, sustain engagement, and support successful learning outcomes.

Several influential theories explain why adults choose to engage in learning and what sustains their commitment over time. Intrinsic and extrinsic motivation highlight the importance of both personal satisfaction and external rewards. Self-Determination Theory demonstrates that autonomy, competence, and relatedness are fundamental psychological needs that underpin motivation. Knowles' theory of andragogy emphasises the importance of relevance, self-direction, and internal motivation, while Expectancy Theory, Goal Theory, Achievement Motivation Theory, Maslow's Hierarchy of Needs, and Transformative Learning each provide complementary perspectives on the diverse factors that influence adult learning behaviour. Collectively, these theories demonstrate that no single explanation adequately accounts for adult motivation and that effective educational practice requires consideration of multiple motivational influences.

The chapter also examined the importance of self-efficacy and confidence. Adults who believe they can successfully complete learning tasks are more likely to participate, invest effort, persist through challenges, and achieve positive learning outcomes. Self-efficacy develops through successful experiences, observing others succeed, constructive feedback, and supportive learning environments. Educators therefore play a critical role in strengthening learners' confidence by providing achievable challenges, timely feedback, appropriate scaffolding, and opportunities for early success.

Persistence and learner retention were explored as essential components of successful adult education. Adult learners frequently encounter interruptions arising from employment, family responsibilities, financial pressures, health concerns, and other life circumstances. Persistence depends upon the interaction between learner characteristics, educational quality, social support, and institutional practices. Flexible delivery, learner support services, relevant learning experiences, and positive educational relationships all contribute to improved retention and completion rates.

The chapter identified a wide range of barriers that may limit adult participation in education. Situational, institutional, dispositional, financial, technological, cultural, workplace, and psychological barriers each influence participation in different ways, and many learners experience several simultaneously. Reducing these barriers requires coordinated action from educators, educational institutions, employers, governments, and communities to improve accessibility, flexibility, equity, and learner support.

Learner engagement was presented as a multidimensional concept involving behavioural, cognitive, emotional, and social participation. Engagement is strengthened when learning is meaningful, authentic, collaborative, learner-centred, and connected to adults' existing knowledge and future goals. Strategies such as active learning, authentic assessment, scaffolding, constructive feedback, learner autonomy, inclusive teaching practices, Universal Design for Learning, culturally responsive teaching, reflective practice, and technology-enhanced learning all contribute to greater participation, deeper learning, and improved learner success.

Finally, the chapter highlighted the importance of learner wellbeing and resilience. Adult learners who experience positive wellbeing and develop resilience are better equipped to manage stress, adapt to change, recover from setbacks, and persist towards their educational goals. Educational environments that foster psychological safety, supportive relationships, realistic goal setting, reflective practice, and opportunities for success not only improve academic performance but also prepare learners for lifelong learning and continuous professional development.

Motivation, engagement, and learner success are central to effective adult education. Although adults participate in learning for diverse reasons, research consistently demonstrates that meaningful, relevant, and learner-centred educational experiences are most likely to sustain motivation and promote long-term participation. Adult learners respond positively when education addresses real-life problems, supports personal and professional goals, and recognises the knowledge and experience they already possess.

No single theory fully explains adult motivation. Instead, adult learning is best understood through the combined insights provided by theories of motivation, self-efficacy, engagement, persistence, and wellbeing. These complementary perspectives demonstrate that successful learning depends upon far more than the delivery of knowledge. Learner success is shaped by the interaction of psychological, social, organisational, and environmental factors that influence whether adults choose to participate, remain engaged, overcome challenges, and continue learning throughout their lives.

The chapter also demonstrates that learner success is a shared responsibility. While adult learners exercise considerable autonomy and personal responsibility, educators, institutions, employers, and policymakers all contribute to creating learning environments that foster motivation, remove barriers, strengthen confidence, and support persistence. Flexible delivery, inclusive curriculum design, supportive teaching practices, authentic learning experiences, constructive feedback, and accessible learner support services all enhance participation and achievement.

Ultimately, successful adult education extends beyond qualification completion or academic achievement. Its broader purpose is to develop capable, confident, resilient, and self-directed lifelong learners who are able to adapt to changing personal, professional, and societal circumstances. By understanding the factors that influence motivation, engagement, persistence, wellbeing, and learner success, educators can create learning environments that not only improve educational outcomes but also empower adults to continue learning, growing, and contributing throughout their lives.

Chapter 7

Diversity, Inclusion, and Global Perspectives

Every learner brings a unique combination of cultural background, personal identity, life experience, language, values, abilities, and social circumstances that shape how they engage with education. As societies become increasingly interconnected through globalisation, migration, technological change, and international collaboration, adult educators are teaching increasingly diverse groups of learners whose educational needs, expectations, and experiences may differ significantly. Recognising and valuing this diversity has become an essential component of effective adult education rather than an optional consideration.

This chapter examines how diversity, inclusion, and global perspectives influence adult learning and educational practice. It explores the ways in which culture shapes learning, compares cross-cultural approaches to adult education, considers Indigenous perspectives on learning, and examines issues of gender, equity, social justice, disability, migration, language and literacy, and lifelong learning from an international perspective. The chapter also discusses how educators can create inclusive, culturally responsive, and accessible learning environments that recognise the strengths and needs of diverse adult learners. By understanding the complex relationship between identity, culture, and education, adult educators can design learning experiences that not only improve educational outcomes but also promote equity, participation, and lifelong learning in an increasingly diverse and global society.

Cultural Influences on Learning

Culture is a fundamental influence on how adults perceive, experience, and engage in learning. It shapes individuals' beliefs about knowledge, communication, authority, collaboration, motivation, and the purpose of education. Adult learners bring to educational settings a wide range of cultural values, traditions, languages, experiences, and worldviews that influence not only what they learn but also how they prefer to learn (Conceição & Biniecki, 2024; Singh, 2019). Understanding these cultural influences enables educators to create more inclusive, respectful, and effective learning environments that recognise and value learner diversity (Erol & Altunay, 2023; Lwanga et al., 2024). As UNESCO has acknowledged, the very notion of "adult" is culturally and socially sensitive in a global society, and the meaning of learning itself is far from obvious (Ranieri & Pachler, 2014). Culture and cultural determinants can prevent cultural minority groups from participating as much as majority groups in adult education programmes (Singh, 2019).

Culture influences learners' expectations of the educational process. In some cultures, teachers are viewed as authoritative experts whose role is to transmit knowledge, and learners may demonstrate respect by listening attentively, avoiding questions, and waiting for explicit instruction (Henschke, 2011b; Wong, 2013). Research on teaching styles in Taiwan, for instance, found that although instructors tended toward andragogical approaches, a significant difference persisted between practiced teaching styles and students' preferences, with researchers surmising that Eastern cultural influence contributes to this gap (Henschke, 2011b). In contrast, other cultures encourage open discussion, questioning, debate, and collaborative knowledge construction, with educators acting as facilitators rather than sole sources of knowledge (Baschiera, 2019; Lutomia & Bello-Bravo, 2024). Adult learners entering educational environments with different expectations may initially find unfamiliar teaching approaches challenging. For example, learners accustomed to teacher-directed instruction may be hesitant to participate in learner-centred activities that require independent inquiry or critical discussion (Lwanga et al., 2024), while learners familiar with collaborative approaches may expect greater opportunities for dialogue and shared decision-making (Chakraborty, 2024). As Lwanga et al. note, the increasing diversity of adult learners in educational settings invites educators to envision and embrace learner diversity in the classroom and reflect on their instructional styles (Lwanga et al., 2024; . The selection of methods, techniques, and tools in adult education depends on the needs, interests, cultural structure, and other characteristics of the adult learners participating (Erol & Altunay, 2023).

Communication styles are also strongly influenced by culture and can affect classroom interactions. High-context cultures often rely on indirect communication, shared understanding, and non-verbal cues, whereas low-context cultures generally favour direct, explicit communication (Economides, 2008). Economides highlights that in collaborative learning environments, learners from diverse cultures require different communication and collaboration tools adapted to their cultural profiles (Economides, 2008). Similarly, cultural norms influence eye contact, turn-taking in conversations, willingness to express disagreement, and attitudes towards questioning instructors (Economides, 2008; Henschke, 2011b). Research has shown, for example, that Hispanic adult learners demonstrated strong preferences for feedback and concrete learning, while the impact of feedback on learning processes was found to be highly dependent on cultural context, with in-class questions producing different effects in Hong Kong compared to the United States (Economides, 2008). Educators who recognise these differences are better positioned to interpret learner behaviour accurately and avoid mistaking cultural communication preferences for disengagement or lack of understanding (Lwanga et al., 2024; Singh, 2019).

Cultural values also influence attitudes towards individual and collective learning. Learners from more individualistic cultures may place greater emphasis on personal achievement, independence, and individual responsibility for learning (Kizilcec & Cohen, 2017; Rothstein-Fisch et al., 2009). Cultures that give priority to the needs of the individual, such as independence, freedom of choice, and self-expression, can be described as individualistic, while those that prioritise family or group needs, social relationships, group success, and shared property can be described as collectivistic (Rothstein-Fisch et al., 2009). In contrast, learners from collectivist cultures may value cooperation, group harmony, shared success, and collaborative problem-solving (Kizilcec & Cohen, 2017; Lutomia & Bello-Bravo, 2024). Brookfield's critique of Knowles's andragogy noted that it centred on learning experiences of those in individualistic, not collectivistic societies, where learning takes place in groups and activities are more frequently done together (Lutomia & Bello-Bravo, 2024). Kizilcec and Cohen demonstrated empirically that a self-regulation intervention was effective for learners in individualist countries but not for those in collectivist cultures, establishing that even highly efficacious educational interventions may be culturally bounded in their effects (Kizilcec & Cohen, 2017). Ubuntu philosophy, for instance, encourages a holistic view of education where individuals are active participants in a broader social and cultural context, emphasising interconnectedness, diverse perspectives, and collaborative learning communities (Lutomia & Bello-Bravo, 2024). These differing perspectives influence participation in group activities, peer learning, assessment preferences, and classroom interactions (Economides, 2008; Ilie, 2019). Although adult learners may prefer group activities, research has found that they sometimes report learning better individually, suggesting that collaborative learning is important but that levels of autonomy and independence in learning remain high (Ilie, 2019). Cross-cultural research on peer-educators in lifelong learning organisations similarly found that U.S. participants emphasised individual enrichment while Portuguese participants stressed collective and relational benefits, reflecting broader cultural orientations (Casanova et al., 2026). Effective adult education recognises that both individual and collaborative approaches have value and provides opportunities for learners to engage in multiple ways (Chakraborty, 2024; Conceição & Biniecki, 2024; Rothstein-Fisch et al., 2009).

Language and literacy represent additional cultural influences on learning. Adult learners studying in a second or additional language may possess extensive professional knowledge and life experience but encounter difficulties expressing that knowledge in the language of instruction (Cantarutti et al., 2023; Lagua et al., 2022). Cultural differences in academic writing conventions, communication styles, and educational terminology may also affect participation and assessment performance (Baidoo-Anu et al., 2023; Núñez, 2024). Providing clear explanations, accessible resources, visual supports, and opportunities for clarification helps reduce unnecessary linguistic barriers while maintaining academic standards (Bedoya et al., 2023; Ferrel-Ballestas et al., 2022).

Previous educational experiences, which are themselves shaped by cultural contexts, also influence adult learning (Baque et al., 2024; Núñez, 2024). Educational systems differ considerably across countries in their teaching methods, assessment practices, classroom relationships, and expectations of learners (Baidoo-Anu et al., 2023; Romero et al., 2020). Adults who have experienced highly structured educational environments may initially struggle with self-directed learning, while those accustomed to independent learning may adapt more readily to learner-centred approaches (Núñez, 2024). Some learners may also have experienced interrupted schooling, educational disadvantage, conflict, migration, or discrimination, all of which can influence confidence, self-efficacy, and willingness to participate in further education (Fernández & Olivencia, 2021; Mejía et al., 2023).

Cultural identity intersects with many other aspects of diversity, including age, gender, religion, socioeconomic background, disability, and professional identity (Totena, 2023; Villafuerte, 2020). Adult learners should therefore not be viewed solely through the lens of their national or ethnic culture. Individuals often belong to multiple cultural communities simultaneously, and their learning preferences are influenced by the interaction of these identities rather than by culture alone (Bedoya et al., 2023; Rodríguez et al., 2024). Avoiding stereotypes and recognising individual differences within cultural groups are essential components of culturally responsive adult education (Baidoo-Anu et al., 2023; Rodríguez et al., 2024).

Culturally responsive teaching seeks to acknowledge, respect, and incorporate learners' cultural knowledge, experiences, and perspectives into the educational process (Onofre et al., 2024; Rodríguez et al., 2024). Rather than expecting learners to adapt completely to a single educational culture, culturally responsive educators build on learners' existing knowledge and create opportunities for multiple perspectives to be explored (Baque et al., 2024; Velarde et al., 2024). Learning materials that include diverse examples, international case studies, and culturally relevant contexts help learners see their own experiences reflected in the curriculum while broadening understanding of other perspectives (Núñez, 2024; Onofre et al., 2024). Encouraging learners to contribute examples from their own professional, cultural, or community experiences enriches discussion and supports collaborative knowledge construction (Bedoya et al., 2023; Velarde et al., 2024).

Inclusive curriculum design also contributes to culturally responsive learning. Universal Design for Learning, differentiated instruction, flexible assessment methods, and accessible learning resources provide multiple ways for learners to access information, engage with learning activities, and demonstrate competence (Bedoya et al., 2023; Coral, 2024; Ferrel-Ballestas et al., 2022). These approaches reduce barriers that may disproportionately affect learners from diverse cultural and linguistic backgrounds while benefiting all learners (Mejía et al., 2023; Totena, 2023). Flexible teaching practices recognise that there is no single way to demonstrate understanding or achieve learning outcomes (Bedoya et al., 2023; Coral, 2024).

Globalisation has further increased the importance of cultural competence in adult education (Cantarutti et al., 2023; Romero et al., 2020). Universities, vocational education providers, and workplaces increasingly serve culturally diverse learner populations, while many educational programs prepare learners for employment in international and multicultural environments (Angarita et al., 2018; Núñez, 2024). Developing intercultural competence, the ability to communicate, collaborate, and work effectively with people from different cultural backgrounds, has therefore become an important educational outcome in its own right (Angarita et al., 2018; Cantarutti et al., 2023). Learning environments that encourage respectful dialogue, critical reflection, and appreciation of diverse perspectives help prepare adults for participation in increasingly interconnected societies and workplaces (Rodríguez et al., 2024; Romero et al., 2020).

Recognising cultural influences on learning strengthens educational quality, equity, and learner engagement (Bedoya et al., 2023; Elizalde, 2020). When educators understand how culture shapes learning experiences, expectations, communication, and participation, they are better able to design learning environments that are inclusive, supportive, and responsive to learner diversity (Baidoo-Anu et al., 2023; Lagua et al., 2022). Such environments not only improve educational outcomes but also foster mutual respect, intercultural understanding, and the lifelong learning capabilities needed in an increasingly diverse and globalised world (Elizalde, 2020; Romero et al., 2020).

As a practical example, consider a flight instructor at a flying school is training an international student pilot who recently relocated to Australia to complete a Commercial Pilot Licence (CPL). The student previously learned in an educational system where instructors were regarded as unquestioned authorities, and students were expected to observe, memorise, and follow instructions precisely rather than ask questions or openly discuss their understanding.

During the first few lessons, the instructor notices that although the student appears attentive, they rarely ask questions during pre-flight briefings and consistently respond with "yes" when asked if they understand the lesson. However, during the flight, the student occasionally demonstrates uncertainty when applying procedures, particularly during simulated emergencies and navigation exercises. Rather than assuming the learner is disengaged or unprepared, the instructor recognises that the student's behaviour may reflect cultural expectations about respecting instructors rather than a lack of interest or ability.

To create a more culturally responsive learning environment, the instructor modifies their teaching approach. Instead of asking, "Do you understand?", they ask the student to explain the procedure in their own words or describe what actions they would take in a particular scenario before entering the aircraft. During debriefings, the instructor reassures the student that asking questions is viewed as a normal and expected part of aviation training and that discussing mistakes openly contributes to safer flying rather than reflecting poorly on the learner. Over time, the instructor gradually encourages the student to take greater responsibility for flight planning, weather interpretation, and decision-making while continuing to provide appropriate guidance and support.

Recognising that English is the student's second language, the instructor supplements verbal explanations with diagrams, aircraft checklists, cockpit photographs, and demonstration videos. Technical aviation terminology is introduced progressively, with opportunities for clarification before each lesson. During radio communication practice, the instructor allows additional time for the student to develop confidence using standard aviation phraseology without feeling rushed or embarrassed. These strategies reduce unnecessary language barriers while maintaining the high standards required for aviation safety.

The instructor also recognises that aviation itself has a distinct professional culture that emphasises communication, teamwork, and the willingness to question unsafe situations regardless of seniority. As the student's confidence develops, the instructor incorporates scenario-based exercises requiring the student to challenge unsafe decisions, communicate concerns clearly, and exercise pilot-in-command authority during simulated flights. These activities not only develop technical flying skills but also help the learner adapt to the safety culture expected within the aviation industry.

By recognising the influence of culture on communication, learning preferences, and classroom expectations, the instructor creates an environment in which the student feels respected, supported, and increasingly confident to participate actively in the learning process. Rather than requiring the learner to abandon their cultural identity, the instructor builds upon the student's strengths while helping them develop the communication, decision-making, and critical thinking skills essential for safe and effective professional aviation practice.

Cross-Cultural Perspectives on Adult Education

Adult education is shaped not only by educational theory but also by the cultural, historical, political, economic, and social environments in which learning takes place. Although many of the theories that have influenced contemporary adult education were developed in Europe and North America (Clair & Käpplinger, 2021; Moll, 2023), adult learning is practised differently across the world in ways that reflect diverse understandings of knowledge, teaching, work, community, and the broader purposes of education (Khattab & Wong, 2018; Yunga, 2019). Cross-cultural perspectives emphasise that adult education cannot be understood through a single universal framework. Instead, effective educational practice requires an appreciation of how cultural values shape learner motivations, expectations, teaching approaches, and educational outcomes (Bishri, 2023; Khattab & Wong, 2018; Moll, 2023).

The purposes of adult education also vary considerably across societies. In many countries, adult learning is closely linked to workforce development, employability, economic productivity, and lifelong skills development, with vocational education and professional learning designed to prepare adults for changing labour markets and technological innovation (Khattab & Wong, 2018; Mbotake, 2016; Moll, 2023; Yunga, 2019). Elsewhere, adult education serves broader social objectives, including community development, civic engagement, cultural preservation, social justice, environmental sustainability, health promotion, and democratic participation (Ajayi, 2019; Clair & Käpplinger, 2021). These differing priorities influence curriculum design, instructional methods, assessment practices, and the roles adopted by educators and learners. For example, Freire's critical pedagogy positioned literacy education as a means of helping adults critically examine their social, political, and economic circumstances, demonstrating that adult education may function as an instrument of social transformation as well as individual development (Moll, 2023).

Cultural values also influence fundamental assumptions about how learning occurs. Educational philosophies that have emerged from many Western traditions often emphasise learner autonomy, critical thinking, self-direction, and individual responsibility for learning (Clair & Käpplinger, 2021; Khattab & Wong, 2018; Moll, 2023). These principles underpin influential approaches such as Knowles' andragogy, self-directed learning, and transformative learning (Clair & Käpplinger, 2021; Moll, 2023; Yunga, 2019). However, scholars have argued that these theories reflect the social and cultural assumptions of the contexts in which they were developed and may not fully represent the educational values of all societies. In particular, andragogy has been criticised for being grounded in the experiences of Western, individualistic, middle-class adults, with assumptions that may inadequately recognise the diversity of cultural traditions, social structures, and educational practices found globally (Clair & Käpplinger, 2021; Moll, 2023).

In many collectivist societies, learning is understood primarily as a shared social activity that emphasises cooperation, mutual responsibility, respect for elders, and maintaining harmonious relationships within families and communities (Bishri, 2023; Khattab & Wong, 2018). Educational traditions in countries such as Sri Lanka, for example, have been influenced by cultural values that prioritise respect for authority, collectivism, and communal harmony, while adult education across the Arab world reflects the combined influence of religion, language, political structures, socioeconomic conditions, and ethnic diversity (Bishri, 2023; Khattab & Wong, 2018). Research has consistently shown that learning is inseparable from learners' cultural identities, prior experiences, and belief systems, suggesting that educational approaches should be responsive to the social and cultural contexts in which learning occurs (Khattab & Wong, 2018).

These differences have important implications for educational practice. Teaching methods that rely heavily on independent learning, individual assessment, or highly self-directed approaches may not align with the expectations or preferences of learners whose educational experiences have emphasised collective responsibility and community participation (Bishri, 2023; Khattab & Wong, 2018; Moll, 2023). Likewise, learners from historically marginalised or culturally diverse communities may approach learning in ways that differ from the assumptions embedded within traditional Western models of adult education. Consequently, many scholars advocate adapting adult education to local cultural contexts rather than applying universal teaching models. Culturally responsive adult education seeks to balance internationally recognised educational principles with local knowledge, values, traditions, and learner experiences, ensuring that learning remains both globally informed and locally relevant (Khattab & Wong, 2018; Moll, 2023).

Cross-cultural research has demonstrated that teaching strategies effective in one cultural context may not produce the same outcomes elsewhere. For example, approaches that encourage open debate, critical questioning of instructors, or individual competition may be highly effective in some educational systems but may create discomfort in cultures where respect for authority, consensus-building, or indirect communication are more highly valued (Hahambu et al., 2012; Khuzwayo, 2022). Similarly, interventions designed to increase learner autonomy or self-regulation may require adaptation when implemented in cultures that emphasise collective decision-making and shared responsibility (Dhungana, 2021; Hahambu et al., 2012). These findings highlight the importance of adapting educational practices to local cultural contexts rather than assuming that one model of teaching is universally applicable (Dhungana, 2021; Hahambu et al., 2012).

Language is another important cross-cultural consideration. Adult learners participating in education through a second or additional language often face challenges that extend beyond vocabulary or grammar (Cooper, 2010; Liu, 2023). Educational terminology, academic writing conventions, classroom interaction styles, and assessment expectations differ across countries and educational systems (Dhungana, 2021; Shohel et al., 2024). Learners may possess substantial professional expertise but require additional support to communicate that knowledge effectively in the language of instruction (Cooper, 2010). Providing multilingual resources, visual supports, opportunities for discussion, and culturally appropriate examples helps create more equitable learning environments for linguistically diverse learners (Dhungana, 2021; Shohel et al., 2024).

Cross-cultural perspectives also emphasise the importance of recognising Indigenous knowledge systems and local ways of knowing. Indigenous approaches to learning often place greater emphasis on relationships, oral traditions, storytelling, observation, participation, connection to land, intergenerational learning, and community responsibility (Hahambu et al., 2012; Khuzwayo, 2022). Rather than viewing these approaches as alternatives to formal education, many educators now recognise their value in broadening understandings of knowledge, learning, and education (Jeannin, 2021; Khuzwayo, 2022). Incorporating Indigenous perspectives enriches adult education by acknowledging multiple forms of knowledge and challenging assumptions that Western educational models represent the only valid approach to learning (Hahambu et al., 2012; Jeannin, 2021; Khuzwayo, 2022).

Globalisation has significantly expanded opportunities for cross-cultural learning. International student mobility, migration, multinational workplaces, online education, and global professional networks increasingly bring together learners from diverse cultural backgrounds (Kuffuor et al., 2024; Lihe, 2018). Adult learners frequently participate in multicultural classrooms where different perspectives, experiences, and problem-solving approaches contribute to richer learning experiences (Jeannin, 2021; Shohel et al., 2024). At the same time, educators must be prepared to facilitate intercultural communication, manage cultural misunderstandings, and create inclusive environments where all learners feel respected and able to participate (Dhungana, 2021; Kuffuor et al., 2024).

Developing intercultural competence has therefore become an important goal of adult education. Intercultural competence involves the ability to communicate effectively, collaborate respectfully, and work productively with people from different cultural backgrounds (Kuffuor et al., 2024; Liu, 2023). It includes cultural awareness, empathy, openness to diverse perspectives, reflective practice, and the ability to adapt communication and behaviour across different cultural contexts (Kuffuor et al., 2024; Wickline et al., 2024). These capabilities are increasingly valued across higher education, vocational education, professional practice, and international workplaces (Kuffuor et al., 2024; Lihe, 2018; Liu, 2023).

Cross-cultural perspectives also encourage educators to examine their own cultural assumptions. Teaching practices, curriculum design, assessment methods, and definitions of learner success are often influenced by educators' own educational experiences and cultural values (Dhungana, 2021; Shohel et al., 2024). Reflective educators recognise that what appears to be "normal" or "best practice" within one educational tradition may not be universally appropriate (Hahambu et al., 2012; Khuzwayo, 2022). By critically examining their own assumptions and remaining open to alternative educational perspectives, educators can design learning experiences that are more inclusive, flexible, and responsive to learner diversity (Dhungana, 2021; Kuffuor et al., 2024; Shohel et al., 2024).

Cross-cultural perspectives broaden our understanding of adult education by recognising that learning is both a universal human activity and a culturally situated experience. Although adults around the world share common needs for learning, personal growth, and adaptation, the ways in which these needs are expressed and supported vary considerably across cultures (Dhungana, 2021; Hahambu et al., 2012). Effective adult education therefore requires not only sound pedagogical knowledge but also cultural humility, intercultural competence, and a willingness to value diverse ways of knowing, teaching, and learning (Jeannin, 2021; Khuzwayo, 2022; Kuffuor et al., 2024). As societies become increasingly interconnected, these cross-cultural perspectives will continue to play a central role in creating equitable, inclusive, and globally relevant adult education systems (Kuffuor et al., 2024; Lihe, 2018).

The following example illustrates how cross-cultural perspectives can be incorporated into adult education practice. It demonstrates how an educator can recognise culturally influenced differences in communication, participation, authority, collaboration, and assessment without stereotyping individual learners. By using flexible and inclusive teaching strategies, the educator enables adults from different cultural backgrounds to contribute their experience while achieving shared learning outcomes.

A university in Singapore delivers an executive leadership programme for experienced managers from across Southeast Asia, including participants from Singapore, Japan, Indonesia, Vietnam, India, and the Philippines. Although all participants have extensive management experience, they bring different cultural values, workplace norms, and educational experiences that influence how they approach learning.

During the first workshop, the facilitator notices clear differences in classroom participation. Managers from Australia and the United Kingdom, who are also attending the programme, readily volunteer opinions, debate case studies, and openly question the facilitator's assumptions. In contrast, several participants from Japan and Vietnam are more reserved during whole-class discussions, preferring to listen carefully before contributing. Rather than interpreting this behaviour as a lack of engagement, the facilitator recognises that many participants come from educational and workplace cultures where careful reflection, respect for seniority, and avoiding public disagreement are considered appropriate professional behaviours.

To create a learning environment that accommodates these differing cultural expectations, the facilitator redesigns the learning activities. Instead of relying primarily on open classroom discussion, participants first analyse case studies individually before discussing them in small multicultural groups. Each group appoints a spokesperson to present its conclusions, ensuring that all members contribute without placing unnecessary pressure on individuals who may be less comfortable speaking in large groups.

The facilitator also encourages participants to compare leadership practices from their own countries. When examining employee motivation, Japanese participants describe consensus-based decision-making and long-term organisational commitment, while Indonesian managers discuss the importance of relationship-building and community values. Participants from India explain how leadership often involves balancing organisational objectives with highly diverse cultural and linguistic workforces. Rather than identifying one approach as superior, the facilitator guides discussion towards understanding how leadership practices are shaped by cultural, historical, and organisational contexts.

Assessment is designed to reflect this diversity. Instead of requiring every participant to complete the same type of assignment, learners may choose between an individual reflective analysis, a workplace improvement project, or a collaborative organisational case study. Regardless of the format selected, each assessment requires participants to analyse how cultural factors influence leadership decisions within their own professional environments and to justify strategies appropriate for their local context.

By recognising that adult learners bring culturally shaped assumptions about leadership, communication, and learning, the facilitator creates an educational experience that values multiple perspectives rather than promoting a single model of effective management. Participants not only develop new leadership knowledge but also strengthen their intercultural competence, gaining a deeper appreciation of how effective leadership must be adapted to different cultural settings. This prepares them to lead diverse teams and collaborate successfully in the increasingly international business environments of the Asia-Pacific region.

Indigenous Approaches to Learning

Indigenous approaches to learning represent diverse educational traditions that have developed over thousands of years within Indigenous communities across the world. While these approaches vary between cultures, many share common characteristics that distinguish them from dominant Western educational models (Kayira et al., 2022; Newman et al., 2022; Peltier, 2021). Rather than viewing learning as primarily an individual process focused on acquiring abstract knowledge, Indigenous perspectives often understand learning as relational, experiential, holistic, and closely connected to community, culture, spirituality, and the natural environment (Elliott-Groves & Fish, 2024; Gaudet & Chilton, 2018; Huambachano, 2019; Newman et al., 2022; Peltier, 2021). Knowledge is viewed not simply as information to be transmitted, but as something that is lived, shared, and applied within the context of relationships and responsibilities (Chiblow & Meighan, 2023; Gaudet & Chilton, 2018; Mani & Rai, 2021) .

A defining feature of many Indigenous learning traditions is their holistic view of education. Learning is commonly understood as involving the intellectual, physical, emotional, cultural, and spiritual dimensions of life rather than concentrating solely on cognitive development (Peltier, 2021; White, 2022; Wilbiks, 2021). Adults are encouraged to develop practical skills, cultural identity, ethical responsibility, and community relationships simultaneously (Gaudet, 2020; Newman et al., 2022). This holistic perspective recognises that personal wellbeing, family, community, Country, and cultural identity are interconnected, and that effective learning supports the development of the whole person rather than isolated competencies (Peltier, 2021; Sebastian & Giovanangeli, 2024; Warburton & Chambers, 2007).

Relationships occupy a central position within Indigenous education (Elliott-Groves & Fish, 2024; Gaudet & Chilton, 2018; Newman et al., 2022). Learning typically occurs through participation in family and community life, with knowledge shared across generations by Elders, family members, and experienced community members (Chiblow & Meighan, 2023; Kayira et al., 2022; Newman et al., 2022). Elders are respected not simply because of their age but because of their accumulated knowledge, cultural authority, wisdom, and responsibility for maintaining and transmitting cultural traditions (Chiblow & Meighan, 2023; Kayira et al., 2022; Newman et al., 2022; Sebastian & Giovanangeli, 2024). Rather than viewing teachers as the sole source of knowledge, Indigenous education often recognises multiple knowledge holders, with learning occurring through observation, participation, discussion, mentoring, and shared experience (Gelderen & Guthadjaka, 2019; Jaggi et al., 2024; Mani & Rai, 2021; Wilbiks, 2021).

Experiential learning is another defining characteristic of many Indigenous educational traditions. Knowledge is developed through direct participation in meaningful activities rather than through abstract instruction alone. Adults learn by observing experienced practitioners, practising skills within authentic contexts, reflecting on their experiences, and gradually assuming greater responsibility within their communities. This close integration of learning with everyday life ensures that knowledge remains relevant, practical, and directly connected to real-world responsibilities. Rather than separating education from daily experience, many Indigenous educational programs embed learning within community life, enabling adults to develop knowledge and skills through active participation in culturally significant activities. Community-based programs often involve learners working alongside Elders, master craftspeople, language custodians, and other community members while participating in cultural events, traditional arts, language revitalisation, music, food preparation, and other cultural practices. Through these experiences, learning becomes inseparable from participation in community life, allowing learners to develop practical competence, cultural understanding, and a deeper sense of identity simultaneously (Townley et al., 2023; Whitinui et al., 2015). Indigenous methodologies similarly emphasise that authentic understanding develops through respectful engagement with Indigenous communities and lived cultural experiences rather than through classroom instruction alone. Experiential learning grounded in Indigenous knowledge systems also supports learners in exploring and strengthening their cultural identities (Townley et al., 2023).

Storytelling has long been a central teaching method within many Indigenous cultures. Stories communicate historical knowledge, cultural values, ethical principles, environmental understanding, social responsibilities, and practical skills by linking learning to people, places, relationships, and lived experiences. Rather than presenting information as isolated facts, storytelling encourages learners to reflect, interpret meaning, and connect new knowledge with existing cultural understandings. Indigenous educational philosophies also emphasise collaborative knowledge creation, recognising that learning emerges through shared experiences, dialogue, and relationships within communities (Angelo et al., 2022). In Australia, for example, educational practices such as Acknowledgement to Country activities frequently incorporate local Indigenous languages, stories, songs, artworks, and connections to the natural environment, helping learners develop an appreciation of local culture while strengthening relationships with place and community (Benton et al., 2021).

Many Indigenous approaches also place considerable emphasis on learning through observation, attentive listening, and gradual participation before assuming independent responsibility. Rather than expecting learners to demonstrate competence immediately, individuals are encouraged to spend time watching experienced community members, reflecting on their actions, and progressively engaging in activities as their confidence, understanding, and skills develop. Learning is therefore viewed as an ongoing developmental journey in which patience, respect, and careful observation are essential components of becoming a knowledgeable and responsible member of the community. This process is closely linked to the role of Elders, who serve as custodians of cultural knowledge and play a central role in preserving and transmitting traditions across generations (Kawano, 2023). Contemporary Indigenous education initiatives continue to reflect these principles by bringing together Elders, Indigenous knowledge holders, community members, and educators to support culturally grounded learning, strengthen relationships, and promote the renewal of Indigenous knowledge through collaborative and intergenerational engagement (Townley et al., 2023).

Connection to land, often referred to as *Country* in many Indigenous Australian contexts, is another defining feature of Indigenous approaches to learning. Country is understood not simply as a physical location but as a living source of identity, history, spirituality, knowledge, and responsibility. Consequently, learning frequently occurs within natural environments where learners engage directly with ecological systems, cultural practices, seasonal cycles, and sustainable resource management. Contemporary Indigenous educational approaches increasingly advocate place-based learning that positions Indigenous knowledge as a central framework for understanding the environment rather than as supplementary content (Benton et al., 2021). Similar approaches are evident in Indigenous communities throughout North America, Aotearoa New Zealand, Hawai'i, Scandinavia, Africa, and other regions, where cultural identity is closely connected to relationships with land and community. Research has also highlighted that the historical dispossession of Indigenous peoples from their traditional lands has significantly disrupted the transmission of Indigenous knowledge and contributed to educational disadvantage. As a result, decolonising educational practices increasingly seek to restore Indigenous worldviews and reconnect learning with the cultural, environmental, and spiritual relationships that underpin Indigenous knowledge systems (Nesterova et al., 2024).

Collective responsibility is likewise fundamental to many Indigenous educational philosophies. Learning is generally understood as serving not only individual development but also the wellbeing of families, communities, and future generations. Knowledge carries both privileges and responsibilities, with learners expected to apply what they have learned in ways that contribute to community life, cultural continuity, and collective wellbeing. Studies of Indigenous communities have shown that educational success is frequently associated with developing responsibility for others, strengthening relationships within the community, and preserving cultural traditions rather than focusing solely on individual achievement or academic performance (Lins & Florêncio, 2023). Family and community support also play a critical role in encouraging participation in education and helping learners pursue personal and professional aspirations (O'Loughlin, 2021). These perspectives differ from educational models that prioritise competition, individual success, and personal advancement. Instead, many Indigenous education programs are designed to strengthen cultural identity, support community development, and honour Indigenous ways of knowing through relational, culturally grounded educational practices that emphasise collaboration, reciprocity, and care for others (Angelo et al., 2022; Townley et al., 2023).

Respect for multiple ways of knowing is another important characteristic. Indigenous knowledge systems encompass scientific observation, environmental knowledge, cultural practices, oral histories, spirituality, language, and lived experience (Bergier & Anderson, 2021; A. Hossain et al., 2023; Webkamigad et al., 2019). These knowledge systems are increasingly recognised as sophisticated and valuable contributions to education, environmental management, healthcare, governance, and sustainable development (Locke, 2022; Peterson et al., 2019). Contemporary adult education increasingly acknowledges that Indigenous knowledge complements rather than competes with Western scientific and educational traditions, offering alternative perspectives for understanding complex social and environmental challenges (Assefa, 2021; Cappiali, 2024; Duncan et al., 2022).

Adult education providers have increasingly sought to incorporate Indigenous perspectives through culturally responsive teaching practices. These may include involving Indigenous community members and Elders in educational programs (Bally et al., 2025; Buxton, 2018; Kim, 2023), incorporating local histories and cultural knowledge into curricula (Peters et al., 2024), recognising Indigenous languages (Bergier & Anderson, 2021; Webkamigad et al., 2019), adopting collaborative learning approaches (O'Shea et al., 2023), using storytelling and experiential activities (Hulko et al., 2021; Ignacio et al., 2023; Locke, 2022), and creating culturally safe learning environments (Buxton, 2018; Kim, 2023; O'Shea et al., 2023). Importantly, authentic incorporation of Indigenous perspectives requires meaningful partnerships with Indigenous communities rather than simply adding Indigenous content to existing curricula (Locke, 2022; McIlduff et al., 2022).

Indigenous approaches to learning also contribute significantly to contemporary educational theory. Their emphasis on experiential learning, relational learning, community participation, lifelong learning, reflection, and holistic development aligns closely with many modern theories of adult education while offering broader understandings of knowledge, identity, and learning (Assefa, 2021; O'Shea et al., 2023; Peterson et al., 2019). These perspectives challenge educators to move beyond narrowly individualistic models of education and recognise that learning is fundamentally embedded within cultural, social, environmental, and historical contexts (Assefa, 2021; Cappiali, 2024; Duncan et al., 2022).

Indigenous approaches to learning demonstrate that education is not simply the acquisition of knowledge or skills but a lifelong process of developing relationships, responsibilities, cultural identity, and practical wisdom (Assefa, 2021; Peterson et al., 2019; Webkamigad et al., 2019). By recognising and valuing Indigenous educational traditions, adult educators can create more inclusive, culturally responsive, and holistic learning environments that respect diverse ways of knowing while enriching educational practice for all learners (Cappiali, 2024; Kim, 2023; O'Shea et al., 2023).

The principles of Indigenous approaches to learning can be applied across a wide range of adult education settings. Rather than replicating Indigenous cultural practices outside their appropriate context, educators can incorporate underlying principles such as relational learning, experiential learning, community participation, storytelling, place-based education, and holistic development to create more inclusive and meaningful learning experiences.

Example 1: Community Health Worker Training

An adult education provider delivers a community health worker program for Indigenous and non-Indigenous learners in northern Canada. Instead of relying solely on classroom lectures, the program includes Elders, local health practitioners, and community members who share traditional knowledge alongside contemporary healthcare practices. Learners participate in community wellness activities, visit local health services, and learn about traditional approaches to nutrition, mental wellbeing, and environmental health. Storytelling sessions allow Elders to share experiences that illustrate the importance of holistic health, while learners reflect on how Indigenous and Western health practices can complement one another. Through this approach, students develop clinical knowledge while strengthening cultural awareness, communication skills, and relationships with the communities they will eventually serve.

Example 2: Environmental Management Program

A university in Aotearoa New Zealand offers a professional development program for environmental managers. As part of the course, participants work alongside Māori knowledge holders to understand traditional approaches to waterway management, biodiversity conservation, and sustainable resource use. Rather than studying these concepts exclusively through textbooks, learners participate in field activities on local whenua (land), observe environmental monitoring techniques, and discuss the cultural significance of ecosystems with community representatives. By combining Indigenous ecological knowledge with contemporary environmental science, learners develop a more comprehensive understanding of sustainable land management while appreciating multiple knowledge systems.

Example 3: Leadership Development

An Indigenous leadership program in the United States prepares emerging tribal leaders through community-based learning rather than conventional classroom instruction alone. Participants spend time observing experienced community leaders, attending tribal council meetings, assisting with community projects, and receiving mentoring from Elders. Storytelling is used to explore historical leadership decisions and cultural values, while learners gradually assume leadership responsibilities within supervised community initiatives. Reflection sessions encourage participants to consider how leadership involves service, cultural responsibility, and stewardship of future generations rather than individual authority or status.

Example 4: Vocational Education and Apprenticeship

An adult vocational training provider offering traditional woodworking partners with Indigenous artisans to deliver part of the program. Learners observe master craftspeople before practising techniques themselves, progressing from simple to more complex tasks as their competence develops. Alongside technical skills, learners explore the cultural significance of traditional designs, sustainable harvesting of materials, and the responsibilities associated with preserving cultural knowledge. Assessment includes not only the quality of the finished product but also demonstrated respect for cultural protocols, collaboration with mentors, and responsible use of natural resources.

Example 5: Adult Language Revitalisation Program

A community-based language revitalisation course supports Indigenous adults seeking to reconnect with their ancestral language. Learning occurs through conversations with fluent speakers, storytelling circles, traditional songs, community events, and participation in cultural activities rather than through grammar exercises alone. Language learning is integrated with local history, cultural practices, and relationships with Country, enabling participants to strengthen both linguistic competence and cultural identity. As learners gain confidence, they begin mentoring younger community members, reinforcing the Indigenous principle that knowledge carries a responsibility to support future generations.

Example 6: Corporate Cultural Awareness Training

A mining company operating on Indigenous land introduces a professional development program for managers and supervisors. Rather than presenting Indigenous culture as a series of facts to memorise, the program is co-designed and co-delivered by local Traditional Owners. Participants spend time on Country learning about cultural heritage, land management, and local history through guided walks, discussions, and storytelling. Managers are encouraged to reflect on how Indigenous perspectives can inform environmental stewardship, stakeholder engagement, and workplace decision-making. The emphasis is on building respectful relationships and understanding Indigenous perspectives through direct engagement rather than passive instruction.

These examples demonstrate that Indigenous approaches to learning are applicable far beyond formal Indigenous education. By incorporating principles such as relational learning, experiential participation, observation, storytelling, community engagement, respect for multiple ways of knowing, and holistic development, adult educators can create learning environments that are more authentic, culturally responsive, and meaningful for diverse groups of learners.

Gender and Adult Education

Gender has long influenced access to education, participation in learning, educational experiences, and the outcomes achieved by adult learners (Chuang, 2015; Dosunmu & Dichaba, 2019, 2020; Saif Juma Salim Al, 2022). Although substantial progress has been made toward educational equality in many parts of the world, gender continues to shape opportunities for lifelong learning through social expectations, cultural norms, economic conditions, family responsibilities, and institutional practices (Chuang, 2015; Saif Juma Salim Al, 2022; Thinley & Gyeltshen, 2024). Contemporary adult education recognises that gender is not simply a biological characteristic but a social construct that influences individuals' experiences, identities, opportunities, and interactions within educational environments (Carter, 2020; Saif Juma Salim Al, 2022). Consequently, effective adult education seeks to provide equitable opportunities for learners of all genders while acknowledging that barriers to participation may differ according to individual circumstances (Chuang, 2015; Tellado, 2017).

Historically, educational opportunities were often distributed unequally according to gender. In many societies, men were more likely to receive formal education, while women were expected to prioritise domestic responsibilities and caregiving roles (Dosunmu & Dichaba, 2019; Hamilton, 2023; Thinley & Gyeltshen, 2024). Adult education frequently emerged as a pathway through which women could acquire literacy, vocational skills, professional qualifications, and civic knowledge that had previously been inaccessible (Amusa, 2022; Dosunmu & Dichaba, 2020; Thinley & Gyeltshen, 2024). Throughout the twentieth century, women's participation in adult education expanded significantly alongside broader movements for gender equality, increased workforce participation, and improved access to higher education (Dosunmu & Dichaba, 2019; Hamilton, 2023; Ignatovich & Walter, 2021). Today, women represent a substantial proportion of adult learners in many countries, particularly in higher education and continuing professional development (Dosunmu & Dichaba, 2019; Ischinger, 2010), although important disparities remain across regions and fields of study (Dosunmu & Dichaba, 2020; Ischinger, 2010; Saif Juma Salim Al, 2022; Slowey et al., 2020).

Despite improvements in participation, gender-related barriers continue to affect adult learning. Women are more likely than men in many societies to balance education alongside unpaid caregiving responsibilities, including caring for children, older family members, or relatives with disabilities (Hanson & Fletcher, 2021). These competing demands can limit opportunities to attend classes, participate in workplace training, or undertake further study (Hanson & Fletcher, 2021). Financial constraints, limited access to childcare, inflexible course schedules, and workplace cultures may further restrict participation (Hanson & Fletcher, 2021). Conversely, men may encounter barriers in educational settings where social expectations discourage participation in literacy programs, health education, or occupations traditionally associated with women, such as nursing, early childhood education, or community services. Gender stereotypes therefore have the potential to constrain educational participation for learners of all genders.

Gender also influences learners' educational experiences. Classroom interactions, teaching materials, assessment practices, and educator expectations may consciously or unconsciously reinforce traditional gender roles. Examples include textbooks that disproportionately portray men in leadership, science, engineering, or technical occupations while depicting women primarily in caregiving or domestic roles. Similarly, assumptions about the abilities, interests, or career aspirations of learners can influence the support and opportunities they receive. Adult educators therefore have a responsibility to critically examine curriculum materials and teaching practices to ensure they promote equity, diversity, and inclusion rather than reinforcing stereotypes, (Zhu et al., 2023).

The relationship between gender and occupational segregation remains an important issue within adult education. Many vocational education and training systems continue to reflect traditional patterns of employment, with women overrepresented in health, education, administration, and community services, while men remain dominant in engineering, construction, information technology, mining, and many skilled trades (Zhu, 2023). These patterns are influenced by socialisation, cultural expectations, role models, workplace cultures, and perceptions of suitable occupations rather than by differences in capability (Hanson & Fletcher, 2021; Zhu, 2023). Adult education plays an important role in encouraging learners to pursue careers based on their interests, abilities, and aspirations rather than gender stereotypes, thereby supporting greater workforce diversity and addressing skill shortages across multiple industries (Zhu, 2023).

Intersectionality provides an important framework for understanding gender within adult education. Coined by legal scholar Kimberlé Crenshaw, intersectionality recognises that gender interacts with other aspects of identity, including age, ethnicity, Indigenous identity, disability, socioeconomic status, sexual orientation, migration background, religion, and geographic location (Gnanadass & Merriweather, 2024; Thomas et al., 2021; Varsik & Gorochovskij, 2023). As a result, individuals who share the same gender may experience education very differently depending on their broader social circumstances (Gnanadass & Merriweather, 2024; Thomas et al., 2021; Varsik & Gorochovskij, 2023). For example, a woman living in a remote community may face different educational barriers from a woman living in a metropolitan area, while an Indigenous man may experience different challenges from a non-Indigenous man with similar educational goals. Adult educators increasingly recognise the importance of considering these intersecting identities when designing inclusive learning environments (Thomas et al., 2021; Varsik & Gorochovskij, 2023; Zhu et al., 2023).

Gender diversity has also become an increasingly important consideration in contemporary adult education. Many educational institutions now recognise that learners may identify as transgender, non-binary, gender diverse, or gender fluid (Carlson, 2024; Palkki & Sauerland, 2018). Inclusive educational practice involves creating respectful learning environments in which all learners feel safe, valued, and able to participate fully (Davis et al., 2024; Horton, 2020). This may include using inclusive language, respecting individuals' names and pronouns (Carlson, 2024; Davis et al., 2024; Vice et al., 2026), providing appropriate facilities where possible Horton, 2020; , addressing discriminatory behaviour promptly (Davis et al., 2024; Vice et al., 2026), and ensuring institutional policies support diversity and inclusion (Horton, 2020; Royaltey-Quandt, 2016) Such practices contribute to psychological safety, learner engagement, and positive educational outcomes for all participants (Carlson, 2024; Horton, 2020; Mobbs & Tashjian, 2022).

Adult educators can promote gender equity through a range of inclusive teaching practices. These include selecting learning materials that represent diverse genders in a variety of professional and leadership roles (Brussino & McBrien, 2022; Cecilia et al., 2020), encouraging equitable participation during discussions (Mobbs & Tashjian, 2022), challenging gender stereotypes when they arise (Brussino & McBrien, 2022; Palkki & Sauerland, 2018), providing flexible learning arrangements that accommodate family and caring responsibilities (Alston et al., 2021), and designing assessments that allow learners to demonstrate competence in different ways (Brussino, 2021). Educators should also reflect on their own assumptions and biases to ensure expectations are based on learners' abilities rather than gender-related assumptions (Brussino & McBrien, 2022; Mobbs & Tashjian, 2022).

Technology has created both opportunities and challenges for gender equity in adult education. Online learning has increased access for many adults who previously faced barriers related to geography, disability, employment, or caring responsibilities (Alston et al., 2021; Mizzi, 2021). Flexible delivery enables learners to study at times that suit their personal circumstances, benefiting many women and men balancing multiple responsibilities (Alston et al., 2021). However, digital inequality persists in some regions, where access to reliable internet, technology, or digital literacy remains uneven. Gender differences in digital participation may therefore continue to influence access to lifelong learning opportunities in certain contexts.

Internationally, gender remains an important focus of educational policy. Sustainable Development Goal 4 seeks to ensure inclusive and equitable quality education and promote lifelong learning opportunities for all, while Sustainable Development Goal 5 aims to achieve gender equality and empower all women and girls (Cecilia et al., 2020). Together, these goals recognise that equitable access to education contributes to improved health, economic participation, social inclusion, democratic engagement, and sustainable development. Adult education plays a central role in achieving these objectives by providing opportunities for individuals to develop new knowledge, skills, confidence, and employment opportunities throughout their lives (Brussino & McBrien, 2022; Cecilia et al., 2020).

The United Nations adopted the 2030 Agenda for Sustainable Development in 2015, establishing 17 Sustainable Development Goals designed to address major global challenges, including poverty, health, education, inequality, environmental sustainability, and economic development. Education is recognised as a key driver of progress because it equips individuals with the knowledge, skills, and opportunities needed to improve their lives and contribute to society.

Sustainable Development Goal 4 (Quality Education) focuses on ensuring that everyone has access to inclusive, equitable, and high-quality education throughout their lives. While much attention is often given to school education, SDG 4 explicitly promotes lifelong learning, recognising that adults also require ongoing opportunities to develop literacy, vocational skills, professional qualifications, digital capabilities, and personal competencies. This is particularly important as technological change, evolving labour markets, and demographic shifts require adults to continually update their knowledge and skills.

Sustainable Development Goal 5 (Gender Equality) aims to eliminate discrimination against women and girls and ensure equal participation in education, employment, leadership, and public life. Within adult education, this means removing barriers that prevent women from participating in learning, such as financial disadvantage, cultural expectations, caregiving responsibilities, limited access to training, or discrimination in educational and workplace settings. It also encourages greater participation by women in fields where they have traditionally been underrepresented, including science, technology, engineering, and leadership.

These two goals are closely interconnected because education is recognised as one of the most effective mechanisms for advancing gender equality. When women and girls have equitable access to education throughout their lives, they are more likely to secure employment, earn higher incomes, participate in civic and political life, make informed health decisions, and contribute to economic and social development. At the same time, improving gender equality increases access to education by reducing discriminatory practices and creating more inclusive learning environments.

Adult education contributes directly to both SDGs by providing opportunities for people to continue learning beyond compulsory schooling. Adult literacy programs, vocational education, university study, workplace training, community education, digital skills programs, and professional development all help individuals improve their employment prospects, adapt to changing industries, participate more fully in society, and engage in lifelong learning. These opportunities not only benefit individual learners but also contribute to broader national goals, including improved workforce productivity, reduced inequality, greater social cohesion, and sustainable economic development (Brussino & McBrien, 2022; Cecilia et al., 2020).

The principles of gender equity can be applied in many different vocational education settings to improve participation, learning experiences, and employment outcomes for learners of all genders. The following examples illustrate how training providers in different countries could incorporate inclusive recruitment, teaching, assessment, learner support, and industry engagement practices into mechanical engineering and fabrication programs. While the examples are situated within the vocational education systems of the United Kingdom and France, the underlying principles are applicable across a wide range of engineering and technical disciplines and demonstrate how adult education can contribute to both high-quality vocational training and greater gender equality.

Example 1: United Kingdom – Mechanical Engineering Apprenticeship

A further education college delivering a Level 3 Mechanical Engineering apprenticeship could embed gender equity throughout the program to encourage greater participation and success among women in engineering trades. Recruitment campaigns could feature female mechanical engineers, CNC machinists, maintenance technicians, and fabrication supervisors working in industries such as aerospace, rail, automotive manufacturing, and renewable energy. Open days could include presentations from current female apprentices and industry mentors to demonstrate that engineering careers are open to people of all genders.

Within the workshop, instructors could ensure that every apprentice develops competence across the full range of practical activities rather than allowing learners to self-select familiar tasks. All apprentices would rotate through machining, CNC programming, welding, fabrication, precision measurement, fault diagnosis, and assembly activities while taking turns leading practical projects. Assessment would be based entirely on the occupational standards for the apprenticeship, ensuring that judgements are made according to technical competence, safe working practices, and problem-solving ability rather than assumptions about physical strength or previous experience. Flexible delivery options, such as evening theory classes, recorded lectures, and additional workshop access, could also support apprentices balancing employment or caring responsibilities. Partnerships with local engineering employers could provide female apprentices with mentors working in advanced manufacturing, helping learners build confidence, expand professional networks, and see clear career pathways within the engineering sector. Such initiatives support greater workforce diversity while helping employers address ongoing engineering skills shortages.

Example 2: France – Metal Fabrication and Welding Program

A vocational lycée and apprentice training centre (*Centre de Formation d'Apprentis*) delivering adult qualifications in chaudronnerie (metal fabrication), welding, and industrial maintenance could implement a range of practices to promote gender equity. Promotional materials could deliberately present women working as welders, sheet metal fabricators, robotic welding operators, quality inspectors, and production managers, helping to challenge traditional perceptions of fabrication as a male occupation. During induction, learners could explore the importance of diversity within modern manufacturing and discuss how inclusive workplaces contribute to innovation, safety, and productivity.

Practical training could be organised so that every learner gains experience with manual and robotic welding, metal forming, fabrication drawings, non-destructive inspection, quality assurance, and computer-controlled manufacturing systems. Rather than allocating physically demanding tasks according to gender, instructors could teach appropriate lifting techniques, ergonomic work practices, and the correct use of mechanical handling equipment to all learners. Industry placements could be arranged with fabrication companies that have established diversity and inclusion policies, allowing apprentices to observe mixed-gender engineering teams and experience contemporary workplace practices. Learners requiring additional flexibility because of family or caring commitments could be offered blended learning opportunities and alternative scheduling for theory components while maintaining the practical competency requirements of the qualification. Program outcomes could be monitored by tracking recruitment, completion, employment, and progression by gender, allowing the training provider to continually improve participation and learner success. Through these measures, vocational education can contribute to more equitable participation in engineering while supporting the objectives of lifelong learning and gender equality promoted through international education policy.

Equity and Social Justice

Equity and social justice are foundational principles in adult education that seek to ensure all individuals have fair opportunities to access, participate in, and benefit from learning throughout their lives (Pernas & Gómez, 2016; Urrutxi et al., 2021). While equality focuses on treating everyone the same, equity recognises that learners begin from different circumstances and may require different forms of support to achieve comparable educational outcomes (Ó. Espinoza, 2020; Santamaría & Vila, 2016). As Bronfenbrenner (1974) noted, equality refers to quantity, whereas equity is linked to social justice in relation to distribution in education (Santamaría & Vila, 2016). The distinction is further clarified by the observation that equity demands acting with justice, since equality of access or opportunities is often insufficient to promote social justice, and therefore it is necessary to do more. Social justice extends this principle by addressing the broader social, economic, cultural, and political conditions that create educational disadvantage, (Ó. Espinoza, 2020; Flores, 2022). Indeed, equity has an individual orientation, while social justice transcends equity and seeks more egalitarian treatment for different social groups to avoid any type of discrimination (Ó. Espinoza, 2020). Together, these concepts promote education as both an individual right and a means of creating a more inclusive, equitable, and democratic society (Pernas & Gómez, 2016; Trujillo et al., 2023; Urrutxi et al., 2021).

Equity in adult education acknowledges that learners differ in their educational backgrounds, financial circumstances, cultural experiences, language proficiency, disabilities, employment commitments, family responsibilities, and geographic location (Alarcón & D'Angelo, 2020; Cortes & Loor, 2024;

Rojas-Botero et al., 2024). These differences influence an individual's ability to participate successfully in education (Arrabal, 2018; Cortes & Loor, 2024). Consequently, equitable educational practice involves identifying and reducing barriers to participation rather than assuming that identical treatment will produce fair outcomes (Santamaría & Vila, 2016). Equity in education must mean considering the differences with which learners arrive at the educational system and giving more opportunities to those who need them most to guarantee acceptable levels of learning (Alarcón & D'Angelo, 2020). Support may include flexible course delivery, financial assistance, accessible learning materials, language support, assistive technologies, recognition of prior learning, academic support services, and culturally responsive teaching practices (Basurto et al., 2024; Cortes & Loor, 2024; Rodríguez, 2024). The elimination of physical, technological, and social barriers that could hinder access is positioned as an essential component of inclusion (Rodríguez, 2024). Furthermore, equitable approaches encompass multiple dimensions including equity in access, equity in participation, equity in permanence, and equity in outcomes (Alarcón & D'Angelo, 2020; Flores, 2022; Worthman, 2020).

Social justice broadens this perspective by recognising that educational disadvantage is often linked to structural inequalities within society (Argüelles, 2023; Flores, 2022). Factors such as poverty, discrimination, unemployment, racism, gender inequality, disability, social exclusion, and historical marginalisation can significantly influence educational opportunities (Flores, 2022; Rojas-Botero et al., 2024; Urrutxi et al., 2021). Three major approaches to social justice have been identified: distributive justice, justice as recognition, and justice as participation, each emphasising different aspects of equality and equity while complementing one another to provide a more integral vision (Worthman, 2020). Adult education therefore has the potential not only to develop knowledge and vocational skills but also to promote greater social inclusion, civic participation, economic mobility, and empowerment (Cortes & Loor, 2024; Murillo, 2007; Rodríguez, 2024). The state bears responsibility for ensuring that all individuals can achieve educational results that allow them to enter the labour market and the social system under equal conditions and rights (Ó. Espinoza, 2020). By improving access to education, individuals are better positioned to participate in employment, contribute to their communities, and exercise their rights as active citizens, (Pernas & Gómez, 2016; Trujillo et al., 2023). As the 2015 World Education Forum declared, inclusion and equity in education are considered the cornerstone of a transformative agenda, ensuring that all persons, regardless of sex, age, race, ethnicity, language, religion, disability, or socioeconomic origin, must have access to inclusive and equitable quality education and opportunities (Pernas & Gómez, 2016; Urrutxi et al., 2021).

The concept of social justice has been influenced by numerous philosophical and educational traditions. Liberal theories emphasise equal rights and equal opportunity, while critical theorists argue that education should challenge systemic inequalities and empower learners to transform society (Bindhani & Gopinath, 2024). Humanistic educators focus on respecting the dignity and potential of every learner (Hue & Karim, 2024), whereas contemporary inclusive education frameworks emphasise participation, diversity, and belonging (Cologon, 2024; Lakkala et al., 2024; W. Zhu et al., 2022). Although these perspectives differ in emphasis, they share a commitment to ensuring that education serves all members of society rather than reinforcing existing social divisions, (Kangas et al., 2023).

Within adult education, equity and social justice influence curriculum design, teaching practice, assessment, and institutional policy (Debasu & Yitayew, 2024). Educators are encouraged to develop inclusive learning environments that recognise learner diversity and value multiple forms of knowledge and experience (Massiah et al., 2024). Teaching materials should reflect diverse cultures, genders, occupations, and perspectives, while assessment methods should allow learners multiple opportunities to demonstrate competence (Mitchell, 2023; Mpofu & Sefotho, 2024). Institutions may also implement policies that support flexible enrolment, reasonable adjustments for learners with disabilities, financial assistance, childcare support, and outreach programs for underrepresented communities (Buli-Holmberg & Kamenopoulou, 2017).

Equity and social justice are particularly important within vocational education and training because access to education directly influences employment opportunities and economic participation (Debasu & Yitayew, 2024). Adults seeking to retrain after redundancy, migrants developing language and occupational skills, Indigenous learners, people with disabilities, older workers adapting to technological change, and individuals from disadvantaged socioeconomic backgrounds may all require different forms of educational support (Buli-Holmberg & Kamenopoulou, 2017). By addressing these barriers, vocational education contributes to workforce participation while reducing social and economic inequalities (Debasu & Yitayew, 2024).

Technological developments have created new opportunities for advancing educational equity through online learning, blended delivery, open educational resources, and digital learning platforms (Manalo & Chua, 2024). These approaches can increase access for learners in rural and remote communities, those with mobility limitations, or adults balancing employment and family responsibilities (Manalo & Chua, 2024). However, digital technologies can also create new inequities where learners lack reliable internet access, appropriate devices, digital literacy, or suitable learning environments (Manalo & Chua, 2024). Consequently, promoting digital inclusion has become an increasingly important aspect of educational equity (Manalo & Chua, 2024).

Disability and Inclusive Education

Disability and inclusive education are central concepts in contemporary adult education, reflecting the principle that all individuals, regardless of physical, sensory, intellectual, cognitive, or psychosocial disability, should have equitable opportunities to access, participate in, and succeed in lifelong learning (Aldhilan et al., 2025; Alterator et al., 2022; Ambady & Sukumaran, 2025; Kunwar & Adhikari, 2023). Inclusive education moves beyond simply allowing learners with disabilities to enrol in educational programs; it seeks to create learning environments in which all learners can participate meaningfully, achieve their educational goals, and feel respected and valued (Kunwar & Adhikari, 2023; Long & Guo, 2023; Nkoma & Hay, 2018; Quick et al., 2021). Rather than viewing disability as a limitation within the individual, inclusive education increasingly recognises that many barriers arise from the design of educational systems, environments, and teaching practices rather than from the disability itself (Aldhilan et al., 2025; Alterator et al., 2022; Muafiah et al., 2025; Sakız, 2016).

Historically, disability was often understood through the medical model, which viewed disability primarily as an impairment that required treatment, rehabilitation, or special education (Aldhilan et al., 2025; Hunt et al., 2021). Educational responses frequently involved segregating learners into specialist institutions or expecting individuals to adapt to mainstream educational systems (Hunt et al., 2021). While medical interventions remain important for many individuals, this approach often focused on deficits rather than capabilities and sometimes resulted in exclusion from mainstream education and employment (Alterator et al., 2022; Hunt et al., 2021; Sakız, 2016).

Contemporary adult education has increasingly adopted the social model of disability, which argues that disability results not solely from an individual's impairment but from physical, social, attitudinal, and institutional barriers that restrict participation (Ross-Gordon, 2024). Under this perspective, inaccessible buildings, inflexible teaching methods, poorly designed assessments, negative attitudes, inaccessible technologies, and discriminatory policies create disability by preventing individuals from participating on equal terms (Kavitha, 2023; Rogers-Shaw et al., 2023). Inclusive education therefore focuses on removing these barriers so that learners can participate according to their abilities (Gedzyk-Nieman & Derouin, 2025; Kavitha, 2023).

Many countries now complement the social model with a rights-based approach, recognising education as a fundamental human right (Biao, 2022). This perspective is reflected in the United Nations Convention on the Rights of Persons with Disabilities (CRPD), which affirms the right of persons with disabilities to inclusive education at all levels, including lifelong learning and vocational education (Biao, 2022; Hinzen, 2020). Rather than requiring learners to fit existing systems, educational institutions have a responsibility to create environments that accommodate learner diversity and ensure equitable participation (Rogers-Shaw et al., 2017; Rogers-Shaw et al., 2023).

Within adult education, disability encompasses a broad range of conditions. These may include physical disabilities affecting mobility or dexterity, sensory disabilities such as blindness or hearing impairment, intellectual disabilities, learning disabilities including dyslexia and dyscalculia, autism spectrum conditions, acquired brain injuries, chronic health conditions, and mental health conditions such as anxiety, depression, or post-traumatic stress disorder (Ross-Gordon, 2024). Many disabilities are not immediately visible, highlighting the importance of creating supportive learning environments without requiring learners to disclose personal information unnecessarily (Gedzyk-Nieman & Derouin, 2025; Rogers-Shaw et al., 2023).

Inclusive adult education is based on the principle that learner diversity should be anticipated rather than treated as an exception. Instead of modifying courses only after barriers arise, educators increasingly design learning experiences that are accessible to the widest possible range of learners from the outset (Bustamante & Mejía, 2017; Collay et al., 2024). This philosophy is reflected in Universal Design for Learning (UDL), which encourages multiple means of presenting information, engaging learners, and allowing learners to demonstrate their knowledge and skills (Temesio-Vizoso, 2016; , Moya et al., 2020), Núñez-Sotelo & López, 2022). By providing flexibility within curriculum design, UDL benefits not only learners with disabilities but also those from diverse linguistic, cultural, educational, and socioeconomic backgrounds (Collay et al., 2024; García et al., 2022; Neira-Pesántez et al., 2025).

Teaching practices play a significant role in creating inclusive learning environments. Educators can improve accessibility by providing learning materials in multiple formats, using clear and consistent language, incorporating captions and transcripts into multimedia resources, allowing sufficient processing time during discussions, and structuring learning activities to accommodate different learning preferences (Barletta, 2023; Gálvez-Palomeque et al., 2023; Willis et al., 2023). Practical demonstrations, visual supports, digital resources, collaborative learning, and scaffolded instruction enable many learners to engage more effectively with course content (Ares & Cuevas, 2022; Idilia del Pilar Reyna et al., 2024; Morrillo et al., 2024).

Assessment practices also require careful consideration. Inclusive assessment does not lower academic or vocational standards; rather, it enables learners to demonstrate the required competencies in ways that minimise unnecessary barriers (Ramírez-Solórzano & Navas, 2024; Willis et al., 2023). Reasonable adjustments may include additional time during assessments, assistive technologies, alternative assessment formats, accessible examination environments, interpreters, readers, scribes, or modified presentation methods, provided that the integrity of the learning outcomes and competency standards is maintained (Gálvez-Palomeque et al., 2023; García et al., 2023; Vera & Gámez, 2024). The objective is to assess what learners know and can do, rather than the effects of inaccessible assessment methods (Ramírez-Solórzano & Navas, 2024; Willis et al., 2023).

Technology has significantly expanded opportunities for inclusive adult education. Screen readers, speech-to-text software, text-to-speech applications, electronic magnification, hearing assistance systems, captioning, alternative input devices, accessible learning management systems, and artificial intelligence-based accessibility tools enable many learners to participate more fully in education (García et al., 2023; Godinho et al., 2021; Rivas, 2024; Treviño et al., 2024). Online and blended learning also provide increased flexibility for adults who experience mobility limitations, chronic illness, fatigue, or mental health conditions (Moya et al., 2020; Temesio-Vizoso, 2016). However, these benefits depend on digital resources being designed according to recognised accessibility standards (Bustamante & Mejía, 2017; Godinho et al., 2021). Poorly designed online learning environments may create new barriers if accessibility is not considered during development (Morrillo et al., 2024; Neira-Pesántez et al., 2025).

Inclusive education is particularly important within vocational education and training because access to skills development directly influences employment opportunities, financial independence, and social participation (Pereira et al., 2024; Sousa et al., 2024). Adults with disabilities have historically experienced lower rates of workforce participation, higher unemployment, and fewer opportunities for career advancement than the general population (Lawrence et al., 2025; Smith et al., 2023; Ximba et al., 2026). Inclusive vocational education can help address these disparities by providing accessible training, workplace learning opportunities, reasonable workplace adjustments, and strong partnerships with employers that promote inclusive recruitment and employment practices (Erickson et al., 2014; Smith et al., 2023; Sousa et al., 2024; Ximba et al., 2026).

Educators also play a vital role in fostering inclusive classroom cultures. Respectful communication, high expectations, learner-centred teaching, collaboration, and recognition of individual strengths contribute to psychological safety and learner engagement (Sousa et al., 2024; Wissell et al., 2022). Rather than focusing on limitations, inclusive educators adopt a strengths-based approach that recognises the knowledge, experiences, resilience, and capabilities that learners with disabilities bring to the educational environment (Lawrence et al., 2025; Smith et al., 2023). Confidentiality, learner autonomy, and collaborative planning are equally important, ensuring that learners participate in decisions about any support or adjustments they may require (Curnow et al., 2025; Ezerins et al., 2023; Lawrence et al., 2025).

As an application example, consider a vocational education provider delivering a certificate course in Mechanical Engineering to a cohort of adult learners that includes individuals with physical disabilities, dyslexia, hearing impairment, autism spectrum disorder, acquired brain injuries, and anxiety disorders. Rather than making adjustments only after learners encounter difficulties, the program is designed from the outset using Universal Design for Learning (UDL) principles so that multiple pathways exist for learners to access information, participate in practical activities, and demonstrate competence. This proactive approach reflects the social model of disability by recognising that barriers often arise from the design of the learning environment rather than from the learners themselves.

Several adult learning theories inform the design of the program. Drawing on Knowles' theory of andragogy, instructors recognise that adult learners bring valuable workplace experience and prefer learning that is relevant to real occupational tasks. Learners are encouraged to discuss previous engineering or trade experiences, identify individual learning goals, and participate in planning aspects of their learning. Flexible delivery options, including online theory modules, recorded demonstrations, and self-paced learning activities, allow learners to balance study with employment, health conditions, and family responsibilities while maintaining responsibility for their own learning.

Constructivist learning principles are applied by engaging learners in authentic engineering problems that require collaboration, investigation, and practical decision-making. During workshop projects, mixed-ability teams work together to manufacture components, interpret engineering drawings, diagnose equipment faults, and develop solutions to production problems. Learners contribute according to their strengths while developing new technical skills through interaction with instructors and peers. Scaffolded instruction, demonstrations, visual supports, and guided questioning enable learners requiring additional assistance to progressively build competence without reducing the expected vocational standards.

The program also incorporates experiential learning by providing repeated opportunities to apply theoretical concepts during practical workshop activities and workplace simulations. After completing machining, welding, assembly, or maintenance tasks, learners participate in structured reflection sessions where they evaluate the effectiveness of their techniques, identify areas for improvement, and relate their experiences to engineering theory. This reflective process enables learners to transform practical experience into deeper professional understanding while developing confidence in their abilities.

Social learning theory further supports inclusion by encouraging peer modelling and collaborative learning. Experienced apprentices and workplace mentors demonstrate engineering tasks, safe work practices, and problem-solving techniques that learners observe before practising independently. Peer mentoring allows learners with different strengths to support one another, promoting both technical competence and social inclusion. Importantly, mentoring relationships are based on mutual learning rather than assumptions about disability, reinforcing a strengths-based approach to education.

Humanistic principles are evident in the respectful, learner-centred culture established throughout the program. Educators work collaboratively with learners to determine appropriate reasonable adjustments while maintaining confidentiality and learner autonomy. Some learners use speech-to-text software, screen readers, enlarged engineering drawings, captioned instructional videos, ergonomic workstations, or additional assessment time. Others may complete oral technical explanations instead of lengthy written responses where literacy difficulties are unrelated to the competency being assessed. Throughout the program, assessment remains focused on whether learners can demonstrate the required engineering competencies rather than on the effects of inaccessible assessment methods.

Technology further enhances accessibility by providing engineering simulations, captioned instructional videos, digital workshop manuals compatible with screen readers, and online collaboration tools that enable learners to review content at their own pace. Practical workshop activities are supported through adaptive equipment, accessible workstations, and mechanical lifting devices, ensuring that learners can safely participate regardless of physical ability.

The effectiveness of the inclusive program is evaluated by monitoring learner participation, retention, competency completion, workplace placement success, and graduate employment outcomes across different learner groups. Learner feedback is regularly used to identify remaining barriers and improve course design. Through the combined application of andragogy, constructivism, experiential learning, social learning theory, and humanistic education, the provider creates an inclusive vocational learning environment that maintains rigorous engineering standards while ensuring that disability does not become a barrier to participation, competence, or future employment.

Migration and Refugee Education

Migration and refugee education has become an increasingly important field within adult education as global migration, forced displacement, humanitarian crises, labour mobility, and international resettlement continue to reshape societies (Aleghfeli et al., 2024; Guo, 2021; Slowey et al., 2020). Adult migrants and refugees often seek education not only to develop new knowledge and vocational skills but also to rebuild their lives, gain employment, participate in their new communities, and achieve social and economic independence (Gravani et al., 2019; Guo, 2021; Καρανικόλα & Παλαιολόγου, 2021). Adult education therefore plays a critical role in supporting integration, lifelong learning, social inclusion, and human development for people who have crossed international borders under a wide variety of circumstances (Gravani et al., 2019; Guo, 2021; Slowey et al., 2020; Καρανικόλα & Παλαιολόγου, 2021).

Although the terms are sometimes used interchangeably, migrants and refugees represent distinct groups with different educational needs (Aleghfeli et al., 2024; Mareci et al., 2023). Migrants generally relocate voluntarily for reasons such as employment, education, family reunification, or improved living conditions (Gravani et al., 2019; Slowey et al., 2020). Refugees, by contrast, are individuals who have been forced to flee their countries because of war, persecution, violence, conflict, or serious human rights violations and are protected under international law, including the 1951 United Nations Convention Relating to the Status of Refugees and its 1967 Protocol (Aleghfeli et al., 2024; Slowey et al., 2020). Asylum seekers are individuals whose refugee claims have not yet been determined (Mareci et al., 2023). These different migration pathways often influence learners' previous educational experiences, legal rights, employment opportunities, and access to education (Aleghfeli et al., 2024; Mareci et al., 2023; Slowey et al., 2020).

Adult migrants and refugees bring significant strengths to educational environments, including multilingual abilities, diverse cultural knowledge, professional expertise, resilience, and extensive life experience (Guo, 2021; Losoncz, 2015). Many have completed higher education or vocational qualifications before migration, while others possess substantial workplace skills acquired through informal or non-formal learning (Losoncz, 2015; Slowey et al., 2020; Terrón-Caro et al., 2022). Adult education increasingly recognises these existing capabilities through strengths-based approaches that value learners' prior experiences rather than focusing solely on educational deficits (Aleghfeli et al., 2024; Kukovetz & Sprung, 2014).

However, many migrant and refugee learners also encounter substantial barriers to educational participation (Aleghfeli et al., 2024; Mareci et al., 2023; Oliver, 2010; Καρανικόλα & Παλαιολόγου, 2021). Language proficiency is often the most immediate challenge, affecting access to vocational training, higher education, employment, and community participation (Aleghfeli et al., 2024; Izquierdo et al., 2021; Jeon, 2019; Mareci et al., 2023; Καρανικόλα & Παλαιολόγου, 2021). Learners may also experience unfamiliar educational systems, different teaching and assessment practices, financial hardship, interrupted schooling, non-recognition of overseas qualifications, limited digital literacy, or responsibilities associated with employment and family settlement (Aleghfeli et al., 2024; Jeon, 2019; Losoncz, 2015; Mareci et al., 2023; Terrón-Caro et al., 2022). Refugees may additionally experience trauma, grief, displacement, prolonged periods without formal education, uncertainty regarding immigration status, or ongoing concern for family members remaining overseas (Losoncz, 2015; Καρανικόλα & Παλαιολόγου, 2021). These factors can significantly influence learners' confidence, motivation, concentration, and educational participation (Izquierdo et al., 2021; Καρανικόλα & Παλαιολόγου, 2021).

Adult education providers therefore frequently adopt holistic approaches that extend beyond classroom instruction. Language learning is commonly integrated with vocational education rather than taught in isolation, allowing learners to develop occupational communication skills while acquiring technical competencies. Settlement support, career guidance, mental health services, academic skills development, digital literacy training, and recognition of prior learning (RPL) are often provided alongside formal education to assist learners in successfully transitioning into employment and community life.

Inclusive teaching practices are particularly important when working with culturally and linguistically diverse learners. Educators are encouraged to use clear language, visual supports, practical demonstrations, collaborative learning activities, and scaffolded instruction while avoiding unnecessary jargon or culturally specific assumptions (H. J. Kim, 2024; Uribe-Zarain et al., 2024). Learning materials should reflect cultural diversity and recognise multiple perspectives rather than presenting knowledge exclusively through the lens of the host country (H. J. Kim, 2024; Lu & Troyan, 2022). Respect for learners' cultural identities, languages, religious practices, and previous educational experiences contributes to psychological safety and learner engagement (H. J. Kim, 2024; Uribe-Zarain et al., 2024; Zhang-Wu, 2017).

Intercultural competence has become an essential capability for adult educators. Effective teachers develop an understanding of cultural differences in communication, classroom participation, authority relationships, assessment expectations, and learning preferences while avoiding stereotypes or deficit-based assumptions (Collazo, 2019; Uribe-Zarain et al., 2024). Rather than expecting learners to abandon their cultural identities, inclusive adult education seeks to create learning environments where cultural diversity is recognised as an educational asset that enriches learning for all participants (Lu & Troyan, 2022; Uribe-Zarain et al., 2024; Zhang-Wu, 2017).

Recognition of prior learning is particularly significant for migrants and refugees because many arrive with substantial professional expertise that may not be formally recognised within their new country. Engineers, teachers, healthcare professionals, tradespeople, and technicians may possess years of experience but require assistance in demonstrating equivalence with local qualification frameworks. Effective RPL processes help reduce unnecessary duplication of training, accelerate workforce participation, and acknowledge the knowledge and competencies learners have already acquired through previous education, employment, and life experience.

Vocational education and training plays a particularly important role in supporting economic integration. Industry-specific language programs, apprenticeships, work placements, bridging qualifications, licensing programs, and partnerships with employers enable adult migrants and refugees to develop both technical competence and workplace confidence. These pathways support employment while helping employers address workforce shortages in sectors such as healthcare, construction, manufacturing, engineering, transport, hospitality, agriculture, and community services.

Technology has further expanded educational opportunities for migrant and refugee learners. Online language learning platforms, translation applications, multilingual learning resources, digital literacy programs, and blended learning environments enable adults to continue studying while balancing employment, family responsibilities, and settlement commitments (Stranger–Johannessen & Damiani, 2024). Open educational resources, including multilingual stories and digital platforms, have been developed specifically for immigrant and refugee language learners, with projects such as LIDA Stories providing country-specific resources for youth and adult immigrants learning the language of the host country (Stranger–Johannessen & Damiani, 2024). Blended learning approaches that combine digital learning with face-to-face activities have been recommended as particularly effective for this population (Stranger–Johannessen & Damiani, 2024), and the provision of specific actions to strengthen migrants' and refugees' digital literacy skills is considered essential for enabling access to online learning opportunities (Stranger–Johannessen & Damiani, 2024). At the same time, educators must recognise that some learners may have limited digital skills or restricted access to technology, requiring additional support to participate effectively in online learning (Kennedy et al., 2026; Stranger–Johannessen & Damiani, 2024). Research on educational technology in crisis-affected contexts has demonstrated that flexible modalities, including online, blended, and offline formats, are necessary to address barriers to digital access among refugee and migrant communities (Kennedy et al., 2026).

Migration and refugee education is closely connected to theories of adult learning. Andragogy recognises that adult migrants and refugees bring extensive life experience that can enrich learning (Mukhalalati & Taylor, 2019; Obashoro-John & Findsen, 2023). Knowles defined andragogy as "the art and science of helping adults learn," with core principles including the role of experience, readiness to learn, and orientation to learning (Moll, 2023; Mukhalalati & Taylor, 2019; Obashoro-John & Findsen, 2023). Adult learners are understood to possess different needs, interests, and experiences that shape their engagement with education (Arghode et al., 2017; Mukhalalati & Taylor, 2019). Transformative learning theory explains how migration often requires individuals to critically reassess previously held assumptions, identities, and ways of understanding the world while adapting to new cultural, social, and professional environments (Lefay et al., 2025; Mukhalalati & Taylor, 2019). Transformative learning stresses the importance of questioning frames of reference among adults (Lefay et al., 2025), and this process of critical reflection generates transformative learning by engaging in higher and more demanding levels of thought (Rubaia'an, 2023). Experiential learning enables learners to connect prior knowledge with new workplace practices (Arghode et al., 2017; Silva, 2018).

Kolb's model of experiential learning describes learning as a holistic process whereby knowledge is created through the transformation of experience (Kumar, 2025; Silva, 2018). Social constructivism highlights the importance of collaborative learning, language development, and participation within supportive learning communities (Lefay et al., 2025; Mukhalalati & Taylor, 2019; Wilson, 2018). Social constructivism emphasises the social and cultural environmental demands as integral dimensions of the learning process (Wilson, 2018) and stresses that learning is not just an individual behaviour but a process of cultural participation (Lin et al., 2024). According to constructivism, individuals construct new knowledge through the interaction between their previous skills and knowledge, the skills and knowledge gained from social interaction with peers and teachers, and social activities (Mukhalalati & Taylor, 2019). Humanistic education emphasises dignity, respect, learner autonomy, and personal growth, recognising that many learners are rebuilding confidence and identity alongside acquiring new skills (Obashoro-John & Findsen, 2023; Parker, 2021). The humanistic strand in adult learning accentuates the locus of control within an individual and gives greater credence to how people change across different environments through social interaction (Obashoro-John & Findsen, 2023). Humanistic learning theory has had a profound effect on adult learning theory, with andragogy, self-directed learning, and transformative learning all having roots in humanistic psychology (Silva, 2018). A humanist approach requires that educators be attentive to their learners' emotional and affective needs in order to facilitate their personal growth and development (Parker, 2021).

Language, Literacy, and Numeracy

Language, literacy, and numeracy (LLN) are fundamental capabilities that underpin learning, employment, community participation, and lifelong education. Within adult education, LLN extends well beyond the ability to read, write, and perform basic calculations. It encompasses the communication, comprehension, reasoning, and problem-solving skills that enable adults to participate effectively in education, the workplace, and society (Santos & McKinney, 2019). As economies become increasingly knowledge-based and technology-driven, strong LLN skills have become essential for vocational competence, digital participation, civic engagement, and personal independence (Churchill et al., 2011; Kyröläinen & Kuperman, 2021). Modern definitions of literacy now include problem-solving ability, numeracy, and the capacity to communicate using modern electronic media (Churchill et al., 2011), while international policy frameworks such as UNESCO's Sustainable Development Goal 4.6 emphasise that ensuring all youth and a substantial proportion of adults achieve literacy and numeracy remains a global challenge. Adults with poor literacy and numeracy skills face multiple sources of disadvantage, including higher unemployment, lower wages, and difficulty exercising their rights in society.

Language refers to the ability to communicate effectively through speaking, listening, reading, writing, and increasingly through digital and multimodal forms of communication (Kral, 2016). Adult learners require language skills to understand instructions, participate in discussions, interpret workplace documentation, complete assessments, collaborate with colleagues, and communicate with clients and supervisors. For many adults, language learning also involves acquiring English or another second language while simultaneously developing vocational knowledge (Más et al., 2012; Young-Scholten, 2013). Consequently, language development is often integrated into adult education rather than taught as an isolated subject (Más et al., 2012). Research on immigrant populations demonstrates that adults with limited schooling face considerable challenges in acquiring the linguistic competence and literacy skills that support participation in the economic and social life of their new communities (Young-Scholten, 2013), and that linguistic distance between a learner's first language and the language of instruction can significantly impede integration and literacy outcomes (Kuperman, 2022).

Literacy has evolved from a narrow focus on reading and writing to a much broader concept involving the ability to locate, understand, evaluate, interpret, create, and communicate information across a variety of contexts (Cairney, 2016; Churchill et al., 2011). Contemporary literacy includes document literacy, digital literacy, media literacy, information literacy, financial literacy, health literacy, and workplace literacy (Choi et al., 2023; Niyibizi et al., 2020; Santos & McKinney, 2019). Adults must interpret manuals, policies, contracts, safety procedures, diagrams, graphs, online information, and technical documentation while critically evaluating the reliability and relevance of information (Choi et al., 2023; Wild et al., 2022). Literacy is therefore viewed as a socially situated practice that varies according to the demands of different workplaces, communities, professions, and cultures (Cairney, 2016; Kral, 2016). The PIAAC framework defines literacy as "understanding, evaluating, using and engaging with written texts to participate in society, to achieve one's goals, and to develop one's knowledge and potential" (Wild et al., 2022, p. 3), and research has demonstrated that literacy proficiency in adulthood is shaped by socioeconomic, sociodemographic, and linguistic factors (Kyröläinen & Kuperman, 2021; Vágvölgyi et al., 2024). Furthermore, the changing nature of communication and the pervasiveness of digital and social media have prompted a shift from limited definitions of literacy toward the richer concept of multiliteracies (Cairney, 2016; Cartner & Hallas, 2020; Kral, 2016). Health literacy research further illustrates the breadth of contemporary literacy demands, encompassing functional, linguistic, and digital dimensions that affect individuals' capacity to act on health information (Choi et al., 2023; Santos & McKinney, 2019).

Numeracy refers to the ability to use mathematical knowledge and quantitative reasoning to solve problems encountered in everyday life, education, and work (Wild et al., 2022). Rather than focusing solely on arithmetic, adult numeracy includes estimating, measuring, interpreting graphs and tables, calculating quantities, analysing data, understanding probability and risk, budgeting, financial decision-making, and applying mathematical reasoning in practical situations (Santos & McKinney, 2019). Within vocational education, numeracy may involve calculating dimensions, tolerances, costs, percentages, volumes, dosages, loads, production rates, or statistical quality measures depending on the occupational context. Research suggests that low literacy and low numeracy should be regarded as independent phenomena with distinct profiles, analogous to the distinction between dyslexia and dyscalculia (Vágvölgyi et al., 2024). The PIAAC international survey assesses numeracy alongside literacy and problem-solving as core information-processing skills in the adult population (Kyröläinen & Kuperman, 2021; Wild et al., 2022), and programmes in adult basic education increasingly integrate numeracy with literacy and digital literacy instruction (Niyibizi et al., 2020; Santos & McKinney, 2019).

Adult learners enter educational programs with widely differing LLN capabilities. Some possess highly developed academic literacy but limited vocational communication skills, while others have extensive practical experience yet limited formal education. Migrants and refugees may possess strong literacy in their first language while developing proficiency in a second language (McCaffery et al., 2019; McCaffery et al., 2016). Learners with dyslexia, dyscalculia, hearing impairment, acquired brain injury, interrupted schooling, or limited educational opportunities may require additional support to develop the LLN skills necessary for successful participation (Anapey et al., 2025; Kunwar, 2024). Adult education therefore recognises that LLN development is highly individual and influenced by learners' previous educational experiences, cultural backgrounds, occupations, confidence, and learning goals (Ahmed, 2019; Windisch, 2015).

The concept of embedded LLN has become central to contemporary vocational education. Rather than teaching language, literacy, and numeracy as separate subjects, educators integrate these capabilities into vocational learning activities (Ahmed, 2019; Bak & O'Maley, 1970). For example, apprentice electricians develop literacy by interpreting wiring regulations and technical manuals, language skills by communicating with clients and supervisors, and numeracy through calculating electrical loads and cable sizes. Nursing students interpret patient records, calculate medication dosages, and communicate clinical information, while engineering apprentices read technical drawings, complete workplace documentation, calculate tolerances, and interpret measurement data. Embedding LLN within authentic vocational contexts increases learner motivation by demonstrating the direct relevance of these skills to workplace performance (Ahmed, 2019; Coysh, 2010; Windisch, 2015).

Adult learning theories provide valuable guidance for LLN education. Knowles' theory of andragogy emphasises that adults learn most effectively when instruction is relevant to their immediate personal or occupational needs (McCaffery et al., 2019). LLN instruction therefore focuses on authentic workplace documents, practical calculations, and real communication tasks rather than abstract academic exercises (Ahmed, 2019; Windisch, 2015). Adults are more likely to engage with literacy and numeracy development when they clearly understand how these skills improve their employment opportunities and everyday functioning (Fletcher et al., 2011; Muscat et al., 2020).

Constructivist learning theory suggests that adults build new LLN knowledge by connecting it with their previous experiences. Learners actively construct understanding through authentic problem-solving, discussion, collaboration, and reflection rather than memorising isolated rules. For example, a construction apprentice develops numeracy by calculating material quantities for an actual building project rather than completing unrelated mathematical worksheets. Similarly, literacy develops through reading workplace documentation, completing job reports, and participating in team meetings.

Experiential learning theory further supports LLN development by integrating practical experience with reflection and application. Adults repeatedly use communication, reading, writing, and mathematical skills during workplace simulations, practical projects, and industry placements before reflecting on their performance and applying improvements in subsequent tasks. This cycle strengthens both vocational competence and LLN capability.

Social learning theory highlights the importance of observing skilled practitioners and participating within communities of practice. Learners acquire workplace language, technical vocabulary, documentation practices, and problem-solving strategies through interaction with instructors, mentors, supervisors, and peers. Collaborative learning activities encourage learners to communicate ideas, explain reasoning, and negotiate solutions, further strengthening language and literacy development.

Transformative learning theory may also influence LLN development. Adults who have previously experienced educational failure or who believe they are "poor at reading" or "bad at mathematics" may critically reassess these long-held assumptions through supportive educational experiences. Successfully mastering workplace literacy or numeracy tasks can increase confidence, self-efficacy, and willingness to engage in further learning.

Assessment of LLN in adult education focuses on identifying learners' existing capabilities and determining the support required for successful participation. Diagnostic assessment enables educators to tailor instruction to individual needs, while ongoing formative assessment monitors learner progress throughout the course. Importantly, LLN assessment is intended to support learning rather than exclude learners from educational opportunities. Many vocational education systems therefore incorporate foundation skills assessment to ensure learners receive appropriate assistance while maintaining the integrity of vocational competency standards.

Technology has significantly expanded opportunities for LLN development. Online learning platforms, interactive literacy software, digital dictionaries, translation tools, speech-to-text applications, text-to-speech software, numeracy applications, artificial intelligence tutors, and adaptive learning systems provide personalised learning experiences that adults can access at their own pace. Digital literacy itself has become an essential component of LLN, as adults increasingly require competence in using computers, mobile devices, online communication platforms, spreadsheets, databases, and digital workplace systems.

International Approaches to Lifelong Learning

Lifelong learning is the principle that learning occurs throughout an individual's entire life and across multiple contexts, extending beyond compulsory schooling and formal qualifications. It recognises that people continually acquire knowledge, skills, attitudes, and competencies through formal education, vocational training, higher education, workplace learning, community participation, family life, and personal experiences. Within adult education, lifelong learning has become one of the most influential international policy frameworks, supporting economic development, social inclusion, active citizenship, personal fulfilment, and sustainable development (Akther, 2020; Biesta, 2021; Špolar & Holford, 2014).

The concept has evolved considerably over the past century. Early adult education movements primarily focused on literacy, citizenship, and community development, with European history tracing the development of adult education to seventeenth-century thinkers such as Comenius and Grundtvig who recognised the problem of illiteracy (Špolar & Holford, 2014). During the second half of the twentieth century, rapid technological change, globalisation, changing labour markets, and increased life expectancy highlighted the need for continuous learning throughout adulthood (Akther, 2020; Biesta, 2021). The contemporary beginnings of lifelong learning can be traced to UNESCO's embrace of the concept in the 1960s, when the organisation declared "lifelong education" as the master concept for all its educational planning, policy-making, and practice (Špolar & Holford, 2014). Rather than viewing education as something completed during youth, governments and international organisations increasingly recognised learning as a lifelong process that enables individuals to adapt to changing social, economic, and technological conditions (Biesta, 2021; Milana, 2012).

The modern international understanding of lifelong learning was strongly influenced by UNESCO's Faure Report, *Learning to Be* (1972), which argued that education should become a continuous process throughout life rather than being confined to childhood and adolescence (Biesta, 2021; Elfert, 2015; Špolar & Holford, 2014). The report proposed the concept of the "learning society," in which education supports both individual development and democratic participation (Biesta, 2021; Singh, 2015d; Špolar & Holford, 2014). As the Faure Report stated, "the concept of lifelong education (all through life as well as during childhood and adolescence) is emerging clearly as a conscious aspiration" (Špolar & Holford, 2014, p. 36). The report emphasised that if learning is the affair of a lifetime and of a whole society, then it is necessary to think beyond the reform of educational systems toward an "educational community" (Špolar & Holford, 2014).

This vision was further expanded in the Delors Report, *Learning: The Treasure Within* (1996), which introduced the four pillars of lifelong learning: learning to know, learning to do, learning to live together, and learning to be (Bajner, 2019; Biesta, 2021; Elfert, 2019; Milana, 2012; Špolar & Holford, 2014). The Delors Report considered learning throughout life as the key of education and the foundation of a learning society (Y. Lyu, 2023), and argued for the need to rethink and broaden the notion of lifelong education so that it constitutes a continuous process of forming whole human beings (Biesta, 2021; Singh, 2015d). These pillars continue to shape educational policy throughout the world by emphasising that learning serves personal, vocational, social, and civic purposes simultaneously (Bajner, 2019; Biesta, 2021; Milana, 2014). The fourth pillar, "learning to be," explicitly referred back to and reaffirmed the core message of the Faure Report (Bajner, 2019; Milana, 2014). However, some scholars note that while the Delors Report preserved a social-democratic liberal approach, it was not immune to neoliberal ideas such as the updating of skills (Akther, 2020; Milana, 2012).

UNESCO continues to promote lifelong learning as a fundamental human right and an essential component of sustainable development (Elfert, 2019; English & Carlsen, 2019). Through the Recommendation on Adult Learning and Education (RALE) and Sustainable Development Goal 4, UNESCO advocates inclusive and equitable quality education and lifelong learning opportunities for all people regardless of age, gender, disability, socioeconomic background, or geographic location (Elfert, 2019; English & Carlsen, 2019; Panitsides, 2022). SDG 4 calls on Member States to ensure inclusive and equitable quality education and promote lifelong learning opportunities for all (English & Carlsen, 2019; Panitsides, 2022). This approach views lifelong learning not merely as preparation for employment but as a means of improving health, reducing poverty, strengthening democracy, promoting cultural participation, supporting environmental sustainability, and enhancing individual well-being (Elfert, 2019; English & Carlsen, 2019; Stanistreet, 2019). The UNESCO Institute for Lifelong Learning (UIL) has long been engaged with the agenda of lifelong learning across the lifespan through its programmes supporting literacy, policy development, and adult learning and education (English & Carlsen, 2019; Stanistreet, 2022).

The Organisation for Economic Co-operation and Development (OECD) approaches lifelong learning from a complementary perspective that places greater emphasis on workforce development, human capital, productivity, and economic competitiveness (Biesta, 2021; Lee, 2022; Olesen & Wildemeersch, 2012). While UNESCO has tried to sustain a more humanistic dimension, the OECD has more markedly emphasised the economic role of lifelong learning (Lee, 2022; Olesen & Wildemeersch, 2012). OECD policy highlights the importance of continuous skills development throughout working life as occupations evolve due to automation, digitalisation, artificial intelligence, demographic change, and global competition (Biesta, 2021; Cummins & Kunkel, 2015). The Programme for the International Assessment of Adult Competencies (PIAAC), set out by the OECD, concentrates on evaluating adults' proficiency in key skills required in the knowledge age (Boeren & Lido, 2023; Y. Lyu, 2023; Panitsides, 2022). PIAAC data are collected across the world, with its main aim being to gather evidence on the state of the art in developing a strong human capital base for ongoing economic competitiveness (Boeren & Lido, 2023). These surveys demonstrate that adults require ongoing opportunities to develop literacy, numeracy, digital skills, and problem-solving abilities to remain employable in rapidly changing labour markets (Boeren & Lido, 2023; Cummins & Kunkel, 2015; English & Carlsen, 2019; Panitsides, 2022).

The European Union has developed one of the world's most comprehensive lifelong learning frameworks. Through the European Lifelong Learning Strategy, member states encourage adults to move flexibly between vocational education, higher education, workplace learning, and continuing professional development throughout their lives (Amoud & Raymer, 2025). The European Qualifications Framework (EQF) facilitates recognition of qualifications across countries by providing a common reference framework for learning outcomes. Complementary initiatives such as Europass, Erasmus+, and the validation of non-formal and informal learning promote learner mobility, lifelong skills development, and international recognition of competencies (Amoud & Raymer, 2025). Europe's experience highlights the tension between skills-focused initiatives and broader competence-based approaches, as policymakers seek to create not just a "Union of Skills" but potentially a more holistic framework for lifelong learning (Amoud & Raymer, 2025).

Many European countries have integrated lifelong learning into national education systems. The Nordic countries, particularly Finland, Sweden, Denmark, and Norway, have long traditions of liberal adult education, community learning, and folk high schools that emphasise democratic participation, civic engagement, and personal development alongside vocational education. Adult education in these countries is viewed as a public good that contributes to social cohesion, equality, and active citizenship. Flexible pathways between vocational education, universities, workplace learning, and community education allow adults to continually update their knowledge and skills throughout their lives.

In contrast, several East Asian countries emphasise lifelong learning as a mechanism for national economic development and technological innovation. Singapore's SkillsFuture initiative provides every citizen with learning credits that can be used throughout adulthood to undertake accredited education and training (Amoud & Raymer, 2025). This approach encourages continuous upskilling and reskilling while recognising that individuals will likely experience multiple career changes during their working lives. As noted in the literature, "the Singapore experience suggests that the skills development agenda is integral to economic transformation and inclusive growth" (Amoud & Raymer, 2025, p. 47). Similarly, South Korea has invested heavily in lifelong learning centres, digital learning platforms, and adult education policies to support workforce adaptability within a highly technological economy.

Japan has traditionally promoted lifelong learning through community learning centres (Kominkan), universities, cultural education, and workplace training. Japanese lifelong learning policy combines vocational development with broader goals of cultural participation, healthy ageing, community engagement, and active citizenship. This reflects the recognition that learning contributes not only to economic productivity but also to quality of life throughout adulthood.

In Australia, lifelong learning is supported through an interconnected education and training system that includes schools, vocational education and training (VET), universities, adult and community education providers, industry training organisations, and workplace learning. The Australian Qualifications Framework (AQF) facilitates movement between different educational sectors, while recognition of prior learning enables adults to receive formal credit for knowledge and skills acquired through employment and life experience (Ahlgren & Engel, 2011). The AQF defines RPL as "an assessment process that involves assessment of the individual's relevant prior learning to determine the credit outcomes of an individual application for credit". RPL has been embedded in the Australian qualifications system since the introduction of the national framework for the recognition of training in 1992. Lifelong learning policies increasingly emphasise reskilling and upskilling in response to technological change, industry transformation, and an ageing workforce (Rogers, 2011).

The United Kingdom similarly supports lifelong learning through further education colleges, universities, apprenticeships, adult community learning, and continuing professional development (Adanu, 2007; Embo & Valcke, 2017). Policies have increasingly focused on improving workforce skills while also promoting social mobility and widening participation in higher education (Crawford, 2007; Embo & Valcke, 2017). The Leitch Review, for instance, has been reflected in lifelong learning strategies in both England and Scotland, emphasising workforce development and the shift toward higher value-added employment (Rogers, 2011). Flexible qualifications, modular learning, and recognition of workplace learning provide adults with multiple pathways into further education throughout their careers (Adanu, 2007). CPD requirements for teachers in the lifelong learning sector, for example, include a minimum of 30 hours of professional development per year (Embo & Valcke, 2017).

Internationally, lifelong learning increasingly recognises three complementary forms of learning. Formal learning occurs within structured educational institutions leading to recognised qualifications. Non-formal learning takes place through organised activities outside formal education, such as workplace training, professional development, community programs, or industry courses. Informal learning occurs through everyday experiences, self-directed study, family life, hobbies, volunteering, and workplace practice (Rogers, 2011). The European lifelong learning agenda has emphasised these distinctions, with the European Commission defining lifelong learning as referring to all learning activities undertaken throughout life. Contemporary lifelong learning policies acknowledge that valuable learning occurs across all three contexts and encourage systems that recognise learning regardless of where or how it is acquired, (Rogers, 2011).

Adult learning theories provide a strong foundation for lifelong learning policies. Knowles' andragogy emphasises self-directed learning and the importance of adults taking increasing responsibility for their educational development throughout life (Avis et al., 2012). Andragogy has been described as the science and art of adult learning" and is cited often as a significant component of lifelong learning. Experiential learning theory recognises that adults continually learn from workplace experiences, reflection, and practical application (Avis et al., 2012). Transformative learning theory explains how changing life circumstances, career transitions, migration, technological disruption, or personal experiences can prompt adults to critically re-evaluate their assumptions and develop new perspectives (Avis et al., 2012). Social constructivism highlights the importance of collaborative learning, professional communities, and social interaction throughout adult life, while connectivism has become increasingly relevant in digitally connected societies where knowledge is distributed across networks and technologies (Avis et al., 2012). Foundational theories including andragogy, self-directed learning, transformative learning, and experiential learning remain relevant today, while the presence of technology has introduced new theoretical perspectives (Avis et al., 2012).

Despite its widespread acceptance, lifelong learning faces several challenges internationally. Access remains uneven across socioeconomic groups, geographic regions, genders, age groups, and people with disabilities (Rogers, 2011). Financial costs, limited employer support, family responsibilities, digital exclusion, and insufficient recognition of prior learning continue to restrict participation for many adults (Rogers, 2011). The current focus of employers on formal educational opportunities appears narrowly focused on job-specific, in-house training, which limits the development of portable skills and qualifications (Rogers, 2011). Rapid technological change also requires continual updating of educational systems to ensure that learning opportunities remain relevant and accessible throughout increasingly diverse and extended working lives (Avis et al., 2012; Rogers, 2011).

Ultimately, international approaches to lifelong learning share a common recognition that education can no longer be confined to childhood or early adulthood. While countries differ in their priorities, some emphasising economic competitiveness, others social inclusion, democratic participation, or personal development, all recognise that continuous learning has become essential in modern societies. Lifelong learning therefore serves as both an educational philosophy and a policy framework that enables individuals to continually develop their knowledge, skills, capabilities, and identities throughout their lives while contributing to more adaptable, inclusive, and sustainable societies.

Individual Differences and Learning Needs

Individual differences and learning needs are fundamental considerations in adult education because no two adult learners enter an educational program with identical backgrounds, experiences, abilities, motivations, or goals. Adult learners differ in their educational histories, cognitive abilities, cultural backgrounds, language proficiency, occupational experiences, personal interests, health, socioeconomic circumstances, family responsibilities, learning preferences, and confidence. These differences influence not only what learners already know but also how they engage with learning, process new information, and apply knowledge in practical contexts (Goodwill & Chen, 2021). Contemporary adult education therefore recognises learner diversity as a normal characteristic of educational practice rather than an exception requiring special treatment (Fourie & Schlebusch, 2024).

Individual differences arise from a complex interaction of biological, psychological, social, cultural, and environmental factors (Goodwill & Chen, 2021). Some adults may return to education after decades in the workforce, while others may have recently completed secondary school. Learners may possess extensive practical expertise but limited formal education, or they may hold advanced academic qualifications while entering an entirely new vocational field (Fourie & Schlebusch, 2024). Previous educational experiences also shape learners' attitudes towards study. Adults who experienced success in earlier education often display greater confidence and self-efficacy, whereas those who previously struggled may approach learning with anxiety or negative expectations. Recognising these varied starting points enables educators to design learning experiences that build upon learners' existing strengths while addressing areas requiring further development (Fourie & Schlebusch, 2024).

Adult education also acknowledges significant differences in cognitive functioning. Individuals vary in areas such as memory, attention, information processing speed, problem-solving ability, reasoning, creativity, and metacognitive skills (Goodwill & Chen, 2021). These differences should not be interpreted as indicators of intelligence or potential but rather as natural variations in how people process, organise, and apply knowledge. Some learners readily understand abstract theoretical concepts, while others learn more effectively through concrete examples and practical application. Similarly, learners differ in the amount of repetition, scaffolding, or guided practice they require before achieving competence.

Language and cultural diversity further contribute to individual learning needs. Adult learners may speak multiple languages, come from different educational traditions, or possess varying levels of proficiency in the language of instruction (Ahrens et al., 2020; Ray, 2019). Cultural backgrounds influence communication styles, attitudes towards authority, classroom participation, collaboration, assessment expectations, and perceptions of knowledge (Ober et al., 2023; Yıldırım, 2019). Inclusive adult education values these differences as educational assets while providing appropriate language support, culturally responsive teaching, and opportunities for learners to connect new knowledge with their own cultural experiences (Ahrens et al., 2020; Jenkins & Alfred, 2017; Ray, 2019; Tisdell et al., 2013).

Personal circumstances also influence educational participation. Employment commitments, shift work, financial pressures, caring responsibilities, health conditions, disability, geographic location, and access to technology all affect adults' capacity to engage with learning (Nielsen et al., 2008). These external factors often determine when, where, and how adults can study (Askari & Baumgartner, 2024). Consequently, flexible delivery options, blended learning, online education, recognition of prior learning, and part-time study pathways have become increasingly important components of adult education systems (Nielsen et al., 2008).

Motivation varies considerably among adult learners. Some adults study to obtain employment or promotion, while others pursue personal interests, career changes, professional registration, social participation, or intellectual fulfilment (Nielsen et al., 2008). These differing motivations influence learner engagement, persistence, and preferred learning activities (Askari & Baumgartner, 2024). Understanding learners' goals enables educators to establish meaningful learning experiences that connect educational content with learners' personal, professional, and community aspirations (Ray, 2019; Tisdell et al., 2013).

Learning needs refer to the knowledge, skills, support, and educational conditions required for individual learners to achieve successful outcomes. These needs may involve developing foundational language, literacy, numeracy, or digital skills; acquiring vocational competencies; improving academic study skills; building confidence; or overcoming barriers created by disability, disadvantage, or interrupted education (Ahrens et al., 2020; Nielsen et al., 2008). Importantly, learning needs should not be viewed as learner deficiencies but as opportunities for educators to provide appropriate instructional support that enables success (Garrett et al., 2021; Ober et al., 2023).

Effective adult education therefore begins with identifying learner needs through diagnostic assessment, learner interviews, skills audits, workplace observations, or recognition of prior learning processes (Ahrens et al., 2020; Yıldırım, 2019). This information enables educators to tailor instruction appropriately while maintaining consistent learning outcomes and competency standards (Ober et al., 2023). Rather than expecting all learners to follow identical educational pathways, contemporary adult education increasingly promotes flexible learning that accommodates individual differences while preserving academic and vocational rigour (Ahrens et al., 2020; Hahambu et al., 2012).

Several adult learning theories provide important insights into responding to individual differences. Knowles' theory of andragogy emphasises that adults are self-directed learners whose previous experiences represent valuable learning resources, (Guzón et al., 2021; Vega & Domínguez, 2022). Educators therefore encourage learners to draw upon their professional and personal experiences while providing opportunities for individual goal setting and self-directed learning (Guzón et al., 2021; Vega & Domínguez, 2022). Instruction is designed to be relevant to learners' immediate needs and occupational contexts, recognising that adults are more motivated when learning addresses real-life challenges (Guzón et al., 2021; Leupin, 2015).

Constructivist learning theory argues that learners actively construct new understanding by connecting new information with their existing knowledge and experiences. Because every learner possesses unique prior knowledge, educators provide opportunities for discussion, reflection, problem-solving, and collaborative learning that allow individuals to develop personal meaning from educational experiences. Scaffolded instruction enables learners with differing backgrounds to progressively build understanding at appropriate levels of challenge.

Experiential learning theory further recognises that adults learn through cycles of concrete experience, reflection, conceptualisation, and active experimentation (Bernal, 2023; Sevillano et al., 2023). Individual differences influence each stage of this process, as learners bring different workplace experiences, interpret events differently, and apply new knowledge within diverse professional contexts (Sevillano et al., 2023). Practical activities, workplace simulations, case studies, and reflective exercises therefore allow learners to connect theory with their own experiences while accommodating diverse learning pathways (Bernal, 2023).

Humanistic learning theory places particular emphasis on recognising each learner as a unique individual with distinct needs, aspirations, values, and potential. Humanistic educators create supportive learning environments characterised by respect, empathy, trust, and learner autonomy (Vega & Domínguez, 2022). Rather than focusing solely on academic achievement, they support learners' personal growth, confidence, self-efficacy, and lifelong capacity for independent learning (Leupin, 2015; Vega & Domínguez, 2022).

Transformative learning theory acknowledges that adults interpret new learning through their existing beliefs, assumptions, and life experiences (Cancelas-Ouviña & Hurtado-Sierra, 2020; Condom & Zellweger, 2019). Individual differences influence how learners respond to challenging ideas, critical reflection, and perspective transformation (Cancelas-Ouviña & Hurtado-Sierra, 2020; Condom & Zellweger, 2019). Educators therefore encourage respectful dialogue and reflective practice while recognising that learners may progress through transformative learning at different rates depending on their personal histories and readiness for change (Cancelas-Ouviña & Hurtado-Sierra, 2020; Condom & Zellweger, 2019).

Responding to individual differences requires inclusive teaching practices rather than separate educational programs for every learner (H. J. Kim et al., 2023; Montgomery, 2022). Universal Design for Learning (UDL) encourages educators to anticipate learner diversity by providing multiple means of engagement, representation, and expression from the outset (Houghton, 2022; Lewitzky & Weaver, 2022; Nijakowska & Guz, 2024). Learning materials may be presented through text, diagrams, videos, demonstrations, and interactive activities (Ciolfi & Howard, 2024; Saenen et al., 2026). Learners may demonstrate competence through written reports, practical demonstrations, presentations, portfolios, workplace projects, or oral assessments where appropriate (Ciolfi & Howard, 2024; Schlünzen et al., 2024). Such flexibility benefits all learners while reducing barriers for those with disabilities, language differences, or varied educational backgrounds (Liasidou, 2022; Nijakowska & Guz, 2024).

Technology has expanded educators' ability to personalise learning. Learning management systems, adaptive learning software, artificial intelligence tutoring systems, online discussion forums, multimedia resources, and digital assessment platforms enable learners to study at different times, progress at different rates, revisit difficult concepts, and access learning materials in formats that best meet their individual needs (Ciolfi & Howard, 2024; Galvin & Geron, 2021; Montgomery, 2022). Technology also supports accessibility through captioning, screen readers, text-to-speech software, speech recognition, adjustable display settings, and multilingual resources (Ciolfi & Howard, 2024; Rogers-Shaw et al., 2017; Saenen et al., 2026).

Assessment practices similarly reflect recognition of individual differences. Diagnostic assessment identifies existing capabilities, formative assessment provides ongoing feedback during learning, and summative assessment evaluates achievement of learning outcomes (Alamri, 2017) Reasonable adjustments may be provided for learners with disabilities or specific learning needs, provided the integrity of the required learning outcomes or competency standards is maintained (Clark & Karvonen, 2021; Laitusis & Karvonen, 2025). The objective is not to lower standards but to ensure that assessment measures learners' knowledge and skills rather than barriers unrelated to the intended outcomes (Ciolfi & Howard, 2024; Laitusis & Karvonen, 2025).

The following example demonstrates how an adult educator teaching a vocational hairdressing qualification can respond effectively to individual differences and learning needs while maintaining consistent industry competency standards. It illustrates how learner diversity can be recognised as a strength through the application of adult learning theories, including andragogy, constructivism, experiential learning, humanistic learning, and transformative learning, together with inclusive teaching practices and Universal Design for Learning (UDL). The example highlights how flexible instructional strategies, personalised support, and authentic workplace learning can enable adults with diverse backgrounds, experiences, abilities, motivations, and learning needs to successfully develop the knowledge, practical skills, and professional behaviours required for employment in the hairdressing industry.

A vocational education provider delivers a trade-level hairdressing program to a diverse group of adult learners. The cohort includes individuals changing careers after redundancy, experienced salon assistants seeking formal qualifications, recent migrants developing language proficiency, parents returning to the workforce, learners with dyslexia, mature-aged learners re-entering education after many years, and several students who already operate informal home-based hair businesses. Before the course begins, the educator conducts learner interviews, practical skills observations, language and literacy screening, and recognition of prior learning (RPL) assessments to identify each learner's existing capabilities, learning goals, and support requirements. Although every learner is working towards the same industry competency standards, the educator recognises that they will require different learning pathways to achieve those outcomes.

The course design reflects Knowles' theory of andragogy by acknowledging that adult learners bring valuable prior experiences to the classroom. Learners who have previously worked in customer service discuss communication and client consultation skills, while those with informal hairdressing experience demonstrate techniques they have developed in practice. Individual learning plans are established, allowing learners to identify personal goals such as gaining employment in a salon, opening their own business, specialising in colour services, or improving confidence in client communication. Learning activities are closely linked to authentic salon practice so that learners immediately recognise the relevance of the knowledge and skills being developed.

Constructivist learning principles are incorporated through collaborative problem-solving activities. Rather than simply memorising haircutting procedures, learners analyse client consultations, evaluate different hairstyles for varying face shapes and hair textures, and discuss appropriate colouring techniques for diverse client needs. Small-group discussions encourage learners from different cultural and professional backgrounds to share haircare practices, styling traditions, and customer service approaches from their own communities. Scaffolded instruction allows learners with limited prior knowledge to progressively build competence while more experienced learners undertake increasingly complex technical challenges.

The program also applies experiential learning theory through repeated practical salon activities. Learners first observe demonstrations, then practise techniques on mannequin heads before progressing to supervised work with volunteer clients. Following each practical session, learners participate in guided reflection, evaluating the effectiveness of their consultation, technical execution, time management, and client communication. They identify improvements before applying these insights during subsequent practical sessions. This cycle of experience, reflection, conceptualisation, and experimentation allows each learner to progress according to their individual level of development while maintaining consistent vocational standards.

Humanistic learning theory is reflected in the supportive learning environment established by the educator. Recognising that some learners may lack confidence due to previous educational experiences, constructive feedback focuses on individual progress and skill development rather than comparison with classmates. Learners are encouraged to seek assistance without fear of embarrassment, set achievable personal milestones, and celebrate improvements as they develop competence. Respect for learner autonomy allows individuals to pursue elective practical activities that align with their professional interests while still achieving the required competencies.

The educator also recognises the importance of transformative learning. Several learners enter the course believing they are "too old to learn a trade" or lacking confidence after previous academic failure. Through successful practical experiences, supportive feedback, and increasing responsibility during client consultations, many begin to reconsider these long-held assumptions. As their technical competence grows, learners develop stronger professional identities and greater confidence in their ability to succeed within the hairdressing industry.

Universal Design for Learning (UDL) principles ensure that learning materials are accessible to all learners from the outset. Haircutting techniques are presented through written guides, diagrams, high-quality instructional videos, live demonstrations, and interactive digital resources. Learners with language support needs receive illustrated glossaries of technical terminology, while multilingual resources and visual demonstrations assist learners whose first language differs from the language of instruction. Those with dyslexia use text-to-speech software and coloured overlays for written materials, while captioned videos support learners with hearing impairments and reinforce technical terminology for all students.

Assessment methods also accommodate individual differences while maintaining identical competency standards. Learners demonstrate competence through practical client services, professional portfolios, oral questioning, reflective journals, workplace observations, and written knowledge assessments. Reasonable adjustments, such as additional assessment time, oral rather than written responses where appropriate, assistive technologies, or alternative formats for theory assessments, are provided without changing the required industry outcomes. Practical assessments continue to evaluate the same standards of client consultation, hygiene, technical accuracy, safety, and professional conduct expected in commercial salons.

Throughout the program, the educator regularly reviews learner progress using formative assessment, observation, learner feedback, and individual coaching sessions. Additional support is provided where required, while learners demonstrating advanced skills are offered extension activities, such as creative colour correction, advanced styling techniques, or mentoring less experienced classmates. By applying the principles of andragogy, constructivism, experiential learning, humanistic education, transformative learning, and Universal Design for Learning, the educator creates an inclusive vocational learning environment that recognises individual differences as strengths rather than barriers. Every learner is supported to achieve the same professional hairdressing competencies, although the pathway taken to reach those competencies reflects each learner's unique experiences, aspirations, and learning needs.

Key Concepts and Conclusions

This chapter has explored how diversity, inclusion, and global perspectives shape contemporary adult education. It has demonstrated that adult learning is not culturally neutral or universally experienced but is influenced by learners' identities, cultural backgrounds, social contexts, life experiences, and access to educational opportunities. Effective adult educators recognise this diversity as a strength and design learning environments that are inclusive, equitable, culturally responsive, and accessible to all learners.

The chapter has shown that culture influences how adults understand knowledge, communicate, participate, collaborate, and relate to authority within educational settings. Consequently, teaching approaches that are effective in one cultural context may require adaptation in another. Cross-cultural perspectives emphasise the importance of intercultural competence, cultural humility, and recognising that multiple educational traditions and ways of knowing can contribute to effective adult learning. Rather than applying a single educational model universally, educators should adapt teaching practices to learners' cultural contexts while maintaining high educational standards.

Indigenous approaches to learning broaden contemporary understandings of adult education by emphasising relationships, community participation, experiential learning, storytelling, respect for Elders, connection to place, and holistic development. These approaches remind educators that learning extends beyond the acquisition of knowledge and skills to include identity, responsibility, relationships, and community wellbeing. Respectfully incorporating Indigenous perspectives enriches educational practice and promotes culturally responsive learning environments for all learners.

The chapter has also examined the continuing influence of gender on educational participation and outcomes. Although significant progress has been made towards gender equality, barriers associated with gender roles, occupational stereotypes, caring responsibilities, and institutional practices continue to influence lifelong learning opportunities. Adult educators play an important role in promoting gender equity through inclusive curriculum design, equitable participation, flexible delivery, and learning environments that value learners of all genders. International frameworks such as the United Nations Sustainable Development Goals reinforce the importance of education as a mechanism for advancing gender equality and social inclusion.

Equity and social justice extend the discussion by recognising that equal treatment alone does not necessarily produce fair educational outcomes. Adult learners begin from different circumstances and may require different forms of support to achieve comparable success. Inclusive educational practice therefore involves identifying and reducing barriers related to socioeconomic status, disability, migration, language, discrimination, and other forms of disadvantage while maintaining consistent educational standards. Adult education contributes not only to workforce development but also to greater social participation, civic engagement, and democratic inclusion.

Disability and inclusive education demonstrate that educational systems should be designed to accommodate learner diversity rather than expecting learners to adapt to inflexible educational structures. Contemporary approaches emphasise removing physical, technological, instructional, and attitudinal barriers through accessible curriculum design, Universal Design for Learning, assistive technologies, inclusive assessment, and strengths-based teaching. These principles benefit not only learners with disabilities but all adults participating in lifelong learning.

Migration and refugee education further illustrate the importance of recognising learners' existing knowledge, cultural identities, resilience, and prior experiences. Adult migrants and refugees often require integrated support that combines language development, vocational education, recognition of prior learning, settlement assistance, and culturally responsive teaching. By valuing learners' strengths while addressing barriers to participation, adult education supports both successful integration and lifelong learning.

Language, literacy, and numeracy underpin every aspect of adult learning. Rather than viewing these capabilities as separate foundation subjects, contemporary adult education increasingly embeds them within authentic vocational, professional, and community contexts. Strong language, literacy, and numeracy skills enable adults to communicate effectively, solve problems, participate confidently in employment, and continue learning throughout their lives.

International approaches to lifelong learning reinforce the understanding that education is a continuous process extending across the lifespan. Although countries differ in their educational priorities, most recognise lifelong learning as essential for personal development, workforce adaptability, social cohesion, active citizenship, and sustainable development. International organisations such as UNESCO, the OECD, and the European Union have established frameworks that encourage adults to continue learning through formal, non-formal, and informal education throughout their lives.

Finally, the chapter has highlighted that recognising individual differences and learning needs is fundamental to effective adult education. Learners differ in their experiences, motivations, cognitive abilities, languages, cultures, personal circumstances, and educational goals. Adult learning theories—including andragogy, constructivism, experiential learning, humanistic learning, and transformative learning—provide educators with complementary approaches for responding to this diversity. Together with inclusive teaching practices, Universal Design for Learning, flexible assessment, and appropriate technologies, these approaches enable educators to create learning environments in which every learner has the opportunity to succeed while achieving consistent educational and vocational standards.

Ultimately, diversity, inclusion, and global perspectives are not peripheral considerations in adult education but fundamental principles that shape educational quality and effectiveness. Adult educators who recognise cultural diversity, promote equity, value multiple ways of knowing, respond to individual learning needs, and create inclusive learning environments are better equipped to prepare adults for participation in increasingly diverse, interconnected, and rapidly changing societies. In doing so, adult education fulfils its broader purpose of supporting lifelong learning, personal growth, workforce development, social justice, and global citizenship.

Richard Skiba

Chapter 8

Theory Informed Teaching

Understanding how adults learn is fundamental to effective teaching, yet knowing what to teach is only one part of successful educational practice. Equally important is understanding how learning occurs and why particular teaching approaches are more effective than others in different educational contexts. Adult learners bring diverse experiences, motivations, prior knowledge, cultural backgrounds, and professional aspirations to the learning environment. Consequently, educators require more than subject expertise and practical experience; they need a sound understanding of educational theory to make informed decisions about curriculum design, instructional strategies, assessment, learner support, and reflective practice. Theory-informed teaching provides the intellectual framework that enables educators to create engaging, inclusive, evidence-based learning experiences that respond to the complexity and diversity of adult learning.

This chapter examines the principles of theory-informed teaching and demonstrates how educational theory can be translated into effective instructional practice. It begins by exploring the concept of theory-informed teaching and the role educational theories play in guiding teaching decisions throughout the learning process. The chapter then examines three learner-centred instructional approaches—performative learning, Problem-Based Learning (PBL), and Project-Based Learning (PjBL)—showing how each draws upon multiple educational theories to promote authentic, active, and meaningful learning. Finally, the chapter explores how educators can apply different theoretical perspectives to meet the needs of individual learners, recognising that no single theory adequately explains all aspects of adult learning. Together, these discussions illustrate how theory-informed teaching enables educators to integrate research, professional judgement, and reflective practice to develop learners who are capable of critical thinking, professional competence, lifelong learning, and effective participation in an increasingly complex world.

Theory-Informed Teaching

Effective adult education is built upon more than practical experience, subject expertise, or intuition. While experienced educators often develop highly effective teaching practices through reflection and professional experience, educational theory provides a systematic foundation for understanding how adults learn and how teaching can be designed to maximise learning outcomes (Carré, 2015; McLean et al., 2014). As (Schunk, 2012) argues, cited in Nifaoui, "practical experience without theory is limited to a given situation that lacks a global approach and a framework for organising the way of teaching and learning" (Nifaoui, 2020). Theory-informed teaching involves applying established educational theories and research to guide instructional planning, curriculum design, learner support, assessment, and reflective practice (Carré, 2015; McLean et al., 2014). Rather than prescribing a single method of teaching, educational theories provide complementary perspectives that help educators make informed decisions about responding to the diverse needs, experiences, motivations, and learning contexts of adult learners (Carré, 2015).

Throughout the development of adult education, numerous theories have emerged to explain different aspects of learning. Behaviourist theories emphasise the role of reinforcement and observable behaviour, while cognitivist theories focus on mental processes such as memory, information processing, and problem-solving (Bocquillon et al., 2020; Nifaoui, 2020). Constructivist approaches highlight the active construction of knowledge through experience and social interaction (Bourgeois, 2018; Nifaoui, 2020), whereas humanistic theories emphasise learner autonomy, personal growth, and self-actualisation (McLean et al., 2014). Experiential learning, transformative learning, social learning, connectivism, and sociocultural perspectives each contribute further insights into how adults acquire knowledge, develop skills, construct meaning, and adapt to changing personal and professional circumstances (Biasin, 2018a; Kandeel, 2020; Turcotte, 2022). Carré identifies the major theoretical systems constructed over the past century in the psychology of learning, including andragogy and newer currents such as self-directed, transformative, and critical learning, as essential to understanding adult education (Carré, 2015). Collectively, these theories provide educators with a rich body of knowledge from which to draw when designing effective learning experiences (McLean et al., 2014; Nifaoui, 2020).

Theory-informed teaching recognises that no single educational theory adequately explains all aspects of adult learning or is appropriate for every educational situation (Carré, 2015; Nifaoui, 2020). Different learning objectives, learner characteristics, subject matter, and educational contexts may require different instructional approaches (McLean et al., 2014). For example, behaviourist principles may be highly effective when teaching procedural safety routines or technical skills requiring accuracy and repetition (Nifaoui, 2020), whereas constructivist or experiential approaches may be better suited to developing problem-solving, professional judgement, or critical thinking (Bocquillon et al., 2020; Bourgeois, 2018). Humanistic approaches may be particularly valuable when supporting learner confidence and motivation, while transformative learning may help adults critically examine long-held assumptions during career transitions or professional development (Biasin, 2018a). Effective adult educators therefore select and combine theoretical approaches according to the educational purpose rather than adhering rigidly to a single teaching philosophy (Carré, 2015; McLean et al., 2014).

Applying educational theory also supports evidence-informed professional practice. Rather than relying solely on personal preference or tradition, educators who understand adult learning theories can justify their instructional decisions, critically evaluate their teaching practices, and adapt their methods in response to learner needs and emerging educational research (Carré, 2015; Nagels et al., 2018). Theory provides a common professional language that enables educators to analyse learning challenges, share effective practices, and engage in ongoing professional development (McLean et al., 2014; Perrault & Levené, 2019). It also supports reflective practice by encouraging educators to examine why particular teaching strategies succeed or require modification within different learning environments (Nagels et al., 2018; Perrault & Levené, 2019).

Theory-informed teaching is particularly important within vocational education, higher education, workplace learning, and professional development, where adult learners bring diverse experiences, expectations, and career goals (Vinatier, 2012). Understanding how adults learn enables educators to create learning environments that are engaging, inclusive, relevant, and responsive to individual differences while maintaining rigorous academic and vocational standards (Carré, 2015; McLean et al., 2014). By integrating educational theory with practical teaching experience, educators can design learning experiences that not only develop knowledge and skills but also foster critical thinking, professional competence, lifelong learning, and personal growth (Biasin, 2018a; Carré, 2015; McLean et al., 2014).

Theory-informed teaching bridges the gap between educational research and classroom practice. It enables educators to move beyond asking *what* should be taught towards understanding *how* adults learn most effectively and *why* particular teaching approaches are appropriate in different contexts. In doing so, theory-informed teaching provides the intellectual foundation for reflective, adaptable, and learner-centred adult education that prepares individuals to learn, work, and thrive throughout their lives.

Theory-informed teaching is implemented by deliberately using educational theories to guide decisions throughout the entire teaching and learning process rather than relying solely on experience, intuition, or tradition. Theories are used to inform how educators analyse learner needs, design curriculum, select teaching strategies, facilitate learning, assess achievement, and evaluate the effectiveness of their practice. In this way, educational theory becomes a practical framework for making informed instructional decisions that improve learning outcomes.

The process commences with developing and understanding the learners. Educators use adult learning theories to identify learners' prior knowledge, experiences, motivations, cultural backgrounds, abilities, and learning goals. For example, an educator drawing on Knowles' theory of andragogy recognises that adults are self-directed learners who bring valuable life experiences to the classroom. Consequently, learner interviews, diagnostic assessments, recognition of prior learning, and discussions about learning goals become important planning tools before instruction begins.

Theory also informs curriculum and learning design. Rather than simply presenting information, educators design learning experiences that reflect how learning occurs according to different theoretical perspectives. Constructivist approaches encourage authentic problem-solving, collaborative learning, and opportunities for learners to build understanding through experience. Experiential learning theory leads educators to incorporate practical activities, workplace simulations, reflective practice, and repeated opportunities to apply new knowledge. Behaviourist principles may guide the sequencing of procedural skills that require demonstration, guided practice, repetition, and reinforcement.

The selection of teaching strategies is similarly informed by theory. Instead of relying on lectures alone, educators choose instructional methods that best suit the intended learning outcomes. Humanistic approaches encourage learner-centred teaching, respect, empathy, and learner autonomy. Social learning theory supports demonstrations, peer mentoring, role modelling, and collaborative learning. Transformative learning encourages critical reflection, discussion of assumptions, and analysis of authentic dilemmas that challenge existing perspectives. Educators often combine multiple theoretical approaches within a single lesson because different aspects of learning require different instructional methods.

Educational theory also shapes the learning environment. Theory-informed educators create environments that encourage psychological safety, active participation, collaboration, and learner engagement. Humanistic theory promotes respectful relationships and learner autonomy, while Universal Design for Learning (UDL) ensures that learning materials, activities, and assessments are accessible to diverse learners. Constructivist and sociocultural theories encourage interaction, discussion, and learning through participation within communities of practice.

Assessment is another area where theory is applied. Theory-informed educators use assessment not only to measure learning but also to support it. Formative assessment provides regular feedback that guides learner improvement, reflecting constructivist and humanistic principles. Experiential learning encourages reflective journals, portfolios, and workplace projects that allow learners to demonstrate learning through authentic performance. Behaviourist approaches may be appropriate when assessing procedural accuracy or compliance with established workplace standards. Regardless of the assessment method, educators ensure that assessment aligns with the intended learning outcomes and the theoretical approach underpinning instruction.

Educational theory also guides learner support. Recognising that adults differ in their experiences, abilities, motivations, and personal circumstances, educators provide appropriate scaffolding, coaching, mentoring, language support, flexible learning pathways, and reasonable adjustments where required. Rather than expecting every learner to progress identically, theory-informed teaching recognises that different learners may require different forms of support while working towards the same educational outcomes.

Finally, theory-informed teaching is implemented through reflective practice and continuous improvement. Educators regularly evaluate the effectiveness of their teaching by analysing learner performance, engagement, feedback, assessment outcomes, and their own observations. Educational theories provide frameworks for interpreting this evidence and making informed improvements to curriculum design, instructional strategies, assessment methods, and learner support. Reflection allows educators to adapt their teaching as learner needs, technologies, workplaces, and educational research continue to evolve.

In practice, theory-informed teaching is rarely based on a single educational theory. Effective educators adopt an eclectic approach, selecting and integrating theories according to the learning context and desired outcomes. For example, a vocational educator teaching hairdressing might use behaviourist principles to teach safe handling of chemicals, constructivist activities to develop client consultation skills, experiential learning during supervised salon practice, social learning through peer observation and mentoring, humanistic principles to build learner confidence, and transformative learning to encourage reflection on professional identity and customer service. Theories therefore function as complementary tools that inform professional judgement rather than rigid prescriptions for teaching.

Performative Learning

Performative learning is an educational approach that emphasises learning through doing, demonstrating, practising, and performing authentic tasks. Rather than viewing learning as simply the acquisition of knowledge, performative learning recognises that genuine understanding develops when learners actively apply knowledge, skills, attitudes, and professional judgement in realistic contexts (Said, 2018; Winberg & Hollis-Turner, 2021). Authentic learning, understood as "learning through the application of knowledge in realistic contexts," makes use of authentic tasks where assignments are drawn from vocational practice (Said, 2018, p. 47). This approach is particularly valuable in vocational education, higher education, and workplace training, where competence is measured by an individual's ability to perform effectively rather than simply recall information (Dahlback et al., 2020; Winberg & Hollis-Turner, 2021). As Winberg and Hollis-Turner note, "competence," interpreted as "professional competence to act," is central to vocational pedagogy and examination systems (Winberg & Hollis-Turner, 2021, p. 8). Effective practical training requires authentic learning tasks that address key vocational competencies (Winberg & Hollis-Turner, 2021), and authentic assessment involves supporting "the job-relevant learning process towards comprehensive professional competence which is based on the competence needed in professional practice" (Dahlback et al., 2020, p. 306).

Within the context of theory-informed teaching, performative learning is not regarded as a learning theory in its own right but as an instructional approach that integrates insights from multiple educational theories to create authentic and meaningful learning experiences. The application of learning theories to educational design provides educators with a coherent framework for developing curricula, teaching strategies, and assessment methods that are aligned with how students learn most effectively (Bajpai et al., 2019). By drawing on these theoretical perspectives, educators gain a deeper understanding of the learning process and can design activities that require learners to demonstrate their knowledge and skills through realistic, practice-based performance (Choi-Lundberg, 2024; Scholtz & Bester, 2018).

This approach is particularly relevant in vocational education, where learners are expected not only to acquire technical knowledge and practical skills but also to develop the attitudes, values, and professional identity required for competent workplace practice (Said, 2018). Accordingly, vocational educators are encouraged to implement pedagogies that enable learners to connect theoretical concepts with practical application while integrating knowledge from multiple sources into their own professional practice (Schaap et al., 2023). The literature identifies authentic learning experiences, work-integrated learning (WIL), and reflective practice as effective strategies for achieving these outcomes because they provide opportunities for learners to apply discipline-specific knowledge in realistic settings while supporting ongoing professional growth (Choi-Lundberg, 2024). In this way, educational theory provides the conceptual foundation for learning design, while performative learning operationalises those theories by engaging learners in authentic tasks that demonstrate competence in practice (Schulz, 2016; Winberg & Hollis-Turner, 2021).

A central assumption of performative learning is that learners develop deeper understanding by actively performing meaningful tasks rather than passively receiving information. Learning becomes more effective when individuals are required to apply knowledge in situations that resemble those they will encounter in professional practice (Dopelt et al., 2023; Ninassi & Burrell, 2023; Osunlaja et al., 2024). Active learning requires the learner to become an active participant in creating their own learning, rather than a passive bystander to the learning process (Wagner & Campbell, 1994). Experiential learning techniques immerse students in collaborative, reflective, and integrative experiences that simulate real-world scenarios (Ninassi & Burrell, 2023), and through such techniques students "learn-by-doing," applying their knowledge to real-world problems (O'Connor & Moran, 2023).

For example, rather than simply attending a lecture on leadership, learners may conduct a team meeting, resolve a workplace conflict, deliver performance feedback to an employee, or lead a project team. Students may take on roles in team projects that simulate real-world leadership challenges, and by actively participating in decision-making, conflict resolution, and team dynamics, they gain a deeper understanding of leadership principles and how to apply them in practice (Ninassi & Burrell, 2023). Simulations and role-plays are frequently employed to allow learners to practice leadership behaviors, such as navigating challenging conversations or managing clinical scenarios (Alhadithi & Almansour, 2025; Dopelt et al., 2023). Leadership-related activities such as giving feedback, delegating duties, building teamwork, managing conflicts, and coping with stress have been incorporated into experiential programs across professional domains (Frich et al., 2014). Similarly, instead of only reading about welding techniques, apprentice welders prepare materials, set up equipment, complete welds, inspect the quality of their work, and evaluate their own performance. In each case, the performance itself becomes both the learning activity and the evidence that learning has occurred. Experiential learning opportunities have been praised for bridging the gap between theory and practice, as students gain a deeper understanding of professional systems, improve their practical skills, and feel more confident in their abilities to lead and manage within professional settings (Khalid et al., 2023; Osunlaja et al., 2024). The performance of meaningful tasks in realistic contexts ensures that the skills and lessons learned can be directly transferred to the workplace, bridging the gap between theory and application (Almeida & Buzády, 2024; O'Connor & Moran, 2023).

Theoretical Foundations of Performative Learning

Performative learning is supported by several complementary educational theories, each contributing a different perspective on how learning occurs (Kitagawa, 2020; Lee & Wong, 2025; Wu et al., 2011).

Behaviourism contributes the principles of repetition, reinforcement, feedback, and skill mastery. Through repeated practice, learners gradually develop automaticity and accuracy. Behaviourism is rooted in the work of Pavlov and Skinner, positing that learning is a change in observable behaviour conditioned through positive or negative reinforcement (Kitagawa, 2020; Omer, 2020b). Positive outcomes, or reinforcers, improve behaviour and learning, whereas adverse outcomes undermine it (Mukhalalati & Taylor, 2019; Rubaia'an, 2023). Three key concepts strengthen learning in this paradigm: contiguity, frequency, and contingency (Omer, 2020b). In AI-driven surgical training, for instance, correct techniques are reinforced with affirmative feedback while incorrect methods receive corrective responses, directly conditioning behavioural change (Lee & Wong, 2025). For example, apprentice electricians repeatedly wire electrical circuits while receiving immediate feedback from their instructor until correct procedures become second nature.

Cognitivism emphasises the mental processes that underpin effective performance. Learners are encouraged to understand not only what to do but also why they are doing it. Cognitivism focuses on inner mental activities, positing that learning is most effective when individuals actively engage in organising and reflecting on new information (Lee & Wong, 2025; Nalliah, 2014). Cognitivism replaced behaviourism in the 1960s as the dominant paradigm, opening the "black box" of the human mind (Nalliah, 2014). Learners are prompted to analyse, reflect, and apply knowledge in various contexts (Tuczyński, 2023). Before performing structural calculations, engineering students learn to analyse loads, identify assumptions, organise information systematically, and apply appropriate mathematical models. Their performance reflects deliberate thinking rather than simple mechanical repetition.

Constructivism views learning as an active process in which learners build their own understanding by solving authentic problems. Constructivists believe that learning is the meaning-making process of discovering complex information and transforming it into internal constructions of reality (Naudé et al., 2014). Rather than following predetermined procedures, learners investigate issues, collaborate with others, make decisions, and develop solutions. Active learning, described as learners doing things and thinking about the things they are doing, is aligned with the constructivism learning philosophy (Bell & Bell, 2020). Examples of constructivist classrooms include problem-based learning, group collaborations, research, and creative projects (Atta et al., 2023). Constructivism focuses on the uniqueness of learners, recognising that their views and values change over time, impacting future knowledge (Hammad et al., 2020). For example, business students may create a comprehensive marketing strategy for a real organisation instead of merely memorising marketing theories. Their completed work demonstrates knowledge that has been actively constructed through experience.

Experiential learning, particularly Kolb's learning cycle, aligns closely with performative learning because it emphasises learning through concrete experience followed by reflection, conceptual understanding, and experimentation. Kolb proposed that experience facilitates learning and knowledge formation, dividing the learning cycle into four phases: concrete experience, reflective observation, abstract conceptualisation, and active experimentation (Chun, 2024; Mukhalalati & Taylor, 2019; Rubaia'an, 2023). Reflection plays an important role by providing a bridge between experience and theoretical conceptualisation (Chen, 2013). Kolb's experiential learning theory proposes a constructivist theory of learning through which "social knowledge is created and re-created in the personal knowledge of the learner" (Bell & Bell, 2020, p. 8). For example, nursing students conduct a simulated patient assessment before reflecting on their clinical decisions, identifying areas for improvement, and applying these insights during subsequent simulations.

Social learning theory also contributes significantly by recognising that people often learn by observing others before attempting tasks independently. Social learning theories integrate the concept of behaviour modelling with cognitive learning so that understanding of task performance is strengthened (Mukhalalati & Taylor, 2019). As Bandura stated, "most human behavior is learned observationally through modeling: from observing others one forms an idea of how new behaviors are performed, and on later occasions this coded information serves as a guide for action" (Nalliah, 2014, p. 51). The role of the instructor is as a model from whom the learner obtains an understanding of how to perform, and methodologies are focused around observation and performance (Kitagawa, 2020). Performative learning commonly follows a sequence in which learners first observe an expert, then practise under supervision, and eventually perform independently. This progression is characteristic of apprenticeships, internships, and workplace mentoring programs.

Humanism complements performative learning by emphasising learner autonomy, personal growth, confidence, and self-direction. Humanism is a paradigm from the 1960s that emphasises human freedom and dignity to achieve full potential, suggesting that learning is self-directed and that adults can plan, manage, and assess their own learning (Mukhalalati & Taylor, 2019; Rubaia'an, 2023). Implications of humanism include critical reflection and feedback, which are the two pillars of self-directed learning (Omer, 2020). AI-enhanced humanistic approaches position feedback as a means of encouraging self-direction and personal agency, supporting learners in making informed decisions rather than dictating learning paths (Lee & Wong, 2025). Learners are encouraged to take responsibility for improving their own performance through reflection, self-assessment, and goal setting, rather than depending solely on instructor evaluation.

Characteristics of Performative Learning

Theory-informed performative learning incorporates a number of common characteristics that distinguish it from more traditional teacher-centred approaches. Learning activities are authentic and reflect real workplace or professional tasks (Baldacci, 2022; Chen & Tang, 2024; Evanick, 2024). Learners actively participate rather than passively receive information (Chen & Tang, 2024; Hwang et al., 2022; Nemakhavhani, 2024; Vindigni, 2023) and are regularly required to solve problems, collaborate with others, reflect on their experiences, and receive constructive feedback (Allan & O'Driscoll, 2025; Nasir et al., 2018; Nemakhavhani, 2024; Routh et al., 2022). Self-assessment and reflection are encouraged (Bajis et al., 2020; Routh et al., 2022), while responsibility for learning is gradually transferred from the educator to the learner as competence develops (Allan & O'Driscoll, 2025). Repeated opportunities for practice allow learners to refine their skills and demonstrate increasing levels of competence (Bajis et al., 2020; Chen & Tang, 2024). Rather than separating theoretical knowledge from practical application, performative learning integrates both throughout the learning process (Casado-Aranda et al., 2021; Nemakhavhani, 2024; Wang, 2024).

Examples in Post-Compulsory Education

Performative learning is widely used across different forms of post-compulsory education. In vocational education, a Certificate III plumbing learner studies the principles of water supply systems before designing and installing a plumbing system within a simulated workplace. Throughout the activity, the educator explains the engineering concepts underlying the installation, supervises practical work, provides ongoing feedback, and encourages learners to reflect on the completed task. Learning is demonstrated through competent performance rather than written knowledge alone.

In higher education, engineering students may investigate the failure of a bridge as part of a structural engineering course. Rather than simply completing a written examination, students analyse engineering drawings, calculate structural loads, identify design failures, develop alternative engineering solutions, and present their recommendations to their peers. This assessment closely mirrors authentic engineering practice by integrating theoretical knowledge with professional problem-solving and communication skills.

Within workplace training, newly appointed warehouse supervisors participate in leadership development by conducting toolbox meetings, allocating work, managing employee performance, resolving workplace conflicts, and receiving coaching from experienced managers. Instead of merely studying leadership theories, learners demonstrate their competence through authentic workplace performance, allowing both themselves and their supervisors to evaluate how effectively theoretical principles have been translated into practice.

The Educator's Role

Within performative learning, the educator's role extends far beyond simply delivering information. Educators become designers of authentic learning experiences, facilitators of learning, coaches, mentors, and providers of constructive feedback (Green, 2020; Shin & Rim, 2023). Theory-informed educators deliberately select learning activities because educational research indicates that these approaches promote deeper understanding and more durable learning (Borders, 2019; Kuiper et al., 2010).

Equally important is explaining the purpose of learning activities. Before commencing a workplace simulation, an educator might explain that the activity has been designed to allow learners to apply a previously introduced decision-making model within a realistic scenario (Jacobs, 2019). By understanding why they are completing the activity and how it supports their professional development, learners become more engaged and develop greater metacognitive awareness of their own learning processes (Borders, 2019; Evanick, 2024).

When informed by educational theory, performative learning offers numerous educational advantages. It creates strong connections between theoretical knowledge and practical application, enabling learners to understand how concepts are used in authentic professional settings (Beer et al., 2020; Kuiper et al., 2010). Because learners actively apply knowledge rather than merely memorising information, performative learning promotes deeper understanding and longer-term retention (Kuiper et al., 2010).

Authentic performance activities also develop workplace competence by providing opportunities to practise professional skills under realistic conditions (Kuiper et al., 2010),. They improve problem-solving ability, encourage critical thinking, increase learner engagement, and build professional confidence through repeated practice and constructive feedback. Reflection and self-assessment support self-regulation (Evanick, 2024; Rens et al., 2020), while the realistic nature of learning activities prepares learners for the demands of professional practice (Beer et al., 2020; Kuiper et al., 2010). These capabilities also foster lifelong learning by enabling individuals to continue developing their knowledge and skills throughout their careers (Borders, 2019; Green, 2020).

Performative learning demonstrates how educational theory can be translated into effective teaching practice. Rather than relying on a single theoretical perspective, it integrates the complementary strengths of multiple learning theories (Borders, 2019; Shin & Rim, 2023). Behaviourism contributes structured practice and reinforcement; cognitivism explains the mental processes that underpin skilled performance; constructivism encourages learners to build understanding through authentic problem-solving (Borders, 2019); experiential learning highlights the value of experience, reflection, and continuous improvement (Jacobs, 2019); social learning theory emphasises observation and modelling (Nugent et al., 2018); and humanism promotes learner autonomy and personal growth (Nugent et al., 2018).

By integrating these theoretical perspectives, theory-informed teaching enables educators to design learning experiences that require learners to perform meaningful tasks, reflect on their performance, receive constructive feedback, and progressively develop the knowledge, skills, and professional judgement required for competent practice (Borders, 2019; Kuiper et al., 2010). Performative learning therefore serves as a bridge between educational theory and authentic workplace performance, ensuring that learners are not only able to explain concepts but are also capable of applying them confidently and effectively in real-world contexts (Beer et al., 2020; Green, 2020; Kuiper et al., 2010).

Problem-Based Learning

Problem-Based Learning (PBL) is a learner-centred instructional approach in which learning begins with an authentic, complex, and often ill-structured problem rather than with the presentation of content (Prieto-López & Ávila, 2021; Velàzquez et al., 2020). Instead of first teaching theory and then asking learners to apply it, PBL reverses this sequence by presenting learners with a realistic problem that requires them to identify what they need to know, locate relevant information, collaborate with others, and develop evidence-based solutions (Rodríguez, 2014; Velàzquez et al., 2020). The educator acts primarily as a facilitator who guides inquiry rather than providing direct instruction (Sardar, 2023)
.

Within theory-informed teaching, Problem-Based Learning is not a learning theory itself but a teaching strategy grounded in several complementary educational theories (Maria Teresa Vargas La et al., 2022). It illustrates how theoretical understandings of learning can be translated into instructional practice that promotes active engagement, critical thinking, and professional competence (Rodríguez, 2014; Sardar, 2023). Rather than relying on lectures as the primary means of instruction, educators use authentic problems to encourage learners to construct knowledge, apply theory, reflect on their learning, and develop the skills required for professional practice (Julca-Asto et al., 2023; Prieto-López & Ávila, 2021).

Problem-Based Learning draws on multiple educational theories, each contributing different principles to its design and implementation. Constructivism provides the strongest theoretical foundation for PBL (Julca-Asto et al., 2023; Rodríguez, 2014). Constructivist theory proposes that learners actively construct knowledge through experience rather than passively receiving information (O'Donnell, 2012; Sardar, 2023). In a PBL environment, learners investigate complex problems, integrate new information with prior knowledge, test possible solutions, and refine their understanding through discussion and reflection (O'Donnell, 2012; Sardar, 2023). Learning occurs because learners actively make sense of authentic situations rather than memorising isolated facts (Baloco & Mendoza, 2022; Rondón-Molina et al., 2024).

PBL is also strongly influenced by experiential learning. Learners engage directly with realistic problems, reflect on their experiences, develop conceptual understanding, and apply that understanding to increasingly complex situations (Maria Teresa Vargas La et al., 2022). This cycle closely mirrors Kolb's model of concrete experience, reflective observation, abstract conceptualisation, and active experimentation. The origins of PBL have been traced to the experiential educational principles of John Dewey during the early decades of the twentieth century (Maria Teresa Vargas La et al., 2022).

Social constructivism, particularly the work of Lev Vygotsky, contributes the emphasis on collaborative learning (Araujo, 2016; Sardar, 2023). Most PBL activities require learners to work in groups where they discuss ideas, challenge assumptions, share expertise, and jointly develop solutions (Julca-Asto et al., 2023). Through dialogue and collaboration, learners often achieve higher levels of understanding than they could independently (Sardar, 2023).

Cognitivism also underpins Problem-Based Learning by recognising that learners must organise information, analyse evidence, evaluate alternatives, and develop structured approaches to solving problems (M, 2014; . PBL encourages learners to develop higher-order cognitive processes such as reasoning, decision-making, and critical thinking rather than simply recalling information (Sardar, 2023; , (M, 2014; .

Elements of humanism are also evident because learners assume responsibility for directing much of their own learning (Rodríguez, 2014). They identify learning needs, determine what information is required, select learning resources, and evaluate their own progress. This emphasis on autonomy supports self-directed learning and lifelong professional development (Baloco & Mendoza, 2022; Rodríguez, 2014).

Characteristics of Problem-Based Learning

Problem-Based Learning differs from traditional instruction in several important ways. Learning begins with a realistic problem rather than a lecture or textbook chapter (Ajani, 2024; Grabinger et al., 2011; Moallem, 2019). The problems are often open-ended and may have multiple acceptable solutions, requiring learners to investigate, analyse, and justify their decisions (Arega & Hunde, 2025; Moallem, 2019; Varadarajan & Ladage, 2024).

Learners work collaboratively to identify what they already know, determine what additional knowledge is required, divide research tasks, and integrate their findings into a coherent solution (Grabinger et al., 2011; Ranmechai & Poonputta, 2023). Throughout the process, educators facilitate discussion by asking questions, encouraging reflection, and guiding inquiry without providing direct answers (Grabinger et al., 2011; Smith et al., 2022; Wang et al., 2016). Reflection is an essential component, enabling learners to evaluate both the quality of their solutions and the effectiveness of their learning strategies (Berenji, 2021; Hajian, 2019; Ranmechai & Poonputta, 2023).

In Problem-Based Learning, the educator's role changes significantly. Rather than acting primarily as a source of information, the educator becomes a facilitator, coach, and guide (Grabinger et al., 2011; Moallem, 2019; Smith et al., 2022; Wang et al., 2016). The educator designs authentic problems (Maoto, 2023), monitors group discussions (Varadarajan & Ladage, 2024), encourages critical questioning (Wang et al., 2016), provides appropriate scaffolding when learners encounter difficulties (Berenji, 2021; , Kim et al., 2018), and helps learners reflect on both the problem-solving process and the knowledge they have developed (Grabinger et al., 2011; Varadarajan & Ladage, 2024).

Theory-informed educators also explain why Problem-Based Learning is being used. Rather than simply assigning a case study, they help learners understand that investigating authentic problems develops analytical thinking, professional judgement, collaboration, and independent learning skills that are essential in the workplace (Ajani, 2024; Chua, 2023). Making the educational purpose explicit also strengthens learners' metacognitive awareness by helping them understand how the activity contributes to their learning (Ranmechai & Poonputta, 2023; Smith et al., 2022).

PBL develops deeper understanding of concepts than passive instruction (Payan-Carreira et al., 2022; Rosário & Raimundo, 2024; Varadarajan & Ladage, 2024). It strengthens critical thinking, analytical reasoning, and decision-making (Alexander et al., 2024; Han, 2024; Rosário & Raimundo, 2024; R. Wang et al., 2025). Collaborative investigation develops communication and teamwork skills (Alexander et al., 2024; Han, 2024; Mawas & Muntean, 2018), while self-directed research promotes independence and lifelong learning (Han, 2024; Rosário & Raimundo, 2024).

By working with realistic professional problems, learners develop professional judgement and confidence, facilitating transfer of learning from educational environments to workplace practice (Alexander et al., 2024; Payan-Carreira et al., 2022; Zenios, 2020). Reflection throughout the process helps learners become more aware of how they learn, strengthening metacognitive skills and their ability to regulate their own learning (Chen & Rattray, 2018; Han, 2024; McMillan et al., 2022).

Examples in Post-Compulsory and Vocational Education

The following examples illustrate how Problem-Based Learning (PBL) can be implemented in different post-compulsory education settings. Although the vocational and higher education contexts differ considerably, both examples demonstrate the same underlying theory-informed teaching principles. Rather than beginning with lectures or presenting theoretical content before practice, learners are first confronted with an authentic professional problem that mirrors those encountered in their future careers. This problem becomes the catalyst for learning, prompting learners to identify what they already know, determine what additional knowledge and skills they require, conduct independent and collaborative research, develop evidence-based solutions, and reflect on both their learning process and their proposed outcomes.

In each example, the educator adopts the role of facilitator rather than information provider, guiding inquiry through questioning, scaffolding, and feedback rather than direct instruction. The learning activities have been deliberately designed to reflect the theoretical foundations of Problem-Based Learning by integrating constructivist, experiential, social constructivist, cognitivist, and humanistic principles. Consequently, learners not only acquire disciplinary knowledge but also develop higher-order thinking, collaboration, professional judgement, self-directed learning, and reflective practice. These examples demonstrate how theory-informed teaching can be translated into authentic educational practice across diverse disciplines while preparing learners for the complex decision-making required in professional environments.

Example 1: Vocational Education and Training – Certificate III in Sport, Aquatics and Recreation

Rather than beginning the lesson with lectures on risk management, activity planning, group leadership, communication, and relevant legislation, the trainer presents learners with an authentic workplace problem.

Scenario: The local council has asked your organisation to deliver a free community "Family Aquatic Fun Day" aimed at increasing physical activity and water safety participation. However, the event has a limited budget, participants will range from toddlers to older adults, one of the lifeguards has become unavailable, and severe weather is forecast for the afternoon. Your team has one week to develop a safe, inclusive and achievable event plan.

Rather than providing the required knowledge upfront, learners work in teams to identify what they need to know before they can solve the problem. Through discussion they recognise gaps in their knowledge relating to aquatic supervision ratios, risk assessments, emergency procedures, activity programming for different age groups, accessibility requirements, weather contingency planning, equipment availability, staffing requirements, and relevant organisational policies.

Each team researches these areas independently using workplace procedures, industry guidelines, legislation, Australian Standards, and sporting organisation resources. Team members divide research responsibilities before sharing their findings with the rest of the group. Throughout the process, the trainer acts as a facilitator by asking questions such as:

- What risks have you identified?
- How will you ensure participant safety?
- What assumptions have you made?
- How will your plan change if the weather deteriorates?
- What additional information do you still need?

The trainer deliberately avoids providing answers, instead encouraging learners to justify their decisions using evidence.

Once sufficient information has been gathered, each team develops a complete event management plan that includes activity schedules, staffing allocations, emergency procedures, risk controls, participant briefings, equipment lists, communication plans and contingency arrangements. Teams then present their proposed solutions to the class, explaining the reasoning behind each decision and responding to questions from their peers.

Following the presentations, learners reflect individually on both their solution and their learning process. They consider what they initially knew, how their understanding developed, how effectively their team collaborated, what additional knowledge they acquired, and how they would improve their approach if faced with a similar workplace situation.

This activity demonstrates theory-informed teaching because the authentic workplace problem becomes the catalyst for learning. Learners construct knowledge through investigation (constructivism), collaborate to solve the problem (social constructivism), learn through authentic experience (experiential learning), organise and evaluate information (cognitivism), and assume responsibility for directing much of their own learning (humanism). Rather than memorising procedures, learners develop the professional judgement required of recreation officers and aquatic facility staff.

Example 2: Associate Degree in Archaeology

Instead of commencing a unit on archaeological excavation methods with lectures covering excavation techniques, stratigraphy, recording systems, ethics, dating methods, artefact identification and heritage legislation, the lecturer introduces students to a realistic archaeological investigation.

Scenario: During preliminary earthworks for a regional highway project, construction workers uncover several stone artefacts, fragments of pottery, charcoal deposits and possible human remains. Construction has been halted while heritage authorities determine whether the site has archaeological significance. Your archaeological team has been engaged to investigate the discovery and provide recommendations regarding excavation, preservation and future management of the site.

Students immediately begin discussing what they already know and what additional information is required before recommendations can be made. They identify numerous knowledge gaps relating to archaeological surveying, excavation methodology, stratigraphic interpretation, dating techniques, cultural heritage legislation, Indigenous consultation requirements, recording standards, ethical responsibilities, conservation techniques and stakeholder management.

Working collaboratively, students allocate research responsibilities across the team. Some investigate excavation methodologies, others examine relevant legislation and ethical obligations, while others research comparable archaeological sites, dating methods or conservation strategies. The lecturer circulates among groups, asking questions that encourage deeper thinking rather than providing direct instruction.

Questions might include:

- What evidence supports your interpretation?
- How reliable are your dating methods?
- What alternative explanations exist?
- How might excavation damage the archaeological record?
- Which stakeholders should be consulted before excavation proceeds?

As students gather information, they progressively develop an archaeological investigation strategy. This includes a site assessment, excavation methodology, recording procedures, sampling strategy, conservation plan, stakeholder consultation process, ethical considerations, and recommendations regarding whether excavation, preservation in situ, or redesign of the construction project represents the most appropriate course of action.

Each group presents its recommendations as though reporting to a heritage authority and project client. Students justify their conclusions using archaeological theory, scientific evidence, legislation and ethical principles. Other groups critically evaluate the proposed solutions and offer alternative interpretations of the available evidence.

The activity concludes with individual reflective journals in which students analyse not only the archaeological decisions they made but also how they approached the investigation. They reflect on how collaboration influenced their thinking, how their interpretation changed as new evidence emerged, which research strategies proved most effective, and what they would do differently in future archaeological investigations.

This example illustrates theory-informed teaching by allowing archaeological knowledge to emerge through investigation of an authentic professional problem. Students actively construct understanding through inquiry (constructivism), engage in collaborative interpretation of evidence (social constructivism), organise complex information into coherent explanations (cognitivism), learn through authentic disciplinary practice (experiential learning), and direct much of their own research and decision-making (humanism). Rather than simply learning archaeological techniques, students begin to think and reason like practising archaeologists.

Although the learning contexts differ, both examples illustrate the same principles of Problem-Based Learning. Learning begins with an authentic professional problem rather than a lecture. Learners determine what they need to know, conduct independent research, collaborate to develop solutions, justify their decisions using evidence, reflect on both their learning and their performance, and receive facilitative guidance rather than direct instruction. In the Sport, Aquatics and Recreation course, the emphasis is on planning safe and inclusive community recreation programs. In the archaeology program, the focus is on investigating and managing a complex heritage site. In both cases, educational theory informs the design of the learning experience, while Problem-Based Learning provides the mechanism through which learners integrate theoretical knowledge with professional practice, preparing them for the complex decision-making required in their future careers.

Relationship to Theory-Informed Teaching

Problem-Based Learning is an excellent example of theory-informed teaching because it translates educational theory into practical instructional design. Constructivism explains why learners build understanding through solving authentic problems (Dillon et al., 2023; Rubaia'an, 2023; Sistermans, 2020; Varadarajan & Ladage, 2024); cognitivism supports the development of reasoning and problem-solving processes (Alhassan, 2020; Bell, 2021; Nalliah, 2014); social constructivism highlights the value of collaboration and guided learning (Alhassan, 2020; Dillon et al., 2023; Foster et al., 2021; Omer, 2020b); experiential learning emphasises learning through experience and reflection (Ajraoui et al., 2019; Bell & Bell, 2020; Varadarajan & Ladage, 2024); and humanism promotes learner autonomy and self-directed inquiry (Alhassan, 2020; Foster et al., 2021; Lee & Wong, 2025; Rubaia'an, 2023). Rather than teaching theory in isolation, Problem-Based Learning integrates these complementary perspectives to create authentic learning experiences that develop both disciplinary knowledge and professional competence (Sheikh & Ahmad, 2015; Sistermans, 2020; Varadarajan & Ladage, 2024). As a result, learners are not only able to explain concepts but can also apply them effectively to analyse, evaluate, and solve complex problems encountered in professional practice (Dillon et al., 2023; Rubaia'an, 2023; Sistermans, 2020).

Project-Based Learning

Project-Based Learning (PjBL) is a learner-centred instructional approach in which learners acquire knowledge and skills by planning, developing, and completing an extended project centred on an authentic question, challenge, or professional task (Dogara, Saud, Kamin, et al., 2020; Louis et al., 2021; Sukackė et al., 2022). Unlike traditional teaching, where knowledge is often presented before learners complete an assignment, Project-Based Learning integrates learning and application throughout the project (Damaiyanti, 2024; Fitri et al., 2024). Learners investigate a real-world issue, apply disciplinary knowledge, collaborate with others, create a meaningful product or solution, and present their work to an authentic audience (Hignasari et al., 2023; Louis et al., 2021; Son & Penry, 2022; Sukackė et al., 2022).

Within theory-informed teaching, Project-Based Learning is not a learning theory but an instructional strategy informed by multiple educational theories. It is characterised as a constructivist pedagogy (Dogara, Saud, Kamin, et al., 2020; Sagita et al., 2023; Sukackė et al., 2022), with social constructivist perspectives highlighting the importance of collaborative interaction and context in knowledge construction (Gao & Chen, 2023). It provides a practical means of translating theoretical understandings of learning into authentic educational experiences that promote engagement, inquiry, collaboration, reflection, and professional competence (Fitri et al., 2024; Son & Penry, 2022; Yazici, 2020). Rather than positioning learners as passive recipients of information, Project-Based Learning encourages them to become active participants who progressively construct knowledge while working towards a clearly defined project outcome (Damaiyanti, 2024; Fitri et al., 2024; Louis et al., 2021; Sukackė et al., 2022).

Project-Based Learning is grounded in several complementary learning theories, each contributing different principles to its design and implementation. Constructivism provides the primary theoretical foundation. Constructivist theory proposes that learners actively construct understanding through meaningful experiences rather than passively absorbing information (Ajraoui et al., 2019; Jeny, 2024). In Project-Based Learning, learners investigate authentic questions, make decisions, solve problems, and integrate new knowledge into existing understandings as they work towards completing their project (Sun & Yahaya, 2024; Woodhouse et al., 2022).

Experiential learning also underpins Project-Based Learning because learners acquire knowledge through experience, reflection, conceptual understanding, and application (Jeny, 2024). As projects progress, learners continually cycle between planning, acting, evaluating outcomes, and modifying their approach, closely reflecting Kolb's experiential learning cycle (Ajraoui et al., 2019; Jeny, 2024).

Social constructivism contributes the emphasis on collaboration. Most projects require learners to work in teams where they negotiate roles, exchange ideas, solve problems collectively, and learn from one another through discussion and shared decision-making (Ajraoui et al., 2019; Sun & Yahaya, 2024). Learning is viewed as both an individual and social process (Ajraoui et al., 2019).

Cognitivism contributes by recognising that successful project work requires learners to organise information, analyse evidence, plan systematically, manage complex tasks, monitor progress, and evaluate outcomes (Ajraoui et al., 2019; Effendi & Hendriyani, 2020). Project-Based Learning develops higher-order cognitive skills such as planning, reasoning, synthesis, and evaluation (Ajraoui et al., 2019).

Humanism is evident through its emphasis on learner autonomy and personal responsibility (Hassan et al., 2022). Learners often make decisions about project direction, allocate responsibilities, manage timelines, evaluate their own progress, and reflect on their learning. This promotes independence, self-direction, and intrinsic motivation (Woodhouse et al., 2022).

Elements of connectivism may also be present, particularly in contemporary education, where learners access digital resources, collaborate online, engage with professional communities, and utilise technology throughout the project (Ajraoui et al., 2019; Jeny, 2024; Woodhouse et al., 2022). Connectivism, as developed by Siemens, emphasises that learning occurs through the process of connecting specialised nodes and information sources within networks, and that digital tools play a critical role in facilitating continuous learning (Barnard-Ashton et al., 2017; Jeny, 2024). In the context of Project-Based Learning, connectivism is evident when students utilise digital platforms, engage with diverse online information sources, and build learning networks that extend beyond the traditional classroom (Jeny, 2024; Woodhouse et al., 2022). Research has identified significant overlaps between on-campus Project-Based Learning pedagogy and online connectivism pedagogy, suggesting these approaches are complementary rather than competing frameworks (Woodhouse et al., 2022).

Characteristics of Project-Based Learning

Project-Based Learning differs from traditional instructional approaches because learning revolves around the completion of a substantial project rather than isolated classroom activities. Projects are typically authentic, interdisciplinary, and designed to address realistic professional or community needs (Bintoro et al., 2024; Kholil et al., 2022). They often extend over several weeks or months, allowing learners to investigate issues in depth and progressively refine their work (Wirzal et al., 2022).

Learners assume significant responsibility for planning their work, identifying information needs, conducting research, solving problems, creating products, and presenting their outcomes (Farizi et al., 2023; Samadun & Dwikoranto, 2022). Throughout the project they receive guidance, feedback, and support from the educator, but much of the learning occurs through active engagement with the project itself (Farizi et al., 2023; Jannah, 2022).

Projects generally culminate in a tangible outcome, such as a report, prototype, presentation, event, design, business proposal, community program, or professional portfolio, which demonstrates both the learning process and the knowledge developed throughout the project (Bakri & Ching, 2021; Wirzal et al., 2022).

Although the two approaches share many similarities, they differ in emphasis (Wirzal et al., 2022). Problem-Based Learning begins with a complex problem that learners investigate to develop understanding and propose solutions (Jufriadi & Andinisari, 2020; Lodge et al., 2018). The learning process itself is the primary focus (Lodge et al., 2018; Mardikaningsih et al., 2024), and the solution may remain conceptual (Jufriadi & Andinisari, 2020).

Project-Based Learning also begins with an authentic challenge but places greater emphasis on designing, creating, implementing, or producing a substantial final product (Jufriadi & Andinisari, 2020). The completed project becomes the demonstration of learning (Jufriadi & Andinisari, 2020).

Within Project-Based Learning, the educator functions as a designer of learning experiences, facilitator, mentor, project manager, and coach (Byrne et al., 2018). The educator develops authentic project briefs (Galaleldin et al., 2018; Indriati et al., 2024), provides milestones (Walters-Williams, 2022), monitors learner progress (Francis et al., 2024; Indriati et al., 2024), facilitates collaboration (Byrne et al., 2018; Indriati et al., 2024), offers formative feedback (Marquis, 2021), and encourages reflection throughout the project (Byrne et al., 2018; Galaleldin et al., 2018; Mitrulescu, 2023).

Rather than directing every aspect of learning, educators provide appropriate scaffolding while progressively transferring responsibility to learners (Byrne et al., 2018; Marquis, 2021). Equally important is explaining why project work is being used. Learners should understand that projects develop professional skills such as collaboration, communication, planning, research, problem-solving, and self-management (Galaleldin et al., 2018; Naseer et al., 2025; Walters-Williams, 2022), all of which are highly valued in the workplace (Francis et al., 2024; Naseer et al., 2025).

Examples of Applying Project-Based Learning

The following examples illustrate how educators could implement Project-Based Learning (PjBL) in different post-compulsory education settings. Although one example is situated within vocational education in Canada and the other within higher education in the United Kingdom, both demonstrate how theory-informed teaching can be translated into authentic educational practice. In each case, learning is organised around an extended, authentic project that reflects the knowledge, skills, and professional practices learners are likely to encounter in their future careers. Rather than presenting knowledge as discrete topics before assessment, educators could progressively introduce concepts, skills, and resources as learners require them throughout the project. In doing so, learners actively construct knowledge, collaborate with others, apply theory to practice, and produce a meaningful outcome that demonstrates both their learning and professional capability.

The first example relates to delivery of a vocational program in photography and digital imaging in Canada. Project-Based Learning could be implemented in a Certificate in Photography and Digital Imaging by organising a substantial portion of the course around a simulated commercial photography commission rather than teaching camera operation, composition, lighting, image editing, client communication, copyright, and portfolio development as isolated topics followed by separate assessments.

For example, the instructor could present learners with a project brief in which a local tourism organisation requires a digital promotional campaign showcasing the cultural heritage, natural landscapes, and recreational opportunities of the surrounding region. Learners could be asked to plan and produce a professional photographic portfolio suitable for use across websites, social media platforms, brochures, and visitor information centres over an eight-week period.

At the commencement of the project, learners could analyse the client's requirements to determine what knowledge and skills would be needed to complete the commission successfully. Through this process, they would identify learning needs relating to photographic composition, lighting techniques, camera settings, image-editing workflows, digital asset management, copyright, model releases, branding requirements, project management, and client communication. Rather than delivering all of this content through sequential lectures, the instructor could introduce these topics as they become relevant during the project.

Learners could work collaboratively in production teams, assuming roles such as project coordinator, lead photographer, lighting specialist, editor, and client liaison. Together they could develop project timelines, organise equipment, identify suitable locations, obtain permissions, prepare shooting schedules, and manage the production workflow. Throughout the project, the instructor would act as a facilitator by reviewing project plans, monitoring progress, asking questions that encourage critical thinking, and providing formative feedback without prescribing solutions.

As learners undertake photographic sessions, they could evaluate lighting conditions, experiment with composition, refine camera settings, apply digital editing techniques, and continually review and improve their work based on feedback from peers and the instructor. The project could culminate in a professional presentation in which learners justify their technical and creative decisions, explain how their photographic portfolio satisfies the client's objectives, discuss copyright and licensing considerations, and evaluate the effectiveness of their project management processes. Individual reflective portfolios could also be completed to encourage learners to critically evaluate their technical development, collaboration, professional judgement, and learning throughout the project.

Such an approach demonstrates theory-informed teaching by enabling learners to construct photographic knowledge through authentic professional practice. Constructivist principles are reflected as learners build understanding through creative problem-solving; experiential learning is evident through repeated photographic practice and reflection; social constructivism supports collaborative production; cognitivism underpins planning, organisation, and decision-making; humanism encourages learner autonomy and responsibility; and connectivism is reflected through the use of online portfolios, digital editing software, collaborative cloud-based workflows, and engagement with professional photography communities.

As a second example, consider delivery of a Bachelor degree level program in Applied Linguistics delivered in the United Kingdom. Project-Based Learning could also be implemented within a Bachelor of Arts in Applied Linguistics by organising a semester-long research project around an authentic language issue encountered within the local community rather than teaching sociolinguistics, discourse analysis, language acquisition, corpus linguistics, research methods, and language policy as separate units of content followed by traditional examinations.

For example, the lecturer could present students with a project brief in which a local council seeks assistance in understanding communication barriers experienced by recently arrived migrant families when accessing public health services. Students could be tasked with investigating the linguistic challenges affecting communication and developing evidence-based recommendations that improve accessibility while respecting linguistic and cultural diversity.

At the beginning of the project, students could analyse the brief and identify the disciplinary knowledge required to undertake the investigation successfully. Through discussion, they would recognise the need to understand language acquisition, multilingualism, intercultural communication, discourse analysis, qualitative interviewing, ethical research practices, language policy, corpus analysis, and community engagement. Rather than introducing each topic in a predetermined sequence, the lecturer could provide learning activities, resources, and guidance as students encounter these knowledge requirements throughout the investigation.

Working collaboratively in research teams, students could prepare research proposals, undertake literature reviews, develop ethics applications, design interview protocols, identify appropriate linguistic analysis methods, and allocate responsibilities among team members. Individual students might assume responsibility for participant recruitment, transcription, corpus development, discourse analysis, statistical analysis, or policy evaluation according to their interests and strengths.

Throughout the semester, the lecturer would facilitate regular research meetings, encourage students to justify methodological decisions, question assumptions, provide formative feedback, and guide inquiry without directing the research itself. As new evidence emerges, students could modify their research questions, refine their analytical approaches, and reconsider their interpretations.

The completed project could require each research team to produce a professional report analysing communication barriers, evaluating existing language support services, and presenting practical recommendations for improving communication between healthcare providers and multilingual communities. Students might also present their findings to an audience of academic staff, community representatives, or language professionals, defending their methodological choices and demonstrating how linguistic theory informed their recommendations. Individual reflective journals could encourage students to evaluate their research processes, collaborative learning, ethical decision-making, and professional development throughout the project.

This approach illustrates theory-informed teaching by embedding linguistic theory within authentic research practice. Constructivism enables students to develop understanding through inquiry and analysis; experiential learning occurs through conducting a realistic research investigation; social constructivism is evident through collaborative inquiry; cognitivism supports research planning, analysis, and interpretation; humanism promotes learner autonomy and professional identity; and connectivism is reflected through the use of digital linguistic corpora, online research databases, collaborative research platforms, and engagement with wider academic communities.

Although these examples are situated within different educational sectors and disciplines, they demonstrate how Project-Based Learning could be implemented using the same theory-informed teaching principles. In both contexts, educators could organise learning around the completion of an authentic project that reflects professional practice rather than around isolated instructional topics. Learners would progressively acquire disciplinary knowledge as it becomes necessary for successful project completion while assuming increasing responsibility for planning, research, collaboration, project management, decision-making, and reflection. Educators would provide scaffolding, milestones, coaching, and formative feedback throughout the project rather than directing every stage of learning. The completed project would serve as both the primary demonstration of learning and a means of preparing learners for professional practice, while structured reflection would help learners evaluate both the quality of their final product and the learning processes they employed. These examples demonstrate how Project-Based Learning can operationalise multiple educational theories to develop disciplinary knowledge, professional competence, critical thinking, collaboration, and lifelong learning across both vocational and higher education contexts.

Applying Theoretical Perspectives to Individual Learning

One of the defining characteristics of theory-informed teaching is that educators recognise there is no single theory that explains how all people learn. Adult learners differ in their backgrounds, prior experiences, motivations, abilities, learning preferences, cultural contexts, career aspirations, and educational needs. Consequently, effective educators draw upon multiple educational theories to understand individual learners and to select teaching approaches that best support their development. Applying theoretical perspectives to individual learning involves interpreting learners' needs through different educational lenses and using those insights to design learning experiences that maximise engagement, understanding, and achievement.

Rather than viewing learning theories as competing or mutually exclusive, theory-informed educators regard them as complementary frameworks that offer different explanations of learning (Matorevhu, 2022; Peterson et al., 2018). Each theory highlights particular aspects of human learning and provides guidance on when certain teaching strategies may be most appropriate (Delahunty & Kimbell, 2021; Matorevhu, 2022; Peterson et al., 2018). Research demonstrates that no single theory individually fully explains what happens when learners learn, and therefore none can be applied universally (Matorevhu, 2022). Each theory has its own merits and limitations, and to maximise learning, a combination of theories should be used (Matorevhu, 2022). For instance, the transformation of educational practice has been shaped by multiple key learning theories, behaviourism, cognitivism, constructivism, and connectivism, each of which offers distinct explanations of how learning occurs (Mahat, 2022).

Students' existing level of mastery should strongly influence whether to use teacher- or learner-centred pedagogies in a specific situation, and fostering competences means adopting sequences and combinations of pedagogies that find a balance between different approaches (Peterson et al., 2018). By understanding these theoretical perspectives, educators can make informed decisions about how to support individual learners in different contexts (Ananga, 2020; Matorevhu, 2022; Peterson et al., 2018) rather than relying on a single instructional approach for every situation. Pedagogy selection must target the subject matter and the needs of the learners, learning theories, objectives, instructional methods, interaction, and assessment (Ananga, 2020). As Hägg and Kurczewska note, pedagogy is regarded as a fundamental instructional entry point for students facing new subjects, but as learners develop their knowledge base, a switch to alternative approaches is needed, illustrating how different theories serve different moments in the learning process (Hägg & Kurczewska, 2020). Similarly, constructivist teaching pedagogy is acknowledged as not effective for all learning environments but rather situational (Mbise & Lekule, 2020), further reinforcing the principle that educators must draw upon multiple theoretical frameworks to address diverse teaching and learning situations (Matorevhu, 2022; Peterson et al., 2018).

Constructivism encourages educators to recognise that learners actively construct knowledge based on their existing experiences and understanding. When applying a constructivist perspective to individual learning, educators seek to identify learners' prior knowledge and experiences before introducing new concepts. Learning activities are then designed to help individuals connect new information with what they already know through discussion, inquiry, authentic tasks, and reflection. For learners with extensive workplace experience, educators may provide opportunities to relate theoretical concepts to their own professional practice, allowing previous experience to become a foundation for further learning.

A cognitivist perspective focuses on the internal mental processes involved in learning, including attention, memory, reasoning, and problem-solving. Applying cognitivism to individual learning involves helping learners organise information into meaningful structures, manage cognitive load, develop effective learning strategies, and strengthen critical thinking. Educators may adapt their teaching by breaking complex concepts into manageable components, using visual organisers, encouraging metacognitive reflection, or explicitly teaching strategies that improve memory, comprehension, and knowledge transfer.

Behaviourist principles may be particularly valuable when learners are developing foundational knowledge, procedural skills, or behaviours that require accuracy and consistency. Educators applying behaviourist approaches provide clear demonstrations, guided practice, immediate feedback, repetition, and positive reinforcement to help learners achieve competence. Individual learners who require additional practice or confidence-building may benefit from structured exercises and regular feedback before progressing to more complex or independent learning activities.

Humanistic perspectives emphasise the learner as a whole person, recognising that motivation, self-esteem, personal goals, emotions, and autonomy significantly influence learning. Applying humanistic principles requires educators to understand learners' aspirations, provide choices where appropriate, encourage self-directed learning, and create supportive learning environments characterised by trust, respect, and psychological safety. Individual learning plans, negotiated assessment options, and opportunities for self-reflection enable learners to take increasing ownership of their educational journey.

Experiential learning highlights the importance of learning through experience, reflection, conceptual understanding, and application. When applying this perspective, educators provide authentic activities that enable learners to practise skills in realistic contexts before reflecting on their experiences and identifying opportunities for improvement. Individual learners may engage in workplace projects, simulations, laboratory activities, fieldwork, or professional practice experiences that allow them to apply theoretical concepts to practical situations and develop professional judgement.

Social constructivist perspectives recognise that learning is influenced by interaction with others. Educators applying this perspective create opportunities for collaboration, discussion, peer learning, mentoring, and cooperative problem-solving. Individual learners benefit from sharing ideas, receiving feedback, observing different perspectives, and participating in communities of practice where knowledge is constructed collectively. For learners who may lack confidence or prior experience, collaboration with more knowledgeable peers or educators can provide appropriate scaffolding that supports learning within their zone of proximal development.

Connectivism has become increasingly relevant in contemporary education because learners operate within digitally connected environments where knowledge is distributed across networks of people, organisations, and technologies. Applying connectivist principles involves helping learners locate, evaluate, organise, and apply information from multiple digital sources while developing professional learning networks that support continuous learning beyond formal education. Educators encourage learners to critically evaluate online information, participate in professional communities, and use digital technologies to collaborate, communicate, and remain current within their disciplines.

Applying theoretical perspectives also requires educators to recognise that learners' needs change over time. A novice learner may initially benefit from behaviourist strategies that provide structured guidance and repeated practice while developing foundational skills. As competence increases, constructivist, experiential, and project-based approaches may become more appropriate because learners possess sufficient knowledge to investigate authentic problems, make independent decisions, and reflect critically on their practice. Similarly, learners who begin with considerable educator support may progressively assume greater responsibility for directing their own learning as their confidence and expertise develop.

Theory-informed teaching therefore requires continual observation, reflection, and adaptation. Educators monitor learner progress, gather evidence about engagement and achievement, and adjust teaching strategies in response to individual learning needs. This may involve providing additional scaffolding for some learners, increasing opportunities for collaboration, modifying learning activities, incorporating technology, or offering alternative ways for learners to demonstrate their understanding. Rather than expecting learners to adapt to a fixed teaching method, educators adapt their teaching in ways that maximise learning for diverse individuals.

Applying theoretical perspectives to individual learning is an exercise in informed educational judgement. Learning theories provide complementary frameworks that help educators understand why learners respond differently to particular teaching approaches and how instruction can be adapted to support their development. By drawing upon multiple theoretical perspectives, educators are better equipped to create learner-centred environments that recognise individual differences, foster engagement, promote deeper understanding, and prepare learners for lifelong learning and professional practice.

The following examples illustrate how educators can apply performative learning, problem-based learning, and project-based learning to support the needs of individual learners. In each case, the educator recognises that learners enter post-compulsory education with different levels of prior knowledge, workplace experience, confidence, motivation, learning preferences, and career aspirations. Rather than expecting every learner to complete identical learning activities in the same manner, the educator adapts learning experiences to suit individual needs while ensuring that all learners achieve the required learning outcomes. These approaches enable educators to personalise learning by providing varying levels of support, selecting authentic contexts that reflect learners' interests and experience, and progressively increasing learner independence as competence develops.

Performative learning can be applied to individual learning by tailoring authentic performance tasks to each learner's current level of competence while progressively increasing the complexity of those tasks as confidence and capability develop. The emphasis is not on altering the required learning outcomes but on adapting the pathway through which individual learners demonstrate competence.

For example, in a Certificate III in Early Childhood Education and Care (Australia)/ New Zealand Certificate in Early Childhood Education and Care/ SVQ Social Services (Children and Young People) (Scotland)/ Higher Certificate in Early Childhood Care and Education (South Africa), an educator could recognise that learners possess varying levels of workplace experience. A learner who has previously worked in a childcare centre might be encouraged to independently plan, deliver, and evaluate a literacy activity for preschool children, drawing upon both existing experience and newly acquired theoretical knowledge. In contrast, a learner with no previous industry experience could initially observe an experienced educator conducting similar activities before leading small sections of the lesson under supervision. As confidence develops, the learner could progressively assume responsibility for delivering the entire activity independently. Although both learners ultimately demonstrate the same competency, the educator adjusts the level of support and performance expectations throughout the learning process to reflect individual learning needs.

A similar approach could be implemented in a Bachelor of Nursing program. A student who demonstrates strong communication skills and clinical confidence could participate in complex patient simulations involving multiple health conditions, competing priorities, and interdisciplinary decision-making. Another student who experiences anxiety during clinical simulations could initially practise specific communication skills with a standardised patient before gradually progressing to more comprehensive scenarios as confidence increases. Throughout both learning pathways, the educator provides feedback tailored to each learner's performance, encouraging reflection on strengths, identifying areas for improvement, and supporting the gradual development of professional competence. In this way, performative learning allows authentic workplace performance to be adapted to the individual while maintaining common professional standards.

Problem-based learning can be individualised by presenting learners with authentic problems that reflect their existing knowledge, workplace experience, interests, or intended career pathways. Although learners may apply similar problem-solving processes, the nature and complexity of the problems can be adapted to maximise engagement and promote meaningful learning.

For example, in a Certificate IV in Information Technology (Australia)/ New Zealand Certificate in Information Technology (Level 5)/ BTEC Level 4 Higher National Certificate (HNC) in Digital Technologies or Computing (UK)/ Associate of Applied Science (AAS) in Information Technology (USA), an educator could differentiate learning by assigning technical support problems appropriate to each learner's capabilities. A learner with previous experience in computer networking might investigate intermittent failures within a small business network, requiring analysis of network architecture, security settings, hardware faults, and system configurations. Another learner with more limited technical experience could investigate why a single workstation cannot access shared network resources, enabling the learner to develop foundational troubleshooting skills before progressing to more complex networking scenarios. In both cases, the educator facilitates inquiry through questioning, guidance, and feedback rather than providing direct solutions, enabling learners to develop independent problem-solving abilities at an appropriate level of challenge.

Similarly, in a Master of Public Health program, students could investigate public health issues that align with their individual professional interests. One student with an interest in Indigenous health could examine barriers to preventive healthcare in remote communities, while another interested in epidemiology might investigate declining childhood vaccination rates in metropolitan areas. Although both students employ the same research and analytical framework, each explores a problem that reflects their own career aspirations and prior knowledge. By facilitating inquiry rather than directing solutions, the educator enables each learner to construct knowledge that is both academically rigorous and personally meaningful.

Project-based learning can also be adapted to support individual learning by allowing learners to undertake substantial projects that reflect their personal interests, professional aspirations, and existing experience while still addressing common learning outcomes. Individualisation occurs through learner choice, project design, and varying degrees of autonomy rather than through different assessment standards.

For example, in a Certificate IV in Photography and Digital Imaging/ WSQ qualifications in Media Production or Photography (Singapore)/ Ausbildung zum Fotografen (Germany)/ CAP Photographe or Bac Professionnel Photographie (Fance), learners could complete a professional photographic portfolio as the major assessment task while selecting project themes that align with their own interests. One learner may choose to produce a wildlife conservation photography portfolio documenting local ecosystems, another may develop a commercial product photography campaign for a small business, while a third may create a portrait exhibition celebrating members of the local community. Although each learner applies the same principles of composition, lighting, image editing, copyright, and client communication, the projects reflect individual creative interests and intended career pathways. Throughout the project, the educator provides milestones, coaching, formative feedback, and technical guidance while progressively encouraging learners to assume greater responsibility for planning, production, and presentation.

A similar approach could be adopted in a Bachelor of Architecture program. Rather than requiring every student to complete an identical design project, the educator could allow learners to select a project that aligns with their professional interests. One student may design a sustainable rural health clinic, another may develop an affordable housing complex, while another designs a culturally responsive community centre. Each project requires learners to apply architectural theory, environmental sustainability principles, planning regulations, digital modelling techniques, and professional communication skills. However, learners investigate different design challenges that reflect their individual interests and career aspirations. Regular design reviews, reflective discussions, and formative feedback enable the lecturer to support each learner's development while maintaining consistent academic expectations across the cohort.

Although performative learning, problem-based learning, and project-based learning all support learner-centred education, they individualise learning in different ways. Performative learning adapts authentic workplace performances to suit each learner's current level of competence, gradually increasing independence as capability develops. Problem-based learning individualises learning by presenting authentic problems that reflect learners' backgrounds, experience, interests, or career goals while encouraging inquiry, critical thinking, and evidence-based decision-making. Project-based learning supports individual learning by allowing learners to undertake substantial projects that align with their personal interests and professional aspirations while demonstrating common learning outcomes.

Across all three approaches, the educator's role extends beyond delivering content to continually observing learner progress, providing appropriate scaffolding, offering formative feedback, encouraging reflection, and progressively transferring responsibility for learning to the individual. Rather than relying on a single instructional strategy for every learner, theory-informed educators apply these complementary approaches flexibly to accommodate learner diversity, promote meaningful engagement, and prepare individuals for the autonomy, adaptability, and professional judgement required throughout their careers

Key Concepts and Conclusions

Theory-informed teaching recognises that effective adult education is grounded in educational theory as well as professional experience. Rather than relying on intuition or a single preferred teaching method, educators draw upon complementary theoretical perspectives to understand how adults learn and to make informed decisions about curriculum design, teaching strategies, learner support, assessment, and reflective practice. Educational theories provide a conceptual framework that enables educators to create learning experiences that are responsive to learner diversity, educational purpose, and professional context.

The chapter demonstrates that educational theories are most valuable when viewed as complementary rather than competing explanations of learning. Behaviourism, cognitivism, constructivism, humanism, experiential learning, social learning, transformative learning, connectivism, and sociocultural perspectives each explain different dimensions of learning and contribute unique insights into teaching practice. Effective educators therefore adopt an eclectic, evidence-informed approach, selecting and combining theories according to learner needs, learning outcomes, and contextual requirements rather than adhering rigidly to a single philosophy.

Three learner-centred instructional approaches, performative learning, Problem-Based Learning, and Project-Based Learning, illustrate how educational theory can be translated into authentic teaching practice. Although none of these approaches constitutes a learning theory in its own right, each operationalises multiple theories by engaging learners in meaningful activities that integrate knowledge, skills, reflection, collaboration, and professional judgement. These approaches demonstrate that learning is most effective when learners actively participate in realistic tasks that mirror the complexity of professional practice rather than passively receiving information.

Performative learning emphasises learning through authentic performance, allowing learners to demonstrate competence by applying knowledge and skills in realistic contexts. Problem-Based Learning develops critical thinking and professional reasoning by placing authentic problems at the centre of the learning process, while Project-Based Learning requires learners to plan, create, and complete substantial projects that integrate disciplinary knowledge with practical application. Although each approach has a distinct focus, all encourage active participation, reflection, collaboration, learner autonomy, and the progressive transfer of responsibility from educator to learner.

A central principle of theory-informed teaching is that learning should be adapted to the individual rather than expecting individuals to adapt to a single teaching method. Adult learners bring diverse experiences, motivations, abilities, confidence levels, cultural backgrounds, and career aspirations to educational settings. By understanding these differences through multiple theoretical perspectives, educators can personalise learning by varying the level of guidance, selecting authentic contexts that reflect learners' interests and experience, providing appropriate scaffolding, and progressively increasing learner independence as competence develops.

The chapter also reinforces the evolving role of the educator. Rather than acting solely as a transmitter of knowledge, the educator becomes a designer of learning experiences, facilitator of inquiry, coach, mentor, assessor, and reflective practitioner. Theory-informed educators continually evaluate learner progress, analyse evidence of learning, provide timely feedback, and refine their teaching practices in response to learner needs and emerging educational research. Reflection and continuous improvement are therefore essential components of effective teaching practice.

Theory-informed teaching bridges the gap between educational research and classroom practice. It provides educators with the professional knowledge required to justify instructional decisions, design engaging and authentic learning experiences, and foster learner-centred environments that promote deep understanding, critical thinking, professional competence, and lifelong learning. By integrating multiple educational theories with practical teaching expertise, educators are better equipped to prepare adult learners for the complex and continually changing demands of contemporary workplaces and society.

Chapter 9

Digital Learning and Emerging Technologies

Technology has transformed the way adults access, experience, and apply learning. The widespread availability of digital devices, high-speed internet, cloud computing, artificial intelligence, and immersive technologies has expanded educational opportunities far beyond the traditional classroom. Adult learners can now engage in flexible, personalised, and collaborative learning experiences across higher education, vocational education and training, workplace learning, and professional development. At the same time, rapid technological change requires educators to continually evaluate how emerging technologies can be used to enhance learning while responding to evolving workforce demands.

However, technology alone does not improve learning. Digital tools are most effective when they are integrated within thoughtfully designed learning experiences that are informed by sound educational theory. Whether learning occurs through online environments, mobile devices, learning management systems, open educational resources, virtual simulations, artificial intelligence, or digital credentials, the quality of learning depends on the pedagogical decisions made by educators. Effective technology-enhanced learning therefore requires educators to understand not only the capabilities of digital technologies but also how these technologies can support constructivist, social constructivist, connectivist, cognitivist, behaviourist, humanistic, and experiential approaches to adult learning.

This chapter examines the role of digital learning and emerging technologies in contemporary adult education. It explores major approaches and technologies, including online and blended learning, mobile learning, learning management systems, open educational resources, artificial intelligence, virtual and augmented reality, learning analytics, adaptive learning, and micro-credentials with digital badges. Throughout the chapter, each technology is examined in terms of its defining characteristics, theoretical foundations, educational applications, benefits, limitations, and relationship to theory-informed teaching. Practical examples drawn from higher education, vocational education and training, workplace learning, and lifelong learning demonstrate how these technologies can be applied to create engaging, accessible, and authentic learning experiences that prepare adults for participation in an increasingly digital and knowledge-driven world

Online and Blended Learning

Online and blended learning have become integral components of contemporary education, providing flexible, learner-centred approaches that accommodate the diverse needs of learners across vocational education, higher education, workplace training, and professional development. Although these approaches utilise digital technologies to support teaching and learning, they differ in how face-to-face and online learning experiences are combined (Raad & Odhabi, 2021). Both approaches are grounded in educational theory and, when carefully designed, can promote active learning, collaboration, self-directed learning, and lifelong learning (Kasperski et al., 2023; Lane & Stephens, 2023).

Online learning refers to education delivered primarily or entirely through digital technologies, with learners and educators interacting remotely via the internet (Lane & Stephens, 2023; Woollard, 2020). Learning may occur synchronously, where educators and learners participate in real-time using video conferencing or virtual classrooms, or asynchronously, where learners access recorded lectures, readings, discussion forums, quizzes, and learning activities at times that suit their individual schedules (Bernard et al., 2004; Woollard, 2020). Many online courses combine both approaches, allowing learners to benefit from scheduled interaction while maintaining flexibility (Mukherjee & Maity, 2022; Shakeeb, 2020).

One of the defining characteristics of online learning is its accessibility. Learners can participate regardless of geographical location, making education available to those who may otherwise face barriers due to work commitments, family responsibilities, disability, or distance (Besa & Parcon, 2021; Woollard, 2020). Digital learning environments also allow learners to revisit learning materials, progress at an individual pace, and engage with a wide range of multimedia resources including videos, simulations, interactive activities, podcasts, and virtual laboratories (Lockey et al., 2022).

Successful online learning requires careful instructional design rather than simply transferring classroom materials into a digital environment (Kasperski et al., 2023; Lane & Stephens, 2023; Shakeeb, 2020). Effective online courses provide clear learning pathways, opportunities for interaction, timely feedback, accessible resources, authentic assessment, and regular communication between educators and learners (Kasperski et al., 2023; Mukherjee & Maity, 2022; Shakeeb, 2020). The educator's role extends beyond delivering content to facilitating online discussions, monitoring learner engagement, providing academic and pastoral support, and creating a sense of community among geographically dispersed learners (Kasperski et al., 2023; Mukherjee & Maity, 2022; Rehn et al., 2016).

Blended learning combines face-to-face teaching with online learning activities in a deliberately integrated manner (Bandara & Jayaweera, 2024; Min & Yu, 2023). Rather than viewing online learning as a replacement for classroom teaching, blended learning seeks to utilise the strengths of both environments (Alammary, 2024; Nkhoma et al., 2019). As Chen notes, "adopting a blended learning approach will lead to creating a more multimodal, engaging and learner centered learning environment" (Chen, 2022, p. 1). Learners may complete online preparatory activities before attending workshops, participate in face-to-face practical sessions, continue collaborative discussions online, and submit assessments through a learning management system (Bandara & Jayaweera, 2024; Nkhoma et al., 2019)

The proportion of online and face-to-face learning varies considerably depending on the educational context (Alammary, 2024). In vocational education, theoretical knowledge may be delivered online while practical skills are developed during workshops, laboratories, or workplace training (Alsaif, 2021; Farmati et al., 2023). In higher education, lectures may be replaced by recorded presentations that learners view before attending seminars focused on discussion, problem-solving, and application of knowledge (Alammary, 2024; Nkhoma et al., 2019). Workplace training frequently combines online compliance modules with supervised practical activities and mentoring (Dahri et al., 2024; Farmati et al., 2023).

Blended learning provides flexibility while retaining opportunities for direct interaction with educators and peers (Alsaif, 2021; Bandara & Jayaweera, 2024; Farmati et al., 2023). Face-to-face sessions can be used for activities that benefit from immediate feedback, practical demonstrations, collaborative problem-solving, and professional discussion (Alammary, 2024; Min & Yu, 2023), while online components support independent study, revision, reflection, and ongoing communication outside scheduled classes (Alsaif, 2021; Bandara & Jayaweera, 2024; Farmati et al., 2023). As Alsaif observes, blended learning's use of digital technology improves the educational process "in terms of degree of flexibility, flexibility in terms of time and place, learning rhythm, learning forms (audio/video, text/pictures, online/offline, individual/team work), and, lastly, adaptability to the learner's needs" (Alsaif, 2021, p. 12).

Theoretical Foundations

Online and blended learning are not learning theories themselves but instructional approaches informed by multiple educational theories. Several scholars have confirmed that effective online instruction draws upon multiple theoretical foundations rather than relying on a single theory (Banihashem & Macfadyen, 2021; Dunaway, 2011; Lee & Wong, 2025). As Lamtara (2023b, p. 18)noted, "pedagogical practices should be based on appropriate learning theories according to lesson objectives and learning outcomes," with educators drawing on behaviourism, cognitivism, constructivism, and connectivism as needed. Similarly, Banihashem and Macfadyen (2021) argued that careful reflection on the learning theory underpinning pedagogical design should guide the selection of learning analytics and instructional approaches.

Constructivism underpins many online and blended learning environments by encouraging learners to actively construct knowledge through authentic activities, inquiry, discussion, and reflection. Constructivist learning environments rest on the premise that learning is individually and socially constructed through interactions with the world. In e-learning contexts, constructivist theory supports effective dialogue, group work, and knowledge construction through online platforms (Bell, 2021). Digital games or game-based simulations that are designed primarily for education and training rather than entertainment in distance education also reflect constructivist theory by encouraging learners to actively construct knowledge through meaningful interaction, inquiry, and reflection within learning environments designed to facilitate understanding rather than simply transmit information.

Social constructivism emphasises the importance of interaction, collaboration, and shared meaning-making in the learning process. From this perspective, conceptual understanding develops through discussion, cooperation, the exchange of ideas, and the negotiation of meaning among learners. Web

2.0 technologies and social media platforms support these principles by providing opportunities for communication, collaboration, and knowledge sharing in online environments (Celestini, 2021). Online discussion forums, collaborative digital workspaces, and shared learning platforms enable learners to engage with peers, construct understanding collectively, and participate in meaningful dialogue despite geographical separation (Bell, 2021; Bonanno et al., 2019).

Connectivism, developed by George Siemens and Stephen Downes, proposes that learning occurs through networks of people, digital resources, organisations, and technologies. Knowledge is viewed as being distributed across these interconnected networks, and learning involves the ability to establish, navigate, and continually expand these connections. Siemens argued that traditional learning theories do not fully address the characteristics of learning in the digital age, where information is abundant, constantly changing, and accessed through technology (Dunaway, 2011). In fields such as surgical education, connectivism highlights the important role that technological and social networks play in supporting learning and professional development (Lee & Wong, 2025). Consequently, connectivism has become a significant pedagogical framework for technology-enhanced learning because it supports learner autonomy, interaction, diversity, and ongoing participation within digital learning communities (Banihashem & Macfadyen, 2021).

Cognitivism contributes to online and blended learning by informing the organisation and presentation of digital learning materials. Cognitivist theory focuses on the mental processes involved in learning, including thinking, information processing, memory, and problem-solving. Within e-learning environments, cognitivist principles support the logical sequencing of content, adaptation of learning materials to learners' levels of understanding, and the use of multiple forms of representation to strengthen knowledge acquisition (Bell, 2021). Digital course design commonly incorporates problem-solving activities, decision-making tasks, advanced organisers, concept maps, and other strategies that help learners organise information into meaningful cognitive structures (Lee & Wong, 2025).

Behaviourism continues to influence aspects of online learning where structured practice, immediate feedback, and reinforcement are beneficial. Behaviourist theory proposes that learning occurs through changes in observable behaviour resulting from conditioning, environmental stimuli, and reinforcement (Lee & Wong, 2025). These principles underpin measurable learning objectives, computer-assisted instruction, automated assessment, and programmed learning activities that are widely used in digital education (Bell, 2021). Automated quizzes, adaptive learning systems, and interactive tutorials provide learners with immediate feedback and repeated opportunities to practise skills, reinforcing correct responses and supporting mastery of foundational knowledge and procedures.

Humanism is reflected in online learning through its emphasis on learner autonomy, flexibility, personal growth, and self-directed learning. Humanistic learning environments prioritise respectful relationships, learner wellbeing, community, and opportunities for individuals to pursue meaningful educational goals. Within digital learning contexts, humanistic principles support learners' ongoing personal development, self-realisation, and continual updating of knowledge and skills (Bell, 2021). By providing choice, encouraging self-regulation, and promoting authentic learning activities, online learning environments foster intrinsic motivation and support adult learners in taking responsibility for their own learning (Lee & Wong, 2025).

Experiential learning is increasingly incorporated into online and blended learning through virtual simulations, workplace projects, scenario-based activities, remote laboratories, and structured reflection. Experiential learning is recognised as an effective educational approach because it enables learners to acquire knowledge through cycles of experience, reflection, conceptualisation, and application. Digital simulations, educational games, virtual experiments, and authentic workplace activities allow learners to apply theoretical concepts in realistic contexts while receiving opportunities to reflect on their performance and refine their understanding. By integrating constructivist and experiential principles, online learning environments encourage learners to connect theory with practice through authentic, reflective, and application-focused learning experiences (Lee & Wong, 2025).

Characteristics of Effective Online and Blended Learning

Well-designed online and blended learning environments typically share several characteristics. Learning outcomes are clearly articulated, learning materials are logically organised, and activities encourage active participation rather than passive content consumption (Baillargeon et al., 2017; Samyan, 2022). Opportunities for interaction with educators and peers are deliberately built into the course through discussions, collaborative tasks, virtual meetings, or peer feedback (Henri & Lundgren-Cayrol, 2001; Sacré et al., 2020). Structured interactions, including forums, peer tutoring, and cooperative learning, have been shown to constitute important support for learning, with research demonstrating a positive link between organised discussions and learning outcomes (Sacré et al., 2020). The collaborative dimension of online learning environments is further reinforced when courses integrate both synchronous and asynchronous modalities, enabling diverse forms of exchange between learners and instructors (Baillargeon et al., 2017; Henri & Lundgren-Cayrol, 2001).

Accessibility is also a central consideration. Learning resources should be compatible with assistive technologies, available across multiple devices, and designed according to principles such as Universal Design for Learning (UDL) (Heilporn et al., 2023; Marceau et al., 2020; Rogers-Shaw et al., 2017). UDL is a framework that conceptualises knowledge through learner-centred foci emphasising accessibility, collaboration, and community, and is widely used to design courses from primary to higher education levels (Rogers-Shaw et al., 2017). In the francophone literature, this framework is referred to as *conception universelle de l'apprentissage* (CUA), and distance education is considered particularly conducive to integrating its principles due to the emphasis on instructional design, technology, and learner support (Baillargeon et al., 2017; Heilporn et al., 2023; Marceau et al., 2020). Accessibility becomes possible when obstacles to learning are removed and all learners can use a given training device (Marceau et al., 2020). Regular formative assessment and timely feedback enable learners to monitor their progress and identify areas requiring further development (Bazine et al., 2020; Sharaf, 2020). Technologies are understood to favour interactions, adapt learning through new modalities, and provide more feedback on the learning process (Bazine et al., 2020). Educators also establish a consistent online presence by communicating regularly, facilitating discussions, and responding promptly to learner questions, thereby reducing the sense of isolation that can accompany remote learning (Baillargeon et al., 2017; Henri & Lundgren-Cayrol, 2001; Sacré et al., 2020). Supervision and mentoring practices, including encouraging regular participation, recognising peer contributions, and synthesising discussions, are essential to sustaining motivation and engagement in distance learning (Baillargeon et al., 2017; Henri & Lundgren-Cayrol, 2001).

Online and blended learning offer numerous educational advantages. They increase access to education by reducing geographical and scheduling barriers, provide flexibility for learners balancing study with employment and family commitments (Khalil & Ebner, 2016), and support personalised learning through self-paced study and adaptive technologies (Kesylė & Melnikova, 2024; Leppan et al., 2018).

Multimedia resources, simulations, and interactive activities can enhance learner engagement and accommodate diverse learning preferences (Mittal et al., 2022; Nuankaew, 2022). These approaches also encourage the development of digital literacy, self-directed learning, collaboration, and lifelong learning skills that are increasingly important in contemporary workplaces (Khalil, 2020; Mittal et al., 2022).

Blended learning, in particular, combines the flexibility of online learning with the social interaction and practical experiences of face-to-face education, often leading to higher learner satisfaction and stronger opportunities for authentic application of knowledge (Mittal et al., 2022; Nuankaew, 2022).

Despite these advantages, online and blended learning also present challenges. Learners require sufficient digital literacy, self-regulation, time management, and motivation to engage successfully with independent study (Elibol & Bozkurt, 2023; Khalil, 2020). Unequal access to reliable internet connections, appropriate devices, or accessible technologies can create barriers for some learners (Alharbi et al., 2020). Educators must also develop competencies in digital pedagogy, online facilitation, and instructional design rather than simply adapting traditional classroom practices (Kesylė & Melnikova, 2024; Lopes & Soares, 2018).

Maintaining learner engagement, fostering a sense of community, ensuring academic integrity, and providing meaningful feedback can also be more challenging in online environments (Chen et al., 2024; Elibol & Bozkurt, 2023; Rajabalee et al., 2019). Consequently, successful implementation requires thoughtful curriculum design, appropriate technological infrastructure, ongoing learner support, and continuous evaluation and improvement (Alfy et al., 2019; Lopes & Soares, 2018).

Relationship to Theory-Informed Teaching

Online and blended learning exemplify theory-informed teaching because their effectiveness depends on the deliberate application of educational theory rather than the use of technology alone (Corbett & Spinello, 2020; Downes, 2020; Thota, 2015). Technology is simply the medium through which learning occurs; meaningful learning results from the instructional decisions made by educators (Chen & Chan, 2024; Jeny, 2024; Thota, 2015). By integrating constructivist inquiry (Thota, 2015), social collaboration, (Joksimovíc et al., 2018), experiential activities (Jeny, 2024), connectivist networking (Corbett & Spinello, 2020; Downes, 2020; Tschofen & Mackness, 2012), cognitivist instructional design (Thota, 2015), behaviourist reinforcement (Thota, 2015), and humanistic support for learner autonomy (Downes, 2020; Tschofen & Mackness, 2012), educators can create online and blended learning environments that are engaging, inclusive, flexible, and educationally effective (Chen & Chan, 2024; Downes, 2020; Jeny, 2024). When grounded in sound pedagogical principles, these approaches enable learners to develop disciplinary knowledge, professional competence (Oddone et al., 2019), critical thinking (Downes, 2020), digital capability (Atay & Şumuer, 2021; Chen & Chan, 2024), and the capacity for lifelong learning (Conceição & Biniecki, 2024; Downes, 2020).

Example application in Post-Compulsory Education

The following example illustrates how an educator could implement blended learning within a vocational Plumbing program. Rather than delivering all instruction through traditional classroom lectures followed by workshop activities, the educator deliberately integrates online and face-to-face learning to support both theoretical understanding and practical skill development.

At the beginning of a unit on installing heated and cold water services, learners could complete an online learning module through the institution's learning management system before attending practical workshops. The module could include short instructional videos explaining water supply principles, interactive diagrams illustrating plumbing system layouts, readings covering relevant plumbing regulations and Australian Standards, digital animations demonstrating water pressure and flow, and self-marking quizzes that provide immediate feedback. Learners would complete these activities at their own pace, allowing them to revisit complex concepts whenever necessary before progressing to practical work.

Online discussion forums could enable learners to compare different installation methods, discuss common plumbing faults encountered in workplaces, and ask questions about the theoretical content before attending class. Learners employed in different plumbing businesses could also share workplace experiences, exposing the class to a wider variety of installation practices and industry contexts than would be possible in a traditional classroom alone.

When learners attend the face-to-face workshop, the educator could assume that foundational knowledge has already been introduced online. Instead of repeating theoretical content, workshop time could focus on practical application. Learners might install a complete water supply system within a simulated domestic bathroom, interpret plumbing drawings, measure and cut pipework, install valves and fittings, pressure-test the completed system, and rectify faults identified during testing. Throughout these activities, the educator provides demonstrations where required, supervises practical work, asks questions that encourage problem-solving, and provides immediate formative feedback as learners develop their practical competence.

Digital technologies could continue to support learning throughout the practical phase. Learners might use tablets or smartphones to access installation guides, instructional videos, manufacturer specifications, and plumbing standards while completing practical tasks. They could upload photographs of completed installations to an online portfolio, record short reflective videos explaining the decisions they made during installation, and receive feedback from both the educator and their peers through the learning management system. Following each workshop, learners could complete an online reflective journal analysing the challenges they encountered, identifying improvements for future installations, and connecting practical experiences with the theoretical principles studied online.

This blended learning approach demonstrates theory-informed teaching by integrating multiple educational perspectives. Constructivist principles are evident as learners actively construct understanding by applying theoretical knowledge during authentic plumbing installations. Social constructivism is reflected through online discussions, peer collaboration, and shared problem-solving during workshop activities. Connectivism is demonstrated as learners access manufacturer resources, plumbing regulations, instructional videos, and professional industry networks to support their learning. Cognitivist principles underpin the logical sequencing of online learning materials and the use of visual resources that help learners organise technical knowledge before applying it in practice. Behaviourist principles are incorporated through repeated practical exercises, immediate feedback, and online quizzes that reinforce correct procedures. Humanistic principles support learner autonomy by allowing learners to progress through online content at an appropriate pace while encouraging self-reflection and responsibility for learning. Experiential learning is realised through authentic plumbing installations, structured reflection, and the continual refinement of practical skills through repeated application.

By combining online theoretical instruction with supervised practical workshops, learners gain the flexibility of self-paced digital learning while continuing to develop the practical competence, professional judgement, and workplace skills that are essential for successful employment as plumbers. This approach enables valuable face-to-face workshop time to be devoted to higher-order learning activities, coaching, troubleshooting, and authentic skill development rather than the delivery of theoretical content alone.

Mobile Learning

Mobile learning (m-learning) is an educational approach that uses portable digital devices such as smartphones, tablets, laptops, and wearable technologies to enable learning at any time and in almost any location (Khozouie & Fotouhi-Ghazvini, 2013; Madjarov, 2012; Mohiuddin et al., 2021). While online learning and distance education generally focus on learning delivered through internet-based platforms, mobile learning emphasises the flexibility and portability of learning made possible through mobile technologies (Adejo et al., 2018). Rather than being confined to classrooms or desktop computers, learners can access educational resources, communicate with educators and peers, complete assessments, and participate in learning activities while travelling, working, or undertaking field-based tasks (Badidi, 2016; Madjarov, 2014).

The growth of smartphones, wireless internet, cloud computing, and mobile applications has transformed mobile learning from a supplementary educational tool into a significant mode of teaching and learning across vocational education, higher education, workplace training, and professional development (Mohiuddin et al., 2021; Wu & Plakhtii, 2021). Mobile devices provide immediate access to multimedia resources, learning management systems, digital textbooks, instructional videos, podcasts, simulations, and communication platforms, enabling learning to occur within authentic real-world contexts (Badidi, 2016; Matthew et al., 2018).

One of the defining characteristics of mobile learning is its flexibility. Learners are no longer restricted by specific times or locations and can engage with learning whenever opportunities arise (Adejo et al., 2018; Khozouie & Fotouhi-Ghazvini, 2013). Short learning activities, often referred to as microlearning, allow learners to complete brief instructional modules during breaks at work, while commuting, or between workplace tasks (Marshall, 2023). Mobile learning also supports just-in-time learning, where learners access information precisely when it is needed to solve workplace problems or perform unfamiliar tasks (Khozouie & Fotouhi-Ghazvini, 2013).

Another important characteristic is contextual learning. Because mobile devices accompany learners into workplaces, laboratories, construction sites, hospitals, and community settings, learning can occur within the environment where knowledge will ultimately be applied (Ahmad et al., 2020). This allows learners to immediately connect theoretical concepts with authentic professional practice, strengthening the transfer of learning to workplace performance (Madjarov, 2014; Mohiuddin et al., 2021).

Constructivism underpins mobile learning by encouraging learners to actively construct knowledge through exploration, inquiry, and authentic problem-solving. Mobile technologies enable learners to investigate real-world situations, collect data, document observations, and reflect on their experiences while learning occurs within authentic contexts (Cornelius & Marston, 2009; Kissinger, 2013). Authentic learning builds on constructivist learning theories, particularly situated learning, and includes features such as providing authentic learning contexts and giving students an active role as they experience information grounded in reality (Mohamed & Al-Hidabi, 2020). Constructivism takes a holistic approach where each learner, individually and socially, constructs their own knowledge through meaning-making processes that depend on previous experiences. Mobile devices support this by enabling learners to engage with real-world phenomena using scientific tools and inquiry-based approaches (Nikou & Economides, 2021).

Social constructivism is reflected in the collaborative capabilities of mobile technologies. Instant messaging, discussion forums, social media platforms, collaborative documents, and video conferencing allow learners to communicate, share experiences, provide peer feedback, and negotiate meaning regardless of their physical location (Shadiev et al., 2017). Social constructivism stresses the need for collaboration and social interaction between learners (Hidayati et al., 2022), and mobile learning research has demonstrated that student performance is enhanced most when learners collaborate tightly with peers (Shadiev et al., 2017). MALL course design has been shown to promote active collaborative learning that helps students construct learning experiences in real-world environments. These technologies enable learning communities to remain connected beyond the classroom, as mobile learning supports human interaction across multiple situations mediated by mobile technology within networked communities of practice.

Connectivism has particular relevance to mobile learning because mobile devices provide continuous access to extensive networks of information, experts, professional organisations, and digital resources (Hidayati et al., 2022). Connectivism involves understanding learning in a digital age and emphasises increased social interaction through digital platforms (Hidayati et al., 2022). Learners develop the ability to locate, evaluate, synthesise, and apply information from multiple sources while participating in professional learning networks that support lifelong learning. However, Siemens's description of connectivism as a successor to behaviourism, cognitivism, and constructivism has been challenged, as learners may struggle in unstructured learning environments and become overwhelmed by peer-generated content.

Cognitivism informs the design of mobile learning resources through the organisation of information into manageable learning segments. Cognitivism focuses on how learners learn and how information is received, organised, and retrieved by the mind (Hidayati et al., 2022). Mobile learning commonly incorporates multimedia, interactive graphics, quizzes, and short instructional videos that reduce cognitive load while supporting memory, attention, and information processing. Cognitivism has led to inventions including adaptive systems that analyse learner responses and direct them to appropriate actions. Notifications and spaced learning activities can also reinforce long-term retention, though negative effects on cognitive load have been reported when mobile learning environments become overly complex (Shadiev et al., 2017).

Behaviourism continues to influence mobile learning through structured practice and immediate feedback. Educational applications frequently incorporate quizzes, adaptive learning activities, progress tracking, badges, and gamified rewards that reinforce learning through repetition and positive feedback (Hidayati et al., 2022). Programmed instruction amplifies the traditional teacher-centred approach based on behaviourism (Hidayati et al., 2022). These approaches are particularly useful for developing procedural knowledge and reinforcing foundational concepts, and the development from behaviourism to cognitivism implies an increase in self-regulation in learning (Hidayati et al., 2022).

Humanism is evident in the learner-centred nature of mobile learning. Learners determine when, where, and how they engage with learning materials, enabling them to personalise learning according to their individual needs, interests, and professional goals. Mobile learning's student-centred, personal, and ubiquitous characteristics can enhance learners' self-regulation and motivation (Солодка et al., 2022). The framework proposed by (Nikou & Economides, 2021) demonstrated that mobile-based formative assessment activities had a significant positive impact on student perceived levels of autonomy, competence, and relatedness, enhancing intrinsic motivation (Солодка et al., 2022). This flexibility supports learner autonomy, intrinsic motivation, and lifelong learning.

Experiential learning is strengthened because mobile technologies allow learning to occur within authentic environments. Learners can apply theoretical knowledge while performing workplace tasks, document experiences through photographs or videos, receive immediate feedback, and reflect on their learning shortly after the experience occurs (Cornelius & Marston, 2009; Shadiev et al., 2017; Shadiev et al., 2022). Learning by doing is a broad paradigm established to support experiential learning theory and situated learning, requiring learners to perform tasks that have to be learnt. Language learners, for example, have been asked to take pictures of objects and describe them using mobile learning systems in authentic environments, with scholars finding higher learning performance compared to traditional approaches (Shadiev et al., 2022). This continuous cycle of experience, reflection, conceptualisation, and application aligns closely with experiential learning principles.

Mobile learning is supported by a wide range of technologies including smartphones, tablets, wearable devices, mobile learning applications, cloud storage, learning management systems, augmented reality (AR), virtual reality (VR), QR codes, podcasts, e-books, instructional videos, and mobile assessment tools (El-Haggar et al., 2023; Mihus & Gupta, 2023). Many learning management systems are specifically designed to function across mobile devices, allowing learners to access course materials, submit assessments, communicate with educators, and receive feedback regardless of location (El-Haggar et al., 2023).

Emerging technologies such as augmented reality further extend mobile learning by overlaying digital information onto physical environments (Bęś et al., 2025; Blattgerste et al., 2019). Learners can view three-dimensional models, equipment instructions, maintenance procedures, or safety information while interacting directly with workplace equipment (Bęś et al., 2025; Blattgerste et al., 2019).

The educator's role in mobile learning extends beyond providing digital content. Educators design learning activities that take advantage of mobility by encouraging learners to collect workplace evidence, participate in field observations, solve authentic problems, collaborate with peers, and reflect on their experiences (El-Haggar et al., 2023). They also ensure that learning materials are suitable for smaller screens, accessible across multiple devices, and organised into concise, manageable learning segments that support learning on the move (El-Haggar et al., 2023).

Mobile learning provides considerable flexibility by allowing learning to occur whenever and wherever opportunities arise (El-Haggar et al., 2023; Mihus & Gupta, 2023). It supports just-in-time learning, enabling learners to access information precisely when required during workplace tasks (El-Haggar et al., 2023). The portability of mobile devices encourages contextual learning, strengthens the connection between theory and practice, and promotes continuous professional development beyond formal educational settings (El-Haggar et al., 2023; Mihus & Gupta, 2023). Mobile learning also supports learner autonomy, personalised learning pathways, collaboration, digital literacy, and lifelong learning (El-Haggar et al., 2023; Mihus & Gupta, 2023).

Despite its advantages, mobile learning also presents challenges. Learners require reliable access to mobile devices, internet connectivity, and sufficient digital literacy to engage effectively with learning activities (El-Haggar et al., 2023; Mihus & Gupta, 2023). Small screen sizes may limit the presentation of complex information, while frequent notifications and competing applications can create distractions. Educators must carefully design learning materials specifically for mobile devices rather than simply adapting desktop-based resources (El-Haggar et al., 2023). Consideration must also be given to data privacy, cybersecurity, accessibility, and ensuring that learners are not disadvantaged by unequal access to technology (El-Haggar et al., 2023; Mihus & Gupta, 2023).

Mobile learning exemplifies theory-informed teaching because its effectiveness depends upon the thoughtful integration of educational theory with mobile technology. The technology itself does not produce meaningful learning; rather, learning occurs through instructional designs that encourage active inquiry, collaboration, authentic application, self-directed learning, and reflection (El-Haggar et al., 2023). By integrating constructivist, social constructivist, connectivist, cognitivist, behaviourist, humanistic, and experiential principles, educators can create mobile learning experiences that extend education beyond traditional classrooms and enable learners to engage with knowledge wherever authentic learning opportunities arise (El-Haggar et al., 2023).

The following examples illustrate how mobile learning could be implemented in adult education within both higher education and vocational education and training. Although the disciplines differ considerably, each example demonstrates how portable digital technologies can extend learning beyond the classroom, support authentic learning experiences, and enable learners to access resources, collaborate with others, and reflect on their learning within real-world contexts. Together, these examples demonstrate that mobile learning is not confined to a particular educational sector but can be adapted to support theoretical understanding, practical skill development, and professional competence across a wide range of adult learning environments.

A Bachelor of Psychology could integrate mobile learning to enable students to engage with learning in both academic and authentic community settings. Rather than restricting learning to lectures and tutorials, the educator could design activities that encourage students to use smartphones or tablets to access learning resources, collect observations, collaborate with peers, and reflect on experiences wherever learning opportunities arise.

Before attending seminars, students could complete short mobile learning modules consisting of brief videos explaining psychological theories, interactive concept maps, podcasts discussing current research, and low-stakes quizzes that reinforce key terminology and concepts. Push notifications from the learning management system could remind students to complete weekly activities, while spaced retrieval quizzes delivered through mobile applications would strengthen long-term retention of foundational knowledge.

During practical classes or community-based learning activities, students could use mobile devices to record behavioural observations, complete field notes, administer approved psychological questionnaires within ethical guidelines, or document environmental influences on behaviour. In research methods units, students could collect survey data using mobile applications, analyse preliminary findings through cloud-based software, and collaborate with classmates using shared digital workspaces. During placements or supervised practice, learners could use mobile devices to access ethical codes, assessment protocols, and evidence-based intervention guidelines while reflecting on professional practice through secure electronic journals. Educators could provide timely feedback on reflective entries, monitor learner progress, and facilitate discussion forums where students compare observations, analyse case studies, and explore multiple interpretations of psychological phenomena.

This approach demonstrates theory-informed teaching by combining constructivist inquiry through authentic investigations, social constructivist collaboration through online discussions and peer feedback, connectivist engagement with current psychological research and professional organisations, cognitivist organisation of multimedia learning resources, behaviourist reinforcement through regular formative quizzes, humanistic support for learner autonomy, and experiential learning through authentic observation, reflection, and application of psychological knowledge.

A vocational qualification in Bituminous Surfacing could similarly utilise mobile learning to support apprentices and experienced workers undertaking training across road construction and maintenance projects. Because much of the learning occurs in dynamic workplace environments rather than classrooms, mobile technologies allow learners to access information precisely when it is required and immediately apply new knowledge to practical tasks.

Before commencing daily work, learners could complete brief mobile learning modules covering topics such as asphalt mix properties, equipment operation, traffic management requirements, workplace health and safety procedures, environmental controls, and quality assurance processes. Short instructional videos, interactive equipment diagrams, and microlearning activities could reinforce theoretical knowledge before learners begin practical work on site.

While undertaking surfacing operations, learners could use smartphones or tablets to access equipment operating manuals, manufacturer specifications, project drawings, weather forecasts, material testing procedures, and workplace documentation. QR codes attached to paving equipment, rollers, spray units, or testing instruments could provide immediate access to maintenance instructions, safe operating procedures, inspection checklists, and troubleshooting guides. Learners could photograph completed work, record videos demonstrating compaction techniques or quality testing procedures, and upload these to an electronic portfolio as evidence of competency.

Mobile devices could also support quality assurance by enabling learners to record pavement temperatures, compaction results, layer thickness measurements, material quantities, weather conditions, and inspection outcomes directly into digital workplace systems. Following each shift, learners could complete short reflective activities evaluating production outcomes, identifying factors affecting pavement quality, and linking workplace experiences to engineering principles and industry best practice. Supervisors and educators could review submitted evidence, provide timely feedback, and identify areas requiring further coaching while learners remain actively engaged on live construction projects.

This application integrates multiple educational theories. Constructivism is reflected as learners actively solve authentic workplace problems and apply technical knowledge during surfacing operations. Social constructivism is supported through collaboration with supervisors, crew members, and fellow learners using mobile communication platforms. Connectivism enables learners to access current technical specifications, industry guidance, manufacturers' information, and professional resources directly from the worksite. Cognitivist principles underpin the delivery of concise multimedia learning resources and structured microlearning modules that minimise cognitive overload. Behaviourist approaches are evident through repeated practice, digital competency checklists, and immediate feedback on workplace performance. Humanism supports learner autonomy by allowing individuals to access learning resources whenever needed, while experiential learning is achieved through the continual cycle of practical application, reflection, feedback, and improvement within authentic road construction environments.

Learning Management Systems

Learning Management Systems (LMSs) are digital platforms that enable educators to design, deliver, manage, and evaluate learning across online, blended, and face-to-face educational environments (Marshall, 2023; Salah & Thabet, 2021; Yang, 2024). They provide a centralised digital space where learners can access learning materials, participate in learning activities, communicate with educators and peers, complete assessments, receive feedback, and monitor their academic progress (Angelova et al., 2015; Salah & Thabet, 2021). Used extensively in higher education, vocational education and training, workplace learning, and professional development, LMSs have become a fundamental component of contemporary education by supporting flexible, accessible, and learner-centred approaches to teaching and learning (Mauro et al., 2018; Wu & Plakhtii, 2021).

Modern Learning Management Systems extend far beyond functioning as repositories for course materials. They integrate content delivery, communication, assessment, collaboration, learner analytics, record keeping, and administrative functions within a single digital environment (Marshall, 2023; Salah & Thabet, 2021; Yang, 2024). Common features include course pages, discussion forums, announcements, assignment submission portals, online quizzes, gradebooks, calendars, multimedia resources, virtual classrooms, and mobile access (Angelova et al., 2015; Salah & Thabet, 2021). Many LMSs also integrate with external technologies such as video conferencing platforms, e-portfolios, digital libraries, plagiarism detection software, simulations, and collaborative productivity applications, creating comprehensive digital learning ecosystems that support both teaching and administration (Malele, 2023; Marshall, 2023).

The evolution of Learning Management Systems has closely paralleled advances in educational technology and internet-based learning. Early systems introduced during the 1990s primarily distributed learning materials and recorded grades. As internet technologies matured, LMSs expanded to incorporate communication tools, multimedia learning resources, online assessment, and collaborative learning environments (Biljon & Merwe, 2018; Gorshenin, 2018). Contemporary cloud-based platforms now support mobile learning, learning analytics, adaptive learning, artificial intelligence, competency-based education, and integration with a wide range of third-party educational technologies (Gorshenin, 2018; Marshall, 2023; Wu & Plakhtii, 2021). Consequently, LMSs have evolved from relatively simple administrative systems into sophisticated environments capable of supporting complex, interactive, and personalised learning experiences (Apoki & Crişan, 2022; Gorshenin, 2018).

One of the defining characteristics of an LMS is its ability to organise learning into structured pathways aligned with clearly defined learning outcomes. Educators can sequence content, activities, and assessments logically while providing learners with consistent access to recorded lectures, readings, videos, podcasts, simulations, digital resources, and external learning materials from a single location (Salah & Thabet, 2021; Yang, 2024). This structured organisation allows learners to navigate courses more effectively while supporting independent study regardless of time or place (Malele, 2023; Wu & Plakhtii, 2021).

Communication is another central feature of Learning Management Systems. Discussion forums, announcements, messaging systems, virtual classrooms, and collaborative workspaces facilitate ongoing interaction between educators and learners throughout a course (Angelova et al., 2015; Salah & Thabet, 2021). These communication tools support both synchronous and asynchronous learning by enabling learners to ask questions, participate in discussions, collaborate on projects, share ideas, and receive academic support regardless of geographical location (Badidi, 2016; Wu & Plakhtii, 2021).

Assessment and feedback are similarly integrated into most Learning Management Systems. Online quizzes, assignment submission systems, marking rubrics, automated assessment, peer review activities, and electronic gradebooks enable educators to evaluate learner performance efficiently while providing timely feedback (Salah & Thabet, 2021; Yang, 2024). Learning analytics further assist educators by identifying patterns of learner engagement, highlighting learners who may require additional support, and informing ongoing improvements to teaching and course design (Marshall, 2023; Макарчук et al., 2018).

Learning Management Systems effectiveness depends not upon the technology itself but upon the quality of the learning experiences designed by educators (Ali, 2024; Alonso et al., 2005; Corbin & Bugden, 2018). Consequently, LMSs can support multiple educational theories simultaneously, with different features reflecting different perspectives on how learning occurs (Ali, 2024; Dakich, 2014; González & Quiroz, 2019).

Constructivism is evident when educators use LMSs to create inquiry-based learning activities that encourage learners to actively construct knowledge through authentic tasks, case studies, simulations, project work, and reflective journals (Guo, 2018; Martucci et al., 2023). Rather than simply presenting information, these learning activities encourage learners to investigate problems, apply knowledge, and reflect on their experiences within meaningful contexts (González & Quiroz, 2019; Lamtara, 2023a).

Social constructivism is reflected through the collaborative tools available within Learning Management Systems (Alonso et al., 2005; Nikolov et al., 2016). Discussion forums, collaborative documents, peer review activities, group projects, and virtual classrooms provide opportunities for learners to exchange ideas, negotiate meaning, and build understanding collectively (Merzouk et al., 2014; Nikolov et al., 2016). These communication tools enable learners to participate in active learning communities even when studying remotely (António Manuel Diogo dos et al., 2018; Merzouk et al., 2014).

Connectivism aligns closely with contemporary LMSs because they provide access to broader digital learning networks extending beyond the course itself (Ali, 2024; Brieger et al., 2020). Educators can integrate professional organisations, research databases, webinars, multimedia repositories, open educational resources, external websites, and social media into learning activities (Ali, 2024; Lamtara, 2023a). These connections enable learners to engage with current knowledge, professional communities, and rapidly evolving information throughout their studies (Ali, 2024).

Cognitivism informs the organisation and presentation of learning materials within an LMS ((Ali, 2024; Corbin & Bugden, 2018; Yeo, 2024). Course modules are typically arranged in logical sequences, while multimedia resources, concept maps, worked examples, scaffolding activities, retrieval practice, and formative assessments support attention, information processing, memory, and long-term knowledge retention (Ali, 2024; Yeo, 2024). These instructional strategies help learners organise complex information into meaningful cognitive structures (Corbin & Bugden, 2018; Yeo, 2024).

Behaviourism continues to influence many LMS features through structured learning pathways, automated quizzes, adaptive learning activities, badges, progress tracking, and immediate feedback (Corbin & Bugden, 2018; Dakich, 2014). These functions encourage repeated practice, reinforce correct responses, and support the progressive mastery of foundational knowledge and procedural skills through ongoing assessment and reinforcement (Bahattab et al., 2023; Dakich, 2014).

Humanism is reflected in the learner-centred flexibility provided by Learning Management Systems Merzouk et al., 2014). Learners often determine the pace and timing of their study, revisit learning resources when required, monitor their own progress, and engage with optional learning activities according to their interests and needs (Bahattab et al., 2023; Merzouk et al., 2014). These opportunities encourage learner autonomy, self-regulation, intrinsic motivation, and responsibility for personal learning (Kumar & Santosh, 2022; Merzouk et al., 2014).

Experiential learning can also be effectively supported through an LMS by incorporating workplace projects, digital portfolios, simulations, scenario-based learning activities, reflective journals, and authentic assessments (Bahattab et al., 2023; Lamtara, 2023a). Learners document practical experiences, receive formative feedback, reflect on their performance, and connect workplace practice with theoretical knowledge through structured learning activities managed within the digital environment (Bahattab et al., 2023; Hashim & Hamidon, 2022).

Within a Learning Management System, the educator's role extends well beyond uploading learning materials (Corbin & Bugden, 2018; González & Quiroz, 2019). Educators design coherent learning pathways, facilitate discussions, monitor learner engagement, analyse learning analytics, provide timely feedback, and create opportunities for authentic collaboration and inquiry (Christopoulos et al., 2020; Corbin & Bugden, 2018; Merzouk et al., 2014). They also ensure that digital resources remain accessible, assessments align with intended learning outcomes, and learners receive appropriate academic and pastoral support throughout their studies (Corbin & Bugden, 2018; Merzouk et al., 2014). Effective use of an LMS therefore requires both technological competence and strong pedagogical expertise to ensure that technology enhances rather than replaces meaningful teaching (Corbin & Bugden, 2018; González & Quiroz, 2019).

Learning Management Systems offer numerous educational advantages. They centralise learning resources, improve access to education regardless of geographical location, support flexible study arrangements, facilitate communication between educators and learners, streamline assessment processes, and maintain comprehensive records of learner progress (Cui & Coleman, 2021; Ifenthaler & Yau, 2020b; P. Li et al., 2026; Oliwa, 2024). LMS platforms such as Blackboard, Canvas, and Moodle allow educators to create and manage online courses, assignments, and assessments, while also providing tools for communication, collaboration, and feedback (Prasetya & Syarif, 2023; Younas et al., 2026). Multimedia resources, collaborative tools, and learning analytics further enhance learner engagement, support personalised learning, and assist educators in identifying learners who may require additional support (Blumenstein et al., 2018; Brice et al., 2023; Chen et al., 2021; Rossiter et al., 2022). Learning analytics can be utilised for improving student retention, supporting informed decision making, helping to understand learning behaviour, providing personalised assistance, and delivering feedback and interventions (Gumede et al., 2023; Tatnall, 2023). LMSs also encourage lifelong learning by enabling learners to revisit learning materials, monitor their own progress, and continue accessing educational resources beyond scheduled class sessions (Zawacki-Richter et al., 2019). Active participation with interactive content has been shown to improve information retention, learning engagement, attitudes, and thinking skills over static content (Ifenthaler & Yau, 2020a), and online learning embraces flexibility, personalised learning, and adaptive learning experiences, thereby shifting the focus to a student-centred approach (Diahyleva et al., 2024; Ifenthaler & Yau, 2020a).

Despite these advantages, Learning Management Systems also present challenges. Poorly designed courses may overwhelm learners through excessive content, inconsistent organisation, or confusing navigation (Qi et al., 2023), while an overreliance on technology may reduce meaningful interaction if instructional design is weak (Vincent-Ruz & Boase, 2022). Research has noted that low retention rates and limited learner engagement remain prevalent issues, especially in online learning contexts (Kian et al., 2022). Successful participation requires learners to possess adequate digital literacy, reliable internet access, and suitable technological devices (Mansor et al., 2020; Yang, 2025). English language proficiency and technological skills can also impact the success of self-directed learning approaches, which in turn affects learner satisfaction and engagement (Aryani et al., 2021). Educators must likewise invest significant time in developing digital learning resources, mastering platform functionality, and designing engaging online learning experiences; indeed, bespoke mobile-learning resources represent a significant investment of skill and time to develop and maintain (Aryani et al., 2021). Issues relating to accessibility, data privacy, cybersecurity, and equitable access must also be carefully considered to ensure that technology supports rather than dictates educational practice (Kasap, 2023; Nkechi et al., 2024; Sato et al., 2023). Privacy concerns are particularly salient as learning systems can autonomously collect personal information such as learning styles and capabilities, and institutions need robust policies to maintain ethical standards (Kasap, 2023; Sato et al., 2023).

Learning Management Systems exemplify theory-informed teaching because they provide the technological infrastructure through which multiple educational theories can be implemented (Younas et al., 2026). The constructivism theory suggests that learners construct knowledge by actively engaging with new information and building upon prior knowledge and experiences (Younas et al., 2026), and LMS features can support constructivist principles including self-regulation, experiential learning, social learning, and learning community development. The platform itself does not determine how learning occurs; instead, meaningful learning results from the pedagogical decisions made by educators (Aryani et al., 2021; Ifenthaler & Yau, 2020a). Blended learning environments, facilitated through LMS platforms, have demonstrated positive impacts on learners' satisfaction, motivation, knowledge retention, and student engagement when pedagogies align with active and engaging learning requirements (Aryani et al., 2021; Ifenthaler & Yau, 2020a). By thoughtfully integrating constructivist inquiry, social collaboration, connectivist networking, cognitivist instructional design, behaviourist reinforcement, humanistic learner support, and experiential learning opportunities, educators can transform an LMS from a simple content repository into a dynamic learning environment that promotes learner engagement, critical thinking, professional competence, and lifelong learning (Blumenstein et al., 2018; Younas et al., 2026). Gamification strategies, adaptive learning technologies, and personalised feedback systems further demonstrate how theoretical frameworks can be operationalised within LMS platforms to motivate learners and create structured, engaging learning experiences (Diahyleva et al., 2024; Kian et al., 2022; Qi et al., 2023).

Open Educational Resources

Open Educational Resources (OER) are teaching, learning, and research materials that are freely available for anyone to access, use, adapt, modify, and redistribute with few or no legal or financial restrictions (Alzahrani, 2023; Clinton-Lisell, 2021; Tang et al., 2020). Unlike traditional educational resources, which are typically protected by restrictive copyright and require payment or institutional subscriptions (Allen & Katz, 2020; Clinton-Lisell, 2021), OER are generally released under open licences, such as Creative Commons licences, that permit varying degrees of reuse while acknowledging the original creator (Allen & Katz, 2020; Atenas & Havemann, 2014; Santos et al., 2023). These resources may include textbooks, lecture notes, videos, assessments, simulations, images, podcasts, datasets, software, laboratory manuals, and entire online courses (Hung, 2013; Saripudin et al., 2020).

The concept of Open Educational Resources emerged from the broader open education movement, which seeks to increase equitable access to high-quality education by reducing barriers associated with cost, geography, and intellectual property (Abeywardena, 2017; Atenas & Havemann, 2014; Clinton-Lisell, 2021). The term Open Educational Resources was formally adopted by UNESCO in 2002 during the Forum on the Impact of Open Courseware for Higher Education in Developing Countries (Clinton-Lisell, 2021; Duval & Wiley, 2010; Gil-Jaurena, 2013; Wahid et al., 2020). Since then, governments, educational institutions, non-profit organisations, and individual educators have contributed to the growth of openly licensed educational materials that support lifelong learning across schools, vocational education and training, higher education, and workplace learning (Abeywardena, 2017; Alzahrani, 2023; Wahid et al., 2020).

A defining characteristic of OER is that they support the principles commonly referred to as the 5Rs of openness (Alzahrani, 2023; Amornrit et al., 2018; Heck et al., 2020; Santos et al., 2023; Tang et al., 2020). These principles allow users to retain copies of educational materials, reuse them in different contexts, revise or adapt them to meet local needs, remix them with other openly licensed resources to create new materials, and redistribute the original or modified versions to others (Santos et al., 2023; Tang et al., 2021; Tang et al., 2020; Wiley & Hilton, 2018). These permissions distinguish OER from free educational resources that may be available at no cost but remain protected by traditional copyright restrictions (Allen & Katz, 2020; Richardson & Sakho, 2022).

Open Educational Resources encompass a diverse range of educational materials. They may include openly licensed textbooks that replace expensive commercial publications (Allen & Katz, 2020; Clinton-Lisell, 2021), instructional videos, interactive simulations, virtual laboratories, assessment banks, lesson plans, presentation slides, digital images, podcasts, case studies, and complete online courses (Hung, 2013; Saripudin et al., 2020). Many universities also publish open courseware that includes lecture recordings, reading lists, assignments, and supporting resources, enabling learners worldwide to benefit from high-quality educational content regardless of enrolment status (Saripudin et al., 2020).

Theoretical Foundations

Constructivism is reflected when educators use OER to support inquiry-based learning, problem-solving, and authentic investigation (Belle, 2024; Werth & Williams, 2021). Rather than relying on a single prescribed textbook, learners can explore multiple open resources, compare perspectives, investigate real-world problems, and construct their own understanding from diverse sources of information (Belle, 2024; Berg & Toit-Brits, 2023).

Social constructivism is supported through the collaborative nature of many OER initiatives (Letseka & Seeletso, 2021; Putra et al., 2024). Educators frequently adapt and improve existing resources, while learners may contribute to the development of openly shared materials through collaborative projects, peer review, and knowledge creation (Gil-Jaurena, 2020; Letseka & Seeletso, 2021). These practices encourage communities of learning that extend beyond individual institutions, as open educational practices "support the (re)use and production of OER through institutional policies, promote innovative pedagogical models, and respect and empower learners as co-producers on their lifelong learning path" (Ehlers, 2011, p. 12).

Connectivism aligns closely with Open Educational Resources because learners gain access to extensive global networks of knowledge (Ossiannilsson, 2022). OER repositories, digital libraries, professional organisations, research publications, and open online courses enable learners to locate current information, evaluate multiple viewpoints, and participate in broader learning communities throughout their educational and professional lives (Bulathwela et al., 2024; Ossiannilsson, 2022). Connectivism emphasises the power of networking, gathering diverse opinions, and understanding that learning occurs not only within an individual but also within and across networks (Ossiannilsson, 2022).

Cognitivism informs the design of many high-quality OER through the careful organisation of learning materials into logical sequences that support attention, memory, and information processing (Santos et al., 2024). Multimedia resources, worked examples, interactive activities, and formative assessments commonly found within OER help learners organise and integrate new knowledge effectively (Bulathwela et al., 2024; Santos et al., 2024).

Behaviourism is evident in openly licensed practice exercises, quizzes, tutorials, and automated assessment resources that provide learners with repeated opportunities for practice and immediate feedback (Akma et al., 2024; Putra et al., 2024). These resources support the development of foundational knowledge and procedural skills through structured reinforcement (Putra et al., 2024).

Humanism is reflected in the learner-centred philosophy underpinning open education (Berg & Toit-Brits, 2023; Werth & Williams, 2021). Learners have considerable freedom to select resources that match their interests, prior knowledge, language, cultural context, and learning goals (Berg & Toit-Brits, 2023; Bonk & Lee, 2017). This flexibility promotes learner autonomy, self-directed learning, intrinsic motivation, and lifelong learning (Berg & Toit-Brits, 2023; Letseka & Seeletso, 2021).

Experiential learning can be supported when educators combine OER with authentic learning activities such as workplace projects, case studies, simulations, field investigations, and reflective practice (Belle, 2024). Openly licensed resources provide the theoretical foundation that learners can immediately apply within practical learning experiences before reflecting upon their outcomes (Belle, 2024; Werth & Williams, 2021).

Application and Implementation

Educators play a critical role that extends well beyond locating freely available learning materials. They are responsible for evaluating the quality, accuracy, relevance, and accessibility of Open Educational Resources (OER), adapting them to suit specific learning contexts, and integrating them into coherent instructional designs. Because OER are openly licensed digital resources, educators can modify and redistribute materials to better meet the diverse needs, backgrounds, abilities, and learning preferences of adult learners. This flexibility enables learning resources to be contextualised for particular industries, professions, cultural settings, and educational levels while reducing the financial and legal barriers often associated with traditional copyrighted materials (Hagedorn et al., 2011; Patel et al., 2023).

Beyond using existing resources, educators also contribute to the broader open education movement by creating and sharing their own openly licensed teaching materials. The Creative Commons (CC) licensing framework has become the primary mechanism through which educators communicate how others may use, adapt, and redistribute their work. By selecting appropriate licences, educators can encourage collaboration while retaining control over the conditions under which their resources are reused. This collaborative approach supports the continual improvement of educational materials and contributes to a global community of educators, institutions, policymakers, and learners committed to expanding access to high-quality education (Howard, 2019; Patel et al., 2023).

Open Educational Resources provide substantial educational, social, and economic benefits. By removing or significantly reducing the cost of textbooks and other learning materials, OER improve access to education for learners regardless of their financial circumstances. Their open licensing also enables educators to adapt, update, and redistribute resources to reflect current knowledge, local contexts, industry practices, and learner needs without relying on commercial publishing cycles. This adaptability allows learning materials to remain current and relevant while supporting more personalised and inclusive educational experiences (Patel et al., 2023).

Research has also demonstrated that OER can contribute to improved learner outcomes, including higher levels of student performance and retention, particularly among learners who have traditionally experienced barriers to higher education. Beyond reducing financial costs, OER support lifelong learning by making educational opportunities more affordable, accessible, and flexible throughout an individual's personal and professional life. They also promote educational equity by expanding access to high-quality learning resources, encouraging collaboration among educators, and fostering a culture of knowledge sharing across institutions and national boundaries (Hagedorn et al., 2011; Howard, 2019; Patel et al., 2023).

Despite their considerable advantages, Open Educational Resources also present several challenges. Because OER are developed by a wide variety of authors, institutions, and organisations, the quality, accuracy, depth, and suitability of available resources can vary considerably. In some specialised disciplines, educators may find that existing OER are limited in scope, insufficiently comprehensive, or developed for broader survey courses rather than advanced or discipline-specific study. Consequently, educators must carefully evaluate resources before incorporating them into teaching programs and may need to supplement or modify materials to ensure they meet learning outcomes and professional standards (Patel et al., 2023).

Licensing also presents challenges because not all Creative Commons licences provide the same permissions. While open licences facilitate sharing and collaboration, conditions such as Non-Commercial (NC) or No Derivatives (ND) may restrict adaptation or reuse, limiting the flexibility typically associated with OER. Educators therefore require a sound understanding of copyright and licensing to ensure that resources are used appropriately and legally. Additional challenges include locating suitable materials, maintaining the currency and accessibility of resources, providing learners with adequate digital access, and establishing institutional policies, governance structures, and leadership that effectively support the implementation and sustainability of open education initiatives (Hagedorn et al., 2011; Howard, 2019; Patel et al., 2023).

Open Educational Resources exemplify theory-informed teaching because they provide adaptable educational materials that can be aligned with a wide range of pedagogical approaches rather than prescribing a single method of instruction. Their openness enables educators to revise, contextualise, combine, and redesign learning resources to support diverse learner needs, educational contexts, and professional disciplines. This flexibility encourages innovation in curriculum design while promoting equitable access, inclusion, lifelong learning, and collaboration across educational systems (Howard, 2019).

The collaborative nature of OER also supports the continuous improvement of teaching resources through the collective contributions of educators, institutions, and learners worldwide. By encouraging openness, knowledge sharing, and ongoing refinement of educational materials, OER contribute to a more accessible and sustainable educational ecosystem. Supported by appropriate Creative Commons licensing, educators can create learning experiences that align with constructivist, social constructivist, connectivist, cognitivist, behaviourist, humanistic, and experiential approaches while responding effectively to the diverse needs of adult learners in higher education, vocational education and training, workplace learning, and lifelong learning contexts (Hagedorn et al., 2011; Howard, 2019; Patel et al., 2023).

Micro-credentials and Digital Badges

Micro-credentials and digital badges are emerging forms of credentialing that recognise the achievement of specific knowledge, skills, or competencies through relatively short, focused learning experiences (Clements et al., 2020; Kiiskilä et al., 2023; McGreal & Olcott, 2022; Porto & Presant, 2023). Unlike traditional qualifications, which typically require learners to complete comprehensive programs over extended periods, micro-credentials recognise mastery of discrete areas of competence that can be acquired independently or accumulated over time (Diab & Zhang, 2023; McGreal & Olcott, 2022; Tamoliūnė et al., 2023). A micro-credential has been defined as "a certification of assessed learning that is additional, alternate, and complementary to or a formal component of a formal qualification" (Oliver, 2019, p. i), while others describe them as "certified documents that provide recognised proofs of the achievement of learning outcomes from shorter, less duration, educational or training activities (McGreal & Olcott, 2022, p. 3). Digital badges provide the visual and verifiable representation of these achievements, allowing learners to demonstrate their competencies to employers, educational institutions, and professional bodies through secure digital platforms (Clements et al., 2020; Francia et al., 2023; Jirgensons & Kapenieks, 2018; Kiiskilä et al., 2023).

The rapid growth of micro-credentials reflects changing workforce demands, technological advancement, and the increasing need for lifelong learning (Ashizawa et al., 2024; Braxton, 2023; Tamoliūnė et al., 2023). As industries evolve more rapidly than traditional qualification systems can often respond, employers increasingly seek evidence of current, job-specific competencies rather than relying solely on formal qualifications (Diab & Zhang, 2023; Kiiskilä et al., 2023; Tamoliūnė et al., 2023). Micro-credentials allow educational institutions, industry organisations, professional associations, and employers to recognise emerging skills in areas such as cybersecurity, artificial intelligence, project management, sustainability, leadership, data analytics, healthcare, advanced manufacturing, and digital literacy (J. Graebe, 2025; Oliver & Brabin, 2026; Tamoliūnė et al., 2023). They are now widely used across higher education, vocational education and training, workplace learning, professional development, and continuing education (Ashizawa et al., 2024; McGreal & Olcott, 2022; Šarčević, 2024; Ratna Selvaratnam & Michael Sankey, 2025). As Tamoliūnė et al. (2023) note, most learners pursue micro-credentials for work-related purposes, leading to increasing expectations for higher education institutions to offer more competence-based and work-related curricula. Similarly, the IDB has issued over 200,000 badges as of January 2023, reflecting the scale of adoption (Porto & Presant, 2023).

Although the terms are often used together, micro-credentials and digital badges represent different concepts (Clements et al., 2020; McGreal & Olcott, 2022). A micro-credential is the formal recognition awarded after successfully demonstrating defined learning outcomes or competencies through assessment (Diab & Zhang, 2023; McGreal & Olcott, 2022; Porto & Presant, 2023). A digital badge is the digital artefact that represents that achievement (Clements et al., 2020; D. Griffiths et al., 2024;

Swanepoel, 2024). As Fields defined, a digital badge is "a digital signifier of accomplishments, skills, qualities, or experiences" (D. Griffiths et al., 2024, p. 91). Digital badges commonly contain embedded metadata that identifies the issuing organisation, the competencies achieved, assessment criteria, date of issue, expiration (if applicable), and methods for verifying authenticity (Bruno & Morgado, 2022; Clements et al., 2020; Swanepoel, 2024). The Open Badges Infrastructure calls for badges to be "formatted as images enriched with metadata (e.g., issuing organization, badge description, badge requirements, submitted evidence, standards, endorsements)" that allow verification and insight into the skills possessed (Clements et al., 2020, p. 155). This embedded information enables employers and other stakeholders to verify credentials electronically, increasing transparency and trust in the recognition of skills (Bruno & Morgado, 2022; Francia et al., 2023; Kiiskilä et al., 2023). Kiiskilä et al. (2023, p. 166) argue that "the inclusion of these metadata in digital credentials will increase their value beyond that of paper credentials" (Kiiskilä et al., 2023), while blockchain-based approaches have been explored to create permanent, secure verification infrastructure that prevents badges from becoming invalid when issuers cease hosting them (Jirgensons & Kapenieks, 2018; Panagiotidis, 2022).

One of the defining characteristics of micro-credentials is their flexibility. They allow learners to undertake targeted learning that directly addresses immediate professional or personal development needs without enrolling in lengthy qualification programs (Oliver & Brabin, 2026; Темјановски et al., 2023). Many micro-credentials are delivered online or through blended learning, enabling learners to study while remaining employed (Kumar et al., 2022; Oliver & Brabin, 2026). Their modular structure also supports the progressive accumulation of competencies, with some educational institutions allowing multiple micro-credentials to contribute towards larger qualifications through formal credit or recognition arrangements (Bean et al., 2023; J. Graebe, 2025; Selvaratnam & Sankey, 2021; West & Cheng, 2022).

Micro-credentials are increasingly aligned with competency-based education because learners are recognised for demonstrating measurable capabilities rather than simply completing a specified period of study (Braxton, 2023; Porto & Presant, 2023). Assessment typically focuses on authentic demonstrations of competence through workplace projects, portfolios, simulations, practical assessments, case studies, or performance tasks (Orman et al., 2023; Porto & Presant, 2023). This emphasis on demonstrated capability strengthens the relationship between education and workplace performance while providing employers with clearer evidence of specific skills (Brown, Mhichíl, et al., 2021; Perkins & Pryor, 2025).

Micro-credentials and digital badges are not learning theories but credentialing approaches that can support a wide range of educational philosophies (Brown, Mhichíl, et al., 2021; West & Cheng, 2022). Their effectiveness depends upon how learning experiences and assessments are designed rather than the credential itself (Orman et al., 2023; Rísquez et al., 2020).

Constructivism is reflected when learners earn micro-credentials by solving authentic problems, completing workplace projects, conducting investigations, or developing practical solutions that require them to actively construct knowledge (Porto & Presant, 2023; West & Cheng, 2022). Rather than simply recalling information, learners demonstrate their understanding through meaningful application in realistic contexts (Porto & Presant, 2023).

Social constructivism is evident when micro-credential programs incorporate collaborative projects, peer assessment, professional discussion, mentoring, or participation in learning communities (Kumar et al., 2022; Rísquez et al., 2020). Learners develop knowledge through interaction with educators, peers, and workplace experts while demonstrating competencies within authentic social and professional contexts (Rísquez et al., 2020).

Connectivism aligns particularly well with micro-credentials because learners often access knowledge through diverse digital networks, online communities, professional organisations, webinars, open educational resources, and industry experts (D. Griffiths et al., 2024; West & Cheng, 2022). The ability to locate, evaluate, and apply current information becomes a competency in itself, particularly within rapidly evolving industries (Brown, Mhichíl, et al., 2021; Темјановски et al., 2023).

Cognitivism informs the design of many micro-credential programs through carefully sequenced learning modules, multimedia instruction, scaffolded activities, retrieval practice, and formative assessment (Kumar et al., 2022; West & Cheng, 2022). Because micro-credentials usually focus on a narrow set of competencies, instructional design can target specific cognitive processes while minimising unnecessary cognitive load (West & Cheng, 2022).

Behaviourism is reflected through clearly defined learning outcomes, competency standards, structured practice, and criterion-referenced assessment (Virkus, 2019). Learners receive regular feedback as they progressively master specific skills, while digital badges provide visible recognition and reinforcement of successful achievement (Virkus, 2019; West & Cheng, 2022).

Humanism is evident in the learner-centred flexibility offered by micro-credentials. Learners choose credentials that align with their career aspirations, personal interests, or professional development goals, allowing them to direct their own learning pathways according to individual needs (West & Cheng, 2022; Темјановски et al., 2023). This autonomy supports self-directed learning, intrinsic motivation, and lifelong learning (Rísquez et al., 2020; Темјановски et al., 2023).

Experiential learning is particularly relevant because many micro-credentials require learners to demonstrate competence through authentic workplace activities, practical projects, simulations, or reflective practice (J. Graebe, 2025; Porto & Presant, 2023). Learning occurs through experience, followed by reflection, conceptualisation, and further application, rather than through theoretical study alone (Porto & Presant, 2023).

The educator's role in micro-credential programs extends beyond delivering content. Educators design learning experiences that align with clearly defined competencies, develop authentic assessment tasks, provide timely feedback, facilitate learner reflection, and verify that learners have demonstrated the required standards before awarding credentials (J. Lang, 2023; Raj et al., 2024). They also ensure that assessments remain valid, reliable, fair, and aligned with industry expectations (Abdullah & Ghazali, 2023; Donnelly & Maguire, 2020; Lau, 2021). In many cases, educators collaborate closely with employers, professional associations, or industry experts to ensure that micro-credentials remain relevant to current workforce needs (M. Brown et al., 2023; J. Lang, 2023; Lau, 2021; Ratna Selvaratnam & Michael Sankey, 2025).

Micro-credentials provide numerous educational and workforce benefits. They offer flexible, targeted learning opportunities that enable learners to develop specific competencies quickly and efficiently while balancing employment and personal commitments (Abdullah & Ghazali, 2023; Brown, Mhichíl, et al., 2021; Orman et al., 2023) . Their shorter duration allows educational institutions to respond rapidly to emerging industry requirements, ensuring that learning remains current and relevant (McGreal & Olcott, 2022; Ward et al., 2023). Because they recognise demonstrated competence, micro-credentials provide employers with more detailed evidence of workforce capabilities than broad qualification titles alone (Alasmari, 2024; Braxton, 2023; Ratna Selvaratnam & Michael Sankey, 2025). Digital badges further enhance portability by allowing learners to securely share verified achievements through professional networking platforms, digital portfolios, and employment applications (Mitchell et al., 2025; Porto & Presant, 2023; Rísquez et al., 2020).

Micro-credentials also support lifelong learning by encouraging individuals to continually update their knowledge and skills throughout their careers (Abdullah & Ghazali, 2023; Brown, Mhichíl, et al., 2021; J. Lang, 2023). Rather than viewing education as a one-time event, learners can progressively build portfolios of verified competencies that reflect changing technologies, professional responsibilities, and career aspirations (Ashizawa et al., 2024; Braxton, 2023; Ratna Selvaratnam & Michael Sankey, 2025).

Despite their growing popularity, micro-credentials also present several challenges. Their quality, assessment standards, and recognition vary considerably between providers, making it difficult for employers and educational institutions to compare credentials issued by different organisations (McGreal et al., 2022; Oliver & Brabin, 2026; Raj et al., 2024). The absence of universally accepted standards in some jurisdictions can create uncertainty regarding the value and transferability of individual micro-credentials (Abdullah & Ghazali, 2023; Lau, 2021; Raj et al., 2024).

Another challenge involves ensuring that narrowly focused credentials do not fragment learning into isolated skills without developing broader conceptual understanding or professional capability (Mitchell et al., 2025; Raj et al., 2024). Educational providers must therefore balance highly specialised competencies with opportunities for learners to integrate knowledge across disciplines (Donnelly & Maguire, 2020; Ward et al., 2023). Questions also remain regarding credit transfer, long-term recognition, quality assurance, assessment validity, credential verification, and the sustainability of digital credentialing platforms (Abdullah & Ghazali, 2023; Kumar et al., 2022; McGreal et al., 2022).

Micro-credentials and digital badges exemplify theory-informed teaching because they enable educators to recognise learning that has been achieved through diverse pedagogical approaches rather than prescribing a single instructional method (Braxton, 2023; West & Cheng, 2022). Whether learners acquire competencies through constructivist projects, collaborative inquiry, workplace experience, online learning, simulations, reflective practice, or structured instruction, the credential recognises demonstrated achievement against clearly defined standards (Mitchell et al., 2025; Porto & Presant, 2023; Ward et al., 2023). By supporting personalised learning pathways, authentic assessment, learner autonomy, collaboration, and lifelong learning, micro-credentials and digital badges provide a flexible mechanism for recognising adult learning across higher education, vocational education and training, workplace learning, and continuing professional development (M. Brown et al., 2023; Brown, Mhichíl, et al., 2021; Orman et al., 2023).

The following examples illustrate how micro-credentials and digital badges could be incorporated into both higher education and adult secondary education. Both examples aim to demonstrate how targeted credentials can recognise the progressive acquisition of knowledge, practical skills, and professional capabilities without requiring learners to wait until the completion of an entire qualification. In each example, learners accumulate a portfolio of verified achievements that document specific competencies while supporting motivation, lifelong learning, and employability. These examples also demonstrate that micro-credentials can recognise both academic achievement and transferable skills across diverse educational settings.

A university offering a Bachelor of Science with a major in Virology could embed micro-credentials throughout the degree to recognise mastery of specialised scientific knowledge and laboratory competencies. Rather than awarding recognition only upon graduation, students could progressively earn digital badges as they demonstrate proficiency in key areas of virological science.

During the early stages of the program, learners might earn micro-credentials in Laboratory Safety and Biosecurity, Scientific Microscopy, Introduction to Cell Biology, Basic Molecular Biology Techniques,

Scientific Data Recording, and Research Ethics. These credentials would recognise foundational competencies required before undertaking more advanced laboratory work.

As students progress, additional micro-credentials could acknowledge competencies such as Virus Classification and Taxonomy, Viral Genetics, Polymerase Chain Reaction (PCR) Techniques, Cell Culture Techniques, Diagnostic Virology, Immunology for Virologists, Electron Microscopy Interpretation, Laboratory Quality Assurance, Genomic Sequencing, and Bioinformatics for Viral Analysis.

Senior students undertaking research projects could earn advanced micro-credentials in Research Project Design, Scientific Literature Review, Statistical Analysis for Biological Research, Laboratory Leadership, Outbreak Investigation, Emerging Infectious Diseases, Vaccine Development Principles, Public Health Surveillance, Scientific Communication, and Research Publication Preparation.

Each completed micro-credential would be represented by a digital badge containing verifiable metadata identifying the issuing university, learning outcomes achieved, assessment methods, date of completion, and evidence demonstrating competence. Graduates would therefore possess not only a university degree but also a detailed digital portfolio documenting the specific laboratory, research, and professional skills developed throughout their studies.

As a second example, an adult secondary education provider preparing learners for senior secondary certification could similarly use micro-credentials to recognise incremental academic achievement and essential employability skills. Because adult learners often balance education with employment, family responsibilities, and other commitments, recognising smaller milestones can provide regular feedback, increase motivation, and acknowledge meaningful progress toward larger educational goals.

Within English, learners could earn micro-credentials in Academic Reading Strategies, Essay Writing, Critical Thinking, Information Literacy, Research Skills, Public Speaking, Digital Communication, and Workplace Writing.

Mathematics micro-credentials might include Numeracy Foundations, Algebra Fundamentals, Financial Mathematics, Statistical Reasoning, Measurement and Geometry, Spreadsheet Calculations, and Data Interpretation. Science learning could be recognised through badges in Scientific Investigation, Environmental Science, Human Biology, Health Literacy, Laboratory Safety, and Evidence-Based Decision Making.

In addition to academic subjects, the program could recognise broader capabilities that support further education and employment. Learners might earn micro-credentials in Digital Literacy, Artificial Intelligence Literacy, Career Planning, Time Management, Problem Solving, Teamwork, Communication Skills, Workplace Health and Safety, Intercultural Communication, Financial Literacy, and Learning to Learn.

Learners preparing for tertiary study could also complete micro-credentials in Academic Integrity, Referencing Skills, Independent Study Skills, Online Learning Readiness, and University Preparation. Those intending to enter the workforce directly could instead pursue credentials in Resume Preparation, Interview Skills, Customer Service, Workplace Communication, and Professional Behaviour.

Each achievement would be recognised through a digital badge that could be displayed within an electronic portfolio or shared with prospective employers and educational institutions. As learners accumulate multiple badges, they progressively build evidence of both academic capability and transferable employability skills, providing a richer and more detailed record of achievement than a final secondary school certificate alone.

Together, these examples demonstrate that micro-credentials and digital badges can recognise learning across very different educational contexts. In higher education they acknowledge progressively advanced disciplinary knowledge and professional competencies, while in adult secondary education they recognise foundational academic achievement, transferable skills, and readiness for further study or employment. In both settings, they complement rather than replace traditional qualifications by providing verified evidence of specific capabilities acquired throughout the learning journey.

Artificial Intelligence in Adult Education

Artificial intelligence (AI) refers to computer systems capable of performing tasks that normally require human intelligence, including recognising patterns, analysing data, generating content, solving problems, understanding natural language, making recommendations, and supporting decision-making (Brustenga et al., 2018; Mathew et al., 2021). In adult education, AI is increasingly being integrated into teaching, learning, assessment, administration, and learner support (Adıgüzel et al., 2023; Pokrivčáková, 2019). Rather than replacing educators, AI has the potential to enhance educational experiences by providing personalised learning, automating routine tasks, supporting accessibility, and delivering timely feedback that enables educators to focus more on teaching, mentoring, and facilitating learning (Vera, 2024).

The rapid adoption of artificial intelligence in education reflects broader technological developments and changing workforce expectations (Roméro et al., 2024). As digital technologies continue to transform industries, adult learners increasingly require both AI literacy and opportunities to learn with AI-enabled tools (Ngo, 2023). Educational institutions, vocational education providers, universities, and workplace training organisations are therefore incorporating AI into curricula while also using AI to improve the design and delivery of education (Adıgüzel et al., 2023; Roméro et al., 2024). Consequently, AI is becoming both a subject of learning and a tool that supports the learning process itself (Ngo, 2023).

Artificial intelligence encompasses a wide range of technologies. Machine learning enables systems to identify patterns within data and improve performance over time. Natural language processing allows AI systems to understand and generate human language, making conversational assistants and intelligent tutoring systems possible (Brustenga et al., 2018; Mathew et al., 2021; Pokrivčáková, 2019). Computer vision enables AI to interpret images and video, while generative AI can create text, images, audio, computer code, and other digital content in response to user prompts, (Adıgüzel et al., 2023). Learning analytics, adaptive learning systems, intelligent tutoring systems, automated assessment, recommendation engines, speech recognition, and predictive analytics all represent practical applications of artificial intelligence within adult education (Adıgüzel et al., 2023; Vera, 2024).

One of the defining characteristics of AI in education is its ability to personalise learning experiences (Vera, 2024). Traditional educational programs often present identical content to all learners regardless of their prior knowledge, learning pace, or individual needs. AI-enabled systems can analyse learner performance and recommend additional resources, adjust the difficulty of learning activities, identify misconceptions, and provide targeted feedback that supports individual learning pathways (Adıgüzel et al., 2023; Mathew et al., 2021). This personalised approach allows learners to progress at an appropriate pace while receiving support tailored to their developing competencies (Vera, 2024).

Artificial intelligence also supports accessibility and inclusion (Roméro et al., 2024). Speech-to-text and text-to-speech technologies assist learners with hearing or vision impairments, while automated translation supports multilingual learning environments (Pokrivčáková, 2019; Roméro et al., 2024). AI-powered summarisation, reading assistance, caption generation, and personalised study recommendations can reduce barriers for learners with diverse educational backgrounds, language abilities, or learning needs (Roméro et al., 2024). These capabilities contribute to more equitable access to educational opportunities across diverse adult learner populations.

Assessment practices are also being influenced by artificial intelligence. AI systems can automatically mark objective assessments, provide immediate formative feedback, analyse learner responses, identify patterns of misunderstanding, and generate performance reports for educators (Allen et al., 2026; Nieto-Cardenas, 2025; Reihanian et al., 2025). Some AI applications assist with creating assessment questions, developing marking rubrics, or identifying opportunities for remediation (Coşgun, 2025; Symeou et al., 2025). However, high-stakes assessment and decisions regarding learner achievement continue to require professional judgement and appropriate human oversight to ensure fairness, validity, transparency, and academic integrity (Bearman & Ajjawi, 2023; Maio, 2026; Reihanian et al., 2025).

Artificial intelligence is transforming the role of educators rather than eliminating it. Many routine administrative tasks, including preparing quizzes, organising learning materials, responding to frequently asked questions, analysing learner engagement, and generating reports, can now be supported by AI (Nieto-Cardenas, 2025; Symeou et al., 2025). This allows educators to devote more time to designing authentic learning experiences, facilitating discussion, mentoring learners, providing pastoral support, and developing higher-order thinking skills (Bearman & Ajjawi, 2023; Mei et al., 2026). At the same time, educators require new competencies in AI literacy, ethical decision-making, prompt design, digital assessment, and critical evaluation of AI-generated outputs ((Butler-Ulrich et al., 2024; Coşgun, 2025; Qian, 2025; Symeou et al., 2025; Woo et al., 2025).

Theoretical Foundations

Constructivism is reflected when learners use AI as a tool to investigate complex problems, explore multiple perspectives, analyse evidence, generate hypotheses, and develop solutions. Rather than accepting AI-generated responses uncritically, learners actively evaluate information, compare alternative viewpoints, and construct their own understanding through inquiry and reflection.

Social constructivism is supported when AI facilitates collaborative learning. Learners may use AI to generate ideas before participating in group discussions, collaborate on projects, critique AI-generated outputs collectively, or engage in peer review activities. Knowledge continues to be constructed through interaction between learners, educators, and digital tools rather than through isolated individual study.

Connectivism aligns particularly closely with artificial intelligence because AI enables learners to navigate increasingly complex digital knowledge networks. AI-powered search tools, recommendation systems, digital libraries, professional communities, and intelligent assistants help learners locate, evaluate, synthesise, and apply information from diverse sources while developing skills in digital information management.

Cognitivism is reflected through AI systems that provide scaffolded learning, adaptive instruction, retrieval practice, personalised feedback, and carefully sequenced learning activities. By analysing learner performance, AI can recommend resources that reinforce memory, support conceptual understanding, and reduce unnecessary cognitive load.

Behaviourism remains evident in AI-supported practice activities that provide immediate feedback, adaptive questioning, automated reinforcement, mastery learning pathways, and progress monitoring. Learners receive repeated opportunities to practise skills while AI systems reinforce correct responses and identify areas requiring additional practice.

Humanism is supported when AI enables greater learner autonomy and self-directed learning. Adult learners can choose when and how to engage with AI tools, seek additional explanations, request alternative examples, and pursue personalised learning pathways that align with their individual goals, interests, and professional aspirations. The educator remains responsible for fostering ethical decision-making, critical reflection, and personal growth rather than simply delivering information.

Experiential learning is strengthened through AI-powered simulations, virtual laboratories, scenario-based learning, workplace problem-solving, and reflective activities. Learners can practise complex tasks within safe virtual environments, receive immediate feedback, reflect upon their decisions, and transfer their learning to authentic workplace contexts.

Artificial intelligence exemplifies theory-informed teaching because it provides educators with powerful technological capabilities that can be adapted to support a wide range of pedagogical approaches rather than prescribing a single method of instruction. Its educational value lies not in automating teaching but in enriching personalised learning, authentic assessment, collaboration, reflection, accessibility, and learner engagement. When integrated thoughtfully with constructivist inquiry, collaborative learning, connectivist networking, cognitivist instructional design, behaviourist reinforcement, humanistic learner-centred practice, and experiential learning, AI can enhance educational experiences while preserving the educator's essential role in fostering critical thinking, ethical judgement, professional competence, and lifelong learning. Through careful pedagogical design, artificial intelligence becomes a tool that complements human expertise, enabling educators to create more responsive, inclusive, and engaging learning environments for adult learners.

Role of the Educator

Within AI-enhanced learning environments, the educator's role remains central to ensuring that technology is used purposefully, ethically, and pedagogically (Gagné et al., 2026; Hefnawi, 2025). Rather than replacing professional expertise, artificial intelligence serves as another instructional tool that educators integrate into well-designed learning experiences (Haidar & Tassis, 2025; Khalid et al., 2026). Educators determine when and how AI should be incorporated into teaching (Haidar & Tassis, 2025; Hefnawi, 2025), evaluate the quality and accuracy of AI-generated content (Gagné et al., 2026;

Qin & Zhou, 2026), design authentic assessments that promote critical thinking (Gagné et al., 2026; Yan & Chau, 2025), establish clear expectations regarding appropriate AI use (Haidar & Tassis, 2025; Yan & Chau, 2025), and guide learners in developing the skills needed to use these technologies responsibly (Gagné et al., 2026; Yan & Chau, 2025). The effectiveness of AI in education therefore depends not on the technology itself but on the educator's ability to integrate it in ways that support meaningful learning and align with intended learning outcomes (Hefnawi, 2025; Khalid et al., 2026; Yu et al., 2025).

As AI becomes increasingly embedded within workplaces and society, educators also have a responsibility to develop learners' AI literacy (Hefnawi, 2025; Samuel, 2025; Yan & Chau, 2025). Adult learners need to understand how AI systems function (Dilek et al., 2025; Silva et al., 2024), recognise their capabilities and limitations (Dilek et al., 2025; Mikeladze et al., 2024), identify potential bias (Yan & Chau, 2025; Zhang & Rhee, 2025), verify AI-generated information using reliable sources (Gagné et al., 2026; Haidar & Tassis, 2025), protect personal and organisational data (Yan & Chau, 2025; Zhang & Rhee, 2025), and make informed ethical decisions when using AI in academic and professional contexts (Samuel, 2025; Yan & Chau, 2025; Zhang & Rhee, 2025). These capabilities are becoming increasingly important transferable skills that contribute to employability and lifelong learning in a rapidly evolving digital economy (Samuel, 2025; Silva et al., 2024; Yeter et al., 2024).

Benefits and Challenges

Artificial intelligence offers numerous benefits for adult education by enhancing both teaching and learning. AI can provide personalised learning experiences that adapt to individual learner needs, deliver immediate feedback, improve accessibility through assistive technologies, and support flexible learning pathways that accommodate diverse educational backgrounds and learning preferences, (Fathahillah et al., 2023; Khatri & Karki, 2023). Intelligent learning systems can teach students using computing systems and provide immediate and personalised feedback, while AI analyses student knowledge and engagement strategies to restructure effective personalised learning pathways. Learners can access explanations, practice activities, language support, and additional learning resources whenever required, extending educational support beyond scheduled class times, (Alman, 2020). For educators, AI can assist with resource development, assessment design, learning analytics, and the identification of learners who may require additional academic support (Khatri & Karki, 2023; Pazieieva et al., 2024), allowing more time to be devoted to facilitation, mentoring, and learner engagement. AI has the potential to transform assessment practices by automating grades and providing personalised feedback, saving teachers time while maintaining fairness and accuracy (Alamin & Sauri, 2024; Pazieieva et al., 2024).

Artificial intelligence also strengthens lifelong learning by providing continuous access to educational support throughout an individual's career (Altınay et al., 2024). As industries evolve and new knowledge emerges, adult learners can use AI-enabled technologies to update existing skills, explore emerging fields, practise new competencies, and receive personalised guidance that supports ongoing professional development (Altınay et al., 2024). AI revolutionises the notion of lifelong learning by providing tailored experiences outside of the classroom, allowing people to continuously learn new skills throughout their lifetimes. This capacity for continuous learning enables individuals to remain adaptable and responsive within increasingly dynamic workplaces (Altınay et al., 2024).

Despite its considerable potential, the integration of artificial intelligence into adult education presents significant challenges. AI systems are capable of generating inaccurate, misleading, or fabricated information, and they may reproduce biases embedded within the data on which they have been trained (Alamin & Sauri, 2024; Khatri & Karki, 2023; Kumar et al., 2024). Because AI-generated responses often appear authoritative regardless of their accuracy, learners must develop strong critical thinking skills and information literacy to evaluate AI outputs rather than accepting them uncritically (Khatri & Karki, 2023).

Artificial intelligence also raises important questions regarding academic integrity, assessment, privacy, and ethics (Khatri & Karki, 2023; Kumar et al., 2024). The integration of AI in higher education raises concerns about academic integrity and ethical issues, as it has the potential to lead to plagiarism, impede critical thinking, suppress creativity, and erode originality in teaching, research, and scholarship (Khatri & Karki, 2023). Educators must design assessments that encourage authentic demonstration of learning while clearly defining appropriate and inappropriate uses of AI (Khatri & Karki, 2023; Kumar et al., 2024). Additional concerns include data privacy, cybersecurity, intellectual property, algorithmic bias, transparency in automated decision-making, unequal access to digital technologies, varying levels of AI literacy, and the environmental impacts associated with the computational resources required to operate large-scale AI systems (Alamin & Sauri, 2024; Kumar et al., 2024). Data privacy and security remain key issues, as AI systems often require access to large amounts of data, including personal and sensitive information (Pazieieva et al., 2024). There is also a risk that excessive dependence on AI may reduce opportunities for independent reasoning, creativity, problem-solving, and interpersonal communication if technology is used without sound educational purpose (Khatri & Karki, 2023). Concerns have been raised that over-reliance on AI may result in a passive learning style, undervaluing students' critical thinking and problem-solving abilities,. Consequently, effective AI integration requires careful instructional design that ensures technology enhances, rather than diminishes, meaningful learning and human interaction (Altınay et al., 2024; Khatri & Karki, 2023; Kumar et al., 2024).

Successfully integrating artificial intelligence into adult education requires a balanced approach that combines sound pedagogy, institutional governance, and ongoing professional development. Rather than treating AI as either a solution to all educational challenges or a technology to be avoided, educators and institutions should adopt evidence-informed practices that place learning outcomes, critical thinking, and learner development at the centre of instructional design.

One of the most effective strategies is to embed AI literacy within adult education programs. Learners should be taught how to use AI critically and responsibly by questioning outputs, evaluating evidence, cross-checking information with authoritative sources, recognising potential bias, and understanding the limitations of AI systems. By treating AI-generated responses as starting points for investigation rather than definitive answers, learners develop the analytical and evaluative skills required for both academic study and professional practice.

Educational institutions should also establish clear policies and guidelines governing the use of AI. These should outline acceptable uses of AI within learning and assessment, expectations regarding transparency and attribution, responsibilities relating to privacy and confidentiality, and procedures for maintaining academic integrity. Consistent institutional guidance helps educators and learners understand how AI can be used ethically while reducing uncertainty and promoting responsible practice.

Assessment practices should evolve to reflect the presence of AI within education and the workplace. Rather than relying heavily on tasks that AI can readily complete, educators can emphasise authentic assessment requiring learners to demonstrate application, analysis, reflection, communication, practical competence, and professional judgement. Activities that involve workplace projects, collaborative problem-solving, presentations, simulations, portfolios, and supervised practical tasks make it more difficult to rely inappropriately on AI while better reflecting the capabilities required in professional environments.

Professional development is equally important for educators. Ongoing training enables teachers to understand emerging AI technologies, evaluate their educational value, identify potential risks, and integrate them effectively into teaching practice. Building educator confidence also supports more consistent implementation across programs and enables educators to model ethical, informed, and reflective use of AI for their learners.

Institutions should adopt robust governance arrangements to manage issues relating to privacy, cybersecurity, intellectual property, and algorithmic transparency. Selecting reputable AI platforms, implementing appropriate data protection measures, providing guidance on the handling of sensitive information, and regularly reviewing institutional AI policies all contribute to creating a safer and more trustworthy learning environment. Governance should also include periodic evaluation of AI systems to identify potential bias or unintended impacts on learners.

Ensuring equitable access is another important consideration. Educational providers can reduce barriers by offering institutionally supported AI tools, providing access to suitable devices and internet connectivity where possible, and ensuring that AI-enhanced learning activities remain accessible to learners with diverse abilities, educational backgrounds, and technological resources. Inclusive implementation helps prevent AI from widening existing educational inequalities.

Artificial intelligence should also be integrated in ways that complement rather than replace human interaction. Learning activities should continue to promote discussion, collaboration, mentoring, reflection, and interpersonal communication alongside the use of AI tools. Maintaining meaningful relationships between educators and learners preserves the social and collaborative dimensions of adult learning that remain essential for developing professional judgement, ethical reasoning, and lifelong learning skills.

Finally, successful AI integration requires continuous monitoring and evaluation. As AI technologies continue to evolve, educational institutions should regularly review their policies, teaching practices, assessment methods, and technological infrastructure to ensure they remain effective, ethical, and aligned with educational objectives. Ongoing consultation with educators, learners, industry, and professional bodies can further support continuous improvement and ensure that artificial intelligence remains a tool that enhances, rather than compromises, the quality of adult education.

Virtual and Augmented Reality in Adult Education

Virtual reality (VR) and augmented reality (AR) are immersive technologies that are transforming the ways adult learners engage with knowledge, develop practical skills, and participate in authentic learning experiences (Ding et al., 2023; Ghosh & Ravichandran, 2024; Ziker et al., 2021). Both technologies extend traditional teaching methods by enabling learners to interact with digital information in ways that would be difficult, expensive, dangerous, or impossible in conventional educational settings (Ding et al., 2023; Xu et al., 2019). Although they are often discussed together, virtual reality and augmented reality differ in how they present digital content and how learners interact with their environments (Joshi & Pramod, 2023; Iatsyshyn et al., 2020).

Virtual reality creates a fully computer-generated, three-dimensional environment that immerses learners within a simulated world using head-mounted displays, motion controllers, and increasingly, haptic technologies (Barteit et al., 2021; Dicheva et al., 2023). When using VR, learners are separated from their physical surroundings and interact entirely within a virtual environment (Joshi & Pramod, 2023; Iatsyshyn et al., 2020). These environments may replicate real workplaces, laboratories, construction sites, hospitals, aircraft cockpits, or industrial facilities, allowing learners to practise procedures, solve problems, and develop technical skills within realistic but controlled settings (Ding et al., 2023; Doolani et al., 2020; Ghosh & Ravichandran, 2024; Xu et al., 2019).

Augmented reality, by contrast, enhances rather than replaces the physical environment (Joshi & Pramod, 2023; Kuleto et al., 2021; Iatsyshyn et al., 2020). Using smartphones, tablets, smart glasses, or head-mounted displays, AR overlays digital information such as text, images, animations, three-dimensional models, or interactive instructions onto real-world objects and locations (Doolani et al., 2020; Ryoo & Winkelmann, 2021; Ward et al., 2024). Learners remain aware of their physical surroundings while receiving additional contextual information that supports learning, problem-solving, and task performance (Ahmad et al., 2024; Krishnan & Poorani, 2022). This integration of digital and physical environments makes AR particularly valuable for workplace training, equipment maintenance, healthcare, engineering, and technical education (Ahmad et al., 2024; Doolani et al., 2020; Farrell, 2018; Ghosh & Ravichandran, 2024).

The development of VR and AR has been driven by advances in computer graphics, mobile computing, artificial intelligence, sensors, networking, and wearable technologies (Anderson et al., 2010; Chhabhaiya et al., 2024; Partarakis & Zabulis, 2024). Early virtual reality systems were prohibitively expensive and largely confined to military, aviation, and research applications (Concannon et al., 2019; Jagatheesaperumal et al., 2022). Improvements in computing power and reductions in hardware costs have made immersive technologies increasingly accessible to educational institutions and industry (Júnior et al., 2023; Tang & Ho, 2020). Today, VR and AR are widely used in higher education (Concannon et al., 2019; Gudonienė & Rutkauskienė, 2019), vocational education and training (He & Li, 2024; Partarakis & Zabulis, 2024), professional development (Jagatheesaperumal et al., 2022), healthcare education (Chhabhaiya et al., 2024; González-Argote et al., 2023), emergency services (González-Argote et al., 2023), defence (Chhabhaiya et al., 2024; Concannon et al., 2019), manufacturing (Tang & Ho, 2020), engineering (Sik-Lányi, 2014), mining, and workplace safety training (Kaarlela et al., 2022; Partarakis & Zabulis, 2024).

One of the defining characteristics of virtual and augmented reality is their capacity to create highly authentic learning environments. Learners can practise complex procedures, operate specialised equipment, respond to emergencies, and develop technical skills without exposing themselves, other people, or expensive equipment to unnecessary risk. Simulated environments can also be repeated multiple times, allowing learners to refine their performance, build confidence, and learn from mistakes in a safe setting before transferring their skills to real workplaces.

Another important characteristic is the ability to visualise concepts that are difficult to observe directly. Three-dimensional anatomical structures, molecular interactions, engineering systems, architectural designs, geological formations, or industrial processes can be explored interactively, allowing learners to manipulate objects, examine relationships, and develop spatial understanding that would be difficult to achieve using textbooks or two-dimensional images alone. In augmented reality, these visualisations can be superimposed directly onto physical equipment or environments, providing contextual guidance during practical tasks.

Virtual and augmented reality also support experiential learning by allowing learners to engage actively with realistic scenarios rather than passively receiving information (Suh et al., 2023), (Λαμπρόπουλος et al., 2024), (Chahartangi et al., 2026). Learners make decisions, observe the consequences of their actions, receive immediate feedback, and refine their understanding through repeated practice (Chahartangi et al., 2026), Calandrino et al., 2022; , Adnan et al., 2023). These immersive experiences encourage deeper engagement, improve knowledge retention, and support the development of both technical and non-technical skills, including communication, teamwork, leadership, and decision-making (Suh et al., 2023), (Λαμπρόπουλος et al., 2024), (Chahartangi et al., 2026), Tene et al., 2024).

Specifically, immersive learning environments involve using virtual and augmented reality technologies to create more engaging and effective learning experiences that can simulate real-world scenarios and provide interactive feedback (Suh et al., 2023). VR and AR can effectively support experiential learning, as their ability to provide safe virtual learning environments in which individuals can practice and develop their skills and knowledge in real-life simulated working settings significantly contributes to learning outcomes (Λαμπρόπουλος et al., 2024). Virtual patients enable students to engage in diagnostic reasoning, clinical decision-making, and self-directed learning through interactive case simulations, assessments, and immediate feedback (Chahartangi et al., 2026). AR-based systems have been shown to strengthen cognitive and psychomotor skills, support knowledge retention, and foster teamwork and decision-making abilities (Chahartangi et al., 2026). Furthermore, metaverse technology facilitates collaboration and small group activities, improves mentor–mentee interactions, provides opportunities for self-directed learning experiences, and helps develop teamwork skills (Suh et al., 2023). VR and AR technologies also support the development of non-technical competencies, as VR can simulate social interactions for training in leadership development and empathy training (Pireddu et al., 2025), and their ability to support social and collaborative learning and offer experiential learning has significantly contributed to learning outcomes in safety training contexts (Λαμπρόπουλος et al., 2024).

Theoretical Foundations

Constructivism is strongly reflected in VR and AR because learners actively construct knowledge through exploration, experimentation, and interaction with realistic environments (Hajirasouli & Banihashemi, 2022; Marougkas et al., 2023; Zhao et al., 2023). Rather than simply reading about concepts, learners investigate virtual scenarios, manipulate digital objects, solve problems, and develop understanding through firsthand experience (Shihab et al., 2023; Vedpathak & Mithari, 2024). As Marougkas et al. (2023, p. 17) note, "constructivism emphasizes the active participation of learners in constructing their own understanding of knowledge through hands-on experiences and interactions with the virtual environment". Hajirasouli and Banihashemi (2022, p. 12) similarly argue that "authentic learning happens when the newly received information is being assimilated with previously established and perceived knowledge".

Social constructivism is supported when immersive technologies enable collaborative learning. Multi-user virtual environments allow learners to work together, communicate, solve problems, and complete shared tasks despite being geographically separated (Oyekunle et al., 2024; Vedpathak & Mithari, 2024). VR "allows students to participate in immersive, interactive environments that foster social interaction and collaboration, providing opportunities for learners to engage in dialogue, negotiate meaning and co-create knowledge in real time" (Marougkas et al., 2023, p. 12). Educators facilitate discussion, reflection, and peer learning as learners negotiate meaning through shared experiences (Bondin et al., 2026; Hajirasouli & Banihashemi, 2022).

Connectivism aligns with VR and AR through their integration with broader digital learning ecosystems. As Wu et al. observe, AR and VR capabilities "resonate not only with Vygotsky's (1978) SCT and zone of proximal development but also with connectivism," emphasising "how technology creates new opportunities for interactive and collaborative learning within networked environments" (J. G. Wu et al., 2025, p. 3). Virtual environments may incorporate online resources, real-time data, expert guidance, digital libraries, collaborative platforms, and intelligent tutoring systems, enabling learners to connect immersive experiences with wider networks of information and professional practice (Marougkas et al., 2023; Vedpathak & Mithari, 2024).

Cognitivism informs the instructional design of immersive learning experiences by organising scenarios into logical sequences that progressively increase in complexity (Marougkas et al., 2023; Oyekunle et al., 2024). Scaffolding, guided practice, visualisation, feedback, and repetition support information processing, memory formation, and the transfer of knowledge from simulated environments to real-world situations (Bondin et al., 2026; Hsu & Liu, 2024). A guiding principle for creating successful XR educational experiences is "the incorporation of learning theories into the design process" (Oyekunle et al., 2024, p. 55).

Behaviourism is evident through repeated practice, immediate feedback, performance monitoring, and reinforcement (Marougkas et al., 2023). Learners develop procedural skills by practising tasks repeatedly within virtual environments, receiving corrective feedback, and refining their performance until competency is achieved (Botha et al., 2025; Oyekunle et al., 2024). AR-based learning environments have incorporated rewards systems and specific objectives for students to achieve, making them powerful tools for skill development (Botha et al., 2025).

Humanism is reflected in the learner-centred flexibility provided by immersive technologies (Marougkas et al., 2023; Vedpathak & Mithari, 2024). Learners can often progress at their own pace, repeat activities as required, explore areas of personal interest, and develop confidence before performing similar tasks in real workplaces (Shihab et al., 2023; Vedpathak & Mithari, 2024). These opportunities promote autonomy, self-directed learning, and personal growth (Hajirasouli & Banihashemi, 2022; J. G. Wu et al., 2025).

Experiential learning is perhaps the theory most closely associated with virtual and augmented reality (Hajirasouli & Banihashemi, 2022; Marougkas et al., 2023; Zhao et al., 2023). Immersive technologies provide concrete experiences that learners subsequently reflect upon, analyse, conceptualise, and apply to future situations, following Kolb's cycle of "concrete experience, reflective observation, abstract conceptualization and active experimentation" (Amanatidis, 2022; Marougkas et al., 2023; Meany et al., 2024). By safely simulating authentic workplace challenges, VR and AR support the complete experiential learning cycle while allowing repeated practice and continuous improvement (Choi-Lundberg et al., 2023; Hajirasouli & Banihashemi, 2022). As Marougkas et al. conclude, experiential learning "is most appropriate for VR" among the learning theories examined (Marougkas et al., 2023, p. 1).

Role of the Educator

The educator's role within immersive learning environments extends beyond operating technology. Educators design authentic scenarios aligned with learning outcomes (Bondin et al., 2026; Zhang et al., 2026), prepare learners before entering virtual or augmented environments (Holovko, 2024), facilitate learning during immersive activities (Mystakidis et al., 2021; Taylor, 2015), encourage reflection following practical experiences (Elendu et al., 2024; Holovko, 2024), and connect simulated learning with theoretical concepts and workplace practice (Cheng, 2025; Zhang et al., 2026). They also evaluate learner performance (Elendu et al., 2024), provide constructive feedback (Elendu et al., 2024; Taylor, 2015), ensure safe and ethical technology use (Mystakidis et al., 2021; Zhang et al., 2026), and determine when immersive technologies offer genuine educational value rather than simply introducing novelty (Mystakidis et al., 2021).

Specifically, educators must competently operate immersive tools, interpret analytics, and conduct structured debriefing (Zhang et al., 2026). The instructor's active involvement during immersive activities encompasses multiple roles including monitoring, moderating, guiding, facilitating, providing hints, encouraging, and debriefing (Makri et al., 2021). The instructor should provide only guidance, instructions, and probing questions to maintain students' attention and encourage reflective thinking throughout the debriefing process (Holovko, 2024). In serious gaming and simulation contexts, the instructor as in-game facilitator adds complexity and dynamics, provides just-in-time information and meaningful feedback, and scaffolds learner performance (Taylor, 2015). VR and AR are not educational panaceas; they should be used for specific pedagogical purposes when their use can deliver tangible learning advantages (Mystakidis et al., 2021).

Educators must also prepare learners to transfer skills developed within virtual environments to real-world contexts (Ravichandran & Mahapatra, 2023; Taylor, 2015). Structured debriefing sessions, reflective discussions, workplace application activities, and guided feedback help learners interpret their experiences and consolidate learning (Elendu et al., 2024; Holovko, 2024; Taylor, 2015). Debriefing involves a guided discussion where learners reflect on their performance, identify areas for improvement, and receive constructive instructor feedback, helping to consolidate learning, reinforce best practices, and address gaps in knowledge or skills (Elendu et al., 2024). Studies have shown that debriefing significantly enhances learning outcomes and improves clinical performance (Elendu et al., 2024). The facilitator plays a critical role in the transfer process by guiding participants towards reflective practice, fostering connections between the simulated experience and workplace practices (Taylor, 2015). Hybrid curricula combining simulation training with structured placements promote progressive competency development (Zhang et al., 2026), while institutions that combine VR, simulation programs, and supervised live clinical experiences supported by adequate staffing and instructor training achieve the strongest educational outcomes (Zhang et al., 2026). Assessment strategies in immersive environments must be adapted to capture both performance and reflection (Dumitrescu, 2025). In this way, the educator remains central to transforming immersive experiences into meaningful educational outcomes.

Technologies

Virtual and augmented reality technologies have expanded rapidly over the past decade (Elbert et al., 2023; Ortega-Rodríguez, 2022; Rodriguez et al., 2023; Zambrano et al., 2023), with a wide variety of hardware and software now available to support adult education across higher education, vocational education and training (VET), workplace learning, and professional development. The most appropriate platform depends on the intended learning outcomes, available resources, level of immersion required, and the complexity of the skills being developed (Freire, 2024). While some applications provide fully immersive virtual environments, others overlay digital information onto real-world settings (Aguirre-Herráez et al., 2020; Ortega-Rodríguez, 2022; Reyes et al., 2022) or create interactive three-dimensional experiences that support learning without requiring specialised equipment (Cabascango-Trávez, 2023).

Meta Quest is one of the most widely adopted virtual reality platforms in education due to its standalone wireless headsets, which operate without requiring connection to a personal computer. Its portability, relatively low cost, and extensive software ecosystem have made it a popular choice for educational institutions and industry training providers. Meta Quest supports immersive simulations, collaborative virtual environments, and practical skills development across a wide range of disciplines (Meta, 2025). Within adult education, the platform could be used to train emergency responders in hazardous incident management, enable engineering learners to inspect virtual industrial facilities, allow healthcare students to practise patient assessment and clinical procedures, or provide construction apprentices with realistic site safety simulations before entering operational workplaces.

Microsoft Mesh, used in conjunction with mixed reality devices such as Microsoft HoloLens, integrates virtual and augmented reality to create shared three-dimensional collaborative environments. Unlike fully immersive virtual reality, mixed reality enables users to interact with holographic digital objects while remaining aware of their physical surroundings (Microsoft, 2025). These capabilities are particularly valuable in engineering, architecture, manufacturing, and healthcare education, where learners could collaboratively inspect machinery, visualise Building Information Models (BIM), explore complex anatomical structures, or receive interactive maintenance instructions superimposed directly onto physical equipment.

ClassVR is an education-specific virtual reality platform designed to simplify the integration of immersive technologies into teaching. It combines standalone VR headsets with an extensive library of curriculum-aligned learning experiences and classroom management tools, enabling educators to deliver structured immersive lessons without requiring advanced technical expertise (Avantis Education, 2025). Adult education providers could employ ClassVR to simulate industrial workplaces, conduct virtual field excursions, recreate historical settings, or deliver workplace safety training in industries such as mining, logistics, agriculture, environmental management, and emergency services.

ENGAGE XR provides collaborative virtual environments in which educators and learners participate as avatars within realistic classrooms, laboratories, workplaces, and conference venues. The platform supports presentations, collaborative projects, practical demonstrations, meetings, and international training programs (Engage XR, 2025). Universities, training organisations, and corporate learning providers could use ENGAGE XR to facilitate remote laboratory classes, leadership development programs, engineering design reviews, international seminars, or collaborative workplace training involving participants located in different geographical regions.

VictoryXR specialises in immersive educational content for schools, universities, and professional education. Its virtual campuses, laboratories, classrooms, and simulated workplaces enable learners to undertake practical experiences that may otherwise require costly facilities or specialised equipment (VictoryXr, 2025). Adult learners studying science, healthcare, engineering, aviation, or technical trades could use VictoryXR to conduct laboratory experiments, explore biological systems, practise aircraft maintenance procedures, or develop technical competencies within safe and highly realistic virtual environments.

Labster has become one of the leading providers of virtual laboratory simulations for higher education, offering interactive learning experiences across biology, chemistry, physics, medicine, biotechnology, environmental science, and related disciplines (Labster, 2025). Rather than replacing physical laboratories, Labster enables learners to familiarise themselves with laboratory procedures, practise experimental techniques, investigate scientific concepts, and build confidence before undertaking practical laboratory work. This approach is particularly beneficial where access to laboratory facilities is constrained by cost, equipment availability, safety considerations, or geographical distance.

Google Arts & Culture provides a diverse collection of virtual reality and augmented reality experiences that enable learners to explore museums, heritage sites, artworks, historical artefacts, and natural environments from around the world (Google Arts and Culture, 2025). Adult learners studying history, archaeology, anthropology, tourism, museum studies, cultural heritage management, or geography could undertake virtual field trips that provide access to internationally significant collections and locations that may otherwise be inaccessible due to travel costs or physical distance.

Adobe Aero enables educators and learners to develop their own augmented reality experiences without requiring advanced programming knowledge. Using smartphones or tablets, digital objects, animations, instructional overlays, and interactive content can be integrated directly into real-world environments (Adobe, 2025). In vocational education, learners could develop augmented reality maintenance procedures, workplace safety demonstrations, equipment operating guides, or customer training resources, while higher education students might create interactive scientific visualisations, architectural models, or engineering prototypes.

ZapWorks and BlippAR are augmented reality authoring platforms that allow educators to create interactive learning materials by overlaying videos, animations, three-dimensional models, and digital instructions onto printed resources or physical equipment (BlippAr, 2025; ZapWorks, 2025). These platforms are particularly well suited to workplace learning and technical training, where learners can scan machinery, engineering drawings, or technical manuals to receive contextual, step-by-step guidance during maintenance, assembly, troubleshooting, or inspection activities. Such contextual learning supports just-in-time training while reducing reliance on printed documentation and improving procedural accuracy.

Across numerous industries, immersive technologies have been adapted to meet highly specialised training requirements by providing realistic, repeatable, and low-risk learning environments. Aviation has long been recognised as a pioneer in the use of simulation-based training, and many of the educational principles developed within aviation have subsequently been adopted by healthcare, defence, manufacturing, and emergency services. The use of simulation in healthcare education, in particular, has evolved from practices established in aviation and military training, where realistic environments allow learners to develop technical competence, decision-making skills, and procedural confidence before performing tasks in operational settings (Randazzo et al., 2023).

Healthcare is now one of the most significant users of virtual reality for education and professional development. Virtual surgical simulators allow medical practitioners to rehearse complex procedures within safe, controlled environments, enabling repeated practice without risk to patients while improving procedural accuracy, technical proficiency, and clinical confidence (Tomita et al., 2025). These technologies are used across a broad range of medical specialties, including laparoscopic surgery, endoscopy, neurosurgery, emergency medicine, and other clinical disciplines, where learners can progressively develop skills before undertaking supervised practice in real clinical environments (González-Argote et al., 2023).

Defence organisations have similarly embraced immersive technologies to prepare personnel for complex and high-risk operational environments. Virtual reality simulations have been shown to improve the transfer of knowledge and practical skills from training environments to real-world situations, particularly in the preparation of combat medics and civilian emergency responders who must perform effectively under pressure (Lembo et al., 2024). The use of realistic scenarios allows learners to develop technical competence, situational awareness, teamwork, and decision-making while avoiding the risks and costs associated with live operational training.

Advances in virtual modelling have also led to the development of digital twins and three-dimensional virtual reconstructions that support planning and rehearsal before undertaking real procedures. In healthcare, patient-specific anatomical models can be generated from medical imaging to allow clinicians to visualise complex anatomy, plan interventions, and practise procedures before surgery. These technologies enhance precision, improve preparation for complex cases, and contribute to safer clinical practice by enabling practitioners to identify potential challenges before entering the operating theatre (Hameed et al., 2022).

Conditions for Effective Integration

The educational effectiveness of virtual and augmented reality depends far more on their pedagogical integration than on the sophistication of the technology itself. Immersive technologies produce the greatest learning benefits when they are embedded within well-designed instructional frameworks that align with intended learning outcomes, provide appropriate guidance, and encourage active learner engagement. Research has shown that real-time immersive graphics and simulations can improve learning outcomes and clinical skill development compared with more traditional instructional approaches, although these benefits depend substantially on the quality of instructional design, the realism of the learning activities, and the way educators facilitate the learning experience (González-Argote et al., 2023).

Much of the educational value of VR and AR is grounded in experiential learning principles, which emphasise learning through authentic experience, repeated practice, immediate feedback, and structured reflection. Immersive technologies are particularly valuable when they provide access to environments or situations that would otherwise be difficult, costly, dangerous, or impossible to replicate in conventional educational settings. For example, healthcare education uses virtual environments to allow learners to practise responding to rare clinical events, emergency situations, and complex medical procedures that cannot be reproduced easily or safely with real patients (González-Argote et al., 2023). By enabling repeated practice within controlled environments, VR allows learners to refine technical skills, build confidence, and learn from mistakes without exposing patients, equipment, or personnel to unnecessary risk (González-Argote et al., 2023).

Virtual and augmented reality also enhance learning by enabling learners to visualise complex concepts that are difficult to understand through traditional teaching methods alone. Interactive three-dimensional models, virtual anatomical reconstructions, engineering prototypes, and digital twins provide opportunities to explore structures and systems from multiple perspectives, improving conceptual understanding and spatial reasoning (Hameed et al., 2022). Collaborative virtual environments further extend these opportunities by allowing geographically dispersed learners to work together on shared tasks, solve problems collectively, and participate in realistic simulations that mirror professional practice (Lembo et al., 2024; Żydowicz et al., 2024).

When combined with effective instructional design and skilled educator facilitation, immersive technologies have been associated with improvements in learner confidence, engagement, and practical competence. Studies have demonstrated that participation in virtual reality training can increase learners' confidence in performing complex tasks while also encouraging greater acceptance of immersive technologies as valuable educational tools (Tomita et al., 2025). However, reflection, mentoring, and guided debriefing remain essential components of the learning process. These activities help learners interpret their experiences, consolidate new knowledge, and transfer skills developed in simulated environments to authentic workplace situations. Research across healthcare and other professional disciplines has shown that immersive simulation can improve both technical performance and the application of skills in real-world practice when integrated into broader educational programs (Lembo et al., 2024; Randazzo et al., 2023).

Despite these benefits, successful implementation continues to depend upon addressing practical challenges. Financial investment in hardware and software, technical infrastructure, educator training, and the development of high-quality learning experiences remain significant considerations for many institutions. Studies have also identified infrastructure limitations, implementation costs, and the need for discipline-specific training as common barriers to wider adoption (Tomita et al., 2025). Consequently, the successful use of virtual and augmented reality depends not only on technological capability but also on careful planning, ongoing professional development, appropriate institutional support, and a clear focus on educational outcomes.

Benefits and Challenges

Virtual and augmented reality offer numerous educational benefits for adult learners. They provide safe environments for practising hazardous or high-risk procedures, reduce the costs associated with specialised equipment and consumable materials, increase learner engagement, and improve opportunities for repeated practice without adverse consequences. Immersive technologies can also support learners who are geographically dispersed by providing consistent training experiences regardless of location.

The ability to visualise complex concepts in three dimensions enhances conceptual understanding, particularly in disciplines requiring strong spatial reasoning, while realistic simulations allow learners to develop decision-making, communication, teamwork, and leadership skills under authentic conditions. VR and AR also support personalised learning by allowing learners to progress at their own pace and repeat scenarios until confidence and competence are achieved.

The successful implementation of virtual and augmented reality requires careful planning and investment. Hardware, software, maintenance, and content development can involve significant financial costs, particularly for highly specialised simulations. Institutions must also consider ongoing technical support, equipment upgrades, staff training, and integration with existing curricula.

Accessibility presents another challenge, as some learners may experience motion sickness, visual discomfort, fatigue, or difficulty using immersive technologies due to physical or cognitive limitations. Learning activities should therefore provide alternative pathways where appropriate and follow inclusive design principles to ensure equitable participation.

The educational value of immersive technologies also depends upon instructional quality. Poorly designed simulations that prioritise technological novelty over learning outcomes may distract learners or create unnecessary cognitive load. Effective implementation requires scenarios that are pedagogically purposeful, appropriately scaffolded, and integrated into broader learning programs rather than being used as isolated experiences.

Finally, educators must recognise that simulated experiences cannot fully replicate the complexity, unpredictability, and interpersonal dimensions of real workplaces. Immersive technologies should therefore complement, rather than replace, authentic workplace learning, supervised practical experience, and human interaction.

Virtual and Augmented Reality and Theory-Informed Teaching

Virtual and augmented reality exemplify theory-informed teaching because they provide educators with powerful technological environments that can support a wide range of pedagogical approaches (Johnston et al., 2017; Lee-Remond & Ballouk, 2026; Nelson & Gabbard, 2023). Their greatest educational value lies not in the technology itself but in their ability to facilitate authentic, immersive, and learner-centred experiences that encourage active participation, reflection, collaboration, and practical application (Amanatidis, 2022; Drigas & Karyotaki, 2014; Hsu & Liu, 2023; Lovo et al., 2024). When integrated thoughtfully with constructivist inquiry (Hajirasouli & Banihashemi, 2022; Nelson & Gabbard, 2023; Salvetti et al., 2023), collaborative learning (Hsu & Liu, 2023; Pappa & Papadopoulos, 2019), connectivist networking, cognitivist instructional design (Johnston et al., 2017; Walker et al., 2020), behaviourist skill development (Nelson & Gabbard, 2023), humanistic learner-centred education (Hsu & Liu, 2023; Schmidt & Stumpe, 2025), and experiential learning (Amanatidis, 2022; Hajirasouli & Banihashemi, 2022; Lee-Remond & Ballouk, 2026; Lovo et al., 2024), VR and AR can create rich educational experiences that develop both technical competence and higher-order thinking (Al-Rahbi et al., 2026; Dermott et al., 2023; Drigas & Karyotaki, 2014). Used in this way, immersive technologies become valuable tools for preparing adult learners to perform confidently, safely, and effectively in increasingly complex professional environments (Lovo et al., 2024; Pappa & Papadopoulos, 2019; Walker et al., 2020).

Learning Analytics

Learning analytics refers to the collection, measurement, analysis, and interpretation of data about learners and their learning environments to better understand, support, and optimise learning (Díaz-Diaz et al., 2020; Lobos et al., 2022; Mendiola et al., 2018; Sánchez, 2022). It involves using data generated through learning management systems (LMSs), online learning platforms, assessments, simulations, digital resources, and other educational technologies to provide insights into learner behaviour, progress, engagement, and achievement (Bravo et al., 2021; Mosquera & Bucheli, 2021; Ruipérez-Valiente, 2020). Rather than simply recording learner activity, learning analytics seeks to transform educational data into meaningful information that supports evidence-informed decision-making by educators, institutions, and learners themselves (Mella-Norambuena et al., 2022; Mella-Norambuena et al., 2023).

The growth of learning analytics has paralleled the rapid expansion of digital education over the past two decades (Ferreira et al., 2019; Mella-Norambuena et al., 2022). As learning increasingly occurs through online platforms and technology-enhanced environments, large volumes of educational data have become available for analysis (Figaredo et al., 2020; Mendiola et al., 2018; Ruipérez-Valiente, 2020). Every interaction—such as logging into an LMS, accessing learning resources, participating in discussion forums, submitting assessments, viewing videos, completing quizzes, or engaging with simulations—creates digital traces that can be analysed to identify learning patterns and inform educational practice (Aguilar, 2015; Campos, 2023). Advances in artificial intelligence, machine learning, cloud computing, and data visualisation have further expanded the capabilities of learning analytics, enabling increasingly sophisticated approaches to monitoring learner progress and supporting personalised education (Bravo et al., 2021; Mendiola et al., 2018; Okia del Rocio Andrade et al., 2024).

One of the defining characteristics of learning analytics is its ability to identify patterns that may not be immediately apparent through traditional observation alone (Ferreira et al., 2019). Analytics can reveal levels of learner engagement, participation, attendance, resource usage, assessment performance, and progression through learning materials (Lobos et al., 2022; Ortega & Gonzalez-Segarra, 2022). These insights allow educators to identify learners who may require additional support, recognise high-performing learners who are ready for greater challenges, evaluate the effectiveness of instructional materials, and identify aspects of courses that may require redesign (Mendiola et al., 2018; Sánchez, 2022; Zambrano, 2022).

Learning analytics is often described as operating at several levels (Lobos et al., 2022). Descriptive analytics summarises what has already occurred by reporting learner participation, attendance, assessment results, or resource usage (Lobos et al., 2022). Diagnostic analytics examines why particular outcomes have occurred by identifying relationships between learner behaviours and performance (Lobos et al., 2022). Predictive analytics uses historical and current data to estimate the likelihood of future outcomes, such as identifying learners who may be at risk of disengagement or non-completion (Lobos et al., 2022; Mella-Norambuena et al., 2022; Mendiola et al., 2018). Prescriptive analytics goes one step further by recommending interventions, resources, or learning activities that may improve learner success based on the available data (Lobos et al., 2022).

Learning analytics systems analyse learner progress, engagement patterns, and interactions with learning materials to support personalised learning pathways. Such systems can recommend resources, identify gaps in understanding, and support adaptive learning by tracking individuals' learning progression, analysing their personality traits and skills, and adapting curriculum, course content, strategy, and support to individual needs (R. Romano et al., 2024). Learning analytics allows students to be informed about necessary and effective actions, including through personalisation of learning and predictive modelling, with previous studies highlighting improvement in academic performance and development of personalised education as positive factors. Personalised feedback enables students to receive data on academic support services that help them develop professional skills and succeed in higher education. Learning analytics and big data have been identified as essential technological supports for individualised learning (Kucirkova, 2018; Liu, 2026), and the use of educational data mining and learning analytics can help educators determine and adjust teaching and intervention strategies (Alqarni, 2025). This capacity enables adult learners to receive timely support reflecting their individual learning needs while enabling educators to make more informed instructional decisions (Li & Wong, 2023; Pappagallo, 2024).

Learning analytics also contributes to institutional quality assurance and continuous improvement. Educational organisations can analyse aggregated data to evaluate course effectiveness, monitor learner satisfaction, and support evidence-based planning. In China, for example, the integration of learning analytics offers opportunities to support personalised learning and improve curriculum design, with teachers using analytics to ensure proper quality of online education and the development of educational courses and curricula. Educational big data and learning analytics help educators better improve their instruction design and the efficiency of learning environments (Pina et al., 2018). Program evaluation, accreditation, curriculum review, and strategic decision-making increasingly rely on learning analytics, as the incorporation of AI within educational administration encompasses learning analytics and predictive capabilities (Yuyun & Suherdi, 2023).

Learning analytics is a technological and analytical approach rather than a learning theory. Its educational value depends on how data is interpreted and used to support pedagogical decision-making. Research has demonstrated a complete centralisation of experiences in technological developments and a lack of an explicit pedagogical perspective in many personalised learning implementations (H. P. Li et al., 2026). Pina et al. (2018) argue that if personalisation of learning is an "unavoidable challenge" to education, it does not seem that some kinds of educational technology try to include pedagogy or educational foundations in that challenge (H. P. Li et al., 2026). The need for AI-enabled educational platforms to be grounded in theory has been emphasised (Yuyun & Suherdi, 2023), and educators must be able to interpret patterns and take meaningful action rather than relying on opaque indicators (Sunar et al., 2016).

Constructivism is supported when analytics helps educators understand how learners interact with learning activities, identify misconceptions, and provide opportunities for reflection. Interactive and adaptive technologies significantly enhance learning outcomes when integrated within constructivist and blended learning models (Kem, 2022). Smart learning environments facilitate learner self-regulation to promote autonomous learning by allowing learners to have control over their own learning process. Within constructivist approaches, teachers take on roles as mediators and knowledge brokers to scaffold learning appropriately, enabling learners to take more responsibility for their own learning (FitzGerald et al., 2018).

Social constructivism is reflected when learning analytics examines patterns of collaboration, discussion participation, peer interaction, and knowledge sharing within learning communities. Learning analytics for social network analysis can enable personalised learning within online environments. Smart learning environments embrace collaborative learning alongside personalised and adaptive approaches, and analytics can help educators facilitate collaborative learning by identifying learners who may require additional encouragement to participate.

Connectivism aligns particularly closely with learning analytics because both recognise the importance of digital networks and information flows. The field of intelligent agent-supported learning has witnessed integration of multimodal learning analytics, synthesising diverse data streams to create comprehensive learner profiles (Addanki et al., 2022). Analytics provides insights into how learners interact with online resources, digital communities, and educational technologies, supporting learning within increasingly connected digital environments (Pina et al., 2018).

Cognitivism is evident when analytics identifies patterns relating to memory, information processing, learning progression, and cognitive development. AI technologies can simulate real-life scenarios and provide personalised feedback, enabling students to strengthen their reasoning and develop analytical skills (Hwang & Fu, 2020). Smart learning environments offer cognitive guiding and support for developing higher-order abilities (Pina et al., 2018), and educators can use these insights to sequence instruction and provide appropriate scaffolding.

Behaviourism is reflected through the monitoring of observable learner behaviours such as participation, quiz completion, attendance, and resource engagement. Learning analytics means collecting, analysing, and evaluating data relevant to learning, including learning behaviours and learning interaction (Pina et al., 2018). Learning analytics reports can track the frequency of activities (Li & Wong, 2023), enabling educators to identify behavioural patterns and provide timely reinforcement or intervention based on measurable learning activities.

Humanism emphasises learner autonomy and self-directed learning. Learning analytics can support these principles by providing learners with personalised dashboards, progress reports, and opportunities for self-monitoring. Learning analytics can be used within smart learning environments to help learners understand their own learning process, and providing feedback serves to inform both instructors and learners. Personalised technology-enhanced learning can engage learners in conversations about learning how to learn, with e-portfolios and shared online progress files aiding reflection and review of learner progress, supplemented with data from learning analytics dashboards (Evans & Waring, 2011).

Experiential learning is supported when analytics captures learner performance within simulations, workplace activities, practical exercises, or authentic projects. AI technologies can play a significant role in designing complex learning environments that challenge students with analytical tasks, including simulating real-life scenarios (Hwang & Fu, 2020). The resulting data can inform reflection, identify strengths and areas for improvement, and support continuous refinement of practical skills through repeated experience, as adaptive systems analyse input in real time and adjust task difficulty and feedback accordingly (Kem, 2022).

Learning analytics does not replace professional judgement; rather, it provides educators with additional evidence to support informed decision-making (Hakimi et al., 2021; Marshall et al., 2022). Educators interpret analytical data within the broader context of each learner's circumstances, recognising that numerical indicators alone cannot fully explain learning behaviour or performance (Hakimi et al., 2021; McNicol et al., 2023). Effective use of learning analytics requires educators to combine data with observation, communication, assessment, and professional expertise when making decisions about learner support and instructional design (McNicol et al., 2023; Memari & Ruggles, 2025).

Educators also play a critical role in communicating learning analytics to learners in meaningful ways (Hakimi et al., 2021; McNicol et al., 2023). Rather than presenting data as performance scores alone, educators can use analytics to facilitate reflective discussions, identify learning strategies, establish realistic goals, and encourage learners to monitor their own progress (Hakimi et al., 2021; Memari & Ruggles, 2025). In this way, analytics becomes a tool for supporting learner development rather than simply measuring achievement (McNicol et al., 2023; Memari & Ruggles, 2025).

Learning analytics offers numerous benefits for adult education. It enables early identification of learners who may require additional support, facilitates timely intervention before learning difficulties become significant (Potgieter, 2020; Qolamani & Mohammed, 2023), supports personalised learning pathways (Iris Huis in 't & Nagenborg, 2019; Popkhadze, 2021), and provides educators with evidence to improve teaching practices (Qolamani & Mohammed, 2023; Slade & Prinsloo, 2013). Analytics can also increase learner engagement by providing regular feedback, visualising progress, and helping learners understand their strengths and areas requiring further development (Ma et al., 2026; West et al., 2020).

At an institutional level, learning analytics supports quality assurance, curriculum evaluation, learner retention, resource planning, and evidence-informed educational improvement (Qolamani & Mohammed, 2023; Slade & Prinsloo, 2013). Aggregated data can help institutions identify successful teaching strategies, evaluate program effectiveness, monitor learner outcomes, and allocate educational resources more effectively (K. M. L. Jones, 2019; Popkhadze, 2021). These insights contribute to continuous improvement across both academic and vocational education (Slade & Prinsloo, 2013).

The successful implementation of learning analytics depends on responsible and ethical data governance. Educational institutions must establish clear policies regarding the collection, storage, analysis, sharing, and use of learner data while complying with relevant privacy legislation and ethical standards ((Che & Perkins, 2026; Hakimi et al., 2021; Popkhadze, 2021). Learners should understand what data is being collected, why it is being collected, how it will be used, and who has access to it (Braunack-Mayer et al., 2020; K. M. L. Jones, 2019). Transparency helps build trust and encourages responsible use of learning analytics (K. M. L. Jones, 2019; Popkhadze, 2021; Slade & Prinsloo, 2013).

Data quality also remains a significant consideration. Learning analytics is only as reliable as the data on which it is based. Incomplete records, inconsistent data collection, inaccurate information, or inappropriate interpretation may result in misleading conclusions (Braunack-Mayer et al., 2020; Popkhadze, 2021). Educators should therefore avoid making decisions solely on the basis of analytical data and instead combine analytics with professional judgement, learner feedback, and other forms of assessment (Hartman-Caverly, 2019; Qolamani & Mohammed, 2023).

Another challenge is the potential for learning analytics to oversimplify complex learning processes. Metrics such as login frequency, time spent online, or resource access do not necessarily indicate meaningful learning or deep understanding (Hartman-Caverly, 2019; Snelling, 2023). Consequently, analytics should be interpreted as indicators that support inquiry rather than definitive measures of learner success (Potgieter, 2020; Slade & Prinsloo, 2013). Educators must also be mindful of algorithmic bias (Alwahaby & Cukurova, 2024; Yan & Chau, 2025), the risk of labelling learners as "at risk" prematurely (Rehak, 2022; Rubel & Jones, 2020), and the possibility that excessive monitoring may reduce learner autonomy or create perceptions of surveillance (Iris Huis in 't & Nagenborg, 2019; Ma et al., 2026; Potgieter, 2020; Qolamani & Mohammed, 2023; Rubel & Jones, 2020).

Learning analytics exemplifies theory-informed teaching by providing educators with evidence that informs pedagogical decision-making rather than prescribing instructional practice. When interpreted thoughtfully, analytics enables educators to personalise learning, identify opportunities for intervention, evaluate instructional effectiveness, and support learner reflection while remaining responsive to individual needs. Its greatest educational value lies in enhancing, not replacing, the educator's professional judgement. When integrated alongside constructivist inquiry, collaborative learning, connectivist networking, cognitivist instructional design, behaviourist feedback, humanistic learner-centred education, and experiential learning, learning analytics supports more responsive, inclusive, and evidence-informed adult education while maintaining the central importance of meaningful human relationships in the learning process.

Future Trends in Technology-Enhanced Learning

Technology-enhanced learning (TEL) continues to evolve rapidly as advances in digital technologies reshape how adults access, experience, and apply education. While earlier developments focused primarily on digitising traditional learning materials and expanding online access, emerging technologies are increasingly creating intelligent, immersive, personalised, and interconnected learning environments (Ajayi & Udeh, 2024; Egunjobi & Adeyeye, 2024; Supriya et al., 2024). These developments are changing not only the tools available to educators but also the ways learning is designed, delivered, assessed, and supported throughout an individual's lifetime (Boland, 2023; Nyathani, 2023). As workplaces become increasingly digital and knowledge-intensive, adult education will need to prepare learners to adapt continuously to new technologies, changing occupations, and evolving professional practices (Ajayi & Udeh, 2024; Supriya et al., 2024).

One of the most significant trends is the continued integration of artificial intelligence (AI) into education. AI is expected to become increasingly sophisticated in supporting personalised learning pathways, intelligent tutoring, automated feedback, adaptive assessment, learning analytics, and educational resource development (Egunjobi & Adeyeye, 2024; Nyathani, 2023; N. Rane et al., 2024; Supriya et al., 2024). Rather than simply generating content, future AI systems are likely to function as collaborative learning partners that help learners investigate complex problems, identify knowledge gaps, recommend learning resources, and support reflection (Ajayi & Udeh, 2024; Nyathani, 2023). At the same time, educators will play an increasingly important role in developing learners' AI literacy, ethical decision-making, and critical evaluation skills to ensure AI is used responsibly and effectively (Ajayi & Udeh, 2024; Önal & Özdemir, 2025).

The continued development of immersive technologies, including virtual reality (VR), augmented reality (AR), and mixed reality (MR), is expected to expand opportunities for authentic experiential learning (Boland, 2023; Bremner et al., 2020; Egunjobi & Adeyeye, 2024; N. Rane et al., 2024). As hardware becomes more affordable and software more sophisticated, immersive learning environments are likely to become increasingly common across higher education, vocational education, healthcare, engineering, defence, manufacturing, and workplace training (Ajayi & Udeh, 2024; Boland, 2023; Bremner et al., 2020; Nyathani, 2023; Supriya et al., 2024). Future applications may incorporate realistic haptic feedback, artificial intelligence, real-time collaboration, and digital twins that allow learners to interact with highly accurate virtual representations of workplaces, equipment, and operational systems before applying their knowledge in real-world settings (Akdere et al., 2021; N. Rane et al., 2024).

Extended reality (XR), which encompasses virtual reality (VR), augmented reality (AR), and mixed reality (MR), is increasingly recognised as one of the most significant emerging technologies in technology-enhanced learning (Alnagrat et al., 2022; Jagatheesaperumal et al., 2024; Zaman et al., 2024). By combining immersive digital environments with interactive technologies, XR enables learners to engage with authentic tasks that closely resemble real-world practice while remaining in safe and controlled learning environments. As these technologies continue to mature, they are expected to play an increasingly important role in supporting practical skills development, collaborative learning, and experiential education across higher education, vocational education, and workplace training.

One of the defining strengths of XR is its capacity to facilitate collaboration among learners regardless of their geographical location. Shared virtual environments enable participants to work together on projects, solve problems, communicate in real time, and participate in realistic simulations without being physically co-located (Zaman et al., 2024). This capability has considerable implications for distance education, international partnerships, and workplace learning by reducing the need for travel while providing learners with access to educational opportunities that may otherwise be unavailable because of location or resource constraints (Oyekunle et al., 2024).

Virtual laboratories represent another significant application of XR technologies. Immersive laboratory environments enable learners to conduct experiments, explore scientific concepts, and practise technical procedures repeatedly without the limitations imposed by laboratory availability, equipment costs, or safety considerations (Alnagrat et al., 2022). Similarly, XR supports realistic workplace simulations that allow learners to experience professional environments before entering actual workplaces. In healthcare education, immersive simulations recreate dynamic clinical situations that support the development of technical competence, decision-making, and teamwork (Fealy et al., 2023). Engineering education has also adopted XR to simulate interdisciplinary industrial projects, enabling learners from different disciplines to collaborate on authentic engineering problems while developing communication, teamwork, and practical problem-solving skills (Bönsch et al., 2024). Comparable approaches are emerging throughout vocational education, where immersive technologies allow learners to develop and refine workplace skills through realistic practice before applying those skills in operational settings (Boland, 2023).

As learning analytics becomes increasingly sophisticated, ethical governance is emerging as a critical priority for educational institutions. The growing capacity to collect, analyse, and interpret large volumes of learner data offers significant opportunities for personalised learning and evidence-informed teaching, but it also introduces important responsibilities relating to privacy, transparency, and accountability. Research consistently highlights the importance of establishing strong privacy protections, clear governance arrangements, and transparent data management practices to maintain learner trust and support the responsible adoption of learning analytics within education (Joksimovíc et al., 2021).

Ethical concerns extend beyond data privacy alone. Researchers have identified issues relating to informed consent, data ownership, data quality, validity of analytical conclusions, fairness in automated decision-making, institutional accountability, and the ethical governance of learner information (Hakimi et al., 2021). Additional challenges include protecting learner privacy throughout data collection, storage, and sharing; avoiding excessive reliance on predictive algorithms that may influence learner autonomy; and ensuring that educational records are managed responsibly throughout their lifecycle (Guan et al., 2023). As learning analytics continues to evolve towards increasingly predictive and prescriptive capabilities, educational institutions will require comprehensive governance frameworks that balance innovation with ethical responsibility while safeguarding learner rights and maintaining public confidence.

Adaptive learning systems represent another important direction in the future of technology-enhanced learning. These systems use learner data, artificial intelligence, and learning analytics to personalise instructional content, learning pathways, feedback, and assessment according to individual learner needs and progress. Rather than requiring all learners to follow identical instructional sequences, adaptive platforms adjust educational experiences to reflect differences in prior knowledge, learning pace, performance, and learning preferences (Oyekunle et al., 2024).

Emerging adaptive systems are increasingly integrating immersive technologies such as virtual reality with intelligent learning algorithms to provide personalised educational experiences. These environments support self-paced learning while adapting instructional activities according to learner responses, progress, and demonstrated understanding (Jagatheesaperumal et al., 2024). Advances in immersive technologies are also enabling systems that incorporate elements of self-regulated learning by using learner interactions, and in some cases neurophysiological data, to modify learning experiences dynamically in response to learner engagement and performance (Mikkonen et al., 2024). These developments suggest that future adaptive learning environments will become increasingly responsive, personalised, and capable of supporting individual learning journeys.

The continued expansion of micro-credentials and competency-based recognition reflects the growing importance of lifelong learning within rapidly changing workplaces. Flexible credentialing systems allow learners to demonstrate specific competencies as they are acquired, supporting continuous professional development and enabling individuals to update their knowledge throughout their careers. Immersive technologies are contributing to this trend by providing accessible, on-demand training environments in which learners can develop, practise, and demonstrate workplace skills more efficiently while generating evidence of competency that may contribute to recognised credentials (Boland, 2023).

The increasing adoption of competency-based education also highlights the importance of developing credentialing systems that reflect the needs of industry and professional practice. Research suggests that effective competency frameworks should be based on empirical evidence of stakeholder requirements rather than generic lists of skills, thereby supporting the development of portable, meaningful credentials that are recognised across educational institutions, industries, and employers (Mazur, 2026). As technology-enhanced learning continues to evolve, micro-credentials are likely to become increasingly integrated with adaptive learning systems, immersive technologies, and digital credentialing platforms, providing learners with flexible pathways for ongoing education, career progression, and lifelong learning.

The development of blockchain technology has the potential to transform the way educational achievements are recorded, verified, and shared. Blockchain-based credentials could provide secure, tamper-resistant records of qualifications, professional certifications, licences, and micro-credentials that remain under the learner's control throughout their career. Such systems may simplify credential verification, improve international recognition of qualifications, and reduce administrative burdens associated with employment and professional registration.

Digital twins represent another emerging technology with significant educational potential. A digital twin is a dynamic virtual representation of a physical object, process, or system that is continuously updated using real-world data. In adult education, digital twins could allow engineering students to monitor industrial equipment remotely, healthcare professionals to rehearse complex procedures using patient-specific models, logistics managers to optimise supply chains, or infrastructure engineers to analyse the performance of transport networks before implementing operational changes. By integrating simulation with real-time operational data, digital twins create opportunities for authentic, data-driven experiential learning.

The rapid development of the Internet of Things (IoT) will further connect learning environments with workplaces. IoT refers to a network of interconnected physical devices, objects, and systems that communicate and share data through the Internet (Behmanesh et al., 2023; Moshawrab et al., 2022). Sensors embedded within machinery, buildings, vehicles, wearable devices, and industrial systems will generate real-time operational data that can be incorporated directly into education. In workplace contexts, IoT involves embedding various sensors and other smart devices into infrastructure to collect and exchange data (Moshawrab et al., 2022), and IoT sensors installed on production machinery and equipment enable real-time monitoring of environmental conditions, equipment performance, and worker safety (Massaro et al., 2025). Adult learners may increasingly analyse live workplace information, monitor equipment performance, investigate operational trends, and solve authentic problems using current industry data rather than relying solely on simulated examples. Indeed, IoT-based technologies have already been deployed in medical education to collect data and provide interactive feedback to students (Behmanesh et al., 2023), and wearable technology can be used to allow students to track their own learning behaviour and improve their interactive experience. The dynamic potential of IoT devices and sensors in educational settings includes areas such as activity monitoring, physiological monitoring, and data collection (Behmanesh et al., 2023).

Wearable technologies are also expected to play an expanding role in adult education. Smart glasses, wearable sensors, biometric devices, and connected safety equipment can provide real-time guidance, monitor learner performance, collect physiological data, and deliver contextual information during workplace training. Wearable devices have been worn on the head (e.g., smart glasses), wrist (e.g., watches), and chest (e.g., electrocardiogram patches), with head-worn devices enabling better student engagement with learning materials, improved attention, and higher spatial and visual awareness (Khosravi et al., 2021). Due to their imperceptibility and direct contact with the human body, wearables can play a significant role in learning and education (Khosravi et al., 2021). In fields such as healthcare, construction, manufacturing, emergency services, and mining, wearable technologies may support competency development while improving learner safety and providing educators with detailed performance information for coaching and feedback.

In hazardous work environments, specialised wearable devices such as smart helmets equipped with sensors can detect harmful gases, monitor environmental conditions, and assess injuries (Moshawrab et al., 2022). Wearable devices equipped with biometric sensors can monitor vital signs, physical activity, and environmental exposures in real-time, enabling continuous health monitoring and early detection of potential safety risks (Massaro et al., 2025). Furthermore, wearable data can enhance the personalisation of learning and development by integrating with AI-driven platforms, where indicators of individual stress and well-being can be used to identify specific contexts where individuals are most likely to benefit from training. In medical education specifically, Google Glass, smart watches, and biometric devices have been employed to support case-based learning approaches, enabling access to relevant information and capturing real-life scenarios (Behmanesh et al., 2023). Wearable sensors also offer students the advantage of continuous monitoring, providing real-time data on physical activity levels and health status throughout the learning process (Behmanesh et al., 2023).

Advances in natural language processing, speech recognition, and multimodal interfaces are making learning technologies increasingly accessible and conversational. Future educational platforms are likely to support seamless interaction through voice, gesture, handwriting, images, and natural conversation, reducing barriers for learners with diverse abilities, languages, and educational backgrounds. These developments also have the potential to improve accessibility by providing more personalised and inclusive learning experiences.

Another significant trend is the increasing emphasis on open, connected, and global learning ecosystems. Educational institutions, industry, professional bodies, and learners are likely to collaborate more closely through shared digital platforms, open educational resources (Lima-Lopes & Biazi, 2021; Ossiannilsson, 2023), international partnerships (Porter & Perris, 2023; Zeng et al., 2020), and global communities of practice (Atenas et al., 2023; Connolly et al., 2020). Learning will increasingly occur across institutional, organisational, and geographical boundaries (McGreal et al., 2023; Wimpenny et al., 2020), allowing adult learners to engage with diverse perspectives, international experts, and multidisciplinary professional networks throughout their careers (Atenas et al., 2023; Yang & Wu, 2024).

Alongside technological innovation, growing attention is being given to the ethical, legal, and social implications of technology-enhanced learning. Issues relating to privacy, cybersecurity, artificial intelligence governance, algorithmic bias, intellectual property, accessibility, digital equity, sustainability, and learner wellbeing are expected to become increasingly prominent (Cheong, 2024; J. Li, 2025; Pechenkina, 2023; Smuha, 2020). Current ethical frameworks for AI in education typically integrate multidisciplinary knowledge from computer science, ethics, law, and pedagogy to address core topics such as data privacy, algorithmic bias, intellectual property, transparency, and educational equity (J. Li, 2025). The European Commission's High-Level Expert Group on AI has articulated seven requirements for Trustworthy AI, including human agency and oversight, technical robustness and safety, privacy and data governance, transparency, diversity and non-discrimination, societal and environmental wellbeing, and accountability (Smuha, 2020). Privacy and data governance concerns are particularly acute, as AI systems in education rely on vast amounts of student data, raising questions about surveillance, data misuse, and the quality of services provided to learners (Hockly, 2023; Pechenkina, 2023; Smuha, 2020). A significant gap in addressing cybersecurity and data protection has been identified even in major institutional frameworks; for instance, only five World Bank education projects during the pandemic mentioned data privacy or cybersecurity laws (Salimi, 2025). The absence of clear global governance structures for educational data has been described as "alarming" (Chilambarasan, 2025).

Algorithmic bias represents a persistent challenge, as AI systems may reinforce existing inequalities when they optimize for narrow metrics while ignoring contextual goals and values (Cheong, 2024; , (Čermáková et al., 2024; Cheong, 2024; Verma & Dhaigude, 2026). At the organizational level, particularly with the increasing volume of AI-based applications, equitable processes for addressing bias are of utmost importance, including identification of bias types and points of risk in algorithmic training and implementation of transparent practices to minimize or eliminate bias in design (Čermáková et al., 2024). Research on algorithmic justice stresses that while AI can broaden access to education, it may also unintentionally perpetuate unfairness (Verma & Dhaigude, 2026). The close association between adaptive learning, personalized learning, data privacy, and algorithm bias indicates that the promotion of personalized learning must simultaneously address data privacy risks and algorithmic fairness issues (J. Li, 2025).

Intellectual property concerns have also emerged as a significant dimension, particularly regarding the ownership of AI-generated educational content, digital copyright, and knowledge ownership in the context of open educational resources (J. Li, 2025; Versteijlen & Groesbeek, 2024). The earnings models of for-profit EdTech companies are focused on obtaining user data and benefit from a lack of transparency in data use, privacy, and intellectual property (Versteijlen & Groesbeek, 2024). Regulations should clearly specify data privacy, algorithmic transparency, intellectual property rights, and provisions for accountability (Coman et al., 2026).

Questions of digital equity and accessibility remain central, as strong inequalities exist between and within countries regarding the level of education and digital literacy that students enjoy (Mariyono, 2026; Smuha, 2020). While AI-powered educational technology promises to make education more broadly accessible, it can just as well exacerbate inequalities and widen the digital literacy gap (Rivera-Vargas, 2023; Smuha, 2020). The digital divide and linguistic biases in AI models are sources of risk in exacerbating existing educational inequities, making culturally responsive AI designs and inclusive governance frameworks indispensable (Coman et al., 2026). Equitable EdTech practices require that necessary supports are provided so that all learners, regardless of background or ability level, have full access to educational tools, resources, and practices (Čermáková et al., 2024).

Sustainability and learner wellbeing constitute additional dimensions requiring attention. Environmentally, the use of EdTech can reduce carbon footprints but device disposal and energy consumption of hardware and data storage must be considered (Versteijlen & Groesbeek, 2024). The benefits of AI in education cannot fulfil the purpose of enhancing learners' capabilities without taking into account not only their educational but also their moral, mental, and physical wellbeing, particularly when minors are involved (Smuha, 2020). Ethical issues such as collecting learner data, surveillance, and privacy must be considered alongside learner wellbeing and the digital literacies that teachers and learners need (Hockly, 2023).

Educational institutions will therefore need comprehensive governance frameworks that balance innovation with ethical responsibility while ensuring that technology serves educational rather than purely technological objectives. The establishment of specialized institutes and advisory groups tasked with producing ethical frameworks for AI governance in education indicates a concerted move toward unifying approaches (Pechenkina, 2023). Institutions should consider governance frameworks that align AI adoption with institutional values, respect human rights, and prioritize the wellbeing of students (Karimi & Khawaja, 2023; Verma & Dhaigude, 2026). Data governance, cybersecurity, and data privacy risks require institutions to focus not only on technology acquisition but also on technology implementation, responding appropriately to inherent risks (Rivera-Vargas, 2023). Governments should adopt AI-in-education policies aligned with ethical frameworks such as UNESCO's Recommendation on the Ethics of AI (Chilambarasan, 2025; Thakur & Mittal, 2025). Ultimately, the question is no longer whether AI will enter education but how it will be integrated—the choice is between profit-driven or publicly accountable systems designed to serve learners equitably and ethically (Chilambarasan, 2025).

Despite rapid technological change, the future of technology-enhanced learning is unlikely to be defined by technology alone. The most significant advances will arise from thoughtful integration of emerging technologies within sound pedagogical practice. Educators will continue to play a central role in designing meaningful learning experiences, fostering critical thinking, supporting collaboration, encouraging reflection, and developing the ethical and professional capabilities that technology cannot replace. Rather than diminishing the importance of educators, future technologies are expected to augment their ability to personalise learning, provide timely support, and create authentic educational experiences that prepare adult learners for increasingly complex and technology-rich workplaces.

Key Concepts and Conclusions

This chapter has explored how digital technologies are reshaping adult education while reinforcing that effective learning depends on sound pedagogy rather than technology alone. Technology-enhanced learning encompasses a diverse range of approaches, including online and blended learning, mobile learning, learning management systems, open educational resources, artificial intelligence, virtual and augmented reality, learning analytics, adaptive learning, and immersive technologies. Each offers unique opportunities to improve access, flexibility, engagement, collaboration, and authentic workplace learning when aligned with clear educational goals.

A central theme throughout the chapter is that digital technologies are educational tools rather than educational theories. Their effectiveness is determined by how educators integrate them within theory-informed teaching. Constructivism supports inquiry and authentic problem-solving; social constructivism promotes collaboration and shared knowledge construction; connectivism emphasises learning through digital networks; cognitivism informs instructional sequencing and information processing; behaviourism supports practice, reinforcement, and feedback; humanism encourages learner autonomy and self-directed learning; and experiential learning connects theory with authentic workplace application. Technology provides the medium through which these pedagogical principles can be enacted rather than replacing them.

The chapter also demonstrates that technology-enhanced learning has expanded opportunities for lifelong learning by increasing educational accessibility and flexibility. Adult learners can engage with learning regardless of location, access educational resources on demand, participate in global learning communities, and develop competencies that respond to rapidly changing workplace requirements. Technologies such as artificial intelligence, learning analytics, immersive simulations, adaptive learning systems, and digital credentials are enabling increasingly personalised and responsive learning experiences while strengthening the relationship between education and professional practice.

At the same time, technological innovation introduces important ethical, legal, and professional responsibilities. Issues relating to privacy, cybersecurity, data governance, algorithmic bias, intellectual property, accessibility, digital equity, learner wellbeing, and sustainability require careful consideration by educators and institutions. Successful implementation therefore depends upon robust governance frameworks, ongoing professional development, and a commitment to ensuring that technological innovation supports rather than compromises educational quality and learner outcomes.

Technology-enhanced learning has become an integral component of contemporary adult education, offering unprecedented opportunities to expand access, personalise learning, strengthen workplace relevance, and support lifelong professional development. Digital technologies have transformed how education is delivered, yet their educational value continues to depend upon thoughtful instructional design and the deliberate application of sound pedagogical principles rather than the technologies themselves.

Throughout this chapter, it has been shown that no single technology provides a universal solution for adult learning. Instead, educators must select and integrate technologies according to the learning outcomes being pursued, the characteristics of adult learners, the learning context, and the theoretical foundations that best support effective learning. Whether designing an online course, implementing mobile learning, using artificial intelligence, developing immersive simulations, or analysing learner data, successful practice requires educators to remain guided by educational theory, evidence-informed decision-making, and authentic learner-centred practice.

Emerging technologies will continue to reshape education through increasingly personalised, intelligent, immersive, and connected learning environments. Artificial intelligence, extended reality, adaptive learning, digital twins, wearable technologies, learning analytics, and competency-based digital credentials are likely to become increasingly embedded within higher education, vocational education, workplace learning, and professional development. These innovations have considerable potential to enhance learner engagement, improve practical skill development, strengthen industry partnerships, and support continuous lifelong learning.

However, the future of adult education will ultimately be determined not by technological innovation alone but by the ability of educators and institutions to integrate technology ethically, inclusively, and pedagogically. Human expertise, critical thinking, professional judgement, collaboration, reflection, and meaningful educator–learner relationships will remain essential components of effective learning. Technology should therefore be viewed as an enabler of educational innovation rather than as a substitute for high-quality teaching.

The central message of this chapter is that technology-enhanced learning is most effective when technological capability is combined with theory-informed teaching, authentic learning experiences, ethical governance, and a commitment to developing adaptable, digitally capable, and lifelong learners who are prepared to thrive in increasingly complex and technology-rich workplaces.

Chapter 10

The Future of Adult Learning and Vocational Education

The world of adult learning and vocational education is undergoing one of the most significant periods of transformation in its history. Advances in artificial intelligence, automation, digital technologies, globalisation, demographic change, and environmental sustainability are reshaping how people work, the skills they require, and the ways in which education is designed and delivered. At the same time, traditional assumptions that individuals can rely on a single qualification throughout their working lives are being replaced by the expectation that learning will continue across an entire lifetime. These developments are redefining the relationship between education, employment, and society, requiring learners, educators, employers, and governments to rethink how knowledge and skills are developed, recognised, and maintained.

Responding to these changes requires more than simply introducing new technologies or updating curriculum content. It demands education systems that are flexible, inclusive, responsive, and capable of supporting individuals through multiple career transitions and lifelong professional development. Increasingly, learning takes place not only within schools, vocational education providers, and universities, but also through workplaces, professional networks, online communities, digital platforms, and everyday experience. As a result, contemporary adult education is becoming more personalised, competency-based, technology-enabled, and closely connected to the evolving needs of industry and society.

This chapter explores the emerging directions that are shaping the future of adult learning and vocational education. It examines how technological innovation, workforce transformation, lifelong learning ecosystems, competency-based education, digital learning, micro-credentials, recognition of prior learning, and sustainability are influencing educational policy and practice. The chapter also considers the evolving roles of educational institutions, employers, governments, professional bodies, and learners in creating responsive lifelong learning systems. Finally, it demonstrates how established educational theories continue to provide an essential foundation for understanding and designing effective adult learning in an increasingly complex, interconnected, and rapidly changing world.

The Future of Work and Skills Development

The future of work is being shaped by rapid technological innovation, demographic change, globalisation, environmental sustainability, and evolving economic conditions (Ashta & Mogha, 2023; Côté & Su, 2022). Advances in artificial intelligence, automation, robotics, digital communication, cloud computing, and data analytics are transforming industries and redefining the knowledge, skills, and capabilities required for employment (Côté & Su, 2022; Zouinar, 2020). Rather than simply replacing existing jobs, these changes are altering the nature of work itself, creating new occupations, changing existing roles, and requiring workers to continually update their competencies throughout their careers (Côté & Su, 2022; Zouinar, 2020).

Historically, workers could often rely on a single qualification or apprenticeship to sustain an entire career. In contrast, contemporary labour markets increasingly require continuous learning as technologies, regulations, business models, and professional practices evolve (Boulahoual & Rafik, 2024; Psyché et al., 2024). Many occupations are experiencing rapid change, with new digital tools becoming integral to everyday work in fields as diverse as healthcare, engineering, construction, manufacturing, education, logistics, finance, and public administration (Hocquelet, 2020; Krichen, 2023). Consequently, lifelong learning is becoming a fundamental expectation rather than an optional aspect of professional development (Marcellis-Warin, 2022; Petit, 2022).

One of the most significant drivers of workplace transformation is artificial intelligence. AI systems are increasingly supporting activities such as data analysis, customer service, scheduling, quality assurance, predictive maintenance, medical diagnosis, financial modelling, content generation, and decision support (Ashta & Mogha, 2023; Côté & Su, 2022; Zouinar, 2020). Automation is similarly replacing many routine and repetitive tasks while increasing demand for workers who can supervise automated systems, interpret complex information, solve novel problems, and make sound professional judgements (Côté & Su, 2022; Zouinar, 2020). As technology assumes responsibility for predictable activities, human capabilities such as critical thinking, creativity, ethical reasoning, communication, collaboration, and leadership become increasingly valuable (Petit, 2022; Psyché et al., 2024).

The future workforce is therefore expected to require a combination of technical and transferable skills. Technical skills remain essential for operating specialised equipment, using digital technologies, interpreting data, and applying discipline-specific knowledge (Boulahoual & Rafik, 2024; Marcellis-Warin, 2022). Equally important are transferable skills that can be applied across occupations and industries, including communication, teamwork, adaptability, problem-solving, digital literacy, information literacy, emotional intelligence, intercultural competence, and lifelong learning skills (Marcellis-Warin, 2022; Psyché et al., 2024). These capabilities enable individuals to respond effectively to technological change, changing workplace roles, and increasingly complex professional environments (Côté & Su, 2022).

Digital literacy has become a foundational competency for the future workforce. Employees are increasingly expected to use digital technologies confidently, evaluate online information critically, collaborate through digital platforms, protect data security and privacy, and understand the capabilities and limitations of emerging technologies such as artificial intelligence (Akingbola et al., 2025; Ramachandran et al., 2024; Semi et al., 2026). These capabilities extend beyond basic technical proficiency and include the ability to make informed decisions about the reliability, relevance, and ethical use of digital information and tools.

AI literacy is increasingly recognised as an important component of broader digital literacy. It enables workers to use AI systems effectively while evaluating outputs critically, identifying potential errors and bias, protecting confidential information, and applying ethical judgement when interacting with intelligent technologies (Chee et al., 2024; Fan, 2024; Semi et al., 2026). This form of literacy is relevant across the workforce rather than being limited to computer science or technical occupations because AI systems are becoming embedded in decision-making, communication, administration, analysis, and professional practice across many industries (Zhang et al., 2024).

AI literacy competency frameworks commonly include an understanding of data, algorithms, model limitations, error detection, AI-assisted decision-making, and the application of specialised technologies within particular occupational fields. Workers also need the capacity to interpret, analyse, and evaluate AI-generated information in relation to professional standards and workplace contexts (Chee et al., 2024). These competencies support more informed use of AI while reducing the risk that workers will accept automated outputs without sufficient scrutiny.

Ethical capability is central to AI literacy. Workers must be able to recognise algorithmic bias, protect personal and organisational data, understand accountability for AI-supported decisions, and apply responsible AI principles in professional practice (Akingbola et al., 2025; Karakas et al., 2025; Sarwar et al., 2024; Semi et al., 2026). Effective AI literacy education should therefore prepare individuals not only to operate AI systems but also to understand their limitations, evaluate their outputs, and determine when human oversight and professional judgement are required (Ng et al., 2023).

Governments and international organisations are increasingly incorporating AI literacy into workforce policy and professional education. The European Union's regulatory approach places responsibility on organisations using AI systems to ensure that relevant personnel possess an appropriate level of understanding, while international health organisations have emphasised the need to integrate AI competencies into health professions education to prepare a digitally capable workforce (Agarwal et al., 2024; Alqaissi et al., 2026). These developments demonstrate that digital and AI literacy are becoming essential elements of workforce readiness, professional responsibility, and lifelong learning.

The future of work is increasingly interdisciplinary because complex workplace problems rarely fall within the boundaries of a single profession or field of knowledge. Engineers may collaborate with environmental scientists, healthcare professionals with information technology specialists, educators with learning designers, and business professionals with data scientists and cybersecurity experts (Adeoye et al., 2024; Frehywot & Vovides, 2023; Ramachandran et al., 2024). These arrangements reflect the growing interdependence of technical, social, organisational, and ethical knowledge in contemporary workplaces.

Educational programs are therefore placing greater emphasis on interdisciplinary learning, systems thinking, communication, and collaborative problem-solving. These capabilities help learners understand how different areas of expertise contribute to complex problems and enable them to work effectively with professionals who use different concepts, methods, and forms of evidence (Adeoye et al., 2024; Ramachandran et al., 2024; Semi et al., 2026). Interdisciplinary preparation is particularly important in emerging fields where technological developments intersect with public policy, health, business, law, security, and ethics.

Digital public health provides one example of this shift. Workforce preparation in this field may draw upon public health, informatics, data science, computer science, biostatistics, health communication, and business, with practice-based learning occurring across multiple disciplinary contexts (Ramachandran et al., 2024). Cyberbiosecurity similarly requires collaboration across biological, cybersecurity, regulatory, organisational, and technical systems, making communication and multisector coordination central professional capabilities (Adeoye et al., 2024).

AI education also benefits from interdisciplinary curriculum design and institutional collaboration. Sustainable integration of AI into professional education may require contributions from subject-matter experts, computer scientists, data specialists, learning designers, technology organisations, social scientists, legal professionals, and AI ethicists (Frehywot & Vovides, 2023; Semi et al., 2026). Bringing these perspectives together can ensure that learners understand both the technical operation of AI and its broader professional, ethical, social, and regulatory implications.

Interdisciplinary capabilities also strengthen career adaptability. Workers who can transfer knowledge across domains, communicate with diverse professional groups, and integrate different forms of expertise are better prepared to move between roles, industries, and emerging areas of practice. AI literacy frameworks therefore increasingly emphasise transferable competencies that support career pathways extending beyond traditional occupational or disciplinary boundaries (Chee et al., 2024).

Globalisation continues to shape workforce development by connecting organisations, labour markets, education systems, and professional networks across national boundaries. Digital communication technologies allow employers to recruit internationally, coordinate distributed teams, provide services across multiple regions, and collaborate with partners located in different countries (Calonge et al., 2025; Makhachashvili et al., 2024). As a result, many workers now operate within professional environments influenced by diverse cultural, linguistic, legal, and organisational contexts.

These conditions increase the importance of intercultural communication, cultural awareness, foreign language capability in relevant occupations, and the ability to collaborate respectfully with colleagues from different backgrounds. Workers must be able to interpret communication within its cultural context, manage different expectations about teamwork and leadership, and contribute effectively to geographically dispersed groups (Makhachashvili et al., 2024). These capabilities are important not only in multinational organisations but also in local workplaces serving culturally diverse communities and clients.

Remote and hybrid work have further expanded the range of competencies required for effective participation in global workplaces. Employees need strong digital communication skills, self-management, virtual teamwork, online meeting capability, and the ability to maintain productive professional relationships without regular face-to-face interaction (Calonge et al., 2025; Makhachashvili et al., 2024). They may also need to coordinate work across time zones, adapt communication for different digital channels, and manage ambiguity when contextual cues are limited.

Digital literacy supports the development and application of these intercultural and collaborative capabilities. In foreign language and communication education, digital competence can strengthen communication, interdisciplinarity, and the transfer of soft skills across different learning and professional contexts (Makhachashvili et al., 2024). Digital communication platforms can also expose learners to international perspectives and provide opportunities to practise collaboration in culturally diverse virtual environments.

National workforce strategies increasingly reflect the competitive pressures associated with global technological change. Investment in AI upskilling, digital capability, and advanced professional education is being used to strengthen workforce competitiveness and prepare workers for participation in technology-intensive global economies (Calonge et al., 2025). The spread of AI education initiatives across North America, Europe, Asia, and the Middle East further demonstrates that workforce transformation is an international development requiring coordinated attention to skills, education, regulation, and access (Akingbola et al., 2025; Semi et al., 2026).

Environmental sustainability is another major force reshaping occupations, industries, and workforce development. Governments, businesses, and communities are adopting renewable energy technologies, circular economy practices, low-carbon production methods, sustainable design principles, and more efficient approaches to resource use (Nedungadi et al., 2024). These changes are creating new occupations while also altering the responsibilities and skill requirements associated with existing roles.

The transition towards more sustainable economies is increasing demand for competencies in areas such as renewable energy, environmental management, sustainable construction, carbon accounting, resource efficiency, climate risk, sustainable procurement, and environmentally responsible business practice. Sustainability is also becoming embedded within occupations that were not traditionally regarded as environmental roles. Engineers, tradespeople, managers, educators, health professionals, logistics workers, and information technology specialists may all be required to consider environmental consequences as part of routine professional decision-making (Fan, 2024; Nedungadi et al., 2024).

Technology can support sustainability education by expanding access to learning, enabling data-informed environmental analysis, and helping educators develop learning resources that address complex sustainability problems. Generative AI and other digital technologies may contribute to educational access and support broader sustainable development objectives, including reducing inequality and improving participation in education (Nedungadi et al., 2024). However, these technologies must themselves be evaluated in relation to energy consumption, resource use, environmental impact, and equitable access.

Workforce reskilling and upskilling will be essential to ensure that workers can participate successfully in the transition to sustainable industries and practices. Training opportunities must be accessible to workers in different regions, industries, and stages of their careers, while curricula need to remain aligned with changing labour-market requirements (Fan, 2024). This may involve revising existing qualifications, developing targeted micro-credentials, expanding workplace learning, and supporting workers whose occupations are significantly affected by technological and environmental change.

Effective workforce transformation will require collaboration among governments, employers, educational institutions, professional bodies, unions, and communities. These stakeholders share responsibility for identifying emerging skill needs, developing relevant education and training, supporting equitable transitions, and ensuring that sustainability goals are integrated into professional practice. By combining environmental knowledge with digital capability, interdisciplinary collaboration, and lifelong learning, workforce development systems can help prepare individuals for employment in economies that are both technologically advanced and environmentally responsible.

These changes have significant implications for education and training systems. Rather than focusing solely on initial qualifications, educational institutions are increasingly supporting lifelong learning through flexible delivery models, micro-credentials, competency-based education, recognition of prior learning, stackable qualifications, and work-integrated learning (Arnold et al., 2020; Passos, 2021; Swanepoel, 2024). Digital learning technologies enable individuals to undertake professional development while remaining employed, allowing workers to update their skills in response to changing occupational requirements without interrupting their careers (Alamuri & Chittipaka, 2024; Asian Development Bank, 2023).

Technology-enhanced learning will play a central role in future workforce development. Online learning, mobile learning, learning management systems, artificial intelligence, virtual and augmented reality, learning analytics, adaptive learning systems, and digital credentials are making education increasingly accessible, personalised, and responsive to industry needs (Arnold et al., 2020; Ghosh & Ravichandran, 2024; Swanepoel, 2024). Immersive simulations allow learners to practise complex procedures safely before entering the workplace (Arnold et al., 2020; Ghosh & Ravichandran, 2024), while artificial intelligence can personalise learning pathways and provide immediate feedback (Arezki et al., 2019; Arnold et al., 2020; Ghosh & Ravichandran, 2024). Learning analytics enable educators to identify learners requiring additional support, and adaptive learning systems tailor instruction according to individual progress and competency development (Arnold et al., 2020; Kamp, 2023; Swanepoel, 2024).

Workplace learning itself is also evolving. Learning is increasingly integrated into everyday work through just-in-time learning resources, mobile technologies, digital performance support systems, virtual mentoring, collaborative knowledge-sharing platforms, and communities of practice (Arnold et al., 2020; Asian Development Bank, 2023; Drigas & Tsolaki, 2015; Lee & Tan, 2023). Employees no longer rely exclusively on formal training programs but continuously acquire knowledge through workplace experience, professional networks, online resources, and collaborative problem-solving (Alamuri & Chittipaka, 2024; Bhatia et al., 2023; Lee & Tan, 2023). This shift reflects growing recognition that learning and work are becoming increasingly interconnected rather than separate activities (Asian Development Bank, 2023; Noor, 2012).

The educator's role is also changing. Rather than acting primarily as transmitters of knowledge, educators are increasingly facilitating inquiry, critical thinking, collaboration, reflection, digital capability, and professional identity development (Choi-Lundberg, 2024; Shouman et al., 2023). They help learners navigate rapidly changing information, evaluate evidence, apply knowledge in authentic contexts, and develop the adaptability required for lifelong employability (Choi-Lundberg, 2024; Shouman et al., 2023). Educators also play an essential role in fostering ethical decision-making, particularly as artificial intelligence, automation, and data-driven technologies become more prevalent across workplaces (Hussain & Tasnim, 2023).

The future of work also presents important challenges. Technological disruption may widen existing inequalities if individuals lack access to education, digital technologies, or opportunities for reskilling (Falco, 2026). Some occupations may experience significant displacement, while new occupations emerge requiring different competencies. Educational systems, employers, governments, and professional bodies therefore share responsibility for ensuring equitable access to lifelong learning opportunities that enable workers to adapt successfully throughout their careers.

From a theory-informed perspective, the future of work reinforces the importance of multiple educational theories. Constructivism supports authentic problem-solving and knowledge construction in complex workplace contexts (Shouman et al., 2023). Social constructivism emphasises collaboration and shared learning within professional communities. Connectivism recognises the importance of digital networks and continual access to current knowledge (Falco, 2026). Cognitivism informs the organisation of increasingly complex learning experiences and supports information processing in technology-rich environments. Behaviourism remains relevant for developing procedural competence and reinforcing safe workplace practices. Humanism promotes learner autonomy, self-directed learning, and personal growth (Choi-Lundberg, 2024), while experiential learning continues to provide the bridge between theoretical understanding and authentic professional practice (Choi-Lundberg, 2024).

Lifelong Learning Ecosystems

A lifelong learning ecosystem is an interconnected network of educational institutions, workplaces, governments, professional bodies, community organisations, technology providers, and learners that collectively supports learning throughout an individual's life. Unlike traditional education systems, which often treat learning as a sequence of separate stages, lifelong learning ecosystems recognise that people continually develop knowledge and skills across formal education, employment, community participation, and everyday life (Schreiber-Barsch, 2015).

Within such an ecosystem, learning does not occur solely in schools, colleges, universities, or training organisations. It may also take place through workplace projects, mentoring, professional networks, online courses, micro-credentials, community programs, industry training, self-directed study, open educational resources, and informal experience (Schreiber-Barsch, 2015). As Schreiber-Barsch notes, adult education "occurs in companies, in educational organizations, at home, in cultural institutions and in the media" (Schreiber-Barsch, 2015, p. 6). Similarly, informal learning in the workplace encompasses on-the-job learning, action-learning, and experiential processes that complement formal education,. These different forms of learning are connected through pathways that allow learners to enter, leave, resume, and combine education according to changing personal, professional, and social needs. The movement of "Lifelong Learning" emerged precisely as a response to the gap between an individual reality of learning throughout life and a formal education offering limited to the first phases of active life.

The concept of an ecosystem emphasises interdependence. No single institution can provide all the learning required across an individual's life and career. Educational providers contribute structured programs and recognised qualifications, employers provide authentic workplace learning and current industry knowledge, (Behringer & Coles, 2003), governments establish policy and funding arrangements, (Bouchereau & Roxin, 2018), professional bodies define occupational standards, technology providers support digital access, and community organisations respond to local learning needs (Schreiber-Barsch, 2015). As Schreiber-Barsch (2015, p. 6) observes, "the interconnectedness of adult education, learning and its socio-political environment (persons and organizations, aims and educational concepts, as well as the labor market) is closer than that of universities and schools". In Singapore, for instance, the government has significantly increased resources allocated to lifelong learning and established comprehensive continuing education systems involving employers, educational institutions, and workforce development agencies (Bouchereau & Roxin, 2018). The recognition and validation of informal learning also inscribe themselves within the logic of lifelong education, enabling mobility of workers and accumulation of credentials across settings. Learners move between these settings while accumulating knowledge, experience, competencies, and credentials, navigating between diverse learning sites throughout their lives.

A defining characteristic of a lifelong learning ecosystem is flexibility. Learners need opportunities to study at different times, in different locations, and at varying levels of intensity. Online learning, blended delivery, mobile learning, evening programs, workplace training, and modular study arrangements allow adults to participate while managing employment, family, caring, and community responsibilities (OECD, 2022a; Ricart et al., 2014). Indeed, adapting locations and schedules to align with the working schedules of professionals and promoting more work-based learning have been identified as among the most important changes implemented in lifelong learning systems (Ricart et al., 2014).

Another important characteristic is permeability between educational sectors. Learners should be able to move between vocational education, higher education, workplace learning, community education, and professional development without unnecessary duplication or barriers (OECD, 2022a, 2023b; Schröder & Kruger, 2019). Recognition of prior learning, credit transfer, articulation agreements, bridging programs, and stackable credentials can help connect these sectors and create more coherent learning pathways (OECD, 2023b, 2023d). Micro-credentials, for instance, can cater flexibly to diverse workforce needs and allow credentials to be accumulated towards traditional educational qualifications, thereby supporting movement across sectors (OECD, 2023d). However, as observed in some national contexts, several providers play a role in lifelong learning, including higher education institutions, secondary vocational schools, and other further education providers, yet each may operate independently, meaning that regular communication and collaboration between these providers is yet to fully develop (OECD, 2023b). Moving beyond the fragmentation of education and lifelong learning towards overarching and connected governance structures has been identified as beneficial for creating coherent ecosystems (Schröder & Kruger, 2019).

Another important characteristic is permeability between educational sectors. Learners should be able to move between vocational education, higher education, workplace learning, community education, and professional development without unnecessary duplication or barriers (OECD, 2022a, 2023b; Schröder & Kruger, 2019). Recognition of prior learning, credit transfer, articulation agreements, bridging programs, and stackable credentials can help connect these sectors and create more coherent learning pathways (OECD, 2023b, 2023d). Micro-credentials, for instance, can cater flexibly to diverse workforce needs and allow credentials to be accumulated towards traditional educational qualifications, thereby supporting movement across sectors (OECD, 2023d). However, as observed in some national contexts, several providers play a role in lifelong learning, including higher education institutions, secondary vocational schools, and other further education providers, yet each may operate independently, meaning that regular communication and collaboration between these providers is yet to fully develop (OECD, 2023b). Moving beyond the fragmentation of education and lifelong learning towards overarching and connected governance structures has been identified as beneficial for creating coherent ecosystems (Schröder & Kruger, 2019).

Personalisation is also central. Adult learners enter education with diverse knowledge, experiences, goals, languages, cultural backgrounds, and professional identities. Lifelong learning ecosystems therefore need to provide multiple entry points, differentiated support, flexible pathways, and opportunities for learners to pursue goals that are relevant to their circumstances (Behrend et al., 2022; King et al., 2017; Schröder & Kruger, 2019). A more learner-oriented approach, rather than one focused solely on institutional improvements, has been advocated as essential for effective lifelong learning (Schröder & Kruger, 2019). Learning analytics and adaptive technologies may assist with personalisation—personalized learning powered by data-driven cognitive technologies can enable individuals to take more responsibility over their skills future and improve outcomes across the ecosystem (King et al., 2017), but professional guidance and human support remain essential (OECD, 2022a).

The ecosystem model also depends on responsiveness. Learning provision must evolve in response to technological change, emerging industries, workforce shortages, demographic change, environmental priorities, and community needs (Graf et al., 2025; Taylor et al., 2021). Education policies and systems need to be sufficiently agile to react to knowledge developments and redesign programmes to incorporate lifelong learning (Taylor et al., 2021). This requires ongoing collaboration between education providers, employers, governments, and professional organisations so that curricula, qualifications, and professional development remain current and meaningful (Graf et al., 2025; OECD, 2022a; Taylor et al., 2021). Given the scale and speed of change associated with transitions to green and digital economies, it is challenging for policymakers, employers, and educational institutions to predict what skills will be in demand and how to create them, necessitating collaborative and experimental approaches (Graf et al., 2025). Local policy makers must find ways to steer and adjust the provision and content of programmes to labour market demands, while engaging private actors in local skills partnerships is a vital tool for dealing with the inherent complexities of adult learning systems (OECD, 2022a).

Formal, Non-formal, and Informal Learning

Lifelong learning ecosystems integrate formal, non-formal, and informal learning (Singh, 2015c; M. Singh, 2022). Formal learning takes place in an organised and structured educational environment, usually leading to a certificate or a diploma (M. Singh, 2022). It usually involves defined learning outcomes, planned instruction, assessment, and quality assurance (Commonwealth Secretariat, 2022; Singh, 2015b).

Non-formal learning is organised and intentional but may not lead to a full qualification. It refers to semi-structured training without resulting in a formal qualification certification (M. Singh, 2022). It includes workshops, short courses, micro-credentials, industry training, professional seminars, community education, and workplace development programs (Commonwealth Secretariat, 2022; Conrad, 2023a; Scott, 2022a). These forms of learning can respond rapidly to emerging needs and may provide targeted development in specific skills (Brown, Mhichíl, et al., 2021; OECD, 2023b).

Informal learning is the acquisition of practical competences, expertise, work practice and attitudes to perform, which leads to no qualification, degree or certification (M. Singh, 2022). It occurs through everyday experience, workplace problem-solving, observation, reading, professional discussion, online participation, and interaction with colleagues or community members (Conrad, 2023a; Mitchell et al., 2025; Scott, 2022a). Although informal learning may not be structured or assessed, it contributes significantly to professional competence, practical judgement, identity development, and personal growth (Cameron, 2012; Scott, 2022a; Singh, 2015g).

An effective lifelong learning ecosystem enables these forms of learning to complement one another (Mason & Carr, 2021; Mikulec et al., 2020; Singh, 2015c). Knowledge transfer occurs between distinctive and different forms of learning from formal, non-formal and informal learning settings (Singh, 2015c). For example, a worker might develop practical expertise informally through employment, complete a non-formal short course in a new technology, receive recognition of prior learning, and then enter a formal qualification with credit for existing competence (Brown, Mhichíl, et al., 2021; Ratna Selvaratnam & Michael Sankey, 2025; Singh, 2015b). In New Zealand, RPL takes into account formal, non-formal and informal learning, and recognition of existing competences may lead to an increased willingness among employees to take part in workplace training or learning (Singh, 2015b). Similarly, South Korea's Academic Credit Bank System facilitates the recognition of non-formal and informal learning, allowing individuals to obtain academic degrees and vocational certificates on the basis of credits (Ashizawa et al., 2024; Singh, 2015g). The Australian Qualification Framework sets a policy framework for the recognition of informal and non-formal learning through both the VET and higher education sectors (Mason & Carr, 2021), enabling learners to transfer between formal, non-formal, and informal learning opportunities (Mikulec et al., 2020).

Digital Technologies and Lifelong Learning

Digital technologies have become important infrastructure within lifelong learning ecosystems. Learning management systems, online platforms, mobile applications, open educational resources, artificial intelligence, virtual simulations, digital portfolios, and professional networking platforms enable learning to continue beyond institutional boundaries (Fidalgo & Thormann, 2024; H. P. Li et al., 2026). These technologies allow learners to access educational resources, collaborate with others, document achievements, and participate in learning regardless of location (Fidalgo & Thormann, 2024; L. G. Ortiz et al., 2021).

Artificial intelligence may support lifelong learning by recommending resources, identifying knowledge gaps, adapting learning activities, and helping learners plan individual development pathways (Fidalgo & Thormann, 2024; H. P. Li et al., 2026). Learning analytics may assist educators and institutions in identifying disengagement, evaluating program effectiveness, and providing targeted support (Fidalgo & Thormann, 2024; L. G. Ortiz et al., 2021; Panda, 2022). Digital portfolios can allow individuals to collect evidence of formal qualifications, micro-credentials, workplace achievements, projects, and professional development across their careers (Ocheja et al., 2022; L. G. Ortiz et al., 2021).

Digital credentialing is also becoming increasingly important (J. Lang, 2023; H. P. Li et al., 2026). Micro-credentials and digital badges can recognise smaller units of assessed learning and allow learners to demonstrate specific competencies (Aithal & Maiya, 2024; Elsayed, 2023; Kumar et al., 2022; Темјановски et al., 2023). When combined with clear standards, reliable assessment, and recognised pathways, these credentials may allow individuals to build broader qualifications progressively (Aithal & Aithal, 2023; J. Lang, 2023; Ratna Selvaratnam & Michael Sankey, 2025). However, their value depends on quality assurance, transparency, portability, and recognition by employers and educational institutions (Brown, Mhichíl, et al., 2021; J. Lang, 2023; H. P. Li et al., 2026; Ratna Selvaratnam & Michael Sankey, 2025).

Role of Educational Institutions

Educational institutions remain central to lifelong learning ecosystems, but their role extends beyond delivering initial qualifications. Universities, vocational education providers, adult education centres, and community organisations increasingly support learners across multiple stages of life and work (Choi-Lundberg et al., 2024; Crosbie et al., 2024; Ngubane et al., 2024). They may provide short courses, professional development, micro-credentials, bridging programs, recognition services, online learning, and customised workforce training in addition to traditional qualifications (Medina et al., 2020; Papier, 2018, 2020).

Institutions also contribute by connecting learners with employers, professional networks, research, community organisations, and public services (Choi-Lundberg et al., 2024; Hardin-Ramanan et al., 2020; Medina et al., 2020). Work-integrated learning, apprenticeships, placements, industry projects, guest lecturers, and collaborative curriculum development help strengthen the relationship between education and professional practice ((Choi-Lundberg et al., 2024; Crosbie et al., 2024; Hardin-Ramanan et al., 2020; Medina et al., 2020; Mesuwini & Mokoena, 2023; Takei, 2023). Universities may employ strategies such as embedding internships within curricula, facilitating industry-based projects, inviting guest lecturers from professional practice, and engaging employers in curriculum design to ensure alignment between education and workforce needs (Choi-Lundberg et al., 2024; Crosbie et al., 2024; Hardin-Ramanan et al., 2020; Medina et al., 2020). Mentorship programs drawing on alumni and industry contacts further reinforce these connections (Choi-Lundberg et al., 2024), while collaborative arrangements between educational institutions and employers, including staff exchanges and joint research, deepen the integration of theory and practice (Hardin-Ramanan et al., 2020; Medina et al., 2020).

To support lifelong participation, institutions need flexible admissions processes, accessible learner support, career guidance, digital services, and pathways that recognise prior learning (Reddy & Maharaj, 2022; Rongmin, 2024; Ruthanam, 2023). Recognition of prior learning (RPL) serves as a mechanism enabling learners who have acquired skills through non-formal or informal means to have those competences certified, thereby facilitating re-entry into formal education (Bijl & Taylor, 2016; Papier, 2018). Career guidance services within universities should integrate placement and counselling functions, helping adults develop professional and personal pathways rather than merely assisting with job searches. They must also design programs that allow learners to return after periods of employment or personal responsibility without being treated as if they are beginning from the outset (Papier, 2018; Rongmin, 2024). Lifelong learning opportunities afford such returning learners a second chance to build on existing skills and competences (Papier, 2018), while flexible program structures, including online and distance learning modalities, accommodate diverse learner circumstances (Rongmin, 2024; Takei, 2023).

Role of Employers

Employers are important learning partners because much adult learning occurs through work. They provide opportunities for employees to develop competence through supervised practice, mentoring, job rotation, project participation, coaching, professional discussion, and engagement with new technologies. Work-based learning is widely recognised as a process through which individuals learn from their community of practice in the workplace, encompassing learning *for*, *through*, and *at* work (Amish et al., 2023; Jalinus et al., 2023; Lemanski et al., 2016). Organisations that treat learning as part of everyday work are better positioned to adapt to change, improve performance, and retain skilled employees. Research confirms that WBL approaches increase professionalism and motivation among learners while addressing persistent skills gaps and meeting business needs (Amish et al., 2023). Workplace learning meets the needs of individuals for personal growth and professional development, enabling employees to attain high levels of knowledge and participate in professional development activities, with businesses aiming to utilise this knowledge to the benefit of their organisations (Vannasy & Sengsouliya, 2023).

A strong workplace learning culture encourages employees to ask questions, share knowledge, report mistakes, reflect on experience, and participate in continuous improvement. Managers and supervisors contribute by providing feedback, allocating time for development, recognising learning, and ensuring that training is connected to real workplace needs. Research has identified that workplace supervisors help students and employees make connections between what is learned at training institutions and how it is applied at work (Arinaitwe, 2021). Vocational teachers' engagement in workplace learning provides a valuable avenue for enhancing curriculum design and supporting transitions into the workplace, ensuring knowledge and skills are tailored to industry-standard practice (Zhao & Ko, 2024). The organisational culture of learning and leadership are key predictors that make workplace learning realistic, alongside the employer-employee relationship (Vannasy & Sengsouliya, 2023).

Employers may also collaborate with education providers to develop programs, provide equipment or expertise, host learners, validate assessment tasks, and identify emerging skill requirements. University-employer partnerships integrate informal work-based activities with formal classroom-based learning and create flexible learning environments to support collaboration (Amish et al., 2023). Strong partnerships between educational institutions and industry stakeholders are crucial for ensuring curricula meet labour market needs, allowing for the exchange of ideas and resources. Industry partners emphasise the importance of time students spend in workplaces and the collaborative design of training (Ziegler et al., 2020). These partnerships help ensure that education remains relevant while avoiding an overly narrow focus on immediate employer needs at the expense of broader capability and career mobility. As noted in the literature, there is an important balance between job-specific learning and obtaining a broader education that may equip learners for employment in other sectors (Lemanski et al., 2016). Collaboration and partnership between education providers and progressive workplaces are core elements of effective vocational education, requiring equal, mutual, and trustful relationships to support sustainable change (Svensson et al., 2009).

Role of Governments and Professional Bodies

Governments influence lifelong learning ecosystems through policy, regulation, funding, qualification frameworks, employment services, digital infrastructure, and social inclusion initiatives (Dadzie et al., 2020; Dhiman & Solkhe, 2025; OECD, 2020a; Osborne et al., 2022). Public policy can encourage participation by reducing financial barriers, supporting disadvantaged learners, funding reskilling programs, and creating pathways between educational sectors (OECD, 2020a, 2021a, 2021c).

Professional and industry bodies contribute by identifying occupational standards, accrediting programs, recognising continuing professional development, and advising on emerging capability requirements (Maulana & Jun, 2018; OECD, 2020b; Poirier & Remsen, 2017). They can help establish consistency and trust across credentials, although care is required to ensure that professional requirements remain accessible and do not create unnecessary barriers to participation (OECD, 2020b).

Coordination across government portfolios is particularly important. Education, employment, industry, migration, regional development, social services, and digital policy are closely connected within a lifelong learning system (Dhiman & Solkhe, 2025; OECD, 2020a, 2021d; Sujito & Bahagijo, 2022). Fragmented policy can create duplication and gaps (Dhiman & Solkhe, 2025; Osborne et al., 2022), while coordinated approaches can create more coherent support for learners and workers (Dadzie et al., 2020; Lewis & Tolgensbakk, 2019; Sujito & Bahagijo, 2022).

Expanding on these points, the evidence across multiple national contexts illustrates the mechanisms through which governments shape lifelong learning. In Ghana, the government established the Council for Technical and Vocational Education and Training (COTVET) to formulate national policies for skills development across pre-tertiary and tertiary education and the formal, informal, and nonformal sectors, with multiple ministries, including the Ministry of Education, Ministry of Youth and Sports, and Ministry of Finance, sharing oversight responsibilities (Dadzie et al., 2020). Similarly, Germany has developed measures integrating migrants through vocational education and training, including the Recognition Act (2012) and the Vocational Language Training programme (2016), while the Digital Pact invested in digital infrastructure for education (OECD, 2020b). VET systems more broadly are recognized as well-placed to equip students with labour market skills, provide reskilling for displaced workers, and create avenues for upskilling (OECD, 2021c).

Regarding professional and industry bodies, the Federal Institute for Vocational Education and Training (BIBB) in Germany serves as a major instrument for cooperation between stakeholders in the VET sector (OECD, 2020a). In Greece, new governance bodies were created including the National VET Committee to monitor framework implementation and the National Apprenticeship Coordinating Body bringing together various stakeholder representatives (OECD, 2020b). In Thailand, Sector Education and Training Alliances develop sector skills plans that inform occupational demand lists (OECD, 2021c). Poirier and Remsen emphasize that TVET educators must develop universally accepted certifications and public-private partnerships to prepare the future workforce (Poirier & Remsen, 2017).

The challenge of coordinating policy and delivery across government portfolios is well established. In India, skill development initiatives have been affected by fragmented programs, unrealistic projections of future skill demand, and limited capacity to adapt to digital change, reducing their overall effectiveness and highlighting the need for more integrated reform and stronger stakeholder coordination (Dhiman & Solkhe, 2025).

Comparable problems have been identified in Indonesia, where apprenticeship arrangements have operated through separate ministerial programs without a single coordinating policy framework. This has contributed to inconsistent certification practices and limited mutual recognition between agencies responsible for different parts of the system (Maulana & Jun, 2018). In Benin, technical and vocational education and training is similarly distributed across six line ministries without a unified policy or common legal framework, creating difficulties in maintaining coherence across the sector (World Bank team, 2022).

Further analysis of Indonesia has also found that cooperation between ministries and the private sector remains underdeveloped. Effective multisector initiatives, including proposed funding mechanisms such as a Social Development Fund, would require coordinated action across agencies responsible for employment, finance, education, culture, and related policy areas, together with stronger engagement from employers and other stakeholders (Sujito & Bahagijo, 2022). Collectively, these examples demonstrate that fragmented governance can weaken skills policy by duplicating responsibilities, creating inconsistent standards, and limiting the capacity of education and training systems to respond coherently to workforce needs.

Conversely, coordinated approaches yield more coherent outcomes. In Norway, a National Skills Policy Strategy was agreed by five government ministries, with trials of interministerial coordination and a focus on cross-sectoral cooperation (Lewis & Tolgensbakk, 2019). Osborne et al. identify multisectoral collaboration and robust governance systems as common pre-conditions for success in smart learning cities promoting lifelong learning (Osborne et al., 2022). The OECD's work on local entrepreneurship ecosystems in Cambridgeshire highlights how devolution of the Adult Education Budget to a Combined Authority, working with central government on apprenticeship policy, can strengthen skills development through coordinated governance (OECD, 2021d).

Role of the Learner

Within a lifelong learning ecosystem, learners are active participants rather than passive recipients of education. They identify goals, make decisions about learning pathways, evaluate resources, seek feedback, reflect on experience, and connect learning with personal and professional development. Self-directed learning, metacognition, digital literacy, and career management skills therefore become increasingly important (Bittner et al., 2021; Evans, 2018; Karataş & Arpacı, 2021; Marra et al., 2018). Research confirms that self-directed learning skills, metacognitive awareness, and 21st century competences, including digital literacy, are positively interrelated and collectively predict learners' readiness for online learning environments (Karataş & Arpacı, 2021). Metacognition provides a "conceptual foundation to support lifelong and self-directed learning" (Evans, 2018, p. 2), while the relationship between self-directed learning and metacognition is well established, with metacognitive skills being "necessary to become a successful self-directed learner" (Marra et al., 2018). At a higher level, educational self-regulation competence extends across the lifespan to encompass workplace learning and career management, with workers increasingly needing to "manage their careers proactively" through goal-setting and self-regulated learning (Bittner et al., 2021, p. 313).

However, emphasising personal responsibility for learning should not obscure the structural factors that influence participation in lifelong learning. An individual's capacity to engage in education is shaped by a wide range of factors, including financial cost, time constraints, employment conditions, digital access, disability, language proficiency, previous educational experiences, and family responsibilities. Effective lifelong learning ecosystems therefore combine learner agency with coordinated support from educational institutions, workplaces, governments, and communities. Although lifelong learning systems have expanded educational opportunities and improved access for many adults, participation remains uneven across different population groups, indicating that increased availability alone does not guarantee equitable engagement (Singh, 2024).

Research also challenges the assumption that all adults are equally motivated, capable, or willing to manage their own learning independently. While self-directed learning is an important goal, learners differ considerably in their readiness for autonomy and the level of support they require (Dunk & Craft, 2025). Self-regulation develops through purposeful teaching and guided practice rather than emerging naturally, and learners with less developed learning strategies often require explicit instruction, scaffolding, and ongoing support to become effective independent learners (Ambinintsoa, 2020). Furthermore, the quality of self-regulation is influenced by the learning environment. High-pressure educational contexts can encourage more superficial approaches to learning, increase stress, and reduce learner autonomy, persistence, and engagement (Neufeld, 2022). These findings highlight the importance of designing learning environments that balance independence with appropriate guidance and support.

Learners also make important contributions to lifelong learning ecosystems through knowledge sharing, mentoring, participation in professional communities, and collaboration with peers. Adult learners bring substantial personal and professional experience to educational settings, enriching discussion, strengthening curriculum relevance, and supporting workplace learning. Self-directed learning can also foster transferable capabilities that extend beyond formal education, including greater independence, confidence, problem-solving ability, and adaptability within professional environments (Marra et al., 2018). Educational approaches that position learners as active partners in the learning process rather than passive recipients of information have been shown to strengthen metacognitive development, learner engagement, and ownership of learning (Matook et al., 2023).

Effective self-regulated learners also recognise the importance of collaboration. Rather than working in isolation, they actively seek assistance, exchange ideas, and learn from others as part of their ongoing development (Ambinintsoa, 2020). Research further indicates that interpersonal communication skills strengthen the relationship between metacognitive capability and self-directed learning, suggesting that peer interaction, professional dialogue, and collaborative knowledge sharing enhance learners' capacity to manage their own learning effectively (Li et al., 2024). Consequently, successful lifelong learning ecosystems support both learner independence and meaningful participation within collaborative learning communities, recognising that personal agency and social learning are complementary rather than competing dimensions of adult education.

Theoretical Foundations

Lifelong learning ecosystems are not grounded in a single educational theory but instead draw upon multiple complementary perspectives that explain different aspects of adult learning. Together, these theories provide a comprehensive understanding of how adults acquire knowledge, develop skills, and continue learning across diverse educational, workplace, and community settings.

Humanism underpins the emphasis on personal growth, learner autonomy, self-direction, and the pursuit of meaningful personal and professional goals. Influenced by theorists such as Abraham Maslow and Carl Rogers, the humanist perspective views learning as a process of supporting individuals to realise their potential rather than simply responding to external stimuli. Humanist approaches recognise learners as capable, autonomous individuals whose emotions, values, experiences, and aspirations influence learning. Consequently, educational practices informed by humanism seek to develop learners' capacity for lifelong learning, independent judgement, decision-making, and personal development while recognising the dignity and potential of every individual (Bigdeli et al., 2023; Dunaway, 2011; Khan, 2012; Muzumdar & Cheemalapati, 2025).

Constructivism provides another important foundation by proposing that learners actively build new understanding through the interpretation of experience. Rather than receiving knowledge passively, learners connect new ideas to their existing knowledge, experiences, and professional practice. Authentic learning activities, workplace experiences, and problem-solving tasks enable learners to construct increasingly sophisticated understanding over time. This perspective aligns closely with lifelong learning pathways that integrate formal education with workplace learning, prior experience, and ongoing professional development, recognising that meaningful learning develops through active engagement with real-world situations (Arghode et al., 2017).

Social constructivism extends this perspective by emphasising that learning is inherently social and culturally situated. Knowledge develops through interaction with others, participation in communities, mentoring relationships, collaborative problem-solving, and engagement within professional and cultural contexts. Learning is therefore understood not simply as an individual cognitive activity but as a process of participating in shared practices and gradually becoming a member of a professional or learning community. Communities of practice, peer learning, workplace mentoring, and collaborative inquiry all reflect this theoretical orientation and highlight the importance of relationships in knowledge construction (Caruso et al., 2020).

Connectivism is particularly relevant to lifelong learning ecosystems because contemporary learning increasingly occurs across networks of people, organisations, digital technologies, and information sources. This perspective views knowledge as distributed across interconnected networks rather than residing solely within individuals. Learning involves locating, evaluating, connecting, and applying information from diverse sources while continually developing and strengthening personal and professional networks. Digital technologies play a central role by enabling learners to access information, collaborate globally, participate in online communities, and maintain professional learning networks throughout their careers. Connectivism also emphasises learner autonomy, adaptability, and collaborative knowledge creation within networked environments (Bigdeli et al., 2023; Lacasse et al., 2019; Lee & Wong, 2025).

Experiential learning reflects the ongoing relationship between concrete experience, critical reflection, conceptual understanding, and informed action. Building on the work of David Kolb and Carl Rogers, this perspective recognises that meaningful learning develops through repeated cycles of experience, reflection, experimentation, and refinement. Workplace learning, simulations, projects, placements, and professional practice provide opportunities for learners to test ideas, reflect upon outcomes, and continuously improve their knowledge and skills. Experiential learning is therefore particularly well suited to lifelong learning because it supports continuous professional development throughout changing careers and workplaces (Khan, 2012; Lee & Wong, 2025).

Cognitivism contributes to lifelong learning ecosystems by informing the design of structured learning pathways, scaffolding, assessment, and instructional support. This perspective focuses on internal mental processes such as attention, memory, perception, problem-solving, metacognition, and information processing. Learning environments informed by cognitivism seek to organise information in ways that support understanding, reduce unnecessary cognitive load, provide meaningful feedback, and progressively build learner capability. Such approaches are particularly valuable when learners encounter increasingly complex knowledge and need structured guidance to develop expertise (Bigdeli et al., 2023; Ilic, 2021; Tsortanidou et al., 2018).

Behaviourism also retains an important role within lifelong learning ecosystems, particularly where learners are developing procedural knowledge, technical skills, or competencies that require repeated practice. Behaviourist approaches emphasise observable performance, structured instruction, timely feedback, reinforcement, and opportunities for repetition until competence is achieved. These principles remain relevant in vocational education, workplace training, simulation-based learning, and digital learning environments where immediate feedback supports skill acquisition and consistent performance (Bigdeli et al., 2023; Khan, 2012; Lacasse et al., 2019; Tsortanidou et al., 2018).

Collectively, these perspectives demonstrate that lifelong learning cannot be adequately explained by any single educational theory. Each theoretical tradition provides valuable insights into different dimensions of adult learning, including motivation, cognition, social participation, experience, technological networking, and behavioural development. Rather than competing with one another, these theories are complementary and may be applied according to the learning context, learner characteristics, educational goals, and workplace requirements (Lacasse et al., 2019; Muzumdar & Cheemalapati, 2025). Lifelong learning ecosystems therefore represent an integrated educational approach that draws upon multiple theoretical foundations to support learning across diverse settings, technologies, and stages of life.

Relationship to Theory-Informed Teaching

Lifelong learning ecosystems exemplify theory-informed education by recognising that adult learning occurs through multiple processes, settings, relationships, and experiences rather than within a single educational institution or learning environment (Amoud & Raymer, 2025; Muzapparovna, 2024; Ülker et al., 2026). Effective ecosystems therefore extend beyond the provision of technology, qualifications, or organisational structures alone, integrating authentic experience, social participation, structured learning, learner agency, digital networks, reflection, and constructive feedback to support continuous development (Kift, 2025; H. P. Li et al., 2026; Muzapparovna, 2024; Osunwusi, 2020; Ülker et al., 2026).

The literature consistently describes lifelong learning as encompassing formal, non-formal, and informal learning opportunities that occur across education, work, community, and everyday life (Boyadjieva & Ilieva-Trichkova, 2017b; Nogueiro et al., 2022; Osunwusi, 2020; Ülker et al., 2026). Learning beyond traditional educational settings contributes to the development of creativity, critical thinking, problem-solving, adaptability, and a sustained commitment to learning throughout life (Ülker et al., 2026). Similarly, continuing education increasingly relies on flexible approaches such as online learning, interactive technologies, experiential learning, and personalised instruction to meet the diverse and changing needs of adult learners (Muzapparovna, 2024). Strong connections between education and employment further enhance lifelong learning by creating more coherent and coordinated pathways in which educational institutions, workplaces, and other stakeholders each contribute to learner development (Amoud & Raymer, 2025). Emerging digital ecosystems are extending these opportunities through technologies such as artificial intelligence, personalised learning systems, and blockchain-supported credentialing, although recent research suggests that existing learning theories may require further development to explain how these technologies influence learner engagement, autonomy, and long-term skill development across the lifespan (H. P. Li et al., 2026).

Within these ecosystems, educators play an important role in helping learners connect prior experience with new knowledge, identify appropriate learning pathways, participate in collaborative learning communities, and transfer learning between educational and workplace settings (Amoud & Raymer, 2025; Muzapparovna, 2024; Osunwusi, 2020). This perspective reflects a broader understanding of education as an ongoing aspect of everyday life rather than an activity confined to formal institutions (B.-Y. Wang, 2025). Self-determination and self-regulation are also recognised as important characteristics of lifelong learners because they encourage individuals to engage actively with learning opportunities while pursuing longer-term personal and professional goals (Entigar, 2021). At the same time, educators have a responsibility to design inclusive learning experiences that recognise the diversity of adult learners and provide the guidance, flexibility, and support necessary for effective self-directed learning (Amoud & Raymer, 2025; Entigar, 2021). Community education initiatives further contribute by creating learner-centred environments that support not only workforce participation but also civic engagement and community involvement (Amoud & Raymer, 2025). Contemporary perspectives on inclusion also emphasise that learners from diverse backgrounds, including migrants and other underrepresented groups, should be recognised as active contributors whose experiences enrich educational communities rather than being viewed solely through the lens of disadvantage (Entigar, 2021).

Ultimately, lifelong learning ecosystems seek to make learning a continuous, connected, and accessible part of adult life (Jing & Yan, 2020; Kift, 2025; Osunwusi, 2020). Contemporary understandings of lifelong learning extend across all stages of life, multiple learning environments, and diverse areas of knowledge, reflecting its growing importance in education, employment, social participation, and public policy (Kift, 2025). Lifelong learning also serves multiple purposes by supporting economic development, personal fulfilment, social inclusion, and democratic participation (Osunwusi, 2020). Achieving these outcomes depends upon collaboration among learners, educators, employers, governments, professional bodies, technology providers, and community organisations (Muzapparovna, 2024; Pyzocha & Boateng, 2024; Ülker et al., 2026). Research consistently indicates that addressing the challenges associated with lifelong learning requires coordinated action across these stakeholders to promote inclusion, quality assurance, sustainability, and equitable access (Muzapparovna, 2024; Pyzocha & Boateng, 2024). Cross-sector partnerships strengthen community capability by providing integrated learning opportunities, shared resources, and coordinated support services that improve access for diverse learner groups (Pyzocha & Boateng, 2024).

When supported by coherent learning pathways, equitable participation, trusted credentials, ethical governance, responsive technologies, and sound educational practice, lifelong learning ecosystems can strengthen workforce adaptability, individual capability, social inclusion, and community resilience (Amoud & Raymer, 2025; Boyadjieva & Ilieva-Trichkova, 2017b; H. P. Li et al., 2026; Ülker et al., 2026). Recent research also highlights the importance of coordinated policy frameworks, effective institutional governance, and ethically informed technology design to ensure that digital lifelong learning remains inclusive and sustainable (H. P. Li et al., 2026). Strengthening learning opportunities beyond formal educational institutions can further enhance digital capability, civic participation, and social inclusion, making lifelong learning more accessible throughout society (Ülker et al., 2026). Nevertheless, significant equity challenges remain. Participation in adult education continues to vary according to educational background, with individuals who have lower levels of prior education remaining substantially less likely to engage in lifelong learning than those with higher educational attainment (Boyadjieva & Ilieva-Trichkova, 2017b). These findings reinforce the importance of designing lifelong learning ecosystems that not only expand opportunities but also actively address barriers to equitable participation.

Benefits and Challenges

Lifelong learning ecosystems can improve access to education by providing multiple entry points and reducing reliance on a single, linear educational pathway. Rather than requiring learners to complete all education at one stage of life, these ecosystems enable individuals to begin, pause, and resume learning as personal, professional, and economic circumstances change. Learners can combine workplace experience with formal qualifications, professional development, and non-formal learning opportunities throughout their careers (Brown, Mhichíl, et al., 2021; J. Lang, 2023; Porto & Presant, 2023). Flexible learning models, including micro-credentials, have emerged in response to the growing demand for shorter, more accessible forms of education that support ongoing participation across the lifespan (Antonaci, Henderikx, et al., 2021; J. Lang, 2023). International policy initiatives have similarly recognised the importance of providing flexible learning opportunities that enable adults to upskill, reskill, and transition between occupations throughout their working lives (Antonaci, Henderikx, et al., 2021). Digital badges and other alternative credentials further expand participation by allowing learning to be recognised through multiple delivery models and professional development pathways, while recognition of prior learning (RPL) enables individuals to receive formal credit for competencies developed through work, community participation, or previous study (Conrad, 2023a; Donnelly & Maguire, 2020; Porto & Presant, 2023).

These ecosystems also strengthen workforce adaptability by supporting continuous skills development in response to technological change, labour market shifts, and evolving occupational requirements. Workers increasingly need to update existing competencies, acquire new skills, and prepare for career transitions throughout increasingly longer and more dynamic working lives (J. Lang, 2023; Sharma et al., 2024). Micro-credentials provide targeted opportunities for upskilling and reskilling, allowing learners to address emerging skill shortages without necessarily undertaking lengthy qualification programs (Brown, Mhichíl, et al., 2021; J. Lang, 2023). Employers also benefit from more responsive education systems when universities, vocational providers, and industry collaborate to develop credentials that reflect current workplace requirements (Rachel Fitzgerald & Henk Huijser, 2021; Ratna Selvaratnam & Michael Sankey, 2025). At a broader level, lifelong learning ecosystems can contribute to stronger workforce participation, improved social inclusion, and enhanced local capability by helping individuals maintain relevant skills and reducing mismatches between workforce capabilities and labour market demand. For example, stackable digital credential systems have supported continuous professional development within the public sector in parts of Latin America and the Caribbean (Passos, 2021).

Another significant benefit is the recognition of learning acquired across different contexts. By connecting formal qualifications, workplace learning, micro-credentials, professional development, and prior experience, lifelong learning ecosystems reduce unnecessary duplication of learning and enable individuals to build progressively on existing competencies (Braxton, 2023; Brown, Mhichíl, et al., 2021; Conrad, 2023a). Recognition of prior learning provides structured processes through which experiential learning can be assessed against established standards, allowing learners to demonstrate competencies developed through employment, volunteering, community engagement, military service, or other life experiences (Conrad, 2023a). Similarly, universities and professional bodies are increasingly recognising micro-credentials as evidence of learning that may contribute towards formal qualifications or broader professional portfolios (Brown, Mhichíl, et al., 2021). Digital badges also support personalised learning pathways by documenting achievements gained through higher education, workplace training, professional certification, and community participation within a single digital record (Braxton, 2023). These approaches can reduce both the time and financial cost associated with further education by ensuring that learners are not required to repeat competencies they have already demonstrated (Conrad, 2023a; Graebe & McIntyre-Hite, 2024).

Lifelong learning ecosystems can also promote greater educational equity when they are designed with appropriate learner support and inclusive policies (Conrad, 2023a; OECD, 2021g). Flexible delivery models, accessible digital technologies, financial assistance, community-based learning opportunities, and targeted learner support can broaden participation among groups that have traditionally experienced barriers to education (J. Graebe, 2025; H. P. Li et al., 2026; OECD, 2021g). Alternative credentials are increasingly being designed to improve accessibility through flexible scheduling, mobile-compatible platforms, multilingual delivery, and learning opportunities that accommodate diverse employment patterns, geographical locations, and caring responsibilities (J. Graebe, 2025). Community organisations, public libraries, and local learning centres also play an important role by providing access to digital technologies, mentoring, and educational guidance for learners who may otherwise experience social or economic disadvantage (H. P. Li et al., 2026). Recognition of prior learning further contributes to equity by valuing learning that occurs in diverse settings and acknowledging competencies developed outside formal education (Conrad, 2023a). Nevertheless, achieving equitable participation remains an ongoing challenge. Research indicates that the benefits associated with micro-credentials and other flexible learning opportunities are not always distributed evenly across the population, highlighting the importance of ensuring that lifelong learning initiatives are intentionally designed to reach learners from all backgrounds and circumstances (OECD, 2021g).

Building a coherent lifelong learning ecosystem is difficult because education, employment, and community services are often organised through separate institutions, funding arrangements, and regulatory systems. Learners may encounter inconsistent admissions processes, poor credit recognition, duplication of learning, and limited information about available pathways. The lack of shared digital standards and common frameworks for credit transfer, online quality assurance, or digital credential recognition makes joint initiatives across institutions and jurisdictions difficult (OECD, 2021g; World Bank, 2022). Micro-credential programmes still lack universally recognised definitions in many jurisdictions and are often offered as less regulated non-formal programmes outside established data reporting systems (Schmoelz, 2023). Efforts to create cohesive ecosystems require integrating different stakeholders, education and training providers, public authorities, industries, and learners, to collaboratively address fragmentation and maximise benefits (Schmoelz, 2023). The importance of interoperability, recognition, and validation of digital credentials between institutions has been identified as a key challenge, requiring shared and standard frameworks whose quality is recognised across jurisdictions (Read & Olcott, 2025). In higher education specifically, many questions have arisen regarding the applicability of micro-credentials, their quality assurance, inclusion in formal educational settings, and how to integrate them into traditional education systems.

Short courses, micro-credentials, workplace programs, and online learning opportunities vary considerably in their design, assessment, and credibility. Clear standards and transparent information are needed so that learners and employers can judge the value of different forms of learning. The number and diversity of micro-credential offerings have expanded substantially, yet key challenges relate to providing a framework for widespread understanding and recognition, and ensuring learners have the information needed to choose micro-credentials that support their individual needs and career objectives. Since both higher education institutions and private companies grant micro-credential certificates without acting according to a common framework or determined standards, efforts have emerged to create common quality assurance procedures. A study of 750 human resources executives in the United States found that only 14% reported hiring an individual with digital badges and 10% had hired someone with micro-credentials" (OECD, 2021g; World Bank, 2022).

Several governments and governmental agencies have established guidelines about micro-credential descriptors to provide foundations for quality assurance and recognition, including detailed information such as programme title, issuing entity, assessment protocol, learning outcomes, and alignment with external standards (Orman et al., 2023). The persistent tension between the rapid proliferation of micro-credentials and their reliable recognition has been identified as a central challenge in digital lifelong learning research (H. P. Li et al., 2026). Quality assurance policies for both education and employment training programmes must be adapted to the offer of micro-credentials, and education and labour authorities need procedures to coordinate policies having an impact on micro-credential offerings and learners (Schmoelz, 2023). As Read and Olcott (2025) note, there are no guarantees that micro-credentials will provide new revenue streams, while the development and refinement of assessment and certification standards may require significant fiscal and human resources.

Although online technologies expand access for many learners, those without reliable connectivity, suitable devices, accessible platforms, or digital literacy may be disadvantaged. Technology-enhanced learning must therefore be accompanied by inclusive design, technical support, and non-digital alternatives where necessary. The pandemic widened the digital divide, and investment in digital infrastructure and technologies in underserved areas is needed to reach disadvantaged students. In the Philippine context, policies must support the monitoring and evaluation of successful online learning while enhancing interoperability and strengthening cybersecurity and data privacy. Both Malaysia and Singapore continue to wrestle with persistent challenges including unequal access to high-quality digital infrastructure, inconsistent instructional design, and the need to ensure quality assurance in digital delivery. In healthcare micro-credential contexts, limited access to technology, particularly in low-income or rural settings, has been identified as a barrier to participation (Oliver & Brabin, 2026). Those with lower IT competency were less satisfied with micro-credential programmes, suggesting that digital literacy is a key consideration in programme design (Oliver & Brabin, 2026). Inclusive and accessible design, use, and occupation of digital space must be enabled and promoted so that all people can participate in digitalisation. The European digital governance system for vocational and adult education emphasises inclusion as a core dimension, yet findings reveal tensions between emancipatory forces and neoliberal pressures regarding who benefits from digital systems.

Other challenges include insufficient time for learning, insecure employment, training costs, limited employer support, and unequal access to career guidance. Workers in lower-paid, casual, regional, or declining industries may have the greatest need for reskilling but the least opportunity to participate. Prior to the pandemic, non-degree programmes were often outside the scope of public funding, and the costs of learning were largely borne by learners or their employers; therefore, learners from more privileged backgrounds have been the major beneficiaries of fee-based learning opportunities (Orman et al., 2023). Cost has been identified as a deterrent when learners are responsible for paying, and programmes need to be contextually relevant and professionally applicable to motivate learners (Oliver & Brabin, 2026). To ensure that micro-credentials remain accessible to all learners beyond the end of temporary funds, governments may need to consider making micro-credential financing more sustainable (Orman et al., 2023). Different types of learners have different support needs, and tailored programme design and support measures are essential (Orman et al., 2023). Adult learners often balance employment, family responsibilities, and study demands, necessitating flexible delivery and supportive learning environments (Wu et al., 2026). Micro-credential initiatives, properly designed and targeted, can promote greater equity of access to higher education, but ensuring that benefits are made available to learners across all groups of the population remains a key challenge. Access to education should be fair and inclusive for everyone, with a learner-centred approach placing the learner's needs at the centre of design, development, and delivery (OECD, 2023d).

Data governance becomes increasingly important as digital platforms collect information about learner progress, credentials, employment, and professional development. Institutions must protect privacy, maintain cybersecurity, ensure transparency, and avoid using predictive systems in ways that unfairly restrict learner opportunities. The exchange of information should respect the data protection rights of all actors, and end-user consent should comply with the General Data Protection Regulation (GDPR), with individuals able to clearly understand how their data is shared and what options they have to restrict access. A technical infrastructure focused on privacy control and data protection, openness, and interoperability should be established, while avoiding over-reliance on a single platform or commercial providers. Governance frameworks must address ethical pitfalls such as algorithmic bias, data privacy, and the transparency of AI-driven assessments (Ratna Selvaratnam & Michael Sankey, 2025). Future studies need to critically address the use of data science in digital governance systems for vocational and adult education, to fully unpack power relationships and whether they reinforce established biases and inequalities. The findings from bibliometric research underscore the need for coordinated policy frameworks, robust institutional governance, and ethically informed technology design to support inclusive and sustainable digital lifelong learning (H. P. Li et al., 2026). Enhanced online education platforms and tools should be designed following cybersecurity and data privacy regulations, and digital learning must safeguard privacy and security of learner profiles while protecting against fraudulent activities (OECD, 2023d).

Reskilling and Upskilling

Reskilling and upskilling are two closely related strategies that enable individuals and organisations to respond to technological innovation, changing labour market demands, economic transformation, and evolving workplace practices. Although the terms are often used together, they refer to different forms of learning. Upskilling involves developing additional knowledge and competencies that enhance performance within an existing occupation or professional field. Reskilling, by contrast, involves acquiring an entirely new set of competencies that prepares an individual to move into a different role, occupation, or industry (Cook, 1987; Sorenson, 1987). Both are fundamental components of lifelong learning and are increasingly recognised as essential for maintaining workforce adaptability in rapidly changing economies (Morrison, 2006).

Historically, workers often remained in a single occupation for much of their working lives, with relatively limited requirements for substantial retraining (Vafa et al., 2022). However, advances in artificial intelligence, automation, robotics, digital technologies, environmental sustainability initiatives, and changing business models have accelerated the pace of occupational change. Many existing jobs are being redesigned, new occupations are emerging, and some traditional roles are declining or disappearing (Frank et al., 2024; A. J. Griffiths et al., 2024). As a result, workers increasingly require opportunities to continually update existing capabilities while also preparing for entirely new career pathways when necessary (Morrison, 2006; Moss-Pech et al., 2021).

Upskilling focuses on expanding existing capabilities so that individuals can perform more effectively within their current occupation. This may involve learning to use new technologies, adopting revised workplace procedures, developing leadership capabilities, improving digital literacy, or gaining specialised technical expertise (A. J. Griffiths et al., 2024; Morrison, 2006). Upskilling enables workers to remain current as professional standards, legislation, equipment, and organisational practices evolve.

For example, a registered nurse may undertake professional development in telehealth technologies, an electrician may learn to install renewable energy systems, a teacher may develop expertise in artificial intelligence-assisted learning, or a civil engineer may undertake training in digital twin technology and Building Information Modelling (BIM). In each case, the individual remains within the same profession while extending their professional capability (A. J. Griffiths et al., 2024).

Upskilling benefits both employees and employers. Workers improve their employability, professional confidence, productivity, and career progression opportunities, while organisations develop a more capable workforce that can respond to technological and organisational change without relying solely on external recruitment (A. J. Griffiths et al., 2024; Morrison, 2006). Upskilling also supports innovation by enabling employees to apply new knowledge directly within their existing workplaces.

Reskilling involves preparing individuals for substantially different occupational roles by developing competencies that differ significantly from their existing qualifications or experience (Cook, 1987; Sorenson, 1987). This approach becomes particularly important when industries experience structural change, technological disruption, economic restructuring, or workforce shortages in emerging sectors (Frank et al., 2024; Moss-Pech et al., 2021). Historical cases of worker dislocation illustrate how displaced workers have been provided with job skill retraining, including technical preparation and long-term new skill vocational training, to facilitate transitions into new occupations (Cook, 1987; Sorenson, 1987). Research on career mobility further demonstrates that transitions between occupations are shaped by the similarity of skill requirements, suggesting that reskilling efforts must account for the distance between workers' existing competencies and those demanded by target occupations (Frank et al., 2024). Moreover, scholarship on adult education has shown that downward mobility and labour market precarity can compel individuals to pursue credentials in entirely new fields, sometimes at lower educational levels than previously attained, as part of broader reskilling processes (Moss-Pech et al., 2021).

For example, a manufacturing worker displaced through automation may retrain as a renewable energy technician. A retail employee may complete qualifications in cybersecurity, while a military veteran may transition into project management or logistics. Similarly, workers from declining fossil fuel industries may undertake reskilling programs to prepare for employment in renewable energy, environmental management, or advanced manufacturing.

Reskilling typically requires broader educational programs than upskilling because learners must acquire new technical knowledge, occupational standards, workplace practices, and professional identities. Depending on the occupation, reskilling may involve vocational qualifications, university programs, apprenticeships, industry certifications, workplace placements, or combinations of these learning approaches.

Drivers of Reskilling and Upskilling

Several interconnected factors are increasing demand for workforce reskilling and upskilling.

Technological innovation is perhaps the most significant driver. Artificial intelligence, automation, robotics, cloud computing, big data, cybersecurity, and digital communication technologies are changing how work is organised across virtually every industry. Rather than replacing all workers, these technologies are transforming job requirements by automating routine tasks while increasing demand for higher-order cognitive, technical, and interpersonal skills.

Economic restructuring also contributes to changing skill requirements. Global competition, changing consumer expectations, demographic shifts, and new business models continually reshape labour markets. Some industries expand rapidly while others contract, requiring workers to adapt by developing new capabilities.

Environmental sustainability is creating additional demand for reskilling. The transition towards renewable energy, low-carbon industries, circular economy practices, sustainable construction, and environmental management requires workers to develop new technical competencies while adapting existing occupations to more sustainable practices.

Demographic changes also influence workforce development. Ageing populations in many countries increase demand for healthcare, aged care, education, and community services, while technological advances continue to reshape manufacturing, finance, transport, agriculture, and professional services. These changes create ongoing requirements for both occupational mobility and continuous professional development.

Role of Education and Training

Educational institutions play a central role in supporting reskilling and upskilling by providing flexible learning pathways that accommodate adult learners with diverse backgrounds and responsibilities. Traditional full-time study is increasingly complemented by online learning, blended delivery, workplace training, micro-credentials, professional certificates, stackable qualifications, and recognition of prior learning.

Vocational education and training providers often deliver industry-specific skills closely aligned with occupational standards, while universities increasingly offer shorter professional development programs alongside traditional degrees. Industry associations, employers, professional bodies, and technology companies also contribute by developing specialised training aligned with emerging technologies and workforce needs.

Flexible learning is particularly important because many adult learners continue working while undertaking further education. Evening classes, online learning, modular programs, and self-paced study allow learners to balance education with employment, family, and community responsibilities.

Workplace Learning

Employers increasingly view learning as an ongoing organisational activity rather than an occasional training event (Anika et al., 2023; Kour & Ashraf, 2018). This shift reflects a broader recognition that a dynamic business environment necessitates continuous learning among employees, with training being important for skill updating and growth within any organisation (Kour & Ashraf, 2018). Organisations that offer effective training and opportunities for learning and development to their employees have been shown to achieve better employee retention outcomes (Abubakar et al., 2022; Kour & Ashraf, 2018). Indeed, human capital development has become part of an overall effort to achieve employee retention, requiring organisations to invest resources to ensure that employees have the knowledge, skills, and competencies they need to work effectively in a rapidly changing and complex environment (Abubakar et al., 2022).

Workplace learning may encompass a wide variety of approaches. These include mentoring, coaching, job rotation, project assignments, communities of practice, professional discussion, digital learning platforms, and supervised practice (Kajwang, 2022; Nahak & Srinivas, 2023; Naim & Lenka, 2018; Sudharshini, 2024). Mentoring, for instance, has been identified as a key construct with great potential to foster competency development, particularly among younger employees (Naim & Lenka, 2018). Training methods employed by organisations include on-the-job training, workshops, seminars, and e-learning initiatives, all of which are used to disseminate industry-specific knowledge and skills (Anika et al., 2023; Sudharshini, 2024). The development of digital skills through digital learning platforms has also emerged as a significant dimension of workplace learning, with motivation identified as a predictor of digital knowledge and skill acquisition that enables employees to become more proficient and valuable to their organisations (Ebnezer & Rajini, 2024).

These approaches enable employees to apply new knowledge directly within authentic workplace contexts while receiving immediate feedback and support (Abubakar et al., 2022; Kajwang, 2022). The process requires creating the necessary environments in which employees can learn better and apply innovative ideas, acquire new competencies, and develop skills, behaviours, and attitudes (Abubakar et al., 2022). Competency development has been described as "an organisational activity to maintain and enhance employees' career, knowledge, and skills and align the employees with strategic goals of an organisation" (Xulu, 2022, p. 9). In-service learning opportunities and continuous professional development have been shown to be associated with positive learning outcomes, including knowledge gained and skills learned, as well as greater career motivation (Morrow et al., 2024).

Knowledge sharing also constitutes an important dimension of workplace learning, as it involves social interaction and transfer of organisational insights and experiences resulting in the expansion of competencies (Naim & Lenka, 2018). Furthermore, the application of models such as the Develop-

Deploy-Connect framework demonstrates how development through training is an important component of HRM practices that enhances the human capital level in the organisation and fosters employee engagement (Kajwang, 2022).

Many organisations now develop workforce capability strategies that identify emerging skill requirements, map existing competencies, and provide structured professional development aligned with organisational objectives (Chethana & Noronha, 2023; OECD, 2021e; Xulu, 2022). Competency mapping has been described as a strategic talent management method that aids firms in developing and nurturing their employees while also accomplishing their goals, providing a framework for improving the skills, knowledge, behaviour, and qualities within an organisation's human capital (Krishna, 2022). A Talent Management Strategy (TMS) is described as a methodical and intentional technique that an organisation uses to recruit, develop, and retain exceptional individuals in order to achieve its corporate goals, entailing the matching of the workforce's talents and competencies with the organisation's objectives (Sudharshini, 2024).

The OECD has similarly identified the importance of building a digital workforce through attracting talents, developing a retention programme, and allocating skills, emphasising the need for a learning culture and appropriate organisational structures to support digital transformation (OECD, 2021e). Such structured approaches to workforce development help organisations remain competitive while supporting employee career development and retention (Kour & Ashraf, 2018; Ojo & Alli, 2024; Thusi & Chauke, 2023; Xulu, 2022). Organisations that adopt career development programmes are more likely to have a high sense of commitment and retention among their workers (Thusi & Chauke, 2023), and the availability of career development opportunities has been identified as the variable showing the highest significant relationship with employee retention (Abdulazeez, 2020). Training and development, alongside career planning, have been shown to be significant positive predictors of employee retention factors including motivation, workplace flexibility, and work-life balance (Abubakar et al., 2022).

Training programmes should be aligned with organisational objectives and employee needs in order to optimise their effectiveness (Anika et al., 2023; Chethana & Noronha, 2023). A successful organisation will have a well-established procedure for the professional development of every employee that is seamlessly integrated with other procedures such as evaluation and performance monitoring (Chethana & Noronha, 2023). Such approaches help organisations remain competitive while supporting employee career development and retention (Kour & Ashraf, 2018; Msacky, 2024; Ray & Sahu, 2025; Sampson) Continuous learning programmes, skill certifications, and career advancement pathways have been shown to significantly enhance job satisfaction and motivation, with career stagnation identified as a driver of dissatisfaction and increased attrition (Ray & Sahu, 2025).

Digital Technologies

Digital technologies have significantly expanded opportunities for reskilling and upskilling. Learning management systems, mobile learning applications, virtual classrooms, artificial intelligence, virtual and augmented reality, adaptive learning systems, and digital credentialing enable learning to occur regardless of time or location.

Artificial intelligence may personalise learning pathways by recommending resources, identifying knowledge gaps, providing immediate feedback, and adapting instruction according to learner progress. Virtual and augmented reality can provide realistic simulations for practising complex procedures safely before entering real workplace environments. Learning analytics help educators monitor learner progress and provide targeted interventions when additional support is required.

Micro-credentials and digital badges also support workforce development by recognising smaller units of learning that can be completed progressively and combined into larger qualifications. These credentials allow learners to demonstrate specific competencies without necessarily completing lengthy qualification programs before applying new skills in the workplace.

Challenges

Despite their benefits, reskilling and upskilling present several challenges for learners, employers, educational institutions, and policymakers. Workers may encounter financial constraints, limited time, family and caring responsibilities, and uncertainty about future employment opportunities, all of which can reduce participation in professional development and lifelong learning (OECD, 2021h; UNESCO, 2022; World Economic, 2025). Employers may also experience difficulties releasing staff for training, particularly within small and medium-sized enterprises, while accurately forecasting future workforce skill requirements remains increasingly difficult given the pace of technological change and labour market transformation (CEDEFOP, 2023; OECD, 2023c). Educational providers face the ongoing challenge of continuously reviewing curricula, learning resources, and assessment practices to ensure they remain aligned with rapidly evolving technologies, industry standards, and occupational requirements (Brown, Nic Giolla Mhichíl, et al., 2021; D. Lang, 2023; R. Selvaratnam & Michael Sankey, 2025).

Unequal access to education also risks reinforcing existing workforce inequalities. Individuals employed in secure, well-paid occupations often have greater access to employer-funded professional development, structured career pathways, and workplace learning than those employed in casual, low-paid, or insecure forms of work (OECD, 2021h; UNESCO, 2022). Additional barriers may affect participation among rural and remote communities, people with disabilities, culturally and linguistically diverse learners, older workers, and individuals with limited digital access or digital literacy, highlighting the importance of inclusive policies and targeted learner support (S. Graebe, 2025; Li & et al., 2026; OECD, 2021h). Research consistently demonstrates that participation in lifelong learning is unevenly distributed across educational and socioeconomic groups, indicating that expanding learning opportunities alone does not necessarily lead to equitable participation (Boyadjieva & Ilieva-Trichkova, 2017a; OECD, 2021h).

Quality assurance presents a further challenge as short courses, micro-credentials, and online learning opportunities become increasingly widespread. Learners, employers, professional bodies, and educational institutions require confidence that alternative credentials represent meaningful, rigorously assessed competencies and that they are recognised consistently across industries and education sectors (Antonaci, Klemke, et al., 2021; Brown, Nic Giolla Mhichíl, et al., 2021; OECD, 2021h). Maintaining trust in these credentials requires transparent quality standards, robust assessment processes, effective accreditation arrangements, and clear articulation pathways between micro-credentials, formal qualifications, and professional recognition (Conrad, 2023b; Robin Fitzgerald & Henk Huijser, 2021; Graebe & McIntyre-Hite, 2024). Without consistent quality assurance and recognition frameworks, the growing diversity of learning opportunities may reduce rather than enhance confidence in lifelong learning pathways.

Relationship to Theory-Informed Teaching

Reskilling and upskilling align with multiple theories of adult learning because they involve cognitive, social, experiential, behavioural, and personal dimensions of learning. Humanism supports learner autonomy, career development, and personal growth throughout working life. Constructivism recognises that adults build new understanding by connecting learning with prior experience and authentic workplace practice. Social constructivism highlights the importance of mentoring, collaboration, and communities of practice in developing professional competence.

Experiential learning is particularly important because adults refine new knowledge through repeated cycles of workplace application, reflection, and improvement. Cognitivism informs the organisation of learning pathways, scaffolding, and feedback that support increasingly complex knowledge acquisition, while behaviourism contributes through structured practice, reinforcement, and competency-based assessment where procedural mastery is required. Connectivism is increasingly relevant as workers use digital technologies, professional networks, online communities, and distributed knowledge sources to support continual professional development.

Ultimately, reskilling and upskilling represent more than workforce training strategies. They are central components of lifelong learning that enable individuals to remain adaptable, employable, and professionally competent throughout changing careers. As technological, economic, environmental, and social change continues to accelerate, the capacity to continually acquire new knowledge and competencies will become an increasingly important characteristic of individuals, organisations, and societies.

Sustainability and Education

Sustainability and education have become increasingly interconnected as societies seek to address complex environmental, social, and economic challenges. Climate change, biodiversity loss, resource depletion, population growth, technological transformation, and widening social inequalities have highlighted the need for education systems that prepare individuals not only for employment but also for responsible citizenship and sustainable development (International Commission on the Futures of Education, 2021; Sterling, 2011; UNESCO, 2020). Consequently, sustainability has evolved from being a specialised topic within environmental education to becoming a broad educational principle that influences curriculum design, teaching practice, institutional governance, and lifelong learning (International Commission on the Futures of Education, 2021; Tilbury, 2011; Wals, 2015).

Education for sustainability extends beyond teaching environmental science or conservation. It seeks to develop the knowledge, skills, values, attitudes, and behaviours that enable individuals and communities to make informed decisions that support environmental protection, social equity, economic prosperity, and intergenerational responsibility (UNESCO, 2020; Wiek et al., 2011). Learners are encouraged to understand the interconnected nature of ecological, economic, political, cultural, and technological systems and to consider the long-term consequences of individual and collective actions (International Commission on the Futures of Education, 2021; Rieckmann, 2018).

The contemporary understanding of sustainability is strongly influenced by the concept of sustainable development, commonly defined as development that meets present needs without compromising the ability of future generations to meet their own needs (World Commission on & Development, 1987).

This perspective recognises that environmental protection, economic development, and social wellbeing are interdependent rather than competing objectives (International Commission on the Futures of Education, 2021; Sachs et al., 2019).

Education plays a central role in achieving sustainable development because lasting change depends upon informed citizens, capable professionals, innovative organisations, and effective public policy (International Commission on the Futures of Education, 2021; United Nations General Assembly, 2015). Educational institutions prepare future engineers, healthcare professionals, teachers, business leaders, scientists, tradespeople, policymakers, and community members who will make decisions affecting environmental and social outcomes throughout their careers (International Commission on the Futures of Education, 2021; Rieckmann, 2018).

The United Nations Sustainable Development Goals (SDGs), particularly Goal 4: Quality Education, recognise education as both a goal in itself and an enabler of all other sustainable development objectives (United Nations General Assembly, 2015). Education contributes to improved health, reduced poverty, gender equality, responsible consumption, climate action, decent work, and stronger institutions by developing knowledge, critical thinking, and civic participation (International Commission on the Futures of Education, 2021; Leicht, 2018).

Education for Sustainable Development (ESD) is an educational approach that integrates sustainability principles throughout teaching, learning, research, and institutional practice. Rather than treating sustainability as a separate subject, ESD encourages learners to consider sustainability issues across disciplines and professional contexts (International Commission on the Futures of Education, 2021; UNESCO, 2020; Wals, 2015).

This approach promotes systems thinking, enabling learners to understand relationships between environmental, economic, social, and technological systems. It also develops futures thinking, encouraging learners to anticipate emerging challenges, evaluate alternative scenarios, and consider the long-term implications of present decisions. Problem-solving, collaboration, ethical reasoning, and active citizenship are equally important because many sustainability challenges require collective rather than individual responses (International Commission on the Futures of Education, 2021; Rieckmann, 2018; Wiek et al., 2011).

Education for Sustainable Development also encourages learners to move beyond acquiring knowledge towards taking informed action. Learners are encouraged to investigate authentic problems, evaluate evidence, consider multiple perspectives, engage with communities, and develop practical solutions that contribute to more sustainable outcomes (International Commission on the Futures of Education, 2021; Sterling, 2011; Wals, 2015).

Sustainability Across Educational Sectors

Sustainability is increasingly embedded throughout education systems, reflecting its growing importance as a cross-disciplinary priority rather than a stand-alone subject (Gouvernement français, 2023; Morin et al., 2021). Educational institutions are recognising that preparing learners for sustainable futures requires the integration of environmental, social, economic, and ethical considerations across curricula, learning activities, and institutional practices (Alqarni et al., 2023; Gouvernement français, 2023). Consequently, sustainability education begins in the earliest years of learning and continues through school, vocational education, higher education, and lifelong learning (Gouvernement français, 2023).

In early childhood education, sustainability is often introduced through practical, age-appropriate experiences that foster curiosity, respect for nature, and environmental stewardship (Bouttemont, 2010). Activities such as outdoor play, gardening, recycling, caring for plants and animals, and exploring local environments help young children develop an early appreciation of natural systems and the importance of caring for their surroundings (Bouttemont, 2010; Damus, 2023). These experiences establish foundational values and behaviours that can support lifelong environmental responsibility (Bouttemont, 2010; Damus, 2023).

As learners progress through school education, sustainability becomes increasingly integrated across a wide range of subjects rather than being confined to environmental studies alone (Gouvernement français, 2023; Morin et al., 2021). Science classes examine ecosystems, climate change, biodiversity, and renewable energy, while geography explores human-environment interactions and resource management (Guertin et al., 2019; Morin et al., 2021). Economics and business studies consider sustainable development and responsible consumption (Morin et al., 2021), technology and design investigate sustainable innovation (Alqarni et al., 2023), and civics examines the social and political dimensions of environmental decision-making (Morin et al., 2021; Torterat et al., 2021). Through these interdisciplinary approaches, students develop an understanding of the complex relationships between environmental, economic, and social systems while strengthening critical thinking and problem-solving skills (Anderson et al., 2022; Kim, 2015).

Within vocational education and training (VET), sustainability is closely aligned with occupational practice and industry requirements. Learners develop competencies that enable them to apply sustainable principles directly within their chosen professions, reflecting the growing demand for a workforce capable of supporting the transition to more sustainable economies (CEDEFOP, 2021; UNESCO-UNEVOC, 2017). Construction apprentices may learn environmentally sustainable building techniques and energy-efficient construction methods, electricians develop expertise in renewable energy technologies, automotive technicians acquire skills related to electric and hybrid vehicles, agricultural workers explore regenerative farming and sustainable land management practices, and hospitality students learn sustainable food sourcing, waste reduction, and resource-efficient operations (Pavlova, 2018; UNESCO, 2021b). By embedding sustainability within occupational competencies and workplace practice, VET prepares graduates to contribute to increasingly sustainable industries while supporting innovation, resource efficiency, and environmental stewardship throughout their careers (CEDEFOP, 2021; UNESCO-UNEVOC, 2017).

In higher education, sustainability has become an important consideration across diverse academic disciplines as universities respond to global sustainability challenges through teaching, research, community engagement, and institutional governance (Leal Filho et al., 2018; UNESCO, 2021b). Engineering programs increasingly incorporate sustainable infrastructure design, life-cycle assessment, renewable energy systems, and resource efficiency, while business and management programs examine corporate sustainability, environmental governance, ethical leadership, and sustainable finance (Leal Filho et al., 2018; Lozano et al., 2017). Health disciplines recognise the relationships between environmental conditions, climate change, and population health, and teacher education programs prepare future educators to integrate sustainability concepts into curriculum and pedagogy (UNESCO, 2021b). Across universities, sustainability is also embedded within institutional operations, research agendas, interdisciplinary collaboration, and community partnerships, reinforcing its role as a guiding principle for both higher education and wider societal development (Leal Filho et al., 2018; Lozano et al., 2015).

Sustainability in Adult and Workplace Learning

Adult education plays an important role in supporting the transition towards more sustainable economies and communities. Many workers require reskilling or upskilling as industries adopt renewable energy technologies, circular economy principles, low-carbon manufacturing, sustainable procurement, and environmentally responsible business practices (Maclean et al., 2018; Nail-Beatty et al., 2024; Strachan et al., 2022). The transition to a low-carbon economy requires skills development that impacts all job levels, and the focus on "green jobs" alone grossly underestimates the broader workforce skills needs (Strachan et al., 2022). Demand for skills specific to expanding industries such as renewable energy is growing, while demand for skills in fossil fuel-based industries is declining, creating a simultaneous need for reskilling and upskilling across transforming jobs (Maclean et al., 2018).

Workplace learning enables employees to develop competencies related to energy efficiency, waste minimisation, sustainable resource management, environmental compliance, carbon accounting, biodiversity protection, and sustainable supply chains (Cholakkathodi & Bikku, 2026; Maclean et al., 2018; Strachan et al., 2022). The movement toward greener practices encompasses broad goals including energy efficiency, circular resource use, and sustainable procurement (Pollard & Rapp-Wright, 2026). Continuous professional development ensures that professionals remain current as environmental regulations, technologies, and industry standards evolve (Beier et al., 2025; Pollard & Rapp-Wright, 2026). Lifelong learning is recognised as central to remaining productive and employed throughout the lifespan, and workplace learning will continue to be essential as industries undergo transformation (Beier et al., 2025). Employers expect to offer reskilling and upskilling to over 70% of their employees by 2025, though employee engagement in such programmes lags behind (Strachan et al., 2022). Education plays a major role in green economy skills development, but in the context of lifelong learning, other stakeholders including employers, individuals, and policymakers must also be engaged (Strachan et al., 2022).

Community education also contributes by increasing public understanding of sustainability through programs focusing on energy conservation, sustainable transport, water management, recycling, food security, financial sustainability, and disaster resilience (Cholakkathodi & Bikku, 2026; Maclean et al., 2018). Basic education, continuing education, and training are important for empowering people, including those in the informal sector, and for reducing poverty among marginalised communities (Cholakkathodi & Bikku, 2026). Green skills in demand encompass areas including energy, waste management, procurement, financial skills, and communication (Cholakkathodi & Bikku, 2026). Education and training institutions, industries, and all stakeholders must collaborate in evaluating and planning for skills development to support the green transition (Maclean et al., 2018).

Digital Technologies and Sustainability Education

Digital technologies are transforming sustainability education by expanding access to learning resources, improving collaboration, and supporting data-informed decision-making. Online learning, mobile applications, virtual laboratories, geographic information systems (GIS), digital twins, remote sensing, and environmental simulation software enable learners to investigate complex sustainability issues that would otherwise be difficult to observe directly.

Artificial intelligence can assist by analysing environmental data, modelling climate scenarios, supporting resource management, and personalising learning pathways (OECD, 2022c). Virtual and augmented reality can immerse learners in ecosystems, industrial facilities, or future urban environments, allowing them to explore sustainability challenges through realistic simulations. Learning analytics may also help educators evaluate learner engagement and identify opportunities to strengthen sustainability education.

However, digital technologies themselves raise sustainability considerations. Data centres, electronic devices, artificial intelligence systems, and digital infrastructure consume energy and generate electronic waste (Ewim et al., 2023; OECD, 2022c). The energy consumption of data centres is expected to continue rising significantly, with projections suggesting they could consume a substantial share of global electricity (Buyya et al., 2018; Ewim et al., 2023; OECD, 2022c). ICT systems, and data centres in particular, have a heavy environmental footprint featuring high consumption of non-renewable energy, waste production, and CO_2 emissions. The carbon footprint of data centres is among the fastest growing across the entire ICT sector, and without intervention, data centre emissions could rival those of major global industries (Ewim et al., 2023). Cloud computing, while offering potential energy efficiency gains over on-premise infrastructure (Columbres & Victoriano, 2024; Yenugula et al., 2024), still relies on energy-intensive data centres that raise concerns regarding carbon emissions and electronic waste (Uddin et al., 2017; Yenugula et al., 2024). Machine learning and cloud computing can help reduce environmental stress in specific domains but simultaneously raise concerns about energy use, natural resource consumption, and waste disposal (OECD, 2022c). Sustainability education therefore increasingly encourages learners to evaluate both the benefits and environmental impacts of emerging technologies (Uddin et al., 2017; Yenugula et al., 2024).

The Role of Educators

Educators play an important role in developing sustainability literacy. Rather than simply transmitting factual knowledge about environmental issues, they encourage learners to think critically, analyse complex systems, evaluate evidence, consider ethical implications, and develop informed responses to real-world challenges (Hogan & O'Flaherty, 2021; Kari & Disegha, 2026; Pratiwi et al., 2025; Shutaleva, 2023) . Critical thinking enables individuals to question assumptions, examine evidence, consider multiple perspectives, and apply logical reasoning to make informed decisions about environmental issues (Shutaleva, 2023).

Educators foster skills such as systems thinking, critical reflection, and ethical engagement, which are necessary for grappling with environmental complexity (Kari & Disegha, 2026; Pratiwi et al., 2025). Science education, in particular, emphasises the development of critical scientific literacy that enables learners to meaningfully navigate multiple perspectives presented in public debates relating to sustainability issues (Hogan & O'Flaherty, 2021). Environmental literacy itself encompasses knowledge, affect, skills including critical thinking and problem-solving, and responsible behaviour (Sindiani-Bsoul et al., 2026), while educators are expected to help students analyse information, evaluate competing perspectives, and construct evidence-based arguments (Sammel, 2014; Shutaleva, 2023). The integration of ethical and moral elements into education is also recognised as essential, with educators encouraging discussion of the social, economic, ethical, and environmental implications of sustainability challenges (Hogan & O'Flaherty, 2021; Li et al., 2022; Salinas et al., 2022).

Effective sustainability education often involves inquiry-based learning, project-based learning, community engagement, interdisciplinary collaboration, fieldwork, and authentic problem-solving

(Akram et al., 2025; Evans et al., 2017; Kari & Disegha, 2026; Kidman & Casinader, 2019; Pratiwi et al., 2025). These pedagogical approaches are rooted in constructivist learning theories and systems thinking frameworks that emphasise learning through active engagement and problem-solving (Akram et al., 2025; Evans et al., 2017). Inquiry-based practices develop the ability to think critically and engage with issues as an ongoing process (Kidman & Casinader, 2019), while problem-based learning gives students the skills, values, and information needed to relate disciplinary knowledge to global challenges such as pollution, biodiversity loss, and climate change (Kari & Disegha, 2026). Interdisciplinary cooperation is recognised as a key educational pillar alongside ecological literacy (Kari & Disegha, 2026), and experiential learning approaches such as permaculture-based education encourage students to actively engage with ecological systems and sustainability practices (Akram et al., 2025). Community-based projects evaluate students' ability to apply knowledge to local environmental challenges (Kari & Disegha, 2026), and SSI (socio-scientific issues) approaches allow learners to consider the complex systems in which issues are embedded (Sadler et al., 2019). These approaches encourage learners to investigate local and global issues while applying knowledge within meaningful contexts (Akram et al., 2025; Gal, 2024; Hogan & O'Flaherty, 2021).

Educators also model sustainable practices through institutional policies, resource management, curriculum design, and professional behaviour (Kwenin, 2026; Pratiwi et al., 2025; Stephens & Stephens, 2019). Teachers need to be equipped not only with environmental content knowledge but also with pedagogical strategies that foster systems thinking and ethical engagement, requiring the incorporation of sustainability principles into pre-service education and ongoing professional development (Dada et al., 2017; Kwenin, 2026; Pratiwi et al., 2025). Sustainable campuses, environmentally responsible procurement, waste reduction initiatives, and community partnerships reinforce the principles being taught in formal learning activities (Classens et al., 2023; Stephens & Stephens, 2019). For instance, institutions have pursued sustainability through energy efficiency, waste and recycling programmes, sustainable procurement, and education and outreach initiatives (Stephens & Stephens, 2019), while campus food systems alternatives demonstrate how procurement and production practices can embody sustainability values (Classens et al., 2023). Teacher training institutions should integrate comprehensive environmental sustainability education courses into preservice training programmes while providing continuous professional development for in-service teachers (Kwenin, 2026), and curriculum evaluations are necessary to determine progress toward sustainability literacy goals (Dada et al., 2017; Pratiwi et al., 2025).

Sustainability Competencies

International sustainability frameworks increasingly identify a core set of competencies that learners require across different educational levels, industries, and occupations (He, 2024; Holdsworth et al., 2018; Macagno et al., 2024). These competencies are intended to help individuals understand complex sustainability challenges, evaluate possible responses, and contribute to environmentally responsible, socially equitable, and economically viable outcomes (Chinedu et al., 2023). Because sustainability issues frequently cross disciplinary and organisational boundaries, these capabilities are relevant not only to environmental professions but also to business, engineering, health, education, government, trades, and community work (Chinedu et al., 2023; Henderson et al., 2025; Holdsworth et al., 2018).

Systems thinking enables learners to recognise relationships between environmental, economic, social, cultural, and technological systems rather than viewing problems in isolation (Décamps et al., 2021; Herliani, 2021; Jeronen, 2021). Critical thinking supports the evaluation of evidence, assumptions, policies, and competing perspectives (Shutaleva, 2023; Sidiropoulos et al., 2013), while futures thinking helps learners anticipate long-term consequences, examine uncertainty, and consider alternative scenarios (Herliani, 2021; Macagno et al., 2024; Merritt et al., 2018). Strategic thinking builds upon these capabilities by enabling learners to design practical and achievable responses to sustainability challenges (Carrascal et al., 2019; Macagno et al., 2024).

Collaboration is equally important because sustainability problems often require coordinated action across disciplines, organisations, sectors, and cultures (He, 2024; Kari & Disegha, 2026; Pettinger et al., 2023). Ethical reasoning helps learners balance environmental protection, economic development, social justice, cultural values, and intergenerational responsibilities when making decisions (Carrascal et al., 2019; Holdsworth et al., 2018). Self-awareness encourages individuals to reflect on their own values, behaviours, assumptions, and responsibilities (He, 2024), while integrated problem-solving supports the development of solutions that address multiple dimensions of sustainability simultaneously (He, 2024; Macagno et al., 2024).

Together, these competencies are highly transferable and support lifelong learning in rapidly changing professional environments (Holdsworth et al., 2018; Sandri et al., 2018). They enable learners and workers to respond more effectively to emerging technologies, evolving regulatory expectations, changing community priorities, and new sustainability demands within their industries and workplaces (Chinedu et al., 2023; Henderson et al., 2025; Weirs & Osborne, 2020).

Challenges

Despite growing recognition of sustainability within education, significant challenges remain. Sustainability concepts are often incorporated unevenly across curricula, with varying levels of institutional commitment and educator confidence (Kari & Disegha, 2026; Parry & Metzger, 2023). Some educators require additional professional development to integrate sustainability effectively within their disciplines (Heidt & Lamberton, 2014; Parry & Metzger, 2023; Tomasella et al., 2024). Many practicing educators feel unprepared to help learners develop the competencies needed for sustainability (Parry & Metzger, 2023), and business academics unfamiliar with sustainability principles would need support from disciplinary specialists as to how to embed sustainability where relevant throughout curriculum (Heidt & Lamberton, 2014).

Curriculum overcrowding can also limit opportunities to introduce new sustainability content (Parry & Metzger, 2023). Sustainability-related instruction is commonly perceived as an add-on that must compete for time within an already crowded curriculum (Parry & Metzger, 2023). Rather than adding separate sustainability units, many institutions now seek to embed sustainability throughout existing programs (Annan-Diab & Molinari, 2017; Edwards et al., 2017; Kari & Disegha, 2026). Microbiology curricula, for example, should be redesigned to embed sustainability concepts across all subfields rather than merely adding environmental topics (Kari & Disegha, 2026), and embedding sustainable development only in environmental courses or creating specific disciplines not connected to the mainstream curriculum will not be sufficient (Annan-Diab & Molinari, 2017). Edwards et al. describe a spectrum ranging from narrow integration in which sustainability is an "add-on" to broad integration in which it serves as a foundation for cross-disciplinary knowledge (Edwards et al., 2017).

Another challenge involves balancing environmental, economic, and social perspectives (Annan-Diab & Molinari, 2017; Romulo et al., 2024). Sustainability issues frequently involve competing priorities and complex trade-offs, requiring learners to engage with uncertainty rather than expecting simple solutions (Badarnah, 2025; Romulo et al., 2024). The Food-Energy-Water Nexus framework, for instance, recognizes the importance of impacts, trade-offs, and competing interests, and students must be educated on how to navigate these complex relations and holistically consider trade-offs (Romulo et al., 2024). In the context of sustainability, education must move beyond traditional linear approaches to encompass the complexity and uncertainty of real-world systems (Badarnah, 2025). Constructivist and experiential pedagogies emphasize engagement with real-world problems where learners must grapple with uncertainty and integrate stakeholder perspectives (Badarnah, 2025).

Access to sustainability education also varies internationally (Apollo & Mbah, 2021; Makrakis, 2023; Parry & Metzger, 2023). Differences in funding, technology, educational infrastructure, government policy, and local priorities influence the extent to which learners can participate in high-quality sustainability education (Parry & Metzger, 2023), (Semi et (Apollo & Mbah, 2021; Parry & Metzger, 2023; Semi et al., 2026). Institutions operating in resource-constrained environments face significant obstacles including unreliable internet connectivity, outdated hardware, and limited budgets (Semi et al., 2026). In East Africa, governments lack coherent approaches to leverage climate change education as a tool in their adaptation and mitigation strategies (Apollo & Mbah, 2021). Teachers' voices from across the world reveal common obstacles including disciplinary silos, a focus on high-stakes assessments, and inadequate professional learning opportunities, and these hurdles persist despite repeated calls from policymakers to better prepare and support educators (Parry & Metzger, 2023). The DREAM methodology has been applied across South East Asian universities to embed sustainability and SDGs in multiple academic disciplines, demonstrating efforts to address these disparities (Makrakis, 2023).

The DREAM methodology is a formative course assessment and continuous improvement framework developed by Vassilios Makrakis to evaluate teaching, learning, and curriculum implementation, particularly within Education for Sustainable Development (ESD) and initiatives supporting the Sustainable Development Goals (SDGs). Unlike traditional evaluation approaches that measure learning outcomes only at the conclusion of a course, DREAM gathers evidence throughout the learning process to inform ongoing improvements to both teaching practice and curriculum design. The methodology is grounded in the principles of transformative learning, critical pedagogy, constructivism, and assessment for learning, making it particularly well suited to learner-centred and sustainability-focused education.

The name DREAM represents five interconnected stages that guide the continuous evaluation and enhancement of learning. The first stage, Diagnosing, focuses on identifying learners' existing knowledge, experiences, expectations, values, strengths, and learning needs before instruction begins. This diagnostic process enables educators to tailor learning activities to the characteristics of the learner cohort rather than assuming a common starting point. Information may be collected through questionnaires, interviews, discussions, surveys, or reflective activities that provide insight into learners' backgrounds and preparedness.

The second stage, Reviewing and Reflecting, occurs throughout the course as learners regularly examine their experiences and learning progress. Reflection encourages learners to consider what has occurred, how their experiences compared with their expectations, what challenges emerged, how they responded, and what they learned from those experiences. Rather than waiting until the conclusion of a course, this ongoing reflection provides continuous feedback that supports both learner development and instructional improvement.

The Explaining stage extends beyond simple description by encouraging learners and educators to analyse why particular experiences, successes, or challenges occurred. Participants examine the relationships between teaching strategies, learning activities, assessment practices, learner engagement, and learning outcomes to identify the underlying factors that influenced learning. This analytical approach supports deeper understanding and informs evidence-based improvements to educational practice.

During the Assessing stage, evidence gathered throughout the learning process is evaluated to determine how effectively intended learning outcomes have been achieved and how well teaching approaches have supported student learning. Unlike traditional summative assessment, assessment within the DREAM methodology incorporates multiple sources of evidence, including learner reflections, qualitative feedback, observations, discussions, focus groups, and other indicators of learner engagement and achievement. This comprehensive approach provides a richer understanding of learning than reliance on examination results alone.

The final stage, Managing, uses the findings generated throughout the evaluation process to improve future teaching and curriculum design. Rather than viewing assessment as the endpoint of learning, DREAM treats evaluation as the foundation for continuous curriculum enhancement. Educators use the evidence collected to revise learning resources, modify teaching strategies, refine assessment methods, and improve course content before subsequent deliveries, thereby establishing an ongoing cycle of quality improvement.

The DREAM methodology is informed by several complementary educational theories. Transformative learning provides the foundation for encouraging critical reflection and changes in learners' perspectives and assumptions. Critical pedagogy positions learners as active participants who contribute to evaluating and improving educational practice rather than passive recipients of instruction. Constructivism recognises that learners actively construct knowledge through experience, reflection, and interaction with others, while assessment for learning emphasises the use of formative evidence to improve teaching and learning during the educational process rather than simply measuring achievement at its conclusion.

Although originally developed to support the integration of sustainability and the Sustainable Development Goals into university curricula, the DREAM methodology has broader applications across numerous educational settings. It has been used in higher education, vocational education and training (VET), professional development, workplace learning, curriculum evaluation, teacher education, instructional design, and other contexts where continuous improvement and learner-centred approaches are valued. Its emphasis on ongoing learner feedback, critical reflection, evidence-informed decision making, and curriculum enhancement makes it particularly valuable for educational programs seeking to develop higher-order thinking, professional competence, and transformative learning rather than relying solely on end-of-course assessment.

Relationship to Theory-Informed Teaching

Sustainability education aligns closely with multiple theories of adult learning. Humanism supports the development of personal responsibility, ethical awareness, self-directed learning, and commitment to social and environmental wellbeing (Budiarto et al., 2023; Chander et al., 2020; Sabri et al., 2022). Constructivism recognises that learners build understanding by connecting sustainability concepts with authentic experiences and prior knowledge (Olaleye et al., 2024; Wahid & Ain, 2023).

Social constructivism emphasises collaborative inquiry, community engagement, and shared problem-solving, recognising that sustainability challenges require collective action (Ekung & Odesola, 2018; Wahid & Ain, 2023). Experiential learning supports fieldwork, workplace projects, simulations, and community-based learning that allow learners to apply theoretical knowledge to authentic sustainability issues (Budiarto et al., 2023; Ekung & Odesola, 2018; Friberg & Hamilton, 2025).

Cognitivism contributes through the structured development of systems thinking, conceptual understanding, and complex problem-solving (Elsey & Sirichoti, 2003). Behaviourism remains relevant where repeated practice and feedback support the development of sustainable workplace procedures and environmentally responsible behaviours Sabri et al., 2022). Connectivism is increasingly important because sustainability knowledge is distributed across global research networks, professional communities, government agencies, industry organisations, and digital information systems (Abedini et al., 2021; Sabri et al., 2022). Learners must therefore develop the ability to locate, evaluate, integrate, and apply knowledge from multiple interconnected sources (Abedini et al., 2021).

Ultimately, sustainability and education share a common purpose: preparing individuals and communities to create resilient, equitable, and environmentally responsible futures (Budiarto et al., 2023; Chander et al., 2020; Ekung & Odesola, 2018). Education equips learners with the knowledge, competencies, values, and capacity for lifelong learning needed to respond to increasingly complex global challenges (Chander et al., 2020; Gandhi, 2022; Kijmongkolvanich et al., 2024). As sustainability becomes an increasingly important consideration across every profession and sector, education will continue to play a central role in supporting innovation, responsible decision-making, workforce transformation, and sustainable development (Budiarto et al., 2023; Chander et al., 2020; Ekung & Odesola, 2018).

Globalisation and Workforce Mobility

Globalisation has fundamentally transformed the nature of work by increasing the movement of people, knowledge, technology, capital, and services across national borders. Advances in transportation, digital communication, international trade, and multinational business operations have created labour markets that are increasingly interconnected (Beňo, 2020), allowing organisations to recruit talent globally while enabling workers to pursue employment opportunities beyond their home countries (Frazier & Ewing, 2009). At the same time, digital technologies have reduced the importance of geographical location for many occupations (Beňo, 2021; Tsapenko & Grishin, 2022), making international collaboration and remote work increasingly common (Aroles et al., 2021; Contreras et al., 2020). As a result, workforce mobility has become a defining characteristic of the modern global economy (Beňo, 2021; Tsapenko & Grishin, 2022).

Workforce mobility refers to the movement of workers between jobs, industries, occupations, regions, and countries in response to employment opportunities, economic conditions, technological change, or personal aspirations. Mobility may occur within a single organisation through career progression, across industries as workers change occupations, or internationally through migration, temporary employment, professional exchanges, and multinational assignments (Sequera, 2024; Tsapenko & Grishin, 2022). Increasingly, mobility also occurs virtually, with employees contributing to organisations located in other countries without relocating physically through remote work and distributed teams (Contreras et al., 2020; Frazier & Ewing, 2009; Tsapenko & Grishin, 2022).

Globalisation has expanded employment opportunities by allowing businesses to access larger markets and specialised talent while providing workers with access to a broader range of careers (Beňo, 2020; Peiris, 2021). Highly skilled professionals, including engineers, healthcare practitioners, information technology specialists, educators, and researchers, frequently work across international borders where their expertise is in demand (Sequera, 2024). Skilled tradespeople also experience increasing mobility as countries seek workers to address labour shortages in construction, manufacturing, transport, energy, agriculture, and infrastructure (Tomić & Vizinger, 2023). These movements contribute to economic development, knowledge exchange, innovation, and the transfer of technical expertise between regions and industries (Beňo, 2020; Tsapenko & Grishin, 2022).

Technological advances have accelerated workforce mobility by enabling digital collaboration regardless of physical location. High-speed internet, cloud computing, videoconferencing, collaborative software, and artificial intelligence have supported the growth of remote work, virtual teams, freelancing, digital entrepreneurship, and cross-border service delivery (Balakrishnan, 2022; Jabeen et al., 2024). Many organisations now operate with geographically dispersed workforces whose members collaborate across multiple time zones, cultures, and regulatory environments (Balakrishnan, 2022). This shift has created new employment opportunities while requiring workers to develop digital communication, self-management, intercultural collaboration, and virtual teamwork skills (Achoki, 2023; Balakrishnan, 2022).

Globalisation has also influenced the skills increasingly valued by employers. While technical expertise remains essential, organisations now seek workers who possess transferable capabilities that enable them to operate effectively in diverse and rapidly changing environments. These include communication, collaboration, cultural competence, adaptability, digital literacy, critical thinking, problem-solving, creativity, ethical decision-making, and lifelong learning (Achoki, 2023; Marlapudi & Lenka, 2024). Workers must often understand international standards, regulatory requirements, sustainability expectations, and emerging technologies while collaborating with colleagues, clients, and stakeholders from different cultural and linguistic backgrounds (Achoki, 2023; Branca et al., 2022). The World Economic Forum has projected that half of all employees would need to be reskilled by 2025, underscoring the scale of competency transformation required (Kertechian & El-Farr, 2024; Zhao & Jakkampudi, 2023).

Intercultural competence has become particularly important as workplaces become more diverse and internationally connected. Employees increasingly interact with people from different cultural traditions, languages, values, and communication styles. Effective collaboration requires cultural awareness, empathy, respect, and the ability to adapt communication and leadership approaches to diverse contexts (Achoki, 2023). Organisations similarly benefit from inclusive workplace cultures that value diversity, encourage multiple perspectives, and support equitable participation. Research consistently demonstrates that culturally diverse teams can improve innovation, creativity, and organisational problem-solving when effectively managed (Achoki, 2023).

Globalisation has also created significant challenges for workers, employers, and governments. Increased international competition may contribute to job displacement in industries exposed to global market pressures, while automation and outsourcing have altered employment patterns in many sectors (Jawaid & Ahmed, 2023; Kertechian & El-Farr, 2024; Li, 2023). Differences in labour standards, wages, employment protections, and professional recognition between countries can create inequalities and complicate workforce mobility (Hwang & Yoon, 2023; Li, 2023). Skilled migration may generate economic opportunities for individuals while contributing to workforce shortages in countries experiencing the loss of highly qualified professionals, a phenomenon commonly described as "brain drain" (Osatis & Asavanirandorn, 2022). Governments and international organisations therefore seek to balance labour mobility with workforce planning, skills development, and policies that promote equitable economic growth (Amoud & Raymer, 2025; Jawaid & Ahmed, 2023; Shi, 2025).

Education and training systems play a central role in preparing individuals for an increasingly global workforce. Schools, vocational education and training (VET) providers, universities, and workplace learning programs are expanding their focus beyond occupation-specific knowledge to include transferable skills that support career mobility and international employability (Fairman et al., 2022; Minali & Manju, 2018; Teik, 2019). Greater emphasis is being placed on digital capability, intercultural communication, language learning, sustainability, entrepreneurship, ethical practice, and global citizenship (Fairman et al., 2022; M. U. Hassan et al., 2024; Trần, 2019). The importance of intercultural approaches in business and management has been identified as critical to operating in a globalised environment, with structured collaborative engagement needed to develop human capability across borders (Fairman et al., 2022).

Faculty in higher education are increasingly recognising their professional responsibilities in preparing students for global labour markets, though interpretations of "global competence" vary, encompassing the capability to engage and perform effectively in globalised contexts (Trần, 2019). Micro-credentials, professional certifications, recognition of prior learning (RPL), and international qualification frameworks are also improving the portability of skills and qualifications across industries and national borders (Antonaci, Henderikx, et al., 2021; Bartley et al., 2016; Teik, 2019). The Common Microcredentials Framework (CMF), for instance, uses Bologna tools such as the European Qualifications Framework (EQF) and the European Credit Transfer and Accumulation System (ECTS) to provide a foundation for mutual trust and recognition through transparency (Antonaci, Henderikx, et al., 2021). Similarly, the IASSW Global Qualifying Standards document recognises the increasing mobility of professions and aims for consistency within and between qualifications to facilitate portability (Bartley et al., 2016). For engineering, mutual recognition of substantial equivalence of engineering education has been identified as a necessary step for mobility of engineering personnel across borders (Teik, 2019).

From a theoretical perspective, globalisation and workforce mobility are supported by several complementary learning theories. Humanism emphasises learner autonomy, personal growth, and self-directed career development throughout increasingly diverse career pathways. Constructivism and social constructivism recognise that professional knowledge is developed through authentic experiences, collaboration, and participation within diverse cultural and workplace communities. The social construction of economic institutions and the embeddedness of economics in the social domain have been recognised as important for understanding workforce development in globalised contexts (Fairman et al., 2022).

Connectivism highlights the importance of digital networks, online communities, and distributed knowledge for continuous professional learning in globally connected environments; the exponential growth of digital platforms and telemedicine during COVID-19 demonstrated how online tools can take clinical or academic expertise to distant parts of the world, making them more accessible and affordable (Khan et al., 2022). Experiential learning explains how workers develop expertise through international placements, workplace projects, cross-cultural collaboration, and reflection on professional experiences (Bartley et al., 2016; Trần, 2019).

Transformative learning recognises that exposure to different cultures and perspectives can reshape individuals' assumptions, identities, and professional practice; new forms of global practice and internationalisation of education are emerging in response to new mobilities, requiring new forms of education and training to account for the increasing complexities created by intensification in border crossings (Bartley et al., 2016). Higher education institutions have been analysed through multiple theoretical lenses including World Systems Theory and Structuration Theory to understand how they operate within a global education system (M. U. Hassan et al., 2024).

The growth of globalisation and workforce mobility reinforces the importance of lifelong learning. Careers are increasingly characterised by multiple transitions between occupations, industries, employers, and geographical locations rather than continuous employment within a single profession or organisation (Fairman et al., 2022; Mazlan & Jambulingam, 2023). The global economy has impacted the business environment, causing the workforce to become more extensive and diversified, with people from every corner of the world able to work at their desired companies (Mazlan & Jambulingam, 2023). Globalisation has accelerated the reformation of industry and expanded its pervasiveness, seriously challenging many nations by commitments toward greater employee mobility (Fairman et al., 2022). Workers therefore require ongoing opportunities to update technical expertise, develop transferable competencies, and respond to changing labour market demands throughout their working lives (Minali & Manju, 2018; Teik, 2019; Tun, 2007).

Education makes an individual capable of adopting changes, whether innovation, new technology, or new culture, and skilled educated workers are demanded more internationally, which increases labour mobility (Minali & Manju, 2018). National education systems must be able to meet the needs of the global market to maintain and enhance the competitiveness of the national economy (Tun, 2007). Education systems that foster adaptability, resilience, continuous learning, and global perspectives are better positioned to prepare individuals for sustainable careers within an increasingly interconnected world (M. U. Hassan et al., 2024; Teik, 2019; Tun, 2007). As emphasised in the engineering context, what is important is for educators to train future graduates who can embrace lifelong learning and possess strong basic fundamentals, producing graduates who are "ready-to-evolve rather than graduates who are just ready-to-market" (Teik, 2019, p. 3).

Education Policy and Reform

Education policy and reform refer to the processes through which governments, educational institutions, industry bodies, and other stakeholders develop, implement, evaluate, and revise the laws, regulations, funding arrangements, curricula, assessment systems, and governance structures that shape education. Education policy establishes the direction and priorities of education systems, while reform involves planned changes intended to improve educational quality, equity, efficiency, accessibility, and responsiveness to social, economic, technological, and environmental change (Gouëdard et al., 2020; Lokshyna et al., 2021). As societies evolve, education systems must continually adapt to prepare learners with the knowledge, skills, and capabilities required for contemporary life and work (Lokshyna et al., 2021; Vinh et al., 2022).

Education policy operates at multiple levels, ranging from international agreements and national legislation to regional regulations, institutional policies, and classroom practice Michel, 2016), (Mikulec & Ermenc, 2016). International organisations such as UNESCO, the OECD, the World Bank, and the European Union influence education through policy recommendations, comparative research, global development agendas, and international benchmarks (Alazmi, 2024; Carvalho et al., 2023; Engel & Rutkowski, 2014; Lokshyna et al., 2021; Michel, 2016; Singh & Ehlers, 2020). As Alazmi (2024) notes, international institutions exert profound influence through the provision of educational schemes and aid for global education reform efforts, with UNESCO and the World Bank devoted to helping nations formulate and execute their educational policies and strategies. The OECD has become one of the leading international organizations in education policy formation, providing influential policy advice and recommendations to varying levels of government (Engel & Rutkowski, 2014), while its PISA programme has had significant impact on national education policies across many countries (Michel, 2016; Morgan, 2007; Murphy, 2014).

National governments establish legislative frameworks, curriculum standards, qualification systems, funding models, and accountability requirements (Karakhanyan, 2018; Lokshyna et al., 2021), while regional authorities and educational institutions adapt these policies to local contexts (Michel, 2016; Mikulec & Ermenc, 2016). Effective policy therefore requires coordination across multiple levels of governance to ensure consistency while allowing flexibility for diverse educational settings. As Michel describes, the notion of governance in education has evolved toward "multi-level governance" to account for the different levels of decision-making influenced by international standards and benchmarks from the EU, OECD, World Bank, and UNESCO (Michel, 2016). This multi-level character is further reflected in the development of qualifications frameworks, where EU member states develop policies not as autonomous countries but through complex multilevel processes that take place above, between, and within national borders (Mikulec & Ermenc, 2016).

Historically, education reforms have reflected changing societal priorities. During the nineteenth and early twentieth centuries, reforms focused on expanding access to basic education, improving literacy, and supporting industrialisation through mass schooling. Following the Second World War, many countries expanded secondary and higher education to support economic growth, scientific advancement, and social mobility (Karakhanyan, 2018; Suleymanov et al., 2023). More recent reforms have responded to globalisation, technological innovation, demographic change, environmental sustainability, and increasingly knowledge-based economies (Alazmi, 2024; Lokshyna et al., 2021; Michel, 2016). Globalisation has fundamentally changed the format of educational policy development, with the national level now complemented by the supranational level through international organizations (Lokshyna et al., 2021). Contemporary education policy places growing emphasis on lifelong learning, digital transformation, inclusion, workforce development, sustainability, and preparing learners for uncertain futures (Y. Lyu, 2023; Singh & Ehlers, 2020; Vinh et al., 2022). UNESCO, the OECD, the EU, and the World Bank have collaborated to make lifelong learning an international trend in reforming national education and training systems (Y. Lyu, 2023), while organisations such as the World Bank have introduced frameworks for performance measurement and benchmarking to drive concrete reforms in education systems (Singh & Ehlers, 2020).

The emphasis on revising curriculum frameworks to incorporate critical thinking, creativity, collaboration, and problem-solving aligns with global reform trends. Indonesia's Merdeka Belajar reform exemplifies this shift, moving toward foundational learning, student well-being, and competency-based approaches, with curriculum features aligning with global trends in curriculum policy (Kioupi et al., 2023). Cambodia's education reform similarly prioritises improving teaching methods, integrating active learning, and conducting regular student assessment as core academic management strategies (Flood & Banks, 2021). The focus on personalised learning and authentic assessment reflects broader movements toward learner-centred education, with reforms emphasising that curriculum and assessment provide the framework that guides teaching and learning in the classroom (Kioupi et al., 2023).

Educational equity has become a central priority in education policy and reform, with governments and institutions seeking to reduce disparities associated with socioeconomic status, disability, gender, ethnicity, language, and geographic location. Inclusive education policies increasingly focus on ensuring that all learners can participate meaningfully in high-quality education by identifying and removing barriers that restrict access, engagement, and achievement (Filiz & Albay, 2025; Maguvhe & Mutambo, 2023; Schwarz & Sollenberger, 2024). Rather than treating diversity as an exception requiring individual accommodation, contemporary reforms recognise learner diversity as a normal characteristic of education systems and seek to embed inclusion throughout policy, curriculum, teaching, and assessment.

A key framework underpinning these reforms is Universal Design for Learning (UDL), which promotes the proactive design of curriculum, learning environments, teaching practices, and assessments that accommodate the diverse needs of learners from the outset (McKenzie & Dalton, 2020; Moodley, 2024). Instead of relying primarily on individual adjustments after barriers emerge, UDL encourages educators to provide multiple means of engagement, representation, and expression so that learners with differing abilities, cultural backgrounds, languages, and learning preferences have equitable opportunities to participate and succeed. The growing international adoption of UDL demonstrates its influence on inclusive education policy, with implementation reported across countries including Ghana, South Africa, Ireland, New Zealand, Canada, and the United States (Aldhilan et al., 2025; Ezeonwumelu et al., 2025; Maguvhe & Mutambo, 2023; Mantey, 2019; Moodley, 2024).

Inclusive education reforms also emphasise the complementary roles of differentiated instruction and culturally responsive teaching. These approaches recognise that effective inclusion requires flexible teaching strategies that respond to learners' diverse strengths, experiences, cultural identities, and educational needs rather than applying uniform instructional methods (Maguvhe & Mutambo, 2023; Mantey, 2019). For example, South Africa's inclusive education reforms position UDL as a practical mechanism for translating policy objectives into classroom practice, while education policy in Nova Scotia, Canada, highlights the importance of providing equitable, culturally responsive, and linguistically inclusive learning environments that support the wellbeing and achievement of every learner (Maguvhe & Mutambo, 2023). Across diverse contexts, researchers consistently argue that identifying and removing barriers to participation is fundamental to improving educational outcomes for all learners rather than only those with identified disabilities.

Curriculum design has likewise become a significant focus of inclusive education reform. Increasingly, scholars argue that curricula should be designed with learner diversity in mind from the outset, as curricula developed around narrow assumptions about learners can unintentionally exclude students with differing abilities, backgrounds, languages, and learning needs (Whitley & Hollweck, 2020). Experiences from countries such as Zambia further demonstrate that insufficient curriculum reform can limit the successful implementation of inclusive education policies, particularly for learners with special educational needs, highlighting the importance of aligning curriculum design with broader policy objectives (Filiz & Albay, 2025).

The COVID-19 pandemic further accelerated digital transformation across education systems and significantly influenced education policy. The rapid transition to remote and blended learning demonstrated both the potential of educational technologies and the challenges associated with ensuring equitable participation. Many countries subsequently invested in digital learning platforms, online assessment systems, and technology-enabled curriculum reforms to strengthen educational resilience. For example, Indonesia introduced digital platforms to support teachers, schools, and regional education authorities in implementing revised curriculum and assessment practices (Kioupi et al., 2023). At the same time, the pandemic exposed persistent inequalities related to digital access, technological infrastructure, learner support, and educator capability, reinforcing the importance of designing inclusive digital learning environments and addressing barriers such as digital skills, equipment availability, financial constraints, and accessibility requirements (Jardinez & Natividad, 2024; Whitley & Hollweck, 2020).

Another increasingly important area of education reform is strengthening the relationship between education and labour market needs. Governments and education providers are placing greater emphasis on entrepreneurship, innovation, employability, and transferable skills to prepare learners for rapidly changing economies. Entrepreneurship education is increasingly viewed as a means of developing capabilities such as creativity, adaptability, problem-solving, initiative, and responsible risk-taking, while inclusive education policies recognise that broad participation in education contributes to stronger communities, more inclusive labour markets, and sustainable economic development. Growing demand for STEM professionals has further reinforced the need for policies that support participation by learners from diverse backgrounds, including those with disabilities and other traditionally underrepresented groups. Similar workforce-focused reforms have been observed in countries such as Cambodia, where curriculum improvement, innovative teaching practices, and stronger community partnerships are intended to better align education with economic development priorities (Flood & Banks, 2021). Collectively, these reforms reflect a growing recognition that inclusive curriculum design not only promotes educational equity but also equips learners with the knowledge, skills, and competencies required to contribute effectively to diverse workplaces and society (Jardinez & Natividad, 2024).

Education policy has also increasingly incorporated sustainability as a guiding principle. International frameworks, including the United Nations Sustainable Development Goals, particularly SDG 4: Quality Education, encourage education systems to promote inclusive, equitable, and lifelong learning opportunities while integrating sustainability across curricula and institutional practices (Adhikari & Shah, 2021; Sever & Tok, 2022; Soomro et al., 2025). SDG 4 aims to "ensure inclusive and equitable quality education and promote lifelong learning opportunities for all" (Soomro et al., 2025). Education for Sustainable Development (ESD) encourages learners to understand the environmental, social, economic, and cultural dimensions of sustainable development while developing the competencies needed to contribute to more sustainable communities and workplaces (Sever & Tok, 2022; Shava, 2020; Yelubayeva et al., 2023). ESD is expected to transform society by empowering people of all ages and generations, and is recognised as a lifelong learning process and an integral part of quality education (Sever & Tok, 2022). Consequently, sustainability is increasingly reflected in curriculum reform, teacher education, institutional governance, and quality assurance processes, (Cembranel et al., 2024; Hou et al., 2025; Sever & Tok, 2022; Yelubayeva et al., 2023).

Effective education reform depends upon evidence-informed policymaking. Policymakers increasingly draw upon educational research, international comparisons, labour market analysis, learner outcomes, and stakeholder consultation when designing reforms (Ciriello, 2025; Faiz, 2023). Large-scale assessments, educational data, programme evaluations, and longitudinal research help identify areas requiring improvement and assess the effectiveness of policy interventions (Ciriello, 2025; Hoang et al., 2020). At the same time, policymakers must balance empirical evidence with social values, political priorities, economic constraints, and the diverse needs of learners, educators, employers, and communities (Ciriello, 2025; Faiz, 2023). Successful reform therefore requires ongoing consultation, implementation support, monitoring, and evaluation rather than relying solely on policy design (Faiz, 2023).

Despite its importance, education reform presents significant challenges. Educational systems are complex, and reforms often require changes to curriculum, assessment, teacher professional development, institutional leadership, funding, infrastructure, and organisational culture simultaneously (Mahere, 2024). Frequent policy changes may create uncertainty for educators, while insufficient resources, inadequate professional learning, or limited stakeholder engagement can reduce the effectiveness of reform initiatives (Faiz, 2023; Mahere, 2024). Furthermore, balancing accountability with professional autonomy, maintaining equity while promoting excellence, and responding to rapidly changing technological and labour market conditions remain continuing policy challenges across many countries (Ciriello, 2025; Hou et al., 2025).

From a theoretical perspective, education policy and reform are informed by a range of educational theories. Behaviourist perspectives have historically influenced standards-based curricula, measurable learning outcomes, and accountability systems. Cognitivism has informed curriculum sequencing, instructional design, and approaches that support cognitive development. Constructivism and social constructivism have contributed to learner-centred curricula, inquiry-based learning, collaboration, and authentic assessment. Humanism has influenced policies promoting learner wellbeing, personal development, autonomy, and inclusive education, while transformative learning has shaped reforms that encourage critical reflection, social responsibility, and active citizenship (Yelubayeva et al., 2023; Yun-bo & Zhang, 2025). Connectivism has become increasingly relevant as digital technologies, online learning, and global knowledge networks reshape educational practice (Shava, 2020).

Education policy and reform seek to ensure that education remains responsive to changing societal needs while maintaining its broader purposes of developing knowledgeable, capable, ethical, and engaged citizens (Ciriello, 2025). Effective reform extends beyond revising curriculum documents or introducing new technologies; it requires sustained investment in teachers, institutions, learners, and communities, together with ongoing evaluation and continuous improvement (Mahere, 2024; Shava, 2020). As societies continue to confront technological disruption, environmental challenges, demographic change, and global interdependence, education policy will remain central to preparing individuals and communities for an increasingly complex and rapidly evolving world (Ciriello, 2025; Sever & Tok, 2022; Yelubayeva et al., 2023).

Industry 4.0 and Society 5.0

Industry 4.0 and Society 5.0 are two complementary concepts that describe the transformation of economies, workplaces, and societies through advanced digital technologies. While both recognise the importance of digital innovation, they differ in their primary focus. Industry 4.0 concentrates on the digital transformation of manufacturing and industrial processes through automation, connectivity, and intelligent systems (Marku, 2024; Németh et al., 2021; Spoettl, 2020; Spoettl & Tūtlys, 2020). Specifically, Industry 4.0 encompasses cyber-physical systems, the Internet of Things, cloud computing, artificial intelligence, robotics, and big data analytics, primarily applied to manufacturing processes across different sectors (Akyazi et al., 2020; Marku, 2024; Naghipour & Rahim, 2024; Spoettl, 2020). Society 5.0 extends this technological transformation beyond industry by positioning technology as a means of improving human wellbeing, addressing social challenges, and creating a more sustainable and inclusive society (M. A. Hassan et al., 2024; Nešić, 2023; Saniuk & Grabowska, 2023).

Industry 5.0, closely aligned with Society 5.0, emphasises a human-centric, sustainable, and resilient approach, highlighting the importance of human involvement and the "rehumanisation" of automation (Ikenga & Sijde, 2024; Nešić, 2023; Saniuk & Grabowska, 2023; Uzoamaka & Sijde, 2024). In particular, Industry 5.0 seeks to put human welfare at the core of manufacturing systems, achieving societal goals beyond employment and growth (M. A. Hassan et al., 2024), and the European Commission has identified challenge areas including human-machine interaction, artificial intelligence, biotechnologies, and technologies for energy efficiency (Saniuk & Grabowska, 2023). Together, these concepts are reshaping education, workforce development, and lifelong learning by changing the knowledge, skills, and competencies required for future work (Adel, 2024; Marku, 2024; Miah et al., 2024; Sima et al., 2020; YÜCeol, 2021). The emergence of new job roles demanding advanced digital skills, the need for continuous reskilling and upskilling, and the imperative for lifelong learning are widely recognised consequences of these transformations (Adel, 2024; Marku, 2024; Nešić, 2023; Puhovichová & Jankelová, 2022).

Industry 4.0

The term Industry 4.0 originated in Germany in 2011 as part of a national strategy to strengthen manufacturing competitiveness through digital technologies (Abdelmajied, 2022; Bakes et al., 2022). It represents the Fourth Industrial Revolution, following the mechanisation of production through steam power, mass production enabled by electricity, and automation through electronics and information technology (Abdelmajied, 2022; Asadollahi-Yazdi et al., 2020; Mhlanga, 2021; Naudé, 2017). Industry 4.0 builds upon these earlier developments by integrating physical production systems with digital technologies to create highly connected, intelligent, and autonomous workplaces (Abdelmajied, 2022; Ajayi et al., 2023).

At the centre of Industry 4.0 are cyber-physical systems, which integrate sensors, computing, communication networks, and physical equipment to enable machines to monitor, analyse, and respond to changing conditions with minimal human intervention (Abdelmajied, 2022; Verma et al., 2022). These systems are supported by technologies such as the Internet of Things (IoT), artificial intelligence (AI), machine learning, robotics, cloud computing, big data analytics, digital twins, blockchain, additive manufacturing (3D printing), augmented reality, and advanced automation (Abdelmajied, 2022; Bongomin et al., 2019; Bongomin et al., 2020). Together, these technologies enable real-time data exchange, predictive maintenance, intelligent decision-making, flexible production, and improved operational efficiency (Abdelmajied, 2022; Asadollahi-Yazdi et al., 2020).

Industry 4.0 has transformed manufacturing, logistics, healthcare, agriculture, mining, construction, energy, transport, and many service industries (Bongomin et al., 2020; Dachyar et al., 2019; Olapoju, 2022; Wang et al., 2022). Smart factories use interconnected equipment that continuously monitors production performance and automatically adjusts processes to improve quality and efficiency (Abdelmajied, 2022; Verma et al., 2022). Autonomous vehicles support transport and logistics (Olapoju, 2022; Park, 2021), precision agriculture uses sensors and AI to optimise crop production (Bongomin et al., 2020; Nkosi et al., 2024), while healthcare increasingly relies on digital diagnostics, wearable technologies, and data-driven clinical decision-making (Ayele et al., 2024; Bongomin et al., 2020). Similar technologies are being applied across construction through Building Information Modelling (BIM), digital engineering, and automated equipment (Bakes et al., 2022).

The widespread adoption of Industry 4.0 technologies has significantly altered workforce requirements. Routine and repetitive tasks are increasingly automated, while demand has grown for workers capable of operating, maintaining, analysing, and improving complex digital systems (Bongomin et al., 2019; Nkosi et al., 2024). Technical knowledge remains important, but employers increasingly value transferable capabilities such as digital literacy, systems thinking, data analysis, critical thinking, creativity, collaboration, adaptability, and continuous learning (Bongomin et al., 2019; Micheni et al., 2021). Consequently, education systems are expanding their focus beyond occupation-specific skills to prepare learners for technology-rich workplaces characterised by continual innovation (Bongomin et al., 2019; Naudé, 2017).

Society 5.0

Society 5.0 was introduced by the Japanese Government in 2016 through the Fifth Science and Technology Basic Plan (Gobinath, 2024; Rybalko et al., 2021). Whereas Industry 4.0 primarily focuses on transforming industry and manufacturing, Society 5.0 seeks to create a human-centred society in which digital technologies are used to improve quality of life, solve complex societal challenges, and support sustainable development (Adel & Alani, 2024; Chandre, 2024; Gobinath, 2024). Rather than viewing technological advancement as an end in itself, Society 5.0 emphasises technology as a tool for enhancing human wellbeing while balancing economic growth with social inclusion and environmental sustainability (Adel & Alani, 2024; Gobinath, 2024; J. Rane et al., 2024).

The concept describes the evolution of human society through five stages. Society 1.0 represents hunter-gatherer communities, Society 2.0 agricultural societies, Society 3.0 industrial societies, Society 4.0 information societies, and Society 5.0 a "super-smart" society in which cyberspace and the physical world are closely integrated (Chandre, 2024; Gobinath, 2024). Artificial intelligence, the Internet of Things, robotics, autonomous systems, big data, and digital platforms continuously collect, analyse, and utilise information to support decision-making and deliver personalised services across healthcare, education, transport, public administration, and environmental management (Adel & Alani, 2024; Chandre, 2024; Gobinath, 2024).

Within Society 5.0, technology is intended to address broad societal issues such as ageing populations, healthcare access, climate change, disaster resilience, energy management, sustainable cities, food security, and social inclusion (Adel & Alani, 2024; Gobinath, 2024; J. Rane et al., 2024). Smart cities use connected infrastructure to improve transportation, energy efficiency, waste management, and public safety (Adel & Alani, 2024; Chandre, 2024; Gobinath, 2024). Healthcare systems employ AI-assisted diagnosis, telemedicine, wearable devices, and personalised treatment (Adel & Alani, 2024; Chandre, 2024; Gobinath, 2024). Education increasingly incorporates adaptive learning systems, intelligent tutoring, learning analytics, and flexible digital learning environments that support lifelong learning across all stages of life (Gobinath, 2024; Pang et al., 2023; J. Rane et al., 2024).

Comparing Industry 4.0 and Society 5.0

Although Industry 4.0 and Society 5.0 rely on many of the same technologies, their objectives differ. Industry 4.0 primarily seeks to improve productivity, efficiency, competitiveness, and innovation within industrial and business settings. Society 5.0 builds upon these technological advances but places greater emphasis on using innovation to enhance human wellbeing, reduce inequality, strengthen communities, and promote sustainable development (Chaerunnisa et al., 2022). In this sense, Society 5.0 extends the technological capabilities developed through Industry 4.0 into broader social, economic, and environmental contexts (Adel, 2024; Mazur & Walczyna, 2022).

Industry 4.0 is largely technology-driven, focusing on digital transformation within organisations and industries (Adel, 2024). Society 5.0 is human-driven, seeking to ensure that technological innovation serves societal needs rather than simply improving economic performance (Adel, 2024; Chaerunnisa et al., 2022; Mazur & Walczyna, 2022). This distinction has important implications for education, as learners require not only technical competence but also ethical reasoning, intercultural understanding, systems thinking, sustainability literacy, and the capacity to evaluate the social consequences of emerging technologies (Çelik et al., 2020; Chaerunnisa et al., 2022; Mazur & Walczyna, 2022).

Implications for Education

Education systems are increasingly adapting to prepare learners for both Industry 4.0 and Society 5.0. Curricula are incorporating digital technologies, computational thinking, artificial intelligence, robotics, data analytics, cybersecurity, automation, and digital engineering across multiple disciplines (Erümit & KeleŞ, 2023; Hermans et al., 2024; Syarova & Toleva-Stoimenova, 2023). At the same time, greater emphasis is being placed on transferable capabilities including communication, collaboration, creativity, critical thinking, entrepreneurship, ethical decision-making, emotional intelligence, and lifelong learning (Dianova & Schultz, 2023; OECD, 2023a; Quraishi et al., 2024; Zhou et al., 2025). These competencies enable graduates to adapt to rapidly changing technologies while contributing positively to society (Dianova & Schultz, 2023; Erümit & KeleŞ, 2023; Zhou et al., 2025).

Vocational education and training (VET) providers are modernising training to include smart manufacturing, renewable energy technologies, digital construction, precision agriculture, advanced logistics, and automated maintenance systems (Arenas et al., 2022; Hermans et al., 2024). Universities are expanding interdisciplinary programs that combine engineering, computer science, business, sustainability, health, and the social sciences to prepare graduates capable of addressing complex real-world challenges (Dianova & Schultz, 2023; Nyale et al., 2025; Xu, 2026). Workplace learning is also evolving, with organisations increasingly relying on micro-credentials, online learning, simulation, virtual reality, and continuous professional development to support ongoing workforce capability (Ayanwale et al., 2024; Zhou et al., 2025).

Challenges

Despite their potential benefits, both Industry 4.0 and Society 5.0 present significant challenges. Automation may displace some occupations while increasing demand for new forms of expertise, requiring substantial investment in reskilling and upskilling (Fan, 2024; Kertechian & El-Farr, 2024; Marku, 2024). The fourth industrial revolution has triggered substantial transformations in business practices, and the set of skills needed for the changing nature of work is significantly different, necessitating rapid reskilling to ensure the future employability of the existing workforce (Kertechian & El-Farr, 2024). As AI-driven automation reshapes industries, workers in sectors relying on routine tasks face displacement, while new roles emerge in technology, data management, and AI maintenance (Fan, 2024). Reskilling and upskilling initiatives are essential to equip the workforce with the skills required in the AI-driven economy, involving not only technical skills but also soft skills that are uniquely human (Ibegbulam et al., 2023). In many developing countries, the workforce is predominantly low-skilled, tailored to traditional industries rather than the high-tech requirements of Industry 4.0, necessitating strategic approaches to workforce development (Marku, 2024).

Unequal access to technology, digital infrastructure, and education may widen existing social and economic inequalities if not addressed through inclusive policy (Effoduh et al., 2024; Gobinath, 2024; Jing-jing et al., 2023; Kertechian & El-Farr, 2024). High unemployment resulting from automation might lead to further social and income inequalities, relegating many to the lower class while placing few in the upper class (Kertechian & El-Farr, 2024). In the context of Society 5.0, access to technology relies on robust digital infrastructure and reliable internet connectivity; however, in many regions, especially rural and remote areas, infrastructure limitations and inadequate broadband coverage hinder access (Gobinath, 2024). The digital divide highlights disparities in access to AI technologies, which could exacerbate existing social inequalities (Jing-jing et al., 2023). In Africa, unequal access to services like electricity and internet connectivity hampers data collection and utilization, and resolving these disparities requires substantial investment in infrastructure development and digital inclusion initiatives (Effoduh et al., 2024).

The increasing collection and use of personal data raises concerns regarding privacy, cybersecurity, algorithmic bias, surveillance, and ethical governance (Al-Kasasbeh, 2024; Fan, 2024; Gobinath, 2024; Jawaid & Ahmed, 2023; Kertechian & El-Farr, 2024; Kusuma, 2024). AI-driven systems often process vast amounts of data, including personal information, raising questions about user privacy and data retention policies (Kertechian & El-Farr, 2024). Algorithmic decisions are only as unbiased as the data they are trained on, and AI systems that depend on extensive datasets may unintentionally perpetuate bias if those datasets reflect historical inequalities (Fan, 2024; Jing-jing et al., 2023). Issues such as digital surveillance, algorithmic biases, and the impact on employment require ethical frameworks and responsible practices (Al-Kasasbeh, 2024). Companies must balance the economic benefits of data utilization with the ethical imperative to protect individual privacy and ensure responsible data governance (Kusuma, 2024). Society 5.0 recognizes these ethical dimensions and emphasizes that privacy, data protection, transparency, and accountability must be given importance to maintain trust and ensure the well-being of individuals (Gobinath, 2024). Collaborative efforts among governments, industry stakeholders, academia, and civil society are essential for developing ethical AI governance frameworks (Jawaid & Ahmed, 2023; Sharma & Sharma, 2025).

Environmental impacts associated with digital infrastructure, electronic waste, and energy consumption also require careful management to ensure that technological advancement supports sustainable development (Jena et al., 2024; Marku, 2024). The proliferation of AI devices and electronics can contribute to a substantial increase in electronic waste and environmental pollution, and proper disposal and recycling of AI-related electronics are essential to mitigate this issue (Jena et al., 2024). Comprehensive e-waste recycling programs and wildlife conservation partnerships between AI researchers and environmental organizations are among the recommended strategies (Jena et al., 2024). More broadly, the deployment of Industry 4.0 technologies necessitates policies and investments that prioritize environmental stewardship, social equity, and economic viability (Marku, 2024).

Relationship to Learning Theory

The educational transformation associated with Industry 4.0 and Society 5.0 is supported by several complementary learning theories. Connectivism explains learning within digitally connected knowledge networks where information is distributed across people, technologies, and organisations (Corbett & Spinello, 2020; Ghoulam et al., 2024; Shrivastava, 2018). Siemens first introduced connectivism as "a learning theory for the digital age," describing it as "the integration of principles explored by chaos, network, and complexity and self-organization theories" (Corbett & Spinello, 2020, p. 2). Connectivism posits that knowledge is not confined to individual minds but occurs within distributed networks of resources, people, and technologies (Ghoulam et al., 2024), and that learning is the process of creating and connecting new connections within these networks ((Aderibigbe et al., 2023). In these knowledge networks, connectivism supports lifelong learning, collaboration, and critical thinking and fosters the ability to synthesize information from multiple sources (Benjamin et al., 2024). Connectivism also allows students to develop key competences at the foundation of 21st-century teaching and learning, namely critical thinking, collaboration, communication, creativity and innovation, self-direction, making global and local connections, and using technology as a tool for learning (Dziubaniuk et al., 2023).

Constructivism and social constructivism emphasise collaborative problem-solving and authentic learning experiences that mirror complex workplace environments. Constructivism posits that learners actively construct their understanding of the world by assimilating new information and experiences into their existing knowledge structures (Ghoulam et al., 2024; Mokaya, 2025). Social constructivism, developed by Vygotsky, holds that learning is best achieved through interactions with others in society, peers, teachers, and members of the broader community (Mat & Jamaludin, 2024; Tombleson, 2023). Social constructivism focuses on the importance of social interactions and cultural context in the learning process, underscoring the value of collaborative learning environments where knowledge is co-constructed (Richter et al., 2024). In AI-enhanced and digitally mediated courses, students can work together on projects, fostering a collaborative learning culture (Richter et al., 2024), while educators design learning using inquiry-based, hands-on, collaborative approaches that allow students to explore, discover, and construct meaning (Mat & Jamaludin, 2024).

Experiential learning supports the use of simulations, workplace projects, laboratories, and industry placements that allow learners to develop practical expertise through experience and reflection. Workplace learning allows students to apply their theoretical knowledge in real-world settings and develop practical skills (Aderibigbe et al., 2023). Simulations, virtual reality, and augmented reality allow students to visualise abstract concepts, engage in experiential learning, and investigate phenomena not possible in typical teaching laboratories (Choi-Lundberg et al., 2023). Work-based learning encompasses cooperative education, internships, apprenticeships, and experiential learning, and is beneficial because it allows students to put into practice the skills they are learning in the classroom, making learning more relevant and meaningful (Garrison, 2018). Studio-based and design education similarly emphasise learning by doing, contextualised problem-solving, and learning through experiential activities (Dreamson, 2020). Teaching methods grounded in action learning, project and problem-based learning, and experiential learning support self-directed learning as an essential foundation (Na-Songkhla & Sujiva, 2022).

Humanism reinforces learner autonomy, self-directed learning, and personal development. Humanism focuses on personal growth and self-direction, aligning with supportive, personalised feedback where learners set goals, stay motivated, and reflect on their progress (Lee & Wong, 2025). The person-centred approach to learning indicates that the individual has within themselves the ability to alter basic attitudes, self-concept, and to self-direct behaviour when there is a climate of acceptance, authenticity, and positive understanding (Garrison, 2018). Connectivist learning environments similarly require autonomy, connectedness, diversity, and openness as foundational elements for learners to thrive (Benjamin et al., 2024), while the high level of autonomy in learning requires students to take initiative, be creative, and be innovative (Dziubaniuk et al., 2023).

Transformative learning encourages learners to critically examine the ethical, social, environmental, and economic implications of technological change. Transformative learning suggests that meaningful learning occurs through experiences that encourage a deep, structural shift in basic premises of thought, feelings, and actions (Conceição & Biniecki, 2024). The learning of sustainability in business is frequently grounded on theories such as transformative learning, alongside constructivism, social learning theory, and experiential learning (Dziubaniuk et al., 2023). Ethical considerations in curricula foster learning environments where students collaboratively explore and navigate the ethical complexities of emerging technologies (Richter et al., 2024), while education must empower students to take responsibility for their learning, cultivate self-control in utilising technology, and enable them to evaluate and synthesise information critically (Richter et al., 2024).

Together, these theoretical perspectives support educational approaches that prepare learners not only to work with advanced technologies but also to shape their responsible and sustainable application within society. Each framework offers distinct benefits: constructivism promotes experiential learning, humanism prioritises individual growth, and connectivism highlights the value of interconnected networks (Lee & Wong, 2025). The integration of these theories creates dynamic, learner-centred environments (Lee & Wong, 2025), while connectivism has forced educators to rethink what is being done in digital education and how each part fits (Corbett & Spinello, 2020).

Industry 4.0 and Society 5.0 represent complementary visions of the future. Industry 4.0 provides the technological foundation for intelligent production and digitally connected workplaces, while Society 5.0 extends these innovations to create a more inclusive, sustainable, and human-centred society. Design education, for instance, has undergone significant development to respond to technological and social changes and to adequately prepare students for future professional practice within the context of the fourth industrial revolution (Petrova, 2021). For educators, policymakers, and employers, the challenge is not simply to develop technological expertise but to equip learners with the knowledge, values, and capabilities required to ensure that technological innovation contributes to economic prosperity, social wellbeing, and sustainable development. The learning of sustainability in business, grounded in constructivism, transformative learning, and experiential learning frameworks (Dziubaniuk et al., 2023), alongside connectivism's emphasis on navigating digital knowledge networks (Corbett & Spinello, 2020; Shrivastava, 2018), collectively support this broader educational mission. Higher education's expanding role in linking universities with external stakeholders highlights the importance of knowledge and skills transfer for society's benefit (Sanyang & Weller, 2026), while partnerships with business and industry create work-based learning opportunities that bridge the gap between academia and professional practice (Garrison, 2018; Petrova, 2021).

Artificial Intelligence and Human Capability

Artificial intelligence (AI) is transforming how people learn, work, solve problems, and make decisions. While early discussions about AI often focused on automation and the replacement of human labour, contemporary perspectives increasingly view AI as a technology that can augment rather than replace human capability (Akuma, 2025; Scott et al., 2026). Human capability refers to the knowledge, skills, creativity, judgement, ethical reasoning, emotional intelligence, and adaptability that enable individuals to function effectively in personal, professional, and societal contexts (Akuma, 2025). The relationship between AI and human capability is therefore not one of simple substitution but of collaboration, in which humans and intelligent technologies work together to achieve outcomes that neither could accomplish as effectively alone (Akuma, 2025; Scott et al., 2026).

Artificial intelligence encompasses computer systems capable of performing tasks that traditionally require human intelligence. These tasks include recognising patterns, interpreting language, generating text and images, analysing large datasets, making predictions, supporting decision-making, and learning from experience through machine learning algorithms. Advances in generative AI have further expanded these capabilities by enabling systems to produce human-like text, software code, visual content, audio, and video in response to user prompts. As AI systems become increasingly integrated into workplaces and educational environments, they are changing both the nature of work and the capabilities required of learners and professionals.

One of the most significant contributions of AI is its ability to enhance human cognitive capability. AI can rapidly analyse vast quantities of information, identify trends that may not be immediately apparent to humans, and provide recommendations that support evidence-informed decision-making (Austine et al., 2024; Ibegbulam et al., 2023; Osasona et al., 2024; Rawas, 2024). Researchers, engineers, healthcare professionals, financial analysts, educators, and policymakers increasingly use AI to process complex information (Ibegbulam et al., 2023; Kolivand et al., 2025; Osorio et al., 2024; Rawas, 2024), allowing them to devote more time to interpretation, innovation, and strategic thinking (Elamin, 2024; Ibegbulam et al., 2023; Osorio et al., 2024). Rather than replacing human expertise, AI often serves as a cognitive partner that extends human analytical capacity while leaving responsibility for judgement and final decision-making with people (Austine et al., 2024; Chen et al., 2026; Iyer, 2025; Yuan et al., 2024).

Artificial intelligence also supports productivity by automating routine, repetitive, and time-consuming tasks. Administrative activities such as scheduling, document preparation, transcription, data entry, inventory management, customer support, and report generation can increasingly be completed or assisted by AI systems. This automation enables workers to concentrate on activities requiring creativity, interpersonal communication, leadership, ethical reasoning, and complex problem-solving. In education, AI can assist with preparing learning resources, providing formative feedback, analysing learner progress, and personalising learning pathways, allowing educators to devote greater attention to mentoring, facilitation, and learner support.

Creativity, once considered uniquely human, has also been influenced by advances in AI. Generative AI tools can assist with brainstorming ideas, drafting written content, generating computer code, composing music, producing visual designs, and developing prototypes. Rather than diminishing human creativity, these technologies can accelerate creative processes by providing multiple alternatives, encouraging experimentation, and reducing the time required to develop initial concepts. Human creativity remains essential, however, because people provide the purpose, context, critical evaluation, and ethical judgement needed to refine AI-generated outputs into meaningful and appropriate solutions.

The growing integration of AI has shifted the capabilities valued in modern workplaces. Technical knowledge remains important, but employers increasingly seek individuals who can work effectively alongside intelligent technologies. Digital literacy is now complemented by AI literacy, which encompasses understanding how AI systems function, recognising their capabilities and limitations, evaluating AI-generated information critically, writing effective prompts, interpreting AI outputs, and using AI responsibly and ethically. Equally important are transferable capabilities such as critical thinking, communication, collaboration, adaptability, systems thinking, and lifelong learning, which enable individuals to respond effectively as technologies continue to evolve.

Despite its considerable potential, AI has important limitations that reinforce the continuing importance of human capability. AI systems operate by identifying statistical patterns within data rather than possessing human understanding, consciousness, or lived experience. Consequently, AI-generated outputs may contain factual inaccuracies, fabricated information, embedded biases, or recommendations that fail to account for ethical, cultural, or contextual considerations. Human oversight remains essential to verify information, evaluate alternative perspectives, exercise professional judgement, and ensure that decisions align with legal, ethical, and organisational requirements. In high-stakes fields such as healthcare, engineering, aviation, education, and law, AI is most appropriately viewed as a decision-support tool rather than an autonomous decision-maker.

The ethical implications of AI have become an increasingly important aspect of human capability. Individuals using AI must understand issues relating to privacy, data security, intellectual property, algorithmic bias, transparency, accountability, and responsible use. Organisations are developing governance frameworks that establish appropriate standards for AI implementation, while educators are increasingly recognising the need to develop ethical AI literacy alongside technical competence. Responsible AI use requires users to understand not only how AI systems operate but also the potential consequences of relying on them without adequate human review.

Artificial intelligence is also reshaping education itself. Intelligent tutoring systems, adaptive learning platforms, learning analytics, automated assessment, and generative AI tools enable increasingly personalised learning experiences (Annuš, 2024b; Lameras & Arnab, 2021; Nguyen et al., 2024; Suvendu & S, 2024; Virvou, 2023; Zawacki-Richter et al., 2019). Learners can receive immediate feedback, access explanations tailored to their current level of understanding, and engage with interactive learning resources that adapt as their knowledge develops (Annuš, 2024b; Govender, 2024; Nguyen et al., 2024). Educators, meanwhile, can use AI to identify learners requiring additional support, monitor engagement, and design more responsive learning experiences (Lameras & Arnab, 2021; Nayak et al., 2023; S. V. Singh, 2022). Nevertheless, effective education continues to depend upon the expertise of educators in fostering motivation, critical thinking, collaboration, ethical reasoning, and meaningful human interaction—capabilities that cannot be fully replicated by technology (Annuš, 2024a; Çelik et al., 2024; Lameras & Arnab, 2021).

The relationship between AI and human capability is particularly significant in the context of lifelong learning. As AI transforms occupations and creates new forms of work, individuals must continually update their knowledge and develop new competencies throughout their careers (Agostini & Picasso, 2024; Annuš, 2024a). Reskilling and upskilling increasingly include not only learning to use AI tools but also developing complementary human capabilities that remain difficult to automate (Çelik et al., 2024; Mitra, 2023). These include creativity, emotional intelligence, leadership, negotiation, intercultural competence, ethical decision-making, resilience, and the capacity to learn continuously in response to technological change (Çelik et al., 2024; Li & Wong, 2023).

Several educational theories provide valuable perspectives for understanding how AI can support human capability. Constructivism emphasises that learners actively construct knowledge through interaction with information, experience, and others, suggesting that AI should support inquiry and exploration rather than replace active learning. Social constructivism highlights the importance of collaboration and dialogue, reinforcing the continuing role of educators and peer interaction even within technology-enhanced learning environments. Connectivism is particularly relevant because it views learning as occurring across networks of people, digital technologies, and information sources, with AI becoming one component of an individual's broader knowledge network. Humanism reminds educators that learning extends beyond cognitive development to include personal growth, autonomy, motivation, and self-actualisation, while experiential learning demonstrates the importance of applying knowledge through authentic practice and reflection rather than relying solely on AI-generated information. Transformative learning further encourages learners to critically examine their assumptions about technology, work, ethics, and society, enabling them to become thoughtful and responsible users of AI rather than passive consumers of technological outputs.

Artificial intelligence should be viewed as a technology that has the potential to amplify human capability rather than replace it (Carrubbo & Megaro, 2024; Kumar & Bargavi, 2024). Intelligence Augmentation refers to the use of technology, particularly AI, to enhance human capabilities rather than replace them, implying that AI can collaborate with humans for amplifying their cognitive and decision-making capabilities in various contexts (Carrubbo & Megaro, 2024). Unlike the common fear of total replacement of humans by machines, intelligence augmentation emphasizes collaboration between people and intelligent systems, where AI is seen as an ally that can improve human performance and solve complexities (Carrubbo & Megaro, 2024). The discourse around AI utilization has shifted from being narrowly focused on automation-based tasks to the augmentation of human capabilities (Ifenthaler et al., 2024), and collaboration between humans and AI represents a paradigm shift in which AI systems enhance human talents (Kumar & Bargavi, 2024).

The greatest benefits arise when AI complements human strengths by increasing access to information, supporting analysis, enhancing creativity, and automating routine activities, while people continue to provide judgement, ethics, empathy, contextual understanding, and innovation. Both human intelligence and artificial intelligence have their own advantages and disadvantages and can complement each other's strengths to shape hybrid intelligent models of human-machine collaboration (Liu & Fu, 2024). AI cannot replace human intelligence completely; while AI can perform certain tasks better than humans, such as analysing vast amounts of data, it cannot replicate human creativity, intuition, and empathy, as AI lacks the emotional intelligence and holistic understanding that humans possess (Virvou, 2023). Intelligent machines are superior to people in computing and memorization, but people are superior to intelligent machines in terms of emotions and innovation (Wang, 2023). Enhanced human-AI collaboration can lead to more creative solutions and better decision-making (Bobiţan et al., 2024), and AI should serve to support and amplify human strengths, fostering a harmonious and productive workplace (Bobiţan et al., 2024). Advances in AI technology require humans to continue to improve their abilities in order to adapt to technological changes, though AI cannot replace the emotional aspects that are unique to humans (Dewi et al., 2024).

As education systems and workplaces continue to evolve, developing the ability to collaborate effectively with AI will become an essential capability. Human-machine collaboration is at the heart of modern education (Wang, 2023), and the reciprocal collaboration of humans and AI is a suitable approach for enhancing the capacities of both (Ifenthaler et al., 2024). For humans working with AI systems in the workplace or in policy making, developing an adequate mental model of the underlying mechanisms of AI is essential, and Intelligence Awareness in humans should be addressed more vigorously (Korteling et al., 2021). The study of how AI systems and humans can coexist is among the most vital questions, including which decisions should be made by AI, which by humans, and which in collaboration (Fulton et al., 2022). Cross-sector collaboration is key in meeting AI-era challenges, with governments, educational institutions, and the private sector working together to ensure that future generations have a strong foundation (Dewi et al., 2024). Strategic reskilling and continuous learning are critical to align employee skills with new business paradigms shaped by AI (Bobiţan et al., 2024).

Success in the age of artificial intelligence will therefore depend not only on technological proficiency but also on cultivating the uniquely human qualities that enable individuals to apply technology responsibly, creatively, and in ways that contribute to individual wellbeing and the broader public good. Responsible Artificial Intelligence is about human responsibility for the development of intelligent systems along fundamental human principles and values, to ensure human flourishing and wellbeing in a sustainable world (Fulton et al., 2022). The importance of maintaining human oversight and developing human-centric skills such as empathy, emotional intelligence, and cultural competence ensures that AI serves to support and amplify human strengths (Bobiţan et al., 2024). The three most important trends identified for deploying AI in education were privacy and ethical use of AI, the importance of trustworthy algorithms, and equity and fairness (Ifenthaler et al., 2024). The iterative and people-oriented nature of computational design thinking ensures that the evolution of intelligent systems is consistent with human needs and desires, embodying the concept of "human-centered AI" (Liu & Fu, 2024).

Building Learning Organisations

The concept of the learning organisation describes an organisation that continuously develops its capacity to create, acquire, share, and apply knowledge in ways that improve performance, innovation, and adaptability (Brătianu et al., 2020; Farrukh & Waheed, 2015; Surijah, 2019). Rather than viewing learning as an activity confined to formal training programs, learning organisations embed learning into everyday work, decision-making, collaboration, and continuous improvement (Qutoshi & Rajbhandari, 2016; Rana et al., 2016). Employees at all levels are encouraged to reflect on their experiences, share knowledge, solve problems collaboratively, and continually develop new capabilities (Binson & Joseph, 2023; Qehaja et al., 2017; Qutoshi & Rajbhandari, 2016). As organisations operate in increasingly complex, technology-driven, and rapidly changing environments, the ability to learn faster than competitors has become a significant source of long-term success (Demaj, 2019; Farrukh & Waheed, 2015; Surijah, 2019).

The idea of the learning organisation gained prominence through the work of Peter Senge in *The Fifth Discipline* (1990) (W. Ahmad et al., 2023; Brătianu et al., 2020; Cavallaro & Nault, 2020; Jeong et al., 2007). Senge (1990) argued that organisations must move beyond traditional hierarchical management structures towards cultures that encourage continuous learning, systems thinking, shared vision, collaboration, and personal growth (Bui & Baruch, 2010; Jeong et al., 2007; Mukti & Dudija, 2016; Rana et al., 2016). Rather than treating learning as an individual responsibility alone, learning organisations recognise that organisational performance depends upon collective learning, knowledge sharing, and the ability to adapt to changing circumstances (Binson & Joseph, 2023; Bui & Baruch, 2011; Qehaja et al., 2017; Qutoshi & Rajbhandari, 2016).

At the heart of a learning organisation is the recognition that knowledge is one of its most valuable strategic resources. Unlike physical assets, knowledge grows when it is shared and applied. Learning organisations therefore establish systems that encourage employees to capture lessons learned, exchange expertise, document successful practices, and apply new knowledge across teams and departments. Formal training remains important, but equal emphasis is placed on workplace learning, mentoring, coaching, communities of practice, collaborative projects, reflective practice, and experiential learning.

Senge identified five disciplines that characterise effective learning organisations (Breiter, 2002; Hutasuhut et al., 2021; Küppers, 2022; Morrison & Rosenthal, 1997). The first is personal mastery, which encourages individuals to pursue continuous learning, professional growth, and self-improvement throughout their careers (Breiter, 2002; Küppers, 2022; Morrison & Rosenthal, 1997). Senge sees personal mastery as relating to "a special level of proficiency achieved through a commitment to lifelong learning" and as "a key characteristic of the development of the learning organization" (Morrison & Rosenthal, 1997, p. 5). Employees are supported to develop new knowledge, refine their skills, and reflect on their own performance while taking responsibility for their ongoing development (Hutasuhut et al., 2021; Morrison & Rosenthal, 1997).

The second discipline involves examining mental models, which are the assumptions, beliefs, and ways of thinking that influence how individuals interpret information and respond to challenges (Breiter, 2002; Mbah & Brohmann, 2021; Morrison & Rosenthal, 1997). Learning organisations encourage employees to question existing assumptions, remain open to alternative perspectives, and critically evaluate established practices rather than accepting them without reflection (Morrison & Rosenthal, 1997). According to Senge (1992, p. 174), cited in Morrison and Rosenthal (1997), "the discipline of managing mental models — surfacing, testing, and improving our internal pictures of how the world works — promises to be a major breakthrough for building learning organizations". Challenging mental models supports innovation by preventing organisations from becoming constrained by outdated ways of thinking, as "untested mental models block people's ability to learn" (Morrison & Rosenthal, 1997, p. 8).

The third discipline is the development of a shared vision (Breiter, 2002; Hutasuhut et al., 2021; Küppers, 2022; Morrison & Rosenthal, 1997). Rather than relying solely on organisational directives imposed from senior management, learning organisations seek to develop a common understanding of organisational goals, values, and aspirations that employees collectively support (Hutasuhut et al., 2021; Morrison & Rosenthal, 1997). A learning organisation requires more than a vision developed by senior management; employees throughout the organisation need to understand, support, and contribute to a common set of goals and aspirations (Hutasuhut et al., 2021). When individuals share a collective understanding of the organisation's purpose and future direction, their personal learning becomes more closely aligned with organisational priorities. This shared commitment strengthens motivation, encourages collaboration, and helps create a culture in which employees work collectively towards common objectives rather than pursuing isolated individual goals (Küppers, 2022; Morrison & Rosenthal, 1997).

The fourth discipline, team learning, recognises that many contemporary organisational challenges are too complex to be addressed by individuals working independently (Breiter, 2002; Morrison & Rosenthal, 1997; Tscherne, 2021). Senge describes team learning as the process through which groups develop their collective capacity to achieve shared goals, arguing that organisational learning depends more on the learning of teams than on the capabilities of individual employees (Morrison & Rosenthal, 1997). Team learning is fostered through dialogue, collaborative problem-solving, shared reflection, open communication, and collective decision-making, enabling members to build upon one another's knowledge and experiences (Morrison & Rosenthal, 1997; Tscherne, 2021). By integrating diverse perspectives and expertise, teams are often able to generate more innovative and effective solutions than individuals working in isolation, making them the fundamental unit through which organisational learning occurs (Morrison & Rosenthal, 1997; Tscherne, 2021).

The fifth and integrating discipline is systems thinking, which provides the conceptual framework that connects and reinforces the other four disciplines (Breiter, 2002; Hutasuhut et al., 2021; Küppers, 2022; Morrison & Rosenthal, 1997). Systems thinking encourages employees and leaders to understand organisations as dynamic, interconnected systems in which people, processes, and structures continually influence one another rather than functioning as isolated departments or activities (Küppers, 2022; Mbah & Brohmann, 2021). It serves as the foundation of the learning organisation because the benefits of personal mastery, mental models, shared vision, and team learning are realised most effectively when these disciplines operate together as an integrated whole (Breiter, 2002; Küppers, 2022). Through systems thinking, decision-makers are encouraged to recognise how actions taken in one area may produce intended and unintended consequences elsewhere in the organisation. This broader perspective enables organisations to identify underlying causes of problems rather than merely addressing their symptoms, supporting more informed decision-making, strategic planning, organisational learning, and continuous improvement (Hutasuhut et al., 2021; Küppers, 2022).

Building a learning organisation requires creating a culture in which learning is valued, encouraged, and rewarded. Psychological safety plays an important role in this process because employees are more likely to ask questions, admit mistakes, experiment with new ideas, and share knowledge when they feel safe to do so without fear of blame or punishment (Barton, 2021; Burrell & Brauner, 2021; Fujiyoshi-Ito et al., 2021; Kumah, 2025; Mazzocato et al., 2025). Organisations that encourage open communication, constructive feedback, and collaborative problem-solving are generally more successful at developing cultures of continuous learning than those characterised by rigid hierarchies and risk avoidance (Ip et al., 2025; Malik et al., 2021; Okon et al., 2024).

Leadership is fundamental to establishing and sustaining a learning organisation. Leaders influence organisational culture by modelling curiosity, supporting professional development, encouraging experimentation, and viewing mistakes as opportunities for learning rather than occasions for punishment (Kumah, 2025; Santos et al., 2025; Sharp et al., 2023). Effective leaders promote reflective practice, provide opportunities for coaching and mentoring, allocate resources for learning, and recognise the contributions of employees who share knowledge and contribute to organisational improvement (Mazzocato et al., 2025; Norman et al., 2025; Okon et al., 2024). Leadership within learning organisations is often distributed across multiple levels rather than concentrated solely among senior managers, allowing expertise and innovation to emerge throughout the organisation (Ip et al., 2025; Kline, 2023).

Knowledge management systems also play an important role in learning organisations by supporting the collection, organisation, sharing, and application of knowledge. Digital technologies such as learning management systems, enterprise knowledge bases, collaborative platforms, artificial intelligence, intranets, and social learning networks enable employees to access information, share expertise, and collaborate regardless of geographical location (Ahmad, 2023; Burrell & Brauner, 2021). These technologies support both formal learning through structured training and informal learning through day-to-day interactions and problem-solving (Mazzocato et al., 2025; Omar et al., 2026).

Continuous improvement is another defining characteristic of learning organisations. Rather than viewing processes as fixed, learning organisations regularly collect evidence, evaluate outcomes, identify opportunities for improvement, and implement changes based on feedback and organisational learning (Demirkesen et al., 2021; Okon et al., 2024; Sherwood, 2015). Approaches such as reflective practice, after-action reviews, quality assurance processes, performance measurement, action research, and continuous improvement methodologies help organisations learn systematically from both successes and failures (Kumah, 2025; Maben et al., 2024; Sherwood, 2015). This commitment to evidence-informed decision-making enables organisations to remain responsive to changing technologies, customer expectations, and external environments (Malik et al., 2021; Okon et al., 2024).

The rapid pace of technological change has further increased the importance of learning organisations. Digital transformation, artificial intelligence, automation, Industry 4.0, and Society 5.0 require organisations to continually update workforce capabilities and organisational knowledge (Achoki, 2023; Cramarenco et al., 2023; Sharma & Kohli, 2023). Learning organisations are generally better positioned to adapt because they invest in reskilling, upskilling, lifelong learning, and innovation (Achoki, 2023; Ekuma, 2023; Gamberini & Pluchino, 2024). Rather than reacting only after change occurs, they proactively anticipate future capability requirements and prepare employees for emerging technologies and new ways of working (Achoki, 2023; Bobițan et al., 2024; Tenakwah & Watson, 2024).

Building a learning organisation also contributes to employee engagement and organisational resilience. Employees who have opportunities to learn, develop new skills, participate in decision-making, and contribute to organisational improvement often experience higher levels of motivation, job satisfaction, and organisational commitment (Ajayi & Udeh, 2024; Edeh et al., 2025; Ekuma, 2023). During periods of disruption, organisations with strong learning cultures are typically more capable of adapting because employees are accustomed to continuous learning, collaborative problem-solving, and organisational change (Edeh et al., 2022; Z. Y. Lim et al., 2024; Nakash, 2024).

Despite these advantages, creating a learning organisation presents several challenges. Research confirms that cultural resistance represents a primary barrier; organisational cultures that do not support knowledge sharing discourage employee participation, as individuals may withhold information fearing it could diminish their importance or job security (Brødsjø et al., 2023). This cultural dimension has been identified as the single biggest challenge in knowledge management implementation, surpassing technical obstacles (Robinson et al., 2025). Similarly, a "culture of resistance to change" has been documented across organisational settings, intensifying pressures on employees and hindering engagement.

Limited leadership support constitutes another critical inhibitor. Studies consistently demonstrate that without visible top management commitment and follow-through, employees are reluctant to adopt knowledge-sharing behaviours (Arnold et al., 2023; Robinson et al., 2025). Research in school settings has shown that knowledge management "cannot" be implemented without effective leadership, and that weak leadership makes launching such initiatives exceedingly difficult (Wooi, 2024). Leadership development, particularly for middle and frontline managers, is frequently insufficient, which hinders learning, fosters resistance to change, and disrupts teamwork (Ip et al., 2025).

Insufficient time for learning and inadequate investment in professional development further constrain organisational learning. Challenges in knowledge sharing across departments and hierarchical levels arise partly due to time constraints (Burrell, 2018). Leadership development and professional development remain low priorities in many sectors, despite evidence that continuous learning and training are essential for organisational adaptation.

Poor knowledge-sharing practices are compounded by technological and structural barriers. The integration of different tools across departments can create compatibility issues, making it difficult for employees to access information seamlessly. Rigid hierarchical structures have been shown to have a significant and negative relationship with the quality of organisational outcomes, as overly rigid hierarchies suppress employee voice and innovation (Al-Azri et al., 2025; Thombare et al., 2026).

Regarding the challenge of information growth, research confirms that the rapid pace of change has led to information overload, making it difficult for organisations to acquire and manage knowledge effectively (Arnold et al., 2023). Information overload affects the quality of individuals' decisions and can decrease productivity (Arnold et al., 2023). Employees struggle to navigate multiple platforms, leading to frustration. Organisations must therefore develop strategies that ensure learning translates into improved performance rather than merely increasing access to information, as early productivity and relevant knowledge are valued over overwhelming volumes of general information (Burrell, 2018). Effective strategies include creating dedicated knowledge-sharing positions, contextualising information to make it more useful, and developing organisational processes that clarify where to find information (Arnold et al., 2023).

Several educational theories provide valuable insights into how learning organisations develop. Experiential learning emphasises learning through experience, reflection, conceptualisation, and experimentation, supporting workplace learning and continuous improvement. Constructivism recognises that employees actively construct knowledge by engaging with authentic workplace challenges, while social constructivism highlights the importance of collaboration, dialogue, and shared meaning-making within teams. Connectivism explains how knowledge increasingly exists across networks of people, digital technologies, and organisational information systems, making the ability to locate, evaluate, and apply knowledge as important as possessing it. Transformative learning encourages employees to critically examine existing assumptions, question established practices, and adopt new perspectives that support organisational innovation and adaptation. Humanism reinforces the importance of personal growth, self-directed learning, intrinsic motivation, and developing individuals' full potential within supportive organisational cultures.

Building a learning organisation is not achieved through isolated training initiatives but through creating an organisational culture in which learning becomes embedded in everyday work. Organisations that encourage continuous learning, knowledge sharing, collaboration, reflective practice, and evidence-informed improvement are better positioned to innovate, adapt to change, and sustain long-term success. As workplaces continue to evolve in response to technological advancement, globalisation, and changing societal expectations, the capacity to learn collectively will remain one of the defining characteristics of successful organisations and an essential component of lifelong learning.

As an example of implementation of building a learning organisation, consider a university that delivers a Master of Business Administration (MBA) with a focus on strategic leadership and organisational change. The business school seeks to operate as a learning organisation by embedding continuous learning throughout the program rather than treating learning as a series of isolated subjects. Faculty members, professional staff, industry partners, and students work collaboratively to improve both the curriculum and the learning experience.

Students are encouraged to pursue personal mastery by developing individual learning plans at the beginning of the program. They identify leadership capabilities they wish to strengthen, regularly reflect on their progress through learning journals, and revise their goals following feedback from lecturers, peers, and workplace mentors. This encourages lifelong learning and professional growth beyond the completion of the degree.

The program also challenges students' mental models by exposing them to complex case studies involving ethical dilemmas, organisational change, artificial intelligence, and sustainability. Students compare multiple perspectives, question their assumptions about leadership and management, and critically evaluate different approaches before proposing evidence-based solutions. This reflective process encourages learners to reconsider established beliefs and develop more sophisticated ways of thinking.

A shared vision is established across the program through a commitment to developing responsible, innovative, and globally aware leaders. This vision is communicated consistently throughout every subject, ensuring that assessments, teaching strategies, and industry projects contribute towards common graduate outcomes. Students understand how each learning activity contributes to their overall professional development and future leadership responsibilities.

Team learning is embedded through multidisciplinary consulting projects in which students from business, information technology, and engineering collaborate to solve authentic organisational problems provided by industry partners. Team members analyse data, negotiate solutions, evaluate alternative strategies, and reflect on both their individual and collective performance. The collaborative process enables students to benefit from diverse knowledge and experiences while developing communication, leadership, and teamwork skills valued by employers.

Finally, the program adopts systems thinking when reviewing graduate outcomes. Instead of evaluating each subject independently, academic staff analyse student progression across the entire degree. When graduates demonstrate weaknesses in strategic decision-making, the review identifies that opportunities to integrate finance, marketing, operations, and leadership occur too late in the curriculum. In response, interdisciplinary projects are introduced earlier in the program so students develop systems thinking throughout their studies rather than only in the final semester.

As a further example, a Registered Training Organisation (RTO) delivers the Certificate III in Commercial Cookery through a combination of classroom learning and workplace apprenticeships. The RTO applies learning organisation principles to create a culture of continuous improvement involving trainers, apprentices, employers, workplace supervisors, and industry representatives.

Each apprentice develops personal mastery by maintaining an electronic skills portfolio that records workplace experiences, photographs of completed dishes, customer feedback, reflective journals, and competency achievements. Trainers review the portfolio during regular coaching sessions, helping apprentices identify strengths, address skill gaps, and establish goals for further professional development. This reinforces the expectation that chefs continue learning throughout their careers rather than only during their apprenticeship.

Apprentices are encouraged to examine their mental models during practical workshops. For example, learners who believe that speed is the primary measure of kitchen performance participate in reflective discussions after service periods, analysing how communication, planning, food safety, teamwork, and customer satisfaction contribute equally to successful commercial kitchen operations. These activities help apprentices replace narrow assumptions with a broader understanding of professional practice.

The RTO establishes a shared vision with employers and apprentices that graduates will become highly skilled chefs who consistently demonstrate food safety, professionalism, sustainability, customer service, and teamwork. This shared vision informs classroom instruction, workplace supervision, assessment, and industry engagement, ensuring that all stakeholders work towards common learning outcomes.

Team learning occurs naturally during commercial kitchen simulations where apprentices rotate through roles such as head chef, sous chef, pastry chef, and kitchen hand. After each service, teams conduct structured debriefings to discuss communication, workflow, quality control, customer feedback, and production efficiency. Apprentices learn not only from their own experiences but also from observing their colleagues and collectively identifying opportunities for improvement.

The RTO also applies systems thinking to improve training quality. When several apprentices struggle to complete assessments on time, trainers do not assume the issue relates solely to learner ability. Instead, they examine the broader training system, including workplace rostering, supervisor availability, assessment scheduling, learning resources, literacy support, and communication between employers and trainers. The review identifies inconsistent workplace mentoring as a major contributing factor. The RTO responds by introducing supervisor training, increasing workplace visits, and improving communication with employers, resulting in higher apprentice completion rates and improved workplace performance.

Future directions for adult learning theory and practice

Adult learning is entering a period of profound transformation driven by technological innovation, demographic change, globalisation, sustainability, evolving labour markets, and new understandings of how adults learn throughout their lives (Aurelio, 2022; Barrios et al., 2022). Traditional models that viewed adult education as preparation for a single occupation are increasingly being replaced by lifelong and life-wide learning approaches that recognise learning as a continuous process extending across careers, communities, and personal development (Estella & Vera, 2015; Ramiro, 2022; Segovia, 2019). Consequently, the future of adult learning theory and practice will involve integrating established educational theories with emerging technologies and evolving societal needs while maintaining a strong commitment to learner-centred education (Canal et al., 2024; Martínez-Alvarez & López, 2024; Monferrato, 2024).

One of the most significant future directions is the continued expansion of lifelong learning as a guiding principle for education systems (Estella & Vera, 2015; Sevillano et al., 2023; Vera, 2010). Rapid technological advancement, automation, artificial intelligence, and changing workforce demands mean that knowledge and skills can become outdated more quickly than in previous generations (Aurelio, 2022; Monferrato, 2024). Adults are therefore increasingly expected to engage in continuous reskilling and upskilling throughout their working lives (Norberto Alonso Arroyave Villa, 2023; Segovia, 2019). Rather than viewing education as a discrete stage completed during early adulthood, governments, employers, and educational institutions are recognising learning as an ongoing process that supports employability, career transitions, personal fulfilment, active citizenship, and social participation across the lifespan (Estella & Vera, 2015; Ramiro, 2022; Sevillano et al., 2023; Vera, 2010).

The increasing integration of artificial intelligence (AI) into education is expected to reshape both adult learning theory and educational practice. AI-powered learning systems can personalise learning pathways, provide immediate feedback, recommend learning resources, identify knowledge gaps, and adapt instruction to individual learner needs (Chen, 2025; Kristiawan et al., 2024; Oke et al., 2023; Y. Wang, 2025). Generative AI tools can support learners by assisting with research, writing, problem-solving, language development, programming, and creative activities (Alzubi, 2024; Chen, 2025; Fang & Han, 2025). However, these technologies also reinforce the continuing importance of uniquely human capabilities such as critical thinking, ethical reasoning, creativity, emotional intelligence, and professional judgement (Kristiawan et al., 2024; Schenck, 2024). Future adult education will increasingly focus on developing AI literacy alongside digital literacy, enabling learners to work effectively with intelligent technologies while critically evaluating their outputs and limitations (Alzubi, 2024; Kristiawan et al., 2024; Schenck, 2024).

Digital transformation is continuing to expand opportunities for more flexible, accessible, and personalised approaches to adult learning. Online learning, blended learning, mobile learning, virtual and augmented reality, simulation technologies, learning analytics, and other immersive digital environments enable adults to participate in education regardless of their geographical location or work commitments (Boltz et al., 2021; Ghobrini & Sarnou, 2022; Joshi, 2021). These technologies support self-paced study (Cheng et al., 2023; Xiong et al., 2021), facilitate collaborative knowledge sharing through digital networks and communities of practice (Kulal et al., 2024), provide authentic workplace simulations that enhance experiential learning (Borgos et al., 2023), and increasingly allow educational experiences to be adapted to the needs, preferences, and progress of individual learners (Kaddoura & Husseiny, 2023; Meaney, 2024). At the same time, the effective use of digital technologies depends upon ensuring equitable access to digital infrastructure (Fortman et al., 2023; Župan, 2023), developing the digital capabilities of both educators and learners (Malik & Hooda, 2023), and designing learning environments that are accessible to people with diverse backgrounds, abilities, and learning needs (Alasmari & Al-Zahrani, 2026; Kasztelewicz et al., 2022).

The growth of digital education has transformed how adult learners engage with formal and informal learning. Online and distance education, including massive open online courses (MOOCs), blended learning, and mobile learning, have substantially reduced the constraints of time and location, enabling learners to participate in education alongside employment and family commitments (Boltz et al., 2021; Ghobrini & Sarnou, 2022). Blended learning combines online and face-to-face instruction to create more flexible, learner-centred educational experiences (Joshi, 2021), while mobile learning extends access through the widespread availability of smartphones and portable digital devices (Cheng et al., 2023). Advances in virtual reality (VR), augmented reality (AR), immersive environments, and emerging metaverse technologies are creating increasingly interactive educational experiences that allow learners to practise complex skills within realistic virtual settings (Borgos et al., 2023; Kaddoura & Husseiny, 2023). Similarly, simulation-based learning and virtual laboratories provide authentic practice environments that are particularly valuable in disciplines where access to specialist equipment, laboratories, or clinical settings may be limited (Kaddoura & Husseiny, 2023).

These technologies provide a range of important pedagogical benefits for adult learners. Self-paced learning enables individuals to progress according to their own schedules, making education more compatible with employment and other personal responsibilities (Boltz et al., 2021; Xiong et al., 2021). Online discussion forums, collaborative digital workspaces, and virtual classrooms promote interaction, peer learning, and knowledge sharing despite geographical separation (Kulal et al., 2024; Meaney, 2024). At the same time, advances in learning analytics and artificial intelligence are making it increasingly possible to personalise education by identifying learner needs, adapting assessments, recommending resources, and creating individual learning pathways that improve engagement and learning outcomes (Alasmari & Al-Zahrani, 2026; Kaddoura & Husseiny, 2023). These developments allow adult learners greater flexibility in balancing study with work and family commitments while supporting more responsive and learner-centred educational experiences (Ghobrini & Sarnou, 2022; Xiong et al., 2021).

Despite these opportunities, digital transformation also presents significant challenges relating to equity and inclusion. Unequal access to reliable internet connections, suitable digital devices, and affordable technology continues to create barriers for learners in rural, remote, and economically disadvantaged communities, contributing to persistent disparities in educational participation and outcomes (Fortman et al., 2023; Župan, 2023). Digital inequality is also influenced by factors such as income, educational background, geographic location, gender, and other demographic characteristics (Župan, 2023). Although online education and MOOCs were initially promoted as mechanisms for widening participation, research suggests that learners who already possess educational, technological, and socioeconomic advantages are often best positioned to benefit from these opportunities, reinforcing existing educational inequalities commonly described as the Matthew Effect (Kasztelewicz et al., 2022). Consequently, the expansion of digital learning must be accompanied by deliberate strategies that address both access to technology and broader issues of educational equity (Malik & Hooda, 2023).

Developing digital capability has therefore become a priority for both educators and learners. Many educators continue to require greater preparation and ongoing professional development to design, facilitate, and evaluate high-quality online learning experiences (Meaney, 2024; Xiong et al., 2021). Effective digital teaching increasingly requires competencies that extend beyond technical proficiency to include online pedagogy, data literacy, digital assessment, learner engagement, and inclusive instructional design (Alasmari & Al-Zahrani, 2026). Likewise, learners require not only access to digital technologies but also the skills and confidence needed to use them effectively, making digital literacy an essential component of educational equity and lifelong learning (Kulal et al., 2024; Malik & Hooda, 2023).

Ensuring accessibility is equally important as digital education expands. Learning management systems, online platforms, and digital learning resources must be designed to accommodate learners with diverse physical, sensory, cognitive, and learning needs (Ghobrini & Sarnou, 2022; Zdravkova et al., 2022). Although many digital platforms now incorporate accessibility features, research indicates that design limitations continue to prevent some learners from participating fully in online education (Zdravkova et al., 2022). Universal Design for Learning (UDL) and other inclusive design frameworks provide practical approaches for creating learning environments that minimise barriers and maximise participation for all learners, particularly in increasingly digital educational contexts (Zdravkova et al., 2022). When digital technologies are implemented thoughtfully, they have considerable potential to improve educational quality, expand participation, and provide more flexible, learner-centred learning experiences for diverse populations (Fortman et al., 2023; Ghobrini & Sarnou, 2022).

International organisations have highlighted digital inclusion as a critical component of achieving equitable education. UNESCO and other global agencies emphasise that technological innovation should support broader educational goals of accessibility, lifelong learning, inclusion, and social equity rather than simply increasing the availability of digital tools (Malik & Hooda, 2023; Ossiannilsson, 2022). These priorities align closely with Sustainable Development Goal 4, which seeks to ensure inclusive and equitable quality education and promote lifelong learning opportunities for all (Malik & Hooda, 2023; Ossiannilsson, 2022). As emerging technologies such as artificial intelligence, learning analytics, and MOOCs continue to reshape education, researchers increasingly argue that equal attention must be given to issues of equity, accessibility, open education, and social justice to ensure that technological advancement benefits all learners rather than widening existing educational disparities (Alasmari & Al-Zahrani, 2026).

Future adult learning is also expected to become increasingly learner-centred and competency-based. Rather than measuring learning primarily through time spent in classrooms or instructional hours, education systems are placing greater emphasis on demonstrated competence, practical capability, and authentic performance. Competency-based education recognises that learners acquire knowledge and skills at different rates, and that progression should be determined by the achievement of clearly defined learning outcomes rather than the duration of study (Porto & Presant, 2023). This shift is evident across multiple sectors, including professional and healthcare education, where competency frameworks are increasingly being integrated into continuing professional development despite the persistence of traditional time-based training models (Braxton, 2023). Similarly, competency verification forms the basis of many contemporary micro-credentialing systems, where learning is recognised once learners demonstrate that they have achieved specified performance standards, regardless of how or where that learning occurred (Diab & Zhang, 2023).

This transition towards competency-based learning is accompanied by the growing adoption of alternative credentialing systems, including micro-credentials, digital badges, stackable qualifications, recognition of prior learning (RPL), workplace assessment, and competency-based education. These approaches provide adults with more flexible opportunities to receive formal recognition for learning acquired through employment, volunteering, military service, professional practice, and other informal or non-formal learning experiences. Digital badges and related credentials enable learners to assemble personalised learning pathways that combine formal qualifications with achievements gained through professional organisations, workplace learning, community participation, and prior experience (Orman et al., 2023). Likewise, contemporary digital credential frameworks increasingly recognise the educational value of experiential and work-based learning, encouraging the formal recognition of authentic workplace capabilities alongside traditional classroom-based education (Kumar et al., 2022). Micro-credentials are therefore emerging as complementary credentials that may stand alone or contribute towards larger qualifications through articulation and recognition of prior learning (Mitchell et al., 2025; Ratna Selvaratnam & Michael Sankey, 2025).

An important characteristic of these alternative credentials is their stackability, which enables learners to accumulate smaller units of verified learning over time and combine them into larger qualifications or professional profiles. Research has identified portability, interoperability, standardisation, security, verifiability, and stackability as essential characteristics of effective micro-credential systems (Braxton, 2023). Stackable credentials support competency-based learning by allowing learners to progressively build specialised knowledge and skills that can later contribute to formal qualifications (D. Lyu, 2023). Within Europe, for example, common micro-credential frameworks have been developed to facilitate the recognition of these credentials across higher education institutions through recognition of prior learning processes (Antonaci, Henderikx, et al., 2021). The effectiveness of such systems depends heavily upon robust recognition of prior learning arrangements, which enable previously acquired knowledge and skills to contribute towards future educational pathways (Tamoliūnė et al., 2023). Similar developments are evident in healthcare education, where competency-based lifelong learning frameworks encourage practitioners to integrate formal education with workplace learning, while recognition of prior learning acknowledges competencies developed through employment, professional practice, volunteering, and alternative credentialing pathways (Braxton, 2023; D. Griffiths et al., 2024).

These developments reflect a broader recognition that valuable learning occurs across many different contexts rather than exclusively within formal educational institutions. Micro-credentials and related recognition systems support lifelong learning by extending opportunities to assess and recognise knowledge gained through non-formal and informal learning while allowing learners to combine previous qualifications, workplace experience, and newly acquired competencies into coherent educational pathways (J. Graebe, 2025). Rather than following a single linear educational journey, adult learners increasingly combine university qualifications, vocational education, professional certifications, digital badges, workplace learning, and community-based experiences throughout their careers (Orman et al., 2023). Micro-credentials also provide opportunities for continuous professional development by recognising specialised knowledge and emerging skills that complement existing qualifications while demonstrating an ongoing commitment to lifelong learning (D. Lyu, 2023). As labour markets increasingly prioritise demonstrated capabilities over traditional qualifications alone, governments, employers, and educational institutions are investing in micro-credential systems to support workforce upskilling, reskilling, and more flexible access to education across the OECD. Collectively, these developments suggest that micro-credentials, recognition of prior learning, and competency-based credentialing will become increasingly important components of the evolving lifelong learning ecosystem and the broader credential landscape of the twenty-first century.

Workplace learning will continue to play a central role in adult education. Organisations increasingly recognise that continuous learning is essential for maintaining competitiveness in rapidly changing environments shaped by Industry 4.0, Society 5.0, digital transformation, and globalisation. Learning is becoming more closely integrated with everyday work through coaching, mentoring, communities of practice, action learning, workplace projects, collaborative problem-solving, and reflective practice. The concept of the learning organisation is therefore expected to become increasingly important as employers seek to create cultures that encourage knowledge sharing, innovation, adaptability, and continuous professional development.

Increasing attention is also being directed towards interdisciplinary learning. Complex global challenges such as climate change, public health, cybersecurity, artificial intelligence, demographic change, and sustainable development cannot be addressed through single disciplines alone. Adult learners increasingly require the ability to integrate knowledge from multiple fields, collaborate across professions, and apply systems thinking to complex problems. Future educational programs are therefore likely to place greater emphasis on interdisciplinary curricula, project-based learning, collaborative inquiry, and authentic real-world problem-solving.

Another significant direction is the growing integration of Education for Sustainable Development (ESD) into adult learning. As societies seek to achieve the United Nations Sustainable Development Goals, adult education is increasingly expected to develop not only occupational competence but also sustainability literacy, ethical decision-making, systems thinking, global citizenship, and environmental responsibility. Adult learners are being prepared to contribute to more sustainable workplaces and communities through learning experiences that address the interconnected environmental, economic, social, and cultural dimensions of sustainability. This reflects an expanding view of adult education as contributing to both individual development and societal wellbeing.

The future of adult learning also places increasing importance on equity, inclusion, and accessibility. Diverse learner populations require educational environments that accommodate differences in culture, language, disability, educational background, age, and life experience. Universal Design for Learning (UDL), culturally responsive teaching, differentiated instruction, accessible technologies, and inclusive assessment practices are expected to become more deeply embedded within educational design. Rather than treating inclusion as an additional consideration, future adult education increasingly views accessibility as a fundamental characteristic of high-quality learning environments.

Globalisation continues to influence adult learning by increasing opportunities for international collaboration, cross-cultural learning, and workforce mobility. Adults increasingly participate in global online learning communities, multinational workplaces, international professional networks, and collaborative research projects. Consequently, adult education is placing greater emphasis on intercultural competence, multilingual communication, global citizenship, ethical leadership, and the ability to collaborate effectively across cultural and disciplinary boundaries.

The role of the educator is also evolving. Rather than functioning primarily as transmitters of knowledge, educators are increasingly becoming facilitators, learning designers, mentors, coaches, and curators of learning experiences. As information becomes more readily available through digital technologies and AI, educators focus more on supporting critical inquiry, reflective practice, collaboration, ethical reasoning, and the application of knowledge to authentic contexts. This shift reinforces the importance of pedagogical expertise while recognising that technology complements rather than replaces effective teaching.

Future developments in adult learning theory are likely to become increasingly interdisciplinary. While foundational theories such as Behaviourism, Cognitivism, Constructivism, Humanism, Experiential Learning, Transformative Learning, and Social Learning Theory remain highly relevant, emerging perspectives are expanding our understanding of learning in digitally connected and complex environments. Connectivism highlights learning across networks of people and technologies, while complexity theory, systems thinking, neuroscience, and learning analytics provide new insights into how adults acquire, apply, and transfer knowledge. Rather than replacing traditional theories, these perspectives are likely to complement them by explaining aspects of learning that earlier theories did not fully address.

Research is also expected to place greater emphasis on understanding the relationship between human learning and intelligent technologies. As artificial intelligence becomes increasingly integrated into education and work, scholars will continue investigating how AI influences cognition, creativity, decision-making, motivation, collaboration, assessment, and professional identity (Mikkonen, 2025; Turvey & Pachler, 2025; Webb et al., 2020). Questions surrounding ethical AI use, data privacy, algorithmic bias, academic integrity, and the appropriate balance between human expertise and machine assistance are likely to become major areas of adult learning research and policy (Kosbar, 2025; Mikkonen, 2025; Mouta et al., 2024; Webb et al., 2020).

The integration of AI into educational contexts raises fundamental questions about the convergence and divergence of human-mediated and algorithmically mediated pedagogy, including the risk that distinctive qualities of human-mediated pedagogy could deteriorate as algorithmic mediation increases (Turvey & Pachler, 2025). Scholars have emphasized that human agency remains critical within AI-mediated educational environments, and that teacher education must develop understanding of both the limitations and potential of AI (Turvey & Pachler, 2025; Webb et al., 2020). The concept of "hybrid intelligence", where human expertise and machine computation operate synergistically, has significant implications for curriculum design, requiring education to address digital ethics, algorithmic bias, and accountability (Mikkonen, 2025; Turvey & Pachler, 2025).

Research highlights that explainability and accountability must be incorporated into machine learning system design from the outset to meet social, ethical, and legislative requirements (Webb et al., 2020). Bibliometric analyses have revealed increasing scholarly examination of tensions surrounding data ethics, academic integrity, and workforce disruption as algorithms permeate educational processes. Systematic reviews of ethical dimensions in AI in education have identified blind spots, noting that global policy often fails to adequately address the ethics of AI in education and that there is a shortage of training opportunities for teachers and the general public on these matters (Mouta et al., 2024). Furthermore, the lack of clear institutional policies on AI use confronts universities with significant challenges, with only a minority of university leaders reporting formal policy frameworks governing AI use (Kosbar, 2025).

The development of policies and codes of conduct for machine learning in education, addressing both opportunities and risks, will be essential as these technologies play an increasingly important role in learning (Kosbar, 2025; Webb et al., 2020). As the post-digital perspective suggests, new understandings are needed regarding what it means for machines to have agency to learn, as distinct from human agency, with implications for both the philosophy of learning and educational policy (Jandrić & Hayes, 2020).

Despite these emerging opportunities, adult education will continue to face significant challenges. Ensuring equitable access to technology, supporting learners through rapid occupational change, maintaining educational quality, protecting privacy, addressing misinformation, and balancing technological innovation with meaningful human interaction will require thoughtful educational leadership and evidence-informed policy. Adult education must also respond to ageing populations, changing workforce participation patterns, increasing cultural diversity, and growing expectations for flexible and personalised learning opportunities.

The future of adult learning theory and practice is likely to be characterised by greater flexibility, personalisation, interdisciplinarity, technological integration, and lifelong engagement with learning. While digital technologies and artificial intelligence will continue to transform educational practice, the fundamental principles of adult learning remain unchanged. Adults continue to learn most effectively when learning is meaningful, relevant, collaborative, reflective, and connected to authentic personal, professional, and societal contexts. Future adult education will therefore build upon established learning theories while embracing new technologies and innovative practices that enhance, rather than replace, the uniquely human dimensions of learning.

Key Concepts and Conclusions

This chapter examined how adult learning and vocational education are evolving in response to rapid technological, economic, social, and environmental change. It demonstrated that the future of work is being shaped by technological innovation, artificial intelligence, automation, globalisation, demographic change, and sustainability, making lifelong learning an essential requirement for career development rather than an optional extension of initial education. Workers are increasingly expected to combine technical expertise with transferable capabilities such as critical thinking, creativity, communication, collaboration, adaptability, ethical judgement, digital literacy, AI literacy, and a commitment to continuous learning.

The chapter highlighted that digital and artificial intelligence literacy have become foundational competencies across all occupations. Individuals must be able to use intelligent technologies responsibly, critically evaluate AI-generated information, and apply ethical principles when making professional decisions. At the same time, the increasing complexity of workplace challenges is driving greater emphasis on interdisciplinary learning, systems thinking, intercultural competence, digital communication, and collaboration across professions, industries, and cultural boundaries. Sustainability has also emerged as a significant influence on workforce capability, embedding environmental responsibility into occupations across all sectors and increasing demand for green skills and sustainable professional practice.

Education systems are responding to these changes by adopting more flexible and learner-centred approaches to education. Technology-enhanced learning, competency-based education, micro-credentials, work-integrated learning, recognition of prior learning, and stackable qualifications are creating more accessible and responsive pathways that enable adults to continue learning throughout their careers. Lifelong learning ecosystems recognise that learning occurs across formal, non-formal, and informal settings and depends upon collaboration between educational institutions, employers, governments, professional bodies, technology providers, communities, and learners. Mechanisms such as recognition of prior learning, articulation agreements, credit transfer, and digital credentialing help connect these diverse learning experiences into coherent educational pathways.

The chapter also demonstrated that digital technologies have become central infrastructure within lifelong learning ecosystems. Learning management systems, artificial intelligence, learning analytics, virtual simulations, digital portfolios, and digital credentials support continuous learning across the lifespan while enabling greater flexibility, accessibility, and personalisation. Educational institutions now extend beyond delivering initial qualifications by providing lifelong learning opportunities, career guidance, industry partnerships, workforce development, and recognition of prior learning. Employers similarly play a critical role by embedding learning into workplace practice through mentoring, coaching, professional development, communities of practice, and collaboration with education providers. Governments and professional bodies support these ecosystems through policy development, funding, qualification frameworks, occupational standards, quality assurance, and coordinated workforce planning.

Finally, the chapter emphasised that adult learners are active participants in their own learning. Successful lifelong learning depends upon self-direction, metacognition, reflection, collaboration, career management, and learner agency, while also recognising that equitable participation requires coordinated support to address structural barriers. Lifelong learning ecosystems are informed by multiple complementary educational theories, including Humanism, Behaviourism, Cognitivism, Constructivism, Social Constructivism, Connectivism, and Experiential Learning. Together, these perspectives provide a comprehensive theoretical foundation for theory-informed teaching that integrates authentic experience, learner autonomy, collaboration, digital technologies, reflection, and inclusive educational practice.

The future of adult learning and vocational education will be defined by continuing technological innovation, changing labour markets, demographic transformation, environmental sustainability, and increasing global interconnectedness. As occupations continue to evolve at an unprecedented pace, lifelong learning has become fundamental to employability, professional competence, career resilience, and active participation in society. The traditional expectation that a single qualification will sustain an entire career has been replaced by the need for continuous learning, adaptation, and professional renewal throughout life.

Preparing learners for this future requires far more than developing technical expertise alone. Adult learners must cultivate transferable capabilities that enable them to respond confidently to continual workplace transformation, including critical thinking, communication, collaboration, adaptability, digital and AI literacy, ethical decision-making, systems thinking, and intercultural competence. These capabilities support career mobility, lifelong employability, and the capacity to work effectively within increasingly diverse, technology-rich, and interdisciplinary environments.

Education systems are responding by becoming more flexible, personalised, and learner-centred. Online learning, work-integrated learning, competency-based education, micro-credentials, recognition of prior learning, stackable qualifications, and technology-enhanced learning provide adults with multiple opportunities to develop new capabilities while balancing employment, family responsibilities, and changing career aspirations. These developments acknowledge that valuable learning occurs throughout life and across multiple contexts rather than solely within formal educational institutions.

The chapter further demonstrated that lifelong learning is most effective when understood as an interconnected ecosystem rather than a sequence of isolated educational experiences. Educational institutions, employers, governments, professional bodies, technology providers, communities, and learners each contribute unique but complementary roles in creating accessible, coherent, and responsive learning opportunities. Strong collaboration among these stakeholders enables education systems to remain relevant to changing workforce requirements while promoting inclusion, learner mobility, and social participation.

Equally important is the recognition that no single educational theory adequately explains the complexity of adult learning in contemporary society. Humanism, Behaviourism, Cognitivism, Constructivism, Social Constructivism, Connectivism, and Experiential Learning each illuminate different dimensions of learning and together provide a robust theoretical foundation for designing educational experiences that support personal growth, workplace competence, collaboration, technological engagement, and continuous professional development.

Ultimately, the future of adult learning will depend not only on advances in technology but also on the ability of education systems to remain inclusive, equitable, learner-centred, and responsive to changing societal needs. Artificial intelligence, digital technologies, and emerging learning environments offer unprecedented opportunities to personalise education and expand participation; however, these opportunities must be accompanied by ethical governance, quality assurance, equitable access, and strong partnerships between education, industry, government, and communities. By integrating theory-informed teaching with lifelong learning ecosystems and maintaining a commitment to continuous improvement, adult learning and vocational education can prepare individuals not only for changing careers but also for meaningful participation in an increasingly complex, interconnected, and sustainable world.

References

Aakernes, N. (2020). Both Novice and Expert? How Apprentices Develop Vocational Competence in Workplaces Where Technology Is Continuously Changing: Examples From the Norwegian Media Graphics Programme. *Nordic Journal of Vocational Education and Training*, 18-42. https://doi.org/10.3384/njvet.2242-458x.2010118

Aaron, C. S. (2014). The Self-Management of Diabetes in Older African American Women Caregivers of Persons With Dementia. https://doi.org/10.17077/etd.tkti4wqj

Abakah, F., & Antwi, A. (2024). Using Experiential Learning in Solving 3-Dimensional Problems in Trigonometry. *Gulf Journal of Mathematics*, *16*(2), 364-379. https://doi.org/10.56947/gjom.v16i2.1817

Abakumova, I., Тельнова, О. В., & Фоменко, В. Т. (2012). Development of Pupils in Interdisciplinary Contexts. *Российский Психологический Журнал*, *9*(4), 41-53. https://doi.org/10.21702/rpj.2012.4.5

Abbajobir, F. A. (2026). The Contribution of Integrated Functional Adult Education to Rural Livelihood Empowerment and Social Cohesion in Ethiopia: Evidence From Jimma Zone. https://doi.org/10.21203/rs.3.rs-9798109/v1

Abblitt, S., Davis, S., D'Rozario, T., Hồ, T., Pariah, Y., & Rasouli, R. (2023). How to Build a Learning Designer. *Ascilite Publications*, 1-11. https://doi.org/10.14742/apubs.2023.617

Abbott, M. L., & Lee, K. (2022). Researcher-Supported Professional Learning and Development for Instructors in Adult L2 Programs: Factors Leading to Ongoing Communities of Practice. *Modern Language Journal*, *106*(3), 617-634. https://doi.org/10.1111/modl.12793

Abdelmajied, F. E. Y. (2022). Industry 4.0 and Its Implications: Concept, Opportunities, and Future Directions. https://doi.org/10.5772/intechopen.102520

Abdulazeez, A. O. (2020). Talent Management Strategies and Employees' Retention in Chi Farms LTD (Ajanla Farm), Oyo State, Nigeria. *World Journal of Business and Management*, *6*(2), 1. https://doi.org/10.5296/wjbm.v6i2.17609

Abdullah, F., & Ghazali, N. (2023). Micro-Credentials at Higher Education Institutions: Towards Smooth Sailing Ahead. 118-126. https://doi.org/10.2991/978-2-38476-196-8_11

Abedini, A., Abedin, B., & Zowghi, D. (2021). Adult Learning in Online Communities of Practice: A Systematic Review. *British journal of educational technology*, *52*(4), 1663-1694. https://doi.org/10.1111/bjet.13120

Abery, E., Drummond, C., & Bevan, N. (2015). Work Integrated Learning: What Do the Students Want? A Qualitative Study of Health Sciences Students' Experiences of a Non-Competency Based Placement. *Student Success*, *6*(2), 87-91. https://doi.org/10.5204/ssj.v6i2.288

Abeysekera, I., Sunga, E., Gonzales, A., & David, R. (2024). The Effect of Cognitive Load on Learning Memory of Online Learning Accounting Students in the Philippines. *Sustainability*, *16*(4), 1686. https://doi.org/10.3390/su16041686

Abeywardena, I. S. (2017). An Empirical Framework for Mainstreaming OER in an Academic Institution. *Asian Association of Open Universities Journal*, *12*(2), 230-242. https://doi.org/10.1108/aaouj-11-2017-0036

Abid, G., Ahmed, A., Naqvi, F. N., Rehmat, M., & Anum, L. (2021). Nomological Network of Thriving at Work. *Academic Journal of Social Sciences (Ajss)*, *5*(1), 73-108. https://doi.org/10.54692/ajss.2021.05011428

Abid, G., Contreras, F., Rank, S., & Ilyas, S. (2023). Sustainable Leadership and Wellbeing of Healthcare Personnel: A Sequential Mediation Model of Procedural Knowledge and Compassion. *Frontiers in Psychology*, *13*. https://doi.org/10.3389/fpsyg.2022.1039456

Abidin, K. R. (2016). The Concept of Language Learning in Behaviorism Perspective. *Register Journal*, *2*(2). https://doi.org/10.18326/rgt.v2i2.434

Abrahams, K., Abdullah, A., Pillay, M., & Kathard, H. (2022). Emerging Professional Identity Formation: Exploring Coloniality in the Rehabilitation Professions. *Professions and Professionalism*, *12*(1). https://doi.org/10.7577/pp.4627

Abrassart, A., & Wolter, S. C. (2019). Investigating the Image Deficit of Vocational Education and Training: Occupational Prestige Ranking Depending on the Educational Requirements and the Skill Content of Occupations. *Journal of European Social Policy*, *30*(2), 225-240. https://doi.org/10.1177/0958928719855298

Abri, H. A., Behforouz, B., & Zargaran, Z. (2024). Dual Perspectives: How English Lecturers Perceive Peer and Supervisor Observations. *International Journal of Learning Teaching and Educational Research*, *23*(7), 102-119. https://doi.org/10.26803/ijlter.23.7.6

Abubakar, B., Oluwade, D. O., & Ibrahim, U. A. (2022). Evaluating the Effects of Human Capital Development on Employee Retention in Nigerian Universities. *Journal of Human Resource and Sustainability Studies*, *10*(03), 617-651. https://doi.org/10.4236/jhrss.2022.103037

Abukari, A. (2005). Conceptualising Lifelong Learning: A Reflection on Lifelong Learning at Lund University (Sweden) and Middlesex University (UK). *European journal of education*, *40*(2), 143-154. https://doi.org/10.1111/j.1465-3435.2004.00216.x

Achille, T. M. (2024). Perception De L'échec Et Motivation Académique Des Etudiants en 2e Année en Faculté De Lettres Et Des Sciences Humaines a L'Université De Douala. *Journal of Advanced Psychology*, *5*(2), 16-32. https://doi.org/10.47941/japsy.1730

Achilleos, J., Douglas, H., & Washbrook, Y. (2021). Educating Informal Educators on Issues of Race and Inequality: Raising Critical Consciousness, Identifying Challenges, and Implementing Change in a Youth and Community Work Programme. *Education Sciences*, *11*(8), 410. https://doi.org/10.3390/educsci11080410

Achoki, P. M. (2023). Upskilling and Reskilling for a VUCA World. *Gile Journal of Skills Development*, *3*(2), 34-52. https://doi.org/10.52398/gjsd.2023.v3.i2.pp34-52

Acosta-Leal, D. A., Gil, M. A., Montañez, J. C. C., & Martínez, E. H. P. (2023). La Andragogía Como Teoría Mediadora Del Aprendizaje. Guía Para Docentes. https://doi.org/10.26620/uniminuto/978-958-763-606-2

Acton, K. (2023). Reinventing EdD Research Methods Courses. *Impacting Education Journal on Transforming Professional Practice*, *8*(4), 3-10. https://doi.org/10.5195/ie.2023.323

Adami, P., Rodrigues, P. B., Woods, P. J., Becerik-Gerber, B., Soibelman, L., Copur-Gencturk, Y., & Lucas, G. (2021). Effectiveness of VR-based Training on Improving Construction Workers' Knowledge, Skills, and Safety Behavior in Robotic Teleoperation. *Advanced Engineering Informatics*, *50*, 101431. https://doi.org/10.1016/j.aei.2021.101431

Adanu, T. (2007). Continuing Professional Development (CPD) in State-owned University Libraries in Ghana. *Library Management*, *28*(6/7), 292-305. https://doi.org/10.1108/01435120710774440

Addanki, K., Holdsworth, J., Hardy, D., & Myers, T. (2022). A Preliminary Study Using Academagogy to Uncover the Problems That Block Adult Online Learner Engagement. 95-104. https://doi.org/10.1145/3511861.3511872

Addleman, R. A., Brazo, C. J., & Cevallos, T. (2011). Transformative Learning Through Cultural Immersion. *Nwjte*, *9*(1). https://doi.org/10.15760/nwjte.2011.9.1.5

Adejo, O., Ewuzie, I., Usoro, A., & Connolly, T. (2018). E-Learning to M-Learning: Framework for Data Protection and Security in Cloud Infrastructure. *International Journal of Information Technology and Computer Science*, *10*(4), 1-9. https://doi.org/10.5815/ijitcs.2018.04.01

Adel, A. (2024). The Convergence of Intelligent Tutoring, Robotics, and IoT in Smart Education for the Transition From Industry 4.0 to 5.0. *Smart Cities*, *7*(1), 325-369. https://doi.org/10.3390/smartcities7010014

Adel, A., & Alani, N. H. (2024). Human-Centric Collaboration and Industry 5.0 Framework in Smart Cities and Communities: Fostering Sustainable Development Goals 3, 4, 9, and 11 in Society 5.0. *Smart Cities*, *7*(4), 1723-1775. https://doi.org/10.3390/smartcities7040068

Adeoye, S., Batarseh, F. A., Brown, A. M., & Kaufman, E. K. (2024). Understanding the Landscape of Cyberbiosecurity for Integrative Educational Programming. *Journal of the Asabe*, *67*(1), 207-217. https://doi.org/10.13031/ja.15739

Aderibigbe, S. A., AbdelRahman, A. R. A., & Othman, H. A. (2023). Using Online Discussion Forums to Enhance and Document Students' Workplace Learning Experiences: A Semi-Private Emirati University's Context. *Education Sciences*, *13*(5), 458. https://doi.org/10.3390/educsci13050458

Aderibigbe, S. A., Dias, J. M., & Abraham, M. S. (2021). Understanding Issues Affecting Students' Commitment to Online Discussion Forums in Undergraduate Courses. *International Journal of Interactive Mobile Technologies (Ijim)*, *15*(01), 4. https://doi.org/10.3991/ijim.v15i01.17939

Adhikari, D. R., & Shah, B. B. (2021). The State of the Art in the Incorporation of Sustainable Development Goals in Nepalese Universities. *International Journal of Sustainability in Higher Education*, *22*(6), 1373-1401. https://doi.org/10.1108/ijshe-11-2020-0460

Adhikari, S., & Kunwar, R. (2023). An Overview of the Conceptualization, Theoretical Foundations and Rationale for Developing a Local Curriculum. *Integrated Journal for Research in Arts and Humanities*, *3*(4), 128-136. https://doi.org/10.55544/ijrah.3.4.17

Adıgüzel, T., Kaya, M. H., & Cansu, F. K. (2023). Revolutionizing Education With AI: Exploring the Transformative Potential of ChatGPT. *Contemporary Educational Technology*, *15*(3), ep429. https://doi.org/10.30935/cedtech/13152

Adobe. (2025). Adobe Aero. https://www.adobe.com/products/aero.html

Adomako, N. O., Nti, K. O., Chepkorir, J., Asante, K. P., Ofori, N., Agordekpe, E. E., Chen, Y., Miller, H., & Himmelfarb, C. D. (2025). Who's Left Behind? Exploring the Digital Divide in Health App Use Among Adults With Hypertension and Diabetes. https://doi.org/10.1101/2025.11.24.25340934

Agarwal, G., Maroo, K., Zomorodian, P., Bhatt, N., Sanli, D., Sharma, A., & Shelmerdine, S. C. (2024). Radiology AI Training and Assessment—challenges, Innovations, and a Path Forward. *Bjr|open*, *7*(1). https://doi.org/10.1093/bjro/tzaf026

Agostini, D., & Picasso, F. (2024). Large Language Models for Sustainable Assessment and Feedback in Higher Education. *Intelligenza Artificiale*, *18*(1), 121-138. https://doi.org/10.3233/ia-240033

Aguiar, D. R. d. C. (2023). The Pedagogical Praxis in Critical Environmental Education: Potentialities in Different Contexts. *Concilium*, *23*(7), 352-359. https://doi.org/10.53660/clm-1255-23e39

Aguilar, D. A. G. (2015). *Analítica Visual en eLearning*

Aguilar, D. L., Pérez, P. R. Á., González-Ramos, J. A., & Delgado, Y. G. (2023). El Desarrollo De Conductas Resilientes en La Lucha Contra El Abandono Académico Universitario. *Educación Xx1*, *26*(2), 91-116. https://doi.org/10.5944/educxx1.35891

Aguirre-Herráez, R. G., García-Herrera, D. G., Guevara-Vizcaíno, C. F., & Erazo-Álvarez, J. C. (2020). Realidad Aumentada Y Educación en El Ecuador. *Revista Arbitrada Interdisciplinaria Koinonía*, *5*(5), 415. https://doi.org/10.35381/r.k.v5i5.1052

Aheisibwe, I., & Kintu, D. (2019). Exploring the Effectiveness of Informal Apprenticeship in a Community of Practice: A Case Study of Katwe, Kampala-Uganda. *African Journal of Teacher Education*, *8*, 238-253. https://doi.org/10.21083/ajote.v8i0.5353

Ahlgren, L., & Engel, L. (2011). Lifelong Learning Through SMEs: Exploring Workplace Learning in the UK. *Journal of Workplace Learning*, *23*(5), 331-348. https://doi.org/10.1108/13665621111141920

Ahmad, A. R., Ali, D. F., Othman, N. F., Wahab, N. A., & Kamaruzaman, N. (2024). Technological Innovations in Welding Education: A Review of Teaching and Learning Approaches. *International Journal of Academic Research in Progressive Education and Development*, *13*(4). https://doi.org/10.6007/ijarped/v13-i4/23801

Ahmad, B. E., Saad, Z. A., Aminuddin, A. S., & Abdullah, M. A. (2023). Self-Directed Learning of Malay Undergraduate Students. *Studies in Self-Access Learning Journal*, *14*(3), 244-266. https://doi.org/10.37237/140302

Ahmad, M. O. (2023). Psychological Safety, Leadership and Non-Technical Debt in Large Scale Agile Software Development. *35*, 327-334. https://doi.org/10.15439/2023f8595

Ahmad, N., Hoda, N., & Alahmari, F. (2020). Developing a Cloud-Based Mobile Learning Adoption Model to Promote Sustainable Education. *Sustainability*, *12*(8), 3126. https://doi.org/10.3390/su12083126

Ahmad, W., Azam, T., Arshad, M., Ahmed, B., & Zaman, H. M. F. (2023). Faculty Members' Perception of Learning Organization: A Case of Higher Education Institutions. *Sage Open*, *13*(1). https://doi.org/10.1177/21582440231154409

Ahmed, M. (2019). A Case Study on the Development of Adult Language, Literacy and Numeracy Skills. *Icst Transactions on E-Education and E-Learning*, *6*(19), e1. https://doi.org/10.4108/eai.30-10-2018.159602

Ahmed, M. M., Rahman, A., Hossain, M. K., & Tambi, F. B. (2022). Ensuring Learner-Centred Pedagogy in an Open and Distance Learning Environment by Applying Scaffolding and Positive Reinforcement. *Asian Association of Open Universities Journal*, *17*(3), 289-304. https://doi.org/10.1108/aaouj-05-2022-0064

Ahrens, A., Zaščerinska, J., Melnikova, J., Jurgaityte, V., Aļeksejeva, L., & Gukovica, O. (2020). Culturally Responsive Teaching of Immigrants in Adult Education: A Case Study in Sweden. *Journal of Regional Economic and Social Development*, *1*(12), 18. https://doi.org/10.17770/jresd2020vol1.12.5390

Ahsanu, M., Purwati, T., & Wardani, E. (2021). Unpacking Reflective Practice in the Praxis of English Language Teaching in Indonesia. https://doi.org/10.31235/osf.io/au4ec

Aishath, Z., Omar, I. M., & Waheeda, A. (2021). Correlation Between Lecturers' Professional Development Activities and Their Competencies in Maldives Higher Education Institutes. *International Journal of Learning Teaching and Educational Research*, *20*(9), 18-37. https://doi.org/10.26803/ijlter.20.9.2

Aithal, P. S., & Aithal, S. (2023). Predictive Analysis on Future Impact of Ubiquitous Education Technology in Higher Education and Research. *International Journal of Applied Engineering and Management Letters*, 88-108. https://doi.org/10.47992/ijaeml.2581.7000.0190

Aithal, P. S., & Maiya, A. K. (2024). Innovations in Higher Education Industry – Shaping the Future. *SSRN Electronic Journal*. https://doi.org/10.2139/ssrn.4770797

Ajani, O. A. (2024). Enhancing Problem-Solving Skills Among Pre-Service Teachers in Higher Education: A Systematic Literature Review. *Journal of Pedagogical Sociology and Psychology*. https://doi.org/10.33902/jpsp.202424002

Ajayi, E. A. (2019). The Role of Traditional Folklore in Facilitating Adult Learning in Nigeria. *International Review of Education*, *65*(6), 859-877. https://doi.org/10.1007/s11159-019-09807-z

Ajayi, F. A., & Udeh, C. A. (2024). Review of Workforce Upskilling Initiatives for Emerging Technologies in It. *International Journal of Management & Entrepreneurship Research*, *6*(4), 1119-1137. https://doi.org/10.51594/ijmer.v6i4.1003

Ajayi, O., Bagula, A., & Maluleke, H. (2023). The Fourth Industrial Revolution: A Technological Wave of Change. https://doi.org/10.5772/intechopen.106209

Ajraoui, T. E., Kaddour, K. B., & Zeriouh, M. (2019). Transversal Skills in Vocational Education and Training: The Case of ENSAM Engineering Students. *European Scientific Journal Esj*, *15*(8). https://doi.org/10.19044/esj.2019.v15n8p237

Akbari, M., Nguyen, H., McClelland, R., & Houdt, K. V. (2021). Design, Implementation and Academic Perspectives on Authentic Assessment for Applied Business Higher Education in a Top Performing Asian Economy. *Education + Training*, *64*(1), 69-88. https://doi.org/10.1108/et-04-2021-0121

Akdere, M., Jiang, Y., & Acheson, K. (2021). To Simulate or Not to Simulate? Comparing the Effectiveness of Video-based Training Versus Virtual Reality-based Simulations on Interpersonal Skills Development. *Human Resource Development Quarterly*, *34*(4), 437-462. https://doi.org/10.1002/hrdq.21470

Akingbola, A., Benson, A. E., Makinde, A. O., Shekoni, M., Animashaun, K., & Fakiyesi, T. (2025). Post-Pandemic Era: Global Trends, Benefits, and Barriers in Integrating Artificial Intelligence Into Public Health Education. *Frontiers in Public Health*, *13*. https://doi.org/10.3389/fpubh.2025.1648970

Akinsooto, T. A., Tino, C., & Fedeli, M. (2020). Fostering Critical Reflection in the Frame of Transformative Learning in Adult Education: Italian and Nigerian Comparative Case Studies. 165-180. https://doi.org/10.36253/978-88-5518-155-6.11

Akintolu, M., & Letseka, M. (2021). The Andragogical Value of Content Knowledge Method: The Case of an Adult Education Programme in Kwa-Zulu Natal Province of South Africa. *Heliyon*, *7*(9), e07929. https://doi.org/10.1016/j.heliyon.2021.e07929

Akma, S., Seman, M. N. A., Jameelah, M., Hashim, M., Zainal, Z. E. A., Nor, I., Nizam, M., Matori, S., Zaini, D., & Zulkifli, J. (2024). *International Journal on E-Learning and Higher Education*, *19*(2). https://doi.org/10.24191/ijelhe.v19n2

Akob, B. (2016). The Development of Multicultural Discourse in the Historical Learning: a Case Study on Samudra University Langsa Aceh. *International Journal of Multicultural and Multireligious Understanding*, *3*(5), 1. https://doi.org/10.18415/ijmmu.v3i5.51

Akram, T., Ahmad, S., Hafeez, R., Nazir, A., & Kirmani, N. S. (2025). Students' Perspectives on Permaculture as a Tool for Environmental Literacy. *Journal of Natural Resources and Life Sciences Education*, *54*(2). https://doi.org/10.1002/nse2.70034

Aksah, S. J., Izni, N. A., Mahmood, N. A., Tajuddin, N. I. I., Zainuddin, N., & Rahmat, N. H. (2023). Affiliation, Achievement and Power in Learning Motivation: How Do They Relate? *International Journal of Academic Research in Business and Social Sciences*, *13*(8). https://doi.org/10.6007/ijarbss/v13-i8/17742

Akther, J. (2020). Influence of UNESCO in the Development of Lifelong Learning. *Open Journal of Social Sciences*, *08*(03), 103-112. https://doi.org/10.4236/jss.2020.83010

Akuma, I. (2025). Human and Artificial Intelligence: A Review of Competencies, Collaboration, and Ethical Implications. https://doi.org/10.5281/zenodo.15292956

Akyazi, T., Goti, A., Oyarbide-Zubillaga, A., Alberdi, E., Carballedo, R., Ibeas, R., & Bringas, P. G. (2020). Skills Requirements for the European Machine Tool Sector Emerging From Its Digitalization. *Metals*, *10*(12), 1665. https://doi.org/10.3390/met10121665

Al-Kasasbeh, O. (2024). The Transformation of the Economy: Exploring the Impacts and Opportunities of the Digital Economy. *Organize Journal of Economics Management and Finance*, *3*(1), 10-17. https://doi.org/10.58355/organize.v3i1.75

Al-Mosawi, A. (2019). Essential Training Knowledge for Physicians Working in the Field of Professional Training and Development, and Medical Schools' Instruction. *Research and Reviews on Healthcare Open Access Journal*, *4*(3). https://doi.org/10.32474/rrhoaj.2019.04.000187

Al-Mosawi, A. (2022). Principles of Training and Development for Physicians. *Medpress Psychiatry and Behavioral Sciences*, *1*(1). https://doi.org/10.33582/mppbs.2022.202209004

Al-Mosawi, A. (2023). Modern Instruction in Medicine: An Overview. *Hospital and Clinical Management*, *2*(1). https://doi.org/10.58489/2836-2292/007

Al-Rahbi, A., Agarwal, H. S., Sirasanagandla, S. R., Azmi, M. O., Al-Badi, Q., Kolekar, A., Rachoori, S., Kumar, R. S., Sakr, H. F., & Joseph, M. A. (2026). Revolutionizing Medical Physiology Education: A Narrative Review of the Transition From Traditional Classrooms to Artificial Intelligence. *Frontiers in Medicine*, *13*. https://doi.org/10.3389/fmed.2026.1781197

Al-Salem, R. E. (2020). A New Link in the Chain? Arabic-Language Citizenship Education Courses and the Integration of Resettled Syrian Refugees in Canada. *Refuge Canada S Journal on Refuge*, *36*(1), 14-29. https://doi.org/10.25071/1920-7336.40535

Al-Azri, H., Al-Harrasi, N. H., & Al-Aufi, A. S. (2025). Exploring Variations in the Experience of Information Literacy Among Decision-Makers of Omani Government Companies: A Phenomenographic Study. *Journal of Librarianship and Information Science*, *58*(2), 786-798. https://doi.org/10.1177/09610006251357035

AlAfnan, M. A., & Dishari, S. (2024). ESD Goals and Soft Skills Competencies Through Constructivist Approaches to Teaching: An Integrative Review. *Journal of Education and Learning (Edulearn)*, *18*(3), 708-718. https://doi.org/10.11591/edulearn.v18i3.21408

Alaka, N. S., Akpoviroro, K. S., Amos, A. O., & Sanusi, A. (2018). *Foreign Trade Economics Finance Law*, *100*(5). https://doi.org/10.31617/zt.knute.2018(100)05

Alamer, A. (2021). Basic Psychological Needs, Motivational Orientations, Effort, and Vocabulary Knowledge. *Studies in Second Language Acquisition*, *44*(1), 164-184. https://doi.org/10.1017/s027226312100005x

Alamin, F., & Sauri, S. (2024). Education in the Era of Artificial Intelligence: Axiological Study. *Progres Pendidikan*, *5*(2), 146-150. https://doi.org/10.29303/prospek.v5i2.423

Alammary, A. (2024). Blended Learning Delivery Methods for a Sustainable Learning Environment: A Delphi Study. *Sustainability*, *16*(8), 3269. https://doi.org/10.3390/su16083269

Alamri, A. S. (2017). *University Faculty and Diverse Students' Self-Reported Attitudes Toward Inclusive Teaching Strategies* University of North Texas].

Alamuri, S., & Chittipaka, V. (2024). Corporate Learning in a Disruptive Digital Era: a New Global Organizational Paradigma (A Case-Study of IBM's Your Learning Platform). *Brazilian Journal of Development, 10*(5), e69568. https://doi.org/10.34117/bjdv10n5-032

Alarcón, E. A. G., & D'Angelo, N. I. (2020). La Cuestión De La Justicia en La Educación Superior en América Latina. Una Revisión De Su Abordaje. *Revista Latinoamericana De Estudios Educativos, 50*(3), 13-46. https://doi.org/10.48102/rlee.2020.50.3.122

Alaryani, M., Alremeithi, S., Ali, F. A., & Ikuesan, R. A. (2024). PentHack: AI-Enabled Penetration Testing Platform for Knowledge Development. *European Conference on Cyber Warfare and Security, 23*(1), 27-36. https://doi.org/10.34190/eccws.23.1.2493

Alasmari, T. (2024). Reshaping Vocational Training: A study on the Recognition of Micro-Credentials in Job Markets. *Education + Training, 66*(2/3), 233-251. https://doi.org/10.1108/et-07-2023-0282

Alasmari, T., & Al-Zahrani, A. (2026). From Crisis to Innovation: Forecasting Educational Technology Research Through Bibliometric Analysis of Pandemic-Era Publications (2020–2024). *Journal of Educational Computing Research, 64*(3), 756-786. https://doi.org/10.1177/07356331251414548

Alazmi, A. A. (2024). Examining the Influence of International Organizations in Globalizing Education Policy in Kuwait: A Qualitative Study. *International Journal of Educational Management, 38*(6), 1649-1666. https://doi.org/10.1108/ijem-09-2023-0469

Albert, K. (2017). The Certification Earnings Premium: An Examination of Young Workers. *Social Science Research, 63*, 138-149. https://doi.org/10.1016/j.ssresearch.2016.09.022

Alcantar, C. M., Hernandez, E., Reyes, A. N., Jauregui-Gomez, M., & Kim, V. (2024). Para Mi Gente: Community Responsive Civic Practices at Hispanic Serving Community Colleges. *New Directions for Community Colleges, 2024*(205), 87-101. https://doi.org/10.1002/cc.20614

Aldawsari, A., Andrusyszyn, M.-A., & Babenko-Mould, Y. (2014). Abstracts, Oral. *International Journal of Qualitative Methods, 13*(1), 484-567. https://doi.org/10.1177/160940691401300128

Aldhilan, D., Rafiq, S., & Afzal, A. (2025). Navigating Barriers and Building Pathways: Inclusive Education for Children With Special Needs. *F1000research.* https://doi.org/10.12688/f1000research.172211.1

Aleghfeli, Y. K., McIntyre, J., Hunt, L., & Stone, C. (2024). Safety, Belonging and Success in Education for Refugees in Europe: A Systematic Review. *European journal of education, 60*(1). https://doi.org/10.1111/ejed.12852

Alexander, E. S., White, A. A., Varol, A., Appel, K., & Lieneck, C. (2024). Team- And Problem-Based Learning in Health Services: A Systematic Literature Review of Recent Initiatives in the United States. *Education Sciences, 14*(5), 515. https://doi.org/10.3390/educsci14050515

Alfy, S. E., Gómez, J. M., & Dani, A. (2019). Exploring the Benefits and Challenges of Learning Analytics in Higher Education Institutions: A Systematic Literature Review. *Information Discovery and Delivery, 47*(1), 25-34. https://doi.org/10.1108/idd-06-2018-0018

Alghowinem, S., Bagiati, A., Salazar-Gómez, A. F., & Breazeal, C. (2024). Leading in the AI Era: An Interactive Experiential Hands-on Learning Approach for Professionals and Leaders. https://doi.org/10.18260/1-2--47719

Alhadeff-Jones, M. (2018). Pour Une Conception Rythmique Des Apprentissages Transformateurs. *Phronesis, 7*(3), 43-52. https://doi.org/10.7202/1054408ar

Alhadithi, A., & Almansour, M. (2025). Leadership and Management Competency Frameworks in Undergraduate Medical Education: A Scoping Review. *BMC Medical Education, 26*(1). https://doi.org/10.1186/s12909-025-08489-6

Alharbi, K., Alrajhi, L., Cristea, A. I., Bittencourt, I. I., Isotani, S., & James, A. (2020). Data-Driven Analysis of Engagement in Gamified Learning Environments: A Methodology for Real-Time Measurement of MOOCs. 142-151. https://doi.org/10.1007/978-3-030-49663-0_18

Alhasov, Y., Verbytska, A., & Kolenichenko, T. (2020). Teaching English to Adult Learners Within Extracurricular Activities at University: Barriers and Motivation Factors. *Advanced Education, 7*(15), 12-19. https://doi.org/10.20535/2410-8286.195696

Alhassan, M. (2020). Informing Health Professions Education Through Modern Educational Theories: A Necessary Process With Potential Pitfalls. *Pan African Medical Journal*, *36*. https://doi.org/10.11604/pamj.2020.36.236.23196

Alheit, P., & Dausien, B. (2002). *The Double Face of Lifelong Learning: Two Analytical Perspectives on a Silent Revolution*. Studies in the Education of Adults.

Ali, A. (2024). Implication of E-Learning and AI Individualized Pedagogical Perspectives. *Journal of Electrical Systems*, *20*(7s), 2510-2515. https://doi.org/10.52783/jes.4074

Aljohani, O. H., & Alajlan, S. M. (2020). Motivating Adult Learners to Learn at Adult-Education Schools in Saudi Arabia. *Adult Learning*, *31*(4), 150-160. https://doi.org/10.1177/1045159519899655

Allan, H., & O'Driscoll, M. (2025). Using Recontextualisation Theory to Understand Learning Across Multiple Sites in Simulation-Based Nurse Education. *Journal of Advanced Nursing*, *82*(5), 5513-5520. https://doi.org/10.1111/jan.70163

Allcock, A. (2018). Iriss ESSS Outline: Mental Health of Young People With Learning Disabilities in Transition. https://doi.org/10.31583/esss.20181217

Allen, B. (2011). Critical Communication Pedagogy as a Framework for Teaching Difference and Organizing. 103-126. https://doi.org/10.4135/9781452230467.n5

Allen, J., & Katz, S. (2020). Teaching With OER During Pandemics and Beyond. *Journal for Multicultural Education*, *14*(3/4), 209-218. https://doi.org/10.1108/jme-04-2020-0027

Allen, Z. v., Forgues-Martel, S., Venables, M. J., Ghanney, Y., Villeneuve, A., Dongmo, J. B., Ahmed, M., Archibald, D., & Jolin-Dahel, K. (2026). Can AI Match Human Experts? Evaluating LLM-Generated Feedback on Resident Scholarly Projects. https://doi.org/10.64898/2026.03.04.26346878

Aller, L., & Almrwani, A. M. (2022). Self-Doubt in Nursing Students. *Advances in Nursing Science*, *47*(2), 153-165. https://doi.org/10.1097/ans.0000000000000475

Almagarbeh, B. (2022). Learners' Aptitude Towards Using Digital Techniques in Learning EFL: Internet Discourse. *Journal of Practical Studies in Education*, *3*(2), 1-6. https://doi.org/10.46809/jpse.v3i2.42

Alman, S. M. (2020). Department of Computer Science and Information Technology, College of Computer Science &Amp; Information Technology, Firat University, Turkey. *Qubahan Academic Journal*, *1*(1), 46-63. https://doi.org/10.48161/qaj.v1n1a8

Almeida, F., & Buzády, Z. (2024). FLIGBY for Graduates' Employability Enhancement in Computer Engineering. *Computer Applications in Engineering Education*, *32*(6). https://doi.org/10.1002/cae.22789

Almengor, R. A. (2018). Reflective Practice and Mediator Learning: A Current Review. *Conflict Resolution Quarterly*, *36*(1), 21-38. https://doi.org/10.1002/crq.21219

Alnagrat, A. J. A., Ismail, R. C., & Idrus, S. Z. S. (2022). The Significant and Challenges of Extended Reality Technologies in Learning and Training During Covid-19 Pandemic. *Journal of Human Centered Technology*, *1*(2), 44-55. https://doi.org/10.11113/humentech.v1n2.21

Alonso, F., López, G., Manrique, D., & Viñes, J. M. (2005). An Instructional Model for Web-based E-learning Education With a Blended Learning Process Approach. *British journal of educational technology*, *36*(2), 217-235. https://doi.org/10.1111/j.1467-8535.2005.00454.x

Alqaissi, N., Qtait, M., Awwad, K., Mesk, Z., Farajalla, F., Ziad, Y., Abu-Salameh, F., Hmedat, A., Halayka, R., Ajlouni, M., & Shareef, Y. (2026). Knowledge, Attitudes, and Practices Toward Artificial Intelligence Among Health Sciences Students: A Cross-Sectional Study in Palestine. *Plos One*, *21*(6), e0352785. https://doi.org/10.1371/journal.pone.0352785

Alqarni, A. M. Z., Afy-Shararah, M., & Salonitis, K. (2023). Integrating Sustainability Into Engineering Curricula: Challenges, Framework, and Impact on Student Competencies in Saudi Arabian Higher Education. https://doi.org/10.3233/atde230921

Alqarni, A. S. (2025). Artificial Intelligence-Critical Pedagogic: Design and Psychologic Validation of a Teacher-Specific Scale for Enhancing Critical Thinking in Classrooms. *Journal of Computer Assisted Learning*, *41*(3). https://doi.org/10.1111/jcal.70039

Alsaif, M. (2021). Cultural Values That Impact Blended Learning Acceptance and Effectiveness. *International Journal of Instructional Technology and Educational Studies*, *2*(2), 9-18. https://doi.org/10.21608/ihites.2021.88651.1046

Alston, G. D., Hansman, C. A., & Freeman, S. A. (2021). Equity-focused Online Facilitation for Adult Learners. *New Directions for Adult and Continuing Education*, *2021*(169), 61-69. https://doi.org/10.1002/ace.20414

Alterator, S., Cleveland, B., & Boys, J. (2022). The Evaluation of Inclusive School Environments: A Scoping Review of the Literature. *Iul Research*, *3*(6). https://doi.org/10.57568/iulres.v3i6.313

Altınay, Z., Altınay, F., Sharma, R. C., Dağlı, G., Shadiev, R., Yıkıcı, B., & Altınay, M. (2024). Capacity Building for Student Teachers in Learning, Teaching Artificial Intelligence for Quality of Education. *Societies*, *14*(8), 148. https://doi.org/10.3390/soc14080148

Aluko, F. R., Krull, G., & Mhlanga, E. (2022). Shaping Open, Distance and E-Learning in Post School Education and Training: A Call for a Revised Agenda. *Perspectives in Education*, *40*(1), 1-17. https://doi.org/10.18820/2519593x/pie.v40.i1.1

Alvarez, M. P. U. (2023). La Andragogía en La Enseñanza Del Turismo en Instituciones Superiores. *Ciencia Latina Revista Científica Multidisciplinar*, *7*(4), 3705-3721. https://doi.org/10.37811/cl_rcm.v7i4.7218

Alves, N., Schmidt-Lauff, S., Doutor, C., & Campos, L. S. (2020). Contexts of Recognition of Prior Learning: A Comparative Study of RPL Initiatives in Brazil, Portugal, and Germany. *Andragoske Studije*(2), 87-110. https://doi.org/10.5937/andstud2002087a

Alvunger, D. (2024). Curriculum Making Across Sites of Activity in Upper Secondary School Vocational Education and Training: A Review of the Research in Sweden. *International Journal for Research in Vocational Education and Training*, *11*(3), 303-333. https://doi.org/10.13152/ijrvet.11.3.1

Alwahaby, H., & Cukurova, M. (2024). Navigating the Ethical Landscape of Multimodal Learning Analytics: A Guiding Framework. https://doi.org/10.35542/osf.io/adxuq

Alzahrani, A. (2023). Embracing the Promise of Open Educational Resources: Faculty Insights and Implications in Higher Education. https://doi.org/10.21203/rs.3.rs-2984517/v1

Alzubi, A. A. F. (2024). Generative Artificial Intelligence in the EFL Writing Context: Students' Literacy in Perspective. *Qubahan Academic Journal*, *4*(2), 59-69. https://doi.org/10.48161/qaj.v4n2a506

Amali, N. A. K., Seng, H. Z., Rahmat, N. H., Mustafa, N. C., & Ibrahim, I. W. (2023). An Investigation of Learners' Motivational Drive to Learn Foreign Language: A Look at Vroom's Expectancy Theory. *International Journal of Academic Research in Business and Social Sciences*, *13*(3). https://doi.org/10.6007/ijarbss/v13-i3/16493

Amanatidis, N. (2022). Augmented Reality in Education and Educational Games-Implementation and Evaluation: A Focused Literature Review. *Computers and Children*, *1*(1), em002. https://doi.org/10.29333/cac/11925

Ambady, K. G., & Sukumaran, P. (2025). Effects of Response Cards Intervention on Academic Engagement and Achievement in Students With Intellectual Disabilities During Inclusive Elementary Mathematics Instruction. https://doi.org/10.21203/rs.3.rs-6338352/v1

Ambinintsoa, D. V. (2020). *Fostering Learner Autonomy in an EFL Malagasy Context* Victoria University of Wellington].

Ambrósio, S., Maria Helena Araújo e, S., Pinto, S., & Simões, A. R. (2014). Perspectives on Educational Language Policy: Institutional and Students' Voices in Higher Education. *European Journal of Language Policy*, *6*(2), 175-194. https://doi.org/10.3828/ejlp.2014.4

Ameri, M. A. (2020). Criticism of the Sociocultural Theory. *Budapest International Research and Critics Institute (Birci-Journal) Humanities and Social Sciences*, *3*(3), 1530-1540. https://doi.org/10.33258/birci.v3i3.1082

America, C., & Mallon, P. (2020). "Connectedness" in Business Studies Pedagogy: Is a Transition Year Alluring? *Perspectives in Education*, *38*(2), 70-85. https://doi.org/10.18820/2519593x/pie.v38.i2.05

Amery, C., & Griffin, A. (2020). Exploring Communities of Practice in the NHS: A Core Medical Trainee Experience. *Future Healthcare Journal*, *7*(2), e1-e5. https://doi.org/10.7861/fhj.2019-0034

Amish, M., Jihan, S., & Wyatt, A. (2023). Models of Work-Based Learning With Examples of Practice in Petroleum Engineering – A University-Employer Partnership. https://doi.org/10.33422/4th.globalet.2023.04.103

Amornrit, P., Na-Songkhla, J., & Wannapiroon, P. (2018). A Study of Use and Supporting Factors to Effective Use of Open Educational Resources Towards Active Learning in the Context of

Higher Education in Thailand. *Suranaree Journal of Social Science*, *12*(1), 17-36. https://doi.org/10.55766/nsjp2368

Amoud, L. E., & Raymer, A. L. (2025). Reimagining Workforce Development: A Critical Analysis of Approaches in Europe, Singapore, and the United States. *New Directions for Adult and Continuing Education*, *2025*(185), 44-53. https://doi.org/10.1002/ace.20556

Amponsah, S. (2020). Exploring the Dominant Learning Styles of Adult Learners in Higher Education. *International Review of Education*, *66*(4), 531-550. https://doi.org/10.1007/s11159-020-09845-y

Amusa, D. R. (2022). Women's Education and the Achievement of Learning Cities in Lagos, Nigeria. *Tudásmenedzsment*, *23*(Special Issue), 36-45. https://doi.org/10.15170/tm.2022.23.k1.4

An, Y. (2020). A History of Instructional Media, Instructional Design, and Theories. *International Journal of Technology in Education*, *4*(1), 1-21. https://doi.org/10.46328/ijte.35

Analyti, E., Schoina, I., Travlou, C., Mitroyanni, E., Drongitis, A., & Kriparopoulou, A. (2024). Learning Object Repository Systems and Their Use in Greek Education. *Technium Education and Humanities*, *10*, 28-40. https://doi.org/10.47577/teh.v10i.11910

Ananga, P. (2020). Pedagogical Considerations of E-Learning in Education for Development in the Face of COVID-19. *International Journal of Technology in Education and Science*, *4*(4), 310-321. https://doi.org/10.46328/ijtes.v4i4.123

Ananmalay, V., Benen, D., Luo, J., Silva, I. M. D., Smith, E. J., & Taniyama, T. (2022). Perspectives and Experiences of International Students at a Swedish University During the COVID-19 Pandemic. Nr 23 (2022): DYSKURSY MŁODYCH. https://doi.org/10.34768/dma.vi23.630

Anapey, G. M., Obeng-Treve, E., Ahwireng, D., & Aheto, S.-P. K. (2025). Neurodiverse Children Learning Needs Assessment for Digital Inclusion in Mainstream Education. *Sage Open*, *15*(4). https://doi.org/10.1177/21582440251395452

Andersen, T. (2018). The Value of Peer Mentorship as an Educational Strategy in Nursing. *Journal of Nursing Education*, *57*(4), 217-224. https://doi.org/10.3928/01484834-20180322-05

Anderson, A. M., & Or, J. (2023). Fostering Intercultural Effectiveness and Cultural Humility in Adult Learners Through Collaborative Online International Learning. *Adult Learning*, *35*(3), 143-155. https://doi.org/10.1177/10451595231182447

Anderson, E. F., McLoughlin, L., Liarokapis, F., Peters, C., Petridis, P., & Freitas, S. d. (2010). Developing Serious Games for Cultural Heritage: A State-of-the-Art Review. *Virtual Reality*, *14*(4), 255-275. https://doi.org/10.1007/s10055-010-0177-3

Anderson, J., Tytler, R., & Williams, G. (2022). Supporting Curriculum Innovation in Integrated STEM for Secondary Teachers in Australia. *Research in Integrated Stem Education*, 1-30. https://doi.org/10.1163/27726673-00101001

Andrew, N., & Ferguson, D. (2008). `Constructing Communities for Learning in Nursing'. *International Journal of Nursing Education Scholarship*, *5*(1). https://doi.org/10.2202/1548-923x.1579

Ang, S. M. (2021). ST. Paul University Philippines' Thrust Towards Education for Sustainable Development. *International Journal of Education and Literacy Studies*, *9*(4), 271. https://doi.org/10.7575/aiac.ijels.v.9n.4p.271

Angarita, V. N., Barreto, C. R., Acevedo, C. M. A., & Barrios, J. C. (2018). Formación Y Desarrollo De Competencias TIC E Intercultural De Educadores Infantiles Para La Convivencia Escolar. *Revista De Paz Y Conflictos*, *11*(1). https://doi.org/10.30827/revpaz.v11i1.6490

Angelo, D., Disbray, S., Singer, R., O'Shannessy, C., Simpson, J., Smith, H., Meek, B., & Wigglesworth, G. (2022). *Learning (In) Indigenous Languages*. OECD.

Angelova, N. V., Kiryakova, G., & Yordanova, L. (2015). Cloud-Based LMS for E-Learning. *TJS*, *13*(Suppl.1), 386-391. https://doi.org/10.15547/tjs.2015.s.01.066

Angulo-Alvarez, M., & López, F. (2024). Observación De Experiencias Educativas en Institutos Tecnológicos: Un Caso De Estudio. *Revista Científica Uisrael*, *11*(2), 31-46. https://doi.org/10.35290/rcui.v11n2.2024.1141

Anika, A., Islam, M., & Ahmed, F. (2023). Factor Influencing Recruitment of Training and Development in Automobile Industry. https://doi.org/10.21203/rs.3.rs-3793823/v1

Ann, O. W., Swanto, S., & AlSaqqaf, A. (2018). Pre-Service ESL Teachers Engaging in Reflective Practice: Current Observations and Perceived Challenges. *Journal of Research Policy & Practice of Teachers & Teacher Education*, *8*(2), 5-18. https://doi.org/10.37134/jrpptte.vol8.no2.2.2018

Ann, T. W. (2023). *Disability Identity, Belongingness, and Persistence in Students With Disabilities in Higher Education* San Francisco State University].

Anna, N. E. V., Kaur, K., & Yanti Idaya Aspura Mohd, K. (2023). Teaching Strategies for Library Instruction: Directions From the Literature. *Malaysian Journal of Library & Information Science*, *28*(2), 63-87. https://doi.org/10.22452/mjlis.vol28no2.4

Annan-Diab, F., & Molinari, C. (2017). Interdisciplinarity: Practical Approach to Advancing Education for Sustainability and for the Sustainable Development Goals. *The International Journal of Management Education*, *15*(2), 73-83. https://doi.org/10.1016/j.ijme.2017.03.006

Annuš, N. (2024a). Education in the Age of Artificial Intelligence. *Tem Journal*, 404-413. https://doi.org/10.18421/tem131-42

Annuš, N. (2024b). Educational Software and Artificial Intelligence: Students' Experiences and Innovative Solutions. *Information Technologies and Learning Tools*, *101*(3), 200-226. https://doi.org/10.33407/itlt.v101i3.5479

Anselmann, S. (2022). Learning Barriers at the Workplace: Development and Validation of a Measurement Instrument. *Frontiers in Education*, *7*. https://doi.org/10.3389/feduc.2022.880778

Antonaci, A., Henderikx, P., & Ubachs, G. (2021). The Common Microcredentials Framework for MOOCs and Short Learning Programmes. *Journal of Innovation in Polytechnic Education*, *3*(1), 5-9. https://doi.org/10.69520/jipe.v3i1.89

Antonaci, A., Klemke, R., & Specht, M. (2021). The Washback Effect of Micro-credentials. *Journal of Learning for Development*, *8*(2), 228-254.

António Manuel Diogo dos, R., Yakovleva, O., Smyrnova-Trybulska, E., & Морзе, Н. (2018). From Socratic Behaviourism to Digital Constructivism. *International Journal of Research in E-Learning*, *4*(1). https://doi.org/10.31261/ijrel.2018.4.1.02

Apoki, U. C., & Crişan, G. C. (2022). A Modular and Semantic Approach to Personalised Adaptive Learning: WASPEC 2.0. *Applied Sciences*, *12*(15), 7690. https://doi.org/10.3390/app12157690

Apollo, A., & Mbah, M. F. (2021). Challenges and Opportunities for Climate Change Education (CCE) in East Africa: A Critical Review. *Climate*, *9*(6), 93. https://doi.org/10.3390/cli9060093

Araujo, E. V. (2016). Modelos Educativos: Estrategias De Actualizaciones Didácticas Y Pedagógicas. *Revista Científica Ciencia Y Tecnología*, *1*(3). https://doi.org/10.47189/rcct.v1i3.89

Ardiansyah, W., & Ujihanti, M. (2017). Constructivism and Its Perspectives Related to Teaching and Learning Process in the Classroom : A Conceptual Framework. *Tarbawy Jurnal Pendidikan Islam*, *4*(1), 117-143. https://doi.org/10.32923/tarbawy.v4i1.815

Ardill, A. (2019). Deep Critique: Critical Pedagogy, Marxism, and Feminist Standpoint Theory in the Corporate Classroom. 143-163. https://doi.org/10.1163/9789004398597_009

Arega, N. T., & Hunde, T. S. (2025). Constructivist Instructional Approaches: A Systematic Review of Evaluation-based Evidence for Effectiveness. *Review of Education*, *13*(1). https://doi.org/10.1002/rev3.70040

Arenas, I. D. R., Giraldo-Martínez, E., Atienza, M., Sánchez, D. G., & Jiménez-Medina, E. (2022). Adapting Education to Industry 4.0: A Comparative Study. *Educación Y Humanismo*, *24*(43). https://doi.org/10.17081/eduhum.24.43.5590

Ares, P. A., & Cuevas, P. G. (2022). EXPERIENCIA DE FORMACIÓN DOCENTE EN TIC Y RECURSOS DIGITALES INCLUSIVOS: Proyecto DIPCE. *Tics & Ead Em Foco*, *8*(2), 45-65. https://doi.org/10.18817/ticseademfoco.v8i2.627

Arezki, R., Belhaj, F., & Shah, P. (2019). Promoting a New Economy for the Middle East and North Africa. https://doi.org/10.1596/31984

Arghode, V., Brieger, E. W., & McLean, G. N. (2017). Adult Learning Theories: Implications for Online Instruction. *European journal of training and development*, *41*(7), 593-609. https://doi.org/10.1108/ejtd-02-2017-0014

Argüelles, M. d. C. Z. (2023). Análisis De Las Desigualdades Y De La Política Social Cubana Desde Una Perspectiva Interseccional. 245-269. https://doi.org/10.54871/ca23p10j

Argüelles, M. J. M., Plana-Erta, D., & Bertran, À. F. (2022). Impact of Using Authentic Online Learning Environments on Students' Perceived Employability. *Educational Technology Research and Development*, *71*(2), 605-627. https://doi.org/10.1007/s11423-022-10171-3

Ariffin, T. F. T., & Hashim, R. A. (2010). Teachers' Engagement in Workplace Learning Activities: Validation of a Measure. *International Journal of Management Studies*, *17*. https://doi.org/10.32890/ijms.17.2.2010.10001

Arinaitwe, D. (2021). Practices and Strategies for Enhancing Learning Through Collaboration Between Vocational Teacher Training Institutions and Workplaces. *Empirical Research in Vocational Education and Training*, *13*(1). https://doi.org/10.1186/s40461-021-00117-z

Arinaitwe, D., Mifsud, L., Kato, H., & Sannerud, A. R. (2022). Learning Through Collaboration Between Vocational Teacher Training Institutions and Workplaces: Barriers and Contradictions. *Nordic Journal of Vocational Education and Training*, *12*(1), 25-50. https://doi.org/10.3384/njvet.2242-458x.2212125

Arinaitwe, D., & Sannerud, A. R. (2019). Analysing the Interplay Between Institutional-Based and Workplace Learning. *Skandinavisk Tidsskrift for Yrker Og Profesjoner I Utvikling*, *4*(1), 109-135. https://doi.org/10.7577/sjvd.3249

Ariwijaya, T., & Ningsih, L. F. Y. (2020). Adult Learners' Barriers of Learning: The World of Married Women. *Couns-Edu | the International Journal of Counseling and Education*, *5*(2), 81-86. https://doi.org/10.23916/0020200526920

Armson, G. (2009). Exploring a Space for Emergent Learning to Occur: Encouraging Creativity and Innovation in the Workplace. *American Journal of Business Education (Ajbe)*, *2*(1), 1-16. https://doi.org/10.19030/ajbe.v2i1.4015

Arniza, A. E. (2024). Revolutionizing Learning: Leveraging Social Media Platforms for Empowering Open Educational Resources. *International Journal on E-Learning and Higher Education*, *19*(1), 83-106. https://doi.org/10.24191/ijelhe.v19n1.1915

Arnold, A. C., Lindsey, A., McCoy, A. P., Khademian, A. M., Lockee, B. B., Adams, C. A., Amelink, C., Glover, C., Harris, C. M., Potts, C., Pike, D., Whittaker, D., Hare, D. L., Tegarden, D. P., Tinapple, D., Ucko, D., Nahapetian, E., & Hou, F. (2020). Adaptive Life-Long Learning for an Inclusive Knowledge Economy. https://doi.org/10.21061/adaptive-learning-report

Arnold, M., Goldschmitt, M., & Rigotti, T. (2023). Dealing With Information Overload: A Comprehensive Review. *Frontiers in Psychology*, *14*. https://doi.org/10.3389/fpsyg.2023.1122200

Aroles, J., Ćećez-Kecmanović, D., Dale, K., Kingma, S., & Mitev, N. (2021). New Ways of Working (NWW): Workplace Transformation in the Digital Age. *Information and Organization*, *31*(4), 100378. https://doi.org/10.1016/j.infoandorg.2021.100378

Arrabal, A. A. (2018). Equidad Y Educación De La Primera Infancia en La Agenda Educativa Mundial. *Revista Latinoamericana De Ciencias Sociales Niñez Y Juventud*, *17*(1), 47-59. https://doi.org/10.11600/1692715x.17102

Artino, A. R., Jr. (2012). Academic self-efficacy: from educational theory to instructional practice. *Perspectives on medical education*, *1*(2), 76-85. https://doi.org/10.1007/s40037-012-0012-5

Aryal, M. (2024). Reflective Teaching and Practices in the Classroom. *Journal of Multidisciplinary Research Advancements*, *2*(1), 26-31. https://doi.org/10.3126/jomra.v2i1.66635

Aryani, O. D., Bharati, D. A. L., & Astuti, P. (2021). Teachersâ€™ Practices in Using Educational Mobile Applications to Teach English. *English Education Journal*, *11*(1), 37-55. https://doi.org/10.15294/eej.v11i1.40710

Aryeequaye, S. A., Corneau, K., & Duchscher, J. E. B. (2025). Professional Role Transition in Nursing: Leveraging Transition Theory to Mitigate the Current Human Resource Crises. *Healthcare*, *13*(6), 671. https://doi.org/10.3390/healthcare13060671

Asadollahi-Yazdi, E., Couzon, P., Nguyen, N. Q., Ouazène, Y., & Yalaoui, F. (2020). Industry 4.0: Revolution or Evolution? *American Journal of Operations Research*, *10*(06), 241-268. https://doi.org/10.4236/ajor.2020.106014

Ashby, S., Ryan, S., Gray, M., & James, C. (2012). Factors That Influence the Professional Resilience of Occupational Therapists in Mental Health Practice. *Australian Occupational Therapy Journal*, *60*(2), 110-119. https://doi.org/10.1111/1440-1630.12012

Ashfia, A., & Septianti, A. P. (2024). Psychological Theories (Stimulus Response and Cognitive) in Arabic Learning Videos (Descriptive Analysis). *International Journal of Social Science and Human Research*, *7*(04). https://doi.org/10.47191/ijsshr/v7-i04-49

Ashizawa, S., Ziguras, C., & Yonezawa, A. (2024). Convergence or Fragmentation? Recent Developments in Recognition of Microcredentials and Their Impact on Higher Education in Asia and the Pacific. *Journal of International Cooperation in Education*, *26*(1), 116-130. https://doi.org/10.1108/jice-11-2023-0031

Ashta, A., & Mogha, V. (2023). Les Risques Liés À L'innovation : Le Cas De L'intelligence Artificielle. *Technologie Et Innovation*, *8*(1). https://doi.org/10.21494/iste.op.2023.0908

Asian Development Bank. (2023). *Reimagine Tech-Inclusive Education*. A. D. Bank.

Askari, M., & Baumgartner, L. M. (2024). The Applicability of Critical Pedagogy in Adult English Language Teaching (ELT). *International Journal of Linguistics*, *16*(5), 189. https://doi.org/10.5296/ijl.v16i5.22236

Askari, M., & Behdarvandirad, S. (2025). Critical Pedagogy in Adult English Language Teaching: A Scoping Literature Review. *New Horizons in Adult Education and Human Resource Development*, *37*(3), 148-165. https://doi.org/10.1177/19394225251328394

Asriati, A. (2025). Psychological Insights Into Human Resource Management Strategy: A Qualitative Exploration and Review of Literature. *Golden Ratio of Human Resource Management*, *5*(1), 198-212. https://doi.org/10.52970/grhrm.v5i1.503

Assefa, Y. (2021). Integration of Indigenous Knowledge Into Adult Learning Material Development and the Facilitation Methodology. *Education Research International*, *2021*, 1-10. https://doi.org/10.1155/2021/2231759

Åström, E., Sörman, D. E., Sörqvist, P., & Ljungberg, J. K. (2025). The Lifelong Learner's Compass: Cultivating Attitudes With Self-Reflection, Openness, Curiosity and Motivation. *Adult education quarterly*, *75*(4), 291-310. https://doi.org/10.1177/07417136241312174

Ataş, U., & Daloğlu, A. (2024). Developing a Comprehensive Profile of Professional Development and Identity for English Language Teacher Educators. *European journal of education*, *59*(4). https://doi.org/10.1111/ejed.12729

Atay, B., & Şumuer, E. (2021). College Students' Readiness for Connectivist Learning: The Development and Validation of a Scale. *International Journal of Information and Learning Technology*, *38*(2), 230-242. https://doi.org/10.1108/ijilt-12-2019-0124

Atehortúa, J. V. (2020). A Decolonial Pedagogy for Teaching Intersectionality. *Nordic Journal of Comparative and International Education (Njcie)*, *4*(1), 156-171. https://doi.org/10.7577/njcie.3555

Atenas, J., & Havemann, L. (2014). Questions of Quality in Repositories of Open Educational Resources: A Literature Review. *Research in Learning Technology*, *22*. https://doi.org/10.3402/rlt.v22.20889

Atenas, J., Nerantzi, C., & Bussu, A. (2023). A Conceptual Approach to Transform and Enhance Academic Mentorship: Through Open Educational Practices. *Open Praxis*, *15*(4), 271-287. https://doi.org/10.55982/openpraxis.15.4.595

Atiku, B. A. (2023). Analytical Overview of Some Motivational Theories and Their Implications in Provision of Continuous Professional Development of Staff of Colleges of Education and Analogous Institutions. *British Journal of Multidisciplinary and Advanced Studies*, *4*(3), 188-201. https://doi.org/10.37745/bjmas.2022.0226

Atkinson, R. C., & Shiffrin, R. M. (1968). Human memory: A proposed system and its control processes. In K. W. Spence & J. T. Spence (Eds.), *The psychology of learning and motivation: Advances in research and theory* (Vol. 2, pp. 89-195). Academic Press.

Atmani, Z. E., Madrane, M., & Janati-Idrissi, R. (2023). The Reflective Practice a Promising Path for Professionalizing Nursing Education. *Jett*, *14*(2). https://doi.org/10.47750/jett.2023.14.02.036

Atta, S. A., Clark, L. J., & Bonyah, E. (2023). Mathematics Learning Experience at the Basic School and Its Influence on One's Program of Choice at the Institution Higher of Learning. *Golden Ratio of Social Science and Education*, *3*(1), 36-49. https://doi.org/10.52970/grsse.v3i1.230

Auclair, D. (2023). Discours Progressiste Et Crises des Institutions Éducatives : L'Horizon 2030 De L'UNESCO Contre Le Sujet. *Cahiers Société*(5), 121. https://doi.org/10.7202/1110124ar

Auerbach, P., & Green, F. (2024). Reformulating the Critique of Human Capital Theory. *Journal of Economic Surveys*, *39*(5), 1839-1851. https://doi.org/10.1111/joes.12675

Aurelio, M. (2022). Propuestas Y Experiencias Para Mejorar La Educación Y El Empleo. https://doi.org/10.36006/16369-1

Austin, R. D. Board 5: Construction Engineering Division: Tailoring Construction Management Instruction to the Emerging Adult Learner. https://doi.org/10.18260/1-2--32363

Austine, U. O., Stanley, O. C., Osariemen, A. T., Olalekan, D. M., Alexander, L., Babatunde, Y. Y., & Mayowa, A. J. (2024). The Impact of AI on US Labor Markets. 33-48. https://doi.org/10.5121/csit.2024.140403

Authors, V. (2011). IRRODL Volume 12, Number 1; Special Issue: Prior, Experiential, and Informal Learning in the Age of Information and Communication Technologies. *The International Review of Research in Open and Distributed Learning*, *12*(1), 1. https://doi.org/10.19173/irrodl.v12i1.1111

Avantis Education. (2025). ClassVR. https://www.classvr.com

Avis, J., Canning, R., Fisher, R., Morgan-Klein, B., & Simmons, R. (2012). Teacher Education for Vocational Education and Training: A Comparative Study of the Scottish and English Systems Set Within a European Context. *Scottish Educational Review*, *44*(2), 14-23. https://doi.org/10.1163/27730840-04402003

Awgichew, S., & Seyoum, Y. (2017). Integrated Functional Adult Literacy: Existing Practices and Challenges in Eastern Ethiopia. *International Journal of Education and Literacy Studies*, *5*(4), 86. https://doi.org/10.7575/aiac.ijels.v.5n.4p.86

Awiphan, R., Ruengorn, C., Phosuya, C., Koyratkoson, K., Kanjanarat, P., Thavorn, K., Wongpakaran, N., Wongpakaran, T., & Nochaiwong, S. (2026). Individual Socioeconomic Position Among the General Population and Economic Downturns-Related Perceived Stress, Psychological Resilience, and Wellbeing in Thailand. *Frontiers in Public Health*, *14*. https://doi.org/10.3389/fpubh.2026.1855202

Ayanwale, M. A., Adelana, O. P., Ishola, A. M., & Adeeko, O. (2024). Education 4.0: Exploring Computer Science Teachers' Readiness. *Eurasia Journal of Mathematics Science and Technology Education*, *20*(8), em2492. https://doi.org/10.29333/ejmste/14918

Ayele, E. D., Gavriel, S., Ferreira, J., Teeuw, W. B., Philimis, P., & Gillani, G. A. (2024). Emerging Industrial Internet of Things Open-Source Platforms and Applications in Diverse Sectors. *Telecom*, *5*(2), 369-399. https://doi.org/10.3390/telecom5020019

Ayuningrum, N. D., Ngazizah, N., & Ratnaningsih, A. (2024). The Effectiveness of Authentic Assessment Instrument Based on Higher Order Thinking Skills Integrated With Character Education. *Journal of Innovation and Research in Primary Education*, *3*(1), 1-10. https://doi.org/10.56916/jirpe.v3i1.548

Azambuja, R., Baudot, L., & Matsubara, S. (2025). Closing the Books or Keeping Them Open? Identity Work in Partner Retirement From Big 4 Accounting Firms. *Contemporary Accounting Research*, *42*(3), 1839-1869. https://doi.org/10.1111/1911-3846.13044

Azaria, A., Valkanos, E., & Dukakis, N. (2017). In-House Experiential Education After the Greek Crisis: Recessionary Effects on Experiential Learning Design and Implementation of an Effective Vocational Training. *Industrial and Commercial Training*, *49*(5), 256-269. https://doi.org/10.1108/ict-04-2017-0029

Baan, N. v. d. (2026). *Facilitating the Education-to-Work Transition* Maastricht University].

Babacan, A., & Babacan, H. (2012). The Transformative Potential of an Internationalised Human Rights Law Curriculum. *Journal of Transformative Education*, *10*(4), 199-218. https://doi.org/10.1177/1541344613478470

Babacan, A., & Babacan, H. (2015). A Transformative Approach to Work Integrated Learning in Legal Education. *Education + Training*, *57*(2), 170-183. https://doi.org/10.1108/et-07-2013-0098

Bachman, C., & Stewart, C. (2011). Self-Determination Theory and Web-Enhanced Course Template Development. *Teaching of Psychology*, *38*(3), 180-188. https://doi.org/10.1177/0098628311411798

Bachmann, V., & Moisio, S. (2022). Reclaiming the Knowledge Society From "Urban Economization": Thinking With the Frankfurt School. *Geographische Zeitschrift*, *110*(2), 62. https://doi.org/10.25162/gz-2022-0001

Bachore, M. M., Dagaga, E. G., & Lerebo, T. D. (2024). Do Teachers Think on Their Feet? The Awareness and Practice of Reflective Approach Among Secondary School Teachers in Ethiopia. *Heliyon*, *10*(14), e34232. https://doi.org/10.1016/j.heliyon.2024.e34232

Badarnah, L. (2025). Holistic Education for a Resilient Future: An Integrated Biomimetic Approach for Architectural Pedagogy. *Biomimetics*, *10*(6), 369. https://doi.org/10.3390/biomimetics10060369

Badidi, E. (2016). A Cloud-Based Framework for Personalized Mobile Learning Provisioning Using Learning Objects Metadata Adaptation. 368-375. https://doi.org/10.5220/0005810603680375

Badyal, D., & Singh, T. (2017). Learning Theories: The Basics to Learn in Medical Education. *International Journal of Applied and Basic Medical Research*, *7*(5), 1. https://doi.org/10.4103/ijabmr.ijabmr_385_17

Bae, B. (2018). In Pursuit of Transformative Learning: Exploring the Stimulation of Curiosity Through Critical Reflection in the College Classroom. https://doi.org/10.22371/05.2018.016

Baetens, F., & Cheah, W. L. (2013). Being an International Law Lecturer in the 21st Century: Where Tradition Meets Innovation. *Cambridge Journal of International and Comparative Law*, *2*(4), 974-1011. https://doi.org/10.7574/cjicl.02.04.140

Bahattab, A., Caviglia, M., Hubloue, I., Corte, F. D., & Ragazzoni, L. (2023). Scenario-Based E-Simulation Design for Global Health Education: Theoretical Foundation and Practical Recommendations. *Journal of Medical Internet Research*, *25*, e46639. https://doi.org/10.2196/46639

Baidoo-Anu, D., Gyamerah, K., Mahama, I., & Sasu, E. O. (2023). Towards Classroom Inclusivity: Exploring K-12 Teachers' Sensitivity to Cultural Diversity (<i>Hacia La Inclusividad en El Aula: Explorando La Sensibilidad a La Diversidad Cultural Del Profesorado De K-12</I>). *Culture and Education*, *35*(4), 938-975. https://doi.org/10.1080/11356405.2023.2200590

Bailey, A., Kooij, D., Reuver, R. d., & Kupper, N. (2026). Barriers and Facilitators to Participation in Interventions for Employee Health and Wellbeing in the University Context. *Stress and Health*, *42*(2). https://doi.org/10.1002/smi.70163

Baillargeon, M., Gérin-Lajoie, S., LaBillois, D., Leduc, D., Parent, S., Parr, M., Plante, P., & Thibault, A. (2017). Formation À Distance en Enseignement Supérieur. https://doi.org/10.14375/np.9782760547179

Baird, M., Bozick, R., & Zaber, M. A. (2022). Beyond Traditional Academic Degrees: The Labor Market Returns to Occupational Credentials in the United States. *Iza Journal of Labor Economics*, *11*(1). https://doi.org/10.2478/izajole-2022-0004

Bajis, D., Chaar, B., & Moles, R. (2020). Rethinking Competence: A Nexus of Educational Models in the Context of Lifelong Learning. *Pharmacy*, *8*(2), 81. https://doi.org/10.3390/pharmacy8020081

Bajner, M. (2019). Lifelong Learning Redefined: From Sustainability to Generational Learning. *Andragoška Spoznanja*, *25*(3), 35-45. https://doi.org/10.4312/as.25.3.35-45

Bajpai, S., Semwal, M., Bajpai, R., Car, J., & Ho, A. H. Y. (2019). Health Professions' Digital Education: Review of Learning Theories in Randomized Controlled Trials by the Digital Health Education Collaboration. *Journal of Medical Internet Research*, *21*(3), e12912. https://doi.org/10.2196/12912

Bak, T., & O'Maley, P. (1970). Towards Professional Responsibility for Language and Literacy: Exploring Vocational Teachers' Emerging Language and Literacy Understandings and Identities. *Literacy and Numeracy Studies*, *23*(1), 50-72. https://doi.org/10.5130/lns.v23i1.4424

Bakare, T. V. (2020). Factors Affecting the Use of Andragogical Principles in Adult Education Programs in Lagos State, Nigeria. *International Journal of Technology and Management Research*, *4*(1), 23-34. https://doi.org/10.47127/ijtmr.v4i1.73

Baker, D. (2019). Potential Implications of Degree Apprenticeships for Healthcare Education. *Higher Education Skills and Work-Based Learning*, *9*(1), 2-17. https://doi.org/10.1108/heswbl-01-2018-0006

Baker, M. A., Robinson, S., & Kolb, D. (2012). Aligning Kolb's Experiential Learning Theory With a Comprehensive Agricultural Education Model. *Journal of Agricultural Education*, *53*(4), 1-16. https://doi.org/10.5032/jae.2012.04001

Bakes, O., Košťálová, J., & Vávra, J. (2022). Awareness of the Terminology of the Industry 4.0 Concept in the Czech Republic. https://doi.org/10.3846/bm.2022.898

Bakki, A., Oubahssi, L., George, S., & Cherkaoui, C. (2020). A Model and Tool to Support Pedagogical Scenario Building for Connectivist MOOC. *Technology Knowledge and Learning*, *25*(4), 899-927. https://doi.org/10.1007/s10758-020-09444-8

Bakri, S. R. A., & Ching, L. S. (2021). Bridging the Gap Between the Derivatives and Graph Sketching in Calculus. *Asian Journal of University Education*, *16*(4), 121. https://doi.org/10.24191/ajue.v16i4.11962

Baktymbet, S. S. (2021). Assessment of Human Capital Development and Its Impact on the Economy of the Country. *Bulletin of the Karaganda University Economy Series*, *103*(3), 4-14. https://doi.org/10.31489/2021ec3/4-14

Balakrishnan, J. (2022). Building Capabilities for Future of Work in the Gig Economy. *NHRD Network Journal*, *15*(1), 56-70. https://doi.org/10.1177/26314541211064726

Balbín, L. G. E., Ávila, M. C., Mori, R. L. G., Barrios, S. T., & Yvonne del Carmen Cruz, C. (2022). Enseñanza De La Resiliencia en Escuelas Rurales: Análisis Sistemático. *Ciencia Latina Revista Científica Multidisciplinar*, *6*(6), 8773-8810. https://doi.org/10.37811/cl_rcm.v6i6.4032

Baldacci, B. (2022). Experiential Learning: Investigating Lecturers' Perceptions of Authentic Learner Activities. *Scope Contemporary Research Topics (Learning and Teaching)*, 30-39. https://doi.org/10.34074/scop.4011008

Baldwin, C. A. (2021). The Alert Collector: Universal Design for Learning. *Reference & User Services Quarterly*, *59*(3/4), 173-176. https://doi.org/10.5860/rusq.59.3/4.7716

Baldwin, C. K. (2020). Implementer Learning in Developmental Evaluation: A Cultural Historical Activity Theory and Expansive Learning Analysis. *Evaluation*, *26*(1), 27-48. https://doi.org/10.1177/1356389019895033

Ballout, S., & DePace, N. (2025). From Curriculum to Consciousness: A Justice-Oriented Pedagogical Model in Undergraduate Community-Based Nursing Education. *Nursing inquiry*, *32*(4). https://doi.org/10.1111/nin.70055

Bally, J., Burles, M., Widyaratne, A., Spurr, V. A., Hodgson-Viden, H., & Sinha, R. (2025). Supporting Indigenous Family Caregivers of Children With Life-Threatening and Life-Limiting Illness in One Canadian Province: Healthcare Providers' Perspectives. *Children*, *12*(7), 895. https://doi.org/10.3390/children12070895

Baloco, C., & Mendoza, O. J. L. (2022). Ambientes Virtuales De Aprendizaje Con Metodología De Aprendizaje Basado en Problemas (ABP): Una Estrategia Didáctica Para El Fortalecimiento De Competencias Matemáticas. *Praxis*, *18*(2), 324-344. https://doi.org/10.21676/23897856.3919

Bamber, J., & Crowther, J. (2011). Speaking Habermas to Gramsci: Implications for the Vocational Preparation of Community Educators. *Studies in Philosophy and Education*, *31*(2), 183-197. https://doi.org/10.1007/s11217-011-9277-0

Bandara, K. M. N. T. K., & Jayaweera, B. P. A. (2024). Commentary on the Applications of Blended Learning in the Teaching and Learning Process – A Review. *Journal of Research in Education and Pedagogy*, *1*(2), 83-97. https://doi.org/10.70232/jrep.v1i2.10

Bandura, A. (1977). *Social Learning Theory*. Prentice-Hall.

Bandura, A. (1986). *Social Foundations of Thought and Action: A Social Cognitive Theory*. Prentice-Hall.

Banegas, D. L. (2022). Ethnic Equity, Mapudungun, and CLIL. *Aila Review*, *35*(2), 275-296. https://doi.org/10.1075/aila.22018.ban

Banihashem, S. K., & Macfadyen, L. P. (2021). Pedagogical Design: Bridging Learning Theory and Learning Analytics. *Canadian Journal of Learning and Technology*, *47*(1). https://doi.org/10.21432/cjlt27959

Banikowski, A. K., & Mehring, T. A. (2017). Strategies to Enhance Memory Based on Brain-Research. *Focus on Exceptional Children*, *32*(2). https://doi.org/10.17161/fec.v32i2.6772

Baque, D. L. Z., Lozano, N. S. M., Villacreses, G. C. C., & Bravo, H. M. T. (2024). Estrategia Didáctica Para El Aprendizaje Desarrollador en La Asignatura Matemática en Los Estudiantes De Cuarto Nivel De La Carrera De Educación. *Ciencia Latina Revista Científica Multidisciplinar*, *8*(4), 3300-3329. https://doi.org/10.37811/cl_rcm.v8i4.12569

Barath, S., & Ross, A. (2024). Conceptualising the Experiences of Continuing Professional Development of Young Private Sector Audiologists as an Attribute of Andragogy. *Health Sa Gesondheid*, *29*(0). https://doi.org/10.4102/hsag.v29i0.2683

Barbosa, P., & Wang, W. (2020). Challenging the Politics of Education: Intertwining Dialogic Pedagogy With a Research-Based Practicum for First-Year Pre-Service Teachers. *Dialogic Pedagogy a Journal for Studies of Dialogic Education*, *8*. https://doi.org/10.5195/dpj.2020.306

Barefield, T., & Nicolaides, A. (2022). A Cultural-Historical Analysis of Marsick and Watkins' Informal and Incidental Learning Model. *Adult education quarterly*, *73*(2), 150-168. https://doi.org/10.1177/07417136221125462

Barger, R. P. (2020). Democratization of Education Through Massive Open Online Courses in Asia. *Iafor Journal of Education*, *8*(2), 29-46. https://doi.org/10.22492/ije.8.2.02

Barjesteh, H., Vijayaratnam, P., Isaee, H., & Manoocherzadeh, M. (2026). Bridging Criticality and Digital Literacy in the Age of AI: A Mixed-Method Comparative Study of Teachers' and Learners' Perceptions of Critical Digital Pedagogy in Iranian EFL Institutes. https://doi.org/10.21203/rs.3.rs-8167945/v1

Barker, C., Madden, G., & Petersen, B. (2023). Welcome Back. *Canadian Journal for the Study of Adult Education*, *35*(02). https://doi.org/10.56105/cjsae.v35i02.5726

Barletta, C. M. (2023). Inclusión Y Accesibilidad en La Educación Superior: Estrategias Para Promover Un Entorno Educativo Inclusivo. *Trayectorias Universitarias*, *9*(16), 129. https://doi.org/10.24215/24690090e129

Barnard-Ashton, P., Rothberg, A., & McInerney, P. (2017). The Integration of Blended Learning Into an Occupational Therapy Curriculum: A Qualitative Reflection. *BMC Medical Education*, *17*(1). https://doi.org/10.1186/s12909-017-0977-1

Barnard, J., Betteney, M., & Lambirth, A. (2022). Beginner Teachers and Classroom Communities: A Thematic Analysis of UK Beginner Teachers' Experiences in Initial Teacher Education and Beyond. *Teaching and teacher education*, *119*, 103871. https://doi.org/10.1016/j.tate.2022.103871

Barnes, P., Morris, M. P., Pierce, A. L., & Shaffer, T. J. (2022). Answering the Call: Offering and Analyzing Civil Discourse Opportunities in Undergraduate Public Affairs Education. *Journal of Public Affairs Education*, *29*(2), 213-231. https://doi.org/10.1080/15236803.2022.2148604

Barnett, R. (1990). *The Idea of Higher Education*. Open University Press.

Barrett, J. L., Denegar, C. R., & Mazerolle, S. M. (2018). Challenges Facing New Educators: Expanding Teaching Strategies for Clinical Reasoning and Evidence-Based Medicine. *Athletic Training Education Journal*, *13*(4), 359-366. https://doi.org/10.4085/1304359

Barrios, M. J. L., Chávez, Z. R. M., Iparraguirre, R. A., García, J. A. R., Espino, A. M. E., & Espino, J. C. E. (2022). Educación Virtual en La Calidad De Los Aprendizajes: Tendencias De Las Herramientas Informáticas. https://doi.org/10.31219/osf.io/9bsvq

Barros, A., Bristow, A., Contu, A., Wanderley, S., & Prasad, A. (2024). Politicizing and Humanizing Management Learning and Education With Paulo Freire. *Management learning*, *55*(1), 3-16. https://doi.org/10.1177/13505076231221792

Barros, R. (2019). The Role of Transnational Bodies in Lifelong Learning and the Politics of Measurement. 53-63. https://doi.org/10.1163/9789004413320_005

Barry, M., Kuijer-Siebelink, W., Nieuwenhuis, L., & Haan, N. S. (2016). Communities of Practice: A Means to Support Occupational Therapists' Continuing Professional Development. A Literature Review. *Australian Occupational Therapy Journal*, *64*(2), 185-193. https://doi.org/10.1111/1440-1630.12334

Barteit, S., Lanfermann, L., Bärnighausen, T., Neuhann, F., & Beiersmann, C. (2021). Augmented, Mixed, and Virtual Reality-Based Head-Mounted Devices for Medical Education: Systematic Review. *Jmir Serious Games*, *9*(3), e29080. https://doi.org/10.2196/29080

Bartlett, D., & Sims, D. (2021). Assessment Strategies and Employability in the Environmental and Engineering Sectors. *Compass Journal of Learning and Teaching in Higher Education*, *14*(2). https://doi.org/10.21100/compass.v14i2.1182

Bartlett, S., & Ross, J. (2022). Mental Health and Wellbeing of Tertiary Learners: What Do We Need to Know. *Scope Contemporary Research Topics (Health and Wellbeing)*(7), 66-71. https://doi.org/10.34074/scop.3007016

Bartley, A., Beddoe, L., Duke, J., Fouché, C., Harington, P., & Shah, R. (2016). Crossing Borders: Key Features of Migrant Social Workers in New Zealand. *Aotearoa New Zealand Social Work, 23*(3), 16-30. https://doi.org/10.11157/anzswj-vol23iss3id157

Barton, C. (2021). Embedding Social Pedagogy and Psychological Safety With Virtual Teams. *International Journal of Social Pedagogy, 10*(1). https://doi.org/10.14324/111.444.ijsp.2021.v10.x.008

Barus, I. R. G., & Bontisesari. (2023). Exploring the Potentials of Microlearning in English Language Teaching in Higher Education. *Asian Journal of Applied Education (Ajae), 2*(4), 557-566. https://doi.org/10.55927/ajae.v2i4.6356

Baschiera, B. (2019). The University of the Third Age in Italy: A Dynamic, Flexible, and Accessible Learning Model. 57-67. https://doi.org/10.1007/978-3-030-21515-6_5

Basma, J. (2015). Enhancing Motivation to Learn Among Arab Pre-Service Teachers in Arab and Mixed Colleges. *Athens Journal of Education, 2*(4), 359-382. https://doi.org/10.30958/aje.2-4-5

Basten, M., Meyer-Ahrens, I., Fries, S., & Wilde, M. (2014). The Effects of Autonomy-Supportive vs. Controlling Guidance on Learners' Motivational and Cognitive Achievement in a Structured Field Trip. *Science Education, 98*(6), 1033-1053. https://doi.org/10.1002/sce.21125

Basurto, E. G. A., Camila Almeida De Andrade, C., & Curimilma, C. R. B. (2024). Educación Inclusiva en Ecuador: Propuesta Desde Las TIC. *Ciencia Latina Revista Científica Multidisciplinar, 8*(5), 9033-9052. https://doi.org/10.37811/cl_rcm.v8i5.14296

Bateman, K. M., Kelly, G. J., Licona, P. R., & Cunningham, C. M. (2025). Examining the Affordances of Engineering and Curricular Supports for Learning Among Emerging Bilingual Elementary Afterschool Students. *Science Education, 110*(2), 357-378. https://doi.org/10.1002/sce.70007

Bates, E. S. (2014). Towards Effective Military Training in International Humanitarian Law. *International Review of the Red Cross, 96*(895-896), 795-816. https://doi.org/10.1017/s1816383115000557

Bates, T., & Glikman, V. (2021). One History of Technology and Media in Distance Education. *Médiations Et Médiatisations*(6), 12-34. https://doi.org/10.52358/mm.vi6.190

Batz, R., Walter, M., Burnham, M. M., & Fiore, L. B. (2025). Holding Up the Mirror: The Role of Teacher Educators and Syllabi in Perpetuating or Disrupting Inequity. *Infant Mental Health Journal Infancy and Early Childhood, 46*(4), 391-405. https://doi.org/10.1002/imhj.22158

Baumgartner, L. M., & Carr-Chellman, D. (2024). Adult Learning: A Multifaceted Endeavor. *New Directions for Adult and Continuing Education, 2024*(184), 92-96. https://doi.org/10.1002/ace.20550

Baytiyeh, H., & Naja, M. (2012). Identifying the challenging factors in the transition from colleges of engineering to employment. . *European Journal of Engineering Education, 37*(1), 3-14.

Bazán, V., Zambrano, A. I., Bone, V. C. C., Guilindro, Y. M. L., & Torres-Ruiz, M. (2024). Estrategias Didácticas en La Enseñanza De La Contabilidad en Bachillerato: Un Estudio Comparativo. *Revista Social Fronteriza, 4*(5), e45471. https://doi.org/10.59814/resofro.2024.4(5)471

Bazine, N., Battistelli, A., & Lagabrielle, C. (2020). Environnement Psycho-Technologique (EPT) Et Comportements D'apprentissage Avec Les Technologies (CAT) : Developpement Et Adaptation Française De Deux Mesures. *Psychologie Du Travail Et Des Organisations, 26*(4), 330-343. https://doi.org/10.1016/j.pto.2020.08.001

Bean, M., Grant, S., Hardaker, G., & Ward, R. (2023). Signals and Noise: Communicating achievement Through Alternative Credentials. *International Journal of Information and Learning Technology, 40*(4), 305-316. https://doi.org/10.1108/ijilt-02-2023-0020

Bearman, M., & Ajjawi, R. (2023). Learning to Work With the Black Box: Pedagogy for a World With Artificial Intelligence. *British journal of educational technology, 54*(5), 1160-1173. https://doi.org/10.1111/bjet.13337

Beatriz, M. H., Magdalena, S. P. M., & Gerardo, L. A. Herramientas Online De Libre Acceso Para El Apoyo a La Comunicación Y Al Diseño De Materiales Didácticos en Los Procesos De Enseñanza-Aprendizaje a Través De La Red. https://doi.org/10.31428/10317/2313

Beaty, J. S., & Cleary, R. K. (2025). Foundational Excellence: Theories for Modern Colorectal Surgery Education—Key Pedagogical Frameworks and Strategies That Every Colorectal Surgeon

Teaching Medical Students, Residents, and Co-Workers Should Know. *Clinics in Colon and Rectal Surgery, 39*(04), 261-266. https://doi.org/10.1055/s-0045-1813668

Beauregard, C., Rousseau, C., & Mustafa, S. (2015). The Use of Video in Knowledge Transfer of Teacher-Led Psychosocial Interventions: Feeling Competent to Adopt a Different Role in the Classroom / L'utilisation De La Vidéo Dans Le Transfert De Connaissances Dans Les Interventions Psychosociales. *Canadian Journal of Learning and Technology, 41*(1). https://doi.org/10.21432/t2s907

Beck, V. S., Boys, S. K., Haas, H. J., & King, K. N. (2017). How Do You Use Experiential Learning to Bridge the Classroom and the Real World? *New Directions for Teaching and Learning, 2017*(151), 97-115. https://doi.org/10.1002/tl.20251

Bednarskaya, L. D. (2022). Principles of Consistency in Methodological Science. *Proceedings of Southern Federal University Philology, 26*(3), 169-181. https://doi.org/10.18522/1995-0640-2022-3-169-181

Bedoya, G. J. A., Rivera, N., Sacón, J. P. R., Rosado, A. M. V., & Delgado, L. R. G. (2023). Inclusión Educativa Y Diversidad Cultural. *Acvenisproh Académico.* https://doi.org/10.47606/acven/aclib0041

Beer, J. d., Walt, M. v. d., & Bunt, B. (2020). The Affordances of Case-Based Teaching That Draws on Drama in Pre-Service Teacher Education. 189-214. https://doi.org/10.4102/aosis.2020.bk215.07

Beer, W. d., Harvey, R., & Hill, A. (2019). The Current Status of RANZCP CPD 2017 Programme: A Shift to Regulation. *Australasian Psychiatry, 27*(2), 192-194. https://doi.org/10.1177/1039856218822728

Behmanesh, A., Davarzani, M. H., Shams, R., & Soleimanpour, S. (2023). Internet of Things in Medical Education: A Systematic Mapping Review. https://doi.org/10.21203/rs.3.rs-3565910/v1

Behrend, C., Maldonado-Mariscal, K., & Schröder, A. (2022). Bridging Social Innovation in Education and Organisational Learning. *European Public & Social Innovation Review, 7*(1), 17-29. https://doi.org/10.31637/epsir.22-1.2

Behringer, F., & Coles, M. (2003). The Role of National Qualifications Systems in Promoting Lifelong Learning. https://doi.org/10.1787/224841854572

Beier, M. E., Saxena, M., Kraiger, K., Costanza, D. P., Rudolph, C. W., Cadiz, D. M., Petery, G. A., & Fisher, G. G. (2025). Workplace Learning and the Future of Work. *Industrial and Organizational Psychology, 18*(1), 84-109. https://doi.org/10.1017/iop.2024.57

Bell, R. (2021). Underpinning the Entrepreneurship Educator's Toolkit: Conceptualising the Influence of Educational Philosophies and Theory. *Entrepreneurship Education, 4*(1), 1-18. https://doi.org/10.1007/s41959-020-00042-4

Bell, R., & Bell, H. (2020). Applying Educational Theory to Develop a Framework to Support the Delivery of Experiential Entrepreneurship Education. *Journal of Small Business and Enterprise Development, 27*(6), 987-1004. https://doi.org/10.1108/jsbed-01-2020-0012

Bell, R., & Liu, P. (2019). Educator Challenges in the Development and Delivery of Constructivist Active and Experiential Entrepreneurship Classrooms in Chinese Vocational Higher Education. *Journal of Small Business and Enterprise Development, 26*(2), 209-227. https://doi.org/10.1108/jsbed-01-2018-0025

Belland, B. R. (2016). Context of Use of Computer-Based Scaffolding. 55-77. https://doi.org/10.1007/978-3-319-02565-0_3

Belle, L. J. (2024). Remote Online Collaborative Learning for Knowledge Co-Creation Through WhatsApp in an ODL Programme. *Futurity Education*, 106-124. https://doi.org/10.57125/fed.2024.06.25.06

Bello, M. B. (2023). An Understanding of Paulo Freire's "Pedagogy of the Oppressed". *Canadian Journal of Family and Youth / Le Journal Canadien De Famille Et De La Jeunesse, 15*(1), 225-231. https://doi.org/10.29173/cjfy29905

Below, S. v., Powell, J. J. W., & Roberts, L. W. (2013). Educational Systems and Rising Inequality. *Sociology of Education, 86*(4), 362-375. https://doi.org/10.1177/0038040713496585

Benati, K., & Fischer, J. (2020). Beyond Human Capital: Student Preparation for Graduate Life. *Education + Training, 63*(1), 151-163. https://doi.org/10.1108/et-10-2019-0244

Beneito-Montagut, R., Begueria, A., & Cassián, N. (2018). «Hago De Todo Y No Sé Hacer Funcionar Nada»: Aprendizaje Afectivo Y Relacional De Tecnologías Digitales en Adultos Mayores. *Aula, 24*(0), 77. https://doi.org/10.14201/aula2018247792

Benjamin, J., Pillow, M. T., MacNeill, H., Masters, K., Agrawal, A., & Mehta, N. (2024). Reflections From the Pandemic: Is Connectivism the Panacea for Clinicians? *Journal of Medical Internet Research, 26*, e53344. https://doi.org/10.2196/53344

Benke, M., & Rachwał, T. (2022). The Evolution of Vocational Education and Training in Hungary and Poland 1989–2035. *The Hungarian Educational Research Journal, 12*(3), 328-356. https://doi.org/10.1556/063.2022.00061

Bennett, E. E., & McWhorter, R. R. (2022). Dancing in the Paradox: Virtual Human Resource Development, Online Teaching, and Learning. *Advances in Developing Human Resources, 24*(2), 99-116. https://doi.org/10.1177/15234223221079440

Bennett, R., Strehlow, K., & Hill, B. (2022). Myth-Busting in an Aboriginal Pre-University Bridging Program: Embedding Transformative Learning Pedagogy. *The Australian Journal of Indigenous Education, 51*(1). https://doi.org/10.55146/ajie.2022.24

Beňo, M. (2020). Four Factors That Will Shape the Future of Work. *Journal on Advances in Theoretical and Applied Informatics, 5*(1). https://doi.org/10.26729/jadi.v5i1.3114

Beňo, M. (2021). Office Evolution From Ancient Age to Digital Age (E-Working). 182-192. https://doi.org/10.1007/978-3-030-72651-5_18

Benoliel, P., & Schechter, C. (2017). Is It Personal? Teacher's Personality and the Principal's Role in Professional Learning Communities. *Improving Schools, 20*(3), 222-235. https://doi.org/10.1177/1365480217703725

Bentley, C., Browman, G. P., & Poole, B. (2010). Conceptual and Practical Challenges for Implementing the Communities of Practice Model on a National Scale - A Canadian Cancer Control Initiative. *BMC Health Services Research, 10*(1). https://doi.org/10.1186/1472-6963-10-3

Benton, M., Hearn, S., & Marmolejo-Ramos, F. (2021). Indigenous Students' Experience and Engagement With Support at University: A Mixed-Method Study. *The Australian Journal of Indigenous Education, 50*(2), 256-264. https://doi.org/10.1017/jie.2021.1

Berenji, S. (2021). Enhancing Metacognitive Scaffolding and Comprehension Ability Through Problem-Based Learning in an EFL Context. *Education Research International, 2021*, 1-9. https://doi.org/10.1155/2021/6766793

Berg, G. v. d., & Toit-Brits, C. d. (2023). Adoption and Development of OERs and Practices for Self-Directed Learning: A South African Perspective. *Teacher Education Through Flexible Learning in Africa (Tetfle), 4*(1). https://doi.org/10.35293/tetfle.v4i1.4247

Berg, S. A., & Chyung, S. Y. (2008). Factors That Influence Informal Learning in the Workplace. *Journal of Workplace Learning, 20*(4), 229-244. https://doi.org/10.1108/13665620810871097

Bergier, A., & Anderson, K. (2021). "Step Into Learning When Ready": Towards a Strength-Based Approach to Indigenous Language Education in a University Setting. *Winhec International Journal of Indigenous Education Scholarship*(1). https://doi.org/10.18357/wj161202120273

Bergkvist, L., Ahlin, K., Humble, N., Mozelius, P., & Johansson, J. O. R. (2023). Redesigning Professional Development on Digital Transformation Using Andragogy as a Theoretical Lens. *European Conference on E-Learning, 22*(1), 25-32. https://doi.org/10.34190/ecel.22.1.1821

Bergman, L., & Hessel, F. (2023). From "Confusion" and "Fear" to "Broadened Horizons." Students' Transformative Experiences in Two Higher Education Contexts: Argentina and Sweden. *Journal of Transformative Education, 22*(2), 137-156. https://doi.org/10.1177/15413446231179896

Bergseng, B. (2019). Percentage of Firms Employing More Than 10 People Providing Work-Based Learning Opportunities to Learners Enrolled in Upper Secondary VET, 2015. https://doi.org/10.1787/97a4ab53-en

Bermea, G. O. (2022). Humanistic Advising: Applying Humanistic Theory to the Practice of Academic Advising. *Nacada Review, 3*(1), 3-20. https://doi.org/10.12930/nacr-20-07

Bermúdez, J. M. M. (2017). El Énfasis en Biotecnología Como Escenario De Inclusión Y Desarrollo De Una Actitud Emprendedora. *Bio-Grafía, 10*(19), 351. https://doi.org/10.17227/bio-grafia.extra2017-7124

Bernal, B. K. M. (2023). Estrategias De Formación Del Espíritu Emprendedor en Adultos Mayores Jubilados. *Revista Latinoamericana Ogmios*, *3*(8), 11-24. https://doi.org/10.53595/rlo.v3.i8.073

Bernard, D. (2024). Reimagining Indigenous Education in Costa Rica Through Paulo Freire's Pedagogical Lens: A Critical Analysis. *Latam Revista Latinoamericana De Ciencias Sociales Y Humanidades*, *5*(4). https://doi.org/10.56712/latam.v5i4.2503

Bernard, R., Abrami, P. C., Lou, Y., Wade, A., Wozney, L., Wallet, P., Fiset, M., & Huang, B. (2004). How Does Distance Education Compare With Classroom Instruction? A Meta-Analysis of the Empirical Literature. *Review of Educational Research*, *74*(3), 379-439. https://doi.org/10.3102/00346543074003379

Berragan, E. (2011). Simulation: An Effective Pedagogical Approach for Nursing? *Nurse Education Today*, *31*(7), 660-663. https://doi.org/10.1016/j.nedt.2011.01.019

Berríos-Arvey, H., Barría, R. S., & Vega, E. C. (2024). Trends in Health Sciences Education: Remote Classroom, Artificial Intelligence and Clinical Field. https://doi.org/10.35542/osf.io/8r7tk

Berry, D. (2021). Athletic Training Education: Educating the Future Athletic Training Professionals. *Athletic Training Education Journal*, *16*(4), 235-236. https://doi.org/10.4085/1947-380x-21-066

Bertoni, G., Giardulli, B., Minozzi, B., Coppola, I., Furri, L., Testa, M., & Battista, S. (2024). The Experience of Pedagogical Training on Postgraduate Rehabilitation Health Professionals: A Qualitative Study. *Plos One*, *19*(12), e0314920. https://doi.org/10.1371/journal.pone.0314920

Bęś, P., Strzałkowski, P., Górniak-Zimroz, J., Szóstak, M., & Janiszewski, M. (2025). Innovative Technologies to Improve Occupational Safety in Mining and Construction Industries—Part II. *Sensors*, *25*(18), 5717. https://doi.org/10.3390/s25185717

Besa, L. M., & Parcon, R. E. (2021). On Instructional Materials: The Pre-Service Teachers Preferences in Time of Pandemic. *Basic and Applied Education Research Journal*, *2*(1), 23-28. https://doi.org/10.11594/baerj.02.01.04

Bezard, C. L., & Shaw, S. (2017). Developing Multicultural Self-Awareness Through a Transformative Learning Experience. *Journal of Research in Technical Careers*, *1*(2), 36. https://doi.org/10.9741/2578-2118.1011

Bezerra, A. C. (2021). From Critical Information Literacy to a Critical Theory of Information. *The International Review of Information Ethics*, *30*(1). https://doi.org/10.29173/irie411

Bhandari, U. (2024). Recognition of Prior Learning in Nepal: A Gateway to Socio-Economic Inclusion. *Journal of Technical and Vocational Education and Training*, *18*(1), 1-9. https://doi.org/10.3126/tvet.v18i1.62706

Bhang, C., Yoon, E., Alejandre, S. M., Lefebvre, É., & Yeh, E. T. (2025). Resilience Among 1.5 Generation Korean American Emerging Adults. *Asian American Journal of Psychology*, *16*(4), 339-352. https://doi.org/10.1037/aap0000383

Bhatia, K. V., Pathak-Shelat, M., & Arora, P. (2023). DIY Education in the Digital Era: Youth-Driven Learning Strategies and Curricula for the Future of Work Opportunities. *Education and Information Technologies*, *29*(2), 1407-1426. https://doi.org/10.1007/s10639-023-11750-4

Bhattarai, D. P., & Basnet, H. B. (2022). Understanding the Nepali Classroom Practices: A Constructivist Perspective. *Journal of Research and Development*, *5*(1), 33-40. https://doi.org/10.3126/jrdn.v5i1.50093

Bi, G. (2023). Promoting Self-Efficacy of Prospective Music Teachers: Theoretical Insights. *Society Integration Education Proceedings of the International Scientific Conference*, *1*, 82-91. https://doi.org/10.17770/sie2023vol1.7073

Biao, I. (2022). Financing Adult Learning and Education (ALE) Now and in Future. *International Review of Education*, *68*(2), 213-231. https://doi.org/10.1007/s11159-022-09950-0

Biasin, C. (2018a). L'apprentissage Transformateur : État Des Lieux Et Portée Heuristique D'un Construit en Développement. *Phronesis*, *7*(3), 1. https://doi.org/10.7202/1054403ar

Biasin, C. (2018b). Transformative Learning: Evolutions of the Adult Learning Theory. *Phronesis*, *7*(3), 5-17. https://doi.org/10.7202/1054404ar

Biavaschi, C., Eichhörst, W., Giulietti, C., Kendzia, M. J., Muravyev, A., Pieters, J., Rodríguez-Planas, N., Schmidl, R., & Zimmermann, K. F. (2012). Youth Unemployment and Vocational Training. *SSRN Electronic Journal*. https://doi.org/10.2139/ssrn.2158300

Biesta, G. (2017). Don't Be Fooled by Ignorant Schoolmasters: On the Role of the Teacher in Emancipatory Education. *Policy Futures in Education*, *15*(1), 52-73. https://doi.org/10.1177/1478210316681202

Biesta, G. (2021). Reclaiming a Future That Has Not Yet Been: The Faure Report, UNESCO's Humanism and the Need for the Emancipation of Education. *International Review of Education*, *68*(5), 655-672. https://doi.org/10.1007/s11159-021-09921-x

Bigdeli, S., Hosseinzadeh, Z., Dehnad, A., Sohrabi, Z., Aalaa, M., Haghani, F., & Atlasi, R. (2023). Underpinning Learning Theories of Medical Educational Games: A Scoping Review. *Medical Journal of the Islamic Republic of Iran*. https://doi.org/10.47176/mjiri.37.26

Bijl, A. V. d., & Taylor, V. (2016). Nature and Dynamics of Industry-Based Workplace Learning for South African TVET Lecturers. *Industry and Higher Education*, *30*(2), 98-108. https://doi.org/10.5367/ihe.2016.0297

Biljon, J. v., & Merwe, R. v. d. (2018). Trends, Drivers and Barriers Influencing Cloud Computing Services for Mobile Interactions in Teaching and Learning. 1-7. https://doi.org/10.1109/ictas.2018.8368738

Billett, S. (2001). *Learning in the Workplace: Strategies for Effective Practice*. Allen & Unwin.

Billett, S. (2011). *Vocational Education: Purposes, Traditions and Prospects*. Springer.

Billsberry, J., & Alony, I. (2023). The MOOC Post-Mortem: Bibliometric and Systematic Analyses of Research on Massive Open Online Courses (MOOCs), 2009 to 2022. *Organizational Behavior Teaching Review*, *48*(4), 634-670. https://doi.org/10.1177/10525629231190840

Bindhani, S., & Gopinath, G. (2024). Inclusive Education Practices: A Review of Challenges and Successes. *International Journal for Multidisciplinary Research*, *6*(2). https://doi.org/10.36948/ijfmr.2024.v06i02.17341

Binson, P. A., & Joseph, M. (2023). Learning Organization: The Cornerstone of Organizational Development in Non Profit Management. *International Journal for Multidisciplinary Research*, *5*(6). https://doi.org/10.36948/ijfmr.2023.v05i06.11338

Bintoro, H. S., Aulya, R., & Wanabuliandari, S. (2024). The Problem-Based Learning Model Assisted With the Ethlaf Application Contains Ethnomathematics to Improve Students' Conceptual Understanding Capability. *Journal of Education Research and Evaluation*, *8*(1), 163-172. https://doi.org/10.23887/jere.v8i1.69805

Birdsall, S. (2022). Exploring Effective Pedagogies in Environmental and Sustainability Education for Teachers: A Story of New Zealand Pre-Service Teachers' Learning Experiences. *Brock Education Journal*, *31*(2), 40-62. https://doi.org/10.26522/brocked.v31i2.941

Bishri, R. (2023). Student-Centered Learning Approaches in the State Universities of Sri Lanka: Exploring Socio-Cultural Impacts Through a Conceptual Analysis. *Colombo Journal of Multi-Disciplinary Research*, *8*(1-2), 75-99. https://doi.org/10.4038/cjmr.v8i1-2.76

Bitrán, M. (2021). Burnout Y Bienestar Estudiantil: Una Mirada Nueva a Un Problema Viejo. *Ars Medica Revista De Ciencias Médicas*, *46*(4), 71-76. https://doi.org/10.11565/arsmed.v46i4.1855

Bittner, J. V., Roßnagel, C. S., & Staudinger, U. M. (2021). Educational Self-Regulation Competence: Toward a Lifespan-Based Concept and Assessment Strategy. *International Journal for Educational and Vocational Guidance*, *22*(2), 307-325. https://doi.org/10.1007/s10775-021-09491-2

Black, K., & Warhurst, R. (2025). Enhancing Employability Through an Authentic Experiential Learning Exercise. *Organizational Behavior Teaching Review*, *50*(2), 230-258. https://doi.org/10.1177/10525629251394311

Blackburne, G. (2024). A Conceptual Instructional Design Model for Executive Education. *Research in Education Curriculum and Pedagogy Global Perspectives*. https://doi.org/10.56395/recap.v1i1.7

Blackford, K., Bona, M. D., & Crawford, G. (2024). A Preliminary Analysis of "Passport to Practice": Investigating Development of Core Competencies in Undergraduate Health Promotion Students. *Global Health Promotion*, *31*(4), 27-36. https://doi.org/10.1177/17579759241230065

Blaschke, L. M. (2021). The Dynamic Mix of Heutagogy and Technology: Preparing Learners for Lifelong Learning. *British journal of educational technology*, *52*(4), 1629-1645. https://doi.org/10.1111/bjet.13105

Blaschke, L. M., & Hase, S. (2019). Heutagogy and Digital Media Networks. *Pacific Journal of Technology Enhanced Learning*, *1*(1), 1-14. https://doi.org/10.24135/pjtel.v1i1.1

Blaschke, L. M., & Marín, V. I. (2020). Applications of Heutagogy in the Educational Use of E-Portfolios. *Revista De Educación a Distancia (Red)*, *20*(64). https://doi.org/10.6018/red.407831

Blattgerste, J., Renner, P., & Pfeiffer, T. (2019). Augmented Reality Action Assistance and Learning for Cognitively Impaired People. 270-279. https://doi.org/10.1145/3316782.3316789

BlippAr. (2025). BlippAR augmented reality platform. https://www.blippar.com

Blumenstein, M., Liu, D., Richards, D., Leichtweis, S., & Stephens, J. M. (2018). Data-Informed Nudges for Student Engagement and Success. https://doi.org/10.31234/osf.io/qjmpn

Bobiţan, N., Dumitrescu, D., Popa, A. F., Sahlian, D.-N., & Turlea, I. C. (2024). Shaping Tomorrow: Anticipating Skills Requirements Based on the Integration of Artificial Intelligence in Business Organizations—A Foresight Analysis Using the Scenario Method. *Electronics*, *13*(11), 2198. https://doi.org/10.3390/electronics13112198

Bocquillon, M., Gauthier, C., Bissonnette, S., & Derobertmasure, A. (2020). Enseignement Explicite Et Développement Decompétences : Antinomie Ou Nécessité? *Formation Et Profession*, *28*(2), 3. https://doi.org/10.18162/fp.2020.513

Boeren, E., & Holford, J. (2016). Vocationalism Varies (A Lot). *Adult education quarterly*, *66*(2), 120-142. https://doi.org/10.1177/0741713615624207

Boeren, E., & Lido, C. (2023). Capturing Lifelong Learning Data Through International Surveys and Novel Innovative Methods. 207-219. https://doi.org/10.36253/979-12-215-0253-4.22

Bøje, R. B., Musaeus, P., Sørensen, D., & Ludvigsen, M. S. (2022). Toward Nurses' Transformative Agency in Transitional Care for Older Adults: A Change-Laboratory Intervention. *Global Qualitative Nursing Research*, *9*. https://doi.org/10.1177/23333936221087622

Bol, T., & Weeden, K. A. (2014). Occupational Closure and Wage Inequality in Germany and the United Kingdom. *European Sociological Review*, *31*(3), 354-369. https://doi.org/10.1093/esr/jcu095

Boland, B. (2023). Extended Reality Vocational Training's Ability to Improve Soft Skills Development and Increase Equity in the Workforce. *Ai Computer Science and Robotics Technology*, *2*. https://doi.org/10.5772/acrt.22

Boltz, L. O., Yadav, A., Dillman, B., & Robertson, C. (2021). Transitioning to Remote Learning: Lessons From Supporting K-12 Teachers Through a MOOC. *British journal of educational technology*, *52*(4), 1377-1393. https://doi.org/10.1111/bjet.13075

Bonanno, P., Klichowski, M., & Lister, P. (2019). A Pedagogical Model for CyberParks. 294-307. https://doi.org/10.1007/978-3-030-13417-4_23

Bond, W. F., Lammers, R. L., Spillane, L., Smith-Coggins, R., Fernandez, R., Reznek, M. A., Vozenilek, J. A., & Gordon, J. A. (2007). The Use of Simulation in Emergency Medicine: A Research Agenda. *Academic Emergency Medicine*, *14*(4), 353-363. https://doi.org/10.1197/j.aem.2006.11.021

Bondin, A., Zammit, J. P., Francalanza, E., Borg, J. C., Dallasega, P., Lanzone, M., Maffei, A., Monetti, F. M., Amza, C. G., Chicioreanu, T. D., Ljubić, S., & Batistić, L. (2026). Immersive Technologies in Engineering Education: A Systematic Review Through the Lens of Constructive Alignment and Bloom's Taxonomy. *Computer Applications in Engineering Education*, *34*(3). https://doi.org/10.1002/cae.70201

Bongomin, O., Ocen, G. G., Nganyi, E. O., Alex, M., & Omara, T. (2019). Exponential Disruptive Technologies and the Required Skills of Industry 4.0: A Review. https://doi.org/10.20944/preprints201910.0240.v1

Bongomin, O., Yemane, A., Kembabazi, B., Malanda, C., Mwape, M. C., Mpofu, N. S., & Tigalana, D. (2020). Industry 4.0 Disruption and Its Neologisms in Major Industrial Sectors: A State of the Art. *Journal of Engineering*, *2020*, 1-45. https://doi.org/10.1155/2020/8090521

Bonk, C. J., & Lee, M. M. (2017). Motivations, Achievements, and Challenges of Self-Directed Informal Learners in Open Educational Environments and MOOCs. *Journal of Learning for Development*, *4*(1). https://doi.org/10.56059/jl4d.v4i1.195

Bono, G., Reil, K., & Hescox, J. (2020). Stress and Wellbeing in Urban College Students in the U.S. During the COVID-19 Pandemic: Can Grit and Gratitude Help? *International Journal of Wellbeing*, *10*(3), 39-57. https://doi.org/10.5502/ijw.v10i3.1331

Bönsch, J., Greif, L., Hauck, S., Kreuzwieser, S., Mayer, A., Michels, F. L., & Ovtcharova, J. (2024). Virtual Engineering: Hands-on Integration of Product Lifecycle Management, Computer-Aided Design, eXtended Reality, and Artificial Intelligence in Engineering Education. *Chemie Ingenieur Technik*, *96*(11), 1460-1474. https://doi.org/10.1002/cite.202300169

Boonyuen, S., Charungkaittikul, S., & Ratana-Ubol, A. (2016). Integrating Transformative Learning and Action Learning Approaches to Enhance Ethical Leadership for Supervisors in the Hotel Business. *SHS Web of Conferences*, *26*, 01136. https://doi.org/10.1051/shsconf/20162601136

Borders, L. D. (2019). Science of Learning: Evidence-Based Teaching in the Clinical Supervision Classroom. *Counselor Education and Supervision*, *58*(1), 64-79. https://doi.org/10.1002/ceas.12124

Borgos, J., Kinser, K., & Kline, L. M. (2023). The Borderless Market for Open, Distance, and Digital Education. 355-369. https://doi.org/10.1007/978-981-19-2080-6_22

Borisova, T. G., & Bekhteneva, R. A. (2015). Certified Public Accountant Is Unmatched Advantage on the Global Job Market. *European Journal of Economic Studies*, *12*(2), 64-68. https://doi.org/10.13187/es.2015.12.64

Botes, V. (2024). Using a Word Game to Revitalise Language and Promote Equity and Inclusion in a Quantitative Subject. https://doi.org/10.18573/conf2.k

Botha, B. S., Hugo, L., Jordaan, H., & Lombard, W. A. (2025). A Conceptual Framework to Implement Extended Reality in Undergraduate Agricultural Sciences. *Journal of Natural Resources and Life Sciences Education*, *54*(2). https://doi.org/10.1002/nse2.70037

Botross, P. (2024). Equipping Disciplers (Part 1): A Pedagogical Framework for Equipping the <i>Whole</I> Person. *Christian Education Journal Research on Educational Ministry*, *21*(3), 290-309. https://doi.org/10.1177/07398913241302103

Bou-Ammar, H., Grosnit, A., Maraval, A., Refinath, S. N., Zhao, Z., Doran, J., Paolo, G., Thomas, A., Kumar, A., Khandelwal, K., Benechehab, A., Cherkaoui, H., El-Hili, Y. A., Shao, K., Hao, J., Yao, J., Kégl, B., & Wang, J. (2025). Kolb-Based Experiential Learning for Generalist Agents With Human-Level Kaggle Data Science Performance. https://doi.org/10.21203/rs.3.rs-7472642/v1

Bou, D., Sran, L., Teu, S., Tep, C., Kheuy, S., Vy, S., & Dul, V. (2024). Examining Self-Assessment of Student Teachers' Self-Directed Learning: A Case Study at Battambang Teacher Education College, Cambodia. https://doi.org/10.31219/osf.io/uq8t3

Bouchereau, A., & Roxin, I. (2018). Internet Des Objets Pour L'apprentissage Humain. *Internet Des Objets*, *18*(1). https://doi.org/10.21494/iste.op.2018.0217

Boulahoual, A., & Rafik, M. (2024). The Role of E-Learning in the Evolution of Professions: a Moroccan Contextualization. *Global Journal of Management and Business Research*, 1-14. https://doi.org/10.34257/gjmbrgvol24is1pg1

Bourgain, M. (2022). L'enseignement Créatif Et Critique Du Management. *Finance Contrôle Stratégie*, *NS-13*. https://doi.org/10.4000/fcs.9567

Bourgeois, É. (2018). Le Développement De L'enfant : La Contribution De Piaget Au Champ De L'éducation. *Revue Internationale D Éducation De Sèvres*, *79*, 99-108. https://doi.org/10.4000/ries.7077

Bouttemont, C. d. (2010). Références Bibliographiques Du Dossier « qualité, Équité Et Diversité Dans Le Préscolaire ». *Revue Internationale D Éducation De Sèvres*(53), 141-150. https://doi.org/10.4000/ries.919

Bouwer, M., Venketsamy, R., & Bipath, K. (2021a). Remodelling Work-Integrated Learning Experiences of Grade R Student Teachers. *South African Journal of Higher Education*, *35*(4). https://doi.org/10.20853/36-3-4331

Bouwer, M., Venketsamy, R., & Bipath, K. (2021b). Remodelling work-integrated learning experiences of Grade R student teachers *South African Journal of Higher Education*, *35*(4). https://doi.org/10.20853/35-4-4331

Bova, B. M., & Kroth, M. (2001). Workplace Learning and Generation X. *Journal of Workplace Learning*, *13*(2), 57-65. https://doi.org/10.1108/13665620110383645

Boverie, P., Grassberger, R., & Law, V. (2013). Leading Individual Development and Organizational Change Around Learning, Meaning, and Nurturing Environment. *Advances in Developing Human Resources*, *15*(4), 382-400. https://doi.org/10.1177/1523422313498564

Bovshik, A. S., & Gaivoronskaya, S. O. (2022). Storytelling in the Inclusive L2 Classroom. *Linguistics & Polyglot Studies*, *8*(1), 80-93. https://doi.org/10.24833/2410-2423-2022-1-30-80-93

Boyadjieva, P., & Ilieva-Trichkova, P. (2017a). Adult Education as a Common Good: Conceptualisation and Measurement. *International Journal of Lifelong Education*, *36*(3), 345-358. https://doi.org/10.1080/02601370.2016.1272888

Boyadjieva, P., & Ilieva-Trichkova, P. (2017b). Between Inclusion and Fairness. *Adult education quarterly*, *67*(2), 97-117. https://doi.org/10.1177/0741713616685398

Boyd, D., Boyd, S. C., & Berry, P. (2009). A Primer for Accounting Certification: Complete Analysis of the Process With Listing of Sources. *American Journal of Business Education (Ajbe)*, *2*(7), 83-96. https://doi.org/10.19030/ajbe.v2i7.4587

Boyd, V., Woods, N. N., Kumagai, A. K., Kawamura, A., Orsino, A., & Ng, S. (2022). Examining the Impact of Dialogic Learning on Critically Reflective Practice. *Academic Medicine*, *97*(11S), S71-S79. https://doi.org/10.1097/acm.0000000000004916

Bozkurt, A. (2022). Resilience, Adaptability, and Sustainability of Higher Education: A Systematic Mapping Study on the Impact of the Coronavirus (Covid-19) Pandemic and the Transition to the New Normal. *Journal of Learning for Development*, *9*(1), 1-16. https://doi.org/10.56059/jl4d.v9i1.590

Brabazon, H., Esmail, J., Locklin, R. B., & Stirling, A. (2020). Beyond Employability: Defamiliarizing Work-Integrated Learning With Community-Engaged Learning. *Engaged Scholar Journal Community-Engaged Research Teaching and Learning*, *5*(3), 21-41. https://doi.org/10.15402/esj.v5i3.70364

Branca, T. A., Fornai, B., Colla, V., Pistelli, M. I., Faraci, E. L., Cirilli, F., & Schröder, A. (2022). Skills Demand in Energy Intensive Industries Targeting Industrial Symbiosis and Energy Efficiency. *Sustainability*, *14*(23), 15615. https://doi.org/10.3390/su142315615

Brătianu, C., Prelipcean, G., & Bejinaru, R. (2020). Exploring the Latent Variables Which Support SMEs to Become Learning Organizations. *Management & Marketing*, *15*(2), 154-171. https://doi.org/10.2478/mmcks-2020-0010

Braunack-Mayer, A., Street, J., Tooher, R., Feng, X., & Scharling-Gamba, K. (2020). Student and Staff Perspectives on the Use of Big Data in the Tertiary Education Sector: A Scoping Review and Reflection on the Ethical Issues. *Review of Educational Research*, *90*(6), 788-823. https://doi.org/10.3102/0034654320960213

Bravo, L. E. C., Tarazona-Bermúdez, G.-M., & Molano, J. I. R. (2021). Tecnología Y Analítica Del Aprendizaje: Una Revisión a La Literatura. *Revista Científica*, *41*(2), 150-168. https://doi.org/10.14483/23448350.17547

Braxton, S. N. (2023). Competency Frameworks, Alternative Credentials and The evolving Relationship of Higher education and Employers In recognizing Skills and Achievements. *International Journal of Information and Learning Technology*, *40*(5), 373-387. https://doi.org/10.1108/ijilt-10-2022-0206

Breiter, A. (2002). Wissensmanagementsysteme in Schulen Oder: Wie Bringe Ich Ordnung Ins Chaos? *Medienpädagogik Zeitschrift Für Theorie Und Praxis Der Medienbildung*, *6*, 1-26. https://doi.org/10.21240/mpaed/06/2002.11.06.x

Bremner, R., Gibbs, A., & Mitchell, A. R. (2020). The Era of Immersive Health Technology. *Emj Innovation*, 40-47. https://doi.org/10.33590/emjinnov/19-00132

Breton, H. (2018). Vers Une Relecture Herméneutique Et Biographique De La Théorie De L'apprentissage Transformateur. *Phronesis*, *7*(3), 35-42. https://doi.org/10.7202/1054407ar

Brewer, S. A., & Klein, J. D. (2006). Type of Positive Interdependence and Affiliation Motive in an Asynchronous, Collaborative Learning Environment. *Educational Technology Research and Development*, *54*(4), 331-354. https://doi.org/10.1007/s11423-006-9603-3

Brian, T. (2021). Educational Equity and the Application of Technology: A Critical Approach. *Scope Contemporary Research Topics (Learning and Teaching)*(10), 83-89. https://doi.org/10.34074/scop.4010009

Brice, L., Harrison, A., & Cadwallader, A. H. (2023). Transforming Learning Support in ODFL: The LESS Model. *Journal of Open Flexible and Distance Learning*, *27*(1), 42-53. https://doi.org/10.61468/jofdl.v27i1.581

Briceño, N. C. B., Mogollón, M. L., & Naranjo, M. E. G. (2019). Andragógica Y Praxeología: Trialéctica Para La Transformación Desde La Comunicación Universitaria. *Innova Research Journal*, *4*(3.2), 63-79. https://doi.org/10.33890/innova.v4.n3.2.2019.1057

Briciu, B., Ste-Marie, L., & Léa-Langlois, M. (2024). Recovering Hope and a Sense of Agency Through Transformative Education for Internationally Educated Healthcare Professionals in Canada. *Journal of Transformative Education*, *23*(4), 523-542. https://doi.org/10.1177/15413446241296079

Bridgman, H., Bird, M. L., Heyworth, K. J., Maine, G., Hardcastle, S., Murray, S., Radford, J., Elmer, S., Norris, K., Dean, T., Marlow, A., Williams, A. D., & Todd, A. (2020). Evaluating an Interprofessional Workshop on Persistent Pain: The Role of Adult Learning and Social Identity Theories. *Journal of Applied Learning & Teaching*, *3*(Sp. Iss. 1), 129-139. https://doi.org/10.37074/jalt.2020.3.s1.11

Brieger, E. W., Arghode, V., & McLean, G. N. (2020). Connecting Theory and Practice: Reviewing Six Learning Theories to Inform Online Instruction. *European journal of training and development*, *44*(4/5), 321-339. https://doi.org/10.1108/ejtd-07-2019-0116

Briggs, A. (2018). *Non-Profit Leadership Learning : A Narrative Research Study Exploring the Transformation of Experience Into Leadership Learning* Northeastern University].

Brítez, M. Á. A. (2023). Las Teorías Del Aprendizaje Y Su Aplicación en La Enseñanza De La Contabilidad. *Revista Científica Estudios E Investigaciones*, *12*(2), 68-88. https://doi.org/10.26885/rcei.12.2.68

Brnich, M. J., Jr., R. L. D., Mallett, L., & Vaught, C. (2002). Strategies for Improving Miners' Training. https://doi.org/10.26616/nioshpub2002156

Brodén, K., Johansson, J. O. R., Bergkvist, L., Humble, N., & Mozelius, P. (2024). Design Principles for a Higher Education Course Focusing on Lifelong Learning. *The Electronic Journal of E-Learning*, *22*(7), 31-39. https://doi.org/10.34190/ejel.22.7.3517

Brødsjø, V., Sandøy, B., & Hustad, E. (2023). Exploring Onboarding Processes for IT Professionals: The Role of Knowledge Management. *European Conference on Knowledge Management*, *24*(1), 148-158. https://doi.org/10.34190/eckm.24.1.1370

Broek, S., Linden, J. v. d., Maria Anna Catharina Theresia, K., & Semeijn, J. (2023). What Makes Adults Choose to Learn: Factors That Stimulate or Prevent Adults From Learning. *Journal of Adult and Continuing Education*, *29*(2), 620-642. https://doi.org/10.1177/14779714231169684

Bronfenbrenner, U. (1974). Developmental Research, Public Policy, and the Ecology of Childhood. *Child Development*, *45*(1), 1-5. https://doi.org/10.2307/1127743

Brookfield, S. (2018). Transformative Learning and the Awareness of White Supremacy. *Phronesis*, *7*(3), 63-69. https://doi.org/10.7202/1054410ar

Brookfield, S. D. (2017). *Becoming a Critically Reflective Teacher* (2nd ed.). Jossey-Bass.

Brown, A., & Bimrose, J. (2014). Model of Learning for Career and Labour Market Transitions. *Research in Comparative and International Education*, *9*(3), 270-286. https://doi.org/10.2304/rcie.2014.9.3.270

Brown, J. (2017). Curating the "Third Place"? Coworking and the Mediation of Creativity. *Geoforum*, *82*, 112-126. https://doi.org/10.1016/j.geoforum.2017.04.006

Brown, J. S., Collins, A., & Duguid, P. (1989). Situated Cognition and the Culture of Learning. *Educational Researcher*, *18*(1), 32-42. https://doi.org/10.3102/0013189X018001032

Brown, K. M. (2004). Leadership for Social Justice and Equity: Weaving a Transformative Framework and Pedagogy. *Educational Administration Quarterly*, *40*(1), 77-108. https://doi.org/10.1177/0013161x03259147

Brown, K. M. (2006). Leadership for Social Justice and Equity: Evaluating a Transformative Framework and Andragogy. *Educational Administration Quarterly*, *42*(5), 700-745. https://doi.org/10.1177/0013161x06290650

Brown, L. R., McCray, P. D., Neal, J. L., & Caldwell-Serges, L. (2023). Corporate Digital Literacy Mandates. *International Journal of Advanced Corporate Learning (Ijac)*, *16*(2), 18-36. https://doi.org/10.3991/ijac.v16i2.35747

Brown, M., McGreal, R., & Peters, M. (2023). A Strategic Institutional Response to Micro-Credentials: Key Questions for Educational Leaders. *Journal of Interactive Media in Education*, *2023*(1). https://doi.org/10.5334/jime.801

Brown, M., Mhichíl, M. N. G., Beirne, E., & Lochlainn, C. M. (2021). The Global Micro-Credential Landscape: Charting a New Credential Ecology for Lifelong Learning. *Journal of Learning for Development*, *8*(2), 228-254. https://doi.org/10.56059/jl4d.v8i2.525

Brown, M., Nic Giolla Mhichíl, M., Beirne, E., & Mac Lochlainn, C. (2021). The Global Micro-credential Landscape: Charting a New Credential Ecology for Lifelong Learning. *Journal of Learning for Development*, *8*(2), 228-254. https://doi.org/10.56059/jl4d.v8i2.525

Brudney, J. J. (2021). Square Pegs and Round Holes: Shrinking Protections for Unpaid Interns Under the Fair Labor Standards Act. https://doi.org/10.4337/9781800885042.00019

Bruillard, É., Ghabara, K., Huguenin, S., Jolicoeur, P.-L., Laferrière, T., Nadeau-Tremblay, S., Papı, C., & Pelletier, M.-A. (2021). L'apprentissage Connecté Des Jeunes en Contexte D'éducation Formelle Francophone. *Canadian Journal of Learning and Technology*, *47*(4). https://doi.org/10.21432/cjlt28060

Bruin, L. d. (2019). The Use of Cognitive Apprenticeship in the Learning and Teaching of Improvisation: Teacher and Student Perspectives. *Research Studies in Music Education*, *41*(3), 261-279. https://doi.org/10.1177/1321103x18773110

Bruner, J. S. (1961). The act of discovery. *Harvard Educational Review*, *31*(1), 21-32.

Bruner, J. S. (1966). *Toward a theory of instruction*. Harvard University Press.

Brunetti, I., & Corsini, L. (2019). School-to-Work Transition and Vocational Education: A Comparison Across Europe. *International Journal of Manpower*, *40*(8), 1411-1437. https://doi.org/10.1108/ijm-02-2018-0061

Bruno, M. L., & Morgado, L. (2022). Lessons Learned From Initiatives to Roll Out Digital Credentials in Europe. *1*, 8215-8224. https://doi.org/10.21125/iceri.2022.2130

Brussino, O. (2021). Building Capacity for Inclusive Teaching. https://doi.org/10.1787/57fe6a38-en

Brussino, O., & McBrien, J. (2022). *Gender Stereotypes in Education*. OECD.

Brustenga, G. G., Alpiste, M. F. i., & Castells, N. M. (2018). Briefing Paper: Chatbots in Education. https://doi.org/10.7238/elc.chatbots.2018

Buckingham, C. R. (2000). Philosophies of Adult Education as Practiced by Agricultural Education Teachers in Pennsylvania, West Virginia, and Virginia. https://doi.org/10.33915/etd.1063

Budd, J. M., Chu, C. M., Dali, K., & O'Brien, H. L. (2015). Making an Impact Through Experiential Learning. *Proceedings of the Association for Information Science and Technology*, *52*(1), 1-4. https://doi.org/10.1002/pra2.2015.14505201007

Budiarto, M. K., Yusuf, M., & Subagya. (2023). Implementation of Pedagogical, Andragogical, and Heutagogical Approaches in Education System Sustainability. *Indonesian Journal of Educational Research and Review*, *6*(2), 281-298. https://doi.org/10.23887/ijerr.v6i2.59889

Budwig, N., & Alexander, A. J. (2021). Exploring the Conceptual Frameworks Guiding Developmental Research and Practice in Higher Education: Some Challenges for Transdisciplinary Work. *Human Development*, *65*(1), 1-18. https://doi.org/10.1159/000514553

Buhrig, R. (2023). The Case for Case-Based Learning in Police Recruit Training. *The Police Journal Theory Practice and Principles*, *97*(2), 336-353. https://doi.org/10.1177/0032258x231181322

Bui, H. (2020). From the Fifth Discipline to the New Revolution: What We Have Learnt From Senge's Ideas Over the Last Three Decades. *The Learning Organization*, *27*(6), 495-504. https://doi.org/10.1108/tlo-04-2020-0062

Bui, H., & Baruch, Y. (2010). Creating Learning Organizations: A Systems Perspective. *The Learning Organization*, *17*(3), 208-227. https://doi.org/10.1108/09696471011034919

Bui, H., & Baruch, Y. (2011). Learning Organizations in Higher Education: An Empirical Evaluation Within an International Context. *Management learning*, *43*(5), 515-544. https://doi.org/10.1177/1350507611431212

Bukko, D., & Dhesi, J. (2021). Doctoral Students Living, Leading, and Learning During a Pandemic. *Impacting Education Journal on Transforming Professional Practice*, *6*(2), 25-33. https://doi.org/10.5195/ie.2021.167

Bulathwela, S., Pérez-Ortiz, M., Holloway, C., Cukurova, M., & Shawe-Taylor, J. (2024). Artificial Intelligence Alone Will Not Democratise Education: On Educational Inequality, Techno-Solutionism and Inclusive Tools. *Sustainability*, *16*(2), 781. https://doi.org/10.3390/su16020781

Buli-Holmberg, J., & Kamenopoulou, L. (2017). Attaining New Knowledge on Inclusive Education: A Case Study of Students' Voices. *Athens Journal of Education*, *4*(4), 363-378. https://doi.org/10.30958/aje.4-4-5

Burgess, A., & McGregor, D. (2018). Peer Teacher Training for Health Professional Students: A Systematic Review of Formal Programs. *BMC Medical Education*, *18*(1). https://doi.org/10.1186/s12909-018-1356-2

Burkšaitienė, N. (2006). Užsienio Kalbos Mokymasis Univeristetinėse Studijose Konstruktyvistinės Paradigmos Kontekste. *Acta Paedagogica Vilnensia*, *16*, 18-26. https://doi.org/10.15388/actpaed.2006.16.9703

Burlakova, I., Budnik, A., & S.Burlakova, E. (2019). Organization of Lifelong Learning in the System of Professional Training of Teachers. https://doi.org/10.3897/ap.1.e0051

Burnham, J. D., & Ponton, M. K. (2021). An Exploration of the Deterrents to Adult Basic Education Among Blue Collar Workers. *Journal of Studies in Education*, *11*(1), 26. https://doi.org/10.5296/jse.v11i1.18169

Burns, E., Silvennoinen, E., Копнов, В. А., Щипанова, Д. Е., Papić-Blagojević, N., & Tomašević, S. (2020). Supporting the Development of Digitally Competent VET Teachers in Serbia and Russia. *The Education and Science Journal*, *22*(9), 174-203. https://doi.org/10.17853/1994-5639-2020-9-174-203

Burrell, D. N. (2018). Exploring Leadership Coaching as a Tool to Improve the People Management Skills of Information Technology and Cybersecurity Project Managers. *Holistica – Journal of Business and Public Administration*, *9*(2), 107-126. https://doi.org/10.2478/hjbpa-2018-0015

Burrell, E., & Brauner, E. (2021). Knowing and Sharing: Transactive Knowledge Systems and Psychological Safety. *Organisationsberatung Supervision Coaching*, *28*(3), 303-317. https://doi.org/10.1007/s11613-021-00722-7

Burt, S. (2024). What Do Coaches Actually Do to Learn and Develop? A Qualitative Exploration of the Development Narratives of Experienced Coaches. https://doi.org/10.24384/a16t-f341

Burwell, K. (2012). Apprenticeship in Music: A Contextual Study for Instrumental Teaching and Learning. *International Journal of Music Education*, *31*(3), 276-291. https://doi.org/10.1177/0255761411434501

Buser, P. (2006). Slowly Forgetting the Pavlovian Adventure? *Comptes Rendus Biologies*, *329*(5-6), 398-405. https://doi.org/10.1016/j.crvi.2006.03.010

Bustamante, J. S. B., & Mejía, C. (2017). Acciones Y Herramientas De Software Para Implementar Entornos De Inclusión Educativa en Colombia. *Teknos Revista Científica*, *17*(2), 38. https://doi.org/10.25044/25392190.903

Butcher, D., & Bruce, A. (2015). Nurses and Lifelong Learning: Creating "Makers and Shapers" or "Users and Choosers"? *Nursing Forum*, *51*(2), 97-104. https://doi.org/10.1111/nuf.12126

Butler-Ulrich, T., Hughes, J., & Morrison, L. (2024). Creativity and Generative AI for Preservice Teachers. https://doi.org/10.5772/intechopen.1007517

Buttigieg, K. R., & Calleja, C. (2020a). Bildung and Transformative Learning Theory: Two Peas in a Pod? *Journal of Transformative Education*, *19*(2), 166-185. https://doi.org/10.1177/1541344620971673

Buttigieg, K. R., & Calleja, C. (2020b). <i>Bildung</I> and Transformative Learning Theory: Two Peas in a Pod? *Journal of Transformative Education*, *19*(2), 166-185. https://doi.org/10.1177/1541344620971673

Buus, N., Ong, B., & McCloughen, A. (2022). Open Dialogue Trainees' Perspectives on Learning Processes and Psychotherapeutic Practice: A Prospective Focus Group Study. *Family Process*, *62*(3), 993-1005. https://doi.org/10.1111/famp.12832

Buxton, L. (2018). Aboriginal Ways of Seeing and Being: Informing Professional Learning for Australian Teachers. *Alternative an International Journal of Indigenous Peoples*, *14*(2), 121-129. https://doi.org/10.1177/1177180118764124

Buyya, R., Srirama, S. N., Casale, G., Calheiros, R. N., Simmhan, Y., Varghese, B., Gelenbe, E., Javadi, B., Vaquero, L. M., Netto, M. A. S., & Toosi, A. N. (2018). A Manifesto for Future Generation Cloud Computing. *Acm Computing Surveys*, *51*(5), 1-38. https://doi.org/10.1145/3241737

Bwegyeme, J., & Munene, J. C. (2021). Experiencial Learning; Learning Abilities That Enhance Problem Solving Ability. *International Journal of Experiential Learning & Case Studies*, *6*(1), 1-14. https://doi.org/10.22555/ijelcs.v6i1.263

Byrne, J. R., Kearney, S., & Sullivan, K. (2018). Technology-Mediated Collaborative Learning: The Bridge21 Activity Model in Theory and Practice. 309-330. https://doi.org/10.1007/978-3-030-01551-0_16

Cabascango-Trávez, G. (2023). El Uso De La Realidad Aumentada en La Enseñanza De Ciencias: Un Enfoque Integrador en Educación Secundaria. *Revista Científica Kosmos*, *2*(1), 39-50. https://doi.org/10.62943/rev.cien.kos.v2i1.43

Cadorette, J., McCurry, M. K., & Chin, E. (2023). Reflective Transition Practice Model: The New Graduate Registered Nurse. *Nursing Science Quarterly*, *36*(3), 282-291. https://doi.org/10.1177/08943184231169763

Cahyo, E. D., Hidayat, S., Mahgribi, H., Prasetyo, F. I., & Kurniawati, D. A. (2023). Internalization of Wasaṭiyyah Educational Values in the Guidelines for Islamic Life of Muhammadiyah Citizens at Muhammadiyah University of Surakarta. 239-251. https://doi.org/10.2991/978-2-38476-102-9_22

Cain, M., Le, A. H., & Billett, S. (2019). Sharing Stories and Building Resilience: Student Preferences and Processes of Post-Practicum Interventions. 27-53. https://doi.org/10.1007/978-3-030-05560-8_2

Cairney, T. (2016). Community Literacy Practices and Education: Australia. 1-21. https://doi.org/10.1007/978-3-319-02321-2_18-1

Calabrese, J. E., & Capraro, R. M. (2021). The Autonomy of Informal STEM and Benefits of Andragogy With Gifted Children. *Journal of Research in Innovative Teaching & Learning*, *15*(2), 207-217. https://doi.org/10.1108/jrit-08-2021-0060

Calaguas, N. P., & Consunji, P. M. P. (2022). A Structural Equation Model Predicting Adults' Online Learning Self-Efficacy. *Education and Information Technologies*, *27*(5), 6233-6249. https://doi.org/10.1007/s10639-021-10871-y

Caldwell, G. (2011). Whatever Happened to Apprenticeship Learning? *The Clinical Teacher*, *8*(4), 272-275. https://doi.org/10.1111/j.1743-498x.2011.00456.x

Callahan, J. L., & Sandlin, J. A. (2007). The Tyranny of Technology: A Critical Assessment of the Social Arena of Online Learning. *New Horizons in Adult Education and Human Resource Development*, *21*(3-4), 5-15. https://doi.org/10.1002/nha3.10286

Calonge, D. S., Kamalov, F., Aguerrebere, P. M., Hassock, L., Smail, L., Yousef, D., Thadani, D. R., Kwong, T., & Abdulla, N. (2025). Upskilling and Reskilling in the United Arab Emirates: Future-Proofing Careers With AI Skills. *Journal of Adult and Continuing Education*, *32*(1), 98-126. https://doi.org/10.1177/14779714251315288

Calvard, T. (2025). Sensemaking And <i>Management Learning</I> : Productive Tensions and Unrealized Possibilities. *Management learning*, *56*(1), 75-82. https://doi.org/10.1177/13505076241279067

Camara-Walker, B. A., & Irizarry, V. (2013). Meaningful Online Interactions and Writing Improvement. *Hets Online Journal*, *4*(1), 163-203. https://doi.org/10.55420/2693.9193.v4.n1.152

Cameron, R. (2012). Recognising Workplace Learning: The Emerging Practices of E-RPL and E-PR. *Journal of Workplace Learning*, *24*(2), 85-104. https://doi.org/10.1108/13665621211201689

Campos, S. (2023). *Analítica Del Aprendizaje: Identificación De La Efectividad De Los Recursos Didácticos en Entornos Virtuales Según Los Estilos De Aprendizaje De Los Estudiantes*

Camps, A., & Blancafort-Masriera, L. (2019). La Simulación Como Metodología Formativa Innovadora en La Mención Internacional Del Grado De Administración Y Dirección De Empresas. 105-134. https://doi.org/10.24275/uam.5916.8947

Camps, J., & Torres-Carballo, F. (2010). El Teatro Como Instrumento Pedagógico Para El Desarrollo De Competencias: Una Experiencia en El Campo Del Management. *Tic Revista D Innovació Educativa*, *0*(4). https://doi.org/10.7203/attic.4.182

Canal, M. N., Ardavin, A. F., Díaz-Marcos, L., & Aguado-Tevar, O. (2024). Aprendizaje Generativo Integral: Un Modelo Para La Educación Superior Ante Los Desafíos De La Inteligencia Artificial. *European Public & Social Innovation Review*, *9*, 1-21. https://doi.org/10.31637/epsir-2024-1685

Cancelas-Ouviña, L.-P., & Hurtado-Sierra, A. (2020). El Aprendizaje Y Servicio (Aps) en La Enseñanza Del Alemán: Una Propuesta Innovadora Para La Escuela Oficial De Idiomas. *Revista De Estudios Socioeducativos Resed*(8), 113-132. https://doi.org/10.25267/rev_estud_socioeducativos.2020.i8.9

Candas, P. (2012). Étude Des Trajectoires d'Apprentissage Autodirigé : Place De La Planification, Des Intentions, Du Hasard. *Carrefours De L Éducation*, *n° 33*(1), 149-162. https://doi.org/10.3917/cdle.033.0149

Cantarutti, C., Véliz, C., Isbej, L., & Fredes, A. F. (2023). Identificación De Contenidos De Pertinencia Intercultural Para La Atención De Salud Para El Plan De Estudios De Odontología. *Revista Médica De Chile*, *151*(5), 649-658. https://doi.org/10.4067/s0034-98872023000500649

Cappiali, T. M. (2024). Promoting Inclusion and Justice in University Teaching. https://doi.org/10.4337/9781035323456

Carcea, I., & Froemke, R. C. (2019). Biological Mechanisms for Observational Learning. *Current Opinion in Neurobiology*, *54*, 178-185. https://doi.org/10.1016/j.conb.2018.11.008

Cárdenas, O. (2021). The Andragogic and Autopoietic in the Educational Model of the University of the 21st Century. *Revista Digital De Investigación Y Postgrado*, *1*(2), 23-50. https://doi.org/10.59654/vgpypq36

Careless, E. J. (2015a). Social Media for Social Justice in Adult Education: A Critical Theoretical Framework. *Journal of Teaching and Learning*, *10*(1). https://doi.org/10.22329/jtl.v10i1.3972

Careless, E. J. (2015b). "Typing Back": Social Media as Space for Critical Discourse. *New Horizons in Adult Education and Human Resource Development*, *27*(3), 50-55. https://doi.org/10.1002/nha3.20111

Carlisle, K. L., & McCloskey, M. A. (2023). Didactic Recommendations for Teaching About Integrated Care in Human Services Addiction Education. *Journal of Human Services*, *42*(1). https://doi.org/10.52678/001c.74363

Carlsen, A., & Haddad, G. (2013). Introduction. *International Review of Education*, *59*(3), 311-318. https://doi.org/10.1007/s11159-013-9371-3

Carlson, R. (2024). Transgender and Gender-Expansive Singers in Choir: A Review of the Literature. *Update Applications of Research in Music Education*, *44*(1), 57-66. https://doi.org/10.1177/87551233241301797

Carnicelli, S., Boluk, K., & Canosa, A. (2022). Paulo Freire 100: A Commentary From Tourism Scholars. *Journal of Hospitality Leisure Sport & Tourism Education*, *30*, 100377. https://doi.org/10.1016/j.jhlste.2022.100377

Carolino, J. L. T., & Rouco, C. (2022). Proficiency Level of Leadership Competences on the Initial Training Course for Firefighters—A Case Study of Lisbon Fire Service. *Fire*, *5*(1), 22. https://doi.org/10.3390/fire5010022

Carolyne, M. M. P. F. Y. D. P. N. (2022). Strategies Used in Teaching English Language Learners: A Case of Adult Literacy Centres in Imenti-North Sub-County, Kenya. *Journal of Languages and Linguistics*, *4*(1). https://doi.org/10.51317/ecjlls.v4i1.304

Carrascal, S., Gutiérrez, M. M., Acero, J. M. A., & Mateos, M. E. (2019). Acquisition of Competences for Sustainable Development Through Visual Thinking. A Study in Rural Schools in Mixco, Guatemala. *Sustainability*, *11*(8), 2317. https://doi.org/10.3390/su11082317

Carrasco-Walburg, C., Valenzuela-Astudillo, E., Maino-Ansaldo, S., Correa-Díaz, M., & Zapata-Torres, D. (2021). Experiential Teaching-Learning Tools: Critical Study of Representational Media and Immersion in Architecture. 475-488. https://doi.org/10.5151/sigradi2021-359

Carré, P. (1993). l'Apprentissage Autodirigé Dans La Recherche Nord-Américaine. *Revue Française De Pédagogie*, *102*(1), 17-22. https://doi.org/10.3406/rfp.1993.1301

Carré, P. (2015). De L'apprentissage À La Formation. Pour Une Nouvelle Psychopédagogie Des Adultes. *Revue Française De Pédagogie*, *190*, 29-40. https://doi.org/10.4000/rfp.4688

Carrubbo, L., & Megaro, A. (2024). Intelligence Augmentation and Capability Co-Elevation in Healthcare Enabled by Reasoned Transparency. *Itm Web of Conferences*, *62*, 02002. https://doi.org/10.1051/itmconf/20246202002

Carter, A. (2020). Preparing for Uganda: Anticipating the Intersections of Gender, Race, and Nationality in Transcultural Leadership Development Research. https://doi.org/10.31124/advance.13268867

Carter, P. L., & Nicolaides, A. (2023). Transformative Learning: An Emotional (R)evolution. *New Directions for Adult and Continuing Education*, *2023*(177), 25-36. https://doi.org/10.1002/ace.20476

Cartner, H., & Hallas, J. (2020). Aligning Assessment, Technology, and Multi-Literacies. *E-Learning and Digital Media*, *17*(2), 131-147. https://doi.org/10.1177/2042753019899732

Caruso, T. J., Qian, J., Lawrence, K. A., Armstrong-Carter, E., & Domingue, B. W. (2020). From Socrates to Virtual Reality: A Historical Review of Learning Theories and Their Influence on the Training of Anesthesiologists. *Journal of Education in Perioperative Medicine*, *22*(2). https://doi.org/10.46374/volxxii-issue2-caruso

Carvalho, A. E., Cosme, A., & Veiga, A. (2023). Inclusive Education Systems: The Struggle for Equity and the Promotion of Autonomy in Portugal. *Education Sciences*, *13*(9), 875. https://doi.org/10.3390/educsci13090875

Casado-Aranda, L. A., Sánchez-Fernández, J., Montoro-Ríos, F., & Horcajadas, M. I. A. (2021). Evaluation of the Work-Integrated Learning Methodology: Teaching Marketing Through Practitioner Experience in the Classroom. *Mathematics*, *9*(17), 2164. https://doi.org/10.3390/math9172164

Casanova, G., Weil, J., & Cerqueira, M. (2026). Experiencing Retirement Through Teaching: Productive Engagement in Later Life Across Two Cultures. *Journal of Cross-Cultural Gerontology*, *41*(1). https://doi.org/10.1007/s10823-026-09560-5

Castanha, C. S. C., Daboin, B. E. G., Zangirolami-Raimundo, J., Tavares, L. F. B., Abreu, L. C. d., & Raimundo, R. D. (2025). A Practice-Oriented Framework for Basic Life Support Training of Brazilian School Teachers: Integrating Law 13,722/2018, Andragogy, and the Health Belief Model. *Frontiers in Public Health*, *13*. https://doi.org/10.3389/fpubh.2025.1616459

Castegnaro, M., Saxe, E. B., & Miranda, X. (1970). Hacia La Transdisciplinariedad en La Docencia en La Universidad De Costa Rica. *Actualidades Investigativas en Educación*, *12*(1). https://doi.org/10.15517/aie.v12i1.10262

Cathro, V., O'Kane, P., & Gilbertson, D. (2017). Assessing Reflection. *Education + Training*, *59*(4), 427-442. https://doi.org/10.1108/et-01-2017-0008

Catyong, C., & Garciano, L. M. (2023). Recognizing the Status, Needs, and Challenges of Alternative Learning System Students in a Rural District. *Jep*. https://doi.org/10.7176/jep/14-17-04

Cavallaro, L., & Nault, W. J. (2020). Cultivating a Learning Culture in the US Navy. *The Learning Organization*, *28*(3), 298-315. https://doi.org/10.1108/tlo-12-2019-0176

Cecilia, R. R., Ojeda, J. R. G., & Marín-Macías, C. (2020). Analysis of Heteronormativity and Gender Roles in EFL Textbooks. *Sustainability*, *13*(1), 220. https://doi.org/10.3390/su13010220

CEDEFOP. (2021). *Vocational Education and Training and the Green Transition: A Compendium of Inspiring Practices*.

CEDEFOP. (2023). *Skills Forecast: Trends and Challenges to 2035*. https://www.cedefop.europa.eu/

Cela-Ranilla, J. M., González, V. E., Esteve-Mon, F. M., Martínez, J. G., & Cervera, M. G. (2017). El Docente en La Sociedad Digital: Una Propuesta Basada en La Pedagogía Transformativa Y en La Tecnología Avanzada. *Profesorado Revista De Currículum Y Formación Del Profesorado*, *21*(1), 403-422. https://doi.org/10.30827/profesorado.v21i1.10371

Celestini, A. (2021). Serious Games in Higher Distance Education. *Canadian Journal of Learning and Technology*, *46*(3). https://doi.org/10.21432/cjlt27965

Çelik, İ., Gedrimiene, E., Siklander, P., & Muukkonen, H. (2024). The Affordances of Artificial Intelligence-Based Tools for Supporting 21st-Century Skills. *Australasian Journal of Educational Technology*. https://doi.org/10.14742/ajet.9069

Çelik, S., Kirjavainen, S., & Björklund, T. (2020). Educating Future Engineers: Student Perceptions of the Societal Linkages of Innovation Opportunities. https://doi.org/10.18260/1-2--34490

Cembranel, P., Luíza Luchi de Paulo, G., Moro, L. D., Fuchs, P. G., Birch, R. S., & José Baltazar Salgueirinho Osório de Andrade Andrade, G. (2024). The Pivotal Role of Higher Education Institutions in Cultivating a Sustainable Development Goals-Centric Culture. *International Journal of Sustainability in Higher Education*, *25*(7), 1385-1411. https://doi.org/10.1108/ijshe-01-2024-0057

Cennamo, I., Kastner, M., & Schlögl, P. (2020). Signposts of Change in the Landscape of Adult Basic Education in Austria: A Telling Case. *European Journal for Research on the Education and Learning of Adults*, *11*(1), 91-107. https://doi.org/10.3384/rela.2000-7426.rela9204

Cerda, G., Sagredo, A. V., & Saadati, F. (2023). Interacción Compleja Del Estilo Atribucional, Autorregulación Y Resiliencia Respecto Del Rendimiento en Matemáticas. *Revista Complutense De Educación*, *34*(1), 35-45. https://doi.org/10.5209/rced.76753

Čermáková, A., Prado, Y., & Kucirkova, N. (2024). Equity in Edtech by Design. https://doi.org/10.31265/usps.277

Černý, M. (2020). Connectivism in Phenomenological-Pragmatist Tradition. *E-Pedagogium*, *20*(2), 7-24. https://doi.org/10.5507/epd.2020.017

Ceschi, A., Perini, M., Scalco, A., Pentassuglia, M., Righetti, E., & Caputo, B. (2021). Foster Employability and Fight Social Exclusion Through the Development of Lifelong Learning (LLL) Key-Competences: Reviewing Twenty Years of LLL Policies. *European journal of training and development*, *45*(6/7), 475-511. https://doi.org/10.1108/ejtd-07-2019-0126

Chaerunnisa, C., Suryana, S., & Wibowo, L. A. (2022). The Effect of Vocational School Leadership and Educational Innovations on the Culture of Innovation. *657*. https://doi.org/10.2991/aebmr.k.220701.095

Chahartangi, F., Zarifsanaiey, N., Ghoochani, B. Z., & Sharifzadeh, N. (2026). Investigating the Effect of Augmented Reality-Based Virtual Patient Training on Occupational Therapy Students' Clinical Decision-Making: A Quasi-Experimental Study. *Plos One*, *21*(2), e0340759. https://doi.org/10.1371/journal.pone.0340759

Chaiyasit, W., Chomsuwan, K., & Chanchalor, S. (2023). Hybrid Teaching Using Problem-Based Learning to Promote Self-Directed Learning Abilities of Students During the COVID-19 Pandemic. *International Journal of Learning Teaching and Educational Research*, *22*(8), 1-22. https://doi.org/10.26803/ijlter.22.8.1

Chakraborty, S. (2024). Empowering and Engaging Adult Learners: Strategies to Facilitate Self-Directed Learning in Adult Education. *International Journal for Multidisciplinary Research*, *6*(3). https://doi.org/10.36948/ijfmr.2024.v06i03.22216

Chambers, D. (2019). Is Freire Incoherent? Reconciling Directiveness and Dialogue in Freirean Pedagogy. *Journal of Philosophy of Education*, *53*(1), 21-47. https://doi.org/10.1111/1467-9752.12340

Chander, N., Siow, M. L., Ramachandran, S., Kunasekaran, P., & Rathakrishnan, T. (2020). Conceptualizing Inclusive Learning and Development: A Framework Towards Entrepreneurial Competency Practices for Sustainability. *Sustainability*, *12*(17), 6905. https://doi.org/10.3390/su12176905

Chandre, P. (2024). Industry 5.0 for Society 5.0 a New Transformation. *PMJ*, *35*(1s), 127-137. https://doi.org/10.52783/pmj.v35.i1s.2188

Chang, B., Krepper, R., & Giraldo-García, R. (2023). Designing a Mock Training Workshop on Poverty —Transformative Learning as Framework. *Psychology in the Schools*, *60*(12), 4793-4809. https://doi.org/10.1002/pits.23005

Chang, M. (2022). Andragogical Assumptions in the Adult Students' Learning Process to Improve Reading Skills. *Ciencia Latina Revista Científica Multidisciplinar*, *6*(1), 3708-3721. https://doi.org/10.37811/cl_rcm.v6i1.1764

Chao, C. A. (2005). Toward Full Participation in Management Consulting Practice. *Education + Training*, *47*(1), 18-30. https://doi.org/10.1108/00400910510580601

Charles, M. (2018). Erziehung: The Critical Theory of Education and Counter-Education. 988-1005. https://doi.org/10.4135/9781526436122.n59

Chasaya, W. (2023). *Entrepreneurship Education and Students' Entrepreneurial Intentions at Selected Tertiary Institutions in Durban* university of Kwazulu-Natal].

Chasserio, S., & Bacha, E. (2023). Women-Only Training Programmes as Tools for Professional Development: Analysis and Outcomes of a Transformative Learning Process. *European journal of training and development*, *48*(3/4), 455-477. https://doi.org/10.1108/ejtd-12-2021-0210

Chatti, M. A., Jarke, M., & Quix, C. (2010). Connectivism: The Network Metaphor of Learning. *International Journal of Learning Technology*, *5*(1), 80. https://doi.org/10.1504/ijlt.2010.031617

Chau, S., & Ren, L. (2024). How Does Educational Travel Promote Lifelong Learning? *International Journal of Tourism Research*, *26*(4). https://doi.org/10.1002/jtr.2731

Chauvin, R., Fenouillet, F., & Brewer, S. S. (2020). An Investigation of the Structure and Role of English as a Foreign Language Self-Efficacy Beliefs in the Workplace. *System*, *91*, 102251. https://doi.org/10.1016/j.system.2020.102251

Che, C., & Perkins, B. (2026). Academic Stewardship Through AI-Enabled Learning Governance: Ethical Systems for Inclusive Learner Support in Higher Education. https://doi.org/10.21203/rs.3.rs-9759894/v1

Chee, H., Ahn, S., & Lee, J. (2024). A Competency Framework for AI Literacy: Variations by Different Learner Groups and an Implied Learning Pathway. *British journal of educational technology*, *56*(5), 2146-2182. https://doi.org/10.1111/bjet.13556

Chen, C. P. (2011). Life-Career Re-Engagement: Considerations and Implications for Counselling People in Retirement Transition—Part 2. *Australian Journal of Career Development*, *20*(2), 32-41. https://doi.org/10.1177/103841621102000205

Chen, C. X. R., Xie, S., Luo, Z., Hu, Z., Du, X., Huang, D., Hu, H., & Peng, S. (2026). Artificial Intelligence in Anesthesiology Education: Transformative Applications, Challenges, and Future Perspectives. *Frontiers in Medicine*, *13*. https://doi.org/10.3389/fmed.2026.1817855

Chen, D.-L., & Rattray, J. (2018). Critical Thinking Threshold Concepts in PBL. 31-47. https://doi.org/10.1163/9789004375123_004

Chen, F. (2025). Scientific Mapping of GenAI in English as a Foreign Language (EFL) Context. *International Journal of Technology and Human Interaction*, *21*(1), 1-23. https://doi.org/10.4018/ijthi.384377

Chen, L., Manwaring, P., Zakaria, G., Wilkie, S., & Loton, D. (2021). Implementing H5P Online Interactive Activities at Scale. 81-92. https://doi.org/10.14742/ascilite2021.0112

Chen, P.-J. (2026). Learning Experiences and Usability Perceptions of an Immersive Virtual Reality Simulation for Cardiotocography Interpretation Education: A Qualitative-Dominant Mixed Methods Inquiry. *BMC Nursing*, *25*(1). https://doi.org/10.1186/s12912-025-04126-1

Chen, R. H. (2022). Effects of Deliberate Practice on Blended Learning Sustainability: A Community of Inquiry Perspective. *Sustainability*, *14*(3), 1785. https://doi.org/10.3390/su14031785

Chen, X., & Chan, S. (2024). Implementing Digital Pedagogy in TVET: A Connectivist Perspective. *Vocat Tech Edu*, *1*(2). https://doi.org/10.54844/vte.2024.0595

Chen, X., & Tang, M. (2024). Constructing Virtual Teaching and Research Centers for Business English Education: A Framework for Application-Oriented Universities. *Journal of Education and Educational Research*, *11*(2), 184-189. https://doi.org/10.54097/v9dj3k61

Chen, X., Zou, D., Xie, H., & Wang, F. L. (2024). Technology-Enhanced Higher Education: Text Mining and Bibliometrics. *Heliyon*, *10*(4), e25776. https://doi.org/10.1016/j.heliyon.2024.e25776

Chen, Y. L. (2013). The Possibility of Applying YouTube to Motivate Learning Autonomy. *Journal of International Education Research (Jier)*, *9*(3), 207-216. https://doi.org/10.19030/jier.v9i3.7877

Chen, Z., & Murphy, I. (2022). Domain and Pedagogical Competency and Currency: A Framework for Capability Development of Adult Educators. *Journal of Adult and Continuing Education*, *29*(1), 273-293. https://doi.org/10.1177/14779714221125494

Cheng, B., Wang, M., Yang, S. J., Kinshuk, K., & Peng, J. (2011). Acceptance of Competency-Based Workplace E-Learning Systems: Effects of Individual and Peer Learning Support. *Computers & Education*, *57*(1), 1317-1333. https://doi.org/10.1016/j.compedu.2011.01.018

Cheng, C.-C. (2014). Situated Learning and Professional Development: A Case Study of Applying Cognitive Apprenticeship and Community of Practices in a Kindergarten. *Problems of Education in the 21st Century*, *59*(1), 15-24. https://doi.org/10.33225/pec/14.59.15

Cheng, L. (2025). AI- And Big Data-Driven Innovation in Vocational Education: A Case Study of "Belt and Road" Language Service Talent Development". https://doi.org/10.21203/rs.3.rs-7901067/v1

Cheng, R., Zhou, X., & Chen, Y. (2023). Unraveling the Impact and Challenges of Online Learning in Chinese Education. *Research and Advances in Education*, *2*(11), 23-36. https://doi.org/10.56397/rae.2023.11.04

Cheong, B. C. (2024). Transparency and Accountability in AI Systems: Safeguarding Wellbeing in the Age of Algorithmic Decision-Making. *Frontiers in Human Dynamics*, *6*. https://doi.org/10.3389/fhumd.2024.1421273

Chethana, K. M., & Noronha, S. D. (2023). Impact of Talent Management Practices in Higher Educational Institutions. *International Journal of Management Technology and Social Sciences*, 17-46. https://doi.org/10.47992/ijmts.2581.6012.0287

Chew, S. L. (2021). An Advance Organizer for Student Learning: Choke Points and Pitfalls in Studying. *Canadian Psychology/Psychologie Canadienne*, *62*(4), 420-427. https://doi.org/10.1037/cap0000290

Chhabhaiya, N., Patle, B. K., & Bhojane, P. K. (2024). Virtual and Augmented Reality Applications: A Broader Perspective Review. *Journal of Data Science and Intelligent Systems*, *4*(1), 13-25. https://doi.org/10.47852/bonviewjdsis32021064

Chiang, T. H. (2021). What Freirean Critical Pedagogy Says and Overlooks From a Durkheimian Perspective. *Social Inclusion*, *9*(4), 1-11. https://doi.org/10.17645/si.v9i4.4157

Chiang, W. S. (2021). Involvement of External Stakeholders in Designing Pedagogy for Experiential Learning at University Level: A Case Study. *Educatum Journal of Social Sciences*, *7*(2), 45-56. https://doi.org/10.37134/ejoss.vol7.2.5.2021

Chiblow, S., & Meighan, P. J. (2023). Anishinaabek Giikendaaswin and Dùthchas Nan Gàidheal: Concepts to (Re)center Place-Based Knowledges, Governance, and Land in Times of Crisis. *Ethnicities*, *24*(4), 617-634. https://doi.org/10.1177/14687968231219022

Chilambarasan, N. R. (2025). AI-Powered Personalized Learning: Ethical and Practical Integration Into Contemporary Education. https://doi.org/10.21203/rs.3.rs-7580533/v1

Chinedu, C. C., Saleem, A., & Wan Hanim Nadrah Wan, M. (2023). Teaching and Learning Approaches: Curriculum Framework for Sustainability Literacy for Technical and Vocational Teacher Training Programmes in Malaysia. *Sustainability*, *15*(3), 2543. https://doi.org/10.3390/su15032543

Chirimuuta, M. (2021). Reflex Theory, Cautionary Tale: Misleading Simplicity in Early Neuroscience. *Synthese*, *199*(5-6), 12731-12751. https://doi.org/10.1007/s11229-021-03351-w

Cho, J. K. (2025). Exploring the Association Between Time Dedicated to Lifelong Learning and Volunteering Among Older Adults: Socioeconomic Status as a Moderator. *Journal of Applied Gerontology*, *45*(2), 279-288. https://doi.org/10.1177/07334648251339513

Choi-Lundberg, D. L., Butler-Henderson, K., Harman, K., & Crawford, J. (2023). A Systematic Review of Digital Innovations in Technology-Enhanced Learning Designs in Higher Education. *Australasian Journal of Educational Technology*, 133-162. https://doi.org/10.14742/ajet.7615

Choi-Lundberg, D. L., Douglas, T., Bird, M. L., Coleman, B., Greenwood, M., Martin, R., Prior, S., Saghafi, F., & Roehrer, E. (2024). Employability Learning and Teaching Research: A Twenty Year Structured Narrative Review. *Journal of University Teaching and Learning Practice*, *21*(05). https://doi.org/10.53761/g8mryt07

Choi-Lundberg, D. L. e. a. (2024). Employability Learning and Teaching Research: A Twenty Year Structured Narrative Review. *Journal of University Teaching and Learning Practice*, *21*(05). https://doi.org/10.53761/g8mryt07

Choi, G. W., Chang, S. J., Kim, H. J., & Jeong, H. N. (2023). Understanding Health Literacy of Deaf Persons With Hypertension in South Korea: A Cross-Sectional Study. *Plos One*, *18*(11), e0294765. https://doi.org/10.1371/journal.pone.0294765

Cholakkathodi, J., & Bikku. (2026). Upskilling and Reskilling of Waste Workers in India's Circular Economy. https://doi.org/10.21203/rs.3.rs-8602865/v1

Choo, C. C., Devakaran, B., Chew, P. K. H., & Zhang, M. (2019). Smartphone Application in Postgraduate Clinical Psychology Training: Trainees' Perspectives. *International Journal of Environmental Research and Public Health*, *16*(21), 4206. https://doi.org/10.3390/ijerph16214206

Chopra, K. (2019). Vroom's Expectancy Theory to Motivate Academic Library Users in India Using Grounded Theory Approach. *Global Knowledge Memory and Communication*, *68*(4/5), 300-322. https://doi.org/10.1108/gkmc-09-2018-0079

Chowdhury, M. K. B., & Behak, F. B. P. (2021). Online Higher Education in Bangladesh: Challenges and Prospects. *Journal of Education & Social Sciences*, *9*, 54-74. https://doi.org/10.20547/jess0912109105

Chrismastianto, I. A. W., Nitiasih, P. K., & Riastin, P. N. (2024). Freedom Learning in University (MBKM): Strengthening Social Studies Curriculum Through Critical Pedagogy. *Kne Social Sciences*. https://doi.org/10.18502/kss.v9i2.14950

Christens, B. D., Winn, L., & Duke, A. M. (2015). Empowerment and Critical Consciousness: A Conceptual Cross-Fertilization. *Adolescent Research Review*, *1*(1), 15-27. https://doi.org/10.1007/s40894-015-0019-3

Christiansen, A., & Bell, A. (2010). Peer Learning Partnerships: Exploring the Experience of Pre-registration Nursing Students. *Journal of Clinical Nursing*, *19*(5-6), 803-810. https://doi.org/10.1111/j.1365-2702.2009.02981.x

Christopoulos, A., Πέλλας, N., & Laakso, M. J. (2020). A Learning Analytics Theoretical Framework for STEM Education Virtual Reality Applications. *Education Sciences*, *10*(11), 317. https://doi.org/10.3390/educsci10110317

Chua, B. L. (2023). Path Analysis: The Predictive Relationships of Problem-Based Learning Processes on Preservice Teachers' Learning Strategies. *Interdisciplinary Journal of Problem-Based Learning*, *17*(2). https://doi.org/10.14434/ijpbl.v17i2.37322

Chuang, S. (2015). Deterrents to Women's Participation in Continuing Professional Development. *New Horizons in Adult Education and Human Resource Development*, *27*(2), 28-37. https://doi.org/10.1002/nha3.20104

Chuang, S. (2019). Exploring Women-Only Training Program for Gender Equality and Women's Continuous Professional Development in the Workplace. *Higher Education Skills and Work-Based Learning*, *9*(3), 359-373. https://doi.org/10.1108/heswbl-01-2018-0001

Chuang, S. (2021). The Applications of Constructivist Learning Theory and Social Learning Theory on Adult Continuous Development. *Performance Improvement*, *60*(3), 6-14. https://doi.org/10.1002/pfi.21963

Chun, K. H. (2024). Cognitive Constructivism in the Development of Medical Education. *Korean Medical Education Review*, *26*(Suppl 1), S22-S30. https://doi.org/10.17496/kmer.24.002

Chundur, S. (2020). Digital Justice. *The Journal of Community Informatics*, *16*, 118-140. https://doi.org/10.15353/joci.v16i0.3485

Churchill, S., Jezak, M., & Lamoureux, S. A. (2011). Introduction to Part 1. *Olbi Journal*, *3*. https://doi.org/10.18192/olbiwp.v3i0.1090

Chürr, C. (2015). Realisation of a Child's Right to a Basic Education in the South African School System: Some Lessons From Germany. *Potchefstroom Electronic Law Journal/Potchefstroomse Elektroniese Regsblad*, *18*(7), 2404-2455. https://doi.org/10.4314/pelj.v18i7.01

Ciolfi, M., & Howard, L. (2024). The Application of UDL and SoTL to Health Care Education. *New Directions for Teaching and Learning*, *2024*(179), 69-77. https://doi.org/10.1002/tl.20598

Ciriello, R. (2025). How Education Quality and Inclusivity Foster Sustainable Development in the Global South: A Structured Narrative Review. *Review of Development Economics*, *30*(3), 1657-1670. https://doi.org/10.1111/rode.70081

Clair, R. S. (2024). Andragogy: Past and Present Potential. *New Directions for Adult and Continuing Education*, *2024*(184), 7-13. https://doi.org/10.1002/ace.20546

Clair, R. S., & Käpplinger, B. (2021). Alley or Autobahn? Assessing 50 Years of the Andragogical Project. *Adult education quarterly*, *71*(3), 272-289. https://doi.org/10.1177/07417136211027879

Clark, A. K., & Karvonen, M. (2021). Instructionally Embedded Assessment: Theory of Action for an Innovative System. *Frontiers in Education*, *6*. https://doi.org/10.3389/feduc.2021.724938

Clark, C. S. (2016). Watson's Human Caring Theory: Pertinent Transpersonal and Humanities Concepts for Educators. *Humanities*, *5*(2), 21. https://doi.org/10.3390/h5020021

Clarke, L. T. (2022). Transforming Consent Conversations. *Canadian Journal for the Study of Adult Education*, *34*(1). https://doi.org/10.56105/cjsae.v34i1.5678

Clarke, P. (2015). Student Nurses on Placement – Collaborators or Challengers. *The Journal of Adult Protection*, *17*(5), 287-295. https://doi.org/10.1108/jap-11-2014-0032

Classens, M., Adam, K., & Srebot, S. (2023). Food Systems Change and the Alternative Campus Foodscape. *Journal of Agriculture Food Systems and Community Development*, 1-22. https://doi.org/10.5304/jafscd.2023.123.010

Clausen, N. R. (2023). Self-Directed Learning in Problem-and Project-Based Learning: A Study of Self-Direction in the Aalborg PBL Model. https://doi.org/10.54337/aau679676715

Clements, K., West, R. E., & Hunsaker, E. (2020). Getting Started With Open Badges and Open Microcredentials. *The International Review of Research in Open and Distributed Learning*, *21*(1), 153-171. https://doi.org/10.19173/irrodl.v21i1.4529

Clinton-Lisell, V. (2021). Open Educational Resources as Tools to Foster Equity. https://doi.org/10.31219/osf.io/3axus

Closs, L. Q., & Antonello, C. S. (2011). Transformative Learning. *Journal of Transformative Education*, *9*(2), 63-88. https://doi.org/10.1177/1541344611429129

Çobanoğlu, C. (2019). The Global Conference on Education and Research: Volume 3. https://doi.org/10.5038/2572-6374-v3

Cocquyt, C., Zhu, C., de, G., & Vanwing, T. (2018). Contribution of Social-Constructivist and Andragogical Principles in Blended Learning to Adult's Social Inclusion and Social Capital. *Andragoske Studije*(1), 47-78. https://doi.org/10.5937/andstud1801047c

Coelho, D., Goyal, R., Swift, K., & Glasgow, S. (2024). Designing a Communication Skills Professional Development Course for Healthcare Professionals in the UAE: Reflections on the Process, Challenges, Successes, and Implications. *Adult Learning*, *36*(3), 188-200. https://doi.org/10.1177/10451595241286915

Coetzer, A., Kock, H., & Wallo, A. (2017). Distinctive Characteristics of Small Businesses as Sites for Informal Learning. *Human Resource Development Review*, *16*(2), 111-134. https://doi.org/10.1177/1534484317704291

Coffey, L., & Davis, A. (2019). The Holistic Approach to Academia: Traditional Classroom Instruction and Experiential Learning of Student-Athletes. *Education Sciences*, *9*(2), 125. https://doi.org/10.3390/educsci9020125

Cohen-Scali, V. (2012). Competence and Competence Development. https://doi.org/10.3224/86649462

Cohen, G. S., & Cohen, A. (2023). Zoom In: Open Educational Resources for Informal Online Learning During COVID-19 and Beyond. *Australasian Journal of Educational Technology*, 39-55. https://doi.org/10.14742/ajet.8044

Coleman, B. M., Bunch, J. C., Israel, G. D., Roberts, T. G., & Wysocki, A. F. (2025). Effects of Experiential Learning on Students' Transfer Skills. *Sage Open*, *15*(4). https://doi.org/10.1177/21582440251385407

Coleman, P. (2022). The Theory of Human Relatedness as a Potential Underlying Causative Mechanism in Nursing Student Placement Experiences: A UK-Based Critical Realist Study. *International Journal for Research in Vocational Education and Training*, *9*(1), 44-65. https://doi.org/10.13152/ijrvet.9.1.3

Collay, A. G. P., Mancheno, F. M. L., & Nancy del Consuelo Jiménez, R. (2024). Implementación Del Diseño Universal Del Aprendizaje (DUA) en El Aula: Estrategias Para Atender La Diversidad De Estilos De Aprendizaje. *Ciencia Latina Revista Científica Multidisciplinar*, *8*(5), 6987-7018. https://doi.org/10.37811/cl_rcm.v8i5.14120

Collazo, J. (2019). Inequitable Perspectives: First-Year ESL Teachers' Perspectives On, Pedagogy For, and Dispositions to Teach Culturally and Linguistically Diverse Students. https://doi.org/10.3102/1438170

Collins, A., Brown, J. S., & Newman, S. E. (1989). Cognitive apprenticeship: Teaching the crafts of reading, writing, and mathematics. In *Knowing, learning, and instruction: Essays in honor of Robert Glaser.* (pp. 453-494). Lawrence Erlbaum Associates, Inc.

Cologon, K. (2024). "But Marley Can't Play Up Here!" Children Designing Inclusive and Sustainable Playspaces Through Practitioner Research. *Sustainability*, *16*(15), 6626. https://doi.org/10.3390/su16156626

Columbres, M. R. C., & Victoriano, J. M. (2024). Cloud Sustainability: An Analysis and Assessment of the Plateau Prediction of 2023 Gartner Hype Cycle for Emerging Technologies. *International Journal of Sustainable Development and Planning*, *19*(8), 2881-2892. https://doi.org/10.18280/ijsdp.190807

Coman, C., Gherheş, V., Bucs, A., Coman, E., Victoria-Rodica, C., Bódi, D.-C., & Şerbănescu, R.-M. (2026). Navigating Opportunities and Challenges of Generative AI in Higher Education: Towards Responsible, Equitable, and Human-Centered Integration. *Frontiers in Artificial Intelligence*, *9*. https://doi.org/10.3389/frai.2026.1750978

Comings, J. P., Parrella, A., & Soricone, L. (1999). Persistence Among Adult Basic Education Students in Pre-Ged Classes. https://doi.org/10.1037/e371022004-001

Commonwealth Secretariat. (2022). Technical and Vocational Education and Training (TVET) Self-Assessment Toolkit: Participant Manual. https://doi.org/10.14217/comsec.1082

Concannon, B., Esmail, S., & Roberts, M. R. (2019). Head-Mounted Display Virtual Reality in Post-Secondary Education and Skill Training. *Frontiers in Education*, *4*. https://doi.org/10.3389/feduc.2019.00080

Conceição, S. C. O., & Biniecki, S. M. Y. (2024). Technology and Adult Learning. *New Directions for Adult and Continuing Education*, *2024*(184), 65-73. https://doi.org/10.1002/ace.20545

Condom, M. F. i., & Zellweger, F. (2019). Formación E Innovación Docente Universitaria Para Una Educación Transformadora: La Investigación E Indagación Reflexiva Como Requisito. *El Guiniguada.* https://doi.org/10.20420/elguiniguada.2019.265

Connolly, C., Hall, T., Ryan, M., McMahon, J., McGann, M., & Egan, A. (2020). A Fusion of Research-Informed Teaching and Teaching-Informed Research: Designing a Scalable Online Ecosystem for New Partnerships in Educational Research. *Australasian Journal of Educational Technology*, 82-95. https://doi.org/10.14742/ajet.6131

Conrad, D. (2008). Revisiting the Recognition of Prior Learning (RPL): A Reflective Inquiry Into RPL Practice in Canada. *Canadian Journal of University Continuing Education*, *34*(2). https://doi.org/10.21225/d5mg6s

Conrad, D. (2011). The Role of Language in Portfolio Learning. *The International Review of Research in Open and Distributed Learning*, *12*(1), 109. https://doi.org/10.19173/irrodl.v12i1.1062

Conrad, D. (2023a). Accreditation and Recognition of Prior Learning in Higher Education. 801-817. https://doi.org/10.1007/978-981-19-2080-6_44

Conrad, D. (2023b). Recognition of Prior Learning in Higher Education: Contemporary Perspectives.

Conrad, D., & McGreal, R. (2012). Flexible Paths to Assessment for OER Learners: A Comparative Study. *Journal of Interactive Media in Education*, *2012*(2), 12. https://doi.org/10.5334/2012-12

Contreras, F., Baykal, E., & Abid, G. (2020). E-Leadership and Teleworking in Times of COVID-19 and Beyond: What We Know and Where Do We Go. *Frontiers in Psychology*, *11*. https://doi.org/10.3389/fpsyg.2020.590271

Contreras, M. E. (2006). Aprender a Desaprender en La Búsqueda De Un Aprendizaje Transformativo: Apuntes Sobre La Capacitación De Gerentes Sociales. https://doi.org/10.18235/0012047

Conway, P. F. (2022). Andragogy in Prison: Higher Education in Prison and the Tenets of Adult Education. *Adult education quarterly*, *72*(4), 361-379. https://doi.org/10.1177/07417136221100481

Cook, R. F. (1987). Worker Dislocation: Case Studies of Causes and Cures. https://doi.org/10.17848/9780880996181

Coombs, P. H., & Ahmed, M. (1974). *Attacking Rural Poverty: How Non-Formal Education Can Help.* Johns Hopkins University Press.

Coon, S. (2019). Theorizing Critical Carceral Pedagogies: Teaching Toward the Purposes of Higher Education in and Against Prisons. https://doi.org/10.3102/1428633

Cooper, L. (2010). Activists Within the Academy: The Role of Prior Experience in Adult Learners' Acquisition of Postgraduate Literacies in a Postapartheid South African University. *Adult education quarterly*, *61*(1), 40-56. https://doi.org/10.1177/0741713610380441

Coral, P. (2024). Diagnóstico Del Diseño Universal De Aprendizaje Educación Básica De La Unidad Educativa Los Alpes. *Mentor Revista De Investigación Educativa Y Deportiva*, *3*(8), 380-398. https://doi.org/10.56200/mried.v3i8.7844

Corbett, F., & Spinello, E. (2020). Connectivism and Leadership: Harnessing a Learning Theory for the Digital Age to Redefine Leadership in the Twenty-First Century. *Heliyon*, *6*(1), e03250. https://doi.org/10.1016/j.heliyon.2020.e03250

Corbetta, L., Arru, L. B., Mereu, C., Pasini, V., & Patelli, M. (2019). Competence and Training in Interventional Pulmonology. *Panminerva Medica*, *61*(3). https://doi.org/10.23736/s0031-0808.18.03562-0

Corbin, L., & Bugden, L. (2018). Online Teaching: The Importance of Pedagogy, Place and Presence in Legal Education. *Legal Education Review*, *28*(1). https://doi.org/10.53300/001c.6313

Córdoba, M. E. (2020). El Constructivismo Sociocultural Lingüístico Como Teoría Pedagógica De Soporte Para Los Estudios Generales. *Revista Nuevo Humanismo*, *8*(1). https://doi.org/10.15359/rnh.8-1.4

Cornelius, S., & Marston, P. (2009). Towards an Understanding of the Virtual Context in Mobile Learning. *Research in Learning Technology*, *17*(3). https://doi.org/10.3402/rlt.v17i3.10874

Cornet, S., Barpanda, S., Guidi, M.-A. D., & Viswanathan, P. K. (2024). Sustainability Education and Community Development in Higher Education Using Participatory and Case Based Approaches in India. *International Journal of Sustainability in Higher Education*, *25*(9), 94-110. https://doi.org/10.1108/ijshe-07-2022-0242

Cortes, G. A. M., & Loor, G. B. P. (2024). Los Desafíos De Los Entornos Educativos Virtuales Para El Aprendizaje De Los Estudiantes Con Escolaridad Inconclusa. *Latam Revista Latinoamericana De Ciencias Sociales Y Humanidades*, *5*(4). https://doi.org/10.56712/latam.v5i4.2226

Cortés, H. S. R., & Polo, C. F. C. (2021). Una Mirada Histórico-Reflexiva Al Acto Educativo. 173-186. https://doi.org/10.22490/9789586517775.3.1

Coryell, J. E. (2013). Collaborative, Comparative Inquiry and Transformative Cross-Cultural Adult Learning and Teaching. *Adult education quarterly*, *63*(4), 299-320. https://doi.org/10.1177/0741713612471420

Coşgun, G. E. (2025). Artificial Intelligence Literacy in Assessment: Empowering Pre-service Teachers to Design Effective Exam Questions for Language Learning. *British educational research journal*, *51*(5), 2340-2357. https://doi.org/10.1002/berj.4177

Cosnefroy, L. (2011). l'Apprentissage Autorégulé : Perspectives en Formation d'Adultes. *Savoirs*, *n° 23*(2), 9-50. https://doi.org/10.3917/savo.023.0009

Costa, P. I. D., Park, J. S., & Wee, L. (2020). Why Linguistic Entrepreneurship? *Multilingua*, *40*(2), 139-153. https://doi.org/10.1515/multi-2020-0037

Côté, A. M., & Su, Z. (2022). Évolutions De L'intelligence Artificielle Au Travail Et Collaborations Humain-Machine. *Ad Machina L Avenir De L Humain Au Travail*(5), 144-160. https://doi.org/10.1522/radm.no5.1413

Coulombe, S. (2012). Une Modélisation Des Processus Et Des Stratégies d'Apprentissage en Situation De Travail. *Savoirs*, *n° 29*(2), 45-58. https://doi.org/10.3917/savo.029.0045

Cowan, N. (2013). Working Memory Underpins Cognitive Development, Learning, and Education. *Educational Psychology Review*, *26*(2), 197-223. https://doi.org/10.1007/s10648-013-9246-y

Cox, A., & John, V. M. (2016). Transformative Learning in Postapartheid South Africa. *Adult education quarterly*, *66*(4), 303-318. https://doi.org/10.1177/0741713616648376

Coysh, L. (2010). Further Developing the Library Curriculum: Skills for Life Delivery. *Health Information & Libraries Journal*, *28*(1), 82-86. https://doi.org/10.1111/j.1471-1842.2010.00927.x

Cramarenco, R. E., Burcă-Voicu, M. I., & Dabija, D. C. (2023). The Impact of Artificial Intelligence (AI) on Employees' Skills and Well-Being in Global Labor Markets: A Systematic Review. *Oeconomia Copernicana*, *14*(3), 731-767. https://doi.org/10.24136/oc.2023.022

Crane, B., & Hartwell, C. J. (2018). Developing Employees' Mental Complexity: Transformational Leadership as a Catalyst in Employee Development. *Human Resource Development Review*, *17*(3), 234-257. https://doi.org/10.1177/1534484318781439

Crane, C., & Sosulski, M. J. (2020). Staging Transformative Learning Across Collegiate Language Curricula: Student Perceptions of Structured Reflection for Language Learning. *Foreign Language Annals*, *53*(1), 69-95. https://doi.org/10.1111/flan.12437

Crans, S.-L. (2023). *Understanding the Social Side of Learning* [Doctoral Thesis,, Maastricht University].

Cranton, P. (2016). *Understanding and Promoting Transformative Learning* (3rd ed.). Stylus Publishing.

Crawford, K. (2007). Continuing Professional Development in Higher Education: Debating the Academic Perspective. *The International Journal of Knowledge Culture and Change Management Annual Review*, *7*(8), 51-58. https://doi.org/10.18848/1447-9524/cgp/v07i08/50415

Cremers, P. H., Wals, A. E. J., Wesselink, R., & Mulder, M. (2016). Design Principles for Hybrid Learning Configurations at the Interface Between School and Workplace. *Learning Environments Research*, *19*(3), 309-334. https://doi.org/10.1007/s10984-016-9209-6

Cretton, X. S., & Méndez, N. (2022). Perfiles De Gratitud, Necesidades Psicológicas Y Su Relación Con La Resiliencia en Estudiantes No Tradicionales. *Estudios Sobre Educación*, *43*, 115-134. https://doi.org/10.15581/004.43.006

Cristol, D., & Müller, A. (2013). Les Apprentissages Informels Dans La Formation Pour Adultes. *Savoirs*, *n° 32*(2), 11-59. https://doi.org/10.3917/savo.032.0011

Crocco, O. S., & Cseh, M. (2020). Learning, Development and Change in a Community-Based Enterprise in Myanmar. *European journal of training and development*, *45*(4/5), 436-448. https://doi.org/10.1108/ejtd-12-2019-0198

Cronin, C., Donevant, S., Hughes, K. A., Kaunonen, M., Marcussen, J., & Wilson, R. (2026). Reflecting Through Menopause: Introducing the MIRROR Cycle to Support Reflection During the Menopause Transition. *Journal of Research in Nursing*. https://doi.org/10.1177/17449871261446828

Crosbie, B., Gerhardt, T., & Montgomery, J. (2024). The Use of Problem-Solving Methodology to Develop Institutional and Curricular Change: Work-Integrated Learning as a Strategy of Differentiation. *Higher Education Skills and Work-Based Learning*. https://doi.org/10.1108/heswbl-01-2024-0020

Cross, K. P. (1981). *Adults as Learners*. Jossey-Bass.

Cruz, A. Y. V., & Cruz, I. M. (2023). Interacción Crucial: Habilidades Comunicativas Y Competencias Lectoras en El Contexto Educativo Presencial Y en Línea. *Revista De Climatología*, *23*, 2538-2545. https://doi.org/10.59427/rcli/2023/v23cs.2538-2545

Cui, T., & Coleman, A. (2021). Investigating Students' Attitudes, Motives, Participation And Performance Regarding Out-of-Class Communication (OCC) in a Flipped Classroom. *The Electronic Journal of E-Learning*, *18*(6). https://doi.org/10.34190/jel.18.6.007

Cummins, P., & Kunkel, S. (2015). A Global Examination of Policies and Practices for Lifelong Learning. *New Horizons in Adult Education and Human Resource Development*, *27*(3), 3-17. https://doi.org/10.1002/nha3.20107

Cunningham, K. M. W., VanGronigen, B. A., Tucker, P. D., & Young, M. D. (2018). Using Powerful Learning Experiences to Prepare School Leaders. *Journal of Research on Leadership Education*, *14*(1), 74-97. https://doi.org/10.1177/1942775118819672

Curnow, E., Maciver, D., Johnston, L., Johnstone-Cooke, V., Utley, I., Jenkins, N., Meff, T., Muggleton, J., & Rutherford, M. (2025). Learning From the Experiences of Autistic Professionals Working in Health and Education. *Autism in Adulthood*, *8*(4), 788-800. https://doi.org/10.1089/aut.2024.0168

Curto-Reverte, A., Carré, M. C. P., Rius, H. C., & Martí, C. (2025). The Role of Work-integrated Learning in the European Higher Education Area: A Systematic Review. *Review of Education*, *13*(3). https://doi.org/10.1002/rev3.70114

Cvetković, D. (2022). MOOC (Massive Open Online Courses). https://doi.org/10.5772/intechopen.95166

Czaja, S. J., Charness, N., Fisk, A. D., Hertzog, C., Nair, S. N., Rogers, W. A., & Sharit, J. (2006). Factors Predicting the Use of Technology: Findings From the Center for Research and Education on Aging and Technology Enhancement (Create). *Psychology and Aging*, *21*(2), 333-352. https://doi.org/10.1037/0882-7974.21.2.333

Dachner, A. M., Miguel, R. F., & Patena, R. A. (2017). Risky Business: Understanding Student Intellectual Risk Taking in Management Education. *Organizational Behavior Teaching Review*, *41*(3), 415-443. https://doi.org/10.1177/1052562917695775

Dachner, A. M., & Polin, B. (2015). A Systematic Approach to Educating the Emerging Adult Learner in Undergraduate Management Courses. *Organizational Behavior Teaching Review*, *40*(2), 121-151. https://doi.org/10.1177/1052562915613589

Dachyar, M., Zagloel, T. Y. M., & Saragih, L. R. (2019). Knowledge Growth and Development: Internet of Things (IoT) Research, 2006–2018. *Heliyon*, *5*(8), e02264. https://doi.org/10.1016/j.heliyon.2019.e02264

Dada, O. D., Eames, C., & Calder, N. (2017). Impact of Environmental Education on Beginning Preservice Teachers' Environmental Literacy. *Australian Journal of Environmental Education*, *33*(3), 201-222. https://doi.org/10.1017/aee.2017.27

Dadzie, C. E., Fumey, M., & Namara, S. (2020). Youth Employment Programs in Ghana: Options for Effective Policy Making and Implementation. https://doi.org/10.1596/978-1-4648-1579-9

Dahl, K. K. B. (2020). Collaborating Alone? Teachers' and Pedagogues' Unequal Cross-Professional Collaboration and What It Means for Their Professional Learning in Danish Schools. *Power and Education*, *12*(3), 246-260. https://doi.org/10.1177/1757743820943226

Dahlback, J., Olstad, H. B., Sylte, A. L., & Wolden, A.-C. (2020). The Importance of Authentic Workplace-Based Assessment: A Study From VET Teacher Education. *International Journal for Research in Vocational Education and Training*, *7*(3), 302-324. https://doi.org/10.13152/ijrvet.7.3.3

Dahri, N. A., Yahaya, N., Al-Rahmi, W. M., Noman, H. A., Alblehai, F., Kamin, Y., Soomro, R. B., Shutaleva, A., & Al-Adwan, A. S. (2024). Investigating the Motivating Factors That Influence the Adoption of Blended Learning for Teachers' Professional Development. *Heliyon*, *10*(15), e34900. https://doi.org/10.1016/j.heliyon.2024.e34900

Dai, W.-A. (2024). Digital Literacy Education: A New Approach to Cultivating 21st Century Key Competencies. *Journal of Basic and Applied Research International*, *30*(5), 48-57. https://doi.org/10.56557/jobari/2024/v30i58906

Dakich, E. (2014). Theoretical and Epistemological Foundations of Integrating Digital Technologies in Education in the Second Half of the 20th Century. 150-163. https://doi.org/10.1007/978-3-642-55119-2_10

Dale, V., Sullivan, M., & May, S. A. (2008). Adult Learning in Veterinary Education: Theory to Practice. *Journal of Veterinary Medical Education*, *35*(4), 581-588. https://doi.org/10.3138/jvme.35.4.581

Dalton, E., McKenzie, J., & Kahonde, C. (2012). The Implementation of Inclusive Education in South Africa: Reflections Arising From a Workshop for Teachers and Therapists to Introduce Universal Design for Learning. *African Journal of Disability*, *1*(1). https://doi.org/10.4102/ajod.v1i1.13

Damaiyanti, M. (2024). The Key Factors of Project-Based Learning Implementation on Higher Education: A Discourse Analysis Point of View. *Gramatika Stkip Pgri Sumatera Barat*, *10*(1). https://doi.org/10.22202/jg.2024.v10i1.7931

Damoah, B. (2024). Constructivist Mentor: Influence of Mentor Teachers on Mentees. *Jett*, *15*(1). https://doi.org/10.47750/jett.2024.15.01.007

Damus, O. (2023). La Pédagogie Régénératrice Et Réparatrice. *Anthropologie Des Savoirs Des Suds*(1), 7-29. https://doi.org/10.17118/11143/20529

Daneil, I. L. (2021). Comparative Analysis of Learning Style Preferences Between Pre-Science and Pre-Business Students. *Jep*. https://doi.org/10.7176/jep/12-2-02

Dang, V. T., & Chou, Y. C. (2019). Extrinsic Motivation, Workplace Learning, Employer Trust, Self-Efficacy and Cross-Cultural Adjustment. *Personnel Review*, *49*(6), 1232-1253. https://doi.org/10.1108/pr-10-2018-0427

Daniels, H., & Warmington, P. (2007). Analysing Third Generation Activity Systems: Labour-power, Subject Position and Personal Transformation. *Journal of Workplace Learning*, *19*(6), 377-391. https://doi.org/10.1108/13665620710777110

Darden, D. C. (2014). Relevance of the Knowles Theory in Distance Education. *Creative Education*, *05*(10), 809-812. https://doi.org/10.4236/ce.2014.510094

Darko-Asumadu, D. A. (2023). Challenges of Student-Workers' in Distance Education in Ghana: A Myth or Reality? *East African Journal of Education and Social Sciences*, *4*(4), 16-28. https://doi.org/10.46606/eajess2023v04i04.0299

Daryoush, Y. (2013). Improving Job Performance: Workplace Learning Is the First Step. *International Journal of Education and Literacy Studies*, *1*(1), 100-107. https://doi.org/10.7575/aiac.ijels.v.1n.1p.100

Darzón, A. L. D., Rojas, D. S. L., & Castillo, A. A. U. (2021). Educación Interprofesional, Fundamentos Y Estrategias Didácticas: Revisión Del Estado Del Arte. *Rediis / Revista De Investigación E Innovación en Salud*, *4*, 34-47. https://doi.org/10.23850/rediis.v4i4.3101

Davey, C. L. (2004). The Impact of Human Factors on <i>Ab Initio</I> Pilot Training. *Gender Work and Organization*, *11*(6), 627-647. https://doi.org/10.1111/j.1468-0432.2004.00252.x

Davis, L. S., Crain, E. E., & Sheridan, M. (2024). LGBTQ+-affirming Graduate Education: Preparing and Supporting Future Family Court Clinicians. *Family Court Review*, *62*(3), 680-700. https://doi.org/10.1111/fcre.12803

Day, M. (2016). Influencing Practice Through the Development of Clear and Transparent Guidelines for the Recognition of Prior Learning (RPL). *Athens Journal of Health*, *4*(1), 27-50. https://doi.org/10.30958/ajh.4-1-2

Day, R., Carpenter, H., Ker, G., & Ker, P. (2020). Capable NZ: Persevering to Success. *Scope Contemporary Research Topics (Work-Based Learning)*(1), 11-21. https://doi.org/10.34074/scop.6001013

Dean, B. J., & Parson, J. (2021). Integrating Play Based Approaches Into Nursing Education: Teachers as Playful Academics. *The Journal of Play in Adulthood*, *3*(1). https://doi.org/10.5920/jpa.873

DeAngelis, L. M. (2021). Enabling the Exploration of Disorienting Dilemma in the Classroom. *Journal of Education*, *202*(4), 585-600. https://doi.org/10.1177/0022057421991865

Debasu, H., & Yitayew, A. (2024). Examining Elements of Designing and Managing of Creating Inclusive Learning Environment: Systematic Literature Review. *International Journal of Special Education (Ijse)*, *39*(1), 33-43. https://doi.org/10.52291/ijse.2024.39.4

Décamps, A., Allal-Chérif, O., & Gombault, A. (2021). Fostering Knowledge of the Sustainable Development Goals in Universities: The Case of Sulitest. *Sustainability*, *13*(23), 13215. https://doi.org/10.3390/su132313215

DeCapua, A., Marshall, H. W., & Frydland, N. (2017). The Transformational Learning Journey of a Novice ESL Teacher of Low-Literate Adults. *Journal of Transformative Education*, *16*(1), 17-38. https://doi.org/10.1177/1541344617704645

deCarvalho, R. J. (1991). The Humanistic Paradigm in Education. *The Humanistic Psychologist*, *19*(1), 88-104. https://doi.org/10.1080/08873267.1991.9986754

Decker, D. S. (2019). Student Perceptions of Higher Education and Apprenticeship Alignment. *Education Sciences*, *9*(2), 86. https://doi.org/10.3390/educsci9020086

Dediwalage, N. M., Kalogeropoulos, P., Russo, J., & Vale, C. (2025). A Systematic Literature Review of Student, Parent and Teacher Experiences of Mathematics Education in Multicultural Classrooms. *Review of Education*, *13*(3). https://doi.org/10.1002/rev3.70109

DeFoor, M. T., Chung, Y., Zadinsky, J. K., Dowling, J., & Sams, R. (2019). An Interprofessional Cohort Analysis of Student Interest in Medical Ethics Education: A Survey-Based Quantitative Study. https://doi.org/10.21203/rs.2.13642/v2

Deißinger, T. (2012). Reforming the VET System via National Qualification Frameworks? A Comparison of Germany and Austria. 305-320. https://doi.org/10.1007/978-3-531-18757-0_18

Delahunty, T., & Kimbell, R. (2021). (Re)framing a Philosophical and Epistemological Framework for Teaching and Learning in STEM: Emerging Pedagogies for Complexity. *British educational research journal*, *47*(3), 742-769. https://doi.org/10.1002/berj.3706

Delors, J., Al Mufti, I. a., Amagi, I., Carneiro, R., Chung, F., Geremek, B., Gorham, W., Kornhauser, A., Manley, M., Padrón Quero, M., Savané, M.-A., Singh, K., Stavenhagen, R., Won Suhr, M., & Nanzhao, Z. (1996). *Learning: The Treasure Within.* UNESCO Publishing.

Demaj, E. (2019). The Learning Organization Understanding in Disruptive Times and Its Impact on Organizational Performance: The Apparel Industry Case. 125-134. https://doi.org/10.31410/eman.s.p.2019.125

Demirkesen, S., Sadikoglu, E., & Jayamanne, E. (2021). Assessing Psychological Safety in Lean Construction Projects in the United States. *Construction Economics and Building, 21*(3). https://doi.org/10.5130/ajceb.v21i3.7657

Deng, X., & Yang, Z. (2021). Digital Proficiency and Psychological Well-Being in Online Learning: Experiences of First-Generation College Students and Their Peers. *Social Sciences, 10*(6), 192. https://doi.org/10.3390/socsci10060192

Dennick, R. (2016). Constructivism: Reflections on Twenty Five Years Teaching the Constructivist Approach in Medical Education. *International Journal of Medical Education, 7*, 200-205. https://doi.org/10.5116/ijme.5763.de11

Denzinger, F., Backes, S., Job, V., & Brandstätter, V. (2016). Age and Gender Differences in Implicit Motives. *Journal of Research in Personality, 65*, 52-61. https://doi.org/10.1016/j.jrp.2016.09.003

Dermott, G. M., Byrne, A., McLaughlin, R., O'Connor, N., & Griselain, S. (2023). Exploring the Use of Immersive Technologies to Enhance the Student Experience. *Ubiquity Proceedings.* https://doi.org/10.5334/uproc.98

Derwik, P., & Hellström, D. (2023). Successful Competence Development for Retail Professionals: Investigation of Key Mechanisms in Informal Learning. *International Journal of Retail & Distribution Management, 51*(13), 33-46. https://doi.org/10.1108/ijrdm-09-2022-0321

Desmarchelier, R., & Cary, L. J. (2022). Toward Just and Equitable Micro-Credentials: An Australian Perspective. *International Journal of Educational Technology in Higher Education, 19*(1). https://doi.org/10.1186/s41239-022-00332-y

Desta, S. Z., & Gugssa, M. A. (2022). The Implementation of Andragogy in the Adult Education Program in Ethiopia. *Education Research International, 2022*, 1-11. https://doi.org/10.1155/2022/2028248

Deulen, A. A. (2013). Social Constructivism and Online Learning Environments: Toward a Theological Model for Christian Educators. *Christian Education Journal Research on Educational Ministry, 10*(1), 90-98. https://doi.org/10.1177/073989131301000107

Dewayani, J., & Ferdinand, A. T. (2019). Motivation to Transfer, Supervisor Support, Proactive Learning, and Training Transfer: Testing Interaction Effects. *International Journal of Economics and Business Administration, VII*(Issue 3), 141-150. https://doi.org/10.35808/ijeba/314

Dewey, J. (1938). *Experience and Education.* Macmillan.

Dewi, N. K., Jamil, A., & Wijaya, F. W. (2024). Artificial Intelligence and the Jobs of the Future: Preparing Young Moderates for Change. *Lentera Multidisciplinary Studies, 2*(3). https://doi.org/10.57096/lentera.v2i3.96

Dhar, U. (2024). Professional Development Experiences Aligned to the Ideal Self: A Positive Psychology Perspective on Individual Development. *The Journal of Applied Behavioral Science, 61*(4), 749-775. https://doi.org/10.1177/00218863241269055

Dhiman, D., & Solkhe, A. (2025). Shaping Skills: A Qualitative Dive Into India's Evolving Skill Development Policies. *Development Policy Review, 43*(3). https://doi.org/10.1111/dpr.70006

Dhungana, J. (2021). Cultural Diversity Responsive Education: Exploration of Concepts and Dimensions. *Education and Development, 31*(01), 48-66. https://doi.org/10.3126/ed.v31i01.62202

Diab, M., & Zhang, M. (2023). Perceptions of Industry Professionals From Diverse Sectors on Post-Secondary Micro-Credentials on Linkedin. *International Journal of Research in Education Humanities and Commerce, 04*(02), 63-115. https://doi.org/10.37602/ijrehc.2023.4211

Diahyleva, O., Yurzhenko, A., & Kononova, O. (2024). Gamification Techniques in Maritime English Online Courses: Motivating Learners in Virtual Environments. *Advances in Mobile Learning Educational Research, 4*(1), 965-972. https://doi.org/10.25082/amler.2024.01.008

Dianova, V., & Schultz, M. D. (2023). Discussing ChatGPT's Implications for Industry and Higher Education: The Case for Transdisciplinarity and Digital Humanities. *Industry and Higher Education, 37*(5), 593-600. https://doi.org/10.1177/09504222231199989

Dias, B., & Boulder, T. C. (2023). Toward Humanizing Online Learning Spaces. https://doi.org/10.59668/806.12879

Dias, J. M., Aderibigbe, S. A., & Abraham, M. S. (2026). Clinical Mentorship in Nursing Practice in the UAE: A Mixed-Methods Study of Mentor Perspectives on Effectiveness, Training, and Support. *Journal of Nursing Management, 2026*(1). https://doi.org/10.1155/jonm/3224685

Díaz-Diaz, O., Díaz, Y. E., & Saldívar-González, S. J. (2020). Análisis Bibliometrico De La Análitica De Datos en La Educación Superior Mediante Los Sistemas De Administración De Aprendizaje. *Vinculatégica Efan, 6*(1), 658-671. https://doi.org/10.29105/vtga6.1-615

Dicheva, N., Rehman, I. U., Anwar, A., Nasralla, M. M., Husamaldin, L., & Aleshaiker, S. (2023). Digital Transformation in Nursing Education: A Systematic Review on Computer-Aided Nursing Education Pedagogies, Recent Advancements and Outlook on the Post-Covid-19 Era. *Ieee Access, 11*, 135659-135695. https://doi.org/10.1109/access.2023.3337669

Diken, F., & Özdemir, N. (2021). The Effects of Culture on Learning Styles: The Sample of Ondokuz Mayıs University. *Journal of Education in Black Sea Region, 6*(2), 54-69. https://doi.org/10.31578/jebs.v6i2.231

Diko, M. (2023). Harnessing the Power of Education for Social Change. *Eureka Social and Humanities*(5), 43-55. https://doi.org/10.21303/2504-5571.2023.003051

Dilek, M., Baran, E., & Aleman, E. (2025). AI Literacy in Teacher Education: Empowering Educators Through Critical Co-Discovery. *Journal of Teacher Education, 76*(3), 294-311. https://doi.org/10.1177/00224871251325083

Dillard, N., Sisco, S., & Collins, J. C. (2024). Expanding Experiential Learning in Contemporary Adult Education: Embracing Technology, Interdisciplinarity, and Cultural Responsiveness. *New Directions for Adult and Continuing Education, 2024*(184), 30-38. https://doi.org/10.1002/ace.20539

Dillon, M., Heap, J., Patel, D., & Khan, S. A. (2023). Capacity Building for Food Safety Practices in Fish Farmers in Bangladesh: A Problem-Based Learning Approach. *International Journal of Applied Engineering and Management Letters*, 98-111. https://doi.org/10.47992/ijaeml.2581.7000.0169

Ding, K., Zhang, X., & Zhuang, M. (2023). Exploration of the Development Path of Future Vocational Education. *Journal of Research in Vocational Education, 5*(3). https://doi.org/10.53469/jrve.2023.05(03).06

Ding, T. (2023). Social Dialogue and Its Role in Germany Vocational Education and Training. *Lecture Notes in Education Psychology and Public Media, 2*(1), 715-722. https://doi.org/10.54254/2753-7048/2/2022xxx

Dipitso, P. O. (2023). Employers' Perspectives on Employability Skills and Attributes of Mining Engineering Undergraduates in South Africa. *The Southern Journal of Engineering Education, 2*, 101-123. https://doi.org/10.15641/sjee.v2i1.1491

Dison, L. (2016). Scrutinising the Role of Reflection in a Postgraduate Diploma in Higher Education. *South African Journal of Higher Education, 30*(6). https://doi.org/10.20853/30-6-718

Diz-Comesaña, M. E., Vázquez, B. G., & López, N. R. (2022). Nivel De Atención, Perfil De Aprendizaje Y Rendimiento Académico. *Visual Review International Visual Culture Review / Revista Internacional De Cultura Visual, 11*(1), 1-11. https://doi.org/10.37467/revvisual.v9.3634

Doan, L. K. (2019). Finding Community: An Exploration Into an Induction Support Pilot Project. *Journal of Childhood Studies*, 68-79. https://doi.org/10.18357/jcs.v44i1.18778

Dobrich, E. (2022). Prospects for Adult Learning and Global Change: A Canadian Perspective With Recommendations for Professional Practice. *European Journal of Teaching and Education, 4*(4), 57-69. https://doi.org/10.33422/ejte.v4i4.867

Dodge, R. B., & McKeough, M. (2003). Internship and the Nova Scotia Government Experience. *Education + Training, 45*(1), 45-55. https://doi.org/10.1108/00400910310459662

Dogara, G., Saud, M., & Kamin, Y. (2020). Work-Based Learning Conceptual Framework for Effective Incorporation of Soft Skills Among Students of Vocational and Technical Institutions. *Ieee Access, 8*, 211642-211652. https://doi.org/10.1109/access.2020.3040043

Dogara, G., Saud, M., Kamin, Y., & Nordin, M. S. (2020). Project-Based Learning Conceptual Framework for Integrating Soft Skills Among Students of Technical Colleges. *Ieee Access*, *8*, 83718-83727. https://doi.org/10.1109/access.2020.2992092

Dolenc, N., Tai, R. H., & Williams, D. (2020). Excessive Mentoring? An Apprenticeship Model on a Robotics Team. *Journal of Research in Stem Education*, *6*(2), 91-114. https://doi.org/10.51355/jstem.2020.81

Donaldson, C., & Villagrasa, J. (2024). Integrative Entrepreneurship Education: The Case of EDEM Business School. *Journal of Management and Business Education*, *7*(3), 419-434. https://doi.org/10.35564/jmbe.2024.0023

Donnelly, R., & Maguire, T. (2020). Building Digital Capacity for Higher Education Teachers: Recognising Professional Development Through a National Peer Triad Digital Badge Ecosystem. *European Journal of Open Distance and E-Learning*, *23*(2), 1-19. https://doi.org/10.2478/eurodl-2020-0007

Doo, M. Y., Bonk, C. J., & Heo, H. (2020). A Meta-Analysis of Scaffolding Effects in Online Learning in Higher Education. *The International Review of Research in Open and Distributed Learning*, *21*(3). https://doi.org/10.19173/irrodl.v21i3.4638

Doolani, S., Wessels, C., Kanal, V., Sevastopoulos, C., Jaiswal, A., Nambiappan, H. R., & Makedon, F. (2020). A Review of Extended Reality (XR) Technologies for Manufacturing Training. *Technologies*, *8*(4), 77. https://doi.org/10.3390/technologies8040077

Doolittle, P. E., Wojdak, K., & Walters, A. L. (2023). Defining Active Learning: A Restricted Systemic Review. *Teaching & Learning Inquiry the Issotl Journal*, *11*. https://doi.org/10.20343/teachlearninqu.11.25

Dopelt, K., Shevach, I., Vardimon, O. E., Czabanowska, K., Nooijer, J. d., Otok, R., Leighton, L., Bashkin, O., Duplaga, M., Levine, H., MacLeod, F., Malowany, M., Okenwa-Emegwa, L., Zelber-Sagi, S., Davidovitch, N., & Barach, P. (2023). Simulation as a Key Training Method for Inculcating Public Health Leadership Skills: A Mixed Methods Study. *Frontiers in Public Health*, *11*. https://doi.org/10.3389/fpubh.2023.1202598

Dorsey, G. N. (2009). Praxis. https://doi.org/10.31979/etd.t9rd-2gky

Dosunmu, A. G., & Dichaba, M. (2019). Access to Higher Education and Career Advancement for Women in the South African Work Environment. *South African Journal of Higher Education*, *33*(5). https://doi.org/10.20853/33-5-3602

Dosunmu, A. G., & Dichaba, M. (2020). Lifelong Learning and Career Progression of Women in Nigeria. 1-20. https://doi.org/10.1007/978-3-030-43042-9_18-1

Douieb, M. M. (2021). La Integración De Las TIC en La Universidad Marroquí: Aproximación Pedagógica. *Communication Papers*, *10*(20), 101. https://doi.org/10.33115/udg_bib/cp.v10i20.22609

Downes, S. (2007). What Connectivism Is. *Half an Hour*. https://halfanhour.blogspot.com/2007/02/what-connectivism-is.html

Downes, S. (2012). *Connectivism and Connective Knowledge: Essays on Meaning and Learning Networks*. National Research Council Canada.

Downes, S. (2020). Recent Work in Connectivism. *European Journal of Open Distance and E-Learning*, *22*(2), 113-132. https://doi.org/10.2478/eurodl-2019-0014

Dreamson, N. (2020). Online Design Education: Meta-Connective Pedagogy. *International Journal of Art & Design Education*, *39*(3), 483-497. https://doi.org/10.1111/jade.12314

Drew, J. (2008). The Study of the Impact of the M.S. In Agronomy Distance Education Program on Student Careers. https://doi.org/10.31274/rtd-180813-16579

Drewery, D., Sproule, R., & Pretti, T. J. (2020). Lifelong Learning Mindset and Career Success: Evidence From the Field of Accounting and Finance. *Higher Education Skills and Work-Based Learning*, *10*(3), 567-580. https://doi.org/10.1108/heswbl-03-2019-0041

Drigas, A., & Karyotaki, M. (2014). Learning Tools and Applications for Cognitive Improvement. *International Journal of Engineering Pedagogy (Ijep)*, *4*(3), 71. https://doi.org/10.3991/ijep.v4i3.3665

Drigas, A., & Tsolaki, V. (2015). Lifelong Learning and ICTs. *International Journal of Recent Contributions From Engineering Science & It (Ijes)*, *3*(2), 15. https://doi.org/10.3991/ijes.v3i2.4353

Drury, S. A. M., Andre, D., Goddard, S., & Wentzel, J. (2016). Assessing Deliberative Pedagogy: Using a Learning Outcomes Rubric to Assess Tradeoffs and Tensions. *Journal of Deliberative Democracy*, *12*(1). https://doi.org/10.16997/jdd.245

Duarte, O. E. C. (2021). Andragogy and Its Interdisciplinary Character in University Education. *Revista Digital De Investigación Y Postgrado*, *1*(1), 77-88. https://doi.org/10.59654/qx1ye451

Dubouloz, C. J. (2014). Discours Commémoratif Muriel Driver 2014. *Canadian Journal of Occupational Therapy*, *81*(4), E1-E10. https://doi.org/10.1177/0008417414559013

Duchatelet, D., Cornelissen, F., & Volman, M. (2023). Features of Experiential Learning Environments in Relation to Generic Learning Outcomes in Higher Education: A Scoping Review. *Journal of Experiential Education*, *47*(3), 400-423. https://doi.org/10.1177/10538259231211537

Duff, M. C. (2019). Perspectives in AE— Adult Black Males and Andragogy: Is There a Goodness of Fit. *New Horizons in Adult Education and Human Resource Development*, *31*(4), 51-58. https://doi.org/10.1002/nha3.20264

Duffy, P. A. (2008). *Holes in the Health Care Safety Net: An Ethnographic Study of a Free Medical Clinic* Iowa State University].

Dujo, Á. G. d., & García, A. V. M. (2009). Caracterización Pedagógica De Los Entornos Virtuales De Aprendizaje. *Teoría De La Educación Revista Interuniversitaria*, *14*. https://doi.org/10.14201/2976

Dumitrescu, C. (2025). Use of Emerging Technologies in Rehabilitation Education and Practice. *Frontiers in Rehabilitation Sciences*, *6*. https://doi.org/10.3389/fresc.2025.1655454

Dunaway, M. (2011). Connectivism. *Reference Services Review*, *39*(4), 675-685. https://doi.org/10.1108/00907321111186686

Duncan, H. E., Smith, J., & Bachewich, L. (2022). Mino-Pimaatisiwin Beginning the Journey Towards Decolonisation and Reconciliation. *Australian and International Journal of Rural Education*, *32*(2), 1-18. https://doi.org/10.47381/aijre.v32i2.332

Dunk, V., & Craft, J. (2025). Beyond Self-Directed Learning: The Role of Self-Regulated Learning in Nursing Education. *Journal of Nursing Education*, *64*(5), 287-293. https://doi.org/10.3928/01484834-20250108-07

Dunn, C. (2004). Industry Training Strategies in Australia and New Zealand. *Labour Employment and Work in New Zealand*. https://doi.org/10.26686/lew.v0i0.1294

Duplàa, E., Pecorini, B. C., Weber, J., & Blouin, M. (2023). L'apprentissage en Ligne Dans Le Contexte De La Quatrième Révolution Industrielle : Le Cas D'un Module Connectiviste en Contexte Universitaire. *Canadian Journal of Learning and Technology*, *49*(1), 1-21. https://doi.org/10.21432/cjlt28411

Durrans, B., Whale, J., & Calais, M. (2020). Benchmarking a Sustainable Energy Engineering Undergraduate Degree Against Curriculum Frameworks and Pedagogy Standards From Industry and Academia. *Energies*, *13*(4), 822. https://doi.org/10.3390/en13040822

Duval, E., & Wiley, D. (2010). Guest Editorial: Open Educational Resources. *Ieee Transactions on Learning Technologies*, *3*(2), 83-84. https://doi.org/10.1109/tlt.2010.11

Dykes, G. (2014). Recognition of Prior Learning (Rpl) and Support: Are the Learning Needs of RPL First-Year Students Different? *Social Work/Maatskaplike Werk*, *45*(3). https://doi.org/10.15270/45-3-207

Dziubaniuk, O., Ivanova-Gongne, M., & Nyholm, M. (2023). Learning and Teaching Sustainable Business in the Digital Era: A Connectivism Theory Approach. *International Journal of Educational Technology in Higher Education*, *20*(1). https://doi.org/10.1186/s41239-023-00390-w

Ebner, C., Graf, L., & Nikolai, R. (2013). New Institutional Linkages Between Dual Vocational Training and Higher Education: A Comparative Analysis of Germany, Austria and Switzerland. 281-298. https://doi.org/10.1007/978-94-007-6119-3_14

Ebnezer, E., & Rajini, G. (2024). Digitalization of Human Resource Management and the Significance of Motivation in Acquiring Digital Skills for Employee Retention. *Salud Ciencia Y Tecnología - Serie De Conferencias*, *3*, 887. https://doi.org/10.56294/sctconf2024887

Eckes, E. J., Gutierrez, D., Peterson, A., Hubbard, M., Barrett, K., & Matlock, A. M. (2016). Educating Health Care Providers in Treatment of Patients With Ebola Virus Disease. *Critical Care Nurse*, *36*(5), e8-e13. https://doi.org/10.4037/ccn2016814

Economides, A. A. (2008). Culture-aware Collaborative Learning. *Multicultural Education & Technology Journal*, *2*(4), 243-267. https://doi.org/10.1108/17504970810911052

Edeh, F. O., Neji, D. O., Irem, C. O., Alum, E. U., Ніценко, В., Owere, G. O., Chukwu, A. U., Oyekezie, K. S. U., & Anyalor, C. M. (2025). Investigating the Role of Perceived Organisational Support in the Relationship Between Human Resource Skill Adjustment and Corporate Sustainability in the Hospitality Industry. https://doi.org/10.21203/rs.3.rs-7114536/v1

Edeh, F. O., Ugboego, A. C., & Adama, L. (2022). Human Resource Skills Adjustment and Organisational Resilience in Times of Global Crisis. *Kelaniya Journal of Human Resource Management*, *17*(1), 60-79. https://doi.org/10.4038/kjhrm.v17i1.70

Edelist, T., Friesen, F., Ng, S., Fernández, N., Bélisle, M., Lechasseur, K., Rochette, A., Vachon, B., & Caty, M. È. (2024). Critical Reflection in Team-based Practice: A Narrative Review. *Medical Education*, *58*(10), 1166-1177. https://doi.org/10.1111/medu.15462

Edgar, D. (2012). Learning Theories and Historical Events Affecting Instructional Design in Education. *Sage Open*, *2*(4). https://doi.org/10.1177/2158244012462707

Edokpolor, J. E., & Omiunu, S. E. (2017). Role of Career Development Services in Developing Students Lifelong Learning Self-Efficacy: Vocational and Adult Education Lecturers' Perceptions. *Baltic Journal of Career Education and Management*, *5*(1), 30-45. https://doi.org/10.33225/bjcem/17.5.30

Edwards, M., Benn, S., & Starik, M. (2017). Business Cases for Sustainability-Integrated Management Education. https://doi.org/10.4337/9781785361241.00010

Efendi, F., Has, E. M. M., Pradipta, R. O., Houghty, G. S., Oda, H., & Tsujita, Y. (2025). Retention of International Nurses in Receiving Country: Voices of Indonesian Nurses in Japanese Healthcare Facilities. *International Journal of Nursing Sciences*, *12*(1), 3-11. https://doi.org/10.1016/j.ijnss.2024.12.008

Effendi, H., & Hendriyani, Y. (2020). The Conceptual and Hypothetical Model of Interactive Blended Problem Based Learning. *Jpi (Jurnal Pendidikan Indonesia)*, *8*(2), 285. https://doi.org/10.23887/jpi-undiksha.v8i2.24162

Effoduh, J. O., Akpudo, U. E., & Kong, J. D. (2024). Toward a Trustworthy and Inclusive Data Governance Policy for the Use of Artificial Intelligence in Africa. *Data & Policy*, *6*. https://doi.org/10.1017/dap.2024.26

Egan-Simon, D. (2022). Active Agents of Change: A Conceptual Framework for Social Justice-Orientated Citizenship Education. *Equity in Education & Society*, *1*(2), 297-310. https://doi.org/10.1177/27526461221089350

Egetenmeyer, R., & Fedeli, M. (2017a). *Adult Education and Work Contexts: International Perspectives and Challenges*. https://doi.org/10.3726/b12114

Egetenmeyer, R., & Fedeli, M. (2017b). Adult Education and Work Contexts: International Perspectives and Challenges. https://doi.org/10.3726/b12114

Egetenmeyer, R., Guimarães, P., & Németh, B. (2017). Joint Modules and Internationalisation in Higher Education. https://doi.org/10.3726/b12110

Egunjobi, D., & Adeyeye, O. (2024). Revolutionizing Learning: The Impact of Augmented Reality (AR) and Artificial Intelligence (AI on Education. *International Journal of Research Publication and Reviews*, *5*(10), 1157-1170. https://doi.org/10.55248/gengpi.5.1024.2734

Ehlers, U. (2011). From Open Educational Resources to Open Educational Practices. *ELearning Papers*. .

Eichhörst, W., Rodríguez-Planas, N., Schmidl, R., & Zimmermann, K. F. (2013). A Roadmap to Vocational Education and Training Systems Around the World. *SSRN Electronic Journal*. https://doi.org/10.2139/ssrn.2199793

Eichhörst, W., Rodríguez-Planas, N., Schmidl, R., & Zimmermann, K. F. (2015). A Road Map to Vocational Education and Training in Industrialized Countries. *Ilr Review*, *68*(2), 314-337. https://doi.org/10.1177/0019793914564963

Eisenstein, A. (2022). Engaging Engineering Students With Engineering Entrepreneurship and the Start-Up Working Environment Through Supervised Entrepreneurial Work-Integrated Learning. *Proceedings of the Canadian Engineering Education Association (Ceea)*. https://doi.org/10.24908/pceea.vi.15845

Ekuma, K. (2023). Rethinking Upskilling and Reskilling in the Age of AI and Automation: A fsQCA Approach. https://doi.org/10.20944/preprints202309.0055.v1

Ekung, S., & Odesola, I. A. (2018). Learning Models for Effective Propagation of Sustainable Construction Practices in the Built Environment. *Malaysian Journal of Sustainable Environment*, *4*(1), 113. https://doi.org/10.24191/myse.v4i1.5610

El-Haggar, N., Amouri, L., Alsumayt, A., Alghamedy, F. H., & Aljameel, S. S. (2023). The Effectiveness and Privacy Preservation of IoT on Ubiquitous Learning: Modern Learning Paradigm to Enhance Higher Education. *Applied Sciences*, *13*(15), 9003. https://doi.org/10.3390/app13159003

El-Kishawi, M., Khalaf, K., & Winning, T. (2021). Understanding Motor Skill Learning as Related to Dentistry. *Dentistry Journal*, *9*(6), 68. https://doi.org/10.3390/dj9060068

Elamin, M. O. I. (2024). AI Through the Ages: Unlocking Key Opportunities and Navigating Challenges in the History and Future of Artificial Intelligence. *International Journal of Religion*, *5*(12), 1152-1166. https://doi.org/10.61707/tarzcj89

Elbert, M. J. P., Mendoza, B. M. Z., Aguirre, K. A. M., & Cárdenas, M. A. V. (2023). Realidad Virtual, Realidad Aumentada Y Realidad Extendida en La Educación. *Recimundo*, *7*(2), 74-88. https://doi.org/10.26820/recimundo/7.(2).jun.2023.74-88

Elder, A. C. (2023). Statistics Attitudes After Using Guided Project-Based Learning as an Andragogical Strategy in a Graduate Statistics Course. *Statistics Education Research Journal*, *22*(3), 4. https://doi.org/10.52041/serj.v22i3.436

Elendu, C., Amaechi, D. C., Okatta, A. U., Amaechi, E. C., Elendu, T. C., Ezeh, C. P., & Elendu, I. D. (2024). The Impact of Simulation-Based Training in Medical Education: A Review. *Medicine*, *103*(27), e38813. https://doi.org/10.1097/md.0000000000038813

Elfert, M. (2015). UNESCO, the Faure Report, the Delors Report, and the Political Utopia of Lifelong Learning. *European journal of education*, *50*(1), 88-100. https://doi.org/10.1111/ejed.12104

Elfert, M. (2019). Revisiting the Faure Report and the Delors Report. 17-25. https://doi.org/10.1163/9789004413320_002

Elibol, S., & Bozkurt, A. (2023). Student Dropout as a Never-Ending Evergreen Phenomenon of Online Distance Education. *European Journal of Investigation in Health Psychology and Education*, *13*(5), 906-918. https://doi.org/10.3390/ejihpe13050069

Elizalde, M. L. M. (2020). Equidad Y Responsabilidad Social Como Bien Común. Desafíos Para La Educación Superior en México. 29-45. https://doi.org/10.48190/9789942241412.2

Elkachradi, R., Boudallaa, I., Hillali, M., & Kadouri, A. (2023). Lifelong Learning Among Healthcare Professionals in Public Hospitals: Historical Analysis and Multiple Case Studies in Morocco. *E3s Web of Conferences*, *412*, 01078. https://doi.org/10.1051/e3sconf/202341201078

Ellinger, A. D., & Cseh, M. (2007). Contextual Factors Influencing the Facilitation of Others' Learning Through Everyday Work Experiences. *Journal of Workplace Learning*, *19*(7), 435-452. https://doi.org/10.1108/13665620710819384

Elliott-Groves, E., & Fish, J. (2024). Overcoming Human Exceptionalism: The Role of Ethical Nature-Culture Relations in the Developmental Contexts of Indigenous Children. *Child Development*, *95*(6), 1894-1905. https://doi.org/10.1111/cdev.14195

Elliott, T. H. (2024). Theory Analysis and Evaluation of Emancipatory Nursing Praxis: A Theory of Social Justice in Nursing. *International Journal of Nursing Knowledge*, *35*(1), 32-39. https://doi.org/10.1111/2047-3095.12414

Elsayed, A. N. (2023). The Use of Blockchain Technology in Education: A Comprehensive Review and Future Prospects. *SSRN Electronic Journal*. https://doi.org/10.2139/ssrn.4523322

Elsayed, N., & Hassan, M. K. (2024). Audit Simulation and Learning Styles: Enhancing Students' Experiential Learning and Performance at a MENA University. *International Journal of Auditing*. https://doi.org/10.1111/ijau.12345

Elsayed, N., Ismael, H. R., & Saadullah, S. M. (2023). Enhancing Students' Understanding and Performance in a Distance-Learning Setting: Evidence From an Audit Simulation at a GCC University. *Accounting Research Journal*, *36*(1), 1-20. https://doi.org/10.1108/arj-12-2021-0359

Elsey, B., & Sirichoti, K. (2003). The Theory and Practice of Workplace Learning in the Adoption of Integrated Pest Management by Tropical Fruit Growers in Thailand. *Journal of Workplace Learning, 15*(2), 53-62. https://doi.org/10.1108/13665620310464094

Eltayb, A., & Valcke, J. (2024). Insights From Glocal Educators. *Journal of Comparative & International Higher Education, 16*(5). https://doi.org/10.32674/kznfxn70

Embo, M., & Valcke, M. (2017). Continuing Midwifery Education Beyond Graduation: Student Midwives' Awareness of Continuous Professional Development. *Nurse Education in Practice, 24*, 118-122. https://doi.org/10.1016/j.nepr.2015.08.013

Ememe, P. I. (2020). Open Distance Learning: An Alternative Strategy for Rural Teachers' Professional Development in Lagos State, Nigeria. *Tetfle, 2*(1). https://doi.org/10.35293/tetfle.v2i1.89

Emmenegger, P., & Seitzl, L. (2020). Social Partner Involvement in Collective Skill Formation Governance. A Comparison of Austria, Denmark, Germany, the Netherlands and Switzerland. *Transfer European Review of Labour and Research, 26*(1), 27-42. https://doi.org/10.1177/1024258919896897

Endersby, L., Jenkins, D., & Priest, K. L. (2019). Leadership Education: Advancing a Community of Practice. *New Directions for Student Leadership, 2019*(164), 141-149. https://doi.org/10.1002/yd.20363

Éneau, J. (2016). Autoformation, Autonomisation Et Émancipation *Recherches & Éducations, 16*, 21-38. https://doi.org/10.4000/rechercheseducations.2489

Engage XR. (2025). ENGAGE XR. https://engagevr.io

Engel, L., & Rutkowski, D. (2014). Global Influences on National Definitions of Quality Education: Examples From Spain and Italy. *Policy Futures in Education, 12*(6), 769-783. https://doi.org/10.2304/pfie.2014.12.6.769

Engeström, Y. (2000). Activity theory as a framework for analyzing and redesigning work. *Ergonomics, 43*(7), 960-974. https://doi.org/10.1080/001401300409143

Engler, J. N., Shedlosky-Shoemaker, R., & Deffler, S. A. (2025). Self-Assessment Using American Psychological Association Guidelines 3.0: A Strategy for College and Career Success. *Scholarship of Teaching and Learning in Psychology, 11*(1), 81-90. https://doi.org/10.1037/stl0000423

English, L. M., & Carlsen, A. (2019). Lifelong Learning and the Sustainable Development Goals (SDGs): Probing the Implications and the Effects. *International Review of Education, 65*(2), 205-211. https://doi.org/10.1007/s11159-019-09773-6

Entigar, K. E. (2021). To Choose Not to Be Included: Critical Perspectives on Practices of Inclusion With Adult Immigrant Students. *International Review of Education, 67*(6), 833-849. https://doi.org/10.1007/s11159-021-09933-7

Eraut, M. (1994). *Developing Professional Knowledge and Competence*. Falmer Press.

Eraut, M. (2004). Informal Learning in the Workplace. *Studies in continuing education, 26*(2), 247-273. https://doi.org/10.1080/158037042000225245

Erbay, A. (2020). Descriptive Analysis of Reasons of Participating Literacy Courses and Effects of Being Literate. *Yaşadıkça Eğitim, 34*(1), 99-110. https://doi.org/10.33308/26674874.2020341169

Erickson, W., Schrader, S. v., Bruyère, S. M., VanLooy, S., & Matteson, D. S. (2014). Disability-Inclusive Employer Practices and Hiring of Individuals With Disabilities. *Rehabilitation Research Policy and Education, 28*(4), 309-328. https://doi.org/10.1891/2168-6653.28.4.309

Erol, E., & Altunay, E. (2023). Investigation of the Challenges of Public Education Centre Instructors. *Anadolu University Faculty of Education Journal, 7*(3), 739-764. https://doi.org/10.34056/aujef.1228945

Erümit, S. F., & KeleŞ, E. (2023). Examining Computer Science Education of Asia-Pacific Countries Successful in the PISA. *Journal of Educational Technology and Online Learning, 6*(1), 82-104. https://doi.org/10.31681/jetol.1154913

Esgandanian, A. (2023). Youth Apprentice Engagement in Ontario: What Can Be Learned From Switzerland's Vocational Education and Training (VET) System? https://doi.org/10.32920/24065922.v1

Eslit, E. R. (2023). Elevating Language Acquisition Through Deep Learning and Meaningful Pedagogy in an AI-Evolving Educational Landscape. https://doi.org/10.20944/preprints202309.0658.v1

Espinoza, L. I. V. (2023). Educación, Bildung Y Desarrollo Humano. *Ducere Revista De Investigación Educativa*, *2*(1), e202301. https://doi.org/10.61303/2735668x.v2i1.3

Espinoza, Ó. (2020). Acceso Al Sistema De Educación Superior en Chile. *Universidades*, *68*(74), 7-30. https://doi.org/10.36888/udual.universidades.2017.74.484

Espinoza, S. L. (2020). El Aprendizaje: Posibilidades Teóricas Para Comprenderlo Más Allá De La Escuela. *Revista Redca*, *2*(6), 22. https://doi.org/10.36677/redca.v2i6.13937

Esquivel, A. C., Gónzales, D. Y. B., & Castillo, G. A. T. (2022). Estilos De Aprendizaje De Estudiantes De Primer Ciclo De La Universidad Andina Del Cusco, 2018. *Yachay - Revista Científico Cultural*, *11*(1), 562-568. https://doi.org/10.36881/yachay.v11i1.550

Essomba, M. À., Lleonart, P., Alfonso, L., & Bin, H. (2022). Education for Sustainable Development in Educating Cities: Towards a Transformative Approach From Informal and Non-Formal Education. *Sustainability*, *14*(7), 4005. https://doi.org/10.3390/su14074005

Estella, A., & Vera, C. S. (2015). 1995-2015: 20 Años De Aprendizaje Permanente en La Unión Europea. *Revista Iberoamericana De Educación*, *69*(3), 93-116. https://doi.org/10.35362/rie693114

Esterhazy, R., Fossland, T., & Stalheim, O.-R. (2020). What Counts as Quality Feedback? Disciplinary Differences in Students' and Teachers' Perceptions of Feedback. 155-174. https://doi.org/10.1007/978-3-030-41757-4_9

Evanick, J. (2024). Transferable Best Practices and Mutual Learning Opportunities Between Instructional Design in Higher Education and the Corporate World. *International Journal of Advanced Corporate Learning (Ijac)*, *17*(2), 63-75. https://doi.org/10.3991/ijac.v17i2.45555

Evans, C., & Waring, M. (2011). How Can an Understanding of Cognitive Style Enable Trainee Teachers to Have a Better Understanding of Differentiation in the Classroom? *Educational Research for Policy and Practice*, *10*(3), 149-169. https://doi.org/10.1007/s10671-011-9101-1

Evans, G. J. (2018). Windmills of Your Mind: Metacognition and Lifelong Learning. *Proceedings of the Canadian Engineering Education Association (Ceea)*. https://doi.org/10.24908/pceea.v0i0.12985

Evans, K. (2015). Apprentissage Tout Au Long De La Vie : Politique Sociale Et Agentivité Individuelle. *Savoirs*, *N° 37*(1), 11-33. https://doi.org/10.3917/savo.037.0011

Evans, L., Thornton, B., & Usinger, J. (2012). Theoretical Frameworks to Guide School Improvement. *Nassp Bulletin*, *96*(2), 154-171. https://doi.org/10.1177/0192636512444714

Evans, L. M., Joseph, T., Jozwik, S., & Bartlett, M. (2024). Preparing Inclusive Early Childhood Educators (<scp>PIECE</Scp>): A Conceptualization of Multilingualism, English Learning, and Inclusivity. *Tesol Journal*, *15*(4). https://doi.org/10.1002/tesj.863

Evans, M., & Boucher, A. R. (2015). Optimizing the Power of Choice: Supporting Student Autonomy to Foster Motivation and Engagement in Learning. *Mind Brain and Education*, *9*(2), 87-91. https://doi.org/10.1111/mbe.12073

Evans, N., Stevenson, R. B., Lasen, M., Ferreira, J. A., & Davis, J. M. (2017). Approaches to Embedding Sustainability in Teacher Education: A synthesis of the Literature. *Teaching and teacher education*, *63*, 405-417. https://doi.org/10.1016/j.tate.2017.01.013

Ewim, D. R. E., Ninduwezuor-Ehiobu, N., Orikpete, O. F., Egbokhaebho, B. A., Fawole, A. A., & Onunka, C. (2023). Impact of Data Centers on Climate Change: A Review of Energy Efficient Strategies. *The Journal of Engineering and Exact Sciences*, *9*(6), 16397-16301e. https://doi.org/10.18540/jcecvl9iss6pp16397-01e

Eyal, O., & Berkovich, I. (2024). Ethics Education in Management in Academia and HRD: An Identity Control Theory-Based Model. *Journal of Workplace Learning*, *36*(8), 658-675. https://doi.org/10.1108/jwl-02-2024-0031

Ezeonwumelu, V. U., Offor, U. I., Nwikpo, M. N., Onunkwo, C. M., & Nwaru, P. E. (2025). Promoting Inclusive Education to Boost Entrepreneurial Intentions: A Catalyst for Attainment of Sustainable Development Goals. *European Journal of Contemporary Education and E-Learning*, *3*(1), 65-76. https://doi.org/10.59324/ejceel.2025.3(1).06

Ezerins, M. E., Simon, L., Vogus, T. J., Gabriel, A. S., Calderwood, C., & Rosen, C. C. (2023). Autism and Employment: A Review of the "New Frontier" of Diversity Research. *Journal of Management*, *50*(3), 1102-1144. https://doi.org/10.1177/01492063231193362

Fairman, B., Voak, A., & Babacan, H. (2022). Human Mobility and Dismantling Cultural Dominance. *Journal of Resilient Economies (Issn 2653-1917)*, *2*(1). https://doi.org/10.25120/jre.2.1.2022.3917

Faiz, H. (2023). Navigating Pakistan's Educational Policy Landscape in Pursuit of Sustainable Development Goal 4: A View of Stakeholders. *Pakistan Languages and Humanities Review*, *7*(III). https://doi.org/10.47205/plhr.2023(7-iii)44

Falco, L. D. (2026). School Counselors and the Future of Work: Preparing P–12 Students for Career Development Amid AI, Technological Change, and Economic Volatility. *Journal of Employment Counseling*, *63*(2), 129-140. https://doi.org/10.1002/joec.70012

Fan, X. (2024). Artificial Intelligence's Impact on Employment: Challenges, Potential Consumers, and Policy Responses Through Automation and Workforce Rehabilitating. *Lecture Notes in Education Psychology and Public Media*, *73*(1), 67-72. https://doi.org/10.54254/2753-7048/73/20241040

Fang, H.-L. (2025). Enhancing Student Learning Confidence and Career Readiness Through Instructor Demonstration and Industry Collaboration: An Empirical Study of Hairdressing Education in Taiwan. https://doi.org/10.21203/rs.3.rs-6318289/v1

Fang, M. (2023). Leadership Coaching to Drive Transformation in the Experience Economy. https://doi.org/10.5772/intechopen.108104

Fang, S., & Han, Z. (2025). On the Nascency of ChatGPT in Foreign Language Teaching and Learning. *Annual Review of Applied Linguistics*, *45*, 148-178. https://doi.org/10.1017/s026719052510010x

Fang, Z., & Sim, N. (2024). Does Lifelong Learning Matter for the Subjective Wellbeing of the Elderly? A Machine Learning Analysis on Singapore Data. *Plos One*, *19*(6), e0303478. https://doi.org/10.1371/journal.pone.0303478

Fanguy, M., Costley, J., Almusharraf, N., & Almusharraf, A. (2023). Online Collaborative Note-Taking and Discussion Forums in Flipped Learning Environments. *Australasian Journal of Educational Technology*, *39*(2), 142-158. https://doi.org/10.14742/ajet.8580

Farashah, A. D., & Blomquist, T. (2021). Work Experiences of Qualified Immigrants: A Review of Theoretical Progress. *Equality Diversity and Inclusion an International Journal*, *41*(7), 1063-1090. https://doi.org/10.1108/edi-01-2019-0046

Farizi, S. F., Umamah, N., & Soepeno, B. (2023). The Effect of the Challenge Based Learning Model on Critical Thinking Skills and Learning Outcomes. *Anatolian Journal of Education*, *8*(1), 191-206. https://doi.org/10.29333/aje.2023.8113a

Farmati, C., Yeou, M., & Benzahaf, B. (2023). Blended Learning in English for Specific Purposes Instruction: A Systematic Review. *Digital Education Review*(44), 114-124. https://doi.org/10.1344/der.2023.44.114-124

Farooqi, R., & Scutt, D. (2026). Practical Considerations for Bedside Teaching: A Critical Application of Kolb's Experiential Learning Cycle. *The Clinical Teacher*, *23*(3). https://doi.org/10.1111/tct.70437

Farrell, W. A. (2018). Learning Becomes Doing: Applying Augmented and Virtual Reality to Improve Performance. *Performance Improvement*, *57*(4), 19-28. https://doi.org/10.1002/pfi.21775

Farrukh, M., & Waheed, A. (2015). Learning Organization and Competitive Advantage-an Integrated Approach. *Journal of Asian Business Strategy*, *5*(4), 73-79. https://doi.org/10.18488/journal.1006/2015.5.4/1006.4.73.79

Fathahillah, F., Fakhri, M. M., & Ahmar, A. S. (2023). Analysis of Artificial Intelligence Literacy in the Blended Learning Model in Higher Education. *Eduline Journal of Education and Learning Innovation*, *3*(4), 566-575. https://doi.org/10.35877/454ri.eduline2049

Fatima, A. (2021). Gig Economy: A Mediating Role of Motivation Between Workplace Learning Activities and Creative Performance. *International Journal of Management Research and Emerging Sciences*, *11*(2). https://doi.org/10.56536/ijmres.v11i2.235

Faure, E., Herrera, F., Kaddoura, A.-R., Lopes, H., Petrovsky, A. V., Rahnema, M., & Ward, F. C. (1972). *Learning to Be: The World of Education Today and Tomorrow*. UNESCO Publishing.

Fayard, A. L., & DeSanctis, G. (2005). Evolution of an Online Forum for Knowledge Management Professionals: A Language Game Analysis. *Journal of Computer-Mediated Communication*, *10*(4), 00-00. https://doi.org/10.1111/j.1083-6101.2005.tb00265.x

Fazio, M. V., Fernández-Coto, R., & Ripani, L. (2016). Apprenticeships for the XXI Century: A Model for Latin America and the Caribbean? https://doi.org/10.18235/0012817

Fealy, S., Irwin, P., Taçgın, Z., See, Z. S., & Jones, D. (2023). Enhancing Nursing Simulation Education: A Case for Extended Reality Innovation. *Virtual Worlds*, *2*(3), 218-230. https://doi.org/10.3390/virtualworlds2030013

Febriansyah, D., & El-Alami, D. (2021). Moderate Islam Vis-a-Vis Salafism in Indonesia: an Ideological Competition. *Walisongo Jurnal Penelitian Sosial Keagamaan*, *29*(1), 55-78. https://doi.org/10.21580/ws.29.1.7212

Fedeli, M., & Olatunji, T. I. (2024). My Theory, My Tool, My Tribe: Unpacking Scholars' Perspectives on the Evolution of Transformative Learning. *Journal of Transformative Education*, *23*(4), 444-466. https://doi.org/10.1177/15413446241286481

Federighi, P. (2018). Non-Linear Paths in Transitions Through the Labour Market. 129-145. https://doi.org/10.36253/978-88-6453-672-9.20

Fee, A., & Gray, S. J. (2020). Expatriates as Catalysts: What and How Vietnamese Locals Learn From Self-Initiated Expatriates. *Cross Cultural & Strategic Management*, *27*(3), 389-416. https://doi.org/10.1108/ccsm-08-2019-0145

Fejes, A., & Nicoll, K. (2013). Editorial: Approaches to Research in the Education and Learning of Adults. *European Journal for Research on the Education and Learning of Adults*, *4*(1), 7-16. https://doi.org/10.3384/rela.2000-7426.relae6

Feldmann, L. (2016). Considerations in the Design of WBL Settings to Enhance Students' Employability. *Higher Education Skills and Work-Based Learning*, *6*(2), 131-145. https://doi.org/10.1108/heswbl-09-2014-0044

Fennimore, T. F. (1988). *The Helping Process Overview Guidebook. Dropout Prevention Series.*

Fenwick, T. (2005). Ethical Dilemmas of Critical Management Education. *Management learning*, *36*(1), 31-48. https://doi.org/10.1177/1350507605049899

Fergusson, L., & Laan, L. v. d. (2021). Work + Learning: Unpacking the Agglomerated Use of Pedagogical Terms. *Journal of Work-Applied Management*, *13*(2), 302-314. https://doi.org/10.1108/jwam-12-2020-0053

Fernández, L. M. A., Jacquemin, V. M., Pianovi, M. R., & Pokolenko, A. A. (2019). Innovación Educativa Y Participación Ciudadana Desde La Experiencia Del IMES en El Presupuesto Participativo Posadas. *La Saeta Universitaria Académica Y De Investigación*, *7*(1), 19-36. https://doi.org/10.56067/saetauniversitaria.v7i1.120

Fernández, M., & Olivencia, J. J. L. (2021). Los Fondos De Conocimiento en El Desarrollo De La Interculturalidad. Dilemas Y Paradojas Ante Los Efectos De La Covid19 en Alumnado Inmigrante en Situación De Vulnerabilidad. *Revista Temas Sociais*, *1*(1), 31-48. https://doi.org/10.53809/2021-01-ts-n.1-31-48

Fernández, T., Zuloeta, E. R. B., & Montalvo, A. G. (2023). Revisión Bibliométrica Sobre Bienestar Psicológico en La Revista De Investigación en Psicología - UNMSM (1998-2022). *Psiquemag/ Revista Científica Digital De Psicología*, *12*(2), 17-26. https://doi.org/10.18050/psiquemag.v12i2.2450

Ferreira, A. C., Santiago, M. A., María de los Ángeles López, O., & González, O. A. G. (2019). Analítica Del Aprendizaje Y Las Neurociencias Educativas: Nuevos Retos en La Integración Tecnológica. *Revista Iberoamericana De Educación*, *80*(1), 31-54. https://doi.org/10.35362/rie8013428

Ferrel-Ballestas, L.-F., Flórez-Gómez, D.-Y., & Rosa, N. G. l. (2022). Diseño De Cartilla Para La Enseñanza De La Biología en Estudiantes De La Cultura Wayuu Bajo Los Principios Del DUA. *Revista Latinoamericana De Educación Inclusiva*, *16*(2), 89-108. https://doi.org/10.4067/s0718-73782022000200089

Fidalgo, P., & Thormann, J. (2024). The Future of Lifelong Learning: The Role of Artificial Intelligence and Distance Education. https://doi.org/10.5772/intechopen.114120

Field, J. (2006). *Lifelong Learning and the New Educational Order* (2nd ed.). Trentham Books.

Fields, L., Dean, B. A., Perkiss, S., & Moroney, T. (2022). Education on the Sustainable Development Goals for Nursing Students: Is Freire the Answer? *Nursing inquiry*, *29*(4). https://doi.org/10.1111/nin.12493

Figaredo, D. D., Reich, J., & Ruipérez-Valiente, J. A. (2020). Analítica Del Aprendizaje Y Educación Basada en Datos: Un Campo en Expansión. *Ried Revista Iberoamericana De Educación a Distancia*, *23*(2), 33. https://doi.org/10.5944/ried.23.2.27105

Figurska, I., & Sokół, A. (2016). The Process of Knowledge Acquisition With the Use of Various Teaching Methods and Its Effect on the Creativity of Employees of the Creative Sector. *Mediterranean Journal of Social Sciences*. https://doi.org/10.5901/mjss.2016.v7n6p143

Filipiak, D., & Caraballo, L. (2019). Growing Together: Literacy and Agency in an Early-College Research Collaborative. *Journal of Adolescent & Adult Literacy*, *63*(3), 337-341. https://doi.org/10.1002/jaal.1006

Filiz, A., & Albay, C. (2025). Global Trends In <scp>STEM</Scp> -based Special Education Research: Bibliometric Mapping of the Literature. *British Journal of Special Education*, *52*(4), 597-612. https://doi.org/10.1111/1467-8578.70056

Findsen, B. (2016). Learning in Later Life: Benefits and Challenges for Volunteers and Agencies. *Current Aging Science*, *9*(3), 203-209. https://doi.org/10.2174/1874609809666160506122422

Finkel, S. E. (2014). The Impact of Adult Civic Education Programmes in Developing Democracies. *Public Administration and Development*, *34*(3), 169-181. https://doi.org/10.1002/pad.1678

Fischbach, S., & Guerrero, V. B. (2020). Brand Stories: Transformative Learning Through Digital Brand Storytelling (DBS). *Journal of Advertising Education*, *24*(2), 133-149. https://doi.org/10.1177/1098048220948515

Fithriani, F., Syabuddin, S., Gunawan, G., Zainuddin, T., & Sulaiman, S. (2021). Teacher as a Role Model in the 2013 Curriculum Development. *Jurnal Ilmiah Islam Futura*, *21*(2), 240. https://doi.org/10.22373/jiif.v21i2.7516

Fitri, R., Lufri, L., Alberida, H., Amran, A., & Fachry, R. (2024). The Project-Based Learning Model and Its Contribution to Student Creativity: A Review. *Jpbi (Jurnal Pendidikan Biologi Indonesia)*, *10*(1), 223-233. https://doi.org/10.22219/jpbi.v10i1.31499

FitzGerald, E., Jones, A., Kucirkova, N., & Scanlon, E. (2018). A Literature Synthesis of Personalised Technology-Enhanced Learning: What Works and Why. *Research in Learning Technology*, *26*(0). https://doi.org/10.25304/rlt.v26.2095

Fitzgerald, R., & Huijser, H. (2021). Co-creating Micro-credentials with Industry: Emerging Models for Higher Education.

Fitzgerald, R., & Huijser, H. (2021). Exploring Industry-University Partnerships in the Creation of Short Courses and Micro-Credentials. 340-344. https://doi.org/10.14742/ascilite2021.0149

Fleming, P. (2017). The Human Capital Hoax: Work, Debt and Insecurity in the Era of Uberization. *Organization Studies*, *38*(5), 691-709. https://doi.org/10.1177/0170840616686129

Fleming, T. (2018). Learning Careers and Transformative Learning. 1-13. https://doi.org/10.1163/9789004375475_001

Fleming, T. (2022a). Mezirow's Theory of Transformative Learning and Freire's Pedagogy. *Adult Education Critical Issues*, *2*(2), 7-19. https://doi.org/10.12681/haea.32302

Fleming, T. (2022b). Transforming Everyday Experience: Transformative Learning, Disorienting Dilemmas and Honneth's Theory of Recognition. *Journal for the Theory of Social Behaviour*, *52*(4), 563-578. https://doi.org/10.1111/jtsb.12355

Fleming, T. (2024). Transformative Learning Transforming Research. *European Journal for Research on the Education and Learning of Adults*, *15*(1), 13-26. https://doi.org/10.3384/rela.2000-7426.4841

Fletcher, J., Nicholas, K., & Davis, N. (2011). Supporting Adults to Address Their Literacy Needs Using E-Learning. *Journal of Open Flexible and Distance Learning*, *15*(1), 17-29. https://doi.org/10.61468/jofdl.v15i1.9

Flood, M., & Banks, J. (2021). Universal Design for Learning: Is It Gaining Momentum in Irish Education? *Education Sciences*, *11*(7), 341. https://doi.org/10.3390/educsci11070341

Flore, J., & Govender, I. (2021). Factors Influencing Secondary School Teachers' Beliefs and Intention to Accept Online Professional Development: An Empirical Study in Mauritius. *Universal Journal of Educational Research*, *9*(4), 880-890. https://doi.org/10.13189/ujer.2021.090422

Flores, G. d. l. C. (2022). Política Educativa Y Equidad: Desafíos en El México Contemporáneo. *Revista Latinoamericana De Estudios Educativos*, *52*(1), 71-92. https://doi.org/10.48102/rlee.2022.52.1.468

Florey, S., & Émery--Bruneau, J. (2023). Innover en Didactique De La Poésie : Quelle Pérennité ? *Tréma, 59*. https://doi.org/10.4000/trema.8289

Fordham, A. J. (2005). Using a Competency Based Approach in Nurse Education. *Nursing Standard, 19*(31), 41-48. https://doi.org/10.7748/ns2005.04.19.31.41.c3841

Formenti, L., & Hoggan-Kloubert, T. (2023). Transformative Learning as Societal Learning. *New Directions for Adult and Continuing Education, 2023*(177), 105-118. https://doi.org/10.1002/ace.20482

Fortman, J., Quintana, R. M., & Aguinaga, J. M. (2023). Conceptions of Time in Educational Technology. *Online Learning, 27*(4). https://doi.org/10.24059/olj.v27i4.4056

Fosnot, C. T. (2013). *Constructivism: Theory, perspectives, and practice* (2 ed.). Teachers College Press.

Foster, K., Jacob, S., Schapiro, S., Gupta, M., Hale, T., & Boardman, J. M. (2025). Transitioning Into Mental Health Nursing: A Scoping Review. *International Journal of Mental Health Nursing, 34*(5). https://doi.org/10.1111/inm.70144

Foster, M. K., Taylor, V., & Walker, J. L. (2021). Conceptualtheoretical Framework. https://doi.org/10.4337/9781789901122.00006

Fourie, E., & Coetzee, E. (2017). The Use of a Therapeutic Jurisprudence Approach to the Teaching and Learning of Law to a New Generation of Law Students in South Africa. *Potchefstroom Electronic Law Journal/Potchefstroomse Elektroniese Regsblad, 15*(1), 367-390. https://doi.org/10.17159/1727-3781/2012/v15i1a2468

Fourie, M., & Schlebusch, G. (2024). Cognitive Psychology and the Classroom: What Teachers Need to Know. *Journal of Educational Sciences & Psychology, 14 (76)*(1), 64-75. https://doi.org/10.51865/jesp.2024.1.08

Fragkos, K. C. (2016). Reflective Practice in Healthcare Education: An Umbrella Review. *Education Sciences, 6*(3), 27. https://doi.org/10.3390/educsci6030027

Fraher, A. L., & Gabriel, Y. (2014). Dreaming of Flying When Grounded: Occupational Identity and Occupational Fantasies of Furloughed Airline Pilots. *Journal of Management Studies, 51*(6), 926-951. https://doi.org/10.1111/joms.12081

Francia, A., Mariani, S., Adduce, G., Vecchiarelli, S., & Zambonelli, F. (2023). Digital Management of Competencies in Web 3.0: The C-Box® Approach. *Future Internet, 15*(11), 350. https://doi.org/10.3390/fi15110350

Francis, N., Pritchard, C., Prytherch, Z. C., & Rutherford, S. (2024). Making Teamwork Work: Enhancing Teamwork and Assessment in Higher Education. *Febs Open Bio, 15*(1), 35-47. https://doi.org/10.1002/2211-5463.13936

Franco-Tantuico, M. (2021). Active Learning: A Concept Analysis With Implications for Nursing Education. *Nursing Education Perspectives, 43*(4), 233-237. https://doi.org/10.1097/01.nep.0000000000000895

Frank, M. R., Moro, E., South, T., Rutherford, A., Pentland, A., Taska, B., & Rahwan, I. (2024). Network Constraints on Worker Mobility. *Nature Cities, 1*(1), 94-104. https://doi.org/10.1038/s44284-023-00009-1

Franky, G. A. (2009). Potencialidades Pedagógicas De Los Entornos De Simulación, Desde La Perspectiva De La Cognición Situada. *Tecné Episteme Y Didaxis Ted*(25). https://doi.org/10.17227/ted.num25-468

Frazier, M. W., & Ewing, S. (2009). Online Free Markets and Entrepreneurship in the Muslim World. *Economic Affairs, 29*(2), 28-35. https://doi.org/10.1111/j.1468-0270.2009.01890.x

Frehywot, S., & Vovides, Y. (2023). An Equitable and Sustainable Community of Practice Framework to Address the Use of Artificial Intelligence for Global Health Workforce Training. *Human Resources for Health, 21*(1). https://doi.org/10.1186/s12960-023-00833-5

Freire, J. L. (2024). Realidad Aumentada vs Realidad Virtual: Un Análisis Comparativo en La Educación Superior. *Reincisol, 3*(6), 6025-6048. https://doi.org/10.59282/reincisol.v3(6)6025-6048

Freire, P. (1970). *Pedagogy of the Oppressed.* Continuum.

Freire, P. (1970). *Pedagogy of the oppressed.* . Herder and Herder.

Freire, Y. M. O., Cueva, M. A. S., & Macías, M. E. P. (2021). Resiliencia Y Neurociencia. Reflexiones en El Contexto Educativo. *Revista Educare - Upel-Ipb - Segunda Nueva Etapa 2 0, 25*(2), 372-386. https://doi.org/10.46498/reduipb.v25i2.1494

French, H., Arias-Shah, A., Gisondo, C., & Gray, M. M. (2020). Perspectives: The Flipped Classroom in Graduate Medical Education. *Neoreviews, 21*(3), e150-e156. https://doi.org/10.1542/neo.21-3-e150

French, P., James-Gallaway, C. D., & Bohonos, J. W. (2021). Examining Intergroup Dialogue's Potential to Promote Social Justice in Adult Education. *Journal of Transformative Education, 20*(1), 44-61. https://doi.org/10.1177/15413446211000037

Friberg, J. C., & Hamilton, M. (2025). Learning and Development Through the Scholarship of Teaching and Learning. *New Directions for Adult and Continuing Education, 2025*(187), 32-39. https://doi.org/10.1002/ace.70005

Frich, J. C., Brewster, A. L., Cherlin, E., & Bradley, E. H. (2014). Leadership Development Programs for Physicians: A Systematic Review. *Journal of General Internal Medicine, 30*(5), 656-674. https://doi.org/10.1007/s11606-014-3141-1

Frkal, R. A. (2018). Integrating Care and Critical Reflection in Women's Leadership Development Programs. Vol. 38 (2018). https://doi.org/10.21423/awlj-v38.a329

Fuchs, A., Nekhendzy, V., & Greif, R. (2025). The Evolving Landscape of Airway Education in Anesthesia for Head and Neck Surgery. *Current Opinion in Anaesthesiology, 38*(6), 817-824. https://doi.org/10.1097/aco.0000000000001572

Fujiyoshi-Ito, A., Sato, K., Yumoto, Y., Sasaki, M., & Ogata, Y. (2021). A Concept Analysis of Psychological Safety: Further Understanding for Application to Health Care. *Nursing Open, 9*(1), 467-489. https://doi.org/10.1002/nop2.1086

Fuller, A., & Unwin, L. (2003). Learning as apprentices in the contemporary UK workplace: creating and managing expansive and restrictive participation. *Journal of Education and Work, 16*(4), 407-426.

Fuller, A., & Unwin, L. (2011). Apprenticeship as an Evolving Model of Learning. *Journal of Vocational Education and Training, 63*(3), 261-266.

Fulton, R., Fulton, D., & Kaplan, S. (2022). The Framework of Artificial Intelligence (FAI): Driving Triggers, State of the Art Over Time and Industry Adoption Influencers. *International Journal of Artificial Intelligence & Applications, 13*(3), 85-108. https://doi.org/10.5121/ijaia.2022.13307

Fute, A., Wan, X., Oubibi, M., & Bulugu, J. B. (2022). Adult Literacy Education and Reduction of Poverty in Tanzania: A Review of Policies and Their Implementation. *Journal of Education, 203*(4), 931-938. https://doi.org/10.1177/00220574221075204

Gagné, J. C. D., Carrejo, D. J., & Hf, H. (2026). Generative AI Ethics Toolkit for Clinical Prioritization in Nursing Education. *Nursing Ethics, 33*(4), 1169-1180. https://doi.org/10.1177/09697330261428627

Gagné, R. M. (1965). *The conditions of learning*. Holt, Rinehart and Winston.

Gagné, R. M., Briggs, L. J., & Wager, W. W. (1992). *Principles of instructional design* (4 ed.). Harcourt Brace Jovanovich.

Gal, A. (2024). A Hackathon as a Promoter of Environmental Citizenship Among Fifth-Grade Students—the Attitudes of Students, Educators, Ornithologists, and People Working in High-Tech. *Education Citizenship and Social Justice, 20*(2), 319-341. https://doi.org/10.1177/17461979241253556

Gal, I., Grotlüschen, A., Tout, D., & Kaiser, G. (2020). Numeracy, Adult Education, and Vulnerable Adults: A Critical View of a Neglected Field. *ZDM, 52*(3), 377-394. https://doi.org/10.1007/s11858-020-01155-9

Galaleldin, M., Anis, H., Dumond, P., & Knox, D. (2018). Scaffolding Strategies for Teaching Engineering Design in a Collaborative Project-Based Learning Environment. *Proceedings of the Canadian Engineering Education Association (Ceea)*. https://doi.org/10.24908/pceea.v0i0.10548

Galanaki, E., & Malafantis, K. D. (2022). Albert Bandura's Experiments on Aggression Modeling in Children: A Psychoanalytic Critique. *Frontiers in Psychology, 13*. https://doi.org/10.3389/fpsyg.2022.988877

Galkienė, A., & Monkevičienė, O. (2021). Preconditions of Transforming the Educational Process by Applying Inclusive Education Strategies: Theoretical Background. 1-21. https://doi.org/10.1007/978-3-030-80658-3_1

Gallo, V., Cotton, S., Theeke, L., Barnhart, C., Marino, L., Messer, J., Minor, A., & Smothers, A. (2022). Evidence of Diversity, Equity, and Inclusion in Nursing Syllabi: A Descriptive Analysis. *Journal of Nursing Education*, *61*(12), 665-671. https://doi.org/10.3928/01484834-20221003-02

Gallup, A. (2023). What We Know About Registered Apprenticeship: A Systematic Review and Synthesis of 30 Years of Empirical Research. *Economic Development Quarterly*, *38*(1), 25-39. https://doi.org/10.1177/08912424231196792

Gálvez-Palomeque, P. A., Chalco–Torres, L. E., & Galarza-Mora, W. G. (2023). Optimizando La Enseñanza en Educación Superior: Una Revisión De Estrategias Para La Inclusión De Estudiantes Con Necesidades Educativas Especiales. *Mqrinvestigar*, *7*(3), 4300-4316. https://doi.org/10.56048/mqr20225.7.3.2023.4300-4316

Galvin, T., & Geron, M. (2021). Building a Community of Practice Across an Institution. 323-344. https://doi.org/10.4018/978-1-7998-7106-4.ch017

Gamberini, L., & Pluchino, P. (2024). Industry 5.0: A Comprehensive Insight Into the Future of Work, Social Sustainability, Sustainable Development, and Career. *Australian Journal of Career Development*, *33*(1), 5-14. https://doi.org/10.1177/10384162241231118

Gandhi, R. (2022). Significance of New Education Policy (NEP) 2020 for Adult Education and Lifelong Learning Program. *International Journal of Management Technology and Social Sciences*, 79-95. https://doi.org/10.47992/ijmts.2581.6012.0179

Ganira, K. L., & Odundo, P. A. (2023). Experiential Learning Style Models on Implementation of Pre-Primary School Social Studies Curriculum: Systematic Review of Related Literature. *Asian Journal of Education and Social Studies*, *42*(1), 9-20. https://doi.org/10.9734/ajess/2023/v42i1905

Gao, P., Zhao, Y., Wang, F., & Lu, M. (2020). A New Teaching Pattern and Evaluation System of Knowledge Point Puzzle in Massive Online Computer Courses. 103-106. https://doi.org/10.1109/iccse49874.2020.9201775

Gao, T., & Chen, Y. (2023). Social and Affection: Design Procedures and Methods Curriculum Reform Based on Project-Based Learning. 341-351. https://doi.org/10.2991/978-2-38476-128-9_41

García-Romero, D., & Lalueza, J. L. (2019). Procesos De Aprendizaje en Aprendizaje-Servicio Universitario: Una Revisión Teórica. *Educación Xx1*, *22*(2). https://doi.org/10.5944/educxx1.22716

García, J. G. (2022). El Lenguaje Como Una Herramienta Psicológica Para El Aprendizaje a Partir Del Trabajo Colaborativo en La Educación Presencial Y a Distancia Dentro De La Zona De Desarrollo Próximo. *Dilemas Contemporáneos Educación Política Y Valores*. https://doi.org/10.46377/dilemas.v10i1.3255

García, J. M. V., Guerrero, W. J. L., & Alvarez, J. E. A. (2023). Enfoque Inclusivo en Lengua Y Literatura: Su Función Socializadora en Los Estudiantes Con Discapacidad. *Ciencia Latina Revista Científica Multidisciplinar*, *7*(5), 4456-4472. https://doi.org/10.37811/cl_rcm.v7i5.8045

García, L. M. P., Leiva, J.-J., & Santos-Villalba, M. J. (2022). El Diseño Universal Para El Aprendizaje en La Formación Digital Del Profesorado Desde Una Mirada Pedagógica Inclusiva. *Revista Latinoamericana De Educación Inclusiva*, *16*(2), 109-123. https://doi.org/10.4067/s0718-73782022000200109

García, M. E. L., Balcázar, P. A. L., & Gallardo, A. (2024). Aprendizaje Basado en Problemas en El Desarrollo De Habilidades De Pensamiento Crítico De Los Estudiantes Del Séptimo Ciclo De La Carrera De Pedagogía De Las Ciencias Experimentales Química Y Biología. *Ciencia Latina Revista Científica Multidisciplinar*, *8*(2), 5856-5879. https://doi.org/10.37811/cl_rcm.v8i2.10991

Gardner, A. C., Maietta, H. N., Gardner, P. D., & Perkins, N. (2021). Postsecondary Adult Learner Motivation: An Analysis of Credentialing Patterns and Decision Making Within Higher Education Programs. *Adult Learning*, *33*(1), 15-31. https://doi.org/10.1177/1045159520988361

Garnett, J. (2015). Recognition of Prior Learning: Opportunities and Challenges for Higher Education. *Journal of Work-Applied Management*, *7*(1), 28-37. https://doi.org/10.1108/jwam-10-2015-001

Garrett, S. D., Miller, V., & Gilbert, C. (2021). The Challenge of Culturally Responsive Teaching in Rural Community Colleges. *New Directions for Community Colleges*, *2021*(195), 107-118. https://doi.org/10.1002/cc.20470

Garrison. (2018). *Technical College Student Attitudes Toward Learning 21st Century Work Skills* Northeastern University].

Garth, B., Kirby, C., Nestel, D., & Brown, J. B. (2024). Becoming a General Practice Supervisor: A Longitudinal Multi-case Study Exploring Key Supportive Factors. *The Clinical Teacher*, *21*(4). https://doi.org/10.1111/tct.13738

Garza, R., & Harter, R. A. (2014). Perspectives From Pre-Service Mathematics and Science Teachers in an Urban Residency Program. *Education and Urban Society*, *48*(4), 403-420. https://doi.org/10.1177/0013124514533989

Gaster, B., & Largent, E. A. (2024). What Makes a Better Life for People Facing Dementia? Toward Dementia-Friendly Health and Social Policy, Medical Care, and Community Support in the United States. *The Hastings Center Report*, *54*(S1). https://doi.org/10.1002/hast.1554

Gaudet, J. C. (2020). Project George: An Indigenous Land-Based Approach to Resilience for Youth. *International Journal of Indigenous Health*, *16*(2). https://doi.org/10.32799/ijih.v16i2.31668

Gaudet, J. C., & Chilton, C. (2018). Milo Pimatisiwin Project. *International Journal of Indigenous Health*, *13*(1), 20-40. https://doi.org/10.32799/ijih.v13i1.30264

Gautam, D. K., & Basnet, D. (2020). Organizational Culture for Training Transfer: The Mediating Role of Motivation. *International Journal of Organizational Analysis*, *29*(3), 769-787. https://doi.org/10.1108/ijoa-04-2020-2147

Gautam, G. (2017). Understanding the Quintessence of Clinical Legal Education: Nepalese Experience. *Kathmandu School of Law Review*, 177-192. https://doi.org/10.46985/jms.v5i1.163

Gauvain, M. (2018). Collaborative Problem Solving: Social and Developmental Considerations. *Psychological Science in the Public Interest*, *19*(2), 53-58. https://doi.org/10.1177/1529100618813370

Gazi, Y. (2014). Issues Surrounding a Heutagogical Approach in Global Engineering Education. https://doi.org/10.18260/1-2--20722

Gedzyk-Nieman, S. A., & Derouin, A. (2025). Ensuring Access for All Nursing Students: Classroom Accommodations and Universal Design for Learning. *Creative Nursing*, *31*(4), 392-398. https://doi.org/10.1177/10784535251392561

Gelderen, B. v., & Guthadjaka, K. (2019). School Is Home, Home as School: Yolngu '˜On Country' and 'Through Country' Place-Based Education From Gäwa Homeland. *Australian and International Journal of Rural Education*, *29*(3), 56-75. https://doi.org/10.47381/aijre.v29i3.235

Gelpi-Trudo, R., Valle, M. d., Pagano, C. M., Andrés, M. L., Coni, A. G., & Juric, L. C. (2021). Adaptación Y Validación De Un Cuestionario De Compromiso Escolar en La Escuela Primaria. *Revista Iberoamericana De Evaluación Educativa*, *14*(1), 173-189. https://doi.org/10.15366/riee2021.14.1.010

Gendera, S., Alexander, J., Burton, J., Fisher, K., & Robinson, S. (2025). People With Disability Influencing Sustained Improvements to Employment. *Social Policy and Administration*, *60*(4), 585-595. https://doi.org/10.1111/spol.70022

George, M. (2010). Ethics and Motivation in Remedial Mathematics Education. *Community College Review*, *38*(1), 82-92. https://doi.org/10.1177/0091552110373385

Gerdin, G., Philpot, R., Smith, W., Schenker, K., Moen, K. M., Larsson, L., Linnér, S., & Westlie, K. (2021). Teaching for Student and Societal Wellbeing in HPE: Nine Pedagogies for Social Justice. *Frontiers in Sports and Active Living*, *3*. https://doi.org/10.3389/fspor.2021.702922

Gerken, M. (2016). *How Do Employees Learn at Work ?* Maastricht University].

Gerlach, J. D., & Reinagel, T. P. (2016). Experiential Learning in MPA Programs: A Case for Complementarity Between Internship and Service Learning Requirements. *Ps Political Science & Politics*, *49*(1), 132-138. https://doi.org/10.1017/s1049096515001158

Gewurtz, R., Coman, L., Dhillon, S., & Jung, B. (2016). Problem-Based Learning and Theories of Teaching and Learning in Health Professional Education. *Journal of Perspectives in Applied Academic Practice*, *4*(1). https://doi.org/10.14297/jpaap.v4i1.194

Ghamrawi, N. A. R., Ghamrawi, N., & Shal, T. (2025). Reimagining Teacher Preparation: An Andragogical Lens on Transforming Preservice Education. *Journal of Adult and Continuing Education*, *32*(2), 365-390. https://doi.org/10.1177/14779714251388277

Ghobrini, R. E. A., & Sarnou, H. (2022). Migrating vs. Decentralizing MOOC-based E-Content to Teaching MA Language and Communication Students. https://doi.org/10.31235/osf.io/tx28r

Ghori, U. (2011). Teaching International Trade Law to the Google Generation: Perspectives on Teaching a Diverse Audience. *SSRN Electronic Journal*. https://doi.org/10.2139/ssrn.2717240

Ghosh, L., & Ravichandran, R. (2024). Emerging Technologies in Vocational Education and Training. *Journal of Digital Learning and Education*, *4*(1), 41-49. https://doi.org/10.52562/jdle.v4i1.975

Ghoulam, K., Bouikhalene, B., Babori, A., & Falih, N. (2024). Exploring the Impact of Mobile Devices in Electronics E-Learning: A Case Study Evaluating the Effectiveness of Mobile Learning Applications in the Field of Electronics and Sensors. *Advances in Mobile Learning Educational Research*, *4*(2), 1058-1072. https://doi.org/10.25082/amler.2024.02.001

Giambo, D. A., & Garrido, L. E. (2024). Cross-Cultural Course and Service-Learning Study-Away: A Means to Transformative Learning. *Journal of Transformative Education*, *23*(1), 98-118. https://doi.org/10.1177/15413446241261970

Gibbs, L., Linden, K., Teakel, S. L., & Dell, E. Y. (2023). Initiatives Supporting Student Retention in Early Childhood Teacher Education Qualifications. *Australasian Journal of Early Childhood*, *49*(1), 17-31. https://doi.org/10.1177/18369391231220050

Gil-Jaurena, I. (2013). Open Praxis, Volumen 5 Issue 2. *Open Praxis*, *5*(2). https://doi.org/10.5944/openpraxis.5.2.77

Gil-Jaurena, I. (2020). Open Praxis Vol. 12 Issue 4. *Open Praxis*, *12*(4), 437. https://doi.org/10.5944/openpraxis.12.4.1208

Gill, R. L. (2016). Bridging the Knowledge Gap Between Study and Professional Practice: An Australian Case Example. *International Journal of Academic Research in Progressive Education and Development*, *5*(1). https://doi.org/10.6007/ijarped/v5-i1/2010

Ginty, C., Killoran, B., & Clinch, G. (2023). Building Lifelong Learning Participation With the ATU Higher Education 4.0 Project: Recognising Prior Learning (RPL) and Providing Career and Learning Pathways With MyCareerPath.ie. *Ubiquity Proceedings*. https://doi.org/10.5334/uproc.93

Girei, M. M., Abdullah, Z., & Hammanjoda, B. (2020). Review of Related Literature on Problem-Based Learning and Cognitive Apprenticeship for Instruction in Agriculture as Vocational Education. *International Journal of Academic Research in Progressive Education and Development*, *9*(2). https://doi.org/10.6007/ijarped/v9-i2/7734

Glavič, P. (2020). Identifying Key Issues of Education for Sustainable Development. *Sustainability*, *12*(16), 6500. https://doi.org/10.3390/su12166500

Glikman, V. (2021). Il Était Une Fois... La Formation À Distance. *Médiations Et Médiatisations*(6), 3-11. https://doi.org/10.52358/mm.vi6.228

Gnanadass, E., & Merriweather, L. R. (2024). To Transgress or Not?: Critical Theory as a Framework for Change. *New Directions for Adult and Continuing Education*, *2024*(184), 56-64. https://doi.org/10.1002/ace.20544

Gobinath, T. (2024). Society 5.0 the Next Revolution in Human Progress. *Shanlax International Journal of Arts Science and Humanities*, *11*(3), 61-68. https://doi.org/10.34293/sijash.v11i3.6898

Goda, Y., Sudo, K., & Yamada, M. (2025). First Steps in the Workplace: Why Do Some New Graduates Grow Quickly While Others Advance Gradually? *Sage Open*, *15*(4). https://doi.org/10.1177/21582440251394373

Godinho, S. D. S., Rivela, C. V., Medrado, S. O., Marmo, J., & Lanuque, A. (2021). Educación Inclusiva Y Accesibilidad Digital. *Revista Científica Arbitrada De La Fundación Menteclara*, *6*. https://doi.org/10.32351/rca.v6.249

Godoy, C. d. S. O., & J., C. M. (2024). The Educational Potential of Cyberfeminisms: An Intersectional Analysis of @Coletivoandorinha, @Feministasemmovimento and @feministas.pt Profiles. *International Conference on Gender Research*, *7*(1), 431-439. https://doi.org/10.34190/icgr.7.1.2173

Goldman, J. D. G., & Bradley, G. L. (2001). Programs for Retaining High School Re-Enterers for Credentialling and Employment. *Curriculum and Teaching*, *16*(2), 81-103. https://doi.org/10.7459/ct/16.2.07

Gonzales, A. M. (2020). Empowering the Lowland Indigenous Community Through Child Health and Nutrition Program in Occidental Mindoro Province, Philippines. *Asean Journal of Community Engagement*, *4*(1). https://doi.org/10.7454/ajce.v4i1.1081

Gonzalez-Nuñez, F. P., Llanos, J. F. S., & Preciado, M. A. C. (2024). Aprendizaje Experiencial Basado en Simuladores De Negocios Sobre La Satisfacción Y El Desempeño Académico. *Revista Docentes 2 0*, *17*(1), 111-122. https://doi.org/10.37843/rted.v17i1.440

González-Argote, J., Lepez, C. O., Castillo-González, W., Bonardi, M. C., Cano, C. A. G., & Vitón-Castillo, A. A. (2023). Use of Real-Time Graphics in Health Education: A Systematic Review. *Eai Endorsed Transactions on Pervasive Health and Technology*, *9*, e3. https://doi.org/10.4108/eetpht.v9i.3209

González, L. F. M., & Quiroz, V. G. (2019). Instructional Design in Online Education: A Systemic Approach. *European journal of education*, *2*(3), 43-52. https://doi.org/10.26417/ejed.v2i3.p64-73

Goodburn, C. (2020). Growing Up in (And Out Of) Shenzhen: The Longer-Term Impacts of Rural-Urban Migration on Education and Labor Market Entry. *The China Journal*, *83*, 129-147. https://doi.org/10.1086/705540

Goodwill, A. M., & Chen, S. H. A. (2021). The Science of Lifelong Learning. https://doi.org/10.31234/osf.io/juefx

Google Arts and Culture. (2025). Google Arts & Culture. https://artsandculture.google.com

Gordon, J. (2015). Glimpsing the Future in the Past: VET in Europe. *European journal of education*, *50*(4), 440-460. https://doi.org/10.1111/ejed.12151

Gorey, C. (2021). How Effectively Is the Cadet School Able to Develop Leaders of Character? *Journal of Military History and Defence Studies*, *2*(1). https://doi.org/10.33232/jmhds.2.1.21

Gormally, S., Coburn, A., & Beggan, E. (2021). Idealistic Assertions or Realistic Possibilities in Community and Youth Work Education. *Education Sciences*, *11*(9), 561. https://doi.org/10.3390/educsci11090561

Gorp, K. V. (2026). Making All Repertoires Count: Re-envisioning TBLT Through Critical Multilingual Language Awareness. *Modern Language Journal*, *110*(2), 657-681. https://doi.org/10.1111/modl.70051

Gorshenin, A. (2018). Toward Modern Educational IT-ecosystems: From Learning Management Systems to Digital Platforms. 1-5. https://doi.org/10.1109/icumt.2018.8631229

Gottron, D., Williams, P., & Jiang, M. (2024). Hope Mediates the Relationship Between Childhood Adversity and Academic Resilience Among Appalachian Young Adults. *Theory & Practice in Rural Education*, *14*(1), 49-72. https://doi.org/10.3776/tpre.2024.v14n1p49-72

Gouëdard, P., Pont, B., Hyttinen, S., & Huang, P. (2020). Curriculum Reform. https://doi.org/10.1787/efe8a48c-en

Gould, D. J., Sawarynski, K. E., & Mohiyeddini, C. (2022). Academic Management in Uncertain Times: Shifting and Expanding the Focus of Cognitive Load Theory During COVID-19 Pandemic Education. *Frontiers in Psychology*, *13*. https://doi.org/10.3389/fpsyg.2022.647904

Gouthro, P., & Holloway, S. M. (2013). Preparing Teachers to Become Lifelong Learners: Exploring the Use of Fiction to Develop Multiliteracies and Critical Thinking. *Language and Literacy*, *15*(3), 50. https://doi.org/10.20360/g2588t

Gouthro, P. A., & Holloway, S. M. (2023). Critical Social Theory, Inclusion, and a Pedagogy of Hope. *European Journal for Research on the Education and Learning of Adults*, *14*(3), 325-341. https://doi.org/10.3384/rela.2000-7426.4662

Gouvernement français. (2023). Bibliographie Indicative. 197-200. https://doi.org/10.4000/books.pressesenssib.17958

Govender, R. (2024). A Reflection on the Use of Artificial Intelligence Tools for the Meaningful Learning of Mathematics. *Pythagoras*, *45*(1). https://doi.org/10.4102/pythagoras.v45i1.831

Govender, R., & Booyse, C. (2024). A Heuristic Determination of Innovative Pedagogical Practices for an Online or Blended Context. https://doi.org/10.5772/intechopen.1005342

Grabarski, M. K., Mouratidou, M., & Fellows, I. (2023). The Evolution of Learning and Technological Innovation. 58-76. https://doi.org/10.4018/978-1-6684-7442-6.ch004

Grabinger, S., Dunlap, J. C., & Duffield, J. A. (2011). Rich Environments for Active Learning in Action: Problem-Based Learning. *Research in Learning Technology*, *5*(2). https://doi.org/10.3402/rlt.v5i2.10558

Graebe, J. (2025). Alternative Credentials in ANCC Nursing Continuing Professional Development Accreditation™: Advancing Workforce Readiness, Equity, and Lifelong Learning. *The Journal of Continuing Education in Nursing*, *56*(9), 353-357. https://doi.org/10.3928/00220124-20250814-02

Graebe, J., & McIntyre-Hite, L. (2024). Call to Action: Leveraging American Nurses Credentialing Center Nursing Continuing Professional Development Credit for Prior Learning—Part 2. *The Journal of Continuing Education in Nursing*, *55*(9), 416-420. https://doi.org/10.3928/00220124-20240821-02

Graebe, S. (2025). Alternative Credentials and Workforce Accessibility.

Graebe, S., & McIntyre-Hite, L. (2024). Recognition of Learning Through Micro-credentials and Alternative Credentials.

Graf, L. (2013). The Hybridization of Vocational Training and Higher Education in Austria, Germany, and Switzerland. https://doi.org/10.2307/j.ctvbkjt2g

Graf, L., Marques, M., & Lambrechts, A. A. (2025). Skills Development for the Twin Transition: Building Transnational Skills Ecosystems Through Experimentalist Governance. *Regulation & Governance*, *20*(2), 509-523. https://doi.org/10.1111/rego.70076

Graham, D. N. A. (2019). Benefits of Online Teaching for Onground Teaching at a Historically Black Colleges and Universities. *Online Learning*, *23*(1). https://doi.org/10.24059/olj.v23i1.1435

Grandy, G., & Mavin, S. (2020). Informal and Socially Situated Learning: Gendered Practices and Becoming Women Church Leaders. *Gender in Management an International Journal*, *35*(1), 61-75. https://doi.org/10.1108/gm-03-2019-0041

Gravani, M. N., Hatzopoulos, P., & Chinas, C. (2019). Adult Education and Migration in Cyprus: A Critical Analysis. *Journal of Adult and Continuing Education*, *27*(1), 25-41. https://doi.org/10.1177/1477971419832896

Gray, J. A. (2018). Leadership Coaching and Mentoring: a Research-Based Model for School Partnerships. *International Journal of Education Policy and Leadership*, *13*(12). https://doi.org/10.22230/ijepl.2018v13n12a844

Gray, J. A., Walker, A., Zimmerman, S. A., & Dickson, J. S. (2022). Leadership-Focused Coaching in Action: An Approach to Continuous Improvement and Supporting Public Schools. *International Journal of Education Policy and Leadership*, *18*(1). https://doi.org/10.22230/ijepl.2022v18n1a1159

Gray, S. W. (2006). Executive Coaching: Towards a Dynamic Alliance of Psychotherapy and Transformative Learning Processes. *Management learning*, *37*(4), 475-497. https://doi.org/10.1177/1350507606070221

Green, W. J. (2020). Foreword. xxxvii-xxxviii. https://doi.org/10.4102/aosis.2020.bk215.0f

Greer, T. W. (2013). Facilitating Successful Re–entries in the United States: Training and Development for Women Returners. *New Horizons in Adult Education and Human Resource Development*, *25*(3), 41-61. https://doi.org/10.1002/nha3.20030

Gregg, N., & Nelson, J. M. (2017). Empirical Studies on the Writing Abilities of Adolescents and Adults With Learning Difficulties. 73-95. https://doi.org/10.1163/9789004346369_006

Griffiths, A. J., Hurley-Hanson, A. E., Giannantonio, C. M., Nash, A. M., Walrod, W., Walrod, P., Torres, R., & Delgado, R. (2024). Identifying Critical Employability Skills for Employment Success of Autistic Individuals: A Content Analysis of Job Postings. *Journal of Research in Special Educational Needs*, *24*(3), 566-577. https://doi.org/10.1111/1471-3802.12651

Griffiths, D., Burgos, D., & Aceto, S. (2024). Open Education and Alternative Digital Credentials in Europe. *The International Review of Research in Open and Distributed Learning*, *25*(1), 89-108. https://doi.org/10.19173/irrodl.v25i1.7412

Griffiths, K. (2015). Personal Coaching: Reflection on a Model for Effective Learning. *Journal of Learning Design*, *8*(3). https://doi.org/10.5204/jld.v8i3.251

Grigg, R., Lewis, H., Morse, M., & Crick, T. (2024). Rethinking Student Teachers' Professional Learning in Wales: Promoting Reflection-in-action. *The Curriculum Journal*, *35*(4), 690-705. https://doi.org/10.1002/curj.271

Grigoroviča, E., Slavinska, A., Jansone-Ratinika, N., & Bahs, G. (2022). Aspects of Human Capital Management of Healthcare Workforce in the Context of Lifelong Learning: A Rapid Review. *Society Integration Education Proceedings of the International Scientific Conference*, *1*, 753-766. https://doi.org/10.17770/sie2022vol1.6865

Grummell, B. (2023). Maintaining Deep Roots. *European Journal for Research on the Education and Learning of Adults*, *14*(1), 145-161. https://doi.org/10.3384/rela.2000-7426.4322

Grus, C. L., Lagbo, V., & Rozensky, R. H. (2025). Applying Principles of Adult Learning and Andragogy to the Design of Continuing Education Programs in Psychology. *Practice Innovations*, *10*(2), 108-119. https://doi.org/10.1037/pri0000268

Grzegorczyk, G. (2018). Dialogue and Language as Factors Contributing to Transformative Learning in Academic Tutoring. *Psychology of Language and Communication*, *22*(1), 164-186. https://doi.org/10.2478/plc-2018-0007

Guan, X., Feng, X., & Islam, A. Y. M. A. (2023). The Dilemma and Countermeasures of Educational Data Ethics in the Age of Intelligence. *Humanities and Social Sciences Communications*, *10*(1). https://doi.org/10.1057/s41599-023-01633-x

Gudonienė, D., & Rutkauskienė, D. (2019). Virtual and Augmented Reality in Education. *Baltic Journal of Modern Computing*, *7*(2). https://doi.org/10.22364/bjmc.2019.7.2.07

Guercio, J. M. (2018). The Importance of a Deeper Knowledge of the History and Theoretical Foundations of Behavior Analysis: 1863–1960. *Behavior Analysis Research and Practice*, *18*(1), 4-15. https://doi.org/10.1037/bar0000123

Guerra, A. (2017). Integration of Sustainability in Engineering Education. *International Journal of Sustainability in Higher Education*, *18*(3), 436-454. https://doi.org/10.1108/ijshe-02-2016-0022

Guerrin, B. (2012). Albert Bandura Et Son Œuvre. *Recherche en Soins Infirmiers*, *N°108*(1), 106-116. https://doi.org/10.3917/rsi.108.0106

Guertin, M.-A., Poirier-Ghys, G., Reed, M. G., Dupras, J., & Sauriol, G. R. (2019). L'éducation Relative À L'environnement Dans La Réserve De Biosphère Du Mont Saint-Hilaire : Observations Et Tensions. *Éducation Relative À L Environnement*, *Volume 15 - 1*. https://doi.org/10.4000/ere.3810

Guile, D., & Lahiff, A. (2016). Apprenticeship for 'Liquid Life': Learning in Contingent Work Conditions for Contingent Employment. *Vocations and Learning*, *10*(2), 157-175. https://doi.org/10.1007/s12186-016-9166-3

Guillemette, F. (2020). Passer Du Modèle Transmissif À Un Modèle De L'apprentissage Guidé. *Enjeux Et Société Approches Transdisciplinaires*, *7*(2), 42-73. https://doi.org/10.7202/1073360ar

Guimarães, P. (2017). The Usefulness of Adult Education: Lifelong Learning in the European Union and the Portuguese Public Policy. *Andragoška Spoznanja*, *23*(4), 35-50. https://doi.org/10.4312/as.23.4.35-50

Guimarães, P., & Mikulec, B. (2020a). The Paradox of Non-formal and Informal Learning Validation: European Policy Perspectives. *European Journal for Research on the Education and Learning of Adults*, *11*(2), 165-180.

Guimarães, P., & Mikulec, B. (2020b). Paradox of Utilitarian Recognition of Prior Learning. *European Journal for Research on the Education and Learning of Adults*, *12*(1), 109-122. https://doi.org/10.3384/rela.2000-7426.ojs1479

Güler, A. T., & Esen, M. (2021). EFL Teachers' Perceptions on Learner Autonomy in Online Instruction During the Lockdown Period. *English as a Foreign Language International Journal*, *25*(6), 6-25. https://doi.org/10.56498/802562021

Güllüpınar, F. (2012). The Unequal Structure of the German Education System: Structural Reasons for Educational Failures of Turkish Youth in Germany. *Spaces and Flows an International Journal of Urban and Extraurban Studies*, *2*(2), 93-112. https://doi.org/10.18848/2154-8676/cgp/v02i02/53849

Gumede, P., Beharry-Ramraj, A., Nzimande, S. B., & Ramson, S. M. (2023). Digital Transformation in South African Public Higher Education Institutions – A Pedagogical Approach. *E-Journal of Humanities Arts and Social Sciences*, 141-152. https://doi.org/10.38159/ehass.202341413

Gunnlaugson, O. (2007). Shedding Light on the Underlying Forms of Transformative Learning Theory. *Journal of Transformative Education*, *5*(2), 134-151. https://doi.org/10.1177/1541344607303526

Guo, H. (2018). Application of a Computer-Assisted Instruction System Based on Constructivism. *International Journal of Emerging Technologies in Learning (Ijet)*, *13*(04), 33. https://doi.org/10.3991/ijet.v13i04.8468

Guo, S. (2021). Transnational Migration, Refugee Studies and Lifelong Learning. *International Review of Education*, *67*(6), 723-731. https://doi.org/10.1007/s11159-022-09936-y

Guo, Z., & Sheffield, J. (2006). Habermasian Inquiring System: Toward a General Framework for Knowledge Management Research. 162c-162c. https://doi.org/10.1109/hicss.2006.201

Gupta, N., Ali, K., Jiang, D., Fink, T., & Du, X. (2024). Learner Agency in Undergraduate Dental Education- A Scoping Review Through Self-Regulated and Self-Directed Learning. https://doi.org/10.21203/rs.3.rs-4947130/v1

Gutiérrez-Gómez, P. (2023). Aprendizaje Desarrollador De Competencias en Ingeniería Civil: Abordando La Complejidad Y La Transversalidad. *593 Digital Publisher Ceit*, *8*(4), 471-498. https://doi.org/10.33386/593dp.2023.4.1906

Gutiérrez, M., Tomás, J. M., & Alberola, S. (2018). Apoyo Docente, Compromiso Académico Y Satisfacción Del Alumnado Universitario. *Estudios Sobre Educación*, *35*, 535-555. https://doi.org/10.15581/004.35.535-555

Guzón, J. L., Alonso, F. G., & Cervantes, A. O. (2021). Aprendizaje Servicio en Educación Superior Entre España Y México. Hacia Los ODS. *Alteridad*, *17*(1), 76-88. https://doi.org/10.17163/alt.v17n1.2022.06

Hachem, H., & Formosa, M. (2023). Barriers to Learning at a U3A in Lebanon: A Structurationist Perspective. *International Journal of Population Studies*, *9*(3), 1. https://doi.org/10.36922/ijps.375

Haddad, G., & O'Connor, D. (2022). Developing Players for Athlete Leadership Groups in Professional Football Teams: Qualitative Insights From Head Coaches and Athlete Leaders. *Plos One*, *17*(8), e0271093. https://doi.org/10.1371/journal.pone.0271093

Hagedorn, G., Mietchen, D., Morris, R., Agosti, D., Penev, L., Berendsohn, W. G., & Hobern, D. (2011). Creative Commons Licenses and the Non-Commercial Condition: Implications for the Re-Use of Biodiversity Information. *Zookeys*, *150*, 127-149. https://doi.org/10.3897/zookeys.150.2189

Hagen, M., & Park, S. (2016). We Knew It All Along! Using Cognitive Science to Explain How Andragogy Works. *European journal of training and development*, *40*(3), 171-190. https://doi.org/10.1108/ejtd-10-2015-0081

Hager, P., & Smith, E. (2004). The Inescapability of Significant Contextual Learning in Work Performance. *London Review of Education*, *2*(1). https://doi.org/10.1080/1474846042000177465

Hägg, G., & Kurczewska, A. (2020). Guiding the Student Entrepreneur – Considering the Emergent Adult Within the Pedagogy–andragogy Continuum in Entrepreneurship Education. *Education + Training*, *62*(7/8), 759-777. https://doi.org/10.1108/et-03-2020-0069

Hahambu, C. M. E., Brownlee, J., & Petriwskyj, E. (2012). Elementary Teacher Education in Papua New Guinea: Towards a Culturally Connected Perspective of Teaching. *Australian Journal of Teacher Education*, *37*(4). https://doi.org/10.14221/ajte.2012v37n4.4

Haidar, R., & Tassis, A. (2025). Power, Pedagogy, and Algorithms: An Exploratory Study of EAP Instructors' Critical Engagement With Generative AI. https://doi.org/10.21203/rs.3.rs-7014667/v1

Haidari, S. M., Karakuş, F., & KoÇOĞLu, A. (2020). Üstün Yeteneklilerin Eğitiminde Güvenli Öğrenme Ortamına İlişkin Öğrenci Ve Öğretmen Görüşleri. *Kuramsal Eğitimbilim*, *13*(2), 311-333. https://doi.org/10.30831/akukeg.581632

Hajian, S. (2019). Transfer of Learning and Teaching: A Review of Transfer Theories and Effective Instructional Practices. *Iafor Journal of Education*, *7*(1). https://doi.org/10.22492/ije.7.1.06

Hajirasouli, A., & Banihashemi, S. (2022). Augmented Reality in Architecture and Construction Education: State of the Field and Opportunities. *International Journal of Educational Technology in Higher Education*, *19*(1). https://doi.org/10.1186/s41239-022-00343-9

Hakimi, L., Eynon, R., & Murphy, V. A. (2021). The Ethics of Using Digital Trace Data in Education: A Thematic Review of the Research Landscape. *Review of Educational Research*, *91*(5), 671-717. https://doi.org/10.3102/00346543211020116

Halder, D., Bastaki, E. M. A., Suleymanova, S., Muhammad, N., & Purushothaman, A. (2024). Utilizing Open Educational Practices to Support Sustainable Higher Education in the United Arab Emirates. *Asian Association of Open Universities Journal*, *19*(2), 117-134. https://doi.org/10.1108/aaouj-07-2023-0086

Hall, L., Maughan, C., Wilkes, M., Thorpe, T., Forrest, J., & Harrison, A. (2015). Swimming Not Drowning – Resilience as a Key Determinant of Success for Aboriginal and Torres Strait Islander Pre-Tertiary Students. *Journal for Multicultural Education*, *9*(3), 159-173. https://doi.org/10.1108/jme-12-2014-0045

Halman, M., Baker, L., & Ng, S. (2017). Using Critical Consciousness to Inform Health Professions Education: A literature Review. *Perspectives on medical education*, *6*(1), 12-20. https://doi.org/10.1007/s40037-016-0324-y

Halpern, R., & Tucker, C. (2015). Leveraging Adult Learning Theory With Online Tutorials. *Reference Services Review*, *43*(1), 112-124. https://doi.org/10.1108/rsr-10-2014-0042

Halupa, C. (2015). Pedagogy, Andragogy, and Heutagogy. 143-158. https://doi.org/10.4018/978-1-4666-8571-0.ch005

Hameed, B. M. Z., Somani, S., Keller, E. X., Balamanigandan, R., Mahapatra, S., Pietropaolo, A., Tonyalı, Ş., Juliebø-Jones, P., Naik, N., Mishra, D. K., Kumar, S., Chlosta, P., & Somani, B. (2022). Application of Virtual Reality, Augmented Reality, and Mixed Reality in Endourology and Urolithiasis: An Update by YAU Endourology and Urolithiasis Working Group. *Frontiers in Surgery*, *9*. https://doi.org/10.3389/fsurg.2022.866946

Hameleers, R. J., Terlouw, A. J., Laan, R., & Pol, M. v. d. (2022). Investing in Student Wellbeing: A Design-Based Educational Research Approach to Develop a Longitudinal Proactive Program to Improve Wellbeing of Medical Students. https://doi.org/10.21203/rs.3.rs-2317703/v1

Hamilton, R. D. (2023). The Fight for Social Justice – Lalage Bown: A Lifetime's Mission. 29-40. https://doi.org/10.36253/979-12-215-0253-4.07

Hamm, L., Maston, M., McLoughlin, J. G., & Smith, J. (2021). Educators Responding to Rapid Demographic Change in New Brunswick: "It Is Not Inclusion if They Are Just Sitting in the Desk". *Alberta Journal of Educational Research*, *67*(2), 178-201. https://doi.org/10.55016/ojs/ajer.v67i2.68630

Hammad, R., Khan, Z., Safieddine, F., & Ahmed, A. (2020). A Review of Learning Theories and Models Underpinning Technology-Enhanced Learning Artefacts. *World Journal of Science Technology and Sustainable Development*, *17*(4), 341-354. https://doi.org/10.1108/wjstsd-06-2020-0062

Hammond, C., & Feinstein, L. (2005). The Effects of Adult Learning on Self-Efficacy. *London Review of Education*, *3*(3). https://doi.org/10.1080/14748460500372754

Han, I. (2024). Evaluating the Quality of Research on Problem-Based Learning in Korean English Classrooms. *Forum for Linguistic Studies*, *6*(3), 813-823. https://doi.org/10.30564/fls.v6i3.6637

Han, S. h., Oh, E., Kang, S., Lee, S., & Park, S. H. (2022). The Mediating Role of Informal Learning on Job Performance: The Work-Learning Dual System in South Korea. *Journal of Workplace Learning*, *34*(5), 437-454. https://doi.org/10.1108/jwl-07-2021-0101

Hanshaw, G., & Hanson, J. (2019). Using Microlearning and Social Learning to Improve Teachers' Instructional Design Skills: A Mixed Methods Study of Technology Integration in Teacher Professional Development. *International Journal of Learning and Development*, *9*(1), 145. https://doi.org/10.5296/ijld.v9i1.13713

Hanson, C. L. (2013). Exploring Dimensions of Critical Reflection in Activist–Facilitator Practice. *Journal of Transformative Education*, *11*(1), 70-89. https://doi.org/10.1177/1541344613488834

Hanson, C. L., & Fletcher, A. J. (2021). Beyond the Trinity of Gender, Race, and Class: Further Exploring Intersectionality in Adult Education. *European Journal for Research on the Education and Learning of Adults*, *12*(2), 135-148. https://doi.org/10.3384/rela.2000-7426.3360

Harden, K., Pflepsen, A., & King, S. (2018). Relationships Between Coach Support and Teachers' Adoption of New Instructional Practices: Findings From the Nigeria Reading and Access Research Activity (RARA). 169-204. https://doi.org/10.3768/rtipress.2018.bk.0022.1809.6

Hardin-Ramanan, S., Gopee, S., Rowtho, V., & Charoux, O. (2020). Graduate Work-Readiness in Mauritius: A Multi-Stakeholder Approach. *Journal of Teaching and Learning for Graduate Employability*, *11*(1), 93-109. https://doi.org/10.21153/jtlge2020vol11no1art937

Hardin-Ramanan, S., Soupramanien, L. D. B., & DeLapeyre, D. (2018). Project #NuKapav: A Mauritian Service-Learning Case Study. *Equality Diversity and Inclusion an International Journal*, *37*(2), 167-181. https://doi.org/10.1108/edi-01-2017-0008

Harnish, D., & Wilke-Schnaufer, J. (1998). ≪p>Work-Based Learning in Occupational Education And Training</P&gt. *The Journal of Technology Studies*, *24*(2), 21-30. https://doi.org/10.21061/jots.v24i2.a.3

Harper, G. J., Cameron, R., & Edwards, C. (2024). A Framework for the Facilitation of Accelerated Leadership and Management Capability Development in the Workplace. *Humanities and Social Sciences Communications*, *11*(1). https://doi.org/10.1057/s41599-024-02758-3

Harreveld, B., & Singh, M. (2009). Contextualising Learning at the Education-training-work Interface. *Education + Training*, *51*(2), 92-107. https://doi.org/10.1108/00400910910941264

Harris, A. (2008). Leading Innovation and Change: Knowledge Creation by Schools for Schools. *European journal of education*, *43*(2), 219-228. https://doi.org/10.1111/j.1465-3435.2008.00343.x

Harris, D. B., & Roter, D. (2024). Profound Love and Dialogue: Paulo Freire and Liberation Education. *HLRP Health Literacy Research and Practice*, *8*(3). https://doi.org/10.3928/24748307-20240613-02

Hart, J., & Bone, E. (2022). Authenticity Ahead of Interdisciplinarity – A Scoping Review of Student Experiences in Interdisciplinary Science Projects. *Journal of Teaching and Learning for Graduate Employability*, *13*(1), 109-126. https://doi.org/10.21153/jtlge2022vol13no1art1487

Hartman-Caverly, S. (2019). Human Nature Is Not a Machine: On Liberty, Attention Engineering, and Learning Analytics. *Library Trends*, *68*(1), 24-53. https://doi.org/10.1353/lib.2019.0029

Hartree, A. (1984). Malcolm Knowles' theory of andragogy: A critique. *International Journal of Lifelong Education*, *3*(3), 203-210.

Hashim, N., & Hamidon, Z. (2022). Blended Learning in Technical and Vocational Education and Training (TVET) Training Institute. *International Journal of Academic Research in Progressive Education and Development*, *11*(1). https://doi.org/10.6007/ijarped/v11-i1/12343

Hashim, N. C., Majid, N. A. A., Arshad, H., Hashim, H., & Alyasseri, Z. A. A. (2022). Mobile Augmented Reality Based on Multimodal Inputs for Experiential Learning. *Ieee Access*, *10*, 78953-78969. https://doi.org/10.1109/access.2022.3193498

Haslam, S. A., Haslam, C., & Steffens, N. K. (2025). A Badge, a Part of My History and Also a Root: A Longitudinal Qualitative Exploration of the Social Identity Model of Identity Change in Retired Elite Athletes. *British Journal of Social Psychology*, *64*(3). https://doi.org/10.1111/bjso.70005

Hassam, Z., Gulzar, N., & Latif, A. (2022). The Role of Educators in Supporting the Mental Well-Being of Postgraduate Pharmacist Distance Learners. *American Journal of Pharmaceutical Education*, *86*(9), ajpe8847. https://doi.org/10.5688/ajpe8847

Hassan, J., Devi, A., & Ray, B. (2022). Virtual Laboratories in Tertiary Education: Case Study Analysis by Learning Theories. *Education Sciences*, *12*(8), 554. https://doi.org/10.3390/educsci12080554

Hassan, M. A., Zardari, S., Farooq, M. U., Alansari, M. M., & Nagro, S. (2024). Systematic Analysis of Risks in Industry 5.0 Architecture. *Applied Sciences*, *14*(4), 1466. https://doi.org/10.3390/app14041466

Hassan, M. U., Murtaza, A., & Rashid, K. (2024). Redefining Higher Education Institutions (<scp>HEIs</Scp>) in the Era of Globalisation and Global Crises: A Proposal for Future Sustainability. *European journal of education*, *60*(1). https://doi.org/10.1111/ejed.12822

Hassi, M.-L., & Laursen, S. L. (2015). Transformative Learning. *Journal of Transformative Education*, *13*(4), 316-340. https://doi.org/10.1177/1541344615587111

Hawse, S., & Wood, L. (2019). Designing Workplace Induction Programs to Support the Transition of New-Career Engineers to Practice. *Higher Education Skills and Work-Based Learning*, *9*(1), 18-29. https://doi.org/10.1108/heswbl-02-2018-0014

He, L., & Chen, Z. L. (2026). Powered Auditory Training and Its Role in Strengthening Musical Memory: A Systematic Review in Higher Music Education. *European journal of education*, *61*(2). https://doi.org/10.1111/ejed.70569

He, M. (2024). Using Large Language Models to Enhance Genre-Based English for Academic Purposes (EAP) Writing: A Path to Sustainable Development. *International Journal of Education and Humanities*, *17*(1), 133-143. https://doi.org/10.54097/f1qfkr93

He, X., & Li, L. (2024). Teaching Optimization of Virtual Reality and Augmented Reality Technology in Vocational Education. *Journal of Computational Methods in Sciences and Engineering*, *24*(6), 3734-3747. https://doi.org/10.1177/14727978241297013

Hearn, S., & Long, M. (2023). Adult Learners' College Access &Amp; Choice Process: A Literature Review. *Student Affairs on Campus*, *6*(1), 37-47. https://doi.org/10.54155/saoc.v6i1.16

Hébert, Y. M. (2016). Youth in Plural Cities, Multiculturalism and Citizenship: Policy Challenges and Opportunities. *Foro De Educación*, *14*(20), 199-230. https://doi.org/10.14516/fde.2016.014.020.011

Hébrard, P. (2013). Ambiguities and Paradoxes in a Competence-Based Approach to Vocational Education and Training in France. *European Journal for Research on the Education and Learning of Adults*, *4*(2), 111-127. https://doi.org/10.3384/rela.2000-7426.rela9002

Heck, T., Kullmann, S., Hiebl, J., Schröder, N., Otto, D., & Sander, P. (2020). Designing Open Informational Ecosystems on the Concept of Open Educational Resources. *Open Education Studies*, *2*(1), 252-646. https://doi.org/10.1515/edu-2020-0130

Hedin, N. (2010). Experiential Learning: Theory and Challenges. *Christian Education Journal Research on Educational Ministry*, *7*(1), 107-117. https://doi.org/10.1177/073989131000700108

Hefnawi, A. (2025). Purposeful AI Integration in Education: Aligning Policy, Literacy, and Practice. *Management in Education*, *40*(2), 138-142. https://doi.org/10.1177/08920206251394517

Heidt, T. v. d., & Lamberton, G. (2014). How Academics in Undergraduate Business Programs at an Australian University View Sustainability. *Australian Journal of Environmental Education*, *30*(2), 215-238. https://doi.org/10.1017/aee.2015.7

Heijden, B. v. d., Boon, J., Klink, M. v. d., & Meijs, E. (2009). Employability Enhancement Through Formal and Informal Learning: An Empirical Study Among Dutch Non-academic University Staff Members. *International Journal of Training and Development*, *13*(1), 19-37. https://doi.org/10.1111/j.1468-2419.2008.00313.x

Heilporn, G., Larose, S., Beaulieu, C., Janosz, M., Cellard, C., Bureau, J. S., & Côté, S. M. (2023). Pratiques Pédagogiques Inclusives en Enseignement Postsecondaire, en Contexte Pandémique. *Revue Internationale De Pédagogie De L'enseignement Supérieur*, *39*(2). https://doi.org/10.4000/ripes.4661

Helskog, G. H. (2014). The Healing Power of Dialogos Dialogues: Transformative Learning Through Dialogical Philosophizing. *Open Journal of Social Sciences*, *02*(11), 79-83. https://doi.org/10.4236/jss.2014.211011

Henderson, T., Carrero, A., & Michel, J. O. (2025). Engineering Students' Sustainability Learning: Examining Opportunities to Learn About Sustainability and Their Role in Student Outcomes. *Journal of Engineering Education*, *114*(3). https://doi.org/10.1002/jee.70010

Henning, T. (2012). Writing Professor as Adult Learner: An Autoethnography of Online Professional Development. *Online Learning*, *16*(2). https://doi.org/10.24059/olj.v16i2.251

Henri, F., & Lundgren-Cayrol, K. (2001). Apprentissage Collaboratif À Distance. https://doi.org/10.1353/book.15576

Henschke, J. A. A Productive Decade of Andragogy's History and Philosophy. 44-71. https://doi.org/10.4018/978-1-61520-745-9.ch004

Henschke, J. A. (2010a). Beginnings of the History and Philosophy of Andragogy 1833-2000. 1-30. https://doi.org/10.4018/978-1-61520-694-0.ch001

Henschke, J. A. (2010b). A Productive Decade of Andragogy's History and Philosophy. https://doi.org/10.4018/9781615207459.ch004

Henschke, J. A. (2011a). Considerations Regarding the Future of Andragogy. *Adult Learning*, *22*(1), 34-37. https://doi.org/10.1177/104515951102200109

Henschke, J. A. (2011b). A Productive Decade of Andragogy's History and Philosophy. 44-71. https://doi.org/10.4018/978-1-61520-745-9.ch004

Herliani, E. (2021). Developing Sustainability Literacy Through STEM Learning: A Review. *Seaqis Journal of Science Education*, *1*(1), 36-43. https://doi.org/10.58249/sjse.v1i1.23

Hermans, S., Neutens, T., wyffels, F., & Petegem, P. V. (2024). Empowering Vocational Students: A Research-Based Framework for Computational Thinking Integration. *Education Sciences*, *14*(2), 206. https://doi.org/10.3390/educsci14020206

Hernández, A., Escobar, S. G., Norma Ivonne González Arratia López, F., & Eguiarte, B. E. B. (2019). Estrés, Autoeficacia, Rendimiento Académico Y Resiliencia en Adultos Emergentes. *Electronic Journal of Research in Educational Psychology*, *17*(47), 129-148. https://doi.org/10.25115/ejrep.v17i47.2226

Herrera, L. D., & Hoz, A. D. l. (2014). El Ciclo De Aprendizaje Experiencial en La Enseñanza. Caso ESUMER-GIZ: Seminario Sobre Cooperación Y Relaciones Comerciales Con Alemania. *Nodos Y Nudos*, *4*(37). https://doi.org/10.17227/01224328.3128

Herrero, P. P. i., Ciraso-Calí, A., & Arnau-Sabatés, L. (2017). PIAAC Results on Opportunities for Informal Learning in the Workplace in Spain /<i>Resultados PIAAC Sobre Las Oportunidades Para El Aprendizaje Informal en El Puesto De Trabajo en España</I>. *Culture and Education*, *29*(1), 151-182. https://doi.org/10.1080/11356405.2017.1278938

Herrington, J., & Oliver, R. (2000). An Instructional Design Framework for Authentic Learning Environments. *Educational Technology Research and Development*, *48*(3), 23-48. https://doi.org/10.1007/BF02319856

Hess, A. N. (2019). Academic Librarians' Educational Factors and Perceptions of Teaching Transformation: An Exploratory Examination. *Evidence Based Library and Information Practice*, *14*(3), 52-76. https://doi.org/10.18438/eblip29526

Heymann, J., Sprague, A., & Raub, A. (2020). The Urgency of Advancing Equality. 1-15. https://doi.org/10.1525/luminos.81.b

Hezlett, S. A. (2016). Enhancing Experience-Driven Leadership Development. *Advances in Developing Human Resources*, *18*(3), 369-389. https://doi.org/10.1177/1523422316645887

Hicks, A. (2020). Negotiating Change: Transition as a Central Concept for Information Literacy. *Journal of Information Science*, *48*(2), 210-222. https://doi.org/10.1177/0165551520949159

Hidayati, A., Bentri, A., & Eldarni. (2022). Supporting Factors for the Implementation of Mobile Learning for Elementary School Students Using an Authentic Approach and Real-World Activities. *International Journal of Interactive Mobile Technologies (Ijim)*, *16*(05), 107-120. https://doi.org/10.3991/ijim.v16i05.26551

Hignasari, L. V., Gusti Ayu Made Arna, P., & Wijaya, I. K. W. B. (2023). Transformative Strategies in Higher Education: SWOT Analysis for Project-Based Learning Models. *Utamax Journal of Ultimate Research and Trends in Education*, *5*(3), 214-230. https://doi.org/10.31849/utamax.v5i3.15748

Hill, B. (2017). Research Into Experiential Learning in Nurse Education. *British Journal of Nursing*, *26*(16), 932-938. https://doi.org/10.12968/bjon.2017.26.16.932

Hill, B., Nilson, C., Uink, B., & Fetherston, C. (2023). Transformation at the Cultural Interface: Exploring the Experiences of Aboriginal and Torres Strait Islander University Students. *The Australian Journal of Indigenous Education*, *52*(2). https://doi.org/10.55146/ajie.v52i2.656

Hill, L. H., Rogers-Shaw, C., & Carr-Chellman, D. (2025). Meeting the Needs Of <i>All</I> Adult Learners: Threats to the Future of Adult Education. *Adult Learning*, *36*(4), 207-214. https://doi.org/10.1177/10451595251380596

Hinck, J. M. (2022). Leadership Coaching as a Transformative Process in the Military. https://doi.org/10.24384/6cf9-v073

Hine, S. F. M., & María Nuria Rodríguez de, M. (2021). Hacer El Pensamiento Visible Para El Fortalecimiento De Las Habilidades Comunicativas en Inglés De Niños De Transición. *Colombian Applied Linguistics Journal*, *23*(1), 49-62. https://doi.org/10.14483/22487085.16818

Hinostroza, E. F. C. (2020). Relación Entre La Resiliencia Y La Calidad De Vida en Estudiantes Universitarios De Cajamarca. *Avances en Psicología*, *28*(1), 59-71. https://doi.org/10.33539/avpsicol.2020.v28n1.2112

Hinton, R., & Kazmi, A. (2022). Afterword. 439-450. https://doi.org/10.11647/obp.0256.18

Hinzen, H. (2020). 100 Years of Volkshochschule – 50 Years of DVV International. Local and Global Perspectives on Adult Education and Lifelong Learning. 199-211. https://doi.org/10.36253/978-88-5518-155-6.14

Hix, J. W. (2013). *Measuring the Effectiveness of Transfer of Learning Constructs and Intent to Transfer in a Simulation-Based Leadership Training Program* University of North Texas].

Ho, C. S. M., Ho, Y. L., & Bai, B. (2026). Toward High-Quality Professional and Vocational Education and Training (PVET) Development: Building Skills-Ecosystem in Chinese Societies. *Beijing International Review of Education*, *8*(2), 139-153. https://doi.org/10.1177/25902547261457683

Hoang, A.-D., Nguyen, C. Y., Nguyen, L.-K.-N., Vuong, Q. H., Dam, M. Q., Tran, T., & Nguyen, T. T. (2020). Introducing a Tool to Gauge Curriculum Quality Under Sustainable Development Goal 4: The Case of Primary Schools in Vietnam. *International Review of Education*, *66*(4), 457-485. https://doi.org/10.1007/s11159-020-09850-1

Hockly, N. (2023). Artificial Intelligence in English Language Teaching: The Good, the Bad and the Ugly. *Relc Journal*, *54*(2), 445-451. https://doi.org/10.1177/00336882231168504

Hocquelet, M. (2020). De L'influence Du Numérique Sur Les Compétences Dans Les Entrepôts De La Logistique. 102-109. https://doi.org/10.4000/books.cereq.2356

Hodge, S. (2007). The origins of competency-based training. *Australian Journal of Adult Learning*, *47*(2), 179-209.

Hodkinson, H. M. (2010). Learning to Work No Longer: Exploring "Retirement". *Journal of Workplace Learning*, *22*(1/2), 94-103. https://doi.org/10.1108/13665621011012889

Hoem, I. B., & Ursin, M. (2025). Transformative Learning and Participatory Approaches With Youth: A Discussion on Distinctions. *Children & Society*, *40*(4), 719-729. https://doi.org/10.1111/chso.70007

Hoffert, M. M., Passalacqua, K. D., Haftka-George, A., Lanfranco, O. A., & Martin, R. A. (2021). A Curriculum for Enhancing Physician Teaching Skills: The Value of Physician-Educator Partnerships. *Journal of Medical Education and Curricular Development*, *8*. https://doi.org/10.1177/23821205211032013

Hogan, D., & O'Flaherty, J. (2021). Addressing Education for Sustainable Development in the Teaching of Science: The Case of a Biological Sciences Teacher Education Program. *Sustainability*, *13*(21), 12028. https://doi.org/10.3390/su132112028

Hoggan-Kloubert, T., & Mabrey, P. E. (2022). Civic Education as Transformative Education. *Journal of Transformative Education*, *20*(3), 167-175. https://doi.org/10.1177/15413446221103173

Hoggan, C., & Higgins, K. (2023). Understanding Transformative Learning Through Metatheory. *New Directions for Adult and Continuing Education*, *2023*(177), 13-23. https://doi.org/10.1002/ace.20475

Hoggan, C., Mälkki, K., & Finnegan, F. (2016). Developing the Theory of Perspective Transformation. *Adult education quarterly*, *67*(1), 48-64. https://doi.org/10.1177/0741713616674076

Holcomb, L. (2025). Designing Strategic and Interactive Signing Instruction to Support Signed Literacy for Deaf Learners. *International Journal of Designs for Learning*, *16*(2), 34-49. https://doi.org/10.14434/ijdl.v16i2.40280

Holdo, M. (2022). Critical Reflection: John Dewey's Relational View of Transformative Learning. *Journal of Transformative Education*, *21*(1), 9-25. https://doi.org/10.1177/15413446221086727

Holdsworth, S., Sandri, O., Hayes, J., & Maslen, S. (2025). Delivering Ongoing Learning to Engineers Designed Using the Professional Learning Design Framework. *Journal of Adult and Continuing Education*, *32*(2), 234-263. https://doi.org/10.1177/14779714251344504

Holdsworth, S., Thomas, I., & Sandri, O. (2018). Assessing Graduate Sustainability Attributes Using a Vignette/Scenario Approach. *Journal of Education for Sustainable Development*, *12*(2), 120-139. https://doi.org/10.1177/0973408218792127

Holladay, P. J. (2024). Introduction to Teaching Agritourism. 1-16. https://doi.org/10.4337/9781802208191.00005

Holmes, A. G. D. (2019). Constructivist Learning in University Undergraduate Programmes. Has Constructivism Been Fully Embraced? Is There Clear Evidence That Constructivist Principles Have Been Applied to All Aspects of Contemporary University Undergraduate Study? *Shanlax International Journal of Education*, *8*(1), 7-15. https://doi.org/10.34293/education.v8i1.819

Holmes, J., & Woodhams, J. M. (2013). Building Interaction: The Role of Talk in Joining a Community of Practice. *Discourse & Communication*, *7*(3), 275-298. https://doi.org/10.1177/1750481313494500

Holovko, D. (2024). Use of Virtual Simulators for Training Specialists for Modern Labor Market Requirements. *Věda a Perspektivy*(11(42)). https://doi.org/10.52058/2695-1592-2024-11(42)-167-174

Hope, E. C., Anyiwo, N., Palmer, G. J. M., Bañales, J., & Smith, C. D. (2023). Sociopolitical Development: A History and Overview of a Black Liberatory Approach to Youth Development. *American psychologist*, *78*(4), 484-495. https://doi.org/10.1037/amp0001119

Horna, J., & Reyes, B. G. A. (2019). Estilos De Aprendizaje De Estudiantes Universitarios De Ciencias Biológicas. *Revista De Estilos De Aprendizaje*, *12*(23), 50-70. https://doi.org/10.55777/rea.v12i23.1210

Horton, C. (2020). Thriving or Surviving? Raising Our Ambition for Trans Children in Primary and Secondary Schools. *Frontiers in Sociology*, *5*. https://doi.org/10.3389/fsoc.2020.00067

Horton, L. B., MacGregor, C. J., Anderson, A. L., Glaessgen, T. A., & Ongaga, K. (2023). Life-Changing Conversations: A Look at How One Civil Discourse Program at a Midwestern State University Impacts the Transformative Learning of Its' Participants. *International Journal of Business and Applied Social Science*, 16-25. https://doi.org/10.33642/ijbass.v9n1p2

Hosoda, Y. (2006). Development and Testing of a Clinical Learning Environment Diagnostic Inventory for Baccalaureate Nursing Students. *Journal of Advanced Nursing*, *56*(5), 480-490. https://doi.org/10.1111/j.1365-2648.2006.04048.x

Hossain, A., Ying, K., & Malik, A. (2023). Remote Learning During COVID-19 and Transformative Learning Theory: Tendency Towards Quadruple Helix Model for Future Post-Secondary Education in Indigenous Contexts. https://doi.org/10.21203/rs.3.rs-3124186/v1

Hossain, K. I., Ahmed, I., & Mahmud, M. S. (2023). English Language Anxiety in Adult Learners: Challenges and Opportunities. *Iubat Review*, *6*(2), 158-183. https://doi.org/10.3329/iubatr.v6i2.71314

Hou, A. Y. C., Ying, C., Lin, A. F. Y., Su, E. H. C., Zhou, K. Z., & Tao, C. H. (2025). Did Global Rankings and National Accreditation Drive Taiwanese Universities to Advance Sustainable Development Goals (SDGs)?—Competitive Advantage, a Moral Symbol or Leading to a Paradigm Shift. *Higher Education Quarterly*, *79*(3). https://doi.org/10.1111/hequ.70031

Hou, X.-b., Li, X.-k., & Zhang, B.-B. (2024). Effect of Poverty Alleviation Relocation on Non-Farm Employment of Household Labor Force and Its Mechanism:Based on the Analysis of 7th-Phase Tracking Data of 857 Farmers in the Poverty-Stricken Areas of Wuling Mountain. *自然资源学报*, *39*(4), 978. https://doi.org/10.31497/zrzyxb.20240414

Houghton, J. (2022). New to Universal Design for Learning? Start Here! *New Directions for Teaching and Learning*, *2022*(172), 11-22. https://doi.org/10.1002/tl.20525

Housel, D. A. (2019). When Co-Occurring Factors Impact Adult Learners: Suggestions for Instruction, Preservice Training, and Professional Development. *Adult Learning*, *31*(1), 6-16. https://doi.org/10.1177/1045159519849910

Houssami, S., & Benattabou, D. (2024). Intersections of Motivation and Autonomous Learning Activities Among Moroccan EFL University Students. *Tesol and Technology Studies*, *5*(2), 1-20. https://doi.org/10.48185/tts.v5i2.1194

Hovington, S., Blanchet-Cohen, N., & Mann-Feder, V. (2020). Innovation in a Capstone Course in Youth Work. *International Journal of Child Youth and Family Studies*, *11*(3), 146-166. https://doi.org/10.18357/ijcyfs113202019713

Howard, J., Cohen, A., Stephenson, C. M., Lamborg, A., Leonard, S. D., Biagini, D., Becks, V., Stirnkorb, A., Rivera, N., Loos, G. P., Schulte, P. A., Baker, C., Felde, D., Barton, K., Cronin, C., Payne, H., Brown, M., Larson, M., & Moir, S. (2004). Report From the 1999 National Conference on Workplace Safety &Amp; Health Training. Putting the Pieces Together &Amp; Planning for the Challenges Ahead. https://doi.org/10.26616/nioshpub2004132

Howard, V. J. (2019). Open Educational Resources in Behavior Analysis. *Behavior Analysis in Practice*, *12*(4), 839-853. https://doi.org/10.1007/s40617-019-00371-4

Howardson, G. N., & Behrend, T. S. (2015). The Relative Importance of Specific Self-efficacy Sources in Pretraining Self-efficacy Beliefs. *International Journal of Training and Development, 19*(4), 233-252. https://doi.org/10.1111/ijtd.12060

Howe, A., Lo, J., Jaswal, S., Bani-Fatemi, A., Chattu, V. K., & Nowrouzi-Kia, B. (2023). Engaging Employers in Apprentice Training: Focus Group Insights From Small-to-Medium-Sized Employers in Ontario, Canada. *International Journal of Environmental Research and Public Health, 20*(3), 2527. https://doi.org/10.3390/ijerph20032527

Hsu, K. C., & Liu, G. Z. (2023). The Construction of a Theory-based Augmented Reality-featured Context-aware Ubiquitous Learning Facilitation Framework for Oral Communication Development. *Journal of Computer Assisted Learning, 39*(3), 883-898. https://doi.org/10.1111/jcal.12792

Hsu, K. C., & Liu, G. Z. (2024). Towards a Design of the Best Practices for <scp>AR</Scp>-Guided Oral Communication Development: A Systematic Review of Selected Research Published From 2000 to 2023. *Journal of Computer Assisted Learning, 41*(1). https://doi.org/10.1111/jcal.13103

Huambachano, M. (2019). Traditional Ecological Knowledge and Indigenous Foodways in the Andes of Peru. *Review of International American Studies, 12*(1), 87-110. https://doi.org/10.31261/rias.6866

Huang, H.-W. (2021). Effects of Smartphone-Based Collaborative Vlog Projects on EFL Learners' Speaking Performance and Learning Engagement. *Australasian Journal of Educational Technology*, 18-40. https://doi.org/10.14742/ajet.6623

Huang, R. (2024). Comparative Analysis of the Differences in the Education System in China vs Germany. *Journal of Education Humanities and Social Sciences, 33*, 48-56. https://doi.org/10.54097/3be6kt31

Huang, X., Mayer, R. E., & Usher, E. L. (2020). Better Together: Effects of Four Self-Efficacy-Building Strategies on Online Statistical Learning. *Contemporary Educational Psychology, 63*, 101924. https://doi.org/10.1016/j.cedpsych.2020.101924

Huda, M. Q., Sulaeman, M., & Marfuah, S. (2023). Inclusivity in Islamic Conservatism: The Moderate Salafi Movement in Kediri, Indonesia. *Religious Jurnal Studi Agama-Agama Dan Lintas Budaya, 7*(1), 77-92. https://doi.org/10.15575/rjsalb.v7i1.22648

Hue, M. T., & Karim, S. (2024). Developing a Sense of Belonging Among Ethnic Minority Youth in Hong Kong: Challenges and Opportunities. *International Journal of Comparative Education and Development, 26*(1), 56-75. https://doi.org/10.1108/ijced-05-2023-0034

Huggins, S. (2017). Practice-Based Learning in Higher Education. *Library Trends, 66*(1), 1-12. https://doi.org/10.1353/lib.2017.0024

Huiyan, Z. (2019). The Strategic Conception of Implementing Co-Op Education in Architecture Major——Take Wuhan Polytechnic University as an Example. https://doi.org/10.2991/mmetss-19.2019.170

Hulko, W., Mahara, M. S., Wilson, D., & Campbell-McArthur, G. (2021). Culturally Safe Dementia Care: Building Nursing Capacity to Care for First Nation Elders With Memory Loss. *International Journal of Older People Nursing, 16*(5). https://doi.org/10.1111/opn.12395

Hung, C. S. (2013). Framework for the Development of OER-based Learning Materials in ODL Environment. *Open Praxis, 5*(4), 315. https://doi.org/10.5944/openpraxis.5.4.99

Hunt, X., Saran, A., White, H., & Kuper, H. (2021). PROTOCOL: Effectiveness of Interventions for Improving Educational Outcomes for People With Disabilities in Low- and Middle-income Countries: A Systematic Review. *Campbell Systematic Reviews Better Evidence for a Better World, 17*(4). https://doi.org/10.1002/cl2.1197

Hunter-Johnson, Y. (2021). A Leap of Academic Faith and Resilience: Nontraditional International Students Pursuing Higher Education in the United States of America. *Journal of International Students, 12*(2). https://doi.org/10.32674/jis.v12i2.1986

Hunter, C. P. (2014). Perspectives in AE—Intentional Incidental Learning in the Workplace: Implications for Adult Learning. *New Horizons in Adult Education and Human Resource Development, 26*(2), 49-53. https://doi.org/10.1002/nha3.20062

Hussain, S. H. M., & Tasnim, R. (2023). Enhancing Undergraduates' Digital Competencies Through a Mentorship Model. *Environment-Behaviour Proceedings Journal, 8*(SI15), 253-259. https://doi.org/10.21834/e-bpj.v8isi15.5096

Hutasuhut, I. J., Adruce, S. A. Z., & Jonathan, V. (2021). How a Learning Organization Cultivates Self-Directed Learning. *Journal of Workplace Learning, 33*(5), 334-347. https://doi.org/10.1108/jwl-05-2020-0074

Hutchins, R. M. (1968). *The Learning Society*. Frederick A. Praeger.

Hutfilter, A. F., Lehmann, S., & Kim, E. Y. (2018). Improving Skills and Their Use in Germany. https://doi.org/10.1787/8a251b1f-en

Hwang, G. J., Chang, C. C., & Chien, S. Y. (2022). A Motivational Model-based Virtual Reality Approach to Prompting Learners' Sense of Presence, Learning Achievements, and Higher-order Thinking in Professional Safety Training. *British journal of educational technology, 53*(5), 1343-1360. https://doi.org/10.1111/bjet.13196

Hwang, G. J., & Fu, Q. K. (2020). Advancement and Research Trends of Smart Learning Environments in the Mobile Era. *International Journal of Mobile Learning and Organisation, 14*(1), 114. https://doi.org/10.1504/ijmlo.2020.10024691

Hwang, J., & Yoon, S. W. (2023). Workplace Learning for the Disadvantaged: Perspectives From Adult Education and Human Resource Development. *New Directions for Adult and Continuing Education, 2023*(179), 91-104. https://doi.org/10.1002/ace.20505

Hyde, B. (2021). Critical Discourse and Critical Reflection in Mezirow's Theory of Transformative Learning: A Dialectic Between Ontology and Epistemology (And a Subtext of Reflexivity Mirroring My Own Onto-Epistemological Movement). *Adult education quarterly, 71*(4), 373-388. https://doi.org/10.1177/07417136211003612

Hyden, B., & Cherrstrom, C. A. (2025). Instructional Design and Facilitation for Multigenerational Online Learning in Higher Education—Baby Boomers, Generation X, Millennials, and Generation Z. *New Directions for Adult and Continuing Education, 2025*(187), 15-22. https://doi.org/10.1002/ace.70001

Ibegbulam, C. M., Olowonubi, J. A., Fatounde, S. A., & Oyegunwa, O. A. (2023). Artificial Intelligence in the Era of 4ir: Drivers, Challenges and Opportunities. *Engineering Science & Technology Journal, 4*(6), 473-488. https://doi.org/10.51594/estj.v4i6.668

Ibrahim, N., Freeman, S. A., & Shelley, M. (2012). Exploring the Impact of Work Experience on Part-Time Students' Academic Success in Malaysian Polytechnics. *Career and Technical Education Research, 37*(1), 57-74. https://doi.org/10.5328/cter37.1.57

Ibrahim, N. F. W., Darus, N. A., & Radzi, A. H. M. (2022). Students' Acceptance and Motivation on Andragogical Approaches Towards Their Academic Performance. *Jurnal Intelek, 17*(2), 74-81. https://doi.org/10.24191/ji.v17i2.18041

Ibsen, M. F. (2023). Domination, Social Norms, and the Idea of an Emancipatory Interest. *Constellations, 31*(2), 160-173. https://doi.org/10.1111/1467-8675.12674

Idilia del Pilar Reyna, R., Álvarez, L., Coello, J. J. B., Gómez, I. A. P., Vélez, Y. P. R., & Méndez, Y. X. O. (2024). Implementación De Modelos Educativos Inclusivos: Retos Y Oportunidades. *South Florida Journal of Development, 5*(2), 755-768. https://doi.org/10.46932/sfjdv5n2-028

Ifenthaler, D., Majumdar, R., Gorissen, P., Judge, M., Mishra, S., Raffaghelli, J. E., & Shimada, A. (2024). Artificial Intelligence in Education: Implications for Policymakers, Researchers, and Practitioners. *Technology Knowledge and Learning, 29*(4), 1693-1710. https://doi.org/10.1007/s10758-024-09747-0

Ifenthaler, D., & Yau, J. Y. (2020a). Reflections on Different Learning Analytics Indicators for Supporting Study Success. *International Journal of Learning Analytics and Artificial Intelligence for Education (Ijai), 2*(2), 4. https://doi.org/10.3991/ijai.v2i2.15639

Ifenthaler, D., & Yau, J. Y. (2020b). Understanding Learning Analytics Indicators for Predicting Study Success. 36-41. https://doi.org/10.14742/ascilite2020.0101

Igira, F. T., & Gregory, J. (2009). Cultural Historical Activity Theory. 434-454. https://doi.org/10.4018/978-1-60566-659-4.ch025

Iglesias-Díaz, P., & Pérez, C. R. (2021). Aulas Afectivas E Inclusivas Y Bienestar Adolescente: Una Revisión Sistemática. *Educación Xx1*, *24*(2). https://doi.org/10.5944/educxx1.28705

Ignacio, J. V. P. (2020). Beyond the Classroom: Reconsidering the Role of Learner-Centered Teaching Using Paulo Freire's Philosophy of Education. *International Journal of Humanities and Social Sciences*, *12*(2), 52-62. https://doi.org/10.26803/ijhss.12.2.4

Ignacio, M., Sense-Wilson, S., Lucero, D. M., Antone, Q., Locklear, T., Crowder, R., Lee, J., & Gavin, A. R. (2023). Establishing Core Content for Culturally Grounded Harm Reduction Intervention for Urban Indigenous After-School Youth in the Pacific Northwest. *International Journal of Indigenous Health*, *18*(1). https://doi.org/10.32799/ijih.v18i1.39510

Ignatovich, E., & Walter, P. (2021). Political Posters, the Soviet Enlightenment and the Construction of a Learning Society, 1917-1928. *European Journal for Research on the Education and Learning of Adults*, *13*(1), 75-96. https://doi.org/10.3384/rela.2000-7426.3561

Ignatyeva, E. V., Козлова, T. A., & Ryabkova, Y. V. (2020). Conditions for the Realization of the Idea of Future Teachers' Lifelong Learning. *SHS Web of Conferences*, *87*, 00098. https://doi.org/10.1051/shsconf/20208700098

Ikenga, U. G., & Sijde, P. v. d. (2024). Twenty-First Century Competencies: About Competencies for Industry 5.0 and the Opportunities for Emerging Economies. https://doi.org/10.20944/preprints202403.0143.v1

Ilic, P. (2021). The Challenge of Information and Communications Technology in Education. *SHS Web of Conferences*, *102*, 01009. https://doi.org/10.1051/shsconf/202110201009

Ilie, V. (2019). Study on Adult Learning Motivation. *International Journal of Human and Behavioral Science*, *5*(1), 11-28. https://doi.org/10.19148/ijhbs.543980

Illeris, K. (2018). *Contemporary Theories of Learning: Learning Theorists ... In Their Own Words* (2nd ed.). Routledge.

Ilomäki, L., Löfgren, S., & Toom, A. (2022). Teachers' Perceptions on Relevant Upper-Secondary Vocational Graduate Competencies and Their Development. *Nordic Journal of Vocational Education and Training*, *12*(2). https://doi.org/10.3384/njvet.2242-458x.2212298

Ilyas, S., Kashif, M. F., & Iqbal, H. S. (2020). An Exploratory Study About the Role of Experiential Learning for Developing Professional Skills at Higher Education. *Global Social Sciences Review*, *V*(III), 341-350. https://doi.org/10.31703/gssr.2020(v-iii).36

İNankul, H. (2016). Behavioral Learning Theories and a Review for Police Basic Training. *Journal of International Social Research*, *9*(42), 1540-1540. https://doi.org/10.17719/jisr.20164216263

Indriati, L., Neo, M., & Tan, H. Y. (2024). Enhancing Authentic Assessment in Large-Class Design Education Through Authentic Project-Based Learning. *International Journal of Learning Teaching and Educational Research*, *23*(9), 432-452. https://doi.org/10.26803/ijlter.23.9.22

Ingles, D. M. (2023). *Examining Reflective Practice in Science With Practice Portfolios: An Experiential Learning Program at Iowa State University* Iowa State University].

International Commission on the Futures of Education. (2021). *Reimagining Our Futures Together: A New Social Contract for Education*. https://www.unesco.org/en/futures-education/report-resources

Ioannou, N. (2024). Exploring Trends and Challenges in Professional Development of Adult Educators Across European Member States: Insights and Recommendations for Improvements. *Social Education Research*, 258-272. https://doi.org/10.37256/ser.5220244450

Ions, K., & Sutcliffe, N. (2019). Barriers to Constructing Experiential Learning Claims Through Reflective Narratives. *Higher Education Skills and Work-Based Learning*, *10*(1), 126-140. https://doi.org/10.1108/heswbl-04-2019-0053

Ip, E., Srivastava, R., Lentz, L., Jasinoski, S. C., & Andérson, G. S. (2025). Antecedents of Workplace Psychological Safety in Public Safety and Frontline Healthcare: A Scoping Review. *International Journal of Environmental Research and Public Health*, *22*(6), 820. https://doi.org/10.3390/ijerph22060820

Iris Huis in 't, V., & Nagenborg, M. (2019). It's Getting Personal: The Ethical and Educational Implications of Personalised Learning Technology. *Journal of Philosophy in Schools*, *6*(1). https://doi.org/10.21913/jps.v6i1.1567

Isah, E. E., & Byström, K. (2015). Physicians' Learning at Work Through Everyday Access to Information. *Journal of the Association for Information Science and Technology*, *67*(2), 318-332. https://doi.org/10.1002/asi.23378

Ischinger, B. (2010). Foreword. 3-4. https://doi.org/10.1787/edu_today-2010-1-en

Ishak, M. A., & Obeng-Ofori, D. (2022). Occupational Health and Safety Issues Among Artisans in the Bibiani Municipality. *10*(2), 1441-1459. https://doi.org/10.53075/ijmsirq/6564453446

Ishaq, I., Degeng, I. N., Praherdhiono, H., & Degeng, M. D. K. (2024). Learning Design Heutagogical Approach in Developing Self-Determined Learning Skills. *Journal of Education Research and Evaluation*, *8*(2), 373-383. https://doi.org/10.23887/jere.v8i2.76371

Isidori, E., Poplavskaya, N., Alonzi, R., Martino, M. D., Leonova, I., & Şandor, I. (2023). What Does It Mean "To Learn" in the Internet Age? , *6*, 330-344. https://doi.org/10.15405/epes.23056.30

Itasanmi, S. A., Ajani, O. A., Mabayoje, D. E., & Odiaka, S. I. (2023). Predictive Influence of Locus of Control on Adult Learners' Academic Attitude and Engagement. *Nurture*, *17*(4), 723-736. https://doi.org/10.55951/nurture.v17i4.503

Iyer, S. S. (2025). Data-Driven Decision Making: The Key to Future Health Care Business Success. *Ra Journal of Applied Research*, *11*(03). https://doi.org/10.47191/rajar/v11i3.06

Izquierdo, C. R., Carrasco, S., & Pedro, L. N. (2021). Language and Social Integration in Times of Increasing Anti-Immigration Discourses. *Migraciones Publicación Del Instituto Universitario De Estudios Sobre Migraciones*(51), 61-91. https://doi.org/10.14422/mig.i51y2021.003

Jabeen, M., Aakif, Z., & Afridi, H. A. (2024). Unlocking Pakistan's Digital Potential: A Roadmap for Workforce Digitalization and Economic Transformation. *Journal of Information Technology Teaching Cases*, *16*(1), 102-113. https://doi.org/10.1177/20438869241280980

Jackson, C., & Manley, K. (2021). Contemporary Challenges of Nursing CPD: Time to Change the Model to Meet Citizens' Needs. *Nursing Open*, *9*(2), 880-891. https://doi.org/10.1002/nop2.941

Jackson, D. (2014). *Academic Self-Efficacy of Adult First-Generation Students Enrolled in Online Undergraduate Courses* University of North Texas].

Jackson, D. (2017). Using Work-Integrated Learning to Enhance Career Planning Among Business Undergraduates. *Australian Journal of Career Development*, *26*(3), 153-164. https://doi.org/10.1177/1038416217727124

Jacobs, B., Babineau, K., & Parker, D. (2024). From Childcare to Educare: Inspiring Change in Early Childhood Education for Rural Tennessee. https://doi.org/10.51388/20.500.12265/222

Jacobs, J. (2019). Client-based Project Work as Experiential Education. *Design Management Review*, *30*(1), 46-54. https://doi.org/10.1111/drev.12161

Jácome, N. E. G., & Martinez, M. E. M. (2024). Desarrollo De Habilidades De Pensamiento Crítico en Adultos: Aplicaciones Andragógicas en El Aula. *Boletín Científico Ideas Y Voces*, *4*(3). https://doi.org/10.60100/bciv.v4i3.168

Jagatheesaperumal, S. K., Ahmad, K., Al-Fuqaha, A., & Qadir, J. (2022). Advancing Education Through Extended Reality and Internet of Everything Enabled Metaverses: Applications, Challenges, and Open Issues. https://doi.org/10.48550/arxiv.2207.01512

Jagatheesaperumal, S. K., Ahmad, K., Al-Fuqaha, A., & Qadir, J. (2024). Advancing Education Through Extended Reality and Internet of Everything Enabled Metaverses: Applications, Challenges, and Open Issues. *Ieee Transactions on Learning Technologies*, *17*, 1120-1139. https://doi.org/10.1109/tlt.2024.3358859

Jaggi, A., Rogers, K. G., Rogers, H. G., Daniels, A. Y., Ens, E., & Pinkcham, S. (2024). "We Need to Run Our Own Communities". *Australian and International Journal of Rural Education*, *34*(1), 122-144. https://doi.org/10.47381/aijre.v34i1.706

Jain, A., & Dutta, D. (2018). Millennials and Gamification: Guerilla Tactics for Making Learning Fun. *South Asian Journal of Human Resources Management*, *6*(1), 29-44. https://doi.org/10.1177/2322093718796303

Jain, C., & Madaan, M. (2020). Students' Perception Regarding Pedagogy, Andragogy, and Heutagogy as Teaching–learning Methods in Undergraduate Medical Education. *Journal of Education and Health Promotion*, *9*. https://doi.org/10.4103/jehp.jehp_221_20

Jaldemark, J., Lindqvist, M. H., Mozelius, P., & Ryberg, T. (2021). Editorial Introduction: Lifelong Learning in the Digital Era. *British journal of educational technology*, *52*(4), 1576-1579. https://doi.org/10.1111/bjet.13128

Jalinus, N., Syahril, S., Haq, S., & Kassymova, G. K. (2023). Work-Based Learning for the Engineering Field in Vocational Education: Understanding Concepts, Principles and Best Practices. *Journal of Engineering Researcher and Lecturer*, *2*(1), 9-17. https://doi.org/10.58712/jerel.v2i1.22

Jameson, M. M. (2019). Time, Time, Time: Perceptions of the Causes of Mathematics Anxiety in Highly Maths Anxious Female Adult Learners. *Adult education quarterly*, *70*(3), 223-239. https://doi.org/10.1177/0741713619896324

Jameson, M. M., & Fusco, B. R. (2014). Math Anxiety, Math Self-Concept, and Math Self-Efficacy in Adult Learners Compared to Traditional Undergraduate Students. *Adult education quarterly*, *64*(4), 306-322. https://doi.org/10.1177/0741713614541461

Jamhari, J., & Asrori, S. (2022). In the Making of Salafi-Based Islamic Schools in Indonesia. *Al-Jami Ah Journal of Islamic Studies*, *60*(1), 227-264. https://doi.org/10.14421/ajis.2022.601.227-264

Jandrić, P., & Hayes, S. (2020). Postdigital We-Learn. *Studies in Philosophy and Education*, *39*(3), 285-297. https://doi.org/10.1007/s11217-020-09711-2

Jannah, N. (2022). Problem-Based Learning Strategies as the Main Concept of Quality Learning. *Falasifa Jurnal Studi Keislaman*, *13*(1), 68-82. https://doi.org/10.62097/falasifa.v13i1.861

Jansen, A., & Wieland, A. (2024). Developing Resilient Leaders: A training for Students. *Journal of Leadership Education*. https://doi.org/10.1108/jole-06-2024-0073

Jansen, F. (2023). Teachers' Learning Experiences While Teaching Refugees in Greece: "it's Pretty Extreme". *Adult Education Discourses*, *24*. https://doi.org/10.34768/dma.vi24.684

Jansen, G., Janssen, S., Levels, M., & Fregin, M.-C. (2025). Why Are Employees Most Susceptible to Automation Least Likely to Retrain? Automation Risks and Inequalities in Learning Intention, Perceived Opportunities, and Learning Participation Among Employee Groups. *Economic and Industrial Democracy*, *47*(2), 382-403. https://doi.org/10.1177/0143831x251331749

Jardinez, M. J., & Natividad, L. (2024). The the Advantages and Challenges of Inclusive Education: Striving for Equity in the Classroom. *Shanlax International Journal of Education*, *12*(2), 57-65. https://doi.org/10.34293/education.v12i2.7182

Jarvis, P. (2004). *Adult Education and Lifelong Learning: Theory and Practice* (3rd ed.). RoutledgeFalmer.

Javeed, L. (2019). Supporting Clinical Practice Through Rehearsals. *Northwest Journal of Teacher Education*, *14*(1). https://doi.org/10.15760/nwjte.2019.14.1.2

Jawaid, S. A., & Ahmed, S. (2023). The Transformative Economic Impact of Artificial Intelligence. https://doi.org/10.20944/preprints202311.1384.v1

Jayaraman, S. (2021). Immigrant Women Workers in the Settlement Sector: Niche Employment or Occupational Segmentation? a Comparative Study of Germany and Canada. https://doi.org/10.32920/ryerson.14664864

Jeannin, L. (2021). Agency Within the Context of Pedagogies, Epistemologies and the Transformative Curricula. 1-10. https://doi.org/10.1163/9789004468443_001

Jena, M. C., Mishra, S. K., & Moharana, H. S. (2024). The Way Forward to Overcome Challenges and Drawbacks of AI. *Computing and Artificial Intelligence*, 1581. https://doi.org/10.59400/cai1581

Jenkins, A. (2021). Ecological Global Citizenship Education: A Reframing. *International Journal of Development Education and Global Learning*, *13*(2). https://doi.org/10.14324/ijdegl.13.2.03

Jenkins, C., & Alfred, M. V. (2017). Understanding the Motivation and Transformation of White Culturally Responsive Professors. *Journal of Adult and Continuing Education*, *24*(1), 81-99. https://doi.org/10.1177/1477971417738793

Jeno, L. M., Raaheim, A., Kristensen, S. M., Kristensen, K. D., Hole, T. N., Haugland, M., & Mæland, S. (2017). The Relative Effect of Team-Based Learning on Motivation and Learning: A Self-Determination Theory Perspective. *Cbe—life Sciences Education*, *16*(4), ar59. https://doi.org/10.1187/cbe.17-03-0055

Jentsch, B., Klein, R., Sandner, E., Bignami, F., & Paulus, M. (2022). Early Life Family Care for and With Migrants: The Development of a Transcultural Transition Mentoring Competence Profile. *Journal of Family Issues*, *44*(11), 2961-2980. https://doi.org/10.1177/0192513x221113851

Jeny, D. P. (2024). Classrooms to Networks: Applying Connectivism Principles in Pedagogy ```. *Nawala Education*, *1*(5), 63-73. https://doi.org/10.62872/0ypqjy52

Jeon, S. (2019). Many OECD Countries Face and Will Face an Ageing Population. https://doi.org/10.1787/a00bd85d-en

Jeong, S., Han, S. J., Lee, J., Sunalai, S., & Yoon, S. W. (2018). Integrative Literature Review on Informal Learning: Antecedents, Conceptualizations, and Future Directions. *Human Resource Development Review*, *17*(2), 128-152. https://doi.org/10.1177/1534484318772242

Jeong, S. H., Lee, T. W., Kim, I. S., Ha, J. S., & Kim, M. J. (2007). The Effect of Nurses' Use of the Principles of Learning Organization on Organizational Effectiveness. *Journal of Advanced Nursing*, *58*(1), 53-62. https://doi.org/10.1111/j.1365-2648.2006.04206.x

Jeronen, E. (2021). Transitioning to Quality Education. https://doi.org/10.3390/books978-3-03897-893-0

Jézégou, A. (2008). Apprentissage Autodirigé Et Formation À Distance. *Distances Et Savoirs*, *6*(3), 343-364. https://doi.org/10.3166/ds.6.343-364

Jimoyiannis, A., & Tsiotakis, P. (2016). Self-Directed Learning in E-Portfolios: Analysing Students' Performance and Learning Presence. *Icst Transactions on E-Education and E-Learning*, *3*(9), 151120. https://doi.org/10.4108/eai.10-3-2016.151120

Jing-jing, W., Mao, W., & Wenjie, W. (2023). The Ethics of Artificial Intelligence: Sociopolitical and Legal Dimensions. *Isslp*, *2*(2), 27-32. https://doi.org/10.61838/kman.isslp.2.2.6

Jing, C., Wu, Y., Huang, L., Wu, Y., & Guan, Y. (2025). Integrating Kolb's Experiential Learning Theory Into Nursing Education: A Four-Stage Intervention With Case Analysis, Mind Maps, Reflective Journals, and Peer Simulations for Advanced Health Assessment. *Frontiers in Medicine*, *12*. https://doi.org/10.3389/fmed.2025.1616392

Jing, X., & Yan, Y. (2020). Challenge, Change and Response: Research Into Education in a Globalized Perspective. *International Dialogues on Education Journal*, *7*(1/2). https://doi.org/10.53308/ide.v7i1/2.30

Jirgensons, M., & Kapenieks, J. (2018). Blockchain and the Future of Digital Learning Credential Assessment and Management. *Journal of Teacher Education for Sustainability*, *20*(1), 145-156. https://doi.org/10.2478/jtes-2018-0009

Joerke, G. A. O., José Leônidas Alves do, N., Souza, L. B. P., Santos, A. A. d., Costa, F. R. d., Silva, P. A. D., Fernando Luiz Cas de Oliveira, F., Paschoal, P. C. G., Paz, J. F. D., Hansel, T. F., Costa, K. M., Ataliba, P. M., & Barreto, M. S. (2023). Paulo Freire E a Educação Popular: Diálogos Para a Transformação Social. *Observatório De La Economía Latinoamericana*, *21*(7), 7103-7120. https://doi.org/10.55905/oelv21n7-069

Jog, S. (2021). *Investigating the Impact of Leadership Development on the Formation of Public Sector Leader Identities* Victoria University of Wellington].

Jõgi, L. (2023). Potovanje Iz Preteklosti v Sedanjost. *Andragoška Spoznanja*, *29*(2), 107-124. https://doi.org/10.4312/as/13473

Jõgi, L., & Gross, M. (2009). The Professionalisation of Adult Educators in the Baltic States. *European journal of education*, *44*(2), 221-242. https://doi.org/10.1111/j.1465-3435.2009.01380.x

Johansen, R. B., Sookermany, A. M., & Isaksen, G. (2021). Twisting the Pedagogy in Military Education – Experiences Drawn From a Problem-Based Teaching Approach at the Norwegian Defence University College. 236-259. https://doi.org/10.31374/book2-k

John, J. G. (2023). Reaching Out to Group Contingencies in the ESL Classroom. *Journal of Higher Education Theory and Practice*, *23*(1). https://doi.org/10.33423/jhetp.v23i1.5794

Johnson, E. S. (2008). Out of My Viewfinder, Yet in the Picture. *Science Technology & Human Values*, *33*(1), 53-76. https://doi.org/10.1177/0162243907306194

Johnson, K., Ennis-Cole, D., & Bonhamgregory, M. (2020). Workplace Success Strategies for Employees With Autism Spectrum Disorder: A New Frontier for Human Resource Development. *Human Resource Development Review*, *19*(2), 122-151. https://doi.org/10.1177/1534484320905910

Johnson, S., Blackman, D., & Buick, F. (2018). The 70:20:10 Framework and the Transfer of Learning. *Human Resource Development Quarterly*, *29*(4), 383-402. https://doi.org/10.1002/hrdq.21330

Johnston, E., Olivas, G. W., Steele, P., Smith, C., & Bailey, L. (2017). Exploring Pedagogical Foundations of Existing Virtual Reality Educational Applications: A Content Analysis Study. *Journal of Educational Technology Systems*, *46*(4), 414-439. https://doi.org/10.1177/0047239517745560

Johrendt, J., Hector, S., Benzinger, K., Salinitri, G., Jaekel, A., Northwood, D. O., & Watters, M. (2010). Learning Outcomes Achievement in Cooperative Education: A Survey of Engineering Students. https://doi.org/10.18260/1-2--16581

Jokic, S., Albrecht, N., & Smith, S. (2019). Mindful Self-Compassion and Adult Learner Retention in Post-Compulsory Education. *Obm Integrative and Complementary Medicine*, *04*(01), 1-30. https://doi.org/10.21926/obm.icm.1901004

Jokikokko, K. (2021). Challenges and Possibilities for Creating Genuinely Intercultural Higher Education Learning Communities. *Journal of Praxis in Higher Education*, *3*(2), 26-51. https://doi.org/10.47989/kpdc111

Joksimovíc, S., Dowell, N., Poquet, O., Kovanović, V., Gašević, D., Dawson, S., & Graesser, A. C. (2018). Exploring Development of Social Capital in a CMOOC Through Language and Discourse. *The Internet and Higher Education*, *36*, 54-64. https://doi.org/10.1016/j.iheduc.2017.09.004

Joksimovíc, S., Marshall, R., Rakotoarivelo, T., Ladjal, D., Zhan, C., & Pardo, A. (2021). Privacy-Driven Learning Analytics. 1-22. https://doi.org/10.1007/978-3-030-86316-6_1

Jones, J. (2006). An Inter-industry Comparison of VET in Australian SMEs. *Education + Training*, *48*(8/9), 584-596. https://doi.org/10.1108/00400910610710029

Jones, J., Johnston, J., Ndiaye, N., Tokar, A., Singla, S., Skinner, N. A., Strehlow, M., & Utunen, H. (2024). Health Care Workers' Motivations for Enrolling in Massive Open Online Courses During a Public Health Emergency: Descriptive Analysis. *Jmir Medical Education*, *10*, e51915-e51915. https://doi.org/10.2196/51915

Jones, K. (2019). From Critical to Transformative Pedagogy in Architectural Education. 124-130. https://doi.org/10.35483/acsa.teach.2019.21

Jones, K. M. L. (2019). Learning Analytics and Higher Education: A Proposed Model for Establishing Informed Consent Mechanisms to Promote Student Privacy and Autonomy. *International Journal of Educational Technology in Higher Education*, *16*(1). https://doi.org/10.1186/s41239-019-0155-0

Jones, T. C. T. (2020). *"Why Have I Never Learned This?": Exploring Health Disparities Through a Critical Consciousness Lens* Iowa State University].

Jooste, K., & Jasper, M. (2010). A Framework for Recognition of Prior Learning Within a Postgraduate Diploma of Nursing Management in South Africa. *Journal of Nursing Management*, *18*(6), 704-714. https://doi.org/10.1111/j.1365-2834.2010.01160.x

Jooste, N., & Shaikh, A. (2026). Reigniting Engagement: A Multidisciplinary, Authentic Case-based Approach to Anatomy and Physiology Education for Medical Imaging Students. *Anatomical Sciences Education*, *19*(7), 1181-1188. https://doi.org/10.1002/ase.70238

Jørgensen, W., & Hadders, H. (2015). The Significance of Communities of Practice: Norwegian Nursing Students' Experience of Clinical Placement in Bangladesh. *Nursing Open*, *2*(1), 36-46. https://doi.org/10.1002/nop2.15

Joshi, S. (2021). Rising Importance of Remote Learning in India in the Wake of COVID-19: Issues, Challenges and Way Forward. *World Journal of Science Technology and Sustainable Development*, *18*(1), 44-63. https://doi.org/10.1108/wjstsd-10-2020-0079

Joshi, S., & Pramod, P. J. (2023). A Collaborative Metaverse Based a-La-Carte Framework for Tertiary Education (CO-MATE). *Heliyon*, *9*(2), e13424. https://doi.org/10.1016/j.heliyon.2023.e13424

Jossou, M. C., Gautier, A., Hernandez, S., & Tiberghien, B. (2022). L'apprentissage Expérientiel en Situation Extrême : Le Cas De L'École Nationale Supérieure Des Officiers De Sapeurs-Pompiers Pour La Gestion De L'accueil Et De L'hébergement Des Rapatriés De Wuhan. *Gestion Et Management Public, Volume 9 / N° 4*(4), 119-126. https://doi.org/10.3917/gmp.094.0119

Jufriadi, A., & Andinisari, R. (2020). JITT With Assessment for Learning: Investigation and Improvement of Students Understanding of Kinematics Concept. *Momentum Physics Education Journal*, 94-101. https://doi.org/10.21067/mpej.v4i2.4669

Julca-Asto, M. J. M., Duran-Llaro, K. L., Álvarez-Medina, G. M., & Donato-Palacios, M. (2023). El Aprendizaje Basado en Problemas Para La Enseñanza De Las Ciencias. *Revista Arbitrada Interdisciplinaria Koinonía*, *8*(2), 390-408. https://doi.org/10.35381/r.k.v8i2.2882

Jumaah, F. M. (2024). Exploring Constructivist Learning Theory and Its Applications in Teaching English. *The American Journal of Social Science and Education Innovations*, *6*(8), 7-19. https://doi.org/10.37547/tajssei/volume06issue08-02

Jun, W. (2021). A Study on Cause Analysis of Digital Divide Among Older People in Korea. *International Journal of Environmental Research and Public Health*, *18*(16), 8586. https://doi.org/10.3390/ijerph18168586

Júnior, E. L. P., Moreira, M. Â. L., Portella, A. G., Azevêdo, C. M. d., Igor Pınheıro de Araújo, C., Fávero, L. P. L., Gomes, C. F. S., & Santos, M. d. (2023). Systematic Literature Review on Virtual Electronics Laboratories in Education: Identifying the Need for an Aeronautical Radar Simulator. *Electronics*, *12*(12), 2573. https://doi.org/10.3390/electronics12122573

Juntunen, J., Tuomikoski, A. M., Pramila-Savukoski, S., Kaarlela, V., Keinänen, A. L., Kääriäinen, M., & Mikkonen, K. (2024). Healthcare Professionals' Experiences of Required Competencies in Mentoring of Interprofessional Students in Clinical Practice: A Systematic Review of Qualitative Studies. *Journal of Advanced Nursing*, *81*(2), 701-729. https://doi.org/10.1111/jan.16347

Jurkova, S., & Guo, S. (2018). Connecting Transculturalism With Transformative Learning: Toward a New Horizon of Adult Education. *Alberta Journal of Educational Research*, *64*(2), 173-187. https://doi.org/10.55016/ojs/ajer.v64i2.56383

Kaarlela, T., Padrão, P., Pitkäaho, T., Pieskä, S., & Bobadilla, L. (2022). Digital Twins Utilizing XR-Technology as Robotic Training Tools. *Machines*, *11*(1), 13. https://doi.org/10.3390/machines11010013

Kaatrakoski, H., Littlejohn, A., & Hood, N. (2016). Learning Challenges in Higher Education: An Analysis of Contradictions Within Open Educational Practice. *Higher Education*, *74*(4), 599-615. https://doi.org/10.1007/s10734-016-0067-z

Kaçar, I. G. (2021). A Case Study of Turkish Pre-Service Teachers of English in an International Exchange Program: ELF and WE Perspectives. *Eurasian Journal of Applied Linguistics*, 435-457. https://doi.org/10.32601/ejal.911484

Kaddoura, S., & Husseiny, F. A. (2023). The Rising Trend of Metaverse in Education: Challenges, Opportunities, and Ethical Considerations. *Peerj Computer Science*, *9*, e1252. https://doi.org/10.7717/peerj-cs.1252

Kahu, E. R. (2013). Framing student engagement in higher education. *Studies in Higher Education*, *38*(5), 758-773. https://doi.org/10.1080/03075079.2011.598505

Kaihlanen, A. M., Gluschkoff, K., Koskinen, S., Salminen, L., Strandell-Laine, C., Fuster, P., Sveinsdóttir, H., Fatkulina, N., Chianáin, L. N., Stubner, J., & Leino-Kilpi, H. (2021). Final Clinical Practicum Shapes the Transition Experience and Occupational Commitment of Newly Graduated Nurses in Europe—A Longitudinal Study. *Journal of Advanced Nursing*, *77*(12), 4782-4792. https://doi.org/10.1111/jan.15060

Kajwang, B. (2022). Application of Develop Deploy Connect Model in Managing Talent in the Insurance Sector. *International Journal of Sociology*, *6*(1), 1-10. https://doi.org/10.47604/ijs.1449

Kalashnyk, L., Ruda, N., Oserska, O., & Nazarenko, O. (2024). "SCHOOLS FOR LADIES AND GENTLEMEN" AS a SPECIFIC FORM OF NON-FORMAL EDUCATIONAL INSTITUTIONS IN FAR EAST COUNTRIES (China, Japan, South Korea and Singapore). *Society Integration Education Proceedings of the International Scientific Conference*, *1*, 665-674. https://doi.org/10.17770/sie2024vol1.7812

Kalenda, J., Kočvarová, I., & Boeren, E. (2023). Impact of <scp>COVID</Scp>-19 on Participation and Barriers in Nonformal Adult Education in the Czech Republic. *European journal of education*, *58*(4), 681-698. https://doi.org/10.1111/ejed.12580

Kalimullina, A. A. (2023). Two-Way Feedback With the Use of Digital Tools in the Classroom: Effect Size. *Education & Self Development*, *18*(4), 91-105. https://doi.org/10.26907/esd.18.4.08

Kallou, S., & Kikilia, A. (2021). A Transformative Educational Framework in Tourism Higher Education Through Digital Technologies During the COVID-19 Pandemic. *Advances in Mobile Learning Educational Research*, *1*(2), 37-47. https://doi.org/10.25082/amler.2021.01.005

Kálmán, C. (2019). Human Resource Policy Makers', Teachers', and Adult Learners' Views on the Impact of Context- And Teacher-Generated Autonomy on Adult Learners' L2 Motivation: Interview Studies in Hungarian Corporate Contexts. *Journal of Adult Learning Knowledge and Innovation*, *3*(2), 49-60. https://doi.org/10.1556/2059.03.2019.08

Kambara, C. (2026). Psychometric Properties of a Workplace Experiential Learning Scale Among Hospital-Based Registered Dietitians in Japan. *Journal of Human Nutrition and Dietetics*, *39*(2). https://doi.org/10.1111/jhn.70251

Kamp, A. (2023). Navigating the Landscape of Higher Engineering Education. https://doi.org/10.59490/mg.72

Kamp, M. v. d. (2003). Still Struggling With the Class Struggle? *Journal of Transformative Education*, *1*(2), 136-140. https://doi.org/10.1177/1541344603001002006

Kandaurova, A. V., & Milovanova, N. G. (2023). Socio-Cultural Conditionality of the Content of Education of Children and Youth. *Bulletin of the Chuvash State Pedagogical University Named After I Y Yakovlev*(3(120)), 097-105. https://doi.org/10.37972/chgpu.2023.120.3.012

Kandeel, R. H. (2020). LMOOC for Developing Teaching and Learning French as a Foreign Language (FFL). *Xlinguae*, *13*(1), 24-36. https://doi.org/10.18355/xl.2020.13.01.02

Kane, A. (2017). *Internationally Qualified Nurses' Perceptions of Patient Safety: New Zealand Case Studies* Victoria University].

Kangas, J., Lastikka, A. L., & Arvola, O. (2023). Inclusive Play: Defining Elements of Playful Teaching and Learning in Culturally and Linguistically Diverse ECEC. *Education Sciences*, *13*(9), 956. https://doi.org/10.3390/educsci13090956

Kanyesigye, S. T., Uwamahoro, J., & Kemeza, I. (2022). The Effect of Professional Training on in-Service Secondary School Physics 'Teachers' Motivation to Use Problem-Based Learning. *International Journal of Learning Teaching and Educational Research*, *21*(8), 271-287. https://doi.org/10.26803/ijlter.21.8.16

Karakas, G., Plooy, D., & Chapman, A. (2025). "Enhance—Yes, Replace—No": Psychology Students' Perceptions of Artificial Intelligence in Psychological Practice and Education. *Scholarship of Teaching and Learning in Psychology*. https://doi.org/10.1037/stl0000466

Karakhanyan, S. (2018). Armenia: Transformational Peculiarities of the Soviet and Post-Soviet Higher Education System. 73-96. https://doi.org/10.1007/978-3-319-52980-6_3

Karataş, K., & Arpacı, İ. (2021). The Role of Self-Directed Learning, Metacognition, and 21st Century Skills Predicting the Readiness for Online Learning. *Contemporary Educational Technology*, *13*(3), ep300. https://doi.org/10.30935/cedtech/10786

Kari, A., & Disegha, G. C. (2026). Integrating Sustainability Thinking Into Microbiology Education: A Conceptual Framework for the Next Generation of Environmental Scientists. *Journal of Microbiology and Biology Education*. https://doi.org/10.1128/jmbe.00289-25

Karimi, H., & Khawaja, S. (2023). The Impact of Artificial Intelligence on Higher Education in England. *Creative Education*, *14*(12), 2405-2415. https://doi.org/10.4236/ce.2023.1412154

Karlsson, T. (2023). Something to Do Something to Be : A Study of Students' Vocational Identity Formation in Adult Education. https://doi.org/10.3384/9789180754255

Karlsson, T., Muhrman, K., & Nyström, S. (2021). A Path Towards a Possible Future – Adult Students' Choice of Vocational Education. *Vocations and Learning*, *15*(1), 111-128. https://doi.org/10.1007/s12186-021-09280-6

Karmelita, C. (2018). Exploring the Experiences of Adult Learners in a Transition Program. *Journal of Adult and Continuing Education*, *24*(2), 141-164. https://doi.org/10.1177/1477971418791587

Kasa, M., Kho, J., Yong, D., Hussain, K., & Lau, P. S. Y. (2020). Competently Skilled Human Capital Through Education for the Hospitality and Tourism Industry. *Worldwide Hospitality and Tourism Themes*, *12*(2), 175-184. https://doi.org/10.1108/whatt-12-2019-0081

Kasap, M. (2023). Hello China and MOOCs: Shaping a New Paradigm in Education. *Eximia*, *12*, 238-251. https://doi.org/10.47577/eximia.v12i1.361

Kasperski, R., Porat, E., & Blau, I. (2023). Analysis of Emergency Remote Teaching in Formal Education: Crosschecking Three Contemporary Techno-Pedagogical Frameworks. *Research in Learning Technology*, *31*. https://doi.org/10.25304/rlt.v31.2982

Kastner, M., & Motschilnig, R. (2021). Interconnectedness of Adult Basic Education, Community-Based Participatory Research, and Transformative Learning. *Adult education quarterly*, *72*(3), 223-241. https://doi.org/10.1177/07417136211044154

Kasztelewicz, A., Tomaszewska, B., Rahner, S., Winter, I., & Voss, V. (2022). MOOC Education for Sustainable Development. The Erasmus+ Project Experience. *Polityka Energetyczna – Energy Policy Journal*, *25*(4), 165-180. https://doi.org/10.33223/epj/156341

Kattari, L. (2025). Transformative Learning to Politicized Collective Identity: How Cisgender Parents and Caregivers of Transgender and Gender Diverse Youth Become Change Makers for <scp>TGD</Scp> Justice. *Journal of Family Theory & Review*, *17*(2), 254-264. https://doi.org/10.1111/jftr.12608

Kavashev, Z. (2024). A Bibliometric Performance Analysis of Publication Productivity Within Pedagogy, Andragogy, and Heutagogy Continuum: Outcomes of SciVal Analytics. *E-Learning and Digital Media*, *22*(6), 534-555. https://doi.org/10.1177/20427530241239406

Kavitha, L. A. (2023). Universal Design: Promoting Equity and Accessibility. *Shanlax International Journal of Arts Science and Humanities*, *11*(S1i2-Nov), 7-10. https://doi.org/10.34293/sijash.v11is1i2-nov.7307

Kawano, K. (2023). Engaging a Kanaka ʻŌiwi Literature Review Methodology Through Research on Native Hawaiian Culture-Based Education. *Review of Educational Research*, *93*(6), 862-900. https://doi.org/10.3102/00346543221149004

Kayi, E. A. (2024). Transitioning to Blended Learning During <scp>COVID</Scp> -19: Exploring Instructors and Adult Learners' Experiences in Three Ghanaian Universities. *British journal of educational technology*, *55*(6), 2760-2786. https://doi.org/10.1111/bjet.13475

Kayira, J., Lobdell, S., Gagnon, N., Healy, J., Hertz, S., McHone, E., & Schuttenberg, E. (2022). Responsibilities to Decolonize Environmental Education: A Co-Learning Journey for Graduate Students and Instructors. *Societies*, *12*(4), 96. https://doi.org/10.3390/soc12040096

Keating, M. (2012). Developing Social Capital in 'Learning Borderlands': Has the Federal Government's Budget Delivered for Low-Paid Australian Workers? *Literacy and Numeracy Studies*, *20*(1), 5-24. https://doi.org/10.5130/lns.v20i1.2617

Keegan, P. (2021). Critical Affective Civic Literacy: A Framework for Attending to Political Emotion in the Social Studies Classroom. *The Journal of Social Studies Research*, *45*(1), 15-24. https://doi.org/10.1016/j.jssr.2020.06.003

Keeler, H. (2024). The Rising Call for Nursing Professional Development to Elevate Preceptor and Clinical Instruction. *Journal for Nurses in Professional Development*, *40*(4), 181-183. https://doi.org/10.1097/nnd.0000000000001069

Kefalaki, M. (2023). Education for Sustainable Development (ESD) in the Greek Education System. *Journal of Applied Learning & Teaching*, *6*(1). https://doi.org/10.37074/jalt.2023.6.1.ss3

Kellenberg, F., Schmidt, J., & Werner, C. (2017). The Adult Learner: Self-Determined, Self-Regulated, and Reflective. *Journal of Pedagogy and Psychology Signum Temporis*, *9*(1), 23-29. https://doi.org/10.1515/sigtem-2017-0001

Keller, D. S., Grossman, R. C., & Winter, D. C. (2020). Choosing the New Normal for Surgical Education Using Alternative Platforms. *Surgery (Oxford)*, *38*(10), 617-622. https://doi.org/10.1016/j.mpsur.2020.07.017

Kem, D. (2022). Personalised and Adaptive Learning: Emerging Learning Platforms in the Era of Digital and Smart Learning. *International Journal of Social Science and Human Research*, *05*(02). https://doi.org/10.47191/ijsshr/v5-i2-02

Kennedy, A. M., Veer, E., & Kemper, J. A. (2022). Social Marketing AS Pedagogy. *Journal of Social Marketing*, *12*(3), 337-353. https://doi.org/10.1108/jsocm-08-2021-0192

Kennedy, E., Chase, E., Pherali, T., Witthaus, G., Laurillard, D., Chan, M. J., Tyrosvoutis, G., Sa Phyo Arkar Myo, H., San, S. S., Win, N., & Charoensukaran, W. (2026). Collaborative Co-design in

Action: Reimagining Teacher Professional Development for Crisis-affected Contexts. *British journal of educational technology*, *57*(3), 797-821. https://doi.org/10.1111/bjet.70054

Kertechian, K. S., & El-Farr, H. (2024). Dissecting the Paradox of Progress: The Socioeconomic Implications of Artificial Intelligence. https://doi.org/10.5772/intechopen.1004872

Kesylė, K., & Melnikova, J. (2024). Unveiling the Potential of Learning Analytics in Learning Design: Theoretical Implications for General Education. *Regional Formation and Development Studies*, 59-71. https://doi.org/10.15181/rfds.v42i1.2612

Ketev, K., Rash, I., Sandeva, M., Marinov, B., Bosheva, M., & Qayumi, K. (2023). Comparison of Knowledge Retention After the Use of a Virtual Patient Versus a High-Fidelity Physical Simulator and Traditional Training. *Creative Education*, *14*(06), 1217-1229. https://doi.org/10.4236/ce.2023.146077

Khabanyane, K. E., Maimane, J. R., & Ramabenyane, M. J. (2014). A Critical Reflection on Transformative Learning as Experienced by Student-Teachers During School-Based Learning. *Mediterranean Journal of Social Sciences*. https://doi.org/10.5901/mjss.2014.v5n27p452

Khadka, J. (2020). Andragogy: Process and Context Based (PCB) Model for Adult Learners. *Jaas*. https://doi.org/10.7176/jaas/63-04

Khalid, A. F., George, M. A., Eggen, C., Sritharan, A., Wali, F., & Viens, A. M. (2023). Using Simulation-Based Experiential Learning to Increase Students' Ability to Analyze Increasingly Complex Global Health Challenges: A Mixed Methods Study. https://doi.org/10.1101/2023.12.19.23300262

Khalid, S., Ifran, S., Huang, X., Tadesse, E., & Dainkun, J. (2026). <scp>AI</Scp> -Integrated Professional Development for Pre-Service Teachers: A Systematic Review of Impacts on Pedagogical Strategies and Technological Self-Efficacy. *European journal of education*, *61*(2). https://doi.org/10.1111/ejed.70547

Khalil, M. (2020). MOLAM: A Mobile Multimodal Learning Analytics Conceptual Framework to Support Student Self-Regulated Learning. https://doi.org/10.35542/osf.io/avq2m

Khalil, M., & Ebner, M. (2016). What Massive Open Online Course (MOOC) Stakeholders Can Learn From Learning Analytics? , 1-30. https://doi.org/10.1007/978-3-319-17727-4_3-1

Khalil, M. K., & Elkhider, I. A. (2016). Applying Learning Theories and Instructional Design Models for Effective Instruction. *Ajp Advances in Physiology Education*, *40*(2), 147-156. https://doi.org/10.1152/advan.00138.2015

Khan, H. M. N. (2024). Assessing the Alignment of Teaching Strategies With Respect to Malcolm Knowles' Model of Andragogy for Prospective Teachers. *Global Educational Studies Review*, *IX*(III), 164-171. https://doi.org/10.31703/gesr.2024(ix-iii).16

Khan, J. S. (2012). Concept Mapping. *The Professional Medical Journal*, *19*(05), 588-596. https://doi.org/10.29309/tpmj/2012.19.05.2405

Khan, N., Gilliar, W., Bamrah, J. S., & Dave, S. (2022). Post-Covid-19: Can Digital Solutions Lead to a More Equitable Global Healthcare Workforce? *Bjpsych International*, *20*(1), 18-23. https://doi.org/10.1192/bji.2022.12

Khandekar, S. (2021). Paulo Freirean Ideas at the Grassroots: From Problem Posing Education to Critical Consciousness. *The International Journal of Community and Social Development*, *3*(1), 30-44. https://doi.org/10.1177/25166026211002937

Khanyase, S. F. (2022). *Mitigating Violence Directed at Teachers: A Narrative Inquiry* UNIVERSITY OF KWAZULU-NATAL].

Kharroubi, S., & ElMediouni, A. (2024). Conceptual Review: Cultivating Learner Autonomy Through Self-Directed Learning &Amp; Self-Regulated Learning: A Socio-Constructivist Exploration. *International Journal of Language and Literary Studies*, *6*(2), 276-296. https://doi.org/10.36892/ijlls.v6i2.1649

Khatib, M., Sarem, S. N., & Hamidi, H. (2013). Humanistic Education: Concerns, Implications and Applications. *Journal of Language Teaching and Research*, *4*(1). https://doi.org/10.4304/jltr.4.1.45-51

Khatri, B. B., & Karki, P. D. (2023). Artificial Intelligence (AI) in Higher Education: Growing Academic Integrity and Ethical Concerns. *Nepalese Journal of Development and Rural Studies*, *20*(01), 1-7. https://doi.org/10.3126/njdrs.v20i01.64134

Khattab, A., & Wong, D. (2018). Integrating Western and Arab Leadership Development Practices: An Example of the Challenge Bridging Global and Local Adult Learning Perspectives. *Andragoske Studije*(2), 65-85. https://doi.org/10.5937/andstud1802065k

Khatter, A., Thalaachawr, K., & Blyth, M. (2024). Student Engagement and Fostering Ownership of Learning. *Journal of Applied Learning & Teaching*, *7*(1). https://doi.org/10.37074/jalt.2024.7.1.38

Khokhotva, O., & Albizuri, I. E. (2019). Student Voice in Lesson Study as a Space for EFL Teachers' Learning: A Case Study in Kazakhstan. *International Journal for Lesson and Learning Studies*, *9*(2), 153-166. https://doi.org/10.1108/ijlls-06-2019-0054

Kholil, A., Supriana, E., & Handayanto, S. K. (2022). Enhancing Senior Students Critical Thinking Skills on Thermodynamics Topic Using Problem-Based Learning Integrated With STEM. *Jurnal Riset Pendidikan Fisika*, *7*(1), 11. https://doi.org/10.17977/um058v7i1p11-15

Khosravi, S., Bailey, S. G., Parvizi, H., & Ghannam, R. (2021). Learning Enhancement in Higher Education With Wearable Technology. https://doi.org/10.48550/arxiv.2111.07365

Khozouie, N., & Fotouhi-Ghazvini, F. (2013). M-CRM on Cloud Computing by (M-Learning). *International Journal of Management & Information Technology*, *6*(1), 738-743. https://doi.org/10.24297/ijmit.v6i1.755

Khurma, O. A., Albahti, F., Ali, N., & Bustanji, A. (2024). AI ChatGPT and Student Engagement: Unraveling Dimensions Through PRISMA Analysis for Enhanced Learning Experiences. *Contemporary Educational Technology*, *16*(2), ep503. https://doi.org/10.30935/cedtech/14334

Khuzwayo, M. E. (2022). Towards the Development of the Decolonized Pedagogy for Higher Education in South Africa: A Students' Perspective. https://doi.org/10.5772/intechopen.101287

Kian, T. W., Sunar, M. S., & Su, G. E. (2022). The Analysis of Intrinsic Game Elements for Undergraduates Gamified Platform Based on Learner Type. *Ieee Access*, *10*, 120659-120679. https://doi.org/10.1109/access.2022.3218625

Kichukova, K., & Taneva, T. (2021). Achievement Motivation and Attitude of Medical Laboratory Assistants to Continuing Education. *The Education and Science Journal*, *23*(6), 185-215. https://doi.org/10.17853/1994-5639-2021-6-185-215

Kidman, G., & Casinader, N. (2019). Developing Teachers' Environmental Literacy Through Inquiry-Based Practices. *Eurasia Journal of Mathematics Science and Technology Education*, *15*(6). https://doi.org/10.29333/ejmste/103065

Kiely, R., Sandmann, L. R., & Truluck, J. (2004). Adult Learning Theory and the Pursuit of Adult Degrees. *New Directions for Adult and Continuing Education*, *2004*(103), 17-30. https://doi.org/10.1002/ace.145

Kift, S. (2025). Foreword. *Journal of Teaching and Learning for Graduate Employability*, *12*(1), i-v. https://doi.org/10.21153/jtlge2021vol12no1art1015

Kiiskilä, P., Kukkonen, A., & Pirkkalainen, H. (2023). Are Micro-Credentials Valuable for Students? Perspective on Verifiable Digital Credentials. *Sn Computer Science*, *4*(4). https://doi.org/10.1007/s42979-023-01797-y

Kijmongkolvanich, S., Sirisawat, S., & Egwutvongsa, S. (2024). Creative Thinking: The Development of Lifelong Learning With the Principle of Andragogy for the Elderly Group in Northern Thailand. *Kuey*. https://doi.org/10.53555/kuey.v30i4.1162

Killam, L. A., & Camargo-Plazas, P. (2021). Revisioning Assessment and Evaluation in Nursing Education Through Critical Caring Pedagogy. *Advances in Nursing Science*, *45*(1), E15-E30. https://doi.org/10.1097/ans.0000000000000382

Killingberg, N. M., Kubberød, E., & Blenker, P. (2020). Preparing for a Future Career Through Entrepreneurship Education: Towards a Research Agenda. *Industry and Higher Education*, *35*(6), 713-724. https://doi.org/10.1177/0950422220969635

Kim, B. (2023). Looking Back and Looking Forward. *Canadian Journal for the Study of Adult Education*, *35*(02). https://doi.org/10.56105/cjsae.v35i02.5597

Kim, G.-j. (2015). Changer Le Paradigme De L'apprentissage Dans La Région Asie-Pacifique. *Revue Internationale D Éducation De Sèvres*, *68*, 37-47. https://doi.org/10.4000/ries.4337

Kim, H., Caytiles, R. D., & Kim, T.-h. (2011). Design of an Effective WSN-Based Interactive U-Learning Model. *International Journal of Distributed Sensor Networks*, *8*(1), 514836. https://doi.org/10.1155/2012/514836

Kim, H. J. (2024). Guidelines for Culturally and Linguistically Responsive Instruction in Higher Education. *Tesol Journal*, *15*(4). https://doi.org/10.1002/tesj.807

Kim, H. J., Kong, Y., & Tirotta-Esposito, R. (2023). Promoting Diversity, Equity, and Inclusion: An Examination of Diversity-Infused Faculty Professional Development Programs. *Journal of Higher Education Theory and Practice*, *23*(11). https://doi.org/10.33423/jhetp.v23i11.6224

Kim, J., Kim, H., Park, E. C., & Jang, S. Y. (2023). The Association Between the Type of Offline Social Participation and Digital Literacy Among Korean Older Adults. https://doi.org/10.21203/rs.3.rs-3309790/v1

Kim, J., Livingston, M. A., Jin, B., & Hwang, J. (2023). Fundamentals of Digital Health Literacy: A Scoping Review of Identifying Core Competencies to Use in Practice. *Adult Learning*, *35*(3), 131-142. https://doi.org/10.1177/10451595231178298

Kim, J., You, J., Yoo, M., & Jang, C. S. (2026). Role of Learning in Human Resource Development: Bibliometrics and Topic Modeling of Articles in AHRD Journals. *Advances in Developing Human Resources*. https://doi.org/10.1177/15234223261426473

Kim, M. K., Kim, S. M., & Bilir, M. K. (2014). Investigation of the Dimensions of Workplace Learning Environments (WLEs): Development of the WLE Measure. *Performance Improvement Quarterly*, *27*(2), 35-57. https://doi.org/10.1002/piq.21170

Kim, Y. (2024). Application of Social Constructivism in Medical Education. *Korean Medical Education Review*, *26*(Suppl 1), S31-S39. https://doi.org/10.17496/kmer.23.044

Kimmel, S. B., Gaylor, K. P., Grubbs, M. R., & Hayes, J. B. (2012). Good Times to Hard Times: An Examination of Adult Learners' Enrollment From 2004–2010. *Journal of Behavioral and Applied Management*, *14*(1). https://doi.org/10.21818/001c.17903

King, M., Marshall, A., & Zaharchuk, D. (2017). Responding to the Global Skills Crisis. *Strategy and Leadership*, *45*(2), 33-41. https://doi.org/10.1108/sl-02-2017-0015

Kinsella, E. A. (2008). Embodied Reflection and the Epistemology of Reflective Practice. *Journal of Philosophy of Education*, *41*(3), 395-409. https://doi.org/10.1111/j.1467-9752.2007.00574.x

Kinsella, E. A. (2010). Professional Knowledge and the Epistemology of Reflective Practice. *Nursing Philosophy*, *11*(1), 3-14. https://doi.org/10.1111/j.1466-769x.2009.00428.x

Kioupi, V., Nawire, A. W., Musungu, S., Nzuve, F., & Giannopoulos, G. A. (2023). Policy and Practice on Inclusive Higher Education in the UK and Kenya: A Theoretical Framework and Recommendations. *Sustainability*, *15*(18), 13540. https://doi.org/10.3390/su151813540

Kiramba, L. K., & Smith, P. H. (2019). "Her Sentence Is Correct, Isn't It?": Regulative Discourse in English Medium Classrooms. *Teaching and teacher education*, *85*, 105-114. https://doi.org/10.1016/j.tate.2019.06.011

Kirch, S. A., & Sadofsky, M. J. (2021). Medical Education From a Theory–Practice–Philosophy Perspective. *Academic Pathology*, *8*, 23742895211010236. https://doi.org/10.1177/23742895211010236

Kissinger, J. (2013). The Social &Amp; Mobile Learning Experiences of Students Using Mobile E-Books. *Online Learning*, *17*(1). https://doi.org/10.24059/olj.v17i1.303

Kitagawa, K. (2020). Disaster Risk Reduction Activities as Learning. *Natural Hazards*, *105*(3), 3099-3118. https://doi.org/10.1007/s11069-020-04443-5

Kittel, A., Kunz, R., & Seufert, T. (2021). Self-Regulation in Informal Workplace Learning: Influence of Organizational Learning Culture and Job Characteristics. *Frontiers in Psychology*, *12*. https://doi.org/10.3389/fpsyg.2021.643748

Kittel, A., Spittle, M., Larkin, P., & Spittle, S. (2023). 360°VR: Application for Exercise and Sport Science Education. *Frontiers in Sports and Active Living*, *5*. https://doi.org/10.3389/fspor.2023.977075

Kiwelu, J. E. M., & Ogbonna, J. E. (2020). Integrating Online Autodidacticism in Lifelong Learning for Adults in Sub-Saharan Africa. *Journal of Applied Science Information and Computing*, *1*(1), 49-56. https://doi.org/10.59568/jasic-2020-1-1-06

Kizilcec, R. F., & Cohen, G. L. (2017). Eight-Minute Self-Regulation Intervention Raises Educational Attainment at Scale in Individualist but Not Collectivist Cultures. *Proceedings of the National Academy of Sciences, 114*(17), 4348-4353. https://doi.org/10.1073/pnas.1611898114

Klassen, M., Reeve, D., Evans, G. J., Rottmann, C., Sheridan, P., & Simpson, A. (2020). Engineering: Moving Leadership From the Periphery to the Core of an Intensely Technical Curriculum. *New Directions for Student Leadership, 2020*(165), 113-124. https://doi.org/10.1002/yd.20373

Kline, R. (2023). Paradigm Lost? Reflections on the Effectiveness of NHS Approaches to Improving Employment Relations. *BMJ Leader, 7*(4), 314-318. https://doi.org/10.1136/leader-2022-000729

Kloubert, T. (2019). Promoting Perspective Transformation in Post-Totalitarian Societies. *Journal of Transformative Education, 18*(1), 8-23. https://doi.org/10.1177/1541344619876274

Knapke, J. M., Hildreth, L., Molano, J. R. V., Schuckman, S., Blackard, J. T., Johnstone, M., Kopras, E. J., Lamkin, M., Lee, R. C., Kues, J. R., & Mendell, A. (2024). Andragogy in Practice: Applying a Theoretical Framework to Team Science Training in Biomedical Research. *British Journal of Biomedical Science, 81*. https://doi.org/10.3389/bjbs.2024.12651

Knowles, M. S. (1973). *The Adult Learner: A Neglected Species*. Gulf Publishing.

Knowles, M. S. (1975). *Self-Directed Learning: A Guide for Learners and Teachers*. Cambridge Adult Education.

Knowles, M. S. (1980). *The Modern Practice of Adult Education: From Pedagogy to Andragogy* (Revised edition ed.). Cambridge Adult Education.

Knowles, M. S. (1984). *Andragogy in Action*. Jossey-Bass.

Knowles, M. S., Holton, E. F., III, & Swanson, R. A. (2015). *The Adult Learner: The Definitive Classic in Adult Education and Human Resource Development* (8th ed.). Routledge.

Knowles, M. S., Holton, E. F., & Swanson, R. A. (2020). *The Adult Learner* (9th ed.). Routledge.

Koca, D., & Ortakaya, Ü. İ. (2023). An Examination of the Role of Vocational Training Centers in Ensuring School-Industry Cooperation From Employer's Perspective: The Ostim Case. *Opus Journal of Society Research*. https://doi.org/10.26466/opusjsr.1349283

Koirala, S. (2019). Toward a Negotiated Autonomy: Culture, Teaching Perceptions, and Participation of Bhutanese Refugees in an Adult Migrant English Program in Australia. *Tesol Journal, 11*(1). https://doi.org/10.1002/tesj.460

Kokkos, A. (2022). Transformation Theory as a Framework for Understanding Transformative Learning. *Adult Education Critical Issues, 2*(2), 20-33. https://doi.org/10.12681/haea.32541

Kolb, D. A. (1984a). *Experiential Learning*. Prentice-Hall.

Kolb, D. A. (1984b). *Experiential Learning: Experience as the Source of Learning and Development*. Prentice Hall.

Kolivand, P., Azari, S., Bakhtiari, A., Namdar, P., Ayyoubzadeh, S. M., Rajaie, S., & Ramezani, M. (2025). AI Applications in Disaster Governance With Health Approach: A Scoping Review. *Archives of Public Health, 83*(1). https://doi.org/10.1186/s13690-025-01712-2

Korte, R., & Lin, S. (2012). Getting on Board: Organizational Socialization and the Contribution of Social Capital. *Human Relations, 66*(3), 407-428. https://doi.org/10.1177/0018726712461927

Korteling, J. E., Boer-Visschedijk, G. C. v. d., Blankendaal, R., Boonekamp, R., & Eikelboom, A. R. (2021). Human- Versus Artificial Intelligence. *Frontiers in Artificial Intelligence, 4*. https://doi.org/10.3389/frai.2021.622364

Kosbar, Y. (2025). TransformED Futures: Towards Human-Centred, Ethical and Inclusive Use and Governance of AI in Higher Education. *World Futures Review, 17*(4), 254-271. https://doi.org/10.1177/19467567251406292

Košmerl, T., Radovan, M., & Radovan, D. M. (2024). The Funding Puzzle in Adult Education for Sustainable Development: Case Study of Slovenia. https://doi.org/10.20944/preprints202408.0302.v1

Kotzee, B., & Palermos, S. O. (2021). The Teacher Bandwidth Problem: MOOCs, Connectivism, and Collaborative Knowledge. *Educational Theory, 71*(4), 497-518. https://doi.org/10.1111/edth.12495

Koukoulidis, N. M., Kotluk, N., & Brown, J. C. (2024). The Status of Culturally Relevant Teacher Education in the European Context: A Systematic Review of Research. *Review of Education, 12*(3). https://doi.org/10.1002/rev3.70002

Koulaouzides, G., & Romano, A. (2022). Una Faccia, Una Razza: Similarities, Differences, and Parallels in Adult Education Policy Development in Greece and Italy. *International Journal of Education Policy and Leadership*, *18*(1). https://doi.org/10.22230/ijepl.2022v18n1a1203

Koulaouzidis, G. A., Eschenbacher, S., Romano, A., & Soeiro, D. (2024). Unveiling and Challenging the Identity, Formation, and Evolution of Transformation Theory. *Adult Education Critical Issues*, *4*(1), 49-56. https://doi.org/10.12681/haea.36292

Kour, R., & Ashraf, M. (2018). The Impact of Women Participative Leadership Style on Employees' Satisfaction and Employees' Performance: Evidence From JandK Education Sector. *Prestige International Journal of Management & It - Sanchayan*, *07*(01), 84-95. https://doi.org/10.37922/pijmit.2018.v07i01.006

Koutsoukos, M. (2022). GR-IDEA: An Innovative Model of Mentoring in Adult Education. *Education Quarterly Reviews*, *5*(3). https://doi.org/10.31014/aior.1993.05.03.540

Kowalski, R., & Russell, C. K. (2020). How We SEE Is How We Learn: Reflection in the Workplace. *Advances in Developing Human Resources*, *22*(3), 239-247. https://doi.org/10.1177/1523422320927294

Kragt, D., & Day, D. V. (2020). Predicting Leadership Competency Development and Promotion Among High-Potential Executives: The Role of Leader Identity. *Frontiers in Psychology*, *11*. https://doi.org/10.3389/fpsyg.2020.01816

Kral, I. (2016). From the Local to the Global. 61-76. https://doi.org/10.1007/978-94-6300-444-2_4

Krichen, M. (2023). How Artificial Intelligence Can Revolutionize Software Testing Techniques. 189-198. https://doi.org/10.1007/978-3-031-27499-2_18

Krichevsky, B. (2021). University-School Divide: The Original Problem in Teacher Education. *Human Arenas*, *6*(2), 264-291. https://doi.org/10.1007/s42087-021-00213-2

Krikorian, M. (2020). Traditional Narratives of Higher Education Cultural Tensions and Ethical Considerations in Adult Learning. *Journal of Critical Thought and Praxis*, *9*(1). https://doi.org/10.31274/jctp.9744

Kriner, B. A., Coffman, K. A., Adkisson, A. C., Putman, P. G., & Monaghan, C. (2015). From Students to Scholars. *Adult Learning*, *26*(2), 73-80. https://doi.org/10.1177/1045159515573021

Krishna, S. (2022). Impact of Talent Management on Employee Performance and Retention. *Interantional Journal of Scientific Research in Engineering and Management*, *06*(06). https://doi.org/10.55041/ijsrem14571

Krishna, S. A., Mythiri, B., & Karthika, V. K. (2024). Reimagining English Language Teaching: Towards Developing a Heutagogical Framework. *Higher Education for the Future*, *11*(1), 76-89. https://doi.org/10.1177/23476311231222964

Krishna, V. (2019). Developing a Competency-Based Credentialing Framework for National Human Resource Development. *International Journal of Human Resource Development Practice Policy & Research*, *4*(1), 1-18. https://doi.org/10.22324/ijhrdppr.4.102

Krishnan, L. R. K., & Poorani, S. (2022). Emerging Trends in Training Knowledge Workers; New Economy. *International Journal of Learning and Development*, *12*(4), 26. https://doi.org/10.5296/ijld.v12i4.20413

Kristiawan, D., Bashar, K., & Pradana, D. A. (2024). Artificial Intelligence in English Language Learning: A Systematic Review of AI Tools, Applications, and Pedagogical Outcomes. *The Art of Teaching English as a Foreign Language (Tatefl)*, *5*(2), 207-218. https://doi.org/10.36663/tatefl.v5i2.912

Kruger, R. (2015). Reflective Practice: A Comparative Dimensional Analysis of the Concept in Nursing and Education Studies. *Nursing Forum*, *51*(2), 139-150. https://doi.org/10.1111/nuf.12129

Kubberød, E., & Pettersen, I. B. (2018). The Role of Peripherality in Students' Entrepreneurial Learning. *Education + Training*, *60*(1), 2-15. https://doi.org/10.1108/et-06-2016-0106

Kuchumova, G., Kuzhabekova, A., Almukhambetova, A., & Nurpeissova, A. (2024). Women's Science, Technology, Engineering, and Mathematics Persistence After University Graduation: Insights From Kazakhstan. *Journal of Career Development*, *51*(3), 408-428. https://doi.org/10.1177/08948453241251466

Kucirkova, N. (2018). Personalised Learning With Digital Technologies at Home and School: Where Is Children's Agency? , 133-153. https://doi.org/10.1108/978-1-78714-879-620181009

Kuczera, M. (2017). Striking the Right Balance. https://doi.org/10.1787/995fff01-en

Kuczera, M., & Field, S. (2018). Share of 25-Year-Olds and Older Among Current Apprentices (2012, 2014). https://doi.org/10.1787/9789264298507-graph2-en

Kuczera, M., T. Bastianić and S. Field. (2018). A Closer Look at the Economics of Training in Israel: Involving Employers Through Youth Apprenticeship and Sectoral Training Levies. 47-62. https://doi.org/10.1787/9789264302051-5-en

Kuechler, W., & Stedham, Y. (2017). Management Education and Transformational Learning: The Integration of Mindfulness in an MBA Course. *Organizational Behavior Teaching Review*, *42*(1), 8-33. https://doi.org/10.1177/1052562917727797

Kuffuor, O., Aggrawal, S., Jaiswal, A., Smith, R. J., & Morris, P. V. (2024). Transformative Pathways: Implementing Intercultural Competence Development in Higher Education Using Kotter's Change Model. *Education Sciences*, *14*(7), 686. https://doi.org/10.3390/educsci14070686

Kuiper, R., Murdock, N. L., & Grant, N. K. (2010). Thinking Strategies of Baccalaureate Nursing Students Prompted by Self-Regulated Learning Strategies. *Journal of Nursing Education*, *49*(8), 429-436. https://doi.org/10.3928/01484834-20100430-01

Kukovetz, B., & Sprung, A. (2014). Is Adult Education a 'White' Business? Professionals With Migrant Backgrounds in Austrian Adult Education. *European Journal for Research on the Education and Learning of Adults*, *5*(2), 161-175. https://doi.org/10.3384/rela.2000-7426.rela9051

Kulal, A., Dinesh, S., Abhishek, N., & Anchan, A. (2024). Digital Access and Learning Outcomes: A Study of Equity and Inclusivity in Distance Education. *International Journal of Educational Management*, *38*(5), 1391-1423. https://doi.org/10.1108/ijem-03-2024-0166

Kuleto, V., Ilić, M., Stănescu, M., Ranković, M., Šević, N. P., Păun, D., & Teodorescu, S. (2021). Extended Reality in Higher Education, a Responsible Innovation Approach for Generation Y and Generation Z. *Sustainability*, *13*(21), 11814. https://doi.org/10.3390/su132111814

Kulo, V., & Cestone, C. (2021). A Bibliometric Analysis of the 100 Most Cited Articles on Problem-Based Learning in Medical Education. https://doi.org/10.31235/osf.io/tnpzs

Kultaieva, M., & Sahuichenko, V. (2019). Transformaciones De Las Prácticas Autodidactas en Los Contextos Culturales Posindustriales. *Dilemas Contemporáneos Educación Política Y Valores*. https://doi.org/10.46377/dilemas.v30i1.1047

Kumah, A. (2025). Adverse Event Reporting and Patient Safety: The Role of a Just Culture. *Frontiers in Health Services*, *5*. https://doi.org/10.3389/frhs.2025.1581516

Kumar, A., Kumar, A., Bhoyar, S., & Mishra, A. K. (2024). Does ChatGPT Foster Academic Misconduct in the Future? *Public Administration and Policy*, *27*(2), 140-153. https://doi.org/10.1108/pap-05-2023-0061

Kumar, J. A., Richard, R. J., Osman, S., & Lowrence, K. (2022). Micro-Credentials in Leveraging Emergency Remote Teaching: The Relationship Between Novice Users' Insights and Identity in Malaysia. *International Journal of Educational Technology in Higher Education*, *19*(1). https://doi.org/10.1186/s41239-022-00323-z

Kumar, N. (2025). Development of a Conceptual Framework for Postgraduate Non-Surgical Aesthetic Curriculum. *International Journal of Aesthetic Plastic Surgery*, *1*(2), 151-160. https://doi.org/10.1177/30499240251344131

Kumar, N., & Santosh, D. (2022). Investigate the Realities and Misconceptions of Multimedia Learning. *Journal of Educational Research and Policies*, *4*(9). https://doi.org/10.53469/jerp.2022.04(09).30

Kumar, S., & Bargavi, S. K. M. (2024). Trust's Significance in Human-Ai Communication and Decision-Making. *Interantional Journal of Scientific Research in Engineering and Management*, *08*(02), 1-10. https://doi.org/10.55041/ijsrem28468

Kunwar, R. (2024). Fostering Student Success: A Prevailing Provision for Dyscalculia and Dyslexia in the School Curriculum and Strategies for Support. *Nepal Journal of Neuroscience*, *21*(1), 29-35. https://doi.org/10.3126/njn.v21i1.63381

Kunwar, R., & Adhikari, S. (2023). An Exploration of the Conceptualization, Guiding Principles, and Theoretical Perspectives of Inclusive Curriculum. *Journal of Contemporary Research in Social Sciences*, *5*(1), 1-13. https://doi.org/10.55214/26410249.v5i1.217

Kuperman, V. (2022). Effects of Linguistic Distance on Cognitive Skills, Health, and Social Outcomes in Canadian Immigrants. *Frontiers in Political Science*, *4*. https://doi.org/10.3389/fpos.2022.874195

Küppers, E. W. U. (2022). „Systemisches Management" – Zukunftsperspektive Jenseits Neoliberaler Ökonomie. *Wirtschaftsinformatik & Management*, *14*(5), 330-337. https://doi.org/10.1365/s35764-022-00426-2

Kusaila, M. M. (2019). Impact of Informal Learning in the Accounting Profession. *Journal of Workplace Learning*, *31*(7), 429-441. https://doi.org/10.1108/jwl-01-2019-0007

Kushch, J., & Korolenko, Y. (2024). «resilience» as a Factor of Health Preservation: Historical Aspect and Modern Understanding. *Вісник Науки Та Освіти*(10(28)). https://doi.org/10.52058/2786-6165-2024-10(28)-617-628

Kušić, S., Vrcelj, S., & Zovko, A. (2020). Sustainability of Adult Education: Andragogues as Change Agents. *International Journal of Social Sciences and Education Research*, *6*(3), 443-449. https://doi.org/10.24289/ijsser.759203

Kusuma, E. A. (2024). The Ethical Perspective of Digital-Technology-Driven Economic Disruption. *International Conference of Business and Social Sciences*, 958-973. https://doi.org/10.24034/icobuss.v4i1.580

Kuzminov, Y., Sorokin, P. S., & Froumin, I. (2019). Generic and Specific Skills as Components of Human Capital: New Challenges for Education Theory and Practice. *Foresight and Sti Governance*, *13*(2), 19-41. https://doi.org/10.17323/2500-2597.2019.2.19.41

Kwenin, I. A. (2026). Teachers' Environmental Sustainability Attitudes and Its Effects on Their Environmental Sustainability Teaching Practices. https://doi.org/10.21203/rs.3.rs-9695642/v1

Kyröläinen, A.-J., & Kuperman, V. (2021). Predictors of Literacy in Adulthood: Evidence From 33 Countries. *Plos One*, *16*(3), e0243763. https://doi.org/10.1371/journal.pone.0243763

Labster. (2025). Virtual labs and simulations. https://www.labster.com

Lacasse, M., Douville, F., Gagnon, J., Simard, C., & Côté, L. (2019). Theories and Models in Health Sciences Education – A Literature Review. *The Canadian Journal for the Scholarship of Teaching and Learning*, *10*(3). https://doi.org/10.5206/cjsotl-rcacea.2019.3.9477

Lacher, S. (2024). Promoting Equal Access in German Adult Education. *Andragoška Spoznanja*, *30*(2), 81-99. https://doi.org/10.4312/as/16259

Lagua, Z. C. G., Lagua, L. A. G., Lagua, S. F. G., & Lagua, S. M. G. (2022). Formación Del Docente Universitario Frente Al Contexto Pluricultural, Pluriétnico Y Plurilingüe. *Concienciadigital*, *5*(4.1), 132-154. https://doi.org/10.33262/concienciadigital.v5i4.1.2406

Lai, C. (2015). Modeling Teachers' Influence on Learners' Self-Directed Use of Technology for Language Learning Outside the Classroom. *Computers & Education*, *82*, 74-83. https://doi.org/10.1016/j.compedu.2014.11.005

Laidlaw, J. M., & Johnson, J. (2025). From 'Pool Noodle Tag' to Building Snow Forts: Experiential Outdoor Education as Pedagogical Job Crafting During the COVID-19 Pandemic. *Journal of Experiential Education*, *49*(1), 216-239. https://doi.org/10.1177/10538259251359572

Laitusis, C. C., & Karvonen, M. (2025). A Case for Reimagining Universal Design of Assessment Systems. *Educational Measurement Issues and Practice*, *44*(3), 5-14. https://doi.org/10.1111/emip.12674

Lajmiri, L. (2016). The Effects of Applying Communication Tools Based on Connectivism Theory on Students' Academic Achievement and the Value of Their Engagement. *The Turkish Online Journal of Design Art and Communication*, *6*(AGSE), 1849-1859. https://doi.org/10.7456/1060agse/067

Lakin, M. B., & Crossman, G. (2004). Utilizing Experiential Learning for Capstone Project Credit. American Society for Engineering Education Annual Conference & Exposition,

Lakkala, S., Beaton, M. C., & Kokko, K. (2024). A Conceptual Analysis of Inclusive Education in Terms of Participation and Agency. *Support for Learning*, *39*(3), 125-136. https://doi.org/10.1111/1467-9604.12477

Lalendle, L. L., Msila, V., & Matlabe, S. M. (2020). Humanising and Decolonising Adult Basic Education and Training (ABET) in South Africa. 103-149. https://doi.org/10.18820/9781928480716/06

Lambini, C. K., Goeschl, A., Wäsch, M., & Wittau, M. (2021). Achieving the Sustainable Development Goals Through Company Staff Vocational Training—The Case of the Federal Institute for

Vocational Education and Training (BIBB) INEBB Project. *Education Sciences*, *11*(4), 179. https://doi.org/10.3390/educsci11040179

Lameras, P., & Arnab, S. (2021). Power to the Teachers: An Exploratory Review on Artificial Intelligence in Education. *Information*, *13*(1), 14. https://doi.org/10.3390/info13010014

LaMonica, S. (2000). ESL CD-ROM Principles and Their Application: A Software Evaluation. https://doi.org/10.15760/etd.7189

Lamtara, S. (2023a). Faculty's Pedagogical Knowledge for Technology-Enhanced Learning in Higher Education. *International Journal of Higher Education Pedagogies*, *4*(2), 1-24. https://doi.org/10.33422/ijhep.v4i2.412

Lamtara, S. (2023b). Faculty's Pedagogical Knowledge for Technology-Enhanced Learning in Higher Education. *International Journal of Higher Education Pedagogies*, *4*, 1-24. https://doi.org/10.33422/ijhep.v4i2.412

Landig, R. C. F., & Landig, A. M. F. (2024). Political Awareness and Civic Consciousness in Fostering Community Engagement of HUMSS Students. *International Journal of Social Science Humanity & Management Research*, *3*(07). https://doi.org/10.58806/ijsshmr.2024.v3i7n20

Landry, J., & Lieres, B. v. (2022). Strengthening Democracy Through Knowledge Mobilization: Participedia – A Citizen-Led Global Platform for Transformative and Democratic Learning. *Journal of Transformative Education*, *20*(3), 206-225. https://doi.org/10.1177/15413446221103191

Lane, A., & Stephens, J. H. (2023). What It Is and Is Not: Pedagogy in Online Nursing Education Delivery. *The International Review of Research in Open and Distributed Learning*, *24*(4), 119-134. https://doi.org/10.19173/irrodl.v24i4.7212

Lang, D. (2023). Lifelong Learning and Micro-credentials: Supporting Workforce Transitions in a Changing Economy.

Lang, J. (2023). Workforce Upskilling: Can universities Meet the Challenges of Lifelong Learning? *International Journal of Information and Learning Technology*, *40*(5), 388-400. https://doi.org/10.1108/ijilt-01-2023-0001

Lang, X., Zeng, T., Ni, S., Jiang, L., Pan, Q., & Wu, M. (2024). Exploring the First-Time Transition to Parenthood in Mainland China: A Qualitative Study on the Experiences of Fathers and Mothers Using the Transition Shock Model. *Frontiers in Psychology*, *15*. https://doi.org/10.3389/fpsyg.2024.1249211

Lange, T. (2012). German Training Revisited: An Appraisal of Corporatist Governance. *Education + Training*, *54*(1), 21-35. https://doi.org/10.1108/00400911211198878

Lassnigg, L. (2024). Perspective Chapter: Sustaining Dual Apprenticeship Systems – Similarities and Differences in Austria, Germany, and Switzerland. https://doi.org/10.5772/intechopen.112561

Lau, J. (2021). Digital Badge Metadata: A Case Study in Quality Assurance. *Journal of Innovation in Polytechnic Education*, *3*(1), 27-36. https://doi.org/10.69520/jipe.v3i1.93

Laubscher, D. (2019). A Teaching–learning Framework for Adaptive Instruction Using Cooperative Learning and Socratic Questioning to Promote Self-Directed Learning: A Systematic Literature Review. 239-282. https://doi.org/10.4102/aosis.2019.bk134.08

Lave, J., & Wenger, E. (1991). *Situated Learning: Legitimate Peripheral Participation*. Cambridge University Press.

Law, H. Y. (2013). Reinventing Teaching in Mathematics Classrooms. *International Journal for Lesson and Learning Studies*, *2*(2), 101-114. https://doi.org/10.1108/20468251311323360

Lawrence, K., Mellifont, D., & Bulkeley, K. (2025). Disability Employment Policy in Australia: Barriers and Facilitators to Workforce Participation. *Social Policy and Administration*, *60*(1), 112-123. https://doi.org/10.1111/spol.13147

Lawrence, K. C., & Maphalala, M. C. (2021). Opportunities for Social Justice in the Curriculum and Pedagogical Practices in Higher Education Institution Spaces in South Africa. *Curriculum Perspectives*, *41*(2), 143-151. https://doi.org/10.1007/s41297-021-00140-w

Lawrence, R. L. (2017). Understanding Collaborative Leadership in Theory and Practice. *New Directions for Adult and Continuing Education*, *2017*(156), 89-96. https://doi.org/10.1002/ace.20262

Lazorak, O., Belkina, O., & Yaroslavova, E. (2021). Changes in Student Autonomy via E-Learning Courses. *International Journal of Emerging Technologies in Learning (Ijet)*, *16*(17), 209. https://doi.org/10.3991/ijet.v16i17.23863

Le, T. D., Dinh, T. D., Do, P. H., Pham, V. D., & Nguyen, N. S. (2026). Beyond the Flag: A Framework for Integrating Cybersecurity Competitions Into K-12 Education for Cognitive Apprenticeship and Ethical Skill Development. https://doi.org/10.48550/arxiv.2602.16921

Leal Filho, W., Raath, S., Lazzarini, B., Vargas, V. R., de Souza, L., Anholon, R., Quelhas, O. L. G., Haddad, R., Klavins, M., & Orlovic Lovren, V. (2018). The Role of Transformation in Learning and Education for Sustainability. *Journal of Cleaner Production*, *199*, 286-295. https://doi.org/10.1016/j.jclepro.2018.07.017

Lecas, J.-C. (2006). Behaviourism and the Mechanization of the Mind. *Comptes Rendus Biologies*, *329*(5-6), 386-397. https://doi.org/10.1016/j.crvi.2006.03.009

Ledezma-Ramírez, D. F. (2024). Aprendizaje Mediante Casos Prácticos Integradores en La Enseñanza De La Ingeniería. Un Caso De Estudio Sobre Dinámica Estructural. https://doi.org/10.1590/scielopreprints.8055

Lee-Remond, S., & Ballouk, H. (2026). Evolving Design Education: Mapping the Integration of Emerging Technologies. *Proceedings of the Design Society*, *6*, 3061-3070. https://doi.org/10.1017/pds.2026.10664

Lee, C. S., Yeo, R. Q. Y., Goh, D. H., Ang, R. P., & Ng, B. (2024). A Transitional Approach to Examine the Influences of Information Practices on Lifelong Learning. *Proceedings of the Association for Information Science and Technology*, *61*(1), 541-546. https://doi.org/10.1002/pra2.1055

Lee, D. (2012). Apprenticeships in England: An Overview of Current Issues. *Higher Education Skills and Work-Based Learning*, *2*(3), 225-239. https://doi.org/10.1108/20423891211271764

Lee, F., & Wong, S. W. (2025). Surgical Education Reimagined: The Convergence of Learning Theories and Artificial Intelligence. *BMC Medical Education*, *26*(1). https://doi.org/10.1186/s12909-025-08039-0

Lee, M.-K. (2016). To Be Autonomous or Not to Be: Issues of Subsuming Self-Determination Theory Into Research on Language Learner Autonomy. *Tesol Quarterly*, *51*(1), 220-228. https://doi.org/10.1002/tesq.343

Lee, M. (2022). Revisiting the Faure Report: Contemporary Legacy and Challenged Legitimacy. *International Review of Education*, *68*(5), 691-707. https://doi.org/10.1007/s11159-022-09979-1

Lee, W. C., & Tan, B. C. Y. (2023). Workplace Learning Strategies, Enablers, and Challenges in the Context of Digital Innovation. *Journal of Workplace Learning*, *35*(2), 192-209. https://doi.org/10.1108/jwl-04-2022-0045

Lefay, G., Briand, C., Sauvageau, A., Drolet, M.-J., Vachon, B., Luconi, F., Thomas, A., & Nadeau-Tremblay, J. (2025). Philosophical Analysis of the Recovery College Learning Model: Characterization and Connections to Learning Theories. *Frontiers in Psychiatry*, *16*. https://doi.org/10.3389/fpsyt.2025.1613074

Lehmann, W. (2000). Is Germany's Dual System Still a Model for Canadian Youth Apprenticeship Initiatives? *Canadian Public Policy*, *26*(2), 225. https://doi.org/10.2307/3552557

Leicht, A., Heiss, J., & Byun, W. J. (Eds.). . (2018). *Issues and Trends in Education for Sustainable Development.* UNESCO.

Lemanski, T., Mewis, R., & Overton, T. (2016). An Introduction to the Recent Literature on Approaches to Work-Based Learning. *New Directions in the Teaching of Physical Sciences*(6), 3-9. https://doi.org/10.29311/ndtps.v0i6.370

Lembo, D., Daga, F. A., Calì, C., Garbossa, D., Manfredi, M., Odetto, L., Ostacoli, L., Paccotti, P., Raimondo, S., Reimondo, G., & Sciascia, S. (2024). Early Introduction of Simulation in the Medical Curriculum: The MedInTo Perspective. *Frontiers in Medicine*, *10*. https://doi.org/10.3389/fmed.2023.1280592

Lemieux, S. L. (2004). *Depreciating Print: Putting Video-Based Instruction to the Test*

Lemmetty, S., & Brem, A. (2023). Editorial Transformative Learning as a Catalyst for Innovation in Times of Uncertainty. *Ieee Engineering Management Review*, *51*(4), 6-9. https://doi.org/10.1109/emr.2023.3335694

Lemmetty, S., & Collin, K. (2020). Self-Direction. 1-7. https://doi.org/10.1007/978-3-319-98390-5_147-1

Lengrand, P. (1975). *An Introduction to Lifelong Education*. UNESCO Institute for Education.

Lensjø, M. (2020). Stories of Learning: A Case Study of Norwegian Plumbers and Apprentices in TVET at the Construction Site and in a Training Agency. *International Journal for Research in Vocational Education and Training*, *7*(2), 148-166. https://doi.org/10.13152/ijrvet.7.2.2

Leong, D. C. P. (2022). Emotional Stability and Motivation of 21st Century Learners: A Comparative Review of Learning Theories. *Quantum Journal of Social Sciences and Humanities*, *3*(6), 68-80. https://doi.org/10.55197/qjssh.v3i6.190

Leong, T. G., & Raja Lailatul Zuraida Raja Maamor, S. (2023). Development and Validation of ConProRet-A Module for the Retention of Algebraic Conceptual and Procedural Knowledge for Form 1 Student. *Educatum Journal of Science Mathematics and Technology*, *10*(2), 58-69. https://doi.org/10.37134/ejsmt.vol10.2.7.2023

Leow, A., Chua, S., Billett, S., & Le, A. H. (2023). Employers' Perspectives of Effective Continuing Education and Training in Singapore. *Higher Education Skills and Work-Based Learning*, *13*(2), 217-232. https://doi.org/10.1108/heswbl-05-2022-0115

Leppan, R. G., Botha, R. A., & Niekerk, J. v. (2018). Process Model for Differentiated Instruction Using Learning Analytics. *South African Computer Journal*, *30*(2). https://doi.org/10.18489/sacj.v30i2.481

Leroy, N., Bressoux, P., Sarrazin, P., & Trouilloud, D. (2013). Un Modèle Sociocognitif Des Apprentissages Scolaires : Style Motivationnel De L'enseignant, Soutien Perçu Des Élèves Et Processus Motivationnels. *Revue Française De Pédagogie*, *182*, 71-92. https://doi.org/10.4000/rfp.4008

Leshem, S., Zion, N., & Friedman, A. (2015). A Dream of a School. *Sage Open*, *5*(4). https://doi.org/10.1177/2158244015621351

Letseka, M., & Seeletso, M. (2021). Open Learning in the African Context: Challenges and Possibilities for Self-Directed Learning. 175-186. https://doi.org/10.1007/978-981-16-4099-5_9

Leupin, R. E. (2015). Aprendizaje Transformativo Y Programas De Investigación en El Desarrollo Docente Universitario. *Redu Revista De Docencia Universitaria*, *13*(3), 309. https://doi.org/10.4995/redu.2015.5431

Levesque, L. L. (2021). Trust in Classroom-as-Organization Simulations: Parallel Experiences of Participants and Facilitators. *Simulation & Gaming*, *52*(3), 333-345. https://doi.org/10.1177/1046878120987583

Levy, D. (2017). Online, Blended and Technology-Enhanced Learning: Tools to Facilitate Community College Student Success in the Digitally-Driven Workplace. *Contemporary Issues in Education Research (Cier)*, *10*(4), 255-262. https://doi.org/10.19030/cier.v10i4.10039

Lewis, A. (2024). Exploring the Impact of Gamified Language Learning Platforms on Adult Second Language Acquisition: A Case Study in Spanish Education. *The Review of Contemporary Scientific and Academic Studies*, *4*(2). https://doi.org/10.55454/rcsas.4.02.2024.003

Lewis, C., & Tolgensbakk, I. (2019). Public Policy on Career Education, Information, Advice and Guidance: Developments in the United Kingdom and Norway. https://doi.org/10.4337/9781788118798.00019

Lewis, K. L., & Novak, A. M. (2023). How Professional Learning Impacts Gifted Educators' Perceptions and Practices: a Phased Analysis of the Four-Zone Equity Driven PL Model. https://doi.org/10.35542/osf.io/2hras

Lewis, L. (2020). Adult Community Learning, Wellbeing, and Mental Health Recovery. 627-658. https://doi.org/10.1007/978-3-030-14625-2_16

Lewitzky, R. A., & Weaver, K. D. (2022). Developing Universal Design for Learning Asynchronous Training in an Academic Library. *Partnership the Canadian Journal of Library and Information Practice and Research*, *16*(2), 1-18. https://doi.org/10.21083/partnership.v16i2.6635

Leyretana, K., & Trinidad, J. E. (2021). Predicting or Preventing Lifelong Learning? The Role of Employment, Time, Cost, and Prior Achievement. *Journal of Adult and Continuing Education*, *28*(2), 658-673. https://doi.org/10.1177/14779714211054555

Li, B., Sjöström, J., Ding, B., & Eilks, I. (2022). Education for Sustainability Meets Confucianism in Science Education. *Science & Education*, *32*(4), 879-908. https://doi.org/10.1007/s11191-022-00349-9

Li, H. (2025). Reflective Practice for Pre-Service Teachers' Professional Development. *Sage Open*, *15*(3). https://doi.org/10.1177/21582440251363136

Li, H. P., Suo, Z., Wu, C., Wider, W., Fauzi, M. A., Wu, H., & Sivaraman, S. (2026). Science Mapping of Lifelong Learning in the Digital Age: Current Trends and Future Prospects. *European journal of education*, *61*(2). https://doi.org/10.1111/ejed.70655

Li, J. (2023). The Diffusion of Automation Technology in Global Supply Chains and Its Heterogeneous Effects on the Labor Market. *Law and Economy*, *2*(9), 15-30. https://doi.org/10.56397/le.2023.09.03

Li, J. (2025). Navigating Ethical Dilemmas in AI-Enhanced Education. *International Journal of Knowledge Management*, *21*(1), 1-21. https://doi.org/10.4018/ijkm.382384

Li, K. C., & Wong, B. T. M. (2023). Artificial Intelligence in Personalised Learning: A Bibliometric Analysis. *Interactive Technology and Smart Education*, *20*(3), 422-445. https://doi.org/10.1108/itse-01-2023-0007

Li, L., Grimshaw, J., Nielsen, C. P., Judd, M., Coyte, P. C., & Graham, I. D. (2009a). Evolution of Wenger's Concept of Community of Practice. *Implementation Science*, *4*(1). https://doi.org/10.1186/1748-5908-4-11

Li, L., Grimshaw, J., Nielsen, C. P., Judd, M., Coyte, P. C., & Graham, I. D. (2009b). Use of Communities of Practice in Business and Health Care Sectors: A Systematic Review. *Implementation Science*, *4*(1). https://doi.org/10.1186/1748-5908-4-27

Li, M., Holstein, J., & Wedekind, V. (2023). The Historical Shifts of in/Formality of Learning Within Craft Skills Ecosystems in the United Kingdom. *International Journal of Training and Development*, *27*(3-4), 405-421. https://doi.org/10.1111/ijtd.12308

Li, M., Mobley, W. H., & Kelly, A. (2013). When Do Global Leaders Learn Best to Develop Cultural Intelligence? An Investigation of the Moderating Role of Experiential Learning Style. *Academy of Management Learning and Education*, *12*(1), 32-50. https://doi.org/10.5465/amle.2011.0014

Li, P., Liu, J., Yao, J., & Luo, L. (2026). <scp>AI</Scp> -Enhanced E-Learning for Intellectual Property Education: Cultivating International Talents Through Accessible and Personalised Digital Training. *European journal of education*, *61*(1). https://doi.org/10.1111/ejed.70480

Li, W. (2018). Research on Training Difficulties and Breakthrough Path of High Skilled Talents in China's Manufacturing Industry. https://doi.org/10.2991/mmetss-18.2018.103

Li, X. (2018). Tailoring T and I Curriculum for Better Employability: An Exploratory Case Study of Using Internship Surveys to Inform Curriculum Modification. *Onomázein*, *40*, 159-182. https://doi.org/10.7764/onomazein.40.10

Li, X., Wang, M., Feng, X., Yin, x., & Liang, J. (2024). An in-Depth Analysis of the Personal Factors and Their Pathways in Shaping Self-Directed Learning Abilities Among Undergraduate Nursing Students. *Frontiers in Psychology*, *15*. https://doi.org/10.3389/fpsyg.2024.1450462

Li, Y., & et al. (2026). Digital Lifelong Learning Ecosystems: A Bibliometric Review.

Liasidou, A. (2022). Inclusive Pedagogies in Digital <scp>post-Covid</Scp> -19 Higher Education. *British Journal of Special Education*, *50*(1), 6-27. https://doi.org/10.1111/1467-8578.12436

Lichtenstein, G., Chen, H. L., Smith, K. A., & Maldonado, T. A. (2014). Retention and Persistence of Women and Minorities Along the Engineering Pathway in the United States. 311-334. https://doi.org/10.1017/cbo9781139013451.021

Liden, R. C., Anand, S., & Vidyarthi, P. R. (2016). Dyadic Relationships. *Annual Review of Organizational Psychology and Organizational Behavior*, *3*(1), 139-166. https://doi.org/10.1146/annurev-orgpsych-041015-062452

Liễu, T. T. B., Duc, N. H. C., Gleason, N. W., Hai, D. T., & Nguyen, T. D. (2018). Approaches in Developing Undergraduate IT Engineering Curriculum for the Fourth Industrial Revolution in Malaysia and Vietnam. *Creative Education*, *09*(16), 2752-2772. https://doi.org/10.4236/ce.2018.916207

Lihe, H. (2018). Extracurricular Activity-Based Intercultural Education at Chinese Universities Rationale, Design and Practice. https://doi.org/10.2991/icelaic-18.2018.30

Lim, J. J., Roberts, C., Graham, O., Coady, S., Topliss, J., & Staton, C. A. (2024). Understanding Medical Students' Intercalation Decisions to Preserve the Clinical Academic Pipeline: A Mixed-Method Survey. https://doi.org/10.21203/rs.3.rs-4592927/v1

Lim, Z. Y., Lai, J. W., Yap, J. H., Mishra, A., Mokhtar, I. A., Yeo, D. J., & Cheong, K. H. (2023). Unraveling the Dynamics of Lifelong Learning in Singapore: A Comparative Study. *Knowledge*, *3*(3), 449-460. https://doi.org/10.3390/knowledge3030030

Lim, Z. Y., Yap, J. H., Lai, J. W., Mokhtar, I. A., Yeo, D. J., & Cheong, K. H. (2024). Advancing Lifelong Learning in the Digital Age: A Narrative Review of Singapore's SkillsFuture Programme. *Social Sciences*, *13*(2), 73. https://doi.org/10.3390/socsci13020073

Lima-Lopes, R. E. d., & Biazi, T. M. D. (2021). Digital Resources and English as an Additional Language in Higher Education: Possibilities for Internationalization. *Ilha Do Desterro a Journal of English Language Literatures in English and Cultural Studies*, *74*(3). https://doi.org/10.5007/2175-8026.2021.e79735

Lima, C. N. (2024). Possibilities of Human Emancipation: Teacher Training Courses in the Department of Human Sciences at UFF. https://doi.org/10.56238/sevened2023.006-136

Lin, B., Zabit, M. N. B., & Huang, S. (2024). A Study of Chinese Danmaku Video Sites for Self-Directed Learning(CDSDL) From an Educational Perspective: A Case Study of Bilibili Website. https://doi.org/10.4108/eai.13-10-2023.2341298

Lin, C., & Tian, Y. (2020). Research on Unemployment Status and Influencing Factors of Floating Population in China. https://doi.org/10.2991/aebmr.k.200402.041

Lin, X. (2019). Self-Regulated Learning Strategies of Adult Learners Regarding Non-Native Status, Gender, and Study Majors. *Journal of Global Education and Research*, *3*(1), 58-70. https://doi.org/10.5038/2577-509x.3.1.1018

Lin, X., Zhao, T., Schmidt, S. W., & Zhou, S. (2025). Using AI as a Learning Tool Through Simulation Interviews to Enhance Adult Learning. *Adult Learning*, *37*(2), 100-112. https://doi.org/10.1177/10451595251345274

Lindeman, E. C. (1926). *The Meaning of Adult Education*. New Republic.

Lins, L. V. S., & Florêncio, R. R. (2023). Tumbalalá People: History, Territory, Cultural Identities and Their Impacts on Indigenous School Education. *RHSS*. https://doi.org/10.7176/rhss/13-6-05

Lipscomb, M., & Ishmael, A. (2009). Humanistic Educational Theory and the Socialization of Preregistration Mental Health Nursing Students. *International Journal of Mental Health Nursing*, *18*(3), 173-178. https://doi.org/10.1111/j.1447-0349.2009.00603.x

Lithoxoidou, A., & Georgiadou, T. (2023). Critical Thinking in Teacher Education: Course Design and Teaching Practicum. *Education Sciences*, *13*(8), 837. https://doi.org/10.3390/educsci13080837

Liu, C. (2023). Research on Promoting Professional Development of Foreign Language Teachers in Higher Education Based on Cross-Cultural Competence in the Context of Deep Learning. *Applied Mathematics and Nonlinear Sciences*, *9*(1). https://doi.org/10.2478/amns.2023.2.00767

Liu, F., & Li, S. (2023). Unveiling the Roles of Adult Learners in Transformative Learning: An English Teacher's Tale. https://doi.org/10.31219/osf.io/j7w8p

Liu, J. (2021). Inclusiveness in Instructional Design &Amp; Development of Informal Learning Experiences: From Cultural Lenses. *Journal of Applied Instructional Design*, *10*(3). https://doi.org/10.51869/103/jcl

Liu, J. T., Breidenthal, A. P., & Schaffer, B. S. (2022). Let's All Go to the Movies, Together: A Social Cognitive Perspective on Diversity, Equity, and Inclusion in Management Education Video Clips. *Management Teaching Review*, *8*(4), 389-399. https://doi.org/10.1177/23792981221140057

Liu, M. (2026). The Rise of Digital Tutors: A Bibliometric Exploration of Intelligent Agent-Supported Language Learning From 2005 to 2025. *European journal of education*, *61*(1). https://doi.org/10.1111/ejed.70529

Liu, S., Lin, G., Peng, X., Wu, X., Wang, W., Lin, J., Lin, Y., & Yang, L. (2026). Exploring Healthcare Professionals' Views on Integrative Chinese–Western Medicine in the Nutritional Management of Cancer Patients: A Qualitative Study. *Frontiers in Nutrition*, *13*. https://doi.org/10.3389/fnut.2026.1623146

Liu, Y. (2022). Experience and Policy Analysis of Singapore's Talent Construction. *Journal of Education Humanities and Social Sciences*, *4*, 162-170. https://doi.org/10.54097/ehss.v4i.2742

Liu, Y., & Fu, Z. (2024). Hybrid Intelligence: Design for Sustainable Multiverse via Integrative Cognitive Creation Model Through Human–Computer Collaboration. *Applied Sciences*, *14*(11), 4662. https://doi.org/10.3390/app14114662

Llorente, P. A., & Iglesias, E. C. (2018). Desarrollo De La Competencia Digital en La Formación Inicial Del Profesorado De Educación Infantil. *Pixel-Bit*(52), 97-110. https://doi.org/10.12795/pixelbit.2018.i52.07

Lobos, K., Mella-Norambuena, J., Bruna, C., & Fernández, C. (2022). Analíticas De Aprendizaje Para La Toma De Decisiones Pedagógicas en Educación Superior. *Formación Universitaria*, *15*(4), 33-48. https://doi.org/10.4067/s0718-50062022000400033

Lock, J., Lakhal, S., Cleveland-Innes, M., Arancibia, P., Dell, D., & Silva, N. D. (2021). Creating Technology-enabled Lifelong Learning: A Heutagogical Approach. *British journal of educational technology*, *52*(4), 1646-1662. https://doi.org/10.1111/bjet.13122

Locke, M. L. (2022). Asking the Experts: Indigenous Educators as Leaders in Early Education and Care Settings in Australia. *The Australian Journal of Indigenous Education*, *51*(2). https://doi.org/10.55146/ajie.v51i2.12

Locke, R. M., & Jacoby, W. (2019). 2. The Dilemmas of Diffusion: Institutional Transfer and the Remaking of Vocational Training Practices in Eastern Germany. 33-68. https://doi.org/10.7591/9781501744891-005

Lockey, A., Bland, A., Stephenson, J., Bray, J., & Astin, F. (2022). Blended Learning in Health Care Education: An Overview and Overarching Meta-Analysis of Systematic Reviews. *Journal of Continuing Education in the Health Professions*, *42*(4), 256-264. https://doi.org/10.1097/ceh.0000000000000455

Lodge, J. M., Kennedy, G., Lockyer, L., Arguel, A., & Pachman, M. (2018). Understanding Difficulties and Resulting Confusion in Learning: An Integrative Review. *Frontiers in Education*, *3*. https://doi.org/10.3389/feduc.2018.00049

Loeng, S. (2020). Self-Directed Learning: A Core Concept in Adult Education. *Education Research International*, *2020*, 1-12. https://doi.org/10.1155/2020/3816132

Loeng, S. (2023). Pedagogy and Andragogy in Comparison – Conceptions and Perspectives. *Andragoška Spoznanja*, *29*(2), 39-52. https://doi.org/10.4312/as/11482

Lofgreen, J. E. (2023). Navigating Between Scylla and Charybdis: SoTL as Its Own Kind of Inquiry. *Teaching & Learning Inquiry the Issotl Journal*, *11*. https://doi.org/10.20343/teachlearninqu.11.28

Lokhtina, I., & Faller, P. (2024). Rethinking Informal Workplace Learning in Times of Complexity. *Journal of Workplace Learning*, *36*(6), 428-442. https://doi.org/10.1108/jwl-11-2023-0181

Lokman, A., Hassan, F., Ustadi, Y. A., Rahman, F. A. A., Zain, Z. M., & Rahmat, N. H. (2022). Investigating Motivation for Learning via Vroom's Theory. *International Journal of Academic Research in Business and Social Sciences*, *12*(1). https://doi.org/10.6007/ijarbss/v12-i1/11749

Lokshyna, O., Dzhurylo, A., Глушко, О. З., & Shparyk, O. (2021). Secondary Education Reform Priorities Under Globalisation: The Case of Ukraine in Comparison With China, Germany and Poland. *SHS Web of Conferences*, *104*, 01004. https://doi.org/10.1051/shsconf/202110401004

Lombaerts, K., Backer, F. D., & Sari, L. K. (2019). Pre-Service Teachers' Resilience Towards School Children's Problems in Remote Areas. 687-699. https://doi.org/10.15405/epsbs.2019.01.67

Long, J. M. (2020). Anywhere-Anytime Engineering Education in a Complete Undergraduate Program. *International Journal on Innovations in Online Education*, *4*(1), 37. https://doi.org/10.1615/intjinnovonlineedu.2020033158

Long, T., & Guo, J. (2023). Moving Beyond Inclusion to Belonging. *International Journal of Environmental Research and Public Health*, *20*(20), 6907. https://doi.org/10.3390/ijerph20206907

Lopes, A. P., & Soares, F. (2018). Using Moodle Analytics for Continuous E-Assessment in a Financial Mathematics Course at Polytechnic of Porto. https://doi.org/10.21125/iceri.2018.2123

Lopes, C. d. S., Costa, A. B. d., & Zoppo, B. M. (2024). A Influência Das Metodologias Ativas De Aprendizagem Na Promoção Da Autonomia E Inovação Em Pesquisadores Em Formação: Um Estudo Transdisciplinar. *Revista Caderno Pedagógico*, *21*(3), e3108. https://doi.org/10.54033/cadpedv21n3-052

Lopes, T. d. S., & Abílio, F. J. P. (2022). A Educação Ambiental Na Formação Inicial De Professores/As: Contribuições Da Pedagogia Crítica. *Praxis Educativa*, *17*, 1-20. https://doi.org/10.5212/praxeduc.v.17.18558.029

Lopes, T. d. S., & Abílio, F. J. P. (2023). Pressupostos Críticos De Uma Educação Ambiental Para Emancipação. *Educação (Ufsm)*. https://doi.org/10.5902/1984644465852

Loquinte, L. (2016). Professional Development and Skills Performance in Two Higher Educational Institutions of Northeastern Mindanao, Philippines. *SMCC Higher Education Research Journal*, *2*(1). https://doi.org/10.18868/sher2j.02.00416.11

Lorenzetti, L., Azulai, A., & Walsh, C. A. (2016). Addressing Power in Conversation. *Journal of Transformative Education*, *14*(3), 200-219. https://doi.org/10.1177/1541344616634889

Losoncz, I. (2015). Goals Without Means: A Mertonian Critique of Australia's Resettlement Policy for South Sudanese Refugees. *Journal of Refugee Studies*, fev017. https://doi.org/10.1093/jrs/fev017

Lou, N. M., Chaffee, K. E., Lascano, D. I. V., Dinçer, A., & Noels, K. A. (2017). Complementary Perspectives on Autonomy in Self-Determination Theory and Language Learner Autonomy. *Tesol Quarterly*, *52*(1), 210-220. https://doi.org/10.1002/tesq.403

Louis, A. T. S., Thompson, P., Sulak, T., Harvill, M. L., & Moore, M. E. (2021). Infusing 21st Century Skill Development Into the Undergraduate Curriculum: The Formation of the iBEARS Network. *Journal of Microbiology and Biology Education*, *22*(2). https://doi.org/10.1128/jmbe.00180-21

Lovo, S., Leidl, D., Usunier, K., Paslawski, T., Wesolowski, M., Puri, A., Caron, V., & Kim, S. H. (2024). Development and Assessment of Learners' Experiences With a Virtual Reality Learning Platform: Constructivist and Experiential Learning Pedagogies in Master of Physical Therapy Curriculum. *Imagining Sotl*, *4*(1), 39-61. https://doi.org/10.29173/isotl682

Loyens, S. M. M., Magda, J., & Remy, M. J. P. R. (2008). Self-Directed Learning in Problem-Based Learning and Its Relationships With Self-Regulated Learning. *Educational Psychology Review*, *20*(4), 411-427. https://doi.org/10.1007/s10648-008-9082-7

Lozano, R., Ceulemans, K., Alonso-Almeida, M., Huisingh, D., Lozano, F. J., Waas, T., Lambrechts, W., Lukman, R., & Hugé, J. (2015). A Review of Commitment and Implementation of Sustainable Development in Higher Education. *Journal of Cleaner Production*, *108*, 1-18. https://doi.org/10.1016/j.jclepro.2014.09.048

Lozano, R., Merrill, M. Y., Sammalisto, K., Ceulemans, K., & Lozano, F. J. (2017). Connecting Competences and Pedagogical Approaches for Sustainable Development in Higher Education. *Journal of Cleaner Production*, *161*, 1393-1405. https://doi.org/10.1016/j.jclepro.2017.05.131

Lu, X., & Troyan, F. J. (2022). Diversity and Inclusion of Culturally and Linguistically Diverse Students in K-12 Chinese Language Education. *Foreign Language Annals*, *55*(3), 684-703. https://doi.org/10.1111/flan.12639

Luaensutthi, A. (2023). Data Analytics of Online Lessons in Social Studies and Buddhism: Enhancing Dhamma Teaching and Tripitaka Understanding Among Teachers and Students. *Journal of Applied Data Sciences*, *4*(3), 200-212. https://doi.org/10.47738/jads.v4i3.125

Lubbe, A., Mentz, E., & Petersen, N. (2016). Student Support Through Cooperative Base Groups and Its Contribution to the Development of Selfdirected Learning Skills. 99-128. https://doi.org/10.4102/aosis.2016.sdlr14.04

Ludovico, C. (2017). Seeing the World Through Adult Eyes. *College & Research Libraries News*, *78*(5), 250. https://doi.org/10.5860/crln.78.5.250

Lūka, I. (2019). Creating a Culture-Based Language Learning Course for Developing Adult Learners' 21st Century Skills. *Journal of Education Culture and Society*, *10*(2), 151-169. https://doi.org/10.15503/jecs20192.151.169

Lukács, A., Simon, N., Kiss-Kondás, D., Szabó, T. L., Bak, M. K., Bibro, M., Göz, B., Biçki, D., Akar, F. D., Sancaktar, S. İ., Pehlivan, S., Gniewek, T., Golec, J., Jankowicz-Szymańska, A., Kiviluoma-

Ylitalo, M., Kreska-Korus, A., Lillo-Crespo, M., Milert, A., Özkoca, G., . . . Yılmaz, D. (2023). Exploring the Pros and Cons of Online Courses for Physiotherapists — A Theoretical Study. *Health Promotion & Physical Activity*, *24*(3), 33-39. https://doi.org/10.55225/hppa.527

Łukasiewicz, J., Stachyra-Sokulska, A., Saran, T., & Mazur, A. (2018). Motivation for Achievement and Family Perception in Emerging Adults. *Family Upbringing*, *18*(2), 307-324. https://doi.org/10.61905/wwr/170543

Lum, L., Bradley, P., & Rasheed, N. (2011). Accommodating Learning Styles in International Bridging Education Programs. *Higher Education Skills and Work-Based Learning*, *1*(2), 147-168. https://doi.org/10.1108/20423891111128917

Luning, C. R., Ledford, A., & Norton, M. I. (2022). A Strategic Organizational Approach to Developing Leadership Developers. *Journal of Character and Leadership Development*, *10*(1), 15-28. https://doi.org/10.58315/jcld.v10.253

Luo, Y., Bi, Y., & Wang, L. (2026). Examining the Impact of Growth Mindset and Lifelong Learning on Teacher Performance Among University Faculty: The Mediating Role of Professional Development. *Plos One*, *21*(2), e0341741. https://doi.org/10.1371/journal.pone.0341741

Luo, Y., Lin, J., & Yang, Y. (2021). Students' Motivation and Continued Intention With Online Self-Regulated Learning: A self-Determination Theory Perspective. *Zeitschrift Für Erziehungswissenschaft*, *24*(6), 1379-1399. https://doi.org/10.1007/s11618-021-01042-3

Lutomia, A. N., & Bello-Bravo, J. (2024). Increasing Lifelong Learning Using Video Animations: The Case of an Agriculture WhatsApp Group in Kenya. https://doi.org/10.5772/intechopen.114379

Luu, H. N. Q. (2021). Learners' Perceptions Towards Online Learning During the Covid-19 Pandemic: a Study at a Center of Foreign Languages in Vietnam. *Vietnam Journal of Education*, *5*(3), 84-90. https://doi.org/10.52296/vje.2021.124

Lwanga, J. B., Ngaka, W., & Openjuru, G. L. (2024). From Their Experience: A Thread of Participation Uncertainties Amongst Adult Returnees at a University in a Developing Context. *East African Journal of Education Studies*, *7*(4), 365-376. https://doi.org/10.37284/eajes.7.4.2324

Lyle, D. E., & Siao, D. (2021). Analysis of the Differences in Perceptions of Safety Reporting Systems Between Collegiate Aviation Students and Airline Pilots. *Collegiate Aviation Review International*, *39*(2). https://doi.org/10.22488/okstate.22.100235

Lynch, J., & Lungrin, T. (2018). Integrating Academic and Career Advising Toward Student Success. *New Directions for Higher Education*, *2018*(184), 69-79. https://doi.org/10.1002/he.20304

Lyons, P., & Bandura, R. P. (2020). Skills Needs, Integrative Pedagogy and Case-Based Instruction. *Journal of Workplace Learning*, *32*(7), 473-487. https://doi.org/10.1108/jwl-12-2019-0140

Lyons, S. D., & Berge, Z. L. (2012). Social Learning Theory. 3116-3118. https://doi.org/10.1007/978-1-4419-1428-6_1257

Lytle, M. C., Foley, P. F., & Cotter, E. W. (2015). Career and Retirement Theories. *Journal of Career Development*, *42*(3), 185-198. https://doi.org/10.1177/0894845314547638

Lyu, D. (2023). Harnessing Micro-Credentials to Innovate Teaching in the Open University of China: Opportunities and Challenges. *Creative Education*, *14*(05), 899-913. https://doi.org/10.4236/ce.2023.145058

Lyu, Y. (2023). Effects of Intergovernmental Organizations on Countries in the Field of Lifelong Learning. *International Education Studies and Sustainability*, *3*(2), p9. https://doi.org/10.22158/iess.v3n2p9

Ma, B., Ren, B., Li, H., Li, G., Chen, L. Y. E., Shimada, A., & Konomi, S. i. (2026). Designing a Meta-Reflective Dashboard for Instructor Insight Into Student-Ai Interactions. https://doi.org/10.48550/arxiv.2603.22674

Ma, C., Zheng, Y., Zhao, W., Yan, G., Zeng, Y., Ning, X.-H., & Jia, Z. (2025). Nurse-Delivered Telehealth in Home-Based Palliative Care: Integrative Systematic Review. *Journal of Medical Internet Research*, *27*, e73024. https://doi.org/10.2196/73024

Maaz, K., Trautwein, U., Lüdtke, O., & Baumert, J. (2008). Educational Transitions and Differential Learning Environments: How Explicit Between-School Tracking Contributes to Social Inequality in Educational Outcomes. *Child Development Perspectives*, *2*(2), 99-106. https://doi.org/10.1111/j.1750-8606.2008.00048.x

Maben, J., Taylor, C., Jagosh, J., Carrieri, D., Briscoe, S., Klepacz, N., & Mattick, K. (2024). Causes and Solutions to Workplace Psychological Ill-Health for Nurses, Midwives and Paramedics: The Care Under Pressure 2 Realist Review. *Health and Social Care Delivery Research*, 1-171. https://doi.org/10.3310/twdu4109

Mabena-Segoe, I. N., & Mabunda, N. F. (2025). The Human Side of PBL: Psychological Well-Being of Health Sciences Students and Academic Staff: An Integrative Review. https://doi.org/10.21203/rs.3.rs-8357368/v1

Macagno, T., Nguyen-Quoc, A., & Jarvis, S. (2024). Nurturing Sustainability Changemakers Through Transformative Learning Using Design Thinking: Evidence From an Exploratory Qualitative Study. *Sustainability*, *16*(3), 1243. https://doi.org/10.3390/su16031243

Macdonald, B. J. (2014). Critical Theory. 1-9. https://doi.org/10.1002/9781118474396.wbept0226

MacDonald, F., Bottrell, D., & Johnson, B. (2018). Socially Transformative Wellbeing Practices in Flexible Learning Environments: Invoking an Education of Hope. *Health Education Journal*, *78*(4), 377-387. https://doi.org/10.1177/0017896918777005

Maclean, R., Jagannathan, S., & Panth, B. (2018). Education and Skills for Inclusive Growth, Green Jobs and the Greening of Economies in Asia. https://doi.org/10.1007/978-981-10-6559-0

MacMahon, S., Corbett, B. J., Hassall, M., Carroll, A., & Boer, R. D. (2024). 'What Good Looks Like': Building a Shared Understanding of Quality Training and Learning in the Mining Industry. *Vocations and Learning*, *17*(3), 509-536. https://doi.org/10.1007/s12186-024-09353-2

Macpherson, A., & Antonacopoulou, E. P. (2013). Translating Strategy Into Practice: The Role of Communities of Practice. *Journal of Strategy and Management*, *6*(3), 265-285. https://doi.org/10.1108/jsma-11-2012-0061

Madjarov, I. (2012). eLSaaS: A FRAMEWORK FOR MOBILE LEARNING CONTENT ADAPTATION. https://doi.org/10.5220/0003958102760281

Madjarov, I. (2014). Cloud-Based Framework for Mobile Learning Content Adaptation. 381-386. https://doi.org/10.1109/educon.2014.6826122

Mage, S., Benton, D., Wilber, K. H., Rodriguez, R. A., Gonzalez, A., Puga, F., & Meyer, K. (2025). Acceptability of a Web-Based Financial Education Intervention for Latino Caregivers: Mixed Methods Evaluation. *Jmir Formative Research*, *9*, e70471-e70471. https://doi.org/10.2196/70471

Magro, K., & Sfeir, G. (2023). Extending Knowledge and Learning Through the Prism of Cosmopolitan Education: International Perspectives. *International Journal for Talent Development and Creativity*, *10*(1-2), 17-29. https://doi.org/10.7202/1099938ar

Maguvhe, M. O., & Mutambo, A. (2023). A Study of Inclusive Education Provision in Zambia: Curriculum Reform. *African Journal of Disability*, *12*. https://doi.org/10.4102/ajod.v12i0.1067

Mahat, M. (2022). Purpose, Process, Place, Pedagogical Affordance and Product (5Ps) of Student Learning in Higher Education: Beyond Beckers. *Facilities*, *41*(1/2), 140-156. https://doi.org/10.1108/f-06-2022-0086

Mahere, S. M. (2024). Contributions of Higher Education Reforms to Sustainable Development Goals: Some Examples and Experiences From the Department of Educational Administration and Leadership, Faculty of Education, University of Zimbabwe. *Journal of Adult and Continuing Education*, *31*(1), 244-264. https://doi.org/10.1177/14779714241261063

Mahlangu, V. P. (2018). The Good, the Bad, and the Ugly of Distance Learning in Higher Education. https://doi.org/10.5772/intechopen.75702

Mahmood, F., Ariza-Montes, A., Saleem, M., & Han, H. (2022). COVID-19 Exposure and Mental Wellbeing of European Male Employees. *Journal of Men S Health*, *18*(7), 1. https://doi.org/10.31083/j.jomh1807145

Maigua, A. C. I., Nazareno, B. G. C., & Ayuy, A. N. C. (2024). Estrategia Metodológica Para El Desarrollo Del Razonamiento Lógico Verbal en Los Niños De Sexto Año De Educación Básica Durante El Año Lectivo 2023-2024. *Latam Revista Latinoamericana De Ciencias Sociales Y Humanidades*, *5*(5). https://doi.org/10.56712/latam.v5i5.2894

Maimunah, I., Haque, A., & Wargadinata, W. (2023). Hybrid Arabic Learning Based on Constructivism Theory for Higher Education. 462-474. https://doi.org/10.2991/978-2-38476-002-2_44

Maio, J. D. (2026). Trends in Teaching Political Science Research: <i>Teaching Political Science in the Age of AI</i>. *International Political Science Abstracts*, *76*(1), 17-34. https://doi.org/10.1177/00208345261430193

Mairs, K., McNeil, H., McLeod, J., Prorok, J., & Stolee, P. (2013). Online Strategies to Facilitate Health-related Knowledge Transfer: A Systematic Search and Review. *Health Information & Libraries Journal*, *30*(4), 261-277. https://doi.org/10.1111/hir.12048

Maison, L. M. (2022). Beyond Western and Indigenous Perspectives on Sustainability: Politicizing Sustainability With the Zapatista Rebellious Education. *Journal of Transformative Education*, *21*(1), 41-58. https://doi.org/10.1177/15413446221079595

Majola, E., Powell, L., & Jordaan, C. (2024). Vocational Education and Graduate Struggles in the Eastern Cape, South Africa: A Freirean Approach. *Education as Change*, *28*. https://doi.org/10.25159/1947-9417/15943

Makhachashvili, R., Semenist, I., Prihodko, G., Prykhodchenko, O., Rudik, I. V., Ter-Grygoryan, M., & Брус, M. (2024). Interdisciplinary, AI-Interoperable, and Universal Skills for Foreign Languages Education in Emergency Digitization. 107-114. https://doi.org/10.54808/imsci2024.01.107

Makrakis, V. (2023). Using the DREAM Methodology for Course Assessment in the Field of ICT-Enabled Education for Sustainability. *European Journal of Investigation in Health Psychology and Education*, *13*(7), 1378-1391. https://doi.org/10.3390/ejihpe13070100

Makri, A., Vlachopoulos, D., & Martina, R. (2021). Digital Escape Rooms as Innovative Pedagogical Tools in Education: A Systematic Literature Review. *Sustainability*, *13*(8), 4587. https://doi.org/10.3390/su13084587

Male, T. (2006). Being an Effective Headteacher. https://doi.org/10.4135/9781446213285

Malele, V. (2023). Cybersecurity Cloud-Based Online Learning: A Literature Review Approach. *Journal of Information Systems and Informatics*, *5*(4), 1623-1632. https://doi.org/10.51519/journalisi.v5i4.583

Malik, H. D., Din, M., & Afzal, S. (2019). Constructivism in the Pedagogical Practices of the University Faculty. *Global Social Sciences Review*, *IV*(I), 172-180. https://doi.org/10.31703/gssr.2019(iv-i).23

Malik, R., Buljac-Samardžić, M., Amajjar, I., Carina, G. J. M. H., & Scheele, F. (2021). Open Organisational Culture: What Does It Entail? Healthcare Stakeholders Reaching Consensus by Means of a Delphi Technique. *BMJ Open*, *11*(9), e045515. https://doi.org/10.1136/bmjopen-2020-045515

Malik, V., & Hooda, M. (2023). MOOCs Revamping Indian Higher Education: Escalating Access, Equity, and Quality. *Journal of Higher Education Theory and Practice*, *23*(12). https://doi.org/10.33423/jhetp.v23i12.6238

Malilang, C. S. (2019). A Multicultural Community of Practice in Creative Writing. *Educare*(2), 6-29. https://doi.org/10.24834/educare.2019.2.2

Malinen, A., & Piirainen, A. (2023). Towards the Essence of Andragogy – A Hermeneutical Reading of Grundtvig, Knowles, Lindeman and Savićević. *Andragoška Spoznanja*, *29*(2), 19-37. https://doi.org/10.4312/as/14421

Malinga, C. L. An Exploration of the Learners' Views on the Efficacy of Information Communication Technology in Improving Work Performance : The Case of KwaZulu-Natal Department of Health Employees Registered at a Community Learning Centre in Durban. https://doi.org/10.29086/10413/23115

Malinga, C. L. (2024). *An Exploration of the Learners' Views on the Efficacy of Information Communication Technology in Improving Work Performance : The Case of KwaZulu-Natal Department of Health Employees Registered at a Community Learning Centre in Durban*

Malinin, L. (2018). Situated Design-Thinking in Architectural Practice: Analyzing and Extending Schön's Epistemology. *Ardeth*, *2*, 53. https://doi.org/10.17454/ardeth02.05

Mälkki, K., & Green, L. (2018). Working With Edge Emotions as a Means for Uncovering Problematic Assumptions: Developing a Practically Sound Theory. *Phronesis*, *7*(3), 26-34. https://doi.org/10.7202/1054406ar

Malureanu, A., Pânişoară, G., & Lazăr, I. (2021). The Relationship Between Self-Confidence, Self-Efficacy, Grit, Usefulness, and Ease of Use of eLearning Platforms in Corporate Training

During the COVID-19 Pandemic. *Sustainability*, *13*(12), 6633. https://doi.org/10.3390/su13126633

Mamani, M. Y. C., Mamani, G. J. C., Díaz-Vilcanqui, Y. D., & Castillo, W. W. C. (2024). Aprendizaje Activo Y Participativo en El Aula. https://doi.org/10.53595/eip.015.2024

Manalo, J. B., & Chua, E. N. (2024). Instructional Approaches and Teachers' Key Competencies on Inclusive Education for an Improved Learners' Opportunities. *International Journal of Research Publications*, *151*(1). https://doi.org/10.47119/ijrp1001511620246826

Manditereza, B., & Chamboko-Mpotaringa, M. (2024). Blended Learning Approach in TVET Colleges: Lecturers and Students Perceptions of Teaching and Learning Practical Subjects Online. *Journal of Education and Learning Technology*, 315-328. https://doi.org/10.38159/jelt.2024585

Mani, I., & Rai, P. (2021). Mangsuk as Indigenous Knowledge Heritage in Yamphu Community: An Estranged Transformative Learning Space. *Journal of Transformative Praxis*, *2*(1), 19-31. https://doi.org/10.51474/jrtp.v2i1.521

Mannarini, T., Rizzo, M., Brodsky, A. E., Buckingham, S. L., Zhao, J., Rochira, A., & Fedi, A. (2021). The Potential of Psychological Connectedness: Mitigating the Impacts of COVID‐19 Through Sense of Community and Community Resilience. *Journal of Community Psychology*, *50*(5), 2273-2289. https://doi.org/10.1002/jcop.22775

Manojan, K. P. (2019). Capturing the Gramscian Project in Critical Pedagogy: Towards a Philosophy of Praxis in Education. *Review of Development and Change*, *24*(1), 123-145. https://doi.org/10.1177/0972266119831133

Mansor, N. A., Abdullah, N., & Rahman, H. A. (2020). Towards Electronic Learning Features in Education 4.0 Environment: Literature Study. *Indonesian Journal of Electrical Engineering and Computer Science*, *19*(1), 442. https://doi.org/10.11591/ijeecs.v19.i1.pp442-450

Mantey, E. E. (2019). Beyond the Politics of Inclusion: The Policy Environment and the Fate of Children With Disability in the Ghanaian Educational System. *Mediterranean Journal of Social Sciences*, *10*(5), 54-61. https://doi.org/10.2478/mjss-2019-0066

Mantuano, M. O. M., Caviedes, E. C. E., Ladines, K. V. O., Rogel, D. R. P., & Yuqui, C. E. P. (2021). Análisis Del Conductismo, Cognitivismo, Constructivismo Y Su Interrelación Con El Conectivismo en La Educación Postpandemia. *South Florida Journal of Development*, *2*(5), 6850-6863. https://doi.org/10.46932/sfjdv2n5-038

Manuti, A., Pastore, S., Scardigno, A. F., Giancaspro, M. L., & Morciano, D. (2015). Formal and Informal Learning in the Workplace: A Research Review. *International Journal of Training and Development*, *19*(1), 1-17. https://doi.org/10.1111/ijtd.12044

Maoto, R. (2023). Mathematics Education for Relevance, Responsiveness, and Viability in Africa Within the Fourth Industrial Revolution Era. *Pythagoras*, *44*(1). https://doi.org/10.4102/pythagoras.v44i1.767

Maphosa, V., & Maphosa, M. (2023). Opportunities and Challenges of Adopting MOOCs in Africa: A Systematic Literature Review. https://doi.org/10.5772/intechopen.1001298

Maralani, F. M., & Pfeiffer, S. I. (2024). The Unique Social-Emotional Challenges That Bright Immigrant Youth Face. *Gifted Education International*, *41*(1), 137-157. https://doi.org/10.1177/02614294241279354

Marangon, F., Bottega, M., Avoni, S., Veronese, M., & Danielis, M. (2025). Nurses' Experience of Redeployment to a New Intermediate Care Unit for Respiratory Patients: A Qualitative Study. *Journal of Advanced Nursing*. https://doi.org/10.1111/jan.70170

Marceau, N., Lakhal, S., & Séguin, A. (2020). Conception D'un Cours Privé en Ligne Sur La Formation Hybride Pour Le Personnel Pédagogique Du Collégial À Partir De La Conception Universelle De L'apprentissage : Étude De Cas Et Considérations Pratiques. *Canadian Journal of Learning and Technology*, *45*(3). https://doi.org/10.21432/cjlt27862

Marcellis-Warin, N. d. (2022). Analyse Comparative D'écosystèmes en IA Dans Le but De Repérer Les Pratiques Innovantes en Matière De Formation Et De Transfert De Connaissances. https://doi.org/10.54932/sxoh3928

Marchi, M. D., Visintini, E., Magro, G., Dussi, G., Chiappinotto, S., Moreale, R., Moreal, C., & Palese, A. (2025). Magnetism in Nursing Education: A Qualitative Embedded Case Study of High-

Applicant Nursing Programs Amid a National Decline. *Journal of Advanced Nursing*, *82*(6), 6586-6601. https://doi.org/10.1111/jan.70295

Marcos, E. B., Fabretti, V., Ordóñez-Carabaño, Á., Herrera, E. R., & Taviani, S. (2023). A Child-Centred Intercultural Approach to the Socio-Educational Inclusion of Migrant and Refugee Children. *Open Research Europe*, *3*, 220. https://doi.org/10.12688/openreseurope.16999.1

Mardikaningsih, A., Kurniasari, P., & Pradipta, A. W. (2024). Qualitative and Quantitative Study on Students' Critical Thinking of Conceptual E-Scaffolding in Blended Learning Context. *Journal International Inspire Education Technology*, *3*(1), 12-18. https://doi.org/10.55849/jiiet.v3i1.545

Mareci, A., Dumitrache, L., Nae, M., Tudoricu, A., & Cioclu, A. (2023). A Qualitative Exploration of Experiences of Asylum Seekers and People With Refugee Backgrounds in Accessing the Education System in Romania. *Sustainability*, *15*(5), 4120. https://doi.org/10.3390/su15054120

Margaryan, A., Littlejohn, A., & Milligan, C. (2013). Self-Regulated Learning in the Workplace: Strategies and Factors in the Attainment of Learning Goals. *International Journal of Training and Development*, *17*(4), 245-259. https://doi.org/10.1111/ijtd.12013

Margeviča-Grīnberga, I., Lake, S., Pīgozne, T., Siirilä, J., & Surikova, S. (2023). Un Marco Pedagógico Para Promover La Competencia Financiera Sostenible en La Educación Y Formación Profesional Secundaria. *Aula Abierta*, *52*(1), 61-70. https://doi.org/10.17811/rifie.52.1.2023.61-70

Marginson, S. (2016). *The Dream Is Over: The Crisis of Clark Kerr's California Idea of Higher Education*. University of California Press.

Margolis, A. A. (2020a). Zone of Proximal Development (ZPD) and Organization of Students Learning Activity. *Psychological Science and Education*, *25*(4), 6-27. https://doi.org/10.17759/pse.2020250402

Margolis, A. A. (2020b). Zone of Proximal Development, Scaffolding and Teaching Practice. *Cultural-Historical Psychology*, *16*(3), 15-26. https://doi.org/10.17759/chp.2020160303

María Amor Barros del, R., Cunha, A., Doroch, A., Mediavilla-Martínez, B., & Milhâilâ, R. (2023). Foreign Language Teacher Education in Europe: Advancing Towards Convergence. https://doi.org/10.36443/9788418465451

Maria Teresa Vargas La, T., Chavez, E. A. A., & Martinez, W. R. D. (2022). Metodologías De Enseñanza Universitaria Un Reto Para El Docente Universitario. *Revista De Ciencia Tecnología E Innovación*, *20*(26), 11-33. https://doi.org/10.56469/rcti.v20i26.704

Mariño, Ó., Hérnandez, C. H., & Flórez, F. B. (2022). Experiencia De Formación Para Profesores De Ingeniería Desde El Aprendizaje Experiencial Y El Trabajo Colaborativo. *Encuentro Internacional De Educación en Ingeniería*. https://doi.org/10.26507/paper.2490

Mariyono, D. (2026). Governing Agentic AI in Global Education: A Justice-Oriented Framework for Alignment, Responsibility, and Distributed Autonomy. https://doi.org/10.21203/rs.3.rs-8890914/v1

Markowitsch, J., & Bjørnåvold, J. (2022). Scenarios for Vocational Education and Training in Europe in the 21st Century. *The Hungarian Educational Research Journal*, *12*(3), 235-247. https://doi.org/10.1556/063.2021.00116

Marku, M. (2024). Navigating the Future of Work: Human Capital in the Age of Industry 4.0. *Interdisciplinary Journal of Research and Development*, *11*(1), 190. https://doi.org/10.56345/ijrdv11n127

Marlapudi, K., & Lenka, U. (2024). Unlocking the Potential: Redefining Talent and Competency Mapping for Industry 4.0. *Management Research Review*, *47*(11), 1805-1832. https://doi.org/10.1108/mrr-07-2023-0496

Marotta, J. A. (2024). Transformative Learning and Professional Advancement During an EdD Program. *Impacting Education Journal on Transforming Professional Practice*, *9*(1), 3-11. https://doi.org/10.5195/ie.2024.402

Marougkas, A., Troussas, C., Krouska, A., & Sgouropoulou, C. (2023). Virtual Reality in Education: A Review of Learning Theories, Approaches and Methodologies for the Last Decade. *Electronics*, *12*(13), 2832. https://doi.org/10.3390/electronics12132832

Marquis, T. L. (2021). Formative Assessment and Scaffolding Online Learning. *New Directions for Adult and Continuing Education*, *2021*(169), 51-60. https://doi.org/10.1002/ace.20413

Marra, R. M., Plumb, C., & Hacker, D. J. (2018). Board 119: Developing Metacognitive Skills in PBL Undergraduate Engineering. https://doi.org/10.18260/1-2--29893

Marshall, R., Pardo, A., Smith, D., & Watson, T. J. (2022). Implementing Next Generation Privacy and Ethics Research in Education Technology. *British journal of educational technology*, *53*(4), 737-755. https://doi.org/10.1111/bjet.13224

Marshall, S. (2023). Perspective Chapter: The Learning Management System of 2028 and How We Start Planning for This Now. https://doi.org/10.5772/intechopen.110120

Marsick, V. J., & Watkins, K. E. (2007). Les Tensions De L'apprentissage Informel Sur Le Lieu De Travail. *Revue Française De Pédagogie*(160), 29-38. https://doi.org/10.4000/rfp.586

Marsick, V. J., & Watkins, K. E. (2018). Introduction to the Special Issue: An Update on Informal and Incidental Learning Theory. *New Directions for Adult and Continuing Education*, *2018*(159), 9-19. https://doi.org/10.1002/ace.20284

Marszowski, R. (2019). Dual Education System in a Knowledge-Based Economy. *Scientific Papers of Silesian University of Technology Organization and Management Series*, *2019*(137), 89-114. https://doi.org/10.29119/1641-3466.2019.137.7

Martin-Thomsen, T. C., Scagnetti, G., McPhee, S., Akenson, A. B., & Hagerman, D. (2021). The Scholarship of Critique and Power. *Teaching & Learning Inquiry the Issotl Journal*, *9*(1), 279-293. https://doi.org/10.20343/teachlearninqu.9.1.19

Martin, J. (2018). Skills for the 21st Century. https://doi.org/10.1787/96e69229-en

Martínez-Alvarez, N., & López, L. M. (2024). Sinergia Piaget, Vygotsky Y La Inteligencia Artificial en La Educación Universitaria. *Vinculatégica Efan*, *10*(4), 70-84. https://doi.org/10.29105/vtga10.4-948

Martínez-Izquierdo, L., & Sánchez, M. T. (2022). Dual Vocational Education and Training Systems' Governance Model and Policy Transfer: The Role of the European Union in Its Diffusion. *Social Sciences*, *11*(9), 403. https://doi.org/10.3390/socsci11090403

Martins, I. (2008). Learning to Live Together: The Contribution of Intercultural Education. *European journal of education*, *43*(2), 197-206. https://doi.org/10.1111/j.1465-3435.2008.00350.x

Martucci, G., Pereira, J., Busa, C., Centeno, C., Csikós, Á., Franchini, L., Elsner, F., Raccichini, M., Mihăilescu-Marin, M.-M., Moşoiu, D., Bernabé, S. R., Scherg, A., & Consorti, F. (2023). Online Learning in Palliative Care Education of Undergraduate Medical Students: A Realist Synthesis. *Palliative Care and Social Practice*, *17*. https://doi.org/10.1177/26323524231218279

Maruatona, T. (2014). Lifelong Learning for Social Inclusion of Ethnic Minorities in Botswana. *Education Citizenship and Social Justice*, *10*(1), 37-54. https://doi.org/10.1177/1746197914558397

Más, F. S., Mein, E., Fuentes, B., Thatcher, B., & Balcázar, H. (2012). Integrating Health Literacy and ESL. *Health Promotion Practice*, *14*(2), 263-273. https://doi.org/10.1177/1524839912452736

Mas'ud, A., Fuad, A. Z., & Zaini, A. (2019). Evolution and Orientation of Islamic Education in Indonesia and Malaysia. *Journal of Indonesian Islam*, *13*(1), 21. https://doi.org/10.15642/jiis.2019.13.1.21-49

Maslow, A. H. (1954). *Motivation and Personality*. Harper & Row.

Maslow, A. H. (1970). *Motivation and Personality* (2nd ed.). Harper & Row.

Mason, J., & Carr, D. (2021). Boosting Lifelong Learning Through Digital Online Education. https://doi.org/10.2991/assehr.k.210508.024

Massaro, S., Yang, T.-C., & Christopoulos, G. I. (2025). Wearable Devices Methodology: Opportunities and Challenges in Human Resource Management. *Human Resource Management Journal*, *36*(1), 158-177. https://doi.org/10.1111/1748-8583.70014

Massiah, A., Shotte, G., Rowe, V., & Minott, C. (2024). Educational Leadership for Social Transformation: An Inclusive Approach for Schools as Places of Belonging. *Power and Education*, *17*(2), 142-158. https://doi.org/10.1177/17577438241297239

Mat, N. C., & Jamaludin, K. A. (2024). Effectiveness of Practices and Applications of Student-Centered Teaching and Learning in Primary Schools: A Systematic Literature Review. *International Journal of Academic Research in Progressive Education and Development*, *13*(3). https://doi.org/10.6007/ijarped/v13-i3/21733

Mathew, A. N., Rohini, V., & Paulose, J. (2021). NLP-based Personal Learning Assistant for School Education. *International Journal of Electrical and Computer Engineering (Ijece)*, *11*(5), 4522. https://doi.org/10.11591/ijece.v11i5.pp4522-4530

Mathews, M. (2007). Constructivist Pedagogy for the Business Communication Classroom. *Journal of College Teaching & Learning (Tlc)*, *4*(11). https://doi.org/10.19030/tlc.v4i11.1529

Matook, S., Wang, Y. M., Koeppel, N., & Guerin, S. (2023). Metacognitive Skills in Low-Code App Development: Work-Integrated Learning in Information Systems Development. *Journal of Information Technology*, *39*(1), 41-70. https://doi.org/10.1177/02683962231170238

Matorevhu, A. (2022). Teacher Educators' Nature of Understanding of Adult Learning Theories Application in Pre-Service Teachers' Classes. *Electronic Journal of Education Social Economics and Technology*, *3*(1), 24-32. https://doi.org/10.33122/ejeset.v3i1.50

Matsuo, M. (2015a). A Framework for Facilitating Experiential Learning. *Human Resource Development Review*, *14*(4), 442-461. https://doi.org/10.1177/1534484315598087

Matsuo, M. (2015b). Human Resource Development to Facilitate Experiential Learning: The Case of <scp>Y</Scp>ahoo <scp>J</Scp>apan. *International Journal of Training and Development*, *19*(3), 199-210. https://doi.org/10.1111/ijtd.12056

Matthew-Maich, N., Ploeg, J., Jack, S. M., & Dobbins, M. (2010). Transformative Learning and Research Utilization in Nursing Practice: A Missing Link? *Worldviews on Evidence-Based Nursing*, *7*(1), 25-35. https://doi.org/10.1111/j.1741-6787.2009.00172.x

Matthew, U. O., Kazaure, J. S., & Okafor, N. U. (2018). Contemporary Development in E-Learning Education, Cloud Computing Technology &Amp; Internet of Things. *Eai Endorsed Transactions on Cloud Systems*, 169173. https://doi.org/10.4108/eai.31-3-2021.169173

Matthews, B. L., Daigle, J., & Cooper, J. (2020). Causative Effects of Motivation to Transfer Learning Among Relational Dyads: The Test of a Model. *European Journal of Management and Business Economics*, *29*(3), 297-314. https://doi.org/10.1108/ejmbe-07-2019-0120

Maulana, A. A., & Jun, G. (2018). Relationship Between Apprenticeship Training With Information and Communication Technologies at Vocational High School in West Java Province. *Innovation of Vocational Technology Education*, *14*(2), 55. https://doi.org/10.17509/invotec.v14i2.14357

Mauro, M. H., Gaspar, M. A., Ohashi, F. K., & Fábio Luí-s Falchi de, M. (2018). Estudo Comparativo Das Tendências De E-Learning E Das Características De Soluções Líderes De Learning Management System. *Renote*, *15*(2). https://doi.org/10.22456/1679-1916.79278

Mawas, N. E., & Muntean, C. H. (2018). Supporting Lifelong Learning Through Development of 21st Century Skills. *1*, 7343-7350. https://doi.org/10.21125/edulearn.2018.1723

Mayombe, C. (2023). Facilitating Non-School-Based Technical and Vocational Training for Disadvantaged Youths in South Africa. *Journal of Technical Education and Training*, *15*(1). https://doi.org/10.30880/jtet.2023.15.01.016

Mazlan, M. R. M., & Jambulingam, M. (2023). Challenges of Talent Retention: A Review of Literature. *Journal of Business Management Review*, *4*(2), 078-091. https://doi.org/10.47153/jbmr42.6302023

Mazur, B., & Walczyna, A. (2022). Sustainable Development Competences of Engineering Students in Light of the Industry 5.0 Concept. *Sustainability*, *14*(12), 7233. https://doi.org/10.3390/su14127233

Mazur, R. (2026). Transversal Competencies for Future Tourism Graduates: Stakeholder Alignment and Curriculum Innovation in European VET. https://doi.org/10.21203/rs.3.rs-10352918/v1

Mazzocato, P., Herber, K., Keita, M., & Roczniewska, M. (2025). Let's Talk About the Elephant in the Room: A Psychological Safety Climate Intervention Among Nursing Teams—A Qualitative Evaluation. *Journal of Advanced Nursing*, *82*(3), 2442-2457. https://doi.org/10.1111/jan.70059

Mbah, M., & Brohmann, B. (2021). Das Lernen in Organisationen. 387-412. https://doi.org/10.14361/9783839456682-018

Mbatha, S. (2024). Evaluating Educationists' Andragogical Strategies for Teaching Numeracy: A Case Study of the Kwazulu-Natal Adult Correctional Centre Classroom. *E-Journal of Humanities Arts and Social Sciences*, 1596-1610. https://doi.org/10.38159/ehass.20245813

Mbise, S., & Lekule, C. (2020). Constructivist Teaching Pedagogy: A Strategy for Innovation, Industrialization and Sustainable Development. *Es*, *6*(10). https://doi.org/10.53555//es.v6i10.1403

Mbokota, G., Myres, P. K., & Stout-Rostron, D. S. (2022). Exploring the Process of Transformative Learning in Executive Coaching. *Advances in Developing Human Resources*, *24*(2), 117-141. https://doi.org/10.1177/15234223221079026

Mbotake, S. G. (2016). Factoring Adult Learners' Generational Mix in Translator Education. *International Journal of Comparative Literature and Translation Studies*, *4*(3). https://doi.org/10.7575/aiac.ijclts.v.4n.3p.59

McBrien, J. (2022). *Social and Emotional Learning (SEL) of Newcomer and Refugee Students.* O. a. U. o. S. Florida.

McCaffery, K., Morony, S., Muscat, D. M., Hayen, A., Shepherd, H. L., Dhillon, H. M., Smith, S. K., Cvejic, E., Meshreky, W., Luxford, K., & Nutbeam, D. (2019). Evaluation of an Australian Health Literacy Program Delivered in Adult Education Settings. *HLRP Health Literacy Research and Practice*, *3*(S1). https://doi.org/10.3928/24748307-20190402-01

McCaffery, K., Morony, S., Muscat, D. M., Smith, S. K., Shepherd, H. L., Dhillon, H. M., Hayen, A., Luxford, K., Meshreky, W., Comings, J. P., & Nutbeam, D. (2016). Evaluation of an Australian Health Literacy Training Program for Socially Disadvantaged Adults Attending Basic Education Classes: Study Protocol for a Cluster Randomised Controlled Trial. *BMC Public Health*, *16*(1). https://doi.org/10.1186/s12889-016-3034-9

McClain, A. L. (2024). New Developments in Transformative Learning. *New Directions for Adult and Continuing Education*, *2024*(184), 20-29. https://doi.org/10.1002/ace.20540

McClain, A. L., & Mallary, K. J. (2025). "When We Design for Adult Learners With Disabilities, Everyone Benefits": Adaptable-Inclusive-Motivated (AIM) Model for Adult Educators to Include Adult Learners With Disabilities. *Adult education quarterly*, *75*(4), 269-290. https://doi.org/10.1177/07417136251316259

McDermott, I., Astbury, J., Jacobs, S., Willis, S., Hindi, A., Em, S., & Schafheutle, E. (2023). To Be or Not to Be: The Identity Work of Pharmacists as Clinicians. *Sociology of Health & Illness*, *45*(3), 623-641. https://doi.org/10.1111/1467-9566.13605

McGill, C. M. (2024). Advising Leaders' Perceptions of Advisor Workplace Learning. *New Horizons in Adult Education and Human Resource Development*, *36*(3), 176-190. https://doi.org/10.1177/19394225241265832

McGinty, J. F., Oest, J., & Divido, J. (2024). Digital Immigrants: Leveraging the Technology Acceptance Model to Increase Digital Literacy in Older Adults. *New Directions for Adult and Continuing Education*, *2024*(181), 9-18. https://doi.org/10.1002/ace.20517

McGrath, S., Ladaah, G. O., Lotz-Sisitka, H., Allais, S., Zeelen, J., Wedekind, V., Ramsarup, P., Monk, D. H., Metelerkamp, L., Russon, J. A., Kyaligonja, B., Robbins, G., Adrupio, S., Ocan, D., Nyeko, K., Primo, A., Molebatsi, P., Tshabalala, T., Muhangi, S., & Openjuru, M. (2023). Transitioning Vocational Education and Training in Africa. https://doi.org/10.47674/9781529224658

McGreal, R., Mackintosh, W., Cox, G., & Olcott, D. (2022). Bridging the Gap: Micro-Credentials for Development. *The International Review of Research in Open and Distributed Learning*, *23*(3), 288-302. https://doi.org/10.19173/irrodl.v23i3.6696

McGreal, R., Mackintosh, W., & Lane, D. (2023). OER-based Online Micro-Courses. *Journal of Open Flexible and Distance Learning*, *27*(1), 5-13. https://doi.org/10.61468/jofdl.v27i1.599

McGreal, R., & Olcott, D. (2022). A Strategic Reset: Micro-Credentials for Higher Education Leaders. *Smart Learning Environments*, *9*(1). https://doi.org/10.1186/s40561-022-00190-1

McIlduff, C., Acharibasam, J. B., Starr, V., & Chapados, M. (2022). Engaging Indigenous Older Adults With Technology Use to Respond to Health and Well-Being Concerns and Needs. *Healthcare Management Forum*, *35*(5), 257-264. https://doi.org/10.1177/08404704221103521

McKenzie, J., & Dalton, E. (2020). Universal Design for Learning in Inclusive Education Policy in South Africa. *African Journal of Disability*, *9*. https://doi.org/10.4102/ajod.v9i0.776

McLean, M., Cilliers, F., & Wyk, J. V. (2014). Développement Professoral : Hier, Aujourd'hui Et Demain Guide AMEE N°36. *Pédagogie Médicale*, *15*(3), 183-237. https://doi.org/10.1051/pmed/2015001

McMillan, M., Little, P., Yoon, S., & Park, M. (2022). Critical Thinking and Metacognition: Processes and Outcomes Within the Learning Cycles. *Journal of Problem-Based Learning*, *9*(2), 110-118. https://doi.org/10.24313/jpbl.2022.00213

McNamara, J., & Ruinard, E. (2016). Evaluation of Collaborative Assessment of Work Integrated Learning. *International Journal of Clinical Legal Education*, *23*(1), 5. https://doi.org/10.19164/ijcle.v23i1.488

McNicol, T., Carthouser, B., Bongiovanni, I., & Abeysooriya, S. (2023). Improving Ethical Usage Of corporate Data in Higher Education: Enhanced Enterprise Data Ethics Framework. *Information Technology and People*, *37*(6), 2247-2278. https://doi.org/10.1108/itp-12-2022-0971

Meadows, S., & Braine, R. D. (2022). The Work Identity of Leaders in the Midst of the COVID-19 Pandemic. *Frontiers in Psychology*, *13*. https://doi.org/10.3389/fpsyg.2022.958679

Meaney, M. (2024). Countering the Matthew Effect in Massive Open Online Courses (MOOCs): Policy Recommendations to Promote Inclusive Design. https://doi.org/10.35542/osf.io/asuyj

Meany, B. C., Fenn, J., Rescober, S., Narang, K., & Dhanda, J. (2024). Communication in Virtual Reality as Applied to Medical Education. *Arts & Humanities Open Access Journal*, *6*(1), 56-59. https://doi.org/10.15406/ahoaj.2024.06.00222

Medhi, M., & Baruah, T. (2025). Disconnected Lives: A Comparative Study of Digital Divide Among Adult Women in Tintukura and Bengenabari Villages, Assam. https://doi.org/10.21203/rs.3.rs-6826243/v1

Medina, A., Hernández, J. C., Muñoz, E., & Rus-Casas, C. (2020). Identification of Educational Models That Encourage Business Participation in Higher Education Institutions. *Sustainability*, *12*(20), 8421. https://doi.org/10.3390/su12208421

Medina, R. M., & Effrosyni, C. (2024). Exploring Refugee School Teachers' Roles in Culturally Diverse Adult Classrooms in Greece. *Education Sciences*, *14*(12), 1315. https://doi.org/10.3390/educsci14121315

Megawati, M., Riady, Y., Firman, F., Fitrah, Y., & Kusmana, A. (2023). Implications of Falsification Theory and the Constructivism Paradigm in the Development of Science Education and Learning. *Etude Journal of Educational Research*, *4*(1), 9-13. https://doi.org/10.56724/etude.v4i1.257

Meghnagi, M., & Tuccio, M. (2022). The Recognition of Prior Learning. https://doi.org/10.1787/2d9fb06a-en

Mehta, R. (2024). Education Investment and Human Capital Development in India. *Journal of Poverty Investment and Development*, *9*(1), 50-63. https://doi.org/10.47604/jpid.2587

Mei, D., Moore, K., & Sayler, B. (2026). Preparing Students for AI-Powered Materials Discovery: A Workflow-Aligned Framework for AI Literacy, Equity, and Scientific Judgment. https://doi.org/10.48550/arxiv.2605.09624

Mei, W. K., Khair, Z., & Othman, M. F. b. (2022). Transformational Learning in Teacher Context. *International Journal of Academic Research in Business and Social Sciences*, *12*(8). https://doi.org/10.6007/ijarbss/v12-i8/14538

Mejía, M. L. F., Peñaranda, L. A. F., Hurtado, M. D. B., Salazar, E. J. A., Cacho, L. P., & Guimaray, J. I. Z. (2023). Universidad Inclusiva E Intercultural en Latinoamérica: Una Visión De Los Marcos Genéricos. https://doi.org/10.31219/osf.io/urfxh

Mella-Norambuena, J., Quintana, M. G. B., & López-Angulo, Y. (2022). Modelos Predictivos Basados en Uso De Analíticas De Aprendizaje en Educación Superior. *Texto Livre Linguagem E Tecnologia*, *15*, e36310. https://doi.org/10.35699/1983-3652.2022.36310

Mella-Norambuena, J., Sáez-Delgado, F., López-Angulo, Y., Sáez-Delgado, Y., & Ron, V. L. (2023). Analíticas De Aprendizaje Y Su Potencial Para Una Educación De Calidad Sostenible. *Ciencia Latina Revista Científica Multidisciplinar*, *7*(1), 5446-5468. https://doi.org/10.37811/cl_rcm.v7i1.4840

Mellard, D. F., Krieshok, T. S., Fall, E., & Woods, K. L. (2012). Dispositional Factors Affecting Motivation During Learning in Adult Basic and Secondary Education Programs. *Reading and Writing*, *26*(4), 515-538. https://doi.org/10.1007/s11145-012-9413-4

Mello, M. I. C. d., Cunha, C. J. C. d. A., d, S. S. M., & A., D. G. (2023). Development of Leaders of Technologically-Based Companies: A Trans-Formative Learning Perspective. *Revista Caribeña De Ciencias Sociales*, *12*(1), 52-84. https://doi.org/10.55905/rcssv12n1-004

Melnychenko, S., Ovchynnyk, V., Hudal, S., & Kulyk, M. (2024). Pedagogical Aspects of the Development of Leadership Qualities in Cadets of Higher Military Educational Institutions. *Scientific Bulletin of Mukachevo State University Series "Pedagogy and Psychology"*, *10*(1), 87-97. https://doi.org/10.52534/msu-pp1.2024.87

Melo, V. (2019). Emancipatory Education and Youth Engagement in Brazil: A Case Study Bridging the Theory and Practice of Education for Social Transformation. *Education Sciences*, *9*(1), 23. https://doi.org/10.3390/educsci9010023

Melrose, S., Miller, J., Gordon, K., & Janzen, K. J. (2012). Becoming socialized into a new professional role: LPN to BN student nurses' experiences with legitimation. *Nursing Research and Practice*(2012).

Melrose, S., Park, C., & Perry, B. (2020). Creative Clinical Teaching in the Health Professions. https://doi.org/10.15215/aupress/9781771993319.01

Memari, M., & Ruggles, K. (2025). Artificial Intelligence in Elementary STEM Education: A Systematic Review of Current Applications and Future Challenges. https://doi.org/10.48550/arxiv.2511.00105

Mendiola, M. S., Moreno, J., Godínez, T. B., & Martínez-González, A. (2018). La Analítica Del Aprendizaje en Educación Médica: ¿punto De Inflexión? *Gaceta Médica De México*, *155*(1). https://doi.org/10.24875/gmm.18004801

Mengting, L. (2025). Investigating the Role of Self-Identity in Adults' Intention for Lifelong Arts Education: An Empirical Study of Dance Learning in China. *Adult education quarterly*, *76*(2), 122-141. https://doi.org/10.1177/07417136251344535

Mensah, E. T., Chen, M., Ntim, S. Y., & Gabrah, A. (2023). Analysing Dewey's Vocational Aspects of Education and Maslow's Theory of Motivation in Support of Vocational Education and Training. *Discover Education*, *2*(1). https://doi.org/10.1007/s44217-023-00042-1

Mentz, E., & Beer, J. d. (2019). The Affordances of Change Laboratories for Improved Sustainability and Impact of in-Service Teacher Professional Development Programmes. https://doi.org/10.20472/tec.2019.008.018

Merhout, J. W., & Cothran, P. E. (2006). Increasing Demand for IT Auditing Creates New Career Options for AIS/MIS/IT Students. *Review of Business Information Systems (Rbis)*, *10*(4), 41-50. https://doi.org/10.19030/rbis.v10i4.4436

Merriam, S. B., & Bierema, L. L. (2014). *Adult Learning: Linking Theory and Practice*. Jossey-Bass.

Merriam, S. B., & Kee, Y. (2014). Promoting Community Wellbeing: The Case for Lifelong Learning for Older Adults. *Adult education quarterly*, *64*(2), 128-144. https://doi.org/10.1177/0741713613513633

Merritt, E. G., Archambault, L., & Hale, A. (2018). Sustainability Education in Elementary Classrooms: Reported Practices of Alumni From a Pre-Service Teacher Course. *Discourse and Communication for Sustainable Education*, *9*(1), 18-35. https://doi.org/10.2478/dcse-2018-0002

Merzliakova, E., Gradovski, M., & Ødegaard, E. E. (2022). Dialogue About "Learning" Across the Northern Norway-Russia Border. *Dialogic Pedagogy a Journal for Studies of Dialogic Education*, *10*, A43-A64. https://doi.org/10.5195/dpj.2022.416

Merzouk, A., Kurosinski, P., & Kostikas, K. (2014). E-Learning for the Medical Team: The Present and Future of ERS Learning Resources. *Breathe*, *10*(4), 296-304. https://doi.org/10.1183/20734735.008814

Mesquita, I., Ribeiro, J., Santos, S., & Morgan, K. (2014). Coach Learning and Coach Education: Portuguese Expert Coaches' Perspective. *The Sport Psychologist*, *28*(2), 124-136. https://doi.org/10.1123/tsp.2011-0117

Mesuwini, J., & Mokoena, S. (2023). TVET Lecturer Work-Integrated Learning: Opportunities and Challenges. *International Journal of Learning Teaching and Educational Research*, *22*(8), 415-440. https://doi.org/10.26803/ijlter.22.8.22

Mesuwini, J., Thaba-Nkadimene, K. L., Mzindle, D., & Makwara, C. (2022). Exploring TVET Lecturer Perceptions About Work-Integrated Learning in South Africa. *Skandinavisk Tidsskrift for Yrker Og Profesjoner I Utvikling*, *7*(1). https://doi.org/10.7577/sjvd.4784

Meta. (2025). Meta Quest. https://www.meta.com/quest

Metsämuuronen, J., & Räsänen, P. (2018). Cognitive–Linguistic and Constructivist Mnemonic Triggers in Teaching Based on Jerome Bruner's Thinking. *Frontiers in Psychology*, *9*. https://doi.org/10.3389/fpsyg.2018.02543

Meylani, R. (2024). Innovations With Schema Theory: Modern Implications for Learning, Memory, and Academic Achievement. *International Journal for Multidisciplinary Research*, *6*(1). https://doi.org/10.36948/ijfmr.2024.v06i01.13785

Meza, I. (2022). Implicaciones De La Teoría Del Procesamiento De Información O Cognitivismo en Aprendices Universitarios. Menciones Al Conductismo Y Constructivismo. *Investigación Y Postgrado*, *37*(2), 217-229. https://doi.org/10.56219/investigacinypostgrado.v37i2.1468

Mezirow, J. (1991). *Transformative Dimensions of Adult Learning*. Jossey-Bass.

Mezirow, J. (2000). Learning to Think like an Adult: Core Concepts of Transformation Theory. In J. Mezirow (Ed.), *Learning as Transformation: Critical Perspectives on a Theory in Progress* (pp. 3-33). Jossey-Bass.

Mhlanga, D. (2021). Artificial Intelligence in the Industry 4.0, and Its Impact on Poverty, Innovation, Infrastructure Development, and the Sustainable Development Goals: Lessons From Emerging Economies? *Sustainability*, *13*(11), 5788. https://doi.org/10.3390/su13115788

Miah, M. T., Késmárki-Gally, S. E., Dancs, A., & Fekete-Farkas, M. (2024). A Systematic Review of Industry 4.0 Technology on Workforce Employability and Skills: Driving Success Factors and Challenges in South Asia. *Economies*, *12*(2), 35. https://doi.org/10.3390/economies12020035

Miao, Q., Eva, N., Newman, A., Nielsen, I., & Herbert, K. (2019). Ethical Leadership and Unethical Pro-Organisational Behaviour: The Mediating Mechanism of Reflective Moral Attentiveness. *Applied Psychology*, *69*(3), 834-853. https://doi.org/10.1111/apps.12210

Michel, A. (2016). Complex Education Systems: From Steering Change to Governance. *European journal of education*, *51*(4), 513-521. https://doi.org/10.1111/ejed.12186

Michelot, F. (2022). Pour Faire Face Aux Défis Informationnels, Numériques Et Médiatiques Du 21e Siècle. *Revue Hybride De L Éducation*, *5*(2), 1-27. https://doi.org/10.1522/rhe.v5i2.1235

Micheni, E., Wechuli, A. N., Murumba, J., & Machii, J. (2021). Fostering the Fourth Industrial Revolution Technologies for Youth and Women Empowerment. *Jiea*. https://doi.org/10.7176/jiea/11-1-05

Microsoft. (2025). Microsoft Mesh and HoloLens. https://www.microsoft.com/mesh

Middleton, L., Hall, H., & Raeside, R. (2018). Applications and Applicability of Social Cognitive Theory in Information Science Research. *Journal of Librarianship and Information Science*, *51*(4), 927-937. https://doi.org/10.1177/0961000618769985

Mihalcea, A. D. (2017). Employer Branding and Talent Management in the Digital Age. *Management Dynamics in the Knowledge Economy*, *5*(2), 289-306. https://doi.org/10.25019/mdke/5.2.07

Mihus, I., & Gupta, S. K. (2023). Exploring the Foundations of the Digital Economy: an in-Depth Overview. 6-37. https://doi.org/10.36690/bm-id-eu-6-37

Mikeladze, T., Meijer, P. C., & Verhoeff, R. P. (2024). A Comprehensive Exploration of Artificial Intelligence Competence Frameworks for Educators: A Critical Review. *European journal of education*, *59*(3). https://doi.org/10.1111/ejed.12663

Mikelli, D. (2020). Pedagogy of Difference 2.0: Interactive Documentary Practices and Participatory Research With Young People. *Convergence the International Journal of Research Into New Media Technologies*, *27*(2), 438-457. https://doi.org/10.1177/1354856520934724

Mikkonen, K. (2025). The Evolution and Future of Nursing Education: Bridging Evidence-Based Practice and Hybrid Intelligence. *Journal of Advanced Nursing*, *82*(S2). https://doi.org/10.1111/jan.70434

Mikkonen, K., Ferdinando, H., Sobocinski, M., Kuivila, H. M., Pramila-Savukoski, S., Vhitehead, T., Ropponen, P., Myllylä, T., Paunonen, J., Halili, E., Koutonen, J., Taikina-Aho, J.-M., Siipo, A., & Järvelä, S. (2024). How Does Human-Centred Extended Reality Support Healthcare Students' Learning in Clinical Conditions? , 181-188. https://doi.org/10.1007/978-3-031-59080-1_13

Mikulec, B. (2017). On Democratic Adult Education: Navigating the Dialogue Between Consensus and Dissensus in Contemporary Multicultural Communities. Nr 18 (2017): DYSKURSY MŁODYCH. https://doi.org/10.34768/dma.vi18.71

Mikulec, B., & Ermenc, K. S. (2016). Qualifications Frameworks Between Global and European Pressures and Local Responses. *Sage Open*, *6*(2). https://doi.org/10.1177/2158244016644948

Mikulec, B., Howells, A., Mihajlović, D., Turiman, P., Ellias, N. N., & Douglas, M. (2020). National Qualifications Frameworks as a Policy Instrument for Lifelong Learning in Ghana, Malaysia and Serbia. 81-98. https://doi.org/10.36253/978-88-5518-155-6.06

Mikulec, B., & Kovšca, M. (2023). Professionalisation of Adult Education in Slovenia From a Multi-Level Perspective. 175-193. https://doi.org/10.36253/979-12-215-0151-3.15

Mikulec, E., & Ramalho, T. (2022). Enacting Critical Pedagogy Online. https://doi.org/10.3726/b19369

Milana, M. (2012). Political Globalization and the Shift From Adult Education to Lifelong Learning. *European Journal for Research on the Education and Learning of Adults*, *3*(2), 103-117. https://doi.org/10.3384/rela.2000-7426.rela0070

Milana, M. (2014). UNESCO, Adult Education and Political Mobilization. *Confero Essays on Education Philosophy and Politics*, *2*(1), 73-107. https://doi.org/10.3384/confero.2001-4562.140604a

Milana, M., & Holford, J. (2014). Adult Education Policy and the European Union: Theoretical and Methodological Perspectives. https://doi.org/10.1163/9789462095489

Millar, R. J., Sahoo, S., Yamashita, T., & Cummins, P. (2020). Literacy Skills, Language Use, and Online Health Information Seeking Among Hispanic Adults in the United States. *Patient Education and Counseling*, *103*(8), 1595-1600. https://doi.org/10.1016/j.pec.2020.02.030

Miller, E., & Konstantinou, I. (2021). Using Reflective, Authentic Assessments to Embed Employability Skills in Higher Education. *Journal of Work-Applied Management*, *14*(1), 4-17. https://doi.org/10.1108/jwam-02-2021-0014

Miller, R. D. (2024). *I Don't Know What I Don't Know: an Action Research Study Regarding the Impact of Youth Travel/Club Sports on High School Athletic Participation* Northeastern University].

Milosheva, M., Hall, H., Robertson, P. K. J., & Cruickshank, P. (2023). Information Literacy Competencies for Career Transitions in the Digital Age. 71-100. https://doi.org/10.29085/9781783305803.007

Min, W., & Yu, Z. (2023). A Systematic Review of Critical Success Factors in Blended Learning. *Education Sciences*, *13*(5), 469. https://doi.org/10.3390/educsci13050469

Minali, B., & Manju, S. (2018). Is Higher Education a Potential Means of Promoting Labour Mobility? *Ijemr*, *2*(1), 27-31. https://doi.org/10.22662/ijemr.2018.2.1.027

Mirzajani, H., Mahmud, R., Ah mad Fauzi Mohd, A., & Wong, S. L. (2015). A Review of Research Literature on Obstacles That Prevent Use of ICT in Pre-Service Teachers' Educational Courses. *International Journal of Education and Literacy Studies*, *3*(2). https://doi.org/10.7575/aiac.ijels.v.3n.2p.25

Mishra, L., & Raina, R. L. (2021). Online Teaching Effectiveness in Higher Education: Faculty Perspectives in India. *Journal of University Teaching and Learning Practice*, *18*(8). https://doi.org/10.53761/1.18.8.5

Mishra, N. R. (2023). Constructivist Approach to Learning: An Analysis of Pedagogical Models of Social Constructivist Learning Theory. *Journal of Research and Development*, *6*(01), 22-29. https://doi.org/10.3126/jrdn.v6i01.55227

Mishra, S., & Ray, S. (2025). Association Between Early Maladaptive Schemas and Developmental Crisis Among Young Adults: Mediating Role of Cognitive Flexibility. *Annals of Neurosciences*. https://doi.org/10.1177/09727531251394926

Mitchell, A. D., Oswald, B., Voon, T., & Larcombe, W. (2011). Education in the Field: A Case Study of Experiential Learning in International Law. *Legal Education Review*, *21*(1). https://doi.org/10.53300/001c.6243

Mitchell, F. (2023). The Implications of a Non-Inclusive Education for Children and Young People With Autism. *Kairaranga*, *24*(1), 33-42. https://doi.org/10.54322/kairaranga.v24i1.160

Mitchell, S., Hart, J. M., Gharaibeh, M., McMahon, G. T., Rhoda, A., Fitzpatrick, S., Wuliji, T., & Janczukowicz, J. (2025). Principles to Award Learning Achievements for Lifelong Learning in Health Using Micro-Credentials: An International Delphi Study. *Human Resources for Health*, *23*(1). https://doi.org/10.1186/s12960-024-00969-y

Mitra, N. K. (2023). Artificial Intelligence: Opportunities and Challenges in Health Professions Education. *International E-Journal of Science Medicine & Education*, *17*(3), 2-8. https://doi.org/10.56026/imu.17.3.2

Mitrulescu, C. M. (2023). Creating EFL Activities for Military Students: A Project-Based Learning Approach. *Land Forces Academy Review*, *28*(4), 296-302. https://doi.org/10.2478/raft-2023-0035

Mittal, P., Kaur, A., & Jain, R. (2022). Online Learning for Enhancing Employability Skills in Higher Education Students: The Mediating Role of Learning Analytics. *Tem Journal*, 1469-1476. https://doi.org/10.18421/tem114-06

Mizzi, R. C. (2021). Teaching for LGBTQ Inclusion in Online Settings. *New Horizons in Adult Education and Human Resource Development*, *33*(3), 70-74. https://doi.org/10.1002/nha3.20339

Mlambo, M., Silén, C., & McGrath, C. (2021). Lifelong Learning and Nurses' Continuing Professional Development, a Metasynthesis of the Literature. *BMC Nursing*, *20*(1). https://doi.org/10.1186/s12912-021-00579-2

Moallem, M. (2019). What Effects Does Problem-Based Learning Have on Student Learning Outcomes? https://doi.org/10.3102/1439964

Mobbs, D., & Tashjian, S. M. (2022). Ten Simple Rules for Unbiased Teaching. *Plos Computational Biology*, *18*(10), e1010344. https://doi.org/10.1371/journal.pcbi.1010344

Močinić, S., & Tatković, S. (2021). Using Kolb's Learning Model in Structured Types of Professional Education and Training for Teachers. *Revija Za Elementarno Izobraževanje*, *4*(4), 409-433. https://doi.org/10.18690/rei.14.4.409-433.2021

Mohajer, N., & Earnest, J. (2009). Youth Empowerment for the Most Vulnerable. *Health Education*, *109*(5), 424-438. https://doi.org/10.1108/09654280910984834

Mohamad, N. S., & Romli, T. R. M. (2021). Application of Theory of Social Constructivism in Teaching Arabic Teachers to Apply Higher Order Thinking Skill. *International Journal of Academic Research in Progressive Education and Development*, *10*(2). https://doi.org/10.6007/ijarped/v10-i2/10130

Mohamed, A. L., & Al-Hidabi, D. A. Y. (2020). Mobile Device-Based Authentic Learning in Educational Environments: A Systematic Review. *Iium Journal of Educational Studies*, *7*(2), 40-55. https://doi.org/10.31436/ijes.v7i2.270

Mohamed, F. S., & Abdallah, D. T. (2021). Applying Lexical Density and Lexical Diversity for Developing Writing Quality and Quantity. 118-88 ,(1)50 ,مجلة كلية التربية بالإسماعيلية. https://doi.org/10.21608/jfes.2021.173721

Mohamed, M., Rashid, R. A., & Alqaryouti, M. H. (2022). Conceptualizing the Complexity of Reflective Practice in Education. *Frontiers in Psychology*, *13*. https://doi.org/10.3389/fpsyg.2022.1008234

Mohammad, M. A. A. (2026). Authentic Workplace Learning in Technology-Enhanced Vocational Education: Examining Situated Learning Principles in Industrial English Training. https://doi.org/10.21203/rs.3.rs-8817222/v1

Mohammadi, F. S., & Asadzadeh, H. (2011). Testing the Mediating Role of Teachers' Self-Efficacy Beliefs in the Relationship Between Sources of Efficacy Information and Students Achievement. *Asia Pacific Education Review*, *13*(3), 427-433. https://doi.org/10.1007/s12564-011-9203-8

Mohammed, D., Abdelrahman, S., Ali, R., & Gabra, S. (2023). Correlation Between Organizational Learning Culture, Goal Orientation, Managerial Effectiveness and Workplace Learning for Nurses. *Minia Scientific Nursing Journal*, *013*(3), 9-16. https://doi.org/10.21608/msnj.2023.206645.1060

Mohammed, I. (2020). Leadership Behavior Analysis: The Case of Three Academic Middle-Level Leaders in Higher Education. *Open Journal of Leadership*, *09*(03), 141-155. https://doi.org/10.4236/ojl.2020.93009

Mohammed, N. F., Bakar, N. A., Mohamed, N., & Fariz, N. M. (2026). Anti-Corruption Education Framework for Higher Education: A Qualitative Analysis. *Sage Open*, *16*(1). https://doi.org/10.1177/21582440251410024

Mohd Amirul Syafiq bin, A., Ma'rof, A. M., Kamalden, T. F. T., & Ahmad, N. A. (2023). Achievement Motivation and Academic Success: A Study of Student-Athletes in a Malaysian Sports School. *International Journal of Academic Research in Progressive Education and Development*, *12*(2). https://doi.org/10.6007/ijarped/v12-i2/16633

Mohiuddin, K., Islam, M. A., Khaleel, M. A., Miladi, M. N., Nasr, O. A., Khan, S. A., Khan, M. A., & Fatima, H. (2021). Determining Mobile Learning Potentialities in Emerging Computing

Paradigms: An Exploratory Approach for Performance Optimization. https://doi.org/10.21203/rs.3.rs-1025389/v1

Mokaya, C. (2025). The Effects of Evaluation in Distance Learning. *Africa Journal of Technical and Vocational Education and Training*, *9*(1). https://doi.org/10.69641/afritvet.2024.91186

Mokotso, R. I. (2024). Enhancing Religious Education Teaching and Learning for Sustainable Development in Lesotho. *HTS Teologiese Studies / Theological Studies*, *80*(1). https://doi.org/10.4102/hts.v80i1.9151

Molina-Montes, A., Pérez-Villamizar, D. I., Domínguez-Angarita, D. D., & Yohaid-Trujillo, Y. L. (2022). Investigación Y Desafíos Educativos en El Contexto Mundial en Torno a La Educación Emocional Bajo La Perspectiva De La Psicología Positiva. Una Revisión Sistemática. *Aibi Revista De Investigación Administración E Ingeniería*, *10*(3), 109-120. https://doi.org/10.15649/2346030x.3099

Molina, G. C. T. (2022). Perspectivas Teóricas Y Metodológicas De Emociones en Aprendizaje en Línea: Revisión Sistemática. *Alpha Centauri*, *3*(1), 43-51. https://doi.org/10.47422/ac.v3i1.73

Moll, I. (2023). A Critique of Andragogy in the South African TVET Context. *Journal of Vocational Adult and Continuing Education and Training*, *6*(1). https://doi.org/10.14426/jovacet.v6i1.318

Moll, I. (2024). A Psychological Critique of Knowles' Andragogy as a Theory of Learning. *Andragoška Spoznanja*, *30*(1), 151-170. https://doi.org/10.4312/as/16396

Møller, B. (2024). Workplace Learning Potential in Uncertain Situations in Care Work. *Journal of Workplace Learning*, *37*(1), 78-92. https://doi.org/10.1108/jwl-12-2023-0187

Monferrato, M. (2024). El Papel De La Universidad en El Desarrollo De Competencias Digitales Y La Reducción Del Desajuste De Habilidades Desde Una Perspectiva Del Capital Humano Para El Siglo XXI. *Debate Universitario*, *14*(23), 45-57. https://doi.org/10.59471/debate2024172

Monk, R. L. (1996). The Motivation of Managers for Training. *Management Development Review*, *9*(3), 26-32. https://doi.org/10.1108/09622519610772148

Montgomery, D. (2022). Integrating Technology With Instructional Frameworks to Support All Learners in Inclusive Classrooms. *The Open/Technology in Education Society and Scholarship Association Journal*, *2*(2), 1-16. https://doi.org/10.18357/otessaj.2022.2.2.31

Moodie, G. (2008). *From Vocational to Higher Education: An International Perspective*. Open University Press.

Moodie, G. (2011). The Interplay Between Lifelong Learning and Vocational Education and Training. 813-827. https://doi.org/10.1007/978-94-007-2360-3_49

Moodley, D. E. (2024). Universal Design for Learning and Writing Centres in South African Higher Education. *Perspectives in Education*, *42*(2), 15-30. https://doi.org/10.38140/pie.v42i2.7830

Moon, H. (2018). Unleashing Apprenticeship: From Onboarding to Professional Development. *European journal of training and development*, *42*(1/2), 110-124. https://doi.org/10.1108/ejtd-06-2017-0056

Moore, S., Hope, E. C., Eisman, A. B., & Zimmerman, M. A. (2016). Predictors of Civic Engagement Among Highly Involved Young Adults: Exploring the Relationship Between Agency and Systems Worldview. *Journal of Community Psychology*, *44*(7), 888-903. https://doi.org/10.1002/jcop.21815

Moosa, M., & Aloka, P. J. O. (2022). Motivational Factors as a Driver for Success for First-Year Students at a Selected Public University in South Africa. *Student Success*. https://doi.org/10.5204/ssj.2415

Moquin, R., Dewey, M., Weinhold, A., Green, O., & Young, A. R. (2023). The Flipped Classroom Approach in a Pediatric Anesthesiology Fellowship Curriculum. *Cureus*. https://doi.org/10.7759/cureus.43979

Morais, J. M. O., Francisco Thiago Chaves de, O., Nóbrega-Therrien, S. M., & Souza, S. G. d. (2023). Paulo Freire's Contributions to the Education of Young People and Adults: A Narrative Review. *Educação Em Revista*, *39*. https://doi.org/10.1590/0102-469840514t

Moreno, J. G. V. (2023). ¿Qué Se Está Haciendo Para Fomentar La Resiliencia en Las Escuelas? Un Acercamiento Al Sistema Educativo en General Y, Especialmente, Al Colombiano. *Ciencia Latina Revista Científica Multidisciplinar*, *7*(2), 7422-7440. https://doi.org/10.37811/cl_rcm.v7i2.5891

Moretti, L., Mayerl, M., Schlögl, P., & Wolter, S. C. (2017). So Similar and Yet So Different: A Comparative Analysis of a Firm's Cost and Benefits of Apprenticeship Training in Austria and Switzerland. *SSRN Electronic Journal*. https://doi.org/10.2139/ssrn.3056619

Morgan, C. (2007). *OECD Programme for International Student Assessment: Unraveling a Knowledge Network* Carleton University].

Morin, O., Langé, J.-M., & Barthes, A. (2021). Enjeux Socio-Cognitifs De L'engagement : La Durabilité en Contexte Scolaire. *Educations*, *4*(1). https://doi.org/10.21494/iste.op.2021.0641

Morley, D. A. (2018). The 'Ebb and Flow' of Student Learning on Placement. 173-190. https://doi.org/10.1007/978-3-319-75166-5_10

Moroz, H. (2020). The Role of Social Protection in Building, Protecting, and Deploying Human Capital in the East Asia and Pacific Region. https://doi.org/10.1596/34783

Morrillo, C. F. S., Palacios, I. A. M., Palacios, G. M. M., Palacios, H. F. M., & Cumbicos, K. M. C. (2024). Desarrollo Y Evaluación De Recursos Educativos Digitales Para La Educación Inclusiva. *Latam Revista Latinoamericana De Ciencias Sociales Y Humanidades*, *5*(2). https://doi.org/10.56712/latam.v5i2.1913

Morris, E. K., Todd, J. T., Midgley, B. D., Schneider, S., & Johnson, L. (1990). The History of Behavior Analysis: Some Historiography and a Bibliography. *The Behavior Analyst*, *13*(2), 131-158. https://doi.org/10.1007/bf03392530

Morris, T. H. (2019). Self-Directed Learning: A Fundamental Competence in a Rapidly Changing World. *International Review of Education*, *65*(4), 633-653. https://doi.org/10.1007/s11159-019-09793-2

Morris, T. H. (2023). Four Dimensions of Self-Directed Learning: A Fundamental Meta-Competence in a Changing World. *Adult education quarterly*, *74*(3), 236-254. https://doi.org/10.1177/07417136231217453

Morrison, D. (2020). Eutagogoy Extended: A Conceptual Framework to Support Informal, Self-Determined Lifelong Learning for Older Adults. *Journal of Studies in Education*, *10*(3), 143. https://doi.org/10.5296/jse.v10i3.17297

Morrison, M., & Rosenthal, A. (1997). Exploring Learning Organizations: Enacting Mental Models - the Power of the Rosenthal Stage. *Journal of Workplace Learning*, *9*(4), 124-129. https://doi.org/10.1108/13665629710177856

Morrison, P. S. (2006). Research Into Labour, Employment and Work in New Zealand: Overview of the Twelfth Conference. *Labour Employment and Work in New Zealand*. https://doi.org/10.26686/lew.v0i0.1303

Morrow, E., Kelly, C., Killeen, C., Naessens, E., & Lynch, M. G. (2024). Exploring a Career Pathway for Home Support Workers in Ireland: A Systematic Scoping Review of the International Evidence. *Frontiers in Health Services*, *4*. https://doi.org/10.3389/frhs.2024.1360920

Morse, S. M. (2006). Assessing the Value. *Journal of European Industrial Training*, *30*(9), 735-755. https://doi.org/10.1108/03090590610715040

Moshawrab, M., Adda, M., Bouzouane, A., Ibrahim, H., & Raad, A. (2022). Smart Wearables for the Detection of Occupational Physical Fatigue: A Literature Review. *Sensors*, *22*(19), 7472. https://doi.org/10.3390/s22197472

Mosquera, J. M. L., & Bucheli, V. (2021). Analítica De Aprendizaje Como Estrategia De Apoyo Al Aula Invertida en Cursos De Programación: Una Revisión Sistemática De Literatura. *Investigación E Innovación en Ingenierías*, *9*(1), 114-135. https://doi.org/10.17081/invinno.9.1.4464

Moss-Pech, C., Lopez, S. H., & Michaels, L. (2021). Educational Downgrading: Adult Education and Downward Mobility. *Sociology of Education*, *94*(2), 143-158. https://doi.org/10.1177/0038040720982890

Mou, T.-Y. (2021). Online Learning in the Time of the COVID-19 Crisis: Implications for the Self-Regulated Learning of University Design Students. *Active Learning in Higher Education*, *24*(2), 185-205. https://doi.org/10.1177/14697874211051226

Mouta, A., Llorente, A. M. P., & Sánchez, E. M. T. (2024). Uncovering Blind Spots in Education Ethics: Insights From a Systematic Literature Review on Artificial Intelligence in Education. *International Journal of Artificial Intelligence in Education*, *34*(3), 1166-1205. https://doi.org/10.1007/s40593-023-00384-9

Moya, E. C., Nieves, L. B. H., & Montes, R. (2020). Educación Virtual Para Todos: Una Revisión Sistemática. *Education in the Knowledge Society (Eks)*, *21*, 13. https://doi.org/10.14201/eks.23448

Mozelius, P., Cleveland-Innes, M., Lindqvist, M. H., & Jaldemark, J. (2024). Critical Aspects of a Higher Education Reform for Continuous Lifelong Learning Opportunities in a Digital Era. *Electronic Journal of Knowledge Management*, *22*(1), 26-39. https://doi.org/10.34190/ejkm.22.1.3258

Mpofu, J., & Sefotho, M. M. (2024). The Relationship Between the Philosophy of Ubuntu and the Principles of Inclusive Education. *Perspectives in Education*, *42*(2), 128-144. https://doi.org/10.38140/pie.v42i2.7338

Msacky, R. (2024). Retention of Human Resources for Health in the Decentralised Health System in Tanzania: Does Training Matter? *Journal of Policy and Development Studies*, *16*(1), 74-91. https://doi.org/10.4314/jpds.v16i1.5

Msiska, F. G. W. (2015). Education, Learning, Training: Critical Issues for Development. *International Review of Education*, *61*(5), 713-716. https://doi.org/10.1007/s11159-015-9510-0

Msonde, S. E. (2021). Revisiting the Idea of Learner-Centered Pedagogy: The Theoretical Perspective. *Journal of Education*, *203*(2), 468-478. https://doi.org/10.1177/00220574211031970

Muafiah, E., Susanto, S., Warsah, I., Puspitasari, D., & Puspita, A. R. (2025). Implementation and Problems of Education Based on Gender Equality, Disability, and Social Inclusion at Schools in Indonesia. *International Journal of Sociology of Education*, 1-20. https://doi.org/10.17583/rise.14279

Mucenski-Keck, L., Hintz, A. F., & Fedoryshyn, M. W. (2012). Planning for Your Second Accounting Job. *American Journal of Business Education (Ajbe)*, *5*(5), 539-546. https://doi.org/10.19030/ajbe.v5i5.7210

Mudau, P. K., & Berg, G. v. d. (2023). Guidelines for Supporting a Community of Inquiry Through Graded Online Discussion Forums in Higher Education. *Education Sciences*, *13*(9), 963. https://doi.org/10.3390/educsci13090963

Mueller, B., Andrew, M., & Connor, M. (2022). Building Belonging in Online WIL Environments – Lessons (Re)learnt in the Pandemic Age: A Collaborative Enquiry. *Journal of University Teaching and Learning Practice*, *19*(4). https://doi.org/10.53761/1.19.4.15

Mueller, S., & Anderson, A. R. (2014). Understanding the Entrepreneurial Learning Process and Its Impact on Students' Personal Development: A European Perspective. *The International Journal of Management Education*, *12*(3), 500-511. https://doi.org/10.1016/j.ijme.2014.05.003

Mugambi, M. (2018). Linking Constructivism Theory to Classroom Practice. *International Journal of Humanities Social Sciences and Education*, *5*(9). https://doi.org/10.20431/2349-0381.0509014

Mugler, J. (2025). Redivisions of Work: Manufacturing Labor and Training in Switzerland's Knowledge Economy. *Anthropology of Work Review*, *46*(2). https://doi.org/10.1111/awr.70011

Muhtar, M., Setiyono, B., Fitriyah, F., & Sardini, N. H. (2022). The Role of Educational Institutions in Countering Political Polarization: Building Inclusive Communities and Civic Engagement. *Qalamuna Jurnal Pendidikan Sosial Dan Agama*, *14*(1), 833-848. https://doi.org/10.37680/qalamuna.v14i1.4919

Mukhalalati, B., & Taylor, A. (2019). Adult Learning Theories in Context: A Quick Guide for Healthcare Professional Educators. *Journal of Medical Education and Curricular Development*, *6*. https://doi.org/10.1177/2382120519840332

Mukherjee, B., Smith, R., & Khalsa, G. (2024). An Experiential Approach to Training Medical Faculty to Coach: "The Total Experience Was Much More Than the Sum of Its Parts". *Journal of Clinical Psychology in Medical Settings*, *31*(4), 769-780. https://doi.org/10.1007/s10880-024-10038-0

Mukherjee, M., & Maity, C. (2022). Emergency Remote Learning (ERL) in the COVID-19 Era: Perceived Experience of Indian Learners of Higher Education. *Asian Association of Open Universities Journal*, *17*(2), 178-193. https://doi.org/10.1108/aaouj-03-2022-0042

Mukti, D. P., & Dudija, N. (2016). Organizational Change Dynamics : A Learning Organization Process Toward World Class Organization. https://doi.org/10.2991/isclo-15.2016.15

Mulcahy, D. (2000). Turning the contradictions of competence: competency-based training and beyond. *Journal of Vocational Education and Training*, *52*(2), 259-280.

Müller, C. V., Scheffer, Â. B. B., & Closs, L. Q. (2020). Volunteer Tourism, Transformative Learning and Its Impacts on Careers: The Case of Brazilian Volunteers. *International Journal of Tourism Research*, *22*(6), 726-738. https://doi.org/10.1002/jtr.2368

Mungania, P. (2004). *Employees' Perceptions of Barriers in E-Learning : The Relationship Among Barriers, Demographics, and E-Learning Self-Efficacy* University of Louisville].

Muñoz, A. P., & Ramírez, M. B. (2015). Teachers' Conceptions of Motivation and Motivating Practices in Second-Language Learning: A Self-Determination Theory Perspective. *Theory and Research in Education*, *13*(2), 198-220. https://doi.org/10.1177/1477878515593885

Murakami, K. (2021). Learning in Retirement: Developing Resilience and Becoming a Resourceful Practitioner of Life. *Learning Culture and Social Interaction*, *28*, 100463. https://doi.org/10.1016/j.lcsi.2020.100463

Murillo, P. O. (2007). El PAEBA: La Alfabetización Y Educación Básica De Adultos en El Marco De La Cooperación Iberoamericana. *Revista Iberoamericana De Educación*, *44*(1), 1-7. https://doi.org/10.35362/rie4412286

Murphy, D. (2014). Issues With PISA's Use of Its Data in the Context of International Education Policy Convergence. *Policy Futures in Education*, *12*(7), 893-916. https://doi.org/10.2304/pfie.2014.12.7.893

Murray, A., Murray, P., Howells, K., Uthmani, N., & McMillan, N. (2025). The Embodiment of Equitable Ways to Develop Agentic Wellbeing Through Movement Maximising Personal and General Spaces: Re-Tooling Affordances as Drivers of Social Justice. *Equity in Education & Society*, *4*(2), 242-262. https://doi.org/10.1177/27526461251324217

Muscat, D. M., Morony, S., Nutbeam, D., Ayre, J., Shepherd, H. L., Smith, S. K., Dhillon, H. M., Hayen, A., Luxford, K., Meshreky, W., & McCaffery, K. (2020). Learners' Experience and Perceived Impact of a Health Literacy Program in Adult Basic Education: A Qualitative Study. *Public Health Research & Practice*, *30*(2). https://doi.org/10.17061/phrp29231909

Musset, P. (2019). Improving Work-Based Learning in Schools. https://doi.org/10.1787/918caba5-en

Mustafaa, F. N., & Martinez, M. N. (2022). A Visionary Working Model for Pursuing Social Justice Praxis Through Educational Psychology Courses. *Teaching of Psychology*, *50*(2), 191-197. https://doi.org/10.1177/00986283211065732

Muthamia, N. M., & Kilika, J. M. (2022). Towards a Theoretical Model for Learning Organization Strategy Towards Improved Service Delivery: A Review of Literature. *International Research Journal of Economics and Management Studies*, *1*(3), 65-76. https://doi.org/10.56472/25835238/irjems-v1i2p110

Muthaura, P. N., Khamis, T., Ahmed, M., & Hussain, S. R. (2015). Perceptions of the Preparedness of Medical Graduates for Internship Responsibilities in District Hospitals in Kenya: A Qualitative Study. *BMC Medical Education*, *15*(1). https://doi.org/10.1186/s12909-015-0463-6

Muzapparovna, N. G. (2024). Advancements in Continuing Education: Exploring Innovative Approaches to Teaching in the Modern Era. *International Journal of Advance Scientific Research*, *4*(4), 89-96. https://doi.org/10.37547/ijasr-04-04-16

Muzhanova, T., Yakymenko, Y., Zaporozhchenko, M., & Tyshchenko, V. (2022). International Vendor-Neutral Certification for Information Security Professionals. *Cybersecurity Education Science Technique*, *4*(16), 129-141. https://doi.org/10.28925/2663-4023.2022.16.129141

Muzumdar, P., & Cheemalapati, S. (2025). Exploring Student Interactions With AI-Powered Learning Tools: A Qualitative Study Connecting Interaction Patterns to Educational Learning Theories. *Journal of Education and Training*, *13*(2), 24. https://doi.org/10.5296/jet.v13i2.23228

Mwikonji, P. (2021). Role of Ethics in Promoting Organization Performance. *Journal of Public Relations*, *1*(1), 29-39. https://doi.org/10.47941/jpr.654

Myers, F., Baxter, J., Selby-Fell, H., & Morales, A. (2024). Navigating From Industry to Higher Education: Practitioner Transitions to Academic Life. *British educational research journal*, *51*(2), 930-948. https://doi.org/10.1002/berj.4109

Mystakidis, S., Fragkaki, M., & Filippousis, G. (2021). Ready Teacher One: Virtual and Augmented Reality Online Professional Development for K-12 School Teachers. *Computers*, *10*(10), 134. https://doi.org/10.3390/computers10100134

Na-Nan, K., Chaiprasit, K., & Pukkeeree, P. (2017). Influences of Workplace Environment Factors on Employees' Training Transfer. *Industrial and Commercial Training*, *49*(6), 303-314. https://doi.org/10.1108/ict-02-2017-0010

Na-Songkhla, J., & Sujiva, S. (2022). A HyFlex-Flipped Class in Action Learning: A Connectivist MOOC for Creative Problem-Solving. *Contemporary Educational Technology*, *14*(4), ep392. https://doi.org/10.30935/cedtech/12554

Nafukho, F. M., Alfred, M. V., Chakraborty, M., Johnson, M., & Cherrstrom, C. A. (2017). Predicting Workplace Transfer of Learning. *European journal of training and development*, *41*(4), 327-353. https://doi.org/10.1108/ejtd-10-2016-0079

Nagels, M., Abel, M. H., & Tali, F. (2018). Focus on the Agency of Learners to Innovate in Pedagogy. 27-40. https://doi.org/10.1108/978-1-78756-555-520181004

Naghipour, M. S., & Rahim, Z. A. (2024). 5G Competencies in Consideration of Industry 4.0 and Beyond: A Review of the Propositions. *International Journal of Academic Research in Business and Social Sciences*, *14*(6). https://doi.org/10.6007/ijarbss/v14-i6/21807

Nahak, N., & Srinivas, S. (2023). Employee Engagement and Retention Strategies on Innovation Performance in Talent Management Can Be Viewed as "Two Sides of the Same Coin". *Ecb*, *12*(si10). https://doi.org/10.48047/ecb/2023.12.si10.00416

Naik, P., & Chattaraj, D. (2025). Exploring Transformative Learning: Perspectives From Indian Educational Philosophies and Practices. *Journal of Transformative Education*. https://doi.org/10.1177/15413446251405182

Nail-Beatty, O., Ibáñez, A., Ayadi, R., Świeboda, P., Njamnshi, A. K., Occhipinti, J. A., Hynes, W., Ikiz, B., Castro-Aldrete, L., O'Brien, K. K., & Platt, M. L. (2024). Brain Health Is Essential for Smooth Economic Transitions: Towards Socio-Economic Sustainability, Productivity and Well-Being. *Brain Communications*, *6*(6). https://doi.org/10.1093/braincomms/fcae360

Naim, M. F., & Lenka, U. (2018). Development and Retention of Generation Y Employees: A Conceptual Framework. *Employee Relations the International Journal*, *40*(2), 433-455. https://doi.org/10.1108/er-09-2016-0172

Nakamura, Y. T. (2025). Professional Identity Development Through Embodied and Reflective Practice. *Journal of Transformative Education*. https://doi.org/10.1177/15413446251405179

Nakash, M. (2024). Reskilling and Upskilling the Workforce: Text Mining to Identify Patterns and Trends in Research. *Proceedings of the International Conference on Research in Human Resource Management*, *1*(1). https://doi.org/10.33422/icrhrm.v1i1.169

Nalliah, S. (2014). Applying the Learning Theories to Medical Education: A Commentary. *International E-Journal of Science Medicine & Education*, *8*(1), 50-57. https://doi.org/10.56026/imu.8.1.50

Nance, K., & Hay, B. (2007). Certifying the Computer Security Professional Using the Project Management Institute's PMP Model. 491-496. https://doi.org/10.1007/978-0-387-72367-9_46

Napari, A. (2020). Using Cognitive Apprenticeship to Enhance the Study of Budgeting and Budgetary Control: An Action Research1. *The Universal Academic Research Journal*, *2*(1), 1-14. https://doi.org/10.17220/tuara.2020.01.1

Narushima, M., Liu, J., & Diestelkamp, N. (2016). Lifelong Learning in Active Ageing Discourse: Its Conserving Effect on Wellbeing, Health and Vulnerability. *Ageing and Society*, *38*(4), 651-675. https://doi.org/10.1017/s0144686x16001136

Naseer, F., Tariq, R., Alshahrani, H. M., Alruwais, N., & Al-Wesabi, F. N. (2025). Project Based Learning Framework Integrating Industry Collaboration to Enhance Student Future Readiness in Higher Education. *Scientific Reports*, *15*(1). https://doi.org/10.1038/s41598-025-10385-4

Nasir, B., Jurd, K., Cook, S., Kwan, M., & Ostini, R. (2018). Developing an Internship Preparedness Program for Final Year Medical Students. *Mededpublish*, *7*, 219. https://doi.org/10.15694/mep.2018.0000219.1

Nassir, S. (2020). Adjustment of the Andragogical Principles to Adults Learning Information and Communication Technologies - Case of the Educational Administration Cycle (Rabat). *International Journal of Advanced Trends in Computer Science and Engineering*, *9*(1), 8-13. https://doi.org/10.30534/ijatcse/2020/02912020

Naudé, L., Bergh, T. J. v. d., & Krüger, I. (2014). "Learning to Like Learning": An Appreciative Inquiry Into Emotions in Education. *Social Psychology of Education*. https://doi.org/10.1007/s11218-014-9247-9

Naudé, W. (2017). Entrepreneurship, Education and the Fourth Industrial Revolution in Africa. *SSRN Electronic Journal*. https://doi.org/10.2139/ssrn.2998964

Naumčiuk, T. (2023). Adult Learners' Motivation to Learn a Foreign Language. *PSTP*, 112-119. https://doi.org/10.56131/pstp.2023.27.1.158

Nayak, K. V. (2026). Contextualizing Critical Consciousness: Community Engagement and Cultural Practices Among <i>Adivasi</I> Adults in Jharkhand, India. *Journal of Social Issues*, *82*(2). https://doi.org/10.1111/josi.70063

Nayak, S., Bhat, S., NI, S., & Hegde, P. (2023). Artificial Intelligence in the Field of Education. *International Research Journal of Modernization in Engineering Technology and Science*. https://doi.org/10.56726/irjmets43309

Nazir, N., Ishaq, M., & Saleem, Z. (2022). Reflection: A Review to Develop Teachers Professionally and Personally. *Pakistan Journal of Social Research*, *04*(04), 974-982. https://doi.org/10.52567/pjsr.v4i04.581

Ndaimani, A. (2026). Overcoming Barriers in Global Nursing Leadership Development: Participant Experiences and a Novel Framework for Equitable Blended Learning. https://doi.org/10.21203/rs.3.rs-9396562/v1

Ndimande, B. S. (2018). The Politics of History Textbooks in South African Classroom in the Era of Curriculum 2005. *Revista Linhas*, *19*(41), 139-159. https://doi.org/10.5965/1984723819412018139

Nebeker, C. (2014). A Proposal for Thinking Strategically About Ethics Education. *The Journal of Philosophy Science & Law*, *14*(1), 32-46. https://doi.org/10.5840/jpsl20141413

Neden, J. (2007). Assessment and Reflexivity in Family Therapy Training. *Journal of Family Therapy*, *29*(4), 373-377. https://doi.org/10.1111/j.1467-6427.2007.00405.x

Nedungadi, P., Tang, K. Y., & Raman, R. (2024). The Transformative Power of Generative Artificial Intelligence for Achieving the Sustainable Development Goal of Quality Education. *Sustainability*, *16*(22), 9779. https://doi.org/10.3390/su16229779

Neilsen-Hewett, C., Lindsay, G., Warren, J., Tonge, K. L., & Cronin, L. (2022). Early Childhood Leadership: Risk and Protective Factors During the COVID-19 Pandemic. *Australasian Journal of Early Childhood*, *47*(3), 219-232. https://doi.org/10.1177/18369391221103286

Neimeyer, G. J., Taylor, J. M., & Wear, D. M. (2010). Continuing Education in Psychology: Patterns of Participation and Aspects of Selection. *Professional Psychology Research and Practice*, *41*(4), 281-287. https://doi.org/10.1037/a0019811

Neira-Pesántez, F., Hulgo-Pullay, J., Molina-Cabrera, S., Sánchez-Tituaña, N., & Núñez-Naranjo, A. (2025). Educación Inclusiva: Recursos Tecnológicos Para Estudiantes Con Necesidades Específicas. *593 Digital Publisher Ceit*, *10*(1-2), 81-96. https://doi.org/10.33386/593dp.2025.2.2957

Nelson, C. R., & Gabbard, J. L. (2023). Pedagogical Design Considerations for Mobile Augmented Reality Serious Games (MARSGs): A Literature Review. *Electronics*, *12*(21), 4524. https://doi.org/10.3390/electronics12214524

Nelson, S. (2012). The Lost Path to Emancipatory Practice: Towards a History of Reflective Practice in Nursing. *Nursing Philosophy*, *13*(3), 202-213. https://doi.org/10.1111/j.1466-769x.2011.00535.x

Nemakhavhani, R. B. (2024). Enhancing Student Engagement Through Problem-Based Learning: A Case of the Built Environment. *International Conference on Education Research*, *1*(1), 238-245. https://doi.org/10.34190/icer.1.1.3053

Németh, E. B., Németh, K., & Szép, K. (2021). Development of Competencies for Employability Digital Competencies. *Gile Journal of Skills Development*, *1*(2), 68-81. https://doi.org/10.52398/gjsd.2021.v1.i2.pp68-81

Nesamany, B., & Subasini, N. S. (2024). Postgraduate Medical Educators' Perception of Continuous Professional Development in Pedagogy and Andragogy: The CME and CPD Crossover. *International Journal of Academic Research in Progressive Education and Development*, *13*(1). https://doi.org/10.6007/ijarped/v13-i1/18365

Nešić, A. (2023). Reshaping the Future of Work: Navigating the Impacts of Lifelong Learning and Digital Competences in the Era of 5.0 Industry. *Social Informatics Journal, 2*(1), 1-6. https://doi.org/10.58898/sij.v2i1.01-06

Nesterova, Y., Couch, D., & Nguyen, H. T. T. (2024). Teachers' Understandings of Barriers to Indigenous Children's Academic Success in Taiwan. *British educational research journal, 50*(6), 2937-2956. https://doi.org/10.1002/berj.4062

Neufeld, A. (2022). On the Challenges of Embedding Assessments of Self-Regulated Learning Into Licensure Activities in Health Professions Education. *Canadian Medical Education Journal, 13*(4), 110-114. https://doi.org/10.36834/cmej.75780

Neufeld, A., & Kassam, A. (2025). Why Respect for Learner Autonomy Is an Ethical Priority. *The Clinical Teacher, 22*(2). https://doi.org/10.1111/tct.70062

Neupane, B. (2021). A Critical Analysis of English Language Teaching Situation in Nepal. *Marsyangdi Journal*, 14-26. https://doi.org/10.3126/mj.v2i1.39962

Newman, J., Rivkin, I., Brooks, C., Turco, K., Bifelt, J., Ekada, L., & Philip, J. (2022). Indigenous Knowledge: Revitalizing Everlasting Relationships Between Alaska Natives and Sled Dogs to Promote Holistic Wellbeing. *International Journal of Environmental Research and Public Health, 20*(1), 244. https://doi.org/10.3390/ijerph20010244

Newman, T. J., Alvarez, M. A. G., & Kim, M. (2017). An Experiential Approach to Sport for Youth Development. *Journal of Experiential Education, 40*(3), 308-322. https://doi.org/10.1177/1053825917696833

Newton, B. (1995). Tune in the Bible: A Computer-Based Adult-Learning System. https://doi.org/10.32597/dmin/610

Ng, E. (2020). Successful Implementation of E-Learning in Self-Financed Higher Education: Experience From Hong Kong. *Journal of Educational Technology and Online Learning, 3*(1), 91-107. https://doi.org/10.31681/jetol.655496

Ng, F. Y. C., Thirunavukarasu, A. J., Cheng, H., Tan, T. F., Gutiérrez, L., Lan, Y., Ong, J. C. L., Chong, Y. S., Ngiam, K. Y., Ho, D., Wong, T. Y., Kwek, K., & Doshi-Velez, F. (2023). Artificial Intelligence Education: An Evidence-Based Medicine Approach for Consumers, Translators, and Developers. *Cell Reports Medicine, 4*(10), 101230. https://doi.org/10.1016/j.xcrm.2023.101230

Ng, H. Z., & Baharom, S. (2018). An Analysis on Adult Learners' Satisfaction in Online Education Programmes. *International Journal of Interactive Mobile Technologies (Ijim), 12*(7), 70. https://doi.org/10.3991/ijim.v12i7.9665

Ng, S., Crukley, J., Brydges, R., Boyd, V., Gavarkovs, A., Kangasjarvi, E., Wright, S., Kulasegaram, K., Friesen, F., & Woods, N. N. (2022). Toward 'Seeing' Critically: A Bayesian Analysis of the Impacts of a Critical Pedagogy. *Advances in Health Sciences Education, 27*(2), 323-354. https://doi.org/10.1007/s10459-021-10087-2

Nganga, C., Jamison, K., & Shepard, S. D. (2025). Transforming Knowledge Into Praxis in Equity and Justice-Oriented Educational Leadership. *Journal of Research on Leadership Education, 20*(3), 347-371. https://doi.org/10.1177/19427751251341295

Ngo, N. H. H., Cherrington, S., & Crabbe, D. (2022). An Integrated Framework of Professional Development for Vietnamese Lecturers of English as a Foreign Language. *Relc Journal, 55*(2), 468-483. https://doi.org/10.1177/00336882221085783

Ngo, T. T. A. (2023). The Perception by University Students of the Use of ChatGPT in Education. *International Journal of Emerging Technologies in Learning (Ijet), 18*(17), 4-19. https://doi.org/10.3991/ijet.v18i17.39019

Ngozwana, N. (2025). Navigating Learning Challenges: Insights From Adult Learners With Visual Impairment During the COVID-19 Pandemic in Eswatini. *Adult education quarterly, 75*(3), 199-215. https://doi.org/10.1177/07417136241290513

Ngubane, N., & Makua, M. (2021). Intersection of Ubuntu Pedagogy and Social Justice: Transforming South African Higher Education. *Transformation in Higher Education, 6.* https://doi.org/10.4102/the.v6i0.113

Ngubane, P. B., Mncube, D. W., & Ajani, O. A. (2024). Exploring Experiences of TVET Graduates' Work-Integrated Learning in a Rural Ecology: The Case of TVET Colleges in KwaZulu-Natal. *International Journal of Research in Business and Social Science (2147-4478)*, *13*(5), 803-812. https://doi.org/10.20525/ijrbs.v13i5.3366

Nguyen, A. (2020). Digital Inclusion. 1-15. https://doi.org/10.1007/978-3-030-48277-0_14-1

Nguyen, A., Kremantzis, M. D., Essien, A., Petrounias, I., & Hosseini, S. (2024). Editorial: Enhancing Student Engagement Through Artificial Intelligence (AI): Understanding the Basics, Opportunities, and Challenges. *Journal of University Teaching and Learning Practice*, *21*(06). https://doi.org/10.53761/caraaq92

Nguyen, H. T., Nguyen, Q. H., & Mậu, C. H. (2020). Guide Pedagogical Students to Design and Organize Experience-Based Learning Activities in Schools. *International Journal of Learning Teaching and Educational Research*, *19*(9), 99-117. https://doi.org/10.26803/ijlter.19.9.6

Nguyen, M. H. (2017). Liberal Education and the Connection With Vygotsky's Theory of the Zone of Proximal Development. *Cultural-Historical Psychology*, *13*(1), 81-88. https://doi.org/10.17759/chp.2017130108

Nichols, M., Choudhary, N., & Standring, D. (2020). Exploring Transformative Learning in Vocational Online and Distance Education. *Journal of Open Flexible and Distance Learning*, *24*(2), 43-55. https://doi.org/10.61468/jofdl.v24i2.435

Nicolaides, A., & Scully-Russ, E. (2018). Connections, Questions, Controversies, and the Potential Paths Forward. *New Directions for Adult and Continuing Education*, *2018*(159), 103-124. https://doi.org/10.1002/ace.20290

Nielsen, W., Nicol, C., & Owuor, J. (2008). Culturally-Responsive Mathematics Pedagogy Through Complexivist Thinking. *Complicity an International Journal of Complexity and Education*, *5*(1). https://doi.org/10.29173/cmplct8780

Nieto-Cardenas, J. (2025). Owlgorithm: Supporting Self-Regulated Learning in Competitive Programming Through LLM-Driven Reflection. https://doi.org/10.48550/arxiv.2511.09969

Nieuwenhove, L. V., & Wever, B. D. (2023). Psychosocial Barriers to Adult Learning and the Role of Prior Learning Experiences: A Comparison Based on Educational Level. *Adult education quarterly*, *74*(1), 62-87. https://doi.org/10.1177/07417136231147491

Nifaoui, A. (2020). Learning Theories and Foreign Language Teaching Methods: A Review of the Literature. *Journal of English Language Teaching and Applied Linguistics*, *2*(3), 28-37. https://doi.org/10.32996/jeltal.2020.2.3.3

Nihishova, A., & Kryvonosova, E. (2020). Design and Implementation of Technologies for Professionally Oriented Adult Education. *International Journal of Current Approaches in Language Education and Social Sciences*, *2*(1). https://doi.org/10.35452/caless.2020.20

Nijakowska, J., & Guz, E. (2024). Inclusion and Accessibility in Teacher Education. *Neofilolog*(62/1), 43-60. https://doi.org/10.14746/n.2024.62.1.4

Nikitara, M., Kalu, A., Latzourakis, E., Constantinou, C. P., & Velonaki, V. S. (2025). Training Nurses for Disasters: A Systematic Review on Self-Efficacy and Preparedness. *Healthcare*, *13*(24), 3323. https://doi.org/10.3390/healthcare13243323

Nikolov, R., Shoikova, E., Krumova, M., Kovatcheva, E., Dimitrov, V., & Chikalanov, A. (2016). On Learning in a Smart City Environment. *Serdica Journal of Computing*, *9*(3-4), 223-240. https://doi.org/10.55630/sjc.2015.9.223-240

Nikou, S. A., & Economides, A. A. (2021). A Framework for Mobile-Assisted Formative Assessment to Promote Students' Self-Determination. *Future Internet*, *13*(5), 116. https://doi.org/10.3390/fi13050116

Ninassi, C., & Burrell, D. N. (2023). Teaching Business Leadership Skills to Professionals in Healthcare Cybersecurity, Biodefense and Biotechnology Through Experiential Learning Methods. *Health Economics and Management Review*, *4*(3), 82-94. https://doi.org/10.61093/hem.2023.3-07

Niyibizi, E., Niyomugabo, C., Ngarukiye, V., Uwizeyemariya, A., Yanzigiye, B., Tabaro, C., Mukamana, H., & Mutarutinya, V. (2020). Eclectic Traditional Value Hub Model or Four-in-One: An Innovative Booster of Community Quadriliteracy Among Adults in Rwandan Rural Communities. *Tetfle*, *2*(1). https://doi.org/10.35293/tetfle.v2i1.93

Njenga, M. (2022). Teacher Participation in Continuing Professional Development: A Theoretical Framework. *Journal of Adult and Continuing Education*, *29*(1), 69-85. https://doi.org/10.1177/14779714221123603

Njiro, E. (2014). Moving Transfer to Transformative Learning: A Curriculum Model for Adult Educators in Open Distance Learning (ODL). *Journal of Educational and Social Research*. https://doi.org/10.5901/jesr.2014.v4n3p479

Nkechi, A. A., Ojo, A. O., & Eneh, O. A. (2024). Impact of Artificial Intelligence in Achieving Quality Education. https://doi.org/10.5772/intechopen.1004871

Nkhoma, M., Nkhoma, C., Thomas, S., & Le, N. Q. (2019). Transforming a First-Year Accounting Course Using a Blended Learning Pathway. https://doi.org/10.28945/4305

Nkoma, E., & Hay, J. F. (2018). Educational Psychologists' Support Roles Regarding the Implementation of Inclusive Education in Zimbabwe. *Psychology in the Schools*, *55*(7), 850-866. https://doi.org/10.1002/pits.22147

Nkosi, M., Agholor, I. A., & Olorunfemi, O. D. (2024). Agro-Investments Among Small Farm Business Entrepreneurs in the Era of the Fourth Industrial Revolution: A Case in the Mpumalanga Province, South Africa. *Administrative Sciences*, *14*(5), 85. https://doi.org/10.3390/admsci14050085

Noel-Weiss, J., Bassett, V., & Cragg, B. (2006). Developing a Prenatal Breastfeeding Workshop to Support Maternal Breastfeeding Self-Efficacy. *Journal of Obstetric Gynecologic & Neonatal Nursing*, *35*(3), 349-357. https://doi.org/10.1111/j.1552-6909.2006.00053.x

Nöel, L.-A., Liu, T. L., & Rider, T. R. (2019). Design Thinking and Empowerment of Students in Trinidad and Tobago. *Cultural and Pedagogical Inquiry*, *11*(3), 52-66. https://doi.org/10.18733/cpi29503

Nogueiro, T., Saraiva, M., Jorge, F., & Chaleta, E. (2022). The Erasmus+ Programme and Sustainable Development Goals—Contribution of Mobility Actions in Higher Education. *Sustainability*, *14*(3), 1628. https://doi.org/10.3390/su14031628

Nohilly, M., & Tynan, F. (2022). The Evolution of Wellbeing in Educational Policy in Ireland: Towards an Interdisciplinary Approach. *International Journal of Wellbeing*, *12*(1), 42-53. https://doi.org/10.5502/ijw.v12i1.1663

Nolan, H. A., & Roberts, L. (2023). Trigger Warnings as Tools for Learning—theorising an Evolving Cultural Concept. *Medical Education*, *58*(2), 185-195. https://doi.org/10.1111/medu.15172

Nomdo, G. (2023). Unpacking the Notion of 'Criticality' in Liberatory Praxis: A Critical Pedagogy Perspective. *Critical Studies in Teaching and Learning*, *11*(SI). https://doi.org/10.14426/cristal.v11isi.644

Nomlomo, V. (2021). Insights Into Student Teachers' Identities and Perspectives on the Role of Language in Science Education. 249-267. https://doi.org/10.52779/9781990995057/14

Noor, A. K. (2012). Training for the Next Wave. *Mechanical Engineering*, *134*(03), 34-37. https://doi.org/10.1115/1.2012-mar-2

Norberto Alonso Arroyave Villa. (2023). El Enfoque De Las Microcredenciales en La Educación Superior. *Rastros Rostros*, *26*(1), 1-40. https://doi.org/10.16925/2382-4921.2024.01.09

Noriega, G., Herrera, L. C., Montenegro, M., & Torres–Lista, V. (2020). Autoestima, Motivación Y Resiliencia en Escuelas Panameñas Con Puntajes Diferenciados en La Prueba TERCE. *Revista De Investigación Educativa*, *38*(2), 533-548. https://doi.org/10.6018/rie.410461

Norman, K., Lewis, A., & Carter, L. (2025). The Role of the Nurse Leader in Facilitating Effective Support and Communication to Raise Concerns. *British Journal of Nursing*, *34*(3), 154-158. https://doi.org/10.12968/bjon.2024.0170

Norris, D. M., & Lefrère, P. (2011). Transformation Through Expeditionary Change Using Online Learning and Competence-□ building Technologies. *Research in Learning Technology*, *19*(1). https://doi.org/10.3402/rlt.v19i1.17105

Nouioua, R. (2018). Attitudes to Learner Autonomy and Learning Strategies in Algerian EFL Context. 97. https://doi.org/10.34174/0079-000-050-039

Noya, A. (2018). Map of Portugal's Adult Learning System. 161-165. https://doi.org/10.1787/9789264298705-9-en

Ntseane, P. G. (2011). Culturally Sensitive Transformational Learning. *Adult education quarterly*, *61*(4), 307-323. https://doi.org/10.1177/0741713610389781

Nuankaew, P. (2022). Self-Regulated Learning Model in Educational Data Mining. *International Journal of Emerging Technologies in Learning (Ijet)*, *17*(17), 4-27. https://doi.org/10.3991/ijet.v17i17.23623

Nugent, A., Lodge, J. M., Carroll, A., Bagraith, R., MacMahon, S., Matthews, K., & Sah, P. (2018). Higher Education Learning Framework: An Evidence Informed Model for University Learning. https://doi.org/10.14264/348c85f

Núñez, C. E. B. (2024). Desarrollo De Competencias Docentes Para La Enseñanza Intercultural en Entornos Virtuales. *Revista Educa Umch*(24), 50-64. https://doi.org/10.35756/educaumch.202424.295

Nurdauletova, B., Артыкбаев, Ж. О., Amirbekova, A., Koshimova, B., Otarova, A., & Zhetkizgenova, A. (2024). Enhancing Cultural Awareness Through Project-Based Learning: A Study on Historical Preservation in Kazakhstan. *Journal of Ethnic and Cultural Studies*, *11*(3), 247-268. https://doi.org/10.29333/ejecs/2214

Nurhidayah, M. I. (2024). Reflective Practice in the Teaching Practicum Program: Pre-Service English Teachers' Reflections. *Journal of Applied Studies in Language*, *8*(1), 11-19. https://doi.org/10.31940/jasl.v8i1.11-19

Nutwell, E., Mokashi, P., & Christy, A. D. (2023). Online Course Design for the Engineering Workforce: Bringing Theory to Practice. *Advances*, *11*(2). https://doi.org/10.18260/3-1-1153-36045

Nweke, C. C., & Owoh, A. T. (2020). On ≪i>pedagogy of the Oppressed</I>: An Appraisal of Paulo Freire's Philosophy of Education. *Ogirisi a New Journal of African Studies*, *16*, 62-75. https://doi.org/10.4314/og.v16i1.4

Nyale, D., Karume, S., Kipkebut, A., & Mukudi, F. (2025). Digital Skills Landscape: A Systematic Review of Current Academic Programs, Industry Demands, and the Digital Divide's Impact on Graduate Competencies. *Industry and Higher Education*, *40*(2), 229-241. https://doi.org/10.1177/09504222251370105

Nyamunda, J., & Westhuizen, T. v. d. (2020). Developing Entrepreneurial Self-Efficacy : A Transformative Learning Theory Approach. *Journal of Contemporary Management*, *17*(se1), 44-66. https://doi.org/10.35683/jcm194e.76

Nyaradi, A., Solomons, T., & McNaught, K. (2025). Beyond the Classroom: The Transformative Experience of Short Rural Immersion Programs for Health Professional Students: A Narrative Review. *Academic Medicine*, *59*(7), 682-693. https://doi.org/10.1111/medu.15612

Nyathani, R. (2023). AI-Enabled Learning and Development: Hr's New Paradigm. *Journal of Marketing & Supply Chain Management*, *2*(2), 1. https://doi.org/10.47363/jmscm/2023(2)117

Nygren, H., Nissinen, K., Rautopuro, J., Mäkitalo, K., & Ullakko, K. (2019). Skills Behind the Robotics – How to Re-Educate Workers for the Future. 197-200. https://doi.org/10.1007/978-3-030-18141-3_16

Nzembayie, K. F., Buckley, A. P., & Talay, I. (2024). Experiential Pedagogies for Cultivating Entrepreneurial Mindsets: Action Design Learning as Tailored Framework. *Entrepreneurship Education and Pedagogy*, *9*(2), 176-207. https://doi.org/10.1177/25151274241292276

O'Leary, P. (2023). New Model to Evaluate Values, Beliefs and Assumptions in the Recognition of Prior Learning. 765-772. https://doi.org/10.4995/head23.2023.16073

O'Loughlin, R. A. (2021). *The Ontario Anti-Bullying Framework and Its Potential Impact on Indigenous Students in Northern Ontario* Carleton University].

O'Shea, C. M., Hudson, D. B., Suwak, K., & Mackey, H. J. (2023). Placating White Fragility for Indigenous Students and Communities. *Journal of Cases in Educational Leadership*, *27*(2), 36-48. https://doi.org/10.1177/15554589231208947

O'Connor, C., & Moran, G. (2023). The Challenge of Skill Development Through "Experiencing" Integrated Marketing Communications. *Higher Education Skills and Work-Based Learning*, *14*(3), 610-624. https://doi.org/10.1108/heswbl-06-2023-0167

O'Connor, E., Cianciotta, J., & Crête, D. (2021). Exploring Student Satisfaction in Experiential Learning at the University of Ottawa. *New Directions for Teaching and Learning*, *2021*(167), 23-32. https://doi.org/10.1002/tl.20456

O'Donnell, A. M. (2012). Constructivism. 61-84. https://doi.org/10.1037/13273-003
O'Flynn, E., Stephens, S., Cunningham, I., Burke, A., & McLaughlin, C. (2023). Experiential Learning and the Entrepreneurial University: An Irish Case Study. *Industry and Higher Education*, *37*(5), 663-672. https://doi.org/10.1177/09504222221151146
O'Neill, S., & Thomson, M. M. (2013). Supporting Academic Persistence in Low-skilled Adult Learners. *Support for Learning*, *28*(4), 162-172. https://doi.org/10.1111/1467-9604.12038
Oates, S. (2019). The Importance of Autonomous, Self-Regulated Learning in Primary Initial Teacher Training. *Frontiers in Education*, *4*. https://doi.org/10.3389/feduc.2019.00102
Obashoro-John, O. A., & Findsen, B. (2023). Learning in Later Years: Perspectives From Nigeria. 191-205. https://doi.org/10.36253/979-12-215-0253-4.21
Ober, T. M., Lehman, B., Gooch, R., Oluwalana, O., Solyst, J., Phelps, G., & Hamilton, L. S. (2023). Culturally Responsive Personalized Learning: Recommendations for a Working Definition and Framework. *Ets Research Report Series*, *2023*(1), 1-14. https://doi.org/10.1002/ets2.12372
Obiagu, A. (2025). Teaching Self-Reliance and Empowerment in Civics Classroom: Towards Sociopolitical Capabilities Approach. *Citizenship Social and Economics Education*. https://doi.org/10.1177/14788047251327548
Obradović-Ratković, S., Woloshyn, V., & Sethi, B. (2020). Reflections on Migration, Resilience, and Graduate Education. *Journal of Comparative & International Higher Education*, *12*(Winter), 81-111. https://doi.org/10.32674/jcihe.v12iwinter.2014
Ocheja, P., Agbo, F. J., Oyelere, S. S., Flanagan, B., & Ogata, H. (2022). Blockchain in Education: A Systematic Review and Practical Case Studies. *Ieee Access*, *10*, 99525-99540. https://doi.org/10.1109/access.2022.3206791
Oddone, K., Hughes, H., & Lupton, M. (2019). Teachers as Connected Professionals. *The International Review of Research in Open and Distributed Learning*, *20*(3). https://doi.org/10.19173/irrodl.v20i4.4082
Odularu, A. T., Khalo, X., Mashiyi, N. F., & Nkohla, M. B. (2023). Exploring COVID-19 Pandemic Impact, Online Engagement, and Digital Divide on Disadvantaged Undergraduate Students in South African Universities. https://doi.org/10.20944/preprints202305.0573.v1
Odutayo, A. O. (2023). Deliberative Dialogue as a Viable Instructional Method for Civic Learning Outcomes. *Studies in Learning and Teaching*, *4*(2), 319-328. https://doi.org/10.46627/silet.v4i2.233
OECD. (2005). Promoting Adult Learning. https://doi.org/10.1787/9789264010932-en
OECD. (2013). Ageing Societies. https://doi.org/10.1787/trends_edu-2013-9-en
OECD. (2017). People on the Move. https://doi.org/10.1787/b3d31c5b-en
OECD. (2018a). *Seven Questions about Apprenticeships: Answers from International Experience*. OECD Publishing.
OECD. (2018b). *Skills Strategy Implementation Guidance for Portugal: Strengthening the Adult-Learning System*. O. Publishing.
OECD. (2019). What Characterises Upper Secondary Vocational Education and Training? https://doi.org/10.1787/a1a7e2f1-en
OECD. (2020a). Education Policy Outlook in Germany. https://doi.org/10.1787/47b795b1-en
OECD. (2020b). Education Policy Outlook in Greece. https://doi.org/10.1787/f10b95cf-en
OECD. (2020c). Les Possibilités De L'apprentissage en Ligne Pour Les Adultes : Premiers Enseignements De La Crise Du COVID-19. https://doi.org/10.1787/0ef7c9bf-fr
OECD. (2021a). Abbreviations and Acronyms. https://doi.org/10.1787/7ab165bf-en
OECD. (2021b). Enhancing Adult Learning in Light of the Consequences and Opportunities of the COVID-19 Pandemic. https://doi.org/10.1787/fd4220f0-en
OECD. (2021c). Improving Access and Quality of Vocational Education and Training in Thailand. https://doi.org/10.1787/da2a3f57-en
OECD. (2021d). Local Entrepreneurship Ecosystems and Emerging Industries. https://doi.org/10.1787/044ffc1d-en
OECD. (2021e). The OECD Framework for Digital Talent and Skills in the Public Sector. https://doi.org/10.1787/4e7c3f58-en
OECD. (2021f). *OECD Skills Outlook 2021: Learning for Life*.

OECD. (2021g). Quality and Value of Micro-Credentials in Higher Education. https://doi.org/10.1787/9c4ad26d-en

OECD. (2021h). *Skills Outlook 2021: Learning for Life.* https://doi.org/10.1787/0ae365b4-en

OECD. (2021i). *Skills Outlook 2021: Learning for Life.* OECD Publishing.

OECD. (2022a). Future-Proofing Adult Learning Systems in Cities and Regions. https://doi.org/10.1787/11fa26cc-en

Oecd. (2022b). How Vocational Education and Training (VET) Systems Can Support Ukraine. https://doi.org/10.1787/e8e86ce2-en

OECD. (2022c). Measuring the Environmental Impacts of Artificial Intelligence Compute and Applications. https://doi.org/10.1787/7babf571-en

OECD. (2023a). Education and Innovation for the Digital and Green Transitions: How Higher Education Can Support Effective Curricula in Schools. https://doi.org/10.1787/3dedf4cb-en

OECD. (2023b). Micro-Credential Policy Implementation in Finland, the Slovak Republic, Slovenia and Spain. https://doi.org/10.1787/c3daa488-en

OECD. (2023c). *OECD Skills Strategy 2023: Skills for a Resilient Green and Digital Transition.* https://www.oecd.org/skills/

OECD. (2023d). Public Policies for Effective Micro-Credential Learning. https://doi.org/10.1787/a41f148b-en

OECD. (2024). *PISA Vocational Education and Training (VET).* OECD Publishing. https://doi.org/10.1787/b0d5aaf9-en

Oh, S.-E., Lee, K. J., Mok, H., & Jo, K. (2025). Bridging the Gap: A Phenomenological Study of Transfer Students' Journey Into Professional Nursing. *Nursing Reports*, *15*(2), 72. https://doi.org/10.3390/nursrep15020072

Ojo, S., & Alli, O. M. (2024). Influence of Talent Management and Organisational Performance on Employee Retention Among First Bank Employees in Osogbo, Osun State. *Quest Journal of Management and Social Sciences*, *6*(1), 15-32. https://doi.org/10.3126/qjmss.v6i1.67146

Oke, O. A., Jamil, D. I., & Çavuş, N. (2023). The Impact of Artificial Intelligence in Foreign Language Learning Using Learning Management Systems: A Systematic Literature Review. *Information Technologies and Learning Tools*, *95*(3), 215-228. https://doi.org/10.33407/itlt.v95i3.5233

Okia del Rocio Andrade, P., Zambrano, M. M. C., Montenegro, S., Chafuelán, S. M. C., & Arias, E. A. R. (2024). La Incidencia De La Inteligencia Artificial en La Educación Secundaria Del Ecuador. *Revista Imaginario Social*, *7*(1). https://doi.org/10.59155/is.v7i1.125

Okochi, C., Gold, A., Christensen, A., & Batchelor, R. L. (2023). Early Access to Science Research Opportunities: Growth Within a Geoscience Summer Research Program for Community College Students. *Plos One*, *18*(12), e0293674. https://doi.org/10.1371/journal.pone.0293674

Okon, E. U., Akpan, C. S., & Akarika, D. C. (2024). Role of Organisational Culture in Enabling Effective Communication and Total Quality Management Practices in Champion Breweries Plc, Uyo, Akwa Ibom State, Nigeria. *Aksu Journal of Social Sciences*, *4*(1), 102-117. https://doi.org/10.61090/aksujoss.2024.008

Okorie, & Uzoaru, C. (2021). Changing Pedagogies in Adult Education Teaching and Learning: Heutagogy an Option. *International Journal of Research and Innovation in Social Science*, *05*(12), 760-769. https://doi.org/10.47772/ijriss.2021.51238

Olaleye, O. A., Adekunle, O. A., Adekunle, A., & Atieno, V. (2024). Bridging Gaps in Farmer Education: The Role of L3F in Fostering Human, Social and Financial Capital for Rural Development. *Journal of Developing Country Studies*, *8*(1), 26-46. https://doi.org/10.47604/jdcs.2434

Olapoju, O. M. (2022). Autonomous Ships, Port Operations, and the Challenges of African Ports. *Maritime Technology and Research*, *5*(1), 260194. https://doi.org/10.33175/mtr.2023.260194

Olatunji, T. I., Bano, S., Dazer, J., Capaccioli, M., & Qiu, R. (2025). From Disorientation to Transformation: Employing Cross-Cultural Dialogues for Understanding Learning in Contexts of Migration. *Journal of Transformative Education*, *24*(1), 56-77. https://doi.org/10.1177/15413446251346169

Olesen, H. S., & Wildemeersch, D. (2012). Editorial: The Effects of Policies for the Education and Learning of Adults - From 'Adult Education' to 'Lifelong Learning', From 'Emancipation' to

'Empowerment'. *European Journal for Research on the Education and Learning of Adults*, *3*(2), 97-101. https://doi.org/10.3384/rela.2000-7426.relae5

Oliveira, G., Lindquist, C., Shiratori, E. S., & Baptaglin, L. A. (2024). Venezuelan Migration to Brazil: Teachers and Students Co-Constructing Multicultural Education Inside Classrooms. *Journal for Multicultural Education*, *18*(4), 540-553. https://doi.org/10.1108/jme-02-2024-0023

Oliver, B. (2019). *Making micro-credentials work for learners, employers and providers*. D. University.

Oliver, E. (2010). Research and Development in Adult Education: Fields and Trends. https://doi.org/10.3224/86649304

Oliver, K., & Brabin, J. (2026). The Fast-Paced World of Micro-Credentials in Healthcare: A Scoping Review. *Journal of Paramedic Practice*, *18*(4), 160-169. https://doi.org/10.12968/jpar.2025.0086

Oliwa, R. (2024). Contextualising the Hyflex Model of Instruction for Language Classes. *Theory and Practice of Second Language Acquisition*, *10*(1), 1-23. https://doi.org/10.31261/tapsla.14757

Omar, A. (2007). Las Perspectivas De Futuro Y Sus Vinculaciones Con El Bienestar Y La Resiliencia en Adolescentes. *Psicodebate*, *7*(0), 141. https://doi.org/10.18682/pd.v7i0.432

Omar, I., Abosheisha, M., & Hafez, A. (2026). Learning From Medical Errors in Invasive Procedures: Current Situation and Perspectives of Improvement: A Semi-Structured Interview. *Journal of Evaluation in Clinical Practice*, *32*(4). https://doi.org/10.1111/jep.70504

Omar, R., & Rashid, R. A. (2019). Challenges for Professional Development of Malaysian ESL Teachers. *Kne Social Sciences*. https://doi.org/10.18502/kss.v3i19.4897

Omer, A. A. A. (2020a). Clinical teaching and learning: From theory and research to application. *Saudi Journal of Medicine and Medical Sciences*, *8*(3), 196-203. https://doi.org/10.4103/sjmms.sjmms_117_20

Omer, A. A. A. (2020b). The Importance of Theory to Inform Practice – Theorizing the Current Trends of Clinical Teaching: A Narrative Review. *Sudan Journal of Medical Sciences*, *15*(4), 383-398. https://doi.org/10.18502/sjms.v15i4.8161

Omodan, B. I. (2022). Analysis of Emancipatory Pedagogy as a Tool for Democratic Classrooms. *International Journal of Research in Business and Social Science (2147-4478)*, *11*(2), 348-354. https://doi.org/10.20525/ijrbs.v11i2.1662

Omodan, B. I., & Mtshatsha, N. D. M. (2022). Reinventing Students' Self-Esteem in Classrooms: The Need for Humanistic Pedagogy. *Jetl (Journal of Education Teaching and Learning)*, *7*(2), 139. https://doi.org/10.26737/jetl.v7i2.3243

Önal, G., & Özdemir, T. E. (2025). Future Trends and Challenges in Occupational Therapy: Insights From a Delphi Study. *British Journal of Occupational Therapy*, *89*(2), 102-115. https://doi.org/10.1177/03080226251368235

Ong, C. H. (2014). Goal Orientation of Adult Students Towards Learning Strategies: The Malaysian Context. *Psychological Thought*, *7*(2), 156-167. https://doi.org/10.5964/psyct.v7i2.114

Onjewu, A. K. E., Godwin, E. S., Azizsafaei, F., & Appiah, D. (2024). The Influence of Technology Use on Learning Skills Among Generation Z: A Gender and Cross-Country Analysis. *Industry and Higher Education*, *39*(2), 139-157. https://doi.org/10.1177/09504222241263227

Onofre, J. M. J., Pilco, R. M. J., Villacis, S. G. P., & Mora, M. (2024). Inclusión De La Historia Y Cultura Afroecuatoriana en El Currículo De Estudios Sociales en Las Escuelas Ecuatorianas. *Ciencia Latina Revista Científica Multidisciplinar*, *8*(2), 2959-2983. https://doi.org/10.37811/cl_rcm.v8i2.10726

Onwuatuegwu, I. N. (2023). An Investigation Into the Philosophical Differences Between Pedagogical and Andragogical Approaches to Education. *Journal of Education Review Provision*, *3*(1), 26-31. https://doi.org/10.55885/jerp.v3i1.181

Oosten, E. B. V. (2014). The Impact of Emotional Intelligence and Executive Coaching on Leader Effectiveness. *Academy of Management Proceedings*, *2014*(1), 12793. https://doi.org/10.5465/ambpp.2014.12793abstract

Oosten, E. B. V., Buse, K., & Bilimoria, D. (2017). The Leadership Lab for Women: Advancing and Retaining Women in STEM Through Professional Development. *Frontiers in Psychology*, *8*. https://doi.org/10.3389/fpsyg.2017.02138

Opoku, M. A., Choi, S. B., & Kang, S. W. (2019). Psychological Safety in Ghana: Empirical Analyses of Antecedents and Consequences. *International Journal of Environmental Research and Public Health, 17*(1), 214. https://doi.org/10.3390/ijerph17010214

Orfanidou, C., & Kedraka, K. (2022). Identity Development as a Learning Outcome: A Case Study. *Έρευνα Στην Εκπαίδευση, 11*(1), 189-196. https://doi.org/10.12681/hjre.31598

Orman, R., Şimşek, E., & Çakır, M. A. K. (2023). Micro-Credentials and Reflections on Higher Education. *Higher Education Evaluation and Development, 17*(2), 96-112. https://doi.org/10.1108/heed-08-2022-0028

Orn, L. (2022). *Effects of Using Self-Directed Learning to Foster Lifelong Learning Attitudes of Functional Literacy Facilitators* Chulalongkorn University].

Orn, L. (2023). Effects of Using Self-Directed Learning to Foster Lifelong Learning Attitudes of Functional Literacy Facilitators. https://doi.org/10.58837/chula.the.2022.264

Ornellas, A., Falkner, K., & Stålbrandt, E. E. (2019). Enhancing Graduates' Employability Skills Through Authentic Learning Approaches. *Higher Education Skills and Work-Based Learning, 9*(1), 107-120. https://doi.org/10.1108/heswbl-04-2018-0049

Orsini, C., Binnie, V. I., Wilson, S. L., & Villegas, M. J. P. (2017). Learning Climate and Feedback as Predictors of Dental Students' Self-determined Motivation: The Mediating Role of Basic Psychological Needs Satisfaction. *European Journal of Dental Education, 22*(2). https://doi.org/10.1111/eje.12277

Ortega-Rodríguez, P. J. (2022). De La Realidad Extendida Al Metaverso: Una Reflexión Crítica Sobre Las Aportaciones a La Educación. *Teoría De La Educación Revista Interuniversitaria, 34*(2), 189-208. https://doi.org/10.14201/teri.27864

Ortega, M. X. P., & Gonzalez-Segarra, A. (2022). Analíticas De Aprendizaje en La Educación Superior Mediante La Aplicación De Herramientas Colaborativas. *Identidad Bolivariana, 6*(2), 154-181. https://doi.org/10.37611/ib6ol2154-181

Ortiz, E. A., Quispe, G. D. M., Chino, R. M., Calderón, I. A. D., Bravo, N. T., & Marina del Águila Gonzales de del, C. (2021). El Cognitivismo: Perspectivas Pedagógicas, Para La Enseñanza Y Aprendizaje Del Idioma Inglés, en Comunidades Hispanohablantes. *Paidagogo, 3*(1), 89-102. https://doi.org/10.52936/p.v3i1.48

Ortiz, L. G., Clougher, D., Anderson, T., & Maina, M. F. (2021). IDEAS for Transforming Higher Education: An Overview of Ongoing Trends and Challenges. *The International Review of Research in Open and Distributed Learning, 22*(2), 166-184. https://doi.org/10.19173/irrodl.v22i2.5206

Ortogero, S. P., Ray, A. B., & Connor, K. E. (2024). Self-Regulated Multimedia Cognitive Learning Model: Enhancing Vocabulary Acquisition in Adult English Learners. *International Journal of Tesol Studies*. https://doi.org/10.58304/ijts.20250103

Osasona, F., Daraojimba, A. I., Atadoga, A., Onwusinkwue, S., & Dawodu, S. O. (2024). Ai Integration in Business Analytics: A Review of Usa and African Trends. *Computer Science & It Research Journal, 5*(2), 432-446. https://doi.org/10.51594/csitrj.v5i2.793

Osatis, C., & Asavanirandorn, C. (2022). An Exploring Human Resource Development in Small and Medium Enterprises in Response to Electric Vehicle Industry Development. *World Electric Vehicle Journal, 13*(6), 98. https://doi.org/10.3390/wevj13060098

Osborne, M. (2017). Industry Challenges for an Under-Skilled Workforce. https://doi.org/10.12794/metadc2137352

Osborne, M., Maitra, S., & Uflewska, A. (2022). Smart Learning Cities Promoting Lifelong Learning Through Working Lives. 376-389. https://doi.org/10.4135/9781529757217.n24

Osorio, C. P., Leucci, F., & Porrini, D. (2024). Analyzing the Relationship Between Agricultural AI Adoption and Government-Subsidized Insurance. *Agriculture, 14*(10), 1804. https://doi.org/10.3390/agriculture14101804

Ossiannilsson, E. (2022). MOOCS for Lifelong Learning, Equity, and Liberation. https://doi.org/10.5772/intechopen.99659

Ossiannilsson, E. (2023). Open Educational Resources (OER) and Some of the United Nations Sustainable Development Goals. *International Journal of Information and Learning Technology, 40*(5), 548-561. https://doi.org/10.1108/ijilt-01-2023-0002

Ostendorff, S. P. The Classroom Is Only Part of the Challenge: Employing Interventions That Help Retain Adult Learners in Higher Education. https://doi.org/10.17760/d20621607

Ostendorff, S. P. (2023). *The Classroom Is Only Part of the Challenge: Employing Interventions That Help Retain Adult Learners in Higher Education* Northeastern University].

Osunlaja, O., Enahoro, A., Maha, C. C., Kolawole, T. O., & Abdul, S. (2024). Healthcare Management Education and Training: Preparing the Next Generation of Leaders - A Review. *International Journal of Applied Research in Social Sciences*, *6*(6), 1178-1192. https://doi.org/10.51594/ijarss.v6i6.1209

Osunwusi, A. O. (2020). The Nigerian Education Milieu in the SDG Era: The Roles, Values, Challenges and Prospects of ODL and Lifelong Learning. *International Journal of Learning and Development*, *9*(2), 140. https://doi.org/10.5296/ijld.v10i1.16184

Otaki, F., Ho, S. B., Nair, B., AlGurg, R., Stanley, A., Khamis, A. H., Paulus, A., & Alsuwaidi, L. (2025). Effects of Building Resilience Skills Among Undergraduate Medical Students in a Multi-Cultural, Multi-Ethnic Setting in the United Arab Emirates: A Convergent Mixed Methods Study. *Plos One*, *20*(2), e0308774. https://doi.org/10.1371/journal.pone.0308774

Ouherrou, N., Auli-Giraldo, M., & Metz, S. M. (2022). Assessing E-Portfolio Acceptability in an Online Course. https://doi.org/10.36315/2022v1end121

Owen, J. (2020). The Use of Practice-Based Learning Days on Psychological Wellbeing Practitioner Training Programmes: A Self-Determination Theory Perspective. *Psychology Teaching Review*, *26*(2), 19-25. https://doi.org/10.53841/bpsptr.2020.26.2.19

Owusu-Agyeman, Y. (2024). How Lifelong Learning Shapes the Professional Development of Staff in Higher Education Institutions. *Quality Education for All*, *1*(2), 134-150. https://doi.org/10.1108/qea-01-2024-0013

Oyekunle, D., Claude, B. E. A., Waliu, A. O., Adekunle, T. S., & Ugochukwu, O. M. (2024). Cloud Based Adaptive Learning System: Virtual Reality and Augmented Reality Assisted Educational Pedagogy Development on Clinical Simulation. *Journal of Digital Health*, 49-62. https://doi.org/10.55976/jdh.32024126849-62

Paavola, S., Lipponen, L., & Hakkarainen, K. (2004). Models of Innovative Knowledge Communities and Three Metaphors of Learning. *Review of Educational Research*, *74*(4), 557-576. https://doi.org/10.3102/00346543074004557

Pacas, G. M. (2021). Aprendizaje Experiencial en La Formación Docente: Un Abordaje Disruptivo. *Eca Estudios Centroamericanos*, *76*(766), 375-387. https://doi.org/10.51378/eca.v76i766.6484

Pacavita, O. S. M., & González, M. C. L. (2018). Teoría Andragógica: Aciertos Y Desaciertos en La Formación Docente en TIC. *Praxis & Saber*, *9*(19), 161-181. https://doi.org/10.19053/22160159.v9.n19.2018.7926

Paja, S. N. (2013). A Phenomenological Study of Life-Changing Adult Learning in a Two-Course Leadership Development Series. https://doi.org/10.31274/etd-180810-3292

Palalas, A., & Doran, M. (2024). Digital Wellness Framework for Online Learning. *Canadian Journal of Learning and Technology*, *49*(3), 1-25. https://doi.org/10.21432/cjlt28581

Palkki, J., & Sauerland, W. (2018). Considering Gender Complexity in Music Teacher Education. *Journal of Music Teacher Education*, *28*(3), 72-84. https://doi.org/10.1177/1057083718814582

Pan, W., Chen, L., & Zhan, W. (2021). Positioning Construction Workers' Vocational Training of Guangdong in the Global Political-Economic Spectrum of Skill Formation. *Engineering Construction & Architectural Management*, *28*(9), 2489-2515. https://doi.org/10.1108/ecam-11-2019-0641

Panagiotidis, P. (2022). Blockchain in Education - The Case of Language Learning. *European journal of education*, *5*(1), 66-82. https://doi.org/10.26417/443gjm83

Panda, S. (2020). Covid-19 Pandemic and Innovations in Institutional Transformation, Technology and Pedagogy. *Journal of Learning for Development*, *7*(3), 264-270. https://doi.org/10.56059/jl4d.v7i3.478

Panda, S. (2022). Book Review: Reimagining Digital Learning for Sustainable Development: How Upskilling, Data Analytics, and Educational Technologies Close the Skill Gap. *The International*

Review of Research in Open and Distributed Learning, *23*(3), 274-279. https://doi.org/10.19173/irrodl.v23i3.6242

Panda, S. (2023). Editorial: Learners as Focus in 'Learning for Development' (L4D). *Journal of Learning for Development*, *10*(3), i-iv. https://doi.org/10.56059/jl4d.v10i3.1259

Pang, T. Y., Lee, T.-K., & Murshed, M. (2023). Towards a New Paradigm for Digital Health Training and Education in Australia: Exploring the Implication of the Fifth Industrial Revolution. *Applied Sciences*, *13*(11), 6854. https://doi.org/10.3390/app13116854

Pang, V., & Lee, P. L. (2013). Motivational Factors in Continuing Education an Academic Achievement of Adult Learners. *Malaysian Journal of Learning and Instruction*, *10*. https://doi.org/10.32890/mjli.10.2013.7651

Panitsides, E. A. (2022). Adult Education: Reaching the Unreached. *Emerald Open Research*, *1*(13). https://doi.org/10.1108/eor-13-2023-0011

Pant, D., & Rastogi, A. (2024). Development of Wellbeing Among School Learners Through Positive Pedagogy. *International Journal of Wellbeing*, *14*(2), 1-27. https://doi.org/10.5502/ijw.v14i2.3203

Papadakis, S. (2023). MOOCs 2012-2022: An Overview. *Advances in Mobile Learning Educational Research*, *3*(1), 682-693. https://doi.org/10.25082/amler.2023.01.017

Papadimitriou, S. T. (2015). Μέσα Επικοινωνίας Και Μάθηση: Η Ελληνική Περίπτωση. *Διεθνές Συνέδριο Για Την Ανοικτή & Εξ Αποστάσεως Εκπαίδευση*, *8*(1A). https://doi.org/10.12681/icodl2015.22

Papier, J. (2018). JOVACET Volume 1 Issue 1. *Journal of Vocational Adult and Continuing Education and Training*, *1*(1). https://doi.org/10.14426/jovacet.v1i1.3

Papier, J. (2020). Journal of Vocational, Adult and Continuing Education and Training, Volume 3, Issue 1 2020. *Journal of Vocational Adult and Continuing Education and Training*, *3*(1), 172. https://doi.org/10.14426/jovacet.v3i1.136

Pappa, D., & Papadopoulos, H. (2019). A Use Case of the Application of Advanced Gaming And Immersion Technologies for Professional Training: The GAMEPHARM Training Environment for Physiotherapists. *The Electronic Journal of E-Learning*, *17*(2). https://doi.org/10.34190/jel.17.2.08

Pappagallo, S. (2024). Chatbots in Education: A Dual Perspective on Innovation and Ethics. *Journal of Digital Pedagogy*, *3*(1), 3-10. https://doi.org/10.61071/jdp.2420

Par, P., Mohamed, M., & Mohamed, M. (2023). Programme Basé Sur Deux Théories Des Actes De Langage Et De L'apprentissage Situationnel Pour Développer Quelques Compétences Pragmatiques en FLE Et L'auto-Efficacité Auprès Des Étudiantes Du Cycle Secondaire. مجلة جامعة الفيوم للعلوم التربوية والنفسية, 17(1), 967-124. https://doi.org/10.21608/jfust.2023.286217

Paraluppi, M. C., Nonato, A. C., & Pio, D. A. M. (2024). Medical Training in the Context of Primary Health Care: Reflexive Portfolios as an Innovative Didactic Proposal. *Cuadernos De Educación Y Desarrollo*, *16*(3), e3573. https://doi.org/10.55905/cuadv16n3-026

Parba, J. (2021). Teaching Critical Vocabulary to Filipino Heritage Language Learners. *Education Sciences*, *11*(6), 260. https://doi.org/10.3390/educsci11060260

Paris, S. G., & Parecki, A. D. (1995). Metacognitive Aspects of Adult Literacy. https://doi.org/10.1037/e402542005-001

Park, B. (2026). Applying Transformative Learning to Faith Identity Formation: Small-Group Ministry Among Emerging Adults in South Korea. *Christian Education Journal Research on Educational Ministry*. https://doi.org/10.1177/07398913261427778

Park, J. H., & Rojewski, J. W. (2006). The Learning Organization Model Across Vocational and Academic Teacher Groups. *Career and Technical Education Research*, *31*(1), 23-48. https://doi.org/10.5328/cter31.1.23

Park, K.-h. (2021). Impact of Digital Convergence Technology on the Logistics Value Chain. *Asia-Pacific Journal of Convergent Research Interchange*, *7*(1), 33-48. https://doi.org/10.47116/apjcri.2020.01.04

Parker, B., & Myrick, F. (2010). Transformative Learning as a Context for Human Patient Simulation. *Journal of Nursing Education*, *49*(6), 326-332. https://doi.org/10.3928/01484834-20100224-02

Parker, N. (2021). A Learning Model on the Readiness to Learn Self-Care for Trauma Survivors. *Journal of Practical Nurse Education and Practice*, *1*(1). https://doi.org/10.29173/jpnep4

Parr, A., Binagwaho, A., Stirling, A., Davies, A., Mbow, C., Hessen, D. O., Nader, H. B., Jamil Salmi, Burkins, M. B., Ramakrishna, S., Serrano, S., Schmelkes, S., Shijun, T., & McMowan, T. (2022). Knowledge-Driven Actions: Transforming Higher Education for Global Sustainability. https://doi.org/10.54675/ybtv1653

Parra, J. C. A., Garzón, B. C. R., Rojas, K. J. F., Vásquez, V. J. C., & Chong, M. D. (2023). Pedagogical Paradigms in Teaching EFL Classroom: Evaluating Approaches of Cognitivism, Constructivism, Pragmatism, and Connectivism. *Ciencia Latina Revista Científica Multidisciplinar*, *6*(6), 14029-14045. https://doi.org/10.37811/cl_rcm.v6i6.4671

Parry, S. E., & Metzger, E. P. (2023). Barriers to Learning for Sustainability: A Teacher Perspective. *Sustainable Earth Reviews*, *6*(1). https://doi.org/10.1186/s42055-022-00050-3

Parsons, D., Sparks, H., Vo, D., & Singh, A. (2023). MOOCS and Micro-Credentials as Launch Pads to Further Education: Challenges and Experiences. https://doi.org/10.5772/intechopen.1001332

Parsons, L. (2016). Competitive Mooting as Clinical Legal Education: Can Real Benefits Be Derived From an Unreal Experience? *Australian Journal of Clinical Education*, *1*(1). https://doi.org/10.53300/001c.5091

Partarakis, N., & Zabulis, X. (2024). A Review of Immersive Technologies, Knowledge Representation, and AI for Human-Centered Digital Experiences. *Electronics*, *13*(2), 269. https://doi.org/10.3390/electronics13020269

Parzen, M., & Janzen, K. (2019). Facilitation of Disorientating Events for the RPN to BScN Learner. *Quality Advancement in Nursing Education - Avancées en Formation Infirmière*, *5*(2). https://doi.org/10.17483/2368-6669.1190

Pascoe, D. (2017). How Can Legal Education Speak to the Discovery-Enriched Curriculum? *Asian Journal of Legal Education*, *4*(1), 17-32. https://doi.org/10.1177/1326365x16667155

Passos, R. (2021). Reimagining Digital Learning for Sustainable Development. *Journal of Learning for Development*, *8*(3), 621-624. https://doi.org/10.56059/jl4d.v8i3.592

Pástor, D. M. L., María de los Ángeles, B., & Benavides, J. P. C. (2021). Dispositivos Tecnológicos: Uso Académico en Estudiantes Universitarios. *Revista Científica Uisrael*, *8*(1e), 23-39. https://doi.org/10.35290/rcui.v8n1e.2021.480

Pata, K., Maslo, I., & Jõgi, L. (2021). Transforming Adult Education From Neo-Liberal to Holistically Inclusive Adult Education in Baltic States. 139-165. https://doi.org/10.1007/978-3-030-65002-5_8

Patchett, M. M. (2017). Historical Geographies of Apprenticeship: Rethinking and Retracing Craft Conveyance Over Time and Place. *Journal of Historical Geography*, *55*, 30-43. https://doi.org/10.1016/j.jhg.2016.11.006

Patel, A., Bulger, A., Jarrett, K., Ginwright, S. S., Chandran, K. B., & Wyss, J. M. (2021). Summer Research Internships Prepare High School Students for 21st Century Biomedical Careers. *The Journal of Stem Outreach*, *4*(1). https://doi.org/10.15695/jstem/v4i1.13

Patel, D., Parsley, S., Cannell, P., & Havemann, L. (2023). Practising Open Education. 293-319. https://doi.org/10.2307/j.ctv2wk7261.25

Patel, S. (2023). Role of Reflection in Education and Practice in Anaesthesia: Purpose, Process, Pitfalls and Promotion. *International Medical Education*, *2*(4), 262-275. https://doi.org/10.3390/ime2040025

Paudel, P. K., & Eberhardt, C. (2023). Approaching Apprenticeship in Nepal: Lessons From Dual-Vet of Germany. *Journal of Technical and Vocational Education and Training*, *17*(1), 1-18. https://doi.org/10.3126/tvet.v17i1.52405

Paul, R., Dedemus, D., Boyce, M., & Johnston, K. L. (2020). The "Engineers Have Feelings" Project: Integrating Mental Wellness and Lifelong Learning Skills in First-Year Undergraduate Engineering Courses. *Proceedings of the Canadian Engineering Education Association (Ceea)*. https://doi.org/10.24908/pceea.vi0.14154

Paulos, C. (2018). Ways of Learning of Adult Educators in Uncertain Professional Contexts. 169-180. https://doi.org/10.1163/9789004375475_013

Pavlova, M. (2018). Fostering Inclusive, Sustainable Economic Growth and Green Skills Development Through Vocational Education and Training. In *Handbook of Vocational Education and Training*. Springer.

Pawlukewicz, J., Ondrus, S., Ellis, D., Kean, L., Freymond, N., Schmid, J., DeGeer, I., Patterson, D., Keefe, R. H., Cox, K., Steiner, S., Lukens, J., Solomon, P., & Gair, S. (2013). *Journal of Social Work Values and Ethics*, *10*(1). https://doi.org/10.55521/10-010-100

Payan-Carreira, R., Rebelo, H., & Sebastião, L. (2022). Perspective Chapter: Active Learning Strategies in the Veterinary Medicine Programme Under the Think4Jobs Project. https://doi.org/10.5772/intechopen.105969

Pazieieva, A., Smolinska, O., Syniuk, O., Turytsya, O., & Leskiv, H. (2024). Modern Economic-Legal Methods of Using Artificial Intelligence in an Educational Company: Information Technology to to Self-Development and Scientific Activity. *International Journal of Religion*, *5*(10), 1170-1180. https://doi.org/10.61707/x3q7t837

Peace, R. J. (2006). Probationer Training for Neighborhood Policing in England and Wales: Fit for Purpose? *Policing an International Journal*, *29*(2), 335-346. https://doi.org/10.1108/13639510610667691

Pechenkina, E. (2023). 9. Artificial Intelligence for Good? , 239-266. https://doi.org/10.11647/obp.0363.09

Peiris, J. M. P. M. (2021). HR Department's Compelling New Role in Workplace Transition in Response to the Coronavirus. *Kelaniya Journal of Human Resource Management*, *16*(1), 58. https://doi.org/10.4038/kjhrm.v16i1.87

Peiser, G., Ambrose, J. F., Burke, B., & Davenport, J. (2018). The Role of the Mentor in Professional Knowledge Development Across Four Professions. *International Journal of Mentoring and Coaching in Education*, *7*(1), 2-18. https://doi.org/10.1108/ijmce-07-2017-0052

Peltier, S. (2021). The Child Is Capable: Anishinaabe Pedagogy of Land and Community. *Frontiers in Education*, *6*. https://doi.org/10.3389/feduc.2021.689445

Pembridge, J. (2014). A Comparison of Adult Learning Characteristics Between First-Year and Senior Capstone Students. https://doi.org/10.18260/1-2--19925

Peng, Z., & Peng, S. (2017). Study on MO Zi's Thoughts of Labor Division. https://doi.org/10.2991/icoss-17.2017.37

Penland, J. L. (2015). A Blended Approach: Persistence and Resiliency in Higher Ed. *Journal of Teaching and Teacher Education*, *3*(1), 43-50. https://doi.org/10.12785/jtte/030104

Pereira, J., Sousa, C., Matos, A., Cobello, S., Milli, E., Güner, G., Yiğit, E. H., Cura, G., & Kızılkum, Ş. (2024). Transnational Legal Frameworks and Context of Employment and Intellectual Disability. 9-35. https://doi.org/10.24140/nobarriers.v1.p01.02

Pérez, M. P., & Andrades, P. J. M. (2020). La Reforma Curricular Y La Praxis Andragógica De Los Docentes en La Modalidad De Jóvenes, Adultas Y Adultos. *Warisata - Revista De Educación*, *2*(5), 99-111. https://doi.org/10.33996/warisata.v2i5.150

Perin, D. (2011). Facilitating Student Learning Through Contextualization. *Community College Review*, *39*(3), 268-295. https://doi.org/10.1177/0091552111416227

Perkins, J., & Pryor, M. (2025). Digital Badges: Pinning Down Employer Challenges. *Journal of Teaching and Learning for Graduate Employability*, *12*(1), 24-38. https://doi.org/10.21153/jtlge2021vol12no1art1027

Perla, L., Vinci, V., & Scarinci, A. (2023). Innovating University Teaching With Micro-Credentials: An Ongoing Research Experimentation. *Research on Education and Media*, *15*(1), 72-79. https://doi.org/10.2478/rem-2023-0010

Pernas, R. G., & Gómez, J. A. C. (2016). La Accesibilidad en Las Realidades De La Vida Cotidiana: La Pedagogía Social en La Construcción Del Derecho a Una Educación Inclusiva. *Education Policy Analysis Archives*, *24*, 91. https://doi.org/10.14507/epaa.24.2458

Perrault, B., & Levené, T. (2019). L'autoévaluation Accompagnée Et Contractualisée :Quelles Retombées Sur Le Développement Professionnel?Ce Qu'en Disent Les Futurs Enseignants Et Formateurs D'adultes. *Revue Internationale De Pédagogie De L'enseignement Supérieur*, *35*(1). https://doi.org/10.4000/ripes.1989

Perry, S. A. B. (2021). Participatory Feeling: Re-Visioning Transformative Learning Theory Through Heron's Whole Person Perspective. *Adult education quarterly*, *71*(4), 338-355. https://doi.org/10.1177/07417136211016779

Peters, H. J., Peterson, T. R., & Community, T. D. W. (2024). A Dakota Cultural Intervention's Influence on Native Students' Sense of Belonging: A CBPR Case Study. *Aera Open*, *10*. https://doi.org/10.1177/23328584231221793

Peterson, A., Dumont, H., Lafuente, M., & Law, N. (2018). Understanding Innovative Pedagogies. https://doi.org/10.1787/9f843a6e-en

Peterson, S. S., Jang, S. Y., Miguel, J. S., Styres, S., & Madsen, A. (2019). Infusing Indigenous Knowledge and Epistemologies: Learning From Teachers in Northern Aboriginal Head Start Classrooms. *McGill Journal of Education / Revue Des Sciences De L Éducation De McGill*, *53*(1), 26-46. https://doi.org/10.7202/1056281ar

Petit, L. (2022). La Rencontre De L'Intelligence Artificielle (IA) Et De L'esprit Critique (EC) : Nouveaux Enjeux ? Nouvelle Formation ? *Communication Technologies Et Développement*, *12*. https://doi.org/10.4000/ctd.8256

Petrescu, M., Stalmeijer, R. E., Rethans, J. J., & Thammasitboon, S. (2020). Learning in Pediatric Emergency Situations: A Qualitative Study of Residents' Perspectives*. *Pediatric Critical Care Medicine*, *21*(10), 886-892. https://doi.org/10.1097/pcc.0000000000002428

Petrova, M. N. (2021). The Design Studio as a Network – Application of Connectivism in the Planning of an Online Learning Space. https://doi.org/10.35199/epde.2021.84

Pettinger, C., Tripathi, S., Shoker, B., & Hodge, G. (2023). Collaborative Leadership to Support Sustainability in Practice for Dietitians as Allied Health Professionals. *Journal of Human Nutrition and Dietetics*, *36*(6), 2323-2335. https://doi.org/10.1111/jhn.13211

Phelan, S. E., & Glackin, C. (2020). A Cautionary Note on Microcredentialing in Entrepreneurship Education. *Entrepreneurship Education and Pedagogy*, *4*(3), 261-281. https://doi.org/10.1177/2515127419899599

Phillips, G. (2011). A Proposed Certification Process for Business Teachers in Jamaica. *College Teaching Methods & Styles Journal (Ctms)*, *3*(1), 23. https://doi.org/10.19030/ctms.v3i1.5272

Phuong, N. T. Y. (2018). Theory of Constructivism and the Orientation of Hanoi Retraining College for Teachers and Educational Managers in Building a Model of Constructive School in Vietnam. *American Journal of Educational Research*, *6*(6), 703-709. https://doi.org/10.12691/education-6-6-18

Piaget, J. (1952). *The origins of intelligence in children*. International Universities Press.

Piaget, J. (1970). *Science of education and the psychology of the child*. Orion Press.

Picca, A., & Reed, S. (2023). Off to Boarding School: Exploring How Physicians Learn in Tumor Board. *Pediatric Blood & Cancer*, *70*(11). https://doi.org/10.1002/pbc.30632

Picón, P. E. C., & Correa, L. J. H. (2022). La Educación Multicultural Desde La Hermenéutica Analógica De Mauricio Beuchot. *Tendencias Pedagógicas*, *39*, 164-177. https://doi.org/10.15366/tp2022.39.013

Piercy, G. (2013). Transformative Learning Theory and Spirituality. *Journal of Instructional Research*, *2*, 30-42. https://doi.org/10.9743/jir.2013.2.14

Pīgozne, T., Lūka, I., & Surikova, S. (2019). Promoting Youth Entrepreneurship and Employability Through Non-Formal and Informal Learning: The Latvia Case. *Center for Educational Policy Studies Journal*, *9*(4), 129-150. https://doi.org/10.26529/cepsj.303

Pilozo, A. G. M., Zapata, E., Salazar, J. A. L., Macias, D. S. P., & Carabajo, M. F. G. (2024). Impacto Del Apoyo Emocional De La Familia en El Desarrollo Educativo. *Ciencia Latina Revista Científica Multidisciplinar*, *8*(4), 2917-2930. https://doi.org/10.37811/cl_rcm.v8i4.12538

Pina, A. R. B., Castañeda, L., & Segura, J. A. (2018). Personalisation in Educational Technology: The Absence of Underlying Pedagogies. *International Journal of Educational Technology in Higher Education*, *15*(1). https://doi.org/10.1186/s41239-018-0095-0

Pireddu, A., Giliberti, C., Innocenti, A., Simeoni, C., & Bonafede, M. (2025). A Novel Macro-Level Model in Evaluating Health and Safety Training Based on Virtual Reality. *International Journal of Environmental Research and Public Health*, *22*(9), 1378. https://doi.org/10.3390/ijerph22091378

Pires, P., Batista, M., Marinho, D. A., Medina, A. A., Mesquita, H., & Ibáñez, S. J. (2021). Training and Profile of Special Olympics Portugal Coaches: Influence of Formal and Non-Formal Learning. *International Journal of Environmental Research and Public Health*, *18*(12), 6491. https://doi.org/10.3390/ijerph18126491

Pittman, J., & Gioia, D. (2020). Transformative Learning and Inclusion in a Global Social Work Course. *Advances in Social Work*, *19*(2), 446-462. https://doi.org/10.18060/22899

Poëllhuber, B., Roy, N., & Moukhachen, M. (2017). Les Stratégies D'apprentissage Et De Collaboration Déployées Par Des Utilisateurs-Clés Dans Le MOOC « Innovations Technopédagogiques en Enseignement Supérieur (ITES) ». *Raisons Éducatives*, *N° 21*(1), 209-228. https://doi.org/10.3917/raised.021.0209

Poirier, S., & Remsen, M. A. (2017). Technical and Vocational Education and Training. 284-310. https://doi.org/10.4018/978-1-5225-1811-2.ch013

Pokrivčáková, S. (2019). Preparing Teachers for the Application of AI-powered Technologies in Foreign Language Education. *Journal of Language and Cultural Education*, *7*(3), 135-153. https://doi.org/10.2478/jolace-2019-0025

Polesel, J. (2010). Vocational Education and Training (VET) and Young People. *Education + Training*, *52*(5), 415-426. https://doi.org/10.1108/00400911011058352

Pollard, B. (2020a). Examining the Pedagogical and Institutional Phenomena That Influence the Facilitation of Critical Consciousness. *Alberta Journal of Educational Research*, *66*(2), 109-129. https://doi.org/10.55016/ojs/ajer.v66i2.56614

Pollard, B. (2020b). Exploring the Impact of Critical Pedagogy: Do Preservice Teachers Develop Critical Consciousness When Taught by Critical Pedagogues? *The Canadian Journal of Action Research*, *20*(3), 74-100. https://doi.org/10.33524/cjar.v20i3.466

Pollard, C., & Rapp-Wright, H. (2026). Sustainable Analytical Chemistry Laboratories: The Critical Role of Education. *RSC Sustainability*, *4*(7), 2849-2866. https://doi.org/10.1039/d5su00921a

Polly, D., & Byker, E. J. (2020). Considering the Role of Zone of Proximal Development and Constructivism in Supporting Teachers' TPACK and Effective Use of Technology. *Revista De Educación a Distancia (Red)*, *20*(64). https://doi.org/10.6018/red.408661

Polly, D., Mraz, M., & Algozzine, R. (2013). Implications for Developing and Researching Elementary School Mathematics Coaches. *School Science and Mathematics*, *113*(6), 297-307. https://doi.org/10.1111/ssm.12029

Popkhadze, N. (2021). The Good, the Bad and the Ugly: AI in the Higher Education. *Hallinnon Tutkimus*, *40*(4), 254-263. https://doi.org/10.37450/ht.107896

Porras-Hernández, L. H., & Salinas-Amescua, B. (2011). Nonparticipation in Adult Education. *Adult education quarterly*, *62*(4), 311-331. https://doi.org/10.1177/0741713611406980

Porter, D., & Perris, K. (2023). Institutional Partnerships and Collaborations in Online Learning. 585-604. https://doi.org/10.1007/978-981-19-2080-6_32

Porto, S. C. S., & Presant, D. (2023). The IDB Digital Credential Framework: Principles and Guidelines for Creating and Issuing Credentials. https://doi.org/10.18235/0004903

Portugal, L. M. (2015). Hiring, Training, and Supporting Online Faculty for Higher Student Retention Efforts. *Journal of Instructional Research*, *4*(2015), 94-107. https://doi.org/10.9743/jir.2015.13

Potgieter, I. (2020). Privacy Concerns in Educational Data Mining and Learning Analytics. *The International Review of Information Ethics*, *28*. https://doi.org/10.29173/irie384

Poudel, M. P. (2024). Empowering Students Through a Self-Directed Learning Pedagogy in Mathematics Education. *Communications on Applied Nonlinear Analysis*, *31*(1), 238-252. https://doi.org/10.52783/cana.v31.409

Pouwels, J. (2019). We Are in Need of Each Other. Paulo Freire and the Role of Conflicts in Education. *International Journal of Social Pedagogy*, *7*(1). https://doi.org/10.14324/111.444.ijsp.2019.v7.1.009

Póvoa, N., Brunstein, J., & Longmore, A.-L. (2025). Exploring the Transformative Power of Literature: Aesthetic Experiences in Business Education. *Journal of Transformative Education*, *24*(3), 426-448. https://doi.org/10.1177/15413446251371058

Prasetya, R. E., & Syarif, A. (2023). English Language Learners' Attitude and Challenge Utilized M-Learning: Comparison Mobile Application Moodle and Google Classroom. *Elt Worldwide Journal of English Language Teaching*, *10*(1), 28. https://doi.org/10.26858/eltww.v10i1.37218

Pratiwi, D. I., Budiaman, B., & Artanti, G. D. (2025). Ecological Future: Designing a Model Through the Integration of ISM and DEMATEL in Ecological Literacy Modeling. https://doi.org/10.21203/rs.3.rs-7785366/v1

Pratt, D. D. (1993). Andragogy after twenty-five years. *New Directions for Adult and Continuing Education*(57), 15-23.

Preece, J. (2003). Education for Transformative Leadership in Southern Africa. *Journal of Transformative Education*, *1*(3), 245-263. https://doi.org/10.1177/1541344603001003005

Preussler, J. M., DeSalvo, A. M., Tweeten, B., Klaphake, J., Cody, M., McGhee, P. M., Dawson, K., Schoeppner, K., & Auletta, J. J. (2025). Proceedings From the Reimagining Caregiver Workshop: Addressing Caregiver Requirements for Hematopoietic Cell Transplant. *Transplantation and Cellular Therapy*, *31*(7), 458.e451-458.e410. https://doi.org/10.1016/j.jtct.2025.04.006

Priadi, A. A., Cahyadi, T., Purba, D., Harini, N. V., & Zuhri, Z. (2019). Recognition of Prior Learning for Master Marine With Indonesian Qualifications Framework. https://doi.org/10.2991/icei-19.2019.62

Price, B. (2020). English Language Teaching Professionals' Trajectories Within the Context of Learning Organisations. *The Hungarian Educational Research Journal*, *10*(1), 47-65. https://doi.org/10.1556/063.2020.00004

Prieto-López, Y., & Ávila, B. A. L. (2021). Estrategia Didáctica Para El Perfeccionamiento Del Proceso De Enseñanza-Aprendizaje De Los Estudiantes De La Universidad Tecnológica Ecotec-Facultad De Marketing Y Comunicación Con Enfoque Constructivista. *South Florida Journal of Development*, *2*(5), 6575-6584. https://doi.org/10.46932/sfjdv2n5-020

Prieto, D. M. G., & Rojas, A. L. D. (2024). Transformaciones en Las Perspectivas De Significado en Estudiantes De Ingeniería. Una Aproximación Desde Las Prácticas Pedagógicas. *Revista Perspectivas*, *9*(1), 29-46. https://doi.org/10.22463/25909215.3997

Prokou, E. (2008). A Comparative Approach to Lifelong Learning Policies in Europe: The Cases of the UK, Sweden and Greece. *European journal of education*, *43*(1), 123-140. https://doi.org/10.1111/j.1465-3435.2007.00338.x

Psyché, V., Tremblay, D. G., & Baptiste, V. P. J. (2024). Les Compétences À Développer Pour La Gestion De Projets en IA : Part De Soi, Part D'autrui. *Médiations Et Médiatisations*(17), 9-28. https://doi.org/10.52358/mm.vi17.386

Puhani, P. A., & Weber, A. (2005). Does the Early Bird Catch the Worm? Instrumental Variable Estimates of Educational Effects of Age of School Entry in Germany. *SSRN Electronic Journal*. https://doi.org/10.2139/ssrn.851191

Puhovichová, D., & Jankelová, N. (2022). Effects of Industry 4.0 on Human Capital and Future of Jobs. 400-412. https://doi.org/10.53465/edamba.2021.9788022549301.400-412

Purvis, J. K. (2025). Influences Shaping Nurses' Continuing Professional Education Choices and Learning Pathways: An Exploratory Case Study. *Canadian Journal of Nursing Research*, *57*(3), 352-363. https://doi.org/10.1177/08445621251322249

Purwati, D., Mardhiah, A., Nurhasanah, E., & Ramli, R. (2022). The Six Characteristics of Andragogy and Future Research Directions in EFL: a Literature Review. *Elsya Journal of English Language Studies*, *4*(1). https://doi.org/10.31849/elsya.v4i1.7473

Putra, F. P. E., Degeng, I. N. S., Ulfa, S., & Kamdi, W. (2024). The Evolution of Quality Education: Impacts and Challenges of Using Open Educational Resources (OER) and Open Educational Practices (OEP) in the Conceive - Design - Implement - Operate (CDIO) Framework. *Tem Journal*, 386-395. https://doi.org/10.18421/tem131-40

Pyzocha, A., & Boateng, S. (2024). Digital Skills Partnerships: Philadelphia Office of Children and Families, Adult Education Division. *Adult Literacy Education the International Journal of Literacy Language and Numeracy*, *6*(1), 30-32. https://doi.org/10.35847/apyzocha.sboateng.6.1.30

Qafisheh, M. M. (2012). The Role of Legal Clinics in Leading Legal Education: The Model From the Middle East. *Legal Education Review*, *22*(1). https://doi.org/10.53300/001c.6259

Qehaja, A. B., Kutllovci, E., & Havolli, Y. (2017). Learning Organization Survey: A Study in Kosovan Enterprises. *Zbornik Ekonomskog Fakulteta U Zagrebu*, *15*(1), 51-64. https://doi.org/10.22598/zefzg.2017.1.51

Qi, X., Sun, G., & Yue, L. (2023). Applying Self-Optimised Feedback to a Learning Management System for Facilitating Personalised Learning Activities on Massive Open Online Courses. *Sustainability*, *15*(16), 12562. https://doi.org/10.3390/su151612562

Qian, Y. (2025). Prompt Engineering in Education: A Systematic Review of Approaches and Educational Applications. *Journal of Educational Computing Research*, *63*(7-8), 1782-1818. https://doi.org/10.1177/07356331251365189

Qin, H., & Zhou, Z. (2026). Feasibility of Implementing a Multicultural Curriculum Through Artificial Intelligence: Perspectives of Educational Science Experts. *Scientific Reports*, *16*(1). https://doi.org/10.1038/s41598-026-48953-x

Qin, J. (2022). A Theoretical Review on the Role of English as a Foreign Language Teachers' Self-Disclosure in Shaping Classroom Climate and Immediacy. *Frontiers in Psychology*, *13*. https://doi.org/10.3389/fpsyg.2022.945046

Qing, A. K. T. (2023). Copping Phenomenon in China: Theoretical Perspective on Vocational Colleges in Shanxi. *Russian Law Journal*, *11*(6s). https://doi.org/10.52783/rlj.v11i6s.907

Qiu, R. (2023). Traditional Gender Roles and Patriarchal Values: Critical Personal Narratives of a Woman From the Chaoshan Region in China. *New Directions for Adult and Continuing Education*, *2023*(180), 51-63. https://doi.org/10.1002/ace.20511

Qolamani, K. I. B., & Mohammed, M. M. (2023). The Digital Revolution in Higher Education: Transforming Teaching and Learning. *Qalamuna Jurnal Pendidikan Sosial Dan Agama*, *15*(2), 837-846. https://doi.org/10.37680/qalamuna.v15i2.3905

Quick, T., Flatley, B., Sellwood, C., Alam, O., & Vukovich, I. (2021). A Model of Inclusion for Children With Disability in NSW Community Preschools. *Australasian Journal of Special and Inclusive Education*, *45*(2), 150-163. https://doi.org/10.1017/jsi.2021.16

Quinot, G., & Tonder, S. V. (2015). The Potential of Capstone Learning Experiences in Addressing Perceived Shortcomings in LLB Training in South Africa. *Potchefstroom Electronic Law Journal/Potchefstroomse Elektroniese Regsblad*, *17*(4), 1350. https://doi.org/10.4314/pelj.v17i4.05

Quraishi, T., Ulusi, H., Muhid, A., Hakimi, M., & Olusi, M. R. (2024). Empowering Students Through Digital Literacy: A Case Study of Successful Integration in a Higher Education Curriculum. *Journal of Digital Learning and Distance Education*, *2*(8), 667-681. https://doi.org/10.56778/jdlde.v2i8.208

Qutoshi, S. B., & Rajbhandari, M. M. S. (2016). A Philosophical View on and a Technical Approach to Leading a Learning Organization. *Journal of Education and Research*, *6*(2), 12-26. https://doi.org/10.3126/jer.v6i2.22145

Raad, M. E. A., & Odhabi, H. (2021). Hybrid Learning Here to Stay! *Frontiers in Education Technology*, *4*(2), p121. https://doi.org/10.22158/fet.v4n2p121

Rabinowitz, L., & Tondreau, A. (2021). Balancing Literacies. 86-120. https://doi.org/10.4018/978-1-7998-4906-3.ch005

Rahim, M. A., Zainuddin, A., Soh, M. C., Kamal, A. M., Razak, S. A., & Rahmat, N. H. (2023). Exploring Motivation to Learn Online Through McLelland's Theory. *International Journal of Academic Research in Business and Social Sciences*, *13*(2). https://doi.org/10.6007/ijarbss/v13-i2/15652

Rahman, N. L. A., Mokhtar, M., & Ali, D. F. (2022). Self-Initiated Professional Development (SI-PD) Elements for Malaysian Vocational Colleges' TVET Teachers Using Fuzzy Delphi Method (FDM). *Journal of Technical Education and Training*, *14*(1). https://doi.org/10.30880/jtet.2022.14.01.009

Raj, R., Singh, A., Kumar, V., & Verma, P. (2024). Achieving Professional Qualifications Using Micro-Credentials: A Case of Small Packages and Big Challenges In higher Education. *International Journal of Educational Management*, *38*(4), 916-947. https://doi.org/10.1108/ijem-01-2023-0028

Rajabalee, Y. B., Santally, M. I., & Rennie, F. (2019). A Study of the Relationship Between Students' Engagement and Their Academic Performances in an eLearning Environment. *E-Learning and Digital Media*, *17*(1), 1-20. https://doi.org/10.1177/2042753019882567

Ralalatiana, M. C., & Vatz-Laaroussi, M. (2016). Quand Projet D'immigration Rime Avec Inscription Dans Les Cours De Francisation. La Trajectoire Langagière De Neuf Immigrantes Scolarisées Dans La Région Montréalaise. *Diversité Urbaine*, *15*(1), 87-107. https://doi.org/10.7202/1037873ar

Ralph, E. G., Walker, K., & Wimmer, R. (2009). Practicum and Clinical Experiences: Postpracticum Students' Views. *Journal of Nursing Education*, *48*(8), 434-440. https://doi.org/10.3928/01484834-20090518-02

Ralph, E. G., Wimmer, R., & Walker, K. (2008). 9. Clinical and Practicum Education in the Professions: The Student Voice. *Collected Essays on Learning and Teaching*, *1*, 55-60. https://doi.org/10.22329/celt.v1i0.3178

Ramachandran, S., Chang, H.-J., Worthington, C., Kushniruk, A., Ibáñez-Carrasco, F., Davies, H., McKee, G., Brown, A., Gilbert, M., & Iyamu, I. (2024). Digital Competencies and Training Approaches to Enhance the Capacity of Practitioners to Support the Digital Transformation of Public Health: Rapid Review of Current Recommendations. *Jmir Public Health and Surveillance*, *10*, e52798-e52798. https://doi.org/10.2196/52798

Rambaree, K., & Nässén, N. (2021). Digitalization of Critical Reflection With ATLAS.ti Software in Social Work Supervision. *Social Sciences*, *10*(3), 95. https://doi.org/10.3390/socsci10030095

Ramírez-Solórzano, F. L., & Navas, C. D. H. (2024). Inclusión Educativa: Desafíos Y Oportunidades Para La Educación De Estudiantes Con Necesidades Especiales. *Revista Científica Zambos*, *3*(3), 44-63. https://doi.org/10.69484/rcz/v3/n3/57

Ramiro, M. R. (2022). La Alienación De La Esperanza: El Aprendizaje Permanente Y La Educación De Adultos en La Política Educativa Europea (2000-2021). *Revista Educación Política Y Sociedad*, *7*(2), 248-273. https://doi.org/10.15366/reps2022.7.2.011

Ramos-Rodríguez, E., Martínez, P. F., & Ponte, J. P. d. (2016). An Approach to the Notion of Reflective Teacher and Its Exemplification on Mathematics Education. *Systemic Practice and Action Research*, *30*(1), 85-102. https://doi.org/10.1007/s11213-016-9383-6

Rana, S., Ardichvili, A., & Polesello, D. (2016). Promoting Self-Directed Learning in a Learning Organization: Tools and Practices. *European journal of training and development*, *40*(7), 470-489. https://doi.org/10.1108/ejtd-10-2015-0076

Ranalli, J., & Chappell, S. (2017). A Model for Development of Employer Engagement at a Small Campus. https://doi.org/10.18260/1-2--27483

Randazzo, G., Reitano, G., Carletti, F., Iafrate, M., Betto, G., Novara, G., Moro, F., & Zattoni, F. (2023). Urology: A Trip Into Metaverse. *World Journal of Urology*, *41*(10), 2647-2657. https://doi.org/10.1007/s00345-023-04560-3

Rane, J., Kaya, Ö., Mallick, S. K., & Rane, N. L. (2024). Automated Machine Learning (AutoML) in Industry 4.0, 5.0, and Society 5.0: Applications, Opportunities, Challenges, and Future Directions. https://doi.org/10.70593/978-81-981271-0-5_5

Rane, N., Choudhary, S., & Rane, J. (2024). Leading-Edge Metaverse in Education: Framework, Applications, Challenges, and Future Development. *Tesol and Technology Studies*, *5*(2). https://doi.org/10.48185/tts.v5i2.1002

Ranieri, M., & Pachler, N. (2014). Inventing and Re-Inventing Identity: Exploring the Potential of Mobile Learning in Adult Education. *Prospects*, *44*(1), 61-79. https://doi.org/10.1007/s11125-014-9294-1

Ranmechai, S., & Poonputta, A. (2023). Problem-Based Learning Approach for Developing Learning Management Skills in Undergraduate Mathematics Education. *International Journal of Learning Teaching and Educational Research*, *22*(9), 296-309. https://doi.org/10.26803/ijlter.22.9.16

Rasmussen, P. (2024). The Implications of Critical Theory for Adult Education Policy. *Journal of Adult and Continuing Education*, *31*(2), 374-390. https://doi.org/10.1177/14779714241266797

Ravichandran, R. (2023). Bridging the Gap: The Role of Apprenticeship Training Programs. *Journal of Vocational Education Studies*, *6*(1), 156-166. https://doi.org/10.12928/joves.v6i1.8006

Ravichandran, R., & Mahapatra, J. (2023). Virtual Reality in Vocational Education and Training: Challenges and Possibilities. *Journal of Digital Learning and Education*, *3*(1), 25-31. https://doi.org/10.52562/jdle.v3i1.602

Rawas, S. (2024). AI: The Future of Humanity. *Discover Artificial Intelligence*, *4*(1). https://doi.org/10.1007/s44163-024-00118-3

Ray, D., & Nelson, T. (2025). They Asked Me if I Wanted to Grow: The Self-Directed, Lifelong Learning of Middle Managers in the Banking Industry. *New Horizons in Adult Education and Human Resource Development*, *37*(1), 46-57. https://doi.org/10.1177/19394225241308671

Ray, K. K., & Sahu, S. (2025). Building a Resilient Healthcare Workforce in India: A Talent Management Perspective. *NHRD Network Journal*, *18*(1), 15-27. https://doi.org/10.1177/26314541251325699

Ray, S. M. (2019). Teaching Case— Applications of Culturally Relevant Pedagogy in a Community College Classroom. *New Horizons in Adult Education and Human Resource Development*, *31*(4), 65-69. https://doi.org/10.1002/nha3.20267

Ray, S. M., & Zarestky, J. (2021). Gender in Vocational Education and Training: An Integrative Review. *European journal of training and development*, *46*(9), 876-893. https://doi.org/10.1108/ejtd-12-2019-0196

Read, T., & Olcott, D. (2025). Mobilizing University Futures: A Blueprint for Change. *Policy Futures in Education*, *23*(7), 1364-1381. https://doi.org/10.1177/14782103251341405

Reddig, J. M., Smith, G. R., Siyahrood, S. A., Morris, W. G., Bae, Y., Crutcher, K., Kos, J., Dass, R. K., Kim, J., Siddiqui, M. N., Weitekamp, D., Thajchayapong, P., Kakar, S., Endert, A., & Crossley, S. (2026). Guidelines for Designing AI Technologies to Support Adult Learning. https://doi.org/10.48550/arxiv.2605.04616

Reddy, P., & Shaw, R. (2019). Becoming a Professional. *Education + Training*, *61*(2), 272-288. https://doi.org/10.1108/et-10-2018-0210

Reddy, T., & Maharaj, M. (2022). Marketing Communication Tools for World of Work Opportunities. *International Journal of Research in Business and Social Science (2147-4478)*, *11*(4), 323-332. https://doi.org/10.20525/ijrbs.v11i4.1793

Rees, C. E., Lee, S., Huang, E., Denniston, C., Edouard, V., Pope, K., Sutton, K., Waller, S., Ward, B., & Palermo, C. (2019). Supervision Training in Healthcare: A Realist Synthesis. *Advances in Health Sciences Education*, *25*(3), 523-561. https://doi.org/10.1007/s10459-019-09937-x

Reghenzani-Kearns, D. (2021). Globalisation: Impacts and Responses. xli-xlii. https://doi.org/10.4102/aosis.2021.bk214.f2

Regmi, K. D. (2020). Lifelong Learning in Nepal: Contexts and Prospects. *Amc Journal*, *1*(1), 121-132. https://doi.org/10.3126/amcj.v1i1.33480

Rehak, K. M. (2022). Privacy and Data Rights for Adult English as a Second Language (ESL) Students. 102-119. https://doi.org/10.4018/978-1-6684-5892-1.ch006

Rehn, N., Maor, D., & McConney, A. (2016). Navigating the Challenges of Delivering Secondary School Courses by Videoconference. *British journal of educational technology*, *48*(3), 802-813. https://doi.org/10.1111/bjet.12460

Reichborn-Kjennerud, K. (2021). Key Insights Into What Makes Public Organizations Learn From Training Programs. *Nordic journal of working life studies*. https://doi.org/10.18291/njwls.129052

Reichenberg, M. (2021). What Predicts Adult Immigrants' Engagement in Host Language Training Programmes: Social Class, Forms of Capital, or Academic Self-Efficacy? *Research Result Pedagogy and Psychology of Education*, *7*(3), 40-52. https://doi.org/10.18413/2313-8971-2021-7-3-0-4

Reid, A., Richards, A., & Willox, D. (2021). Connecting Experiences to Employability Through a Meaning-Making Approach to Learning. *Journal of Teaching and Learning for Graduate Employability*, *12*(2), 99-113. https://doi.org/10.21153/jtlge2021vol12no2art1013

Reihanian, I., Hou, Y., & Sun, Q. (2025). From Pilots to Practices: A Scoping Review of GenAI-Enabled Personalization in Computer Science Education. *Ai*, *7*(1), 6. https://doi.org/10.3390/ai7010006

Reisel, L., Hegna, K., & Imdorf, C. (2015). Gender Segregation in Vocational Education: Introduction. 1-22. https://doi.org/10.1108/s0195-631020150000031023

Reitz, C. (2009). Marcuse in America—Exile as Educator. *Fast Capitalism*, *5*(2), 41-48. https://doi.org/10.32855/fcapital.200902.005

Remer, D. S., & Martin, M. (2009). Project and Engineering Management Certification. *Leadership and Management in Engineering*, *9*(4), 177-190. https://doi.org/10.1061/(asce)lm.1943-5630.0000031

Remington, T. F. (2017). Closing the Skills-Jobs Gap: Russia and China Compared. *SSRN Electronic Journal.* https://doi.org/10.2139/ssrn.3078505

Rennick, J. B. (2015). Learning That Makes a Difference: Pedagogy and Practice for Learning Abroad. *Teaching & Learning Inquiry the Issotl Journal, 3*(2), 71-88. https://doi.org/10.20343/teachlearninqu.3.2.71

Rens, J., White, L., & Botha, L. (2020). The Role of Reflection as a Vehicle for Self-Directed Learning During Work-Integrated Learning of Student Teachers. 247-278. https://doi.org/10.4102/aosis.2020.bk215.09

Resch, K. (2022). Diversity Skills for Future Teachers: How Transformative Learning Prepares Pre-Service Teachers for Diversity in Austrian Schools. *Higher Education Skills and Work-Based Learning, 13*(1), 66-79. https://doi.org/10.1108/heswbl-05-2022-0096

Revuelta, L., Rodríguez-Fernández, A., & Izar-de-la-Fuente, I. (2024). Ajuste Psicosocial: Qué Es, Qué Variables Lo Determinan Y Cuál Es Su Modificabilidad Educativa en El Alumnado Con Y Sin NEAE. *Congreso Internacional Ideice, 14*, 126-148. https://doi.org/10.47554/cii.vol14.2023.pp126-148

Reyes, L. A. S., Rodríguez-Cruz, A., Cruz, R. O., Mendiola-Precoma, J., Guevara, J. E. R., & Yañez, Z. Y. C. (2022). Diagnóstico Del Uso E Implementación De La Realidad Aumentada en La Escuela De Bachilleres De La Universidad Autónoma De Querétaro. *Revista Tecnología Ciencia Y Educación*, 69-94. https://doi.org/10.51302/tce.2022.2812

Rezvanova, L. A., Rezvanov, S. V., & Титова, E. B. (2019). Education in the Context of the Personality's Development Genesis. *Manuscript, 12*(1), 107-111. https://doi.org/10.30853/manuscript.2019.1.23

Rhodes, C., & Fletcher, S. (2013). Coaching and Mentoring for Self-efficacious Leadership in Schools. *International Journal of Mentoring and Coaching in Education, 2*(1), 47-63. https://doi.org/10.1108/20466851311323087

Ricart, C., Morán, T., & Kappaz, C. (2014). Toward a National Framework of Lifelong Learning in Mexico. https://doi.org/10.18235/0012801

Richardson, S. D., & Sakho, J. R. (2022). Creating Equitable Access: Using OER for Socially Just Educational Leaders. *Journal for Multicultural Education, 16*(5), 443-455. https://doi.org/10.1108/jme-12-2021-0224

Richter, S., Giroux, M., Piven, I., Sima, H., & Dodd, P. (2024). A Constructivist Approach to Integrating AI in Marketing Education: Bridging Theory and Practice. *Journal of Marketing Education, 47*(2), 94-111. https://doi.org/10.1177/02734753241288876

Rico, R. L. A. (2022). Creencias De Autoeficacia Estudiantiles Y Docentes: ¿Un Vínculo Inseparable? Propuestas Para La Intervención. *Revista Ucmaule*(62), 106-131. https://doi.org/10.29035/ucmaule.62.106

Rieckmann, M. (2018). Learning to Transform the World: Key Competencies in Education for Sustainable Development. In A. Leicht, J. Heiss, & W. J. Byun (Eds.), *Issues and Trends in Education for Sustainable Development* (pp. 39-59). UNESCO.

Ries, E., Dunham, H., Corriston, K., Alexander, K., Taylor, V., & Wetzel, M. M. (2023). Look to the Side: Peer Partnerships in Preservice Literacy Teacher Education. *Literacy Research Theory Method and Practice, 72*(1), 258-277. https://doi.org/10.1177/23813377231182096

Rijn, M. B. v., Yang, H., & Sanders, K. (2013). Understanding Employees' Informal Workplace Learning. *Career Development International, 18*(6), 610-628. https://doi.org/10.1108/cdi-12-2012-0124

Rintala, H., & Nokelainen, P. (2019). Vocational Education and Learners' Experienced Workplace Curriculum. *Vocations and Learning, 13*(1), 113-130. https://doi.org/10.1007/s12186-019-09229-w

Ríos, C. Y. Q., & Rodríguez, E. B. (2022). Enseñanza De Las Matemáticas Durante El Confinamiento Por COVID-19. *Boletín Científico Investigium De La Escuela Superior De Tizayuca, 8*(Especial), 77-83. https://doi.org/10.29057/est.v8iespecial.9986

Rismiyanto. (2017). Andragogical Orientation: An Anti-Mainstream in Teaching English to Young Learners. 157-165. https://doi.org/10.24176/03.3201.19

Risner, M. E., & Kumar, S. (2016). Graduate Student Perceptions of a Globally Networked Course. *Journal of Applied Research in Higher Education*, *8*(3), 287-301. https://doi.org/10.1108/jarhe-01-2015-0009

Rísquez, A., Cassidy, D., & O'Suilleabhain, G. (2020). Badge of Honour? An Exploration of the Use of Digital Badges to Support a Partnership Approach to Faculty Development. *Australasian Journal of Educational Technology*, *36*(5), 18-29. https://doi.org/10.14742/ajet.6112

Rius, H. C., Boumeggoti, D., Areny, J. F., & Bernadó, B. S. (2021). Cross-Border Previous Learning Recognition: Enhancing Lifelong Learning and Social Inclusion. https://doi.org/10.4995/head21.2021.12972

Rivas, N. (2024). Tecnología Asistiva Para La Inclusión Educativa en Ecuador. *Ciencia Latina Revista Científica Multidisciplinar*, *8*(3), 417-433. https://doi.org/10.37811/cl_rcm.v8i3.11228

Rivera-Vargas, P. (2023). What Is "Algorithmic Education" and Why Do Education Institutions Need to Consolidate New Capacities? https://doi.org/10.31235/osf.io/zwpfk

Robinson, E., Bhandari, D., Donohue, R., & Lokmic-Tomkins, Z. (2025). Addressing Healthcare Professionals' Fatigue for Effective Climate Action Engagement. *The Journal of Climate Change and Health*, *21*, 100366. https://doi.org/10.1016/j.joclim.2024.100366

Rocco, F. D., Romagnoli, C., Ciaccioni, S., Demarie, S., Doupona, M., Capranica, L., Padua, E., & Guidotti, F. (2026). Sustainable Athletes' Career Pathways and Mental Health Support: An Integrative Umbrella Review. *Sports*, *14*(6), 251. https://doi.org/10.3390/sports14060251

Rodger Roberto Alves de, S. (2024). Pedagogy and Andragogy: Project-Based Education to Promote Active and Meaningful Learning. *Revista Interseção*, *6*(1), 231-250. https://doi.org/10.48178/intersecao.v6i1.485

Rodrigo, C., Iniesto, F., & García-Serrano, A. (2024). Applying Andragogy for Integrating a MOOC Into a Formal Online Learning Experience in Computer Engineering. *Heliyon*, *10*(1), e23493. https://doi.org/10.1016/j.heliyon.2023.e23493

Rodrigues, R. L. (2024). A Ética Na Práxis Freireana: Educação Para a Liberdade E Emancipação Humana. *Revista Katálysis*, *27*. https://doi.org/10.1590/1982-0259.2024.e96359

Rodríguez, A. B. P. (2021). Conectivismo Y Diseño Instruccional: Ecología De Aprendizaje Para La Universidad Del Siglo XXI en México. *Márgenes Revista De Educación De La Universidad De Málaga*, *2*(1), 4-20. https://doi.org/10.24310/mgnmar.v2i1.9349

Rodríguez, D. H. G. (2024). Políticas Y Estrategias De Inclusión Educativa en La Educación Superior. *Ciencia Latina Revista Científica Multidisciplinar*, *8*(1), 8823-8840. https://doi.org/10.37811/cl_rcm.v8i1.10211

Rodríguez, R. G., Figueroa, R. A. A., Soledispa, L. M. F., & Figureoa, J. A. A. (2024). El Peligro Del Olvido: La Importancia Del Otro, La Otredad Y El Respeto a La Cultura De Los Pueblos en La Formación De Educadores. *Revista Científica De Innovación Educativa Y Sociedad Actual Alcon*, *4*(4), 195-204. https://doi.org/10.62305/alcon.v4i4.220

Rodríguez, S. L. (2014). El Aprendizaje Basado en Problemas Para La Educación Médica: Sus Raíces Epistemológicas Y Pedagógicas. *Revista Med*, *22*(2), 32. https://doi.org/10.18359/rmed.1168

Rodriguez, W. J. M., Andrade-Girón, D., Zúñiga-Rojas, M. G., Ramírez, E. T. S., Rivera, I. P. C., Sánchez, J., & Soto, F. G. C. (2023). Artificial Intelligence and Augmented Reality in Higher Education: A Systematic Review. *Data & Metadata*, *2*, 121. https://doi.org/10.56294/dm2023121

Roessger, K., Roumell, E., & Weese, J. (2018). *Testing Andragogical Assumptions Across Countries Using Multi-Level Analyses of PIAAC Data.*

Roffey, S., & Quinlan, D. M. (2021). Positive Education With Disadvantaged Students. 645-674. https://doi.org/10.1007/978-3-030-64537-3_25

Rogers-Shaw, C., Carr-Chellman, D., & Choi, J. (2017). Universal Design for Learning: Guidelines for Accessible Online Instruction. *Adult Learning*, *29*(1), 20-31. https://doi.org/10.1177/1045159517735530

Rogers-Shaw, C., Park, T. W., & Mohney, K. D. (2023). Strategies for Supporting Learners With Disabilities in Postsecondary Education. *New Horizons in Adult Education and Human Resource Development*, *35*(4), 199-207. https://doi.org/10.1177/19394225231189136

Rogers, C. R. (1969). *Freedom to Learn.* Merrill.

Rogers, L. (2011). Lifelong Learning: Links Between Policy and Practice. *London Review of Education*, *9*(2). https://doi.org/10.1080/14748460.2011.585894

Rojas-Botero, M. L., Mújica, Ó. J., Avellaneda, X., Cáceres, C. F., Castro, A., Cox, A. L., Galvão, L. A. C., Duarte, I. G., Más, P., Pino, S. d., Rojas, K., Sojo, A., Fuentes, M. U., Sáenz, R., & García-Saisó, S. (2024). No Seguir Ciegos a La Equidad: Lecciones Por Aprender De La Pandemia De COVID-19 en Las Américas. *Revista Panamericana De Salud Pública*, *48*, 1. https://doi.org/10.26633/rpsp.2024.71

Roka, J. (2023). Use of Freire's Conscientization in Mathematics Learning. *Academic Journal of Mathematics Education*, *6*(1), 54-57. https://doi.org/10.3126/ajme.v6i1.63798

Rolle, D. S. (2023). *A Descriptive Analysis of the Relationship Between Student Career Support Services Efficacy and on-Time Graduation in Online Education* Northeastern University].

Romano, A., Kramlich, D. J., Marsick, V. J., Zakrzewski, T. C., Anderson-Sathe, L., Brunstein, J., Cunliff, E., & Longmore, A.-L. (2024). Practicing Embodied Listening for Transformation: The Transformative Listening Protocol. *Journal of Transformative Education*, *23*(1), 31-58. https://doi.org/10.1177/15413446241258681

Romano, R., Lauría, A., Naldi, E., & Mezzapesa, C. (2024). EDULEARN24 Proceedings. *1*. https://doi.org/10.21125/edulearn.2024

Romero, B. C., Acuña, J. E. C., & Caicedo–Quiroz, R. (2020). Multiculturalidad E Interculturalidad en La Escuela Ecuatoriana: Expresión De Identidad Cultural. *Dilemas Contemporáneos Educación Política Y Valores*. https://doi.org/10.46377/dilemas.v8i1.2404

Roméro, M., Reyes, J., & Kostakos, P. (2024). Generative Artificial Intelligence in Higher Education. 129-143. https://doi.org/10.1007/978-3-031-55272-4_10

Romulo, C., Venkataraman, B., Caplow, S., Ajgaonkar, S., Allen, C. R., Anandhi, A., Anderson, S. W., Azzarello, C. B., & Brundiers, K. (2024). Implementing Interdisciplinary Sustainability Education With the Food-Energy-Water (FEW) Nexus. *Humanities and Social Sciences Communications*, *11*(1). https://doi.org/10.1057/s41599-024-03332-7

Rondini, A. C. (2015). Observations of Critical Consciousness Development in the Context of Service Learning. *Teaching Sociology*, *43*(2), 137-145. https://doi.org/10.1177/0092055x15573028

Rondón-Molina, J. A., Torres-Chinga, Á. R., Maliza-Cruz, W. I., & Alba–Castellanos, O. (2024). Estrategia Didáctica Sustentada en El Aprendizaje Basado en Proyectos en La Asignatura Matemática Del Bachillerato Técnico en Informática. *Mqrinvestigar*, *8*(3), 3942-3965. https://doi.org/10.56048/mqr20225.8.3.2024.3942-3965

Rongmin, L. (2024). Rethinking Education and Training: Creating High-Quality Employment Opportunities for Vocational Students. *Advances in Vocational and Technical Education*, *6*(2). https://doi.org/10.23977/avte.2024.060214

Rosário, A. T., & Raimundo, R. (2024). Enhancing Business Higher Education Through Simulation-Based Learning, Problem-Based Learning, and Challenge-Based Learning. https://doi.org/10.20944/preprints202407.0747.v1

Rose, J., & Chmutina, K. (2021). Developing Disaster Risk Reduction Skills Among Informal Construction Workers in Nepal. *Disasters*, *45*(3), 627-646. https://doi.org/10.1111/disa.12435

Ross-Gordon, J. M. (2024). Learning Disabilities and the Adult Learner. *New Directions for Adult and Continuing Education*, *2024*(184), 46-55. https://doi.org/10.1002/ace.20542

Ross-Gordon, J. M., Gordon, S. P., Alston, G. D., Dawson, K., & Aacken, C. V. (2015). Efforts to Transform Learning and Learners: The First Decade of an Innovative Doctoral Program. *Journal of Thought*, *49*(1-2), 52. https://doi.org/10.2307/jthought.49.1-2.52

Rossiter, E., Thomson, T. J., & Fitzgerald, R. (2022). Supporting University Students' Learning Across Time and Space: A From-Scratch, Personalised and Mobile-Friendly Approach. *Interactive Technology and Smart Education*, *21*(1), 108-130. https://doi.org/10.1108/itse-07-2022-0082

Roth, G., Assor, A., Kanat-Maymon, Y., & Kaplan, H. (2007). Autonomous Motivation for Teaching: How Self-Determined Teaching May Lead to Self-Determined Learning. *Journal of Educational Psychology*, *99*(4), 761-774. https://doi.org/10.1037/0022-0663.99.4.761

Roth, W. M., & Lee, Y. J. (2007). "Vygotsky's Neglected Legacy": Cultural-Historical Activity Theory. *Review of Educational Research*, *77*(2), 186-232. https://doi.org/10.3102/0034654306298273

Roth, W. M., Mavin, T. J., & Dekker, S. (2014). The Theory-Practice Gap: Epistemology, Identity, and Education. *Education + Training*, *56*(6), 521-536. https://doi.org/10.1108/et-11-2012-0117

Rothes, A., Lemos, M. S. d., & Gonçalves, T. (2016). Motivational Profiles of Adult Learners. *Adult education quarterly*, *67*(1), 3-29. https://doi.org/10.1177/0741713616669588

Rothstein-Fisch, C., Trumbull, E., & Garcia, S. G. (2009). Making the Implicit Explicit: Supporting Teachers to Bridge Cultures. *Early Childhood Research Quarterly*, *24*(4), 474-486. https://doi.org/10.1016/j.ecresq.2009.08.006

Roumell, E. A., & Jabarkhail, S. (2023). Supporting Adult Working Learners. *New Directions for Adult and Continuing Education*, *2023*(179), 5-16. https://doi.org/10.1002/ace.20498

Roumell, E. A., Salajan, F. D., & Todoran, C. (2018). A Survey of U.S. Education Policy Regarding the Education of Adults. *Educational Policy*, *34*(5), 785-815. https://doi.org/10.1177/0895904818802416

Routh, J., Paramasivam, S. J., Cockcroft, P., Nadarajah, V. D., & Jeevaratnam, K. (2022). Using Learning Theories to Develop a Veterinary Student Preparedness Toolkit for Workplace Clinical Training. *Frontiers in Veterinary Science*, *9*. https://doi.org/10.3389/fvets.2022.833034

Roux, R. (2024). Truth and Transformation: The Test of Truth in the Development of New or Revised Ideas, and Related Values and Beliefs. *Journal of Transformative Education*, *23*(1), 9-30. https://doi.org/10.1177/15413446241268550

Rovai, A. P., Ponton, M. K., Derrick, M. G., Wiggers, N. R., & Flannagan, J. S. (2020). An Exploration of Attitudinal and Situational Factors Related to Intrinsic Motivation and Autonomy in Teacher Education Students. *International Journal of Education*, *12*(1), 99. https://doi.org/10.5296/ije.v12i1.16160

Rowe, L. (2019). Educating for the Modern World: A Report Review. *Journal of Work-Applied Management*, *11*(1), 5-16. https://doi.org/10.1108/jwam-06-2019-0014

Roxas, M. Y. D., Aguirre, J. M., Fias-ilon, L. L., & Alumno, M. C. (2023). Educational Philosophies of Selected Colleges and Universities in Baguio City and Benguet Province Philippines. *Cognizance Journal of Multidisciplinary Studies*, *3*(6), 294-303. https://doi.org/10.47760/cognizance.2023.v03i06.018

Roy, É. L. (2003). Au-Delà Du Discours. https://doi.org/10.1787/9789264299443-fr

Roy, R., & Al-Absy, M. S. M. (2022). Impact of Critical Factors on the Effectiveness of Online Learning. *Sustainability*, *14*(21), 14073. https://doi.org/10.3390/su142114073

Roy, R., & Marsafawy, H. E. (2021). Bridging Recognition of Prior Learning (RPL) and Corporate Social Responsibility (CSR): Circular Flow of Interaction Among the University, Industry, and People. *Sustainability*, *13*(8), 4532. https://doi.org/10.3390/su13084532

Royaltey-Quandt, V. (2016). *Preparing Educational Leaders for Queer Inclusion in Schools* San Francisco State University].

Różewski, P., Jankowski, J., Bródka, P., & Michalski, R. (2015). Knowledge Workers' Collaborative Learning Behavior Modeling in an Organizational Social Network. *Computers in Human Behavior*, *51*, 1248-1260. https://doi.org/10.1016/j.chb.2014.12.014

Rubaia'an, M. A. B. (2023). Dental Education: A Guide for Novice Tutors. *Cureus*. https://doi.org/10.7759/cureus.43227

Rubel, A., & Jones, K. M. L. (2020). The Temptation of Data-Enabled Surveillance: Are Universities the Next Cautionary Tale? *SSRN Electronic Journal*. https://doi.org/10.2139/ssrn.3559429

Rubić, I., & Matijević, D. (2019). Autonomy and Motivation: A Self-Determination Theory Perspective on Esp Motivation. *Journal of Teaching English for Specific and Academic Purposes*, 147. https://doi.org/10.22190/jtesap1902147r

Ruipérez-Valiente, J. A. (2020). El Proceso De Implementación De Analíticas De Aprendizaje. *Ried Revista Iberoamericana De Educación a Distancia*, *23*(2), 85. https://doi.org/10.5944/ried.23.2.26283

Ruiz, M. J. U., García, L. P., & Verdú, F. J. M. (2018). Fact Checking: Un Nuevo Desafío Del Periodismo. *Profesional De La Información*, *27*(4), 733. https://doi.org/10.3145/epi.2018.jul.02

Ruslin, R. (2017). The Learning Experience of Automotive Students at a Vocational School in Indonesia: Perspectives of School Stakeholders. *SSRN Electronic Journal*. https://doi.org/10.2139/ssrn.4242972

Ruthanam, M. (2023). Facing the Unknown: Managing Work Integrated Learning in the Midst of Covid-19. 99-109. https://doi.org/10.2991/978-2-38476-006-0_9

Rutherford-Hemming, T. (2012). Simulation Methodology in Nursing Education and Adult Learning Theory. *Adult Learning*, *23*(3), 129-137. https://doi.org/10.1177/1045159512452848

Rybalko, L., Horoshko, V., & Onishchuk, L. M. (2021). Health Care Education: Realities and Prospects. *Zeszyty Naukowe Wyższej Szkoły Technicznej W Katowicach*(13), 173-180. https://doi.org/10.54264/0016

Ryoo, J., & Winkelmann, K. (2021). Introduction. 1-16. https://doi.org/10.1007/978-3-030-58948-6_1

Ryu, G., & Moon, S.-G. (2019). The Effect of Actual Workplace Learning on Job Satisfaction and Organizational Commitment. *Journal of Workplace Learning*, *31*(8), 481-497. https://doi.org/10.1108/jwl-05-2019-0061

Sa'adi, S. a. (2016). Introduction to Views of Connectivism Theory of Learning. *Register Journal*, *3*(2), 201. https://doi.org/10.18326/rgt.v3i2.201-214

Saal, L. K., & Dowell, M.-M. S. (2014). A Literacy Lesson From an Adult "Burgeoning" Reader. *Journal of Adolescent & Adult Literacy*, *58*(2), 135-145. https://doi.org/10.1002/jaal.325

Saar, E., Ure, O. B., & Holford, J. (2013). Introduction. https://doi.org/10.4337/9780857937360.00007

Sabri, S., Gani, A., Yadegaridehkordi, E., Eke, C. I., & Shuib, L. (2022). A Survey on Mobile Learning for Adult Learners: State-of-the-Art, Taxonomy, and Challenges. *Ieee Access*, *10*, 83884-83897. https://doi.org/10.1109/access.2022.3195285

Sachs, J. D., Schmidt-Traub, G., Mazzucato, M., Messner, D., Nakicenovic, N., & Rockström, J. (2019). Six Transformations to Achieve the Sustainable Development Goals. *Nature Sustainability*, *2*(9), 805-814. https://doi.org/10.1038/s41893-019-0352-9

Sacré, M., Lafontaine, D., & Toczek, M. C. (2020). Liens Entre Les Composantes Des Dispositifs D'enseignement Hybride Et Les Performances Des Étudiants De L'enseignement Supérieur : Une Revue Systématique. *Mesure Et Évaluation en Éducation*, *42*(3), 109-152. https://doi.org/10.7202/1074105ar

Sadler, T. D., Friedrichsen, P., & Zangori, L. (2019). Uma Abordagem Para O Ensino Através De Questões Sociocientíficas E Aprendizagem Baseada Em Modelos (SIMBL). *Educação E Fronteiras*, *9*(25), 08-26. https://doi.org/10.30612/eduf.v9i25.11006

Saenen, L., Stevens, R., Hermans, K., Struyven, K., & Emmers, E. (2026). From Principles to Practice: Universal Design in Higher Education—A Systematic Review. *Review of Education*, *14*(1). https://doi.org/10.1002/rev3.70129

Safourcade, S., & Alava, S. (2009). S'auto Évaluer Pour Agir : Rôle Du Sentiment D'efficacité Personnelle Dans Les Pratiques D'enseignement. *Questions Vives Recherches en Éducation*, *Vol.6 n°12*, 109-123. https://doi.org/10.4000/questionsvives.444

Sagita, S., Rahmat, A., Priyandoko, D., & Sriyati, S. (2023). Sustainability of Project-Based Learning: Challenge and Obstacles From Students Perception Point of View. *Jurnal Penelitian Pendidikan Ipa*, *9*(2), 810-816. https://doi.org/10.29303/jppipa.v9i2.3044

Sahin, G. E., Bayram, G. A., Sierra, A., Akdemir, S., Tarakçı, D., & Tunali, A. N. (2025). Effects of Artificial Intelligence Based Physiotherapy Educational Approach in Developing Clinical Reasoning Skills: A Randomized Controlled Trial. *BMC Medical Education*, *25*(1). https://doi.org/10.1186/s12909-025-07926-w

Şahin, M., Erişen, Y., & ÇELİKÖZ, N. (2016). The Transformational Learning of Three Adult Academicians. *Mediterranean Journal of Humanities*, *6*(1), 299-299. https://doi.org/10.13114/mjh.2016119306

Sahoo, S., Millar, R. J., Yamashita, T., & Cummins, P. (2021). Educational Attainment, Literacy Skills, Nativity, and Motivation to Learn Among Middle-Aged Adults in the United States. *Adult Learning*, *34*(1), 15-29. https://doi.org/10.1177/10451595211048554

Said, A. (2018). Vocational Teaching-Learning Through the Eyes of Undergraduate Vocational Students in Malta: A Qualitative Exploratory Study. *International Journal for Research in Vocational Education and Training*, *5*(1), 42-63. https://doi.org/10.13152/ijrvet.5.1.3

Said, S., & Abdallah, W. (2024). Enhancing Lifelong Learning and Professional Growth: Exploring the Role of Self-Directed Learning for University Educators. *Journal of Adult and Continuing Education*, *30*(2), 439-462. https://doi.org/10.1177/14779714241236282

Saif Juma Salim Al, W. (2022). Factors Influencing Omani Women's Autonomy to Study Overseas: A Synthesis of the Literature. *Open Journal of Social Sciences*, *10*(13), 217-226. https://doi.org/10.4236/jss.2022.1013017

Saint-Luc, F. (2018). De La Confrontation Coopérative Interculturelle À L'autoformation Coopérative. Le Cas Des Formateurs D'enseignants Au Sein Du Mouvement International De L'Ecole Moderne. *Revista Eletrônica De Educação*, *12*(2), 408-430. https://doi.org/10.14244/198271992586

Sakız, H. (2016). Ability, Examination and Inclusive Education: Stretching the Hard Lines of the Educational System. *Educational Process International Journal*, *5*(1), 65-75. https://doi.org/10.12973/edupij.2016.51.5

Salah, S., & Thabet, M. (2021). E-Learning Management Systems- A Feature-Based Comparative Analysis. *Journal of Information Systems and Technology Management*, *18*. https://doi.org/10.4301/s1807-1775202118003

Salajan, F. D., & Roumell, E. A. (2021). Tracing the Historical Construction of a Vocational Training, Adult Education and Lifelong Learning Policy Space in the European Union From 1951 to Present. *European Educational Research Journal*, *22*(3), 347-367. https://doi.org/10.1177/14749041211065324

Salas, D. P. (2017). Challenges of Adult Education in the Contemporary Age: A Conceptual Approach. *Revista Republicana*, *22*, 47-64. https://doi.org/10.21017/rev.repub.2017.v22.a20

Saldivar, J. M. N. (2025). Crafting Minds: A Constructivist Blueprint for Philippine Education Reform. *Cognizance Journal of Multidisciplinary Studies*, *5*(1), 247-259. https://doi.org/10.47760/cognizance.2025.v05i01.020

Salimi, F. (2025). Aligning Policy and Practice: The World Bank's Approach to EdTech in Sub-Saharan Africa. *Policy Futures in Education*, *23*(6), 1134-1156. https://doi.org/10.1177/14782103251324275

Salinas-Atausinchi, Y., & Huaman-Lucana, R. (2021). Critical Pedagogy: An Emancipatory Alternative in the Neoliberal Context. *Revista Innova Educación*, *3*(4), 146-161. https://doi.org/10.35622/j.rie.2021.04.011.en

Salinas-Navarro, D. E., Vilalta-perdomo, E. L., Michel-Villarreal, R., & Montesinos, L. (2024). Designing Experiential Learning Activities With Generative Artificial Intelligence Tools for Authentic Assessment. *Interactive Technology and Smart Education*, *21*(4), 708-734. https://doi.org/10.1108/itse-12-2023-0236

Salinas, I., Guerrero, G., Satlov, M., & Hidalgo, P. (2022). Climate Change in Chile's School Science Curriculum. *Sustainability*, *14*(22), 15212. https://doi.org/10.3390/su142215212

Salvetti, F., Capshaw, T. L., Zanin, L., O'Connor, K. C., Zeng, Q., & Bertagni, B. (2023). The GW Mobile Learning Center: Mixed-Reality Within an Immersive and Interactive Learning Setting. *International Journal of Advanced Corporate Learning (Ijac)*, *16*(2), 93-108. https://doi.org/10.3991/ijac.v16i2.35737

Salvo-Garrido, S., Praihuén, S. S. M., Urra, O. V., & Cabrera, J. (2021). Desarrollo De Factores Protectores Y Obstaculizadores De La Resiliencia Socioeducativa De Niños Y Niñas, Según La Perspectiva Docente. *Revista Colombiana De Educación*, *1*(83). https://doi.org/10.17227/rce.num83-11026

Samadun, S., & Dwikoranto, D. (2022). Improvement of Student's Critical Thinking Ability Sin Physics Materials Through the Application of Problem-Based Learning. *Ijorer International Journal of Recent Educational Research*, *3*(5), 534-545. https://doi.org/10.46245/ijorer.v3i5.247

Samaroo, S., Cooper, E., & Green, T. C. (2013). Pedandragogy: A Way Forward to Self–engaged Learning. *New Horizons in Adult Education and Human Resource Development*, *25*(3), 76-90. https://doi.org/10.1002/nha3.20032

Samin, S. M. (2026). Curriculum, Teaching, and Assessment Gaps in Arabic Teacher Education: A Transformative Learning Integration Model (TLIM). https://doi.org/10.21203/rs.3.rs-10234093/v1

Sammel, A. J. (2014). Science as a Human Endeavour: Outlining Scientific Literacy and Rethinking Why We Teach Science. *Creative Education*, *05*(10), 849-857. https://doi.org/10.4236/ce.2014.510098

Sampson, D. Talent Management Practices in the State Health Services Sector in Ghana: A Case Study of Nurses in Three (3) Health Institutions. https://doi.org/10.23889/suthesis.63644

Samuel, A. (2025). Empowering the Future: Advancing Adult Education for a Responsible Future. *New Directions for Adult and Continuing Education*, *2025*(188), 7-14. https://doi.org/10.1002/ace.70010

Samyan, K. A. (2022). Digital Learning : Challenges and Issues. *Journal of Tikrit University for Humanities*, *29*(3, 1), 34-60. https://doi.org/10.25130/jtuh.29.3.1.2022.24

Sanatullova-Allison, E. (2014). Memory Retention in Second Language Acquisition and Instruction: Insights From Literature and Research. *Iafor Journal of Language Learning*, *1*(1). https://doi.org/10.22492/ijll.1.1.02

Sanchez-Ballado, T. (2025). Mind Over Chatter: A Metacognitive Checklist to Fine-Tune Your Teaching. *American String Teacher*, *75*(4), 33-37. https://doi.org/10.1177/00031313251371659

Sanchez, B. (2020). History, Research, and Theory to Practice: Scaffolding Framework for Graduate Online Courses. *Frontiers in Education Technology*, *3*(3), p9. https://doi.org/10.22158/fet.v3n3p9

Sánchez, C. J. R. (2016). Ambientes De Aprendizaje Colaborativo Y Creación Colectiva en 3 Weeks Bside Project Experience. *Revista Sobre*, *2*, 09-24. https://doi.org/10.30827/5046

Sánchez, E. A., & Fernández, M. C. (2019). Aplicación De Una Metodología De Enseñanza Que Usa El Modelamiento Matemático Enmarcado en La Teoría Del Ciclo De Kolb, Para Abordar El Contenido De Función Cuadrática. *Paideia Revista De Educación*(65), 133-160. https://doi.org/10.29393/pa65-14eaam20014

Sánchez, E. Y. P. (2022). La Analítica Del Aprendizaje Y Las Expectativas De Los Estudiantes Universitarios. *Academo Revista De Investigación en Ciencias Sociales Y Humanidades*, *9*(2), 151-164. https://doi.org/10.30545/academo.2022.jul-dic.4

Sandberg, B. (2024). Outcomes of Art-Based Leadership Development: A Qualitative Metasummary. *Behavioral Sciences*, *14*(8), 714. https://doi.org/10.3390/bs14080714

Sanders, J., Damen, M. A., & Dam, K. v. (2015). Are Positive Learning Experiences Levers for Lifelong Learning Among Low Educated Workers? *Evidence-Based HRM a Global Forum for Empirical Scholarship*, *3*(3), 244-257. https://doi.org/10.1108/ebhrm-01-2014-0002

Sandlin, J. A., Wright, R. R., & Clark, C. A. (2011). Reexamining Theories of Adult Learning and Adult Development Through the Lenses of Public Pedagogy. *Adult education quarterly*, *63*(1), 3-23. https://doi.org/10.1177/0741713611415836

Sandri, O., Holdsworth, S., & Thomas, I. (2018). Assessing Graduate Sustainability Capability Post-Degree Completion. *International Journal of Sustainability in Higher Education*, *19*(1), 2-14. https://doi.org/10.1108/ijshe-08-2016-0160

Sani, L. b. A., & Mohamad, M. (2023). Exploring the Theory, Benefits, and Challenges of Mobile Web-Based Learning (MWBL) for English as a Second Language (ESL) Learners. *International Journal of Academic Research in Progressive Education and Development*, *12*(3). https://doi.org/10.6007/ijarped/v12-i3/19003

Saniuk, S., & Grabowska, S. (2023). Knowledge and Skills Development for Implementing the Industry 5.0 Concept. *European Conference on Knowledge Management*, *24*(2), 1188-1195. https://doi.org/10.34190/eckm.24.2.1329

Sanséau, P. Y., & Ansart, S. (2013). Accreditation of Prior Experiential Learning as a Catalyst for Lifelong Learning: Analysis and Proposals Based on French Experiments. *Journal of International Education Research (Jier)*, *9*(4), 317-328. https://doi.org/10.19030/jier.v9i4.8083

Santamaría, J. S., & Vila, M. G. B. (2016). Desarrollando El Éxito Educativo Para Todos: Reflexiones, Propuestas Y Retos Conceptuales en Torno a La Equidad Educativa. *Reice Revista Iberoamericana Sobre Calidad Eficacia Y Cambio en Educación*, *12*(2). https://doi.org/10.15366/reice2014.12.2.005

Santhi, V., & Malathi, R. D. (2024). Activity-Based Learning for Literacy Development Among Non-Teaching Staff Strategies and Outcomes. *International Journal for Multidisciplinary Research*, *6*(6). https://doi.org/10.36948/ijfmr.2024.v06i06.31989

Santiago, C. d. S., Costa, A. C. F., & Lima, T. B. d. (2024). Qual O Estilo De Aprendizagem Predominante Nos Estudantes De Secretariado Executivo Bilíngue Da Universidade Federal Da Paraíba, No

Brasil? *Revista Tempos E Espaços Em Educação*, *17*(36), e20233. https://doi.org/10.20952/revtee.v17i36.20233

Santibáñez, G. S. (2016). Comprensión Evolutiva en El Análisis De Casos Desde El Modelo De R. Kegan [An Evolutive Comprehension in Cases Analysis From the Model of R. Kegan]. *Clínica Contemporánea*, *7*(1), 23-36. https://doi.org/10.5093/cc2016a2

Santos, L., Goodwin, W., & McArthur, M. (2025). Impact of Humble Leadership on Attitudes to Error Disclosure: The Mediating Role of Psychological Safety and Safety Climate in Australian Veterinary Practices. *BMJ Leader*, *9*(4), 394-399. https://doi.org/10.1136/leader-2024-001149

Santos, L. C. B., Franqueira, A. d. S., Lima, M. W. H., Schäfer, A. G., Francisco Luiz, G. d. C., & Dayse Karoline, S. S. d. C. (2024). Instructional Design and Its Contribution to the Effectiveness of Self-Directed Learning. https://doi.org/10.56238/sevened2023.008-018

Santos, L. M. D. (2019). Rural Public Health Workforce Training and Development: The Performance of an Undergraduate Internship Programme in a Rural Hospital and Healthcare Centre. *International Journal of Environmental Research and Public Health*, *16*(7), 1259. https://doi.org/10.3390/ijerph16071259

Santos, M. G., & McKinney, J. (2019). Putting the Literacy Back Into Health Literacy: Interventions in US Adult Literacy and English Language Programmes. 385-398. https://doi.org/10.51952/9781447344520.030

Santos, R. R. d., Santos, M. T. P., & Ciferri, R. R. (2023). ProvOER Model: A Provenance Model for Open Educational Resources. *Heliyon*, *9*(2), e13311. https://doi.org/10.1016/j.heliyon.2023.e13311

Sanyang, S., & Weller, G. (2026). Integrating Knowledge and Skills Transfer to Support Practice Readiness in Blended Learning Healthcare Management Education. https://doi.org/10.21203/rs.3.rs-8491437/v1

Saragih, S., Tjakraatmadja, J. H., & Pratama, A. P. (2023). Urgency of Managing Digitalization in Higher Education. *International Journal of Management Entrepreneurship Social Sciences and Humanities*, *6*(1), 14-24. https://doi.org/10.31098/ijmesh.v6i1.1339

Saraiva, M., Nogueiro, T., Jorge, F., & Chaleta, E. (2022). Erasmus and International Credit Mobility in a Contribution to the Sustainable Development Goals Associated With Quality Education and Reducing Inequalities. *Ijaedu- International E-Journal of Advances in Education*, 149-162. https://doi.org/10.18768/ijaedu.1141654

Šarčević, I. (2024). A New Agenda on Micro-Credentials: Filling the Gaps in the European Approach. *Journal of Learning for Development*, *11*(1), 181-186. https://doi.org/10.56059/jl4d.v11i1.1143

Sardar, A. (2023). Constructivism in Education: A Comprehensive Review of Theory and Research. *International Research Journal of Modernization in Engineering Technology and Science*. https://doi.org/10.56726/irjmets43553

Sarfo, F. K. (2016). Psychological Framework for Quality Technical and Vocational Education and Training in the Twenty-First Century. 1-26. https://doi.org/10.1007/978-3-319-17727-4_65-1

Saripudin, S., Djohar, A. a., Rohendi, D., & Abdullah, A. G. (2020). Developing Information Technology in OpenCourseWare: From Movements to Opportunities in Asia. *Indonesian Journal of Science and Technology*, *5*(3), 308-320. https://doi.org/10.17509/ijost.v5i3.24886

Sartoni, M., Marconi, F., Torracca, B., Pedonese, F., Nuvoloni, R., & Гуиди, A. (2025). Redesigning Food Handler Training: A Gamified Approach Tested in Italy's Large-Scale Retail. *Foods*, *14*(16), 2803. https://doi.org/10.3390/foods14162803

Sarwar, M. A., Saima, M., & Afshan, G. (2024). The Role of Artificial Intelligence in Shaping the Future of Education at Higher Secondary Level. *Journal of Education and Social Studies*, *5*(1), 34-45. https://doi.org/10.52223/jess.2024.5104

Sasere, O. B., & Makhasane, S. D. (2023). Adult Learning Theory Tenets: A Panacea to ICT Skills Gaps Among Educators in South Africa. *Journal of Education*(92), 58-75. https://doi.org/10.17159/2520-9868/i92a04

Šatienė, S. (2017). Self-Directed Learning in Later Life: Motives and Learning Competences of Retired Teachers. *E-Pedagogium*, *17*(2), 50-61. https://doi.org/10.5507/epd.2017.024

Sato, S. N., Condés, E., Rubio-Zarapuz, A., Dalamitros, A. A., Yañéz-Sepúlveda, R., Tornero-Aguilera, J. F., & Clemente-Suárez, V. J. (2023). Navigating the New Normal: Adapting Online and Distance Learning in the Post-Pandemic Era. *Education Sciences*, *14*(1), 19. https://doi.org/10.3390/educsci14010019

Savolainen, R. (2012). Expectancy-value Beliefs and Information Needs as Motivators for Task-based Information Seeking. *Journal of Documentation*, *68*(4), 492-511. https://doi.org/10.1108/00220411211239075

Sawchuk, P. H. (2007). Understanding Diverse Outcomes for Working-Class Learning: Conceptualising Class Consciousness as Knowledge Activity. *The Economic and Labour Relations Review*, *17*(2), 199-216. https://doi.org/10.1177/103530460701700212

Sayer, L. (2013). Communities of Practice, a Phenomenon to Explain Student Development In Community Nursing. *Primary Health Care Research & Development*, *15*(4), 430-440. https://doi.org/10.1017/s1463423613000455

Schaap, H., Khaled, A., Faber, M., Bijlsma, N., & Bruijn, E. d. (2023). Vocational Knowledge Development During Teacher-Student Interactions: Exploring Vocational Conversations. *Pedagogische Studiën*, *100*(1), 35-66. https://doi.org/10.59302/ps.v100i1.13990

Schalkwyk, S. v., Hafler, J. P., Brewer, T. F., Maley, M., Margolis, C. Z., McNamee, L., Meyer, I. S., Peluso, M. J., Schmutz, A. M., Spak, J. M., & Davies, D. (2019). Transformative Learning as Pedagogy for the Health Professions: A Scoping Review. *Academic Medicine*, *53*(6), 547-558. https://doi.org/10.1111/medu.13804

Schedlitzki, D. (2019). Developing Apprentice Leaders Through Critical Reflection. *Higher Education Skills and Work-Based Learning*, *9*(2), 237-247. https://doi.org/10.1108/heswbl-09-2018-0095

Schenck, A. (2024). Examining Relationships Between Technology and Critical Thinking: A Study of South Korean EFL Learners. *Education Sciences*, *14*(6), 652. https://doi.org/10.3390/educsci14060652

Schiller, E. (2022a). Developing Senior Language Learners' Autonomy With One-to-One Counselling for Autonomous Learning. https://doi.org/10.15476/elte.2022.080

Schiller, E. (2022b). *Developing Senior Language Learners' Autonomy With One-to-One Counselling for Autonomous Learning* EÖTVÖS LORÁND UNIVERSITY].

Schjetne, E. B. C., & Мельникова, Н. М. (2020). Community Work in the North — Merging Economy and Psychology. *Vestnik of Saint Petersburg University Psychology*, *10*(1), 46-60. https://doi.org/10.21638/spbu16.2020.104

Schlünzen, E. T. M., Bracken, S., & Schlünzen, K. (2024). Lesson Study and the Constructionist, Contextualized, and Meaningful Approach to Facilitate Universal Design for Learning Implementations. https://doi.org/10.31235/osf.io/423nj

Schmid, E., Scharnhorst, U., & Kammermann, M. (2020). Developing Two-Year Apprenticeships in Norway and Switzerland. *Vocations and Learning*, *14*(1), 55-74. https://doi.org/10.1007/s12186-020-09254-0

Schmidt, R., & Stumpe, B. (2025). Systematic Review of Mobile Augmented Reality Applications in Geography Education. *Review of Education*, *13*(1). https://doi.org/10.1002/rev3.70042

Schmidt, T. (2020). Reformed and Reduced: Vocational Education and Structural Oppression. *Power and Education*, *12*(3), 276-291. https://doi.org/10.1177/1757743820967027

Schmidt, T., Hodge, S., Byrne, A. L., Wirihana, L., Connor, J., Cole, R., Heidke, P., & Bradshaw, J. (2025). One Profession, Two Ways of Thinking: Challenges in Developing Australia's Nursing Workforce. *Nursing inquiry*, *32*(2). https://doi.org/10.1111/nin.70026

Schmoelz, A. (2023). Digital Humanism, Progressive Neoliberalism and the European Digital Governance System for Vocational and Adult Education. *Journal of Adult and Continuing Education*, *29*(2), 735-759. https://doi.org/10.1177/14779714231161449

Schnepfleitner, F. M., & Ferreira, M. (2021). Transformative Learning Theory – Is It Tıme to Add a Fourth Core Element? *Journal of Educational Studies and Multidisciplinary Approaches*, *1*(1), 40-49. https://doi.org/10.51383/jesma.2021.9

Scholtz, D., & Bester, M. (2018). Project-Based Learning: Panacea for Change or Old Wine in New Bottles? *South African Journal of Higher Education*, *32*(6). https://doi.org/10.20853/32-6-2967

Schön, D. (1983). *The Reflective Practitioner. How professionals think in action*. Temple Smith.

Schön, D. A. (1992). *The Reflective Practitioner: How Professionals Think in Action*. Routledge.

Schönwetter, D. J. (1993). Attributes of Effective Lecturing in the College Classroom. *Canadian Journal of Higher Education*, *23*(2), 1-18. https://doi.org/10.47678/cjhe.v23i2.183159

Schraw, G., Crippen, K. J., & Hartley, K. (2006). Promoting Self-Regulation in Science Education: Metacognition as Part of a Broader Perspective on Learning. *Research in Science Education*, *36*(1-2), 111-139. https://doi.org/10.1007/s11165-005-3917-8

Schreiber-Barsch, S. (2015). Adult and Continuing Education in France. https://doi.org/10.3278/37/0577w

Schreiber-Barsch, S., Bulugu, J. B., & Eble, L. (2023). Encountering Works by Nyerere and Freire. *European Journal for Research on the Education and Learning of Adults*, *14*(1), 79-96. https://doi.org/10.3384/rela.2000-7426.4420

Schröder, A., & Kruger, D. H. (2019). Social Innovation as a Driver for New Educational Practices: Modernising, Repairing and Transforming the Education System. *Sustainability*, *11*(4), 1070. https://doi.org/10.3390/su11041070

Schugurensky, D. (2011). *Paulo Freire*. Continuum.

Schugurensky, D., & Myers, J. P. (2008). Citizenship Education: Theory, Research and Practice. *Encounters in Theory and History of Education*, *4*. https://doi.org/10.24908/eoe-ese-rse.v4i0.655

Schüll, A. (2024). Design Elements for Gamified E-Learning: On Fueling Intrinsic Motivation by Digital Storytelling and Challenges. https://doi.org/10.5772/intechopen.1006548

Schulz, C. (2016). Applied Learning Educators Searching for a Pedagogical Model. 605-631. https://doi.org/10.4018/978-1-4666-9634-1.ch030

Schumann, M., & Baraka, A. A. E. (2024). Using REFLECT Rubric for Assessing the Reflective Use of E-Portfolios, a Qualitative Pilot Study at Alexandria Faculty of Medicine. *Journal of Health Professions Education and Innovation*, *1*(2), 21-35. https://doi.org/10.21608/jhpei.2024.251191.1005

Schumann, M., Lehmann, M., & Peters, H. (2025). Beyond Carrots and Sticks. Exploring Faculty Motivation to Join a Digital Health Professions Educator Program. *Frontiers in Medicine*, *12*. https://doi.org/10.3389/fmed.2025.1554011

Schunk, D. H. (1987). Peer Models and Children's Behavioral Change. *Review of Educational Research*, *57*(2), 149. https://doi.org/10.2307/1170234

Schunk, D. H. (2012). *Learning Theories, an Educational Perspective* (6th ed. ed.). Pearson Education Inc.

Schunk, D. H. (2020). *Learning Theories: An Educational Perspective* (8th ed.). Pearson.

Schürmann, E., & Beausaert, S. (2016). What Are Drivers for Informal Learning? *European journal of training and development*, *40*(3), 130-154. https://doi.org/10.1108/ejtd-06-2015-0044

Schwarz, V. S., & Sollenberger, L. (2024). Integrating Disability Into Equity-Focused General Education Teacher Preparation. *Frontiers in Education*, *9*. https://doi.org/10.3389/feduc.2024.1372380

Schweder, S., & Raufelder, D. (2021). Needs Satisfaction and Motivation Among Adolescent Boys and Girls During Self-directed Learning Intervention[☆]. *Journal of Adolescence*, *88*(1), 1-13. https://doi.org/10.1016/j.adolescence.2021.01.007

Scoppio, G., & Covell, L. (2016). Mapping Trends in Pedagogical Approaches and Learning Technologies: Perspectives From the Canadian, International, and Military Education Contexts. *Canadian Journal of Higher Education*, *46*(2), 127-147. https://doi.org/10.47678/cjhe.v46i2.186043

Scott, C. (2012). Collaborating With the Real World: Opportunities for Developing Skills and Values in Law Teaching. *Indiana Health Law Review*, *9*(2). https://doi.org/10.18060/16614

Scott, C. L. (2005). Multicultural Immersion Programming: An Adult Learning Context. *New Horizons in Adult Education and Human Resource Development*, *19*(4), 12-20. https://doi.org/10.1002/nha3.10230

Scott, I., Vegt, A. v. d., & Douglas, P. (2026). How Can We Prevent Clinical Deskilling When Using AI? *BMJ Quality & Safety*, bmjqs-2026-020130. https://doi.org/10.1136/bmjqs-2026-020130

Scott, J. L. (2022a). *Assessing New Dimensions of an Organization's Learning Culture* University of North Texas].

Scott, J. L. (2022b). Assessing New Dimensions of an Organization's Learning Culture. https://doi.org/10.12794/metadc1944371

Sebastian, T., & Giovanangeli, A. (2024). Reimagining Kinship Systems and Networks: Interconnectedness of Aboriginal Ecologies in Australia (Human and Land Rights). *Junctures the Journal for Thematic Dialogue*(24), 63-79. https://doi.org/10.34074/junc.24063

Sebastien, N. (2017). Effective Educational Leadership Training for Improving Student Learning Outcomes. *Journal of Interdisciplinary Teacher Leadership*, *2*(2), 27-31. https://doi.org/10.46767/kfp.2016-0017

Sebova, D. (2020). *Understanding the "What" and the "How" of Developing Future Leaders in Global Organizations : An Instrumental Case Study of a First-Level Leader Development Program in a Large Global Organization* Northeastern University].

Sebti, L., & Iqtadar, S. (2025). "I Still Want to Challenge My Special Ed Kids, but I Don't Know How!": Secondary Teachers' Perceptions and Practices of Inclusive Math and Science Education in the United States. *Journal of Research in Special Educational Needs*, *25*(4), 930-944. https://doi.org/10.1111/1471-3802.70021

Secretariat, C. (2022). Technical and Vocational Education and Training (TVET) Self-Assessment Toolkit: Participant Manual. https://doi.org/10.14217/comsec.1082

Seevaratnam, V., Gannaway, D., & Lodge, J. M. (2023). Design Thinking-Learning and Lifelong Learning for Employability in the 21st Century. *Journal of Teaching and Learning for Graduate Employability*, *14*(1), 167-186. https://doi.org/10.21153/jtlge2023vol14no1art1631

Segovia, J. D. (2019). Empleo Y Aprendizaje Permanente en España: Implicaciones en La Formación De Las Personas Adultas. *Curriculo Sem Fronteiras*, *19*(3). https://doi.org/10.35786/1645-1384.v19.n3.25

Selçuk, E., Bozak, A., Özkaya, A., Doğan, Y., & SarigÖZ, O. (2026). Mapping Lifelong Learning Research: A Holistic Text Mining Analysis of Educational Trends and Gaps (1980–2024). *Sage Open*, *16*(2). https://doi.org/10.1177/21582440261455217

Selvaratnam, R., & Sankey, M. (2021). The State of Micro-Credentials Implementation and Practice In Australasian Higher Education. *Open Praxis*, *13*(2), 228. https://doi.org/10.5944/openpraxis.13.2.130

Selvaratnam, R., & Sankey, M. (2025). Industry–University Partnerships for Micro-credentials and Workforce Development.

Selvaratnam, R., & Sankey, M. (2025). An Integrative Literature Review of the Implementation of Micro-Credentials in Higher Education: Implications for Practice in Australasia. *Journal of Teaching and Learning for Graduate Employability*, *12*(1), 1-17. https://doi.org/10.21153/jtlge2021vol12no1art942

Semi, M. M. A., Utsho, M. R., Kamal, M. B., Sizan, A. A., Tasnim, A. F., Yeasmin, S., & Parvin, M. R. (2026). Artificial Intelligence in Public Health Education: A Scoping Review of Workforce Competency Development. *Health Science Reports*, *9*(3). https://doi.org/10.1002/hsr2.72066

Senge, P. M. (1990). *The fifth discipline: The art and practice of the learning organization*. Doubleday/Currency.

Senge, P. M. (1992). *The Fifth Discipline Field Book*. Doubleday/Currency.

Sequera, J. (2024). <scp>STRUGGLING WITH THE DIGITAL NOMAD</scp>: Transnational Teleworkers as the New 'Creative Class' in the Urban Marketplace? *International Journal of Urban and Regional Research*, *49*(1), 204-213. https://doi.org/10.1111/1468-2427.13293

Serim, A. B. Ö., & Bilgili, A. (2023). Digitized Higher Education: Digital Transformation in Education From a Bibliometric Perspective. *The Journal of International Scientific Researches*, *8*(3), 531-549. https://doi.org/10.23834/isrjournal.1359200

Sethi, A., Schofield, S., McAleer, S., & Ajjawi, R. (2018). The Influence of Postgraduate Qualifications on Educational Identity Formation of Healthcare Professionals. *Advances in Health Sciences Education*, *23*(3), 567-585. https://doi.org/10.1007/s10459-018-9814-5

Setiyo, S., Riyanto, Y., & Nurlaila, L. (2018). Comparison of Self Efficacy on Cognitive Apprenticeship and on the Job Training. https://doi.org/10.2991/icei-18.2018.139

Sever, S. D., & Tok, E. (2022). Education for Sustainable Development in Qatar. 329-347. https://doi.org/10.1007/978-981-19-7398-7_17

Sevillano, M. Á. P., Cadavieco, J. F., Rodríguez, M. S. G., & Castro, S. S. (2023). Formación Docente en Educación De Adultos. Un Estudio De Casos Desde Un Enfoque Multicontextual. *Aula Abierta, 52*(3), 213-220. https://doi.org/10.17811/rifie.52.3.2023.213-220

Seyi-Oderinde, D. R. (2021). Rethinking Mental Health Literacy Programmes for Enhanced Help-Seeking Behaviour Among Young Male Adults. *Interdisciplinary Journal of Rural and Community Studies, 3*(2), 41-50. https://doi.org/10.51986/ijrcs-2021.vol3.02.05

Shadiev, R., Hwang, W. Y., Huang, Y. M., & Liu, T. Y. (2017). Facilitating Application of Language Skills in Authentic Environments With a Mobile Learning System. *Journal of Computer Assisted Learning, 34*(1), 42-52. https://doi.org/10.1111/jcal.12212

Shadiev, R., Wang, X., Halubitskaya, Y., & Huang, Y. M. (2022). Enhancing Foreign Language Learning Outcomes and Mitigating Cultural Attributes Inherent in Asian Culture in a Mobile-Assisted Language Learning Environment. *Sustainability, 14*(14), 8428. https://doi.org/10.3390/su14148428

Shah, M. A. (2022). Teachers as Reflective Practitioners: From Individualism to Vygotskian Social Constructivism. *Alberta Journal of Educational Research, 68*(3), 297-307. https://doi.org/10.55016/ojs/ajer.v68i3.68598

Shah, R. K. (2019). Effective Constructivist Teaching Learning in the Classroom. *Shanlax International Journal of Education, 7*(4), 1-13. https://doi.org/10.34293/education.v7i4.600

Shah, S. T. H., Shah, S. M. A., & El-Gohary, H. (2022). Nurturing Innovative Work Behaviour Through Workplace Learning Among Knowledge Workers of Small and Medium Businesses. *Journal of the Knowledge Economy, 14*(4), 3616-3639. https://doi.org/10.1007/s13132-022-01019-5

Shahbaz, S. (2021). Schools as Learning Organizations: Impact of Age &Amp; Experience of Preschool Teachers Understanding About Disciplines of Learning Organization in Peter Senge's Model. *Journal for Educators Teachers and Trainers, 12*(01). https://doi.org/10.47750/jett.2021.12.01.005

Shahid, W., Rauf, S., & Arshad, M. U. (2023). The Need and Importance of Field Work Practice in Field Settings for Social Workers. *RHSS*. https://doi.org/10.7176/rhss/13-10-04

Shaik, N. (2022). Transforming Teaching and Learning in Early Childhood Care and Education During COVID-19 in a Poor Community of the Cape Flats, South Africa. *Early Childhood Education Journal, 51*(5), 791-800. https://doi.org/10.1007/s10643-022-01329-y

Shakeeb, S. (2020). Remote Online Teaching (ROT) in Response to COVID-19 Pandemic: Exploring Instructors' Experience. *The Maldives National Journal of Research, 8*(1). https://doi.org/10.62338/dvfnpj73

Shaketange, L. (2018). Challenges and Opportunities for Implementing Recognition of Prior Learning at the University of Namibia. *Creative Education, 09*(13), 2070-2087. https://doi.org/10.4236/ce.2018.913150

Shanks, R. (2023). Informal Coaching and Mentoring and an "Informal Turn" in Teacher Professional Learning. *International Journal of Mentoring and Coaching in Education, 12*(4), 440-452. https://doi.org/10.1108/ijmce-10-2022-0091

Shanthi, A., & Purwarno, P. (2022). Communicative Guanxi: Making Online Discussion Forums Come Alive. *International Journal of Social Science Research, 10*(2), 179. https://doi.org/10.5296/ijssr.v10i2.19953

Sharaf, E. A. E.-S. i. d. (2021). Utilisation De L'apprentissage Expérientiel Pour Développer Quelques Compétences De La Lecture Active en Français Et L'engagement Cognitif Chez Les Étudiants Du Cycle Secondaire. بحوث فى تدريس اللغات, 17(17), 328-453. https://doi.org/10.21608/ssl.2021.84894.1087

Sharaf, E. A. E. (2020). Conception d'Un Environnement d'Apprentissage Ubiquitaire Pour Développer La Compétence De Production Écrite Du Texte Argumentatif Auprès Des Étudiants De La Section De Français À La Faculté De Pédagogie De Kafr El-Cheikh. المجلة التربوية لكلية التربية بسوهاج, 79(79), 27-81. https://doi.org/10.21608/edusohag.2020.116723

Sharma, A. K., & Sharma, R. (2025). Governance in the Age of Artificial Intelligence: A Comparative Analysis of Policy Framework in BRICS Nations. *Ai Magazine, 46*(2). https://doi.org/10.1002/aaai.70010

Sharma, H., Jain, V., Mogaji, E., & Babbilid, A. S. (2024). Blended Learning and Augmented Employability: A Multi-Stakeholder Perspective of the Micro-Credentialing Ecosystem in Higher Education. *International Journal of Educational Management*, *38*(4), 1021-1044. https://doi.org/10.1108/ijem-12-2022-0497

Sharma, N., & Sharma, P. (2021). Constructivism: Implication for Research and Effective Learning. *BSSS Journal of Education*, *10*(1), 28-35. https://doi.org/10.51767/je1003

Sharma, R. C., & Kohli, D. (2023). Embracing Digital Transformation for Sustainable Development: The Synergy Between Learning Organizations, Human Resource Management, and Service Quality. *Journal of Chinese Human Resource Management*, *14*(3), 3-17. https://doi.org/10.47297/wspchrmwsp2040-800501.20231403

Sharp, S., Founder, S. S., Speaker, S. S., Facilitator, Speaker, M., Founder, S., & Speaker, S. (2023). Discretionary Effort Is the Holy Grail in the Workplace, but How Do You Cultivate It? *Journal of Psychology and Neuroscience*. https://doi.org/10.47485/2693-2490.1074

Sharpe, D., Rajabi, M., Chileshe, C., Joseph, S. M., Sesay, I., Williams, J. H., & Sait, S. (2021). Mental Health and Wellbeing Implications of the COVID-19 Quarantine for Disabled and Disadvantaged Children and Young People: Evidence From a Cross-Cultural Study in Zambia and Sierra Leone. *BMC Psychology*, *9*(1). https://doi.org/10.1186/s40359-021-00583-w

Shava, G. N. (2020). Quality Education for Sustainable Development in Zimbabwean Higher Education. 1-26. https://doi.org/10.1007/978-3-030-43042-9_6-1

Shay, S., Wolff, K., & Clarence-Fincham, J. (2016). Curriculum Reform in South Africa: More Time for What? *Critical Studies in Teaching and Learning*. https://doi.org/10.14426/cristal.v4i1.1971

Sheikh, A., & Ahmad, F. (2015). Forensic Medicine. *The Professional Medical Journal*, *22*(07), 838-843. https://doi.org/10.29309/tpmj/2015.22.07.1169

Sheinman, N., & Russo-Netzer, P. (2021). Mindfulness in Education: Insights Towards an Integrative Paradigm. 609-642. https://doi.org/10.1007/978-3-030-64537-3_24

Shelton-Strong, S. J. (2020). Advising in Language Learning and the Support of Learners' Basic Psychological Needs: A Self-Determination Theory Perspective. *Language Teaching Research*, *26*(5), 963-985. https://doi.org/10.1177/1362168820912355

Shelton-Strong, S. J. (2022). Sustaining Language Learner Well-Being and Flourishing: A Mixed-Methods Study Exploring Advising in Language Learning and Basic Psychological Need Support. *Psychology of Language and Communication*, *26*(1), 415-449. https://doi.org/10.2478/plc-2022-0020

Shepherd, J. (2023). Unlocking the Future of Nursing Education and Continuing Professional Development by Embracing Generative Artificial Intelligence and Advanced Language Models. *Interdisciplinary Journal of Partnership Studies*, *10*(2), 4. https://doi.org/10.24926/ijps.v10i2.5614

Sherlock, J. J., & Nathan, M. L. (2008). How Power Dynamics Impact the Content and Process of Nonprofit CEO Learning. *Management learning*, *39*(3), 245-269. https://doi.org/10.1177/1350507608090876

Sherwood, G. (2015). Perspectives: Nurses' Expanding Role in Developing Safety Culture: Quality and Safety Education for Nurses – Competencies in Action. *Journal of Research in Nursing*, *20*(8), 734-740. https://doi.org/10.1177/1744987115621142

Shi, L. (2025). Global Perspectives on AI Competence Development: Analyzing National AI Strategies in Education and Workforce Policies. *Human Resource Development Review*, *24*(4), 447-476. https://doi.org/10.1177/15344843251332360

Shi, Y., & Lin, X. (2021). Exploring the Characteristics of Adults' Online Learning Activities: A Case Study of EdX Online Institute. *Research in Learning Technology*, *29*. https://doi.org/10.25304/rlt.v29.2622

Shi, Y., Ma, D., Zhang, J., & Chen, B. (2021). In the Digital Age: A Systematic Literature Review of the E-Health Literacy and Influencing Factors Among Chinese Older Adults. *Journal of Public Health*, *31*(5), 679-687. https://doi.org/10.1007/s10389-021-01604-z

Shihab, S. R., Sultana, N., & Samad, A. (2023). Pedagogy Designing With Augmented Reality: A Paradigm Shift in Educational Approaches. *Interdiciplinary Journal and Hummanity (Injurity)*, *2*(11), 878-892. https://doi.org/10.58631/injurity.v2i11.136

Shikuku, D. N., Bar-Zeev, S., Ladur, A. N., Allott, H., Mwaura, C., Nandikove, P., Uyara, A., Tallam, E., Ndirangu-Mugo, E., Waweru, L., Nyaga, L., Bashir, I., Bedwell, C., & Ameh, C. (2024). Experiences, Barriers and Perspectives of Midwifery Educators, Mentors and Students Implementing the Updated Emergency Obstetric and Newborn Care- Enhanced Pre-Service Midwifery Curriculum in Kenya: A Nested Qualitative Study. https://doi.org/10.21203/rs.3.rs-4320477/v1

Shileche, A., VanLeeuwen, C., VanLeeuwen, J., Walton, C., & Ogola, L. K. (1970). Impact of Agricultural Development Project on Participating Women Farmers in Naari, Kenya. *Cradle of Knowledge African Journal of Educational and Social Science Research (The)*, *10*(3), 112-120. https://doi.org/10.4314/ajessr.v10i3.2

Shin, H., & Rim, D. (2023). Development and Assessment of a Curriculum Model for Virtual Simulation in Nursing: Curriculum Development and Pilot-Evaluation. *BMC Medical Education*, *23*(1). https://doi.org/10.1186/s12909-023-04283-4

Shoaib, L. A., Rohani, M. M., Safii, S. H., Idris, N., & Hussin, R. (2025). A Peek Under the Mask: Exploring Dental Students' Experiences Through focus Group Discussions. *BMC Medical Education*, *25*(1). https://doi.org/10.1186/s12909-025-07902-4

Shohel, M. M. C., Ashrafuzzaman, M., Ahmed, S. B., Tasnim, N., Akter, T., Islam, G. M. R., Siddik, M. A. B., & Mitu, S. R. (2024). Participatory Pedagogical Approaches in Higher Education: Understanding From the Practices in Bangladesh, the UK and the USA. https://doi.org/10.5772/intechopen.114070

Shonfeld, M., Cotnam-Kappel, M., Judge, M., Ng, C., Ntebutse, J. G., Williamson-Leadley, S., & Yildiz, M. N. (2021). Learning in Digital Environments: A Model for Cross-Cultural Alignment. *Educational Technology Research and Development*, *69*(4), 2151-2170. https://doi.org/10.1007/s11423-021-09967-6

Shore, A., & Dinning, T. (2023). Developing Student's Skills and Work Readiness: An Experiential Learning Framework. *Journal of Work-Applied Management*, *15*(2), 188-199. https://doi.org/10.1108/jwam-02-2023-0016

Shouman, D., Itani, A., & Kawtharani, A. (2023). Preparing Students With Twenty First Century Skills for the Future Post-Pandemic Era. https://doi.org/10.5772/intechopen.1001993

Shousha, A. I. (2021a). Language Difficulties Faced by Saudi Diploma Students at King Abdulaziz University: A Case Study. https://doi.org/10.31235/osf.io/bgc63

Shousha, A. I. (2021b). Language Difficulties Faced by Saudi Diploma Students at King Abdulaziz University: A Case Study. *Arab World English Journal*, *12*(2), 142-157. https://doi.org/10.24093/awej/vol12no2.10

Shrivastava, A. (2018). Using Connectivism Theory and Technology for Knowledge Creation in Cross-Cultural Communication. *Research in Learning Technology*, *26*(0). https://doi.org/10.25304/rlt.v26.2061

Shujahat, M., & Wang, M. (2022). The Online Communities of Practices and Faculty Members' Professional Development: A Critical Literature Review. *Multidisciplinary Business Review*, *15*(2), 54-64. https://doi.org/10.35692/07183992.15.2.6

Shutaleva, A. (2023). Ecological Culture and Critical Thinking: Building of a Sustainable Future. *Sustainability*, *15*(18), 13492. https://doi.org/10.3390/su151813492

Siacor, K. H., Ng, B., & Liu, W. C. (2024). Fostering Student Motivation and Engagement Through Teacher Autonomy Support: A Self-Determination Theory Perspective. *International Journal of Instruction*, *17*(2), 583-598. https://doi.org/10.29333/iji.2024.17232a

Sidiropoulos, L., Wex, I., & Sibley, J. (2013). Supporting the Sustainability Journey of Tertiary International Students in Australia. *Australian Journal of Environmental Education*, *29*(1), 52-79. https://doi.org/10.1017/aee.2013.15

Siemens, G. (2005a). Connectivism: A Learning Theory for the Digital Age. *International Journal of Instructional Technology and Distance Learning*, *2*(1), 3-10. http://www.itdl.org/Journal/Jan_05/article01.htm

Siemens, G. (2005b). Connectivism: A Learning Theory for the Digital Age. http://www.elearnspace.org/Articles/connectivism.htm

Sifakis, N. (2007). The Education of Teachers of English as a Lingua Franca: A Transformative Perspective. *International Journal of Applied Linguistics*, *17*(3), 355-375. https://doi.org/10.1111/j.1473-4192.2007.00174.x

Sik-Lányi, C. (2014). The Thousand Faces of Virtual Reality. https://doi.org/10.5772/58637

Sikora, A., Titova-Kosturkova, T., Janevska, G., Kostov, M., Halunga, S., Suciu, G., & Georgiev, G. (2023). EREMI: An Innovative Interdisciplinary Approach for Higher Education in Resource Efficient Manufacturing Environments. *Sustainability*, *15*(13), 10244. https://doi.org/10.3390/su151310244

Silva, D. D., Jayatilleke, S., El-Ayoubi, M., Issadeen, Z., Moraliyage, H., & Mills, N. (2024). The Human-Centred Design of a Universal Module for Artificial Intelligence Literacy in Tertiary Education Institutions. *Machine Learning and Knowledge Extraction*, *6*(2), 1114-1125. https://doi.org/10.3390/make6020051

Silva, I. (2018). How Learning Theories Can Be Applied in Accounting Education. *Millenium - Journal of Education Technologies and Health*(7), 13-20. https://doi.org/10.29352/mill0207.01.00181

Silva, N. D., Tetrick, L. E., Slack, K. J., Etchegaray, J. M., Latting, J. K., Beck, M. H., & Jones, A. P. (2003). Is There a Relationship Between Employee Perceptions of Organizational Learning Practices and Employee Performance? *The Psychologist-Manager Journal*, *6*(1), 104-116. https://doi.org/10.1037/h0095921

Sim, E., & Bierema, L. L. (2025). Reexamining Learning Transfer Through a Critical Lens: A Conceptual Study on Key Factors Shaping the Transfer of "Critical" Adult Learning. *Advances in Developing Human Resources*, *27*(4), 136-165. https://doi.org/10.1177/15234223251382407

Sima, V., Gheorghe, I. G., Subić, J., & Nancu, D. (2020). Influences of the Industry 4.0 Revolution on the Human Capital Development and Consumer Behavior: A Systematic Review. *Sustainability*, *12*(10), 4035. https://doi.org/10.3390/su12104035

Simanjuntak, F. N. (2019). Facilitating the Need for Self-Esteem to Approach the Adolescents' Chemistry Learning Style. *People International Journal of Social Sciences*, *5*(2), 765-777. https://doi.org/10.20319/pijss.2019.52.765777

Simel, S. (2020). Education for a Positive Self-Image in a Contemporary School. *Journal of Education Culture and Society*, *4*(2), 108-115. https://doi.org/10.15503/jecs20132.108.115

Singh, A. K. (2025). From Policy to Practice: Challenges and Pathway for Advancing Multilingual Education Through India's National Education Policy 2020. *Policy Futures in Education*, *23*(4), 746-765. https://doi.org/10.1177/14782103251320532

Singh, M. (2015a). Coordination and Stakeholder Interests and Motives. 89-102. https://doi.org/10.1007/978-3-319-15278-3_5

Singh, M. (2015b). Features of Best Practice From Country Examples. 103-158. https://doi.org/10.1007/978-3-319-15278-3_6

Singh, M. (2015c). Global Perspectives on Recognising Non-Formal and Informal Learning. https://doi.org/10.1007/978-3-319-15278-3

Singh, M. (2015d). Introduction. 1-16. https://doi.org/10.1007/978-3-319-15278-3_1

Singh, M. (2015e). Policy and Legislative Environment. 47-61. https://doi.org/10.1007/978-3-319-15278-3_3

Singh, M. (2015f). Recognition, Validation and Accreditation of Non-formal and Informal Learning: International Perspectives and Trends. In. UNESCO Institute for Lifelong Learning.

Singh, M. (2015g). RVA's Role in Education, Working Life and Society. 63-88. https://doi.org/10.1007/978-3-319-15278-3_4

Singh, M. (2015h). Sharing Learning: Cross-Country Observations. 159-191. https://doi.org/10.1007/978-3-319-15278-3_7

Singh, M. (2022). RVA in Asia and the Pacific Region. 39-61. https://doi.org/10.1007/978-981-19-2072-1_3

Singh, R. (2016). The Impact of Intrinsic and Extrinsic Motivators on Employee Engagement in Information Organizations. *Journal of Education for Library and Information Science*, *57*(2), 197-206. https://doi.org/10.3138/jelis.57.2.197

Singh, R. K. (2021). Clinical Legal Education—A Robust Instrument for Attainment of Justice: An Indian Perspective. *Asian Journal of Legal Education*, *9*(1), 7-22. https://doi.org/10.1177/09717218211030784

Singh, S. (2024). Onkraj Izobraževanja. *Andragoška Spoznanja*, *30*(2), 39-61. https://doi.org/10.4312/as/16287

Singh, S., & Ehlers, S. (2020). Employability as a Global Norm: Comparing Transnational Employability Policies of OECD, ILO, World Bank Group, and UNESCO. 131-147. https://doi.org/10.36253/978-88-5518-155-6.09

Singh, S., Molina-Naar, M., & Ehlers, S. (2020). Policies for Professionalisation in Adult Learning and Education: A Comparative Study From India, Colombia and Denmark. *Andragoske Studije*(2), 33-61. https://doi.org/10.5937/andstud2002033s

Singh, S. S. (2019). Adult Education Efforts for Refugees: A Case of a Border City in Turkey. Nr 19 (2018): DYSKURSY MŁODYCH. https://doi.org/10.34768/dma.vi19.45

Singh, S. V. (2022). The Impact of AI on Teaching and Learning in Higher Education Technology. *Journal of Higher Education Theory and Practice*, *22*(13). https://doi.org/10.33423/jhetp.v22i13.5514

Siok, T. H., Sim, M. S., & Rahmat, N. H. (2023). Motivation to Learn Online: An Analysis From McClelland's Theory of Needs. *International Journal of Academic Research in Business and Social Sciences*, *13*(3). https://doi.org/10.6007/ijarbss/v13-i3/16471

Sioukas, A. (2022). Constructivism and the Student-Centered Entrepreneurship Classroom: Learning Avenues and Challenges for US College Students. *Industry and Higher Education*, *37*(4), 473-484. https://doi.org/10.1177/09504222221135311

Siregar, R. A., Lengkanawati, N. S., & Damayanti, I. L. (2024). Metacognitive Strategies in Mediating Efl Adult Learners' Self-Regulated Language Learning: a Systematic Review. *LLT Journal a Journal on Language and Language Teaching*, *27*(1), 252-272. https://doi.org/10.24071/llt.v27i1.5502

Sirko, R., Bezverkhnia, H., Zaverukha, O., Чупахіна, C., & Kyrsta, N. (2020). Motivation to Get a Second Higher Education: Psychological and Pedagogical Aspect. *International Journal of Learning Teaching and Educational Research*, *19*(2), 188-202. https://doi.org/10.26803/ijlter.19.2.11

Sistermans, I. J. (2020). Integrating Competency-Based Education With a Case-Based or Problem-Based Learning Approach in Online Health Sciences. *Asia Pacific Education Review*, *21*(4), 683-696. https://doi.org/10.1007/s12564-020-09658-6

Siswanto, S., & Ngadri, N. (2022). Al-Azhar and Moderate: Exploring the Role of Alumni Al-Azhar of Egypt in Strengthening Ummah Unity in Indonesia Through the Concept of "Wasathiyyah". *Religio Education*, *2*(2), 116-132. https://doi.org/10.17509/re.v2i2.52054

Sithole, M. S., & Shokane, A. L. (2023). Workplace Learning in the Contemporary Supervision Landscape: The Case of Supervision in a Social Service Organisation. *Social Work/Maatskaplike Werk*, *59*(2). https://doi.org/10.15270/59-2-1125

Sjöberg, S., Többe-Schukalla, M., Singh, S., & Martinss, K.-H. (2016). Community Work in Germany and Sweden in Context of Changing Welfare Models. *International Social Work*, *61*(4), 553-570. https://doi.org/10.1177/0020872816666622

Skrefsrud, T. A. (2024). Intercultural Dialogue in Diverse Classrooms: Debating the Socratic Dialogue Method From a Postcolonial Perspective. *Religions*, *15*(1), 98. https://doi.org/10.3390/rel15010098

Slade, B. (2013). Professional Learning in Rural Practice: A Sociomaterial Analysis. *Journal of Workplace Learning*, *25*(2), 114-124. https://doi.org/10.1108/13665621311299799

Slade, S., & Prinsloo, P. (2013). Learning Analytics. *American Behavioral Scientist*, *57*(10), 1510-1529. https://doi.org/10.1177/0002764213479366

Slettli, V. (2019). Developing Entrepreneurial Mindset for Transformational Entrepreneurship: The Case of Nordic Transformative Learning Circles. *Journal of Entrepreneurship Management and Innovation*, *15*(4), 77-105. https://doi.org/10.7341/20191543

Sligo, F., Tilley, E., & Murray, N. (2011). Do Apprentices' Communities of Practice Block Unwelcome Knowledge? *Education + Training*, *53*(4), 297-309. https://doi.org/10.1108/00400911111138460

Slowey, M., Schuetze, H. G., & Zubrzycki, T. (2020). Implications of Migration and Ageing Populations for Inclusion and Equality in Higher Education and Lifelong Learning. 3-21. https://doi.org/10.1007/978-3-030-28227-1_1

Smak, M. (2022). Models of Private Sector Involvement in Vocational Education and Training. *European Journal of Teaching and Education*, *4*(1), 37-47. https://doi.org/10.33422/ejte.v4i1.746

Smith, A. K., Jaussen, C., & Nestor, K. R. (2020). The Seminar Goes to Work: Towards a Theory of Collaborative Literacy. *Industry and Higher Education*, *34*(4), 263-271. https://doi.org/10.1177/0950422220912183

Smith, C. F., Hall, S., Border, S., Adds, P. J., & Finn, G. M. (2015). Interprofessional Anatomy Education in the <scp>U</Scp>nited <scp>K</Scp>ingdom and <scp>I</Scp>reland: Perspectives From Students and Teachers. *Anatomical Sciences Education*, *8*(4), 360-370. https://doi.org/10.1002/ase.1548

Smith, E. (2010). A Review of Twenty Years of Competency-based Training in the Australian Vocational Education and Training System. *International Journal of Training and Development*, *14*(1), 54-64. https://doi.org/10.1111/j.1468-2419.2009.00340.x

Smith, J. (2019). The Intersection of Learning and Work: An Examination of an Administrative License Program. https://doi.org/10.3102/1439745

Smith, K., Maynard, N., Berry, A., Stephenson, T., Spiteri, T., Corrigan, D., Mansfield, J., Ellerton, P., & Smith, T. (2022). Principles of Problem-Based Learning (PBL) in STEM Education: Using Expert Wisdom and Research to Frame Educational Practice. *Education Sciences*, *12*(10), 728. https://doi.org/10.3390/educsci12100728

Smith, M. K. (2002). Malcolm Knowles, informal adult education, self-direction and andragogy. *The Encyclopedia of Informal Education*.

Smith, P. J. (2003). Workplace Learning and Flexible Delivery. *Review of Educational Research*, *73*(1), 53-88. https://doi.org/10.3102/00346543073001053

Smith, T. J., Hugh, C., & Fontechia, S. A. (2023). Unemployment and Underemployment of People With Disabilities: An Untapped Resource Within the Global Economy. https://doi.org/10.5772/intechopen.1003706

Smuha, N. A. (2020). Trustworthy Artificial Intelligence in Education: Pitfalls and Pathways. *SSRN Electronic Journal*. https://doi.org/10.2139/ssrn.3742421

Smythe, F. (2020). Language Inclusiveness in Education: Implications for Immigrant Students in France and New Zealand. *New Zealand Journal of Educational Studies*, *55*(1), 215-246. https://doi.org/10.1007/s40841-020-00164-2

Snelling, P. (2023). Am I My Students' Nurse? Reflections on the Nursing Ethics of Nursing Education. *Nursing Ethics*, *31*(1), 52-64. https://doi.org/10.1177/09697330231193858

Snyman, M., & Berg, G. v. d. (2017). The Significance of the Learner Profile in Recognition of Prior Learning. *Adult education quarterly*, *68*(1), 24-40. https://doi.org/10.1177/0741713617731809

Sofo, F. (2011). Developing Communities: Serving ACE Through Tertiary Education. *Education + Training*, *53*(6), 484-498. https://doi.org/10.1108/00400911111159458

Şoitu, C. T., Grecu, S.-P., & Asiminei, R. (2022). Health Security, Quality of Life and Democracy During the COVID-19 Pandemic: Comparative Approach in the EU-27 Countries. *International Journal of Environmental Research and Public Health*, *19*(21), 14436. https://doi.org/10.3390/ijerph192114436

Sommer, M., Andvig, E., Riis, R., & Bongaardt, R. (2022). Becoming Relationally Spacious: Mental Healthcare Students' Experiences With the Immunity to Change Workshop. *Journal of Transformative Education*, *21*(3), 354-370. https://doi.org/10.1177/15413446221129871

Son, E. H., & Penry, T. (2022). Variations in Project-Based Course Design. *Journal of Problem Based Learning in Higher Education*. https://doi.org/10.54337/ojs.jpblhe.v10i1.6821

Song, Y.-J., Chang, W., & Chang, J. (2018). The Effective Factors of Professional Learning : Study on Accounting Firms in Korea. *Journal of Asian Finance Economics and Business*, *5*(2), 81-94. https://doi.org/10.13106/jafeb.2018.vol5.no2.81

Soomro, R. B. K., Soomro, A. B., Sahito, Z., & Memon, I. (2025). Human Rights Education &Amp;amp; Sustainable Development Goal 4: Highlighting Intersections &Amp;amp; Synergies Through Mixed Method Research. https://doi.org/10.21203/rs.3.rs-6365885/v1

Sørensen, K. (2024). SDG4, Education: Education as a Lever for Sustainable Development. 88-103. https://doi.org/10.1017/9781009467766.006

Sorenson, D. L. (1987). The Cummins Engine Company Dislocated Worker Project. 15-30. https://doi.org/10.17848/9780880996181.ch2

Soroa, M., Aizpurua, A., & Lameirinhas, J. (2023). Resiliencia Del Alumnado De Formación Profesional: Relación Con El Apoyo Del Profesorado, El Apoyo Social Percibido Y El Autoconcepto. *Escritos De Psicología / Psychological Writings*, *16*(2), 123-131. https://doi.org/10.24310/escpsi.16.2.2023.16322

Sorte, P. B., & Kim, N. J. (2023). Integrating Augmented Reality and Problem-Based Learning Into English Language Teaching Through Instructional Design. *Revista Tempos E Espaços Em Educação*, *16*(35), e19073. https://doi.org/10.20952/revtee.v16i35.19073

Sousa, C., Pereira, J., & Casimiro, C. (Eds.). (2024). *No Barriers: Strategies and Best Practices for the Employment of Individuals With Intellectual Disability*. https://doi.org/10.24140/nobarriers.v1.

Soussi, S., Naceur, A., Berger, D., Ghérissi, A., Belgacem, A., & Zairi, I. (2021). Impact De L'apprentissage Systémique Sur La Motivation À Apprendre Des Étudiants Dans Un Contexte D'apprentissage en Milieu Clinique. *Revue Internationale De Pédagogie De L'enseignement Supérieur*, *37*(2). https://doi.org/10.4000/ripes.2900

Spatafora, N. (2021). Education and Health for Inclusiveness. *Imf Working Paper*, *2021*(060), 1. https://doi.org/10.5089/9781513571539.001

Spiel, C., Schwartzman, S., Busemeyer, M. R., Cloete, N., Drori, G. S., Lassnigg, L., Schober, B., Schweisfurth, M., Verma, S., Bakarat, B., Maassen, P., & Reich, R. (2018). The Contribution of Education to Social Progress*. 753-778. https://doi.org/10.1017/9781108399661.006

Splan, R. K., Porr, C. A. S., & Broyles, T. W. (2011). Undergraduate Research in Agriculture: Constructivism and the Scholarship of Discovery. *Journal of Agricultural Education*, *52*(4), 56-54. https://doi.org/10.5032/jae.2011.04056

Spoettl, G. (2020). AUTO 4.0: Anticipation of Skills for Employees Due to Digitalization - Identification of "Occupational Profiles". https://doi.org/10.5772/intechopen.90185

Spoettl, G., & Tūtlys, V. (2020). Education and Training for the Fourth Industrial Revolution. *Jurnal Pendidikan Teknologi Dan Kejuruan*, *26*(1), 83-93. https://doi.org/10.21831/jptk.v26i1.29848

Špolar, V. A. M., & Holford, J. (2014). Adult Learning: From the Margins to the Mainstream. 35-50. https://doi.org/10.1163/9789462095489_004

Spooner, V. S., & John, V. M. (2020). Transformative Learning in a Context of Deep Division: The Importance of Learning About Self for Leadership. *South African Journal of Higher Education*, *34*(6). https://doi.org/10.20853/34-6-4077

Staddon, J. E. R. (2024). A Broken Thread: A Review of <i>Pavlov's Legacy: How and What Animals Learn</I>. By Robert A. Boakes. *Journal of the Experimental Analysis of Behavior*, *122*(2), 240-244. https://doi.org/10.1002/jeab.4203

Stafeniak, J. E., & Scott-Brown, J. A. (2016). The Design of a Cognitive Apprenticeship to Facilitate Storytime Programming for Librarians. *Contemporary Educational Technology*, *7*(4). https://doi.org/10.30935/cedtech/6180

Stafford, V. (2021). Using Google Shared Files to Facilitate Successful Online Student Group Collaboration. *Journal of Applied Learning & Teaching*, *4*(1), 129-133. https://doi.org/10.37074/jalt.2021.4.1.21

Stanistreet, P. (2019). Leave No One Behind: The Contribution of Adult Learning and Education. *International Review of Education*, *65*(4), 509-514. https://doi.org/10.1007/s11159-019-09798-x

Stanistreet, P. (2022). The Golden Thread and the Common Treasury. *International Review of Education*, *68*(4), 501-510. https://doi.org/10.1007/s11159-022-09973-7

Stanley, T. (2017). Situated Learning in Accountancy: An Employer Perspective. *Accounting Research Journal*, *30*(3), 265-282. https://doi.org/10.1108/arj-02-2015-0014

Stapleton, M. (2020). Enacting Education. *Phenomenology and the Cognitive Sciences*, *20*(5), 887-913. https://doi.org/10.1007/s11097-020-09672-4

Stathakarou, N., Kononowicz, A. A., Mattsson, E., & Karlgren, K. (2024). Gamification in the Design of Virtual Patients for Swedish Military Medics to Support Trauma Training: Interaction Analysis and Semistructured Interview Study. *Jmir Serious Games*, *12*, e63390. https://doi.org/10.2196/63390

Staudinger, U. M., & Bowen, C. E. (2011). A Systemic Approach to Aging in the Work Context. *Journal for Labour Market Research*, *44*(4), 295-306. https://doi.org/10.1007/s12651-011-0086-2

Stawnychko, L. (2025). Leadership Training for Department Chairs: Integrating Formal Training, Experiential Learning, and Mentorship. *Journal of Leadership Studies*, *18*(4), 6-17. https://doi.org/10.1002/jls.21916

Stenberg, L. C. (2022). Do International Business Students Who Are Studying in Australia Have Similar Academic Outcomes? A Comparison of Undergraduate and Postgraduate Students. *Journal of Education and Training*, *10*(1), 58. https://doi.org/10.5296/jet.v10i1.20217

Stephen, O. O., & Festus, O. O. (2022). Utilization of Work-Based Learning Program to Develop Employability Skill of Workforce (Craftsmen) in Construction Industry Towards Industrial Development. *Indonesian Journal of Educational Research and Technology*, *2*(3), 179-188. https://doi.org/10.17509/ijert.v2i3.43970

Stephens, K. N. R., & Stephens, A. E. (2019). Comparing and Contrasting Sustainability Plans for Taking Their Campuses Green: A Tale of Two Institutions. *2*, 238-242. https://doi.org/10.36315/2019v2end050

Stephens, M., Kramer, C., & Eugene, L. (2023). COVID Faces: Unmasking the Emotional Impact of the Pandemic. *Journal of Regional Medical Campuses*, *6*(1). https://doi.org/10.24926/jrmc.v6i1.4396

Stephens, M. L., Lowney, J., Ybarra, C. M., & Stephens, S. (2022). Adult Learning Facilitators in Any Setting, Instructional Designers' Roles as the Adult Learner and Educator, and Their Perceived Knowledge Construction When Preparing Graduate Faculty for Teaching Online: An Interpretive Study. *New Horizons in Adult Education and Human Resource Development*, *34*(1), 3-15. https://doi.org/10.1002/nha3.20319

Štěrba, R. (2018). Not Too Well-Known "Predecessors" of C. R. Rogers's Humanistic Pedagogy (J. A. Comenius, J. I. Felbiger, J. F. Herbart, O. Chlup). *Universal Journal of Educational Research*, *6*(4), 619-628. https://doi.org/10.13189/ujer.2018.060405

Sterling, S. (2011). Transformative Learning and Sustainability: Sketching the Conceptual Ground. *Learning and Teaching in Higher Education*, *5*, 17-33.

Stokes, N., & Berlacher, K. (2022). The Science of Learning and Art of Education in Cardiology Fellowship. *Methodist Debakey Cardiovascular Journal*, *18*(3), 4-13. https://doi.org/10.14797/mdcvj.1088

Storey, R. J., Lyndgaard, S. F., & Kanfer, R. (2025). The Future of Work and Learning Hinges on Social and Relational Processes. *Industrial and Organizational Psychology*, *18*(1), 128-133. https://doi.org/10.1017/iop.2024.68

Storey, V. A., & Wang, V. C. X. (2016). Critical Friends Protocol: Andragogy and Learning in a Graduate Classroom. *Adult Learning*, *28*(3), 107-114. https://doi.org/10.1177/1045159516674705

Stoten, D. W. (2021a). Building Adaptive Management Capability: The Contribution of Heutagogy to Management Development in Turbulent Times. *Journal of Management Development*, *40*(2), 121-137. https://doi.org/10.1108/jmd-10-2019-0448

Stoten, D. W. (2021b). Navigating Heutagogic Learning: Mapping the Learning Journey in Management Education Through the OEPA Model. *Journal of Research in Innovative Teaching & Learning*, *15*(1), 83-97. https://doi.org/10.1108/jrit-07-2020-0038

Strachan, S., Greig, A., & Jones, A. (2022). Going Green Post COVID-19: Employer Perspectives on Skills Needs. *Local Economy the Journal of the Local Economy Policy Unit*, *37*(6), 481-506. https://doi.org/10.1177/02690942231151638

Stranger–Johannessen, E., & Damiani, V. (2024). Multilingual Stories for Immigrants and Refugees: A Language-as-Resource Approach. 133-149. https://doi.org/10.1007/978-981-99-7196-1_10

Studdard, N. L., Dawson, M., & Jackson, N. (2013). Fostering Entrepreneurship and Building Entrepreneurial Self-Efficacy in Primary and Secondary Education. *Creative and Knowledge Society*, *3*(2), 1-14. https://doi.org/10.2478/v10212-011-0033-1

Styles, M., Middleton, H., Schafheutle, E., & Shaw, M. (2022). Educational Supervision to Support Pharmacy Professionals' Learning and Practice of Advanced Roles. *International Journal of Clinical Pharmacy*, *44*(3), 781-786. https://doi.org/10.1007/s11096-022-01421-8

Sualehi, S. H. (2023). (Some) Adults Left Behind: Digital Literacy and the Working Learner. *New Directions for Adult and Continuing Education*, *2023*(179), 67-77. https://doi.org/10.1002/ace.20503

Suarez-Balcazar, Y. (2020). Meaningful Engagement in Research: Community Residents as Co-creators of Knowledge. *American Journal of Community Psychology*, *65*(3-4), 261-271. https://doi.org/10.1002/ajcp.12414

Suárez, A. E. G., Amor del Rosario Lalama, F., Prieto, A. G. V., Reyna, M. J. M., Guijarro, M. J. A., & Cobos, M. L. P. (2024). Estrategias Psicopedagógicas Para La Mejora De La Resiliencia Académica en Estudiantes De Secundaria. *South Florida Journal of Development*, *5*(6), e3983. https://doi.org/10.46932/sfjdv5n6-002

Sudharshini, F. M. (2024). Optimizing Workforce Stability: Unraveling the Influence of Talent Management Strategies on Employee Retention. *Shanlax International Journal of Management*, *11*(iS1-Jan), 138-144. https://doi.org/10.34293/management.v11iis1-jan.7155

Sufirmansyah, S. (2019). Actualization of Andragogical Learning Strategies for Higher Education in Disruption Era. *Didaktika Religia*, *6*(2), 351-370. https://doi.org/10.30762/didaktika.v6i2.1111

Suh, I. H., McKinney, T., & Siu, K. C. (2023). Current Perspective of Metaverse Application in Medical Education, Research and Patient Care. *Virtual Worlds*, *2*(2), 115-128. https://doi.org/10.3390/virtualworlds2020007

Sujito, A., & Bahagijo, S. (2022). Pioneering the Road to Prosperity for the Future Workforce: Building Social Development Fund (SDF) Roadmap in Indonesia. *Jkap (Jurnal Kebijakan Dan Administrasi Publik)*, *26*(2), 162. https://doi.org/10.22146/jkap.78219

Sukackė, V., Guerra, A., Ellinger, D., Carlos, V., Petronienė, S., Gaižiūnienė, L., Gelabert, S. B., Tallada, A. M., & Brose, A. (2022). Towards Active Evidence-Based Learning in Engineering Education: A Systematic Literature Review of PBL, PjBL, and CBL. *Sustainability*, *14*(21), 13955. https://doi.org/10.3390/su142113955

Sukri, U. (2024). Self-Transformation: The Importance of Education for Adults in the Modern Era. *Global International Journal of Innovative Research*, *2*(10), 2477-2487. https://doi.org/10.59613/global.v2i10.343

Sule, S. Y., Lare, O. J., & Oguntokun, J. E. (2024). Exploring the Efficacy of Vocational Training Programs in Motor Vehicle and Mechanic Works Trade: A Case Study of Science and Technical Colleges in Northeast Nigeria. *International Journal of Education Management and Technology*, *2*(1), 63-80. https://doi.org/10.58578/ijemt.v2i1.2859

Suleymanov, E., Alirzayev, E., & Valehov, J. (2023). Transferring Resource Revenues Into Human Capital in Post-Soviet Countries. *SSRN Electronic Journal.* https://doi.org/10.2139/ssrn.4445847

Sulindra, E., Cendra, A. N., & Hartani, T. (2024). Utilization of Instructional Technology in English Language Teaching (ELT) Based on Constructivism: A Literature Review. *English Education and Literature Journal*, *4*(02), 141-153. https://doi.org/10.53863/ejou.v4i02.1150

Sulistyowati, T. (2019). Bottom-Up and Top-Down Listening Processes Within Cognitive Constructivist Learning Theory. *Prominent*, *2*(1). https://doi.org/10.24176/pro.v2i1.2962

Sultana, R. G. (2013). Career Management Skills: Assessing for Learning. *Australian Journal of Career Development*, *22*(2), 82-90. https://doi.org/10.1177/1038416213496759

Sumithran, S., Chowdhury, R., & Barnes, M. (2022). Synecdochising Student Identities: EAL Teachers' Positioning of Adult EAL Students in Australia. *Qualitative research journal*, *23*(1), 13-26. https://doi.org/10.1108/qrj-05-2022-0064

Sumner, J. (2018). Community-Supported Learning: Practicums, Adult Education and Post-Secondary Institutions. *Alberta Journal of Educational Research*, *64*(2), 97-108. https://doi.org/10.55016/ojs/ajer.v64i2.56384

Sun, D., & Yahaya, N. (2024). The Framework of Online Learning Based on PBL-PL for College English Courses. *International Journal of Academic Research in Progressive Education and Development*, *13*(3). https://doi.org/10.6007/ijarped/v13-i3/21687

Sun, L. (2025). Integrating Culturally Sustaining Pedagogies With Social Justice Language Education: Insights From a Field-Based English Teacher Education Course. *International Journal of Applied Linguistics*, *36*(1), 701-713. https://doi.org/10.1111/ijal.12807

Sun, M. (2023). Exploring the Opportunities and Challenges of AI Technology in College English Teaching. 1532-1541. https://doi.org/10.2991/978-2-38476-126-5_171

Sun, Y. (2024). Power Dynamics in Translingual Practices for Chinese as a Second Language Writing Education. *Language Teaching*, *58*(1), 116-132. https://doi.org/10.1017/s0261444824000089

Sunar, A. S., Abdullah, N. A., White, S., & Davis, H. (2016). Personalisation in MOOCs: A Critical Literature Review. 152-168. https://doi.org/10.1007/978-3-319-29585-5_9

Sung, J., Sheng, Y. Z., Liau, A. K., Xinhui, A. C., Liu, L., & Coates, H. (2022). Augmenting the Role of Higher Education Institutions in Lifelong Learning: Designing an Indicator Framework for Policy Application. *International Journal of Chinese Education*, *11*(1). https://doi.org/10.1177/22125868211072931

Sunggingwati, D., & Haviluddin, H. (2019). Learning Style Preferred by English and Computer Students in Indonesia Context. *International Journal of Emerging Technologies in Learning (Ijet)*, *14*(10), 46. https://doi.org/10.3991/ijet.v14i10.9997

Supriya, Y., Bhulakshmi, D., Bhattacharya, S., Gadekallu, T. R., Vyas, P., Kaluri, R., Koppu, S., Brown, D. J., & Mahmud, M. (2024). Industry 5.0 in Smart Education: Concepts, Applications, Challenges, Opportunities, and Future Directions. *Ieee Access*, *12*, 81938-81967. https://doi.org/10.1109/access.2024.3401473

Surijah, A. B. (2019). Learning Organization in Practices: A Study Case of the Indonesian Companies. https://doi.org/10.2991/icoi-19.2019.19

Surikova, S. (2024). Towards the Effective Professional Upskilling of Low-Skilled Employees: Key Influencing Factors. *Problems of Education in the 21st Century*, *82*(5), 737-757. https://doi.org/10.33225/pec/24.82.737

Susilo, H., & Nugroho, R. (2019). Training of Andragogic Learning Through Model Experiential Learning for Equality Education. https://doi.org/10.2991/icei-19.2019.42

Susler, O., & Babacan, A. (2021). Embedding Critical Reflection in Legal Education. *Law in Context a Socio-Legal Journal*, *37*(3). https://doi.org/10.26826/law-in-context.v37i3.154

Suvendu, R., & S, D. P. (2024). AI-Driven Flipped Classroom: Revolutionizing Education Through Digital Pedagogy. *British Journal of Education Learning and Development Psychology*, *7*(2), 169-179. https://doi.org/10.52589/bjeldp-ltdjflih

Svensson, L., Randle, H., & Bennich, M. (2009). Organising Workplace Learning: An Inter-organisational Perspective. *Journal of European Industrial Training*, *33*(8/9), 771-786. https://doi.org/10.1108/03090590910993625

Swain, J. E., Kumlien, K., & Bond, A. (2020). An Experiential Exercise for Teaching Theories of Work Motivation: Using a Game to Teach Equity and Expectancy Theories. *Organization Management Journal*, *17*(3), 119-132. https://doi.org/10.1108/omj-06-2019-0742

Swanepoel, E. (2021). Shifting Economies and the Need for New Skills. 127-144. https://doi.org/10.4102/aosis.2021.bk214.06

Swanepoel, J. A. (2024). Cohesive Online Education Model Using Emergent Technologies to Improve Accessibility and Impact. *Education Sciences*, *14*(5), 522. https://doi.org/10.3390/educsci14050522

Swartzell, K. L., Fulton, J. S., & Gaudecker, J. R. v. (2023). Older Adult Caregivers' Perspectives on Wound Care Resources: A Qualitative Study. *Research in Gerontological Nursing*, *16*(4), 194-201. https://doi.org/10.3928/19404921-20230503-03

Sweet, S. F. (2022). Using Guided Critical Reflection to Discover Deepened and Transformative Learning in Leadership Education. *Journal of Research on Leadership Education*, *18*(4), 600-621. https://doi.org/10.1177/19427751221118951

Sweetman, M. M. (2018). A Qualitative Exploration of Transformative Learning Within an Online Leadership Course. *Journal of Occupational Therapy Education*, *2*(2). https://doi.org/10.26681/jote.2018.020206

Syaifar, B., & S.Pd, M. P. L. E. R. (2017). Application of Concept and Andragogy Methods on the Civil Servant. *International Journal of Humanities Social Sciences and Education*, *4*(12). https://doi.org/10.20431/2349-0381.0412009

Syarova, S., & Toleva-Stoimenova, S. (2023). Cybersecurity Issues in the Secondary and Higher Education Systems' Curricula. 003. https://doi.org/10.28945/5114

Sykes, S., & Wills, J. (2019). Critical Health Literacy for the Marginalised: Empirical Findings. 167-182. https://doi.org/10.51952/9781447344520.014

Sylte, A. L. (2020). Predicting the Future Competence Needs in Working Life: Didactical Implications for VET. *International Journal for Research in Vocational Education and Training*, *7*(2), 167-192. https://doi.org/10.13152/ijrvet.7.2.3

Symeou, L., Louca, L. T., Kavadella, A., Mackay, J., Danidou, Y., & Raffay, V. (2025). Development of Evidence-Based Guidelines for the Integration of Generative <scp>AI</Scp> in University Education Through a Multidisciplinary, Consensus-Based Approach. *European Journal of Dental Education*, *29*(2), 285-303. https://doi.org/10.1111/eje.13069

Symonds, J., D'Urso, G., & Schoon, I. (2023). The Long-Term Benefits of Adolescent School Engagement for Adult Educational and Employment Outcomes. *Developmental psychology*, *59*(3), 503-514. https://doi.org/10.1037/dev0001458

Taheri, M., Motealleh, S., & Younesi, J. (2021). Workplace Fun and Informal Learning: The Mediating Role of Motivation to Learn, Learning Opportunities and Management Support. *Journal of Workplace Learning*, *34*(3), 229-241. https://doi.org/10.1108/jwl-05-2021-0062

Takei, H. (2023). A Comparative Study of Cooperative Education Programs in Germany, the United Kingdom, Italy, and Finland. *Global Journal of Educational Studies*, *9*(1), 50. https://doi.org/10.5296/gjes.v9i1.20817

Tamashita, T., Kim, G., Liou, C.-l., Ando, T., Bardo, A. R., & Liu, D. (2024). The Roles of Education and Literacy in the Digital Divide Among Middle-Aged Adults: Cross- National Evidence From the United States, Japan, and South Korea. *Adult Literacy Education the International Journal of Literacy Language and Numeracy*, *6*(1), 4-16. https://doi.org/10.35847/tyamashita.gkim.cliou.tando.abardo.dliu.6.1.4

Tamoliūnė, G., Greenspon, R., Teresevičienė, M., Volungevičienė, A., Trepulė, E., & Daukšienė, E. (2023). Exploring the Potential of Micro-Credentials: A Systematic Literature Review. *Frontiers in Education*, *7*. https://doi.org/10.3389/feduc.2022.1006811

Tan, X., Choi, H., & Kang, S. W. (2026). The Resource Conversion Mechanism: Trust, Leader's Vision of Talent, and Informal Training as Pathways to Organizational Commitment. *Behavioral Sciences*, *16*(6), 944. https://doi.org/10.3390/bs16060944

Tanaka, S., & Takeuchi, O. (2024). Sociocultural Influences on Young Japanese English Learners: The Impact of Parents' Beliefs on Learning Motivation. *Journal for the Psychology of Language Learning*, *6*(1), 1-15. https://doi.org/10.52598/jpll/6/1/3

Tang, H., Lin, Y.-J., & Qian, Y. (2021). Improving K-12 Teachers' Acceptance of Open Educational Resources by Open Educational Practices: A Mixed Methods Inquiry. *Educational Technology Research and Development*, *69*(6), 3209-3232. https://doi.org/10.1007/s11423-021-10046-z

Tang, H., Lin, Y. J., & Qian, Y. (2020). Understanding K-12 Teachers' Intention to Adopt Open Educational Resources: A Mixed Methods Inquiry. *British journal of educational technology*, *51*(6), 2558-2572. https://doi.org/10.1111/bjet.12937

Tang, Y. M., & Ho, H. L. (2020). 3D Modeling and Computer Graphics in Virtual Reality. https://doi.org/10.5772/intechopen.91443

Tangen, J. L., & Borders, L. D. (2017). Applying Information Processing Theory to Supervision: An Initial Exploration. *Counselor Education and Supervision*, *56*(2), 98-111. https://doi.org/10.1002/ceas.12065

Tannenbaum, S. I., & Wolfson, M. A. (2022). Informal (Field-Based) Learning. *Annual Review of Organizational Psychology and Organizational Behavior*, *9*(1), 391-414. https://doi.org/10.1146/annurev-orgpsych-012420-083050

Tanyu, M., Spier, E., Pulizzi, S., & Rooney, M. (2020). Improving Education Outcomes for Students Who Have Experienced Trauma and/or Adversity. https://doi.org/10.1787/54d45980-en

Tapis, G. P., Church, K. S., & Webb, T. Z. (2020). Preparing for the Hybridization of the Accounting Profession: A CISA Boot Camp Case Study. *Ais Educator Journal*, *15*(1), 25-58. https://doi.org/10.3194/1935-8156-15.1.25

Tarhouny, S. E., Hassouna, A., Khalifa, M., & Mansour, T. (2021). Etiquette of Virtual Learning; Tips to Ensure Effective Learning for Medical Students. *Mededpublish*, *10*(1). https://doi.org/10.15694/mep.2021.000067.1

Tatarinceva, A., & Gode, I. (2016). Lifelong Education as a Key Factor of the Transformation and Development of the Human Capital's Potential. *Society Integration Education Proceedings of the International Scientific Conference*, *4*, 163. https://doi.org/10.17770/sie2016vol4.1537

Tatnall, A. (2023). Editorial for EAIT Issue 1, 2023. *Education and Information Technologies*, *28*(1), 1-12. https://doi.org/10.1007/s10639-023-11596-w

Taylor, A.-S. A. (2015). The Active Instructor: Benefits and Barriers to Instructor-Led Serious Gaming. 1-8. https://doi.org/10.1109/vs-games.2015.7295787

Taylor, A. (2014). Community Service-Learning and Cultural-Historical Activity Theory. *Canadian Journal of Higher Education*, *44*(1), 95-107. https://doi.org/10.47678/cjhe.v44i1.183605

Taylor, D. C. M., & Hamdy, H. (2013). Adult learning theories: Implications for learning and teaching in medical education. *Medical teacher*, *35*(11), e1561-e1572.

Taylor, S., Mthimkhulu, S., & Mpanza, S. (2021). The Role of Local Economic Development Agencies and Nonprofit Organisations in Local Economic Development: South African Case Studies. 53-74. https://doi.org/10.4102/aosis.2021.bk214.03

Teele, T., & Nkoane, M. M. (2024). Critical Reflections of Emerging Farmers' Individual Perspective Towards Assessments in Recognition of Prior Learning for Unit Standards Credit of Agricultural Courses. *Journal of Adult and Continuing Education*, *31*(3), 645-663. https://doi.org/10.1177/14779714241287327

Teemant, A., Upton, T. A., & Sherman, B. (2024). Inclusive Learning Communities: An English Learner Framework for All Educators. *Tesol Journal*, *15*(3). https://doi.org/10.1002/tesj.798

Tehlan, A. (2023). A Study on Different Modes of E- Learning. *Interantional Journal of Scientific Research in Engineering and Management*, *07*(12), 1-11. https://doi.org/10.55041/ijsrem27436

Teik, C. H. (2019). Developing Engineering Workforce, Designing a Better Future. *Iem Journal*, *80*(1). https://doi.org/10.54552/v80i1.16

Tellado, I. (2017). Bridges Between Individuals and Communities: Dialogic Participation Fueling Meaningful Social Engagement. *Research on Ageing and Social Policy*, *5*(1), 8-31. https://doi.org/10.17583/rasp.2017.2389

Temesio-Vizoso, S. (2016). Educación Inclusiva: Retos Y Oportunidades. *Revista De Educación a Distancia (Red)*(51). https://doi.org/10.6018/red/51/9

Tenakwah, E. S., & Watson, C. (2024). Embracing the AI/automation Age: Preparing Your Workforce for Humans and Machines Working Together. *Strategy & Leadership*, *53*(1), 32-48. https://doi.org/10.1108/sl-05-2024-0040

Teoh, G. B. S., & Liau, A. W. L. (2023). Distance Learners' Emotional Intelligence and Perceptions of Their Situational Barriers in Learning English. *Turkish Online Journal of Distance Education*, *24*(2), 120-144. https://doi.org/10.17718/tojde.1048366

Terrell, M. (2006). Anatomy of Learning: Instructional Design Principles for the Anatomical Sciences. *The Anatomical Record Part B the New Anatomist*, *289B*(6), 252-260. https://doi.org/10.1002/ar.b.20116

Terrón-Caro, M. T., Cárdenas-Rodríguez, R., Ortega-de-Mora, F., Aleksic, K., Bergano, S., Biligha, P., Chiappelli, T., Grigoli, A. R. D., Díaz-Jiménez, R. M., Fouskas, T., Frelih, M., Girón, T., Guo, W., Koulierakis, G., Lapov, Z., & Lyberopoulou, L. (2022). Policy Recommendations Ebook. Migrations, Gender and Inclusion From an International Perspective. https://doi.org/10.46661/rio.20220727_1

Teszenyi, E., Devecchi, C., & Richardson, T. (2019). Enhancing Learning for Early Years Foundation Degree Students: Empowerment Through Heutagogy and Reflecting on the Notion of Knowledgeable Others. *Polish Journal of Educational Studies*, *72*(1), 24-44. https://doi.org/10.2478/poljes-2019-0003

Thakur, B., & Mittal, N. (2025). Role of the United Nations Educational, Scientific and Cultural Organization (UNESCO), the Organization for Economic Cooperation and Development (OECD) and Global EdTech Standards in Shaping Artificial Intelligence and Cyber Law Curriculum Policies: A Comparative Study of Finland, Singapore and India. *International Annals of Criminology*, 1-14. https://doi.org/10.1017/cri.2025.10108

Thapaliya, P., & Luitel, B. C. (2024). Reflection-on-/in-/for-Actions: Deconstructing Hegemonic Pedagogical Culture in Science Education. *Cultures of Science*, *7*(1), 64-76. https://doi.org/10.1177/20966083241241351

Thiel, R., Bowling, A., Rumble, J. N., McFadden, B. R., Stofer, K. A., & Folta, K. M. (2022). Impact of Teaching Methods on Learner Preferences and Knowledge Gained When Informing Adults About Gene Editing. *Advancements in Agricultural Development*, *3*(1), 70-86. https://doi.org/10.37433/aad.v3i1.133

Thinley, R., & Gyeltshen, G. (2024). A Theoretical Review on Non-Formal Education and Women Empowerment in Bhutan. *International Journal of Social Science and Human Research*, *07*(10). https://doi.org/10.47191/ijsshr/v7-i10-91

Thistlethwaite, J. (2013). Practice-Based Learning Across and Between the Health Professions: A Conceptual Exploration of Definitions and Diversity and Their Impact on Interprofessional Education. *International Journal of Practice-Based Learning in Health and Social Care*, *1*(1), 15-28. https://doi.org/10.11120/pblh.2013.00003

Thomas, A., Menon, A., Boruff, J., Rodríguez, A. M., & Ahmed, S. (2014). Applications of Social Constructivist Learning Theories in Knowledge Translation for Healthcare Professionals: A Scoping Review. *Implementation Science*, *9*(1). https://doi.org/10.1186/1748-5908-9-54

Thomas, C., MacMillan, C. P., McKinnon, M., Torabi, H., Osmond-McLeod, M. J., Swavley, E., Armer, T., & Doyle, K. (2021). Seeing and Overcoming the Complexities of Intersectionality. https://doi.org/10.20944/preprints202101.0172.v1

Thomas, S., & Glazzard, J. (2025). Volatile Learning Environments: Exploring the Barriers to Inclusion in Prison Education Through Collaborative Writing. *Policy Futures in Education*. https://doi.org/10.1177/14782103251341407

Thombare, D. V., Mailapalli, J., Adeyemi, E., & Bommer, S. (2026). Conceptual Framework to Advancing Healthcare System Through Agile Methodologies, Human Factors Engineering and Adaptive Leadership. *Journal of Health Organization and Management*, *40*(9), 96-109. https://doi.org/10.1108/jhom-06-2025-0316

Thompson, T. L. (2010). Self-employed and Online: (Re)negotiating Work-learning Practices. *Journal of Workplace Learning*, *22*(6), 360-375. https://doi.org/10.1108/13665621011063478

Thomson, L. K. (1994). *Writing a Needs Assessment to Determine the Needs of the Adult Graduate Student* Iowa State University].

Thoresen, P., Fielding, A., Gillieatt, S., & Thoresen, S. (2016). Identifying the Needs of Refugee and Asylum-Seeking Children in Thailand: A Focus on the Perspectives of Children. *Journal of Refugee Studies*, few028. https://doi.org/10.1093/jrs/few028

Thoron, A. C., & Myers, B. E. (2012). Effects of Inquiry–based Agriscience Instruction and Subject Matter–based Instruction on Student Argumentation Skills. *Journal of Agricultural Education*, *53*(2), 58-69. https://doi.org/10.5032/jae.2012.02058

Thorpe, M., Healy, C., & Olsen, D. (2024). Bridging Workforce Skills Gaps in Rural America Amid 'The Great Resignation' by Leveraging a Triple Helix Collaboration. *Industry and Higher Education*, *39*(2), 198-211. https://doi.org/10.1177/09504222241277409

Thota, N. (2015). Connectivism and the Use of Technology/Media in Collaborative Teaching and Learning. *New Directions for Teaching and Learning*, *2015*(142), 81-96. https://doi.org/10.1002/tl.20131

Thunqvist, D. P., Tønder, A. H., & Reegård, K. (2019). A Tale of Two Reforms: Institutional Change in Vocational Education and Training in Norway and Sweden in the 1990s. *European Educational Research Journal*, *18*(3), 298-313. https://doi.org/10.1177/1474904118823104

Thusi, X., & Chauke, R. (2023). Strategies for Retaining Scarce Skills and Reducing Turnover in the South African Public Sector. *International Journal of Management Entrepreneurship Social Sciences and Humanities*, *6*(1), 120-134. https://doi.org/10.31098/ijmesh.v6i1.1242

Tian, J., & Ng, S. B. (2024). Learning Style and English Argumentative Writing Proficiency – How They Complement Each Other. *Environment-Behaviour Proceedings Journal*, *9*(28), 139-145. https://doi.org/10.21834/e-bpj.v9i28.5909

Tilbury, D. (2011). *Education for Sustainable Development: An Expert Review of Processes and Learning.*

Tınmaz, H. (2012). Social Networking Websites as an Innovative Framework for Connectivism. *Contemporary Educational Technology*, *3*(3). https://doi.org/10.30935/cedtech/6080

Tinoco, E. C., Lazo, J. L. G., Sierra, G. L. Á., & Carhuallanqui, K. R. P. (2022). Aprendizaje Experiencial De Kolb en Estudiantes De La Facultad De Ingeniería Metalúrgica Y De Materiales De La Universidad Nacional Del Centro Del Perú. *Prospectiva Universitaria*, *18*(1), 99-109. https://doi.org/10.26490/uncp.prospectivauniversitaria.2021.18.1634

Tisdell, E. J., Taylor, E. W., & Forté, K. S. (2013). Community-Based Financial Literacy Education in a Cultural Context. *Adult education quarterly*, *63*(4), 338-356. https://doi.org/10.1177/0741713613477186

Tkachenko, O., & Ardichvili, A. (2017). Cultural-Historical Activity Theory's Relevance to HRD: A Review and Application. *Human Resource Development Review*, *16*(2), 135-157. https://doi.org/10.1177/1534484317696717

Tobbell, J., O'Donnell, V., & Zammit, M. (2010). Exploring Transition to Postgraduate Study: Shifting Identities in Interaction With Communities, Practice and Participation. *British educational research journal*, *36*(2), 261-278. https://doi.org/10.1080/01411920902836360

Todić, J., & Christensen, M. C. (2022). Integrating Critical, Engaged, and Abolitionist Pedagogies to Advance Antiracist Social Work Education. *Advances in Social Work*, *22*(2), 389-415. https://doi.org/10.18060/24972

Tolosa-Guzman, I.-A., Nieto, S. L. F., Vargas-Pinilla, O.-L., & Rodríguez-Ibagué, L. F. (2021). Reflexiones Pedagógicas 26. https://doi.org/10.12804/issne.2500-6150_10336.31127_ceatp

Tomasella, B., Akbar, B., Lawson, A., Howarth, R., & Bedford, R. (2024). Embedding the Sustainable Development Goals Into Higher Education Institutions' Marketing Curriculum. *Journal of Marketing Education*, *46*(2), 155-174. https://doi.org/10.1177/02734753241231182

Tombleson, B. (2023). Building a Model for Transmedia Learning. *Ascilite Publications*, 244-253. https://doi.org/10.14742/apubs.2023.686

Tomić, D., & Vizinger, K. (2023). Effects of Remote Business During the Covid-19 Pandemic - A Literature Review. *American Journal of Economics and Business Innovation*, *2*(1), 1-13. https://doi.org/10.54536/ajebi.v2i1.1088

Tomita, D., Abdelhakim, M., Bartková, J., Gulsum, A., Sato, A., Alotaibi, N. H., Aboulhassan, M., Dongcai, L., & Tomita, Y. (2025). From Innovation to Integration: A Global Mixed-Methods Study of VR, Metaverse, and 3D Simulation in Healthcare Training and Clinical Setting. *Frontiers in Digital Health*, *7*. https://doi.org/10.3389/fdgth.2025.1632528

Tong-Battung, G. (2021). Framing the Processes in Asynchronous CPD Online for Accounting Teachers in the Philippines. *Liceo Journal of Higher Education Research*, *17*(2). https://doi.org/10.7828/ljher.v17i2.1541

Tonnsen, S., Alexander, A. E., Topolka-Jorissen, K., & Jacobs, J. E. (2017). Influences of Early Life Experiences on the Generative Leadership Roles of Four Female Professors of Educational Leadership. Vol. 31 (2011). https://doi.org/10.21423/awlj-v31.a82

Toro, E., & Kisi, A. (2022). Digital Literacy, a Necessary Skill for Training Language Teachers in Albania. *Xlinguae*, *15*(4), 3-17. https://doi.org/10.18355/xl.2022.15.04.01

Torre, D., Groce, V. A., Gunderman, R. B., Kanter, J. H., Durning, S. J., & Kanter, S. L. (2017). Freire's View of a Progressive and Humanistic Education: Implications for Medical Education. *Mededpublish*, *6*, 119. https://doi.org/10.15694/mep.2017.000119

Torre, N. O., Vidal, Ó. F., & Ferran, A. P. (2021). Constructivist Learning Models in Training Programs. https://doi.org/10.3926/oms.407

Torres, D., Pulukuri, S., & Abrams, B. (2023). Step Back, Translate, and Extend: An Instructional Framework for Enhancing Knowledge Transfer and Self-Efficacy Across Chemistry Courses. *Journal of Chemical Education*, *100*(12), 4696-4706. https://doi.org/10.1021/acs.jchemed.3c00964

Torres, E. M. A., Pérez, N. Y. M., Cardenas, K. J. A., & Cardenas, B. S. A. (2021). Implicancias De La Nueva Normalidad Por El Covid-19 en El Sistema Educativo Peruano. *South Florida Journal of Development*, *2*(5), 6260-6277. https://doi.org/10.46932/sfjdv2n5-001

Torres, E. M. A., Pérez, N. Y. M., Cardenas, K. J. A., & Cardenas, B. S. A. (2022). El Conectivismo, Un Nuevo Paradigma Para La Educación. *South Florida Journal of Development*, *3*(1), 361-379. https://doi.org/10.46932/sfjdv3n1-028

Torresan, S., & Hinterhuber, A. (2023). Continuous Learning at Work: The power of Gamification. *Management Decision*, *61*(13), 386-412. https://doi.org/10.1108/md-12-2020-1669

Torterat, F., Jesiorski, A., & Langé, J.-M. (2021). Récits D'élèves Sur Les Enjeux De Durabilité / Soutenabilité. Quelles Manifestations De L'engagement ? *Educations*, *4*(1). https://doi.org/10.21494/iste.op.2021.0639

Totena, M. P. (2023). Inclusión De Personas Con Discapacidad en Instituciones Educativas Indígenas. *Sciencevolution*, *3*(7), 60-71. https://doi.org/10.61325/ser.v3i7.77

Tough, A. (1979). *The Adult's Learning Projects: A Fresh Approach to Theory and Practice in Adult Learning* (2nd ed.). Ontario Institute for Studies in Education.

Townley, C., Grace, R., Woodrow, C., Baker, E., Staples, K., Locke, M. L., & Kaplun, C. (2023). Educator Perspectives on Embedding Acknowledgement to Country Practices in Early Learning Centres in Australia. *Australasian Journal of Early Childhood*, *48*(2), 86-98. https://doi.org/10.1177/18369391231173182

Tracy, J. F. (2015). The Impact of Choice: Exercise Motivation and Physical Activity in College Students Enrolled in Fitness for Life. https://doi.org/10.33915/etd.6825

Trần, L. T. (2019). Australian Faculty's Perceptions of Their Professional Responsibilities in Preparing Students for Global Labor Market. https://doi.org/10.3102/1446387

Treviño, G. F. F., Mallozzi, Z. L. T., Galán, J. L. J., & Córdova, C. V. I. (2024). Sistema Digital De Enseñanza Y Aprendizaje Para Las Personas Sordas Aplicando Inteligencia Artificial. 39-90. https://doi.org/10.37572/edart_3107241854

Trewartha, G., Janssens, L., & Harrison, S. (2026). Digital Exclusion Faced by People Living With Chronic Respiratory Disease: Challenges, Implications and Solutions. *Chronic Respiratory Disease*, *23*. https://doi.org/10.1177/14799731261441307

Trujillo, A. L., Gómez, J. A. C., & Merino, D. S. (2023). Políticas Educativas, Ocio Inclusivo Y Equidad Social Ante Los Nuevos Futuros De La Educación. *Pedagogia Social Revista Interuniversitaria*(43), 15-29. https://doi.org/10.7179/psri_2023.43.01

Tsapenko, I. P., & Grishin, I. (2022). Virtualization of Cross-Border Labor Migration. *Herald of the Russian Academy of Sciences*, *92*(5), 580-589. https://doi.org/10.1134/s1019331622050069

Tscherne, M. (2021). Orientierung Durch Werte. *Schule Verantworten | Führungskultur_innovation_autonomie*(2), 57-63. https://doi.org/10.53349/sv.2021.i2.a84

Tschofen, C., & Mackness, J. (2012). Connectivism and Dimensions of Individual Experience. *The International Review of Research in Open and Distributed Learning*, *13*(1), 124. https://doi.org/10.19173/irrodl.v13i1.1143

Tsortanidou, X., Karagiannidis, C., & Koumpis, A. (2018). Role and Value of Learning Theories in the Area of Adaptive Educational Hypermedia Systems Incorporating Learning Styles. *International Journal of Emerging Technologies in Learning (Ijet)*, *13*(01), 93. https://doi.org/10.3991/ijet.v13i01.7614

Tsukube, T., & Matsuo, M. (2020). The Impact of Cognitive Apprenticeship on the Perceived Growth of Junior Doctors. *Journal of Workplace Learning*, *32*(7), 489-499. https://doi.org/10.1108/jwl-04-2020-0055

Tuczyński, K. (2023). Distance Education: An Analysis of Learning Theories in an E-Learning Environment. *Kultura I Edukacja*, *2023*(4 (142)), 67-79. https://doi.org/10.15804/kie.2023.04.04

Tully, N., & Avramenko, A. (2015). Au Fait Law Placements. *Higher Education Skills and Work-Based Learning*, *5*(2), 152-167. https://doi.org/10.1108/heswbl-05-2014-0016

Tun, K. M. (2007). Convocation Address I - Globalisation and Its Implication on Education and Human Resource Development. *Ousl Journal*, *4*(0), 78. https://doi.org/10.4038/ouslj.v4i0.340

Tundaan, M. D. (2023). Examining the Extent of Implementation of Inclusive Education and Teachers' Essential Practices in Handling Learners With Special Educational Needs. *International Journal of Research Publications*, *138*(1). https://doi.org/10.47119/ijrp1001381122023 5732

Tunno, P. (2023). Engineering Global Competencies Through Study Abroad. https://doi.org/10.18687/laccei2023.1.1.1676

Turcotte, J. (2022). Les Interactions Entre La Pédagogie Critique Et La Théorie Socioculturelle De l'Activité Comme Cadre d'Analyse De l'Innovation Des Pratiques Enseignantes. *Revue Internationale Du Crires Innover Dans La Tradition De Vygotsky*, *6*(2), 56-65. https://doi.org/10.51657/ric.v6i3.51687

Turner, J. R., Baker, R. M., Schroeder, J., Johnson, K., & Chung, C. H. (2018). Leadership Development Techniques. *European journal of training and development*, *42*(9), 538-557. https://doi.org/10.1108/ejtd-03-2018-0022

Turner, K. (2019). One-to-One Learning and Self-Determination Theory. *International Journal of Instruction*, *12*(2), 1-16. https://doi.org/10.29333/iji.2019.1221a

Turns, J., Sattler, B., Yasuhara, K., Borgford-Parnell, J., & Atman, C. J. (2014). Integrating Reflection Into Engineering Education. 24.776.771-724.776.716. https://doi.org/10.18260/1-2--20668

Turvey, K., & Pachler, N. (2025). A Topological Exploration of Convergence/Divergence of Human-mediated and Algorithmically Mediated Pedagogy. *British journal of educational technology*, *57*(3), 619-638. https://doi.org/10.1111/bjet.70007

Tūtlys, V., & Spöttl, G. (2017). From the Analysis of Work-Processes to Designing Competence-Based Occupational Standards and Vocational Curricula. *European journal of training and development*, *41*(1), 50-66. https://doi.org/10.1108/ejtd-10-2015-0078

Tymchuk, L., Grytsyk, N., Yahupov, V., Syvokhop, Y., Hrinchenko, T., & Svystun, V. (2021). Andragogy: Theory and Practice of Adult Education Development in Ukraine. *Revista Romaneasca Pentru Educatie Multidimensionala*, *13*(2), 185-205. https://doi.org/10.18662/rrem/13.2/417

Tynjälä, P., & Häkkinen, P. (2005). E-learning at Work: Theoretical Underpinnings and Pedagogical Challenges. *Journal of Workplace Learning*, *17*(5/6), 318-336. https://doi.org/10.1108/13665620510606742

Tzovla, E., & Kedraka, K. (2023). Follow-Up: A Neglected Experiential Process Within a Transformational Learning Community for Teachers as Adult Learners During the Pandemic. *Adult Education Critical Issues*, *3*(1), 39-46. https://doi.org/10.12681/haea.34822

Uddin, M., Okai-Ugbaje, S., & Saba, T. (2017). Green ICT Framework to Reduce Carbon Footprints in Universities. *Advances in Energy Research*, *5*(1), 1-12. https://doi.org/10.12989/eri.2017.5.1.001

Uddin, M. S. (2019). Critical Pedagogy and Its Implication in the Classroom. *Journal of Underrepresented & Minority Progress*, *3*(2), 109-119. https://doi.org/10.32674/jump.v3i2.1788

Ugheoke, S. O. (2024). Predictive Power of Self-Efficacy, Familiarity With Training Content on Job Performance: Moderating Effect of Training Effectiveness in Oman's Higher Education Institutions. *Malaysian Management Journal*, *28*, 245-274. https://doi.org/10.32890/mmj2024.28.9

Ugurel-Kamisli, M. (2022). Language and Culture Learning Needs of Adult Refugees. *New Directions for Adult and Continuing Education*, *2022*(175-176), 85-93. https://doi.org/10.1002/ace.20470

Ülker, Ü., Karyń, A., & Perzyca-Borowska, E. (2026). Out-of-School Learning Practices Within the Scope of Lifelong Learning: Comparison Between Türkiye and Poland. *European journal of education*, *61*(1). https://doi.org/10.1111/ejed.70535

Umbreş, R. (2021). Buckets of Steam and Left-Handed Hammers. The Fool's Errand as a Signal of Epistemic and Coalitional Dominance - Journal of Cognition and Culture 2022. https://doi.org/10.31234/osf.io/vkc42

UNESCO-UNEVOC. (2017). *Greening Technical and Vocational Education and Training: A Practical Guide for Institutions*. UNESCO-UNEVOC International Centre.

UNESCO. (2016). *Recommendation on Adult Learning and Education.*

UNESCO. (2020). *Education for Sustainable Development: A Roadmap.* https://www.unesco.org/en/articles/education-sustainable-development-roadmap

UNESCO. (2021a). *Reimagining Our Futures Together: A New Social Contract for Education.* UNESCO.

UNESCO. (2021b). *Reimagining Our Futures Together: A New Social Contract for Education.*

UNESCO. (2022). *Futures of Education: Learning to Become.* https://www.unesco.org/futuresofeducation

Unger, M., Ernstzen, D., Statham, S., & Titus, A. (2025). Embedding Social Accountability Through Transformative Pedagogies: A Case Study From a South African Physiotherapy Curriculum. https://doi.org/10.21203/rs.3.rs-7327666/v1

United Nations General Assembly. (2015). *Transforming Our World: The 2030 Agenda for Sustainable Development.* https://sdgs.un.org/2030agenda

Upenieks, L. (2021). Uncertainty in Faith, Fear of Death? Transitions in Religious Doubt and Death Anxiety in Later Life. *Omega - Journal of Death and Dying*, *87*(3), 814-837. https://doi.org/10.1177/00302228211029475

Uribe-Canónigo, R. D. (2017). El Aprendizaje en La Era Digital. Perspectivas Desde Las Principales Teorías. *Aibi Revista De Investigación Administración E Ingeniería*, *5*(2), 29-33. https://doi.org/10.15649/2346030x.439

Uribe-Zarain, X., Hellman, A. B., & Bell, A. B. (2024). Transforming Teacher Attitudes and Practices: The Efficacy of Sustained Professional Development in Culturally Responsive Education. *Tesol Journal*, *16*(1). https://doi.org/10.1002/tesj.896

Uribe, A. C. R. (2022). No Good-Bye and No Thanks: A Case Study on the Experience of Mexican University Professors Retiring in Times of Pandemic. *Journal of Behavior Health & Social Issues*, *14*(2), 11-22. https://doi.org/10.22201/fesi.20070780e.2022.14.2.81310

Uriyo, A. Cases of Resiliency : Exploring Experiences of Refugee Women Participating in U.S. Textile and Apparel Nonprofit Organizations' Skills Training Programs. https://doi.org/10.32469/10355/69878

Urrutxı, L. D., Álvarez, M. a. Á., Sáez, I. A., & Arizti, N. B. (2021). Voluntad Política a Favor De La Educación Inclusiva Y Equitativa: Inicios Y Desarrollo Analizando Su Significado. *Education Policy Analysis Archives*, *29*(January - July), 64. https://doi.org/10.14507/epaa.29.4976

Ursula, P. A. (2024). Application of Humanistic Learning Theory in Increasing Student Learning Motivation. *International Journal of Sustainable Social Science (Ijsss)*, *2*(5), 323-334. https://doi.org/10.59890/ijsss.v2i5.2625

Useche, A. C., Galvis, Á., Arceo, F. D. B., Rivera, A. E. P., & Muñoz-Reyes, C. (2022). Reflexive Pedagogy at the Heart of Educational Digital Transformation in Latin American Higher Education Institutions. *International Journal of Educational Technology in Higher Education*, *19*(1). https://doi.org/10.1186/s41239-022-00365-3

Usman, J., Syabuddin, S., & Zakaria, F. (2021). <i>Teungku</I> Identity Development: The Role of <i>Dayah</I> Community of Practice. *Sage Open*, *11*(3). https://doi.org/10.1177/21582440211031532

Usman, M. B. (2026). When Education Fails to Translate Into Work: Structural Mismatch, Graduate Unemployment, and MSME Labour Demand in Ghana. https://doi.org/10.21203/rs.3.rs-9017405/v1

Utami, B. N. P. (2021). The Behaviorism Perspectives on English Learning: Its Theorical and Implications for Foreign English Learner in Online Learning Era. *Proceeding of International Conference on Islamic Education*, 29-41. https://doi.org/10.51425/icie.vi.23

Uthman, Y. O. O. (2023). How Transformative Education Empowers Social Justice: Bridging the Gap Between Knowledge and Action. *Competitive Journal of Education*, *2*(4), 276-283. https://doi.org/10.58355/competitive.v2i4.81

Uysal, D. (2023). Academic Motivation Types and Learning Styles of Students in a Faculty of Education. *Anadolu University Faculty of Education Journal*, *7*(4), 931-944. https://doi.org/10.34056/aujef.1302734

Uzoamaka, I. G., & Sijde, P. v. d. (2024). Twenty-First Century Competencies; About Competencies for Industry 5.0 and the Opportunities for Emerging Economies. https://doi.org/10.20944/preprints202407.0429.v1

Uzorka, A. (2021). Faculty Members' Communication Needs With Regards to Technology. *Journal of Learning for Development*, *8*(1), 111-128. https://doi.org/10.56059/jl4d.v8i1.444

Uzorka, A., Makeri, Y. A., Osigwe, M. U., & Ben, I. N. (2020). An Investigation of the Teaching Needs of Faculty Members With Regard to Technology. *International Journal of Technology in Education and Science*, *5*(1), 70-107. https://doi.org/10.46328/ijtes.152

ÜZÜM, B., & Duruhan, K. (2020). Alman Eğitim Sisteminin Sosyal Ve Kültürel Yapı, İstihdam Ve Ekonomi Bağlamında Etkisinin İncelenmesi. *Opus Toplum Araştırmaları Dergisi*, 1-1. https://doi.org/10.26466/opus.596969

Vaca, C. A. M. (2017). Personalidad Y Liderazgo: Análisis Teórico De Su Relación. *Innova Research Journal*, *2*(5), 106-114. https://doi.org/10.33890/innova.v2.n5.2017.173

Vafa, K., Du, T., Kanodia, A., Athey, S., & Blei, D. M. (2022). CAREER: A Foundation Model for Labor Sequence Data. https://doi.org/10.48550/arxiv.2202.08370

Vágvölgyi, R., Sahlender, M., Schröter, H., Nagengast, B., Dresler, T., Schrader, J., & Nuerk, H. C. (2024). Low Literacy Skills in Adults Can Be Largely Explained by Basic Linguistic and Domain-General Predictors. *Frontiers in Psychology*, *15*. https://doi.org/10.3389/fpsyg.2024.1422896

Vaishali, V., & Mısra, P. K. (2020). Implications of Constructivist Approaches in the Classrooms: The Role of the Teachers. *Asian Journal of Education and Social Studies*, 17-25. https://doi.org/10.9734/ajess/2020/v7i430205

Valadas, S. T., Vilhena, C., & Fragoso, A. (2019). Transitions to Retirement: Perceptions of Portuguese Older Men. *Andragoška Spoznanja*, *25*(2), 37-51. https://doi.org/10.4312/as.25.2.37-51

Valentino, L. (2010). At the Intersection of Work and School: Internships and the Liberal Arts. https://doi.org/10.14418/wes01.1.515

Valenzano, N. (2021). Marxist and Personalist Influences in Paulo Freire's Pedagogical Anthropology. *Tendencias Pedagógicas*, *38*, 68-82. https://doi.org/10.15366/tp2021.38.007

Vanhulle, S. (2005). How Future Teachers Develop Professional Knowledge Through Reflective Writing in a Dialogical Frame. *L1 Educational Studies in Language and Literature*, *5*(3), 287-314. https://doi.org/10.1007/s10674-005-8561-8

Vannasy, V., & Sengsouliya, S. (2023). Key Predictors of the Implementation of Workplace Learning in Higher Education. *Andragoška Spoznanja*, *29*(1), 81-97. https://doi.org/10.4312/as/10447

Varadarajan, S., & Ladage, S. (2024). Problem-Based Learning (PBL): A Literature Review of Theory and Practice in Undergraduate Chemistry Laboratories. *Journal of Chemical Education*, *101*(8), 3027-3038. https://doi.org/10.1021/acs.jchemed.3c01335

Vargas, M. G. (2024). Una Educación Intercultural Basada en Prácticas Cognitivas Situadas. *Andamios*, *21*(54), 83-109. https://doi.org/10.29092/uacm.v21i54.1060

Varma, C., & Malik, S. B. (2024). Perspective Chapter: TVET in the 21st Century – A Focus on Innovative Teaching and Competency Indicators. https://doi.org/10.5772/intechopen.112516

Varsik, S., & Gorochovskij, J. (2023). Intersectionality in Education. https://doi.org/10.1787/dbb1e821-en

Vásquez, G. C., Pareja, E. M. D., & Tudela, J. M. O. (2023). Diseño Didáctico De Aprendizaje Basado en La Web Para Entrenamiento Docente en Incorporación De Tecnología a La Clase De Inglés. *Praxis*, *19*(2), 287-305. https://doi.org/10.21676/23897856.4979

Vaughn, L. M., Farrah, J., & C, B. R. (2009). Cultural Health Attributions, Beliefs, and Practices: Effects on Healthcare and Medical Education. *The Open Medical Education Journal*, *2*(1), 64-74. https://doi.org/10.2174/1876519x00902010064

Vaughter, P., Noguchi, F., & Li, S. (2022). Minding the GAP: An Overview of Five Years of Education for Sustainable Development (ESD) Projects Under the Global Action Programme (GAP) Within Regional Centres of Expertise (RCEs) on ESD. *Frontiers in Sustainability*, *3*. https://doi.org/10.3389/frsus.2022.978938

Vedpathak, M., & Mithari, P. (2024). Harnessing Information Technology in Learning: Exploring Emerging Trends and Innovative Approaches. *Journal of Digital Learning and Distance Education*, *2*(10), 753-759. https://doi.org/10.56778/jdlde.v2i9.221

Vega, N. V., & Domínguez, L. C. (2022). ¿Se Deben Confiar Actividades Profesionales en Cirugía Mínimamente Invasiva Al Médico General? Una Propuesta Basada en El Aprendizaje Experiencial. *Revista Colombiana De Cirugía*. https://doi.org/10.30944/20117582.1445

Velarde, I. L. A., Grünebaum, A. A. C., Bode, S. M., & Aguilar, J. (2024). Papel De Los Fondos De Conocimiento en La Educación. Enfoque Educativo Que Promueve La Paz. *Revista Latinoamericana De Estudios Educativos*, *54*(2), 41-62. https://doi.org/10.48102/rlee.2024.54.2.625

Velásquez, R. E. S. (2021). El Aprendizaje Basado en Problemas en El Desarrollo De Competencias. *Journal Latin American Science*, *5*(2), 792-821. https://doi.org/10.46785/lasjournal.v5i2.106

Velàzquez, R. V., Piguave, C. C., Valdés, I. E., & Zúñiga, K. M. (2020). Metodologías De Enseñanza-Aprendizaje Constructivista Aplicadas a La Educación Superior. *Revista Científica Sinapsis*, *3*(18). https://doi.org/10.37117/s.v3i18.399

Venkatraman, S., Souza-Daw, T. d., & Kaspi, S. (2018). Improving Employment Outcomes of Career and Technical Education Students. *Higher Education Skills and Work-Based Learning*, *8*(4), 469-483. https://doi.org/10.1108/heswbl-01-2018-0003

Vera, C. S. (2010). «Educación Permanente» Y «aprendizaje Permanente»: Dos Modelos Teórico-Aplicativos Diferentes. *Revista Iberoamericana De Educación*, *52*, 203-230. https://doi.org/10.35362/rie520617

Vera, F. S. (2024). Developing Effective Educational Chatbots With GPT: Insights From a Pilot Study in a University Subject. *Trends in Higher Education*, *3*(1), 155-168. https://doi.org/10.3390/higheredu3010009

Vera, G. C. C., & Gámez, M. R. (2024). Tecnología Adaptativa Y Personalización Del Aprendizaje Para Personas Con Discapacidad Visual De La Básica Superior. *Boletín Científico Ideas Y Voces*, *4*(3). https://doi.org/10.60100/bciv.v4i3.170

Vera, L. d. (2023). Reflection-in-Action: Insights From High School Writing Classes. *PCP*, *2022*(1), 112. https://doi.org/10.37546/jaltpcp2022-13

Verenikina, I. (2004). From Theory to Practice: What Does the Metaphor of Scaffolding Mean to Educators Today? *Outlines Critical Practice Studies*, *6*(2), 5-16. https://doi.org/10.7146/ocps.v6i2.2140

Verma, A., Bhattacharya, P., Madhani, N., Trivedi, C., Bhushan, B., Tanwar, S., Sharma, G., Bokoro, P. N., & Sharma, R. (2022). Blockchain for Industry 5.0: Vision, Opportunities, Key Enablers, and Future Directions. *Ieee Access*, *10*, 69160-69199. https://doi.org/10.1109/access.2022.3186892

Verma, A., & Dhaigude, A. S. (2026). Bridging Theory and Practice: A Mixed-Method Exploration of AI's Role in Higher Education. *Sage Open*, *16*(1). https://doi.org/10.1177/21582440261417610

Vermeulen, L. P. (2009). Flight Instructors' Perceptions of Pilot Behaviour Related to Gender. *Sa Journal of Industrial Psychology*, *35*(1). https://doi.org/10.4102/sajip.v35i1.819

Versteeg, M., & Kappe, R. (2021). Resilience and Higher Education Support as Protective Factors for Student Academic Stress and Depression During Covid-19 in the Netherlands. *Frontiers in Public Health*, *9*. https://doi.org/10.3389/fpubh.2021.737223

Versteijlen, M., & Groesbeek, M. J. (2024). Perspective Chapter: Educational Technology Under Scrutiny in Higher Education – A Framework for Balancing Environmental, Economic and Social Aspects in a Blended Design. https://doi.org/10.5772/intechopen.1005117

Verzat, C., Jore, M., Toutain, O., & Silberzahn, P. (2016). Apprendre Par Soi-Même L'entrepreneuriat <i>via</I> Un MOOC. *Revue Française De Gestion*, *42*(257), 33-52. https://doi.org/10.3166/rfg.2016.00050

Vice, J., Beckett, K., Finamore, D., O'Neal, C., Bahle, M., & Velky, S. (2026). Opening Doors and Minds. *International Journal of Bias Identity and Diversities in Education*, *10*(1), 1-23. https://doi.org/10.4018/ijbide.405238

Vicente, R. S., Reomero, J. I., Flores, L. C., Amora, M. S., & Almagro, R. E. (2024). Comparative Analysis of Two Educational Systems in Both Basic and Higher Education Between Japan and Singapore. *Asian Journal of Education and Social Studies*, *50*(1), 114-135. https://doi.org/10.9734/ajess/2024/v50i11243

VictoryXr. (2025). VictoryXR. https://victoryxr.com

Villafuerte, F. S. U. (2020). Interacciones Docentes-Estudiantes Y Prácticas Sexistas en El Aula Del Sistema De Educación Intercultural Bilingüe. *Revista Scientific*, *5*(15), 129-149. https://doi.org/10.29394/scientific.issn.2542-2987.2020.5.15.6.129-149

Villarroel, V., Suárez, M. P. G., Bruna, D., & Castillo-Rabanal, I. (2021). Aplicación De La Metodología De Aprendizaje Experiencial en Educación Superior. *Podium*(40), 41-58. https://doi.org/10.31095/podium.2021.40.3

Vinatier, I. (2012). PASTRÉ Pierre. La Didactique Professionnelle. Approche Anthropologique Du Développement Chez Les Adultes. *Revue Française De Pédagogie*, *179*, 138-141. https://doi.org/10.4000/rfp.3730

Vincent-Ruz, P., & Boase, N. R. B. (2022). Activating Discipline Specific Thinking With Adaptive Learning: A Digital Tool to Enhance Learning in Chemistry. *Plos One*, *17*(11), e0276086. https://doi.org/10.1371/journal.pone.0276086

Vindigni, G. (2023). Adaptive and Re-Adaptive Pedagogies in Higher Education: A Comparative, Longitudinal Study of Their Impact on Professional Competence Development Across Diverse Curricula. *European Journal of Theoretical and Applied Sciences*, *1*(4), 718-743. https://doi.org/10.59324/ejtas.2023.1(4).66

Vinh, L. A., Han, P., Khaing, M. M., & Farrar, O. (2022). An Emerging Dragon: Vietnamese Education After Resolution 29. 99-123. https://doi.org/10.1007/978-3-030-93951-9_5

Vipler, B. (2021). Transformative Learning in Graduate Medical Education: A Scoping Review. *Journal of graduate medical education*, *13*(6), 801-814. https://doi.org/10.4300/jgme-d-21-00065.1

Virkus, S. (2019). The Use of Open Badges in Library and Information Science Education in Estonia. *Education for Information*, *35*(2), 155-172. https://doi.org/10.3233/efi-190257

Virvou, M. (2023). Artificial Intelligence and User Experience in Reciprocity: Contributions and State of the Art. *Intelligent Decision Technologies*, *17*(1), 73-125. https://doi.org/10.3233/idt-230092

Vivar, R. L., Micco, P. D., & Guraziu, E. (2023). Proyecto Self-E: Desarrollo De Competencias De Emprendimiento en Estudiantes Secundarios en El Entorno Europeo, a Través De La Simulación. *Apuntes De Ciencia & Sociedad*, *11*(01), 23-30. https://doi.org/10.18259/acs.2023004

Vlieghe, J. (2013). Alphabetization as Emancipatory Practice: Freire, Rancière, and Critical Pedagogy. *Philosophy of Education*, *69*, 185-193. https://doi.org/10.47925/2013.185

Vogelsang, M. R. (1993). *Transformative Experiences of Female Adult Students* Iowa State University].

Volchenkova, K. (2015). An Andragogical Approach to Forming Faculty Communication Competence in Foreign Language for International Activities. *Bulletin of the South Ural State University Series Education Education Sciences*, *7*(4), 18-25. https://doi.org/10.14529/ped150403

Voon, X. P., Wong, S. L., Wong, L. H., Khambari, M. N. M., & Syed-Abdullah, S. I. S. (2022). Developing Pre-Service Teachers' Computational Thinking Through Experiential Learning: Hybridisation of Plugged and Unplugged Approaches. *Research and Practice in Technology Enhanced Learning*, *18*, 006. https://doi.org/10.58459/rptel.2023.18006

Vries, M. d., Land-Zandstra, A. M., & Smeets, I. (2019). Citizen Scientists' Preferences for Communication of Scientific Output: A Literature Review. *Citizen Science Theory and Practice*, *4*(1). https://doi.org/10.5334/cstp.136

Vujkov, A. (2024). Vocational Training in German-Speaking Countries. https://doi.org/10.5772/intechopen.112109

Vygotsky, L. S. (1978). *Mind in Society: The Development of Higher Psychological Processes*. Harvard University Press.

Wade, S. E., Fauske, J. R., & Thompson, A. (2008). Prospective Teachers' Problem Solving in Online Peer-Led Dialogues. *American Educational Research Journal*, *45*(2), 398-442. https://doi.org/10.3102/0002831207308224

Wagner, R. J., & Campbell, J. (1994). Outdoor-based Experiential Training: Improving Transfer of Training Using Virtual Reality. *Journal of Management Development*, *13*(7), 4-11. https://doi.org/10.1108/02621719410063350

Wahid, M. E., & Ain, R. (2023). Theoritical Implementation as Folklore Digital Reconstruction in Malay Literature Education. *Idealogy Journal*. https://doi.org/10.24191/idealogy.v8i2.447

Wahid, R., Saidin, K., & Arif, N. A. (2020). A Clog on the Right to Quality Education Through OER: A Case of Japan and Malaysia. *Journal International Studies*, *14*. https://doi.org/10.32890/jis.14.2018.8004

Wahlström, N. (2022). School and Democratic Hope: The School as a Space for Civic Literacy. *European Educational Research Journal, 21*(6), 994-1008. https://doi.org/10.1177/14749041221086721

Wakahiu, J. (2017). Mentoring: A Model for Cultivating Leadership Competencies in Kenyan Women Religious. Vol. 33 (2013). https://doi.org/10.21423/awlj-v33.a102

Walker, J., Towey, D., Pike, M., Kapogiannis, G., Elamin, A., & Wei, R. (2020). Developing a Pedagogical Photoreal Virtual Environment to Teach Civil Engineering. *Interactive Technology and Smart Education, 17*(3), 303-321. https://doi.org/10.1108/itse-10-2019-0069

Wallace, D. (2020). Identifying With Borders and Boundaries: The Place of Critical Pedagogy as Social Responsibility Education. 139-154. https://doi.org/10.1108/s2055-364120200000023010

Wals, A. E. J. (2015). *Beyond Unreasonable Doubt: Education and Learning for Socio-Ecological Sustainability in the Anthropocene.* Wageningen University. https://edepot.wur.nl/365312

Walters-Williams, J. (2022). H-Cup: Increasing Higher Order Thinking Skills Levels Through a Framework Based on Cognitive Apprenticeship, Universal Design and Project Based Learning. *Creative Education, 13*(09), 2878-2902. https://doi.org/10.4236/ce.2022.139181

Walters, C., Charles, J. P., & Bingham, S. (2016). Impact of Short-Term Study Abroad Experiences on Transformative Learning. *Journal of Transformative Education, 15*(2), 103-121. https://doi.org/10.1177/1541344616670034

Wang, B.-Y. (2025). Creating Just Learning Spaces: A Path to an Equitable Hub for Lifelong Learning. *New Horizons in Adult Education and Human Resource Development, 38*(3), 260-277. https://doi.org/10.1177/19394225251353581

Wang, F., Hao, S., Tsang, K. T., Wong, J. P., Fung, K., Li, A. T., Jia, C.-X., & Cheng, S. (2024). Empowering Chinese University Health Service Providers to Become Mental Health Champions: Insights From the ACE-LYNX Intervention. *Frontiers in Psychiatry, 15.* https://doi.org/10.3389/fpsyt.2024.1349476

Wang, G., Zhang, T., Wang, C., Pang, Y., Yang, T., Sun, C., Hu, Y., & Zhang, P. (2022). Development of Energy and Mining Governance System Based on New-Generation Information Technology. *Strategic Study of Cae, 24*(1), 176. https://doi.org/10.15302/j-sscae-2022.01.019

Wang, H. (2024). Exploration of Financial Management Curriculum Reform in Vocational College. *Proceedings of Business and Economic Studies, 7*(2), 114-119. https://doi.org/10.26689/pbes.v7i2.6765

Wang, J. (2023). Developing Intelligent Education in China. 514-518. https://doi.org/10.2991/978-2-38476-068-8_64

Wang, M., & Shan, W. (2023). A Research on the Adaptation Problems of Students Returning From Military Service in Higher Education Institutions Based on Subject Knowledge Mapping. *Applied Mathematics and Nonlinear Sciences, 9*(1). https://doi.org/10.2478/amns.2023.2.01010

Wang, Q., Li, H., Pang, W., Liang, S., & Su, Y. (2016). Developing an Integrated Framework of Problem-Based Learning and Coaching Psychology for Medical Education: A Participatory Research. *BMC Medical Education, 16*(1). https://doi.org/10.1186/s12909-015-0516-x

Wáng, Q., & Millward, I. (2014). Developing a Unified Psychological Model of Coaching and Mentoring in Supporting the Learning and Development of Adolescents. *International Journal of Evidence Based Coaching and Mentoring, 12*(2), 91-108. https://doi.org/10.1177/174183051401200208

Wang, R., Liu, C., & Yi, F. (2025). Reconfiguring Tutor–peer Roles to Foster Critical Thinking in Medical PBL Assisted by Tutor-managed chatGPT. *Medical Education, 60*(8), 894-907. https://doi.org/10.1111/medu.70100

Wang, S., Shi, X., & Cai, C. (2024). Advancing Digital Transformation in Chinese Education: A Comprehensive Analysis of Current Trends, Opportunities, Challenges, and Strategic Imperatives. *Journal of Education and Culture Studies, 8*(1), p81. https://doi.org/10.22158/jecs.v8n1p81

Wang, V., Torrisi-Steele, G., & Reinsfield, E. (2020). Transformative Learning, Epistemology and Technology in Adult Education. *Journal of Adult and Continuing Education, 27*(2), 324-340. https://doi.org/10.1177/1477971420918602

Wang, V. C. X., & Sarbo, L. (2004). Philosophy, Role of Adult Educators, and Learning. *Journal of Transformative Education, 2*(3), 204-214. https://doi.org/10.1177/1541344604265105

Wang, V. X., Henschke, J. A., & Fay, K. M. (2013). A Critical Review of Reflectivity, Andragogy, and Confucianism. 356-376. https://doi.org/10.4018/978-1-4666-4249-2.ch021

Wang, V. X., Torrisi-Steele, G., & Hansman, C. A. (2019). Critical Theory and Transformative Learning: Some Insights. *Journal of Adult and Continuing Education*, *25*(2), 234-251. https://doi.org/10.1177/1477971419850837

Wang, W., Hussin, M., & Majid, M. Z. A. (2024). Employment Skills in Tertiary Work-Based Learning: A Multiple-Stakeholder Investigation Into China's "3 + 1" Programs. *International Journal of Learning Teaching and Educational Research*, *23*(9), 198-219. https://doi.org/10.26803/ijlter.23.9.11

Wang, W., Liu, H., & Wu, Y. J. (2024). How Do Reward Personalization Options Influence the Public's Willingness to Participate in Innovation Projects? Insights From Crowdfunding in Industry 5.0. *European Journal of Innovation Management*, *28*(1), 85-112. https://doi.org/10.1108/ejim-07-2023-0584

Wang, W. H., Khasawneh, M. A. S., Sarabani, L., & Ismail, S. M. (2025). Investigating the Influence of Positive Classroom Management Strategies on Academic Buoyancy, Academic Engagement, Psychological Well-Being and Language Achievement: Visions From <scp>EFL</Scp> Learners. *European journal of education*, *60*(3). https://doi.org/10.1111/ejed.70149

Wang, X., & Poon, C. H. (2023). Intrinsic Factors Influencing Career Development Among College Music Teachers in Establishment Stage From Fujian, China. *Malaysian Journal of Social Sciences and Humanities (Mjssh)*, *8*(12), e002641. https://doi.org/10.47405/mjssh.v8i12.2641

Wang, X., Wang, M., Zeng, X., & Zhang, P. (2021). Learning by Teaching and Research: Practice and Exploration of the Construction of First-Class Chemistry Specialty at Hangzhou Normal University. *University Chemistry*, *0*(0), 2010051-2010050. https://doi.org/10.3866/pku.dxhx202010051

Wang, Y. (2025). <scp>L2</Scp> Students' Speaking Skills in Robot-Assisted Language Learning: A Meta-Analysis. *European journal of education*, *61*(1). https://doi.org/10.1111/ejed.70416

Wang, Z., Chen, L., & Anderson, T. (2014). A Framework for Interaction and Cognitive Engagement in Connectivist Learning Contexts. *The International Review of Research in Open and Distributed Learning*, *15*(2). https://doi.org/10.19173/irrodl.v15i2.1709

Wangding, S., Lingard, L., Haidet, P., Vipler, B., Sukhera, J., & Moniz, T. (2024). Disorienting or Transforming? Using the Arts and Humanities to Foster Social Advocacy. *Perspectives on medical education*, *13*(1), 192-200. https://doi.org/10.5334/pme.1213

Warburton, J., & Chambers, B. (2007). Older Indigenous Australians: Their Integral Role in Culture and Community. *Australasian Journal on Ageing*, *26*(1), 3-7. https://doi.org/10.1111/j.1741-6612.2007.00213.x

Ward, L., Gordon, A., & Kirkman, A. (2024). Innovative and Effective Education Strategies for Adult Learners in the Perioperative Setting. *Aorn Journal*, *119*(2), 120-133. https://doi.org/10.1002/aorn.14079

Ward, R., Crick, T., Davenport, J. H., Hanna, P. R., Hayes, A., Irons, A., Miller, K., Moller, F., Prickett, T., & Walters, J. (2023). Using Skills Profiling to Enable Badges and Micro-Credentials to Be Incorporated Into Higher Education Courses. *Journal of Interactive Media in Education*, *2023*(1). https://doi.org/10.5334/jime.807

Wareing, M. (2014). Using Student Lived Experience to Test the Theoretical Basis of Work-Based Learning. *International Journal of Practice-Based Learning in Health and Social Care*, *2*(1), 35-50. https://doi.org/10.11120/pblh.2013.00022

Wargent, E. T., & Stocker, C. J. (2021). Successful Instillation of Professionalism in Our Future Doctors. *Mededpublish*, *10*(1). https://doi.org/10.15694/mep.2021.000173.1

Warhurst, R., & Black, K. (2024). Theorising Later-Career as a Basis for Enhancing Inclusion and Extending Working Lives Through Human Resource Development. *Human Resource Development Review*, *23*(4), 522-544. https://doi.org/10.1177/15344843241269166

Warne, L. (2000). Understanding Organisational Learning in Military Headquarters: Findings From a Pilot Study. *Australasian Journal of Information Systems*, *7*(2). https://doi.org/10.3127/ajis.v7i2.279

Warren, F. M. (2019). Spotlight on Technical and Vocational Education and Training (TVET). https://doi.org/10.58261/xpdk1549

Waseem, T., & Aslam, F. (2020). Educational Learning Theories &Amp; Their Implications in Modern Instructional Designs. *Health Professions Educator Journal, 3*(2), 25-31. https://doi.org/10.53708/hpej.v3i2.9

Wasim, J., Youssef, M. H., Christodoulou, I., & Reinhardt, R. (2023). The Path to Entrepreneurship: The Role of Social Networks in Driving Entrepreneurial Learning and Education. *Organizational Behavior Teaching Review, 48*(3), 459-493. https://doi.org/10.1177/10525629231219235

Wastesson, K. (2024). Learning Managerial Work : First-Line Managers' Learning in Everyday Work Within Swedish Elderly Care. https://doi.org/10.3384/9789180755092

Wati, A., & Prawiro, I. Y. (2019). Pre-Service Efl Teachers' Perceptions and Actual Practices of Giving Oral Feedback in Apprenticeship Program: Exploratory Study. *Wiralodra English Journal, 3*(2), 303-313. https://doi.org/10.31943/wej.v3i2.63

Watkins, K. E., & Marsick, V. J. (2014). Adult Education &Amp; Human Resource Development: Overlapping and Disparate Fields. *New Horizons in Adult Education and Human Resource Development, 26*(1), 42-54. https://doi.org/10.1002/nha3.20052

Watkins, K. E., Marsick, V. J., Wofford, M. G., & Ellinger, A. D. (2018). The Evolving Marsick and Watkins (1990) Theory of Informal and Incidental Learning. *New Directions for Adult and Continuing Education, 2018*(159), 21-36. https://doi.org/10.1002/ace.20285

Watson, J. B. (1913). Psychology as the Behaviorist Views It. *Psychological Review, 20*(2), 158-177.

Watts, R. J., Diemer, M. A., & Voight, A. M. (2011). Critical consciousness: Current status and future directions. *New directions for child and adolescent development, 2011*(134), 43-57.

Webb, M., Fluck, A., Magenheim, J., Malyn-Smith, J., Waters, J., Deschênes, M., & Zagami, J. (2020). Machine Learning for Human Learners: Opportunities, Issues, Tensions and Threats. *Educational Technology Research and Development, 69*(4), 2109-2130. https://doi.org/10.1007/s11423-020-09858-2

Webkamigad, S., Cote-Meek, S., Pianosi, B., & Jacklin, K. (2019). Exploring the Appropriateness of Culturally Safe Dementia Information With Indigenous People in an Urban Northern Ontario Community. *Canadian Journal on Aging / La Revue Canadienne Du Vieillissement, 39*(2), 235-246. https://doi.org/10.1017/s0714980819000606

Webster-Wright, A. (2009). Reframing Professional Development Through Understanding Authentic Professional Learning. *Review of Educational Research, 79*(2), 702-739. https://doi.org/10.3102/0034654308330970

Wehmeyer, M. L., Cheon, S. H., Lee, Y., & Silver, M. (2021). Self-Determination in Positive Education. 225-249. https://doi.org/10.1007/978-3-030-64537-3_9

Weirs, J., & Osborne, A. (2020). Refocusing Sustainability Education: Using Students' Reflections on Their Carbon Footprint to Reinforce the Importance of Considering CO2 Production in the Construction Industry. *Frontiers in Built Environment, 6*. https://doi.org/10.3389/fbuil.2020.00023

Weissova, L., Gregersen-Hermans, J., & Pantelić, D. (2024). Academic Voices: Continuing Professional Development for Teaching in Internationalized Classrooms. *Journal of Comparative & International Higher Education, 16*(5). https://doi.org/10.32674/31dr7314

Wenger-Trayner, E., & Wenger-Trayner, B. (2015). *Learning in landscapes of practice: Boundaries, identity, and knowledgeability in practice-based learning*. Routledge.

Wenger, E. (1998). *Communities of Practice: Learning, Meaning, and Identity*. Cambridge University Press. https://doi.org/10.1017/CBO9780511803932

Wenger, É. (1999). *Communities of Practice. Learning, Meaning and Identity*. Cambridge University Press.

Werquin, P. (2021). Decolonising Recognition of Prior Learning – The Drawbacks of Policy Mimicking. *Research in Comparative and International Education, 16*(4), 488-502. https://doi.org/10.1177/17454999211063480

Werth, E., & Williams, K. (2021). What Motivates Students About Open Pedagogy? Motivational Regulation Through the Lens of Self-Determination Theory. *The International Review of Research in Open and Distributed Learning, 22*(3), 34-54. https://doi.org/10.19173/irrodl.v22i3.5373

West, D., Luzeckyj, A., Toohey, D., Vanderlelie, J., & Searle, B. (2020). Do Academics and University Administrators Really Know Better? The Ethics of Positioning Student Perspectives in Learning Analytics. *Australasian Journal of Educational Technology*. https://doi.org/10.14742/ajet.4653

West, R. (2023). Critical Pedagogy in the Undergraduate Music Therapy Curriculum: A Grounded Theory Study of Music Therapy Educators. *The Arts in Psychotherapy*, *85*, 102068. https://doi.org/10.1016/j.aip.2023.102068

West, R. E., & Cheng, Z. (2022). Digital Credential Evolution. 1-20. https://doi.org/10.1007/978-981-19-0351-9_71-1

Wheelahan, L. (2015). Not just skills: what a focus on knowledge means for vocational education. *Journal of curriculum studies*, *47*(6), 750-762.

Wheelahan, L., & Carter, R. (2001). National Training Packages: A New Curriculum Framework for Vocational Education and Training in Australia. *Education + Training*, *43*(6), 303-316. https://doi.org/10.1108/eum0000000005755

White, L. (2022). "Momma, Today We Were Indian Chiefs!" Pathways to Kan'nikonhrí:io† Through Indigenous Holistic Education. *Frontiers in Education*, *7*. https://doi.org/10.3389/feduc.2022.699627

Whitinui, P., McIvor, O., Robertson, B. G., Morcom, L., Cashman, K. A., & Arbon, V. (2015). The World Indigenous Research Alliance (WIRA): Mediating and Mobilizing Indigenous Peoples' Educational Knowledge and Aspirations. *Education Policy Analysis Archives*, *23*, 120. https://doi.org/10.14507/epaa.v23.2052

Whitley, J., & Hollweck, T. (2020). Inclusion and Equity in Education: Current Policy Reform in Nova Scotia, Canada. *Prospects*, *49*(3-4), 297-312. https://doi.org/10.1007/s11125-020-09503-z

Wickline, V. B., Wiese, D. L., & Aggarwal, P. (2024). Increasing Intercultural Competence Among Psychology Students Using Experiential Learning Activities With International Student Partners. *Scholarship of Teaching and Learning in Psychology*, *10*(3), 272-290. https://doi.org/10.1037/stl0000288

Wiek, A., Withycombe, L., & Redman, C. L. (2011). Key Competencies in Sustainability: A Reference Framework for Academic Program Development. *Sustainability Science*, *6*(2), 203-218. https://doi.org/10.1007/s11625-011-0132-6

Wight, K. (2021). Fighting for Social Lives. *Sfu Educational Review*, *14*(1), 8-16. https://doi.org/10.21810/sfuer.v14i1.1227

Wilbiks, J. M. P. (2021). Indigenizing the Introduction to Psychology Course: Initial Course Content Suggestions and Call for Collaboration. *Canadian Psychology/Psychologie Canadienne*, *62*(4), 391-399. https://doi.org/10.1037/cap0000284

Wilbur, K., & Taylor, A. D. J. (2018). Does a Blended Learning Environment Suit Advanced Practice Training for Pharmacists in a Middle East Setting? *International Journal of Pharmacy Practice*, *26*(6), 560-567. https://doi.org/10.1111/ijpp.12437

Wild, H. A., Kyröläinen, A.-J., & Kuperman, V. (2022). How Representative Are Student Convenience Samples? A Study of Literacy and Numeracy Skills in 32 Countries. *Plos One*, *17*(7), e0271191. https://doi.org/10.1371/journal.pone.0271191

Wildemeersch, D., Guimarães, P., & Fejes, A. (2023). Editorial: Inclusion, Adult Education and Social Justice. *European Journal for Research on the Education and Learning of Adults*, *14*(3), 303-306. https://doi.org/10.3384/rela.2000-7426.5028

Wiley, D., & Hilton, J. (2018). Defining OER-Enabled Pedagogy. *The International Review of Research in Open and Distributed Learning*, *19*(4). https://doi.org/10.19173/irrodl.v19i4.3601

Wilhelmson, L. (2006). Dialogue Meetings as Nonformal Adult Education in a Municipal Context. *Journal of Transformative Education*, *4*(3), 243-256. https://doi.org/10.1177/1541344606290439

Wilkes, G. (2021). Ethics, Religion and Human/E Limits in War. *Religija I Tolerancija*, *19*(36), 369-380. https://doi.org/10.18485/rit.2021.19.36.9

Williams, C. (2010). Understanding the Essential Elements of Work-Based Learning and Its Relevance to Everyday Clinical Practice. *Journal of Nursing Management*, *18*(6), 624-632. https://doi.org/10.1111/j.1365-2834.2010.01141.x

Williams, L. (2024). Indigenous Intergenerational Resilience and Lifelong Learning: Critical Leverage Points for Deep Sustainability Transformation in Turbulent Times. *Sustainability*, *16*(11), 4494. https://doi.org/10.3390/su16114494

Williamson, G. R. (1999). Teaching Sociology to Nurses: Exploring the Debate. *Journal of Clinical Nursing*, *8*(3), 269-274. https://doi.org/10.1046/j.1365-2702.1999.00230.x

Williamson, M., & Watson, R. L. (2006). Learning Styles Research: Understanding How Teaching Should Be Impacted by the Way Learners Learn Part II: Understanding How Learners Prefer to Receive Information. *Christian Education Journal Research on Educational Ministry*, *3*(2), 343-361. https://doi.org/10.1177/073989130600300209

Willis, J., Arnold, J., & DeLuca, C. (2023). Accessibility in Assessment for Learning: Sharing Criteria for Success. *Frontiers in Education*, *8*. https://doi.org/10.3389/feduc.2023.1170454

Willocks, K. (2022a). Developing Leaders Using on-Line Action Learning: An Activity Theory Analysis. *Advances in Developing Human Resources*, *25*(1), 27-44. https://doi.org/10.1177/15234223221139215

Willocks, K. (2022b). Making Leadership as Practice Development Visible: Learning From Activity Theory. *Management learning*, *54*(5), 705-726. https://doi.org/10.1177/13505076221105870

Wilson, S., Tolley, C., Ardle, R. M., Lawson, L., Beswick, E., Hassan, N., Slight, R., & Slight, S. P. (2024). Recommendations to Advance Digital Health Equity: A Systematic Review of Qualitative Studies. *NPJ Digital Medicine*, *7*(1). https://doi.org/10.1038/s41746-024-01177-7

Wilson, S. D. (2018). Implementing Co-Creation and Multiple Intelligence Practices to Transform the Classroom Experience. *Contemporary Issues in Education Research (Cier)*, *11*(4), 127-132. https://doi.org/10.19030/cier.v11i4.10206

Wimpenny, K., Stefanelli, C., Affouneh, S., Almakari, A., Tweissi, A., Naby, H. A., & Tawfik, S. (2020). Breaking Boundaries and Building Bridges Across Knowledge-Sharing Communities. 253-276. https://doi.org/10.1163/9789004422988_012

Winberg, C., & Hollis-Turner, S. (2021). Practical Subjects in the Vocational Curriculum: A Critical Review of the Literature. *Journal of Education*(85), 1-22. https://doi.org/10.17159/2520-9868/i85a01

Windisch, H. C. (2015). Adults With Low Literacy and Numeracy Skills. https://doi.org/10.1787/5jrxnjdd3r5k-en

Winslow, B. W. (2003). Family Caregivers' Experiences With Community Services: A Qualitative Analysis. *Public Health Nursing*, *20*(5), 341-348. https://doi.org/10.1046/j.1525-1446.2003.20502.x

Wirzal, M. D. H., Halim, N. S. A., Nik Abdul Hadi Md, N., & Bustam, M. A. (2022). Metacognition in Science Learning: Bibliometric Analysis of Last Two Decades. *Jurnal Penelitian Dan Pengkajian Ilmu Pendidikan E-Saintika*, *6*(1), 43-60. https://doi.org/10.36312/esaintika.v6i1.665

Wisniewski, B., & Zierer, K. (2021). Functions and Success Conditions of Student Feedback in the Development of Teaching and Teachers. 125-138. https://doi.org/10.1007/978-3-030-75150-0_8

Wissell, S., Karimi, L., Serry, T., Furlong, L., & Hudson, J. (2022). Leading Diverse Workforces: Perspectives From Managers and Employers About Dyslexic Employees in Australian Workplaces. *International Journal of Environmental Research and Public Health*, *19*(19), 11991. https://doi.org/10.3390/ijerph191911991

Woerkom, M. v. (2010). Critical Reflection as a Rationalistic Ideal. *Adult education quarterly*, *60*(4), 339-356. https://doi.org/10.1177/0741713609358446

Wofford, M. G., Ellinger, A. D., & Watkins, K. E. (2013). Learning on the Fly: Exploring the Informal Learning Process of Aviation Instructors. *Journal of Workplace Learning*, *25*(2), 79-97. https://doi.org/10.1108/13665621311299771

Wolcott, M. D., Kornegay, E. C., & Brame, J. L. (2022). Safe to Speak: Fostering Psychological Safety Among Incoming Predoctoral Dental Students. *Journal of Dental Education*, *86*(7), 863-873. https://doi.org/10.1002/jdd.12891

Wolek, F. W. (1999). The Managerial Principles Behind Guild Craftsmanship. *Journal of Management History (Archive)*, *5*(7), 401-413. https://doi.org/10.1108/13552529910297460

Wolmarans, H. (2007). Business Simulations in Financial Management Courses: Implications for Higher Education. *South African Journal of Higher Education*, *20*(2). https://doi.org/10.4314/sajhe.v20i2.25578

Wong, K. F. (2013). A Study of the Universities of the Third Age in Hong Kong : An Evaluation of Elder Academy Model for the Life Long Learning of Older People. https://doi.org/10.14793/soc_etd.31

Woo, D. J., Wang, D., Yung, T., & Guo, K. (2025). Effects of a Prompt Engineering Intervention on Undergraduate Students' <scp>AI</Scp> Self-efficacy, <scp>AI</Scp> Knowledge and Prompt Engineering Ability: A Mixed Methods Study. *British educational research journal, 52*(2), 1442-1469. https://doi.org/10.1002/berj.70087

Woo, Y., & Reeves, T. C. (2019). Interaction in Asynchronous Web- Based Learning Environments: Strategies Supported by Educational Research. *Online Learning, 12*(3-4). https://doi.org/10.24059/olj.v12i3-4.1691

Wood, D., Bruner, J. S., & Ross, G. (1976). The role of tutoring in problem solving. *Journal of Child Psychology and Psychiatry, 17*(2), 89-100.

Woodhouse, A., Ellwood, S., & Lynch, T. (2022). Pivoting Culinary Education During COVID-19 Part Two. *Journal of Innovation in Polytechnic Education, 4*(1), 104-108. https://doi.org/10.69520/jipe.v4i1.109

Woodward, D. R., Booth, S., Allen, E., Forbes, A., Donnelly, C., & Taylor, J. (2021). Constructivism and Reflective Practice: An Evidence-Based Approach to Tertiary Teaching. *Scope Contemporary Research Topics (Learning and Teaching)*(10), 145-153. https://doi.org/10.34074/scop.4010014

Woodward, D. R., Booth, S., Allen, E., Forbes, A., & Morton, C. (2023). Reflective Practice for Educators and Learners, and the Benefits of Being a Reflective Practitioner. *Scope Contemporary Research Topics (Learning and Teaching)*(12), 96-103. https://doi.org/10.34074/scop.4012007

Woodward, D. R., Booth, S., Redfearn, S., Allen, E., & Forbes, A. (2020). Developing Evidence-Based Teaching Practice. *Scope Contemporary Research Topics (Work-Based Learning)*(1), 68-77. https://doi.org/10.34074/scop.6001007

Wooi, C. T. (2024). Knowledge Management in the 21st Century: Trends, Developments, and Strategies. *International Journal of Multidisciplinary Applied Business and Education Research, 5*(10), 4234-4257. https://doi.org/10.11594/ijmaber.05.10.29

Woollard, J. (2020). Pedagogical Approaches of Intermediate Schoolteachers for Teaching Students by Building Online Classroom Communities During the COVID-19 Pandemic. *Jep*. https://doi.org/10.7176/jep/11-30-02

World Bank. (2022). Digital Transformation of Philippine Higher Education. https://doi.org/10.1596/38481

World Bank team. (2022). *Building a Responsive and Resilient Vocational Education and Training System in Benin*. W. B. Group.

World Commission on, E., & Development. (1987). *Our Common Future*. Oxford University Press.

World Economic, F. (2025). *The Future of Jobs Report 2025*. https://www.weforum.org/reports/the-future-of-jobs-report-2025

Worthman, S. S. (2020). Las Políticas De Equidad Educativa a Nivel Superior: Una Propuesta De Marco De Análisis. *Tla-Melaua Revista De Ciencias Sociales, 14*(48). https://doi.org/10.32399/rtla.14.48.507

Wright, A. J., Williams, N. J., Starling, T., Reynolds, A. C., & Garcia-Lavin, B. (2023). Deep-Structure Curriculum Liberation for Social Responsiveness in Graduate Health Service Psychology Training. *Training and Education in Professional Psychology, 17*(1), 22-30. https://doi.org/10.1037/tep0000434

Wu, J. G., Zhang, D., Lee, S. M., & Xian, J. (2025). Game On! Digital Gaming and Augmented Reality/Virtual Reality in Language Learning. *International Journal of Computer-Assisted Language Learning and Teaching, 15*(1), 1-15. https://doi.org/10.4018/ijcallt.367601

Wu, M., Yang, Y., Lin, Y. Y., Chen, M., & Chen, Y. (2025). Effectiveness and Influencing Factors of Standardised Training for Nursing Assistants: A Cross-Sectional Study. *BMJ Open, 15*(2), e088644. https://doi.org/10.1136/bmjopen-2024-088644

Wu, S., Firdaus, R. B. R., Wang, Y., & Kamsi, N. S. (2026). Mapping Higher Education Research in Malaysia and Singapore (2014–2024): A Bibliometric Analysis Through the Research–Policy–Practice Lens. *Sage Open, 16*(1). https://doi.org/10.1177/21582440261423292

Wu, W., & Plakhtii, A. (2021). E-Learning Based on Cloud Computing. *International Journal of Emerging Technologies in Learning (Ijet)*, *16*(10), 4. https://doi.org/10.3991/ijet.v16i10.18579

Wu, W. H., Hsiao, H. C., Wu, P. L., Lin, C. H., & Huang, S. H. (2011). Investigating the Learning-theory Foundations of Game-based Learning: A Meta-analysis. *Journal of Computer Assisted Learning*, *28*(3), 265-279. https://doi.org/10.1111/j.1365-2729.2011.00437.x

Wu, Y., & Crocco, O. S. (2019). Critical Reflection in Leadership Development. *Industrial and Commercial Training*, *51*(7/8), 409-420. https://doi.org/10.1108/ict-03-2019-0022

Wuestewald, T. (2015). Adult Learning in Executive Development Programs. *Adult Learning*, *27*(2), 68-75. https://doi.org/10.1177/1045159515602256

Wylie, A., Wong, J., & Bowen-Jones, E. (2020). Transformative Learning and Medical Students Global Health Essays – A Qualitative Study. *Mededpublish*, *9*, 146. https://doi.org/10.15694/mep.2020.000146.1

Wyn, J. (2014). Conceptualizing Transitions to Adulthood. *New Directions for Adult and Continuing Education*, *2014*(143), 5-16. https://doi.org/10.1002/ace.20100

Xhomaki, B., Maletic, A., Mitri, D. D., & Szücs, A. (2019). Reimagining Education for the Digital Age Position Paper of the Lifelong Learning Platform. *Opus Et Educatio*, *5*(4). https://doi.org/10.3311/ope.282

Ximba, T. M., Biljon, H. M. v., Adams, F., & Niekerk, L. v. (2026). Employment of Persons With Disabilities in the Open Labour Market: A Scoping Review. *African Journal of Disability*, *15*. https://doi.org/10.4102/ajod.v15i0.1873

Xiong, Y., Que, L., & Xiao-li, L. I. (2021). Ubiquitous E-Teaching and E-Learning: China's Massive Adoption of Online Education and Launching MOOCs Internationally During the COVID-19 Outbreak. *Wireless Communications and Mobile Computing*, *2021*(1). https://doi.org/10.1155/2021/6358976

Xu, C., Wang, M., Luo, Q., & Zhang, Y. (2023). A Study on Experiential Learning Model of Tourism Management in Higher Education in China: A Case Study of Chongqing Normal University. *Open Journal of Social Sciences*, *11*(03), 232-242. https://doi.org/10.4236/jss.2023.113016

Xu, S. J., Ni, Q., & Du, Q. (2019). The Effectiveness of Virtual Reality in Safety Training: Measurement of Emotional Arousal With Electromyography. https://doi.org/10.22260/isarc2019/0003

Xu, W., & Knijnik, J. (2021). Critical Chinese as an Additional Language Education in Australia: A Journey to Voices, Courage and Hope. *British educational research journal*, *47*(6), 1635-1651. https://doi.org/10.1002/berj.3747

Xu, Z. (2026). Bridging Language and Data: Transforming Agricultural Curricula for Data Analytics Through Linguistic Insights. *Plos One*, *21*(5), e0348935. https://doi.org/10.1371/journal.pone.0348935

Xulu, L. (2022). *The Importance of Talent Management in a Changing Retail Business Banking Work Environment* University of Kwazulu-Natal].

Yadav, A. (2017). Problem Based Learning as Tool of Self-Reflective Learning Among Medical Undergraduate Student. *International Journal of Advanced Biochemistry Research*, *1*(2), 29-32. https://doi.org/10.33545/26174693.2022.v6.i2a.112

Yalmeh, Z. J., Zainalipour, H., & Zarei, E. (2020). Designing a Higher Education Curriculum Framework Based on Connectivism Approach. *Iranian Evolutionary Educational Psychology Journal*, *2*(1), 1-12. https://doi.org/10.29252/ieepj.2.1.1

Yan, L., & Gašević, D. (2026). Agentivism: A Learning Theory for the Age of Artificial Intelligence. https://doi.org/10.48550/arxiv.2604.07813

Yan, Y., & Chau, T. (2025). A Systematic Review of AI Ethics in Education. *Journal of Global Information Management*, *33*(1), 1-50. https://doi.org/10.4018/jgim.386381

Yang, G. (2025). Conceptual Model to Examine Students' Use Of <scp>H5P</Scp> Technology in Online Learning Environments: The Integration of the Unified Theory of Acceptance and Use of Technology Model and Perceived Pedagogical Value Model. *European journal of education*, *60*(4). https://doi.org/10.1111/ejed.70286

Yang, X., Jiang, M., & Zhao, Y. S. (2017). Effects of Noise on English Listening Comprehension Among Chinese College Students With Different Learning Styles. *Frontiers in Psychology*, *8*. https://doi.org/10.3389/fpsyg.2017.01764

Yang, X., & Wu, W. (2024). Advancing Digital Transformation in TVET Through International Cooperation: Approaches by the UNESCO Chair on Digitalization in TVET. *Vocat Tech Edu*, *1*(2). https://doi.org/10.54844/vte.2024.0585

Yang, Y. (2024). Design and Implementation of SaaS Online Education Management System for Expanding LMS System. *Scalable Computing Practice and Experience*, *25*(4), 3048-3061. https://doi.org/10.12694/scpe.v25i4.2983

Yang, Y., Majumdar, R., Li, H., Akçapınar, G., Flanagan, B., & Ogata, H. (2021). A Framework to Foster Analysis Skill for Self-Directed Activities in Data-Rich Environment. *Research and Practice in Technology Enhanced Learning*, *16*(1). https://doi.org/10.1186/s41039-021-00170-y

Yanow, D., & Tsoukas, H. (2009). What Is Reflection-In-Action? A Phenomenological Account. *Journal of Management Studies*, *46*(8), 1339-1364. https://doi.org/10.1111/j.1467-6486.2009.00859.x

Yaşar, O. (2016). Cognitive Aspects of Computational Modeling and Simulation in Teaching and Learning. *The Journal of Computational Science Education*, *7*(1), 2-14. https://doi.org/10.22369/issn.2153-4136/7/1/1

Yasmiatti, L. W. (2024). Forming an Ethical Mindset: Applying Constructivist Learning to Teach Anti-Corruption Education in Junior High Schools. *Nawala Education*, *1*(5), 74-88. https://doi.org/10.62872/yw3yz151

Yasmin, F., & Islam, M. I. (2021). Theoretical Framework: an Investigation Into Transversal Competencies Among Pakistani Undergraduates. *International Journal of Linguistics and Culture*, *2*(2), 101-116. https://doi.org/10.52700/ijlc.v2i2.30

Yazdani, N., Murad, H. S., & Shuja, A. (2017). Wholistic Management Education (WME): Theorizing the Contextualized Applicability of Transformative Learning in Management Education Discourse. *Sukkur Iba Journal of Management and Business*, *4*(1), 42-63. https://doi.org/10.30537/sijmb.v4i1.103

Yazici, H. J. (2020). Project-Based Learning for Teaching Business Analytics in the Undergraduate Curriculum*. *Decision Sciences Journal of Innovative Education*, *18*(4), 589-611. https://doi.org/10.1111/dsji.12219

Yeheyis, M. B., Reza, B., Hewage, K., Ruwanpura, J. Y., & Sadiq, R. (2016). Evaluating Motivation of Construction Workers: A Comparison of Fuzzy Rule-Based Model With the Traditional Expectancy Theory. *Journal of Civil Engineering and Management*, *22*(7), 862-873. https://doi.org/10.3846/13923730.2014.914103

Yelubayeva, P., Tashkyn, E., & Berkinbayeva, G. (2023). Addressing Challenges in Kazakh Education for Sustainable Development. *Sustainability*, *15*(19), 14311. https://doi.org/10.3390/su151914311

Yenugula, M., Sahoo, S. K., & Goswami, S. S. (2024). Cloud Computing for Sustainable Development: An Analysis of Environmental, Economic and Social Benefits. *Journal of Future Sustainability*, *4*(1), 59-66. https://doi.org/10.5267/j.jfs.2024.1.005

Yeo, S. (2024). The Application of Cognitive Teaching and Learning Strategies to Instruction in Medical Education. *Korean Medical Education Review*, *26*(Suppl 1), S1-S12. https://doi.org/10.17496/kmer.24.006

Yeter, I. H., Yang, W., & Sturgess, J. B. (2024). Global Initiatives and Challenges in Integrating Artificial Intelligence Literacy in Elementary Education: Mapping Policies and Empirical Literature. *Future in Educational Research*, *2*(4), 382-402. https://doi.org/10.1002/fer3.59

Yi-Cline, F. (2020). Learning to Teach Emergent Bilinguals: Mainstream Preservice Secondary Teachers in Student Teaching. https://doi.org/10.3102/1573603

Yıldırım, R. (2019). Teaching in Culturally and Linguistically Diverse Classrooms: Turkish EFL Instructors' Experience. *Journal of Language and Linguistic Studies*, *15*(3), 1155-1170. https://doi.org/10.17263/jlls.631557

Yilmaz, M., & Yılmaz, U. (2019). The Relation Between Social Learning and Visual Culture. *Lnternational Electronic Journal of Elementary Education*, *11*(4), 421-427. https://doi.org/10.26822/iejee.2019450837

Yin, A. Examining Problem Solving Skills Between Students With and Without Engineering Work Experience. https://doi.org/10.18260/1-2--15741

Yoon, D.-Y., Han, S., Sung, M., & Cho, J. (2018). Informal Learning, Organizational Commitment and Self-Efficacy. *Journal of Workplace Learning*, *30*(8), 640-657. https://doi.org/10.1108/jwl-02-2018-0034

Younas, M., Noor, U., Menhas, R., Jiang, Y., & El-Dakhs, D. A. S. (2026). Exploring University Students' Attitudes Toward Gamified Learning Management Systems: An Extended Technology Acceptance Model Approach. *European journal of education*, *61*(1). https://doi.org/10.1111/ejed.70531

Young-Scholten, M. (2013). Low-Educated Immigrants and the Social Relevance of Second Language Acquisition Research. *Second Language Research*, *29*(4), 441-454. https://doi.org/10.1177/0267658313491266

Young, D. G., & Bunting, B. D. (2024). Rethinking College Transitions: Legitimate Peripheral Participation as a Pathway to Becoming. *Aera Open*, *10*. https://doi.org/10.1177/23328584241255631

Young, D. R., Dorn, L., & Karic, I. (2024). Pathways of Hope, Breaking Barriers: Refugee and Immigrant Women Leading Change in Early Childhood Workforce. https://doi.org/10.5772/intechopen.1007371

Young, M., & Allais, S. (2013). *Implementing National Qualifications Frameworks Across Five Continents*. Routledge.

Young, M. D., Mountford, M., & Skrla, L. (2006). Infusing Gender and Diversity Issues Into Educational Leadership Programs. *Journal of Educational Administration*, *44*(3), 264-277. https://doi.org/10.1108/09578230610664850

Yousaf, A., & Masood, S. (2024). Balancing Virtual Boundaries: Investigating the Impact of Online Learning on Socialization and Social Skill Development. *Asian Journal of University Education*, *20*(2), 439-453. https://doi.org/10.24191/ajue.v20i2.27031

Yousef, A. M. F., & Sumner, T. (2020). Reflections on the Last Decade of MOOC Research. *Computer Applications in Engineering Education*, *29*(4), 648-665. https://doi.org/10.1002/cae.22334

Yoyo, Y., Hussin, M., Rahman, A., Zaini, A. R., & Sutipyo, R. (2023). Cultural Perceptions of Santri on Arabic Language Learning. *Jurnal Pendidikan Islam*, *9*(2), 187-198. https://doi.org/10.15575/jpi.v9i2.24103

Yu, H. (2023). Adult Education in Scotland a Critical Analysis of Strategic and Policy. *Lecture Notes in Education Psychology and Public Media*, *2*(1), 858-867. https://doi.org/10.54254/2753-7048/2/2022539

Yu, H., Sun, P., Wanru, G., & Baoshan, L. (2025). Transforming Engineering Education Through Enhancing AI Literacy of College Teachers. *Journal of Integrated Design and Process Science*, *29*(2-3), 156-173. https://doi.org/10.1177/10920617251405473

Yu, S., Chen, L., & Wang, Z. (2023). Trends of Future Development. 379-409. https://doi.org/10.1007/978-981-99-6269-3_7

Yuan, C., Tang, J., & Cao, Y. (2024). The Impact of Artificial Intelligence on Economic Development: A Systematic Review. *International Theory and Practice in Humanities and Social Sciences*, *1*(1), 130-143. https://doi.org/10.70693/itphss.v1i1.57

Yuan, R., Wang, K., & Li, M. (2025). Incorporating "Assessment as Learning" (<scp>AaL</Scp>) in Language Teacher Education: An Exploratory Practice. *Tesol Quarterly*, *60*(1), 392-405. https://doi.org/10.1002/tesq.70027

YÜCeol, N. (2021). The Steps to Be Taken in Higher Education for Successful Adaptation to Industry 4.0. *Yükseköğretim Dergisi*, *11*(3), 563-577. https://doi.org/10.2399/yod.21.617715

Yun-bo, L. I., & Zhang, H. (2025). Practical Legal Education as Transformative Learning in China: Advancing SDG 4 and SDG 16. *European journal of education*, *60*(4). https://doi.org/10.1111/ejed.70272

Yunga, D. (2019). Transformative Learning: The Richness of Formality and Informality. *Andragoske Studije*(2), 69-91. https://doi.org/10.5937/andstud1902069y

Yurt, E. (2025). The Role of Learning Style in the Relationship Between Task Value, Self-Efficacy, and Math Achievement: A Moderated Mediation Analysis. *European journal of education*, *60*(4). https://doi.org/10.1111/ejed.70268

Yusof, F. M. (2025). Mapping the Negative Emotional Trajectories of Chinese University Novice English Foreign Language Teachers Across Developmental Stages: Insights From the Cognitive Appraisal Theory. *European journal of education*, *60*(2). https://doi.org/10.1111/ejed.70136

Yusuf, M. A. (2021). Adult Learning Principles a Strategy for Improving Instructional Delivery in Higher Institutions of Learning in Nigeria. *Asian Journal of Education and Social Studies*, 38-45. https://doi.org/10.9734/ajess/2021/v15i130369

Yuyun, I., & Suherdi, D. (2023). Components and Strategies for Personalized Learning in Higher Education: A Systematic Review. 271-290. https://doi.org/10.2991/978-2-38476-054-1_23

Zaber, M. A., Karoly, L. A., & Whipkey, K. (2019). Reimagining the Workforce Development and Employment System for the 21st Century and Beyond. https://doi.org/10.7249/rr2768

Zainuddin, A., Latiff, Z. I. A., & Bakar, Z. A. (2024). Exploring the Influence of Long-Term Memory in Information Processing in Learning Process. *International Journal of Academic Research in Progressive Education and Development*, *13*(4). https://doi.org/10.6007/ijarped/v13-i4/22901

Zaman, A., Abir, M. R., & Mursalin, S. (2024). Extended Reality in Education and Training: Enhancing Trustworthiness. *International Journal of Science and Research Archive*, *11*(1), 1705-1720. https://doi.org/10.30574/ijsra.2024.11.1.0206

Zambrano, K. A. M. (2022). La Aplicación De La Analítica Del Aprendizaje Para Mejorar La Toma De Decisiones en Educación: Casos De Estudio en Diferentes Contextos Educativos. *Rig*, *1*(1), 26-37. https://doi.org/10.62943/rig.v1n1.2022.54

Zambrano, R. L. C., Romero, M. E. Y., Dávila, K. E. D., & Balarezo, C. E. B. (2023). Realidad Virtual Y Aumentada en La Educación Superior: Experiencias Inmersivas Para El Aprendizaje Profundo. *Religación*, *8*(37), e2301088. https://doi.org/10.46652/rgn.v8i37.1088

Żammit, J. (2024). Sailing or Stumbling: How Do Adult Learners Navigate the Realm of Online Maltese Language Learning? *International Journal of Linguistics and Translation Studies*, *5*(3), 29-44. https://doi.org/10.36892/ijlts.v5i3.464

Zanchetta, M. S., Stevenson, M., Nenadovic, V., Perreault, M., Henry, C. J., & Leong, N. (2016). *Aporia*, *8*(2). https://doi.org/10.18192/aporia.v8i2.2792

ZapWorks. (2025). ZapWorks. https://zap.works

Zavadskienė, L., & Bubnys, R. (2018). The Roles of University Teachers in Building Teacher Students' Capacity of Reflective Practice Through the Process of Experiential Learning. *Society Integration Education Proceedings of the International Scientific Conference*, *1*, 600-610. https://doi.org/10.17770/sie2018vol1.3239

Zawacki-Richter, O., Marín, V. I., Bond, M., & Gouverneur, F. (2019). Systematic Review of Research on Artificial Intelligence Applications in Higher Education – Where Are the Educators? *International Journal of Educational Technology in Higher Education*, *16*(1). https://doi.org/10.1186/s41239-019-0171-0

Zboja, J. J., Jackson, R., & Grimes-Rose, M. (2020). An Expectancy Theory Perspective of Volunteerism: The Roles of Powerlessness, Attitude Toward Charitable Organizations, and Attitude Toward Helping Others. *International Review on Public and Nonprofit Marketing*, *17*(4), 493-507. https://doi.org/10.1007/s12208-020-00260-5

Zdravkova, K., Dalipi, F., & Krasniqi, V. (2022). Remote Education Trajectories for Learners With Special Needs During the Covid-19 Outbreak. *International Journal of Emerging Technologies in Learning (Ijet)*, *17*(21), 89-122. https://doi.org/10.3991/ijet.v17i21.32401

Zeng, M. Q., Chen, H., Shrestha, A., Crowley, C., Ng, E., & Wang, G. (2020). International Collaboration on a Sustainable Forestry Management OER Online Program – A Case Study. https://doi.org/10.4995/head20.2020.11242

Zenios, M. (2020). Educational Theory in Technology Enhanced Learning Revisited: A Model for Simulation-Based Learning in Higher Education. https://doi.org/10.21428/8c225f6e.1cf4dde8

Zepke, N. (2013). Student Engagement: A Complex Business Supporting the First Year Experience in Tertiary Education. *The International Journal of the First Year in Higher Education*, *4*(2). https://doi.org/10.5204/intjfyhe.v4i2.183

Zhan, H., Hou, M., & Tan, F. (2022). Influence of Intelligentization on Enterprise Green Innovation: Evidence From Listed Companies of New Energy Industry in China. *资源科学*, *44*(5), 984-993. https://doi.org/10.18402/resci.2022.05.09

Zhang-Wu, Q. (2017). Culturally and Linguistically Responsive Teaching in Practice: A Case Study of a Fourth-Grade Mainstream Classroom Teacher. *Journal of Education*, *197*(1), 33-40. https://doi.org/10.1177/002205741719700105

Zhang, B., Lee, I., & Moore, K. (2024). An Effectiveness Study of Teacher-Led AI Literacy Curriculum in K-12 Classrooms. *Proceedings of the Aaai Conference on Artificial Intelligence*, *38*(21), 23318-23325. https://doi.org/10.1609/aaai.v38i21.30380

Zhang, H.-Z., & Wang, Q. G. (2025). Comprehensive Education and Flexible Learning: Merging Chinese Wisdom With Contemporary Pedagogical Practices. *European journal of education*, *61*(1). https://doi.org/10.1111/ejed.70382

Zhang, H. (2021). An Institutional Dilemma in China's Skills-Development System: Evidence From Two Apprenticeship Reforms. *The China Quarterly*, *248*(1), 1103-1128. https://doi.org/10.1017/s030574102100059x

Zhang, H., Li, A., Li, Y. J., Nabi, F., Dong, H., & Ma, Y. (2026). Technological Innovations in Veterinary Medical Education: A Narrative Review of Simulation, AI, and Competency-Based Training. *Journal of Veterinary Science*, *27*. https://doi.org/10.4142/jvs.25194

Zhang, J., & Rhee, G. S. (2025). AI Ethics Knowledge Among Chinese Teachers: An Empirical Assessment. https://doi.org/10.21203/rs.3.rs-7042877/v1

Zhang, L. (2024). An Analysis of the Problems and Effectiveness of Children's English Education From the Perspective of Behaviorism. *Journal of Education Humanities and Social Sciences*, *27*, 495-500. https://doi.org/10.54097/2jqfvw93

Zhang, Z., Li, H., & Zhou, J. (2023). Teaching With Social Context in Instructional Video Facilitates Second Language Vocabulary Learning. *Heliyon*, *9*(3), e14540. https://doi.org/10.1016/j.heliyon.2023.e14540

Zhao, J. (2013). Confucius as a Critical Educator: Towards Educational Thoughts of Confucius. *Frontiers of Education in China*, *8*(1), 9-27. https://doi.org/10.1007/bf03396959

Zhao, L., Jiang, M., & Gao, H. (2024). A Thematic Analysis on the Factors Influencing Lifelong Learning Among Medical Students in Jiangxi Province, China. *International Journal of Religion*, *5*(9), 717-732. https://doi.org/10.61707/652zk290

Zhao, X., Ren, Y. Z., & Lee, K. C. S. (2023). Leading Virtual Reality (VR) and Augmented Reality (AR) in Education: Bibliometric and Content Analysis From the Web of Science (2018–2022). *Sage Open*, *13*(3). https://doi.org/10.1177/21582440231190821

Zhao, Y., & Jakkampudi, K. (2023). Assessing Policy Measures Safeguarding Workers From Artificial Intelligence in the United States. *Journal of Computer and Communications*, *11*(11), 149-166. https://doi.org/10.4236/jcc.2023.1111008

Zhao, Y., & Ko, J. (2018). Workplace Learning in the Professional Development of Vocational Education Teachers. *Studia Paedagogica*, *23*(2), 43. https://doi.org/10.5817/sp2018-2-4

Zhao, Y., & Ko, J. (2024). Vocational Teachers' Perceptions on Workplace Learning in Facilitating Students' Professional Engagement in the Context of Industry-University Collaboration in China. *Journal of Workplace Learning*, *36*(4), 282-297. https://doi.org/10.1108/jwl-12-2023-0197

Zheng, Z., Mohd Hafizuddin Syah Bangaan, A., Zaki, H. O., & Tan, Q. L. (2025). Advancing Marketing Science Through Vroom's Expectancy Theory: A Review and Research Agenda. *International Journal of Consumer Studies*, *49*(5). https://doi.org/10.1111/ijcs.70104

Zhong, Q., & Norton, H. (2019). Exploring the Roles and Facilitation Strategies of Online Peer Moderators. *Studies in Self-Access Learning Journal*, 379-400. https://doi.org/10.37237/100405

Zhou, X., Wolstencroft, P., Schofield, L., & Fang, L. (2025). Are Graduates Digitally Unprepared?—A Digital Technology Gap Analysis From Alumni and Employer's Perspectives. *Journal of Computer Assisted Learning*, *41*(4). https://doi.org/10.1111/jcal.70046

Zhu, B., Chau, K. T., & Mokmin, N. A. M. (2024). Optimizing Cognitive Load and Learning Adaptability With Adaptive Microlearning for in-Service Personnel. *Scientific Reports*, *14*(1). https://doi.org/10.1038/s41598-024-77122-1

Zhu, C. (2019). Self-Efficacy and Self-Esteem in Online Learning Environments of Adult Learners. *International Journal of Learning Technology*, *14*(1), 4. https://doi.org/10.1504/ijlt.2019.10022288

Zhu, M., Bonk, C. J., & Berri, S. (2022). Fostering Self-Directed Learning in MOOCs: Motivation, Learning Strategies, and Instruction. *Online Learning*, *26*(1). https://doi.org/10.24059/olj.v26i1.2629

Zhu, W., Hua, Y., Zhu, G., & Wang, L. (2022). Share and Embrace Demographic and Location Diversity: Creating an Instagram-based Inclusive Online Learning Community. *British journal of educational technology*, *53*(6), 1530-1548. https://doi.org/10.1111/bjet.13272

Zhu, Y. (2023). Transforming Adult Education Through Critical Feminist Theory: Towards a Critical Consciousness Learning Model. *New Directions for Adult and Continuing Education*, *2023*(180), 101-111. https://doi.org/10.1002/ace.20515

Zhu, Y., Sun, Q., & Glowacki-Dudka, M. (2023). Women, Race, and Lifelong Learning: Navigating Feminist Approaches in Adult Education. *New Directions for Adult and Continuing Education*, *2023*(180), 5-9. https://doi.org/10.1002/ace.20507

Ziani, M., Codello, P., & Mesny, A. (2021). L'éducation Des Jeunes À L'« Entrepreneuriat Collectif » : Défis Et Tensions Au Cœur Des Coopératives D'initiation À L'entrepreneuriat Collectif Au Québec. *Canadian Journal of Nonprofit and Social Economy Research*, *12*(2). https://doi.org/10.29173/cjnser.2021v12n2a398

Ziegler, R., Chipanga, T., & Magoda, C. M. (2020). Workplace-Based Learning: An Industry Perspective. *South African Journal of Higher Education*, *34*(1). https://doi.org/10.20853/34-1-3395

Ziker, C., Truman, B., & Dodds, H. (2021). Cross Reality (XR): Challenges and Opportunities Across the Spectrum. 55-77. https://doi.org/10.1007/978-3-030-58948-6_4

Zouinar, M. (2020). Évolutions De L'Intelligence Artificielle : Quels enjeux Pour L'activité Humaine Et La relation Humain-Machine Au travail ? *Activites*, *17-1*. https://doi.org/10.4000/activites.4941

Zuhairi, A., Hsueh, A. C. T., & Chiang, I. C. N. (2020). Empowering Lifelong Learning Through Open Universities in Taiwan and Indonesia. *Asian Association of Open Universities Journal*, *15*(2), 167-188. https://doi.org/10.1108/aaouj-12-2019-0059

Župan, M. (2023). Feasibility of Moocs for Legal Education. https://doi.org/10.25234/eclic/28270

Żydowicz, W. M., Skokowski, J., Marano, L., & Połom, K. (2024). Current Trends and Beyond Conventional Approaches: Advancements in Breast Cancer Surgery Through Three-Dimensional Imaging, Virtual Reality, Augmented Reality, and the Emerging Metaverse. *Journal of Clinical Medicine*, *13*(3), 915. https://doi.org/10.3390/jcm13030915

Καρανικόλα, Ζ., & Παλαιολόγου, Ν. (2021). Refugees and Adult Education: A Thematic Analysis on UNESCO's Latest Recommendations. *Andragoske Studije*(1), 57-74. https://doi.org/10.5937/andstud2101057k

Κατσικοπούλου, Ι. (2016). Finding One's Place in Life. https://doi.org/10.12681/eadd/43077

Λαμπρόπουλος, Γ., Fernández-Arias, P., Antón-Sancho, Á., & Vergara, D. (2024). Examining the Role of Augmented Reality and Virtual Reality in Safety Training. *Electronics*, *13*(19), 3952. https://doi.org/10.3390/electronics13193952

Παπαθανασίου, Μ. (2023). Teachers as Adult Learners: Their Need to Transform. *Adult Education Critical Issues*, *3*(1), 26-38. https://doi.org/10.12681/haea.32495

Афанасенко, О. В., Nizhenkovska, I., Holovchenko, O., & Glushachenko, O. O. (2022). Technology-Enhanced Constructivist Learning Environment for Pharmacy Students. *Pharmacy Education*, *22*(1), 778-787. https://doi.org/10.46542/pe.2022.221.778787

Iatsyshyn, A., Ковач, В., Liubchak, V., Zuban, Y. O., Piven, A. G., Sokolyuk, O. M., Iatsyshyn, A., Popov, O., Artemchuk, V., & Shyshkina, M. P. (2020). Application of Augmented Reality Technologies for Education Projects Preparation. *Cte Workshop Proceedings*, *7*, 134-160. https://doi.org/10.55056/cte.318

Кубрушко, П. Ф., & Kozlenkova, E. N. (2019). Continuing Engineering Education: Background and Development Vectors. https://doi.org/10.2991/ismge-19.2019.76

Макарчук, Т. А., Федорович, М. В., Trofimov, V., Лобанов, О. С., Demchenko, S., & Barabanova, M. I. (2018). Quality Management of E-Learning in Information Technology Management Training. 742-748. https://doi.org/10.15405/epsbs.2018.12.91

Осьодло, В., Рахманов, В., Krykun, V., Tarasenko, N., & Aristarkhova, M. (2022). <i>Retracted:</I> Officers' Foreign Language Training in Educational and Information Environment of the Higher Military Educational Institution. *Review of Education*, *10*(1). https://doi.org/10.1002/rev3.3317

Роляк, А., Матієнко, О., Koliadych, Y., Yatsyshin, O., & Dakaliuk, O. (2020). Adults as Non-Traditional Students in the Tertiary Education of Denmark and Ukraine: Comparative Discourse of Structural, Psychologic and Pedagogic Peculiarities. *Independent Journal of Management & Production*, *11*(9), 2215-2234. https://doi.org/10.14807/ijmp.v11i9.1411

Солодка, А., Рускуліс, Л., Demianenko, O., & Zaskaleta, S. (2022). MALL Instructional Course Design: Constructing Out-of-Class Experience. *Arab World English Journal*(8), 40-55. https://doi.org/10.24093/awej/call8.3

Темјановски, Р., Chabukovski, V., Zlatkovski, D., & Todevski, D. (2023). The Need, Practice, and Values of Micro-Credentials in the Academic and Business Sector. *Journal of Economics*, *8*(2), 75-86. https://doi.org/10.46763/joe2382075t

ر. ج. م, إبراهيم. (2024). برنامج قائم على النظرية الترابطية لتنمية مهارات القراءة الرقمية باللغة الفرنسية كلغة أجنبية والوعي المعلوماتي لدى طلاب كلية التربية جامعة المنيا. مجلة البحث في التربية وعلم النفس, 39(1), 442-401. https://doi.org/10.21608/mathj.2024.267143.1416

吴, 蕴. (2017). Analysis on the Talents Cultivation Mode and Employment Effect of Service Outsourcing. *Advances in Education*, *07*(06), 348-353. https://doi.org/10.12677/ae.2017.76054

张, 旭. (2018). The Study of Tech-Based SMEs Brain Drain Under the Theory Perspective of Per Capita Efficiency. *Service Science and Management*, *07*(01), 1-6. https://doi.org/10.12677/ssem.2018.71001

王, 钰. (2016). Study on the External Environment Constitution of Cultivating New-Type Professional Farmers. *Advances in Social Sciences*, *05*(04), 537-542. https://doi.org/10.12677/ass.2016.54075

Index

A

D

E

F

G

H

I

J

K

L

M

N

O

P

Q

R

S

T

www.ingramcontent.com/pod-product-compliance
Lightning Source LLC
LaVergne TN
LVHW082003200926
848481LV00010B/2798
9781764798723

"Brilliant and insightful! *8-Circuit Ascension* explores the psychotherapeutic potential—as well as the shamanic and psychonautic applications—of Leary's 8-Circuit Model of Consciousness, the most inclusive and comprehensive model of human mind states ever devised. *8-Circuit Ascension* advances the theory, and brings it up-to-date with the latest developments in psychology, neuroscience, and psychedelic medicine research. It's just what our wayward world needs. Highly recommended!"

— David Jay Brown, author, *The Illustrated Field Guide to DMT Entities*

"*8-Circuit Ascension* updates Leary's 8-Circuit Model, and makes it accessible for a new generation, especially for mind-body medicine and its many applications."

— Stefan Kasian, N.D., Ph.D., American Psychological Association

"Relevant, applicable, and evolving. It would be wise for all practitioners and students in the fields of psychonautic/psychospiritual/psychedelic therapies and guidance to be well aware of this book."

— Danielle M. Herrera, LMFT, Psychedelic-Assisted Psychotherapist

"A thorough and fascinating review of the theory and application of the 8-Circuit Model, with many new insights drawn from recent trends in research and practice. A must-read for everyone interested in the promise of psychedelic-assisted therapy and the inner journey in all its forms. Highly recommended!"

— David Shoemaker, PsyD, Clinical Psychologist; author, *The Way of the Will*

"*8-Circuit Ascension* brings the foundational concepts into the present, synthesizing Leary and Wilson's groundbreaking concepts with updated frameworks rooted in contemporary research and psychology. A powerful and vital resource for those seeking healing, transformation, and the evolution of human consciousness, and for the therapists, facilitators, and guides who support them in this process."

— Michael Patterson, Grammy nominated record producer and cultural curator

"Wingate and Turetzky unpack Leary's ideas to bring us as thorough and authoritative a sourcebook as any seeker, scholar or researcher could hope for. *8-Circuit Ascension* incorporates updated research, insight and commentary that casts fresh light on Leary's influential work. An invaluable and timely guide to understanding ourselves and why we do the things we do, complete with practical information on how to reprogramme and debug our systems. It will change how you think."

— Grant Morrison, author, producer, musician, counter-culture icon

"Fascinating, in-depth and accessible."

— Michael Mithoefer, MD, MDMA-Assisted Therapy Researcher, *Multidisciplinary Association for Psychedelic Studies* (MAPS)

"*8-Circuit Ascension* updates Leary's model with contemporary advances in neuroscience and psychology. More than just a theoretical framework, it presents an evidence-based approach that has the potential to unlock new pathways for exploring and expanding our understanding of consciousness. A must-read!"

— Heather Berlin, Ph.D., MPH, Associate Prof. of Neuroscience & Psychiatry

"Leary's work on the 8-Circuit Model of Consciousness didn't just change my life—it gave me a framework for understanding human potential in ways I never thought possible. This book takes Leary's groundbreaking concepts and modernizes them, offering a comprehensive guide that could very well be the most important thing you ever read. This isn't just an academic discussion of consciousness—it's a playbook for transformation. It provides a clear structure to navigate and expand our inner worlds. For those on the frontier of mental health, personal growth, and psychedelic exploration, this book is an essential guide. Whether you're a seeker, healer, or adventurer, this is the map you've been waiting for."

— Zappy Zapolin, Psychedelic Concierge, Filmmaker

"An important and essential work. Building upon the foundations of the 8-Circuit Model as developed by Leary, Wilson and Alli, the authors present a lucid and exciting analysis of the model, with in-depth practical discussions, and tools for integrating and working directly with the circuits. Required reading for all explorers of consciousness."

— Gregory Peters, author of *Yogini Magic* and *New Aeon Tantra*

"Having spent years exploring consciousness, I truly believe this book opens new doors of understanding—not just for personal growth, but for grasping the deeper nature of human consciousness itself. I highly recommend this book!"

— Baal Kadmon, author, occultist, historian, philologist

"*8-Circuit Ascension* is an important reference text for psychedelic therapy, and provides in-depth processes for people working in a clinical setting."

— Coby Michael, author of *The Poison Path Herbal*